WAR *and* PEACE

ALSO BY NIGEL HAMILTON

Royal Greenwich: A Guide and History to London's Most Historic Borough (with Olive Hamilton)

Nigel Hamilton's Guide to Greenwich: A Personal Guide to the Buildings and Walks of One of England's Most Beautiful and Historic Areas

The Brothers Mann: The Lives of Heinrich and Thomas Mann, 1871–1950 and 1875–1955

Monty: The Making of a General, 1887–1942

Master of the Battlefield: Monty's War Years, 1942–1944

Monty: Final Years of the Field-Marshal, 1944–1976

Monty: The Man Behind the Legend

JFK: Reckless Youth

Monty: The Battles of Field Marshal Bernard Law Montgomery

The Full Monty: Montgomery of Alamein, 1887–1942

Bill Clinton, An American Journey: Great Expectations

Montgomery: D-Day Commander

Bill Clinton: Mastering the Presidency

Biography: A Brief History

How to Do Biography: A Primer

American Caesars: Lives of the Presidents from Franklin D. Roosevelt to George W. Bush

The Mantle of Command: FDR at War, 1941–1942

Commander in Chief: FDR's Battle with Churchill, 1943

The ABC of Modern Biography (with Hans Renders)

WAR

and

PEACE

FDR'S FINAL ODYSSEY
D-DAY TO YALTA,
1943–1945

Nigel Hamilton

Houghton Mifflin Harcourt
BOSTON · NEW YORK

Copyright © 2019 by Nigel Hamilton

All rights reserved

For information about permission to reproduce selections from this book, write to trade.permissions@hmhco.com or to Permissions, Houghton Mifflin Harcourt Publishing Company, 3 Park Avenue, 19th Floor, New York, New York 10016.

hmhco.com

Library of Congress Cataloging-in-Publication Data
Names: Hamilton, Nigel, author.
Title: War and peace : FDR's final odyssey, D-Day to Yalta, 1943–1945 / Nigel Hamilton.
Description: Boston : Houghton Mifflin Harcourt, [2019] | Series: FDR at war; volume 3 | Includes bibliographical references and index.
Identifiers: LCCN 2018043601 (print) | LCCN 2018051359 (ebook) | ISBN 9780544868540 (ebook) | ISBN 9780544876804 | ISBN 9780544876804 (hardcover)
Subjects: LCSH: Roosevelt, Franklin D. (Franklin Delano), 1882–1945. | World War, 1939–1945 — United States. | World War, 1939–1945 — Diplomatic history.
Classification: LCC D753 (ebook) | LCC D753 .H259 2019 (print) | DDC 940.53/2273–dc23
LC record available at https://lccn.loc.gov/2018043601

Printed in the United States of America
DOH 10 9 8 7 6 5 4 3
4500773126

Photo credits appear on page 503.

Maps by Mapping Specialists, Ltd.

Contents

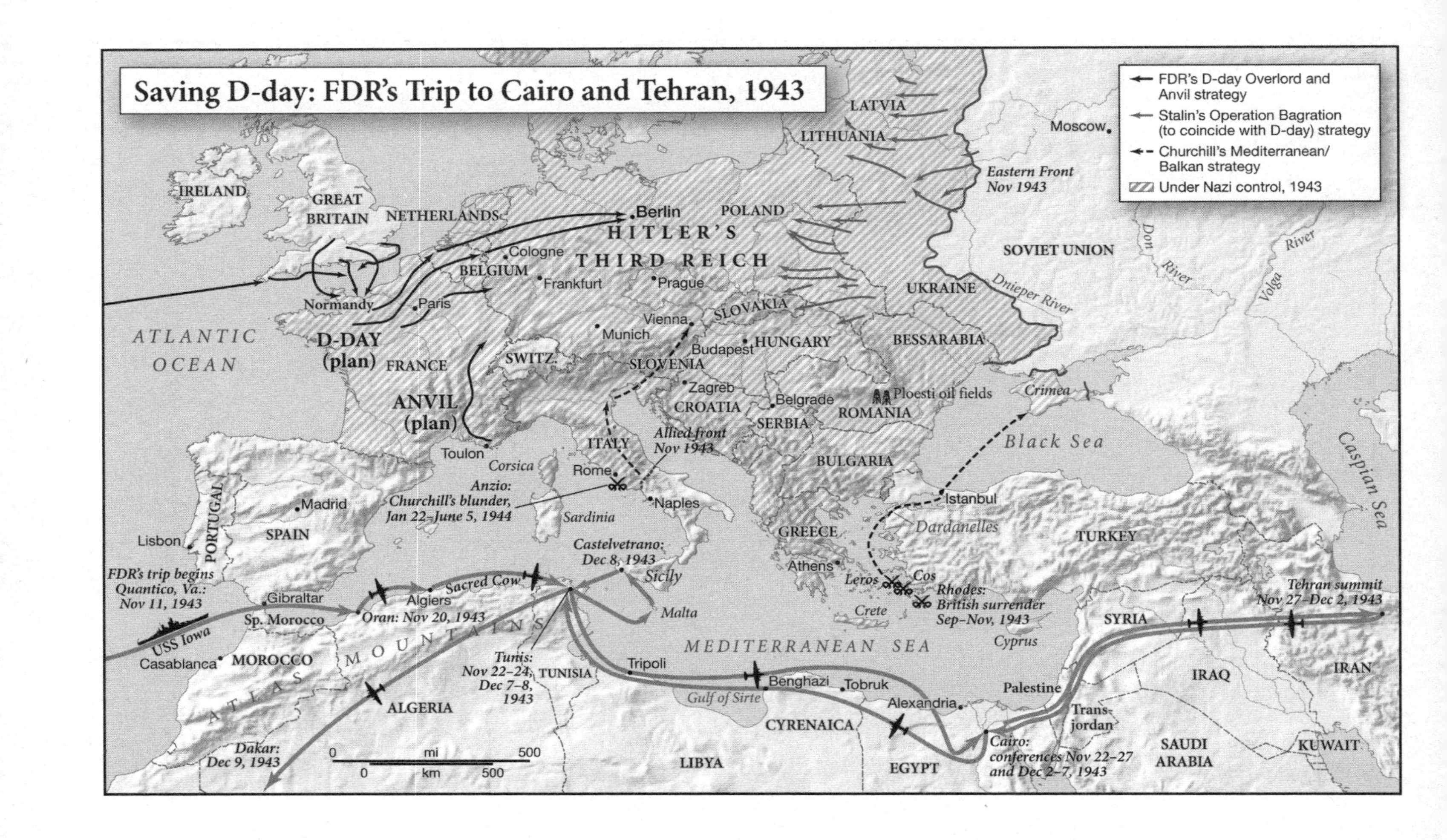

Saving D-day: FDR's Trip to Cairo and Tehran, 1943
FDR's D-day Overlord and Anvil strategy
Stalin's Operation Bagration (to coincide with D-day) strategy
Churchill's Mediterranean/ Balkan strategy
Under Nazi control, 1943
IRELAND
GREAT BRITAIN
NETHERLANDS
Berlin
POLAND
LATVIA
LITHUANIA
Moscow
Eastern Front Nov 1943
HITLER'S THIRD REICH
Cologne
BELGIUM
Frankfurt
Prague
SOVIET UNION
Don River
Volga River
Dnieper River
UKRAINE
Normandy
Paris
SLOVAKIA
Vienna
Munich
ATLANTIC OCEAN
D-DAY (plan)
FRANCE
SWITZ.
Budapest
HUNGARY
BESSARABIA
SLOVENIA
Zagreb
CROATIA
Belgrade
SERBIA
Ploesti oil fields
ROMANIA
Crimea
ANVIL (plan)
Toulon
Corsica
ITALY
Allied front Nov 1943
Rome
BULGARIA
Black Sea
Caspian Sea
Anzio: Churchill's blunder, Jan 22–June 5, 1944
Naples
Sardinia
Istanbul
Dardanelles
Madrid
SPAIN
PORTUGAL
Lisbon
GREECE
TURKEY
Castelvetrano: Dec 8, 1943
Sicily
Athens
Leros
Cos
Rhodes: British surrender Sep–Nov, 1943
FDR's trip begins Quantico, Va.: Nov 11, 1943
Gibraltar
Sp. Morocco
Oran: Nov 20, 1943
Algiers
Sacred Cow
Malta
Crete
Cyprus
Tehran summit Nov 27–Dec 2, 1943
SYRIA
USS Iowa
Casablanca
MOROCCO
ATLAS MOUNTAINS
Tunis: Nov 22–24, Dec 7–8, 1943
TUNISIA
MEDITERRANEAN SEA
Tripoli
Benghazi
Tobruk
Gulf of Sirte
IRAQ
IRAN
Palestine
Alexandria
Trans-jordan
ALGERIA
CYRENAICA
Dakar: Dec 9, 1943
0 mi 500
0 km 500
LIBYA
EGYPT
Cairo: conferences Nov 22–27 and Dec 2–7, 1943
SAUDI ARABIA
KUWAIT

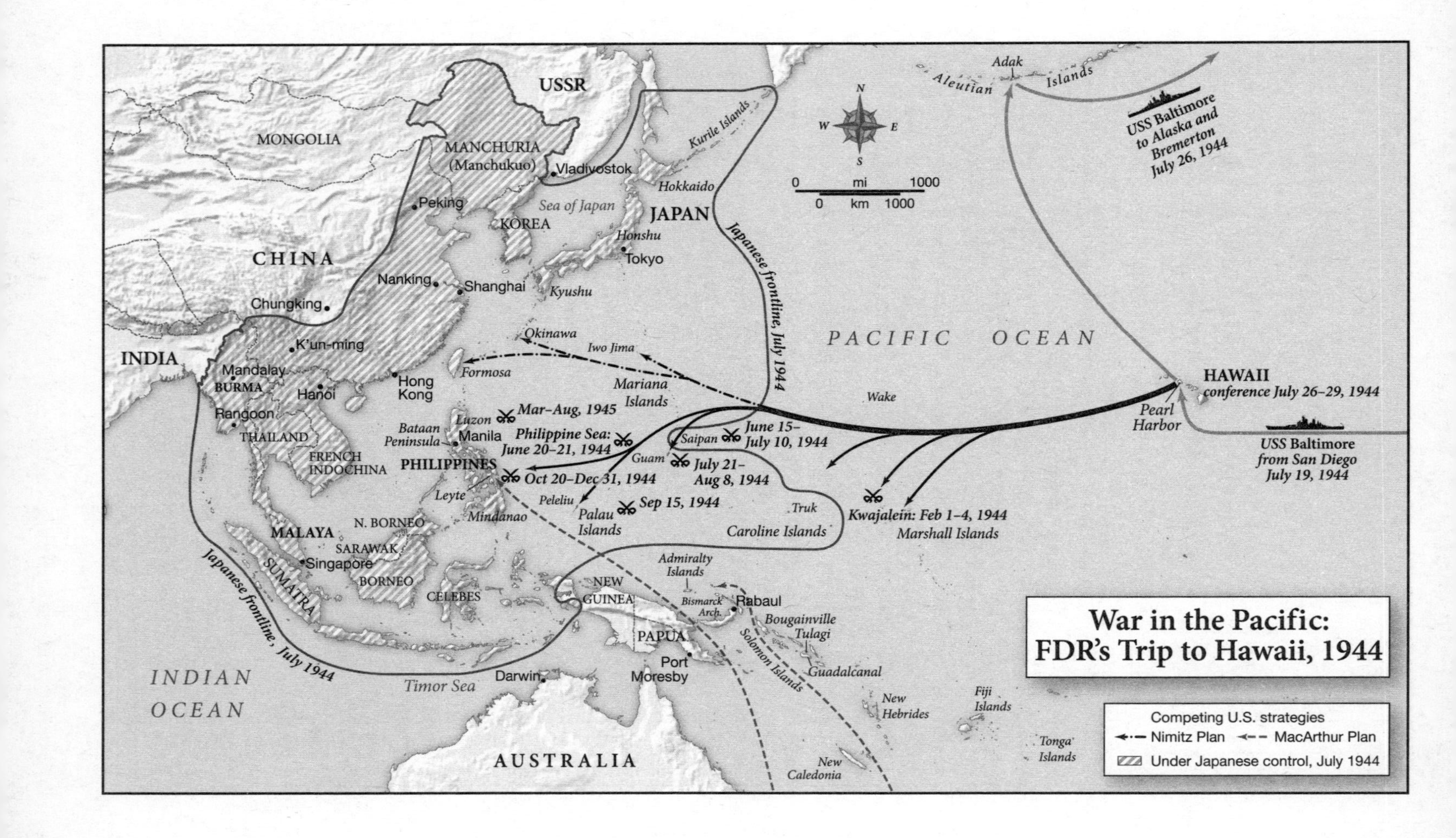
War in the Pacific:
FDR's Trip to Hawaii, 1944
Competing U.S. strategies
Nimitz Plan
MacArthur Plan
Under Japanese control, July 1944
N
S
E
W
0 mi 1000
0 km 1000
USSR
MONGOLIA
MANCHURIA
(Manchukuo)
Vladivostok
Peking
KOREA
Sea of Japan
JAPAN
Hokkaido
Honshu
Tokyo
Kyushu
Kurile Islands
CHINA
Nanking
Shanghai
Chungking
K'un-ming
INDIA
Mandalay
BURMA
Hanoi
Hong Kong
Rangoon
THAILAND
FRENCH INDOCHINA
Okinawa
Iwo Jima
Formosa
Mariana Islands
Luzon
Mar–Aug, 1945
Bataan Peninsula
Manila
Philippine Sea:
June 20–21, 1944
PHILIPPINES
Oct 20–Dec 31, 1944
Leyte
Mindanao
Saipan
June 15–
July 10, 1944
Guam
July 21–
Aug 8, 1944
Peleliu
Palau Islands
Sep 15, 1944
Truk
Caroline Islands
Kwajalein: Feb 1–4, 1944
Marshall Islands
Wake
PACIFIC OCEAN
Japanese frontline, July 1944
Aleutian Islands
Adak
USS Baltimore
to Alaska and
Bremerton
July 26, 1944
HAWAII
conference July 26–29, 1944
Pearl Harbor
USS Baltimore
from San Diego
July 19, 1944
MALAYA
N. BORNEO
SARAWAK
Singapore
BORNEO
SUMATRA
CELEBES
NEW GUINEA
PAPUA
Port Moresby
Admiralty Islands
Bismarck Arch.
Rabaul
Bougainville
Tulagi
Solomon Islands
Guadalcanal
New Hebrides
Fiji Islands
Tonga Islands
New Caledonia
Darwin
Timor Sea
INDIAN OCEAN
AUSTRALIA

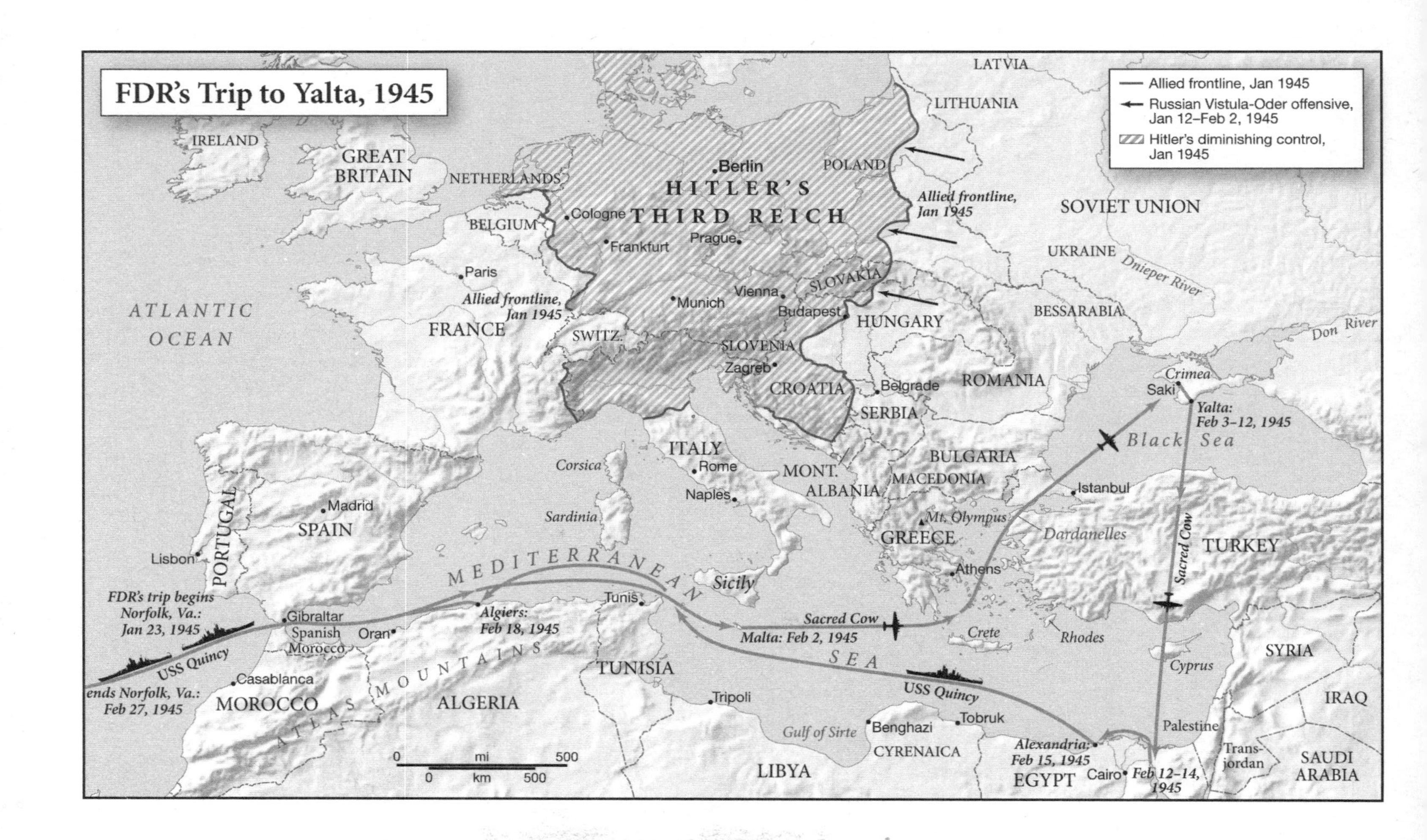
FDR's Trip to Yalta, 1945
Allied frontline, Jan 1945
Russian Vistula-Oder offensive, Jan 12–Feb 2, 1945
Hitler's diminishing control, Jan 1945
IRELAND
GREAT BRITAIN
NETHERLANDS
BELGIUM
Paris
FRANCE
Allied frontline, Jan 1945
ATLANTIC OCEAN
HITLER'S THIRD REICH
Berlin
Cologne
Frankfurt
Prague
Munich
Vienna
Budapest
SWITZ.
SLOVENIA
Zagreb
CROATIA
POLAND
SLOVAKIA
HUNGARY
Allied frontline, Jan 1945
LATVIA
LITHUANIA
SOVIET UNION
UKRAINE
Dnieper River
BESSARABIA
Don River
ROMANIA
Belgrade
SERBIA
BULGARIA
MACEDONIA
MONT.
ALBANIA
GREECE
Mt. Olympus
Athens
Dardanelles
Istanbul
Crimea
Saki
Yalta: Feb 3–12, 1945
Black Sea
Sacred Cow
TURKEY
SYRIA
IRAQ
Cyprus
Palestine
Trans-jordan
SAUDI ARABIA
Crete
Rhodes
Alexandria: Feb 15, 1945
Cairo
Feb 12–14, 1945
EGYPT
Tobruk
Benghazi
CYRENAICA
Gulf of Sirte
LIBYA
Tripoli
USS Quincy
SEA
Sacred Cow
Malta: Feb 2, 1945
Sicily
ITALY
Rome
Naples
Corsica
Sardinia
MEDITERRANEAN
Tunis
TUNISIA
Algiers: Feb 18, 1945
ALGERIA
ATLAS MOUNTAINS
Oran
Gibraltar
Spanish Morocco
Casablanca
MOROCCO
Madrid
SPAIN
PORTUGAL
Lisbon
FDR's trip begins Norfolk, Va.: Jan 23, 1945
ends Norfolk, Va.: Feb 27, 1945
USS Quincy
0 mi 500
0 km 500

Prologue

IN THE SUMMER OF 1940, following evacuation of British and French troops at Dunkirk, the new British prime minister feared an invasion of England by the German Wehrmacht: a prospect that roused him to some of the greatest oratory in history. "We shall fight on the beaches," Mr. Churchill declared before Parliament, "we shall fight on the landing grounds, we shall fight in the fields and in the streets, we shall fight in the hills; we shall never surrender."[1]

It was his greatest speech, and a singular example of leadership of a nation at war.

Three years later, thanks to America's leading contribution to the struggle against Adolf Hitler's Third Reich, the situation was reversed. Hitler was aware the Western Allies were planning an invasion somewhere along the Atlantic Wall, probably on the coast of northwest France, defended by almost a million German troops. "If they attack in the West," he warned his generals in December 1943, "this attack will decide the war." It would come in the spring of 1944, at the end of winter weather, the Führer was certain — though he had the feeling, he said, the British didn't "have their — shall we say — whole heart in this attack."[2]

Hitler was right. Churchill was fearful, and did everything he could as prime minister and quasi commander in chief of the forces of the British Empire to postpone the landings, or sabotage the invasion by embarking on foolish enterprises elsewhere. The story of how President Roosevelt held the feet of the British to the D-day fire is thus a fascinating one. And historic in that Operation Overlord *was* mounted in the spring of 1944. And, as Hitler had warned, it *did* decide the war.

Seventy-five years after the triumphant Allied landings in Normandy it seems a shame, therefore, that FDR's role as commander in chief should largely have been forgotten. For, more than any other single individual, it was Franklin Delano Roosevelt, commanding the Armed Forces of the United States, who *made D-day happen*—moreover who made the fateful decision, in person, in North Africa, over who was to lead the historic landings.

A kind reviewer, Evan Thomas, called *The Mantle of Command,* the first volume of my FDR at War trilogy, "the memoir Roosevelt didn't get to write."[3] While the trilogy is of course a biography rather than a memoir, Thomas's point was that, unlike Winston Churchill, who went on to write his war memoirs, in six grand volumes, FDR did not live to do so. Dying of a cerebral hemorrhage two weeks before Hitler committed suicide, Roosevelt was unable to give his own account of the trials and tribulations he'd faced as U.S. commander in chief since Pearl Harbor, almost three and a half years earlier.

Like Winston Churchill, Franklin Roosevelt had, however, begun to assemble the necessary papers he would need to write his own war story—papers his secretaries had filed in the Oval Office, as well as the secret communications kept under strict security in the White House Map Room. And those, too, that he'd amassed at the presidential library he'd had built beside his home at Hyde Park, and which he proposed to donate to the nation. Four months after the D-day landings, in October 1944, he had even begun to use a young Harvard graduate, Lieutenant George Elsey, the officer in charge of the Map Room team staff, to help him prepare the library to safely house his secret papers from the Oval Office. As Elsey later recalled, "It was remotely possible that [Roosevelt] might lose the [1944 presidential] election to [Thomas] Dewey. He considered it a possibility and he wanted to know what should be done at the library to ensure the safety of his wartime records and papers, which would automatically go with him to Hyde Park. What should be done to protect them"—since, in terms of "the top-secret stuff," the library "was not built with that in mind. So he took me up there, to the library, to make recommendations."

Elsey was later told by the President's naval aide, Admiral Wilson Brown, that there was a reason Elsey had been recommended for the job: that Lieutenant Elsey, "with his background, would be very, very valuable and helpful in doing [FDR's] memoirs." In retrospect, Elsey—who had just prepared his own eyewitness account of the D-day landings and several long reports for the President on U.S. allies such as China—realized "part of the reason why [Roosevelt] spent a good deal of the time with me, talking with me, I think, was that he was just getting better acquainted: sizing me up" for the job of literary assistant.[4]

At the same time, Winston Churchill, too, was contemplating the removal from 10 Downing Street of all the wartime papers he would need to document his leadership in the course of the war—a task in which he would employ a bevy of assistants (military subordinates and the secretary to his wartime cabinet, a group later known as the Syndicate) to help him research and draft his account. Thus when, at the military conference held in Quebec in September 1944, the Canadian prime minister had said he hoped Mr. Churchill would, after war's end, record the saga from his special vantage point as British quasi commander in chief, Churchill had "replied that all of this was practically in his papers already except to give it a certain turn" — assuring Mackenzie King that, in his narrative, he would, in due course, be completely frank, omitting "nothing."[5]

Had Roosevelt survived and published his own account — "He will 'write,'" his cousin Daisy Suckley had noted in September 1944, "and can make a lot of money that way"[6] — the world would have been treated to an interesting postwar literary duel, for FDR's account would have differed markedly from that of Sir Winston (as Churchill became in 1953). The highlight of the tourney would undoubtedly have been their competing accounts of D-day: its inception; the long battle between the two great leaders over its mounting; and its ultimately decisive consequences for the successful prosecution of the war against Hitler.

That literary duel, of course, did not take place. It was the President who died prematurely, at sixty-three, while the aging former prime minister — forced from high office at seventy in the 1945 British general election — devoted himself in his multivolume memoirs to a magisterial if also self-lauding account of how he, rather than Roosevelt, had directed and won World War II.

Given that Sir Winston — whom I deeply admired, and with whom as a college student I proudly stayed for a weekend at Chartwell, his home in Kent, before he died — won the Nobel Prize for Literature in 1953 on the basis of his sextet, and *was* able to "make a lot of money" publishing it, I do not think it unfair to his memory, sixty-five years later, to correct the record regarding his version, and as a historian and biographer to give my own account of the war's military prosecution from President Roosevelt's perspective as U.S. commander in chief.

War and Peace: FDR's Final Odyssey, D-Day to Yalta, 1944–1945, thus commences with the President's voyage abroad to save D-day: averting Churchill's last-ditch efforts to delay and sabotage the landings.

Following that triumphantly successful trip to Tehran, the American Odysseus — a pseudonym FDR had once used at school at Groton — returned to the United States "in the pink": D-day finally and irrevocably set to be mounted in the spring of 1944, under an American general he had personally chosen "on the spot." The invasion, code-named Overlord, would be backed, Stalin had promised him, by a major Russian offensive on the Eastern Front — thus making it impossible for Hitler to reinforce his Wehrmacht armies along the Atlantic Wall. It was the highest point of FDR's career as U.S. commander in chief in World War II.

The Normandy landings, beginning on June 6, 1944, would more than fulfill the President's hopes, and his historic "D-day Prayer," as broadcast to the nation. General Eisenhower's success disproved all Churchill's fears and earlier warnings of disaster; the Allied invasion became, as Hitler had predicted, the "deciding" battle of the war.

Tragically, however, the man who had been ultimately responsible for the Allies' war-winning strategy was unable to take full credit for it. The U.S. commander in chief had fallen gravely ill with flu on his return from the Middle East. And he never got better—a reality his White House doctor, his press secretary, and his staff did their best to hide from snooping reporters and cameramen. For all that he might dream, following the greatest amphibious invasion in history, of setting foot on the shores of liberated Normandy as his U.S. chiefs of staff were able to do, it was not to be. Instead he grew more and more sick with heart disease—more seriously ill, in fact, than any but his closest confidants knew. As Lieutenant Elsey later recalled, the President's health simply plummeted—"he lay in bed, unable to focus or to take an interest, often—just too ill to bother. This was obvious before D-day, and before the fourth election," but it had to be kept from the public lest it affect war morale.[7]

The President never recovered—though he did, miraculously, survive to see at least that the war would be won, and how. As a result, though, his last year as commander in chief was the very opposite of the manner in which he had led the Allies since Pearl Harbor.

Sadly, many historians—too many—have judged Franklin Roosevelt's military role in World War II on the basis of his final year in the Oval Office, thus entirely misconstruing his singular, overarching contribution to victory over the Third Reich and the Empire of Japan. My aim in *War and Peace: FDR's Final Odyssey,* as the culmination of my FDR at War trilogy, therefore, has been not only to chart with fresh clarity how dire was his affliction, but how exactly it affected his decisions and once masterly performance as commander in chief of the Western Allies in World War II after Pearl Harbor. It was a record that Winston Churchill—who had been a great national leader, but a deeply flawed commander in chief of British Empire forces—was only too happy to take from him in literary retrospect.[8]

From triumph to personal tragedy is thus the theme undergirding this volume—for the fact is, from being the victorious captain of the Allied vessel in the momentous struggle against Germany, symbolized in the great D-day landings, the President became thereafter a virtual passenger—with major ramifications not only on the prosecution of the war, but the postwar.

That the President lived long enough to see the approach of unconditional Nazi surrender was, at least, a consolation to those intimates who knew Roosevelt best, and who had seen him *at* his best. As was the sight of him preparing to inaugurate his great contribution to postwar peace: the United Nations. Moreover the fact, too, that he had been able, at least, to die deeply contented in his personal life, despite his cascading ill health. I hope readers of these three volumes will feel the same in coming to the end of this trilogy, seventy-five years after the triumph of his greatest military achievement: D-day.

Book One

PART ONE

Going to See Stalin

1

A Trip to the Mediterranean

THE WEATHER OFF ARGENTIA, Newfoundland, in the fall of 1943 was "boisterous," Captain John McCrea later remembered, recalling his command of the U.S. Navy's most powerful new battleship, the USS *Iowa.*[1] The huge "battlewagon" was alternately resting at anchor in the northern Atlantic and carrying out oceangoing exercises for its assigned role: guarding U.S. convoy routes to and from the war in Europe and Russia.[2]

Launched as the first of its class in January that year and drawing some forty feet in depth, the monster vessel had hit a submerged rock in July, while entering Casco Bay, Maine, at low tide. It had been only its second shakedown cruise, and a fifty-foot gash had been torn in its hull, requiring weeks of repair. Not a good omen.[3]

McCrea, as the ship's captain, had been reprimanded but not dismissed. He had, after all, been naval aide to the President and Commander in Chief until taking command of the huge vessel, and the gash was swiftly repaired. In October 1943 the *Iowa* had finally assumed its active role in North Atlantic duties. Manned by more than two and a half thousand sailors and Marines, weighing fifty-five thousand tons when fully loaded, the vessel could make more than thirty-two knots and boasted sixteen-inch guns that could hit targets twenty miles distant. Among the crew there was hope of an imminent naval engagement — perhaps even a sea duel with the vessel's equivalent: the Bismarck-class battleship and biggest warship of Hitler's Third Reich, the *Tirpitz.*

Such a sea battle was not improbable. Armed with eight fifteen-inch guns, the *Tirpitz* had constantly threatened U.S. and British convoys to Murmansk in aid of the Russian war effort and menaced the North Atlantic — in fact the *Tirpitz* had only weeks before slipped its own moorings on the coast of Norway and made a daring, Viking-like raid. Leading a German armada on September 8, 1943, the battleship had carried out an amphibious operation, executed with total surprise: Zitronella, as it was code-named, an attack from the sea that completely destroyed the Allied weather station at Spitsbergen and a refueling base there, and captured many of its defenders.

In bright sunshine and an unusually gentle wind the USS *Iowa* had just weighed anchor, therefore, to undertake further exercises out at sea when the executive officer brought Captain McCrea a message, marked SECRET. It was from Admiral Ernest J. King, commander in chief, U.S. Fleet. The *Iowa* was ordered not to steam toward Norway and possible action, but directly south, back to United States waters.

Captain McCrea had not the faintest idea what was afoot, or afloat. On McCrea's arrival at Hampton Roads naval base, in Virginia, Admiral Royal Ingersoll, commander of the U.S. Atlantic Fleet, greeted him tartly: "I suppose you know why you are here?"[4]

McCrea shook his head.

"You are going to take the President and the Joint Chiefs of Staff in *Iowa* to the Mediterranean."

The entire American high command? To the Mediterranean?

Captain McCrea took a deep breath. He had, to his credit, considerable experience in handling VIPs. As naval aide to the President he'd even accompanied Mr. Roosevelt on the President's historic seaplane journey to North Africa earlier that very year.[5] There, at Casablanca on the shores of Morocco, the military strategy of the Western Allies had been laid down by the President as U.S. commander in chief, overruling the advice of his chiefs of staff.[6] American forces, he had instructed, would first learn *how* to defeat the Wehrmacht in battle, in the Mediterranean, and only then launch a cross-Channel attack, probably in the spring of 1944—aimed at Berlin.[7]

By October 1943 the year of live-action learning was now almost over. More than a quarter of a million Axis troops had surrendered to General Eisenhower in May 1943, and by July Allied troops were ashore in strength on the island of Sicily—stepping-stone to southern Europe. As Italian troops capitulated or ran away, Wehrmacht units were unable to stem the Allied tide of advance toward Palermo and Reggio, barely three miles from the Italian mainland. The Italian leader, Benito Mussolini, had thereupon been toppled as the Fascist dictator of Italy by his own colleagues; having successfully deposed him, the new Italian government under Marshal Badoglio had surrendered unconditionally.

The war Hitler had declared on America on December 11, 1941, was thus moving toward its climax. What exactly the President had in mind in crossing the Atlantic for a second time that year was unknown to Captain McCrea, as it was to everyone else. But to find out more about the *Iowa*'s role in the President's latest mission, McCrea was ordered to go to the Navy Department in Washington, D.C., and report to Admiral King in person, "tomorrow morning."[8]

Thus began secret preparations for the Commander in Chief of the Armed Forces of the United States to meet, for the first time, the dictator and commander in chief of the Soviet armies, Joseph Stalin.

2

The Meeting Is On

ALL YEAR PRESIDENT ROOSEVELT had pressed Stalin for a meeting — even sending the former ambassador to the Soviet Union, Joseph E. Davies, to Moscow to negotiate a possible rendezvous.[1]

Stalin had several times indicated his willingness to parley. In the end he had always backed out, claiming his responsibilities as commander in chief of all Soviet forces in action against the Wehrmacht were too demanding, and that the bitter battles on the Eastern Front simply precluded him from leaving Russia.

The President had not believed him (Stalin was known to have visited the battlefront only a single time, and only for a day), but had persevered. Hitler's massive invasion of the Soviet Union in 1941 had failed to reach Moscow, and though Wehrmacht troops had neared the Urals in the summer of 1942, that offensive, too, had come to grief. Stalingrad, in February 1943, had marked the end of German advances on the Eastern Front — with Hitler calling off his final massed armored offensive at Kursk in the summer of 1943 when hearing that the Western Allies were ashore in Sicily — threatening to bring down the Pact of Steel and forcing Hitler to divert armored divisions to Italy, not Russia.[2]

The Soviet Union, facing two-thirds of the German army and beginning to defeat the Wehrmacht decisively in battle, would be a major player in the postwar world, the President was well aware. Finding a modus vivendi with Russia would be in America's best interests.[3] It would also be in the interests of the postwar world, too, if a better security system than was set up after World War I could be devised and agreed.[4] Neither Russia nor the United States had been a member of the League of Nations, which Hitler had effectively neutered when announcing Germany's withdrawal in 1933. This time, however, the United States and the Soviet Union could together take a leadership role in guaranteeing postwar peace.

Whether the Soviets would agree to work with the United States and other capitalist nations in the aftermath of the Second World War was another matter. Despite Stalin's radical efforts to industrialize the Soviet Union and mechanize Soviet agriculture, the USSR remained a largely agrarian society, its people ruled by communist dogma, dread of the secret police, fear of forcible resettlement or execution, and paranoia regarding foreign countries. Such paranoia was not only of Stalin's making, however. Hitler's Barbarossa invasion of June 22, 1941, fielding 4 million men drawn from Germany and many other European countries (Austria, Italy, Romania, Hungary, Slovakia, Croatia, Bulgaria, France, and Finland), had resulted in more than 3.5 million casualties by the end of that first year alone. The failure of the United States and Britain to mount a Second Front in 1942, or the year after that, had only increased Russian xenophobia and suspicion. Distrust of those nations bordering the Soviet Union that had aided Hitler in Barbarossa would probably affect relations with Russia for a generation or more, former ambassador Davies had reported to the President on his return from his personal mission to Moscow on behalf of the President in June 1943.[5]

Distrust was one thing. Need, however, was another. Aware the Soviet Union could not continue the war without U.S. Lend-Lease — American exports that were currently supplying more than 10 percent of Russia's war needs — Stalin had closed down the Comintern (the communist, ideological arm of the Soviet Union). In a further, somewhat dubious, demonstration of commitment to the principles of President Roosevelt's Four Freedoms and the Atlantic Charter, the Soviet dictator had even opened Russian churches in the summer of 1943.

But Lend-Lease alone would not be enough to defeat Hitler's legions. Facing no fewer than 260 Axis divisions on the Eastern Front — the majority of them German — the Soviets would never be able to defeat the Third Reich without a Second Front in the west, which would force Hitler to fight major campaigns front and back. Stalin had therefore every interest in Allied military strategy being concerted, which a meeting with President Roosevelt might achieve. At the same time there must be other reasons why it seemed advisable for Stalin *not* to meet with the President, Mr. Roosevelt had mused when discussing the conundrum with his advisers and the people around him. Perhaps his Russian counterpart suffered an "inferiority complex," the President had suggested when talking to his cousin Daisy Suckley.[6] And the Marshal was reported to be deathly afraid of assassination — a fear reflected in his unwillingness to leave the Kremlin. Given his ruthless dictatorship of the USSR since Lenin's death, this was certainly wise. At any rate, whatever the underlying rationale, it had left the coalition of Allies with a sort of vacuum at the very top, in terms of combined military and political strategy — Stalin constantly agreeing to meet and then backing off, offering to send representatives instead.

Finally, in October 1943, Stalin had agreed to a two-week-long international conference of foreign ministers in Moscow. This had produced surprisingly positive results, including provisional agreements between U.S., Soviet, and

British diplomats. Insisting on flying via Africa to Moscow, despite his ill health, Mr. Cordell Hull, the U.S. secretary of state, was even able to get quasi-formal Russian consent to the establishment of an eventual United Nations organization, as well as a private indication from Stalin that the Soviets would join the United States' war against Japan, once Hitler surrendered.

Why, in this case, was it actually *necessary* for Stalin to leave the security of Moscow for a personal meeting with the President of the United States? Why did the President still place so much store by a personal meeting? Why was the President pressing for him to leave Moscow?

Historians, later, tended to see the summit in Tehran—a venue that was ultimately agreed upon only at the last moment, three weeks in advance—as the final compromise in a game of statesmanship, or brinkmanship: the location a matter of competing national honor or pride. To a large extent this was true. In Russian eyes, the U.S. secretary of state had, after all, flown all the way to Moscow—why not, Soviet officials had calculated, the President?

Behind the competing arguments over the venue for this first meeting of the two world leaders, however, there was another reason that would largely elude historians—and certainly eluded the Prime Minister of Britain, the "third wheel" in the proposed conference, at the time.

It did not elude Stalin, though—indeed was, in the end, the reason the Russian premier overcame his reluctance to leave the Soviet Union, or even Moscow: the matter of D-day, as the Second Front landings became known. For without a Second Front the war could not be won.

The Marshal, to be sure, hated the idea of leaving the motherland, where he was secure as an absolute dictator. He was also afraid of flying. But if the British were plotting to back out of D-day, as he'd been given reason to believe, then he had better meet with his two fellow warlords in the struggle to defeat the Third Reich. The Russian premier had thus decided he would go meet the President at a halfway house. The capital of Iran, a "neutral" country across the Caspian Sea, was as far as he would countenance, however. It was currently under Russian and British occupation; it was also a vast American military supply entrepôt—in fact the primary conduit for U.S. Lend-Lease supplies by sea and rail to the Soviet Union.

The three U.S., Russian, and British legations or embassies in Tehran, then, could provide an ideal location for the ultimate high-level, coalition military conference held to decide the Allies' military and political strategy in 1944, in the midst of the most violent war in military history.

The President had accepted Stalin's reasoning. Churchill had reluctantly agreed to the location, too. As prime minister of America's closest military ally he'd insisted, however, that he first meet with President Roosevelt in North Africa or somewhere in the Mediterranean. There, *before* the Tehran summit, he hoped to dissuade the President from following the current plan, agreed at Quebec in August 1943, for the Allied D-day invasion to be mounted in the spring of 1944. Instead Churchill wanted the United States to exploit his latest brainwave: an alternative Mediterranean strategy.

Thus was the stage set for one of the most consequential showdowns in military history—with the President the only man powerful enough to stop Churchill from wrecking Overlord, the agreed battle plan to defeat Nazi Germany.

3

Maximum Secrecy

FOR DAYS SHIPS' CARPENTERS labored on the *Iowa,* as they prepared the huge vessel for its top-secret mission. The crew were given three days' leave.

In the nation's capital McCrea paid a visit to 1600 Pennsylvania Avenue, and saw his old boss at the White House. "I suppose Ernie [Admiral King] has told you about the coming trip?" the President asked McCrea as they sat together in the Oval Office.[1]

Admiral King had — to McCrea's memorable discomfort. At the Navy Department on the Mall (the Navy still refusing to join the Army in the recently completed Pentagon building), the *Iowa*'s captain had innocently remarked that, after delivering the President to the war zone, the battleship would presumably leave North Africa in order to return to combat duties. There was "a moment of silence — ominous silence," McCrea remembered.[2]

Reputed to "shave with a blowtorch," the balding, bullet-headed commander in chief of the U.S. Navy was "strictly business," a man "of few words" — each enunciated with "precision," McCrea later recalled in admiration. There was "never any doubt as to what he said or what he wanted. I can't imagine anyone ever slapping him on the back by way of greeting. He was austere. He was ramrod straight and carried himself with dignity. In my book he was everything a naval officer should have been," McCrea lauded the admiral's memory — discreetly passing over King's well-known penchant for other men's wives.[3] What counted, in McCrea's view, had been the admiral's complete and unequivocal loyalty to the Commander in Chief of the Armed Forces of the United States: the President.

"I shall not subject the President to a Westward, December crossing of the Atlantic in a Heavy Cruiser," King had snapped — which had been the plan McCrea had been informed of earlier. "*Iowa* will remain available for that duty," King ordered, speaking "with a firmness and coldness" that made McCrea regret he'd opened his mouth.[4]

McCrea would thus transport the President to North Africa—and collect him, after he was done.

At the White House the President seemed "in fine fettle," McCrea later recalled. "He looked to be in good health. He had a great zest for living which was contagious," and seemed full of beans. "I am looking forward to it greatly," Mr. Roosevelt remarked—assuring McCrea that in his view there was "nothing so wonderful as a sea voyage by way of relaxation."[5]

A sea voyage in the midst of a world war—in spite of a resurgence of German U-boat attacks on Allied shipping? McCrea was not so sure, but said nothing. As to the departure date, Admiral King had said the President wished first to attend the annual ceremonies of the Unknown Soldier at Arlington Cemetery on Armistice Day, November 11, 1943—a "must." Immediately afterwards, however, he would take his special armored train, the *Ferdinand Magellan,* to Newport News, and stay on it overnight, in order to board the *Iowa* early the next morning, November 12. Admiral King would travel with the President—as would the other chiefs of staff: Admiral William D. Leahy, the President's White House chief of staff as well as chairman of the U.S. Joint Chiefs of Staff; General George C. Marshall, chief of staff of the U.S. Army; and General Henry H. Arnold, chief of staff of the U.S. Army Air Forces. Also General Brehon Somervell, chief of U.S. Army Service Forces, plus dozens of the chiefs' staffs.

McCrea had taken a deep breath. The entire high command of the United States military, to an active theater of war. On one ship!

Secrecy would thus be of paramount importance.

In Hampton Roads the USS *Iowa* was repainted Atlantic-gray. It was also provisioned for the transatlantic journey. Unusual features were added, too. On Sunday afternoon, November 7, "a derrick lighter came alongside and hoisted two very heavy steel battleship-grey elevators aboard. These were welded at once to the *Iowa*'s starboard main and superstructure decks, almost amidships. One elevator rose from the main to the first superstructure deck and its companion was placed inboard, terminating at the flag and signal bridge level."[6]

As one of the officers remarked, "Anyone with half an eye can guess the nature of our next assignment."[7]

Finally, on November 11, 1943, the *Iowa,* having discharged much of its fuel and drawing significantly less water, anchored a few miles up the Potomac River—for McCrea had persuaded Admiral King and the President to repeat the Potomac ruse of 1941. On that occasion, still in peacetime, the President had pretended to go out fishing on his presidential yacht in order to shake off journalists; once at sea he'd secretly transferred to a warship, to meet with

Churchill at Argentia, where he'd gotten Churchill to sign on to the Atlantic Charter.[8]

The *Iowa* positioned off Cherry Point, Virginia, Captain McCrea now ordered the entire ship's crew to assemble on the main aft deck, and explained what for the most part they had intuited: that they'd been selected "by the President and our Commander in Chief to participate in an important mission." That evening, Admiral King's 258-foot yacht and flagship, the USS *Dauntless* (formerly the privately owned yacht *Delphine*), came alongside, bearing the U.S. chiefs of staff and their staffs — though not the U.S. commander in chief, or Admiral Leahy. One by one the officers were taken to their cabins, and aboard the great battleship they spent the night at anchor.

Roosevelt, aboard the 168-foot USS *Potomac,* meantime sailed all night from the Washington Navy Yard — with no prying journalists following. He'd received Stalin's confirmation, cabled on November 10; the meeting in Tehran was on.[9]

"It was a grey morning and cold" at Cherry Point, the *Iowa* ship's chronicler afterwards recorded. The river looked distinctly uninviting. "A chill wind whipped the surface of the Potomac into fluttering white caps. At fifteen minutes before nine, the tiny, white Presidential Yacht, *Potomac,* hove into distant view. As she drew near the *Iowa*'s starboard side the seal of the Chief Executive could be observed on the bridge. A gangway was rigged between the *Iowa*'s side and the *Potomac*'s main decks aft. The President of the United States boarded the *Iowa* at sixteen minutes past nine to be welcomed by Captain McCrea. Mr. Roosevelt had requested that honors be dispensed with. He wore a soft, brown hat and his famed sea cape over a tan sharkskin suit." Sitting in his wheelchair the President was then taken to the captain's quarters "on the starboard side, forward, of the first superstructure deck."[10]

The President was not alone. This time he was accompanied by both his military aides: Major General Edwin "Pa" Watson and naval aide Rear Admiral Wilson Brown. Also his chief of staff, Admiral William Leahy. His dedicated White House counselor, Harry Hopkins, boarded too. And Admiral Ross McIntire, the President's personal physician. They were all now present — with the President occupying the captain's quarters, and McCrea moving to his seagoing cabin in the conning tower above.

McCrea was fully aware of his responsibility. On the shoulders of the President and his senior military subordinates rested the strategy for the ultimate prosecution of the war in Europe: the endgame. In this sense the journey promised to be more historic even than the President's Atlantic Charter meeting in 1941, or his flight to Casablanca in January 1943. Stringent secrecy remained McCrea's major concern, but there were other worries also. Admiral McIntire was not keen on the onward plane journeys that would necessarily follow, perhaps at high altitudes, once the ship reached North Africa — a concern because of the President's heart. Assuming the *Iowa* even got there! Safety was hardly guaranteed. Neither of the President's military aides had been eager for Mr. Roosevelt to travel by sea, since a transatlantic voyage would be subject to U-boat interception. Once in North Africa air travel would, moreover, be subject to Luftwaffe attack. The latter was no idle threat, either; American pilots had, after all, ambushed and shot down the commander in chief of the Japanese navy in the Pacific, Admiral Yamamoto, *four hundred miles* behind Japanese frontlines earlier in April that very year, killing him.[11]

The President's proposed flight route would be from Oran to Cairo, along the shores of North Africa. From there he would fly to southern Iran, and from there the plan was to travel north to Tehran either by air or by train. "Those of us who had to do with the planning for this expedition were very conscious that the President was running grave personal risk," Admiral Brown later wrote candidly, "because we believed that if the enemy could learn of his whereabouts they would spare no effort to attack by air, submarine or assassin. Even with the strictest censorship, rumors of his activities and whereabouts," he explained, "were almost certain to leak out."[12]

By the early summer of 1943 the U-boat menace had been all but extinguished, thanks to U.S. and British air patrols across the Atlantic. But thanks to new German "snorkels," which permitted U-boats to remain submerged and avoid Allied air reconnaissance, the Atlantic battle had now resurged with wolf-pack vengeance. The undersea menace was not all. Above the waves a "new destructive glider-bomb" had been put into combat, "raising havoc" against Allied shipping in the Mediterranean.[13]

"Havoc" was no exaggeration. Often forgotten after the war, the new German guided missile, the "Fritz X" bomb, had been employed for the first time against Allied naval forces during the U.S. invasion of Italy in September 1943 — the latest example of German engineering. It was the world's first effective remote-controlled combat missile, able to penetrate battleships and heavy cruisers. The powerful projectile was steered by *Funkgerät,* or radio remote-control, operated by a technician aboard a motherplane at eighteen thousand feet behind the flying bomb, aiming it directly toward the target. During the Salerno landings on September 9 two Fritz X bombs had sunk the Italian battleship *Roma* as it attempted to abscond and join the Allies — killing the majority of its 1,850-man crew in the process. Another Fritz X had disabled *Roma*'s sister battleship *Italia.* Hits were also scored on the light cruiser USS *Savannah,* killing two hundred American sailors off Salerno — forcing the cruiser out of action and requiring it to return to the United States for twelve months. The USS *Philadelphia* (on which the President had cruised in 1938) was narrowly missed by a Fritz X the same day, September 11. It was fortunate in being only lightly damaged, but HMS *Uganda,* a British light cruiser, was not so lucky on September 13. It received a direct hit by a Fritz X that penetrated through to its keel—a fate that befell HMS *Warspite,* too, a British battleship that was crippled on September 16 along with a number of Allied supply vessels.

New low-level Henschel Hs 293 guided missiles, also steered from accompanying Luftwaffe motherplanes, were sinking yet more Allied naval vessels in the Mediterranean. With one hit on the USS *Iowa,* once it entered Mediterranean waters through the Gibraltar Straits, the Germans could, theoretically, destroy the entire U.S. high command.

For his part Mr. Roosevelt had refused to listen to such fears. His hands were a trifle shaky, his feet swelled when he was tired, and he was "keyed up and couldn't relax," his cousin and loyal companion Daisy Suckley noted in her diary several days before the President's departure,[14] but nothing would stop him now that Stalin had agreed, definitively, to meet in Iran; there was, after all, simply too much at stake.

4

Setting Sail

DAISY WAS YET another person who hated the idea of a second presidential trip abroad that year. Not only was it dangerous for the partially paralyzed president to travel to an active theater of war, in terms of possible enemy interception, but a flight to Tehran — which was as far as Stalin would agree to travel — involved an ascent "over the mountains of up to 15,000 feet," traversing the Zagros range, and inevitably taxing the President's suspect heart, as she'd noted in her diary on October 30, 1943.[1]

Roosevelt's hands may have developed a noticeable tremor, but mentally the President seemed full of ginger and anticipation — pulling down a map of Africa and the "Bible lands" of the Middle East, in his White House study, and showing Daisy his itinerary for the "Long Trip." From Oran, in Algeria, to Cairo, in Egypt, by air; then again by air over Palestine and Iraq, landing in Iran to take the train. He seemed excited at the thought of being far from Washington, with its endless political backbiting and the further loss of Democratic congressional seats suffered in last year's November elections. Also a possible union coal strike to contend with. His adoring cousin had therefore felt torn — the more so since Eleanor, the President's loyal but exceedingly busy spouse, seemed to pay too little attention, in Daisy's opinion, to his precarious health.

Sitting together, they talked about domestic matters — both private and public. Eleanor had become, in a sense, his eyes and ears with respect to national issues, given her many travels across the country: a pleader of social, economic, and political causes, she would often spend an hour or more making her case for action late at night, when she said goodnight to him. Daisy said nothing, but was concerned. "He is just *too* tired, *too* often. I can't help worrying about him," she confided to her diary. On November 8, for example, he'd had some twenty-three appointments in the Oval Office — even before being subjected to Eleanor's near-midnight entreaties.[2]

Worry or no worry over safety, the Long Trip to Tehran was now beginning.

The USS *Potomac* having cast off, the President greeted his Joint Chiefs of Staff and his White House staff in his comfortably large *Iowa* cabin. Whether the "fishing trip" ruse would really work was anybody's guess, but once the battleship weighed anchor no one would know where they were, Admiral Leahy reflected. "Every effort has been made to prevent the leak of any information in regard to the expedition," he noted in his diary that evening; "we will have no communication whatever with the shore, and it is hoped that the President with his staff can succeed in reaching the port of Oran in Africa before the enemy learns of his whereabouts and his intentions."[3]

The *Iowa* thus duly departed from Cherry Point the morning of November 12, 1943. Arriving at Hampton Roads in the evening, it then anchored once more, in darkness, to be refueled in deeper water by two tankers at Berth B, since the ship had pumped out much of its oil in order to have the shallower draft necessary for a Cherry Point rendezvous. Once the tanks were refilled the Commander in Chief asked the ship's captain to wait a few more hours. "You know John," he said to McCrea, "today is Friday. We are about to start on an important mission. Before it is over, many important decisions must be made. I am sailor enough to share the sailors' superstition that Friday is an unlucky day. Do you suppose you could delay getting underway until Saturday—this, of course without interrupting your plans too much?"[4]

Marveling at the Commander in Chief's attention to nautical lore, McCrea therefore ordered the battleship to remain anchored—but with its chains shortened. Its course was designed to take them toward Bermuda, then the Azores, then to continue across the remaining waters of the Atlantic until they reached the narrow Straits of Gibraltar, escorted by successive groups of three destroyers (which carried less fuel), watching for U-boats.

Assuming all went well, the last American destroyer squadron, approaching North Africa, would be "beefed up" with additional British destroyers as well as a U.S. aircraft carrier, or flattop. In order to maintain absolute secrecy regarding the *Iowa*'s passengers, however, and the destination of the voyage, "no information" had been given even to the commander of the first destroyer escort, other than the charge to protect the *Iowa*, and then be ready for "distant service."[5]

Copious amounts of presidential china and silverware had been brought along for use on the ship, to be unloaded and taken by air with the presidential party, once the passengers were transferred to land. Liquor was also aboard—the President skirting the prohibition against alcohol on U.S. naval vessels by claiming that, as U.S. commander in chief, he could do as he wished. This meant he could have his ritual martini "cocktail hour" at the end of the day. He'd also arranged with Captain McCrea that he would have exclusive use of the outdoor first superstructure deck, port and starboard, where he would spend every afternoon thinking, reading, or napping.

Or making notes in the new journal Daisy had given him for his trip — one of the few diary records Roosevelt would keep during the war. As the *Iowa* thus made its way down the Hampton Roads channel toward the ocean on November 13 — the huge warship piloted by an experienced Coast Guard lieutenant commander — the President inked his first entry in his tall, forward-tilting, emphatic script, with its characteristically high-crossed *t*'s and bold capitals.

"This will be another Odyssey," Franklin Delano Roosevelt penned, thinking of Homer's Odysseus the Cunning — or Ulysses, as the Romans called him. The long journey would take place "much further afield & afloat than the hardy Trojan whose name I used to take at Groton when I was competing for school prizes. But it will be filled with surprises," he predicted — aware that any number of things could go wrong.[6]

The prospect of a week's leisure at sea certainly buoyed the President, however. "Enough motion-picture films for a show every night" had been brought with them, according to the signals lieutenant tasked with drawing up a checklist, plus a "well-stocked library of pocket guides, whodunits, and other reading matter."[7]

"I'm safely on my way," the President wrote Daisy in a last letter he left for her at Hampton Roads. "UJ [Uncle Joe, i.e., Stalin] wired he will meet us, so you know all," he confided — thanking her for her gift of a special chemically induced hot-water maker for the high-altitude flight to Tehran — which now looked like a better option than the train. "It's a lovely day but cold — I am being showered with every security and every comfort — & I am optimistic about results. I much wonder when the cat will get out of the bag!"[8]

Time would tell. He turned in sometime after 10:00 p.m.

They were off.

5

Sheer Madness

THANKS TO LAX British censorship in Cairo, the world was to find out all too soon where the leaders of the Western world were meeting.

In the meantime, however, the President attempted to assure his staff and advisers that all would be well. As he noted in his diary, "We are offshore—escorted by destroyers & planes—very luxurious in the *Iowa,* which with her sister ship the N.J. [USS *New Jersey*] are the largest battleships in the world . . . Capt. McCrea is very proud of the *Iowa*—with its population of 2,700 officers & men." He added, "I have his cabin. In my mess are Ads Leahy, W. Brown, McIntire, Gen. Watson & Harry Hopkins. The Joint Staffs are one deck up—Gens. Marshall, Arnold, Somervell, & Adm. King—And below are 100 members of the Planning Staff."[1] In short, he was proud to be traveling as U.S. commander in chief, taking with him his entire high command to the current theater of battle: the Mediterranean. And absolutely determined that the next battlefield would be—as formally agreed at Quebec—northern France, in the spring of 1944.

"The President has been very straight lately and has stood up to Churchill better than at any time heretofore," the elderly war secretary had recorded after seeing the President on November 4—though Henry L. Stimson still distrusted the British prime minister. "With all his lip service Churchill is against the Channel operation and instead is determined to push forward new diversions into the Balkans, Turkey, and possibly other places which will inevitably drain off ships and men from the final decisive conflict which we are trying to ensure in France."[2]

Stimson was right to be suspicious of Churchill—who had been acting strangely in recent weeks, according to reports coming back to Washington, as well as long, plaintive cables the Prime Minister had been sending the President.

The initial cause of Churchill's distress had been the disasters that had befallen his unilateral British attempts to seize the Aegean islands of Rhodes,

Kos, Samos, and Leros, near Turkey, in tandem with the Allied landings on the southern coasts of Italy in September. The Aegean landings were fatally misconceived, and execrably executed: launched exclusively by British Empire forces from the Middle East without U.S. help or even knowledge. Instead of accepting their embarrassing defeat by the Wehrmacht as part of the fortunes of war, however, the Prime Minister had in mid-October 1943 begun a new plot to delay and if possible sabotage Overlord, in favor of redoubled efforts in the Mediterranean and Aegean.

Churchill had always been a "meddler," as one of his senior battlefield generals put it[3] — unable to stop himself from trying to influence and if possible direct operations in the field. Had the Prime Minister been a great military commander, this might have made sense. Ever since his tragicomedy when attempting to command the defense of Antwerp in 1914,[4] and masterminding the assault on Gallipoli in 1915,[5] his attempts at generalship had proven fatal for the ill-prepared troops committed by him to combat. His opportunism demonstrated the force of his vivid imagination and sheer courage; it had rarely, if ever, been based on a realistic appreciation of what he could expect his own soldiers to accomplish against a tough or ruthless enemy. This flaw, sadly, was now leading him to the most consequential mistake of his long life, in terms of military strategy and Allied unity: his third and most egregious attempt to subvert the D-day invasion, and instead reinforce failure in the Mediterranean.

Having assembled his chiefs of staff in London the Prime Minister had on the evening of October 19 asked them not to pursue their work on plans for the Overlord invasion — but, instead, to focus all their attention on covert operational plans to invade the Balkans!

"Pray let this enquiry be conducted in a most secret manner," he'd instructed them, "and on the assumption that commitments into which we have already entered [at Quebec] with the Americans, particularly as regards 'Overlord,' could be modified by agreement to meet the exigencies of a changing situation."[6]

By "secret" he meant: *without American knowledge.*

By "modified" he meant: postponement, if possible to 1945.

By "agreement" he meant: using legerdemain with the President, whom he would try to see on his own in Cairo, and convert, before the proposed meeting with Stalin.

General Sir Alan Brooke, as the head of the British Army, was initially mystified by the Prime Minister's new directive. After all, the American president had recently made clear he would tolerate no postponement of Overlord.

Aware from Field Marshal Sir John Dill, the British liaison to the U.S. chiefs of staff, that in London the Prime Minister was plotting further "adventures" in the Mediterranean and Aegean, the President had summarily dismissed Churchill's plea that General Marshall be asked or ordered to attend a conference with Eisenhower and the British air, ground, and naval commanders in North Africa — a conference that Churchill, ominously, wished to chair.[7] Likewise the President had seen no merit in the Prime Minister's cabled recommendation that the United States should provide American support, despite the recent failures in the Aegean,[8] to a new British attempt to recapture Rhodes "as the key" to a new Mediterranean/Aegean/Balkan strategy. In a memo to Admiral Leahy, the President had rejected Churchill's latest petition in three simple words, to be dispatched to Churchill: "OVERLORD is paramount."[9] Later that same day, the President had sent Churchill a more explanatory personal cable, drafted by Leahy, which ran: "I am opposed to any diversion, which will in Eisenhower's opinion jeopardize the security of his current situation in Italy, the buildup of which is extremely slow considering the well known characteristics of his opponent who enjoys a marked superiority in ground troops and panzer divisions."[10] In other words: focus on Italy, not Rhodes, while preparing for Overlord.

To any normal person, the President's response would have closed the matter. Churchill, however, was no ordinary mortal. Unable to let go of his growing obsession with operations in the Mediterranean after the British fiasco in the Aegean, the Prime Minister had brushed aside the President's negative reply. "I am sure that the omission to take Rhodes at this stage and the ignoring of the whole position in the Eastern Mediterranean," he'd cabled back, "would constitute a cardinal error in strategy."[11]

"I am convinced," Churchill continued, in an attempt to explain such strong language, "that if we were round the table together that this operation could be fitted into our plan without detriment either to the advance in Italy of which you know I have always been an advocate, or to the build-up of OVERLORD" — an operation of which he was not an advocate, but which, he assured the President, "I am prepared faithfully to support."[12]

"Faithful" was not a word commonly attributed to Churchill, as a consummate politician who had so often changed parties and allegiances — switching from the Liberal Party to the Conservatives, and from blind fealty to King Edward VIII to King George VI after King Edward, ignoring his advice, had vacated the British throne to marry Mrs. Simpson.

To buttress his argument for immediate Allied operations to retake Rhodes or hang on in the Dodecanese Islands, Churchill had pooh-poohed any idea the Germans would continue to fight for southern Italy. The Allies now had some "15 divisions ashore," with 12 of them already in combat there, he pointed out to the President.[13] Besides, he was convinced from Ultra decrypts of high-grade German signals that German troops were not intending to offer serious resistance south of the Tuscan mountains, in the north of Italy ("the top of the leg of Italy"),[14] to which they were supposedly ordered by the Führer to retreat if pressed.[15]

Churchill's own chiefs of staff had been as puzzled as the President by such obsession with the island of Rhodes. If the Germans withdrew to the north of Italy, all well and good. But what was the advantage in pursuing — in fact re-pursuing — operations in the eastern Mediterranean and Aegean? Why, above all, change the war's higher strategy to a new, unplanned and unhinged campaign in the Balkans when Overlord, the cross-Channel attack, was now barely six months away? How would a small island like Rhodes be critical, when the Wehrmacht still occupied Crete, the largest of the Greek islands, on which the Luftwaffe had good airfields? Nevertheless, the Prime Minister, imagining himself a great strategist, had "worked himself into a frenzy of excitement about the Rhodes attack," Brooke had noted in his diary; in fact Churchill had "magnified its importance so that he can no longer see anything else and has set his heart on capturing this one island even at the expense of endangering his relations with the President, and also the whole future of the Italian campaign."[16] As Brooke had confessed the next day in the diary he kept both for his own sanity and for his wife to read, "I can control him no more."[17] Even Roosevelt's "very cold reply asking him not to influence operations in the Mediterranean" had met with no success, Brooke recorded.[18] Churchill had merely "wired back again asking President to reconsider the matter. The whole thing is sheer madness," Brooke had bemoaned, "and [Churchill] is placing himself quite unnecessarily in a very false position! The Americans are already desperately suspicious of him, and this will make matters far worse."[19]

The realization that such a revision of Allied strategy would lead to yet another transatlantic battle between the British and American high commands had become more and more alarming to General Brooke. Over the next weeks, however, the Prime Minister's obsession had come closer still to insanity, or an inter-Allied parting of the ways. The Prime Minister's note to the British chiefs of staff on the evening of October 19, "wishing to swing round the strategy back to the Mediterranean at the expense of the cross Channel operation," shocked Brooke. "I am in many ways entirely with him," Brooke noted, given his own lack of faith in Overlord's chances of success, and his wish to focus on dealing the Wehrmacht a major blow in Italy, "but God knows where that may lead us to as regards clashes with the Americans."[20]

Yet Churchill was a master of creating great sagas over names he selected, such as Rhodes and Rome, and inflating dangers where it suited him. At the subsequent 10:30 p.m. War Cabinet meeting in London — a conclave to which Prime Minister Jan Smuts, visiting from South Africa, was also invited — Churchill had described a worsening military situation in Italy that he'd not forecast, but which now *necessitated* that the Allies put more of their eggs into the Mediterranean basket. This would, he claimed, mean making a decision not to switch battle-hardened troops to England, in preparation for Overlord, as currently agreed, provisioned, and planned under the Quebec agreement. Instead, by keeping such veteran forces in the Mediterranean, he would be in a position to throw seasoned British and American troops into further operations in the Aegean and the Balkans, with grand consequences, if he could only get the Americans to agree.

Where, though, was this secret new Churchillian strategy supposed to *lead* the Allies? the President and his American chiefs had wondered in Washington. Who in London was subjecting Churchill's opportunism to critical examination? Possessed of amazing bouts of energy, Churchill had seemed determined to make gold out of straw — with no one in London daring or willing to oppose his new strategy. "I shudder at the thought of another meeting with the American Chiefs of Staff," even Brooke noted despairingly after the meeting on the nineteenth, "and wonder if I can face up to the strain of it"[21] — yet he'd not dared contest the issue with his prime minister, instead confining his real thoughts to his diary.

In this way, in the wake of his late-night meetings in London, Churchill had thrust himself into a new strategic campaign, giving rise to new excitement: excitement that would allow him to marshal his considerable talents of persuasion and rhetoric in a renewed, third major bid to switch Allied forces away from Overlord. Once effected, it would make the cross-Channel gamble, shorn of battle-hardened troops, even less credible as an operation of war than before. Undeterred by the President's insistence on maintaining course, the Prime Minister had thus followed his cabinet meeting, several days later, by giving new notice to the President of the United States that he was definitely not happy with the Overlord plan, or timetable, agreed at Quebec.

"Our present plans for 1944 seem open to very grave defects," Churchill had cabled frankly on October 23 — searching for fresh, apocalyptic rhetoric that would awaken Mr. Roosevelt and his military advisers to the supposed disaster lying ahead in the Mediterranean unless Overlord was put on the back burner. Adolf Hitler, "lying in the center of the best communications in the world, can concentrate at least 40 to 50 days" against either Overlord or the Allies in Italy, Churchill had maintained in a long message, "while holding the other" front (the Eastern Front) from interfering.[22]

Churchill's own hopes of a quick advance to Rome, Pisa, and the Tuscan mountains were meanwhile running into tough opposition, the Prime Minister now acknowledged — though he expressed neither responsibility for nor shame at having ignored the difficulties of fighting in the mountainous eastern Mediterranean, where island by island Wehrmacht forces were methodically liquidating the surviving meager British forces that had landed in the Aegean. It was the disastrous Battle of Crete, which the British had ignominiously lost in 1941, all over again — yet the Prime Minister seemed oblivious to the lessons. Instead of questioning his aims and objectives in the Mediterranean theater, Churchill simply used the dashing of his hopes as an argument for reviving them — this time with American arms and blood.

The new situation threatened the Western Allies with "stalemate" in the war's strategy, Churchill now claimed to the President, unless they doubled down on their efforts, not only in Italy but in the Aegean. He himself had coined the term "soft underbelly" for the Mediterranean — but it was a belly that was in fact far from flabby, he now argued — and would require much sharper harpoons. This left Overlord, in Churchill's mind, as a dangerous distraction, not the other way around.

Allied commitment to Overlord, the cross-Channel invasion, was the problem, Churchill argued. Slavish deference to its timetable would hamstring the war in the Mediterranean, he claimed, and deprive the Allies of great prizes. And for this he blamed the President's insistence on Overlord as the Allied number one priority, with its launch date of May 1, 1944 — barely six months away. "The disposition of our forces between the Italian and the Channel theatres has not been settled by strategic needs but by the march of events, by shipping possibilities, and by arbitrary compromises between the British and Americans," the Prime Minister asserted — ignoring the agreed primacy of Overlord as the main plank in Allied strategy, and deliberately failing to acknowledge the fact that it was he himself who had forced "compromises" over the projected timetable. For at Quebec he had personally insisted upon the addition of special lawyer-style escape clauses, in writing, to the agreement, namely to the effect that D-day must be canceled if the Germans brought more Wehrmacht divisions to France to oppose the landings than the current estimated number. Even the projected date, he now claimed, was a phony one. "The date of OVERLORD itself was fixed by splitting the difference between the American and British view," he asserted, ridiculing the military staffs at Quebec — deliberately oblivious of the paramount need, expressed by the Overlord planning team, for a D-day date as early as possible in 1944 to ensure sufficient summer weather for the subsequent land campaign in France.

Overlord, in short, was a theoretical undertaking, in Churchill's eyes, while the Mediterranean was real — a theater that demanded more American and British forces if it was to be kept going or exploited. "It is arguable that neither the forces building up in Italy nor those available for a May OVERLORD are strong enough for the tasks set them," he'd warned the President[23] — urging him to put all Allied eggs in the Italian basket, not the cross-Channel French one.

The frankness, bordering on cheek, of this cable had been amazing to all who saw it on arrival at the Map Room of the White House.[24] Yet Churchill's latest warning hadn't ended there. What he wanted, the Prime Minister pleaded in his cable, was that there be a halt to current D-day preparations for Overlord. In fact he wanted Mr. Roosevelt to order an immediate standstill in the move of all landing craft and American and British troops currently due to be transferred to Britain. Instead, the vessels and men could be put to better use, the Prime Minister urged the President, in combat in southern Italy, lest their absence "cripple Mediterranean operations without the said craft influencing events elsewhere for many months."[25]

Landing craft? The President, for his part, had not been impressed by Churchill's rhetoric; he was even less impressed by the unstated objectives the Prime Minister had in mind for American landing craft. For what, exactly, were they to be used in the Mediterranean? Roosevelt had asked.

Churchill's answer had been revealing. *Rome!*

As a politician Churchill had always been a dealer in symbols — and symbolism in wartime was certainly crucial, as Joseph Goebbels, Hitler's master propagandist, best knew. But why *Rome,* when the Fascist Italian government and Italian forces had already surrendered to General Eisenhower in September?

The little island of Rhodes had seemed to be a typical Churchill bee-in-the-bonnet when the Prime Minister was haranguing his British chiefs of staff. Now, suddenly, it was Rome. All at once, seizing the Eternal City was a matter of highest priority — overriding preparations for D-day and the timetable of Allied strategy settled at Quebec. *Why* exactly? the President wondered. The major port of Naples, after all, had been successfully cleared of German demolitions, and was now safely in Allied hands. As were the great Foggia airfields. Rome itself was irrelevant, from a strictly military point of view — in fact a red herring.

Churchill certainly offered no military argument — yet insisted Italy's capital city now have priority in all Allied planning, operations, and logistics. "We stand by what was agreed" at Quebec, he maintained in his cable of October 23, but "we do not feel that such agreements should be interpreted rigidly and without review in the swiftly changing situations of war."[26]

The accusatory implication — namely that D-day was hampering great opportunities in the Mediterranean — had led Churchill to his direst new prognostication. Unless the President of the United States were to make the Mediterranean and Aegean the immediate new priority of Allied strategic planning, the Prime Minister warned, there would be a disaster: namely, that "if we make serious mistakes in the campaign of 1944, we might give Hitler the chance of a startling come back."[27]

The argument for Churchill's alternative strategy in Italy sounded all too emotional, speculative — and suspicious. Logic had never been the Prime Minister's strongest suit, but there seemed to be a deeper motive behind his warnings of disaster unless he got his way. British invasion forces on Rhodes had fallen to Wehrmacht forces in September, on Kos on October 4, and seemed likely to be captured or flee from Leros and Samos in November. To bolster his argument for focusing on Rome, however, Churchill had added a new plea: that surviving British troops on Leros, in the eastern Aegean, should now be ordered not only to hold out "at all costs," but be immediately reinforced by General Eisenhower's reserve American forces—including those slated for Overlord! There, on Leros, they could form in the Aegean a great stepping-stone to the Balkans, once Turkey was brought into the alliance . . .[28]

Suffering from influenza at the time, the President had been disturbed by the Prime Minister's telegraphic barrage. Churchill remained, in his eyes, a strange and wonderful individual: a creature of moods and passions, blessed with a gift for metaphor that had no equal in the world; a politician personifying great moral leadership, in both the free world and in the occupied countries. Yet what a *calamity* he was as leader of the forces of the British Empire!

Almost every British military operation Churchill had mounted or touched in the war had failed, the President was aware: from Norway in the spring of 1940 to the Aegean in late 1943, where his folly was now approaching its inevitable end.

Churchill had always been prone to switch from extremes of excitement to despair, the President knew: it was part and parcel of his character. His latest cables had been, Roosevelt thus judged, par for Winston's erratic course. Citing his flu the President had thus continued to politely dismiss them, and hold firm to the military strategy formally agreed at Quebec: Overlord, to be launched on or before May 1, 1944.

Until, that is, the President had learned of Churchill's latest mischief in London—and Moscow.

6

Churchill's Improper Act

AT FIRST THE PRESIDENT had been disbelieving. Was it really possible Winston Spencer Churchill would do such a thing? Yet the facts, when presented, were incontrovertible: Churchill had deliberately ordered that Stalin be shown part of a top-secret Allied military cable from General Eisenhower to the Combined Chiefs — without reference or permission from the President. Moreover, the Prime Minister had doctored the message to give the opposite meaning to the one intended!

Of all the machinations Churchill had employed to sabotage Overlord throughout 1943, including his subversive appeals to congressional leaders on his visit to Washington in May that year,[1] this was without question the most deplorable, the President felt. Clearly, having failed to make any headway in his communications with the President, the Prime Minister had decided to work directly on Stalin, behind the President's back!

Roosevelt's supposition was, unfortunately, correct. "Stalin," Churchill confided to his physician, Dr. Charles Wilson, "seems obsessed by this bloody Second Front," and "ought to be told bluntly that OVERLORD might have to be postponed," as the dutiful doctor had noted in his diary in London on October 24. The Prime Minister was digging in — willfully refusing to take into account how his objections to Overlord could upend the whole Allied alliance. "I will not allow the great and fruitful campaign in Italy to be cast away and end in a frightful disaster, for the sake of crossing the Channel in May," he'd given his reasoning to his physician.[2] It was on that obstinate basis, without telling the President, that he'd deliberately cabled to Anthony Eden, the British foreign secretary who was in Moscow at the foreign ministers' conference, the attachment to Eisenhower's top-secret report to the Combined Chiefs. And had instructed Eden to show it personally to Marshal Stalin.

The attachment was a copy of British general Sir Harold Alexander's latest pessimistic battlefield report from Italy, which General Eisenhower had appended to his message to the Combined Chiefs of Staff. Churchill's cable to Eden, however, had deliberately omitted Eisenhower's commentary on the report, as Alexander's superior officer, in which Eisenhower had stated that he saw no reason for pessimism or alarm in Italy.

Eisenhower had expressed himself, in fact, as very *confident* over the Allies' situation in southern Italy, in his cable to the Combined Chiefs. As Allied commander in chief of all forces in the Mediterranean theater he had merely forwarded General Alexander's report as an attachment, for their information, in explaining why he disagreed with his subordinate. Far from believing Overlord should be postponed, Eisenhower felt the current battle in Italy was a perfect way of keeping the Wehrmacht locked in combat in southern Italy — unable to "withdraw divisions from our front in time to oppose OVERLORD. If we can keep him on his heels until early spring, then the more divisions he uses in a counteroffensive against us, the better it will be for OVERLORD and it then makes little difference what happens to us if OVERLORD is a success."[3]

In other words, General Eisenhower would *welcome* a German counterattack, with the forces he had. There was no mention of a "frightful disaster" such as Churchill had warned the British cabinet to expect if current Allied strategy was not altered. Nor did Eisenhower see any need to halt the current move of landing craft to England, let alone the now battle-hardened combat troops and commanders being transferred for a successful Overlord in the spring . . .

Que faire? News that the Prime Minister of Great Britain had, without telling the President, forwarded a secret, questionable British combat report to Marshal Stalin — with a message to be given to the Russian commander in chief that Overlord would, in all likelihood, have to be postponed as a result of the dire situation in Italy — was, in the view of the U.S. secretary of war, Henry Stimson, rank treachery. "Jerusalem! this made me angry," Stimson had expostulated in his diary, after hearing of it on October 28.[4]

Stimson had fumed all day at the Pentagon. Thanks to Churchill's duplicity (which the Pentagon had heard about indirectly, via General John R. Deane, head of the U.S. Military Mission in Moscow), "Stalin would not have the counter comment of Eisenhower," Stimson bewailed in his diary, "showing that he was not pessimistic at all."

Fortunately General Marshall had immediately — and on behalf of his fellow Joint Chiefs of Staff — "sent a message to Deane," in Moscow, to make sure that Stalin and the Russian high command were told the *actual* truth: that General Eisenhower, the Allied commander in chief in the Mediterranean, was perfectly happy with the battlefield situation in Italy, and that Eisenhower and the Joint Chiefs of Staff of the United States did not think there was any purpose or reason to delay Overlord. "It was perfectly ridiculous," Stimson noted in his diary. "But this shows how determined Churchill is with all his lip service to stick a knife in the back of Overlord and I feel more bitterly about it than I ever have before."[5]

British soldiers could be magnificent when well led, but their leaders were often — too often — manipulative cowards, Stimson felt. And Churchill the worst of them, he rued. So upset was Stimson, in fact, that the next morning he'd gone to the White House without appointment to see the President's counselor, Harry Hopkins. The President's flu had made Roosevelt "almost inaccessible. He has been sick three or four or five days, and now when he'd come out" of hibernation and was working again in the Oval Office, he had so much work on his desk that "it is as difficult to get at him as it would be to get at Mohammed."[6]

Hopkins had been as concerned as Stimson when hearing of Churchill's new perfidy, however — and had insisted the war secretary see the President in person.

Stimson had thus been ushered into the Oval Office, at 11:40 a.m. on October 29. There he ascertained the President had already gotten word of Churchill's deceitful conduct. Stimson explained how General Marshall, on behalf of the Combined Chiefs, had immediately sent a riposte to be given by General Deane to the Soviet high command in Moscow. To Stimson's relief Roosevelt had been in complete "accord with our views in respect to that," Stimson noted in his diary that night — in fact the President "intervened to tell me what his views were in regard to the Balkans. He said he wouldn't think of touching the Balkans," unless the Russians found themselves struggling in southern central Europe and asked for joint operations — which seemed unlikely. What the Russians needed was a Second Front that would force the Wehrmacht to fight both in the east *and* the west. Overlord in the spring of 1944 was, and would remain, official and paramount American strategy for winning the war against Hitler.

Stimson had been relieved. But if Churchill was willing to go behind the President's back in such a way — "dirty baseball" as Stimson termed it — *what else* might the British prime minister be capable of? Deeply concerned lest Churchill seek to sideline General Marshall once appointed to take command of Overlord in northern Europe — where he would no longer be able to argue against Churchillian idiocies in the Aegean — Secretary Stimson and General Marshall had proposed a cable for the President to send to Churchill, saying he could not "make Marshall available immediately" to take command of Overlord. As their draft cable ran, the President should say to Churchill, "I am none the less anxious that preparations proceed on schedule agreed at Quadrant [code-name of the August 1943 Quebec conference] with target date May 1," 1944. It was the British who should, immediately, appoint a "Deputy Supreme Commander for Overlord" — an officer who would receive the "same measure of support as will eventually be accorded to Marshall," and could "well carry the work forward": namely Field Marshal Dill, Air Marshal Portal, or General Alan Brooke.[7]

Though the President held off sending the signal lest it exacerbate the quarrel, the "unpleasant incident"[8] of Churchill's duplicity continued to rankle in the days after: a harbinger, Stimson worried, of further problems, now that planning for the war's end — the occupation of Germany, the defeat of Japan, and the postwar setup for international security — was gathering pace.

Stimson would never forgive Churchill for his Machiavellian machinations. For President Roosevelt, however, the Prime Minister's latest maneuver was, if anything, even more worrying. It was, after all, not Stimson who would have to deal directly with Stalin's probable reaction — which was not difficult to predict, after two years of Allied failure to mount a promised Second Front. There had already been rumors of serious peace-feelers between Nazi Germany and Russia in August and September, when it became clear to Stalin that the Western Allies might never mount a Second Front. Given that the Soviets were still struggling against three-quarters of the German Wehrmacht, Stalin would inevitably lose respect not only for Britain, but for the United States and its word. How then would Roosevelt be able to negotiate from strength with the Russian dictator over the war's end, and over Soviet participation in the war against Japan? Also over the establishment of a postwar system of security, and a United Nations organization the President wanted Russia to join? Moreover, how would the American public — voters, press, institutions, and Jewish Americans especially, who were hearing more and more about Nazi extermination programs in Poland and eastern Europe — react to the postponement or cancellation of Overlord? Would they not see it as a betrayal of their long, often reluctant support for a "Germany First" policy after Pearl Harbor, when the majority of Americans had wanted the President to focus first on defeating Japan?

The President was as disappointed in Churchill as Stimson was, the war secretary found — the Prime Minister's deceit still rankling a week later, when he saw the President in the Oval Office: Roosevelt describing it as "an improper act"[9] akin to treachery. It posed a threat not only to Allied unity with the Red Army in defeating the Third Reich, but to Anglo-American unity and trust, too, at the very moment when unity among the Allies would be vital in ensuring a peaceful transition to a postwar world. Strategic dissension between the primary Western Allies could only cause Stalin to see how *divided* were the U.S. and British governments — and perhaps take advantage of it, just as the potential defeat of Nazi Germany approached. Would Stalin even agree to meet up with the President, in the circumstances?

Deciding it was best to pretend he did not know of Churchill's deviousness, the President had, in the end, merely continued explaining to the Prime Minister that he had *not* changed his mind over Overlord — and would tolerate no

British backsliding on its timetable, if it was to succeed. Moreover, to make extra sure Stalin had not been taken in by Churchill's recent ruse, the President had gone out of his way, before leaving for North Africa, to check with his trusted new ambassador to Moscow, Averell Harriman — cabling him to find out, preferably by return signal, what were the current strategic views of senior Russian military officials — including Stalin himself, if possible.

Harriman had done so — and the American ambassador had been crystal clear in his direct report to the President from Moscow on November 4, 1943. "It is impossible to over-estimate the importance they place strategically," Harriman had signaled (the cable received and decoded at the White House on November 6), "on the initiation of the so-called 'Second Front' next spring."[10]

7

Torpedo!

SUCH WAS THE SITUATION when the USS *Iowa* had left Hampton Roads. As his own first diary entry demonstrated, the President was hopeful — yet was bracing himself for an odyssey full of surprises. That the first one would come on day two of the voyage was on no one's radar, though.

A lifelong mariner, the President had insisted he be shown each morning the daily navigation charts, marked with positions not only of nearby Allied vessels but also of known or suspected U-boats. In addition to their latest snorkels, German U-boats sported new radar, he'd been told by Captain McCrea, as well as an acoustic homing torpedo: the electric *Zaunkönig*, or Winter Wren, traveling at twenty-five knots, which could hit a target at 5,750 meters, or 3.5 miles.

The *Iowa* was currently traveling at the same speed. By midday on November 14, 1943, the battleship and its escort destroyers were 533 miles across the Atlantic. Wanting his crew to be equally prepared for aerial attack once they approached North Africa and the Mediterranean, Captain McCrea had arranged a demonstration of the battleship's armaments for the President and Joint Chiefs of Staff. As he later recalled, the "weather was good," but there was also "a brisk breeze blowing. The state of the sea was choppy, and a succession of white caps was evident on every hand. It had been arranged to exercise the *Iowa*'s 40-mm anti-aircraft batteries that afternoon at 1400 hours, so that the President and the distinguished passengers might see for themselves the efficacy of this particular segment of the *Iowa*'s anti-aircraft defense."[1]

The special presidential elevator made it easy for Mr. Roosevelt to go from one deck to another in his wheelchair, and be wheeled across "ramps built over the coamings and deck obstructions."[2] Thus, at 2:00 p.m., with the President's wheelchair held fast by Petty Officer Arthur Prettyman, his steward, "the escorting destroyers were informed of *Iowa*'s intentions and the drill got underway as scheduled."[3] Weather balloons were inflated and released as targets — the ship's gunners having been ordered to deliver a veritable "curtain of fire to stop enemy planes."[4]

The guns made such a deafening noise the distinguished passengers had to stuff cotton wool in their ears, and for a while the exercise went well. A number of the balloons were shot down: a creditable performance, General Hap Arnold, the chief of staff of the Army Air Forces, noted in his diary.[5] It would be, Captain McCrea later thought, a nice "surprise to those members of the Army Air Force who probably witnessed for the first time the partial capabilities of a ship's air defenses. The drill had been in progress for, I should say, about twenty minutes when in an urgent tone of voice over the T.B.S. [Talk Between Ships] there was a somewhat panic stricken cry of 'Lion, Lion [*Iowa*'s code name]! Torpedo headed your way'—this from the [USS] *Wm D Porter* [destroyer], which at that moment was about 45° on our starboard bow—some 3500 yards distant."[6]

"It should be remarked at this point," McCrea later recalled, "that during the [antiaircraft] firing *Iowa* was on a relatively steady course because the target balloons were going down wind. The *Wm D Porter* was the center ship in the screen. On receipt of the torpedo warning 'Full Speed' was rung up. 'Battle Stations' was immediately sounded. The general alarm gong was clanging—the alarm Boatswains use to prepare the crew to battle stations. Followed by the statement," McCrea recorded, that "'this is not repeat not a drill.'"[7]

"Had there been a leak aboard this trip?" McCrea remembered wondering. "Had we been ambushed?"[8]

"All this took place in seconds. And then a bit late over the T.B.S. from the *Porter* came the statement: 'the torpedo may be ours.'"[9]

Remembering the incident with an embarrassment bordering on shame years later, McCrea allowed that the message had been some "comfort to be sure," in that it meant they were not under enemy attack. This was not much help to the *Iowa,* however; there was still "a torpedo heading our way,"[10] whether German or American.

As ship's captain McCrea now knew, at least, the probable direction from which the torpedo was coming and that there would likely be but one underwater missile. "I immediately altered the course of the ship to the right to head for [USS] *Porter*—this in order to present as narrow a target as possible," he remembered—facing a torpedo, that was, he judged from a professional naval officer's perspective, probably pretty "well aimed."[11]

Would it hit, though? The ship had less than six minutes to respond.

It was just as well McCrea switched course and headed toward the torpedo, for the *Iowa* was 888 feet in length—"12 feet shorter than three football fields laid end to end," as McCrea liked to say.[12]

The captain may have been pedantic in prose, but his swift, instinctive judgment certainly saved the USS *Iowa* from a calamity in the Atlantic Ocean. Below him, down on the first superstructure deck, "President Roosevelt, Admiral Leahy, General Marshall, General Arnold, Harry Hopkins, General Watson,

Admiral Brown and Admiral McIntire had gathered to watch the shooting,"[13] McCrea afterwards recalled.[14] Their admiration soon changed to near-panic.

Harry Hopkins penned his own account that evening. In Hopkins's version, an officer from the bridge two decks above leaned over and yelled, "It's the real thing! It's the real thing!" but the President initially didn't hear. "The president doesn't hear well anyway and with his ears stuffed with cotton he had a hard time getting the officer's words" — words "which I repeated to him several times before he understood. I asked him whether he wanted to go inside — he said, 'No — where is it?'"[15]

Suddenly it seemed as if every gun was trying to hit the racing torpedo, not the barrage balloons — even though, in the choppy ocean water, the torpedo could not clearly be seen, some twenty-five feet below the surface. "More commands from everywhere. Whistles, flags, code signals. The din aboard the ship was terrific," General Arnold later wrote.[16]

Lieutenant William Rigdon, who was part of the President's Map Room signals staff, later described how "President Roosevelt had been watching the gunnery exercise from one of the upper decks when he heard the warning. He called to Arthur Prettyman, his valet: 'Arthur! Arthur! Take me over to the starboard rail! I want to watch the torpedo!'" As Rigdon added, "Prettyman, as he confessed later, was 'shaking all over,'" but he "wheeled the President quickly to a vantage point."[17]

This was scarcely advisable, to say the least, but the President was the President — and Commander in Chief.

Given that the wind "was considerable," and "white caps were everywhere," it was fortunate that at least "the trace" or "wake of the torpedo was visible to us in *Iowa,*" McCrea remembered. "Just about the time we had really hit full speed there was a tremendous explosion on our starboard quarter. The ship shuddered mildly but sufficiently to cause me to turn to *Iowa*'s Executive Officer [Cmdr Thos. J. Casey] and remark, 'Tom, do you think we have been hit?'"[18]

Others certainly thought so.[19] Casey thought not, however, or the impact would have caused even more of a shudder — leaving McCrea, in later years, to ponder the reason. "It is my belief that the turbulence caused by *Iowa* at full speed and executing a turn was sufficient to detonate the torpedo's firing mechanism which resulted in its explosion."[20]

General Arnold, for his part, noted in his diary that night what "a grand rest" he'd been enjoying on the voyage, with "not a worry in the world." Then the "whole character of the maneuver and practice changed," once the alarm sounded, and the ship suddenly swung to starboard. "The wake of the torpedo became quite clear," barely six hundred yards away; "a depth charge went off, all the guns started shooting": 40mm, 20mm, and .50-caliber weapons.[21] As Arnold subsequently concluded, however, it was in vain, since "nothing hit the torpedo."[22]

Whatever caused the torpedo to explode, the underwater missile hadn't actually hit the battleship, mercifully. Yet McCrea was not congratulated. "About this time when, of course, my attention was focused ahead there was purred into my ear a low tone, 'Captain McCrea, what is the interlude?'"[23]

It was Admiral King — who had once told McCrea he was not enough of a "sonofabitch" to make a successful senior combat officer. McCrea explained the situation, upon which King, as U.S. Navy fleet commander, rushed down and joined the President's party on the deck below to report.

General Arnold thought the torpedo had missed the stern of the ship by barely twenty yards. As he noted in his diary that night, "No hit, a miss" — which caused "a thousand sighs of relief."[24] They had almost literally dodged a bullet — and a big one.

"In everyone's mind," Arnold later wrote, "was the question: 'Suppose the torpedo had hit, and it had become necessary to take the President and all the high rank off the *Iowa* in those heavy seas'?"[25]

In a letter he intended to send his cousin Daisy once they reached Oran the President thought it best not to mention the near-miss. "I wish I could tell about things each day," he apologized, "but I dare not."[26]

In the new diary his cousin had given him, by contrast, the President *did* record the ominous event, for later reading. "On Monday last at gun drill," he penned, "our escorting destroyer fired a torpedo at us by mistake. We saw it and missed it by less than 1000 feet." This laconic description left, however, a host of questions — and rereading his entry several months later, back in Washington, Roosevelt dictated a note to be attached to the diary, which he intended to use in writing his memoirs, once the war was over:

"The destroyer in the escort was holding torpedo drill, using the *Iowa* as the spotting target. The firing charge was left in the tube, contrary to regulations. The torpedo was fired and the aim was luckily bad. Admiral King was of course much upset and will I fear take rather drastic disciplinary action. We fired the secondary battery to try to divert the torpedo. Finally we saw it explode a mile or two astern."[27]

"Eventually things calmed down and further shooting was called off," McCrea recalled.[28] By visual communication with USS *Porter* — to avoid the use of radio — it became clear it had been a "torpedo all right — but not from a German submarine," as Hopkins put it.[29] As General Arnold noted later that night, "Practice being over, everyone went back to normal duties and pursuits."[30] It had been an accident, thankfully, not an assassination attempt. "It must have come from some damned Republican!" Hopkins quipped at dinner in the President's mess, attempting to make light of it.[31]

The mishap had been serious, though. As Lieutenant Rigdon later wrote, they had been more than fortunate; it had "just missed being a day of tragedy."

8

A Pretty Serious Set-to

THE PRESIDENT WAS NOT the only one heading to Tehran — and by sea. Prime Minister Churchill had left Plymouth on November 12 — the same day as the President — aboard HMS *Renown*, a World War I battle cruiser. Tellingly, the Prime Minister was going without the current plans for Overlord. He was not interested in Overlord. Instead, General Brooke and his colleagues on the British chiefs of staff were taking with them only plans for the Anglo-American forces in Italy to advance to the Pisa-Rimini line in northern Italy, and for other operations in the Aegean if possible. These plans would, Churchill now insisted, necessitate delaying the D-day assault by "some 2 months" at least, as Brooke noted in his diary.[1] He was clear this would entail a "pretty stiff contest!" with his U.S. counterparts, once they met up in Cairo, en route to Tehran, yet in certain ways Brooke had come to welcome the impending clash.[2] It would be a "pretty serious set to," he acknowledged, one that would "strain our relations with the Americans, but I am tired of seeing our strategy warped by their short-sightedness, and their incompetency!"[3] The British chiefs' formal recommendations to the Prime Minister the previous day had thus stated that, in terms of Overlord, "we do not attach vital importance to any particular date or to any particular number of divisions in the assault and follow up" — perhaps the least impressive admission of the entire war. The chiefs seemed to have no idea how important it would be to launch the assault in spring, to give the Allies a full summer to develop their campaign; British strategy should be, instead, to "stretch the German forces to the utmost," in the hope the Wehrmacht would crack somehow or somewhere. They were therefore "firmly opposed to allowing this date to become our master and to prevent us from taking full advantage of all opportunities that occur to us to follow what we believe to be the correct strategy."[4]

It was a lamentable performance.

From his Wolf's Lair headquarters at Rastenburg, in East Prussia, the Führer had also left on a trip. His was by armored train, traveling to southern Germany: a visit designed to buck up German morale now that the Third Reich had lost its primary partner, Italy.

Given the menacing situation on the Eastern Front, where Russian forces had crossed the Dnieper and Wehrmacht forces were being forced to retreat farther and farther toward the Reich, Dr. Joseph Goebbels, the Reichsminister für Propaganda, had thought the Führer wouldn't have time to make such a "political" trip. Hitler, however, had his own ideas. On November 7, 1943, he'd therefore arrived in Munich, in order to address the party faithful at the Löwenbräukeller, where the Nazi Party had been founded twenty years before. His long speech, broadcast on German radio at 8:15 p.m. that night, had awed even Goebbels, who with the Führer's permission had edited it only lightly.[5]

Addressing himself both to his "party comrades" and *Volksgenossen*, or fellow citizens, the Führer had sounded amazingly confident and rational—indulging in his trademark sneering sarcasm, as well as his passionate belief in Germany's historic destiny in fighting Bolshevism and the Jews. What, he asked those present and those listening on their radios, "would have become of Europe and above all, our German Reich and our beloved homeland, had there not been the faith and the willingness of the [party] individual to risk everything for the movement? Germany would still be what it was at the time: the democratic and impotent state of Weimar origin."[6]

The Nazi Party had, the Führer claimed, saved Europe from "world conquest" by the forces of Bolshevism—a political movement he described as a "Bolshevik-Asian colossus" bent upon subjugating Europe until "it is finally broken and defeated." The colossus was one impelled by Jewry, for the "Jewish democracy of the west will sooner or later lead to Bolshevism," he predicted—the "Jewish-plutocratic west" joining the "Jewish-Bolshevik" east, crushing Germany between them . . .

Instead of submitting to such a dark fate, however, the Nazi Party had successfully set about resisting it. "While in the First World War the German Volk went to pieces at home almost without enemy action," the Führer claimed, "it will not lose the power of its resistance even under the most difficult circumstances."[7] And he warned: "let nobody doubt this or delude himself." Only "criminals" could believe anything was to be gained by an Allied victory. Moreover, "we will deal with these criminals!" he snarled. "What happened in 1918 will not repeat itself in Germany a second time . . . If tens of thousands of our best men, our dearest *Volksgenossen*, fall at the front, then we will not shrink from killing a few hundred criminals at home without much ado . . . If we catch one, he will lose his head. Rest assured, it is much more difficult for me to order a small operation at the front in the realization that perhaps hundreds of thousands will fall, than to sign a sentence that will result in the execution of a few dozen rascals, criminals or gangsters."[8] Allied attempts to bomb Germany into submission he ridiculed, for despite the terrible suffering imposed on "women and children," the "damage done to our industry is largely insignificant." Within two or three years of the war's end two to three million apartments would be constructed to house those made homeless by Allied bombing. Thus, while the "Americans and English are right now planning the rebuilding of the world" through a new United Nations agency, he mocked, "I am right now planning the rebuilding of Germany! There will, however, be a difference: while the rebuilding of the world through the Americans and the English will not take place, the rebuilding of Germany through National Socialism will be carried out with precision and according to a plan."[9]

For this grand German reconstruction project the Führer would bring, he explained, not only his vaunted Todt organization, the civil and military construction force headed by Fritz Todt, but "war criminals roped in for the job. For the first time in their lives, the war criminals will do something useful," he stated.[10]

Which thought led Hitler, in turn, to his next great threat: the preparation and use of *Vergeltungswaffen,* or weapons of revenge. The Allies would reap the whirlwind for having bombed German cities. "Whether or not the gentlemen believe it," Hitler warned, "the hour of retribution will come!" And "if we cannot reach America at the moment, there is one state that is within our reach," he declared, "and we shall fasten onto it" — using France as the launchpad for such weapons of mass destruction, aimed at England.[11]

Allied bombing, currently incurring terrible German civilian casualties, could only harden the determination of a typical German civilian or *Volksgenosse* to support the war fought to the bitter end, "since only a victorious war can help him get his belongings back. And so the hundreds of thousands of the bombed-out are the vanguard of revenge," Hitler declared.[12] The enemy "may hope to wear us out by heavy blood sacrifice. This time, however, the blood sacrifice will consist of two, three, or four enemy sacrifices for every German one. No matter how hard it is for us to bear these sacrifices, they simply oblige us to go further. Never again will it come to pass — as in the [First] World War, when we lost two million and this loss was pointless in the end — that we will today pointlessly sacrifice even a single human being," for there would be no capitulation, only victory, he predicted. "Germany will never capitulate. Never will we repeat the mistake of 1918, namely to lay down our arms at a quarter to twelve. You can rest assured of this: the very last party to lay down its arms will be Germany, and this at five minutes after twelve."[13]

"In conclusion," the Führer said, he wanted to add "one more" thing. "Every week I read at least three or four times that I have either suffered a nervous breakdown, or I have dismissed my friend Göring and Göring has gone to Sweden, or again Göring has dismissed me, or the Wehrmacht has dismissed the party, or the party has by contrast dismissed the Wehrmacht, and then again, the generals have revolted against me, and then again, I have arrested the generals and have had them locked up. You can rest assured: everything is possible," he allowed, "but that I lose my nerve is completely out of the question!"

With this display of black humor, the Führer had signed off. He gave thanks to the "Almighty," who had blessed Germany and had allowed it to "take this fight successfully far beyond the borders of the Reich," rather than to have to fight it "on German soil"—moreover had "helped it to overcome nearly hopeless positions such as the Italian collapse!"[14] He therefore asked the party faithful to leave the beer hall "with fanatical confidence and fanatical faith" that would guarantee Nazi victory, and the domination of Europe. "We fight for this. Many have already fallen for this, and many will still have to make the same sacrifice . . . The blood we spill will one day bring rich rewards for our Volk. Millions of human beings will be granted an existence in new homes," to be built by slave labor, supplied by the Todt organization. And war criminals . . . "Sieg Heil!"[15]

Goebbels was delighted, noting in his diary what a propaganda triumph the speech had been, and what a media defeat for the Allies it would be—especially in the neutral countries. Even the rate of suicide went down in Germany, he noted afterwards. Hitler had confided to him that in the coming two weeks he was hoping for "great things" on the Eastern Front, where German counteroffensive moves were being prepared. Despite their own propaganda boasts, after all, the Russians had been unable to retake the Crimea.

Goebbels was of a mind to believe his supreme leader. After all, German civilian morale was recovering. The British had been trounced in the Aegean, and were being held back in southern Italy. In London Churchill was having to warn Parliament of darker, longer days ahead than he'd once predicted. The recent conference of foreign ministers in Moscow, in which preliminary discussions had been held as to how Germany should be carved up and its industry and manpower distributed to the victors, suddenly looked childishly premature.

"There is only one way for our enemies to win total victory in this war," Goebbels summarized in his diary on November 14, 1943—unaware that President Roosevelt was at sea, halfway across the Atlantic to insist on this very point—"and that is a successful [Allied] invasion of the West."[16]

9

Marshall: Commander in Chief Against Germany

THREE DAYS OUT AT SEA, at 2:00 p.m. on November 15, 1943, the President summoned his Joint Chiefs of Staff to a formal meeting in his cabin, where they could discuss a paper he'd asked them to draw up to counter the British plea for a new, alternative Allied military strategy in the Mediterranean and Aegean.

At the meeting the U.S. chiefs voiced their adamant opposition to any departure from Overlord and its timetable. "The President said as far as he was concerned — Amen," the minutes of the meeting duly recorded.[1]

How to bring the British back into line, though?

Whitewashed by generations of subsequent historians, this was the great tragedy of the war in late 1943: that at a moment when Hitler and Goebbels had no real idea how they could win the war in Europe beyond stoicism and the use of *Vergeltungswaffen,* and recognized the only way they could lose the war was if the Allies launched a Second Front, the Allied coalition faced the danger of being split apart by Winston Churchill and the British.

Churchill's obsession with the Mediterranean and the Aegean, of course, was nothing new. His Dardanelles operation in World War I had resulted, after all, in a fiasco, entailing terrible loss of life and his own resignation as First Sea Lord, or minister responsible for the Royal Navy. Thereafter he'd insisted upon serving as a battalion commander on the Western Front. This battlefield service had only reinforced his opposition to trench warfare, however, and — ironically — fueled his belief in the use of tanks: the very weapon the Wehrmacht had used in 1940 to smash its way across France, backed by modern tactical air support — yet which Churchill declined to believe the Allies could employ in northern France in 1944, targeting Berlin.

Dreading the inevitable casualties the amphibious invasion would entail, and the subsequent land battle if the Allies even managed to get ashore, Churchill had continued to hope British forces could avoid having to face the main armies of the Wehrmacht in the west. Thus, where the President had seen North Africa and the Mediterranean as a proving ground for U.S. military leadership, interservice cooperation, and combat effectiveness in the run-up to the mounting of a successful cross-Channel invasion, Churchill had drawn the opposite lesson. The complete defeat of the British Expeditionary Force in the retreat to Dunkirk in May 1940, and then the fiasco of the Dieppe raid in August 1942, had sapped his confidence in British Empire troops and their commanders. For him, the Mediterranean was not only a crucial lifeline of colonial empire, but also a means to use traditional British *sea power* to surround German-occupied Europe, not attack it, save with bombers. In this way the Soviets would, if possible, be tasked with defeating Hitler, leaving Britain still intact at the heart of a colonial empire that could be reconstructed after the end of hostilities. The agreement over Overlord, timed to take place on or before May 1, 1944, had in his own mind never been more than a piece of lawyer's paper, as he'd put it: a contract which Britain could simply decline to observe, or could keep asking to defer, each moment the bill came due.

That moment, however, was now coming — and in traveling with his small army of advisers and clerks aboard HMS *Renown* toward the Middle East, the Prime Minister began to recognize the enormity of what he was proposing: namely a complete recasting of Allied war strategy, barely six months before the agreed launch date of Overlord.

Afterwards Churchill would drag his literary coat over the British revolt — his subsequent history of the war completely concealing the way in which he had behaved in London in October 1943, as he plotted with his chiefs of staff to subvert the Quebec agreements made only two months earlier: "one of the most blatant pieces of distortion in his six volumes of memoirs," as the distinguished Cambridge historian David Reynolds would write. Having effective command of all British imperial forces in 1943 as prime minister and minister of defense, Churchill had in September of that year begun a "maverick campaign over the Dodecanese." Then, when this had turned to disaster, he had "come close to throwing Overlord overboard," while subsequently hiding his egregious "strategic machinations" in a "willfully inaccurate account."[2] Composing his war memoirs, Reynolds recounted, Churchill had been persuaded by his writing team to suppress the incriminating "key pieces of evidence" against him: the secret instruction to his chiefs of staff, on October 19, to plan an alternative strategy to the one agreed at Quebec; the duplicitous cable to Eden, to explain the "need" to delay Overlord; and his negative subsequent memo to his chiefs while traveling to Cairo.[3] Moreover, the team had wanted him also to delete his draft explanation for his alternative strategy: namely that his own scheme was "to bring Turkey into the war by dominating the Aegean islands, and thereafter enter the Black Sea with the British Fleet and aid the Russians in all their recovery of its northern coast, the Crimea, etc, as well as debouchements near the mouth of the Danube. No British or American army was to be employed in this. Naval and air forces might have sufficed," as Churchill's draft chapter had it.[4] As Reynolds commented: "Sending a British fleet into the Black Sea was hardly a minor operation. And it sounded a little too close for comfort to Gallipoli in 1915" to be included in the published book. Fortunately, it was removed.[5]

The historical truth remained, however — concealed from the public and later historians, who did not have access to the documents Churchill had purloined from 10 Downing Street when unseated as prime minister in 1945: namely that, in the fall of 1943, Winston Churchill had vowed to do everything in his power in London to sabotage Overlord, at the risk of splitting the Allied alliance. What Reynolds left unexamined was the further question, *why?* What could possibly explain Churchill's willingness to risk the transatlantic partnership — in fact the whole Allied coalition against Hitler? Had he, as the British Army chief of staff, Sir Alan Brooke, put it in his diary, simply gone "mad"?

Scratching their heads, historians would never quite make sense of Churchill's fierce determination to get his own way — even to the point of threatening to resign as prime minister on October 29, 1943, as he did, following the War Cabinet meeting that night.[6]

Certainly, like everyone involved in Overlord, Churchill felt a genuine concern at the casualties the D-day invasion would entail. Yet such an explanation did not really hold water, since operations in the Mediterranean, judging by recent Allied experience in southern Italy and in the Dodecanese Islands, would hardly involve fewer casualties, were they to be mounted as the major Allied campaign of 1944 — opposed, undoubtedly, by do-or-die Wehrmacht forces. And for no strategic purpose, except one, if the aim of the Allies was to defeat Nazi Germany: prolonging the British Empire! Besides, the fear of cross-Channel casualties — as the Prime Minister's doctor noted — had never held back Churchill from undertaking an operation that he himself wished to tackle. Gallipoli alone had caused more than 160,000 British Empire casualties — as well as almost 30,000 French.[7]

What, then, in retrospect, could really explain Churchill's scheming and "machinations," as Professor Reynolds described them,[8] at such a juncture in the war? It was, after all, the *third time* that year he was proposing to ditch the agreed Allied strategy — a Peter-like denial, despite having twice formally signed up for a spring 1944 priority for Overlord. What magic, moreover, had Churchill managed to employ to persuade his entire British chiefs of staff to follow him down this hole, in direct opposition to the President? And how had the Prime Minister gotten the entire British War Cabinet to support him in his latest October crusade, and his dire predictions of "disaster" if they didn't? Was it solely Churchill's fabled way with words, and his stature as prime minister? Or was there some other explanation?

Certainly the U.S. president and his advisers, meeting together aboard the USS *Iowa,* seemed unable to understand Churchill's mounting obstructionism toward Overlord in the run-up to Tehran. What was the Prime Minister *playing* at? Given that the United States had rescued Britain from defeat in 1942, and had deliberately blooded its own forces in the Mediterranean in 1943 in preparation of troops and commanders for Overlord, the latest British rebellion was galling. Why were they *doing* it — and how could they be stopped?

It was this question the President put to his advisers in his cabin on November 15, 1943 — aware that, at the end of the day, there was no actual way to enforce the Quebec agreement, if Churchill resigned or withdrew British commitment to military participation in the May 1944 cross-Channel undertaking. The war against Hitler would, effectively, be lost. As it might equally be lost, too, if the President accepted Churchill's alternative, and the Soviets decided to negotiate with Hitler rather than suffer more casualties in assuming the main burden of combat against the Wehrmacht.

The U.S. military was thus between a rock and a hard place — with no obvious way of breaking the impasse.

By suggesting that a Russian observer might attend the Combined Chiefs of Staff meetings in Cairo, prior to Tehran, the President had first hoped, he told his chiefs, the British could perhaps be shamed in front of Stalin's representative. This tactic had failed, however, the President explained in his cabin — and not only because Prime Minister Churchill was "emphatically against the [Russian observer] proposal."[9] Stalin, too, had wrecked the plan, if inadvertently — for once the Soviet premier heard that Generalissimo Chiang Kai-shek would also be present in Cairo, he didn't dare send a senior Soviet official. The Soviet Union was still not at war with Japan — and had no wish to give the Japanese a pretext for declaring war on the Soviet Union, which would in turn force Stalin to withdraw Russian forces from the Eastern Front.

There was therefore probably no alternative, the President felt, but to stonewall once the American team reached Cairo. They would have to try to defer any serious strategic debate over Overlord to "the big conference" in Tehran, where the United States and Russia could together shoot down the British fantasia — even though this would mean facing Stalin, thanks to British intransigence, with a weakened and divided Western Allied hand.

After much discussion, this game plan, then, was decided upon. Regrettably it was the only way for the Americans to proceed, the President made clear. The chiefs would have to ensure, in drawing up an agenda for the preliminary meeting in Cairo, that "the meeting with the [Chinese] Generalissimo and himself and the Joint Chiefs of Staff [was] to be separate from and *precede* any meeting with the British"[10] — i.e., leaving as little time as possible for the British to raise their expected objections to the Quebec agreement before leaving for Tehran.

The only other alternative the chiefs could suggest was a kind of preemptive move: to put on the Cairo Conference agenda a renewed insistence that the yet-to-be-appointed supreme commander of Overlord — an American — should be made supreme commander both of Overlord *and* the Mediterranean — and thus be able to quash any operations in the Mediterranean that might delay Overlord.

It was in the context of this two-part stratagem — namely to stonewall any discussion of changes to Overlord's timetable, while asking the British to agree in Cairo to the merging of both the Second Front *and* Mediterranean operations under one American commander, if necessary, to achieve this objective — that the President finally went on record as stating for the first time, in an official meeting, that the supreme commander of Overlord would be General George Catlett Marshall. As the minutes of the meeting recorded, it was the President's "idea that General Marshall should be the Commander in chief against Germany" — a role in which, in Europe, he would "command all the British, French, Italian and U.S. troops involved in this effort": the defeat of the Third Reich.[11]

10

A Witches' Brew

ABOARD HMS *RENOWN* Winston Churchill was dead set against any such merger of the two theaters, the idea of which he'd learned just before he left England.

The Prime Minister was suffering from a heavy cold, and his mood had darkened aboard the gray British battleship as it made its way south toward the Mediterranean. It didn't help that, on November 15, weather conditions at Gibraltar precluded him from switching to a plane for the onward journey to Malta. Harold Macmillan, the British political adviser to General Eisenhower, thus visited Churchill in his cabin early on November 16, 1943, to witness a "fascinating performance" by the Prime Minister — the "greater part" of which was "a rehearsal of what he is to say at the Military Conference; and he is *terribly* worried and excited about this."[1]

It would be, Churchill explained at Gibraltar, a grand Anglo-American showdown. Far from seeing the Mediterranean as a holding pattern by the Allies while full-scale preparations were made for the spring cross-Channel assault, Churchill was openly caustic about the Allies' failure to be more offensive and flexible in southern Europe — a failure he attributed not only to the "Combined Chiefs of Staff system" but to "our American allies generally." The Americans, he told Macmillan, had all along restrained his military genius, causing the Prime Minister to feel "that all through the war he is fighting like a man with his hands tied behind his back" — and this when he was, ironically, the only man alive who could have "enticed the Americans into the European war at all"[2] — wooing President Roosevelt, as he once put it, "as a man might woo a maid."[3]

The approaching "Sextant" meeting in Cairo, prior to Tehran, was thus, in Churchill's mind, critical to the course of World War II. It would be, Churchill predicted, the war's "'real turning point,'" as the effete, debonair Macmillan recorded Churchill's words in his diary that night, "and the hardest job he [Churchill] has encountered."[4]

The Prime Minister's way with words, of course, was both a mark of his genius and a curse. Macmillan sided with the Prime Minister. Macmillan had been severely wounded on the Western Front, at Loos in 1915, and again in 1916 at the Battle of the Somme — offensive campaigns that had devastated the British Army for no tactical or strategic gain. Like Churchill, Macmillan had been scarred for life by this experience of trench warfare. He'd studied Greek and Latin at Oxford, rather than French or German, and in his role as current British political/diplomatic adviser to General Eisenhower, ever since the Torch invasion of Morocco and Algeria, he'd felt deeply at home in the Mediterranean. He was happy where he was — and hated the very idea of Overlord and a vast Allied campaign in northern France, which might require his transfer to northern Europe. He was, in short, the very last person to point out to the Prime Minister the folly of what he was proposing.

The irony, moreover, was that Macmillan, like Churchill, was half-American — his mother having come from Indiana — and given his daily meetings with General Eisenhower, the seasoned diplomat ought to have foreseen the likely American response to the Prime Minister's obsession with the Mediterranean aimed at Rome. Macmillan found himself too entranced, however, by the brilliance of Churchill's rhetoric to caution him — recording with delight, instead, how Churchill derided his military advisers' pusillanimity: men crushed, Churchill claimed, by the British chiefs of staff system that brought together air, army, and navy strategists. "It leads to weak and faltering decisions — or rather indecisions," the Prime Minister scoffed at their concerns over his Mediterranean plans. "'Why, you may take the most gallant sailor, the most intrepid airman, or the most audacious soldier, put them at a table together — what do you get? *The sum total of their fears!*'"[5]

This was a Churchillian bon mot of a very high order — spoken "with frightful sibilant emphasis," as the political adviser noted in his diary, thinking of the hissing/spitting sound the Prime Minister made when delivering such ex cathedra judgments. Yet it was dangerous talk — with potentially dangerous consequences.

Like Churchill, Harold Macmillan was a bon vivant, and something of an English social and intellectual snob, which also played a part in the gathering saga. Despite his New World heritage Macmillan saw American officers as being, for the most part, brash and often ill-educated in comparison with his own classical Eton and Oxford education — and, to be sure, most were. Like other Englishmen he'd thus forgotten the sorry failures of British military forces in Norway, France, Greece, Crete, Libya, and Egypt, and had begun to feel not gratitude for the way the United States had helped save Britain in 1942 by mounting the Torch invasion, but instead a discernible resentment at growing American economic and military might in the Mediterranean. He thus failed to see that, by encouraging Churchill to defy the President and oppose the formally agreed Quebec strategy, the Prime Minister might thereby drive the President of the United States to "make nice" with Stalin in order to put down Churchill's insurrection — with huge potential political as well as military consequences for the Western Allies.

Would it have helped if Macmillan had been instructed to report to Eden, the British foreign minister, rather than directly to Churchill?

It is doubtful. The British foreign secretary was even more culpable than Macmillan in his failure to challenge the Prime Minister's recalcitrance. Anthony Eden knew *in person* how much the Soviets were counting on a spring '44 invasion to help defeat Hitler, yet had not dared to bring the Prime Minister to his senses. Instead, he had simply passed on the Prime Minister's deceitful cable to Stalin, dutifully telling Stalin that "the British 'would do our very best for "Overlord" but it is no use planning for defeat in the field in order to give temporary political satisfaction.'"[6]

The Second Front a "temporary political satisfaction"? Churchill had instructed Eden to tell the Russian premier that all previous assurances given to Stalin "about May 'Overlord'" in 1944 would now be "modified by the exigencies of the battle in Italy." Nothing, the Prime Minister had cabled, would "alter my determination not to throw away the battle in Italy at this juncture. Eisenhower and Alexander must have what they need to win the battle, no matter what effect is produced on subsequent operations. This may certainly affect the date of 'Overlord.'"[7]

Eden had loyally conveyed this warning to the Russian marshal. "I again emphasized," the foreign secretary afterwards reported back to Churchill, "your anxiety that Stalin should have the latest account of the situation in Italy" — unaware Churchill had deliberately deceived both Eden and Stalin by forwarding General Alexander's battlefield report without Eisenhower's accompanying, contrary assessment. As Eden had summarized his audience with the Russian dictator, he'd told Stalin he "should know not only that you were anxious about it but also that you were insistent that battle in Italy had to be nourished and fought out to victory whatever implications on OVERLORD" — his cable sent to Churchill the very night the Prime Minister was threatening his cabinet colleagues he would resign if he did not get his way over the strategic switch to the Mediterranean, Aegean, and Balkans.

It was small wonder Marshal Stalin had been unimpressed. Eden's assertion that, "in view of the vitally important decisions now confronting the Allies," it was "all the more necessary that the three heads of government should meet as soon as possible" had met with Stalin's disbelief, after looking quizzically at Alexander's top-secret negative account of operations in Italy. "Stalin observed with a smile that if there were not sufficient divisions" in Italy, then "a meeting of the heads of government could not create them." Eden had reported Stalin's response word for word to Churchill — as well as Stalin's remark that it was not for the world's leaders to furnish troops, but for the British and American chiefs of staff to do that. "He then asked point blank whether the [Alexander] telegram which he had just read meant a postponement of Overlord."[8]

For his part Anthony Eden had been deeply embarrassed — and could only quote Churchill's message to him that there was "no use planning for defeat" in Italy.

This was the first Stalin had heard of such a possibility, either from General Deane, heading the U.S. Military Mission in Moscow, or Lieutenant General Hastings Ismay, Churchill's military chief of staff, currently in Moscow for the foreign ministers' conference. Nor had Russian observers in the Mediterranean, or Russian military intelligence, reported such a dire military situation necessitating the "postponement" of Overlord, some five months away. Eden had cited the shortage of landing craft. But why were *landing craft* necessary to defend Allied positions from German counterattack in the mountains of southern Italy, rather than being assembled in Britain for the cross-Channel assault in the coming spring? *To evacuate?* Moreover, why was Churchill proposing to halt the planned move of some seven battle-hardened Allied divisions to Britain for Overlord, as agreed at Quebec, and instead to keep them on the battlefield in Italy, if those divisions were considered essential, experienced troops in ensuring the success of Overlord?

Stalin had thereupon given Eden and his companions a brief and revealing master class on military strategy.

"As he saw it there were two courses open to us," Eden afterwards reported to Churchill.

The first was: "To take up a defensive position north of Rome and use the rest of our forces for OVERLORD."

The second was "to push through Italy into Germany."

Which was it to be? Stalin had asked. To which Eden, in all honesty, had only been able to answer, the first. Stalin had agreed — since it would be "very difficult to get through the Alps," he pointed out, "and that it would suit the Germans well to fight us there." Surely, Stalin remarked, the British had sufficient prestige as a military power "to permit us to pass over to the defensive in Italy," rather than pursue red herrings in Italy or the Aegean?

In view of the *two years'* delay by the Allies before they felt ready to mount a Second Front, Anthony Eden had been surprised at how gently he'd been treated, with "no recrimination of the past. It is clear however," Eden had warned his prime minister, "that M. Stalin expects us to make every effort to stage OVERLORD at the earliest possible moment," moreover the "confidence he is placing in our word is to me most striking."[9]

"*Our word.*" Like Eden, Harold Macmillan knew that the story about a possible "disaster" in southern Italy, unless Overlord was postponed or canceled, was a complete fabrication by the Prime Minister.

Opposing Mr. Churchill, however, was not something Harold Macmillan was willing to hazard. Nor was there anyone else on the British side who would confront him. And this, even though the Prime Minister was becoming "more and more dogmatic," as Macmillan himself recorded in his diary.[10]

Macmillan was a diplomat. But why had Britain's military chiefs of staff—General Alan Brooke, Air Marshal Charles Portal, and Admiral Andrew Cunningham—not restrained the Prime Minister from embarking on a divisive, looming confrontation with their American counterparts?

British military standing in the world had, after all, been tarnished once again by Churchill's recent fiasco in the Aegean: disastrous, unilaterally mounted operations, using British forces from the Middle East that were now nearing their final, humiliating end on the island of Leros—to the delight of Goebbels's propaganda machine. Would such British military standing not be still more tainted, in the eyes of the Russians as well as other nations, were it to be found Britain was going back on its "word," as Stalin put it, and was threatening to pull out of its part in the Overlord invasion, in five months' time?

Day after day Churchill lay in bed composing his "indictment" of the current military situation—and his argument that putting *more* forces into battle in the Mediterranean, instead of preparing them to cross the English Channel, would make better war policy. "He was not at his best, and I feel nervous as to the line he may adopt at this conference," Brooke confided in the small leather-bound journal he was keeping for his wife to read, once he returned home. "He is inclined to say to the Americans, all right if you won't play with us in the Mediterranean we won't play with you in the English Channel. And if they say, 'all right well then we shall direct our main effort in the Pacific,' to reply: 'you are welcome to do so if you wish!'"

The prospect of such a scenario "filled me with gloom!" Brooke added later that evening, after dinner with Churchill and the "long military discussions" that had continued till midnight.[11] "There are times when I feel that it is all a horrid nightmare," Brooke confided[12]—a nightmare out of which he knew he must waken, though, since the very outcome of the war against Hitler was now at stake.

Churchill, in Brooke's opinion, was now "floundering about," with no "clear vision"—in fact exhibiting only "lack of vision": a self-styled strategist who saw "war by theatres and without perspective," a man who, moreover, lacked a "clear appreciation of the influence of one theatre on another!"[13] A man who "discusses Command and Commanders," yet who "has never gained a true grasp of Higher Command organization and what it means."[14] For instance,

Churchill's admiration for the Honorable Sir Harold Alexander, Eisenhower's British field deputy, was all too typical: General Alexander a brave but shallow aristocrat, yet a man the Prime Minister now hoped to appoint supreme Allied commander in the Mediterranean and Middle East, combining both theaters for the first time, once Eisenhower was — as Churchill was insisting — removed from the post . . .

The thought made Brooke squirm. Yet no one in Churchill's court had the wit or courage to protest.

In Algiers, at Eisenhower's main Allied headquarters, there was also little doubt as to the magnitude of the looming showdown. Eisenhower's chief of staff, Lieutenant General Bedell Smith, had just returned from Malta. He'd also been invited to a personal conference with the Prime Minister on HMS *Renown,* which had arrived there safely in the harbor, he reported. "The PM and the British are still unconvinced," Smith warned, "as to the wisdom of Overlord." The British were simply "persistent in their desire to pursue our advantages in the Mediterranean, especially through the Balkans" — no matter what had been agreed at Quebec in August.

The approaching pre–Tehran Conference between the American and British military teams would thus be, General Smith predicted, "the hottest one yet."[15]

11

Fullest Guidance

NOT SINCE CAESAR had approached Pompey's forces on the shores of Egypt, close to Cairo, had there been such a threatening confrontation between two hitherto close allies in the midst of war.

Becoming more and more nervous about the seriousness of the feud—especially when receiving a signal on board the *Iowa* on November 17, approaching North Africa, that the British censor in Egypt had permitted a leak of the forthcoming high-level conference, and even its location, at the Mena House Hotel, close to the famed pyramids—the President now convened a second meeting of his Joint Chiefs of Staff in his stateroom.

The *Iowa* was now closing on the Straits of Gibraltar and the Mediterranean. It was time, he felt, to discuss with his military advisers not only the impending clash with the British, but the larger strategy of the war, the postwar—and relations with the Russians.

Given that he'd arranged to meet with Chiang Kai-shek in Cairo before flying to Tehran, the President was displeased by the British leak, to say the least. One of the *Iowa*'s escort vessels had therefore been summoned alongside by the President, to take an important message. When well away from the battleship's true course, the destroyer radioed London in most secret code to communicate the President's displeasure "that meeting place is known to enemy through press and radio." The President wanted more information concerning the "seriousness of the leak," and possible alternative venues for the meetings with Chiang—and also with the Turkish president, should he agree to come. Perhaps Khartoum, in the Sudan, Mr. Roosevelt posited—a thousand miles from Cairo?

Churchill, chastened by the laxity of the British censor in Egypt, had hastily consulted with his British military advisers. Malta would have suited his plans for a showdown with the American team, since he was already there, but it was too close to German airfields for safety. Or comfort. Amenities for hundreds of U.S. and British staff officers, clerks, and diplomats were required, unless they all remained on board their respective battleships — which in turn would present an even bigger target for guided German missiles. As Churchill's secretary noted in a letter home, "It would have meant housing the American staff in a place with a single tap and a single lavatory."[1]

Understandably the Prime Minister attempted to make light of the news. In a new cable to the President he regretted "the fact that it has leaked a few days earlier" than otherwise it would have done, but he did not consider it catastrophic.[2] Khartoum would involve hurried and unsatisfactory preparations, while Malta was simply too dangerous given the number of attendees. In the end the Prime Minister felt that, if security in Egypt was increased by placing more antiaircraft guns around the Mena House Hotel and adjacent Giza pyramids, and if Cairo city itself was made off-limits to all international personnel in order to guard against assassination, the Sino-U.S.-British conference would best be held as per the program.

Leakage of the locale of the impending military conference, however, left something of a sour taste aboard the *Iowa,* which was now joined by an escort carrier, the USS *Santee* — a former oiler bearing a new deck and carrying twenty-five Grumman TBF Avenger aircraft. Distrust of the British had risen several further degrees.

But it was not only the lapse in security that weighed on Roosevelt and his advisers, or the imminent contest with the British. The looming "battle royal" was depressing enough as they approached the Mediterranean, but the widening gap between American and British military officials betokened further, perhaps inevitable, disagreements in the future. For there was no getting away from the fact that, if the Allied coalition managed to hold together and defeat Nazi Germany, followed by Japan, there would be vast postwar problems — economic, social, refugee, diplomatic — to confront. The United States would no longer be able to revert to isolationism, as it had after World War I. It would be the world's richest and most productive economy, besides possessing the most powerful navy and air forces.

It was time, in other words, for the President to hold a sort of conference-within-a-conference with his U.S. chiefs of staff.

The President had already shared with his advisers his vision of a United Nations authority, the postwar peace to be guaranteed by "Four Policemen": the United States, Britain, the Soviet Union, and China. This vision was all very well in theory. But in the real world, recovering from a destructive global war?

The devil would be in the details. How, for example, should Germany be treated, in defeat? How should the European nations, currently occupied by German forces or allies of the Third Reich, be reconstituted? How far would the communist Soviet Union be willing to cooperate with capitalist democracies? What would be the fate of Czechoslovakia, Danzig, and Poland—the tinderboxes Hitler had used to ignite his war of conquest? What about Japan and the European colonies in the Far East that the Japanese had overrun? Should the white European colonial powers—Britain, France, Belgium, the Netherlands—be pressed to grant sovereignty and assist their former colonies rather than exploit them, if postwar revolution and anticolonial wars were to be avoided?

The matter was already becoming urgent. Only days before, the new French governor of two territories mandated to the French after World War I, Syria and the Lebanon, had simply ignored the results of elections there. To worldwide condemnation the French governor had imprisoned the winning candidates, who had dared call for full sovereignty and the end of the French Mandate. In India, there was a full-scale famine in Bengal—with Jawaharlal Nehru and Mahatma Gandhi still under British arrest, and no sign the British would currently discuss Indian self-government, let alone sovereignty, anytime soon. Similarly the French in Indochina, the Dutch in Indonesia . . . Were American sons to fight and die to resurrect such continued European colonial imperialism?

The three-hour meeting the President convened in his captain's quarters with the Joint Chiefs of Staff aboard the *Iowa* in the early afternoon of November 19, 1943, due west of Gibraltar, thus indicated, for the first time, a growing awareness of the *political* as well as military minefields ahead—minefields that in some ways explained the British fear of likely D-day casualties in Overlord. Given the need for occupation troops in order to reinstate their besieged colonial empire and then to police the vital sea lanes from Britain—their traditional pathways to Palestine, the Middle East, India, and Southeast Asia—the imminence of Overlord had caused much soul-searching in the corridors of Westminster, as Field Marshal Dill had already explained in Washington. In this context Mr. Churchill's passionate, unrelenting espousal of major operations in the Mediterranean and Middle East, even at the risk of destroying Allied unity and provoking an American switch to the Pacific, could be understood as issuing from a concern for Britain's colonial and imperial postwar interests, rather than the desire to bring a swift end to Hitler's mastery of all Europe. "In the course of the discussions the President gave the staff a clearer indication of the direction of his thinking and the fullest guidance on politico-military issues he had given them since America's entry into the war," the official historian of U.S. strategy in World War II, Maurice Matloff, would write.[3]

It was certainly high time. These U.S. admirals and generals, after all, would henceforth be commanding the world's most powerful postwar military, for good or ill.

• • •

"Fullest guidance" was something of a misnomer, however. Very few of Roosevelt's conference discussions were ever recorded or transcribed. Understandably the President had a deep-seated concern over leaks to the media, after years of combating the right-wing isolationist agendas of rich American Republican newspaper owners such as William Randolph Hearst, Robert McCormick, and Cissy Patterson — media moguls who not only opposed the President's policies, but had deliberately published top-secret information and made mischievous assertions — such as revealing the President's Victory Plan, two days before Pearl Harbor, or publishing revelations about U.S. intelligence decryption efforts immediately after Midway, military revelations that would have been considered treason in less liberal countries.

Roosevelt himself, of course, had become a master of press and public relations — and manipulation. By the fall of 1943 he'd held more than *nine hundred* presidential press conferences, as well as giving some twenty-six extended "Fireside Chats" on American radio, each averaging thirty minutes. Still and all, he had remained wary of keeping written minutes — knowing the chiefs of staff were often required to testify before Congress, where the political ramifications of transcribed statements could quickly become footballs. Keeping politics out of the purely military direction of the war had therefore been a primary objective for the President — his appointment of two Republicans as his secretary of war and secretary of the navy intended to demonstrate his political evenhandedness as U.S. commander in chief.

With the endgame of the war approaching, however, the President's unwillingness to be pinned down in writing was becoming more difficult to excuse in terms of clear decision-making. It was also potentially counterproductive. For more than a decade he'd masked his intentions with his characteristic charm, humor, intelligence, and patrician goodwill: listening, airing provisional opinions, testing his own and others' prejudices, exploring ideas, before coming to decisions. At cabinet meetings and other formal committee sessions this had often disappointed members and administration officials like Henry Stimson, who preferred clear-cut, lawyerly presentation and then decisions made categorically in front of witnesses, rather than afterwards, *en privé,* in the Oval Office or the President's study upstairs in the mansion. Like others, Stimson felt the President possessed an imperfect sense of intellectual orderliness, which the war secretary thought a distinct failing for a chief executive. Such frustration had considerable merit, but missed, to be sure, the very essence of Roosevelt's style of leadership: namely his wish to remain open-minded and hear from a wide number of people in seeking an eventual consensus of opinion or direction — consensus that would allow him then to proceed with his overall vision or objective with the least resistance.

This approach to leadership had proven amazingly successful, after all, in terms of national unity after Pearl Harbor — as even Joseph Goebbels, constantly seeking a chance to foment or exploit dissension in America, acknowledged in his diary.

In this sense, though never having worn the uniform, the President had proven himself a better military strategist, on a global scale, than either Hitler or the Japanese high command, if grand strategy be understood as the directing of all possible resources of a nation to attain its political objectives. After Pearl Harbor he'd also led his country's military high command with remarkable skill — quietly dropping Admiral Harold Stark from the Joint Chiefs of Staff, promoting Admiral King to run the U.S. Navy, and using the trio of King, Marshall, and Arnold to put his global strategy into action, monitored by Admiral Leahy, whom he'd made his White House chief of staff in the early summer of 1942. By personally conceiving and quietly promoting government-sponsored use of mass production techniques in the manufacturing of modern arms and weaponry, moreover, he had challenged U.S. industry to reach production targets that had flabbergasted both Hitler and Goebbels. Such output had not only provided arms for the nation, but for the forces of America's allies, including Britain and Russia; it had also allowed the President to pursue a patient two-ocean defensive strategy, followed by the offensive in the Pacific after Midway — yet prioritizing the defeat of Nazi Germany, as the linchpin of his "Germany First" strategy. He'd had to overrule his Joint Chiefs of Staff when they attempted to sell him on a premature cross-Channel invasion in 1942, with untested American troops, just as he'd had to overrule them when they called for a Pacific First approach if the British did not cooperate in defeating the Third Reich. And he'd had to do so once again at the start of 1943, when recognizing how important it was for U.S. forces to gain the necessary combat and command skills — involving army, navy, and air force components — before the Wehrmacht could realistically be defeated in a Second Front invasion of France. In his choice of theater commanders, as in his convictions about modern American air power — from bombers to transports, fighters to aircraft carriers — he had, by and large, shown wisdom, resilience, and steel when necessary. And all the time he had, in addition, maintained unity among the United Nations confronting Hitler and Hirohito, becoming their de facto commander in chief. He had kept the public morale in America high, moreover, with a mixture of realism, pride — and positive, forward-looking rhetoric. The fact that he had achieved this in a global war without resorting to dictatorial methods or inciting great or divisive controversy was, given the fractious nature of American politics and individualist society, almost miraculous. But as the war in Europe moved toward its climax, could this last? Would the inevitable political, cultural, and other differences between the coalition partners begin to militate against Allied endgame success — and the peace that was to follow?

The précis of the President's three-and-a-half-hour November 19 meeting with his Joint Chiefs of Staff, as noted in sparse minutes drawn up by Captain F. B. Royal of the U.S. Navy, reads today like selective notes of an overheard conversation on a train. The notes remain, nevertheless, our most intimate glimpse into America's top military conclave in the run-up to the decisive moment in the prosecution of the war against Hitler, as the British prepared to mount in Cairo a final showdown with their own allies: a showdown intended to force the postponement or even abandonment of Overlord, despite D-day (as it universally became known) being the only practical way, as the U.S. team saw it, to achieve unconditional surrender of the Third Reich. Nor was this a matter of mere theoretical discussion. Millions of lives were at stake in extermination camps run by the Third Reich. Yet Churchill was adamant. As Major General John Kennedy, General Brooke's director of military operations at the British War Office, would later confess, "Had we had our way, I think there is little doubt that the invasion of France would not have been done in 1944."[4]

12

On Board the *Iowa*

IRONICALLY THE AMERICAN chiefs of staff aboard the USS *Iowa* were more divided than their British counterparts — not over the paramount need to mount Overlord in the spring of 1944, but how exactly to deal with the looming British insurrection. On this latter point the U.S. chiefs were in frank disagreement with one another.

Admiral Leahy, as chairman of the Joint Chiefs of Staff Committee, favored a complete refusal to countenance the British proposal for a merging of Mediterranean and Middle East theaters under a new British "Supreme Commander South" — at least until Britain got back into line. The best way to force the British to stand by their commitment to Overlord, Leahy thus declared, was "Proposal 'A,'" which he'd drawn up in writing for the President on behalf of the U.S. chiefs: namely the appointment of an Allied supreme commander to direct both north *and* south theaters in Europe. This, he claimed, would be the only sure way to stop Churchill from pushing once again his Aegean nonsense at the expense of a spring 1944 invasion of France.

General Marshall, unfortunately, begged to differ. Marshall was an officer of the highest integrity and principle, but like General Brooke, he could be naive about political warfare. He knew Churchill had cabled Field Marshal Dill in Washington, and that the British would never, ever agree to such a draconian overall command solution. Moreover Marshall agreed with the British, at least to the extent that the current situation in the Mediterranean was unsatisfactory. General Eisenhower currently commanded all Allied forces in the central Mediterranean, but had no authority over the many (mostly British) forces available in the Middle East (Egypt, Palestine, Mesopotamia, Iran). This undoubtedly weakened the potential application of maximum military force where and when it was needed. Speaking as a military man, General Marshall explained he sympathized with the British in their view that their efforts in the Dodecanese had failed not so much because of local German superiority, but because General Eisenhower had refused to help with major forces from his

Mediterranean reserves. As Marshall put it, "The British doubtless feel, and perhaps rightly so," that a supreme commander both of the Mediterranean *and* Middle East forces "would have influenced the attitude of [Air Marshal] Tedder and [General] Spaatz towards additional air support in the Dodecanese," the consequences of which might then "have been different."[1]

Admiral Leahy disagreed with such hypothesizing. He felt air support would not have turned the tide — not only because the Germans were ruthless and professional in defense, but because the Aegean, as he reminded Marshall, was a graveyard, not a stepping-stone. To get distracted by an invasion of the Aegean islands, when the Allies needed every effort to be put into southern Italy after Salerno, had been a grave strategic error by the British. Above all, it would not have been made under an American supreme commander of Allied forces preparing to invade France, while the Wehrmacht was being held fast in southern Italy in a combined North-and-South American command.

General Arnold kept silent, while Admiral King — who had strong political views — said he was unclear whether it was Mr. Churchill on his own who was, in essence, the bull in the china shop. The Prime Minister had always been a maverick rather than a team player. King therefore wondered whether "the British Cabinet" could be prevailed upon to overrule him.

The meeting aboard the *Iowa* was then spent batting the European command question from corner to corner.

For his part the President, as commander in chief, had no illusions. He felt certain the British were pressing their case for a British-led, expanded Allied theater in the south, embracing both Mediterranean and Middle Eastern forces, for a simple reason: namely in order to divert priority of Allied offensive strategy away from Overlord. As Captain Royal noted, the President suggested that the "over-all Mediterranean command proposed by the British might have resulted from an idea in the back of their heads to create a situation in which they could push our troops into Turkey and the Balkans" instead of northern France — where there were too many German divisions, in their view, to avoid a bloodbath.[2]

Admiral King pointed out that, if that proved to be the case, the Combined Chiefs would still have to give "approval" or denial for Balkan operations, whoever was the supposed supreme commander — prompting the President to observe that it was not only the Combined Chiefs committee who had oversight of major Allied operations. In the end he himself was the Commander in Chief of the Armed Forces of the United States — and he was not going near the Aegean. As he put it, "Even if General Alexander" — Churchill's pet nominee — "should become commander in chief" of the enlarged Mediterranean–Middle East theater, and should "desire to use U.S. troops and landing craft against the Dodecanese," then he, the President, "could say no."[3]

The President did not therefore agree with General Marshall — for although Marshall's argument about applying maximum force had merit, it missed the major point. The Aegean fiasco was an illustration not only of poor independent British performance in combat unless backed by American power, but of Churchill's misguided strategic opportunism and avoidance of head-to-head battle with Hitler's Wehrmacht in France. "The President asked, why Leros, why Cos?" Captain Royal recorded. Roosevelt acknowledged that "the Prime Minister had been upset as regards the United States attitude regarding the Dodecanese," but where did those islands *lead,* for heaven's sakes? If they were considered so strategically important to the Allied coalition, why hadn't the British told the United States in advance of the invasions they were launching there? General Marshall said the most he'd been aware of was a "pink dispatch" slip, or cable — and that the "British always regarded the Dodecanese as of greater importance than we in the United States" . . .

Round and round the discussion went.

In the end the President said that, although the notion of pushing Proposal "A" for an all-Europe commander was the better approach in theory, in order to get Overlord mounted on time, it would in practice only lead to the British digging in and refusing to accept it — which they were within their rights as a major ally to do. With regard to "the matter of the Supreme Allied Commander" for "all Europe," as Leahy had called it, the President was thus doubtful. In demanding an all-Europe commander rather than presenting the idea as a negotiating tactic, Marshall was right; the British would refuse, and the American team would be on a hiding to nothing.

As a temporary means to hold the British to the Quebec agreement on Overlord, however, the President's notion of an American "all Europe" — i.e., north and south — was worth pursuing in Cairo. The President himself would therefore aim to stonewall when they all met. They could deliberately decline to reconsider the matter of Overlord and its timetable till they got to Tehran and met with Stalin — where the Russians could help put iron back into the soul of the faltering British. The President said he would like another meeting with the Joint Chiefs before they went into oral battle with the British on the issue in Cairo, though. "He said he would take up the matter [of Proposal "A," an all-Europe supreme commander] with the Prime Minister at the earliest time" in Egypt, but "we may have difficulty," he warned the chiefs.[4] They might well have to agree to a compromise, or quid pro quo — for without British forces, Overlord simply couldn't be mounted. This reality gave the British veto power, which they were quite capable of using. The matter was very delicate, and would require a mix of patience, determination, and dexterity.

To provide himself with ammunition for the Cairo meeting the President had need of a gentle way of reminding the British who was boss. He had therefore asked his chiefs to give him, in writing, a table of the "total forces that the U.S. and United Kingdom would have at home and abroad by the first of January 1944." The figures they now presented at the meeting on the *Iowa* were telling:

TOTAL MILITARY FORCES
U.S. — 11,000,000
U.K. — 4,500,000[5]

Eleven million men versus four and a half million! What better illustration could there be of the tail trying to wag the dog? American armed forces were now contributing more than *twice* the number of troops as the British (by which was understood British Empire and Commonwealth contingents, including those of Canada, Australia, New Zealand, India, and South Africa) were furnishing. Yet even here the outcome of the impending Cairo deliberations was not guaranteed. The vast majority of U.S. troops were still in training, at home, yet to be dispatched to theaters of war. Thus the number of troops on active service abroad was currently about equal between the two powers — the U.S. Army boasting 2.6 million abroad, the U.S. Navy another million, while total British Empire forces amounted to about the same. Churchill would be bound to argue, then, against having an overall supreme Allied commander of "all Europe" — which he would claim was too big a theater for one man to direct. The Prime Minister would undoubtedly insist upon a British supreme commander in the Mediterranean/Middle East, if the supreme commander of the cross-Channel invasion was, as agreed at Quebec, to be an American, once appointed. When that happened, a British commander in the Mediterranean would inevitably be putty in Churchill's wild hands: unable to resist the Prime Minister's pressure for renewed Allied ventures in Italy, in the Aegean, in the Dardanelles and Balkans.

The more the President and his chiefs aboard the *Iowa* surveyed the looming set-to in Cairo, the more ominous the coming conference appeared, in terms of the Western alliance. Thanks to Churchill's latest opposition, the formally drawn-up Quebec agreements were now, effectively, defunct — and there were worse things to come.

Disagreement between the two nations' militaries was brewing over the disposition of U.S. and British military forces in Europe even *after* the war was won. Draft planning proposals, under the code-name Rankin, had been drawn up for the occupation of Germany and its satellite countries, if and when the Germans collapsed prior to Overlord taking place, or immediately after. These posited a division of Germany into three areas: north, south, and east. The President, leading the discussion, thought "Marshal Stalin might 'okay' such a division," on behalf of the Soviet government: a carve-up in which the Russians would get to occupy all of Prussia, the very seedbed of German militarism. But the British had said they wanted the industrial and Atlantic northern sector of Germany, with the United States having to take the south, below the Moselle River. The President "said he did not like this arrangement. We do not want to be concerned with reconstituting France. France is a British 'baby,'" he told his chiefs. The United States "is not popular in France at the present time," he averred — in part because he had not licensed General de Gaulle to masquerade as an elected French leader before elections could actually take place, once France was liberated. "The British should have France, Luxembourg, Belgium, Baden, Bavaria and Wurtenburg," Mr. Roosevelt therefore argued. "The occupation of these places should be British. The United States should take northwest Germany. We can get our ships into such ports as Bremen and Hamburg, also Norway and Denmark, and we should go as far as Berlin. The Soviets could then take the territory to the east thereof." In his view, the "British plan for the United States to have southern Germany" was anathema, "and he (the President) didn't like it."[6]

The President might have to stomach it, however. As General Marshall pointed out aboard the *Iowa*, if Overlord went ahead — as they all hoped and prayed it would — the plan was for vast numbers of U.S. troops to be fed into France across the Normandy beaches and through floating harbors and French ports, *on the south side of the British*. And this for the simplest reason: namely that the staging areas for U.S. forces were to the southwest or right side of British forces in England. Having subsequently to reverse that communications axis would be a logistical nightmare.

Here was yet another example of the fortunes of war being dependent on geography and supply: the proverbial wagon train.

Military historians would later rue the fact that, in terms of striking across the Rhine into Germany, U.S. forces were located on the right-hand side of the slow and methodical British, rather than on their left. But the effort this redirection would entail, General Marshall pointed out — supported by Admiral King — was just too daunting to consider, however much the President railed against the British plan (even proposing that U.S. occupation forces be sent from the United States via Scotland, in order not to have to cross British lines of communication with the continent).

Such thoughts, in turn, led to the *political* headache the occupation of France would pose — for, as the President predicted, General de Gaulle would "be one mile behind the troops in taking over the government." Mr. Roosevelt therefore "felt that we should get out of France and Italy as soon as possible," once the countries were liberated, "letting the British and French handle their own problem together." There could well be civil war in France, given the ill-feeling between Gaullists and Vichyites, with savage scores to settle. Moreover, with regard to the German capital, if the Third Reich collapsed before, or soon after, the Overlord invasion, there "would definitely be a race for Berlin," Mr. Roosevelt predicted — in competition with the Russians. "We may have to put the United States divisions into Berlin as soon as possible," using Germany's excellent rail system. Ironically, occupying Germany would be a whole lot simpler than France; the "Germans are easier to handle than would be the French under the chaotic conditions that could be expected in France," Admiral Leahy — who had been the President's ambassador to Vichy until the spring of 1942 — opined.[7]

The President thought a million U.S. troops would be needed in Europe "for at least one year, maybe two," to keep order; but as U.S. president he disliked the notion of being "roped in" to any "European sphere of influence." As regards action to be taken on behalf of the United Nations by the "Four Policemen" after the war, there should be no question, he maintained, of using U.S. ground forces to settle "local squabbles in such a place as Yugoslavia. We could use the Army and Navy as an economic blockade and preclude ingress or egress to any area where disorder prevailed." There was even talk of a "buffer state" like Alsace-Lorraine between France and Germany.[8]

There was then talk of China and Korea; of Burma; of Formosa; of bases across the Far East; of the bombing of Japan; of a supreme commander in the Pacific (where currently Admiral Nimitz and General MacArthur commanded separate theaters); of Turkey and the Balkans; and of Churchill's notion of a "European economic federation" — each chief encouraged to voice his opinion.

None of this was of much significance, however, *unless the war against Hitler was won.* And as swiftly as possible. As General Marshall now emphasized, agreeing with the President, the Balkans were a dangerous decoy. Having been asked by the President to reexamine the implications of a Balkan strategy, the U.S. Army chief of staff was clarity itself. "We do not believe that the Balkans are necessary," Captain Royal quoted him saying. "To undertake operations in this region would result in prolonging the war and also lengthening the war in the Pacific. We have now over a million tons of supplies in England for Overlord. It would be going into reverse to undertake the Balkans," and it would "prolong the war materially. It would certainly reduce United States potentialities by two-thirds," he pointed out — adding that "commitments and preparations for Overlord extend as far as the Rocky Mountains in the United States. The British might like to 'ditch' Overlord at this time in order to undertake operations in a country with practically no communications," but for his part he was appalled by such faintheartedness, breach of faith, abrogation of a formal agreement, and deliberate strategic distraction — moreover one that was being pursued purely for British imperial interests. The bottom line was an American no.

"If they insist on any such proposal," Marshall went on, "we could say that if they propose to do that we will pull out and go into the Pacific with all our forces."[9]

A switch to the Pacific? Ironically this was exactly the American response that Churchill was threatening to trigger if he didn't get his way in the Mediterranean, as General Brooke was noting in his diary.[10] This "way" would be accompanied by fantastical Churchillian promises, the President and his chiefs of staff were well aware. The Prime Minister would undoubtedly claim, once in Cairo, that if the Allies moved fast the Danube could be made the key to Berlin. Soviet forces, after all, were now only forty miles from Bessarabia. If in the next two weeks they crossed the Bug River, they'd be "on the point of entering Rumania," the U.S. chiefs envisioned him saying. Churchill's argument would doubtless be that "if someone would now come up from the Adriatic to the Danube, we could readily defeat Germany forthwith."[11] What then?

Dealing with such Churchillian fantasies was, Marshall knew from experience, a constant scourge. For someone who was an experienced, educated soldier in his youth, Churchill had a sense of geography that was often flawed, and it was nothing short of amazing how little account he took of terrain—as well as Wehrmacht opposition, when German forces occupied high ground. Even if the Russians were to ask, at Tehran, for help in striking through Romania toward the Danube—which Marshall doubted—"we will have to be ready to explain to the Soviets the implications of any such move." Air force—i.e., bomber—help, yes; an American army, no. With Overlord only five months away, what was the point, moreover, of dreaming of such wild alternatives? A major Allied campaign in the mountainous Balkans could make projected casualties in Overlord look minuscule! The fact was: the only practical way to force the Germans to surrender unconditionally was to "force the issue from England," shattering the Wehrmacht between Western and Eastern Fronts—Hitler's nightmare. The important thing, in Marshall's view, was that the United States should not even ask the Russians what *they* might like, in terms of military help, whether in Bessarabia or Romania; after all, the war was not being conducted to help the Russians conquer eastern Europe. The U.S. delegation to Tehran should therefore "not bring up the matter of asking the Soviets for their plans until we are committed to our own plans."[12] American plans, pivoting on Overlord . . .

All agreed with General Marshall. But if the British wouldn't commit to Overlord in May?

The chiefs turned to their commander in chief, the President. All now rested upon him.

The task was clear: how to bring an unwilling British bride to the American altar. Only the President, as the bride's brother-in-arms, could get her to the church on time, so to speak, if the war against Hitler was to be won — in fact, if the Western military coalition was to be saved from being sundered by the British.

The next few days, they were all aware, would determine whether Mr. Roosevelt would be successful.

13

In the Footsteps of Scipio and Hannibal

After its race through the Straits of Gibraltar by night, the USS *Iowa* safely disembarked its top-secret passengers at Mers-el-Kébir, the Mediterranean harbor outside Oran, early the next day, November 20, 1943.

Wearing his familiar cape and sporting his famed cigarette holder, the President was lifted into a motorized whaleboat hanging beneath davits over the deck of the battleship, and was lowered into the translucent waters of the Mediterranean. From there he was taken to the shore — to be greeted "on the dock" by his sons Elliott and Franklin Jr., as the President recorded happily in his new leather-bound diary.[1] And by General Eisenhower, too, in person. "Roosevelt weather!" the President shouted to Ike, as the first sun for several days illuminated the November waterfront.[2] "The sea voyage had done father good," Elliott recalled; "he looked fit; and he was filled with excited anticipation of the days ahead."[3]

From Oran the President was driven with General Eisenhower some fifty miles to La Sénia airport, "where we boarded a big C-54, commanded by my old pilot of Casablanca days, Maj. Otis Bryan," the President noted proudly in his diary. "Then a 3½ (hour) flight past Algiers to Tunis & we saw where much of the fighting took place. A drive took us through the ruins of Carthage (but of a Roman date) to my villa — the Casa Blanca — most apt!"[4]

Carthage was a sort of time capsule of ancient and modern empire. In the distance behind them rose the white-peaked caps of the Atlas Mountains. To the north shimmered the Gulf of Tunis and the blue of the Mediterranean.

Prior to General Eisenhower's tenure, guest villa no. 1 — Maison Blanche, or White House, as it was called — had, to the President's great amusement, formerly housed the Wehrmacht's commander in Tunisia. After lunching with his Joint Chiefs, the President reviewed almost three thousand men of Elliott's six-thousand-strong Northwest African Photographic Reconnaissance Wing at nearby La Marsa airfield. But he was equally curious to see the pre-Roman and Roman remains of Carthage and its surrounding region, if possible. Thus, when Eisenhower dissuaded him from flying the next morning to Cairo, as planned, in broad daylight — a flight which could be visually intercepted by enemy fighter aircraft, as Admiral Yamamoto's had been earlier that year in the Pacific — the President gladly agreed to delay his departure until nighttime, on the understanding that Eisenhower would take him "on a personally conducted tour of the battlefields — ancient and modern."[5]

Eisenhower — a student of classical military history — was only too glad to oblige. Sunday afternoon, November 21, 1943, was thus spent in Eisenhower's armored Cadillac, driven by his British female driver, Kay Summersby, and escorted by a phalanx of U.S. Secret Service men in armored vehicles as the President and his Mediterranean theater commander in chief toured a Tunisian battlefield littered with the detritus of recent war, from burned-out tanks to antitank fortifications, and from temporary cemeteries to still-roped-off minefields.

In the presence of his son Franklin Jr., the President "grilled Ike closely, not only on the war that had been fought to a breakthrough by the Allies at Medjez-el-Bab" — where almost three thousand men lay buried, and the famed Hill 609, where American soldiers came of age — but also "on the wars that the Carthaginians had fought in antiquity," as Franklin Jr. related to his brother that night.[6] The President "seemed in good health and was optimistic and confident," Eisenhower himself recalled. "The President's liking for history and his frequent reference to it always gave an added flavor to conversation with him on military subjects."[7] There was, in short, no hint of illness or even fatigue.

As a great maritime empire, the President knew, Phoenicia had dominated the Mediterranean for centuries. The Punic Wars between the Carthaginians and the Greeks, and then the Romans, had led finally to a months-long, historic siege of Carthage — ending in a great naval as well as land assault on the fabled metropolis by Scipio in 146 BC. Thereafter, on the orders of the Roman Senate, Scipio had sacked and demolished the entire city, invoking a curse on anyone who should ever try to rebuild it — or build on it.

In light of the President's discussion of the future of Germany and Berlin with his chiefs of staff, Scipio's actions certainly seemed to have been a salutary example of unconditional enemy surrender. According to the Roman historian Livy, the Carthaginian commander, Hasdrubal, had begged for mercy before Scipio — much to the disgust of Hasdrubal's wife, who threw herself into the flames of a burning building with her children, while her captive husband was taken to Rome in chains to be displayed in Scipio's victory procession, the *triumphus*, along with thousands of chained Carthaginian prisoners, and the spoils of war. And now, several thousand years later, the same proximate city of Tunis had recently been captured from Axis forces in North Africa by the Western Allies, with no escape and after very heavy fighting — yielding on May 12, 1943, more than *a quarter million* Axis prisoners . . .

Berlin, however, was still more than a thousand miles away. Watching a flight of more than fifty U.S. Air Force medium bombers "returning from a tactical mission over the European Continent," the presidential party saw that "some of the 'V's' were not complete, indicating that this particular flight suffered combat losses," the log of the President's trip recorded.[8] There was, clearly, a lot more fighting to be done.

For his part, though, the President seemed visibly moved that, as chief executive and commander in chief of a growing global power, he'd been able to follow in the almost literal footsteps of past conquerors — humbled to be looking at what remained of earlier battlefields. "While traveling through them he speculated upon the possible identity of our battlefields with those of ancient days, particularly with that of Zama," about eighty miles from Tunis, Eisenhower afterwards recalled. There, in October 202 BC, Scipio Africanus, invading from Sicily, had routed the eighty war elephants and forty thousand men of Hannibal's army, using his Roman cavalry and infantry. "So far as either the President or I knew, that battlefield had never been identified by historians, but we were certain, because of the use of elephants by the Carthaginians, that it was located on the level plains rather than in the mountains, where so much of our own fighting," such as Kasserine, took place."[9] Eisenhower therewith recounted for the President the "late battle, the terrain, difficulties encountered, and some of the command personalities"[10] as, in the back of the armored Cadillac, they toured the ground.

Kay Summersby, Eisenhower's driver, found herself starstruck by Mr. Roosevelt, who called her "child," and insisted she dine with him and her boss that evening, before he boarded his plane for the onward journey to Cairo. She was impressed by the President's genuine curiosity — not just about ancient history, but about people. "Child, won't you come here and have lunch with a dull old man?" he'd asked as they sat under a grove of trees, halfway across the battlefield, to enjoy a chicken-sandwich picnic — the sandwiches passed to the President, under the awning, by General Eisenhower. She later wrote she'd "felt as though I had known this vibrant man all my life, as though he were a distant uncle I hadn't seen since babyhood. Mr. Roosevelt had that enviable touch of natural intimacy," she described. He "asked all about my family, about England," as well as posing "keen questions about the role of British women in the war, queries about factory workers, service girls, air raid wardens, and bus 'clippies.'" And "life along the Mediterranean," as she'd experienced it during combat.[11]

How different this was from her meetings with Winston Churchill, who, as his doctor noted, "is not interested in people and cannot follow their thoughts,"

Dr. Wilson reflected in his own diary, as he accompanied the Prime Minister aboard his battleship and ashore at Malta, en route to Cairo. Churchill certainly "likes talking," Dr. Wilson recorded, but was, at best, "an indifferent listener" — one who "has visited nearly all the battlefields and . . . can pick out, in a particular battle, the decisive move that turned the day. But he has never given a thought to what was happening in the soldier's mind, and he has not tried to share his fears. If a soldier does not do his duty, the P.M. says he ought to be shot. It is as simple as that."[12]

14

Two Pieces in a Chess Game

For the President's military advisers the looming clash with the British was worrying. It also troubled General Eisenhower, for reasons both military and personal.

As Eisenhower explained to the President as they toured the Carthaginian battlefields, he'd been summoned to confer in person with the Prime Minister at Malta on November 18. Before flying there, however, he'd gained some idea of what might be afoot, from Lieutenant General Walter Bedell Smith. Smith had warned his boss that Churchill was spoiling for a fight in Cairo. Eisenhower's strategy of fighting in Italy to tie down German forces so that Overlord would face less opposition was, under Churchill's proposed new plan, to be abandoned in favor of a wild new British strategy in the Mediterranean, Aegean, and Balkans.

Smith's prediction of the impending conference at Cairo being the "hottest one yet" therefore had profound personal implications for General Eisenhower. Ike, as he was known to all, had already been informed by the U.S. Navy secretary, Frank Knox, during Knox's recent visit to the theater, that he would undoubtedly be recalled to Washington to replace General Marshall as U.S. Army chief of staff, once Marshall was appointed to the supreme command of Overlord — for the British would undoubtedly insist on the enlarged Mediterranean command being given to a Brit. This was far from a happy thought for Eisenhower. Not only was he loath to leave an active theater of war, but the top Pentagon post was not one, given his relatively junior status in the U.S. Army hierarchy, that he'd ever coveted or in which he would feel comfortable.

Mr. Churchill had made no bones about the new strategy he was going to fight for in Cairo, Eisenhower now told the President. The Prime Minister had said he was looking "forward with great enthusiasm to his meeting with the President, from whom, he said, he always drew inspiration for tackling the problems of war and of the later peace," Eisenhower later recalled. Such rhetoric was, however, an example of the Prime Minister's manner of first circling his prey—a polite preamble before explaining the reason he'd summoned Eisenhower to Malta: namely to press upon the hapless American theater commander the need to postpone or abandon Overlord. The Prime Minister had thus moved on to address "at length," as Eisenhower remembered several years later, "one of his favorite subjects—the importance of assailing Germany through the 'soft underbelly,' of keeping up the tempo of our Italian attack and extending its scope to include much of the northern shore of the Mediterranean"—the Balkans. "He seemed always to see great and decisive possibilities in the Mediterranean, while the project of invasion across the English Channel left him cold."[1]

The President had listened carefully to Eisenhower's report. A casual visitor might have assumed that the President, a charming and genial conversationalist, was merely putting his subordinate at ease. However, most always the President's conversation had a deeper motive. The time he spent traveling with Eisenhower and inspecting the ruins of Carthage certainly proved no exception. In reviewing Eisenhower's conversation and his bearing the President was surreptitiously measuring how well he thought the general would do in the role of U.S. Army chief of staff back in Washington. Or, alternatively, as Allied supreme commander of Overlord. Or even Overlord *and* the Mediterranean—"all Europe"—if the British would agree to this. And how well Ike could handle Churchill.

Would the British agree to the combination of the two theaters, north and south, in Eisenhower's opinion? the President had asked.

Eisenhower had confirmed they probably wouldn't agree to such an expanded supreme command. In fact at Malta Churchill had stated categorically it wasn't an option. The cross-Channel invasion and the Mediterranean *must,* Churchill had emphasized, remain separate entities, separate theaters. The Prime Minister had loyally accepted, he said—though with great embarrassment, given that he had initially promised the job to General Brooke[2]—the President's recent decision, in mid-August that year, that the supreme command of Overlord go to an American officer. Moreover he'd welcomed, he'd said to Ike, the choice of General Marshall for the post, which would ensure Overlord be backed by "an abundance of American power"—and that Britain would therefore not have to shoulder the majority of casualties in the assault. However as prime minister of Great Britain Mr. Churchill had made equally clear he would not be willing to accept an American as supreme commander of *both* theaters combined. Nor would he even accept Eisenhower remaining as commander in chief in the Mediterranean. "I'm sure you will realize, my dear General, that we are quite happy with you," Churchill had stated, "but it would obviously be unfair to us [British] to be foreclosed from both major commands in Europe."[3]

The President had had cause to wince at this—certain that, once Marshall was appointed to supreme command in Britain, Winston would do everything in his power to downplay or delay Overlord, via a British supreme commander in the Mediterranean: promoting yet more audacious southern schemes. These would involve, it seemed, a new invasion of Rhodes, and even Crete—just when they were about to meet Stalin and ought to be presenting a unified commitment to Overlord and the Second Front to be launched in the coming spring!

It was important, though, President Roosevelt was aware, not to lose sight of the positive side of the equation.

"Ike, if one year ago you had offered to bet that on this day the President of the United States would be having his lunch on a Tunisian roadside, what odds could you have demanded?" the President had said with a typical smile during the picnic near Carthage—and the "thought apparently directed his mind to the extraordinary events of the year just past," Eisenhower vividly recalled. "He told me, first, what a disappointment it had been to him that our African invasion came just after, instead of just before, the 1942 elections"—leaving him with a somewhat contrarian Congress to deal with. "He spoke of Darlan, of Boisson and Giraud. He talked of Italy and Mussolini and of the uneasiness he had felt during the Kasserine affair. He told of instances of disagreement with Mr. Churchill, but earnestly and almost emotionally said: 'No one could have a better or sturdier ally than the old Tory!'"[4]

Compared with French leaders at the time, Winston Churchill had certainly been sturdier. But what would happen if Churchill now gained full control of the Mediterranean and Aegean, via a British supreme commander? If Turkey were to join the United Nations struggle against Hitler, "it would result in drawing away support from other operations," as Admiral King and General Marshall had pointed out on board the *Iowa*[5]—turning Turkish requirements into a veritable sump, draining the Allies of significant military resources. Who could say what schemes Churchill would then seek to perpetrate, via General Alexander as his stooge—schemes that would, inevitably, have to be bailed out by the application of yet more American forces once they met serious resistance, leading inevitably to even *more* delay in mounting Overlord?

Alongside the matter of Churchill's impending insurrection, as the President had been driven around the recent Tunisian battlefield, was also the matter of Eisenhower himself. Would Ike, given his relative youth, be able to prevail in the topmost U.S. Army and Air Force hierarchy, if moved back to Washington? He would have nine million soldiers and airmen under his umbrella. But in the weekly meetings of the Joint Chiefs of Staff, how would he be able to handle his exalted older colleagues—especially "Blowtorch" Admiral King?

Certainly, Eisenhower could bring modern battle-experience and a coalition perspective to the U.S. Army and Navy Departments in Washington, the

President could imagine. Studying the young commander in chief, the President could see, moreover, that Eisenhower was, in his genial, friendly way, not unlike himself. Despite his relative youth Eisenhower stood out as a team leader, at once highly intelligent and energetic, a man whose agreeable personality drew people — especially allies — to serve him, not oppose him. In coalition warfare, as in coalition politics, such a man was indispensable; why, then, incarcerate him with fellow American chiefs of staff in Washington, when he could instead be marshaling the disparate Allied coalition forces on the battlefield, leading to the defeat of Hitler in Europe?

As if to voice such second thoughts the President had proceeded to rehearse with Eisenhower, on the historic battlefield, the reasons George Marshall deserved the Overlord post. A post which he had confirmed would go to Marshall, as recorded aboard the *Iowa*.[6] As Eisenhower recalled, "The President spoke briefly to me about the future Overlord commander and I came to realize, finally, that it was a point of intense official and public interest back home."[7]

Eisenhower might be embarrassed, but he was relieved the matter of Overlord's commander was now at least in the open. Admiral King, the night before, had said he would welcome Ike as a new colleague on the Joint Chiefs of Staff Committee — if that was what the President, as commander in chief, so decided. King had made quite clear, nevertheless, that in his view it would be a mistake.

"We now have a winning combination. Why do we want to make a radical change?" King had asked aloud, in front of both Marshall and Eisenhower. "Each of us knows his own role; each of us has learned how to work with the others. Why doesn't the President send you up to Overlord and keep General Marshall in Washington?" Eisenhower recalled Admiral King's words. "'Marshall is the truly indispensable man of this war; Congress and the public trust him, the President trusts him, his associates in the Combined Chiefs of Staff trust him and the Commanders in the field trust him. Why do we change?' He then said that General Arnold, commanding the United States Army Air Forces, as well as a number of others agreed with him."[8]

General Marshall, although present at the discussion, had declined to comment on such "future command arrangements," however — contenting himself with the prim comment that the "President has to make his own decisions. I shall personally have nothing to do with it."[9]

Marshall's refusal to state either a personal opinion or a military view on the matter bespoke his upright, deeply principled character. He could not bear the idea of being accused of putting ambition before country.

Dwight Eisenhower, who admired General Marshall deeply and owed his promotions of the past several years to the U.S. Army chief of staff, had nevertheless been incredulous the general would not be willing to advise the U.S. Commander in Chief as to the pros and cons of such a vital appointment—despite being one of the President's chief military advisers. As Eisenhower later noted in his unpublished memoir, the President's "problem" in Carthage was not simply the looming insurrection of the British, or the command appointment for Overlord that needed to be made, urgently, if the crucial operation was to succeed. It was also the problem of George Marshall himself! "Marshall's sense of duty," Eisenhower later reflected, "was such that he refused even to hint at a preference concerning a future assignment, either to his friends or his superior."[10]

This was, to be sure, of little help to the President, who felt honor-bound to stand by his commitment to make Marshall the supreme Allied commander, but had no idea where Marshall's own wishes or preference lay. George Marshall was, after all, the principal backer and longtime architect of the operation—which would very likely determine the outcome of the war in Europe, for good or ill: victory or defeat. "I just do not know," the President had thus confided to Eisenhower, "what Marshall would like to do because he will not tell me." As president, Mr. Roosevelt would hate to lose the tall, proud, upright Virginian from the Pentagon, "but of one thing I am sure, in military history it is only the name of Field Commanders that are remembered, not that of a Chief of Staff, even though he be the official superior," Eisenhower quoted him.[11] And to illustrate his point the President had said: "You and I know the name of the Chief of Staff in the Civil War [Henry Halleck], but few Americans outside the professional services do." Despite this, as Eisenhower recalled, the President had added "as if thinking aloud: But it is dangerous to monkey with a winning team."[12]

Clearly the President was of two minds—leaving Eisenhower in something of a bind. "I answered nothing," Eisenhower afterwards recorded, "except to state that I would do my best wherever the government might find use for me."[13]

The President had good reasons, however, for holding off on a decision. So long as General Eisenhower remained Allied commander in chief in the Mediterranean, there was no way Churchill could go ahead with crazy capers, or bend Eisenhower to a different strategy from that which had been agreed and formally confirmed by the Combined Chiefs of Staff at Quebec in August. Moreover, what advantage would there be in the President appointing, at this juncture, General Marshall to be supreme commander of an invasion the British were seeking to delay or abandon? Why lose Marshall's great prestige, authority, and advice in Washington for an operation that, if the British had their way, might never take place? Ergo, as Admiral Leahy had argued on the *Iowa*, General Eisenhower should stay as Allied commander in chief in the Mediterranean for as long as it took to get the primacy of the Overlord invasion reestablished, if necessary through a supreme commander for all Europe.[14]

For Eisenhower the wait would be interminable — indeed in his unpublished memoirs he confessed that he "could not escape the feeling that both Marshall and I were something like two pieces in a chess game, each compelled to await the pleasure of the Players."[15]

In his diary the President would note how the fields around Carthage contained "many tanks & trucks destroyed by gun-fire & much barbed wire," as well as "many little cemeteries" — and how different this was to his tour of battlefields in "France in 1918," when he'd been assistant secretary of the Navy. Not only the terrain but also the technology and combination of all arms were reflective, he'd seen, of a new kind of "open warfare," one that was "over in a very short time in comparison."[16] Overlord, he hoped, would be the same: applying massive air power in support of the infantry and armored troops of coalition armies, led by an American, as they relentlessly advanced, and finally made for Berlin.

Back at the White House in Carthage the President, meantime, bade Eisenhower farewell — thanking the young general for taking the time to look after him since his arrival in North Africa. "Father was beaming by the time they got back to the villa," Elliott later recalled — as well as recalling how, according to his brother Franklin Jr., the President, as a seasoned jester, had played a trick on Ike. The President had put a "restraining hand on his arm," before Eisenhower departed, telling him he was afraid he was "going to have to do something to you that you won't like."[17]

Eisenhower had been as mystified as was Franklin Jr. "What was it," the President's son had wondered to himself. "Calling him [Ike] from his theater command? Or sarcasm: promoting him on the spot to some new and bigger command?"[18]

Instead, according to Franklin Jr., it was merely a prelude to the President telling Eisenhower he might need Ike's naval aide, Lieutenant Commander Butcher, to help run the government's Office of War Information back in Washington, when Elmer Davis relinquished the post.

Eisenhower was hugely relieved — assuring the President that would be fine.

The President then disappeared into Casa Blanca for a last brief supper before flying overnight to Cairo.

Against all blandishments by his father, Franklin Roosevelt Jr. refused to accompany him to the Egyptian capital. Instead he insisted on returning to his battle-damaged destroyer, which had to be sailed back to the United States for repairs. The President therefore ate briefly with just his son Elliott, Admiral

Leahy, and his immediate staff. Then at 10:00 p.m., November 21, 1943, he was driven to El Aouina airport to board, once again, the "Sacred Cow," as the big Douglas C-54 was affectionately termed: Air Transport plane No. 950, piloted by Major Otis Bryan.

As president, Mr. Roosevelt was entitled to one of the two special bunks that had been built at the back of the airplane. White House counselor Harry Hopkins — so gaunt that Eisenhower's driver, Kay Summersby, wondered "just how he remained alive"[19] — occupied the other bunk for the overnight flight. The rest would have to make do in their seats.

Without a bed for the nearly two-thousand-mile journey,[20] Admiral Leahy —"dour and quiet," in Elliott's description[21] — resigned himself to a long and fitful night. In his diary the aging former chief of U.S. naval operations would permit himself only a laconic entry, namely that he was "looking forward to a very busy and to a probably very controversial visit in Cairo."[22]

PART TWO

Stonewall Roosevelt

15

Airy Visions

In Berlin the Reichsminister for Propaganda had been tracking the latest press reports from England.

Since the fall of Mussolini and the surrender of Italy to the Allies it had become increasingly clear the Third Reich's best hope of ultimate survival would be a negotiated peace or armistice — either with the Soviets or with the Western Allies. The latter seemed to Goebbels quite doable, thanks to the fine defensive performance of the Wehrmacht in southern Italy and the brilliance with which German forces had rubbed out British attempts to seize Rhodes, Kos, Samos, and now Leros. Press reports from London, moreover, had suggested growing strains in the once-watertight alliance between the United States and Great Britain.

Churchill's speech at the Mansion House in London, before departing for Cairo, had caused Dr. Goebbels to wonder what exactly might be percolating in the highest echelons of Allied war-making — in fact whether the British were getting cold feet over the long-promised cross-Channel invasion. They seemed more worried, in particular, about their colonies. The British prime minister had extolled Russian successes on the Eastern Front, but with regard to British interests had repeated that, as he'd declared the year before, he "did not consider it any part of my duty to liquidate the British Empire" once the war was won. "I do not conceal from you that I hold the same opinion today," Churchill had stated, openly reaffirming his stance that, in terms of the Atlantic Charter and the postwar "Four-Power Agreement" of the United Nations, just agreed at the conference of foreign ministers in Moscow, his aim would only be to provide "food, work, homes" for the British people. He would not, he'd declared before Parliament, get embroiled in "airy visions" or follow "party doctrines" or "party prejudices" or feed "political appetites" or seek to satisfy "vested interests." The war had still to be won, he'd said — and it wouldn't be easy. Churchill had also sounded worried about Hitler's warning of *Vergeltungswaffen,* or weapons of revenge, that would "certainly call for the utmost efficiency and devotion in our firewatchers and Home Guard," as well as require the "fortitude for which the British nation has won renown." In referring to British efforts in the coming phase of the war he had given nothing away, however, save to say "1944 will see the greatest sacrifice of life by the British armies, and battles far larger and more costly than Waterloo or Gettysburg."[1]

At the end of his speech Churchill had openly appealed for unity between the partners in the Allied coalition, and among their loyal supporters in the press and in ordinary homes. The "good will that now exists throughout the English-speaking world" should be used "to aid our armies in their grim and heavy task. Even if things are said in one country or the other which are untrue, which are provocative, which are clumsy, which are indiscreet, or even malicious, let them be stated without heat or bitterness. We have to give our men in the field the best chance. That is the thought," he'd declared, "which must dominate all speech and action."[2]

In reading the transcript of Churchill's speech, Goebbels had been intrigued. By a huge majority the United States Senate had, on November 5, passed the Connally Resolution, recording Congress's determination to pursue the war "until complete victory is achieved." Congress had, moreover, given full Senate support to "the necessity of there being established at the earliest practicable date a general organization, based on the sovereign equality of all peace-loving states, and open to membership by all such states, large and small, for the maintenance of international peace and security": a United Nations authority.[3]

There seemed, therefore, to be a growing disconnect between British and American attitudes to the war. "In [U.S] government circles the isolationists will clearly be only an insignificant minority in the foreseeable future," Goebbels had reluctantly noted[4] — but in England the opposite appeared to be the case: a rejection of "airy visions." Churchill's latest objection to possible decolonization policies — the "liquidation" of the British Empire — was evidence, perhaps, of a kind of "last-stand" imperial mentality in England. The British military fiasco in the Aegean and the Führer's latest threats to use revenge weapons, in response to the carpet bombing of German cities, were clearly prompting much soul-searching in London, Dr. Goebbels could see.

Here, surely, there was a chance for Germany to exploit the difference between the two English-speaking allies: a split that seemed to be widening, according to newspaper articles in Britain, concerning the conduct and strategy of the war itself. Churchill's appeal for unity in England was surely a sign of this — in fact the wily Goebbels detected a potential propaganda lever he might use, if the Führer agreed. From German intelligence there was as yet no actual evidence of strategic disagreement at the highest levels, behind the scenes, but that did not mean it wasn't happening. It was certainly happening in print. "As far as the Second Front is concerned, the English are doing their best to shift the weight of the war onto the Americans," Goebbels reflected.

The British were getting cold feet. "Their press is full of reports and statistical demonstrations to the effect that the country is simply not able to face the sacrifice in blood of a Second Front. Having got numerous European countries to shed their blood, the English, ruled by the wealthy, would now like to make the Americans take the heat, so that as far as possible the English themselves will end the war with their population intact. But the Americans won't let themselves be taken for such a ride."[5]

In his demonic, endlessly exploitative way, the Reichsminister remained a discerning analyst. The British were, it seemed to him, losing heart. The recent surrender of the last British forces on Leros had only intensified, in Goebbels's account, the recognition in Britain that Germany was by no means beaten—and that a cross-Channel invasion could be suicidal.[6] This would make the English more than ever dependent on Stalin, the communist "mass-murderer," to fight Germany, while the British watched out for themselves and their empire.

News that Churchill and the President were on their way to Cairo—to be joined there by Stalin, it was presumed[7]—thus stunned Goebbels. To his chagrin it suggested that, far from being riven by "malicious" tongues and disagreement, the Allied leaders were, in fact, united in their determination to demand the unconditional surrender of Nazi Germany.

16

The American Sphinx

A LIFELONG, PASSIONATE PHILATELIST, Franklin Roosevelt had always been enamored of geography. As soon as they were airborne over Tunis he'd gone to bed in the rear of the Sacred Cow, but had asked to be wakened early, in order to see the sun rise over Cairo as they came in from the Mediterranean.

Once awake, however, the President had asked Major Bryan whether, as pilot, he could turn the plane first south, over the Egyptian desert, and reach the Nile at a higher point. From there the pilot could bank north, and follow the river's twisting course so that they might see the fabled waterway in the dawn light as it reached the ancient city of Cairo — and the pyramids.

Major Bryan had said yes. A new course had therefore been set. Until then they had followed the long Mediterranean shore, in the dark, but now, reaching the Egyptian border, the pilot banked right, flew south, then east across the arid desert until they reached the famous river, then swung north.

As the day began to break on November 22, 1943, the Sacred Cow followed the snaking course of the Nile — the President treated to a magical approach to the Egyptian capital as the plane swooped low over ancient-looking felucca boats, rigged with lateen sails. For Roosevelt, whose love of sailing went back to his childhood, the picture was breathtaking. "We came in the 'back way,'" the President recorded in his diary, "so as to avoid German planes & from the desert saw the Nile 100 miles So.[uth] of Cairo — an amazing scene — & followed the narrow belt of fertile fields, with many villages, until we came to the great pyramids."[1]

They were in no hurry. "Otis Bryan circled the plane constantly to give the Boss a good view," the White House security chief, Mike Reilly, recalled — recording as well the President's somewhat sad remark, looking down at the pyramids: "Man's desire to be remembered is colossal."[2]

In Egypt, due to strict radio silence, the President's decision to take his own romantic route to the capital had caused something of a "flap," however. "Our unreportable sight-seeing jaunt set the Cairo air headquarters on its ear," Reilly recalled, enjoying in retrospect the idea of the anxiety it had caused the British. It was a good two and a half hours after "plane number two of our party had arrived from Tunis" before the Sacred Cow finally touched down in Egyptian pastures — there to find its late arrival had "caused some concern at the field as to the President's safety. Two different groups of fighter-planes had been at the appointed rendez-vous at the scheduled times but each failed to make contact and eventually had to return to their base for refueling."[3]

The relief of the airfield staff was palpable. From RAF Cairo West the President was driven by limousine straight to the personal residence of the U.S. minister to Egypt, Alexander Kirk, where it had been arranged that Mr. Roosevelt would stay. It was "7 m[iles] from Cairo — very comfortable," the President jotted in his diary — and with regard to the pyramids was wonderfully "close to them," he added, happily.[4]

Prime Minister Churchill had arrived two days before, and was nearby, in a similar villa, as was Generalissimo Chiang Kai-shek, who also arrived that morning. Near the pyramids was the Mena House Hotel, almost a mile away — the whole area reminding participants and visitors of the complex of buildings and villas in which they'd gathered at Casablanca, Morocco, earlier that year. "As at Anfa, the whole area is surrounded by barbed-wire fences, guards, guns, tanks, etc.," Harold Macmillan noted in his diary. "Owing to the leakage in the *Daily Mail* and elsewhere, the security arrangements are fantastic. It is impossible to move anywhere near the 'great' without passing two or three barriers and showing innumerable 'passes.' The military conferences, etc., take place in the hotel (as at Anfa) — the secretariat and the officers (other than the highest who are in villas) are lodged there."[5]

The President's Filipino cooks and stewards had followed from Oran, and at the Kirk villa, as it was called, they prepared all the President's meals. The house was "of medium size and is beautifully furnished," Lieutenant Rigdon recorded in the President's log. "It also has a lovely flower garden in the rear with an overlooking patio," where Mr. Roosevelt could spend his "leisure moments" — assuming there were any.[6] Determined that the President be spared "gippy tummy" — a common stomach complaint in Egypt — large amounts of bottled water, milk, and even a Thanksgiving turkey were unloaded, while other supplies were kept ready for the onward flight to Tehran, if the President opted to fly there, over the northern Iranian mountains.

The President's doctor, Admiral McIntire, was against the idea, in view of the President's cardiac history, his relatively high blood pressure, and the medical dangers of high-altitude flying. There was also the possibility of enemy interception, or mechanical malfunction; Major Bryan was thus asked to fly to Tehran and back ahead of the President's trip, taking Agent Mike Reilly with him to check out security conditions when they reached the Iranian capital.[7]

In the meantime Ambassador Averell Harriman had come straight to see the President on his arrival in Cairo, at 10:30 that morning—bringing the latest news from Moscow. The summit in Tehran would start on November 27, in five days' time.

The President was relieved. Five days, though! Aware that Churchill and his team would be armed with countless arguments for a new military strategy, Roosevelt first lunched, then donned his metaphorical armor to meet the Generalissimo, the bewitching Mrs. Chiang, and Mr. Churchill and Churchill's daughter Sarah, who was accompanying the Prime Minister. A kind of benevolent bonhomie would be the President's approach—similar to the "extreme hospitality" that had worked so well in putting down Churchill's attempted insurrection back in May, that year, when Churchill had arrived in Washington with 160 advisers and staff to argue against the notion of a Second Front.[8] He would simply not allow himself or his team to be provoked or enticed into any major confrontation before meeting the Russian "Tsar" in Tehran. Instead, as agreed aboard the *Iowa,* they would simply stonewall.

Harriman, for his part, confirmed his earlier cables from Moscow: namely that Marshal Stalin was counting on Overlord. Any notion the British were spreading that the Russians would prefer Allied operations in the Mediterranean in order to draw Wehrmacht forces away from the Eastern Front were false. Stalin was, Harriman reported, as suspicious of Allied Aegean and Balkan diversions that would delay a Second Front as was the President—and in Ambassador Harriman's view Stalin would happily act as the President's "enforcer" in Tehran. In a three-person summit where Churchill would be in the minority, the President calculated, the British would *have to* back down.

If this sounded unkind or underhanded toward the British, it was. It was nevertheless better than permitting a breakdown in the Anglo-American military coalition, even before they got to Tehran. The plan had to be carried out, then, whatever taste of deceit it might leave. It meant keeping Generalissimo Chiang Kai-shek and Mrs. Chiang at hand while dealing with Churchill over the next several days. As the President noted almost facetiously in his diary that night, "This p.m. the Prime Minister & his daughter Sara [*sic*] and the Chiangs came to call. Then we drove out to see the Pyramids & my old friend the Sphinx. We all dined (not the Sphynx) at my villa."[9]

For Churchill, who had spent so many days penning his indictment of American strategy and operations, the President's obvious attempts at diversion were galling. Yet without such a strategy, the President was aware, the war against Hitler might still be lost.

17

Churchill's "Indictment"

CODE-NAMED SEXTANT, the Cairo Conference, prior to Tehran, would now prove for Churchill one of the most frustrating experiences of the entire war — as it would for his entourage, few of whom appeared to understand the President's maneuver.

As Dr. Wilson had observed, Churchill had spent considerable time working on his formal censure of Allied strategy on the journey from England,[1] to be delivered as a grand "indictment of our mismanagement of our operations in the Mediterranean" before a full plenary session of the Chinese, American, and British leaders and their chiefs of staff. "The shadow of 'Overlord,'" as the Prime Minister had written, ominously,[2] haunted them all — and for this shadow he proposed to blame the American chiefs. The military policy laid down at Quebec had been maintained by the Allies "with inflexible rigidity and without regard to the loss and injury to the allied cause thereby," as he'd put it in the latest draft he'd drawn up in Alexandria's harbor, on his arrival there on November 20, before the President reached Egypt. The "fixed date" for Overlord "will continue to wreck and ruin the Mediterranean campaign," he complained, moreover would doom "our affairs" in "the Balkans," and leave the Aegean islands "in German hands." It was "common knowledge in the Armies that the [Mediterranean] theater is to be bled as much as necessary for the sake of an operation elsewhere in the Spring": Overlord.[3] He therefore wanted all further movement of British troops from the Mediterranean to England to be stopped,[4] and told his British chiefs of staff he wanted "the capture of Rome" at the beginning of January and also "the capture of Rhodes at the end" of that month: January 1944.[5]

General Brooke — though he largely agreed with the Prime Minister regarding the chances of Overlord being a potential disaster — found himself deeply worried. "I wish our conference was over," he'd penned in his own diary on arrival in Cairo, before the conference even began. "It will be a most unpleasant one, the most unpleasant one we have had yet, and that is saying a great deal. I despair of ever getting our American friends to have any sort of strategic vision. Their drag on us has seriously affected our Mediterranean strategy and the whole conduct of the war. If they had come wholeheartedly into the Mediterranean with us we should by now have Rome securely, the Balkans would be ablaze, the Dardanelles would be open, and we should be over the high way to getting Rumania and Bulgaria out of the war."[6] He blamed himself for not having resigned his post rather than "compromise," yet he was too sensible not to realize the risks the Prime Minister and the British team were taking in demanding a change of strategy at such a late date, only five months before Overlord—and a few days before the Tehran Conference began. The next day, he noted, the "PM kept us up till after 1 am. He was in a very excitable mood, I am not happy at the line he proposes to take in approaching the conference."[7]

The President, however, seemed to handle Churchill with his customary charm on November 22—prompting Churchill to tell Brooke he was "very pleased with results of his talks with the President, and thinks we shall not have so very much difficulty. Personally, I doubt this," Brooke added that night.[8]

Neither Churchill nor Brooke seemed to recognize what the President was up to, at the time. Or even afterwards. Writing a decade later, when annotating his diary for partial publication, Brooke would complain that the "whole conference had been thrown out of gear by Chiang Kai-shek arriving here too soon. We should *never* have started our conference with Chiang; by doing so we were putting the cart before the horse. He had nothing to contribute towards the defeat of the Germans . . . Why the Americans attached such importance to Chiang I have never discovered."[9]

Even Churchill's shrewd military chief of staff, General Ismay, was no wiser. For his part he would later lament how, thanks to whole days spent discussing China and the Southeast Asian theater with Chiang Kai-shek, as well as with Admiral Louis Mountbatten, the Allied supreme commander in Southeast Asia, and General Joseph Stilwell, Chiang's American chief of staff, there was little or "no time left to reach agreement as to the exact line which should be taken with the Russians about a Second Front in Europe."[10]

"We should have started this conference by thrashing out thoroughly with the Americans the policy and strategy for the defeat of Germany. We could then have shown a united front to Stalin," Brooke bemoaned in retrospect—ignoring the fact that "the policy and strategy" had already been thrashed out at Quebec.[11] Churchill, though, was of like mind. As he later narrated, what he'd feared on hearing the Chinese generalissimo would arrive in Cairo *before* the Tehran summit, instead of afterwards, "now in fact occurred. The talks of the British and American Staffs were sadly distracted by the Chinese story, which was lengthy, complicated, and minor." The President, to Churchill's chagrin, "was soon closeted in long conferences with the Generalissimo. All hope of persuading Chiang and his wife to go and see the Pyramids and enjoy themselves till we returned from Teheran fell to the ground, with the result that Chinese business occupied first instead of last place at Cairo."[12]

Thus had commenced Franklin Roosevelt's most devious ruse of the war. As the President explained to his son Elliott, who arrived on the afternoon of November 23 — day two of the President's stay in Cairo — "Believe it or not, Elliott, the British are raising questions and doubts again about that western front."[13]

"About OVERLORD?" Elliott responded, amazed. "But I thought that was all settled at Quebec!"[14]

"So did we all," Roosevelt sighed — satisfied, at least, that General Marshall would keep the British at bay in the Combined Chiefs of Staff meetings. These, the President had insisted, must be confined for the moment to the topics of China and Southeast Asia, to prevent the British from tabling a change of Overlord strategy before they all left for Tehran.[15]

To Elliott the President also confided his recognition that Chiang Kai-shek was not fighting the Japanese with much determination, and was not really interested in using Chinese forces to liberate Burma. Why should he, after all, when it was ultimately the British who would then seek to recolonize the country? "Chiang's troops aren't fighting at all — despite the reports that get printed in the papers," the President shared his personal view. "He claims his troops aren't trained, and have no equipment — and that's easy to believe. But it doesn't explain why he's been trying so hard to keep Stilwell from training Chinese troops. And it doesn't explain why he keeps thousands and thousands of his best men up in the northwest — up on the borders of Red China,"[16] where Mao Tse-tung's communist Chinese troops were located.

Political considerations were thus part and parcel of deciding military strategy now, the President accepted. Churchill was seeking to preserve the British Empire and the troops to police it, from Palestine to India and Hong Kong; Chiang was preparing to reopen his civil war with Mao's forces. FDR alone had his eye on the prize. Without Overlord how could the Third Reich be defeated? Only if the Wehrmacht were to be forced to fight hard on two fronts, east and west, could Hitler be crushed.

In an attempt to gain better insight into Churchill's thinking the President had also asked the U.S. ambassador to London, Guy Winant, to come to Cairo. Winant had traveled with Churchill's party. He duly explained to the President how Churchill had gone around London claiming it was Stalin himself who wanted more operations in the Mediterranean, rather than a spring Overlord: the Prime Minister claiming "that Marshal Stalin is chiefly interested at the present moment in stretching German resources and his interest in a second front was not nearly so great as it had been."[17] This simply did not square, however, with what Ambassador Harriman was reporting from his dealings with Stalin in Moscow — another reason for waiting to hear from Stalin's lips, as commander in chief of the Soviet Armies, what was the truth.

In the meantime it was crucial for the Western alliance that there be no meltdown in Cairo. For all that the President had asked his chiefs to keep their cool, tempers in the Combined Chiefs of Staff meetings had quickly become frayed. In discussing Admiral Mountbatten's plans for an amphibious attack on the Andaman Islands as Allied supreme commander in Southeast Asia, for example, the Combined Chiefs of Staff almost came to fisticuffs. General Brooke — skeptical, he claimed, of the project's feasibility — had suddenly suggested that the U.S. landing craft would be better used in a renewed amphibious invasion of Rhodes and the Aegean islands. This suggestion really scattered the pigeons, the President afterwards heard. "Became quite open with all cards on the table, face up at times," General Arnold recorded in his journal.[18] General Stilwell, in his own diary, was blunter. "Brooke got nasty and King got good and sore," he recorded. "King almost climbed over the table at Brooke. God, was he mad. I wished he had socked him."[19]

Churchill's military chief of staff, General Ismay, was of little help in reporting the Combined Chiefs' discussions to his prime minister. Like many staff officers, the affable and rotund Ismay tended to tell his boss what he thought the Prime Minister wanted to hear, rather than the truth. Thus the British team, forced to spend the first days of the conference considering strategy in China and Southeast Asia rather than the Mediterranean, tried desperately to link the two — refusing to back any projected operations in the Far East that might take away from the Prime Minister's intended operations in the Aegean and Mediterranean, despite knowing these had *not* been agreed by the Combined Chiefs of Staff, and that such arguments were infuriating their American counterparts.

By day three of the conference, however, the President's policy of patient stonewalling could be extended no further. At 11:00 on the morning of November 24, the second plenary session began, under the President's chairmanship — and Winston Churchill was finally able to read out his "indictment." It was the Prime Minister's chance to have his threatened showdown, in the President's villa.

18

Showdown

THE PRESIDENT OPENED the session in his most friendly and avuncular way by saying that, having discussed China and the Southeast Asian theater in some detail, they would now have the opportunity to discuss, very briefly, Europe. "Discuss" — but not decide. "Final decisions would depend on the way things went at the conference shortly to be held with Premier Stalin," the President made clear — ruling out, in a few words, British hopes that they could overturn the Quebec strategy *before* going to Tehran. Churchill had blanched.

"There were some reports," the President explained, that Premier Stalin was "only concerned" with Overlord, "to which he attached the highest importance as being the only operation worth considering." Other reports, however, claimed "that Premier Stalin was anxious that, in addition to Overlord in 1944, the Germans should be given no respite throughout the winter, and there should be no idle hands between now and Overlord. The logistic problem was whether we could retain Overlord in all its integrity and, at the same time, keep the Mediterranean ablaze." Lest there be any doubt, the President gave his own view of what they had to expect in Tehran. "Premier Stalin," he predicted, "would be almost certain to demand both the continuation of action in the Mediterranean, and Overlord" — but with Overlord as the primary goal.

The President had asked his ambassador to Moscow to address the U.S. Joint Chiefs of Staff in person earlier that morning, after breakfast, at his villa, in order to put them too in the latest Moscow picture. Harriman had duly warned that, if Overlord "were to be abandoned," then in his opinion the Western Allies had better produce a genuine alternative — i.e., "an operation equally offensive in nature" — if they really expected Russia to go on fighting the Germans. Especially if — as he'd been assured in Moscow was the Soviet "intention" — the Western Allies wanted the Soviets to join "the U.S. and British in the war against Japan as soon as Germany capitulated."[1]

General Deane, for his part, had disagreed, feeling the Soviets wanted more "immediate pressure" to be placed on the Germans in the Mediterranean, to relieve current Wehrmacht pressure on the Eastern Front. In Deane's view the Russians were less concerned about a date for the Overlord invasion than with immediate help.

Once again, the President was open to different views — but chose in this case to reject such supposedly professional military advice. Deane was a good man — former secretary of the Combined Chiefs of Staff — but he was a pen-pusher, not a fighting soldier. As head of the U.S. Military Mission in Moscow he had not even met Stalin — the one individual who decided everything there.

The President's judgment was here at issue, at a critical moment of the war. He trusted Harriman's insight rather than that of General Deane — for in a dictatorship like the Soviet Union, only the dictator's view ultimately counted. As the President explained to the plenary meeting, operations in the eastern Mediterranean would not affect the outcome of the war, in terms of defeating the Third Reich — whereas Overlord manifestly would. To the question where the Germans would go if they were evicted from the Aegean islands, the answer was "nowhere," he pointed out — and this was equally true of the Allies. Unless Turkey agreed to become a belligerent, the eastern Mediterranean was a dead end. Even if Turkey *were* to join the Allies, its inclusion among belligerents would be a logistic drain on Allied resources at best; at worst it would be a quagmire if, as was likely, the Germans persisted in the way they were fighting in Italy and the Balkans.

The President then turned to Mr. Churchill to offer *his* views.

This was Churchill's moment. He had spent the past two weeks dictating, revising, and honing his great "indictment" — and with it on the table before him, he announced his opposition to current Allied policy.

In a long, lawyerly presentation, the Prime Minister ridiculed Allied efforts in the Mediterranean since the invasion at Salerno; criticized the buildup of U.S. air forces in Italy rather than the reinforcement of Allied ground forces there; and insisted upon the capture of the Eternal City of Rome — "for 'whoever holds Rome holds the title deeds of Italy,'" he declared with his characteristic appeal to emotion. The Aegean islands, including Rhodes, should be retaken in the coming weeks and months, he insisted. By sending two divisions, with their landing craft and strong Allied naval forces, the Dardanelles could be used to put pressure on Germany's allies Hungary, Romania, and Bulgaria to change sides, like the Italians . . .[2]

General Brooke, in his diary, thought this a "masterly statement": a brilliant denunciation of the "tyranny" of Overlord, which would "help us in our deliberations" — deliberations to delay and if possible abandon D-day in 1944.[3]

The President and the U.S. chiefs thought the opposite. What on earth, the American participants wondered, was the point of making it a primary Allied objective to "liberate" the rest of Italy, not to mention those nations, like

Austria, Hungary, Romania, and Bulgaria, that were *still* fighting as allies of the Third Reich? Why such a determined British attempt to delay the planned invasion of France, and sabotage the crucial campaign to reach Berlin and end the war in Europe as swiftly as possible, before moving on to Japan?

Perhaps the most unfortunate part of Churchill's peroration, as he read out his long indictment, was not the Prime Minister's obsession with Rhodes, Turkey, and the Dardanelles so much as the growing implication the British were simply afraid of head-to-head confrontation with the Wehrmacht in battle. Behind Churchill's oratory the U.S. chiefs now began to sense a somewhat shameful appeal *not* to fight the Germans in open, major combat, but instead to run away, and fiddle around the periphery of Europe, imagining they would be less costly ventures. And this, even though it was clear to the American team that, to judge by rising casualties in Italy, a Mediterranean strategy might well end up costing *more* lives in the mountainous regions of southern Europe than a straightforward, decisive campaign in the largely flat terrain in northern France, using ever-growing numbers of U.S. tanks, artillery, and armor, and backed by overwhelming Allied air power.

Given that the Germans had already murdered or starved to death some two *million* Russian prisoners, and that Russian losses in battle were worsening on the Eastern Front — where the German Army had recently launched a counteroffensive — it seemed unlikely Marshal Stalin would be impressed by Churchill's alternative Aegean strategy. "Mr. Churchill made a long unconvincing talk," Admiral Leahy noted in his own diary that night, "about the advantage of operations in the Aegean Sea and against the Island of Rhodes" — the admiral contemptuous of what he considered the Prime Minister's disloyalty to the Quebec accords.[4]

In the end President Roosevelt, as chairman of the plenary meeting, had attempted to emphasize the positive — while putting the British team in its place. From his briefcase at the table he thus took out the figures of comparative contributions of the United States and Great Britain to the war effort, and laid them in front of the Prime Minister and the British chiefs of staff.[5] With the Russians putting two *hundred* divisions in the field on the Eastern Front, and an American army of up to ninety divisions to be put into the field in France in the wake of Overlord, the Western and Eastern Fronts could crush the Wehrmacht between them — the same two-front strategy which, after all, had brought slow but absolute victory in North Africa. The Aegean, by contrast, was not only a red herring, but unworthy of great allies.

Sadly, the President's firm yet polite insistence that they stick with the Quebec agenda yielded no change in Churchill's views, or those of his British military advisers. The British team simply stuck to its guns over its appeal for more immediate operations in the eastern Mediterranean. "I put forward our counter proposals for continuing active operations in the Mediterranean at the expense of a postponement of Overlord date," Brooke noted proudly in his private journal—surprised that the American team, looking to the President for guidance, simply declined to discuss the matter further. The showdown had come—but with the President determined still not to permit a revision of Quebec before they met Stalin, or the breakdown of the conference in the meantime.

The next day—with two more days of talks to go—the President attempted to keep U.S.-British "community relations" at least on an even keel. There was a cathedral service in Cairo—with security cast to the four winds, and "everyone present at the service but the cat and the dog," as General Arnold noted humorously in his diary. "Camels and caravans, little ones and big ones, all heading toward the Pyramids or away from them; donkeys and sheep and goats, Arabs and more Arabs."[6] In the evening, at his villa, the President held a Thanksgiving meal, to which he invited the Prime Minister and his entourage—his daughter Sarah, Anthony Eden, Dr. Wilson, the Prime Minister's secretary, and his security officer, Commander Tommy Thompson. The President even made a point of saying, in his toast, that "large families are usually closer united than are small families; and that, this year, with the United Kingdom in our family, we are a large family and more united than ever before."[7]

Not even the President's charm, however, could conceal the fact that relations between the U.S. and British chiefs of staff had become fraught—prompting Admiral Leahy to note in his diary that in the "afternoon session of the Combined Chiefs of Staff held in camera" (so that the secretarial staff and other attendees not be witness to the extent of dissension) "we discussed at length and without any agreement a British proposal to delay cross-channel operations in order to put forth more effort in the Aegean and in Turkey."[8] To disconcert the British, the American chiefs had tabled, for their part, "Proposal A," as a countermemorandum. This asked for the "immediate" appointment of a single supreme commander for all Europe—a commander who would "exercise command over the Allied force commanders in the Mediterranean, in northwest Europe, and the strategic air forces": a commander in chief, Europe, in other words, who would report to the Combined Chiefs of Staff and the national leaders of the U.S. and Great Britain, in order to effect the "common, over-all objective—Defeat of Germany."

It was to no avail.

As the President had feared, the American counterproposal met with a furious rebuttal from Mr. Churchill, who gave his confutation in writing to the President that day.[9]

The standoff was now formalized, ending all hope of the two teams traveling to Tehran other than as still-feuding allies. The British chiefs of staff simply continued to insist that recent "major developments" in the Mediterranean warranted the abandonment of a "fixed date" for Overlord, and claimed "departures" from the Quebec agreements were "not only justified but positively essential." The notion of Overlord "shortening the war" was "entirely illusory," they declared in writing. Overlord would "inevitably paralyze action in other theaters without any guarantee of action across the Channel."[10]

Thus had ended the Allied standoff on Thanksgiving Day, November 25, 1943.

The following day, Friday, November 26, proved even worse — open insults now being traded in the final Combined Chiefs of Staff meeting in the afternoon, called to discuss operations in Burma.

"At 2.30 met Americans," Brooke recorded. "It was not long before Marshall and I had the father and mother of a row!" — this time in front of some forty-four witnesses. Once again all staff officers and stenographers were expelled from the room, leaving just the main service chiefs of the two countries to pursue an "off the record meeting."[11]

Admiral Leahy was beside himself with fury that his British counterparts were willing to back their prime minister to the hilt in canceling amphibious operation plans off Burma — the better to use the landing craft to attack Rhodes, of all places! "The Prime Minister seems determined to remove his landing ships from that effort," Admiral Leahy noted, "and in a discussion that became almost acrimonious at times, I informed our British colleagues that the American staff declined to recede from our present planned operations without orders from the President."[12]

The President, for his part, didn't waver, or change his orders. Overlord as planned and scheduled would go ahead, he insisted; his stonewalling had kept the peace — barely, but effectively. "Today I wound up all that could be accomplished," he noted in his diary for November 26 — "a really successful meeting & a good announcement to be given out in 4 days," when they were no longer there. Overlord was still on track. Meanwhile "we are off to Teheran in the morning" — to meet the Russian dictator and decide the future of the world.[13]

PART THREE

Triumph in Tehran

19

A Vision of the Postwar World

THE TEHRAN SUMMIT would be the most consequential meeting of the war. In the ancient capital of the Persian empire the President hoped at least to nudge his two allies — the one a model of colonial exploitation, the other a model of ruthless communist dictatorship — toward a postwar democratic vision. And this, while maintaining the unity of a United Nations military coalition in struggling to defeat Nazi Germany and the Empire of Japan, and ensuring the majority of people in his own country were willing to back him in a new American role: seeking and guaranteeing postwar peace and economic development on a global stage.

Could such an idealistic vision possibly be made real on the anvil of what was already the most destructive war in human history? In terms of the future, Stalin had never evinced any real interest in the sovereignty or development of noncommunist countries. For his part Churchill had zero interest, as he'd recently said in his speech in London, in decolonizing or dissolving parts of the British Empire. Roosevelt, however, saw himself as a visionary — or practical idealist, as he had once described himself in a letter to the South African premier, Field Marshal Smuts.[1]

Despite the wordy communiqués of the recent Moscow Conference of Foreign Ministers there was as yet no warranty that the summit at Tehran would lead, in reality, to a postwar security system more effective than the League of Nations had been. Yet Roosevelt, as president, while still in office, wanted to give the leader of the Soviet Union and his colleagues an opportunity, at least, to willingly take partial responsibility for the security of the postwar world. Similarly, he hoped Britain could be urged to see its future in a genuine commonwealth of English-speaking nations, rather than continued British exploitation of an oppressive colonial empire that would, he told his son Elliott, only incite resistance and revolution in the coming years. He could not compel either Stalin or Churchill to agree on, let alone work for, such a future — but he could encourage them to give their personal support to his postwar concept of the United Nations authority; a Security Council of the United Nations; and to his notion of the "Four Policemen," who would cauterize and put out fires and flare-ups that threatened world peace: namely the United States, Russia, Great Britain, and China.

Given its burgeoning economic potency and its growing military power, the United States would be able to point the way, the President felt, following defeat of the Axis powers. And the only thing that could hinder that defeat would be if American forces were to become bogged down in the Mediterranean, Aegean, Balkans, and Dardanelles instead of striking across the English Channel in May.

The May 1944 timing of Overlord was crucial, the President was aware. The war would soon be entering its third year in the eyes of American voters. There was still great idealism and hope at home — but would it survive if hostilities, so far from American shores, were prolonged, while the British deliberately postponed Overlord in favor of vague and ill-considered "opportunities" elsewhere?

In Cairo the President had summoned his assistant secretary of war, John McCloy, to his villa. At the President's request McCloy had produced a memorandum on "posthostility policy" in which he noted two "tendencies" that were bound to grow, the closer the Allies came to an end to the war. "Stimulated on the part of our soldiers by their wish to get home," the first tendency, McCloy had reported, was already a growing desire to simply "liquidate the European involvement" — i.e., to win and leave. Or just leave! The other "tendency" was the inevitable hostility to the United States that would increase in Britain and elsewhere, "now that the war is on its way to being won and the invader is no longer at the door." For in McCloy's view the current "dependence on the U.S." — economic and military — would produce an unavoidable adverse reaction.[2] In short, McCloy argued, Americans would want to go home — and Europeans would be glad to get rid of them. As had been the case after World War I.

It was in this context that the assistant secretary of war saw the crucial importance of Winston Churchill — in America. Having helped get the United States to focus its primary efforts on Europe, the Prime Minister now surely needed to help "convince America that she must enter the administration of the peace."[3] Churchill had given a beautiful speech on this subject at Harvard University in September, after the Quebec Conference and agreements.[4] But his latest attempts to sabotage D-day threatened to trigger a tragic American turn *away* from Europe, once again, as after 1918. In McCloy's view it was crucial that the American public should see the liberation of Europe as a great *American* achievement — one that would make the maintaining of postwar peace in Europe a worthwhile *American* objective, after the expenditure of so much American blood and treasure.

Churchill's current antics were, in McCloy's eyes, not simply the airing of an honest disagreement between allies, but a profoundly dangerous twist, threatening American willingness to prosecute the war in Europe — and with potentially tragic consequences. It was vital to convince not only U.S. "leaders, but its citizens, that the United States has a major part in directing the war," McCoy contended[5] — in fact, *the* major part.

In a nutshell, Americans at home should see U.S. forces not simply as auxiliary firemen from a different city, helping European democracies put out the Nazi fire, in the analogy the President had used to promote his Lend-Lease policy before Pearl Harbor, but as the captains of the current war — and also the postwar. "It is vitally necessary to indoctrinate the American people to a recognition of the national responsibility of the country in world affairs," McCloy wrote. "It is essential that the people of America become used to decisions being made in the United States," he continued. "On every cracker barrel in every country store in the U.S.," he'd added, memorably, "there is someone sitting who is convinced that we get hornswoggled every time we attend a European conference."[6]

The President liked the word. Moreover, he agreed to his marrow with McCloy's presentation. Churchill simply could not be allowed to delay or sabotage D-day, nor could he be permitted to use American forces in wild ventures in the Aegean and Balkans that would only convince American voters they were being "hornswoggled." The summit in Tehran was thus a chance to show *Americans* that the U.S. was now in charge: taking full military responsibility for the struggle against the Axis powers, in particular the swiftest possible defeat of Nazi Germany.

McCloy's Cairo memorandum — which closely reflected the views of the secretary of war, Henry Stimson — thus served to stiffen the President's resolve at a critical moment in world conflict.

Churchill should not be permitted, all felt, to ditch his Quebec undertaking. By the same token, however, it was important to avoid causing him to resign as prime minister, as he'd reputedly threatened, or to turn the British cabinet against American postwar policy, as was all too possible. The President would have great need of the Prime Minister, not only as America's junior military ally — his "active and ardent lieutenant," as Churchill had publicly called himself[7] — but as a revered British statesman and leader in the eyes of Americans: encouraging the American people to step up to the postwar plate. Also, if at all possible, to encourage Churchill to see decolonization as a great postwar *ideal* to be embraced, not rejected. Simultaneously, the President must offer the Soviet Union, with its four hundred army divisions under Stalin's direct command, an opportunity also, like America, to grow into a new potential role of international responsibility, in order to help guarantee postwar world peace.

This, in short, was the challenge.

The daytime flight on Saturday, November 27, 1943, at least, allowed the President to look down at the fabled cities of Palestine and the Middle East. At the President's request Major Bryan had circled the plane several times over "Bethlehem & Jerusalem & the Red Sea," Roosevelt recorded in his diary. From the air, however, he wasn't persuaded it was a particularly attractive land: "everything very bare looking — & I don't want Palestine as my homeland," he'd penned with feeling — aware how volatile were the politics of the region, and the competing, often fanatical claims of different religions upon it. "Then hundreds of miles of Arabian desert, then a green ribbon & Bagdad and the Tigris, with another green ribbon, the Euphrates — then bare mountains & we followed the highway over which so much lend lease goes to Russia."

Flying through the mountain passes that Major Bryan had successfully reconnoitered on his exploratory flight, they finally landed at 3:00 p.m., at the Russian air base of Gale Morghe, five miles south of the city. No better symbol of Soviet dependence on American military help to the Soviet Union could there have been than the aircraft they'd seen parked by the runway. "This is a modern airfield," Lieutenant Rigdon wrote in the log, "and on it were noted a large number of [American] lend-lease planes now bearing the Red Star of Russia."[8] From Gale Morghe the President was driven to the U.S. Legation in Tehran — and the historic summit began.

20

In the Russian Compound

ONCE INSTALLED as the guest of Mr. Louis Dreyfus, the American minister to Iran, the President dispatched his naval aide, Admiral Brown, across the city to the Russian Embassy to thank the Soviet government for its invitation to stay there, but to politely turn down the suggestion for the moment. Admiral Brown was tasked, instead, with inviting Marshal Stalin to dinner at the U.S. Legation residence that evening — where, it was assumed, the first of the conference's meetings would take place the next day.

Stalin had left Moscow in a slow, specially camouflaged armored train on November 22 — the same day Roosevelt had arrived in Cairo — and had reached Baku, on the Caspian Sea, on November 26. There — overcoming his mortal fear of flying — the Marshal had boarded an American-manufactured lend-leased C-47 on the morning of November 27 for the three-hundred-mile journey across the Caspian to Tehran — clinging "to his armrests with an expression of utter terror on his face," according to one account.[1] Via his chargé d'affaires, Mr. M. A. Maximov, Stalin duly responded to the President's invitation, saying he was grateful for it, but was fatigued by his long journey. He thus declined.

It was clear that, as in a powwow between two tribal chiefs, honor must first be satisfied before they actually sat down together.

By cable the President had asked Stalin five days earlier where exactly he thought it best he should stay in Tehran — a city located, after all, in the northern, Russian-occupied half of Iran. Stalin, in transit, had not replied. Yet the more the President had thought about the matter, the more he'd recognized the advantages of staying at the Russian or British Embassy — and his arrival at Dreyfus's official U.S. residence, with the time it took to send Admiral Brown through the teeming streets of the city to the Russian Embassy and back, only served to make him more amenable to the idea of a move.

Such an arrangement would spare daily journeys by the Russian and British leaders through the insecure, poorly policed streets of Tehran to and from the U.S. Legation. Which of the two embassies, though? The British Legation was already crowded with personnel, in fact was said to be a somewhat "ramshackle" compound: a former cavalry barracks of the Indian Public Works Department.[2] The Soviet Embassy, by contrast, was reported to be less crowded (its personnel had mostly been moved out, prior to Stalin's arrival) and more comfortable. But also bugged with hidden microphones, in the usual, paranoid Soviet style.

Why not stay at the Russian Embassy, bugged or unbugged, though? Would it not show an American willingness to work with the Russians on both war strategy and postwar security arrangements? Moreover, the very act of choosing a Russian roof over the President's head rather than a British one would symbolize the President's determination not to listen further to Churchill's continuing efforts to press his alternative Mediterranean/Aegean strategy outside the formal plenary meetings. The President had therefore hoped to broach the possibility of a move of living quarters to the Russian Embassy when Stalin came to dinner.

But Stalin's turndown left the President guestless on his first night in Tehran. Belatedly the President had decided he'd better invite Mr. Churchill, who had also arrived in Tehran that morning by plane.

Churchill, however, was not feeling at all well — suffering a heavy cold, again. Flying that morning from the Egyptian capital, his flight had been bumpy and the descent to the ground even bumpier. The Prime Minister, like an irritated headmaster, had whacked his pilot across the ankles with his walking stick when the crew lined up to say goodbye at the airport, complaining of a "bloody bad landing."[3] He'd also lost his voice — yet still hoped he could outmaneuver the President, before the summit-proper began. Banking on Mr. Roosevelt remaining across the city at the U.S. Legation, he'd seen a chance to work on Stalin personally: hoping to convince the Russian dictator in person of his strategy. The Prime Minister thus told his staff he wanted to "start there and then" with a meeting with Stalin, next door, if it could be arranged.

Given his sore throat, his loss of voice, and his aggressive, overwrought mood, however, Dr. Wilson and Churchill's own daughter Sarah dissuaded the Prime Minister. Nor would they allow him to accept the President's tardy invitation to dine that evening at the U.S. Legation, across the city. Instead, on doctor's orders, "he had dinner in bed like a sulky little boy," Sarah recorded.[4]

Day one of the Tehran Summit had thus come to a close with the "Big Three" leaders of the world failing to even sit down with one another. In fact the situation would have been comic, had the summit not been so important in terms of the prosecution of the war — and postwar.

But the real problem was that Joseph Stalin was Russian — and had apparently misunderstood Western protocol: namely the need for him to personally invite the President, if he really wanted Mr. Roosevelt to move quarters to the Russian Embassy, where part of the main building was being converted in great haste into a special guest apartment, complete with a new bathroom. As a result, in typical Soviet fashion, a pedantic subterfuge rather than a simple personal invitation was felt necessary: one in which a German "plot" would suddenly be "discovered," that night, after the President — having dined with Leahy, Hopkins, Brown, Watson, and the two ambassadors he'd brought with him, Winant and Harriman — had gone to bed.

The purpose of the supposed plot was explained late at night by the Russian foreign minister, Vyacheslav Molotov, who'd accompanied Stalin on his train and then plane, to Ambassador Harriman and to the British ambassador to Moscow, A. Clark Kerr. German spies were reported to be plotting an assassination of the President, Molotov asserted with a straight face, having summoned the two diplomats to the Soviet Embassy. The German agents were planning to kill the President at the U.S. Legation, or to attack the great leaders as they moved between the several compounds — the American Legation being over a mile away, through largely unpoliced streets. Marshal Stalin was therefore of the opinion, Mr. Molotov confided to the ambassadors, that the President should move either to the British Legation or to the Russian Embassy, and that the summit meetings should then be held in either the one or other building, to maximize security during the coming days.

Since it was so late the suggestion, based on the fake plot, was only communicated to the President after breakfast at 9:30 a.m. on Sunday, November 28, 1943.

The Russian "discovery" of the plot sounded almost silly, coming after so many weeks of intelligence checks, inspections, and reports. But the Russians were Russians — inscrutable in their inferiority complexes, as FDR had told his cousin Daisy.[5] Churchill once again pressed the President, via his ambassador to Russia, to stay at the British Embassy. However, the Russian offer — in which the President would be housed in a "part of their Embassy that would be under a separate roof and we would have complete independence"[6] — was the more tempting to the President, who, unlike Churchill, had still never met Stalin. Moreover, since Churchill was holding out so adamantly for a change in military strategy, a move to the Russian Embassy would insulate him from the Prime Minister's unending exhortations. Ambassador Harriman was therefore asked to tell Mr. Molotov the President was "delighted with the prospect" of staying with Marshal Stalin[7] — and would move to the Russian compound, he declared, after lunch that very day, at 3:00 p.m., along with his White House staff.

Summoning his U.S. chiefs of staff to the legation residence at 11:30 a.m., the President meanwhile rehearsed with them one last time how to deal with the possibility that Churchill was right, and that the Russians "really need assisting operations" in the Mediterranean to draw off Wehrmacht forces from the Eastern Front.[8]

For ninety minutes the President and his chiefs of staff, together with Harry Hopkins, thus rehearsed the situation. No matter how it was sliced and diced, none of the chiefs could see how Overlord could be mounted in the coming spring, in full measure, if they now allowed Allied forces to be "sucked in" to open-ended Mediterranean and Aegean operations.[9] How, in realistic terms, could Eisenhower or his successor be expected to fight his way to Fiume (annexed by Italy in 1924), on the northeast coast of the Adriatic, in only a matter of weeks, when he had still not got much beyond *Naples*? And when U.S. troops, according to General Mark Clark's reports, were finding the campaign in the mountains of Italy a more and more forbidding challenge, even with Allied air power? Even if the Turks were persuaded to enter the war, they could not be counted on to aid the Allies offensively, the President noted.[10] Could the Allies seriously imagine they could logistically achieve a *working* passage through the Dardanelles in less than "six to eight months"? General Somervell pointed out.[11] And how could the amphibious invasion of Burma, to help free up the transit of supplies to Chiang Kai-shek and provide a U.S. bomber base in the Andaman Islands, proceed, if the necessary landing craft were used in the Mediterranean?

The Germans were "already" aware of the "build-up in the U.K. in preparation for Overlord," and could be expected to toughen their Atlantic Wall still further in the next months — thus *increasing*, not decreasing, the need for the Allies to focus on Overlord, not be distracted elsewhere. Commandos helping guerrillas in the northern Adriatic, or southern France, would be more likely to draw off Wehrmacht divisions — perhaps as many as two — than major, battle-hardened Allied formations getting embroiled in the Dodecanese Islands: islands that would draw away no German forces from the Eastern Front, as the President indicated.[12] "Commando group operations," the President repeated, yes — but "on a small scale."[13] And with regard to Overlord? No major operations elsewhere should be countenanced — especially if they risked losing crucial landing craft needed for the cross-Channel invasion.[14]

It was clear there were simply not enough men or weapons or supplies to do everything simultaneously. In a global conflict logistics ultimately determined what was feasible, and what was not.

Amalgamating the largely British theater of the Middle East with that of the Allied Mediterranean theater might, all agreed, be a sensible idea logistically and in tactical terms, thereby permitting the application of the greatest Allied power at a chosen point. Yet if Marshall thought Churchill would accept the idea of General Eisenhower remaining in supreme command of all Allied forces in the Mediterranean as well as those in the Middle East then Marshall — as presumed commander of Overlord, in the north — was being naive. As the President put it, "We must realize that the British look upon the Mediterranean as an area under British domination" — and the addition of British forces from the Middle East would only make Churchill still *more* determined to exercise his dominion via a British supreme commander — with dire consequences for Overlord. After Rhodes was taken — if it was taken — he would want yet more operations, claiming, "'Now we will have to take Greece.'"[15] This, in turn, would involve yet *more* delay to Overlord — perhaps even its cancellation.[16]

The clock struck one at the American Legation, and the President called an end to the meeting. "The Soviets definitely want something," General Marshall had stated, "and we should find out what it is."[17]

They would find out soon enough — the first plenary session of the summit was due to start at four o'clock that afternoon.

21

The Grand Debate

ENTERING THE GATES of the Russian Embassy the President arrived in front of a "square building of light-brown stone set in a small park," boasting an "imposing portico with white Doric columns," as one historian described it — the park itself "surrounded on all sides by a high stone wall," within which there were fountains, a small lake, villas, and apartments for embassy staff personnel.[1]

The President's assigned quarters were in a house attached to the square main building. "It had three or four large downstairs rooms, as well as quarters for the President's Filipino servants and numerous Secret Service men," the President's interpreter, Charles Bohlen, also recalled — having been asked to join the Commander in Chief there, prior to the expected visit of their host, Marshal Stalin.[2]

A career diplomat, Bohlen had seen the President only a few times in his life — and those at a distance. At thirty-nine, he was understandably nervous to be acting suddenly as personal interpreter to the President, rather than as an attending junior U.S. diplomat. "In the few minutes I had with President Roosevelt before his first meeting with Stalin, I outlined certain considerations regarding interpreting," Bohlen later recounted. "The first and most important was to ask if he would try to remember to break up his comments into short periods." These should best comprise "two or three minutes of conversation," which "would hold their [the Russians'] attention and make my job infinitely easier. Roosevelt understood, and I must say he was an excellent speaker to interpret for, breaking up his statements into short lengths and in a variety of ways showing consideration for my travails. Churchill was much too carried away by his own eloquence to pay much attention to his pleasant and excellent interpreter, short, baldish Major Arthur H. Birse. There were occasions," Bohlen reflected, "when Churchill would speak for five, six, or seven minutes, while poor Major Birse dashed his pencil desperately over the paper, trying to capture enough words to convey the eloquence into Russian."[3]

The President, Bohlen recalled, "seemed to be in excellent health, never showing any signs of fatigue, and holding his magnificent leonine head high. He clearly was the dominating figure at the conference"[4] — the prologue to which now began with Stalin's arrival at the President's quarters at 3:15 p.m.

Dressed in a "simple khaki tunic (he was a Marshal of the Soviet Union) with the Order of Lenin on his chest," and escorted into the President's sitting room by a young American army officer, the five-foot-six-inch Russian dictator entered. The six-foot-three-inch president stretched out his arm from his wheelchair and the two men shook hands.[5] "I am glad to see you," the President — dressed in a blue business suit — said sincerely. "I have tried for a long time to bring this about."[6]

He had — in fact his attempts to arrange a meeting went back more than a year. Stalin, for his part, apologized, citing his "preoccupation with military matters."[7] Then, after that brief introduction, Stalin sat down for forty-five minutes while the two men conversed, seeking to get a measure of each other's personality as they ran through a surprising range of subjects openly and informally.

Bohlen was amazed at how deftly the President put the Russian dictator at ease; likewise, how frank, straightforward, and personal the Marshal was in responding. Warming to his task, Bohlen interpreted for both men as they compared notes on the war and the future.

Stalin flatly admitted that the situation on the Eastern Front was "not too good" — in fact it was "so bad that only in the Ukraine was it possible to take offensive operations," and even there, several important cities such as Zhitomir had recently fallen to German counterattacks.[8] How best to draw off significant numbers of German divisions and reinforcements, the President responded, was in part why he had come to Tehran. Stalin thanked him. The President then switched to postwar reconstruction, offering to share with the Soviet Union some of the Allies' merchant fleet that would become redundant, after hostilities ended — prompting Stalin to say, not only would the Soviet Union be grateful, but how he much hoped "the development of relations between the Soviet Union and the United States" would be "greatly expanded. In return for American equipment, the Soviets, for their part, would like to make available raw materials to the U.S."[9]

Allowing for diplomatic niceties, the sheer optimism about postwar relations was a tremendous relief to Roosevelt — so much so that he launched into a very frank discussion of the Far East, explaining how General Stilwell hoped to supply and train up to sixty Chinese divisions, while Allied forces under Admiral Mountbatten hit Burma from the north, and farther south from the Indian Ocean, to open supply routes to "link up with China."

Burma, however, raised the question of postwar British, French, and Dutch decolonization — a problem already rearing its ugly head in the Lebanon, where the French had refused to grant independence, as promised, despite recent elections — the President blaming de Gaulle and his Free French Committee. "Marshal Stalin said he did not know General De Gaulle personally, but frankly, in his opinion, he was very unreal in his political activities. He explained that General De Gaulle represented the very soul of sympathetic [i.e., anti-Fascist] France, whereas the real physical France" was unfortunately "engaged under Petain in helping our common enemy Germany, by making available French ports, materials, machines, etc., for the German war effort. He said the trouble with De Gaulle was his [Free French] movement had no communication with the physical France, which, in his opinion, should be punished for its attitude during this war. De Gaulle acts as though he were the head of a great state, whereas, in fact, it actually commands very little power."[10]

This was — especially after de Gaulle's personal meeting with Roosevelt at Casablanca in January, earlier that year — music to the President's ears. He "agreed" wholeheartedly — indeed posited, more rhetorically than seriously, that "in the future, no Frenchman over 40, and particularly no Frenchman who had ever taken part in the present French [Vichy] Government should be allowed to return to positions after the war. He said that General Giraud was a good old military type, but with no administrative or political sense, whatsoever." This augured poorly for the eleven French divisions, comprising mostly African soldiers, currently being trained in North Africa. The President's remark in turn prompted Stalin to "expiate" at length "on the French ruling classes," which, the dictator remarked, "should not be entitled to share in any of the benefits of the peace, in view of their past record of collaboration with Germany." Which led the President to declare that he disagreed with Churchill's view that France be "reconstructed as a strong nation," for he felt it ought to be, essentially, punished by hard labor "for many years" before it was reestablished as a worthy nation — not only its government but "the people as well," who should, by honest labor, become "honest citizens."[11]

The President was tiptoeing around the matter of imperialism: the idea that people of color owed the white European nations a living, for which the white colonists were not required to do more than wear smart uniforms, brush their proverbial teeth — and make nice with Hitler. "Marshal Stalin agreed," Bohlen recorded — indeed said he thought it wrong the Allies should be expected by the French to "shed blood to restore Indo-China." Recent "events in the Lebanon" showed how important it would be to train formerly colonized peoples in law and government, ready for "independence." This would be especially important in Southeast Asia and the Pacific, once the Japanese were defeated and removed.

Again Stalin agreed. The "political" challenge of decolonization was, Stalin felt, just as important as the military in "certain colonial areas. He repeated that France should not get back Indochina and that the French must pay for their criminal collaboration with Germany" — views with which the President said he was "100% in agreement with Marshal Stalin." Judging from reports he'd received, the President "remarked that after 100 years of French rule in Indochina, the inhabitants were worse off than they had been before." For his part Chiang Kai-shek had assured him the Chinese had "no designs" on Indochina. With Chiang in Cairo the President had therefore discussed, instead, the idea of a United Nations trusteeship. This would prepare the people of Indochina for complete independence "within a definite period of time, perhaps 20 to 30 years" — and he instanced how this had been the task of U.S. policy in the Philippines, which was due to be given independence immediately after the war with Japan was over. "Marshal Stalin completely agreed with this view" — and with the President's notion of an international fact-finding committee to visit, every year, "the colonies of all nations" and seek to "correct any abuse that they find."[12]

Which led the two world leaders to discuss the largest colonized nation in the world: India. Had millions of Russians and tens of thousands of Americans already died, or been maimed, merely so that Britain could maintain imperial domination over four hundred million people? How could the British keep up a steadfast refusal to grant India self-government, let alone independence? Gandhi and Nehru were still under British arrest, and since August 1943, Churchill had refused to release shipping to send food to Bengal, where by October 1943 serious famine was threatening millions of Indians with starvation and death unless the British acted swiftly.

India, however, was a sacred cow for the Prime Minister, the President warned the Marshal — for Churchill "had no solution to that question, and merely proposed to defer the entire question to the end of the war." Stalin "agreed that this was a sore spot with the British" — to which Roosevelt said he would like to discuss the matter of India again with the Soviet premier "at some future date." Might the answer lie in revolution, as had been the case in Soviet Russia? the President wondered — prompting Stalin, ironically, to caution that India was a "complicated" case, "with different levels of culture," religion, and a caste system that precluded, or would make difficult, an easy transition to sovereign, self-governing nationhood.

For Bohlen the conversation was stunningly informal — each leader seeking to show an openness to the other's personal views and ideas. Given the total suppression of free speech in the Soviet Union, and the complete contrast between their own forms of government — the one proudly capitalistic and democratic, if profoundly racist, the other fiercely communist and a brutal dictatorship — the meeting suggested that the two countries might well find common ground in guiding peaceful development of the world, once the barbaric, expansionist German and Japanese empires were forced to surrender.

All too soon the meeting came to an end, however — for at four o'clock the first plenary session of the summit was due to begin next door, in the conference room of the Soviet Embassy. The Russian dictator exited, the President freshened up, and there now began the long-awaited battle over how best to defeat Hitler.

The President began the meeting, ironically, without his two top military advisers. Owing to a misunderstanding over the time of the first plenary meeting neither General Marshall nor General Arnold was present, having gone instead for a hunting "trip through the mountains," as Arnold noted in embarrassment in his diary.[13] In their absence, at the big circular table at the center of the conference room, the President thus sat with Harriman on his right and Bohlen and Hopkins on his left, while Admiral Leahy, Admiral King, and Major General Deane perched on chairs by the walls of the big room, which was guarded by Soviet secret policemen in plain yet bulky clothes, concealing their pistols.

How would Churchill behave? Warned by the President that there were "storm signals flying in the British legation," the President had sent Ambassador Harriman to see the Prime Minister in the British camp prior to the meeting. Harriman had reported Mr. Churchill happily "waived all claims" to be chairman of the conference, despite being the eldest of the three leaders. He had, however, insisted on "one thing."[14]

The President had asked what it was. To his amusement Harriman explained the Prime Minister wanted to "be allowed to give a dinner party on the 30th," in two days' time, to celebrate his sixty-ninth birthday. Churchill intended to "get thoroughly drunk," as he'd warned the American ambassador, and would "leave the following day."[15]

Clearly Churchill had not lost his sense of humor. Meantime, as the agreed chairman of the conference (the President being the only head of state, not simply a premier or first minister), Mr. Roosevelt opened the proceedings. The British interpreter, Major Birse, later recalled how "Roosevelt sat in his wheeled chair which had been pushed up to the conference table. In that position, with his broad shoulders and fine head, he had the appearance of a tall strong man, and it was only his chair which gave away his infirmity. He beamed on all around the table and looked very much like the kind, rich uncle paying a visit to his poorer relations" — both Stalin and Churchill a foot shorter than the President.[16]

Once again Bohlen was struck by the historic, yet "relaxed" nature of the meeting since "it did not seem possible that the three most powerful men in the world were about to make decisions involving the lives of millions."[17] As the youngest of the leaders the President welcomed his "elders." Although minutes or summaries would be kept for later reference, nothing would be made public for the moment without common consent, so that the leaders could talk with "complete frankness," ensuring their three great nations could work together to prosecute the war, and in the future, when their countries would hopefully continue to enjoy close relations, via similar summits, "for generations to come."[18]

Mr. Churchill seconded the President's introductory remarks.

For his part Stalin welcomed those present to his embassy, saying "history had given to us here a great opportunity," which it was up to them to "use wisely" on behalf of their peoples.[19]

With that the President began his overview of the war — beginning with the Pacific, where American forces were carrying virtually the entire burden of the fight against Japan, together with Australian troops — and in the north, with the Chinese.

"While speaking," Major Birse described the President, "he would frequently take off his pince-nez and use it to emphasize a point."[20] He "summed up the aims" of the operations currently being pursued in the Far East as "(1) to open the road to China and supply that country in order to keep it in the war," and (2) by opening the road to China and through increased use of transport planes to put ourselves in position to bomb Japan proper." In the meantime United States forces would continue to advance in the central and southwest Pacific, moving forward from island to island as they turned back the Japanese rampage. Japan's days of military conquest were numbered, but the "most important theater of the war," he emphasized, was Europe.

For more than eighteen months, the President explained, his high-level conferences with Prime Minister Churchill had been dominated by the challenge of "relieving the German pressure on the Soviet front." Largely because of logistical challenges it hadn't been possible until the Quebec Conference in August that year to "set a date for the cross-channel operations. He pointed out," however, as Bohlen recorded in his dictated notes that evening, "that the English Channel was a disagreeable body of water." As such it "was unsafe for military operations prior to the month of May, and that the plan adopted at Quebec involved an immense expedition and had been set at that time for May 1, 1944."[21]

This was the first time the date of the launching of the Second Front had formally been given to the Russians — for in Moscow, in October, General Ismay, Churchill's chief of staff, had done his best, while he and General Deane shared the Overlord plan with Russian generals, *not* to give a target date to which the British could be held. As if to draw attention away from the date, Churchill now "interceded" to remark how thankful were the British for such a "disagreeable body of water."[22]

Resuming his overview the President explained that, while waiting to mount Overlord, there was the question of how best the American and British forces could help keep or even draw away significant forces of the Wehrmacht from the Russian front. In the spirit of candidness and openness, the President admitted there was currently concern about "what use could be made of allied forces in the Mediterranean in such a way as to bring the maximum aid to the Soviet armies on the Eastern front." Some of these operations, it was warned, might delay the Overlord invasion by "one, two or three months," namely by pursuing schemes "in Italy, the Adriatic and Aegean Seas and Turkey." For his own part, however, the President "emphasized the fact," as he put it, "that in his opinion the large cross-channel operation should not be delayed by secondary operations."[23]

The President's reference to Overlord, his revelation of its target date, and his emphatic declaration that he himself wanted no "delay" or "diversions" left Brooke feeling miserable — a "poor and not very helpful speech," as he sneered in his diary. If so, it was a sign of things to come. "From then on," Brooke added, "the conference went from bad to worse!" — for Stalin *agreed* with the President's preference for Overlord priority, indeed proceeded to advocate "cross Channel operations at the expense of all else"![24]

This was not, in truth, quite what happened — at least as observers other than Brooke saw it. Stalin, in their view, had followed the President's opening resumé by giving a quiet, measured address that impressed everyone. The Marshal began by congratulating the Western Allies on their successes in the Pacific, and assured all those present — American, British, and Russian — that the Soviet Union would definitely join the war against Japan; in fact it would begin sending major forces to Siberia for the task as soon as Nazi Germany was defeated. The Soviet commander in chief then proceeded to give a frank and detailed account of Russian operations since the summer of that year.

As Stalin succinctly put it, the Soviet armies were facing some 260 Wehrmacht divisions, including 10 Hungarian, 20 Finnish, and an estimated 18 Romanian. The Soviets could field 330 divisions, but in offensive warfare this was not enough to guarantee success; moreover, even the "numerical superiority the Soviets possessed" was gradually being "evened out" as Hitler switched more Wehrmacht divisions to the Eastern Front. Winter weather meant that operations had slowed down; the Germans were not only counterattacking but seeking to retake Kiev with 8 panzer divisions — 5 of them fresh — and 23 infantry divisions.

The war, in other words, was by no means won — and if the Allies abandoned their methodical advance and chose half-baked alternatives, it could still be lost. Freeing the Mediterranean for Allied shipping had been a signal success, but to imagine Italy or the northern Adriatic was the proper place to bring down the Third Reich was nonsensical. The Alps, Stalin pointed out, "constituted an almost insuperable barrier" — something "the famous Russian General [Alexander] Suvorov had discovered in his time."[25]

This was a telling admission, since Suvorov had been probably the greatest commander in Russian history: a general who himself had led an Austro-Russian army in Italy; a general who had captured Milan, and who had driven the French out of Italy. He had not subsequently been able to cross the Alps, however, and had been forced to retreat to Russia.

Italy, then, was a futile theater of war, beyond pinning down a limited number of Wehrmacht divisions. The real truth was, it was the Germans who were pinning down significant Allied forces in Italy, rather than the reverse. In the "opinion of the Soviet military leaders," Stalin explained, "Hitler was endeavoring to retain as many allied Divisions as possible where no decision could be reached," in Italy. Whereas, he said, "the best method" — at least in the "Soviet opinion" — "was getting at the heart of Germany with an attack through northern or northwestern France." Even, possibly, an additional attack "through southern France. He admitted that this would be a very difficult operation since the Germans would fight like devils to prevent it" — but better, surely, than wasted efforts in Italy. Or, Stalin added, the Aegean, which was an equally fatuous alternative. For although it would be "helpful" to inveigle Turkey into entering the war on the Allied side, "the Balkans were far from the heart of Germany." As the Marshal concluded, "northern France," or Overlord, "was best."[26]

Churchill, listening to the interpreter's version of Stalin's address, and aware of the twenty other persons in the room, was shocked. Shocked, however, into final reality: the reality that, by seeking to alter the strategy agreed at Quebec, he had taken the British team on a wild goose chase — wasting vital preparatory time for the Overlord invasion, upsetting his most important allies, and pursuing reckless fantasies, the consequences of which had not been thought through and could only shame the British contingent. As cowards, moreover.

All Churchill could do, in the circumstances, was to assure everyone the British "were determined to carry it [Overlord] out in the late spring or early summer of 1944." The Overlord plan envisaged an "initial assault of 16 British [including Canadian] and 19 U.S. Divisions, a total of 35."[27]

Brooke was not impressed by this. "Winston replied and was not at his best," Brooke recorded in his diary. "President chipped in and made matters worse. We finished up with a suggestion partly sponsored by the President that we should close operations in Italy before taking Rome."[28]

Before taking Rome? After all that Brooke and his fellow British chiefs of staff had done over the past two months, on Churchill's instructions, to plot and argue for more offensive action in Italy . . . ? This was the very opposite of music to Brooke's smarting ears. Moreover, the Russian marshal's opinion that Turkey was "beyond hope," as he put it, and his realistic prediction that "nothing could induce her to come into the war on any account," had poured ice-cold water on the Prime Minister's dreams. "Dardanelles were apparently not worth opening," Brooke added. "We sat for 3½ hours and finished up this conference," the British Army head fumed, "by confusing plans more than they ever have before!"[29]

For Churchill the opening meeting had been three and half hours of sheer torture. When, "after the plenary session," Dr. Wilson saw the Prime Minister, he found Churchill "so dispirited" that, as the Prime Minister's personal physician, he "departed from my prudent habit and asked him outright whether anything had gone wrong."

Churchill was nothing if not pithy. "A bloody lot," he declared, "has gone wrong."[30]

So dispirited by the plenary was Churchill, in fact, that Stalin had felt it necessary to assure the Prime Minister he hadn't meant to "belittle" the importance of what had been achieved that year in the Mediterranean, since those operations had been "of very real value." Churchill had "thanked the Marshal for his courtesy," insisting that neither he nor the President had ever considered the Mediterranean "as anything more than a stepping-stone for the main cross-Channel invasion." Britain had a population of only forty-six million, however, and with commitments in "the Middle East, India," et cetera, could not be expected to do too much.[31]

It was a somewhat pathetic confession.

As Stalin had suggested, instead of wasting battle-hardened troops in Italy to take Rome, would it not make more sense, if they really wanted to help guarantee the success of Overlord, to launch a secondary landing in southern France that would draw German divisions away from the English Channel? As a strategist Stalin thus agreed with the President: he would be more "inclined to leave 10 divisions in Italy and postpone the capture of Rome in order to launch the attack in southern France two months in advance of Overlord."[32]

Churchill had been stunned — and had only been able to reply that he hoped "Marshal Stalin would permit him to develop arguments to demonstrate why it was necessary for the allied forces to capture Rome, otherwise it would have the appearance of a great Allied defeat in Italy."[33]

A defeat, when the Italian government had already surrendered?

Stalin had been visibly unimpressed by the value Churchill placed upon such "appearances." Undeterred, Churchill had gone on to make an even more specious argument, claiming that "without the fighter cover which would be possible only from the north Italian fields it would be impossible to invade northern France."

D-day impossible without fighters *operating from northern Italy*?

It was the first the President or any of his chiefs had heard of such a claim. Perhaps the Prime Minister had meant southern France? Whatever he had intended to say, however, it had sounded lame — and the President had put him out of his misery by interceding. The question of "relative timing was very important," Mr. Roosevelt had said — in fact, "nothing should be done to delay the carrying out of Overlord," he'd emphasized. And delay was, clearly, what would happen "if any operations in the eastern Mediterranean were undertaken. He proposed, therefore, that the staffs work out tomorrow morning a plan of operations for striking at southern France," either before, during, or soon after the cross-Channel invasion.[34]

Stalin had concurred, pointing out that "the Russian experience had shown that an attack from one direction was not effective," given the defensive skills of the Wehrmacht. Instead, "the Soviet armies now launched an offensive from two sides at once which forced the enemy to move his reserve back and forth. He added that he thought such a two way operation in France would be very successful."[35]

Still Churchill had resisted, however: loath to surrender his dreams of "victory" in the Mediterranean and Aegean — or to accept a May 1944 launch of an invasion in whose success he had frankly never truly believed. The Prime Minister had therefore countered that "it would be difficult for him to leave idle the British forces in the eastern Mediterranean which numbered some 20 divisions, British controlled, which could not be used outside of that area, merely for the purpose of avoiding any insignificant delay in Overlord."[36]

Even Churchill had realized he wasn't making sense, and that the tide of the plenary meeting was running fiercely against him. He had therefore said that "if such was the decision" to concentrate wholly on Overlord, then the British "would, of course, agree, but they could not wholeheartedly agree to postpone operations in the Mediterranean" on which he'd set his heart — especially an amphibious landing north of Naples that would, he was certain, force the Germans to cede Rome. If Turkey did decline to join the Allies, it would make further Aegean operations pointless, he granted, but "he personally favored some flexibility in the exact date of Overlord. He proposed that the matter be considered overnight and have the staffs examine the various possibilities in the morning."[37]

Stalin could only scoff at such endless obstructionism. As the Russian dictator remarked, he "had not expected to discuss technical military questions" at Tehran, and he had "no military staff" with him. However, if more detailed analysis of the Western powers' own competing plans was required, "Marshal Voroshilov would do his best."[38]

The situation would, in sum, have been risible, had it not been so disheartening. By opposing Overlord's May 1944 date, in defiance of the Quebec agreement, Churchill had not only split the Anglo-American alliance, but had made the British military team look ridiculous — as General Brooke was painfully aware.

Years later, when annotating his diary, Brooke admitted he had had, until then, no idea Stalin was a "strategist." The plenary meeting at the Russian Embassy in Tehran had, however, put this misapprehension to rest. "I rapidly grew to appreciate," he wrote, "the fact that he had a military brain of the very highest calibre. Never once in any of his statements did he make any strategic error, nor did he ever fail to appreciate all the implications of a situation with a quick and unerring eye."[39]

By contrast "Roosevelt never made any great pretence at being a strategist and left Marshall or Leahy to talk for him," Brooke reflected[40] — completely forgetting that General Marshall had not even been present at the meeting. Or that Leahy and King, sitting behind the President, never once spoke. Neither Brooke nor Churchill, in fact, would ever acknowledge their military bungle — or the President's defining role, behind his mask of "country gentleman," in putting down the British revolt.

As for Stalin, he was unequivocal in agreeing with President Roosevelt: a chairman who "filled the part most effectively," as even General Ismay, Churchill's chief of staff, later admitted, recalling that the President was "the picture of health and was at his best throughout the conference — wise conciliatory and paternal," whereas the Prime Minister "was suffering from a feverish cold and loss of voice."[41]

By the evening of November 28, 1943, at Tehran, then, the die was cast, as even General Brooke acknowledged. "This Conference is over when it has only just begun," he complained that night to Churchill's doctor — adding that Stalin "has got the President in his pocket."[42]

Brooke's obtuseness about the President, as about Overlord, was disappointingly representative of the British team supporting the Prime Minister at Tehran, unfortunately: a refusal to face facts prior to Tehran, in Tehran, and after Tehran — indeed even after the war.

Marshal Stalin, however, fascinated them. The Marshal would have made "a fine poker player," General Ismay later recalled. "He did not speak much, but his interventions made in a quiet voice and without any gestures, were direct and decided. Sometimes they were so abrupt as to be rude" to British ears, for he "left no doubt in anyone's mind that he was master in his own house. He saw no point, for example in the proposal that the military experts of the three countries should meet the next morning, at Churchill's request, to examine the implications of Churchill's alternative strategy. 'The decisions are our business,' he said. 'That is what we have come for.'"[43]

Like Brooke, General Ismay would later see Stalin's unwillingness to discuss Churchill's alternative Mediterranean strategy as an example of the dictator's secret plans for postwar Russian hegemony. "It is doubtful if many of those who listened to the discussion grasped the significance of Stalin's determination to keep Anglo-American forces as far as possible away from the Balkans. It was not until later that we realised that his ambitions were just as imperialistic as those of the Czars, whose power and property he now enjoyed," Ismay wrote, "but that he was capable of looking much further ahead than they had ever been."[44]

Such retrospective justifications of British alternative strategy testified to Britain's loyalty to Churchill's political genius and foresight. But in truth they were as silly as Brooke's claim the President was in "Stalin's pocket."

Looking back, Admiral King recalled "the long struggle" it took to get the cross-Channel operation mounted. "You see, the British always shook their heads over what developed as 'Overlord.' They pointed to the highways, the railroads that ran east and west and the ability of the Germans to shift forces up against whatever landing we made before we could get the beachheads established," King later recalled. The "British felt we had posted [connived with] the Russians, so that the Russians and ourselves were of one mind" in relation to Overlord as "the second front"—the Russian front being the first. But such accusations of prior collusion were nonsense, King sniffed. "There was a meeting of our minds," he acknowledged of the Tehran Conference, "but I don't think it was concerted at all. I don't think we prompted the Russians. They had their own ideas"—which simply mirrored those of the United States in terms of clear military strategy: namely how best to defeat Nazi Germany. The British "wanted to recover the Dodecanese," from which they'd been ignominiously expelled in recent weeks; "they wanted to recover Rhodes; they wanted to get into Crete; above all they wanted to go up into Yugoslavia."[45] Stalin had been contemptuous. "Marshal Stalin waved aside all those proposals as side issues. That wasn't what he meant as 'second front.'" Moreover, the Russian marshal was by this time well experienced in the only way to defeat the Wehrmacht in battle. If the Western Allies had so many forces in the Mediterranean, why not use them to reinforce Overlord with landings "in Southern France"? King recalled Stalin's question. The Russian dictator's recommendation of landings in southern France had come as a complete but profoundly welcome surprise to the President and the American team at Tehran, after so much British opposition and backpedaling since Quebec. "I don't think he [Stalin] was prompted by anybody at all in the American side," with regard to southern France. "I don't think Mr. Roosevelt prompted that," King reflected[46]—his memory borne out by the minutes of the meeting.[47]

The Russian marshal, in other words, had spoken as a military strategist who had learned his lesson, after a desperately poor start, in *how to defeat the Wehrmacht*—whereas Churchill and the British chiefs hadn't: forever imagining they could pursue peripheral avenues, from Norway to the Balkans, that took no account of the terrain. Or the enemy.

King had no personal animosity toward the Prime Minister. "King likes

Churchill very much indeed," an American journalist had noted during the summer of 1943, before the Quebec Conference, "although he laughingly said that he always had his hand on his watch when Mr. Churchill was trying to 'sell' a point. Mr. Churchill, said [Admiral] King, is first, last and always for the British Empire and you have to always remember that when dealing with him. This, remarked the Admiral, is as it should be and Churchill is respected for it."[48]

The President, likewise, was fond of Churchill—while holding fast to his pocket watch. Above all Mr. Roosevelt was, as Ismay noted, an experienced chairman: a conciliator who understood that the aging prime minister should if possible be brought down gently. The British could not be bullied into submission—for their full-scale cooperation would be crucial in opening a successful Second Front. Moreover their military presence would be required for years to come in ensuring postwar security, the President reasoned.

Since the U.S. chiefs of staff had already prepared a paper on the pincer-like invasion of southern France ("Anvil") for the Cairo Conference, the President suggested they share it at a Combined Chiefs meeting the next morning, together with Marshal Voroshilov, before the second, afternoon plenary meeting—but should *not* discuss Aegean or other diversions a moment longer.

It was eight o'clock, and Hopkins, for his part, was hungry. "I thought we were going to be late for dinner, when the President suggested an adjournment," Hopkins afterwards told Churchill's doctor.[49]

The President's suggestion had been met with relief all round. The first, three-and-a-half-hour plenary meeting of the world's three most powerful Allied leaders had thus come to an end—Roosevelt retiring to his quarters next door, and as president of the United States inviting Churchill and Stalin to be his guests and dine with him at 8:30 that very evening.

22

A Real Scare

WINSTON CHURCHILL MIGHT be angry, as he admitted to his doctor, but he was too wise not to see he was now outnumbered. He had threatened in front of the British War Cabinet he would resign if he did not get his way with the United States. He had boasted to Harold Macmillan, his political minister at Eisenhower's headquarters, and to his British chiefs of staff, that he was willing to provoke the Americans into switching their forces to the Pacific if he did not get his way over furthering his Mediterranean schemes and delaying Overlord to 1945. It had always been a risky maneuver, however — and the President's stonewalling and refusal to countenance any change in the Quebec agreement in Cairo had effectively disabled the British insurrection.

The fact was, the Prime Minister now recognized, he could not make good on his threats. It was unlikely the deputy prime minister or the British cabinet would support him in walking away from the Tehran Summit, or in deliberately invoking a breakdown in United Nations military strategy, merely to invade Rhodes, a small Greek island. He would have to abide by majority decision-making, lest the "grand alliance" he himself had done so much to create be sundered.

Stalin's open evisceration of Churchill's Mediterranean/Aegean "diversions," however, had cut the Prime Minister to the quick. Tired by the hours-long debate over Overlord, its timetable and its alternatives, the President nevertheless tried to encourage a positive atmosphere in his sitting room, before they sat down to eat. "Roosevelt mixed the pre-dinner cocktails himself, which were unlike anything I have ever tasted," Bohlen remembered. "He put a large quantity of vermouth, both sweet and dry, into a pitcher of ice, added a smaller amount of gin, stirred the concoction rapidly, and poured it out. Stalin accepted the glass and drank but made no comment until Roosevelt asked him how he liked it. 'Well, all right, but it is cold on the stomach,' the dictator said."[1]

It was an "interesting" start to an all-American meal of "steak and baked-potato dinner prepared by his Filipino mess boys," Bohlen recalled.[2] But one that, unfortunately, quickly became as potentially querulous as the plenary; indeed the loose, alcohol-fueled, no-holds-barred discussion at the President's dining table became, in retrospect, shameful: a free-flowing conversation, laced with drink and conducted in front of the leaders' political and diplomatic advisers — Harry Hopkins and Ambassador Harriman, Anthony Eden and Ambassador Kerr, and Russian foreign commissar Vyacheslav Molotov — who, for the most part, listened in horror.

The President's attempts to lighten the tone of the dinner merely led to competitive assertions, claims, and opinions that in hindsight were unworthy of the three leaders, who disparaged not only individuals but whole nations. The French, especially, did not come out well, given that all three leaders detested General de Gaulle, for all that he stood so courageously for French antifascism. Once again, the President and Marshal Stalin voiced an even worse opinion of the French people than of de Gaulle personally. Stalin felt the French deserved no "considerate treatment" once liberated; in his view the "entire French ruling class was rotten to the core." It had "delivered France to the Germans," and was "actively helping our enemies," the Nazis, both in industrial output and manpower supplied to the Third Reich. The dictator "therefore felt that it would be not only unjust but dangerous to leave in French hands any important strategic points after the war."[3]

Roosevelt, who abhorred the idea of the French seeking to reconstitute their colonial empire after the war, agreed about military bases — saying New Caledonia, in the South Pacific, and Dakar, on the west coast of Africa, should not be returned to French rule. Which led Churchill to assure the President and the Premier that for its part, Britain was different from France. Britain was not seeking any territory after victory. Yet "for the future peace of the world" he agreed that the United States, the Soviet Union, Great Britain, and China should definitely have and maintain military bases from which, as the "Four Policemen" of the United Nations, they could nip nascent problems in the bud. Stalin, however, was unwilling to leave it at that. The restoration of Europe's Victorian empires was not, he felt, the purpose in defeating Nazi Germany. "France could not be trusted with any strategic possessions outside her own border," he asserted — claiming that Gaston Bergery, Vichy ambassador to Moscow and then Turkey, was typical: the French more willing to negotiate with their former enemy, the Germans, once the war was over, than with their liberators, the Americans and British.

This was far from an exaggeration — but in its flippant way it offered no constructive idea of how, shorn of its colonial territories, the French were miraculously to be reconditioned as antifascists, let alone anti-Germans. The President's idea of disallowing anyone over the age of forty to stand for election after the war, which he repeated at dinner, was worthless as genuine potential policy. It skirted, moreover, the key to postwar European peace and reconstruction: Germany.

Once again it was the President and Stalin who held court on the moral abomination of Nazism — but with only the vaguest notion, still, of how to deal with it at war's end. "The very word 'Reich,' or 'empire,' should be stricken from the language," the President opined, prompting Stalin not only to agree, wholeheartedly, but to go further, saying that unless the "victorious allies" made sure to "prevent any recrudescence of German militarism, they would have failed in their duty."[4]

Churchill said little, still smarting over the afternoon. Yet the aging prime minister was the only one of the three national leaders who had hands-on knowledge and decades-long experience of dealing with Germany and France. His role in the end-of-war deliberations and decisions yet to be made would be an essential one — especially when considering how the Third Reich should, or should not, be dismembered or reformed.

Fortified by more liquor the conversation then moved on to Poland — another matter of huge political significance, since Poland lay between Germany and the Soviet Union. The country was immensely important for Stalin as a buffer state, as all were aware. In the fall of 1939, following Hitler's invasion and military conquest, it had been split between the two signatories to the Hitler-Stalin Pact. Even the Russian occupation of eastern Poland, however, had failed to provide an effective buffer against Nazi dreams of further conquest. Barbarossa — Hitler's massive, Napoleon-like invasion from German-occupied Poland — had carried the Wehrmacht almost to the brink of the Caspian Sea, and resulted in the death of millions of Russians, as well as the devastating destruction of whole cities such as Stalingrad. Now the boot was on the other foot — and with more than three hundred divisions on the Eastern Front, almost within artillery range of the old Polish border, Stalin proposed — as he had at the Moscow Conference — that the permanent Russian border with Poland be moved west, farther away from Moscow.

With four million Polish American voters in America, the President declined to comment. Nor would he discuss the future of the Baltic states of Lithuania, Latvia, and Estonia: traditionally Russian-controlled territories which had become independent after World War I and also had their stalwart supporters in the United States, especially among émigrés. By favoring an international guarantee of shipping access to and from the Baltic Sea through the Danish-Swedish straits the President hoped to urge the conferees to think positively — to consider open trade and the revival of world commerce rather than defensive positions and possessions. In fact, from this he moved on to float the idea of the future United Nations.[5]

"Roosevelt was about to say something," Bohlen later recalled, "when suddenly, in the flick of an eye, he turned green and great drops of sweat began to bead off his face; he put a shaky hand to his forehead. We were all caught by surprise."[6]

Of the three leaders — despite a long, 7,775-mile journey to Tehran — the President had seemed until that moment the least careworn. Stalin's hair had turned positively gray, according to both Hopkins[7] and the British interpreter,[8] while the Prime Minister, who had very little hair, was still suffering from his bad cold and was tired and out of sorts. The ailing Harry Hopkins, so often looking at death's door himself, now leapt to his feet and wheeled the President to his room, next door. There he was lifted onto the bed and Dr. McIntire examined him.

Poison? Admiral Leahy, in his diary, acknowledged this had been the "immediate assumption" — one so serious in its international implications that for several years there would be no mention of the President's sudden collapse in official minutes, records, memoirs, or histories of the Tehran Summit. Admiral Leahy's diary recorded, however, that "at dinner tonight the President suffered an acute digestive attack which alarmed us because of the possibility that poison had been given to him."[9] Lieutenant Rigdon, in charge of all communications to and from the White House, years later recalled that the "attack threw a real scare into all of us. Our first thought was that his food had been poisoned."[10]

Poison seemed unlikely, if only because the meal had been personally prepared by the President's Filipino staff. Even the cooking range was an unlikely source of the attack, since it had been trucked to the President's quarters that very afternoon, with no advance warning, from General Connolly's Lend-Lease Amirabad compound, when it was realized the Russians had either removed or failed to reinstate all cooking equipment in the President's quarters. A German assassination attempt was, of course, possible — but scarcely a Russian one, given the Soviet Union's dependence upon American supplies, arms, and support in the war.

Was it a *health* scare, then?

The President did not mention the occurrence in his diary. Nor did he in a letter he sent afterwards to Daisy Suckley from Tehran. Neither was it mentioned by Elliott Roosevelt in his detailed postwar memoir covering the conference, published only a year after hostilities ended — perhaps because Elliott hadn't actually been in Tehran that day, or the next (his aircraft having experienced engine trouble in Palestine). Nor did Dr. McIntire mention the incident in his memoirs, also published immediately after the war. The same was the case with Hopkins's biographer, Robert Sherwood, and the White House Secret Service detail commander, Mike Reilly — the President's family and staff by then anxious there be no suggestion, in retrospect, that the President had been ill-advised to have stood for reelection the following year, in 1944.

Rumors of a health scare at Tehran would eventually surface, though. Some would later surmise the President had had a heart attack,[11] or been suffering cancer of the stomach,[12] or gallbladder disease,[13] even a tumor on his liver.[14] Since no further symptoms of a serious medical condition appeared until the following year, however, these would all amount to idle speculation.

In the meantime, assuring himself, his patient, and the President's colleagues that it was merely an attack of acute indigestion, Admiral McIntire insisted the President remain in bed, in his room. For his part Hopkins returned to the dining room, where he reported the doctor's diagnosis to the guests. By this point dinner was over.

Relieved at the news the President's collapse was probably indigestion, the Russian marshal and the Prime Minister retired to a smaller room in the Russian Embassy. There they continued talking into the night—carving up the world, at least in their cups.

For President Roosevelt, Commander in Chief of his nation's armed forces, the historic day was over. There would be at most three more days in Tehran in order to get what he wanted, assuming that he quickly recovered from his gastric attack. Also, that he was able to put down any renewed British attempt to sabotage Overlord.

23

Impasse

MERCIFULLY THE PRESIDENT awoke on Monday, November 29, 1943, feeling much better.

"The President this morning has entirely recovered from his indisposition," Admiral Leahy noted with relief in his diary.[1] "At breakfast in the morning FDR seemed to be completely recovered," Lieutenant Rigdon also recalled,[2] while the President's interpreter, Charles Bohlen, noted the President seemed completely restored to good health and "was alert as ever" when the time came to see Stalin again, privately, at 2:45 p.m., in his quarters.[3]

Mr. Churchill had hoped to have lunch with the President first, but the President had really had enough of Winston's continuing effort to derail Overlord — especially after hearing what had transpired at the special get-together of top American and British chiefs of staff and Russian military officials in the Board Room of the Russian Embassy at 10:30 that morning.

Leahy's report to the President was disturbing. "At a small meeting Marshal Voroshilov, General Brooke, Air Marshal Portal, General Marshall, and I discussed questions to come before the conference," Leahy noted in his diary, but "made little progress toward an agreement because of British desires for postponement of the cross-channel operation that has long been scheduled for next May."[4]

It was really too bad, Leahy sighed. Again and again Marshal Voroshilov had pressed General Brooke to answer why *Turkey,* of all countries, should be allowed to delay or abandon Overlord — for Brooke acknowledged that, by keeping Overlord's essential landing craft in the Mediterranean and Aegean, "the retarding of the date set for Overlord" would definitely be necessary, and that, in pursuance of further Mediterranean "operations he had outlined" — which would involve "the capture of the Dodecanese Islands, beginning with Rhodes" in order to "open sea communications to the Dardanelles," and the establishment of airfields in Turkey — "we should be able to hold and destroy the German forces now in the Mediterranean area while awaiting the date for Overlord."[5]

To Voroshilov — and to the American team — this was pie in the sky, and unworthy of a general who considered himself a great strategist. There were now seven U.S. infantry and two U.S. armored divisions in Britain, training for D-day, with some sixty divisions in the United States to follow them onto the beaches of northern France in 1944. The U.S. chiefs then attempted to discuss Overlord air cover and ports — especially man-made floating ports — but General Brooke's insistence they switch to discussing more operations in the Mediterranean led Voroshilov finally to question whether Brooke really believed in Overlord.

Everything Brooke had hitherto said at the meeting sounded defeatist — from his claim that only three or four divisions could be landed on D-day to his assertion that the May 1, 1944, target date for Overlord could only be met by removing all landing craft "from the Mediterranean now" — a removal that was, in the British view, too perilous to contemplate. It would "bring the Italian operations almost to a standstill," Brooke warned — and to scale down the current fighting in Italy was anathema to the British team. "The British wished, during the preparations for an eventual Overlord, to keep fighting the Germans in the Mediterranean to the maximum degree possible," Brooke had stated. Overlord's landing craft should therefore be kept in the Mediterranean, he insisted. As he argued, "Such operations are necessary not only to hold the Germans in Italy but to create the situation in Northern France which will make Overlord possible."[6]

General Marshall had heard all this before. He emphasized that if further amphibious operations were undertaken in the Mediterranean, "Overlord will inevitably be delayed."[7] Marshal Voroshilov kept pointing out that, from the point of view of coalition warfare — since Soviet forces were anxious to launch a massive simultaneous assault from the east — Overlord *must* have priority, with everything else considered "ancillary": "all the other operations, such as Rome, Rhodes and what not, must be planned to assist Overlord and certainly not hinder it." Brooke's "additional operations" threatened to "hurt Overlord," Voroshilov complained — prompting him to emphasize "that this must not be so. These operations must be planned to secure Overlord, which is the most important operation, and not to hurt it." With Brooke stonewalling, however — the British Army chief predicting that Overlord was "bound to fail" unless his alternatives and provisos were accepted[8] — it became clear to Admiral Leahy that only the three world leaders could break the impasse.

In which case, Leahy reported to the "Big Boss," it would be up to the President of the United States, as chairman of the conference, to engineer a solution that kept the Allies together.

24

Pricking Churchill's Bubble

CHURCHILL WAS NO HELP, the President found when the conferees reassembled. Despite knowing from his British chiefs that their meeting with Marshal Voroshilov and the U.S. Joint Chiefs that morning had gone badly, the Prime Minister simply restated British insistence Overlord be delayed to permit more ventures in the Mediterranean and Aegean.

The President, a paraplegic, had traveled almost eight thousand miles across the world to meet Stalin and rehearse some of the issues their countries would face, once the war was won — prompting Stalin, at dinner the night before, to have his interpreter "tell the President I now understand what it has meant for him to make the effort to come on such a long journey — Tell him that the next time I will go to him."[1] Yet Churchill was relentless. Despite hearing from Stalin's own lips how uninterested were the Russians in any operational plan other than a spring Overlord — if possible with a pincer attack on southern France, together with the planned Russian offensive — to ensure the defeat of Germany in 1944, the British prime minister seemed intent on doing battle *yet again* for his alternative Mediterranean and Aegean strategy.

The President, furious at the British resurrection of their Mediterranean plans that morning, had therefore rejected Churchill's invitation to join him for lunch at the British Legation. Spurned, the Prime Minister was "plainly put out," his doctor noted in his diary, deaf to the reason. "It is not like him," Churchill had murmured, puzzled[2] — and hurt by the excuse Harry Hopkins, as the President's emissary, had attempted to give: that Roosevelt didn't want Stalin to think the Westerners were "ganging up" against him.[3]

For the President, getting a firm decision on the date and absolute priority of Overlord was now more important than Churchill's wounded pride. The Prime Minister's endless efforts to postpone the invasion had begun to sound like a broken, jarring gramophone record. It had taken the President simply too long to effect the encounter with Stalin — a meeting he had always wanted to hold without the Prime Minister, in order that Churchill's attempts to subvert or postpone Overlord should not poison the summit—to now squander the chance of endgame and postwar agreements with the USSR.

It was now day three of the Tehran Summit—and crucial that Churchill's opposition to D-day be put down definitively, so that the leaders could move on to discuss the war's endgame, including the defeat of Japan, as well as the President's postwar plans for a United Nations body. The summit with Stalin had never been intended to be a contentious military confrontation between allies; now, thanks to British intransigence, it was. Instead of Churchill being the President's ardent lieutenant, the Prime Minister had become, at Tehran, the President's most ardent military opponent. He seemed willing not only to risk the Allies losing the war against Hitler, but to vitiate also the possibility of a unified coalition of Western allies, strong and willing to check Soviet communist aspirations, should they proved inimical to Western wishes in the war's endgame—and beyond. It was deplorable—and exactly what, in his heart of hearts, the President had feared when setting off from Hampton Roads with his advisers on November 13.

Aware that Stalin was due to visit him in his quarters at 2:45 p.m., the President meantime thanked Admiral Leahy and General Marshall, who'd come to his rooms at 2:30 to rehearse how best to proceed.

Stalin arrived with Molotov. "Punctual to the minute the Soviet leaders arrived," Elliott Roosevelt, who had arrived almost immediately afterwards, recalled. "I was introduced. We pulled up chairs in front of Father's couch, and I sat back to collect my thoughts."[4] These were largely focused on the face of the small, yet "tremendously dynamic" figure of the dictator, who showed "great reserves of patience and of reassurance" as the President proceeded to sketch out, using a piece of paper, his concept of a postwar peace and security organization that, in contrast to the League of Nations after the Versailles Conference, both of their countries could join this time. It would be, the President outlined, a "large organization composed of some 35 members of the United Nations," which would meet and "make recommendations to a smaller body," or executive committee.

Worldwide or European? Stalin asked.

"World-wide," Roosevelt replied.[5]

Explaining his idea of a UN "executive committee," or security council, the President said it would comprise "the Soviet Union, the United States, United Kingdom and China, together with two additional European states, one South American, one Near East, one Far Eastern country, and one British Dominion." It would deal with "all non-military questions such as agriculture, food, health, and economic questions."[6]

Would it have the power to make binding decisions? Stalin asked. Yes and no, the President responded; it could "make recommendations for settling disputes with the hope that the nations concerned would be guided thereby," but he accepted the U.S. Congress would never agree to be bound, militarily, by "a decision of such a body." Which had led him to his concept of the "Four Policemen," namely the Soviet Union, United States, Great Britain, and China: an ad hoc group that would "have the power to deal immediately with any threat to the peace," as Charles Bohlen noted in his minutes of the conversation, "and any sudden emergency which requires this action."[7]

In later years Bohlen was amazed, as a career diplomat, by Roosevelt's presentation — especially the example the President offered Stalin of a threat "arising out of a revolution or of developments in a small country" that threatened to get out of hand. "This bit of prescience by Roosevelt forecast many of the problems that the United States has had to deal with in the postwar period as a result of communist actions," Bohlen commented in retrospect.

"Stalin did not question Roosevelt's idea on this point," Bohlen recalled — in fact the dictator "never showed any antagonism to the general idea of a world body. It was quite obvious," he reflected, "that Stalin felt it would be much more dangerous to be outside any world organization than to be in it" — providing the Soviet Union could "block actions it did not like."[8] Moreover, when the President argued for a world organization rather than merely a European body — which Churchill favored — and Stalin posited that the United States might, as one of the "Four Policemen," then have to send "American troops to Europe," the President did not demur, though he thought U.S. "planes and ships" rather than land armies would probably be sufficient, especially in enforcing a quarantine rather than embarking on hostilities . . .[9]

The discussion was almost mesmerizing in its visionary nature. It left unexamined, as Stalin pointed out, the question of how to deal with Germany after the Third Reich was defeated. Moreover, Stalin was "dubious about Chinese participation,"[10] despite the President's explanation that this would be an investment in the future, given the sheer size and population of China. For all the myriad details still to be worked out, however, it was at least a blueprint for the postwar world: rough and imprecise, but an ideal all the Allies could *fight* for — if the British would only join with the Overlord program as they had promised at Quebec.

At almost 3:30 p.m. General Watson "looked in the door and announced that everything was ready. We got up and moved into the board room."[11] There the Prime Minister of Great Britain, ever conscious of the magical powers of ceremony, had arranged for a beautiful two-handed sword to be presented as a gift to the citizens of Stalingrad, in recognition of Soviet heroism in defending the crucial city in the winter of 1942–43. The ceremony was accompanied by national anthems played by the Red Army Band.

It was a touching gesture — and Stalin seemed genuinely "moved by this simple act of friendship," Churchill's doctor noted; "he bent over and kissed the sword." His armored train had passed through the ruins of Stalingrad on the way to Tehran, and he planned to visit the city on his way back to Moscow. "Roosevelt said there were tears in his eyes." Ruthless, and unarguably a psychopath in relation to his own people, he appeared, for a moment, actually "human,"[12] Churchill's doctor recorded.[13]

Photographs were duly taken to mark the occasion. Within minutes, however, the second plenary meeting commenced in the conference room—and the fireworks, once more, began.

To open the session the President, as chairman, asked that the British general, Sir Alan Brooke, commence by telling them what had transpired in their military discussions that morning.

Loath to admit the truth—namely, that the military representatives of the three countries had failed utterly to agree on a clear military policy to defeat Nazi Germany—Brooke attempted to paper over the disagreement, hoping that either Churchill would do better in convincing the British and Russians they must postpone the Overlord invasion, or that in yet another meeting of the three nations' generals he could somehow keep badgering the Americans until they, at least, gave in. In which case, he remained confident, the Russians would have to accept the decision, since Overlord was not, in the end, a Russian operation.

If Brooke was really banking on this, however, he was in for a school beating. Which Stalin now administered without turning a hair.

The British were still claiming they believed in Overlord. In that case, he asked: "Who will command Overlord?"[14]

The participants had now come to the critical moment of the Tehran Conference, in Bohlen's later view: "a crisis," as he put it. Moreover one that, sadly, rested entirely on the continued British "objections to fixing a date for Overlord."[15] The Russian dictator's question now cut through to the very core of the deadlock, however. Roosevelt admitted no commander had been appointed.

No commander? For an Allied amphibious assault operation involving perhaps a million men, slated to take place in barely *five months'* time? Could they be serious?

Stalin clearly thought not—for, as Churchill's doctor noted, "he would not believe we meant business until we had decided on the man to command the operation."[16]

For his part the President knew the real reason no commander had been appointed: namely, that he himself had consistently refused to assign his best general, George Catlett Marshall, unless the British agreed to the mounting of the operation on May 1, 1944—a date which would give it the best chance to succeed. Relishing the embarrassing position in which Stalin's question put the British, the President thus turned to the man responsible for the calamity: Winston Churchill.

Mortified and ashamed, Churchill attempted to explain that a British "chief of staff" to a future commander had been appointed long ago, in March that year: General Frederick Morgan, a staff officer who had served as a headquarters officer in the retreat to Dunkirk. But he was forced to admit that, no. No Allied commander of the war's most critical operation had yet been appointed.

Attempting to disclaim responsibility, Churchill went on to explain the eventual commander would, however, be an American general, with a British commander then taking charge of an enlarged Mediterranean theater. Rather than endure further humiliation in front of so many witnesses Churchill then suggested, blushing, that the question would "best be discussed between the three of them" — the President, Marshal Stalin, and himself — "rather than in the large meeting."[17]

Yet the President was as determined as Marshal Stalin to get the matter of Overlord finally settled — in fact more so. He thus countered that an American supreme commander could only be appointed once the matter of Overlord's priority was resolved right there, at Tehran. This prompted Stalin to assure his allies "the Russians do not expect to have a voice in the selection of the [Overlord] Commander-in-Chief; they merely want to know who he is and to have him appointed as soon as possible" — for "nothing good would come out of the operation unless one man was made responsible not only for the preparation but for the execution of the operation."[18]

Churchill thereupon twisted in the wind — saying, off the top of his head, he "thought the appointment could be announced in a fortnight."[19]

"Winston was not good," Brooke noted frankly in his diary, thinking of the Prime Minister's long, vacillating, and digressive speech. "Bad from beginning to end."[20]

Mr. Roosevelt, in Brooke's eyes, hadn't helped — or rather, had refused to do so. Worse, he'd agreed completely with the Russian dictator's lacerating interrogation of the Prime Minister and his reasons for wanting to delay Overlord. Stalin's clear contempt for Churchill's floundering explanations of British pusillanimity seemed to Brooke, as a professional soldier, to be extremely well-directed. "Stalin meticulous with only two arguments," Brooke scribbled in his diary that night with sneaking admiration. "Cross Channel operation on May 1st, and also offensive in Southern France! Americans supported this view," he lamented.[21]

It was clear to almost everyone, however, that although the President might still be gentle with Churchill, Marshal Stalin, for his part, was losing patience with the Prime Minister. Churchill had wasted *yet another day* of plenary meetings. The Russian commander in chief wanted an end to Winston's ramblings, and to achieve it Stalin proposed they there and then agree to a three-point directive, which all three nations could then adhere to, namely: (1) "a date should be set and the [Overlord] operation should not be postponed," in order that the Soviets could time a simultaneous offensive "from the east"; (2) that a more or less simultaneous invasion of southern France should be mounted by the Western Allies, or immediately after Overlord, to help guarantee its success; and (3) that a commander in chief should be appointed forthwith — if possible at Tehran.[22]

Churchill refused. He pleaded that the Allies hadn't "studied" a South of France invasion; nor had they even contemplated dovetailing Overlord with a simultaneous Russian offensive — both of them untrue statements. The planning for Overlord had been progressing since the beginning of 1943, *eleven months* before. A simultaneous Russian offensive had been considered a sine qua non, and the very reason for General Deane's military mission in Moscow — since it was imperative Hitler not be able to reinforce the west, once Overlord began. Moreover American planners had spent a great deal of time exploring a simultaneous invasion of southern France.

Instead of expressing embarrassment, however, the Prime Minister appealed to his fellow leaders to change the subject, and discuss the possibility of neutral Turkey being enjoined to declare war on Germany. In order to put pressure on the Turks to do so, Churchill maintained, it would be necessary to retain landing craft in the Mediterranean — not for an invasion of southern France, but for operations in Italy or Rhodes. As he put it, using his memorable verbal artistry, "Now is the time to reap the crop if we will pay the small price of this reaping."[23]

As Brooke noted in his diary that night, however, this was familiar Churchillian flimflam: a typical invocation, larded with grand and memorable phrases invented on the spur of the moment, but reeking with sentiment rather than dispassionate military analysis. Redolent, moreover, with wolf cries: such as the claim that, unless landing craft were taken away from operations in the Pacific, there could be "no action" at all in the Mediterranean. In addition, he warned, if there were more than "12 mobile divisions" facing the Western Allies in France, Overlord would not be permitted by the British to take place at all . . .[24]

Given that the Soviet armies were facing *260 German divisions* in the field, this sounded pathetic. But the longer Churchill struggled to make his case for postponing Overlord, the more embarrassed all those present became. "I feel more like entering a lunatic asylum or a nursing home than continuing my present job," Brooke confessed in shame in his diary. "I am absolutely disgusted with the politicians' methods of waging war!! Why will they imagine that they are experts at a job they know nothing about! It is lamentable to listen to them!"[25]

Admiral Leahy, by contrast, was proud of his commander in chief — the

President saying unequivocally as chairman that, having heard out the Prime Minister, he was nevertheless "in favor of adhering to the original date for Overlord set at Quebec, namely, the first part of May."[26]

For his part Stalin, after three hours of Churchill's filibustering, came close to losing his temper. In the aftermath Hopkins related to Dr. Wilson how, when "the P.M. began once more to stress the strategic importance of Turkey and Rhodes, no one was surprised when the President intervened. 'We are all agreed,' he said, 'that Overlord is the dominating operation, and that any operation which might delay it cannot be considered by us.'" As Dr. Wilson quoted Hopkins's account that night, the Soviet dictator had been relieved. "Stalin looked at Winston as much as to say: 'Well, what about that?'"

Stalin, who did not suffer fools gladly, had hitherto shown unusual self-control — but it was clearly running out. He disputed Churchill's estimate of the number of Wehrmacht divisions in the Balkans, and the Prime Minister's claim that "with a minimum effort these divisions might be placed in a position where they could no longer be of any value," i.e., smashed.[27] There was no such thing as "minimum effort" in combat against the Wehrmacht.

The President was of like mind — considering it an attitude that merely led to distressing loss of Allied lives, often to no purpose. Defeating Wehrmacht forces entrenched in terrain suitable for good defense was no simple matter, even with modern Allied air power. The President had said that, in Yugoslavia, "commando raids" should certainly "be undertaken in the Balkans and that we should send all possible supplies to Tito in order to require [Axis forces in] Yugoslavia to keep their divisions there,"[28] but "without making any particular commitment which would interfere with Overlord." With this the Soviet dictator completely agreed. Stalin, moreover, doubted frankly whether Turkey would "enter the war"[29] — and even if it did, *it would not lead to the defeat of Nazi Germany.* Only Overlord, in tandem with a Russian offensive from the Eastern Front, could effect this. The rest were "diversions."[30]

Which led the President to sum up the plenary meeting by saying "we should therefore work out plans to contain" the German divisions in southern Europe, lest they be switched to reinforce those in France, but "only on such a scale as not to divert Overlord at the agreed time."[31]

Whenever the President mentioned the May deadline for Overlord, Churchill's heart sank further — for Stalin was in complete accord with the President, saying: "'You are right' — 'You are right.'"[32]

Still the Prime Minister resisted, though. When the President thus moved to get their formal agreement as to "the timing of Overlord," declaring that "it would be good for Overlord to take place about 1 May, or certainly not later than 15 or 20 May, if possible," to which Stalin agreed, saying "there would be suitable weather in May," Churchill interceded yet again, saying no, "he could not agree to that"![33]

The President had thus far been amazingly diplomatic, but Churchill's refusal to agree to D-day finally brought his legendary patience to an end. The President looked toward Stalin — and acting as the President's quasi-enforcer, Stalin duly took up the cudgel. The Marshal "turned on Churchill," Bohlen's minutes recorded. The Russian dictator "said he would like to ask him a simple question: 'Do the English believe in Overlord, or do they not?'"[34]

Churchill was understandably abashed — unused to being challenged so directly. Once again he asserted he was in support of Overlord, but "did not think that the many great possibilities in the Mediterranean should be ruthlessly cast aside as valueless merely on the question of a month's delay in Overlord."

A month's delay? The assertion sounded jejune, and unworthy of a commander in chief of Britain's military forces since 1940. Stalin clarified that the Russians were not asking them to "do nothing" in the Mediterranean; he simply wanted them to hold fast there, in order to set a real target date for Overlord. Still, however, Churchill objected, continuing to insist upon further "operations in the Eastern Mediterranean" — including a new invasion of Rhodes, and more pressure put on Turkey to enter the war — which would "create conditions indispensable to the success of Overlord."[35]

Rhodes indispensable to Overlord? The more the Prime Minister protested against a firm date in May for Overlord, the more obvious it was to everyone in the committee room that he was deliberately prevaricating. His talk of Turkish air bases, and "operations to drive and starve all German divisions out of the Aegean and open the Dardanelles" — operations that "could not be considered as military commitments of an indefinite character" — and comments that "our future will suffer great misfortune if we do not get Turkey into the war" since it would leave British troops "idle" — were a charade. British idleness in the Mediterranean had nothing to do with a target date for Overlord; it was British idleness in England, and in particular at 10 Downing Street, preparing for D-day, that had everything to do with the success of Overlord — now only months away. When Churchill suggested the subject be thrown back to the "ad hoc" committee of the Russian, British, and American military representatives, requiring yet more days of argument, Stalin pointed out he had only agreed to a summit lasting until December 1, though would extend this to leave on December 2, if necessary.[36]

It was at this point that President Roosevelt, as chairman of the gathering, decided to read out a proposed directive he'd penned on a piece of paper in front of him. It was damning.

"The Committee of the Chiefs of Staff," the President's draft read, "will assume that Overlord is the dominating operation." In considering "subsidiary operations" in the Mediterranean area they must take "into consideration that any delay should not affect Overlord."

Stalin agreed — though pointed out there was no mention of an actual date for D-day. To which the President responded, unequivocally, "that the date for Overlord had been fixed at Quebec," namely May 1, 1944, "and that only some much more important matter could possibly affect that date."[37]

Churchill's long, three-and-a-half-month struggle to delay or ditch Overlord was coming to an end — as Churchill himself at last recognized.

In perhaps the worst display of military leadership of his life he had wasted everyone's time in a vain attempt to avoid the cost of British casualties in mounting a successful Overlord. Yet British and American casualties in his alternative, peripheral strategy would be just as daunting if the Western Allies poured new men into the Mediterranean, Aegean, and Balkans — and to no common purpose. Nazi Germany would not be defeated by such operations. Only by launching Overlord, in tandem with a Russian offensive, could the Third Reich be brought down. Churchill's unending pleas for alternative "cheap" operations in the Mediterranean and Aegean were neither strategically sound, nor realistic in any military sense. As the famous British Eighth Army commander, General Bernard Montgomery, facing stalwart German defense in southern Italy in winter conditions, had written to the director of military operations only a few days before, "Why we start frigging about in the Dodecanese, and dispersing our efforts, beats me."[38]

It beat everyone else. Churchill's "frigging about," however, had finally been exposed at Tehran for what it was, in front of his country's two senior partners in the United Nations coalition. Humiliated, Churchill reluctantly caved in — proposing that the chiefs of staff translate the President's directive into clear tripartite military strategy that the leaders could sign off on, the next day.[39]

It had been a historic meeting. To spare Churchill further shame, the President "observed that within an hour a very good dinner would be awaiting all of them, with Marshal Stalin as their host, and that he for one would have a large appetite for it."

With that, at 7:15 p.m. on Monday, November 29, 1943, the three-hour plenary session broke up.

25

War and Peace

SITTING SILENTLY BEHIND the President during the plenary meeting, Admiral Leahy had found himself somewhat awed. "I am very favorably impressed by the Soviets' direct methods and by their plain speaking," the admiral confessed that night in his journal, writing as a professional naval officer. Though the Russian dictator "appears old and worn," he was "soft-spoken and inflexible in his purpose."

But the hero of the day had been Mr. Roosevelt. The President had ensured Churchill was given a fair chance to have his full say but in the end, they had been able to reach the brave and fateful tripartite decision the President wanted. "The meeting was conducted by the President with skill and a high order of diplomacy," Leahy observed.[1] The three leaders would discuss politics over dinner, without the chiefs of staff—for the strategy and timetable of the war against Hitler was now set. Churchill's strategic insurrection had been defeated, and the Allies could now work together to effect the defeat of Germany, then move on to defeat Japan.

Winston Churchill, however, was deeply distraught—aware he'd failed. Back at the British Legation, before dinner, he asked his doctor to syringe his aching throat—snarling at the humiliation of it all. "Nothing more can be done here," Dr. Wilson heard him muttering. "Bloody," he swore as he "stumped out" to join the President and Marshal Stalin at the Soviet Embassy for dinner.[2]

Left to dine in the British Legation, General Brooke, Air Marshal Portal, and Admiral Cunningham presented the mien of defeated men. "They are always the same, quiet and equable," Dr. Wilson noted in his diary, "but tonight they seemed put out. It had been a bad day for our people."[3]

It had.

"To our unity—war and peace!" the President raised a toast at dinner. To which Stalin had responded, in all seriousness, with the words: "I want to tell you, from the Russian point of view, what the President and the United States have done to win this war. The most important thing in this war are machines.

The United States has proven it can turn out from 8,000 to 10,000 airplanes per month. Russia can only turn out, at most, 3,000 airplanes a month. England turns out 3,000 to 3,500, which are principally heavy bombers. The United States, therefore, is a country of machines. Without the use of these machines, through Lend-Lease, we would lose this war."[4]

It had been an unusual wartime acknowledgment from the Soviet communist dictator, but it was sincere. In fact it had prompted the President to sum up the achievement of Tehran, as he saw it, in his response. "We have differing customs and philosophies and ways of life," he remarked. "Each of us works out our scheme of things according to the desires and ideas of our own peoples. But we have proved here at Tehran that the varying ideals of our nations can come together in a harmonious whole, moving unitedly for the common good of ourselves and of the world." Looking outside, he'd therefore been disposed to see, as he put it metaphorically, "that traditional symbol of hope, the rainbow."[5]

At the end of the long day night had fallen. The decision over Overlord, so endlessly contested, was now a done deal—with only the choice of a supreme commander for D-day to be made by the President, as U.S. commander in chief.

In contrast to the President's sense of achievement, General Sir Alan Brooke felt nothing but despair—unable and unwilling to appreciate the gigantic stride the Allies had made toward winning World War II. All the head of the British Army could see was the British showdown over strategy having failed to ignite. The alternative, evasive British military strategy now seemed in ruins. They would have to get behind Overlord. "May God help us in the future prosecution of this war" were the last words he scribbled in his diary in his slashing green hand before going to sleep, for "we have every hope of making an unholy mess of it."[6]

PART FOUR

Who Will Command Overlord?

26

A Commander for Overlord

AT 2:35 P.M. on Thursday, December 2, 1943, the Sacred Cow, piloted by Major Ryan, touched down in Cairo at the end of its thirteen-hundred-mile journey back from Iran.

A "Declaration of the Three Powers" had been jointly signed the night before, and was to be announced to the world once the three leaders had departed. The document, drafted by the President, reaffirmed "our determination that our nations shall work together in war and in the peace that will follow.

"As to war," the declaration had continued, "our military staffs have joined in our round table discussions, and we have concerted our plans for the destruction of the German forces. We have reached complete agreement as to the scope and timing of the operations to be undertaken from the east, west and south."[1]

Three theaters — no longer two! And their "timing" agreed.

Those who'd feared the Allies were bogged down in Italy, or were worried the Soviets were being pushed back in Russia, could now breathe more easily. There could be no rumor or suggestion the Allies were disunited, or no longer working to an agreed inter-Allied plan, or did not mean business in pursuing the unconditional surrender of Nazi Germany and the Empire of Japan. The formal photographs of the three Allied leaders, seated on the front porch of the Russian Embassy in Tehran, said it all: instantly, once published, the most potent image of the war up to that time. (Leahy later likened the scene to that of King Henry VIII meeting François Premier, Roi de France, on the Field of the Cloth of Gold — if only one could ignore, Leahy qualified, the "suffering and squalor" of the golden fields of Tehran.)[2]

"No power on earth can prevent our destroying the German armies by land, their U-boats by sea, and their war planes from the air," the President's text ran. "Our attack will be relentless and increasing. Emerging from these cordial conferences we look with confidence to the day when all peoples of the world may live free lives, untouched by tyranny, and according to their varying desires and their own consciences. We came here with hope and determination. We leave here, friends in fact, in spirit and in purpose."[3]

The document was signed — at Stalin's insistence — in the order "Roosevelt, Stalin, Churchill," reflecting their importance on the world stage. Regarding the postwar world, it recognized "the supreme responsibility resting upon us and all the United Nations to make a peace which will command the good will of the overwhelming mass of the people of the world and banish the scourge and terror of war for many generations. With our diplomatic advisers we have surveyed the problems of the future. We shall seek the cooperation and the active participation of all nations, large and small, whose peoples in heart and mind are dedicated, as our own peoples, to the elimination of tyranny and slavery, oppression and intolerance, into a world family of democratic nations."[4]

Given the non-role the United States had played in international relations in the aftermath of World War I, and then during the Great Depression, the tone and content of such a communiqué struck Dr. Goebbels, in Berlin, as a significant turnaround on the part of the United States. The propaganda minister was frankly puzzled. And alarmed.

The first official announcement that the Allied leaders had met in Cairo had already been made on December 1 — with claims that, joined by Stalin, the Big Three had then met in Tehran, there to prepare a "Propaganda-Manifesto" to be broadcast to the German people. The Reich propaganda minister had duly scorned the idea of such an initiative, thinking it would be a "complete fiasco." The Allied "terror" bombing of German cities had only hardened the determination of the German Volk to resist the "plutocracies," Goebbels recorded in his diary — with some justification. "We're not talking 1918 here," he reflected, "in terms of leadership," since he was confident, thanks to the Führer's political and military stewardship, the German Volk had changed completely in the years since the humiliation of Versailles. Proud of his insight into the German psyche he seemed not to take seriously the three-front warning or timetable, issued by the President and his two allies after the Tehran meeting. Instead, he took heart from the Führer's assurances that with the help of the Reich's growing technical arsenal of new weapons, including *Vergeltungswaffen,* the Allies would be prevented from winning the war, even if Germany could no longer achieve the offensive victory that had seemed so tantalizingly within its grasp in the summer of 1942, the year before.

On the telephone from his military headquarters in East Prussia, Hitler had confided to Goebbels he'd spoken to Goering, the field marshal responsible for the Luftwaffe and aerial warfare. He'd asked the increasingly portly airman to press for "our revenge weapons to be completed as quickly as possible"; in the meantime he himself, as Führer, would hold back the Soviet tide.[5]

For his part, as Gauleiter of Berlin as well as Reich minister for propaganda,

Goebbels assured the Führer the Reich capital would cope with anything the Allies could throw at it from the air. Both men therefore saw the Allied summit in Tehran not as a demonstration of Allied unity, but as part of a "war of nerves" to intimidate the German Volk and break German morale. It wouldn't succeed, both Hitler and Goebbels were confident — and in this respect history would prove them right.

The Allied communiqué was not directed at Germany, though. It was written and released to the Allied press, to be read by soldiers and civilians of the United Nations. Hitler's Third Reich was unlikely to surrender in the same manner as the Italians had done in September, the President felt. Judging by recent combat both on the Eastern Front and in Italy, the Germans had simply become too militarized, too wedded to hard power, and too content with tyranny over other peoples — even fellow German citizens if they were in the least part Jewish, or objected to Nazi doctrine. So effective was Hitler's leadership of the majority of Germans, in fact, Hitler no longer even needed to appear in public; the troops of the vaunted Wehrmacht would take care of the front, and German civilians would take care of the *Vaterland.* Nor was this surprising, in terms of the history of peoples — the record of German prowess and endurance on the field of battle going back thousands of years, to Roman times. The fact was, only military victory on the battlefield, won by Allied soldiers, could now free Europe from the Wehrmacht's iron grip and go on to defeat the Third Reich decisively, followed by Japan. It was a task that would involve significant casualties, the President was well aware, in Europe — and again in the Pacific, if the conquests of the Empire of the Rising Sun in China, Southeast Asia, and the Southwest Pacific were to be rolled back, and a new start to world order be made. In the darkness of a continuing global war such communiqués were thus essential to maintain public faith in the outcome. Moreover, they were an important reminder of what the Allies were fighting *for,* the President insisted. Though the communiqué was signed by all three leaders, the vision of the future was clearly that of the American president, Goebbels recognized. As such it aroused his contempt. He dismissed it as cynical claptrap.

It wasn't, however — even if it smacked of innocent idealism and its goals might not necessarily be realized anytime soon. As the President had recently confided to Walter Lippmann, the syndicated columnist, the Moscow Conference had been a "real success," but the devil would be in the details — and its aims far from easy to achieve. "Sometimes," he'd confided to Lippmann before leaving for Tehran, "I feel that the world will be mighty lucky if it gets 50% of what it seeks out of the war as a permanent success. That might be a high average."[6]

Whatever his inherent skepticism, though, the President understood the power of ideals — and was determined the United States would lead the democratic way, once the war was over. The Soviet Union, which had never known democracy, would have to be encouraged to participate on a global stage, as would the primary colonial empires, which would need to be brought along if wars of independence, such as America's in the eighteenth century, were to be avoided.

In the meantime the British prime minister had shown a welcome "inclination to accept the American point of view on matters that have heretofore been in controversy between the American and British Chiefs of Staff," Leahy had noted with relief in his diary.[7] And in an effort to conciliate the Prime Minister, in return, the President had said he was sending his own plane, the Sacred Cow, to Ankara to bring President Inonu — who'd expressed his willingness through the U.S. ambassador to meet with the President and the Prime Minister — to Cairo, to discuss a possible Turkish entry into the war.

Churchill was touched — though aware, like most members of the delegations, that there was little real hope the Turks would ever fight the Germans, whose ruthlessness was well known to them; they had, after all, been Germany's allies in World War I. Only fools rush in; the Turks would surely wait out the course of the war, and declare their interest only when they saw German defeat in sight.

The more important decision to be made in Cairo before the President and Mr. Churchill went home was thus not the matter of Turkey but of who would command the D-day invasion and subsequent drive to Berlin.

27

A Momentous Decision

LIKE MOST PEOPLE — even Stalin himself — General George Marshall had assumed, before leaving American shores, that he would be made supreme commander of Overlord. So certain had he been, in fact, that his wife, Katherine, had begun sending furniture from the general's official residence, Fort Myer, just outside Washington, into storage, assuming that her husband would be moving to England. A number of Marshall's subordinates in Washington, too, had been notified their services would soon be required abroad.[1] After all, had not the tall, austere, no-nonsense chief of staff of the U.S. Army been assured by the President himself — and on a number of occasions — he would be appointed to the coveted post, most recently during his meeting with the Joint Chiefs aboard the USS *Iowa*?

What, Marshall's colleagues and subordinates later wondered, changed the President's mind? And *when,* exactly, did he change it?

Given the historic importance of the appointment, it would be nothing short of extraordinary that in subsequent years no historian would get even the date right. For the most part historical writers merely glided over it, assigning it little significance — another example, it was generally considered, of the President's easygoing, somewhat hands-off approach to military decision-making compared with Churchill's meddling, sometimes disastrous but always energetic, performance as British commander in chief.

In truth the President took the matter of supreme command of Overlord deeply seriously — perhaps more seriously than any appointment he would make in his entire life. Far from having ignored or underestimated the import of the command, as Stalin assumed, the President had been taking advice for quite some time on the matter, and from many quarters. Back in September 1943, in Washington, he'd written to General John Pershing, for example, to solicit the opinion of the five-star World War I general.

General Pershing knew that the U.S. secretary of war, Henry Stimson, was in favor of Marshall being appointed. As Stimson would put it on November 10, even as the President prepared to leave for Tehran, "Marshall's command of Overlord is imperative for its success."[2] In a tough, uncompromising response in September to the President, however, General Pershing—a professional soldier to his bootstraps, and the legendary commander of U.S. forces in Europe in the previous war—had already warned that, in his opinion, to transfer Marshall to "a tactical command in a limited area, no matter how seemingly important, is to deprive ourselves of the benefit of his outstanding strategical ability and experience. I know of no one at all comparable to replace him as Chief of Staff." In sum, General Pershing had judged it "would be a fundamental and very grave error in our military policy."[3]

The President had been somewhat taken aback by Pershing's vehemence—and surprised, too, given that George Marshall had been director of planning in the First U.S. Division for the Battle of Cantigny, on the Somme, in 1918, and had then served on General Pershing's own headquarters staff in France that year. Marshall was conversant therefore with the battlefields on which the Allies would soon be fighting—battlefields that would be greater even than Pershing's command had been in World War I, for the United States would now be fielding by far the dominant Allied military force in France.

"You are absolutely right about George Marshall," the President had responded with his usual tact, then added, "and yet, I think, you are wrong, too!" General Marshall was by "far and away" the best man to be chief of staff of the U.S. Army at the Pentagon, he agreed. However, the supreme command he was proposing was much more than a "tactical command in a limited area." The "command will include the whole European theater," he'd emphasized.[4]

The whole European theater.

It was this misunderstanding that would confuse later historians—for it was never the President's intention to make Marshall a battlefield commander, when Marshall had never commanded in battle, and when to do so would deprive him of Marshall's superlative talents as the head of the U.S. Army in Washington. His aim, since the spring of 1943, when Churchill came to Washington to plead against the notion of a cross-Channel invasion and beaches that might be littered with British dead, was to hold the feet of the British to the fire: which could be done by ensuring an American commanded "the whole European theater," and could thereby guard against Mediterranean diversions that would imperil Overlord. Marshall was to be, he had thus reaffirmed aboard the USS *Iowa,* in command of "all Europe." Assuming, of course, that the British would agree to such a huge command area (essentially the one commanded by a single NATO supreme commander, beginning in 1952).

They wouldn't, though—and Marshall, aboard the *Iowa*, had ironically agreed with their probable objections, deeming it to be too large a military arena to command effectively. For their part, behind a facade of similar concern over the sheer size of such a fiefdom, the British had no intention of allowing both the Mediterranean and Overlord commands to come under a single supreme commander — unless that commander was British and could be relied upon to do Churchill's bidding. For Winston Churchill was no fool: recognizing that, if the Second Front and Mediterranean theaters were merged under an American commander, his chances of "exploitation" in the Aegean and Balkans would be nullified. The idea of an "all Europe" supreme command had therefore become a bone of Anglo-American contention, rather than a serious command possibility — as Marshall himself had been aware, when sharing with Eisenhower, in Algiers on the way to Tehran, his frustration that the two generals were, as Eisenhower put it, mere "chess pieces" on the President's board.

Churchill had thus claimed, at Cairo, to be perfectly content to accept Marshall in command of British D-day forces — but only on the understanding that Overlord not be permitted to have priority over British schemes in the Mediterranean. Tehran had squelched that possibility, however. The cross-Channel invasion was now cast in stone, as the number one Allied undertaking in the spring of 1944, backed by a simultaneous offensive on the Eastern Front by the Soviets. And with Overlord's primacy and timetable irrevocably established between the three major Allied war powers, the President was now faced with an equally difficult task, from a personal point of view: how to tell Marshall he was not going to go to England, and would not command the historic amphibious invasion.

On the evening of his first full day back in Cairo, Friday, December 3, 1943, Roosevelt thus sent his White House counselor, Harry Hopkins, to Marshall's villa, to sound out how this news would be received.

As the modest, intensely private General Marshall noted in a letter to Harry Hopkins's biographer, Robert Sherwood, after the war, he had not been aware his fellow chiefs of staff had, in previous weeks, "gone to the president opposing my transfer" to the D-day command — nor even that the President had sought General Pershing's opinion on the matter. He was thus astonished when, as he recounted, "Harry Hopkins came to see me Saturday night, before dinner, and told me the President was in some concern of mind over my appointment as Supreme Commander."[5]

Marshall's memory was mistaken, for it was Friday evening, not Saturday, following their return from Tehran. Nevertheless, the general was clear about one thing, in retrospect: namely that the question of his not being appointed supreme commander of Overlord "never came to a head in any way until Hopkins came to see me at Cairo, and told me that the President was very much concerned, because he felt he had to make a decision."[6]

Marshall might have gotten the date wrong, but his recollection as to the

President's urgency was undoubtedly correct. The President had given his assurance to Stalin that he would make the decision on Overlord's command within a couple of days of getting back to Cairo. Moreover, from a purely logistical point of view, the decision on leadership was now critical, since thanks to Churchill the whole Overlord project had been put on virtual hold by the British. If D-day was to be a success, there was not a day to be wasted, Marshall knew. Overlord was the invasion he himself had pressed for, studied carefully, planned for, and had backed for almost two years, even when its chances of success were zero, and the President, as commander in chief, had had to overrule him, and put him in his place.[7] Now, however, the President had gotten Churchill to back off, and D-day to be put on the front burner — with Marshall the clear candidate for the post of supreme commander. Hopkins's visit, in person, therefore threw him. As he himself confessed later, he had no idea how to respond: "I was pulled from many directions and I wouldn't express myself on any of them."[8] For Hopkins's appearance at his villa that evening could in fact mean only one thing: that the President was now having second thoughts.[9]

Embarrassed, Marshall declined to indicate to Hopkins whether he would be disappointed if he was not chosen. He was certainly aware that whoever was selected for the post would go down in history. "I merely endeavored to make it clear that I would go along wholeheartedly with whatever decision the President made. He need have no fears regarding my personal reaction. I declined to state my opinion," he later told Robert Sherwood.[10]

For his part Harry Hopkins was dismayed by Marshall's reaction — for Hopkins had always pressed for Marshall to command the D-day invasion, as had many others — even Churchill,[11] and Stalin. Why didn't Marshall say he was *counting* on getting the historic command; that he *wanted* it; that he felt he was the *right man* to take it?

Marshall, however, was no Patton or MacArthur in terms of putting personal ambition above duty. Besides, at some deeper level he knew it would not help sway or change the President's mind, if the President was not keen to appoint him, for whatever reason.

Hopkins, as go-between, duly brought back the information to Mr. Roosevelt, the Commander in Chief. Marshall would not make "a scene," or be difficult about it, he explained. This was all the President needed to know. Or almost all.

Since so many hundreds of thousands of British and Canadian forces would come under the supreme commander who was appointed, Churchill would have to be consulted — and disarmed, lest he once again attempt to sabotage the Overlord timetable. Before making the appointment official, then, the President made sure he had Churchill's solemn word he would not prejudice Overlord's timing. The next morning at 11:00, on December 4, 1943, in the

President's villa and in front of the assembled Combined Chiefs of Staff, the President thus got Churchill to state formally his "conversion" to Overlord and its timetable: his words to be taken down in minutes of the meeting and typed up by the secretary of the Combined Chiefs of Staff Committee. Churchill, duly but reluctantly, obliged. Overlord was "a task transcending all others," the Prime Minister stated in a characteristic turn of phrase, and it would be launched "during May." He himself would have "preferred the July date," but he was "determined nevertheless to do all in his power to make the May date a complete success."[12]

The President was relieved, and the plenary meeting had subsequently addressed how to support Overlord with the strongest possible pincer landing in southern France—Operation Anvil.

The essential deal, however, was now struck, and in writing. The British would partner with the United States in launching D-day in May 1944, under an American commander. With that cast-iron understanding, the President asked to see Marshall afterwards, on his own. As Marshall recounted, "The President had me call at his Villa, either immediately before or immediately after lunch."

In actuality the President's interview with Marshall was over lunch itself. There, in the comfortable Kirk villa overlooking the pyramids, "in response to his questions, I made virtually the same reply I made to Hopkins," Marshall later explained to Robert Sherwood—namely that "I would not attempt to estimate my capabilities; the President would have to do that. I merely wished to make clear that whatever the decision, I would go along with it wholeheartedly; that the issue was too great for any personal feeling to be considered."[13]

Writing in 1948, Sherwood rightly called the appointment a "momentous decision"—in fact Sherwood considered it "one of the loneliest decisions" the President "ever had to make."[14]

Was it? Sherwood was a playwright in his professional life, but a wartime speechwriter for the President and a man who deeply admired General Marshall. By conveying the loneliness of the decision Sherwood was without doubt trying to soften his account. He wanted to make clear how admired Marshall was by his colleagues, and how much they wanted to see him in the role of supreme commander of the greatest amphibious invasion in military history. As Sherwood wrote, if the President were to choose someone other than Marshall it might even be contested within military and White House circles, for it would not only be against "the impassioned advice of Hopkins," who was not a military man, but against that of the secretary of war, Henry Stimson, who was. It would even go counter to "the known preference of both Stalin and Churchill." In fact it would go against the President's "own proclaimed inclination to give George Marshall the historic opportunity which he so greatly desired and so amply deserved."[15]

While this was true, none of it reflected the real story: namely that the President had never intended to let Marshall leave Washington, unless Marshall himself insisted he wanted and deserved the battlefield command! The President had only spoken of General Marshall as the prospective supreme commander in order to stop Churchill from delaying D-day, without actually making the appointment until Churchill backed down, and the primacy and date of the battle were set in stone — no longer subject to Churchillian sabotage in wild Aegean and Balkan diversions. Once Churchill had formally surrendered over Overlord, there was no need to continue the charade. Unless Marshall insisted on the post — which would complicate matters — the President felt free to appoint the man he now felt certain would be the right individual for the job.

Sherwood's version of the historic decision would become iconic, even being quoted by Churchill in his memoirs,[16] but it was plainly wrong, including the date. The President had never wished to lose Marshall from Washington; he had merely used him as a chess piece in his battle with Churchill, exactly as Marshall had described.

What Sherwood went on to record, however, was certainly true: namely that Roosevelt would never knowingly hurt a friend, or someone who had worked loyally for him. Thus the President's final words, as he lunched with Marshall in Cairo — words that would become famous in the study of command decision-making — were typical of the President's *savoir faire* in complex human relations: "I feel I could not sleep at night with you out of the country."[17]

Of all the President's many tributes and compliments this was perhaps the kindest: a generous recognition of Marshall's extraordinary stature in Washington. Yet even Marshall later recognized there had been much more to it than that: that the decision was not sudden, for the President had in fact been reconsidering the appointment, in the light of General Pershing's advice, for months; that he had also sought the views of Marshall's colleagues on the Joint Chiefs of Staff Committee — and even that of General Eisenhower, on the way to the Egyptian capital.

Marshall had been the President's bishop to Churchill's knight in the long, difficult struggle to get the Prime Minister and the British to conform to a May 1944 invasion date, and fold their Mediterranean/Aegean tents. Now that Churchill had given in, the President had been too much of a gentleman to deny Marshall the command, *if* he said he really wanted it. Sensing the President's underlying intent, however, George Marshall had nobly offered to fall on his sword — denying himself the glory of commanding one of the greatest battles in history. For the fact was, the chess match with Churchill was done. The President wanted someone else for Overlord.

28

A Bad Telegram

WHEN THE SECRETARY OF WAR, Henry Stimson, heard the news at the Pentagon, four days later, he blanched. "I had bad news by telegram today on the United Command of Overlord," Stimson noted in his diary. "Apparently there has been a curious shift there which I just don't understand."[1]

One by one, as the U.S. generals and staff officers flew back from Cairo, Stimson tackled them, urging them to tell him about the "most curious question as to the Commandership."[2] But not even General Somervell, the deputy chief of staff of the U.S. Army, who'd traveled out with the President and had taken part in the Combined Chiefs of Staff meetings in Cairo, could tell him how it had happened, or what was behind the decision.

General Dwight David Eisenhower was also surprised, since he had been under the impression the post was definitely going to Marshall.

The President, however, was the President—and would never regret his decision. Most of his adult life he had had to make choices about subordinates—not least those in the military. Article II, Section 2, of the U.S. Constitution stated that the "President shall be Commander in Chief of the Army and Navy of the United States, and of the Militia of the several States, when called into the actual Service of the United States," and would thereby have the right to appoint all "Officers of the United States." Franklin Roosevelt himself had never served *in* the military; he had, though, served *with* military officers, men and contingents, and for almost a decade, long before he was elected president. Appointed by President Wilson to be assistant secretary of the U.S. Navy in 1913, he'd had almost eight years of daily experience dealing with U.S. naval officers and men—before, during, and in the aftermath of World War I. Becoming U.S. president and commander in chief in 1933, moreover, Roosevelt had amassed a further decade of dealing with army, air, and naval officers and men—and since 1941, the ultimate command of his nation's forces in directing a global war. Unlike the Prime Minister—who *had,* ironically, served in the military—the President was thus for the most part a seasoned, successful chooser of men. Behind his mask of charm and affability, he'd honed over the years an almost unique ability, for a politician, to judge the merits of military officers, according to the context and role in which he wanted them to serve. Admiral Harold Stark, for example, had been a wise counselor to the President as chief of naval operations in peacetime — but not in wartime. The President had quickly but quietly removed him from that role, after Pearl Harbor, sending him as senior U.S. naval liaison to England. Others, too, had had to be retired, sidelined, or transferred.

War, like peace, demanded different talents of different men, he understood. There were those who proved able administrators under stress. There were men who proved efficient and effective planners, but who failed in combat. Officers, too, who proved fine commanders in the field of battle — whether in the air, on ground, or at sea — but poor administrators, or higher commanders. But there were still very few who possessed modern battlefield experience commanding all three services, and international coalition forces. It was a problem the British themselves would fail to resolve in subsequent days, as the Prime Minister and General Brooke argued over who should be appointed supreme commander in the Mediterranean.

Overlord, if successful, would lead to the defeat of the Third Reich, as even Hitler accepted. Its supreme command was therefore a role of vast consequence, both for the individual selected and for the free world. In the American armed forces General MacArthur and Admiral Nimitz were both already in the senior "supreme command" category, directing the army, air, and naval forces of the United States as well as other nations in battle. Douglas MacArthur had, however, declined the President's invitation to fly to Cairo to meet him, sending his chief of staff instead. This was a mistake on MacArthur's part, for in the President's view MacArthur was still a mercurial figure at best: one who might be difficult for the President or Joint Chiefs of Staff to control — a reputation MacArthur could only have countered by a personal appearance, advancing with Mr. Roosevelt the argument that, in battle after battle in the Pacific since his defeat at Bataan, he had proved he was a master of amphibious assault operations, moreover had led Australian commanders and their forces with surprising tact and skill. Equally, Chester Nimitz had shown himself to be a master of amphibious operations and the use of multinational naval, ground, and air combat forces. Of intelligence, too — as at Midway in 1942 and the interception and elimination of his main opponent, Admiral Yamamoto, in 1943. But neither commander had faced the modern Wehrmacht — which they would need to do in the looming do-or-die "face to face" battle with the "remaining masses of the German troops," as Secretary Stimson put it in his weekly survey at the Pentagon: a potentially decisive battle between great armies in northern France.[3] This was a region MacArthur, at least, knew intimately from his heroic World War I experience; Nimitz did not.

By contrast young General Dwight D. Eisenhower was *the man on the spot.* He was a general of now-proven experience in commanding multinational forces of all three services, a commander of high intelligence, popular, and with good political instincts. A man whom other men — whether staff officers or commanders — were proud to serve under. None of the U.S. chiefs of staff had been present when the President spent two days with Eisenhower at Oran and Tunis, before the pre–Tehran Conference in Cairo. Ike had then given his presentation of current strategy and plans in the Mediterranean to the Combined Chiefs before they all left to meet Stalin, which had further impressed the President. In fact the President's mind was by then made up as to whom he would like to appoint — if he could get the British to sign up to D-day, incontrovertibly and definitively, to take place in May 1944. They had, leaving only Marshall's feelings to take into account. Once Hopkins had reported to him that Marshall would naturally be disappointed, but would accept the President's decision with good grace, the President had had no further qualms. This was war. Hundreds of thousands of lives were at stake. He had not hesitated — and, as Hopkins had predicted, Marshall had shown no emotion or disappointment.

General Hap Arnold, chief of staff for the Army Air Forces and Marshall's closest colleague at the Pentagon, thus recorded the historic decision with memorable frankness in his diary that evening, December 4, 1943. "Marshall," the air chief wrote, "had lunch with President; he doesn't get Overlord, Ike does."[4]

Churchill and the British chiefs of staff were informed of the President's decision later that afternoon. That night General Alan Brooke noted in his diary the same as General Arnold. "A difficult day!" Brooke penned as Marshall's mirror figure: chief of staff of his nation's army — and, ironically, the very man who, until the Quebec Conference three months earlier, had not only been promised the supreme command but had counted on it. "Finally asked to dine alone with Winston to discuss questions of command," Brooke recorded, for "the President had today decided that Eisenhower was to command Overlord while Marshall remained as Chief of Staff."[5]

To all intents and purposes the deed, then, was done on December 4, 1943 — not December 5, as Churchill would claim in his memoirs, quoting, as would all historians in the following seven decades, Sherwood's book. By nightfall of December 4, not only were the U.S. and British chiefs of staff aware of the historic decision in the Egyptian capital, but the British prime minister at his own villa near the pyramids, too.[6]

And yet it was not immediately announced, or conveyed to Stalin. Why?

29

Perfidious Albion Redux

SEVEN DECADES LATER it is impossible to know for sure. The personal diaries of those present in Cairo, however, give a clue — for the British contingent, worsted over Overlord at Tehran, appeared determined to have its revenge.

The U.S. chiefs of staff were stunned; in fact could not at first credit what the British were doing. On Friday afternoon, December 3, instead of discussing plans for Anvil, the pincer invasion of southern France to elide with Overlord, the American chiefs had found themselves in "locked horns over Rhodes, Dodecanese, Dardanelles, etc," with the British, General Arnold had noted in his diary.[1]

Rhodes, Dodecanese, *Dardanelles*? Not Anvil?

For almost two hours the U.S. chiefs had afterwards sat with the President and with Hopkins, complaining — hoping, however, it was only a temporary venting of British disappointment. There was clearly "lots to do" in terms of Anvil before they left — but the next day, December 4, had proven even harder. Presiding over the quasi-plenary meeting of the Combined Chiefs of Staff together with the Prime Minister for two hours, the President had wrung out of Churchill his acceptance of Overlord's primacy and timetable — but there were warnings of storms ahead over the logistics necessary to mount Anvil. The British were yet again determined to tie them to operational needs in the Far East. This had entailed a "long involved discussion over war: invasion France," foremost — but also "Burma campaign, Andamans, Aegean Sea, principles to our strategy,"[2] as General Arnold jotted in his diary. For his part Churchill was now insisting on the cancellation of the Andaman Islands invasion, despite the President having promised Generalissimo Chiang Kai-shek it would take place to help the Chinese.

The U.S. chiefs had then gone off to lunch — leaving Marshall alone with the President to be told the President's decision regarding command of Overlord. At 2:30 p.m. the Joint Chiefs had been back again with their opposite numbers, this time at Mena House. There the British declared virtual war on their coalition partners, demanding outright the cancellation of "Buccaneer," the Andaman Islands invasion, while the U.S. chiefs fought to keep it in play. Leahy, King, Marshall, and Arnold were all certain that the British were being deceitful in claiming they needed Buccaneer's landing craft for the Anvil invasion of southern France; they strongly (and correctly) suspected the British were actually aiming at the Aegean islands.

Behind his facade of seemingly endless goodwill, the President, in truth, was becoming more and more disenchanted with the British team. He made clear at the meeting that they only had two days in which to get the Anvil act together. Brooke, who was once again leading the charge, recorded in his diary his fury when he'd "discovered" that morning that his prime minister "had been queering our pitch by suggesting to Leahy that if we did not attack Rhodes we might at any rate starve the place out. We then lunched with the PM and at 2.30 went back for our meeting with the Combined Chiefs of Staff. We were dumbfounded by being informed that the meeting must finish on Sunday at the latest (in 48 hours) as the President was off [to Tunis]! No apologies, nothing. They have completely upset the whole [Cairo and Tehran] meeting by wasting our time with Chiang Kai-shek and Stalin before we had settled any points with them. And now with nothing settled they propose to disappear into the blue and leave all the main points connected with the Mediterranean unsettled. It all looks like some of the worst sharp practice that I have seen for some time," Brooke lamented — insisting yet again that Rhodes be taken by military force *before* Overlord could be tackled.[3]

Why Rhodes? Had it not been agreed at Tehran that the Aegean was not to delay, let alone derail, the Overlord timetable?

In the circumstances, by the end of December 4, 1943, the President had good reason to delay public announcement of his decision to appoint Eisenhower, even to Stalin — for as long as General Eisenhower remained the commander in chief of all Allied forces in the Mediterranean he would be empowered to thwart British plans to invade Rhodes and other diversions. Diversions which, if they met the sort of opposition the Wehrmacht had recently mounted in the Aegean and were still showing in southern Italy, could prejudice the launching of Overlord — the last thing the President wanted Stalin to learn, when Overlord's success depended so much on a massive simultaneous Soviet offensive from the east to preclude reinforcement of their forces in France.

Leahy, for his part, was distraught. By a margin of more than two to one the United States was now the dominant partner in the Western coalition — and would become all the more so in the months ahead. Why did the President have to give in to Churchill?

"Second Cairo," as it became called, would later be whitewashed by Churchill,[4] and be largely overlooked by historians after the war. But as Leahy would note, the U.S. chiefs bravely fought to the very end in the Egyptian capital, on behalf of the President, to avoid having to cancel Buccaneer and risk the whole relationship with the Chinese. It proved a losing battle with the British, however.

At the start of their talks the President "didn't budge" over Buccaneer, Leahy proudly remembered, since Mr. Roosevelt had given Generalissimo Chiang Kai-shek, only the week before — and in person, no less — his solemn promise the operation would take place. He thus had no wish to give the British landing craft for further misadventures in the Mediterranean or Aegean. But as the bitter days of Second Cairo went by, it became obvious the British were not going to give up. "No decision has been made as to the controversy in regard to the Mediterranean versus the Andaman operation," Leahy noted in his diary on the night of December 4, 1943.[5] The next day things were no better. "No progress was made toward a solution of the problem," he recorded[6] — though the President, fearing the worst, began to ask if there were any alternatives, and warned Chiang there might not be enough landing craft for Overlord *and* Buccaneer — though knowing this was not the real reason.[7]

"Neither side would yield," Leahy later recalled. "It was the same story up to 5 p.m. on December 6. At no time in previous or later conferences had the British shown such determined opposition. When the American Chiefs met with Roosevelt at 5 o'clock" that day, it was all over, however. The President was leaving the next morning, and refused to stay to duke it out. He therefore sorrowfully "informed us that in order to bring discussions to an end, he had reluctantly agreed to abandon the Andaman plan and would propose some substitute to Chiang. He was Commander-in-Chief and that ended the argument."[8]

The British had won. Tens of thousands of futile and unnecessary American, British, and Canadian casualties would be the result, at the little Mediterranean town of Anzio, but Churchill was Churchill — and rather than put Overlord in jeopardy, once again, the President had given in.

Admiral Leahy never quite forgave Churchill — not only on behalf of American lives subsequently lost in Italy, but for the effect on Chiang Kai-shek. "It must have been a sad disappointment to Chiang," he reflected later, for the "Chinese leader had every right to feel we had failed to keep a promise." Not only was American honor affected, but the war in Asia. The cancellation of the Andaman plans was bound to affect Chinese willingness to go on fighting the Japanese, rather than turn on their compatriots — Chinese communist troops under Mao Tse-tung. This would leave a huge Japanese army occupying a vast swath of coastal China. Chiang "never had indicated much faith in British intentions, but had relied on the United States. If," therefore, "the Chinese quit, the tasks of MacArthur and Nimitz in the Pacific, already difficult, would be much harder," Leahy lamented. "Japanese man power in great numbers would be released to oppose our advance to the mainland of Japan," he wrote: Chinese territory being the best proposed launching pad for the U.S. air, naval, and army assault.[9]

Admiral King, for his part, was appalled. Marshall had already had a late-night session with Churchill in which the Prime Minister was "red hot" for retaking Rhodes. "All the British were against me," Marshall recalled. "It got hotter and hotter." Finally Churchill had grabbed Marshall's lapels and said, "His Majesty's Government can't have its troops staying idle. Muskets must flame," and "more things like that," Marshall later recounted. "I said, 'God forbid, if I should try to dictate, but' I said, 'not one American soldier is going to die on the goddamned beach.'"[10]

The British would not give way, however — insisting that the landing craft released from Buccaneer should be made available for Churchill's immediate plans in the Mediterranean. The final day of the conference had been the worst. "After the matter had been discussed, and Mr. Churchill had pressed his points," King recalled, Leahy and Arnold had both reluctantly surrendered, as did Marshall — "but King remained obdurate, and would not give an inch," he narrated (using the third person for his account), "because he knew that the Chinese, headed by Chiang Kai-shek, would feel they had been sold out — which was the case — and consequently would not do anything to aid Stilwell" in the Far East theater.[11]

If King was furious, General Stilwell was disbelieving. Summoned to see the President at 4:30 p.m. to hear the bad news, he was the most aggrieved. Understandably incensed, he fumed that night in his diary: "God-awful is no word for it" — Roosevelt "a flighty fool," in his vinegary judgment. "Christ but he's terrible."[12]

King, in retrospect, also blamed the President for going "against the advice of his Joint Chiefs of Staff."[13] This was certainly justified, if unfair. For the truth was, the President simply had no arrows left in his quiver, if he wanted the British to land alongside American troops on D-day. He could only swallow his disappointment and let Stalin know, at least, the *good* news: a commander of Overlord had been appointed.

It was a bittersweet moment for Marshall. "The appointment of General Eisenhower to command of Overlord operation has been decided on," he wrote in his draft message for the President to sign and send to Stalin, via the White House in Washington. The President added the word "immediate" before "appointment," then appended his signature.[14] It was December 6, 1943 — the eve of the anniversary of Pearl Harbor — with the British determined to sabotage operations in the Far East now in a desperate attempt to slip Rhodes into the agreed D-day timetable, even at the risk of delaying it. As a result,

Marshall decided he would not go back to the United States directly, but would fly to the Pacific, and see for himself how best to deal with the consequences of the latest British pusillanimity.

The President, for his part, tried to look on the bright side. Overlord — the "D-day Invasion," as it would come to be called in world history — was at least, finally, "on," in tandem with a massive Soviet offensive that would bring the Third Reich to its knees. Moreover the President had made the most momentous decision of his life in deciding on Overlord's supreme commander — an appointment to which the British, who would be fielding half the forces for D-day, had at least agreed.

With that the President, taking a deep breath, set off for Tunis, where he intended to inform Eisenhower of the appointment in person. And if possible visit his combat commanders in the field: Patton, Clark, and others.

30

In the Field with Eisenhower

ALL IS WELL and I'm on my way home," the President penned in a note to his cousin Daisy Suckley, to be sent by airmail. "The trip was almost a complete success — especially the Russians," he recorded, proudly.[1]

Almost. In war, as in peace, one had often to accept less than what one would ideally wish. And if the British had proven halfhearted in support of Overlord, even perfidious, the Russians at least had shown they were determined to support the American landings. The President thus chose to look upon the trip, thus far, as successful. When Daisy heard on the radio his plane "'flew over Tunis'" on his way back to the United States, she was beside herself with anticipation. "*I wish he would get home!*" she scribbled anxiously in her diary.[2]

The President had not flown over Tunis, but in fact *to* Tunis. Whatever he had told Churchill about needing to get back to Washington, he was certainly not rushing. In the biographies written after his death, his return to America did not figure as more than a footnote. And in the ledger of Roosevelt's wartime trips, the trip home perhaps did not merit more than that. Yet the journey home would be important to him personally, as he "wound down" in the company of General Eisenhower, the man he'd just appointed to command the D-day invasion. For the President was now able to spend two further days with the young general: intent upon encouraging his new supreme commander in how best to approach his new responsibility. Also, if possible, to fly with Ike to the Allied battlefield in Italy, where American troops were fighting hard to contain Wehrmacht forces. He'd be able to see some of the men who had proved themselves in combat, and who would now be transferred to England for the great Overlord invasion. Also some of those who would remain in the Mediterranean theater and perhaps take part in the southern pincer invasion of France, Operation Anvil, if he could get the British to cease trying to wreck it.

Shortly after breakfast on December 7, 1943, the Sacred Cow thus lifted off the runway at Cairo West airfield for its eight-hour flight to Tunis — by day, this time.

Given that he'd had to undertake several weeks earlier the same leg, in the opposite direction, in total darkness, the President had been determined the reverse trip be done in good light, and should follow if possible "the whole length of the British advance last winter—over El Alamein & Benghasi and Tripoli," as he described in his diary. "Most interesting to see it all from the air—endless desert, but much of it broken country with a good lot of tanks & other equipment not yet picked up," owing to unexploded mines and drifting sands. "In another year or two there will hardly be a trace—for even the shell holes & fox holes will be filled up. What a march that was! Over 1000 miles—with fighting practically all the way. But at the end, with the advance of Montgomery into Tunis from the South," he noted, "we and the British struck from the west—and all of W. Africa was in our hands together with 300,000 prisoners."[3]

How proud he was, as the de facto commander in chief of the Western democracies—and curious to see the scars. Admiral McIntire, the President's doctor, later remembered the enthusiasm with which the President had insisted on the exact route—and the low altitude. "A dangerous business even with a fighter escort, for the Germans still maintained active airfields on Crete," McIntire recalled—airfields that were "no more than two hundred miles from Tobruk." The President, however, "could not be dissuaded. All morning we skimmed over the scenes of Montgomery's stand and Rommel's rout, but while the rest of us scanned the skies for German planes, the President had eyes only for the battlefields. After luncheon, to everybody's relief, he decided on a nap, and we took advantage of it to steer a swift, straight course for Tunis."[4]

In retrospect, at least, the physician was amused. In the President's rapt curiosity the doctor could see what others missed, namely the fascination felt by a paraplegic: a would-be warrior who could not walk. Also just how much the sight of what would become a legendary North African battlefield had meant to the President who, as the United States commander in chief, had made it all possible—not only by sending, on his personal orders, the critical Sherman tanks Montgomery had needed in order to smash Rommel, but the vast U.S. contingent required at the other extremity of North Africa for the simultaneous pincer attack: Operation Torch.

The President's daylight flight across Egypt, Libya, and Tunisia was, in other words, not only giving Mr. Roosevelt a chance to see with his own eyes the magnitude of the Allied achievement, but in terms of terrain, what still lay ahead for the Allies. Which made him doubly anxious, regarding D-day, to see for himself evidence of the great American amphibious assault landings on the island of Sicily, at the very center of the Mediterranean, which had been in many ways a rehearsal for Overlord. And even to see, if it could be arranged, Salerno, south of Naples, as he proposed to discuss with General Eisenhower, who met him at El Aouina airport, Tunis, that afternoon.

"Well, Ike, you are to command Overlord," the President first announced with a confident smile as they sat back in Eisenhower's armored limousine.

Eisenhower later recalled how relieved he was to have the President tell him the news in person — and how grateful he was. "Mr. President," he responded with great tact, "I realize that such an appointment involved difficult decisions. I hope you will not be disappointed."[5]

It was an iconic moment: the Commander in Chief sitting with the man he'd selected to carry out the biggest seaborne assault landing ever undertaken. Amazingly, Eisenhower had never seen the current Overlord plans, or been asked to comment on them, despite having been the commander in chief of the Allied invasion of Sicily, six months before, which had involved 3,000 Allied ships, 4,000 aircraft, and 150,000 ground troops. Overlord would be far more forbidding, however; thanks to Allied deception measures, the Allies had initially faced only two German divisions in Operation Husky, the invasion of Sicily; Overlord would face at least forty.

Driven with the President to the "White House," in Carthage, Eisenhower listened as the Commander in Chief recounted some of what had transpired in Tehran — as well as what had happened in Cairo, on his return from the summit. The wily British prime minister would surely seek to mount more operations in the Mediterranean and Aegean, under Eisenhower's successor — operations that could still affect the Overlord timetable, now that the indefatigable British had gotten their way in canceling Buccaneer. These were the fortunes, or misfortunes, of coalition war. Which Eisenhower, having commanded American, British, French, and other forces, as well as dealing with political representatives of those countries, could appreciate.

The President's desire to fly to the battlefield in Italy the next day, however, put the grateful new Overlord commander on the spot.

Having to tell his Commander in Chief no, so soon after hearing of his new appointment, was tough. But it had to be done, Eisenhower recalled. Three days earlier, Luftwaffe planes had penetrated British air defenses at Bari, in southern Italy, and had sunk no fewer than *seventeen* Allied ships. Among them had been an ammunition vessel, and a fuel tanker that exploded and spread "fiery catastrophe" to other vessels — including one that was carrying World War I nerve gas to be used in retaliation against German forces, if the Wehrmacht resorted to the use of such weapons. The wind had been mercifully offshore, and the "escaping gas caused no casualties," Eisenhower explained to the President.[6] The moral, unfortunately, was that the President was too important a life to risk; the Italian mainland was simply not safe against German air attack.

By contrast the planned visit to Sicily *would* be possible, Eisenhower was able to reassure the President. U.S. air forces would maintain their protection, not British. And he would order General Clark to fly back from the front in

Italy, as well as assembling a contingent of combat medal award-winners, to meet the President at Palermo, on Sicily's north coast, after he and the President had made a quick stop in Malta.

With that assurance the President had gone to bed in the "really lovely villa," as the President had earlier described the house in his letter to Cousin Daisy, "just outside the ruins of ancient Carthage"[7] — thrilled at the prospect of flying at least to the two famed Mediterranean islands the next day.

He nearly didn't make it, however.

31

A Flap at Malta

WHAT A DAY!" Roosevelt noted with relief, pride, and excitement in his diary, late on December 8, 1943—having survived something of an aerial drama.

At 8:00 a.m. the Sacred Cow, piloted by Major Bryan, had risen in bright sunlight over the Mediterranean. Passing over Cape Bon and Pantelleria it had made for Malta, the fortress-island that had withstood all that the Luftwaffe and Kriegsmarine could throw at it since the spring of 1941. The distance there was only 310 miles—the President's plane duly escorted by a dozen P-38 fighters. But as Major Bryan lowered the landing flaps, it was found there was a problem. The flaps would not go down.

Even with its recent extension, the runway at Valletta was distinctly short for a C-54 plane flying without landing flaps. Parachuting, moreover, was not really an option for the President. There had been a number of air accidents that year, involving transport planes. Wladyslaw Sikorski, prime minister of the Polish government in exile, had drowned when his Liberator had crashed after takeoff at Gibraltar in July—none of the passengers or crew, except for the pilot, having worn life vests. Partly as a result of the Sikorski accident, all those aboard the Sacred Cow, on orders of the pilot, *were* wearing Mae Wests—but the prospects of saving the President in an emergency ocean landing would be slim.

General Eisenhower's heart was understandably in his boots. As the President's host and guide, he had earlier "worried about German fighters still based in Italy." McIntire recalled the general's concern amid the "shaking of heads" that had taken place as to the merits of such a visit by the President to an active war zone. A mechanical malfunction had not, however, figured in such anxieties. McIntire had "stewed over the prospect of another fatiguing day for the President," he related after the war—his concerns, naturally, having been for the President's health. Now, aboard the Sacred Cow, Admiral McIntire found himself anxious for his own health, too—and perhaps life—when "it was discovered that the landing flaps were out of commission."[1]

There was, McIntire narrated later, "great commotion" aboard the aircraft, "but F.D.R. never turned a hair." The situation was almost comic as the plane carrying the President and the new supreme commander of Overlord circled the small island for twenty minutes, losing height, speed, and fuel. "Fine," the President said. "Here's where we see just how good our pilots are."[2]

Luckily the President's faith was not misplaced. "Major Bryan, praise be, lived up to the best traditions of the air service, for he landed us with a skill," McIntire later recorded, "that brought a permanent grin to the face of [U.S. Air Force] General Spaatz."[3]

For his part, Lieutenant Rigdon could scarcely believe how effortlessly Bryan landed the huge four-engined plane. "It was a tense moment — the landing strip we were to use ran right to the water's edge — but we landed with little more than the usual bump and bounce and stopped in plenty of time."[4]

At the foot of the ramp the presidential party was met by the governor, Lord Gort — whose courage and British phlegm were legendary, despite losing an entire army at Dunkirk.[5] "I got into a jeep & reviewed a mixed British Battalion and Navy Marines," the President happily recorded in his diary that night, as well as troops from the Malta defenses regiment.[6] "Photographers in great numbers made pictures of this ceremony," Admiral Leahy recounted in his own diary, "which may sometime in the future be an event in the history studied in our schools of that period": namely a president of the United States inspecting British troops under overall American command, on a historic island that had withstood years of Nazi siege.[7]

"I spoke and presented the illuminated citation I had written before leaving home — Gort received it & made an excellent little speech," the President jotted in his own diary. "We had intended to leave [for Sicily], but the plane had broken the hydraulic pump & it took 2 hours to repair — so with Lord Gort I had a very interesting drive through the harbor part of Valetta, where most of the damage was done to this heavily bombed little island. Nearly all the houses are demolished or bombed but the people stay. The dockyard is running ¾ repaired."[8]

The tour of Valletta was the President's first firsthand witness of the sheer destructiveness of modern urban bombing. Yet it testified also to the deeply questionable results of such a strategy, given the resilience of the civil population. "This is reputed to be the most bombed spot in the world during the present war," Lieutenant Rigdon noted in the log of the President's trip — the area "still generally a mass of shambles."[9] Admiral Leahy, too, noted that the Navy Yard's "buildings and those adjacent thereto were completely destroyed by the enemy."[10] Yet, thanks to "underground workshops," the dockyard itself was "operating at near normal efficiency," Rigdon recounted — and the same seemed true of the inhabitants. Though Churchill continued to put great faith in the bombing of German cities — in fact had given the President a special stereoscopic viewer to look at images of the destruction of the city of Cologne, for example — there was little evidence in Malta that such a tactic achieved much more, in the long run, than the hardening of civilian morale.

On the flight from Tunis to Malta, Hopkins later told his biographer, "Roosevelt talked at great length to Eisenhower about the prodigious difficulties he would confront during the next few months at his new headquarters in London, where he would be surrounded by the majesty of the British Government," Robert Sherwood recorded, recalling Hopkins's mocking account, "and the powerful personality of Winston Churchill, who still believed, in Roosevelt's opinion, that only through failure of a frontal attack across the Channel into France could the United Nations lose the war."[11]

For his part the President believed the opposite: that only by failing to invade France could the Allies lose the war, now they had the men, the means, and the commander to carry out the operation. Ironically Overlord remained a risk that Churchill, for all that he was an inveterate, unashamed gambler, still feared to take, whatever he had agreed to in Tehran. It would be Eisenhower's great task, as supreme commander in England, to restore the Prime Minister's British courage, such as he'd shown in 1940. And if he couldn't, to restrain him from sabotaging the invasion timetable.

Meanwhile, after lunching on board the President's plane, Major Bryan reported at 1:00 p.m. he thought the aircraft was fit to fly to Sicily. Without incident it duly took off at 1:10 p.m. and "Eisenhower listened attentively" to more of the President's advice, Sherwood recounted, while "the 'Sacred Cow' droned over the Mediterranean waters."[12]

At the Tehran Conference, the President related, Marshal Stalin had shown himself to have a sharp, incisive mind, yet also to be surprisingly human. Whether talking of Poland or the Baltic states he had made no bones about Russian "borderland" demands in ending a war that Hitler had waged so mercilessly against the Soviet Union. Beyond that, however, Stalin had appeared to have no territorial or ideological ambitions; client-states, bowing to Russian wishes around the Soviet Union's periphery and probably under Russian military control, seemed far more important to him than the spread of communist ideas. In the aftermath of Tehran Churchill had expressed continuing anxiety about Russian intentions — yet the Prime Minister had evinced no real or practical idea of how to thwart or shrink them. Without consulting any Poles, Churchill had in fact privately suggested to Stalin at Tehran — using three matches to demonstrate — that the Allies "give" the Soviet Union the eastern part of Poland, and compensate the Poles with an equivalent area of eastern Germany.[13] Such shades of Munich had been uncomfortable for the President, who'd not been present, and had declined to "go there," in the absence of actual Poles at this moment in the war, and given likely public and congressional feelings in America.

It wasn't that such things should not be discussed, in closed session, by world statesmen—especially as a sounding board. But it was important to project *power* to the Russians, he felt—*real* power. The Russian dictator had seemed well aware how dependent the Soviet Union was on American largesse, as his toasts on the last night in Tehran had demonstrated. The Soviets, in other words, were still respectful of American economic and industrial power, from which they hoped to benefit after the war. Yet it was vital, Roosevelt explained to Eisenhower, to show the Soviets that the United States could and would use its military power in guaranteeing world peace. He'd deliberately reminded Stalin that the United States, unlike the Soviet Union, was fighting a *global* war, with a vast military commitment in the Pacific. If potential Soviet expansionism was to be discouraged in Europe, it was crucial America strike across the Channel as soon as possible, and defeat the German armies in northern France.

Eisenhower listened, and agreed—reflecting how, in the end, he did not really understand why Churchill, a man of such personal courage and capacity for risk, was so opposed to the cross-Channel assault. "How often I heard him say, in speaking of Overlord's prospects," Eisenhower would later recall, "'We must take care that the tides do not run red with the blood of American and British youth, or the beaches be choked with their bodies.'"[14] Such a fear, though, never quite squared with Churchill's openness to American and British blood being copiously spilled in the Mediterranean or Aegean, Eisenhower noted. Casualties there were mounting inexorably—and might well escalate even more dramatically once a British supreme commander took on Eisenhower's mantle in the Mediterranean at the end of the month . . .

As Eisenhower later recalled, it was difficult to "escape a feeling that Mr. Churchill's views were colored" by considerations "outside the scope of the immediate military problem": that the Prime Minister was all too happy to disregard the military challenges involved when it suited him, preferring to focus on British political, even personal, prizes or "fruits" dangling before him in his capacious mind. When "fired up about a strategic project, logistics did not exist for him," Eisenhower reflected, "the combat troops just floated forward over and around obstacles—nothing was difficult. Once I charged him with this habit, saying, 'Prime Minister, when you want to do something you dismiss logistics with a wave of your hand,'" but when disliking a proposal, he would list so many "'logistic difficulties'" he would "effectively discourage any unwary listener." The Prime Minister "looked at me with a twinkle in his eye," Eisenhower remembered, replying candidly: "'It does make a difference whether your heart is in a project, doesn't it?'"[15]

Overlord was not just a "project," however. Two million men would take part in the invasion and subsequent battle. It would decide the war against Hitler. Everyone's heart *had* to be in it now—especially those of America's best combat commanders. Among them a soldier still under a dark cloud: General George S. Patton.

32

Homeward Bound!

THIRTY MINUTES AFTER TAKEOFF "I saw Mt Aetna, its top white with snow," the President jotted in his diary. "We skirted the So.[uth] West of Sicily, seeing all the American landing places" Patton's Seventh U.S. Army forces had assaulted in July that year, then came down "at a field outside of Palermo."[1]

This was Castelvetrano airfield, where, on descending the special ramp, the President was met by a phalanx of proud American combat officers. Waiting on the tarmac also was General Arnold, his U.S. Army Air Forces chief of staff, who'd followed him in his own plane from Cairo. In pride of place, though, was General Mark Clark, the U.S. Fifth Army commander who'd been summoned from the battlefield in Italy. And George Patton, whose forces had conquered Sicily but whose headquarters were now being disbanded, in preparation for Overlord and Anvil.

Accompanied by photographers, the President "entered a jeep and departed on a tour of the airfield," Lieutenant Rigdon recorded. Infantry, tank, and airfield defense units were "drawn up for inspection by the President," and honors were rendered by the Thirty-Sixth Engineers' Band, after which "the President proceeded to inspect the troops, driving down the ranks in his jeep. He then took a position in the center and at the front of the troops and, while still in the jeep, decorated a number of the officers with the Distinguished Service Cross—including General Clark, whose courage on the beaches at Salerno had tipped the scales during the invasion in September. "General Clark's decoration came as a complete surprise to him, he told us. He had been given no idea of why he had been called down from the front in Italy to Sicily." In bright sun the "assembled troops" then "passed in review before the President," including Company B, 908th Infantry—"a Colored outfit and a unit of the 7th Army," all of whom had taken "an active part in the recent Sicilian campaign."[2] More photographs were taken.

It was at this point that, although General Patton was not being decorated, the President called him over, and had a special word with him.

News that, at the height of the campaign in Sicily, the general had slapped and threatened shell-shocked soldiers in a field hospital had recently swept the press in America. Some members of Congress were demanding the cavalryman be recalled and demoted for conduct unbecoming of an officer and a gentleman.

Admiral Leahy related later how, in "conversation with the President, General Patton brought up the widely publicized incident of his indiscretion of slapping a soldier whom he believed to be a shirker. Apparently the General still worried about possible repercussions and their effect on his own future." To Patton's huge relief, however, the President "indicated that the matter was a closed incident as far as he was concerned."[3]

He did more, in fact. General Clark later recalled how the President, taking Patton's hand, held it for some time. "General Patton," he murmured from his jeep, "you will have an army command in the great Normandy operation."[4]

Patton — a highly emotional as well as brilliantly aggressive commander — almost fainted. In previous days his mood had swung between despair at the thought of being recalled to the United States and faith in divine intervention. The President's arrival out of the skies above Palermo appeared to be divine. As the President's jeep drove on, according to Mike Reilly of the Secret Service, Patton "burst out sobbing."[5]

For his part the President looked proud and elated. Photographs of President Lincoln visiting his generals on or near the battlefield had become iconic — and here was the thirty-second president, more than a century later, inspecting combat troops and their commanders halfway across the world. Barely two hundred miles from Palermo, American troops were fighting a tough battle against the Wehrmacht on the Rapido River, with American planes supporting them. General Spaatz, whom he'd had a chance to get to know better, would be going with Eisenhower to England for Overlord, he'd decided, after discussing the matter with General Arnold and with Eisenhower. General Eaker, the current U.S. commander of the Eighth Air Force in Britain, would be switched to the Mediterranean, while General Doolittle — whom the President had decorated with the Congressional Medal of Honor on his return to the United States after the bombing of Tokyo the year before — would leave Italy and take Eaker's command in England.

The President, in short, was getting his Overlord ducks in a row, after the seemingly endless, acrimonious "strategical" battle with the British. There were early cocktails and a snack at the Thirty-Second Squadron Officers' Club; then, at 3:30 p.m., the President's party once again reembarked on the Sacred Cow. The lumbering C-54 Skymaster sped down the runway and took off — not to fly to Marrakesh, as had earlier been planned, but now that it was so late, only to Carthage, where the President wanted to spend a final night before the long flight to Dakar, on the west coast of Africa. There, as per Admiral King's instructions, the USS *Iowa* would be waiting. Generals Spaatz and Arnold were meanwhile to fly to Italy.

Ever nervous about assassination — Admiral Darlan, after all, had been assassinated the previous December, in Algiers — the Secret Service had attempted to stop the President from staying another night in Carthage, but they had failed. "The Secret Service men were irritated and fearful," Eisenhower later recalled, "but the President confided to me that he had made up his mind to stay at Carthage an extra night and if a legitimate reason for the delay had not been forthcoming he would have invented one." The delay caused by the wing-flap hydraulics had thus been timely. "I remarked that I assumed the President of the United States would not be questioned in dictating the details of his own travel. He replied with considerable emphasis, 'You haven't had to argue with the Secret Service!'"[6]

An hour before sunrise the next day, December 9, 1943, General Eisenhower, Colonel Elliott Roosevelt, and several other senior officers gathered at El Aouina to say goodbye to the President at the airport. "Mr. Roosevelt was just as friendly and natural as before," Kay Summersby, Ike's chauffeur, later said of his departure. "Mr. Roosevelt complimented me on my driving, thanked me for 'taking care' of him, and then smiled. 'I hope you come to the United States, child. If you do, please be sure to come and see me,'" she remembered him saying. "It was the last time I ever saw him."[7]

Flying across the edge of the Atlas Mountains at eight thousand feet — in thick cloud, but warned by the President's doctor not to go any higher — and then over the Sahara, the Great Desert, the 2,425-mile journey seemed to take forever: Tunisia, Algeria, French Sudan, Mauritania, and finally Senegal.

The port of Dakar had played a menacing role before Operation Torch, given the possibility of German occupation of the Vichy-held navy and air base. But since the Torch invasion, it had been in Allied hands — its U.S.-run airfield, Rufisque, as important as the harbor. "We landed on the field about sunset," the President recorded, greeted there by the general commanding American air units — "a very important point," the President noted, since an "average of 60 planes from the U.S. pass through here every day on their way to the front."[8]

Welcoming the President aboard the USS *Iowa,* Captain McCrea, for his part, was suffering from a scraped leg that had become infected. He'd refused to go to the sick bay, however; instead he showed the Commander in Chief to his captain's quarters, and in great pain climbed to his own cabin on the bridge.

It was as well he did. An incompetent French tugboat pilot almost ran the giant warship aground in the dark,[9] but the President, below, was blissfully content, with cocktails and dinner. "An hour ago," the President noted with relief in his diary, when at last he went to bed, "we weighed anchor & so—homeward bound!"[10]

33

The Odyssey Is Over

ACROSS AMERICA THE TRIUMPH of Tehran had been met with rejoicing —and most of all because the patriarch, by December 17, 1943, was reported in the press to be safely back on American shores.

U.S. TROOPS INVADE NEW BRITAIN, WIN FOOTHOLD; RAF BOMBS BERLIN; "FORTS" ALSO ATTACK REICH; ROOSEVELT IS BACK; CHURCHILL HAS PNEUMONIA ran the *New York Times* banner headline. PRESIDENT ARRIVES. SAFE RETURN DISCLOSED BUT NOT WHEN HE WILL REACH CAPITAL. ABSENT FOR 35 DAYS, the front-page article was headed. "The White House announced today that President Roosevelt had safely returned to this country after his journey to the historic conferences in the Middle East."

The *Times* estimated the President had traveled "over 25,000 miles" — an astonishing figure if it was true. (It wasn't.)

The trip, however, was still shrouded in secrecy. No reporters had accompanied the President to Tehran; the *New York Times* really knew no more than anyone else — and nothing whatever of the historic showdown with Churchill that was to decide the war.

"A completely uneventful voyage," the President meantime summarized the voyage home, writing in his arching, looping hand in the leather-bound diary his cousin had given him. He felt in great form. The USS *Iowa* had been "under escort of destroyers all the way, & of aircraft also for the last 3 days," he penned on December 16, 1943.[1] He'd relaxed, had read books, had sat outside, and had worked on his stamp collection. His White House staff had relaxed, too. They'd passed south of the Cape Verde Islands; Admiral Leahy had had three cavities in his teeth filled, and the ship had made twenty-three knots in "warm trade wind weather," as Leahy noted in his own diary.[2] "Marvellous warm weather & no sea," the President scribbled — "up to yesterday noon when we hit a sudden storm from the coast, & from 60 it dropped last night to 20° Fahrenheit." The crew had changed from khakis into blues, and "this morning early we came in through the Capes. Now we are steaming up Chesapeake Bay. We are all packed up—I am writing in the big room & the boys are having a final game of Gin Rummy."[3]

In truth, while the President was asleep during the storm, and with "visibility zero" outside, the *Iowa* had narrowly missed a merchant vessel standing north of them. "Had it not been for our radar contact, a collision would certainly have occurred," Lieutenant Rigdon subsequently recorded in the President's log.[4]

The President's luck had held, however, and at 4:00 p.m. on December 16 the *Iowa* had anchored off Cherry Point, where its journey had begun. "The little *Potomac* has loomed 6 miles ahead at the mouth of the river and at 4.30 I will transfer to her, after first making a speech to the crew," Roosevelt wrote in high, widely separated words, with *t*'s crossed at the apex. He would stay overnight on his presidential yacht, "and tomorrow morning we should get to the Navy Yard in Washington & soon afterwards I will be at the W.[hite] H.[ouse] & using the telephone. So will end a new Odyssey."[5]

Before disembarking the President made a "short interesting talk to the crew of the U.S.S. *Iowa*" on the ship's quarterdeck, Leahy recorded.[6] In this the President tried to convey something of the import of the conferences in the Middle East. "One of the reasons I went abroad, as you know, was to try by conversations with other nations, to see that this war that we are all engaged in shall not happen again," he explained in his easy, folksy, unmistakable tenor lilt. "We have an idea—all of us, I think—that hereafter we have got to eliminate from the human race nations like Germany and Japan; eliminate them," that was, "from the possibility of ruining the lives of a whole lot of other nations." He'd held "talks in North Africa, Egypt and Persia, with the Chinese, the Russians, Turks and others," and felt "real progress" had been made in looking ahead to an international world order beyond war. "Obviously it will be necessary when we win the war to make the possibility of a future upsetting of our civilization an impossible thing," he'd stated. "I don't say forever," he'd cautioned, however. "None of us can look that far ahead. But I do say as long as any Americans and others who are alive today are still alive. That objective is worth fighting for."[7]

He, Mr. Churchill, and Marshal Stalin had "the same fundamental aims," the President asserted: "stopping what has been going on in these past four years, and that is why I believe from the viewpoint of people—just plain people," the trip had been worthwhile.[8] Despite differing beliefs, they were all "engaged in a common struggle"—a coalition in which the three leaders, as heads of their governments, represented "between two-thirds and three-quarters of the entire population of the world."[9]

Once again Captain McCrea—who was not a natural speaker—was delighted at how the President could deliver such "off-the-cuff talks" that tied the concerns of ordinary men and women to the ideals of a whole nation, in fact all democratic nations.

"After he had embarked on the Yacht *Potomac*," using the special gangway, and once the vessel "pulled away from alongside *Iowa*, with hat in his hand he waved goodbye to us," McCrea recalled.[10]

It would be the last time McCrea saw him, for the *Iowa* was thereafter sent to the Pacific, to take part in many battles and win many battle stars. "The crew," for their part, "responded with a spontaneous cheer."[11]

At 9:15 the next morning, after a final night in the *Potomac* as it made its way upriver, the President was greeted at the Washington Navy Yard by his wife, Eleanor. Following "a month in warmer latitudes," Leahy noted, it was a distinctly bracing return to land.[12] In the bitter cold they were driven to the White House, where at 9:30 a.m. a "large delegation of his friends" were on hand "to welcome him back home," Lieutenant Rigdon recorded.

The cabinet and the quasi cabinet had assembled in the Green Room, before going down, en masse, to the "entrance to the South Door of the White House," Secretary Stimson recorded that night in his own diary—joined by a throng of congressmen who'd come "for the same purpose of greeting him." The President was "wheeled in from his car," Stimson described. "He was in his traveling suit, looked very well, and greeted all of us with very great cheeriness and good humor and kindness. He was at his best. Republicans were mixed with Democrats and they all seemed very glad to have him back safe and sound. We stood around a few minutes." Then the President "went upstairs with Mrs. Roosevelt and the family to their apartment and the rest of us left for our offices."[13]

That afternoon at 2:00, having unpacked and eaten a light lunch, the President then convened the cabinet for the first time since November 5.

The interior secretary, Harold Ickes, had been one of those most deeply worried about the prospect of the trip, and the President's safety. Weeks before, on December 3, Grace Tully, the President's office secretary, had told Ickes that "extraordinary precautions" had been ordered "to protect the parties to the conference," as she reassured him over lunch. His anxiety hadn't, however, been assuaged. "I am more nervous than ever about the President and all of his Army and Navy experts being out of the country where a successful attack might mean the lives of all or at least a large number of them," Ickes had confessed in his diary. It was, in its way, the great weakness of the American constitutional system, he'd reflected. "We would suffer from such an event much more than Great Britain or Russia where the systems of government are so much more flexible, so much so as to permit the strongest leader in each country to be selected" to replace him.[14]

This was debatable—as well as being a slur on his colleague, the vice president, Henry Wallace. It certainly explained, though, why the safe return of the Chief Executive was greeted with almost literal fanfare and jubilation. "He looked very well," Ickes noted after the cabinet meeting. "A lot of us feel easier now that he is back in this country." The "President spoke of his recent conference in Cairo and Tehran. Evidently he was very much taken with Stalin, and this was confirmed later by Ross McIntire," the President's physician. "He likes Stalin because he is open and frank. In discussing Japan, Stalin indicated that he did not care how far the United States and Great Britain went in punishment" — moreover the President "felt the same about Germany, too," after the atrocities the Japanese and Nazis had committed.[15] In Tehran he'd seen the young shah, and had gotten Stalin to sign "an agreement guaranteeing the present and future independence of Iran." The President had suggested to Stalin "that he also sign. Stalin demurred. He said that Russia needed a warm-water port and would like to have one on the Persian Gulf. The President suggested that he believed the government of Iran would be willing to allow Russia to ship in bond through some port and that this would be to the advantage of Iran as well as of Russia. This reservation was made and Stalin signed the agreement."[16]

It was clear that the President saw the Soviet Union as a postwar trading equal — a position that would be far more effective in drawing it out of its repressive, paranoid communist shell than exclusion. "As to Hong Kong, it was suggested that this ought to be given back to China," as part of the inevitable decolonization program the United Nations could oversee, "although Churchill was not very strong for that." He'd advocated "breaking Germany up into five independent states and he also suggested that the industrialized Ruhr," which had fed the Nazi war machine, "and adjacent areas be internationalized."[17]

"He discussed his trip with a very interesting and lithesome touch for the first hour of our meeting," Henry Stimson, the secretary of war, acknowledged. "His narrative, so far as it went, completely confirmed the impressions that I have got from the Minutes concerning Joe Stalin and the part he had taken in it, particularly the scraps that he had had with the Prime Minister. The President said that Joe teased the P.M. like a boy and it was very amusing."[18]

The second hour, however, was devoted to the "terrible mess going on" in America — the MANY MAJOR PROBLEMS [THAT] AWAIT HIM HERE, as the *New York Times*'s headline had noted: rising inflation, a looming countrywide railroad strike, an impending steelworkers strike, and other issues.[19] Nothing had seemed to dampen the President's mood, though, and from the Cabinet Room he was wheeled into the Oval Office shortly after 4:00 p.m. to hold his first press conference since his return — the 927th, almost incredibly, since taking office. He said he planned to give a special radio broadcast to the officers and men of the armed forces, all over the world, in a week's time, on Christmas Eve; then, after the new year, to give his annual State of the Union address to Congress. In the meantime, he fielded off-the-record questions—especially ones about Marshal Stalin and the Second Front—which he batted like flies. When a reporter asked, for example, if "there is anything you can say at this point about the possibility of General Marshall's going to Europe?" (i.e., to command Overlord), the President simply answered "No."[20]

The President confirmed he'd flown by plane to Tehran. Also that he'd recently been "through" Dakar. For security reasons, though, he dodged further questions regarding his itinerary, save for a few tidbits—such as admitting he'd stayed in the Russian Embassy in Tehran, where he'd drunk "up to three hundred and sixty-five toasts."[21] Also that he'd visited with General Eisenhower in the Mediterranean, and been to Sicily, where he'd met General Clark. And General Patton.

The journalists pricked up their ears. Drew Pearson and other members of the press were still stirring the Patton scandalpot—SENATORS HOLD UP PATTON PROMOTION: CHANDLER SAYS SUBCOMMITTEE SUSPECTS OTHER INCIDENTS BESIDES THOSE REVEALED, the *New York Times* headline had run the day before[22]—but the Commander in Chief squelched any question regarding his own views on the matter. With a broad, disarming smile, he quoted the well-worn story of a "former President"—clearly Lincoln—who'd had "a good deal of trouble in finding a successful commander for the armies of the United States." And when he finally did find such a commander, Mr. Roosevelt reminded the assembled correspondents, it was only for the President to be told by "some very good citizens": "You can't keep him. He drinks."

"It must be a good brand of liquor,"[23] the President had memorably responded—and the same held true now, he indicated with a smile.

With that, the journalists were dismissed. The Sun King had seemed as sunny as ever: jovial, charming, confident—and masking with his customary savvy the sheer weight of his myriad responsibilities. There was no mention of his medical drama—his indigestion attack in Tehran—or of the seriousness of the British insurrection with which he'd had to deal. The way forward for the Allies was now, at last, clear. *Overlord.* That was the only thing that really mattered: a matter of sticking to the invasion and its timetable.

On leaving Tunis the President had warned Eisenhower to say nothing yet about his new appointment, or even the title that he would be given, but to keep it "strictly secret." The President himself "would do this from Washington," as Eisenhower later recalled the discussion they'd had at Carthage on the subject. "He toyed with the word 'Supreme' in his conversation but made no decision at the moment. He merely said that he must devise some designation that would imply the importance the Allies attached to the new venture."[24]

Miraculously—perhaps as a result of a homily the President gave to the White House correspondents (and thus their editors) concerning the often fatal consequences of "leaks," which only served to help the enemy—

Roosevelt was thus able to keep secret his decision regarding Overlord and its new commander for another week—by which time General Marshall would be back from a wide-ranging tour he was making of American commands in the Pacific, including a meeting with General MacArthur.[25]

In the meantime, Secretary Stimson—one of the few who *did* know the President's decision to appoint Eisenhower to command Overlord—burned with anxiety. He yearned to know the reason General Marshall had not been chosen.

Stimson would have to wait, however. The President was looking forward to a good night's sleep in his old bed, on the second floor of the White House mansion: his first night home after the most grueling yet historic trip of his life. And, in terms of his good health, the last.

PART FIVE

In Sickness and in Health

34

Churchill's Resurrection

In WASHINGTON, D.C., the British ambassador, Lord Halifax, having come down with flu, had been unable to join the throng welcoming the President home at the White House. In his stead his deputy, Ronald Campbell, had attended the gathering. Campbell had reported back that "the President was in good form"[1] — the view of all who saw or spoke to Mr. Roosevelt on his return from Cairo.

Not so the reports that Lord Halifax was receiving from London. "Reports circulated in the afternoon that Winston had died," Halifax noted in his diary — the Prime Minister gravely ill with pneumonia in Carthage, still.[2] "Harry Hopkins rang me up to know if we had any news. I felt pretty sure that this was not likely to be true," he noted of reports of Churchill's death, "but was none the less relieved when the B.B.C. gave a good report in the evening. I imagine the dangerous time though will be after a few days"[3] — for the Prime Minister, staying in the "White House" villa in quick succession to the President, had contracted suspected pleurisy, too, raising fears it could lead (as it did) to heart trouble.

Would Churchill's ill health cause him to resign? many wondered. Churchill's wife, Clemmie, was summoned urgently to Carthage to be at his bedside. For his part King George VI wrote to the Prime Minister, via his royal scribe Sir Alan Lascelles, to say how "cruel" it was that "the P.M.'s triumphant journey should end in this way." As to Churchill's compelled recuperation in North Africa, the monarch hoped that the "comparative rest may be a blessing in disguise." In short, as Lascelles expressed the sovereign's faith, "good may come of evil."[4]

The trip, for Churchill, had been far from triumphant, however, and the "comparative rest" would prove lethal for tens of thousands of unwitting Allied servicemen in Italy, once Churchill emerged from his Lazarus-like bed. Meanwhile, agreeing to see Secretary Stimson privately at the White House on

December 18, 1943, the President confided to Stimson what had actually taken place in Tehran.

"The President and I were alone and he devoted himself to telling of his recent accomplishments at Tehran," Stimson dictated that night. "He said that when he first met Churchill at these meetings he was surprised at the change in him. He seemed unwell, was peevish and had prejudices against people in a way that was quite unusual to him. He came to the first conference [in Cairo] telling the President that he hoped they could now rearrange some of the things that they had taken up before," at Quebec, and "during the discussions he tried to reopen Overlord and the Eastern Mediterranean matters, like the Dodecanese and Rhodes, and finally concentrated on Turkey."[5]

The President had been having none of that, he explained to Stimson, and "said that he himself had fought hard for Overlord and with the aid of Stalin finally won out, and in his charming way he said: 'I have thus brought Overlord back to you safe and sound on the ways for accomplishment.'"[6]

Stimson, having read the top-secret Tehran minutes, was glad to have the President's personal confirmation. "As he put it, the conference had been successful in all military strategic matters except one," namely "the Burma affair" — a typical Churchill maneuver "where as he described it, Churchill had insisted on halting the program which had been approved at the preceding meeting with Chiang Kai-shek by taking away the necessary landing boats for the amphibian Burma operation to use in the Eastern Mediterranean. Roosevelt had opposed this; and had reminded Churchill of the promises made to Chiang, but Churchill had insisted." Running out of time in Cairo they had finally "compromised on the postponement of the Burma operation"[7] — for without British participation, the landings off Burma could not take place. As a result of British objections, the United States would have to try and make things up to Chiang by sending more air supplies over the Himalayan "hump."

"Now I come to the last matter," the President had continued, "and that is the one of Command." As Stimson recorded, "he described his luncheon with Marshall after the conference was over and their return to Cairo. He let drop the fact, which I had supposed to be true, that Churchill wanted Marshall for the Commander and had assumed that it was settled as, in fact, it had been agreed on in Quebec. The President described, however, how he reopened this matter with Marshall at their solitary luncheon together and tried to get Marshall to tell him whether he preferred to hold the Command of Overlord (now that a General Supreme Commander [of all Europe, north and south] was not feasible) or whether he preferred to remain as Chief of Staff."[8]

Having heard McCloy's version the previous evening, Stimson recognized the President was probably telling him the truth — which was not always the case. "He was very explicit in telling me that he urged Marshall to tell him which one of the two he personally preferred, intimating that he would be very glad to give him the one that he did. He said that Marshall stubbornly refused, saying that it was for the President to decide, and that he, Marshall, would do with equal cheerfulness whichever one he was selected for." It had thus been up to Marshall to make his own preference known—and he hadn't. "The President said that he got the impression that Marshall was not only impartial between the two but perhaps really preferred to remain as Chief of Staff. Finally, having been unable to get him to tell his preference, the President said that he decided on a mathematical basis that if Marshall took Overlord it would mean that Eisenhower would become [U.S. Army] Chief of Staff, but, while Eisenhower was a very good soldier and familiar with the European theater, he was unfamiliar with what had been going on in the Pacific and he also would be far less able than Marshall to handle the Congress; that, therefore, he, the President, decided that he would be more comfortable if he kept Marshall at his elbow and turned over Overlord to Eisenhower."[9]

The truth had thus acquired a certain lacquer in the telling—for it was, of course, the President himself who had never really wanted to send Marshall to Europe, unless to command, literally, *all* Europe. The varnish, however, was in a good cause. Stimson's support, as a first-class lawyer and U.S. secretary of war, would be necessary if the army was to take over the railroads across the nation, should the union strike proceed in a few days' time. Moreover Stimson, on behalf of the President, was meeting almost every day with Dr. Vannevar Bush and others about "S-1"—code-name for the atomic bomb project at Los Alamos. There, thanks to Senator Harry S. Truman's investigating committee on government waste and/or corruption in war manufacturing, it was becoming more and more difficult to maintain secrecy—the "installations" getting "so numerous and so big that they are attracting attention in Congress and people are beginning to talk about it," as Stimson had himself recently noted.[10] Above all Stimson was a Republican, and therefore a vital component in the math of the 1944 presidential election—if the President decided to stand for a fourth term.

To the President's delight Stimson—who thought the world of Marshall, and had been unable to understand why the Overlord appointment "seems to have gone to Ike instead of to George"—seemed mollified, though he openly told the President he'd been "staggered when I heard of the change" in the expected appointment of Marshall. As Stimson confided to the President, he'd known "that in the bottom of his heart it was Marshall's secret desire above all things to command this invasion force into Europe." Being the man he was, the general had nobly concealed it, not wishing to put personal ambition before the good of the country, so that Stimson had found it "very hard work to wring out of Marshall that this was so but I had done so finally beyond the possibility of misunderstanding." It was a thousand pities that he, Henry Stimson, had not accompanied Odysseus on his epic journey. "I wish I had been along with you in Cairo. I could have made that point clear," Stimson stated — admitting that he'd warned Marshall, before the trip, not to be diffident when the matter came up for a decision in Cairo, or Tehran.[11] Hearing the President's argument for keeping Marshall in Washington, now that he'd gotten D-day "on the ways," Stimson was relieved, however — for he, too, would benefit from Marshall's steady hand and uncontested authority over the U.S. military.

The President and his war secretary were then joined, after lunch, by Admiral Leahy and General Arnold to discuss the command arrangements for U.S. Army Air Forces in Overlord and Europe. The matter of Marshall's literal dis-appointment was thus left there; it was over. The announcement would be made by the President in his Christmas Eve broadcast, he told them, just as soon as General Marshall returned from the Pacific, where he was meeting with MacArthur, Nimitz, and air force commanders. The President would, he said, make sure in his broadcast that adequate tribute be paid, by name, to General Marshall and his great responsibilities as U.S. Army chief of staff.

With that, the meeting came to an end, and the waiting for D-day — the "deciding" battle of the war, as Hitler put it in a conference with his own generals two days later,[12] began.

35

In the Pink at Hyde Park

ON THE EVENING of December 23, 1943, leaving "behind a day that would have floored many a rugged man,"[1] the President left Washington for Hyde Park. He felt in top form. He'd given orders for the U.S. military to take over striking railroad companies, and was taking with him this time no fewer than nine White House correspondents aboard the *Ferdinand Magellan* — the "first time they have been with us on the homebound train since Pearl Harbor," William Hassett noted in his diary.[2] The next night, Christmas Eve, he would speak to the nation, and to American soldiers, sailors, and airmen across the globe, from his new library building. Even the reporters, Admiral McIntire later wrote, "agreed that he looked 'in the pink.'"[3]

It would be the President's twenty-seventh Fireside Chat. "The Boss, in good humor, joked with photographers, radio men, and newsreel men while waiting to begin," Hassett recorded, and then at 3:00 p.m. on December 24, the broadcast went live.[4] As his family — his wife and thirteen grandchildren — sat on the floor the President began speaking in his characteristically bold but avuncular tone. Sam Rosenman and Robert Sherwood, his speechwriters, had worked night and day to condense his notes, searching for an order in which to best convey his message to American forces.

The result was perhaps the most intimate and human, almost colloquial, account the President had ever given of the war — and the decisions he was making to save humanity. Also America's new role in facing, as he hoped his country would, the great postwar challenge. He did not therefore mention the impending railroad strike, or the threatened steel strike — only the war the United States was fighting abroad, and what would follow on the global stage.

"On this Christmas Eve," he began, "there are over 10 million men in the armed forces of the United States alone. One year ago 1,700,000 were serving overseas. Today, this figure has been more than doubled to 3,800,000 on duty overseas. By next July one that number overseas will rise to over 5,000,000 men and women." Timewise it was midafternoon "here in the United States, and in the Caribbean and on the Northeast Coast of South America," but in "Alaska and in Hawaii and the mid-Pacific, it is still morning," he pointed out — asking listeners and viewers to recognize therein the global nature of the conflict. "In Iceland, in Great Britain, in North Africa, in Italy and the Middle East, it is now evening. In the Southwest Pacific, in Australia, in China and Burma and India, it is already Christmas Day. So we can correctly say that at this moment, in those Far Eastern parts where Americans are fighting, today is tomorrow. But everywhere throughout the world — throughout this war that covers the world — there is a special spirit that has warmed our hearts since our earliest childhood — a spirit that brings us close to our homes, our families, our friends and neighbors: the Christmas spirit of 'peace on earth, good will toward men.' It is an unquenchable spirit.

"During the past years of international gangsterism and brutal aggression in Europe and in Asia, our Christmas celebrations have been darkened with apprehension for the future. We have said, 'Merry Christmas — a Happy New Year,' but we have known in our hearts that the clouds which have hung over our world have prevented us from saying it with full sincerity and conviction. And even this year, we still have much to face in the way of further suffering, and sacrifice, and personal tragedy. Our men, who have been through the fierce battles in the Solomons, and the Gilberts, and Tunisia and Italy know, from their own experience and knowledge of modern war, that many bigger and costlier battles are still to be fought.

"But — on Christmas Eve this year — I can say to you that at last we may look forward into the future with real, substantial confidence that, however great the cost, 'peace on earth, good will toward men' can be and will be realized and ensured . . .

"A great beginning was made in the Moscow conference last October by Mr. Molotov, Mr. Eden, and our own Mr. Hull. There and then the way was paved for the later meetings" — the summit in Iran from which he had just returned.

"At Cairo and Teheran we devoted ourselves not only to military matters; we devoted ourselves also to consideration of the future — to plans for the kind of world which alone can justify all the sacrifices of this war. Of course, as you all know, Mr. Churchill and I have happily met many times before, and we know and understand each other very well. Indeed, Mr. Churchill has become known and beloved by many millions of Americans, and the heartfelt prayers of all of us have been with this great citizen of the world in his recent serious illness.

"The Cairo and Teheran conferences, however, gave me my first opportunity to meet the Generalissimo, Chiang Kai-shek, and Marshal Stalin — and to sit down at the table with these unconquerable men and talk with them face to face. We had planned to talk to each other across the table at Cairo and Teheran; but we soon found that we were all on the same side of the table. We came to the conferences with faith in each other. But we needed the personal contact. And now we have supplemented faith with definite knowledge.

"It was well worth traveling thousands of miles over land and sea to bring about this personal meeting, and to gain the heartening assurance that we are absolutely agreed with one another on all the major objectives—and on the military means of attaining them.

"At Cairo, Prime Minister Churchill and I spent four days with the Generalissimo, Chiang Kai-shek. It was the first time that we had an opportunity to go over the complex situation in the Far East with him personally. We were able not only to settle upon definite military strategy, but also to discuss certain long-range principles which we believe can assure peace in the Far East for many generations to come.

"Those principles are as simple as they are fundamental. They involve the restoration of stolen property to its rightful owners, and the recognition of the rights of millions of people in the Far East to build up their own forms of self-government without molestation. Essential to all peace and security in the Pacific and in the rest of the world is the permanent elimination of the Empire of Japan as a potential force of aggression. Never again must our soldiers and sailors and marines—and other soldiers, sailors, and marines—be compelled to fight from island to island as they are fighting so gallantly and so successfully today.

"Increasingly powerful forces are now hammering at the Japanese at many points over an enormous arc which curves down through the Pacific from the Aleutians to the jungles of Burma. Our own Army and Navy, our Air Forces, the Australians and New Zealanders, the Dutch, and the British land, air, and sea forces are all forming a band of steel which is slowly but surely closing in on Japan.

"On the mainland of Asia, under the Generalissimo's leadership, the Chinese ground and air forces augmented by American air forces are playing a vital part in starting the drive which will push the invaders into the sea.

"Following out the military decisions at Cairo, General Marshall has just flown around the world and has had conferences with General MacArthur and Admiral Nimitz—conferences which will spell plenty of bad news for the Japs in the not too far distant future.

"I met in the Generalissimo a man of great vision, great courage, and a remarkably keen understanding of the problems of today and tomorrow. We discussed all the manifold military plans for striking at Japan with decisive force from many directions, and I believe I can say that he returned to Chungking with the positive assurance of total victory over our common enemy. Today we and the Republic of China are closer together than ever before in deep friendship and in unity of purpose.

"After the Cairo conference, Mr. Churchill and I went by airplane to Teheran. There we met with Marshal Stalin. We talked with complete frankness on every conceivable subject connected with the winning of the war and the establishment of a durable peace after the war. Within three days of intense and consistently amicable discussions, we agreed on every point concerned with the launching of a gigantic attack upon Germany": Overlord.

"The Russian Army will continue its stern offensives on Germany's eastern front, the Allied armies in Italy and Africa will bring relentless pressure on Germany from the south, and now the encirclement will be complete as great American and British forces attack from other points of the compass."

It was at this point in his broadcast that the President at last announced publicly who would command the "gigantic attack." "The Commander selected to lead the combined attack from these other points is General Dwight D. Eisenhower," the President revealed. "His performances in Africa, in Sicily and in Italy have been brilliant. He knows by practical and successful experience the way to coordinate air, sea, and land power. All of these will be under his control. Lieutenant General Carl Spaatz will command the entire American strategic bombing force operating against Germany.

"General Eisenhower gives up his command in the Mediterranean to a British officer whose name is being announced by Mr. Churchill. We now pledge that new Commander that our powerful ground, sea, and air forces in the vital Mediterranean area will stand by his side until every objective in that bitter theater is attained.

"Both of these new Commanders will have American and British subordinate Commanders whose names will be announced in a few days.

"During the last two days [at] Teheran, Marshal Stalin, Mr. Churchill, and I looked ahead, ahead to the days and months and years that will follow Germany's defeat. We were united in determination that Germany must be stripped of her military might and be given no opportunity within the foreseeable future to regain that might.

"The United Nations have no intention to enslave the German people. We wish them to have a normal chance to develop, in peace, as useful and respectable members of the European family. But we most certainly emphasize that word 'respectable' for we intend to rid them once and for all of Nazism and Prussian militarism and the fantastic and disastrous notion that they constitute the 'Master Race.'

"We did discuss international relationships from the point of view of big, broad objectives, rather than details. But on the basis of what we did discuss, I can say even today that I do not think any insoluble differences will arise among Russia, Great Britain, and the United States. In these conferences we were concerned with basic principles — principles which involve the security and the welfare and the standard of living of human beings in countries large and small. To use an American and somewhat ungrammatical colloquialism, I may say that I 'got along fine' with Marshal Stalin. He is a man who combines a tremendous, relentless determination with a stalwart good humor. I believe he is truly representative of the heart and soul of Russia; and I believe that we are going to get along very well with him and the Russian people — very well indeed.

"Britain, Russia, China and the United States and their allies represent more than three-quarters of the total population of the earth. As long as these four nations with great military power stick together in determination to keep the peace there will be no possibility of an aggressor nation arising to start another world war.

"But those four powers must be united with and cooperate with all the freedom-loving peoples of Europe, and Asia, and Africa, and the Americas. The rights of every nation, large or small, must be respected and guarded as jealously as are the rights of every individual within our own republic.

"The doctrine that the strong shall dominate the weak is the doctrine of our enemies — and we reject it. But, at the same time, we are agreed that if force is necessary to keep international peace, international force will be applied for as long as it may be necessary.

"It has been our steady policy — and it is certainly a common sense policy — that the right of each nation to freedom must be measured by the willingness of that nation to fight for freedom. And today we salute our unseen allies in occupied countries — the underground resistance groups and the armies of liberation. They will provide potent forces against our enemies, when the day of the counter-invasion comes" — D-day!

D-day led to the theme Roosevelt most wanted to continue seeding: the end of isolationism in America. "Through the development of science the world has become so much smaller," he pointed out, "that we have had to discard the geographical yardsticks of the past. For instance, through our early history the Atlantic and Pacific Oceans were believed to be walls of safety for the United States," he acknowledged. "Until recently very few people, even military experts, thought that the day would ever come when we might have to defend our Pacific coast against Japanese threats of invasion. At the outbreak of the first World War relatively few people thought that our ships and shipping would be menaced by German submarines on the high seas or that the German militarists would ever attempt to dominate any nation outside of central Europe." Yet that day had come, in World War I, with unrestricted German U-boat warfare — requiring vast resources and determination before Germany pleaded for an armistice.

"After the Armistice in 1918, we thought and hoped that the militaristic philosophy of Germany had been crushed; and being full of the milk of human kindness we spent the next twenty years disarming, while the Germans whined so pathetically that the other nations permitted them — and even helped them — to rearm. For too many years we lived on pious hopes that aggressor and warlike nations would learn and understand and carry out the doctrine of purely voluntary peace."

The result had been tragically violent. "The well-intentioned but ill-fated experiments of former years did not work. It is my hope that we will not try them again. No — that is putting it too weakly — it is my intention to do all that I humanly can as President and Commander-in-Chief to see to it that these tragic mistakes shall *not* be made again.

"There have always been cheerful idiots in this country who believed that there would be no more war for us if everybody in America would only return into their homes and lock their front doors behind them. Assuming that their motives were of the highest, events have shown how unwilling they were to face the facts.

"The overwhelming majority of all the people in the world want peace. Most of them are fighting for the attainment of peace — not just a truce, not just an armistice — but peace that is as strongly enforced and as durable as mortal man can make it. If we are willing to fight for peace now, is it not good logic that we should use force if necessary, in the future, to keep the peace?"

He was coming to his deepest conviction: namely that America's destiny in the coming years would have to be the safeguarding of the peace that would follow American victory. "I believe, and I think I can say, that the other three great nations who are fighting so magnificently to gain peace are in complete agreement that we must be prepared to keep the peace by force. If the people of Germany and Japan are made to realize thoroughly that the world is not going to let them break out again, it is possible, and, I hope, probable, that they will abandon the philosophy of aggression — the belief that they can gain the whole world even at the risk of losing their own souls."

Lest there be press speculation over his choice of Eisenhower over Marshall, the President wisely included in his broadcast a further mention of his faithful army chief of staff.

"To the members of our armed forces, to their wives, mothers, and fathers, I want to affirm the great faith and confidence that we have in General Marshall and in Admiral King, who direct all of our armed might throughout the world," he added. "Upon them falls the great responsibility of planning the strategy of determining where and when we shall fight. Both of these men have already gained high places in American history, which will record many evidences of their military genius that cannot be published today."[5]

As FDR ended his broadcast, all were energized by the President's clarity, conviction, and confidence — a commander in chief determined to lead America to victory and beyond. If all went well.

• • •

For a while it did. No sooner had the President returned to the White House after Christmas than he gave another scintillating peroration to correspondents in the Oval Office — a press conference that would become known as the President's "Dr. Win-the-War" talk.

In this the President listed the major progressive accomplishments of his New Deal program that he wanted people to remember, or not take for granted in the hurly-burly of war: saving the banking system, preserving farms from foreclosure, establishing the Securities and Exchange Commission and old-age insurance, creating unemployment insurance, instituting bank deposit insurance, providing federal aid for the blind and the crippled — things he thought no one in their right mind would want to go back on after the war, if Republicans had their way.

It was this liberal agenda which, like so many millions of Americans, he wanted to improve upon rather than repeal "when victory comes" — an agenda that should be seen as international. The "program of the past," he declared, "has got to be carried on, in my judgment, with what is going on in other countries . . . We can't go into an economic isolationism, any more than it would pay to go into a military isolationism. This is not just a question of dollars and cents, although some people think it is. It is a question of the long range, which ties in human beings with dollars, to the benefit of the dollars and the benefit of the human beings as a part of this postwar program, which of course hasn't been settled on at all, except in generalities. But, as I said about the meeting in Teheran and the meeting in Cairo, we are still in the generality stage, not in the detail stage, because we are talking about principles. Later on we will come down to the detail stage, and we can take up anything at all and discuss it then." As he put it, in the meantime "it seems pretty clear that we must plan for, and help to bring about, an expanded economy which will result in more security, in more employment, in more recreation, in more education, in more health, in better housing for all of our citizens, so that the conditions of 1932 and the beginning of 1933 won't come back again."[6]

The newspaper correspondents were, for the most part, agog — amazed by the paraplegic president's vigor and energy. "The public works program, the direction of federal funds to starving people. The principle of a minimum wage and maximum hours. The Civilian Conservation Corps" and its work on "Reforestation. The N.Y.A. [National Youth Administration], for thousands of literally underprivileged young people. Abolishing child labor," which "was not thought to be constitutional in the old days, but . . . turned out to be. Reciprocal trade agreements, which of course do have a tremendous effect on internal [economic] diseases. Stimulation of private home building through the F.H.A [Federal Housing Administration]. The protection of consumers from extortionate rates by utilities. The breaking up of utility monopolies, through Sam Rayburn's law. The resettlement of farmers from marginal lands that ought not to be cultivated; regional physical developments, such as T.V.A. [the Tennessee Valley Authority]; getting electricity out to the farmers through the R.E.A. [Rural Electrification Act]; flood control; and water conservation; drought control — remember the years we went through that! — and drought relief; crop insurance, and the ever normal granary; and assistance to farm cooperatives," plus the "conservation of natural resources." . . .

"Well, my list just totaled up to thirty," the President stated, summarizing the achievements of the New Deal — "and I probably left out half of them." In that context his allegory of Dr. New Deal and Dr. Win-the-War — an orthopedic surgeon called upon to minister to the nation following a "bad accident" (at Pearl Harbor) — was typical Roosevelt: whimsical, but at heart deeply serious, not evasive. The postwar beckoned — and unlike certain unnamed doctors, he was not afraid to discuss or embrace the latest therapy, now "that the patient is back on his feet. He has given up his crutches. He isn't wholly well yet, and he won't be until he wins the war" — victory which was not imminent, but was at last within sight. And the good news was that, although, "at the present time, obviously, the principal emphasis, the overwhelming first emphasis should be on winning the war," the nation nevertheless was able to discuss a forward-looking, progressive agenda. One that was based upon prescriptions that had addressed the "disease" of economic boom and catastrophe in the Great Depression and had, in the New Deal, treated it successfully, without the nation resorting to fascism or dictatorship or aggression. And to which the nation could soon return, and build upon, in a new international environment, or world order.

"In other words, we are suffering from that bad accident," the President ended, "not from an internal disease."

It was the afternoon of December 28, and for the most part the journalists in the Oval Office were struck by how positive, humorous, compassionate, idealistic, and forward-looking was the President, as the eve of the New Year, 1944, approached. As one reporter broke the silence: "Does that all add up to a fourth-term declaration?"[7]

The question caused the room to erupt in laughter.

That night, however, the President admitted he was "feeling a *little* miserably," his cousin Daisy noted in her diary at the White House, where she was staying.[8] She herself had been suffering flu, which seemed to be endemic at that time.

All too soon the President himself seemed to be coming down with the virus. By December 30, he had a fever running almost 101 degrees. He was "a little hectic and flushed," and "at loose ends," Daisy described, after he'd had dinner brought in on a tray, sitting with his daughter Anna, who was visiting from California.[9]

Daisy was not a nurse, but since it seemed there was no one else taking care of the President, she administered aspirin and cough medicine, and called the doctor. He told her "we have to expect the increased temp. for the next 48 hrs. as part of the flu. The P. must not catch cold during the night; he would probably be in a perspiration & should have dry clothes," she recorded in her diary.[10]

He would get over it; he would bounce back; he would be fine, the doctor said.

But he wasn't.

36

Sick

THE PRESIDENT'S COUGH quickly developed into bronchitis; his hands trembled more and more; he suffered more headaches. "So it went on; one day up and one day down," Admiral McIntire, his White House physician, later recalled.[1] Tough domestic problems — strikes on the railroads, threatened strikes in the all-important steel industry — seemed to follow him implacably, "and just to make things worse, he had the bad luck to contract influenza. The attack hung on and finally left behind a nagging inflammation of the bronchial tubes," McIntire chronicled. "Coughing spells racked him by day and broke his rest at night."[2]

The flu — and the bronchitis that accompanied it — left the President feeling like a proverbial wet rag, or worse. He felt constantly tired, but he'd always been a fighter. He'd disliked ever giving way to sickness — concerned lest any sign of ill-health, after fighting his way to be able at least to stand on his steel-braced feet after poliomyelitis, should become ammunition for his Republican enemies. In the aftermath of his long, Odyssean journey to the Middle East, however, he was minded to acknowledge the illness this time publicly, in a sort of Fireside Chat. He'd be able to say, in all honesty, he was too sick to go to Congress to deliver his annual State of the Union address but that, instead, he was broadcasting it live from the White House — even filming a portion of it for newsreel, to be shown in movie houses. For it was now time, as the elected president of the United States in his final year of office, for him to set out a social and economic vision of the future: ideals that Americans could pursue in a world in which they were the new standard-bearers of democracy.

The President's political vision, in the midst of a world war, was an aspect of his leadership that his friend Prime Minister Mackenzie King found perhaps the most extraordinary thing of all, when comparing the President and the Prime Minister of Great Britain after Tehran.

"Churchill has been 'raised up' to meet the need of this day in the realm of war, to fight, with the power of the sword, the brute beasts that would devour their fellow men in their lust for power," the Canadian prime minister noted in his diary, a trifle grandiosely. Roosevelt might not be as great a man as an orator, or military "genius"; nevertheless he was undoubtedly a greater man, the Canadian felt, "in his love for his fellow men and in his very sincere desire to improve their lot."

Churchill's biggest problem, King felt, was drink. "I greatly fear that demon may claim him as its own, before he sees the fruits of victory. I pray it may not be so, and that he may be spared to enjoy some of the fruits of victory, which he more than any other single man deserves. It is clear, however, that already it has him 'down', and however much he may recover, his strength & endurance will be greatly lessened for all time, and at any moment he may suffer an attack which may take him off," King confided in his dictated journal — thoughts that led him to reflect once again on the health, both physical and psychic, of his dearer friend, Franklin Roosevelt.

"The President has overtaxed his strength in other ways. He has had a harder battle in many ways than Churchill. His fight for the people has made him many and bitter enemies" — not least conservatives who worried that Roosevelt was spending too much money, both to win the war and to bind the nation to his social and political vision of the future. "He has done too much, I fear, for purely political reasons — the vast expenditures totally regardless of consequences, & which may leave the United States in an appalling condition some day. He has used public office to ensure continuance of power" and keep it away from Republican special interests. Such government expenses "can scarcely be justified — but I believe he has been sincere in his determination to better the conditions of the masses," King had judged. "He is more human than Churchill; each desire to be at the top: Churchill would like to be the ruler of an Empire (Conservative). Roosevelt the head of a Commonwealth (democrat)."

Mackenzie King wondered which man's vision would prevail — the imperialist's or the postimperialist's? "I wonder if his [Roosevelt's] ambition to figure too largely on a world stage may not be his undoing & the undoing of his strength & of his political power? We shall see," the longtime Canadian prime minister and spiritualist had noted — observing, as he dictated this, that the two hands of the clock were "exactly together at 5 past one."[3]

The two men had thereafter exchanged touching Christmas greetings via telegraph, but once the President was wheeled into the White House Diplomatic Reception Room shortly before 8:00 p.m. on January 11, 1944, to broadcast his New Year's message to Congress, it had become clear Prime Minister King was right about hubris at the highest level: Roosevelt was clearly on a new, domestic warpath, while Churchill remained entirely focused on military glory.

As the President cleared his throat and surveyed the bank of microphones and film cameras in the White House that would take his "message" way beyond the fractured, often regressive Southern Democrats ensuring his control of the Congress, he began by saying he'd wanted to follow his normal custom of appearing in person, but "like a great many other people I have had the flu, and although I am practically recovered, my doctor simply would not let me leave the White House to go up to the Capitol."[4]

Delivered in the midst of a continuing world war, the broadcast would, in terms of domestic public policy, be one of the most significant addresses of the twentieth century in America: containing not only a National Service Act recommendation, but a "Second Bill of Rights," as the President called it — the text shown to no one before transmission, lest anyone attempt to dissuade him from his Luther-like propositions. These he intended to be metaphorically nailed on the door of Congress — a set of theses in which he would articulate his vision of a new, postwar democratic society.

Judge Sam Rosenman and Robert Sherwood, the President's speechwriters, had not even been permitted to go home after helping him with his Christmas broadcast, Rosenman later related. Instead they'd once again been pressed into rhetorical service for a speech so outspoken that its many drafts, on instructions of the President, were typed by Grace Tully alone, and were not mimeographed. "Sherwood and I took all possible precautions to prevent a leak. That is not easy in Washington, as anyone with experience there can testify," Rosenman would recall with amusement[5] — not even Harry Hopkins informed, or any member of the cabinet.

The "Second Bill of Rights" came straight from the President's heart, Rosenman believed. "He had seen our fighting men at close hand, their hardships and danger and sufferings — and those neat but crowded American cemeteries. He came back determined to see that the people back home did their share too," in the form of a national service bill, to help win the war, but also to propose they be assured, once the war was won, of a better economic and social structure than that which had produced the Great Depression after World War I.[6]

"This Republic had its beginning, and grew to its present strength, under the protection of certain inalienable political rights — among them the right of free speech, free press, free worship, trial by jury, freedom from unreasonable searches and seizures," the President introduced his theme on CBS, NBC, and other radio stations that were transmitting his voice, live. "They were our rights to life and liberty," the President went on. "As our Nation has grown in size and stature, however — as our industrial economy expanded — these political rights proved inadequate to assure us *equality* in the pursuit of happiness. We have come to a clear realization of the fact that true individual freedom cannot exist without economic security and independence. 'Necessitous men are not free men.' People who are hungry and out of a job are the stuff of which dictatorships are made," he remarked.

"In our day these economic truths have become accepted as self-evident," he asserted—addressing directly the challenge of inequity. In people's minds, if not yet in law, poorer Americans expected a better deal, thanks to a better economy—an improved economy which had now been achieved. "We have accepted, so to speak, [the need for] a second Bill of Rights under which a new basis of security and prosperity can be established for all regardless of station, race, or creed." Among these were:

> The right to a useful and remunerative job in the industries or shops of farms or mines of the Nation;
> The right to earn enough to provide adequate food and clothing and recreation;
> The right of every farmer to raise and sell his products at a return which will give him and his family a decent living;
> The right of every businessman, large and small, to trade in an atmosphere of freedom from unfair competition and domination by monopolies at home or abroad;
> The right of every family to a decent home;
> The right to adequate medical care and the opportunity to achieve and enjoy good health;
> The right to adequate protection from the economic fears of old age, sickness, accident, and unemployment;
> The right to a good education.

"All of these rights spell security," the President insisted. "And after this war is won we must be prepared to move forward, in the implementation of these rights, to new goals of human happiness and well-being."[7]

Inequality—or gross inequality—was the scourge of mankind, the President declared: not only because it was morally wrong, but for practical reasons, because in the end economic inequality, if allowed to grow flagrant, led to economic crises and tyranny when, inevitably, financial bubbles burst. This was self-evident in terms of the rise of fascism in his lifetime. "America's own rightful place in the world depends in large part upon how fully these and similar rights have been carried into practice for our citizens," he maintained. "For unless there is security here at home there cannot be lasting peace in the world. One of the great American industrialists of our day—a man who has rendered yeoman service to his country in this crisis—recently emphasized the grave dangers of 'rightist reaction' in this Nation. All clear-thinking businessmen share his concern. Indeed, if such reaction should develop—if history were to repeat itself and we were to return to the so-called 'normalcy' of the 1920's—then it is certain that even though we shall have conquered our enemies on the battlefields abroad, we shall have yielded to the spirit of Fascism here at home," he warned.[8]

"I ask the Congress to explore the means for implementing this economic bill of rights—for it is definitely the responsibility of the Congress so to do. Many of these problems are already before committees of the Congress in the form of proposed legislation. I shall from time to time communicate with the Congress with respect to these and further proposals. In the event that no adequate program of progress is evolved, I am certain that the Nation will be conscious of the fact."[9]

"The Nation" meant its voters—another reason the President had chosen to broadcast his address over the radio rather than delivering it in Congress in person.

"Our fighting men abroad—and their families at home—expect such a program and have the right to insist upon it. It is to their demands that this Government should pay heed rather than to the whining demands of selfish pressure groups who seek to feather their nests while young Americans are dying.

"The foreign policy that we have been following—the policy that guided us at Moscow, Cairo, and Teheran—is based on the common sense principle which was best expressed by Benjamin Franklin on July 4, 1776: 'We must all hang together, or assuredly we shall all hang separately.'"[10]

"All told, the State of the Union Message was unusually bellicose," Rosenman afterwards admitted, for the President "was in a fighting mood," despite his flu—"and in short order got into some bitter fights with the Congress: one on soldier voting, one on national service, and one on taxes"[11]—the President soon having to veto Congress's budget bill as "relief not for the needy but for the greedy," as he memorably declared (only to have his veto overridden).

Rosenman had been well aware the President would not win all, or indeed any, of his social and economic measures, since he was inevitably facing the rising forces of reaction, following more than a decade of "progressive" Democratic administration. "The fights showed that on domestic, civilian issues the President had lost control of the Congress, and indeed of his own party in Congress"—especially in terms of race, Rosenman would recall. "The small reactionary wing of the Democratic party, principally the Southern members, was working in coalition with the Republican party"—the white supremacists, or "last straws" of slavery and the Civil War: men whose obstructionism convinced Roosevelt there would have to be "a new alignment of political forces in the United States" in the future, in order to head them off.[12]

The President seemed nevertheless determined that, if he undertook to run for a fourth term but didn't win the Democratic Party nomination in the summer of 1944, he would at least go down fighting. Assuming his health recovered.

37

Anzio

In CARTHAGE, CHURCHILL, too, seemed to have no intention of slowing down, let alone resigning from office, despite pneumonia and atrial fibrillation. "Oh, yes, he's very glad I've come, but in five minutes he'll forget I'm here," his wife, Clemmie — who'd been flown in to be at her possibly dying husband's bedside — had been heard to say.[1]

Mrs. Churchill knew her husband better than anyone. The Prime Minister was frustrated but no longer abashed by his defeat over Overlord. He had, after all, "triumphed" in forcing the cancellation of Buccaneer, the Burmese operation — thus releasing crucial landing craft for his pet schemes in the Mediterranean, once General Eisenhower left to command the cross-Channel invasion. As prime minister and de facto commander in chief of British Empire forces, Churchill had also won out in insisting a British officer replace the departing American. At Sir Alan Brooke's urging, the Prime Minister named General Sir Henry Maitland Wilson as Allied supreme commander in the Mediterranean and Middle East, not General Alexander — persuaded by Harold Macmillan, among others, that General Alexander, if left in situ in Italy, would remain in field command of the armies there — and could thus be browbeaten into seizing Rome by *coup de main,* at the Prime Minister's urging.

Recovering rapidly from his pneumonia, the Prime Minister had a new gleam in his eyes, as everyone who had seen him in Carthage had become aware. With Eisenhower slated to depart the theater, the Prime Minister announced he wished to mount an immediate grand "scoop" in Italy, as he called it, in both senses of the word. It was to be, he said, an amphibious invasion only thirty miles in distance from Rome: one that would be recognized by the whole world as a brilliant military "end-run," as well as a great political coup de théâtre — and carried out under the new British supreme command in the Mediterranean. Overnight Churchill's new brainwave, an "amphibious scoop"[2] or "cat-claw" or "end-run,"[3] would, he declared, force the Germans to retreat from their defensive positions north of Naples, on the Rapido River, and give the Allies the greatest Italian prize of all: possession of the Eternal City.

Like Overlord, Anzio would be a gamble — but a British gamble this time. Gambling, in any case, was Churchill's great love — an addiction of which he was completely unashamed, as of his alcohol intake. Neither the aftereffects of pneumonia, nor the danger of once again splitting the Allies over the issue, had thus seemed to have any effect on Churchill's mood as he recuperated in Eisenhower's guest villa, at Carthage, and plotted his tour de force. Clad in silk pajamas and a florid Japanese dressing gown embroidered with dragons, he had risen from his bed not as Lazarus did but to reign as the god Neptune, warrior lord of the Mediterranean: a trident-wielding leader who began to see himself — not his appointee-to-be, General Wilson — as the ultimate military genius or generalissimo directing the Allied forces in the Mediterranean and Middle East in 1944.

The picture would have been comical had it not been so serious. And tragic, in terms of the largely futile loss of life — especially American life — that ensued as a result. Churchill's recent fiasco in the Aegean islands in October and November had taught him only that he needed American forces to succeed. At Anzio — target of the Prime Minister's gamble — tens of thousands of American lives would be on the line, but with the President and U.S. chiefs of staff unable to stop him under the impending new command arrangements.

Thus the Anzio tragedy had begun to unfold in the wake of Churchill's "resurrection" — while at the same time the President, who had returned to the United States the conquering hero of Tehran, had fallen ill again with influenza, yet more seriously, in fact, than he or anyone around him recognized. Debilitated by this and bronchitis, he had found himself suddenly too exhausted to embark on another struggle with Winston. Obsessed by the lure of the Eternal City, Churchill had proceeded to revive, in a matter of days, an earlier contingency plan for a small-scale, outflanking amphibious assault at Anzio — one that General Clark had long since dropped as too diminutive, too risky — and had inflated the project into a massive amphibious assault landing, dwarfing all other operations in the Mediterranean combined.

Equally tragically Sir Alan Brooke — promoted to the rank of field marshal — supported his prime minister! The normally dour and critical artilleryman had found himself delighted by Churchill's bounce back to good health, after the British defeat at Tehran. Moreover Churchill's focus on Italy, in the first instance, rather than the Aegean, had met with Brooke's strong backing — completely ignoring the likely Wehrmacht response. He had therefore agreed to the Prime Minister's request that he should fly back to London on December 19, 1943, to tell his colleagues at the War Office of the new plan, and if possible win them over — thus leaving Churchill without a minder willing, or able, to challenge him.

In this way one of the war's most unnecessary and unfortunate disasters had gathered pace, just when the Allies seemed on the cusp of unified victory, thanks to the President's patient, tenacious strategy. As Churchill had resurrected the plan for an amphibious assault at Dieppe in the summer of 1942, the Prime Minister now invested the abandoned plans for an end-run at Anzio with fresh energy—but also with the landing craft intended for the President's Anvil operation. The new project, the Prime Minister had assured everyone in Carthage, would be a *coup de foudre,* in fact a *coup de grâce*: a grand operation of war, put together and mounted in the next several weeks with almost no rehearsal—and without first submitting the plan to the President or his U.S. military team in Washington.

Day by day Churchill's blunder had thereafter grown bigger. Convinced of its merits in London—far from the realities of the battlefield—the British chiefs of staff had cabled the Prime Minister on December 22 to say they were "in full agreement" that the "present stagnation" in Italy was one which, as the Prime Minister had correctly declared, "cannot be allowed to continue"—though why, they would not say. In such circumstances, they all concurred, the necessary landing craft—many of them due to be sailed to Britain by mid-January—should therefore now be withheld and be readied in the Mediterranean as swiftly as possible "to strike round the enemy's flank and open up the way for a rapid advance to Rome."[4]

Under the new British supreme commander–designate, General Wilson, shortly to be in overall control of the whole theater, stretching from Gibraltar to the Middle East, the Prime Minister had insisted that General Sir Harold Alexander, Wilson's ground forces commander in Italy, take personal responsibility for planning the amphibious assault landing at Anzio, and command it once launched. For the moment the assault proved successful, it could be trumpeted in the press as a great British victory, causing the Prime Minister—who saw himself as a "sort of super Commander-in-Chief," in the eyes of General Mark Clark—to insist at least half of the invasion troops be British, and be recorded as such in media reports.[5] It would, in essence, be a British-led Mediterranean version of Overlord, *before* Overlord.

It was in this wildly overoptimistic, cavalier mood that both Brooke and Churchill had ignored the warnings even of British combat commanders that "Shingle," the code-name for the Prime Minister's Anzio assault, would not necessarily lead to a "rapid advance to Rome." General Montgomery—who was facing heavy German opposition on the right flank of the Allied forces in Italy, had warned Alan Brooke, when Brooke visited the front in Italy in December, there was no chance of reaching Rome before the spring,[6] and that the Prime Minister's supposed stroke of genius at Anzio was ridiculous. Neither Brooke nor Churchill had proved willing to listen, and Montgomery, in any case, had subsequently been chosen to be Eisenhower's ground forces deputy for Overlord—the British War Cabinet insisting on a more dynamic ground force commander in chief for the cross-Channel invasion, when Churchill had fallen ill with pneumonia.[7]

Summoning Eisenhower, the outgoing American commander in chief of Allied Mediterranean forces, to his villa on December 23, 1943, the recovering Prime Minister had nevertheless made clear to Ike, his host at Carthage, that he himself was taking *personal* charge of the planning of Shingle—and nothing General Eisenhower could say in cautioning him had seemed to have any effect.

The weather in Italy was abysmal, the Allied air forces could not provide much tactical air support, the troops were finding the Italian terrain forbidding in winter, and even Montgomery's Eighth Army, on the supposedly easier Adriatic side of Italy, near Ortona, was making no progress, as Eisenhower noted. Casualties were mounting alarmingly. Bitter experience had shown him that the Wehrmacht, so famed for offensive operations (Blitzkrieg), were even greater masters of defense. As he pointed out to the Prime Minister, the "Nazis had not instantly withdrawn from Africa or Sicily merely because of threats to their rear. On the contrary, they had reinforced and fought the battle out to the end."[8]

It had been no use. Vainly, Eisenhower had warned the necessary landing craft for the operation would probably have to be kept in the Mediterranean "long after the agreed-upon date for their release" for Overlord, or of its pincer-assault, Anvil—thus prejudicing the very operation the President had just appointed him to command.

The Prime Minister had refused to listen to such warnings—Churchill vowing he would work on President Roosevelt to agree to retention of the necessary landing craft in the Mediterranean. More craft could surely be built, or converted before D-day's launch in May, Churchill claimed, as the smoke of his cigar billowed and then dispersed. All would be well, he'd assured Ike: the Allies would win a great military victory. Rome would be theirs! And with that the Prime Minister had assembled, on Christmas Day at Carthage, his British team-to-be, who would take over from Eisenhower on January 1, 1944, or thereabouts: General Sir Maitland Wilson, General Sir Harold Alexander, Admiral Sir John Cunningham, and Air Marshal Sir Arthur Tedder: Knights of the Round Table.

The President had warned Ike, on board the Sacred Cow, that he would be under merciless meddling pressure from the Prime Minister once he got to London as supreme commander of Overlord. That this would happen before Eisenhower even got to London was unfortunate, indeed tragic—Eisenhower finding himself, as the outgoing commander in chief in the Mediterranean, an impotent observer. He could only witness, not chair, the proceedings, which the Prime Minister had conducted in his dressing gown like a Japanese shogun.

"It would be folly to allow the campaign in Italy to drag on," Churchill declared — though without explaining why. Instead, pouring forth a torrent of emotionally larded words to damn the notion of southern Italy being a "mere" holding front, as the President — and Marshal Stalin — had portrayed it, he called for action on a grand scale, in the middle of winter. The Allies should not even think of mounting "the supreme operations 'Overlord' and 'Anvil'" in the coming spring, the Prime Minister had asserted, "with our task in Italy half-finished."[9]

What exactly this strategic military "task" was, beyond the glory of reaching Rome, the Prime Minister did not define. Nor did he explain why it would in any way help, let alone be essential to, the agreed primacy of Overlord in defeating the Third Reich. And why in winter, with no time to prepare or rehearse the formations? What was the hurry?

The amphibious attack, the Prime Minister had insisted, should be launched in three weeks' time: on January 20, 1944. It should land two Allied divisions in the first assault, instead of one. These should then be followed up by yet more Allied divisions in subsequent waves, like breakers rolling onto a beach. Without question it would "decide the battle of Rome, and possibly the destruction of a substantial part of the enemy's army."[10]

The sheer amateurishness of the Prime Minister's concept of modern war would, in retrospect, be mind-boggling — even criminal in its folly. But Churchill was Churchill: a force of nature. With the plan agreed by his British subordinates at Carthage he had thus cabled the President, the day after Christmas, 1943, to appeal for the fifty-six landing craft to be held back in the Mediterranean for the assault landing rather than be assigned to Overlord and Anvil, claiming there could be nothing more dangerous "than to let the Italian battle stagnate and fester on for another three months thus certainly gnawing into all preparation for and thus again affecting Overlord. We cannot afford to go forward" with Overlord and Anvil "leaving a vast half-finished job behind us," he stated categorically. "If this opportunity is not grasped," he claimed, "we must expect the ruin of the Mediterranean campaign of 1944." And if the prospect of "ruin" in Italy was not credible in the President's eyes, he'd added, all the senior generals and admirals present at his special Carthage conference — including General Eisenhower — were agreed "that every effort should be made to bring off 'Shingle' on a two-division basis around January 20th, and orders have been issued to General Alexander to prepare accordingly."[11]

Orders already issued? Before the British were even formally vested with supreme command in the Mediterranean?

It was all, *en bref*, Churchillian bunkum: the same notion, even same language Churchill had used back in October 1943 when once again trying to halt or postpone Overlord. Yet to Churchill's own astonishment, then and later, neither the President, nor Marshall (who had arrived back in Washington on December 20), nor even the U.S. chiefs of staff, tried to stop him.

Apart from insisting that those landing craft released from potential duty in Buccaneer be sailed directly to Britain for Overlord, and that any further plans to invade Rhodes or other Aegean shores "must be sidetracked," the U.S. Joint Chiefs of Staff had, it appeared, decided not to contest the plan for Italy—which they had not seen—either to the President or to Mr. Churchill via the Combined Chiefs of Staff. Instead they had contented themselves with a signal to Churchill, which they drafted for the President on December 26, 1943, to be sent the next day, stating that Overlord "must remain the paramount operation and will be carried out on the date agreed to at Cairo and Tehran."[12]

Only wise Admiral Leahy had smelled a proverbial rat—worried at the White House by Churchill's latest cable. In his diary on December 27 Leahy recorded his skepticism. "Messages from the British Prime Minister bring up for consideration the use of the landing craft and men in the Italian campaign," he had jotted with concern, "with a possible delay in the planned landings in France. This is probably a first attempt by the British to extend operations in the Mediterranean, even at the cost of prolonging the war with Germany."[13]

After the triumph of Tehran, in terms of unified Allied strategy to defeat the Third Reich, the Anzio operation was like shooting oneself in the foot. The support of the British in carrying out Overlord had been considered by the Joint Chiefs of Staff, however, to be so crucial to the successful course of the war—especially after their contentious meetings in Cairo—that they had simply given way, not daring to get into a new fight with the Prime Minister.

Barely able to credit Churchill's new madness, and only beginning to reassemble in Washington, oversee final preparations for Overlord, and also focus their attention on plans for the war's endgame in the Pacific, the American military high command team at the Pentagon and Navy Department had thus simply washed their hands of Shingle—tired of battling with the British after so many months of extended indictments, threats, and showdowns. They had thus fatefully declined to recommend the President—who was entirely focused on the political agenda he wished to put before the nation, and suffering from flu—take issue with Churchill over the new British Mediterranean scheme.

Thus, to his own astonishment, Winston Spencer Churchill got away with his martial coup, almost without American objection.

Even Churchill's official biographer was later astonished. "Churchill was delighted and a little surprised that the Americans had accepted that the Anzio landing was to take place," Sir Martin Gilbert would write—the Prime Minister feeling as if he was living a second life, following his close brush with death. "What better place could I die [in] than here?" he'd asked his police security chief, Inspector Walter Thompson, at Carthage. And to his daughter Sarah he'd confided: "If I die, don't worry — the war is won."[14] But the pneumonia *hadn't* carried him away, mercifully. He hadn't died; he was alive — doubly alive. And on the warpath, once again. Under British supreme command in the Mediterranean theater he would *personally* plan and win a great battle, he was resolved — something he had longed to do, ever since becoming prime minister in 1940.

38

The President's Unpleasant Attitude

DAY AFTER DAY, by contrast, the President attempted to deal with affairs of state in the White House, feeling "rotten," coughing incessantly, and sleeping badly. The morning after his State of the Union broadcast, on January 5, 1944, he saw General Eisenhower, who had been ordered home for a few days' rest, before he took up his post in London.

"Eisenhower is being hailed as the great genius of this war," Goebbels sneered in his diary. "He's being showered with laurels after laurels in advance. In his broadcast Roosevelt explains the U.S. now has 10 million soldiers under arms, of which 3.5 million are already overseas. But he has to admit in his speech that there will be terrible suffering ahead. Fortress Europe will be hit from many sides. The idea is not to enslave Germany — this is said for us — just liquidate the master-race. Which are one and the same," Goebbels pointed out. "But Roosevelt has no chance of making good on such Jewish plans; after all, we are here and in the way."[1]

This was true: the Wehrmacht still the world's most formidable military force in battle. Even when he met with the President in the Oval Office, Ike was under the impression the Anzio operation would not go ahead, since he had not been informed, when he left, it was going operational. Moreover he himself was still, on paper, the Allied commander in chief, Mediterranean — for he had not authorized British general Maitland Wilson to take over from his staff, in Algiers, before January 8. As the new supreme commander of Overlord, Eisenhower wanted all possible landing craft to be "gathered up" and reconditioned for D-day, "so as to produce the maximum number in May. This would mean the abandonment of" the Anzio assault, Operation Shingle, Eisenhower signaled to General Bedell Smith, his chief of staff in Algiers, on January 5. But in any case, he added, "that operation is open to grave objections under present conditions."[2]

Poor Smith had been compelled to notify Eisenhower on January 9, four days later, that none of the British commanders in the Mediterranean had voiced "grave objections" to the Prime Minister's scheme. In fact he'd just attended a meeting, he reported, with General Sir Maitland Wilson, the day before. "It is not, repeat not, pleasant to be the guest where you have been the master," he'd lamented — and informed Eisenhower the British had now made a cast-iron decision, at Prime Minister Churchill's insistence, to undertake Shingle on January 22, in less than two weeks' time.[3]

When Eisenhower thus paid a second visit to the White House on January 12, the day after the President's broadcast to Congress and the nation, it was to tell the flu-ridden President his concerns about the Anzio project and its implications for Overlord. As the general explained to Secretary Stimson twice that same day, he was concerned "about the coming offensive in Italy," and his "fears as to the strain it would make on the number of our landing craft."[4]

Stimson was alarmed, noting that night how Eisenhower had "told me of a talk he had had with the Prime Minister who is dead set on making this offensive for political reasons. He [Churchill] said he would not dare go back before his people with the present offensive stalled and himself not able to tell about the secret plans for Overlord and so forth. So apparently this effort has been decided upon by the British who now have command of the whole Mediterranean."[5]

Churchill's latest reasoning, according to Eisenhower, sounded not only tactically but strategically unsound to Stimson. It was also potentially criminal in terms of the likely American casualties. Stimson was therefore filled with "ill-foreboding because it is almost certain to get a force tied up," he predicted in his diary, "which will be obliged to use a large number of landing craft after it has landed" in order to keep open "its line of communications."[6]

When Stimson had another talk later the same day with Eisenhower, his heart sank still further. "I know too well that there will be delays" in returning the landing craft, he recorded, for the boats would be needed to support the stranded troops at Anzio, if they found themselves ringed by the Germans. Churchill was "banking on pulling off this operation quickly," Stimson noted, but he'd heard such talk before. As things stood, given the likelihood Anzio would either fail or be a prolonged disaster, this "would mean that the loss will have to be taken out of Anvil or Overlord, and Overlord is already down to its lowest limit in landing craft."[7]

Secretary Stimson's fears, expressed ten days before the Anzio landings took place, would prove all too prescient. Given Churchill's personal authorship of the operation, the President was the only person with sufficient authority to stop the offensive.

The President, however, was "ill," Stimson noted — and like the U.S. chiefs of staff simply not up to a new battle with Churchill. Mr. Roosevelt had, as even Goebbels had noted, "further, terrible problems with strikes in his own country. He's being forced to federalize the railways, because he's unable to control [union] calls for a strike." The day the President had seen Eisenhower, on Ike's first visit, he'd also conferred with Frances Perkins, his secretary of labor, despite his fever, and he'd even chaired the Pacific War Council — visibly ailing.

And so it had gone on in early January, 1944 — the President, for all that he felt "rotten," meeting in the Oval Office on Pennsylvania Avenue with the men and women who most counted for governance, administration, economy, legislation, and diplomacy in America, at the ultimate apex of political power. He gave press conference after press conference — and watched documentary films from various theaters of combat that gave a visceral picture of what the airmen, sailors, and ground forces were confronting. General Arnold took him to Washington Airport to see the new B-29 bomber that would change the military face of the skies. He saw emissaries from military headquarters across the world: India, the Pacific, Africa, the Mediterranean.

Yet the President's flu symptoms persisted. The budget director, Harold Smith, saw Roosevelt on January 7, for example, and was surprised the President did not give the matter of finance his usual attention. In his bedroom at the White House, "he seemed worried and worn out. I have never seen him so listless," Smith had noted in his diary. "He is not his acute usual self. In fact I was quite startled, at one stage when he was about two-thirds through the Message" — a document otherwise known as the draft budget to Congress. "As he sat up in bed, I saw his head nod. I could not see into his eyes, but it seemed to me that they were completely shut. Yet, he said something to the effect that 'this paragraph is good.'" Smith was stunned. "I have seen the President before when he was ill in his bedroom," he wrote in his diary that night, "but never so groggy."[8]

The President's attention span simply never improved — thus leading chroniclers, later, to misjudge just how dominant his authority and control over his administration and especially his chiefs of staff had been in years past, prior to his sudden ill health. Appointments now became shorter, the President's energy level grew fainter, his blood pressure higher. He virtually never left the White House, save to take "rests" at Hyde Park — away from prying, or merely watching, journalists.

Meanwhile on January 22, 1944, Shingle — Churchill's much-vaunted amphibious invasion at Anzio — went in over the beaches. It proved, just as Eisenhower and Stimson had feared, a calamity. The *forty-three thousand* Allied casualties suffered on the beaches of Anzio over the following four and a half months — including seven thousand who died there — would be a terrible indictment of Brooke's support for the Prime Minister's "resuscitation," but most of all to Churchill's impetuosity and shallowness.

The fact was, the Prime Minister, when fired up with a fantasy, was almost impossible to control, his energy and conviction illustrating what the philosopher-king Lord Francis Bacon had noted in 1620: namely a mind which, having "once adopted an opinion," was wont to draw "all things else to support and agree with it. And though there be a greater number and weight of instances to be found on the other side, yet these it either neglects and despises, or else by some distinction sets aside and rejects, in order that by this great and pernicious predetermination the authority of its former conclusions may remain inviolate."[9]

In this respect the Prime Minister's autocratic and often wild behavior seemed to Brooke, despite Brooke's own approval of the Shingle plan, to be substantially worse even than in November 1943, when Brooke had despaired of having to work under such an impossible commander in chief. Already the day after Churchill's return from Marrakesh to London on January 18, 1944, Brooke was recording he could not "stand much more of it," after four hours of meetings with him. "In all his plans he lives from hand to mouth. He can never grasp a whole plan, either in its width (ie all fronts) or its depth (long term projects). His method is entirely opportunist, gathering one flower here another there! My God how tired I am of working with him!"[10]

A good "bag" of 172 pheasants he shot with three colleagues at Glemham Hall in the Suffolk countryside, on January 22 — the day the Anzio invasion began — temporarily lightened Brooke's mood, as did the first reports from the War Office in London recording "the landing south of Rome had been a complete surprise" to the Wehrmacht — a "wonderful relief!" as the newly minted field marshal noted in his diary.[11] But the ensuing days proved less and less hopeful. All too soon the Wehrmacht's inevitable reaction put the whole concept of a swift "advance on Rome" in peril. Eventually, a week later, the penny began to drop. Anzio was *not* going to lead to the Eternal City anytime soon — in fact it would take many months, if it could be done at all — and threatened to make Overlord and Anvil impossible to mount successfully.

"Hitler has reacted very strongly and is sending reinforcements fast," Brooke noted in his diary on January 31.[12] As Leahy had forecast, Brooke and his colleagues were soon cabling Washington to "convince them that with the turn operations have taken in the Mediterranean, the only thing to do is to go on fighting the war in Italy," and — more ominously — "give up any idea of a weak landing in Southern France."[13]

The campaign in Italy thereafter went from bad to worse — no less than three consecutive and bloody battles being fought at Cassino to try and link Clark's frontline troops with the stranded Allied forces on the beaches of Anzio, without success.

Italy was, as Admiral Leahy had predicted, a disaster.

In private Winston Churchill himself became worried lest he be publicly denounced, and incriminated for his homicidal meddling. "Anzio was my worst moment in the war," he later confided to his doctor. "I had most to do with it. I didn't want two Suvla Bays [i.e., Gallipolis] in one lifetime,"[14] he confessed —having admitted to Eisenhower, Bedell Smith, and Brooke in late February, 1944, as British and American casualties mounted on the beaches, that he'd hoped "to land a wildcat that would tear out the bowels of the Boche. Instead we have stranded a vast whale with its tail flopping about in the water."[15]

Such clever phrases were memorable enough for Brooke to record them, instantly, in his diary. But they could not save the lives of the men who'd been given an impossible task.

Sick at heart and in body, the President was both distressed and full of sympathy for the troops, since the majority of them—and the casualties—were in fact Americans. Was it all a form of revenge for the British not getting their way in delaying or halting Overlord at Tehran?

Neither the President nor his chiefs of staff were willing to let the fiasco prejudice the launch date of Overlord or its supporting operation, Anvil. They soon had to, though: Churchill and the British chiefs now begging for more Allied forces to be committed in Italy to save Anzio. In the circumstances Overlord would, the U.S. chiefs agreed, probably have to be delayed by a month, to June 1944; the Anvil assault would probably have to be abandoned: the price of Churchill's folly.

Field Marshal Brooke, as coconspirator in resurrecting the Anzio scheme, was meanwhile required not only to bear the brunt of Churchill's frustration, anger, and blame, but the inevitable consequence: the Prime Minister switching his abortive energies to a new campaign, this time the war in the Pacific.

"I am quite exhausted after spending 7½ hours today with Winston, and most of that time engaged in heavy argument," Brooke penned on February 25 —aware that the President and his team in Washington would simply accept no change in overall European strategy, let alone in the Pacific.[16]

Anzio was, in short, a catastrophe. Rome was as far away as ever—causing the despairing prime minister to insist he must have an emergency meeting with the President and the American chiefs of staff, even if it meant his flying to the United States or the Bahamas.

Fortunately the President could, at least, use the lingering effects of his flu and bronchitis as a way of deflecting another Churchill visit—which would undoubtedly involve new British appeals to change agreed Allied plans for Overlord, the Second Front, and the Pacific.

Brooke might complain of the "President's unpleasant attitude lately," as he recorded in his diary on February 25,[17] but in the circumstances it was remarkable the President had continued to maintain a polite tone in his responses to Churchill's ever more strident one, when the Prime Minister was making major difficulties now over the best way to defeat Japan, by once again pressing for an invasion of the northern tip of Sumatra. The President simply but firmly made clear he had no intention of meeting with Churchill, nor would he countenance Churchill's plea to keep Overlord's landing craft in the Mediterranean.[18] There was not to be, he cabled, any change in the Overlord and Anvil plans.

For Churchill the shame, embarrassment, and displaced anger were impossible to swallow. The Prime Minister's residence at 10 Downing Street became a war zone. As Brooke would later reflect, "We were just at the beginning of the most difficult period I had with Winston during the whole of the war" — which was certainly saying something.[19] By March Brooke was recording the entire British chiefs of staff were on the point of resignation, and that he himself was "shattered by the present condition of the PM. He has lost all balance and is in a very dangerous mood" — a mood that had impelled the Prime Minister to ignore the President's cables, and again demand a meeting with Roosevelt and the U.S. chiefs of staff, this time "on the 25th of this month!" as Brooke noted in despair.[20] In Brooke's eyes it would be fatuous to ask for a change to Pacific strategy at such a conference, when the Americans were clearly in complete — and hitherto successful — charge of naval, air, and army operations there. It would also be dangerous in terms of Churchill's fragile health — a trip that could result in yet another bout of pneumonia, followed by heart problems, Lord Moran warned again and again throughout March 1944, as American and British troops fought and died to hang on to their toehold at Anzio.[21]

It was not Churchill's heart that was the problem. It was the President's.

39

Crimes Against Humanity

SINCE ADMIRAL MCINTIRE would afterwards remove and destroy (it is believed) all the President's White House medical records, the history of Franklin Roosevelt's health would be, in retrospect, spotty. But in essence the problem at the White House and at Hyde Park in the winter and spring of 1944 was not the severity of the President's influenza or bronchitis. It was their persistence.

"More disturbing than anything else," McIntire wrote later of the President at this time, "there was the definite loss of his usual ability to come back quickly"—at least, in the way Churchill had.[1] The burden of the office—domestic, military, diplomatic—was unrelenting. These three aspects, moreover, were becoming impossible to keep apart: pitting cabinet members, administration officials, and military personnel against one another over a thousand issues, which only the President, as chief executive and commander in chief, could decide. For instance, if the U.S. Army was to meet its goal of ten million servicemen, General Marshall and Secretary Stimson pressed the President, it was a matter of urgency to end current deferments of young men working in agriculture and industry. This naturally inflamed cabinet members and officials responsible for agriculture and industry. There was also the question of whether de Gaulle's Comité Français de la Libération Nationale should be allowed to pose as the lawful French government and administer those areas of France that would—hopefully—be liberated by U.S. and Allied troops in the aftermath of D-day, rather than leaving responsibility to General Eisenhower as Allied supreme commander. This was a prospect the President—who had come to despise de Gaulle for his autocratic methods, intense nationalism, and renewed colonialist aspirations—abhorred. As did the secretary of state, Cordell Hull.

There was also the matter of "Manhattan," the development of an atom bomb, which the President was funding, in secret, via his loyalists in Congress—and the question of how to keep at bay Senator Harry Truman, chair of an investigative committee on the national defense program, who had been only vaguely informed of the project, along with senior members of Congress.

In the end Secretary Stimson and General Marshall had gone to see the President in person at the White House on March 13, 1944, arguing for a new executive order to deal with the threat to "our own Army manpower and the danger that it is to Overlord." The President had assured them he would get them the servicemen they needed—but wanted time to work on the Selective Service officials to ensure a consensus that only he, the legendary Magician or Juggler of the White House, could obtain. As for Truman's threat of "dire consequences" if Stimson didn't come clean on the atomic bomb project, the President wholeheartedly backed Stimson's refusal to bow to Truman's pressure. "Truman is a nuisance and a pretty untrustworthy man," the Republican secretary noted in his diary. "He talks smoothly but he acts meanly," he remarked.[2]

By mid-March, however, those closest to the President had become more and more worried about the burdens he was carrying—and the state of his health. Following Secretary Stimson and Marshall's visit on March 13, Admiral McIntire was finally forced to respond to the growing fears.

The Prime Minister's appeal for a "staff meeting on the Teheran scale in Bermuda about the fifth of April" was out of the question. It was all getting too much for a relentlessly sick president. On McIntire's advice Roosevelt cabled Churchill some days later, saying he was not going to be able to oblige. "The old attack of grippe having hung on and on, leaving me with an intermittent temperature, Ross decided a week ago that it is necessary for me to take a complete rest of about two to three weeks in a suitable climate, which I am definitely planning to do at the end of the month," as the President finally explained on March 20. "I see no way out and I am furious."[3]

He wasn't really. True, he was deeply disappointed that the debacle at Anzio had now forced Eisenhower to delay the D-day landings to the end of May or early June, given the paucity of landing craft. But he said nothing of that—indeed his forbearance, in contrast to that of Churchill, was remarkable. He had attended the annual White House Correspondents' Dinner on March 4, and had given no fewer than three press conferences in the week thereafter. His lapses in focus, hearing, and concentration, however, were becoming all too noticeable. By March 14, the President's daughter Anna had finally decided to confront Admiral McIntire. The President's temperature was fluctuating up and down; he had abdominal distress, his hands were trembling almost uncontrollably, he could not sleep at night, was coughing, and was finding it hard to breathe normally.

Responding, McIntire had belatedly insisted on rest and a forthcoming vacation in a warm climate; also that the President immediately go on a diet, both to lose weight (relieving strain on his heart) and improve his digestion.

For the next ten days the President duly followed this prescription. He did not come down to the Oval Office before noon, and was served his dinner on a tray either in his bedroom or his study, upstairs at the White House, with only his cousin Daisy to keep him company, since Eleanor was traveling.

It didn't help. The President was falling asleep sometimes "bolt upright," Daisy noted on March 23.[4] He was in recurrent pain from headaches, fever, and abdominal discomfort. Grace Tully, Roosevelt's loyal office secretary and stenographer, appealed to Anna to *do* something. For three months now the President had been ailing — with no sign of amelioration, in fact the very opposite.

Anna thus insisted Admiral McIntire bring in specialist consultants to examine the President. And with that the final odyssey of the President's life began.

Admiral McIntire — who had no specialist knowledge outside his area of expertise: ear, nose, and throat — duly began making inquiries. Loyalty to his commander in chief trumped all other considerations, however; McIntire insisted no one outside the family circle, and of course no one in the press, should know. An appointment with a navy cardiologist at the Naval Hospital, away from prying eyes, seemed the safest, surely the most secret, way to proceed. A medical exam was thus arranged to take place after the weekend — its scheduling to be kept strictly under wraps.

First, though, the President announced he wished to hold another press conference, in order to make an important statement. He planned to issue a presidential proclamation, he'd decided, on one of the most egregious aspects of the Nazi conquest of Europe: the deliberate extermination of millions of Jews. His proclamation, "promising to help rescue the Jews from Nazi brutality in Europe," his private secretary, Bill Hassett, noted on Friday, March 24, 1944, would make sure no one could claim they were unaware of what was going on behind the Wehrmacht's frontlines. To be certain it got the broadest possible coverage, moreover, he told his secretary, he was going to read it aloud, word by word, at a press conference in the Oval Office, however ill he currently appeared — and felt. He was "not looking so well in his bedroom, nor later when he held a press and radio conference," Hassett recorded in his diary: "voice husky and out of pitch. This latest cold has taken lots out of him. Every morning, in response to inquiry as to how he felt, a characteristic reply has been 'Rotten' or 'Like hell.'"[5] But it had to be done — and was.

One sentence, in particular, he'd thought summed up his own philosophy, and that of most decent people. "The United Nations are fighting to make a world in which tyranny and aggression cannot exist: a world based upon freedom, equality and justice; a world in which all persons regardless of race, color or creed may live in peace, honor and dignity."[6]

A Trip to Tehran

After laying a wreath in Arlington Cemetery on November 11, 1943, FDR sets off with his Joint Chiefs of Staff to Tehran to meet Churchill and Stalin.

Aboard the USS *Iowa,* the latest American battleship, the President works out a plan of how to deal with Churchill's threatened "indictment" of U.S. strategy.

Met at Oran by General Eisenhower, FDR flies with Ike to his Tunis headquarters. Over two days of talks, FDR weighs whether Marshall or Ike should be the supreme commander of the D-day invasion, Operation Overlord.

Arriving in Cairo for a "showdown," Churchill hopes to get D-day postponed in favor of a Balkan strategy. But FDR avoids a battle with the PM by deliberate sightseeing and long talks with Generalissimo Chiang Kai-shek.

Flying on to Tehran, FDR is finally able, with Stalin's help, to overcome Churchill's objections to D-day in 1944. He even gets Stalin to promise to launch a massive Russian offensive on the Eastern Front to help Overlord succeed.

The Tehran summit marks FDR's greatest achievement as strategist and U.S. commander in chief in World War II: keeping both Britain and the USSR cooperating as allies.

A triumphant FDR appoints General Eisenhower to be supreme commander of Overlord. He spends further days with Ike in North Africa, taking the general with him to Malta and Sicily to review troops. FDR also decorates General Clark and tells Patton (in background) he will be needed for the cross-Channel invasion.

Triumphant Return

Returning home on the USS *Iowa,* the Commander in Chief is feted at the White House by the cabinet and congressional leaders, including (below) Treasury Secretary Henry Morgenthau (left), Secretary of State Cordell Hull, and Director of War Mobilization Judge Byrnes.

Surrounded by Eleanor and his family at Hyde Park, FDR gives a Christmas Eve broadcast announcing his appointment of General Eisenhower, not General Marshall, to lead the "next blow" against Nazi Germany: D-day. He has never looked or sounded more confident. Even Hitler predicts the Allied invasion of France will "decide the war."

Taking off his pince-nez, the President had turned to the assembled reporters and smiled. "Some of you people who are wandering around asking the bellhop whether we have a foreign policy or not" might think about the statement, the President had reflected aloud, for "I think it's a pretty good paragraph." The words went to the heart of what America now stood for, in a world wracked by violence on a hitherto unimaginable scale. "*We have a foreign policy,*" he'd reminded members of the press. "Some people may not know it, but we have." And with that he'd read out the next paragraph. This, in turn, explained what the United Nations were up against: namely that "in most of Europe and in parts of Asia the systematic torture and murder of civilians — men, women and children — by the Nazis and the Japanese continue unabated."[7]

This was not mere rhetoric, he'd emphasized. "In areas subjugated by the aggressors innocent Poles, Czechs, Norwegians, Dutch, Danes, French, Greeks, Russians, Chinese, Filipinos — many others — are being starved or frozen to death or murdered in cold blood in a campaign of savagery. The slaughters of Warsaw, Lidice, Kharkov and Nanking" — "sometimes people forget about Nanking," he'd paused to comment — "the brutal torture and murder by the Japanese, not only of civilians but of our own gallant American soldiers and fliers — these are startling examples of what goes on day by day, year in and year out, wherever the Nazis and the Japs are in military control — free to follow their barbaric purpose."[8]

Cruelty without conscience: this seemed the mantra of Germany and Japan. The Wehrmacht's military occupation of its ally Hungary on March 19, had alarmed the entire free world — not only because it demonstrated the German intention of fighting to the bitter end, but because it had exposed the sheer evil of Hitler's regime, as news poured in to Washington of Wehrmacht and SS troops rounding up perhaps three-quarters of a million Hungarian Jews for deportation and execution.

Three-quarters of a million! "In one of the blackest crimes of all history — begun by the Nazis in the day of peace and multiplied by them a hundred times in time of war," the President continued, "the wholesale systematic murder of the Jews of Europe goes on unabated every hour. As a result of the events of the last few days, hundreds of thousands of Jews, who while living under persecution have at least found a haven from death in Hungary and the Balkans, are now threatened with annihilation as Hitler's forces descend more heavily upon these lands. That these innocent people, who have already survived a decade of Hitler's fury, should perish on the very eve of triumph over the barbarism which their persecution symbolizes, would be a major tragedy."[9]

In a United States where the press tended to focus on matters of domestic policy or lack of progress in Italy, it was important, he'd emphasized, not to forget what Americans were fighting *for* — apologizing for mentioning "more foreign policy" than the journalists probably wanted to hear. He was not ashamed, however. In the proclamation, which was to be broadcast that night, he'd wished to make clear it was not only Hitler who was guilty of such mass murder, but many thousands of Germans — "functionaries and their subordinates" both in Germany and its satellite countries — who were assisting in the annihilation of completely innocent civilians. "All who knowingly take part in the deportation of Jews to their death in Poland or Norwegians and French to their death in Germany are equally guilty with the executioner himself. All who share the guilt shall share the punishment."[10]

It was at that point the President had read out a phrase in the proclamation that would soon go down in the history of jurisprudence. "Hitler," he pronounced, "is committing these crimes against humanity." And was doing so "in the name of the German people."[11]

Crimes against humanity.

The phrase had once been used, in the late nineteenth century, to protest the Belgian king Leopold II's ravaging of African civilians in the Congo.[12] Now it applied to the fate of millions of innocents in present-day Europe. Such crimes — wholly intentional and deliberate, and trumpeted proudly by Hitler in his recent broadcast — were beyond anything imaginable in a supposedly civilized country. "I am asking," the President had continued, "every German and every person of any other nationality everywhere under Nazi domination to show the world by his action that in his heart he does not share these insane criminal desires." Whether in Hungary or anywhere else in Europe, "let him hide these pursued victims, help them to get over their borders, and do what he can to save them from the Nazi hangman. I am asking him also to keep watch, and to record the evidence that will one day be used to convict the guilty" — for such crimes against humanity would not, he'd warned, go unpunished or ever be forgotten.[13]

"In the meantime," the President stated, coming to the climax of his proclamation, "until the victory that is now assured is won, the United States will persevere in its efforts to rescue the victims of brutality of the Nazis and the Japs." He was speaking therefore not only on his own behalf, but on behalf of his administration. "In so far as the necessity of military operations permit, this Government will use all means at its command to aid the escape of all intended victims of the Nazi and Jap executioner — regardless of race or religion or color. We call upon the free peoples of Europe and Asia temporarily to open their frontiers to all victims of oppression. We shall find havens of refuge for them, and we shall find the means for their maintenance and support until the tyrant is driven from their homelands and they may return. In the name of justice and humanity let all freedom-loving people rally to this righteous undertaking.

"Finis," he'd ended.[14]

The journalists departed to file their stories. Exhausted and unwell, the President was determined to leave the White House for a weekend at Hyde Park. There, despite his condition and before his impending medical exam at Bethesda, he was expecting an important visitor. Her name was Mrs. Lucy Rutherfurd.

40

Late Love

THE VISIT OF ROOSEVELT's former mistress marked, as fate would have it, a critical moment in the President's life. For by the strangest coincidence, as the President's health plummeted, causing his doctor to reluctantly, and in secret, contact a cardiologist, Mrs. Rutherfurd's husband had died, making Lucy a widow. And free to see the President without shame, if that was what he wished.

He did. At a time when the very life force, energy, and joie de vivre had seemed to be visibly sapping from the President, Lucy Mercer Rutherfurd's visit to Hyde Park promised him new vitality, at least.

By this time the affair between Franklin Roosevelt and Lucy Mercer was largely forgotten in the nation's capital—though a quarter century earlier it had caused a veritable scandal in Washington social circles, as everyone who knew the parties was aware.[1]

Some blamed Franklin, some blamed Eleanor, some Lucy.

Elliott Roosevelt—who was himself married five times in the course of his life—later wrote that the affair had begun because of Eleanor's decision to no longer have sex with her husband. After bearing six children over the course of a decade, Eleanor had, in 1916, insisted on separate bedrooms, Elliott claimed. Thereafter, he asserted, "my parents never lived together as husband and wife."[2]

Elliott's older brother James, expressing more sympathy for his mother, later recounted how Eleanor confided to his sister Anna that "sex was an ordeal to be borne" and had never been a pleasure for her. Whatever the case, when Franklin, as assistant secretary of the navy during World War I, had returned from Europe on a stretcher in the midst of a global flu epidemic, suffering double pneumonia, and was admitted to hospital on September 20, 1918, Eleanor accidentally came across her former social secretary's perfumed love letters in her husband's briefcase—and had been mortified.

It was not sexual jealousy that had given rise to the specter of suicide and divorce on Eleanor's part; it had been the sheer depth of Lucy's love for Franklin. Moreover the evidence, in writing, that this love was shared. The realization had devastated Eleanor — an orphan who had grown up rich yet unloved and lonely. The affection, attention, respect, and shared pride in their five surviving children had been everything to her — a blessing that had seemed suddenly worthless in her husband's eyes, she'd felt.

Eleanor had not been mistaken. Franklin's feelings for Miss Mercer had been far more than a temporary infatuation. As Lucy's cousin later recalled, Lucy and Franklin had been "very much in love with each other" — and the affair had been no secret to others, however much the couple had sought to observe decorum in Washington, where Franklin was, after all, assistant secretary of the U.S. Navy. To distract attention from the budding scandal the couple had spawned stories in the gossip press involving other courtiers chasing Lucy. Even if they hadn't convinced their friends, these had served to deceive Eleanor — at least, up to the moment she found the letters.[3] The fact that, after two years of furtive adultery, her tall, dynamic, hugely handsome husband was still passionately in love with a beautiful, quiet, graceful woman, nine years younger than herself, had been for Eleanor a bombshell, in spite of the many rumors she'd heard (and dismissed) all that year and the year before.[4]

"The bottom dropped out of my own particular world and I faced myself, my surroundings, my world, honestly for the first time," Eleanor later admitted, sadly.[5] Overnight the affair had threatened to sunder her privileged domestic and social world, as well as her ever-vulnerable self-confidence in it. Worst of all, her husband's pursuit of Lucy Mercer had clearly not been one of his brief or overnight flirtations, as Eleanor looked back. It was serious. After war was declared in the spring of 1917, Eleanor had told Lucy she intended to economize, and would no longer need her services as social secretary. Franklin, however, had promptly found Lucy a job as a yeoman in his office at the Navy Department. This had been throwing caution to the four winds — with serious potential political consequences for the Wilson administration. Franklin's boss, Navy Secretary Josephus Daniels, had deplored the business; in fact Secretary Daniels had had Miss Mercer summarily fired after six months, owing to mounting "gossip" in the capital. This had not, apparently, dampened the couple's ardor for each other — Franklin and Lucy merely using "safe houses" provided by their friends and relatives, and trying wherever possible to conduct their romance out of town.

A whole network of relatives and friends had necessarily and inevitably become parties to the affair, as well as its duplicity. "I saw you 20 miles out in the country," Alice Longworth Roosevelt, Eleanor's sharp-tongued cousin, for example, had written to taunt — and warn — Franklin. "You didn't see me. Your hands were on the wheel but your eyes were on that perfectly lovely lady," she'd teased him — describing Lucy as a "beautiful, charming, and an absolutely delightful creature."[6] To others Alice Longworth would later snidely remark that Franklin "deserved a good time," given that he "was married to Eleanor" — a less than beautiful woman with a receding chin and big teeth who, in Alice's malicious eyes, lacked any sense of humor or fun.

In short, by the time Eleanor had belatedly learned of the affair, all Washington already knew of it — with no shortage of evidence that could be used in court, including "a register from a motel in Virginia Beach showing that father had checked in as man and wife," James Roosevelt later recorded, and had "spent the night" with Lucy.[7]

Shaken and hurt to her core, Eleanor had favored a divorce. Lucy, for her part, had been convinced that a church annulment of Franklin and Eleanor's marriage could be successfully sought, so that she and Franklin could marry without the necessity of his converting to Catholicism or vice versa. It had been Franklin's mother, Sara, however, who had refused to countenance a social scandal such as the divorce of her beloved and only son would pose. His career would be ruined, she had warned, as well as the family name. Sara had even threatened to disinherit Franklin, it was widely rumored, if he went ahead.

As Alice Longworth later reflected — following her own divorce and those of every single one of Franklin's children (often multiple times) in later years — the Victorian and Edwardian eras were unimaginably different from those that followed. "I don't think one can have any idea how horrendous even the idea of divorce was in those days . . . In those days people just didn't go around divorcing one another. Not done, they said. Emphatically."[8]

In the end it was the beautiful, statuesque Lucy Mercer who'd recognized she couldn't do this to Franklin, whom she adored. She'd therefore backed out, honorably. All had been scarred, though.

Reeling emotionally but determined to do her best, Eleanor had agreed to continue acting as her husband's loyal consort in public, and to remain a devoted fellow parent to their five surviving children. For herself, however, Eleanor had vowed to make, if she could, a new, more independent life — both as a woman and a citizen — whatever anybody thought of her. She thus helped Franklin gain the Democratic nomination for the vice presidency in 1920; she helped nurse him through his poliomyelitis and lower-limb paralysis (sadly without improvement) in 1921; and later, when he'd stood successfully for the governorship of New York in 1928, she had become — and remained — his loyal political advocate, supporting him steadfastly when he stood for the presidency itself, in 1932. But in terms of intimacy, although not mutual respect, the air had gone out of their relationship. "Franklin and Eleanor" had become once more — as, to an extent, they had always been — devoted cousins more than husband and wife. Cousins who not only shared the same name but to a large extent the same compassionate political and social ideals, despite their inherited wealth, however much this caused them to be hated as renegades or "traitors to their class" by conservatives and rich Republicans.

The "Franklin and Eleanor" political combination, at least, had been completely sincere. But in her personal life Eleanor had become determined to be a "new woman"—whatever others might say or sneer. Franklin had had his coterie of male advisers as well as female assistants and supporters, especially his personal assistant, Marguerite "Missy" LeHand, whom Eleanor called his "office wife." For her part, Eleanor sought to broaden her life experience, and to share deeper feelings with her *own* selected colleagues, helpers, and admirers—including a succession of women and younger men she could take under her wing as intimate companions over the years. There were Nancy Cook and Marion Dickerman, a lesbian couple she loved, and with whom she built Val-Kill, a cottage on the grounds of Hyde Park, with Franklin's blessing and even architectural advice; Lorena "Hick" Hickok, a lesbian reporter who became her bosom companion, perhaps even lover, in the White House; Earl Miller, the chauffeur whom her husband insisted she employ as driver and security detail, once she became First Lady; and finally Joseph "Joe" Lash, a young, left-wing student and son of Russian Jewish émigrés, whom she "adopted" in 1939 to the point of obsession, and an FBI investigation—writing almost daily to him and even insisting on being allowed to visit him at Guadalcanal on her Red Cross visit to the Pacific in 1943, where Lash had been posted after being drafted into the U.S. Army.

Eleanor's "infatuations," alongside her tireless work for social and political causes (not least over civil rights, which constantly threatened to alienate Southern Democrats, whose legislative support in Congress was vital to FDR), had been brave, open, even reckless—yet they had rarely caused the President to turn a hair, or even to counsel prudence. His loyal, avuncular-style support of his wife and members of her inner circle had been, in retrospect, amazing. He had granted the ladies a lease-for-life on Val-Kill, next to a stream that the women soon dammed in order to make a summer swimming pool. He had even presented Marion Dickerman with a children's book he'd found, *Little Marion's Pilgrimage,* not only signing it "from her affectionate Uncle Franklin" but writing that the gift was being made on the "occasion of the opening of the Love Nest on the Val-Kill."[9]

The Love Nest.

The President's words had not been sarcastic, nor were they a sign of Louis XIV–like largesse; they simply reflected Franklin's genuine generosity of spirit. Perhaps they betrayed a certain guilt, too: that his love affair with Miss Mercer during World War I had hurt Eleanor's confidence in herself as a woman profoundly. And that, in consequence, he owed it to Eleanor to make amends in kind, by encouraging her friendships, if not in their own romantic hearts—both of which had been broken by the affair and its outcome. For however hard she had tried, Eleanor had never been able to reconcile herself to the "act of being physically unfaithful" that had marked Franklin's relationship with Lucy—an aspect which she attributed to the lesser self-discipline of the "average man."[10]

Certainly the matter was far from simple. Franklin's generosity of feeling toward Eleanor, her friends, devotees, and adoptees, was not out of character—it was something that had first drawn Eleanor to her cousin. It also undoubtedly stemmed from sincere gratitude: his acknowledgment that, less than three years after the bust-up over Lucy, it had been Eleanor who had coped with his devastating polio diagnosis, causing Eleanor—not Lucy—to have to deal thereafter with a grave affliction that could never be cured, despite the seemingly endless treatments, rehabilitation, therapies, and fruitless remedies he'd tried.

Franklin Roosevelt would not have been Franklin Roosevelt, known for his loyalty and goodwill, had he broken off *all* contact with Miss Mercer, though. Or with their mutual friends. When Franklin heard Lucy had married, on the rebound, Mr. Winthrop Rutherfurd, a rich widower twice her age, in February 1920, just as Franklin prepared to embark on his bid to win the Democratic Party's vice presidential nomination, he was said to have "started like a horse in fear of a hornet,"[11] so upset was he at hearing the news. Yet he could scarcely have expected Lucy, single and by then aged twenty-eight, to wait in the wings to see whether Eleanor and Franklin would, despite their renewed vows, fail to mend their marriage. And once Franklin had fallen ill with polio the following year, he was, so to speak, a twice-broken man: an invalid to be pitied, who would need all the care and help his family could give him, not a second upheaval.

For the most part Franklin's letters to Lucy were burned after Franklin's death, to avoid posthumous scandal,[12] while nearly all her own letters to him would vanish after his death, presumably destroyed. The handful of extant letters demonstrate incontrovertibly that Lucy and Franklin maintained at least distant contact over the years of her marriage.

As Franklin's star had risen in the political firmament of the Great Depression and New Deal, the President had continued to correspond with Lucy, even to provide for her stepchildren and her daughter, Barbara—to the point, in fact, where he had become a sort of honorary uncle, even "godfather," as he called himself, to Barbara: his daughter manqué. In moments of special pride, moreover, in his rise to the presidency—unique in history for a paraplegic—he'd wanted Lucy Rutherfurd to be near him, if it could be arranged discreetly. He'd thus made sure that Lucy—whose sister had a house in Georgetown — was fetched by White House limousine to attend his first inauguration as thirty-second president of the United States in March 1933; then his second, in January 1937; and his third in 1941. Thereafter, once her husband, Winty Rutherfurd, had become infirm, housebound, bedbound, and in need of a live-in nurse, the President had begun inviting Lucy to visit him briefly at the White House, when she was in Washington, under an assumed name: "Mrs. Johnson."

The relationship with Lucy had thus always had its special character, at once nostalgic and joyful. When on March 19, 1944, Lucy's husband had finally died at age eighty-two, the President had felt he need not prevaricate or be secretive any longer. He would be free to see Lucy — no longer as Mrs. Johnson, but as Mrs. Rutherfurd, a widow. And at Hyde Park, on March 26, before he was to be examined, secretly, by naval doctors tasked with finding out why he had seemed unable to bounce back to good health after his triumphant trip to Tehran.

For his part the President had told Daisy he was feeling like death warmed over, lying down after his press conference on March 24. But as the *Ferdinand Magellan* had pulled out of the special platform beneath the Bureau of Engraving and Printing near the White House that night, he'd felt almost lightheaded. For Lucy had said she would drive to see him at Springwood, his Hyde Park house, for the first time in her life, on Sunday, March 26. And for the first time in months he'd felt wonderfully alive.

41

In the Last Stages of Consumption

LOVELY MILD WEATHER just made for the P.," Daisy Suckley noted in her diary on the morning of Saturday, March 25, 1944, as the President was taken by car from Highland railway station, across the Hudson River from Poughkeepsie, to his home. "Mrs. Rutherfurd is coming up to see him from New York tomorrow, and he hopes to show her around the place, the library, the cottage etc., so he took things easy—had a good nap before lunch, & after lunch sat in a deck chair in the sun."[1]

"In the afternoon sat out on the terrace in the sun for some time—no visitors," Bill Hassett also noted in his diary—all of them retiring "to bed soon after ten o'clock," to be ready for the big day.[2]

The President on the eve of Lucy's visit had felt both indolent and excited. "I've never done such a thing in my life before," Daisy quoted him saying, as he'd relaxed. He was "Robert Louis Stevenson, in the last stages of consumption," he'd described himself—Daisy leaving him on the veranda reading the *London Daily News,* together with his secretary Grace Tully, who was taking dictation. But he purposely hadn't invited Daisy, or Hassett, or Grace to join him when Lucy arrived the next day for lunch—after which he drove Lucy around the whole estate personally, in his Ford Phaeton convertible, with its special hand controls. He insisted on showing Lucy the library and his cottage. After six hours of reunion he finally said goodbye to her at 6:30 p.m.

He'd overdone it—in fact, no sooner had Lucy left Hyde Park than he "felt fever coming on & went to bed," as he afterwards admitted to Daisy, on the phone. He'd arranged, he'd told her, to take the *Magellan* back to Washington early the next morning. On arrival at the White House he would have dinner served on a bed tray, and a good night's sleep before the medical examination at Bethesda Naval Hospital scheduled for Tuesday. There, hopefully, the cause of his medical misery could be cleared up. It would be a full day in Washington, since Eleanor would be returning from the Caribbean. He would be

"X-rayed etc," Daisy wrote, having remained at her house, Wilderstein, near Hyde Park. "I pray they do the right thing by him."[3]

They did. The result, however, would not be the one either the President or Daisy — who had an obsession with diet, on which she blamed the President's poor condition — was quite expecting.

The medical examination that Admiral McIntire had arranged to take place at 11:30 a.m. on March 28 would change President Roosevelt's life forever. And none too soon.

Three months earlier, the President had invented the cognomen "Dr. Win-the-War." Now, however, it was the President himself who needed a doctor — a better doctor — if he was to win both the war and his battle with the approaching Reaper. This turned out to be Lieutenant Commander Howard Bruenn, a thirty-nine-year-old officer in the U.S. Naval Medical Corps (Reserve), currently stationed at the Bethesda Naval Hospital in Maryland, where he was chief of the electrocardiograph department.

Dr. Bruenn later confessed to being "pretty shocked" when the President was wheeled into his examination room to be lifted onto a so-called Gatch bed, the top end of which could be tilted up. Admiral McIntire had painted for him a wholly fictitious picture of the President, saying merely that Roosevelt "was not himself" and was "thought to have had an upper respiratory infection and had not quite regained his strength," Bruenn later recounted.

An upper respiratory infection?

Bruenn found it hard to conceal his alarm. In newsreels at Christmastime, on his triumphant return from Cairo and Tehran, the President had appeared to Bruenn and millions of others to be fighting fit — in speech, mind, and body — or at least body language. That much had been genuine. What was true three months later, however, was the opposite. Roosevelt's condition, in Bruenn's subsequent account, was "God awful."[4] The President, Bruenn found, was "in acute congestive heart failure," with possibly only weeks to live, at most only months.

The President's aspect, close-up, stunned Bruenn. "He appeared to be very tired, and his face was very gray. Moving caused considerable breathlessness."[5] The President was coughing repeatedly, but could bring up no sputum. "I suspected something was terribly wrong as soon as I looked at him," Bruenn later confessed. "His face was pallid and there was a bluish discoloration of his skin, lips and nail beds."[6] He was clearly not getting enough oxygen.

Bruenn began his formal examination by measuring the President's temperature — slightly high — his pulse and respiration rates, as well as his blood pressure. At 186 over 108 this was not good, but not startling for a man in his sixties carrying huge responsibilities and suddenly in a medical examination room. Readings from earlier visits, after escalating in 1940, showed this number to be consistently high for years now—178/88 in November 1940, 188/105 in 1941. But no readings thereafter. Suspicious, Bruenn asked to see the President's more recent numbers. Admiral McIntire, who took the President's blood pressure twice daily, seemed reluctant to show them, claiming they were back in his office, next to the Map Room at the White House. (Bruenn insisted, and they were eventually fetched. All, when the cardiologist saw them, showed systolic readings above 200.) Yet it was only when Bruenn examined the President's chest more closely that he truly appreciated the implications of what he was discovering. Electrocardiograms and X-rays were taken, as well as blood samples for further analysis.

Dr. Bruenn's main finding was incontrovertible: the President was dying. His heart had ballooned in size; it had also moved position. In fact, according to Bruenn's notes, it was "enormous,"[7] with the consequence that even shifting position on the special examination bed "caused [the President] considerable breathlessness" and puffing. Bruenn was now afraid. "My diagnosis was a bombshell," he later recalled.[8] Speaking as an experienced cardiologist, he told Admiral McIntire that the President was "in left ventricular failure" or "acute congestive heart failure,"[9] a condition in which treatment, even if begun immediately, could not prolong the President's life for very long, since he was suffering fatal "hypertensive heart disease."[10] And that, as Bruenn later recalled, "put a different aspect on the whole situation."[11]

It certainly did. Despite his pretense, Admiral McIntire had known what really lay behind the President's worsening condition, but had not dared confront it. He asked Bruenn to "write out what I thought should be done." When he saw the report, however, the admiral exploded. As Bruenn recalled, "When I gave him my recommendations for bedrest, diet, etc., he said, 'You can't do that. This is the President of the United States!'"[12]

McIntire was rattled. The President's heart had, after all, been his constant, secret concern—especially whenever Roosevelt was required to travel at high altitude, first on the flight to Tehran and subsequently, when flying back around the Atlas Mountains to Dakar. But now the truth could not be evaded. The President's symptoms over the three months since Christmas stemmed not from a case of lingering flu, as McIntire had pretended, or a respiratory infection, as he'd prayed it might be. The excuse that McIntire later gave for not having summoned expert medical advice earlier, since the President's condition was "completely unsuspected up to this time,"[13] would be a complete lie, like the reports he made to the press at the time. It was a lie he'd felt compelled to tell, since he himself would be blamed, as Roosevelt's White House physician, if it was found the President should have been receiving treatment, or medical advice and recommendations, at least, for his worsening heart condition. For if Bruenn was right, the President was suffering from a mortal malady, for which there was no treatment at that time beyond sedation, bed rest, diet — and the administering of digitalis, a potentially toxic extract of purple foxglove, as a palliative.

Unable to accept his mistake, McIntire thus accepted Bruenn's medical findings, but not the severity of the President's condition or his treatment strategy. McIntire's response to the latter was "somewhat unprintable," Bruenn recalled[14] — the cardiologist recommending the "patient" cease all activity, physical or mental, and submit to weeks of absolute bed rest "with nursing care." Also completely refrain from smoking and begin a diet to lose weight. As well as take raw codeine for his other ailment, his "acute bronchitis," and be fully sedated at night so he could actually sleep. Also a digitalis regimen — to begin immediately.

"Summarily rejected," McIntire declared, panicking at the thought of the effect on the nation, in the midst of a global war, if word got out the President was mortally ill.[15] What was Bruenn thinking? "The President can't take time off to go to bed," McIntire snapped, as if Bruenn had gone mad. He was the nation's commander in chief. "You can't simply say to him, 'Do this or do that.'"[16]

As for the administering of toxic digitalis, McIntire was even more contemptuous — the medication was simply too radical a treatment for him to contemplate, especially given Bruenn's suggestion that the President begin treatment immediately, lest he die in the next days or weeks.[17]

Bruenn must have held his ground, however, for McIntire — perhaps to save his own career — thought better of his first reaction. At the end of their discussion McIntire told the young cardiologist he would seek further advice.

Using his authority as an admiral, McIntire duly convened a secret meeting with senior medical staff of the Bethesda hospital. An interim compromise was reached among them, whereby "limitation of daily activity must be emphasized," an hour's rest to be taken "after meals," no more swimming, and laxatives given to avoid "straining" of the bowels.[18]

No straining of the bowels? Bruenn was dismayed — but sworn to silence by Admiral McIntire, and told not to tell the President anything. "Appalled at what I found," Bruenn later noted,[19] he considered the President had not long to live unless digitalis was administered immediately. Nothing was said for the moment to the President himself. "He said they took X-rays & all sorts of tests," but they "found nothing drastically wrong," Daisy innocently recorded in relief in her diary that night, after the President had called her, apart from "one sinus clogged up."[20]

It was clear that the initial medical response to the President's health crisis was risible, indeed culpable — in fact, seldom in the history of medical treatment at that high a level can such willful incompetence have been displayed. However, "they are going to put him on a strict diet," Daisy added, hoping that "lemon juice in hot water before breakfast" might be the answer.[21]

• • •

How much the President truly believed Admiral McIntire or Lieutenant Commander Bruenn, who came up to see him at the White House the next morning, is unclear. He said nothing further to Daisy for five days. Nor did he confide anything to Eleanor, who, back from her Caribbean tour, merely told her children their father was suffering prolonged flu.

Meanwhile Bruenn, for his part, was unwilling to watch his new patient die. "Despite everything else," Bruenn recalled, "the need for digitalization, I thought was overriding. Said so to Admiral McIntire. Told him I literally didn't want to have anything to do with the situation unless" it was administered.[22]

The threat of such a resignation, and possible scandal if the reason got out, certainly won the admiral's attention — and that of Captain John Harper, the Bethesda Naval Hospital's commandant. Harper warned Bruenn that his bleak prognosis — especially his insistence on immediate, aggressive treatment — could end their careers.[23] Digitalis, he reminded him, slows the heart rate and allows the heart muscle to contract more efficiently, but it can have serious side effects — ones that could harm or kill the serving president, beginning with heart palpitations, hallucinations, and blurred vision.

The outcome of *not* giving digitalis, however, could be equally fatal, Bruenn protested: leading inexorably and perhaps imminently to complete "congestive heart failure."[24]

It was in this quandary that McIntire decided he must bring in bigger guns — even if they were not, strictly speaking, naval ones. He therefore called for medical advice from two "honorary consultants" to the U.S. Navy, whose discretion could be trusted: Dr. James Paullin, the former president of the American Medical Association, and Dr. Frank Lahey, founder of the famous Lahey Clinic.

Once Dr. Paullin and Dr. Lahey arrived in the nation's capital two days later, however, the two renowned physicians disagreed with each other on how best to proceed.

Together with Admiral McIntire, Captain Harper, the radiologist Dr. Charles Behrens, and poor Lieutenant Commander Bruenn, the cardiologist, the doctors reviewed the results of the "X rays, electrocardiograms, and the other laboratory data concerning the President" — but did not reexamine the President, lest they alarm him — and give rise to dark rumors.

"There was much discussion," Bruenn later recalled,[25] sitting around the table at the Bethesda Naval Hospital — for Dr. Lahey, as a gastrointestinal expert, thought there could be other problems with the patient's digestive system that might be even worse than his heart, especially in view of the President's abdominal discomfort. The collapse of the President at dinner with Stalin and Churchill in Tehran, an account of which McIntire related, was especially troubling. There was even a question about the President's prostate, which Bruenn insisted should also be examined.[26] Most of all, though, Bruenn pleaded for digitalis to be administered as a matter of urgency. Dr. Paullin felt the opposite—that digitalis might make a critical condition even worse, given how ill the President was said to be. "They thought that was all too drastic and extensive," Bruenn later recalled[27]—Paullin recommending they give no treatment for the moment, and hope for the best.

Without seeing the patient in person, though, they had nothing firm to go on, beyond the exam results in front of them. Paullin and Lahey therefore asked to see the President in person—a request McIntire, with a heavy heart, was compelled to agree to.

It was the "emergency" visit of Paullin and Lahey to the White House in the afternoon of March 31, and then again the next morning, on April 1, which escalated the growing crisis—for with the arrival of such eminent physicians the President could hardly fail to recognize the gravity of the situation. His symptoms had remained "essentially unchanged" since Bruenn's first examination—his heart so "grossly enlarged"[28] that he was still finding it difficult to move or breathe.

The evidence, moreover, could no longer be concealed: namely the President's grim appearance.

One visitor to the White House who saw the President at this time was a top newspaper reporter, Turner Catledge. He later recalled an off-the-record interview with the President he'd been granted by Roosevelt's press secretary, Steve Early. The bearing of the nation's chief executive, once proud and leonine, was diminished, the shadows under his eyes had deepened, his face more lined, the skin slack. "When I entered the President's office," recalled Catledge, "and had my first glimpse of him in several months, I was shocked and horrified—so much so that my impulse was to turn around and leave. I felt that I was seeing something I shouldn't see. He had lost a great deal of weight. His shirt collar hung so loose on his neck that you could have put your hand inside it. He was sitting there with a vague, glassy-eyed expression on his face and his mouth was hanging open. Repeatedly he would lose his train of thought, stop and stare blankly at me. It was an agonizing experience."[29] Grace Tully, taking dictation, had reported the same to the President's daughter Anna: that she had found he had "momentarily lost consciousness while he was signing a document," and that "they were all concerned."[30]

The President's condition seemed to be worsening—"he did not appear as well as he had even before. His color was poor, and he looked tired," moreover suffered "several paroxysms of nonproductive cough," Bruenn recalled. Again, "there was much discussion." The President's blood pressure had now reached 200/106. Lahey was still "particularly interested in the gastrointestinal tract but submitted that no surgical procedure" could or should currently be undertaken—for the President might well not survive it. Together with Bruenn and McIntire there was "much beating around the bush," with Dr. Paullin, the most distinguished doctor present, expressing "much skepticism," as Bruenn recalled, "despite the overwhelming evidence."[31] Lahey, however, felt "that the situation to his mind was serious enough to warrant acquainting the President with the full facts in order to assure his full cooperation"[32] — and it was thus agreed, at long last, the President would take digitalis, to save or end his life.

It was, in the event, a tribute to Dr. Bruenn's strength of character that, despite being the youngest and most junior member of the medical cabal, his insistence on administering digitalis, along with frequent monitoring by electrocardiogram to see if it was working, had finally been accepted.

It was not a moment too soon — the young doctor may well have saved the President's life. Time would tell. Meanwhile, however, McIntire deliberately sought to bamboozle the press — and the nation. The admiral's press conference on March 28, given to tamp down speculation about a president whose own recent press conferences had raised concerns about his health, was a model of what would later be called White House "spin."

How, in fact, Admiral McIntire managed to arrange the series of high-level medical consultations at the White House and at Bethesda Naval Hospital without the press getting to the literal heart of the matter, or recognizing its seriousness, would be a testament to McIntire's effectiveness as a guardian of the truth. As "the country was filled with every variety of wild and reckless lie about his physical condition," it was simply monstrous that anyone should question the need for discretion, McIntire later reflected.[33] After all, "for better or worse, the war was being run by just three men" — Mr. Roosevelt, Marshal Stalin, and Winston Churchill — "and it was not over yet."[34]

This was true — D-day still two months away, thanks to Churchill's machinations in Italy. As Daisy would note in her diary a few weeks later, however, declining to tell the President how sick he was, was futile. The President, "when he found out that they were not telling him the *whole* truth & that he was evidently more sick than they said," was not impressed. It was *his* body — and it was failing. "It is foolish of them to attempt to put anything over on *him!*" Daisy sniffed.[35]

In any event, the President wasn't fooled. Heart disease ran in the Roosevelt family. The President's own father had died of heart disease — in fact "Mr. James," as he was known, had had to resort to a sedentary, no-stress existence for the entire final decade of his life — managing to survive that way to the relatively august age of seventy-two. But in a world war? As U.S. president and commander in chief?

Four weeks after the President's revelatory examination at Bethesda Naval Hospital, Frank Knox, the navy secretary, would die of a heart attack, on April 28. Pa Watson, the President's military attaché and appointments secretary, who'd accompanied him to Tehran, had a serious heart condition; he would die of a cerebral hemorrhage early the following year, at age sixty-one. The President's youngest son, John, would die of a heart attack at age sixty-five, in 1981.

"The greatest criticism we can have," McIntire meantime told the press, "is that we have not been able to provide him with enough exercise and sunshine."[36]

A few weeks in the Caribbean, away from the political stresses and strains of Washington, were duly insisted upon. Preliminary preparations for a Caribbean journey were even made. But it would be demanding, in terms of physical movement when the President's breathing was so labored, and an air flight — the simplest form of travel — was considered a no-no in his current condition, given his failing heart. Thus when the President's friend Bernard Baruch offered the use of Hobcaw, his huge baronial estate in South Carolina, as a place of respite, the President accepted with alacrity.

It was not ideal — a vast estate, but mosquito-ridden and awkward in terms of Secret Service security, let alone military security. In its favor, however, was that it could be reached by train.

And it had another attraction. Hobcaw was a mere hundred miles or so from Widow Rutherfurd's house at Aiken, South Carolina. Given circumstances in which the President might not live very long, it would certainly have been churlish for anyone to oppose Mr. Roosevelt's personal wishes. Hobcaw's very isolation, after all, was a great advantage, in terms of recuperation. As was its distance from prying eyes, and the Washington press corps.

Thus was the President's trip to the Hobcaw Barony arranged. There, in the peace and serenity of South Carolina, he would await Overlord: its target date delayed, Eisenhower had reported to him, till the end of May or early June, owing to the shortage of landing craft.

D-day, as it was now being called: the great invasion plan he, like Hitler, was certain would "decide" the war.

PART SIX

D-day

42

"This Attack Will Decide the War"

HIGH UP IN the Bavarian Alps, at the Berghof, near Berchtesgaden, Hitler was also recuperating—suffering elevated blood pressure, cardiac hypertension, and stomach problems.

The Führer had left the Wolf's Lair, his military headquarters in East Prussia, on February 23, 1944. As the Red Army got closer, the camp had become vulnerable to air attack; it therefore required thicker blastwalls, roofs, and shelters. The reinforcing would take months. In the meantime the Führer would go, he announced, to the peace and tranquility of the mountains he loved.

Back in December 1943, Hitler had predicted "the attack in the West will come in the spring; it is beyond all doubt"—the Allied invasion accompanied by diversionary attacks anywhere between Norway and the Balkans. "There is no doubt about it, they have committed themselves. After mid-February, early March on, the attack will take place in the West. I don't have the feeling," the Führer told his senior staff, "that the British have their—shall we say—whole heart in this attack," since they wanted to keep their "divisions intact" to run their empire after the war. "If you look at India, Africa, the Far East, Australia," the Führer reckoned they must have at least "50% of their armed forces out there," and "they want to maintain that, of course, in order not to lose any territory at the last minute." The empire came first. But the signs were the Americans were insisting on launching the Second Front, however unenthusiastic the British. Assuming the cross-Channel invasion went ahead, then, it would thus be the critical moment of the war: one that Hitler claimed to welcome. "If they attack in the West, then this attack will decide the war. If this attack is driven back, the whole affair will be over. Then we can also take forces out very quickly," from France, he assured his high command staff—switching the Wehrmacht and Luftwaffe to deal with the Soviets on the Eastern Front, as well as holding any remaining Western Allied forces in the southern Mediterranean, even evicting them.[1]

A month later, on January 30, 1944, delighted by the way Churchill's invasion forces at Anzio seemed to have failed even to get off the beachhead, Hitler had given a national radio broadcast from the Wolf's Lair to remind Germans at home and in the occupied countries what they were fighting for. The Jews, he'd claimed, were responsible for the war, and they were behind the Bolsheviks, as he called them. "One thing is certain," he'd declared: "there can be only one victor in this fight, and this will either be Germany or the Soviet Union! A victory by Germany means the preservation of Europe; a victory by the Soviet Union means its destruction."[2]

The Jews, the Führer claimed, were responsible for all the ills of the world, including unemployment and the Great Depression.

The toughest struggle was already over, the Führer said in his broadcast —National Socialism having successfully prevailed in the struggle against internal and international Jewry. "Germany's and Europe's victory over the criminal invaders from the west and the east" was not only "an expression of faith for every National Socialist, but also, at the end of this entire fight, his inner conviction." The *Volkstaat* was expressed by and through the Wehrmacht— and this "front will never lose heart. Even in the hardest days, it will remain strong"—part of a "tremendous, world-shaking process" entailing "suffering and pain," yes, but which was "the eternal law of destiny, which states not only that everything great is gained by fighting, but also that every mortal comes into this world by causing pain."[3]

Hitler's ever-growing obsession with Jewry, even as the thousand-year Third Reich faced destruction after only ten years, was extraordinary. The Jewish ratio of the German population in the 1930s had been less than 1 percent. Moreover the Soviet Union, tyrannous though it might be as a political regime, had not attacked Germany—in fact had signed a nonaggression pact with the nation in 1939 in vain hopes of staving off a German invasion. Historians would therefore be at pains to understand the roots of Hitler's fanatical determination.

Allied soldiers fighting the Wehrmacht in close combat, meantime, were confronted by a conundrum. If the Führer's anti-Semitic claims were so specious and so paranoid, how was it possible that he enjoyed the support of almost eighty million Germans?[4] If he was mad, why were millions of Wehrmacht officers and men still defending their ill-gotten conquests abroad? This enigma baffled not only Allied intelligence officers interrogating German POWs, but would puzzle historians both in Germany and outside for many decades thereafter.[5]

Moving his whole high command headquarters staff with him to Berchtesgaden, the Führer was certain the Allied invasion would come in the spring, as was confirmed by German intelligence as well as Eisenhower's promotion to Allied supreme command in London. Hitler seemed to look forward to the prospect, confident the Wehrmacht in the east would continue to carry out orders from above without question, regardless of the casualties. The German army was, after all, already too deeply implicated in mass murder in Poland and then in Russia during its days of heady, ruthless conquest, along with the SS, to question the radical racial views expressed in his broadcast.

Hitler no longer even needed to appear in public, despite Goebbels's repeated pleas he should do so. He had thus not visited a single one of the cities the Allies had bombed over the past six months—including Berlin. Arriving in Munich on his special armored train with its blinds drawn, on February 24, 1944, he had, however, agreed to address the Nazi Party faithful. In the Festsaal of the sixteenth-century Hofbräuhaus, his personal appearance had been greeted as if he were the messiah.

There had been rumors swirling that he was unwell, explaining his long disappearance from public view. In Munich that evening, however, there had been no sign of ill health. "He looks wonderful," Goebbels noted in his diary, "and both physically and mentally is in great form. He gives an extraordinarily vibrant speech, in fact more so than one has heard him speak for ages." He'd given his comrades a brief overview of the war situation on the Eastern Front and in Italy, and had pointed to ever-improving German defensive success on both fronts. "He stresses that Germany's eventual victory, in which he believes more unshakably than ever, depends on Germany's toughness." If the British claimed they were paying Germany back in kind for the Blitz of 1940 and 1941, they would shortly learn the meaning of *real* revenge: for in April, hopefully, the first V-bombs or missiles would be launched on London, as soon as certain "technical difficulties" had been overcome. Above all, though, it was in northern France that the Führer saw Germany's "great new chance" in the prosecution of the war: the Führer describing to them how the invading forces of the Western Allies would be rubbed out in the same way as at Anzio, in Italy. The years leading up to the Nazi seizure of power in 1933 had been far worse, the Führer claimed, than the present situation: a time when the nation was now so clearly unified behind him. He was going to pursue his aims "remorselessly." As the Jews had been defeated in Germany, they would be defeated in every corner of the world. "The Jews in England and America" would be the next ones to face the music.[6]

Arriving at the Berghof, his once-modest mountain villa, the German messiah had thereafter been greeted by a snowstorm on February 25. And by his blond, younger mistress, Eva Braun, who was shocked by Hitler's appearance: stooped, old, his tremors worse than ever.

Standing in the midst of six square miles of military installations, fences, and checkpoints, the Berghof was shrouded in camouflage netting, but also, often, fog. And smoke from machines employed to conceal the mountain from American bombers—a technique that made Hitler's personal physician, Dr.

Theodor Morell, so sick he had asked to be allowed to stay below, in Berchtesgaden, driving up to the beleaguered Führer twice a day to administer injections of amphetamines.

To Dr. Goebbels, on March 4, Hitler had confided he was going to have to occupy Hungary, Germany's hitherto loyal ally, with Wehrmacht troops, as a result of Hungary's *Verrat*—treason, in daring to explore surrender negotiations with the enemy—which "must be punished," as Goebbels recorded in his diary. The Hungarian regent, Miklós Horthy, would be taken into custody when he came to visit the Führer, and the Hungarian army would thereupon be disarmed as an unreliable ally. Once that was done, the "question" of the Hungarian aristocracy and "above all the Jews in Budapest" could be addressed—by deportation and liquidation. "For as long as the Jews sit in Budapest," Goebbels noted, one could do nothing with "the city and the country." There were plenty of willing Wehrmacht troops on hand for the task—and the thought that a German officer would question such orders, let alone refuse to carry them out, was unthinkable.

Goebbels's forecast had proven correct. Admiral Horthy had arrived at the Berghof on March 18. He was threatened, placed in "protective custody," and then forced to agree to German occupation and the roundup of more than half a million Jews—with Romania next on Hitler's list.

The Soviets had meantime reopened a huge new offensive in the Ukraine, on an eleven-hundred-kilometer front, as the winter snows melted. But it was in the West the war would be decided, Hitler had rightly predicted—giving rise to a "war of nerves," or propaganda, that tormented Goebbels. What was the Allies' real strategy? They were clinging to the beaches at Anzio, certainly—but why, he wondered, when the campaign had "no clear strategic purpose" in Italy?

As a master propagandist Dr. Goebbels wished to fan the burning embers of anti-Semitism in Britain and America, and thus prepare the way for a negotiated end to the war, leaving Germany in control of most of the continent. For the moment, however, the propaganda minister accepted the Führer's authority and wishes as supreme commander of the Wehrmacht. Germany *had* to continue to go about its deadly business—most importantly by holding fast in the east while preparing to crush the cross-Channel attack when it finally came. Quick defeat of the invasion might even enable Hitler to mount his next offensive on the Eastern Front that very summer, for which "he needs some forty divisions," Goebbels noted in his diary—namely those in France, currently waiting for the expected Allied invasion. Those forces, backed by the Luftwaffe, he'd be able to transfer, just as soon as "we've smashed the invasion"—though where exactly the invasion would come in northern Europe neither he nor the Führer knew for certain.

There was no question but that the Wehrmacht would succeed in its assignment. It was "*absolut sicher*" — absolutely certain. "He describes to me in detail the forces we have at hand to wipe out the invasion — in fact more than enough." There were, to be sure, "a number of new divisions, inexperienced in battle; however they are filled with fantastic human material, so that we need have no concerns on that account. For example, once the SS Hitlerjugend division goes into combat, we can rest assured it'll do its duty. Moreover the Führer intends to put units of his Leibstandarte SS Adolf Hitler into the division, men who'll definitely inspire the Hitlerjugend with their energy. Other forces in France are also first-class: young, educated, well-trained for the task — guaranteeing the outcome for us in their toughness and fortitude. We're even superior to the enemy in our weaponry — especially our tanks. We have the new 'Panther' and 'Tiger' panzers, which are way better than their predecessors; even though we don't have enough of them yet, we can sprinkle them in amongst the others to produce colossal firepower and defensive capability."[7]

Though he was careful not to dictate anything in his diary that might possibly be used against him in the future, Goebbels did add one rider: namely, that he hoped the Führer's optimism would be borne out by the coming battle. "We've been so often disappointed recently that one feels a certain skepticism; but the stakes are so high, I'm sure our soldiers will acquit themselves well."[8]

For Hitler himself, the waiting for D-day at the Berghof seemed to be exhilarating, even if it made him nervous — affecting his digestion and robbing him of sleep.

Only a successful Allied invasion could save the Jews — and as Field Marshal Rommel, commanding all Wehrmacht forces defending the Atlantic Wall from Holland to the Spanish border, told the Führer, his troops would be completely ready by May 1. "The Führer is certain the invasion will fail, in fact that he can give it a drubbing," Goebbels noted proudly when he saw Hitler again in Munich for the funeral of Adolf Wagner, veteran Nazi Gauleiter of Bavaria who had recently died of a stroke. Rommel's optimism had been a blessing to him, the Führer said, for he remained "convinced" the outcome of the war hung on the impending cross-Channel assault. If it failed, President Roosevelt would not get reelected to a fourth term, he assured his propaganda minister (who didn't share this view) — in fact the recent withdrawal of the 1940 contender, Wendell Willkie, from the Republican nomination campaign proved this, in the Führer's view, for Willkie had backed Roosevelt's foreign policy.[9] No, the Americans were simply not enthusiastic about the war in Europe. Failure on D-day would thus empower the eventual Republican winner of the election to negotiate an armistice with the Reich. Meanwhile in England there was so much war-weariness that a massive German V-bomb campaign, launched in tandem with the slaughter of British troops on the beaches of northern France, as at Dieppe in 1942, would surely bring Churchill's downfall. In fact the Führer had decided, he informed Goebbels, not to launch any of the V-1 rockets that were ready that month, but to wait instead for D-day and then fire them in a concentrated attack on London.

It is easy, in the aftermath of war, to ridicule such predictions — but in the context of the time, Hitler's confidence did not seem misplaced. There might be growing resistance to German occupation across western Europe, but there was also anxiety in the Allied camp. If the Western Allies succeeded in defeating the Wehrmacht in northern France and racing to Berlin, Western democracy would be saved. But if D-day did *not* succeed? Stalin's troops could be held at bay on the Eastern Front, or even pushed back; Hitler's forces would then remain masters of Europe — able either to counterattack the Soviets or come to a negotiated nonaggression pact, as in 1939. There were 158,000 German and non-German Todt-organization workers finishing more than twenty thousand fortified posts, planting six million beach mines, and erecting half a million obstacles along the French beaches of the Atlantic Wall. Behind them were some 468,000 Wehrmacht troops in Army Group B area in northern France to immediately repel the invaders as in some medieval siege.

Hitler's military instincts *as a warrior* were thus not flawed. In his battlefield strategy, however, he continually found himself torn between competing advice from his "experts." Field Marshals Rundstedt and Rommel had distinguished themselves in smashing the French and British armies in Belgium and France in 1940. Now the shoe was on the other foot. Led by Americans, the Allied armies were about to assault the Atlantic Wall: the German version of the Maginot Line. Rommel had insisted the invasion must be defeated instantly, on the beaches — pivoting his counterinvasion forces on bombproof strongpoints, lest the Allies advance inland, at which point it might be too late to prevent a buildup of enemy infantry and armored forces, protected by massive Allied air power. But it was still unclear whether the Allies would target the Pas-de-Calais area — as they had done when assaulting Dieppe — or would invade farther southwest, in Normandy or the Cotentin Peninsula around Cherbourg. In this dilemma Hitler was faced by a conundrum: namely where exactly to position his crucial German armored or panzer divisions in order to erase in good time any weakly armed Allied beachheads before the invading forces could expand and break out.

Field Marshal Rundstedt, the commander in chief West, was wisely given control of the Fifth Panzer Army, or Panzer Group West, as well as all reserve divisions. Concerned lest the Allies open their invasion with a diversionary attack designed to draw off his main armored divisions, the Führer insisted on retaining a veto over the Panzer Army's employment, however. As the German official historian, Detlef Vogel, noted, this "panzer controversy" continued right up "until the Allied landing."[10] A compromise was nevertheless reached.

Some armored divisions were allowed to set up camps close to the coasts, but the main body of German armor was held farther inland, near Paris, to be sent into battle once the *Schwerpunkt*, the main impact, of the Allied invasion became certain.

With extensive Allied deception measures being taken (including a dummy army under Patton in eastern England, suggesting to German military intelligence that the Pas-de-Calais was the true target area), no one could be sure, whether at the various German headquarters in the field, or at Berchtesgaden and the Berghof. Only time would tell — Hitler returning by train from Munich on April 18, 1944. Once at the Berghof, he complained to his doctor of stomach pains and headaches, and turned down Goebbels's request that he make a public speech on May 1, saying his nerves were simply not up to it.[11]

Benito Mussolini, who'd been rescued from detention at Gran Sasso in a daring mission carried out by German paratroopers and SS troops, had come to see the Führer at Klessheim Castle in Berchtesgaden on March 22, leaving his mistress Clara Petacci behind while he stayed there for three days of talks.

As head of the new Fascist Republic of Salò (on Lake Garda), the Duce had appealed for better treatment of Italian POWs and forced laborers in Germany, as well as better weapons for the four new Italian Fascist divisions being trained in the Reich. He'd also attempted to interest Hitler in his ideas for the "socialization of businesses" — i.e., to give workers more nationalized power against Italian industry leaders, "the majority of whom were secretly favorable to the British," as the German ambassador to the republic, Rudolf Rahn, noted.[12]

Hitler had listened, but gave short shrift to any hopes of amelioration of Tripartite Pact relations. The Führer "has no interest in the Duce's social-economic measures," Rahn had noted — quoting Hitler saying: "We Germans must get over the habit of thinking we are the doctors of Europe."[13] Rather, they were its executioners — the Führer, with Field Marshal Wilhelm Keitel beside him, assuring Mussolini the Germans were about to launch a new U-boat campaign in the Atlantic; had new jet fighters that would blast Allied bombers and fighters out of the sky; would soon launch "revenge" missiles on London; and would crush the expected cross-Channel landings with a mixture of panzers and concrete. Besides, he'd added, he'd been rereading histories of Frederick the Great and Frederick's father; all coalitions broke apart within five years — and the Allies' time was now up.

How accurate was Hitler's prediction only the next few months would tell.

43

Simplicity of Purpose

WHAT WAS AMAZING to all at the White House, meanwhile, was that the President's spirit or morale seemed unfazed by "the trouble" with his heart, as he described it to Daisy.[1]

Dr. Bruenn's digitalis regimen had served to rescue him miraculously from the brink of heart failure. But he was still liable to lose consciousness at times, and found himself perspiring profusely for no apparent reason. Moreover his digestion, despite his new diet, had been erratic — upset in part by the digitalis, perhaps — giving him chronic abdominal pain.

Strangely, it didn't seem to matter to him. He'd found sudden, unexpected emotional happiness in his life, however fleeting it might prove. Every moment he and Lucy could be together therefore seemed a godsend. If it was not to last more than a few weeks or months, given his condition, then so be it — for this was the elixir that, together with Dr. Bruenn's medical remedy, revived him in the spring of 1944, shortly before D-day.

Churchill had cabled on March 18 saying that, after more than a year of opposition, doubt, and dire predictions if the President's insistence on carrying out D-day was maintained, he was at last "hardening for Overlord as the time gets nearer."[2] This was welcome news. It was not an apology, to be sure, but it was a confession of sorts: enough, at least, that the ailing president had been able to congratulate the Prime Minister on his change of heart — reminding him that Overlord would be not only the greatest Allied amphibious operation of the war but, if all went well, "synchronized with a real Russian breakthrough" on the Eastern Front.[3]

The President's heart remained the problem, though — one that hadn't changed. At certain times of the day he seemed to be a little closer to his normal self, but at others, far, far removed. He was still only sixty-two — the world leader with the lilting voice and jaunty smile who had not only brought the nation through the Great Depression without resort to tyranny, but had guided the country to its current status in the world: the leader of the democracies. Moreover, where Churchill offered no social, economic, or political vision beyond a deeply moral one — freedom and dignity — the President had managed to give his country a positive political and economic *vision* of the future, to be pursued in a postwar peace which the United States, undertaking a new leadership role, would help guarantee. The rights of U.S. servicemen to vote in the forthcoming presidential election would be upheld, too: the necessary "soldier voting bill" having recently been passed by Congress, and a bipartisan War Ballot Commission being currently established to ensure the necessary ballots reached millions of Americans in uniform, serving away from their home constituencies, especially overseas.[4]

The secretary of war, a Republican, had agreed to sit on the commission. The President's enveloping smile, his unfailing confidence in humanity, his humor and bigheartedness were a kind of dynamo that had kept America humming, Stimson felt — even when things had looked dire, or daunting. "The easy-going confidence of a short and quick victory which was so prevalent last autumn and winter has faded out," Stimson had nevertheless acknowledged in his diary on March 22, with the result that people "now realize that they are in for a long war and a very hard fight for the invasion."[5] This would require their president to be even more inspiring than ever. Fixed now for the end of May 1944, Overlord would be a crucial moment in the prosecution of the war, fought under a supreme commander whose appointment Stimson had initially believed was a mistake, but was now beginning to see as a stroke of presidential genius — not least because it left Marshall in Washington, by Stimson's side, and thus able to take a more and more commanding role in the Joint Chiefs of Staff Committee meetings.

D-day remained, as it had always been, the only method of defeating the Wehrmacht and the Third Reich. Italy was a dead end. Moreover, Allied bombing of Germany had resulted in such high losses in aircrew (up to 85 percent not expected to complete their allotted twenty-five missions)[6] that General Arnold worried lest American military morale might break. For all that the Allied air forces targeted military and war-industry installations, their efforts had failed to cripple German output. Instead they seemed to have intensified the enemy's willingness to "stick it out."

However frightening the predictable casualties, then, D-day and the battle for France were the only way to defeat the Nazis. After attending a rehearsal of plans at the headquarters of General Montgomery, Eisenhower's ground forces commander for the invasion, Churchill cabled the President to say for a *second* time he was becoming "very hard set upon Overlord."[7]

So the big question, as the President prepared to take the recuperative vacation his doctors felt was crucial, was this: Should he run again for the presidency, given the party convention that was fast approaching to choose a nominee? Would it be right to do so, when he was so ill? Or should he await the outcome of his R & R in South Carolina, and then of D-day itself?

The President still felt wretched most of the day, but the sedatives he was receiving from Dr. Bruenn were assuring him ten long hours of sleep each night. Even after waking he was not getting up before noon, and was dealing thereafter only with correspondence, not visitors.

Mercifully his brain, the President found, seemed still to be in good shape for part of the day—though he noticed a huge change in himself. Once the most elliptical of thinkers—one who liked to achieve compromise and consensus through friendly discussion by all parties, right up until the moment when momentum seemed to carry his decisions as on a flood tide—he now demanded simplicity of purpose. The details, he felt, would then take care of themselves.

On April 1, 1944, for example, having just seen Dr. Paullin and Dr. Lahey for a second time at the White House, he'd dictated a letter to Cordell Hull telling him he wanted no further State Department attempts to define the meaning of "unconditional surrender" of the Axis powers. "Unconditional surrender" was targeted at America's real "enemies," he said, and should not alarm those lesser actors who, as satellites, had been under the "duress" of Nazi Germany. "Italy surrendered unconditionally but was at the same time given many privileges. This should be so in the event of the surrender of Bulgaria or Rumania or Hungary or Finland," he told Hull. "Lee surrendered unconditionally to Grant but immediately Grant told him that his officers should take their horses home for the Spring plowing. That is the spirit I want to see abroad—but it does not apply to Germany. Germany understands," he cautioned, "only one language."[8]

Four days later the President had stuck to his guns, despite State Department blowback and his own diminishing strength. "From time to time there will have to be exceptions not to the surrender principle but to the application of it in specific cases. That is a very different thing from changing the principle," the President—who had trained as a lawyer—instructed the secretary of state.

Stalin, the President knew, had never been happy with "unconditional surrender." Nevertheless in Moscow the Marshal had subscribed to it, in the name of Allied unity—and he had not opposed it in Tehran. Insofar as the Russian dictator was prepared to back the policy, it was to ensure the Germans made no separate, negotiated settlement with the Western Allies. Unconditional surrender of the Third Reich had thus remained Allied policy since Casablanca—but with a less absolute adherence to the principle in the case of those countries who had allied themselves with the Third Reich, out of fear or special circumstances. The Finns, for example, had taken part in Hitler's Barbarossa offensive solely to get back the territory in Karelia and Saimaa which the Russians had seized in the Winter War of 1939–40. The United States was not even at war with Finland; the President thus welcomed word of Soviet negotiations with the Finnish government — especially since an immediate negotiated settlement would allow Stalin to turn the overwhelming weight of his armies to support the invasion of France — advancing westward, toward Berlin, not north to Helsinki, where there were virtually no Germans. By the same token, however, Roosevelt did not wish to open the door to a World War I–type outcome, which Germans of a subsequent generation might contest. "If we start making exceptions to the general principle [of unconditional surrender] before a specific case arises," he'd therefore argued on April 5, there would be problems once disaffected Germans began secretly seeking a negotiated peace — as had happened in Ankara recently.[9] The same situation as in 1918 would then arise: a future German leader or military claiming the country had never actually been defeated, but had been stabbed in the back by politicians.

Such presidential communication by memos had spoken volumes — the Oval Office becoming less and less inhabited, and less accessible. Moreover the President was currently holding out on his wife in several significant respects. Not only in the matter of his friend, the widow Mrs. Rutherfurd, but the sheer gravity of his medical condition, which Eleanor took to be just an extension of his everlasting flu — or even, possibly, his sadness at the news that their brave but erratic son Elliott was planning to get divorced, once again. "I think the constant tension must tell," Eleanor wrote to her "adoptee," Joe Lash, whom she'd visited in Guadalcanal. She put it down to "the long burden of responsibility" her husband carried, which "has a share in the physical condition I am sure."[10]

The President's "physical condition," was far worse, however, than Eleanor seemed able or willing to confront directly. "FDR is not well," she'd acknowledged, but "I think we can keep him in good health," she'd assured her young friend, who admired Roosevelt immensely. It was just a matter of cutting down his workload — meaning that "he'll have to be more careful" in future.[11] She'd shown no emotion when told by Admiral McIntire about Dr. Bruenn's examination, and the visit of Drs. Paullin and Lahey — the first time in his twelve years as president that outside specialists had entered the White House[12] — but simply accepted McIntire's assurance there was nothing too serious to worry about.

Eleanor, to be sure, was not alone in such willful unconcern. In some ways it was as if, in the first few months of 1944, almost every close aide and subordinate who gathered around the Commander in Chief was pretending to himself or herself that the President's condition was nothing too serious. Thus, even as the President's blood pressure rose steadily into the 200s, and stayed there, reaching 226/118 despite digitalis and Bruenn's ministrations, no alarm bells were rung — the implications of the President's plummeting health simply attributed to flu, bronchitis, walking pneumonia, and the tensions and burdens of the office.

Not all were as lacking in perspicacity, however. Marquis Childs, a well-known columnist, had been allowed to visit with the President at the White House on April 7. In his notes that night the journalist recorded the President speaking "in a firm voice without hesitation. His face was sallow but he appeared in good health," despite "puffiness about the eyes." He was obviously ignoring his doctor's recommendation that he should quit or cut down on his smoking, though, for he smoked "two and perhaps three cigarettes" in the single hour that Childs was with him. On the other hand, in their conversation, the President seemed now "quite deaf," Childs recorded — or perhaps was not listening. "I had the impression of the man's curious aloneness," Childs added — as if Mr. Roosevelt, the father of the nation, had been left to his own devices in the White House.

In a sense, the President had been. In the previous four days he'd accepted only a single formal appointment! When Childs, at their interview, asked about the burden of his office, the President had been stunningly honest. "I wouldn't say burden. You see I don't work so hard any more. I've got this thing simplified . . . I imagine I don't work as many hours as you do."[13]

As a columnist? The President, in other words, was deliberately changing gears. At least for the moment he seemed content to coast along: allowing his trusted administration chiefs to run the country, and only interceding when he felt, or was advised, it was absolutely necessary.

For the President — and for his doctors — this was doubtless the most sensible way forward. But the implications for others were enormous — especially the military. The White House — the powerhouse and apex of American military strategy and command since Pearl Harbor — was going quiet. The Map Room was becoming more and more of an archive, a reference library — not the busy call center of civilization. The President hardly went into the room anymore, recalled the Map Room officer, Lieutenant Elsey[14] — who had asked if, given the relative inactivity, he might join the D-day naval invasion force currently assembling in Britain, and report back on the Allied landings once they took place.

With the D-day invasion now delayed to early June, the President's condition thus left the government of the United States strangely opaque — and vulnerable — at its highest echelon. There were few people who credited the U.S. vice president, Henry Wallace, with the necessary leadership qualities to take over direction of the war, if the President died of a heart attack or was felled by a stroke. And, given current polls showing a major swing of public opinion toward a Republican candidate in the looming presidential election if the President chose not to stand again, there appeared to be little or no chance of a Democrat nominee winning. Inevitably Franklin Delano Roosevelt, the magician of the Western Allies, would be begged to stand for reelection and guide the nation to victory in 1945 or 1946; he would not be permitted to back out. Nor, at the deepest level, was the President really prepared to do so. As Marquis Childs put it in his notes, "The habit of power had grown on him. He wanted to remain as commander [in chief] until the war was won."[15]

The same day as Child's visit, Dr. Paullin and Dr. Lahey had visited the President again. The effect of digitalis had been "spectacular," Bruenn later recalled.[16] Relieved to find the President looking a bit better and sounding better than he had the week before, though still appearing deathly ill, they now agreed — but with no public mention to be made of the decision — that Lieutenant Commander Bruenn should be transferred part-time from his post at Bethesda Naval Hospital to become, in essence, the President's new White House physician: accompanying him wherever he went, and administering the toxic yet effective treatment, while taking great care by daily electrocardiographic monitoring to ensure that it did not kill or incapacitate the President.

Surprisingly, Admiral McIntire did not seem put out by this change of roles. "He knew his shortcomings," Bruenn later reflected; "he wasn't particularly interested in internal medicine or cardiology. This was out of his field. He was perfectly willing to let somebody else take it over."[17]

The President also seemed pleased with the transfer of medical responsibility, for he *did* feel somewhat better, even though he had no illusions. With virtually no visitors allowed at the White House, so few hours spent working, and even meals taken on a tray in bed, it was uncomfortably like quarantine. When speaking with Frank Lahey, after Lahey and Dr. Paullin had both looked at the President's latest medical exam results, he'd greeted the famous surgeon with the words: "You have good news for me, Dr. Lahey?" Lahey had responded that he did indeed have news. But "Mr. President," he'd continued, "you may not care for what I have to say," Lahey later recalled. "That will be all, Dr. Lahey," the President had cut him off, before the eminent surgeon could say another word. The principle was enough; the rest were details.

The President must have talked things over with his widowed friend, however, for twice that week he left the White House secretly in the afternoon with his secretary Grace Tully, to go "motoring" for an hour or more — a euphemism for collecting Lucy from Georgetown in his car, outside her sister's house, and driving incognito in the area.

If Lucy was saddened by the President's dire condition, she was not put off. She'd nursed her husband through his last years of declining health; she was familiar with the approach of death. On the other hand she could see plainly that her very presence in the car, beside him, made the President feel as if he was walking on air — so happy was he to be in her company. And to be, at least for an hour, almost literally careless.

It was thus with Lucy's emotional and moral support — and the understanding they would visit each other in person while in South Carolina — that the ailing president left Washington by train at 9:30 in the evening on April 8, destination Hobcaw Barony, where he would stay, rest, and fish for a period of two weeks at least.

If he died in the interim he would at least die happy. If he survived, he would hopefully see the summation of his war strategy: the invasion he'd nurtured for two long years, parrying everything his own chiefs of staff had done to launch it prematurely, and the British had done to stop it, or delay it to 1945. An invasion he'd gotten the Russians to promise to support with a synchronized offensive on the Eastern Front that would make it impossible for Hitler to transfer German reserves to the west. An invasion whose Supreme allied commander he had personally chosen, despite the disappointment of General Marshall and Secretary Stimson. An invasion to be launched with the might of American air power, armor, artillery, and weapons that would enable Allied forces to fight their way on to Berlin.

44

The Hobcaw Barony

HOBCAW PROMISED WARM WEATHER, but most of all an "oasis of serenity," as Baruch put it: an almost seventeen-thousand-acre former plantation.[1] Its history had tickled Roosevelt's fancy, moreover. The "Barony" had been granted by King George II to Lord Carteret, in an area situated alongside the King's Highway, or main road, between Wilmington, in North Carolina, and Charleston, South Carolina. It even boasted a British fort from the time of the Revolutionary War, which Baruch wanted to show the President.

The President was not the first chief executive of the nation to visit Hobcaw. Grover Cleveland, a governor of New York (like Roosevelt) and twice elected to the U.S. presidency (like Roosevelt), had twice stayed at the property and shot waterfowl there. The President was not up to holding a gun, though—even had he been keen on the sport. He'd been wheeled off the *Magellan* and driven along back roads by the Secret Service less to ensure his security than to prevent him from being seen in his debilitated state. But a child by the plantation gate had seemed to recognize the ancient figure in the car. As Baruch later recalled, the boy had shouted: "Gee! It's George Washington!'"[2]

Accompanying the President and his chief of staff had been only his White House physician, Admiral McIntire; the masseur of his paralyzed legs, Lieutenant Commander Fox; his naval and army aides, Vice Admiral Brown and General Pa Watson; his communications officer, Lieutenant Rigdon; his valet, Arthur Prettyman; and his bodyguard, Agent Fredericks. Also Fala, his little black terrier. Dr. Bruenn would follow later.

Before leaving Washington the President had called his cousin Daisy, confiding to her he'd changed his plans; he would now stay three or four weeks at Hobcaw, given the one-month delay the British had caused in the launching of D-day. He felt as if he had "sleeping-sickness of some sort," he was having to take so many medications and was so little disposed to work, Daisy had noted.[3]

• • •

The whole medical business "depressed and bored" the President; unfortunately things proved no better at Hobcaw, on arrival, than they had been in Washington. The more he lost weight to ease the pressure on his heart, the more "haggard" he looked, as even his doctor admitted.[4] The weather remained cool. It rained. He had breakfast in his room at 9:30, sat in the sun, if conditions were right, at 11:00 a.m. Took lunch and then a nap, followed by some desultory fishing on the dock by the river. Only a single cocktail before dinner with his staff, then early to bed. In a scratchy hand he wrote to Daisy that he was "really feeling no good."[5]

For the commander in chief of more than ten million American soldiers, sailors, and airmen, and at a moment when the Western Allies were making their final preparations for the deciding battle and campaign of the war, it was a strange, almost otherworldy experience: as if he had already died and was now watching from afar. "Every day the bag would come in from the White House with the mail," Dr. Bruenn later described, "and all the papers would be signed. That took about half an hour or so and that was all the business that was done."[6]

Given that the President had, until then, been the energetic, dynamic conductor of an orchestra embracing the entire U.S. administration, as well as its military, the change was momentous, Dr. Bruenn later reflected. "You see, the President *was* the government," he told an interviewer. "He was his own Secretary of the Treasury, his own Secretary of State. He was running the works, including the war."[7]

It was a retrospective exaggeration, of course, yet it did convey a measure of the President's overriding authority as America's "boss," after three terms in office. Whether he could ever recover his magnetic, inspirational skills was doubtful, however — his doctors either forbidden to look into the crystal ball, or say too much about his current condition. In conversation with the President, Dr. Paullin had been more circumspect than Dr. Lahey. Searching for a metaphor, he likened his new patient to the driver of a once-fast car. It was the President's body that was showing "definite signs of wear and tear," after such a long career on the road. The car's engine was knocking. As Paullin had warned, "If you want to finish the journey, traveling the last ten thousand miles without mishap, you can't keep up any seventy-miles-an-hour clip . . . In plain words, you *must live within your reserve*." To this the President had responded, with his trademark laugh: "Well, I'll agree to quit burning up the road."

And he had. He'd reluctantly but dutifully agreed to Dr. McIntire's new regimen, in which he was to see virtually no visitors and do no more than four hours of work a day.[8]

Half an hour, resting at Hobcaw, was not even that.

• • •

It was perhaps a tribute to Roosevelt's choice of administration officials and selection of his military chiefs of staff that the President's absence from Washington did not result in a collapse of government, or a military mishap, other than the failure at Anzio, for which neither the President nor his chiefs of staff bore responsibility. D-day, though delayed as a result, was still on track—and arousing unprecedented hope and determination among the troops who were training to carry out the great operation. Thus, although men like Secretary Stimson had often cursed the managerial weakness of the Roosevelt administration owing to the President's uniquely personal approach to presidential leadership—keeping the ultimate reins of every department in his own hands, but refusing to have clear chains of command—the period before D-day passed off remarkably uneventfully in Washington. It seemed, in fact, as if the administration was running on a kind of autopilot.

With a presidential election looming later that year, however, autopilot was no guarantee of American victory. Could Vice President Henry Wallace take the reins? As the former secretary of agriculture and a former Republican, Wallace had deliberately been put forward (indeed insisted upon) by Roosevelt as his vice presidential nominee in 1940, thus garnering important constituencies in the election. But once world war had commenced for the United States following the attack on Pearl Harbor, Wallace, as chairman of the Board of Economic Warfare as well as the Supply Priorities and Allocation Board, had not prospered. He'd feuded with competing officials. In terms of the anti-Axis alliance, as an outspoken anticolonialist he hadn't been popular with Churchill, either; moreover he'd had the wool pulled over his eyes by the NKVD, Stalin's secret police, on a tour of Russia, it had transpired. And he was certainly not admired by his cabinet colleague Henry Stimson, whose role as secretary of war was critical, not least in bringing development of the atom bomb to combat readiness before the enemy did.

The likely Republican presidential nominee would be New York governor Tom Dewey. If he won the election in November, Dewey, at only forty-two, would become president in January 1945—a man not regarded as the sort of national leader who, if the war was not over by then, could be trusted to head an international coalition of United Nations fighting the Axis powers. Thanks to the inevitable swing toward a Republican after three full terms of a Democrat in the White House, however, Dewey's chances of winning were growing stronger by the day—assuming, that is, Mr. Roosevelt declined to run. Ergo, Roosevelt, it was argued by many, would simply have to soldier on, despite his health issues.

Should—could—the President stand for a fourth term? This was the abiding, Hamletian question. The President, unfortunately, was still without advice from his loyal White House counselor and companion, Harry Hopkins, who, since Tehran, had been out of action, and out of state, passing the winter mostly in Florida—in and out of hospital with a suspected recurrence of stomach cancer.

As usual, however, the acerbic Hopkins put the conundrum most vividly. When his doctors finally decided by majority vote (5 to 3) to perform abdominal surgery on him once again, in April, Hopkins was heard to say, as he was taken into the Rochester Clinic operating theater: "O.K. boys, move right along. Open me up; maybe you will find the answer to the Fourth Term, or maybe not!"[9]

Maybe not.

Mercifully, after five days in early April the President began to feel somewhat better, even though he still looked a sight—his systolic blood pressure remaining over 200—and continued to suffer stomach pain. On "the fifth evening he spent a while in the drawing room after dinner" instead of going to bed, Lieutenant Rigdon recorded.[10]

Admiral Leahy, on behalf of the President, stoically continued to deal with what he called in his diary the "unnecessary number of telegraphic messages from Washington and from the dispersed war areas"—messages to which, after discussing the more important ones with the President, "I send replies via the White House." These at least could be dispatched unsigned.[11]

Where the President's genuine signature was required, however, Mr. Roosevelt had to dictate replies to Lieutenant Rigdon, who was a stenographer, "whether he felt like working or not," as Rigdon noted with compassion.[12] It was clearly an unsatisfactory way of running a world war, yet the President was the President; it was up to him to discover how much he could really manage, despite the strain on his heart. Meantime, it was important he should not be seen by anyone in his current state of physical dilapidation. The press—a group of White House correspondents staying at an inn in nearby Georgetown, South Carolina—were therefore kept strictly away, to their disappointment and no little skeptical curiosity about not being permitted anywhere close to Hobcaw. As Admiral McIntire later admitted, though, this deception necessarily came at the price of open democracy. It didn't really work, however—for the President's very absence from Washington, and word of his health issues rippling through the city's press corps, merely gave rise to wild speculation. It was "during the South Carolina stay, when bronchitis was the President's one and only trouble apart from fatigue," Admiral McIntire later wrote, still bluffing, "that the country filled with every variety of wild and reckless lie about his physical condition."[13]

Dr. Bruenn, who'd had to reorganize his department at Bethesda Naval Hospital so that he could join the President, arrived on April 17. The President's "blood pressure remained elevated," he wrote in his clinical notes; in fact it exceeded 230 over 126. His "heart remained enlarged," Bruenn openly admitted.[14] This necessitated giving still-higher doses of digitalis, which in turn affected the President's digestion: a seesaw of ills between his chest and stomach. The President nevertheless forced himself to sit with General Mark Clark on April 18, on a trip home the exhausted general had made at General Marshall's insistence — Marshall wanting both to give moral support to his top American combat commander in the Mediterranean, in the wake of the Anzio fiasco, and for the President to hear, firsthand, "the present military situation in his area," as Leahy recorded in his diary.[15]

Writing after the war, Clark recalled how the President "showed a surprising knowledge of details, and was quick to offer ideas as I explained our plans for reaching Rome"[16] — for Roosevelt well remembered the Anzio stretch of coast south of the capital from his childhood. It was not the same president who'd awarded Clark the DSC in the field in Sicily, only four months earlier. Nor could the President offer any encouragement in terms of an advance north of Rome, once the city was reached — for Roosevelt and Leahy both felt that mounting Anvil in southern France, in order to help support D-day and the march across France, was the only real way to defeat the Third Reich. Italy would essentially be a holding front, to keep Wehrmacht forces away from the main theater.

By April 21, however, the President did feel physically well enough to go on an excursion to the nearby Arcadia Plantation. On the twenty-second the President's little fishing party cast their rods from a patrol boat, some fifteen miles into the Atlantic. The catch of bluefish and bonito was excellent, and the "President thoroughly enjoyed" the afternoon, Leahy recorded, as of an invalid — which, in effect, the President now was.[17] A trip to Myrtle Beach followed, by car, then Belle Isle Gardens, and on April 25[18] Mrs. Roosevelt, her daughter Anna, and Anna's husband, John Boettiger, as well as John Curtin, the Australian prime minister, and his wife, and the president of Costa Rica, all arrived by air from Washington — staying only for the afternoon. It was Eleanor's first and only visit to see her husband in South Carolina — and all the failing president could manage, for he was husbanding his modestly reviving strength for another visitor: Lucy Rutherfurd.

Together with her stepdaughter, stepdaughter-in-law, and her stepgrandson, Mrs. Rutherfurd arrived on April 28 from her own estate at Aiken, 140 miles away — Bernard Baruch having even surrendered his gas ration so they could make the drive. For, as Baruch saw it as the President's friend and financial adviser over many years, it was Lucy Rutherfurd's job to try and cheer up the President at the luncheon. Which, gracefully, Lucy did immediately, seated at his right — Franklin having personally sketched the seating arrangement before the party even arrived.[19]

Baruch, handsome and genial at seventy-three, sat at the other end of the table. Even the President's long-awaited tryst did not last long, for at 1:30, while they were having dessert, news came that the secretary of the U.S. Navy, Frank Knox, had died of a heart attack.

Neither Admiral McIntire nor Lieutenant Commander Bruenn was in favor of the President traveling back to Washington for the funeral. Especially when, that afternoon, the President had another gastric attack, involving "stomach pains, nausea and tremors."[20] Dr. Bruenn was able to give the President an injection of codeine, so that he could at least meet briefly with three of the reporters staying in nearby Georgetown.

In the sitting room at Hobcaw the President was able to give them, in person, a statement they could publish, honoring Knox's service to his country. Moreover he related to the reporters how, when "Mr. Knox came to the White House and announced to him that the Japanese had attacked Pearl Harbor" on December 7, 1941, the navy secretary had said immediately: "With your permission I'm leaving in the morning." The President had asked Knox where on earth he intended to go. When the navy secretary had said Hawaii, the President "asked him what he could do there." Mr. Knox had replied: "At least I can find out a great deal more than here." Within two days the secretary had reached Honolulu, the President told the reporters, and on the third day he'd telephoned "to report to the President, and to suggest to him that he organize an investigating group right away—not experts but common sense people who have the confidence of the country. The President said he followed the suggestion of the secretary in naming [Supreme Court] Justice [Owen] Roberts to head the Pearl Harbor investigating board. The episode, the President related, was typical of Knox."[21]

It was also typical of the President—his story encapsulating one of Knox's finest hours. The President sounded unusually emotional. Had his own altercation with the Reaper, the month before, turned out differently, he might have preceded Colonel Knox—who had suffered a major heart attack—to the grave. "I've been told people of superior breeding never let their emotions come to the surface publicly," one of the journalists, Merriman Smith of United Press, later recalled of the President's closeness to tears as he told the story—his grief unusually visible.[22]

Meantime Dr. Bruenn, alarmed by the severity of the President's abdominal pain, wanted an exam performed to check for gallstones, in Washington, as soon as possible. Even this he recognized was simply out of the question, however, until the President's heart condition improved.

How, then, could the President *think* of running for reelection—something that would require strenuous travel, public appearances, and the whole panoply of a presidential campaign? It seemed inconceivable. And unnecessary, in terms of public sympathy for his plight; his doctors (other than Admiral McIntire) would have happily explained to the world why it was impossible, lest the President keel over like Secretary Knox, and die on the hustings.

Daisy Suckley arrived at Hobcaw on May 4, to help take care of the President and provide feminine company. Under "his tan" he "looks thin & drawn & not a bit well," she noted in her diary—adding he "feels good-for-nothing, had just had some sort of an 'attack 'which seems to be in the upper part of the abdomen. He says they don't know what is the matter with him"—or that they didn't want "to tell him."[23]

But the President already knew.

45

A Dual-Purpose Plan

ROOSEVELT STAYED AT HOBCAW till May 6, 1944. In his final weeks there his health had seemed to improve a little — the President "in a much better physical condition than at the time of his arrival," Admiral Leahy summarized in his diary the day they departed by train back to Washington.[1]

The next morning, May 7, the President was in the White House once again — free to see Lucy. His blood pressure remained high, but his stomach seemed temporarily at peace.

In Washington there was no shortage of people now urging Roosevelt to commit to a fourth term, however — including Robert Hannegan, chair of the Democratic National Committee. In his capacity as the President's physician, Lieutenant Commander Bruenn was appalled, yet fatalistic. He later recalled not only the President's high blood pressure, but the "pressure" that was "put on him to run for the fourth term. All those people around him depended on the President exclusively for their jobs, for their reputations, everything. I'm not only talking about the secretaries but such people as Steve Early, the press secretary, everybody. The President was the center pole, no question about it. He wasn't particularly anxious to run," Bruenn recalled with compassion. Roosevelt felt he'd done his job — that he'd deserved a rest, now that the war was on clear course to victory, hopefully by the end of the year in Europe, at least. Yet the issue of running for reelection remained a moral one for him, rather than medical, and "he never asked any questions" or the advice of his doctor.[2]

To maintain the public fiction of bronchitis, if he did decide to run, McIntire would not permit Bruenn to be photographed with the President, let alone be seen visiting his patient. The lieutenant commander was expected to continue running his electrocardiology department at the Bethesda Naval Hospital, without his colleagues even being aware of his "temporary additional duty."[3] None of the President's advisers Bruenn did meet at the White House, indeed no one there, "asked me whether he should run or not," he recalled. "I was Mr. Anonymous most of the time" — though he was given a car so that he could "drive down in the morning" from Bethesda to the White House "and see him, then go back to work at the hospital."[4]

The President "never asked me a question about the medications I was giving him, what his blood pressure was, nothing. He was not interested," Bruenn recalled — the President seemingly content to leave that aspect of his life in others' hands.[5] And though Bruenn had undoubtedly saved the President's life in March, not all of his efforts had a salutary effect, at least in terms of side effects. Overdigitalization wreaked havoc with the President's sense of taste and digestion, while Bruenn's urging him to lose weight was a double-edged sword — the President's weight being all in his upper body, leaving "nothing" in his paralyzed legs, which resembled sticks more than ever. The treatment had certainly taken stress off the President's heart, but "some of it came off his face" also, Bruenn admitted in retrospect. As a result "he began to look haggard. And then we had a job trying to get him to eat again."[6]

Privately Bruenn was of two minds about whether the President should run. "I must say, in all honesty, if I had been asked what my opinion or judgment was," he later reflected, "I would have been greatly swayed by the circumstances. Here we were in the middle of a great war, which had been conducted fortunately or unfortunately on an almost personal basis between Stalin, Churchill, and Roosevelt."[7] To take Roosevelt out of the equation seemed . . . unwise.

The President, for his part, remained equally unsure. Eleanor was of little help — not for lack of compassion, but because she was herself under constant stress. Her attitude toward his plummeting health thus remained one of distant concern. As First Lady and mother of a large family, she had never had time for sickness in the family. "She had a toughness about physical ailments, always minimizing her own," her friend Joe Lash would later write — quoting a letter she'd written, a few days after her visit to Hobcaw, in which she'd claimed "F. looks well but said he still has no 'Pep.'" Yet Eleanor, uniquely among those close to the President, had seen no cause for alarm — it was more tiredness and stress than anything, she thought. "He ought soon to get well," she predicted, if he would only follow her advice and go to Hyde Park, where he could rest, spending only "two or three days a month" in the White House "during the summer months."[8]

In the midst of a world war?

By constrast Mrs. Rutherfurd, who had nursed her husband through his last years, could read the dark signs in her visits with the President. Yet she, too, saw no way out of his standing again for election. Declining to serve, in war, seemed somehow like cowardice.

The President, in any event, spent as little time as possible in Washington. D-day was scheduled for early June, to take advantage of the tides. He had no wish to interfere — as reports from London indicated Churchill was doing, yet again. The Prime Minister was said to be a bundle of nerves, alternately excited and depressed — in fact the ground forces commander, General Montgomery, threatened to resign unless the Prime Minister stopped trying to meddle in matters over which he not only had no idea, but threatened to break the chain of command.[9]

D-day was now in the hands of the combat commanders and their men. Field Marshal Brooke might express contempt, in his diary, for the President's choice of supreme commander — a post earlier promised to Brooke. General Eisenhower was "a swinger and no real director of thought, plans, energy or direction! Just a coordinator — a good mixer, a champion of inter-allied cooperation," Brooke noted sourly after attending the final rehearsal of D-day plans in London on May 15.[10] It was a monstrously obtuse comment, since with almost electric energy Eisenhower and Montgomery had in four months transformed a plan over which Brooke had backpedaled since its inception, failing to give it the priority and support that he, the head of the British Army, should have provided. By the time Eisenhower and Montgomery had taken charge, even Overlord's own Anglo-American planning staffs were predicting defeat.[11]

Fortunately Ike and Monty, not Brooke — or Churchill — would decide the battle.

In Washington, thousands of miles away, there was meanwhile increasing anticipation and nerves — which the President did his best to ignore. In some respects his heart condition and reduced workload rendered him mercifully immune. He now had the utmost confidence in Eisenhower, in his combat generals, and in his American forces. Ever since the reversal at Kasserine they had been learning what worked, and what didn't, against a formidable enemy. In Normandy there were no mountains or high ground, mercifully, to contend with; American armor would be free, once ashore, to confront the Wehrmacht, backed by huge air power and naval guns. He was sure they would acquit themselves well, and he had no wish to provide anything but encouragement to the supreme commander he had personally appointed.

On May 11, 1944, the President motored to Shangri-la, his retreat in Maryland's Catoctin Mountains, with Anna and Colonel John Boettiger, her husband, and only returned on May 15, full of confidence in the outcome of the invasion — telling Admiral Leahy, in fact, to look into the possibility of his crossing the Atlantic once the D-day landings had, hopefully, secured a firm bridgehead in northern France. He also wanted to go on an "inspection cruise to Alaska and Honolulu leaving Washington July 23 for Seattle, thence by destroyers to Alaska, by Cruiser to Pearl Harbor and return via San Diego, Los Angeles and San Francisco, reaching Washington about 20 August."[12]

France? Hawaii? Alaska?

Admiral Leahy was both puzzled and amazed at the President's new plans. Normandy posed too many dangers for a presidential trip. But with the U.S.

Joint Chiefs in undisguised disagreement over future strategy in the Pacific (currently divided into two theaters: the Central Pacific under Admiral Nimitz, and the Southwest Pacific under General MacArthur), only the President, as commander in chief, could resolve the impasse. And in theory the best way for the President to cut the Gordian knot would be to go out to the Pacific and meet with his two top commanders there. If his health allowed.

The President was never simple to decipher, however, as Leahy knew. The upcoming Democratic convention to choose a presidential candidate would hold its ballot of delegates in Chicago on July 20. Was the proposed trip along the West Coast, before and after the President's return from Honolulu, for military reasons — or *political*? Sitting with the President in the Oval Office, Leahy delicately asked the projected trip's "bearing on the approaching political campaign."

"The President replied with much feeling," Leahy noted in his diary: "'Bill, I just hate to run again for election. Perhaps the war will by that time have progressed to a point that will make it unnecessary for me to be a candidate.'"[13] But if not, he would run for a fourth term.

Leahy, who'd accompanied the President to Hobcaw and back home, was stunned. "While I have long been sure that the President would like to retire from his present office," Roosevelt's chief of staff noted, "this is the first time he has expressed himself to me clearly in regard to his attitude toward renomination."[14]

The next day the President went on to clarify his plans. He dropped all notion of flying or sailing to Britain — unwilling to face Churchill's likely demands that, with the impending breakout of Mark Clark's forces from Cassino toward Rome, the President's Anvil landings should be scrapped and more Allied punch be put into Italy. Not wanting to confront yet another Churchillian machination, the President decided he would stay put in the United States. As soon as the D-day landings had been launched, and were successful, he would go in the opposite direction. He would take the *Magellan* to Chicago with Admiral Leahy, then continue across the country to San Diego. From there he would sail to Pearl Harbor, returning via the Aleutians.

The President had, in other words, an itinerary in mind. A dual-purpose plan of how, in very limited health, he would not only decide Allied strategy in the Pacific "on the spot" — deciding which should be the main axis of advance to ensure the defeat of Japan — but would, if asked, agree to stand for the presidency a fourth time: though only as current U.S. commander in chief in war, without time to commit to campaigning. He would thus be — like Hitler — largely out of public sight.

Assuming, of course, D-day proved a success. And his health held.

46

D-day

D-DAY — DELAYED UNTIL the necessary landing craft were *in situ* — now loomed larger and more fatefully than ever. "How that event hangs over us — has been hanging over us, for months," Daisy Suckley noted in her diary on May 19 at Hyde Park. The President seemed relatively better — but "relative" was a big word. He "seems pretty well but not right yet," she added truthfully,[1] and almost every day she worried about his demeanor. He looked "tired, and his color is not good" on May 22; "very tired at dinner" on May 26; "seems tired" on June 4.[2]

Fewer than eight days in the lead-up to the D-day invasion were spent at the White House of the twenty-two — the President staying either at Hyde Park, or Shangri-la, or at the hundred-acre estate of his longtime military aide, General Pa Watson, at Charlottesville, Virginia. The tan he'd acquired at Hobcaw had long faded. "From my observations of these last 2–3 weeks," Daisy summed up, "it looks to me this way: Two or three or four days in Wash.[ington], when he is rather keyed up," and which "get him over-tired. He gets away, & for the first 2 or 3 days it is a question of getting relaxed. The next day or two he is getting rested, & then the whole process is repeated — Just how long this is to continue only time can tell," she penned. If the good spells were to outweigh the bad, "he will be all right, otherwise he will get sick again."[3]

Why run again for the presidency, then? When Daisy asked whom he would choose as his vice president, Roosevelt was coy. "I haven't even decided if I will run myself," he asserted — suggesting that a gung ho, no-nonsense character like the shipbuilder Henry Kaiser would be a better choice as president, for he'd be tougher and more like "the Churchills, Stalins, etc" of the world.[4]

This was typical Franklin Roosevelt — two steps forward, one back. But when Daisy pressed him and asked what, then, "is going to decide you," given that he was pretty much nominated by common consent of the Democratic Party, he admitted it was his health that would decide the matter. For, as he put it, "it wouldn't be fair to the American people to run for another four years" if he knew he couldn't "carry on" for a full term.[5]

Fair? For a full term?

This, too, was typical FDR—for he knew he had no hope of actually recovering, given his heart condition. In fact it was highly questionable whether he would make it to the end of his *current* term, let alone be well enough to undertake a new one. Two hours of appointments in the morning, and that only when working—which was only a third part of every month? Half an hour's correspondence in the afternoon? Could such a part-time position, by any stretch of the imagination, be considered a valid U.S. presidency—and in wartime? Compared with the responsibilities he'd carried over the past eleven years, the notion was a joke. And yet: as Daisy herself noted, by taking so much time off and limiting his workload to bare essentials, "he is getting slowly but steadily better," she felt (or convinced herself)—something that caused her to "hope and fear, with millions throughout the country, that he will run for a fourth term. The world needs him."[6]

Eleanor felt the same way. She spent only two days with him that month, and at Hyde Park—Daisy noting they were "the first two days they have been alone together for years." But it was Lucy who really counted.

The President had seen Mrs. Rutherfurd again, collecting her from her sister's house in Georgetown and taking her motoring. It was Lucy who persuaded him he must stand. He had become a symbol of hope and purpose, across the world. Referring to the endless burdens of his presidential office in an undated letter (the only letter that would survive posthumous attempts to conceal the relationship, after the war),[7] Lucy had addressed Franklin as "poor darling." She'd ended her missive with a sad reflection. "I know one should be proud—very proud of your greatness instead of wishing for the soft life, with the world shut out," she admitted. "One is proud and thankful for what you have given the world and realizes how much more must still be given this greedy world—which never asks in vain."[8]

This was now a destiny, in the midst of a global struggle, he could simply not walk away from. "You have breathed new life into its spirit—and the fate of all that is good is in your dear & capable hands."[9]

Lucy, in short, expected him to do the right thing by the nation, indeed by the world. And thus, despite all his qualms, the President accepted his fate. In the meantime, D-day grew closer. After meeting with a group of forty congressmen in the Oval Office, and dining quietly with his daughter Anna and her husband, John, as well as Daisy, the President was wheeled into the Diplomatic Reception Room on June 5, 1944, to address the nation.

Not about D-day, however—which was being launched at that very moment, he knew, via the Map Room—but about Italy.

"Yesterday, June 4th, Rome fell to American and Allied troops," the President began in a firm voice — his face, on film, looking lined, the skin beneath his eyes dark, and his neck unusually scrawny as he sat hunched forward at the microphones placed on his desk. Behind him stretched shelves filled with leather-bound tomes.

The Eternal City had been liberated, he announced, "by the armed forces of many Nations" — American and British armies foremost, but "gallant" Canadians, too, as well as "fighting" New Zealanders, "courageous French," and "French Moroccans, South Africans, Poles and East Indians" also — all of them fighting "with us on the bloody approaches to the city of Rome." Rome, the "great symbol" not only of the Roman empire in classical times, but latterly the "seat of Fascism." And of "Christianity, which has reached into almost every part of the world." Italy was, as the President put it, "a great mother nation," one which across the centuries had furnished "leaders in the arts and sciences" — "Gallileo and Marconi, Michelangelo and Dante" — thus "enriching the lives of all mankind." He was convinced the Italian people, following the unconditional surrender of their government the year before, would soon be a "peace-loving nation," and were capable of democratic "self-government" — in fact, some Italian forces, after the surrender of their government to the Allies, were now contributing to the "battles against the German trespassers on their soil." [10]

This brought Roosevelt to military strategy in Europe. "From a military standpoint," the President explained as America's commander in chief, "we had long ago accomplished certain of the main objectives of our Italian campaign": "the control of the sea lanes of the Mediterranean to shorten our combat and supply lines, and the capture of the airports of Foggia from which we have struck telling blows on the continent." Italy, in other words, was not the objective.

Reading the transcript in London, Churchill would, the President knew, be made uncomfortable. The advance on Rome, the "first of the Axis capitals" to fall, was a significant achievement, certainly, in that it had forced Hitler to send more forces south "at great cost of men and materials to their crumbling Eastern line and their Western front." But it "would be unwise," the President cautioned listeners, "to inflate in our own minds the military importance of the capture of Rome."

The fact was, the forces of the United Nations had not, in all candor, inflicted sufficient losses on the Third Reich "to cause collapse" of Hitler's Third Reich. Unlike the Italians, who had folded their hand upon Allied invasion, "Germany has not been driven to surrender. Germany has not yet been driven to the point where she will be unable to recommence world conquest a generation hence. Therefore, the victory still lies some distance ahead," the President warned. "We shall have to push through a long period of greater effort and fiercer fighting before we get to Germany itself. The distance will be covered in due time — have no fear of that. But it will be tough and it will be costly, as I have told you many, many times."[11]

At that very moment tens of thousands of Allied paratroopers were already in the air above the English Channel, and even more troops — almost 150,000 — were embarked in vessels crossing the 120-mile expanse of stormy water. Upon that landing, he was aware, hinged the fate of Europe and the world.

He went to bed shortly after 11:00 p.m., but did not sleep. "No word yet of the invasion," noted his cousin Daisy — who was staying at the White House — "which the P. says is starting tonight."[12]

Forty-five minutes after midnight an announcement came on the radio "from Germany" — monitored in Washington — that the invasion had begun, Daisy began her diary entry for Tuesday, June 6, 1944.

A few minutes later, "German radio says landing forces are battling at Le Havre, that German warships are fighting Allied landing craft. No Allied confirmation." There were reports, too, that Calais and Dunkirk, in the Pas-de-Calais, were being attacked by strong Allied bomber formations — but that no seaborne troops had yet been landed there. "'They say' it might be," Daisy recorded, "an Allied feint."[13]

It was. Around 3:00 a.m. Eastern Time in America, Supreme Allied Headquarters in Portsmouth, England, gave out a statement that was broadcast on radio. "Under the command of General Eisenhower, Allied naval forces, supported by strong air forces, began landing Allied armies this morning on the northern coast of France."[14]

It was now official.

Whether the invasion would succeed — especially in view of the almost gale-strength weather over the English Channel — was uncertain. In his pocket, when walking from his sleeping trailer to Southwick House, his forward headquarters near Portsmouth, General Eisenhower had a folded sheet of paper announcing — if it came to that — the failure of the invasion.

The President, receiving reports sent up by the Map Room and listening to the radio in his bedroom, could do nothing. Eventually he too went to sleep, knowing the coming day would be strenuous — General Marshall and his fellow Joint Chiefs of Staff coming to see him, as arranged, at the White House at 11:30 a.m.

Had he made the right decision in appointing young General Dwight Eisenhower to command the operation?

The President knew from his own sources there had been nasty arguments over the bombing of the bridges, railways, and coastal areas of northern France, given the inevitable French civilian casualties this would entail. The British, for their part, were understandably concerned over the reverse: namely the concrete launch ramps that had appeared in profusion near the French coast, ready to unleash Hitler's dreaded revenge weapons, or V-1 flying bombs. To Eisenhower's frustration the RAF had in fact recently been directed to divert major resources to put them out of action—almost ruining the Allied "Transportation Plan" to interdict Wehrmacht reinforcements that would be sent to meet the landings. The President had turned down Churchill's requests to intervene, however, and had backed Eisenhower to the hilt.[15] Churchill had been compelled to accede—still desperately hoping that, despite his cable to the President only two weeks earlier about his "hardening" to the Overlord project,[16] matters in Italy would crown his months of hope, four and a half months since launching Anzio; and that he could, at the very least, get the Anvil invasion definitively canceled, thereby releasing landing craft for further ventures in the Mediterranean.

All the President could meantime say to newspaper and radio correspondents as they collected in the Oval Office at 4:10 that historic afternoon was: "I think the arrangements seem to be going all right . . . Up to schedule."[17]

The Oval Office was crammed tight with a multitude of reporters, secretaries, and assistants sitting cross-legged around his desk—some 180 of them, "jammed" in, as the White House press secretary remarked.[18] It was the President's 954th such meeting—one he began with the very latest dispatch from General Eisenhower, reporting the loss of just two destroyers and one LST (tank landing ship), and barely 1 percent air losses.

Asked how long he had known the date of the planned invasion, the President confided the approximate date had been settled at Tehran, but the exact date "just within the past few days," owing to the awful weather. He had, however, been informed, even as he made his broadcast the night before on the fall of Rome, "that the troops were actually in the boats—in the vessels—on the way across" to France. Asked why it had required six months since Tehran to mount the landings, he questioned his questioner. Had he ever seen the English Channel?

On hearing he hadn't, the President explained how "roughness" was the simple answer—"considered by passengers one of the greatest trials of life, to have to cross the English Channel." It was therefore "one of the greatly desirable and absolutely essential things" to be sure—like Julius Caesar and later Duke William, the Norman conqueror of Britain, who had abandoned his first attempt—of "relatively small-boat weather, as we call it, to get people actually onto the beach. And such weather doesn't begin much before May."[19] With that he finally confided to them that D-day had actually been postponed by a further day, precisely because of the unpredictable weather.

Gently, the President also explained why "we didn't institute a second front a year ago" when "politicians and others" began "clamoring for it." It was "because their plea for an immediate Second Front last year reminds me a good deal of that famous editor and statesman who said years ago, before most of you were born, during the Wilson administration, 'I am not worried about the defense of America. If we are threatened, a million men will spring to arms overnight.' And of course, somebody said, 'What kind of arms? If you can't arm them, then what's the good of their springing to something that ain't there'?"[20]

He had, in other words, delayed the invasion until the United States was ready — in arms, and in battle-hardened men and commanders. The military strategy of a Second Front, he shared with his audience, went back to December 1941 — in fact to the Victory Plan — a leaked copy of which the isolationist Republican publisher of the *Chicago Tribune,* Robert McCormick, had, in an act of rank treason, deliberately revealed in order to try and embarrass the administration. "But there were so many other things that had to be done, and so little in the way of trained troops and munitions to do it with, we have had to wait to do it," he explained, off the record — the matter coming "to a head — the final determination — in Cairo and Teheran. I think it's safe to say that." Indeed he even proceeded to share with them how, although Stalin had been "yelling for a second front" (as one questioner phrased it), the Russian premier's "mind was entirely cleared up at Teheran, when he understood the problem of going across the Channel; and when this particular time" — originally intended to be May 1, 1944 — "was arrived at and agreed on at Teheran, he was entirely satisfied."[21]

Which brought the President back to what he had warned against in his broadcast the previous evening: overconfidence. D-day had now begun; the "whole country is tremendously thrilled" — "a very reasonable thrill," certainly, but "I hope very much that there will not be again too much over-confidence, because over-confidence destroys the war effort." It was "the thing we have got to avoid in this country. The war isn't over by any means. This operation isn't over by any means. This operation isn't over. You don't just land on a beach and walk through to Berlin. And the quicker this country understands it the better."[22]

The President was asked how he was feeling. Since he had been up most of the night, he said "I'm a little sleepy."[23] Amid laughter this ended the historic press conference.

He wasn't going to be able to rest, however — for there was one more thing he needed to do that night: give his own version of King Henry V's Saint Crispin's Day speech in Shakespeare's famous play, but this time as a prayer.

Wisely, while at the Watson estate at Charlottesville, Virginia, the previous weekend, together with his daughter Anna and her husband, the President had asked the general, his host, to help him draft a radio address.[24] Since General

Eisenhower would probably make a military announcement as soon as the landings took place but before the outcome was clear, the Boettigers had suggested the broadcast take the form of a prayer.

It was an inspired idea — one the President had never tried before. True, he had often pronounced his Christian faith in public, and without inhibition. A proud Episcopalian by birth and upbringing, he saw Catholics very much as fellow Christians. He respected the pope as a spiritual leader, and had been a good friend to the archbishop of New York, Cardinal Spellman, for many years. In February 1940 he'd openly and "heartily deprecated" Nazi and Soviet "banishment" of religion — knowing, though, "that some day Russia would return to religion for the simple reason that four or five thousand years of recorded history have proven that mankind had always believed in God in spite of many abortive attempts to banish God."[25] Later that same year he'd written that, "in teaching this democratic faith to American children, we need the sustaining, buttressing aid of those great ethical religious teachings which are the heritage of our modern civilization. For 'not upon strength nor upon power, but upon the spirit of God' shall our democracy be founded."[26]

"Without any questions, *he* writes his speeches," Daisy had noted in her diary — "Whoever 'helps him' is doing the 'mechanical' part, possibly suggesting a helpful phrase, reminding about a small point, etc. The P. is just as ready to accept a suggestion that sounds right, as he is to reject it if he doesn't agree with the other person. His clarity of mind is amazing, as is his open-mindedness."[27]

Having made his final revisions, and with Anna at his side and Daisy present, the President delivered his latest radio address from the Diplomatic Reception Room at 8:30 p.m. on June 6, 1944, as troops of the United Nations hunkered down for the night ten miles inland from the shores of Normandy. Leaving the dead to be buried behind them, the survivors had battled their way across five main beaches and established a thin but firm series of beachheads between Ouistreham and Grandcamp — noncontiguous beachheads stretching sixty miles along the coast, which needed to be stitched together before the great battle could be fought to victory. Or stalemate; perhaps defeat. Either way, it would decide the war in Europe.

"My fellow Americans," the President began — his voice slow, measured, a trifle sad. "Last night when I spoke with you about the fall of Rome, I knew that at that moment troops of the United States and our Allies were crossing the Channel in another and greater operation."[28] There were long gaps in his delivery, as if the President were not reading but speaking from the very depth of his ailing heart, and as a fellow parent. "In this poignant hour I ask you to join me in a prayer," he asked — a prayer he proceeded to read even more slowly: thoughtfully, compassionately, humbly. Several sentences he'd deleted, some words he'd added at the last moment, and a number of phrases he'd compressed, during the day: "our religion" had been inserted before "our civilization," and "a suffering humanity" had taken the place of "millions of other human beings." "Rent by noise and flames" he'd used to replace "split by fire of many cannon." "Terrible violences" had been cut to just "violences."

The effect was spellbinding in its almost funereal resignation to the prospect, and necessity, of killing on such a vast scale, in order to prevent more killing on an even vaster one:

> Almighty God: Our sons, pride of our Nation, this day have set upon a mighty endeavor, a struggle to preserve our Republic, our religion, and our civilization, and to set free a suffering humanity.
>
> Lead them straight and true; give strength to their arms, stoutness to their hearts, steadfastness in their faith.
>
> They will need Thy blessings. Their road will be long and hard. For the enemy is strong. He may hurl back our forces. Success may not come with rushing speed, but we shall return again and again; and we know that by Thy grace, and by the righteousness of our cause, our sons will triumph.
>
> They will be sore tried, by night and by day, without rest—until the victory is won. The darkness will be rent by noise and flame. Men's souls will be shaken with the violences of war.
>
> For these men are lately drawn from the ways of peace. They fight not for the lust of conquest. They fight to end conquest. They fight to liberate. They fight to let justice arise, and tolerance and good will among all Thy people. They yearn but for the end of battle, for their return to the haven of home.
>
> Some will never return. Embrace these, Father, and receive them, Thy heroic servants, into Thy kingdom.
>
> And for us at home—fathers, mothers, children, wives, sisters, and brothers of brave men overseas—whose thoughts and prayers are ever with them—help us, Almighty God, to rededicate ourselves in renewed faith in Thee in this hour of great sacrifice.

The President's D-day Prayer, as it would become known, ended not with the first-drafted phrase "So be it," but with another. Words this time from the Lord's Prayer: "Thy will be done, Almighty God. Amen."[29]

47

The Deciding Dice of War

AT THE BERGHOF there had been no prayers, no press conferences or radio addresses. Hitler had slept until midday on June 6, after staying up half the night talking to Dr. Goebbels.

Goebbels had arrived at Berchtesgaden the day before. He'd come straight from Nuremberg — the great gathering place of the Nazi Party in the 1930s, as filmed by Leni Riefenstahl in her documentary peon to Hitler, *Triumph of the Will.* There, in the medieval city, Goebbels had given a one-hour speech at the Adolf-Hitler-Platz, before an ecstatic crowd estimated at sixty to seventy thousand. The propaganda minister was now looking forward to seeing the Führer.

Hitler looked "radiant," Goebbels noted in his diary, after arriving, moreover seemed in great good humor — unfazed by the fall of Rome, which he thought completely insignificant in terms of the war, which he believed would be "decided in the West," i.e., in France. "I'm so pleased the Führer sees things so realistically and down-to-earth," Goebbels went on. "Were he to become discouraged, it would have a calamitous effect on his staff, and on the whole nation. Thank God that's not the case. In any case, the [Italian] Fascists have given away their spiritual and political center. People are saying Mussolini's authority has reached bottom with the fall of Rome." Had the Italians fought to defend Sicily, the summer before, none of this would have happened, of course. "We need to be clear," Goebbels concluded: "we Germans alone have the power to defend Europe against the plutocracies and Bolshevism."[1]

As far as the impending Allied invasion of France was concerned, though, the Führer was "full of confidence." Field Marshal Rommel, who had telephoned several days before and was hoping to visit the Führer from his home in Bavaria the next day, had "filled him with high hopes"[2] — such high hopes, in fact, that Goebbels admitted to being a trifle skeptical. For in the mountains of the Obersalzberg on June 5, the Führer seemed eerily remote from conditions in northern France. "Up here one only sees the war in its highest direction; the middle and lower echelons are virtually absent," Dr. Goebbels noted.[3]

Their talk, over lunch that day, had wandered over art and theater, as well as the postwar, "showing the Führer's bigness of mind and extraordinarily deep imagination. He is now convinced it will be impossible to come to an arrangement with Britain. He thinks England is doomed and he has decided, if he has the opportunity, to give it the kiss of death. I'm not clear how, but in the past he's put forward thousands of plans that seemed absurd, but which came to pass." The two men walked down to the Berghof teahouse, alone. Goebbels was driven back to Berchtesgaden, but had returned for dinner in the Berghof, after which he and the Führer had watched the latest official newsreel, and had talked more about movies and theater — Hitler's mistress, Fräulein Braun, impressing Goebbels with her insight and critical judgment. They had sat together till 2:00 a.m. in the lounge, exchanging memories, the Führer asking "about this and that. In short, it was just like the old days."[4]

But these were not the old days. At 10:00 p.m. on June 5 reports had already been received from Wehrmacht headquarters in Berchtesgaden that Allied radio transmissions indicated the great invasion might at last be beginning. "I didn't at first take them seriously," Goebbels noted frankly in his diary. When he'd finally left the Berghof and reached Berchtesgaden — after staying several hours at the home of Martin Bormann, head of the Nazi Party Chancellery and secretary of the Führer, on the way — he found the reports were being confirmed. If so, then "the deciding day of this war has dawned," he recorded in his diary. It was June 6, 1944. "I snatch a couple of hours of sleep — for I expect the coming day will bring plenty of cares and problems."[5]

Goebbels was not wrong in his assessment of Hitler's mood, and improved health. The months of rest and mountain air seemed to have done wonders, both for the Führer's body and for his morale. When Goebbels joined him on the afternoon of June 6 at Klessheim Castle for a visit by the Hungarian prime minister, Döme Sztójay, the Führer seemed delighted the invasion had begun in earnest. The fact that, according to monitoring of BBC radio, Mr. Churchill had already that afternoon in Parliament boasted of a victory was considered a sign of weakness. After all, the Prime Minister had congratulated himself and his armies at the start of the great battle for Crete in the spring of 1941 — and had lived to regret it, when the Wehrmacht had trounced the British within days. Since then he had tended to boast only after a battle was over, not before, Goebbels jeered.

The "great battle that will decide the war is finally at hand," the propaganda minister reflected. "I notice something I've observed many times in the past during big moments of crisis: that the Führer's on edge as the crisis unfolds, but once it reaches its climax, it's as if a huge weight has been lifted from his back."[6] Now that the great day had come, Hitler seemed "utterly exhilarated"[7] — as was Field Marshal Göring, who joined them at Klessheim. The vast armies would now be locked in combat, as the deciding battle was joined: the Wehrmacht having a four-to-one advantage in initial numbers, and a plethora of first-class tanks, including Tigers and Panthers. It was, to be sure, uncertain whether the landings in Normandy were a feint, designed to draw away the German panzer reserves from a possible main assault in the Pas-de-Calais area; the Führer therefore authorized Field Marshal von Rundstedt, headquartered at Saint-Germain, outside Paris, to release only two of his top armored divisions to make for the Normandy beaches that afternoon, holding back the rest.[8]

If the Americans could be stopped from reaching Cherbourg, they'd be cornered in the fields of Normandy, feint or no feint. The Wehrmacht would then have no trouble "rubbing out" the lightly armed paratroopers who'd dropped inland. Low cloud and nasty weather would make RAF bomber sights "unusable" below a hundred meters.

"Göring has almost won the battle already," Goebbels noted of the airman's bravado in his diary; "by contrast I feel we need to be careful, in fact extra careful" with regard to what he called "the politics of news." "We don't want to talk of the scalding soup we're going to serve up to the English, but rather, that this is a serious, historically decisive confrontation, a matter of life or death." The Führer agreed with this more sober approach. "If we defeat the invasion, the whole picture of the war will change," Goebbels predicted, the Führer having "no doubts about the outcome."[9] Rommel had called from his home, and was on his way to his headquarters at La Roche-Guyon, on the Seine; von Rundstedt had never left his own headquarters in Saint-Germain. At the "Little Chancellery" in Berchtesgaden, the chief of operations, General Alfred Jodl, was "convinced" they would crush the invasion. Himmler had great hopes for his SS divisions — one already speeding toward the battle, and more to follow. All would be well.

Suppose President Roosevelt had arranged with Stalin that the Russians mount a similar attack on the Eastern Front, though — stabbing the Wehrmacht in the back? "Will there be an offensive in the East?" Goebbels wondered. "If Stalin wants to coordinate his operations with those of the Americans and British, this would be the time," he remarked.[10]

It was a wise concern. At the castle Hitler bade farewell to Dr. Goebbels, expressing kind words. The propaganda minister was touched. "It's amazing with what certainty," Goebbels noted, "the Führer believes in his mission."[11]

Outside, the valley was covered in thick fog, and there was driving rain. "As I leave Salzburg the situation in the West looks no clearer. Our Panzer forces will soon be in action. But they're not there yet. I'm waiting on tenterhooks for our reserves to be committed. In Salzburg I learn from our propaganda department that the whole of Germany is waiting feverishly. One is aware the deciding dice of the whole war are rolling."[12]

"We've suffered so much bad luck these past two years we surely deserve a bit of good luck," Goebbels mused as he made his way back to Berlin. "To be weighed down by so many cares! A major battle like this grinds at one's nerves."[13] And in a kind of forlorn hope he added: "How nice it would be if, for once, Lady Luck would smile again."

Lady Luck did not oblige. Goebbels meanwhile disdained reports the President of the United States had read a prayer on radio, as night fell over Normandy. "To what depths of shame and sham will this pet poodle of the Jews not sink?" he asked — astonished that the "Jewish press" in America would print, on their leader pages, words from the Lord's Prayer — something inconceivable in Germany. "I find myself shaking with disgust when I read such things."[14]

The President, however, felt the prayers of the American nation were being answered — or could, at a minimum, help citizens in the United States find comfort as American troops poured into Normandy to reinforce those who'd breached the vaunted Atlantic Wall. Places like Pointe du Hoc would become synonymous with heroism. The struggle to get ashore on Omaha Beach had been especially bloody, yet casualties on D-day, it transpired, had not exceeded twenty-five hundred killed — far below what had been projected. All were aware, however, the numbers would go up inexorably in succeeding days, as Wehrmacht forces overcame their surprise at the magnitude and location of the assault.

The numbers of Allied troops debouching across the five beaches also rose — the units striking immediately inland, just as Rommel had feared. Moreover they were backed by massive Allied air support, as well as heavy naval guns firing from vessels offshore, using radio-equipped spotters on land. Bayeux was reached on June 7; Saint-Mère-Église on June 9; and on June 10 the U.S. Second Armored Division came ashore across Omaha Beach. Carentan fell to U.S. forces on June 11, and by June 12, six days into the battle, some 326,000 Allied troops were in Normandy, with four of the Allied beachheads joined — the day Generals Marshall and Arnold and Admiral King came ashore with General Eisenhower to assess, with pride and gratitude, American progress.

To the President, Marshall radioed a brief report — and his own view of what was to come. "Morale of all our troops and particularly higher commanders, is high. Replacements of men and materiel are being promptly executed throughout the US beachhead. I was very much impressed by the calm competence of 1st Army Commander Bradley, and by the aggressive tactics of his corps commanders," especially General "Lightning Joe" Lawton Collins. "Our new divisions, as well as those which have been battle tested, are doing splendidly and the Airborne Divisions have been magnificent" — as were beach personnel, and the Navy's temporary, portable floating Mulberry harbor. About the man the President had chosen to be supreme commander, he had nothing but praise. "Eisenhower and his Staff," he radioed, "are cool and confident, carrying out an affair of incredible magnitude and complication with superlative efficiency. I think we have these Huns at the top of the toboggan slide, and the full crash of the Russian offensive should put the skids under them. There will be hard fighting and the enemy will seize every opportunity for a skillful counter stroke," he predicted, "but I think he faces a grim prospect. Releases and estimates from General Eisenhower's Headquarters have been and should continue to be conservative in tone. The foregoing is my personal and confidential estimate."[15]

This was exactly what the President was hoping to hear — aware that the Wehrmacht could well become even more fanatical when facing defeat than in victory. In Oradour-sur-Glane, for example, a detachment of Hitler's SS Das Reich Division, ordered north to take part in the battle of Normandy, had already massacred and burned alive some 642 of the village's inhabitants — 200 women and children deliberately locked and incinerated in the local church on June 10: a bitter warning to the people of France of atrocities to follow, as Waffen-SS units sought retribution among the weakest and most defenseless civilians for the gathering Allied success on the battlefield.

Even Stalin was impressed by accounts of the Allied assault. On June 13, at his weekly press conference, the President was able to read to reporters in the Oval Office a personal signal from the Russian marshal, via the U.S. ambassador in Moscow: "The history of war has never witnessed such a grandiose operation, an operation Napoleon himself had never even attempted."[16]

The President was also able to share with them a cable from General Eisenhower, which "came in yesterday." In it Eisenhower paid tribute to the courage and training of American, British, and Canadian troops, many of whom had been "committed to battle for the first time," and who had "conducted themselves in a manner worthy of their more experienced comrades" in the invasion — men who had learned how to "conquer the German in Africa, Sicily, Italy."[17]

The President's dogged strategy for engaging and defeating the Third Reich was working: seasoned troops alongside virgin soldiers, fighting in Normandy under a young supreme commander Mr. Roosevelt had personally appointed, and who was proving himself to be almost Rooseveltian in his ability to harness individuals of different talents, nationalities, and personalities under his leadership.

48

Architect of Victory

OVER THE ENSUING DAYS the Battle of Normandy now became, as had been predicted, a trial by fire. It was a battle General Montgomery, commanding the Allied armies, was well equipped to fight, with his British and Canadian troops holding back Hitler's massed Panzer Army forces on the left flank of the bridgehead, while General Bradley's U.S. First Army pressed farther inland, orchard by orchard, as well as striking westward to cut off the Cotentin Peninsula, in order to gain possession of the crucial port of Cherbourg — with General Patton held in readiness, as the President had promised, to be unleashed with American armor, once the Allies were ready.

By June 22, 1944, Cherbourg was ready to fall to the Allies — which, despite orders from the Führer to fight to the last man and put up suicidal resistance, it finally did, four days later. Reading the battlefield reports, Secretary Stimson finally confided to his diary that he'd been wrong: that keeping Marshall in Washington in the "position of Chief of Staff rather than to take command in the invasion" had been the right decision by the President, after all. "Now that Eisenhower, who was appointed to the commandership, is doing so well," Stimson recognized, belatedly, "we can afford to have Marshall in supreme command at home where he can see the whole field and throw his influence in every part of the global warfare."[1]

With the Normandy beachhead secure and the Allied armies advancing inland, backed by armor, artillery, and close air support, Stimson excoriated Churchill for having been so "strongly against" Overlord.[2] In doing so, however, the war secretary conveniently erased from his memory how hard he himself had pressed the President for the assault to be mounted in 1942,[3] when it would have been crushed, and how, yet again in early 1943, he had pleaded that an invasion of France be mounted that year, before U.S. forces had scarcely fired a shot against the Germans, only against Vichy French.[4] Nor did he mention how he had bet the President that Torch, the President's "great pet scheme" to invade Vichy-held Morocco and Algeria to crush the Germans in Northwest

Africa, while the British advanced from Egypt in November 1942, would fail.[5] In the flush of victory on June 22, Stimson was thus of a mind to see himself, not the President—who had "wobbled over the lot at different times"[6]—as the great architect of Allied victory.[7]

More than any individual alive President Roosevelt had in truth been responsible for ensuring that Overlord would be mounted successfully in the spring of 1944—an achievement, sadly, that would not be recognized in the President's lifetime. Or indeed afterwards, as others claimed the glory.

Recognition, however, was not something the President sought, nor was he in a state to think about it a great deal in June 1944. It was the soldiers, the airmen, and the brave naval forces who deserved the greatest credit, the President felt—for they had proven they could defeat the vaunted Wehrmacht in combat, on the level playing field of Normandy—laying down their lives to end the Nazi nightmare in Europe for all time.

In a moment of concern lest his "unconditional surrender" policy be an obstacle to German capitulation, the President had suggested before D-day a possible message to the German people, to be broadcast by the leaders of the United States, the Soviet Union, and the United Kingdom, once the battle in Normandy and in Russia was joined. Such a message could point out the futility of further bloodshed, when Germans were opposed by the majority of the world's population of nearly two billion people. "Every German knows in his heart, Germany and Japan have made a terrible mistake," he'd suggested as its logic; they should be warned to "abandon the teachings of evil," since the "more quickly the end in the fighting and the slaughter the more quickly shall we come to a decent civilization in the whole world."[8]

Churchill, often sleepless at the prospect of a Dieppe-style fiasco in Normandy, had rightly been unimpressed. Evoking noble intentions on the one hand while wielding the sword of Damocles with the other? Despite his vaunted rhetorical skills the Prime Minister was reluctant to turn such a suggestion into an Allied appeal anytime soon. It would look like a "peace feeler," he warned the President, rather than an attempt to convince ordinary Germans of the folly of fighting to the last man in defense of tyranny. Moreover it raised weighty questions, such as what was to become of Germany after such a surrender—questions that had only been cursorily addressed at Tehran, in terms of reparations, borders, and the integrity or division of Germany itself. Besides, as Churchill had added in one of his characteristically tart turns of phrase, "nothing of this document would get down to the German pillboxes and front line to affect the fighting troops."[9]

Time and again Churchill had been infamously wrong on strategy. But in this case, the President had concluded, the Prime Minister was probably right:

no good would be served by being "nice" in advance of complete victory. Every yard would be contested by a Wehrmacht not only guilty of evil, but wedded to evil in the many countries it had so ruthlessly conquered or occupied, such as Hungary earlier that spring, and even in Germany itself—especially with regard to Jews, gypsies, Russian prisoners, and political opponents. Appeals to "ordinary Germans" to convince them of Allied moral sincerity, shortly after launching a massive, historic invasion, would do no good, and seemed somewhat strange, given the President's implacable insistence upon unconditional surrender. A mark, possibly, of the President's declining health, about which Churchill had been warned by John Curtin, the visiting Australian prime minister. And the absence of trusted political advisers close to the President, like Harry Hopkins, who was still convalescing.

In any event the idea of such an appeal to save further bloodshed had been quietly dropped. The killing would have to go on in Normandy, mano a mano, until the battle was won. In terms of higher strategy, what was important in the President's mind, in any case, was to ensure that victory in France, when it came, did not result in the Allies becoming mired in new ventures that the impetuous prime minister was once again pressing to advance, behind the scenes—especially in the Mediterranean.

At the end of May, as the liberation of Rome had finally approached, Churchill had admitted to the President that his Anzio operation to seize the city had finally paid a "dividend six months later than I hoped."[10]

It was probably the nearest Churchill would ever get to a confession of error, the President considered. Moreover the belated dividend was something of a Pyrrhic victory, as the President gently pointed out in his Rome broadcast, for Field Marshal Kesselring's forces had merely retreated to a new defensive Trasimene Line along the Orcia River north of Rome—leaving the question of how best to use Allied strength in the Mediterranean. It was a new battle the U.S. chiefs of staff would be facing with their opposite numbers in London: landings in the Bay of Biscay or the south coast of France to support the Overlord battle (Anvil), or further offensives in Italy and Yugoslavia, even the eastern Mediterranean.

For the moment, the President simply withdrew from decision-making. The truth was, he was hors de combat—physically and mentally. He felt he'd done his part in ensuring D-day was launched, together with Stalin's impending offensive in the East. Whether he could undertake more than an hour or two's work a day, let alone accept Churchill's invitation to come to Britain, rather than Normandy, was doubtful. "How I wish I could be with you to see our war machine in operation!" the President had cabled Churchill on June 6[11]—knowing Winston himself was desperate to cross the Channel and witness at least some of the action in a battle he had done so much, for so long, to prevent, but which was now real. Yet the President knew he couldn't manage it. The American "war machine" would have to work on its own. Allowing Roosevelt, in the meantime, to focus his own limited energies on the next "deciding battle" of the world war: Allied strategy in the Pacific.

And whether, in fact, to run for a fourth term.

49

To Be, or Not to Be

LIKE THE PRIME MINISTER, the Führer had for his part become determined to get closer to the battle, given how much hinged upon its outcome. On June 13, 1944, he had given orders, as he'd promised Goebbels he would, for the first massive barrage of *Vergeltungswaffen,* or revenge weapons, to be loosed: targeted indiscriminately on London. Either they would force the Allies to attempt a second, perhaps larger, invasion, which he could with ease "rub out" this time in the Pas-de-Calais with the panzer forces he'd held back, or the bombs would discourage the Allies from attempting such a direct Channel crossing.

The weapons proved frightening and bloody — but inconsequential in military terms. No such actual landings in the Pas-de-Calais had been contemplated by the Allies, in any case, beyond deception measures — Operation Fortitude — which had made the threat of an assault on that coast seem real, and had mercifully kept Hitler's reserve of armored and infantry divisions away from Normandy. The Wehrmacht had thus wasted vital days waiting for landings that never came. Furious, four days later the Führer flew to Wolf's Glen, his heavily protected advance headquarters at Neuville-sur-Margival, near Soissons. In an angry meeting with Field Marshals Rommel and Rundstedt on June 17 he'd called for urgent counterattacks to be launched — just as General Marshall had predicted. These, too, failed — as Marshall had also predicted.

Then, on June 22, came the blow that Stalin had formally promised the President and the Prime Minister at Tehran, if the Western Allies would only commit to a spring Overlord: Operation Bagration, the Belorussian Strategic Offensive Operation, employing more than half a million Russian troops, five thousand tanks, and five thousand aircraft. The nightmare Hitler and Goebbels had always feared — major offensive war against the Wehrmacht on two converging fronts — had finally come to pass.

Now, unable to draw reinforcements from Germany or the Eastern Front, the Wehrmacht armies in France were doomed. By the end of June, the German line in Normandy was "stretched almost to breaking point," the deputy chief of the operations staff, General Walter Warlimont, later described. The Allied "invasion had succeeded," he chronicled. "The 'second front' — or rather the third — was established. Hitler might now have thought back to his statement that an Allied success in the West would decide the war and have drawn the necessary conclusions from it; but instead he was to be seen in front of the assembled company at a briefing conference, using ruler and compass to work out the small number of square miles occupied by the enemy in Normandy and compare them to the great area of France still in German hands."[1]

It was an extraordinary image — the Führer trying to convince himself and his military staff that Germany still occupied more territory in France than the Allies. "Was this really all he was capable of as a military leader? Or did he think that this elementary method" — a ruler and compass — "would have some propaganda effect on his audience? It was a sight," Warlimont wrote with bitterness, "I shall not readily forget."[2]

The President, by contrast, declined to interfere in the battle — which Eisenhower and Montgomery were conducting with steely skill. Montgomery's three battle phases — "the Break-in," "the Dogfight," and "the Break-out" — were now backed by tactical air support on a scale never seen before: an Allied version of Blitzkrieg. But for his part, the President's primary attention had now switched to the Pacific — where the first major bombing raid on the Japanese islands had been carried out on June 15 by B-29 Superfortress planes. On June 19, in waters east of the Philippine Islands, the Japanese navy had, moreover, lost three of its aircraft carriers (the *Taiho, Shokaku,* and *Hiyo*) and more than *six hundred* planes, in a new, large-scale naval battle: a spectacular example of modern naval combat by air and submarine forces that became known as the Battle of the Philippine Sea — and was soon nicknamed "the Great Marianas Turkey Shoot."[3]

Looking back over recent months, Churchill, in a cable to the President on June 23, admitted he could not "think of any moment when the burden of the war has laid more heavily upon me or when I have felt so unequal to its ever-more entangled problems" — problems that now caused him to "greatly admire the strength and courage with which you face your difficulties, especially in a year when you have, what I may venture to call, other preoccupations" — namely the looming party conventions to choose presidential candidates.[4]

But the President's "strength and courage" — as well as his relative silence — masked a much simpler, direr truth: that his days were numbered.

Aboard the *Ferdinand Magellan* the President had gone to Hyde Park on June 15, 1944. There he had stayed at his home for a week, accepting no appointments or undertaking any work, beyond minimal correspondence.

None of the President's close advisers were with him, not even Admiral Leahy. It was as if the President was, not to put too fine a point on it, in a funk. On June 20 he openly confessed to his cousin Daisy he did not know what to do — or rather, what he would choose to do, in terms of reelection. He knew he was dying, but not how long he might have — and this the doctors would not say. Visiting Daisy and members of her family and friends for tea, Daisy's friend Renee Chrisment found the President, at Daisy's baronial house, Wilderstein, "looking much older than last year, but otherwise pretty well," if without "his usual energy & vitality & effervescence." Daisy was more observant, however. "I notice," she confided in her diary, that for all his show of normality and his still-blue eyes, he looked lonely in a corner of the Wilderstein porch where "he sits rather tiredly on his chair, and you can see his heart thumping beneath his shirt."[5]

There was party trouble in Texas, the President confided to Daisy — the Democratic Party tearing itself apart over civil rights in a nasty sign of their chances in the November elections, unless they could pull together behind the President's vision of the future. If the Democratic Party were to split, thanks to the threatened walkout of the Dixiecrats, the President said to her, he simply "will not run." "There would be no justification for a fourth term if he wasn't unanimously demanded," the President's cousin noted. Which brought her thoughts back to an earlier president of the United States, before a similarly important election, in 1920.

"I pray the P. does not have to go through a period of illness and disappointment like Mr. [Woodrow] Wilson," she noted, thinking of the twenty-ninth president, who had suffered a massive, paralyzing stroke. The prospect, for her beloved thirty-second president, seemed ominously similar. "Here we are, only a month away from the De.[mocratic] Convention & the P. doesn't know if he will run, or not."[6]

Book Two

PART SEVEN

The July Plot

50

A Soldier of Mankind

FROM SAN DIEGO to Pearl Harbor is a distance of 3,022 nautical miles—longer, of course, on a ship zigzagging to avoid detection by Japanese submarines.

The President had not visited Hawaii since 1934, when he'd made the voyage also by heavy cruiser. This time his task was to decide, as commander in chief, between two competing strategies for attacking and defeating Japan: that of Admiral Nimitz and that of General MacArthur.

On July 8, 1944—the day the Allies began carpet bombing German forces in Normandy in order to break the cordon of Wehrmacht and SS divisions defending Caen and the direct route to Paris—the President had made a reluctant but momentous decision as to whether he should run for a fourth term. But not about whom he should take as his running mate.

The matter was far from straightforward. On July 5 he'd shared with Daisy the reports he was getting from Democratic Party stalwarts that Vice President Henry Wallace did not have enough national support to stand successfully either as the presidential or vice presidential nominee. "The opposition to Wallace has become very strong & active in the last 6 weeks," Daisy had noted of his remarks, "and something has to be done about it."[1]

Was the President himself moving toward an irrevocable decision? Daisy had wondered. He remained, she thought, "rather uncertain in his mind about himself: whether he is strong enough to take on another term—whether he ought to try it when the future is so difficult—On the other hand, whether it is not his *duty* to carry on, as long as he is able . . . Terrible decisions to have to make . . . for he *has* to make them, himself. Sometimes, he looks fine, other times he looks thin, & pale and old. If one could only *do* something to help him. The doctors are so *half*-efficient with their modern methods."[2]

It was one of the President's doctors, in the end, who had the courage to speak up—though not to the President himself.

Dr. Frank Lahey saw Roosevelt at the White House for a "private consultation" on the afternoon of July 5 — and told Admiral McIntire, afterwards, that he felt strongly the President should *not* run for a fourth term.

Lahey was well aware of the gravity of what he was recommending — in fact he would draw up a formal memorandum several days later, written in the first person, to document what had transpired. Lahey was one of the most distinguished medical professionals in America. He'd not only personally examined the President but had reviewed his medical files, his X-rays, and lab results going back several years. To Admiral McIntire, a fellow doctor, he'd therefore stated frankly "that I did not believe . . . if Mr. Roosevelt was elected again, he had the physical capacity to complete a term." As he recapped in the memorandum he dictated for the record, ever since Roosevelt's return from Tehran, the President had been "in a state which was, if not in heart failure, at least on the verge of it." The President's high blood pressure, and the undoubted coronary damage he had suffered, now made it certain he would soon be felled — in fact "he would again have heart failure," like the episode in March, and would therefore "be unable to complete" another term as president.[3]

"Admiral McIntire was in agreement with this," Dr. Lahey's memorandum recorded, and the admiral assured Lahey he'd conveyed his negative counsel to the President, as well as Lahey's warning that "if he does accept another term, he had a very serious responsibility concerning who is the Vice President."[4]

The President, then, had been officially — or professionally, at least — warned *not* to run. And the President, according to McIntire, fully understood the advice and the implications of not heeding it.

Dr. Lahey's warning was significant, Admiral McIntire had been aware, not only because it confirmed what he and the President already knew, but because it would be increasingly hard, if not impossible, to keep the matter of Roosevelt's health status quiet if he did choose to run — especially with Governor Dewey openly campaigning on the issue of "the age of the members of the cabinet" and "F.D.R.'s health, etc.,"[5] as Daisy noted, having listened with the President and his daughter Anna to candidate Dewey speak in a radio broadcast. A final decision, therefore, must clearly be made.

At 6:20 p.m. on July 8, 1944, after a rare swim in the White House pool, the President was thus driven to 2238 Q Street in Georgetown, where Lucy Rutherfurd was staying with her sister Violetta. Collecting Lucy in person, he brought her back in his car to dinner at the White House — his daughter Anna and her husband, John, joining them there for the first time.

Watching her father as "he talked" with Lucy, Anna "realized this woman had all the qualities of giving a man her undivided attention. She certainly had innate dignity."[6] After dinner they'd all watched a movie, then Anna and John had retired, leaving the President and his former lover alone together until Lucy left, shortly after midnight.

This was not what the doctor had ordered — but may well have continued to preserve, if not save, the President's life. He was now phoning Lucy, according to the chronicler of Roosevelt's last year in office, every day to talk. Despite his veritable death sentence he seemed happy[7] — which was not something his daughter Anna could object to. Or even felt she wished to. He was changing before her eyes: not only in how he looked, but in how he acted. General de Gaulle had visited the President on July 6, at FDR's reluctant invitation, and been given a formal state dinner on July 7 at the White House — where, overcoming his long animosity, the President had treated the proud, arrogant French leader with extraordinary politeness and charm, so that de Gaulle had departed in good spirits.

Some profound shift, then, had taken place in the President's mind, or spirit. Fate had dealt the harshest of blows, at age sixty-two, to his physical heart — but in other ways he currently seemed the happiest man alive. He was up early (for him), the next day, July 9, collecting Lucy once again from Q Street, and with her passed the entire day at Shangri-la, his beloved mountain retreat, returning only at 10:30 p.m.

Thus was the fateful decision discussed and reached, with Lucy's blessing — and against Dr. Lacey's firm advice, for good or ill.

Two days later, at eleven o'clock on the morning of July 11, the President told Steve Early to lock the doors of the Oval Office just as soon as the White House correspondents and other journalists were safely in. It was to be his 961st press and radio conference — and one of the most consequential.

To reporters the President seemed in a surprisingly jovial mood compared with earlier conferences, at the time of D-day. Given that there were close to a hundred people present, he said he himself had asked for the doors to be secured in order to avoid a stampede to the exit once he made his announcement. For he had, he'd stated with a smile, some important news to share with them.

As the President explained to reporters, he'd recently spoken with Robert Hannegan, the chairman of the Democratic Party, at the White House. From Mr. Hannegan he'd received an official request to stand for reelection as president of the United States, since the "clear majority of the delegates to the National Convention" were, "by certified numbers in primaries," now requesting him to run for a fourth term as the Democratic nominee.

Reporters' pens raced across their notepads as they took down the President's words. "This action," he said, reading aloud Hannegan's written request, "is a reflection of the vast majority of the American people that you continue as President in this crucial period in the nation's history." The chairman was therefore tendering "to you the nomination of the Party," the President continued, using his pince-nez, "as it is the solemn belief of the rank and file of Democrats, as well as many other Americans, that the nation and the world need the continuation of your leadership." In an oblique reference to rumors of ill health, Hannegan had expressed his confidence "that the people recognize the tremendous burdens of your office, but I am equally confident that you must continue until the war is won and a firm basis for an abiding peace among men is established."[8]

The President then read out to reporters the reply he'd sent Hannegan the previous night — a message that, he admitted, was "hurried." He'd felt he owed Mr. Hannegan, as he put it, "a simple statement of my position. If the Convention should carry this out" and vote to nominate him for reelection, "I shall accept," he'd written. "If the people elect me, I will serve. Every one of our sons serving in this war has officers from whom he takes his orders. The President is the Commander-in-Chief and he, too, has his superior officer — the people of the United States."[9]

The President added one qualification, however — namely that, "for myself, I do not want to run. By next spring I shall have been President and Commander-in-Chief of the Armed Forces for twelve years." He paused there. "All that is within me cries out to go back to my home on the Hudson River, to avoid public responsibilities, and to avoid also the publicity which in our democracy follows every step of the Nation's Chief Executive."

He paused again. "Such would be my choice. But we of this generation chance to live in a day and hour when our Nation has been attacked, and when its future existence and the future existence of our chosen method of government are at stake.

"To win this war wholeheartedly, unequivocally and as quickly as we can is our task of the first importance. To win this war in such a way that there will be — that there will be no further world wars in the foreseeable future is our second objective. To provide occupations, and to provide a decent standard of living for our men in the Armed Forces after the war, and for all Americans, are the final objectives. Therefore, reluctantly, but as a good soldier, I repeat that I will accept and serve in this office, if I am so ordered by the Commander-in-Chief of us all — the sovereign people of the United States."[10]

In her diary, at Wilderstein, Daisy Suckley penned her disappointment — not at the decision itself, which she'd felt was inevitable in the circumstances, but that the President had not had time to call her personally beforehand. Or afterwards. "He announced to his press conference this morning, that he will run for the presidency if he is wanted — that settles that, at least, & his wondering & pondering is over."

Further, Daisy knew just "how mixed his feelings are on the subject." She could therefore "commiserate & congratulate with equal truthfulness." What was clear, however, was that, in contrast to his mood all spring, the President had now decided. He was going to do his best, as a soldier of mankind.

"He evidently feels well enough," she jotted thoughtfully, "to carry on"—knowing it would kill him.[11]

51

Missouri Compromise

LATE ON JULY 13, 1944, the President set off—destination Hawaii, but via Hyde Park and Chicago, where the *Ferdinand Magellan* was serviced and became a maelstrom of visits, phone calls, and political meetings in the run-up to the Democratic National Convention.

With the prospect of a deeply contested ballot splitting Southern and Northern Democrats, the party chairman, Robert Hannegan, came aboard in person to have the President sign and postdate a secret letter of support for two compromise vice presidential candidates, should no clear winner emerge between Wallace and James F. Byrnes. For Wallace, in two meetings with the President at the White House, had refused to withdraw his candidacy for vice president, and Judge Byrnes, the director of the War Mobilization Office and former Supreme Court justice, was intent on running for the position.

"The P. says Wallace is much nearer to the President's thoughts & view of things" in terms of the postwar peace—an anticolonialist and genuine supporter of the New Deal and civil rights—"but [he] is a poor administrator," despite being the rich owner and CEO of an expanding agricultural feed company, Daisy Suckley recorded in her diary.[1] Owing to Wallace's internecine squabbles with Jesse Jones, the commerce secretary, the President had in fact been compelled to remove Wallace from his chairmanship of the boards of Economic Warfare and Supply Priorities and Allocation, and to send him overseas, to Russia and China—a trip that would add to Wallace's global perspective and reputation. In the bare-bones business of election campaigning across America, however, the vice president was known to have erratic religious beliefs, having begun life as a Presbyterian, become a Theosophist, and then an Episcopalian. This would make him vulnerable to Republican attack if, as expected, the election campaign became bitter.

For his part Jim Byrnes, the diminutive former congressman, former senator, and former U.S. Supreme Court justice, was an extremely effective, if widely disliked, administrator: a Southerner who had loyally backed Roosevelt's New Deal in South Carolina. But Byrnes had also attacked the proposed antilynching law in Congress, as well as the Fair Standards Act—thus making him an enemy both of Northerners *and* labor. In an open, national vote Byrnes stood little chance of winning the Democratic Party nomination for vice president—at least, not without a tremendous and potentially divisive fight that might well sink the party's chances, especially when contrasted with the first-ballot choice of the Republican contender, the young New York governor Thomas Dewey and *his* vice presidential nominee, Ohio governor John Bricker. Byrnes, too, could be vulnerable to religious scrutiny, for he had been baptized a Catholic and had only switched to Protestantism on his marriage. "Well, by next Friday, that question will have been decided," Daisy had ended her diary entry on July 14, after the President had left Washington for Hawaii, "& the P. can relax on the high seas, with no newspapers and no telephones."[2]

The drama in Chicago, however, suggested it would not be so simple. Hannegan favored a dark horse candidate, his former political boss in Missouri: Senator Harry S. Truman, chairman of the Senate subcommittee investigating U.S. war production, and senior member of several other Senate committees. Yet another alternative was an outlier: the still-young (age forty-six) associate justice of the Supreme Court, William O. Douglas, who had worked for the Securities and Exchange Commission from its inception during the Depression to clean up Wall Street.

None of these were inspiring choices, given the somewhat dire situation: namely that anyone in politics with half a brain was aware, by this time, that the rumors of the President's ill health were correct. The President was hardly ever to be seen in Washington, and when he was, he seemed a ghost of his former self. He would, widespread speculation now ran, be unlikely to live out a fourth term. He might not even reach the beginning of such a term next January, 1945. Certainly every delegate who pressed Truman, the reluctant senator from Missouri, to run for the vice presidential nomination in Chicago had made clear to the senator it would inevitably mean the burden of the presidency itself, ere long. Thus when a reporter pointed out that, if he threw his hat into the ring, he might well soon "succeed to the throne," Truman had shown no surprise. In fact he had shaken his broad bespectacled head, saying, "Hell, I don't want to be President"[3]—at least not that way, via "the back door."

Truman had not counted, however, on the President's byzantine way of coming to a firm, ultimate decision. The President's pre-Hawaii trip itinerary would allow him to remain absent for the further backroom shenanigans of the nominating process in Chicago once the train left the city on July 15; making a leisurely cross-country rail journey to California that very afternoon would allow him to let the Democratic Party bigwigs and delegates duke it out in the Windy City, so that he could not be accused of interfering. And his heart condition was a boon, for it caused him to not really care at a certain level: to be, in short, uncharacteristically fatalistic.

Had the President put his presidential weight behind Henry Wallace's candidacy, Truman biographer David McCullough later noted, the vice president would undoubtedly have been nominated by the convention, as in 1940 — indeed, Wallace very nearly was, once the 1944 convention unfolded. But the convention was not the American electorate; moreover the President was strangely unwilling to intercede — absolved by his medical condition, which made stress potentially fatal.

In Missouri, Senator Truman's elderly mother had told reporters she knew the rumors about the President. "They keep predicting that Roosevelt will die in office if he's elected," Mary Ellen Truman was quoted. "The Republicans hope he will," she'd said disparagingly. "They keep saying that I'll die, too, and I'm almost 92. I hope Roosevelt fools 'em."[4]

With Harry Hopkins still convalescing, the President had meantime asked Judge Sam Rosenman to travel with him aboard the *Ferdinand Magellan* as speechwriter and political adviser — first across America, then across the Pacific Ocean to Honolulu. As Rosenman afterwards recalled, the President simply did not have the energy to fight for Wallace, as he had in 1940. Nor did he have the conviction that Wallace would make a strong "national" president, if and when required to take the reins. In Rosenman's words the President "saw as the big task for his fourth term the creation and successful functioning of the United Nations Organization. This would require the complete co-operation of his own party, but the co-operation of the Republican party as well. He was determined not to repeat the mistake of Woodrow Wilson by framing the future peace of the world by himself or with his own party alone." It would therefore have to be a collaborative, "bipartisan approach."[5]

By the time the *Ferdinand Magellan* reached California on July 19 — mobbed by crowds at railway halts throughout Iowa, Missouri, Kansas, Oklahoma, Texas, New Mexico, and Arizona, despite attempts to keep the President's train trip secret — the bartering and excitement among the twenty thousand delegates and party stalwarts back in Chicago had reached fever pitch. Byrnes had by then decided he couldn't win, and had withdrawn his candidacy for the vice presidential spot. Wallace had not, however: leaving him, for a while, the sole surviving contender. His supporters thus planned to stampede the convention into voting for him as the party's vice presidential nominee, once the President had accepted the presidential nomination.

But Wallace's backers had not reckoned with Roosevelt's legendary legerdemain, even when a shell of his former self. At a conference of party bigwigs upstairs at the White House on July 11, the consensus had been unequivocal. "When all the names had been fully canvassed," the President had announced "with an air of finality," if without particular enthusiasm, "'It's Truman,'" Rosenman later wrote[6]—and the President had duly told Wallace this was the consensus. In fact the President had also told Wallace that political experts were warning him the vice president, if renominated, could be responsible for losing up to *three million votes* across the country, given the rising Republican tide. Wallace, however, had produced his own statistics, showing he had 65 percent support in the Democratic Party. In vain the President had noted that this was not the point: such delegates could guarantee him the VP nomination, but lose the party the whole election.

Wallace's obstinacy had made the matter distinctly unpleasant for the ailing president. Truman was, in the end, "the only one with no enemies," the President had explained to Wallace; moreover the Missouri senator would "add a little independent"—i.e., bipartisan—"strength"[7] to the ticket. But Wallace had dug in his heels. He wasn't listening. And the President had been too ill—and by nature too nonconfrontational—to insist.

How the President eventually managed to persuade Truman to overcome his diffidence—as well as his puzzlement as to why the President had not discussed the matter directly with him, before the decision became acute—was to become one of the legendary stories of American political history, with numerous versions. That the President did, however, intercede was never contested. Truman, unwilling to potentially enter the White House by the back door, and having backed Judge Byrnes for the vice presidential nomination, initially declared he would not allow his name to go forward, and he refused to back down even when Hannegan assured him he was the choice not only of the Democratic Committee's board, but also of the President—showing him Mr. Roosevelt's postdated letter regarding two compromise candidates he would be pleased to run with, if the convention became locked: Harry Truman or Bill Douglas.

Mr. Roosevelt's wishes, as President and thus the senior Democrat in the nation, were obviously paramount, Truman conceded. But what in all truth, he countered, were these wishes, when the President was thousands of miles away, en route to more distant climes?

Only when the President's train reached the Marine Corps base siding at San Diego was a telephone call from Hannegan put through to the President from the Blackstone Hotel in Chicago on July 19, 1944. A clear, unmistakable presidential preference, expressed in his actual voice, was now essential in order to avoid bedlam in Chicago, Hannegan explained.

The President's train had arrived in the middle of the night. He was not well. In addition to his heart condition the President had almost been suffocated by the heat in Yuma, Arizona, the previous afternoon, where "the temperature on the station platform was 125 F. at 4 p.m.," Admiral Leahy had noted in his diary.[8] Nevertheless the President agreed to take Hannegan's call. Whether he was aware Truman was also in Hannegan's room is unclear, but the metaphorical temperature in Chicago, it seemed, was now higher than in Yuma. It was the eleventh hour. "I was sitting on one twin bed, and Bob was on the other," Truman himself later recalled. The first convention ballots would be cast that night. "Roosevelt said, 'Bob, have you got that guy lined up yet on that Vice Presidency?'"[9]

Given the seesawing arguments that had colored deliberations in Chicago since the President's brief presence in the city, four days before, as well as the President's failure to provide firm leadership over the matter — a matter of potentially world-historical consequence — this was both a shambles and shameless. Hannegan, who had asked Truman to listen in on the call, and if necessary speak with the President directly, took a deep breath. "'No,'" Hannegan responded. "'He's the contrariest goddamn mule from Missouri I ever saw.'

"'Well,' Roosevelt said, 'you tell him if he wants to break up the Democratic Party in the middle of the war and maybe lose that war that's up to him.'"[10]

If these were the President's words, they were well chosen. Hannegan promised the President, on the phone, he would make it happen.

Truman was completely flummoxed. Why had the President not discussed the matter with him in the run-up to the nomination? "He never said a word to me about the situation. And I said to Bob, 'Why in hell didn't he tell me that when he was here [in Chicago] or before I left Washington?"

Hannegan shrugged. If Truman was the contrariest mule, the President was the most enigmatic.

Reluctantly, Truman had then agreed to stand for the vice presidency: the vote slated to take place at the convention the next evening, July 20.[11]

52

The July Plot

ELEANOR ROOSEVELT HAD TRAVELED to the West Coast on the *Ferdinand Magellan* with the President. She wanted to have the chance to see two of her sons there: Elliott, the airman, and Jimmy, the Marine — both of them combat colonels — and their families.

Once the Chicago decision was made on July 19, the President had felt well enough to visit patients at the San Diego Naval Hospital, and after that to have a family dinner, hosted by Jimmy Roosevelt in nearby Coronado, by the bay. It was also arranged that Jimmy, who was on the staff of the Fifth Marine Division stationed at San Diego, would collect his father from the train the next morning and take him to observe a live-fire Marine beach assault training exercise, prior to the President shipping out to the Pacific.

Jimmy arrived early. "Before the exercise began, I was alone with Father in his private railroad car," he later recalled. "We talked of many things — the war, family, and politics . . . I was struck by Father's irritability over what was happening in Chicago and by his apparent indifference as to whom the convention selected as his fourth-term running mate. He made it clear that he was resigned to the dumping of Vice-President Henry A. Wallace" if the Democrats were to have any real chance of winning, for "he felt that Wallace had become a political liability" in the country, not only in the party. The nomination voting was due to take place later that day. The President had "professed not to 'give a damn'" about whom the delegates came up with once the process came to its climax. "His mind was on the war; the fourth-term race was simply a job that had to be accomplished, and his attitude toward the coming political campaign was one of 'let's get on with it,'" if it had to be done.[1]

At this juncture, just before "we were to leave for the exercise, Father suddenly turned white, his face took an agonized look, and he said to me: 'Jimmy, I don't know if I can make it — I have horrible pains!' It was a struggle for him to get the words out."

"I was so scared," the Marine colonel candidly recalled, "I did not know what to think or do."[2]

At his father's insistence they did not call Dr. Bruenn. Canceling the President's presence at the ten-thousand-man exercise, some forty miles away, was not the major issue. It was the intended radio broadcast that evening that was the problem, for it had been set up with all national radio networks, as well as a swarm of reporters, film cameramen, and photographers. Broadcasting from a special additional railway car, the President would that night deliver his formal acceptance of the 1944 Democratic nomination on national live radio. He would make it in the form of a speech—one on which he'd worked with Rosenman and Elmer Davis, the head of the Office of War Information, for several days. It would be historic, too: the first such nomination acceptance address made during hostilities since the Civil War. And an opportunity to lay out, once again, the President's vision of the future.

But if the acceptance speech had to be canceled?

Given the rumors surrounding the President's health, not being well enough to speak would probably end his father's reelection chances, Colonel James Roosevelt was aware. Convinced, however, that the stomach pain mirrored the one he'd suffered in Tehran while dining with Stalin and Churchill, the President asked only that his son help him out of his chair "and let me stretch out flat on the deck for a while—that may help."[3]

Eleanor was not informed, nor Anna. For "perhaps ten minutes, while I kept as quiet as possible, Father lay on the floor of the railroad car, his eyes closed, his face drawn, his powerful torso occasionally convulsed as the waves of pain stabbed him. Never in all my life," Jimmy later wrote, "had I felt so alone with him—and so helpless."[4]

By the strangest of coincidences, an even bigger crisis was taking place in the Third Reich that same day.

Hitler had left his home in the Obersalzberg at the same time the President had left Hyde Park. He was now at his headquarters in Rastenburg, in East Prussia—having flown there in his Condor—the last journey he would ever make by plane. He'd had another stormy meeting on June 29 with Field Marshals Rundstedt and Rommel at the Berghof. The battle in Normandy was going badly—in fact German defeat was, frankly, now inevitable, given Stalin's simultaneous offensive on the Eastern Front. Cherbourg had fallen, despite the Führer's injunction against surrender. Negotiation was the only option "regarding Germany," in order to save the nation further futile bloodshed, Rommel had begun to argue, when Hitler, in cold fury, had interrupted him and in a rage commanded Rommel to leave the room for his insolence in attempting to speak on "political" matters, about which he knew nothing.[5]

The Führer had been beside himself with anger—unable to accept that a

German general, particularly a popular and decorated general, should dare make such a suggestion. To negotiate except from a strong position was anathema to Hitler. He had seen no future in the proposition — not, at least, until and unless the Wehrmacht turned the tables on the Western Allies. In his own mind this was still achievable. A tremendous storm in the English Channel had destroyed the Mulberry artificial harbor in Normandy's American sector, putting back the Allied reinforcement timetable by five days. The first of Hitler's *Vergeltungswaffen,* the Fieseler Fi.103 (known to the Allies as the buzz bomb), was causing considerable alarm, even panic, in southern England. In addition to the V-1, the Führer still had a second *Vergeltungswaffe* up his sleeve, namely ballistic missiles that could not be shot down by antiaircraft fire or fast fighters; also the new Messerschmitt 262 fighters, powered by jet engines — another miracle of German engineering. It was crucial to hold back the Allies in Normandy and stabilize the Eastern Front, following the massive Russian offensive. Only then could one think of negotiation — especially if the Allies could be split apart politically. These were matters that generals were utterly incompetent to recognize, let alone handle.

Having arrived at the now reinforced Wolf's Lair near the Eastern Front, Hitler had moved into the guest bunker, since his own main accommodation was still in the process of renovation. He'd then summoned Field Marshal Kesselring from Italy on July 19 and awarded him the Knight's Cross of the Iron Cross with Oak Leaves, Swords, and Diamonds, for his sterling performance on the Mediterranean front.

The Eastern Front was still of great concern, though, and at midday on July 20 the Führer was just chairing an operations conference of his senior military staff — some twenty-four officers — in an outdoor hut used as a meeting room, when at 12:42 p.m. there was a terrific explosion. The blast blew the air and debris through the open windows, just as the Führer leaned over the map-covered table on his elbows, poring over air-reconnaissance charts of Russian positions. There was mayhem inside — blood, splinters, dust, and rubble.

"Linge, someone has tried to kill me," the Führer cried to his valet,[6] having been flung against a doorpost and having then made his way from the smoking havoc to the guest bunker — his gray jacket torn and his white underwear showing under his shredded black trousers, like a medieval court jester.

Someone had: Colonel Claus von Stauffenberg, chief of staff to General Fromm, commander of the Reserve, or Replacement, Army. After three failed attempts to set up an assassination at the Berghof, Stauffenberg had managed to prime one of two explosive devices in the Wolf's Lair toilet, which he intended to put in his briefcase and place on the floor near the Führer. He had not quite had time to prime the second one, however, before being called back into the conference hut. Had he been able to put both devices in his briefcase, even with one unprimed, they would have detonated together, killing Hitler.

Alternatively, had the meeting been held in the normal concrete bunker, the ricochet of the blast would have been enough to erase the Führer and obliterate his entire senior staff, including Field Marshal Wilhelm Keitel, head of the Wehrmacht Armed Forces high command, and General Alfred Jodl, chief of staff to Keitel (both later executed at Nuremberg). Instead it merely grazed the Führer and ruptured his eardrums — though it did also puncture the flesh of his right forearm, causing nerve damage.

His hair singed, the bandaged Führer was thus able to change clothes and greet the visiting Benito Mussolini around 3:30 p.m. — though, with his right arm in a sling, he was unable to return Mussolini's Sieg Heil salute.

Amazed that so many had survived the blast, considering the utterly devastated interior of the hut to which Hitler took him, and impressed by the Führer's sangfroid, Mussolini declared: "After all I have seen here, I agree with you completely. This is a sign from Heaven!"[7]

A number of the staff officers had been gravely wounded, however; four of them died over subsequent days. Thinking it was the work of impressed laborers still reinforcing the bunker, Hitler was initially unaware that the attempt had been part of a putsch, or coup d'état, to be mounted in Berlin, Paris, and other cities, just as soon as word of Hitler's death was communicated by Stauffenberg to his fellow conspirators. Ironically, Stauffenberg, who had left the hut after placing his briefcase under Hitler's table, assumed from the huge blast the Führer had been killed; from the nearby airfield he thus telephoned Berlin with the secret code word "Valkyrie," indicating success.

This was another fateful error, for it led inevitably to the plotters emerging, being swiftly rounded up, and then executed, once Hitler's survival was affirmed. From Berlin Dr. Goebbels was soon able to assure the Führer by phone that the capital of the Reich was under full Nazi control. Moreover he was able to persuade the reluctant Führer to give a major radio broadcast that very night. In this the Führer would be able to make clear both to the world and to any others who were still conspiring against him that, contrary to rumor, he had survived the assassination attempt. And that the Third Reich was still under his command.

"I speak to you today for two reasons," Hitler began his broadcast, in an unusually somber, sober, almost listless tone — first "so that you can hear my voice and know that I was not injured and am in good health." Second, "so that you learn about the details of this crime, which is without equal in German history."[8]

For decades, night and day, he had worked, Hitler claimed, "only for my Volk!" Any claims he was dead, or that the Wehrmacht was involved, were specious, he lied — his Austrian accent enough to persuade listeners the broadcast was authentic. No civilian was to take instructions from any department that had been "appropriated by the usurpers," he ordered, announcing that Heinrich Himmler would take command of the Replacement Army from General Fromm, while General Guderian would replace General Schmundt, the operations chief of the general staff, who had been severely wounded — perhaps mortally.

"I am convinced that by crushing this very small clique of traitors and conspirators," the Führer assured listeners, an atmosphere could be reestablished at home to reflect that of the many courageous German soldiers fighting at the front. "After all, it is not right that hundreds of thousands and millions of brave men give everything, while a very small coterie of ambitious, pitiful creatures at home constantly tries to undermine this attitude. This time we will settle accounts in the way we are used to as National Socialists," Hitler warned. (Almost five thousand political opponents were ultimately executed.) There were several mentions of gratitude to "Providence and my Creator" for having saved his life — not on his account, but so that he could "persevere in my work." Orders, moreover, had already gone out to "all troops. They will execute them in blind faith and in accordance with the type of obedience which the German army knows."[9]

Fortuitously the Führer was not alone in Rastenburg. Field Marshal Goering, Foreign Minister Ribbentrop, Martin Bormann, and other dignitaries had also gathered that afternoon, having been invited to be present for the Duce's visit. Following the Führer's broadcast they now listened as Hitler that night gave way to a rage more venomous, if such were possible, than they had ever heard before.

In September 1943 it had been the Italians who'd betrayed their noble allies; this time, however — particularly with Mussolini present — Hitler needed new scapegoats, and he found them in the senior German officer class he had always suspected of despising him as a mere corporal with a low-class Austrian accent. There would have to be an immediate investigation — a military court of honor, set up and charged with executing the guilty. Not by firing squad, either, as befitting commissioned officers. "They must hang immediately," the Führer raved, "without mercy," and in their prison clothes, as common criminals. "He is absolutely determined to set a bloody example and to eradicate a freemasons' lodge which has been opposed to us all the time," Goebbels recorded in his diary, one that "has only awaited the moment to stab us in the back in the most critical hour. The punishment which must now be meted out must have historic dimensions."[10]

Six thousand miles away the President of the United States meantime faced the task of his own radio address that same evening.

Jimmy Roosevelt was unsure whether his father could pull it off. Lifting him from the floor of the *Ferdinand Magellan,* he had that morning "helped him get ready, and the Commander in Chief went to review the exercises," Jimmy recalled—the two of them being driven at 9:10 a.m. to the bluffs to watch the "colossal" mock invasion by Marines.

The day before, the President had listened as senior Marine and naval officers, visiting him on his train, had filled him in on the scope and purpose of the training exercise, which would be on the same lines as the current invasion of the Marianas. How much the President was really able to take in of the exercise was unclear. "In the many photographs taken of him that day, Pa looked tired," Jimmy remembered. "In most of them, however, he wore a big smile; no one would have guessed what he had just been through."[11]

The President had claimed he was feeling better, but in actuality he'd asked to go back to the train less than an hour later. "I had a grand view of the landing operation at Camp Pendleton," he wrote afterwards to Eleanor—who had meantime departed for Los Angeles—but he'd unfortunately "got the colly-wobbles," he recounted, making light of the episode.[12] In truth he'd had to cancel lunch at the home of his daughter-in-law, John's wife, in Coronado, instead spending "the afternoon in his private car resting and listening to the Democratic Convention news as it was broadcast over the radio"—culminating in his official renomination request.[13] Jimmy and his wife then had a quiet dinner with him on board the *Ferdinand Magellan* before his broadcast.

The whole saga—the President having another health emergency at the very moment when twenty thousand delegates and supporters were assembling in Chicago to hear Senator Samuel Jackson, the convention chair, declare FDR to be in good health, and as the delegates prepared to listen to him accept his nomination by radio from San Diego—was almost surreal.

The President was determined to fulfill his commitment, however—the more so since word had come through that there had been a putsch in Germany, with an attempted assassination of the Führer. With Hitler dead, the war might well be over soon. If the attempt had been successful.

At 8:30 p.m. the President was finally wheeled to the extra car to give his speech.

The broadcast took fifteen long minutes to read on air—something the President did in such a firm, authoritative voice that no one listening could have imagined his cardiologist, Dr. Bruenn, was all the time sitting close by in the crowded car, concerned about the President's astronomic blood pressure—or that his son Jimmy, next to him, had only hours earlier wondered if his father were dying at his feet.

What had caused the sudden pains and collapse?

The real reason would never be identified, given the President's recovery that very evening. He had clearly been a near-wreck, though, the rest of that day, July 20—somewhat surprised, like Hitler, to be alive.

• • •

The President's broadcast from San Diego was the very opposite of the one Hitler had given from Rastenburg.

Where Hitler had delivered his broadcast in dry, punitive, snarling diction, Roosevelt's easy tenor voice sought to embrace his listeners in a serious yet inspiring personal conversation: at once intimate and rational. Sam Rosenman had traveled with him, working each day on the script; this the President managed to deliver with his trademark cadences: his voice soaring and then resuming its proud, determined rhetoric as he bade listeners understand, once again, that he would prefer not to have to serve another term, but would rather "retire to the quiet of private life." He was being called to duty by the Democratic Party convention—and by another convention, namely the "sense of obligation to serve if called upon to do so by the people of the United States" in November that year.

"I shall not campaign, in the usual sense, for the office. In these days of tragic sorrow, I do not consider it fitting. And besides," he added—economizing with the truth—"in these days of global warfare, I shall not be able to find the time." He'd just traversed, he explained, "the whole width of the continent," and was now at "a naval base where I am speaking to you. As I was crossing the fertile lands and the wide plains and the Great Divide, I could not fail to think of the relationship between the people of our farms and cities and villages and the people of the rest of the world overseas—on the islands in the Pacific, in the Far East, and in other Americas, in Britain and Normandy and Germany and Poland and Russia itself." For the states through which he'd traveled "are becoming a part of all these distant spots."

Current battles being fought "in Normandy and on Saipan" affected the "security and well-being of every human being in Oklahoma and California," the President pointed out. A new world beckoned Americans, rousing the country from its long sleep. "Mankind changes the scope and breadth of its thought and vision slowly indeed," he reflected—but change was coming, indeed had already taken place.

America's mission—its bipartisan government's mission—was thus: "First, to win the war—to win the war fast, to win it overpoweringly. Second to form worldwide international organizations, and to arrange the armed forces of the sovereign Nations of the world to make another war impossible within the foreseeable future. And third to build an economy for our returning veterans and for all Americans—which will provide employment and provide decent standards of living." This was not a mission that could be handed to "inexperienced or immature hands"—men "who opposed lend-lease and international cooperation against the forces of tyranny," and who would return the country to "breadlines and apple-selling." With this in mind, he quoted, as the climax of his oration, "the greatest wartime President in our history," Abraham Lincoln, who in 1865 had set the goals of the nation—"as God gives us the right"—to "strive on to finish the work we are in; to bind up the nation's wounds; to care for him who shall have borne the battle, and for his widow, and his orphan—to do all which may achieve a just and lasting peace among ourselves, and with all Nations."[14]

It was an exemplary radio performance—one of the President's best ever.

It was not quite enough, however. The President was then asked to reread passages from the speech for the newsreel film cameras, to be shown around the world, and to pose for photographs.

This proved a big mistake. The President's speech had exhausted him, as much mentally as physically. It had capped an extraordinary day, in terms of his health—but the medical crisis had been kept quiet. "One of the pictures was a most unfortunate one," recalled Judge Rosenman later. "It was snapped while the President, with his head bowed over the printed page, was pronouncing a broad vowel, so that his mouth was wide open at the click of the camera." Taken by an Associated Press pool photographer, the photo told the true story, which even a thousand words could not, of "a tragic-looking figure," Rosenman candidly admitted. His "face appeared to be very emaciated because of the downward angle and open mouth; it looked weary, sick, discouraged, and exhausted."[15]

In Washington, the secretary of the interior, Harold Ickes, was appalled. "It was a terrible picture," Ickes recorded in his diary. "The President looked like a sitting ghost and the picture has created a bad impression"—one that might have serious consequences on the November election.[16]

Since the White House press secretary, Steve Early, was not aboard the *Magellan,* there had been no media supervision; the photograph, late that night, was carelessly transmitted to the wire services for general distribution—giving a dreadful yet accurate portrayal of the President's ill health. After more than six months in which the President had avoided the glare of publicity, with only carefully monitored coverage that deliberately disguised the true seriousness of his condition, the photo was like a bombshell.

Panic ensued in the Democratic Party. The picture gave the impression, as one distinguished newspaper editor later wrote, "of a failing elder, a candidate for a nursing home, gasping out his last words, the opposite of the impression his confident, sonorous voice left with the party faithful in distant Chicago," or even those who saw the newsreel film.[17] As Rosenman recorded, the photo "was later published with great glee by enemies of the President, who urged it as proof of their charge that he was no longer physically or mentally competent to manage his office."[18]

Rightly or wrongly, it was the first real indication to the public, behind the many rumors, that the President *was* seriously ill—no longer the same man who had addressed Congress after Japan's attack on Pearl Harbor, or in the many speeches and Fireside Chats he'd given over the years since then.

In the controversy that followed, the hapless photographer was banned from the train and the White House. Meanwhile the FBI, under its director, J. Edgar Hoover, was asked to track down rumors about the President's plummeting health: stories that the President was receiving secret treatment for a fatal illness, and would not survive a fourth term, if elected.

Ironically, the same was the case in Germany. Hitler had survived the *attentat*, miraculously. However, rumors the Führer was now only a shell of his former self soon abounded — and in certain respects were true. He was by now sleeping uninterruptedly for only two hours a night.

Close-up, the Nazi leader often looked a shuffling wreck. Summoned to diagnose the damage to Hitler's ears, Major Erwin Giesing, an ENT specialist at the nearby Lötzen field hospital in East Prussia, found the once-strutting Führer a sorry sight. "He looked to me like an ageing man — almost burned out and exhausted," Giesing would afterwards recall, "like somebody husbanding every last ounce of his strength."[19]

Because of the damage to his eardrums, Hitler was no longer able to fly. "I would so much have liked," he told Dr. Giesing, "to get over to the west," where the Allied armies now numbered more than a million men ashore, and a breakout could happen any day. He would still do so, he claimed, even "as gunner in a single-engined plane" if "the dams burst" in Normandy and the Allies broke out of the bridgehead.[20] Field Marshal Rommel had already reported to him, moreover, what was inevitable. "The troops are everywhere fighting heroically, but the unequal struggle is approaching its end" — the same language the so-called Desert Fox had used at the climax of the Battle of Alamein.

Alamein had been near Cairo — more than seventeen hundred miles from Berlin. Now the "deciding" battle of the war was being fought in northern France, with the Allied frontline barely a hundred miles from Paris: gateway to the borders of the Reich. "It is urgently necessary," Rommel had pleaded, "for the proper conclusion to be drawn from this situation."[21]

The Führer had ignored him. Rommel, in any case, had been gravely wounded on July 17 — the day General Bradley's First U.S. Army forces captured Saint-Lô. The Brittany ports and the Loire Valley now beckoned as the general massed his infantry and armor for a concentrated, massive American attack with heavy U.S. Air Force support, while the British and Canadians tied down four-fifths of German panzers around Caen.

Recuperating at his headquarters, Hitler not only never spoke again in public, but would rarely allow photographs to be taken, save when posed. More than ever he was convinced he was the only person left alive with the iron will and experience of World War I to fight the war to a satisfactory German conclusion, whatever that might be. Retreat, let alone negotiating the surrender of the Third Reich, was not an option.

In San Diego, receiving via the White House Map Room confirmation that Hitler had survived the assassination attempt, the ailing president felt likewise. There would be military reverses to come, in all likelihood, and further attempts to split the Allies. There would be disagreements about the best way to end the war, and about the global structure of the postwar. But, in the struggle with Adolf Hitler, commanding the relentlessly disciplined and blinkered forces of the Third Reich, only he, Franklin Delano Roosevelt, had the necessary experience, strategic vision, as well as leadership ability to direct the United States and its allies to victory. And lay the foundations for successful postwar peace — no matter how sick he was.

It would be, in effect, Roosevelt versus Hitler. A struggle to the bitter end.

PART EIGHT

Hawaii

53

War in the Pacific

Once again observing the sailors' superstition about Friday departures, the USS *Baltimore* — a heavy cruiser weighing more than fourteen thousand tons — slipped out of San Diego harbor in the early minutes of Saturday, July 22, 1944, escorted by four destroyers. Destination: Pearl Harbor. Passenger: Mr. Franklin Roosevelt, President of the United States.

Bristling with big eight-inch guns — nine in total, as well as twelve five-inch thirty-eight-caliber dual-purpose cannons, some seventy-two antiaircraft guns, and two floatplanes — the giant warship was only a year old but already heavily battle-scarred, having participated in attacks on Eniwetok, Kwajalein, Tinian, Palau, Hollandia, Truk, Satawan, Marcus Island, Wake Island, Guam, and Rota. It had just returned, in fact, from the invasion of Saipan in the Marianas, as well as participating in a carrier task force raid on Iwo Jima — 760 miles short of Tokyo. As such the *Baltimore* was a symbol of the U.S. Navy's revolution in conducting modern offensive war: heavy cruisers being used as offshore artillery or bombardment platforms, and antiaircraft screening vessels; U.S. submarines decimating enemy supply vessels; U.S. aircraft carriers supporting assault landings — while also tempting the enemy into major fleet action.

What a contrast this made with the disaster at Pearl Harbor, in barely two and a half years! The ailing President, as a lifelong "navy man," could take personal pride in such a military renaissance — a transformation that reflected on the one hand the miracle of American war production, for which he could take the ultimate credit, and on the other his consistent military strategy since 1941 of a two-ocean war, for the most part under American supreme command.

Installed in the captain's quarters on the starboard side, the President did not, however, appear on deck for several days. Nor did he hold any meetings on board with his Joint Chiefs of Staff — for he was traveling without them, other than his White House chief of staff, Admiral Leahy.

It was the first time in the war, in fact, that the President was traveling outside the United States without his top military advisers — and this for a very good reason, namely that they themselves were at war with one another over strategy in the Pacific. Not only strategy, moreover, but with the problem of how to manage the personal dispute between the two supreme commanders there: General Douglas MacArthur, in the Southwest Pacific, and Admiral Chester Nimitz, in the Central Pacific.

Not one of his Joint Chiefs of Staff — General Marshall, Admiral King, and General Arnold — knew how to handle MacArthur, who had his own views on how best to defeat Japan. Dealing with MacArthur, whom the President had summoned by cable to meet with him in person in Hawaii, where Admiral Nimitz had his Central Pacific command headquarters, would be a trial for the President in his current state of health. But it had to be done. And besides, as U.S. commander in chief it would take the President away from the glare of American media, following his acceptance of the Democratic mantle for the coming presidential election. The six-day, three-thousand-mile voyage across the Pacific would hopefully assure the President enough rest to grapple with MacArthur, and then Governor Dewey. And besides: he would have a chance to inspect his army, navy, and air force contingents on Oahu — thereby debunking, by his very presence, General MacArthur's constant cry that the Pacific was the forgotten theater of the war.

Though he had not been invited to accompany the President, Admiral King was responsible for arranging the naval details of the voyage. He was, moreover, all too well acquainted with what was at stake — in fact had decided to go out to Honolulu for a week to prepare the ground, so to speak. He had stayed in Hawaii for ten days, in the end — anxious to ensure that Admiral Nimitz, who he thought too complaisant in army-navy relations, did not allow MacArthur to bend the President's ear when the general arrived there. MacArthur, King knew, would continue to argue for an immediate and unnecessary invasion of the Philippines, as he had been doing all spring and summer.

King himself was utterly opposed to another Philippines campaign, other than perhaps to establish a military base in Mindanao. The admiral figured that U.S. air and naval forces from the Central Pacific could be better used to target Formosa and the Chinese mainland directly and leave the Japanese-occupied Philippine Islands to wither on the vine. During his ten-day mission King had thus reviewed current and future strategic plans with Nimitz, whom he treated as his avatar at the forthcoming meeting with the President. They had even sailed together to the Marianas to see, in person, some of the islands and atolls that U.S. Marines, playing the dominant part, had recently assaulted with extraordinary heroism. King had still been in Honolulu on July 22, the day the President began his own journey to Hawaii aboard the USS *Baltimore,* though the naval operations chief had left Hickam Field that night — having fully prepped Admiral Nimitz.

Above all, King had urged Nimitz, no effort or expense should be spared in hosting the President, whose role as U.S. commander in chief was to be emphasized during the trip, to avoid accusations at home of the Navy aiding his election campaign in the immediate aftermath of the Democratic convention. This would also play to the President's vanity as commander in chief — King having heard from Admiral Leahy that FDR wasn't keen on the rather lax way King referred to himself as "Commander in Chief, U.S. Fleet," and his constant pressure to create a new five-star title, such as "Grand Admiral," for himself.

Impressed by the U.S. naval and military installations and suitably entertained — perhaps with a fishing expedition arranged, and an evening of Hawaiian music — the President would, King thought, be more partial to Nimitz's Central Pacific strategy for defeating Japan directly, over General MacArthur's laborious, costly, and circuitous route around the Western Pacific perimeter and the Philippines.

Would Admiral Nimitz be *assertive* enough, though? King had wondered as he'd left. The fair-haired admiral was a wonderful navy officer, and popular with his subordinates in Hawaii. But he was not a showman like MacArthur. This worried King.

For himself the President was well aware what was brewing. Ever since U.S. forces proved successful in stemming the Japanese rampage in the Pacific and assumed the offensive at Guadalcanal and Midway, the great strategic question had haunted military planners: What was the best way to advance on Tokyo?

At the high-level summit and Combined Chiefs conference at Quebec in August 1943, it had been agreed that a multipronged advance was best: including an offensive war against Japan waged by the British in northern India and Burma, and in China itself by Chiang Kai-shek, whose forces were based at Chungking, in the Chinese interior. In the Pacific, meanwhile, the war against Japan would go on as before, on two axes.

In the Central Pacific Admiral Nimitz would continue to contain the still-powerful Japanese fleet, while pushing amphibious forces across the ocean toward the Marshalls, Marianas, and Guam. This would put the Army Air Force's new B-29 long-range high-altitude bombers within striking distance of the Japanese islands. Meanwhile MacArthur, backed by smaller naval forces but more ground troops and army air support, would continue his series of "leapfrogging" assaults along the northern coast of New Guinea — bypassing heavily defended Japanese bases to avoid heavy casualties, and putting himself within striking distance of Mindanao and the Philippines.

By following this dual strategy, the United States would have two possible pincers with which to approach and cauterize the main Japanese islands — in fact three pincers, if British forces could open an overland supply route to

China, and U.S. air and naval bases could be built up on the southern Chinese mainland or Formosa.

Since the Quebec Conference, however, this overall Allied strategy had proven a disappointment in India and Burma, where British commanders, in the view of General Stilwell, had indulged in "endless walla-walla and very little fighting."[1] Stilwell himself had been unable to do much better in China, since Chiang Kai-shek seemed at permanent personal loggerheads with "Vinegar Joe" — an ardent warrior who wanted Chinese nationalist forces to fight the Japanese, not be held back to deal with communist troops under Mao.

In the Pacific, by contrast, the U.S. double-axis strategy had worked faster and more successfully than had been anticipated at Quebec — so much so that with the Marshalls falling early to American forces and MacArthur's forces establishing control of the northern coast of New Guinea, it had become imperative to decide which axis was the better one to focus on next. And it was over this conundrum, unfortunately, that the U.S. Joint Chiefs of Staff had become irremediably divided.

All spring and early summer General Arnold had favored the seizure of atolls and air bases in the Central Pacific. General Marshall, however, had wanted a more cautious advance around the Pacific Rim, under MacArthur. Admiral King had opposed this idea, since Nimitz's ships and carriers would, he pointed out, be unable to support MacArthur's army troops ashore when under threat from Japanese land-based bombers in the Philippines. It was a reality the British had learned to their cost when the *Prince of Wales* and HMS *Repulse* had been sunk by Japanese land-based aircraft in December 1941, off the coast of Malaya. King did not intend the U.S. Navy to make the same mistake.

But General MacArthur, the man on the spot, was gung ho to invade the Philippines — in fact had long seen himself as the obvious man to command the entire Pacific theater. Pressing his cause via every visitor to Australia, as well as via correspondence with senators and congressmen, MacArthur had been delighted by a growing movement in Congress and in the anti-Roosevelt press in the late fall of 1943 to have him made the single supreme commander in the Pacific, similar to pressure at that time to make General Marshall the single supreme commander for "all Europe."

The "Navy fails to understand the strategy of the Pacific," MacArthur had complained in writing to the secretary of war in January 1944, pointing out that all operations against Japanese land forces in the Pacific *had* to be backed by land-based American air power, otherwise the "attacks by the [U.S.] Navy, as at Tarawa," would lead to casualties the people of the United States would never tolerate. He hoped Stimson would make this clear to the President.[2]

Statistics had certainly been on MacArthur's side. Tarawa — an atoll in the Gilbert Islands that had been attacked by Nimitz's forces in November 1943

—had resulted in almost four thousand U.S. casualties: 1,696 killed. Similar attacks would meet the same result, MacArthur had prophesied: "tragic and unnecessary massacres of American lives." Ergo, the President must appoint him supremo in the Pacific, with all the backing the U.S. Navy could bring to bear to support his efforts. "Mr. Roosevelt is Navy minded," he'd argued, and Mr. Stimson "must persuade him. Give me central direction of the war in the Pacific," he'd begged, "and I will be in the Philippines in ten months . . . Don't let the Navy's pride of position and ignorance continue this great tragedy to our country."[3]

The President, however, had heard such wild claims from MacArthur before—in fact repeatedly, ever since MacArthur's forces had been attacked in the Philippines in 1941. He was thus unwilling to give the general carte blanche, when Admiral Nimitz had been directing the war in the Central Pacific with even greater naval success. Wisely the President had therefore declined to interfere—the more so since General Arnold had promised him, at the White House, that increasing numbers of the new B-29 Superfortress bombers would be available in the spring of 1944. They would be able, once airfields were established, to bomb Japan from China *and* the Central Pacific, if the Navy managed to seize the islands of the Marianas: Saipan, Tinian, and Guam.

The question of casualties in invading the Marianas, however, had aroused fierce arguments similar to those over Overlord—MacArthur continuing to disparage the idea of seizing the islands, and pressing for a more methodical, leapfrogging peripheral advance.

For a while MacArthur's strategic view, supported by Marshall, had prevailed. The Japanese were clearly going to defend to the last man the territories they had conquered in their 1941–42 rampage—in fact the closer the Allied forces advanced toward the home islands, the more suicidal the Japanese would become, as a matter of martial honor. A mere seventeen men out of almost five thousand soldiers had surrendered at Tarawa. As one historian later wrote, "collective death" seemed to be the preference of the Japanese military rather than "the ignominy of surrender."[4] Even Admiral Nimitz had been of a mind to favor MacArthur's plan of advancing via the Philippines and merely blockading the Japanese islands, until overruled by Admiral King in Washington.

King had been appalled. As he'd pointed out, the Japanese were going to defend *all* their conquests, bar none—like the Germans—whether in the Philippines or other Pacific islands. Retreat was dishonorable in their eyes. It was therefore important to strike before the Japanese had time to reinforce their positions—especially those positions that would be strategically helpful to the Allies. Establishing bases in the Marianas for U.S. strategic bomber forces would be important not only for the bombing of Japan's war industries, but also for protecting U.S. naval forces by providing airfields for land-based

U.S. planes—thus allowing the United States to dominate both the Western and Central Pacific. Occupying the Mariana Islands promised to provide not only airfields, but harbors and submarine facilities. Such a direct, D-day-like assault on the Marianas might even lure the Japanese Combined Fleet into leaving the security of the Sulu Sea, north of Borneo, and accepting large-scale naval battle. Proving his point, successful seizure of the Marshalls had been achieved at relatively little cost, and had been followed by the destruction of the crucial Japanese air and naval base at Truk, which was hammered from the air and largely left by the U.S. Navy to die a slow death. Nimitz's success had thus transformed the war and the balance of power between himself and MacArthur, in King's eyes. In fact King had taken Nimitz to see the ailing president in person at the White House—two weeks before the President's diagnosis of heart failure—to recognize his achievement.

Nimitz had found the President "obviously not well," the admiral's biographer later described, using interviews and Nimitz's own papers. "His face was ashen and his hands trembled. Yet he smiled and turned on the Roosevelt charm for his visitors." He'd also seemed satisfied with the emergency Joint Chiefs' recommendations that morning for "immediate operations in the Pacific Area," after listening to MacArthur's surly deputy, Richard Sutherland, as well as Nimitz himself: namely that Nimitz should help MacArthur vault his way along the New Guinea coast to Hollandia, in April 1944, but that Nimitz should thereafter go ahead and invade the Marianas, in the Central Pacific, on June 15—to be followed by the Palaus in September, from which the bombing of Japan proper could begin. In November MacArthur would then invade Mindanao, the second largest of the Philippine Islands—leaving open the question of whether to then move on to Luzon, the largest and northernmost island, in February 1945, or to ignore the Philippines and assault directly Japanese-held positions in southern China or Formosa. The President had "listened with attention to the briefing and approved the strategy"—glad, he'd said, "to see that the drives were directed toward the China coast, for he was determined to keep China in the war."[5]

The strategic plans hadn't addressed, however, how Japan was actually to be *defeated*—which was, in the end, "his objective." Japan's actual defeat and surrender, the President had emphasized, was a project he wanted undertaken just "as soon as the Allies had enough forces" for the task, for it was only then that his United Nations vision could be implemented. After that cautionary remark, the President's concentration had appeared to fail. "He began asking irrelevant questions and making random comments," Nimitz's biographer recorded.

It had been clear to Nimitz that Roosevelt was "getting tired," and when the President had asked the admiral why, after neutralizing Truk in recent weeks, he'd sent his carriers to "raid the Marianas" in advance of the assault landings

— thus giving the Japanese advance warning of future invasion — Nimitz had taken a deep breath. He'd told the President the story of a famous surgeon who'd added an extra procedure to an appendectomy "as an encore" for his colleagues who'd come to witness the operation. "So you see Mr. President, that was the way it was. We just hit Tinian and Saipan for an encore."[6]

The President had thrown back his head and laughed — but hadn't had the energy or focus to belabor the point. Or his lingering concern.

Nimitz or MacArthur? MacArthur's brief foray into the 1944 presidential nomination campaign had certainly not helped his cause. Asking MacArthur simply to sit still in Australia, Senator Vandenberg — the President's fiercest isolationist foe in the Senate — had plotted to get a silent MacArthur drafted by the Republican Party by acclamation at its convention.[7]

Asking MacArthur to be patient, however, had been naive of Vandenberg. MacArthur would not have been MacArthur had he been willing to hold his tongue, let alone allow others to shepherd his ascent to the presidential throne on the assumption — or presumption — that they would then have control over his actions. MacArthur had thus taken time out from his military duties to respond to an insignificant Nebraska congressman's letter — praising, in a written response of his own, the congressman's recent diatribe against the President's New Deal.

Once published by Congressman Arthur L. Miller, MacArthur's letter had overnight ruined Vandenberg's plan — "crucifying the whole MacArthur movement in one inane moment," as Vandenberg noted in his diary.[8] "MacArthur would have been incomparably the greatest Commander in Chief we could have had in this war," Vandenberg had afterwards rued, and "our most eligible President" at the ultimate "peace table"[9] — completely unaware how unbalanced MacArthur could be at times, how incapable he was of working with any but subordinates, and how often loopy was his military judgment.

Publication of the general's correspondence had, in any event, forced MacArthur to declare he would not accept the Republican nomination even if it were offered to him, acknowledging the "widespread public opinion that it is detrimental to our war effort to have an officer in high position on active service at the front considered for President." As Vandenberg had noted in the privacy of his diary, however, "That is not the *real* reason."[10]

MacArthur's lack of political judgment, in Vandenberg's eyes, had been compounded by hearing stories of "veterans returning from the South Pacific" who were not "enthusiastic about our friend," as the senator had earlier confided to retired general Robert Wood, founder of the America First movement and chairman of Sears, Roebuck and Company. A "frank canvass of the men out there" in the Pacific, from a soldier he trusted, had revealed MacArthur's "growing unpopularity. What does this mean?" Vandenberg had asked Wood, deeply "disturbed."[11] Governor Dewey, meantime, had swept the Republican primaries and ended MacArthur's dream.

As he now sailed to Pearl Harbor, the President was aware, after summoning MacArthur to meet him there, that the very failure of his political aspirations would only make the general more difficult to handle — MacArthur determined to have his way at least over military strategy in the Pacific.

Certainly none of the U.S. chiefs of staff dared to do battle with "Dug-out Doug," as MacArthur was widely known by his army troops. For his part Admiral King deeply — and rightly — resented MacArthur's repeated claim that the U.S. Navy could have "saved" the Philippines after Pearl Harbor, if only Admiral King had been more offensive-minded. Yet King, for all his reputation as a "blowtorch" admiral toward his subordinates in the Navy, had never met MacArthur in person — and remained disinclined to do so. A lamentable public speaker — fearful of giving a talk in front of others unless he had a script before him — King was often so nasty toward army and air colleagues in small meetings that General Marshall, on one occasion, had "finally said to him, thumping the table, 'I will not have the meetings of the Joint Chiefs of Staff dominated by a policy of hatred. I will not have any meetings carried on with this hatred'"[12] — an eruption that had, apparently, "shut up King."[13]

King might be a difficult colleague, but MacArthur was in a league of his own — a veritable nightmare to deal with. Once rescued from Corregidor on the President's orders in March 1942, MacArthur had behaved in an almost megalomaniacal manner in Brisbane — yet had also become irreplaceable as an American leader when restoring Australian self-confidence, following the fall of the Philippines. Moreover as a combat general, MacArthur was no man of straw. Provided he had effective, selfless, and subordinate army, air, and navy combat commanders to do his bidding, he was highly professional, paternalistic, and supportive — as long as they did not detract from "his" limelight.

Irrespective of MacArthur's deplorable vanity, then, the achievements of the general's Southwest Pacific forces had, in short, been commendable. It was in the realm *beyond* his own command orbit that the real problems arose — especially with the U.S. Navy.

As General Marshall would confide later, behind the scenes, the war in the Pacific was a disaster area: "a war of personalities — a very vicious war." The "feeling was so bitter, the prejudice so great," in terms of interservice, intercommand hostility, that "the main thing was to get [them] in agreement."[14]

In the end only one person could achieve it: Mr. Roosevelt, the man who'd appointed both men to their current posts.

Hawaii, in short, would reveal whether the dying commander in chief could still command.

54

Deus ex Machina

THE CLOSER the USS *Baltimore* drew to the Hawaiian Islands on the morning of July 26, 1944, the more obvious it became that the President's arrival would be nothing like his entry into the harbor at Oran, in the Mediterranean, eight months before. This time the warship found itself escorted by some eighteen patrol planes from Oahu, then a further two *hundred* aircraft "operating from carriers at sea off Oahu."[1]

The President was in fact steaming toward a huge Hawaiian welcome.

"I was up on the bridge," Lieutenant Commander Bruenn later recalled. "The President and a couple admirals were there too. We were all very casually dressed in khakis and no neckties. As we came into Pearl Harbor a little boat came out":[2] an admiral's launch bearing the naval district commander, Admiral Robert Ghormley, and the supreme commander in the Central Pacific, Admiral Nimitz, coming to greet the President in person, together with General Robert Richardson, the army officer commanding U.S. Army troops in the Central Pacific. Also Territorial Governor Ingram Stainback (Hawaii was still a U.S. territory, not a state), and other notables. The sun was shining, the temperature was still in the midseventies with only a light wind.

"We assumed, as we neared our destination, that our expected arrival had been kept a secret. Imagine our surprise, therefore, when we steamed into the harbor and up toward the docks, to see all the Navy ships with the men at attention at the rails," recalled Judge Rosenman.[3] "The President's flag," Lieutenant Rigdon recorded in his "Log of the President's Inspection Trip to the Pacific," was "hoisted at the main in the *Baltimore* in recognition of honors rendered" — a "violation of sound security measures in time of war," he noted, "but it was found that the news of the President's visit," in disregard of censorship, "had become common knowledge in Honolulu two days before."[4] "As we entered the harbor there must have been a hundred ships there with sailors manning the rails in whites,"[5] Bruenn described.

A hundred ships? It was a deeply satisfying moment for the President — if not for Mike Reilly and the Secret Service detail, who had flown ahead to Hawaii. In a matter of two and a half years the United States had become the most powerful global military power in the world, bar none—and for the third time in sixteen months the U.S. Commander in Chief was inspecting his forces abroad in an active theater of war.

As the USS *Baltimore* docked at Pier 22-B in the Navy Yard, immediately behind the USS *Enterprise* carrier, the mood on shore was one of understandable excitement. "We finally docked and at least 20 or 24 flag officers came aboard to pay their respects," Dr. Bruenn recalled—"Navy, Army, Marines, Air Corps, the works." As well as the USS *Enterprise* there were no fewer than six more U.S. aircraft carriers moored in the harbor, and three huge battleships, nineteen submarines, more than thirty destroyers, and almost two hundred landing craft and auxiliary vessels.

This was Pearl Harbor, 1944: epicenter of modern American military might in the Pacific, and a reason for all to celebrate. All, to be sure, save General MacArthur, whose plane had reportedly landed from Brisbane, but who was conspicuous by his absence at the pier.

Afterwards, in serious military historiography, it would be FDR's arbitration in the great strategy debate that would be considered the primary feature of the President's trip to Pearl Harbor. And to a large extent this was so—Mr. Roosevelt's personal mediation between Nimitz and MacArthur proving crucial to the conduct of the war in the Pacific. Yet the Commander in Chief was going out to the Pacific as president, too—and Judge Rosenman, his speechwriter and White House counselor, was with him throughout, judging the temper of the times in terms of the President's reelection chances.

Rosenman, in particular, was interested to see the people of Hawaii, not only dignitaries, gathering in the tens of thousands to celebrate the President's arrival. No stone was to be left unturned in hosting the President, Admiral King had instructed—and it was clear Nimitz had followed his orders to the letter, as the huge crowds of civilians and military personnel at the quayside showed.

Though he still often felt desperately ill, the President was touched. As always he was energized by public support, but he nevertheless had his own agenda. By winding MacArthur into his official inspection of the island, he hoped to *educate* the wayward general. Hawaii remained the primary Pacific military headquarters—one that MacArthur had still not visited a single time in all the years he'd been in Australia! Somehow the President must show the prickly egoist there were no hard feelings about his brief foray into presidential politics that spring, and no bias toward one service or command over another. They were all now part of one team and must therefore work to defeat Japan together, not apart—this would be his mantra.

MacArthur's absence from the quayside thus came as something of an insult, for there was no sign of the general, or word from him. "Everybody sat around for maybe half an hour. All of a sudden over the loud speaker: 'General MacArthur is coming aboard,'" Dr. Bruenn remembered. "And here he came wearing a leather jacket and his soft hat, making his appearance. He was that kind of guy."[6]

The general's late appearance certainly stole the show. As Rosenman put it, MacArthur "could be dramatic — at dramatic moments."[7]

Whether it was the President or Admiral Leahy who remarked to MacArthur that wearing a heavy red leather flying jacket was perhaps de trop in summer weather would be disputed later. Either way, MacArthur was, all concurred, completely unembarrassed. He'd failed to get Republican acclamation as presidential nominee, but had now made a spectacular entrance, in front of a vast audience on the quayside, and had no reason to feel shame at his tardiness — or his attire. "Well, you haven't been where I came from, and it's cold up there in the sky"[8] was his rejoinder — deliberately reminding the President how far he'd had to fly to get to the meeting.

There were photographs taken on deck, after which the President asked Nimitz, MacArthur, Ghormley, and Richardson to come to his cabin. There it was agreed they'd all meet the next morning at the residence in Waikiki where the President was going to stay. Together they would tour the island's installations — thus ensuring MacArthur would gain a better idea of the Central Pacific theater's main base. After this they would dine at Waikiki as the President's guests. Only the following day, Friday, July 28, would the President, Nimitz, and MacArthur get together, with Admiral Leahy, at the Waikiki house, to discuss future strategy.

In other words, the President was hoping to "soften up" his two supreme commanders before, like gladiators, they faced off against each other in front of him.

With that agreed, they went briefly back on deck for more photographs and newsreel filming and then went their separate ways.

MacArthur was furious.

55

Slow Torture

THE HOLMES ESTATE on Kalakaua Avenue, Waikiki Beach, was palatial. "Since Mr. Holmes' death a short while ago," wrote Lieutenant Rigdon, "his home has been used by our naval aviators attached to carriers of the Pacific Fleet as a rest home between missions. It is an ideal spot for this purpose as we found it very comfortable and quiet there"[1] — the fliers less happy at having to rest elsewhere, however, during the President's stay.

The grand home fronted the famous beach, with rolling white breakers rippling toward the shore. Behind high walls it possessed grounds "studded with shrubs and tall Royal Hawaiian palm trees," and spouting water fountains. The President occupied a large suite on the third floor, accessible via an outside elevator. Dr. Bruenn was lodged there, too, to be on hand if there was any sign of new heart problems.

General MacArthur was surprised by the President's unmistakable physical decline, despite the restorative effects of the sea voyage to Hawaii. Indeed he would tell his own doctor, on returning to Brisbane: "Doc, the mark of death is upon him! In six months he'll be in his grave."[2]

In his memoirs, two decades later, MacArthur remembered how "shocked" he'd been by the President's "appearance. I had not seen him for a number of years, and physically he was just a shell of the man I had known. It was clearly evident that his days were numbered."[3] Yet, as MacArthur noted, it was more a physical change than a mental deterioration: the President still as focused and determined, in the few hours a day that he was able to work, as ever. Even physically the President's cascading health was not evident save close-up. In photographs and newsreels, taken at a distance, he still looked the leonine figure who'd steered his country in peace and war for a record twelve years: his demeanor as authoritative as ever.

That the President had a shrewd agenda was immediately apparent to MacArthur, who was no fool. There had been no meeting of minds or even social gathering the first evening, following the President's arrival, leaving MacArthur to dine alone with General Richardson and contemplate the next day's "political junket," as MacArthur had disparaged it on the flight from Brisbane. His friend Frazier Hunt, a toadying print and radio journalist who was taking care of MacArthur's correspondence at his headquarters in Australia, later recreated MacArthur's mood at General Richardson's house. "As he walked the floor of his bedroom here in Richardson's quarters this late July night of 1944, he talked without restraint to a trusted member of his staff [probably his press officer and former journalist, Lloyd Lehrbas, who had traveled with him] regarding his long years of struggle and his many defeats and frustrations: and he spoke of his country's inadequate leadership, the terrible mistakes made in the war and America's uncertain future. He seemed to unburden himself in a way he had seldom if ever done before in all his life."[4]

The general had every reason to feel he'd been snookered. He had brought with him his press officer — who censored on MacArthur's behalf all communiqués in and out of the Southwest Pacific until they became, as one biographer wrote, "as lush as the New Guinea jungle" — but the President had also brought his *own* contingent of press officers and journalists: the Office of War Information director, Elmer Davis, and three White House press agency correspondents, Merriman Smith, Howard Fleiger, and Robert Nixon, as well as three radio pool correspondents. Further war reporters and photographers, moreover, were on tap in Hawaii — none of whom would be subject to MacArthur's publicity edicts and manipulation. The general would thus have to travel the breadth of Oahu the next day in the company and shadow of the President of the United States ahead of the following day's strategic duel with Nimitz before the Commander in Chief, at the Holmes estate.

MacArthur was understandably frustrated. He had not even brought his deputy, General Sutherland, only Lehrbas and Bonner Fellers, his military secretary — without maps, even. As MacArthur later wrote, he'd planned to fly straight back to Australia after the military discussion — and had had no interest in touring military camps containing ordinary soldiers, sailors, and airmen. It was no small wonder he felt "as depressed and frustrated" as he had on Corregidor.[5]

The President, however, saw matters very differently.

Admiral Ghormley had come to his suite at Waikiki after dinner on Wednesday to "arrange the schedule of the three-day visit" in detail.[6] MacArthur's showy appearance at the quayside had not unduly upset the President; rather, it had spurred his competitive spirit. He had therefore asked Ghormley for a comfortable open limousine to be sequestered. A famous red one belonging to a local madam was turned down for obvious reasons, but the big black one belonging to the local fire chief, when suggested, had been considered perfect — for the President wanted, he said, to sit three abreast with his two supreme commanders in the theater, literally, of war. The photos and newsreels would play extremely well at home, once he'd departed: the equivalent of those memorable photos of himself with Joseph Stalin and Winston Churchill in Tehran. In this case, though, instead of showing him at the Russian Embassy in Iran, he would be at the heart of the war in the Pacific, together with his two top commanders. Moreover, seen live by servicemen and the people of Oahu, they would present a vivid picture that, contrary to reports of intertheater jealousies, there was a Nimitz-MacArthur unity of strategic and tactical direction of the war in the Pacific, under the aegis of the nation's commander in chief. The President would sit in his usual place on the curb side (for ease of egress in emergency) of the vehicle, with Nimitz on the offside — and MacArthur squashed between them, unable to escape!

The only question was: would the President be able to manage *six hours* of inspection in different venues, when he had barely survived one hour in San Diego? His doctors were ambivalent, but the President insisted. He had rested for almost a week on the USS *Baltimore*; it was time to see whether, standing for reelection in the fall, he could possibly manage one half day.

With two big Stars and Stripes mounted over the limo's front wheels, the flags fluttering in the breeze, they rode past "Base Hospital Number 8 (at McGrew Point, Pearl Harbor), past the many activities at Aiea; past the sugar mill at Waipahu; and on to our first stop — the Marine Corps Air Station." The "Naval Air Station at nearby Barbers Point" followed, and the Naval Ammunition Depot at Lualualei, as they drove through the "many scattered magazines and ammunition stowage facilities," and on to the Navy's "main radio transmitting station for Hawaii," which was "high atop the Waianai Mountains," Lieutenant Rigdon recorded.[7]

As supreme commander, MacArthur had earned an unfortunate reputation for ignoring modern naval warfare techniques — in fact he was known to have set foot aboard a navy vessel only once since being rescued by PT boat from the Philippines. The tour of the island's naval facilities was thus an education. At the Schofield Barracks, forty thousand airmen and army troops had the chance to see the President as he was driven past "long lines of tanks" and "the hangars and plane-covered aprons of Wheeler Field," where wounded men from the battles on Saipan and Guam were being lifted down on stretchers to be taken to the post hospital. With the President's car driven up onto a special raised stand, the troops marched past — even Japanese American troops, since the President had emphatically overridden his Secret Service advice against exposing himself to "any Japanese fanatic in the ranks" who might "shoot the President at point blank range."[8]

After lunch at the officers' mess, the President explained he had not come to make speeches, but to see "with my own eyes" the change since his last visit in 1934, "ten years to the day today," as he said in brief remarks. Of twelve tanks at the review in 1934, "seven broke down before they could get past," and the air review was similarly deficient. He was thus delighted by the transformation. "It is being felt all through this area. All the way down to General MacArthur's area, which, thank the Lord, is coming a little closer towards us, and automatically closer to the enemy than it was two years ago. It is good to see the three services together, because I think this morning I have seen not only the Marine Corps Air working together, but the Navy Air and the Army Air working together in all their component parts. I wish everybody back home could see and understand a little more of what's going on out here."[9]

For MacArthur this was a form of slow torture: a testament to modern amphibious warfare under Admiral Nimitz, but one that left MacArthur looking somewhat tortoise-like. General Richardson marched his entire Seventh Division before the President after lunch, but MacArthur could see his old army friend now only as a sort of prisoner of the Navy: "an unhappy man. He lives like a prince with fine cars and a fine home," MacArthur described, "but he has no authority." There had been an unholy row after the commanding officer of the U.S. Twenty-Seventh Division, fighting alongside U.S. Marines, had had to be relieved by a Marine general during the battle on Saipan as insufficiently aggressive, and Richardson was still smarting. "He is a fine, courteous gentleman, so the Navy have him licked," MacArthur railed at what he considered an insult to the Army. But, as MacArthur put it, "they have beaten him so many times there is nothing more he can do."[10]

The afternoon was no less exhausting than the morning: inspection of the Naval Construction Battalions camp at Moanalua Ridge (the "Seabees," as the President said in his remarks there, "now known on every ocean and every continent" — in fact a part of the armed forces that had "come forward as an institution more quickly than any one I know of in the whole of our history"); then Camp Catlin, home of a unit of the Fleet Marine Force, Pacific; and finally the Royal Hawaiian Hotel, which was partly used "as a rest center for U.S. submarine crews" — and where the President himself had stayed in 1934.[11]

Poor MacArthur had been forced not only to eat camp fare all day, but humble pie, in his own opinion — making only small talk as the presidential party toured the military installations and the President spoke to his Marines, soldiers, sailors, submariners, and airmen. And after six hours of that, there was still dinner to come — for MacArthur was expected to return and dine two hours later, at the Holmes mansion, along with Admiral Nimitz, Admiral Leahy, and Admiral Halsey, commander of the Third Fleet, who was flying in to Honolulu from the United States that afternoon.

General MacArthur could later remember nothing of the meal, only that he was completely without support staff — and at the next morning's strategy conference would have to "go it alone," as he put it.[12]

He would certainly never forget what happened after the meal. The President, who had undertaken more that day in Hawaii than on any single day since his trip to see Stalin eight months earlier, seemed to have accessed some hidden reservoir of energy. As the staff removed the dessert plates, he announced that the commanders were to follow him into the conference room next door, where military and naval maps covered the tables. From one wall hung a ten-foot-high map of the Pacific — and settling into bamboo chairs before it, the two supreme commanders in the Pacific were each asked, without preparation, to make a presentation.

Taking a long bamboo pointer, the President pointed at the Marianas and New Guinea. "Well, Doug," he shot at MacArthur, turning to face the general. "Where do we go from here?"[13]

56

In the Examination Room

As one MacArthur biographer noted, "Even a dying FDR was formidable."[1]

MacArthur was first up. "'Mindaneo,' Mr. President, then Leyte,'" the general answered, using the bamboo pole to locate the islands, "and then Luzon.'"

Luzon was the northernmost of the Philippine Islands. Was it really necessary, in a war to defeat Japan? The parallel with Italy, in terms of the defeat of Hitler's Third Reich, was unavoidable. MacArthur's argument, however, was that possession of the islands would allow American forces to cut Japan off from its vital sources of oil and other necessities of war. It would have major political implications, he added. The liberation of the islands, and the granting of immediate independence to the Philippines — as mandated under the Tydings-McDuffie Act, which the President himself had signed in 1934 — would demonstrate to the world the United States was *not* pursuing an imperialist agenda.

Admiral Nimitz then presented his alternative strategy: bypassing the Philippines, where there were more than three hundred thousand Japanese troops defending the islands, and advancing instead straight across the Pacific to Formosa by the summer of 1945, from which the main Japanese islands could either be blockaded or invaded.

MacArthur, in private, was "sure" Nimitz's argument was Admiral King's "and not his own." In front of the President, though, MacArthur did not see, as he later described his performance, how Nimitz could leave hundreds of thousands of Japanese troops "in his rear in the Philippines," where there were substantial numbers of POWs and civilians. Who knew what the Japanese were capable of, in terms of atrocities? And the casualties involved in a direct line of attack across the Pacific would be, he claimed, daunting.

MacArthur had come to the meeting without staff, but he had done his homework. He was convinced that the President already knew from his Joint

Chiefs of Staff in Washington "the general concept of the [King/Nimitz] plan" but was "doubtful of it." He supposed it was the President's very doubts that had led him to Hawaii to make a decision that he could otherwise have communicated from Washington. The President was nevertheless "entirely neutral in handling the discussion," MacArthur remembered gratefully.[2]

Although no minutes were kept of the meeting, Admiral Leahy's diary entry that night confirmed MacArthur's memory. Having witnessed how the President had run the Tehran meetings, Leahy was no stranger to his skills as a chairman who allowed both sides to present their case, before coming to a decision. In his journal Leahy noted: "General MacArthur is convinced that an occupation of the Philippines is essential before any serious attack on Japanese-held territory north of Luzon. He stated that he now has in his command sufficient ground and air forces to take the Philippine Islands, and his additional needs are only landing craft and naval support."[3]

Nimitz contested MacArthur's claim. The Japanese possessed substantial, well-embedded military forces in the Philippines, but they also had significant naval and air strength in the shallow and contorted island waters, meaning that an American assault landing could not be backed by U.S. naval forces, vulnerable to land-based air attack, but only by U.S. submarines. MacArthur's preferred Philippine strategy would thus not allow for the use of the very weapon Nimitz's forces had so brilliantly developed over previous months: their ability to mount D-day-like amphibious assaults on any chosen island or atoll, while bypassing and neutralizing those that were too heavily defended.

Backwards and forwards the arguments were batted—with the President alive to the strengths and weaknesses of each position.

Formosa was close to the Chinese mainland, yes—and certainly closer to Japan than Luzon—but there were still a million Japanese troops in China, and powerful Japanese defense forces on Formosa itself. Above all, Formosa was nearly two thousand miles from the Marianas. Eisenhower's D-day assault on the beaches of Normandy had been mounted across eighty miles of the English Channel, already an immense undertaking that Churchill had feared would fail. This proposal was of an entirely different order, even with U.S. carrier support.

Under questioning Nimitz agreed, however, that some, at least, of the Philippine Islands would have to be seized by MacArthur, if only to protect Nimitz's left flank, and in order to interdict Japanese reinforcement elsewhere. MacArthur, for his part, was made to face up to the enormity of what he was proposing: namely attacking a huge Japanese army, in well-established defensive positions, in the difficult jungle terrain of the Philippines—islands the Japanese were already reinforcing. Military intelligence was estimating almost ninety thousand Japanese troops on Luzon, with a further twenty-four thousand on the island of Leyte, and perhaps sixty thousand on Mindanao. How did he propose to defeat those forces with only the forces he currently had, as he claimed? And what of U.S. casualties? "But Douglas," the President said, "to take Luzon would demand heavier losses than we can stand. It seems to me we *must* bypass it."[4]

Challenged by the President, MacArthur said that his "losses would not be heavy, any more than they have been in the past. The days of frontal assault are over. Modern infantry weapons are too deadly, and direct assault is no longer feasible. Only mediocre commanders still use it. Your good commanders do not turn in heavy losses." Filipinos, too, would probably assist the U.S. liberators, like the Maquis in France, whereas on Formosa, which had been under Japanese control for half a century, there was no promise of aid.

As Leahy saw it, this was one of the President's finest hours as U.S. commander in chief: questioning and maintaining focus so that, in a calmly professional manner, the two supreme commanders each had their say—and took account of each other's positions. Certainly in the history of the war on the Allied side there had never been anything like it. "Roosevelt was at his best as he tactfully steered the discussion from one point to another and narrowed down the area of disagreement between MacArthur and Nimitz," Leahy later recalled. "The discussion remained on a friendly basis the entire time."[5]

Decades of military historians would later underplay the significance of the parley. Yet to Leahy it was a landmark in the prosecution of the war, a "much more peaceful" argument "than I had been hearing in Washington. Here in Honolulu we were working with facts, not with emotional reactions of politicians." The reported antipathy between the commanders was not evident, either, despite MacArthur's reputation even in his own service. "It was no secret that in the Pentagon Building in Washington there were men who disliked him, to state the matter mildly," Leahy noted with his customary understatement.[6] Before the President in Hawaii, however, the two supreme commanders had exemplified as nowhere else the vast changes in modern warfare—technological, tactical, and logistical—while they pushed their pointers across the great wall map of the Pacific.

For all their military savvy in fighting the Japanese, both MacArthur and Nimitz knew the war was gradually moving into a new phase in the Pacific: one where politics could no longer be excluded. Just as had been the case in Cairo and Tehran, the closer the Allies moved to victory over the once-omnipotent military empires of the Third Reich and Japan, the more did the endgame pose political questions. Admiral King had left a memo with Nimitz to give to MacArthur, detailing British aspirations in the Pacific: namely British occupation of the Dutch East Indies, Admiral Mountbatten to assume responsibility for Australia also, while MacArthur's forces moved into the Philippines, and the transfer of Royal Navy vessels, currently in the Indian Ocean, to the Pacific. Clearly the British, having had to be rescued by the United States military, now wished to reestablish their colonial empire and territories in the Far East, from India to Hong Kong, on the coast of China—and possibly garner more.

The future in Asia—given the war-within-a-war between Chiang Kai-shek's forces and those of Mao Tse-tung in China—was equally uncertain. Even the military challenge would have political consequences, not least at home. Evicting the Japanese from the territories they had overrun promised to be immensely costly, in blood and treasure, before U.S. forces even got to Japan proper. If the Japanese continued to fight to the last man standing, could the war possibly be won without incurring casualties unacceptable to the people of America? Would not Russian help be required, not only in providing air bases for U.S. bombers and harbors for U.S. military supplies, but troops as well, in Manchuria, where a three-quarters-of-a-million-strong Japanese army was stationed?

It was testament to the President's acumen that, listening to his two supreme commanders, he recognized their two unique merits both as commanders and strategists. Nimitz, quiet and determined, had proven MacArthur utterly wrong in the general's adamant opposition to Operation Forager, the recent seizure of the Marianas. Nimitz's brilliant amphibious invasions of Saipan, Tinian, and now Guam had not only permitted the United States to acquire inviolable bases from which to bomb Japan and dominate the Central Pacific with naval forces, including submarines, but would now allow it to establish and maintain such naval bases across the Pacific *after the war was won,* without needing to become a colonial empire like the British—or the Dutch or French.

By the same token, however, MacArthur's operations along the northern coast of New Guinea had been exemplary, confounding Nimitz's predictions. By advancing in leaps and bounds—"leapfrogging"—and by bypassing Japanese strongholds, MacArthur had minimized casualties while pushing his forces within striking distance of the southern Philippines.

Yet for all that, it was MacArthur's political vision that most impressed the President. The general had, it seemed, learned his lesson. In February 1942 the President had had to spur MacArthur into refusing to allow the president of the Philippines to negotiate with the Japanese, arguing that the United States must show the world its commitment to democratic principles, not Japanese militarism.[7] Now, two and a half years later, here was Douglas MacArthur arguing that, in addition to the importance of defeating the Japanese, he and the United States had a "sacred" obligation to the people of the Philippines to liberate them—and demonstrate to the world the genuineness of American principles by returning sovereignty to the islands, as per the Tydings-McDuffie Act of 1934. Fulfilling that goal, MacArthur reasoned, would probably do more to persuade postwar Asia of America's good intentions, and at less ultimate cost, than anything else the United States could do.

The President was of the same strategic mind. If, of course, it could be done successfully, without too great a slaughter.

The specter of slaughter also hung over Admiral Nimitz's proposals to invade Iwo Jima, the Bonin Islands, and Formosa.

Iwo Jima and Formosa, under Nimitz's outline plan, could be used as launchpads for an invasion of Okinawa, the Japanese island halfway between Formosa and the Japanese home islands. The President acknowledged such a strategy to be by far the most direct route to Tokyo. But at what cost in blood, if the Japanese continued to contest, suicidally, every inch and every atoll they had conquered?

The President certainly did not buy MacArthur's claim that the Philippines could have been "relieved" by counteraggressive naval action in early 1942. Nor did he agree that failing to "relieve" the islands as a matter of priority in 1944–45 would not be forgiven by the people of America, let alone the Philippines. War was war. But so was postwar!

MacArthur was surely right in seeing the liberation of the Philippines as a symbolic and political duty of the United States, for it was in the Philippines where the difference between Japanese militarism and American democracy could best be seen by the world. Defeating the Japanese military as swiftly as possible was one thing, but showing that difference in the Philippines — where Japanese occupying forces were committing atrocities on a frightening scale — was arguably more important. Who knew what the Japanese would do to American POWs and civilians in captivity if the Japanese soldiers in the Philippines saw their own homeland being bombed and invaded?

It was a sickening thought for all of them. At midnight the President called it a day, and sent the supreme commanders back to their quarters, "with the President making no final decision," as MacArthur related.[8] They were asked, though, to return at 10:15 a.m. to continue the discussion. That the President had conducted such an intense strategy session at the end of a long day inspecting troops was nothing less than a miracle, given his state of health. Or ill health.

By the next morning both supreme commanders seemed to better understand and respect each other's views, at least. MacArthur argued that, once in possession of the Philippines, he could "sweep down" on the Japanese forces in the Dutch East Indies "from the rear." Moreover, bypassing the Philippines, and thereby forfeiting logistical backup to Nimitz's plans for direct assault, would not make military sense, he claimed, given the gamble Nimitz would be taking in mounting invasions of Formosa and Okinawa, which MacArthur accepted as the most direct way of defeating Japan. However tough the task, the northernmost Philippine island, Luzon, would simply have to be taken. And with this view Admiral Nimitz, departing from King's script, felt compelled to agree.

Thus the consensus emerged—beginning with the decision that the British would be denied a role in the Pacific lest they cause the same kind of problems they had caused in the Mediterranean. With the British considered irrelevant, the two supreme commanders in the Pacific should pursue a biaxial strategy: MacArthur would be responsible for liberating the Philippines and doubling back on the Dutch East Indies, with as much naval and carrier support as Nimitz could give, while Nimitz, for his part, would continue to deal with the Japanese fleet, and establish major strategic bomber airfields on Guam and the Marianas. He would set up naval bases and a new Central Pacific forward headquarters there also, and plan the invasions of Iwo Jima, Formosa, and Okinawa for the next year. Whatever happened in China, this strategy would hopefully be enough to force the Japanese to sue for surrender, unconditionally.

The President had achieved his aim: the two theater commanders were now allies, not rivals. MacArthur would completely rethink his plan of starting his assault on the Philippines by an invasion of Mindanao, where there were more than sixty thousand Japanese troops; Nimitz would rethink his idea of invading Formosa. "I personally am convinced that they are together the best qualified officers in our service for this tremendous task," Leahy considered, "and that they will work together in full agreement toward the common end of defeating Japan."[9]

Could any other commander in chief have managed this? In newsreel film shot that morning, the President looked well—certainly no worse than bald Admiral Leahy, or even MacArthur, whose thin dark hair was pasted across his balding head and who looked pinched and reserved, while Nimitz looked grandfatherly under his thick white hair. Unfortunately this was, of course, an optical illusion—as Admiral Charles Lockwood, commanding the submarine base whose mariners the President had addressed at the Royal Hawaiian Hotel the day before, recalled. The "President's appearance was distressing," he later related, noting that his "skin had that grayish tinge one often sees in the very ill."[10]

Nonetheless the President had seemed masterful in mind and spirit—nursing his limited energy for when it counted, and then pulling out all the stops. There was a brief lunch with MacArthur and Nimitz, both of whom then took their leave of the President: Nimitz to return to his headquarters, MacArthur to fly back from Hickam Field to Brisbane.[11]

The President, however, carried on with a full afternoon of inspections, beginning with the Army Jungle Training Center in Kahana Bay, conducted from the fire chief's open limousine. There he was witness to every aspect of jungle warfare, from enfilading to pillbox assault, bridging to hand-to-hand fighting, with live ammunition; thence to the Naval Air Station at Kaneohe Bay, to inspect Fleet Air Wing Two's carrier pilot-and-aircrew training station. And from there to Kailua, to the amphibious warfare station at Wailupe, followed by the Coast Guard base at Diamond Head. Finally, back at the Holmes estate, there was a "small" dinner party for the President, with music provided by the famous Hawaiian composer and bandleader Bill Akamuhou.

Admiral Leahy was impressed, as he confided to his diary — the performance given "between the house and the sea and under the palm trees, the leaves of which were alternately black and silver in the bright light of a half moon, a very lovely setting and a beautiful sight."[12]

In the midst of the most violent war in human history, it was an idyllic end to an extraordinary day — the last such night the President would experience in his dwindling life.

An even more exhausting schedule of inspections had been set for the next day before the President was to board the USS *Baltimore* in the evening for the voyage home via the Aleutian Islands. Whether this was wise or not, it was a schedule the President declared he was going to keep — and did.

Above all, by continuing his grueling itinerary Roosevelt wanted to prove he could still manage not only the tough business of being president and commander in chief in war, but the physical and mental challenges of standing for reelection a fourth time, with all that would involve — including facing the press.

And so, at 4:45 p.m. on July 29, 1944, two hours before leaving Waikiki, the President addressed a gathering of reporters and radio journalists — telling them, with some emotion, what a "pipe dream" his visit to Hawaii had been. He'd seen America's former "outpost" in the Pacific become its centerpiece — the "main distributing point" for America's prosecution of the war against Japan, which would be pursued relentlessly until the "unconditional surrender" of that empire, just as that of the Third Reich. A willingness to discuss terms in advance might theoretically lead to swifter surrender, he acknowledged — but would only produce the same result as after World War I. "Practically every German denies the fact they surrendered in the last war, but this time they are going to know it. And so are the Japs," he declared. His meetings with Admiral Nimitz and MacArthur over the past two days had been "very successful," and though he obviously could not reveal what had been agreed, the discussions had "involved new offensives against Japan." No meeting with Winston Churchill was planned, since the war in the Pacific was being conducted under American supreme commanders — and would, yes, include the liberation of the Philippines. He did not wish "to possibly give the enemy an inkling as to which way we are going," he said, but "we are going to get the

Philippines back, and without question General MacArthur will take part in it. Whether he goes direct or not, I can't say." As to whether he had merely confirmed existing strategy, or set a new one, the President was typically coy. As he put it, "You review or re-establish strategy about once a week as you go along, it's just normal procedure. But it was very useful, this particular conference. I think it was one of the most important we have held in some time."[13]

It was. But at a greater cost to the President's remaining strength than he would acknowledge. Sam Rosenman and Elmer Davis had been especially concerned lest the sheer intensity of his program of military talks and inspections result in an exhaustion, even physical breakdown, of the President that could not be concealed from public view. Miraculously the opposite had been the case. At Hickam Field the President had spoken to wounded men on stretchers, airlifted from Guam. "What surprise and cheer those boys showed on unexpectedly facing their President!" Lieutenant Rigdon noted.[14]

In the new Naval Hospital at Aiea, above Pearl Harbor, the President had left his car and without self-consciousness toured some of the wards in his wheelchair, wheeled by one of his Secret Service detail.

Many of the patients who had lost arms or legs had had no idea their President was himself disabled. As Rosenman, who accompanied Mr. Roosevelt, later recalled, the President had "insisted on going past each individual bed. He had known for twenty-three years what it was to be deprived of the use of both legs. He wanted to display himself and his useless legs to those boys who would have to face the same bitterness. This crippled man on the little wheel chair wanted to show them that it was possible to rise above such physical handicaps. With a cheery smile to each of them, and a pleasant word at the bedside of a score or more, this man who had risen from a bed of helplessness ultimately to become President of the United States and leader of the free world was living proof of what the human spirit could do to conquer the incapacities of the human body . . . The expressions on the faces on the pillows, as he slowly passed by and smiled, showed how effective was this self-display of crippled helplessness. I never saw Roosevelt with tears in his eyes," Rosenman added; but "that day as he was wheeled out of the hospital he was close to them."[15]

When asked by a reporter, several weeks later, what he made of Republicans' criticism that "the whole trip is political," the President could only smile, and say — to laughter — "Well, they must know better than I do."[16]

Rosenman, for his part, had left the President's group to fly home and work on the coming political campaign. He was also taking a note with him for Eleanor.

"Dearest Babs," the President had scribbled. "Just off — hectic 3 days — very good results. All is well. Ever so much love. Devotedly, F."[17]

57

A Terrible Mistake

LEAVING PEARL HARBOR, the President had every reason to be proud of his trip — the results as good and as historic, he thought, as those he'd gotten at Tehran.

Meantime the press were excited by the President's royal reception in Hawaii — and evidence of his good health. Though their reports would have to be held up until the President was safely back on American soil, their delayed stories from Honolulu carried photos of the President with his supreme commanders and of a host of inspection tours he had made, too — a powerful visual demonstration of America's growing war effort in the Pacific. "New and crushing blows to bring the enemy to his knees and unconditional surrender were planned at a conference by the President," the *Boston Globe*'s war correspondent, Martin Sheridan, reported,[1] together with accounts of Roosevelt's breakneck visit to military installations across Oahu, while the *New York Times* carried the texts of his speeches and talks to troops, Seabees, and wounded soldiers at Aiea.

Elmer Davis, directing the Office of War Information, and Steve Early, the President's press secretary, were both delighted. Moreover when a newspaper in Texas printed a letter from a soldier in Hawaii saying the President of the United States had been there and was going to the Aleutians, the decision was made to release on August 11 all delayed war correspondents' reports from Honolulu.

Davis's and Early's jubilation turned to horror the next day, however.

The President, having reached the Aleutians, had made his two-day tour of the islands, aboard a U.S. destroyer, in miserable weather. At Bremerton he was then scheduled to make a speech to several thousand navy yard workers gathered at the dock, which would also be carried as a radio broadcast to the nation. For reasons no one could ever quite explain, the President was expected to give his address — thirty-five minutes long — standing. Not only standing, but alone, at a lectern, before a crowd that had, by the time the moment arrived, swollen to an estimated ten thousand—and millions more, listening on radio, with further tens of thousands watching newsreel of him that would subsequently be shown in theaters.

The President had not actually stood upright, using his steel leg braces, for more than eight months, however. He had lost so much weight—some twenty pounds—that the braces no longer fit around his waist, it was found, and the steel dug into his skin, beneath his dark suit. Clearly Steve Early's political intent was to show the nation and the world a president returning from the sort of military inspection tour abroad that bespoke his role as U.S. commander in chief, standing on the deck of a U.S. Navy warship returning from Pearl Harbor. But if so, Early had no idea what he was asking of the ailing president.

Whether Roosevelt realized it was not going to work would never be clear. At 5:00 p.m. the President did manage to reach the lectern—holding on to it for dear life, while also having to turn the pages of his script.

Crouching behind the gun turret on the USS *Cummings*, his daughter watched with desperation as her father, in obvious physical distress, did his best. A searing pain, as at San Diego three weeks earlier, ripped his chest, but higher this time: symptoms of the heart attack his doctors had always dreaded.

"He began his address at 5 pm and spoke for 35 min.," Dr. Bruenn chronicled a quarter century after the President's death, when he could no longer be accused of unpatriotic revelation. "During the early part of the speech the President for the first time experienced substernal oppression with radiation to both shoulders. The discomfort lasted about 15 min, gradually subsiding at the end of the period."[2]

A further twenty years later Bruenn disclosed that the heart attack was almost inevitable, following the "heart failure" the President had suffered in the spring. The miracle was that, giving his speech, standing alone at the lectern, Roosevelt hadn't flinched. "He kept on with the speech," Bruenn told an interviewer, recalling the President's sheer determination, then "came below and said, 'I had a helluva pain.' We stripped him down in the cabin of the ship, took a cardiogram"—using equipment transferred from the *Baltimore*—"some blood and so forth." And collectively held their breath.

"Fortunately it was a transient episode, a so-called angina, not a myocardial infarction," which could well have led to death. "But that was really a very disturbing situation," Bruenn reflected—"the first time under my observation that he had something like this. He had denied any pain before"—the President concealing from Bruenn his attack aboard the *Magellan* on July 20. "It was, nevertheless proof positive that he had coronary disease, no question about it."[3]

Roosevelt's courage, unfortunately, was for naught. It certainly went unappreciated by the crowd attending the speech, who were deeply disappointed. With General Patton's breakout toward Paris hitting the headlines, shipyard workers on the West Coast wanted an upbeat speech about the war in the Pacific, the likely defeat of Japan, the kind of peace to come. What they got instead was a dying man's halting account of what he had seen on his trip, as if he'd been a tourist, not the U.S. commander in chief — the President not even mentioning Pearl Harbor until halfway through his peroration.

"No very great enthusiasm was shown by the thousands gathered about the Bremerton dock," Admiral Leahy recalled later.[4]

It was small wonder. As Judge Rosenman noted, having listened in Washington, D.C., to the President's account on the radio, the "report of his trip" was execrable. "Whilst this was not an important speech, it was a major one," Rosenman acknowledged, "and deserved much more attention"[5] than the few afternoon hours the President had spent dictating the draft to Lieutenant Rigdon on board the USS *Cummings* — dictation without input from another soul, or a sounding board to measure how it would come across to a local audience in Washington State. Or, more importantly, to its national audience, which was said to have exceeded forty million listeners.

"The people had not heard the President since the acceptance speech of July 20; he had in the meantime journeyed out to Hawaii and Alaska; Dewey had been out strenuously campaigning" — and accusing the Roosevelt administration of being led by tired old men. "The American people expected that the President had something to say and would say it," Rosenman chronicled, adding, "but the speech had nothing to say, and said it poorly." At best, it was a "rambling account of his journeys and experiences during the past month, and he adlibbed a great deal in a very ineffective manner. It was a dismal failure. That speech, together with the unfortunate photograph of the President in San Diego a few weeks earlier, started tongues wagging — friendly and unfriendly tongues — all through the United States. His enemies concluded that 'the old man is through, finished.' His friends and supporters of many years shook their heads sorrowfully and said, 'It looks like the old master had lost his touch. His campaigning days must be over. It's going to look mighty sad when he begins to trade punches with young Dewey.'"[6]

The fact was, the trip to Hawaii (eventually totaling 13,912 miles) had done the President in; he would, as it transpired, never be the same again.

PART NINE

Quebec

58

A Redundant Conference

A MONTH LATER, at 9:00 a.m. on September 11, 1944, the *Ferdinand Magellan* pulled into Wolfe's Cove, Quebec, on the banks of the St. Lawrence River — the very spot where, during the French and Indian War, General James Wolfe had landed and, after scaling the cliffs and storming the Plains of Abraham on September 13, 1759, been mortally wounded in hand-to-hand combat, as was the defending general Louis-Joseph de Montcalm. In the subsequent Treaty of Paris, the French colonies of North America had been formally ceded to Great Britain, and had become Canada.

Under the Statute of Westminster in 1931, after a century and a half of British rule, the colony or Dominion of Canada had finally become coequal with Britain. Now, in anticipation of the arrival of the President of the United States and the Prime Minister of Great Britain, the Canadian prime minister had made his way by train from Ottawa to Quebec, where he had spent the night at the Citadel, the residence of the governor-general, the Earl of Athlone. There Mr. Roosevelt would also stay, as would Mr. Churchill. For Quebec was once again the site for a major Allied war conference or summit — the second time in barely two years.

Why it was being held at all was a mystery.

The venue, at least, had been the idea of Canadian prime minister Mackenzie King. A Liberal politician now in his ninth consecutive year in office, King had offered to host the conference, following the great triumph in Normandy. The arrival of visitors as august as FDR and Churchill, for a second year in succession, would improve his own chances of reelection, the prime minister felt — knowing he would probably have to go to the polls as the end of the war in Europe approached. With an entire First Canadian Army fighting in France and still more Canadian divisions fighting in Italy, in addition to massive aid being given to Britain, Canada was now a major military coalition ally. Representatives of some forty nations were already preparing to assemble in Montreal for a United Nations Relief and Rehabilitation Administration (UNRRA) conference; the city of Quebec, though, was the more imposing, and would be available for the warrior caste and their film crews. "Canada has emerged in every sense of the word into a world power," King congratulated himself. It was a remarkable improvement over the dark days of 1940, only four years earlier, when he might well, as Canadian prime minister, have found himself hosting refugees and exiles from a defeated England.[1] Now, however, he would be hosting two of the three primary military powers that still counted in Allied military decision-making: the United States, Russia, and Britain, with the end of the war in Europe believed to be at hand.

But not by Mr. Roosevelt. The closer to Germany the Allies advanced, the harder the Germans, like the Japanese, would fight, the President felt—especially after the survival of their beloved Führer on July 20. Roosevelt nonetheless had complete confidence in General Eisenhower as supreme commander in northern Europe. The Allies, after all, had vaulted the Seine by August 20. Paris had fallen to Allied forces on August 24. Despite their precipitate retreat, there seemed no sign of Wehrmacht collapse. German garrisons in the Channel ports were resisting to the death, and few Germans were surrendering unless surrounded; they could be expected to fight even more tenaciously on German soil, the President lamented. The end, in other words, could be bloodier even than the beginning.

Churchill had for months added to the President's strategic concerns: the U.S. Joint Chiefs of Staff telling him that, according to their London liaison staff, the Prime Minister had been an ogre more than an ally in arguing to be allowed to open new Allied fronts in the Mediterranean at the expense of Anvil—the invasion of France from the south that Churchill claimed was now irrelevant. Instead the Prime Minister, it appeared, had again become infatuated with the wacky notion of Blitzkrieg in the forbidding mountains of northern Italy and Yugoslavia—fixating on the so-called Ljubljana Gap, or gateway to Vienna. It was a notion the U.S. chiefs and the President considered ludicrously naive—indeed reminiscent of the British Charge of the Light Brigade at Balaclava in the Crimean War. When Churchill's seemingly unending entreaties had failed, and the Anvil (renamed "Dragoon") assault was launched on August 15, 1944, it had proved, contrary to all Churchill's predictions of ghastly failure, an American triumph.

Once more Churchill had been proven completely wrong as a strategist and tactician. The great port of Marseilles was overrun and liberated in a week, providing major new logistical backup to the Allies in northern France. Within ten days of landing, General Jacob Devers's Sixth Army had joined forces with Eisenhower's troops. Churchill had, in short, egg on his face.

In such circumstances Churchill's urgent pleas for yet another military conference, moreover one to be held to recast Allied military strategy, had seemed fatuous. Marshal Stalin, whose forces were now capitalizing on their great success in Operation Bagration, had certainly shown zero interest in the idea. The war was simply moving too fast; his Russian armies were approaching Warsaw in the north, and in the south Soviet forces were advancing on Bucharest — followed by Soviet troops threatening to overrun Bulgaria in the days after that. Given such progress on the battlefield, what was there to discuss, militarily? Stalin had therefore declined an invitation to meet with his American and British allies in Scotland, the first suggested venue.

The President had not been surprised. In terms of influence in the eventual end-of-war arrangements, as in all property disputes, *possession,* not Map Room charts, would be nine-tenths of the law — and Eisenhower's forces were advancing as fast as possible across northern Europe. It was thus vital not to be distracted by extraneous Churchillian schemes in Italy, the Mediterranean, and the Aegean, which the Germans would in any case inevitably vacate, once the Allies breached the Siegfried Line, protecting their homeland. Instead Eisenhower should be given every conceivable military assistance — including the First Allied Airborne Army — in driving his forces to Berlin, not Florence, Rhodes, or Yugoslavia. This airborne army, the President was assured by General Marshall, would help Eisenhower and Montgomery leapfrog German resistance — and vault, if possible, the Rhine, not the Arno. The paratroopers, tankers, and infantrymen would have to win this battle on the ground: a challenge that a new international conference over military strategy could only distract from.

A summit addressing political or diplomatic issues was equally premature at that moment. A major high-level conference had, after all, already been convened; it was meeting in earnest at Dumbarton Oaks in Washington, D.C. It would prepare the way for a postwar international organization and security setup, with its first meeting to be attended by senior representatives of the Soviet Union, Great Britain, China, and the United States.

Quebec, then, was simply not necessary in the summer of 1944. Moreover it promised a hiding to nothing if Churchill, as was rumored, intended to come to North America with new schemes for the Mediterranean or even the Far East.

Despite this, the more the ailing president had pondered the matter, the more he had begun to warm to Churchill's idea of a military conference. Assembling the U.S. Chiefs of Staff together with senior officers of Great Britain would give him an authentic reason to leave Washington. With only weeks to go before the November election, it would, if held in September, permit him to continue his charade: namely that he was too busy with his duties as U.S. commander in chief to take time to campaign against Governor Dewey.

It was in this less than virtuous way the President had found the notion of a second military conference in Quebec to be a worthwhile project, even without

Stalin's participation. It would placate the ever-argumentative Mr. Churchill. In their meetings with their British opposite numbers, the U.S. chiefs of staff would help the President keep the Prime Minister's wild schemes in check. There would be photographs and film taken — though always from a distance — of the President in his commanding role, as in 1943. In the meantime, back in the United States, his new vice presidential nominee, Senator Harry Truman, could barnstorm the country addressing domestic issues on behalf of the Democratic Party.

Thus had the dubious plot been concocted for the President to go to Quebec, rather than to Scotland, "where he will have a meeting with a distinguished Englishman," Vice President Wallace had noted in his diary after lunching with the President at the White House.[2]

59

The Complete Setting for a Novel

THE NEED FOR THE PRESIDENT not to be seen up close or in public, it was felt by the President's political advisers such as Steve Early, was critical, given the state of the President's health — and morale. Eleven days earlier, immediately on his return from San Diego, the President had asked his new running mate, Harry Truman, to lunch with him at the White House. Truman had come — and been appalled.

It had been a hot day, and the two men had eaten outside, under the famous magnolia tree planted by President Jackson. The President's daughter Anna, who was now acting as his personal assistant, was also there. Photographers had taken pictures of the two men, sitting in shirtsleeves on the South Lawn — pictures the White House press secretary policed lest there be any repetition of Bremerton or San Diego. "The President looked fine and ate a bigger lunch than I did," Truman claimed to reporters afterwards.

Unhappily, this wasn't true. The President was *not* fine.

At his weekly press conference in the Oval Office that morning the President's initial answers to questions from journalists had been almost incoherent — so quiet, also, that reporters had had to ask him to speak louder. He'd seemed distracted, becoming energized only when talking of the possibility of a national work service program for young people after the war, rather than military training.

In the West Wing, before lunch, he'd shown Truman silent movies of his Hawaii trip to meet with MacArthur and Nimitz. Over lunch itself, however, they'd spoken only of the election campaign, with no mention of military matters. Truman had been shocked by the President's appearance — and by the fact that his hand shook so much he was unable to pour cream into his coffee.

"I want you to do some campaigning," Truman later recalled the President saying.

"I'll make some plane reservations to go around over the country," the senator had assured "the Boss" — in fact, "anywhere you want me to go."[1] Truman would never forget the President's response. "Don't fly. Ride the trains," Roosevelt had ordered. "Can't both of us afford to take chances."[2]

The President, it was clear, was running against the clock of life. On August 22, as they'd had tea at Top Cottage on his Hyde Park estate, Roosevelt had confided to Daisy Suckley that he might soon be meeting Churchill in Quebec, where he would stay a week, followed by a day or two together with the Prime Minister at Springwood.[3] He hadn't sounded confident about winning the election, though — or even caring too much if he didn't, to Daisy's astonishment.

This was precisely what worried those who felt the anti-Axis alliance would fall apart, thanks to inter-Allied squabbles, if the President didn't win. Even Vice President Wallace was frank in his diary about the reports he'd been hearing of Roosevelt's health.

"It is curious how many people think the President is completely washed up physically," the vice president had noted on August 16. "In most cases the judgment seems to be based on his appearance at the time of his broadcast at San Diego and the manner of his broadcast from Seattle." But there were others, too, who'd met with the President, or furnished similar reports of other appearances by him. Men such as Captain Maurice Sheehy, a navy chaplain and loyal Roosevelt supporter, who had just returned from Hawaii. Sheehy had reported to Wallace the President "was in terrible shape" there, and "that his hands shook so he could scarcely lift his food to his mouth. He thinks the President will be unable to campaign and that he will have to resign from the nomination" — causing Wallace to reflect: "In that case Dewey will win easily."[4]

Dewey win?

A sort of fatalism seemed to overwhelm the President. As Robert Sherwood, Roosevelt's speechwriter, recalled, "It was not what Dewey was doing or saying that provided the present cause for worry to Hopkins and the others in the White House" in August 1944; "it was the indifferent attitude of Roosevelt himself. He seemed to feel that he had done his duty by allowing his name to be placed before the American people, and if they did not want to re-elect him, that would be perfectly all right with him." As his military aide and appointments secretary General Pa Watson put it to Sherwood, "He just doesn't seem to give a damn."[5]

As Sherwood recalled, "the main problem" in Washington had therefore become how "to persuade the President to descend from his position of dignified eminence as Commander-in-Chief and get into the dusty political arena where he was still undisputed champion."[6] Eleanor, the President's wife, was no help in this regard; she seemed just as fatalistic, perhaps even more so than the President — having long ago renounced her right to persuade him one way

Anzio

Both FDR and Churchill fall ill after Tehran. FDR never recovers, but Churchill does. With FDR too sick to stop him, the PM ignores Ike's warnings. Under new British command in the Mediterranean, Churchill demands an instant Allied invasion at Anzio to reach Rome. It is a disaster, incurring 43,000 casualties over four months, to no purpose.

The Triumph of D-day

PRAYER READ BY THE PRESIDENT ON THE RADIO

THE WHITE HOUSE

JUNE 6, 1944

MY FELLOW AMERICANS:

In this poignant hour, I ask you to join me in prayer:

Almighty God: Our sons, pride of our nation, this day have set upon a mighty endeavor, a struggle to preserve our Republic, our religion, and our civilization, and to set free a suffering humanity.

Lead them straight and true; give strength to their arms, stoutness to their hearts, steadfastness to their faith.

They will need Thy blessings. Their road will be long and hard. The enemy is strong. He may hurl back our forces. Success may not come with rushing speed, but we shall return again and again; and we know that by Thy grace,

In contrast to Anzio, FDR's Overlord strategy works brilliantly. His appointee, Eisenhower, performs splendidly, with Montgomery as Allied ground forces commander. The triumphant landings answer FDR's famous D-day Prayer, and the subsequent campaign does decide the war, as Hitler feared.

The Bomb Plot

July 20, 1944, is a fateful day. In San Diego FDR watches a Marine Corps rehearsal for Pacific landings, while in East Prussia Hitler welcomes the deposed Mussolini to his headquarters. But FDR has suffered a medical crisis, and Hitler is almost assassinated. Both men broadcast that night: FDR to accept the Democratic nomination for President, Hitler to prove he has survived as Führer.

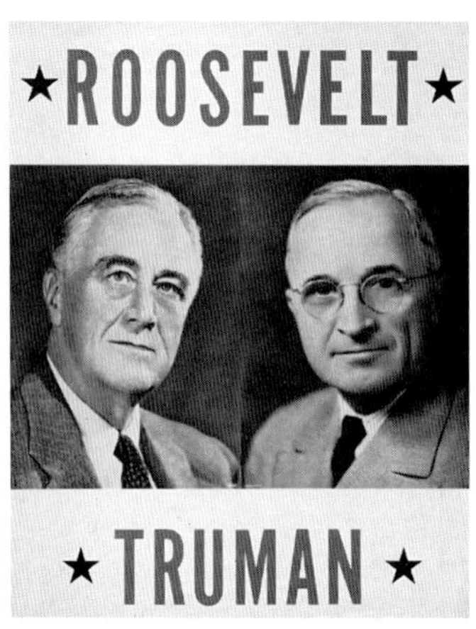

The 1944 Democratic convention in Chicago is a confidence trick; Dr. Frank Lahey warned FDR he could never survive a fourth term. But with no other Democrat able to beat the Republican nominee, Governor Thomas Dewey, FDR is forced to run. He belatedly backs Senator Harry Truman as his running mate, then sails to Pearl Harbor aboard the USS *Baltimore*.

Hawaii

General MacArthur had hoped, vainly, to be the 1944 Republican presidential nominee. Instead, summoned to Pearl Harbor by FDR, he is forced to meet with Admiral Nimitz and Admiral Leahy to tour the navy, air, and army installations there for the first time. Also to debate, before the Commander in Chief, the best way to defeat Japan.

Returning from Pearl Harbor, FDR tries to insulate himself from election campaigning—and the cameras recording his plummeting health—but fails. His speech aboard the USS *Cummings* at Bremerton shipyard is a disaster. Backed by Senator Truman, though, he wins a fourth term and is inaugurated on the White House portico on January 20, 1945.

Aboard the USS *Quincy,* FDR is hailed as a hero at Malta (above), but Yalta is torture for him, given his failing health. Yet no one else can keep the Allied coalition together to defeat the Third Reich and Japan. The summit at least fulfills FDR's dream of a UN organization backed by the Russians and ensures Soviet help in defeating Japan.

Reporting to Congress, the President can no longer stand. The war is almost won; still, Hitler attempts to divide the Allies. Withdrawing to the Little White House, Warm Springs, FDR appeals for common purpose: the defeat of Nazi Germany, then Japan. On April 12, 1945, he dies beside the love of his life, the widowed Lucy Rutherfurd.

EXTRA

San Francisco Chronicle

The City's Only Home-Owned Newspaper

F.D.R. DIES!

War in The West

Tanks of Ninth Cross the Elbe; —It's 57 Miles

The President

Cerebral Hemorrhage Caus[e] Of Death at Warm Spring[s] Truman Is Sworn Into Offic[e]

Cabinet Will Be Retained

Sudden Passing 2½ Hours After The First Attack

FRANKLIN DELANO ROOSEVELT

Eastern Front

Great Battles Rage Along Oder—Final Berlin Drive Is Near

Franklin Delano Roosevelt (Editorial)

Spanish-Jap Rift Leaves U. S. Cool

S. F. Conference Will Go On, Says Truman

or another in his decisions, while she proceeded independently with her own life and causes.

The President had seen Ambassador Halifax on August 19 — causing Halifax to note in his diary he was "not sure how well I thought the President looked. He seemed rather fine drawn, and looked older. He also seemed to be snuffling a bit, with a handkerchief on the table!" He'd said he was delighted with Eisenhower's progress in France, but had sounded dubious about the Germans suing for surrender, even with the Soviets pressing from the east. It was, he thought, "unlikely that any German official would sign unconditional surrender," and the war would continue "rather longer than some optimists" — such as Halifax, who was betting it would all be over in November — "thought." The President had also told Halifax that not only the war but "the election was going to be tough, and that he would suffer from the non-recording of much of the Army vote" — despite the latest decision to let soldiers from some states, at least, vote — "and also of the labour vote"[7] given how many workers were employed in factories away from their home constituencies . . .

Thereafter the President had set off from Hyde Park — Eleanor remaining there, at her Val-Kill cottage, while the President returned to Washington to see the secretary of war, Henry Stimson, after which he was slated to greet the dozens of national representatives participating in the Dumbarton Oaks Conference.

Would relying on his role as U.S. commander in chief be enough, however, to win a presidential election — especially if the President didn't really *want* to win? No one knew.

In the event, only one person could, arguably, inspire the President to fight for the presidency, given how deathly ill he often felt: Lucy Rutherfurd. Without her intercession it seemed doubtful he could make the effort, *or even possessed the will to do so.*

It is unclear if anyone spoke to Lucy about the problem, but perhaps no one needed to — for Mrs. Rutherfurd could see it clearly for herself. Immediately after the Dumbarton Oaks delegates had left the White House on August 23, at any rate, she'd come for tea with him on the lawn by the South Portico, together with her daughter Barbara — whom the President called his "goddaughter" — and Lucy's stepson John Rutherfurd, a U.S. Navy officer. There they'd hatched a plot for the President to spend a day with them at Lucy's Allamuchy estate in New Jersey, on his way back to Hyde Park for Labor Day weekend.

With this lure the President's spirits had seemed miraculously to rekindle: for Lucy, more than any person alive, knew not only how to flatter Franklin by her deep affection, but how to play on his pride as a man: as a father and, in many ways, the father of the nation in war. It was a knack the president's cousin Daisy, who had no children, indeed had never married, simply did not possess.

Military service, moreover, was part and parcel of Lucy Rutherfurd's life. The President had arranged for all three of her stepsons to obtain sea duty in the U.S. Navy at their (and her) insistent request. The Rutherfurd boys were all determined to serve — and there was simply no way, as the idol of the Rutherfurd family, that the President, for his part, would fail to do his part as their commander in chief, however debilitated he felt.

As Lucy later confided to his Russian-language interpreter, Charles Bohlen, who was a friend of the Rutherfurds, the President "kept no wartime secrets or burdens from her."[8] He would have preferred, in his condition, to retire gracefully, at the end of his third term as president. But he had met Stalin in person, now — and had Stalin's respect and cooperation in more ways than anyone could have imagined earlier in the war. Why abandon that personal, highest-level relationship?

True, the Russians were still an enigma, as allies. At Dumbarton Oaks the Russian representative was still insisting on having no fewer than *sixteen* other delegates in the nascent United Nations: one for each Soviet republic, plus the Union. This was a distinct political stumbling block. Moreover despite a personal appeal, the President had completely failed thus far to persuade Stalin to intercede in the struggle for Warsaw, where the Soviets were refusing to lift a finger to help the Polish national uprising that had begun on August 1, only forty miles from the Russian front. The Soviets would thus be a continuing nightmare to deal with: culturally, politically, militarily. Yet without them the war against Hitler might still not be won — and without them the war against Japan might take years to win, as well as incur terrible American bloodshed, given Japanese determination not to surrender an inch of their conquests, let alone their home islands.

In spite of his ill health the President had therefore remained confident he could probably keep Stalin to his promise, at Tehran, to enter the war against Japan, and could eventually persuade Stalin to drop Soviet demands for exaggerated representation at the UN, without wrecking the UN's chances of success. Likewise, he'd shared with Lucy, he could probably handle Winston Churchill, whom he now knew so well, in Quebec. And seeing in her eyes such admiration, any notion of defeatism in terms of the looming election had simply evaporated.

It was in this context the President had hatched his secret plot to visit Lucy — a very risky idea.

In the end, as things developed, it was decided to make Mr. Roosevelt's visit to Mrs. Rutherfurd "above board." It would require, in any case, considerable preparation.

First off, Mike Reilly of the White House Secret Service detail would have to be suborned, since until now the Secret Service had objected to the President taking the train across the bowstring-truss "Hell Gate" bridge in New York, which had been the target of German agents, landed by U-boat, back in 1942. Though the would-be saboteurs had been captured and executed,[9] the Secret Service veto on the President's train using the bridge had been observed ever since.

Reilly was soon dealt with. German sabotage, in September 1944, seemed pretty ridiculous. At the President's insistence, the *Ferdinand Magellan*'s route was therefore reconfigured to follow the Pennsylvania-LeHigh line to Allamuchy, New Jersey, and from there to Hyde Park. The trip to Lucy was on.

Soon after 10:00 p.m. on August 31, 1944, *Magellan* had steamed out of the Bureau of Engraving in darkness, bearing the President, the President's private secretary Bill Hassett, his office secretaries Grace Tully and Dorothy Brady, and Daisy Suckley, too.

Daisy had already found the President "looking awfully tired" at the White House, and having lost yet more weight since his trip to the Pacific, despite daily extra eggnogs.[10] Hassett recorded the same in his own diary, describing the President as still "tense and nervous — not yet rested from his five weeks' trip into the Pacific area via land and sea." The stress of being president was simply too much for him: "Too many visitors at mealtimes — all ages, sexes, and previous conditions of servitude — hardly relaxing for a tired man."[11] However once the train pulled up to a railway crossing on the Rutherfurd estate at 8:30 a.m. on September 1, the next morning, the President had seemed transformed — met there by Mrs. Rutherfurd and her stepson John, John's wife, and their children.

"She is a lovely person," Daisy noted in her diary that morning, "full of charm and with beauty of character shining in her face; no wonder the Pres. has cherished her friendship all these years —. She came on the train hatless & stockingless, in a black figured dress, & black gloves — She is tall & good-looking rather than beautiful or even pretty — Though I remember her as tall & calm & pretty & sad, when her daughter was 6 mos. old, at Aiken," more than twenty years before. Lucy was older now — and by no means triste, Daisy noted, in fact the reverse was the case. "She does not look sad now," the President's cousin described Lucy's happy face — acknowledging "how little one knows about the inner life of others" — especially their love lives.[12]

Rutherfurd Hall was vastly bigger than the President's own house at Hyde Park — an eighteen-thousand-square-foot Elizabethan-style mansion built in 1902, with thirty-eight rooms, a thousand-acre deer park, and farms stretching over five thousand acres. Daisy was asked to play with the children and Fala, while the President — whose car had been driven off the train by its special ramp — "is going to look over" what were known as the Tranquility Estate farms, Daisy recorded — "woods, etc." — and "advise them what to do" in his self-described profession as forester.[13]

Sitting by Allamuchy Pond, "on a cushion on the grass at the water's edge" while Fala and the children played, Daisy had found herself almost in disbelief. "Another adventure!" she'd jotted. It had not ended there, however. Back at the Allamuchy house Lucy "had a lovely room all ready for the P. to take his rest, even to turning down the best linen sheets." But the President had demurred; "he wasn't going to miss any of his visit," Daisy noted.[14]

Lunch for a dozen guests was to have been lobster, but since the President had dined on lobster the night before, and was slated to have it again with Eleanor at Val-Kill that evening, Lucy had arranged for squab or young pigeon to be served with vegetables and salad, and afterwards ice cream. The President "did not rest until we got back on the train." Finally at 3:35 they all "dashed off" to the railway halt, "& got started on schedule," Daisy noted that night. "It was a really lovely day, centering around Mrs. Rutherfurd, who becomes more lovely as one thinks about her — The whole thing was out of a book — a complete setting for a novel, with all the characters at that lunch," including Winty Rutherfurd's former sister-in-law, who'd married a prince, and other Rutherfurds and their wives and husbands, as well as "absent husbands and wives etc."[15]

One of those absent wives, of course, was Eleanor, who would be at Highland station, waiting to take the President and his traveling companions to Val-Kill, "for supper" — a special meal to feature not only lobster, Daisy recorded, but Eleanor's friend Mrs. Pratt's "speciality": German "'appelkuchen.'"[16]

60

Two Sick Men

IF THE PRESIDENT'S ENTOURAGE had become more and more concerned in recent weeks about Roosevelt's declining health, the same could be said of Churchill's courtiers. The Prime Minister's transatlantic voyage aboard the *Queen Mary* had begun at Greenock, Scotland, on September 5, 1944. His doctor, Lord Moran, was taking no chances, and had insisted on taking with him an eminent bacteriologist and nurse, as there were shadows on the Prime Minister's lung X-rays that suggested a possible recurrence of pneumonia.

The intent of the conference, the President had made clear via his ambassador in London, John Winant, was to address military matters only. For Churchill this had been frustrating, given his desire to map out a political future in terms of British imperial interests in the postwar world, but it was a restriction the Prime Minister was forced to accept. The President, now in the midst of an election campaign, naturally wished to avoid accusations of secret political discussions or agreements being made. Such a limitation, Churchill felt, was inherently silly, however. Military strategy at this stage of the war was bound to carry major political implications; he had wisely, therefore, invited both Lord Leathers, his minister in charge of shipping and transport (and an expert on Lend-Lease), and Lord Cherwell, paymaster-general and scientific adviser to the Prime Minister.

Suffering suspected pneumonia, Churchill was neither well nor well-disposed. As far as military strategy was concerned, his agenda for Quebec was at odds with even that of his chiefs of staff, who would be sitting down with their American colleagues on the Combined Chiefs of Staff Committee in the Château de Frontenac. Throughout the voyage the Prime Minister was, his chiefs complained, "quite impossible" to "argue with," as Field Marshal Brooke would later recall. "It was a ghastly time from which I have carried away the bitterest of memories."[1]

• • •

Churchill, for his part, felt the same about Brooke, and even Andrew B. Cunningham, Admiral "ABC," head of the Royal Navy—officers who were collectively "framing up against him," he growled.[2] It "takes little to rouse his vengeful temper," Cunningham noted in his own diary. The admiral was concerned lest this lead to a further souring of relations with their U.S. counterparts, for "he will do anything then to get the better of our allies."[3]

Brooke, who did not tolerate fools gladly, was the most despairing among the Prime Minister's cohort. He found Churchill "old. Unwell, and depressed" during the journey. "He gave me the feeling of a man who is finished, can no longer keep a grip of things, and is beginning to realize it," Brooke confided. "Here we are within 72 hours of meeting the Americans and there is not a single point that we are in agreement with!" he noted with near-despair in his diary. Contradicting the views of his military chiefs, Brooke observed, the Prime Minister was adamant the looming conference should have but a single purpose: namely to get approval for a new Allied bid to reach Vienna from the Adriatic, and that "we were coming to Quebec solely to obtain 20 landing ships out of the Americans to carry out an operation against Istria to seize Trieste."

Vienna?

Even if the idea were feasible militarily—which it was not—Brooke recorded on September 8, "with the rate at which events were moving, Istria might be of no value."[4] At a moment when the Wehrmacht was fighting even harder in the mountains of Italy, an attempt to land a whole army at Istria and strike through the Alps to Vienna was asking for trouble—Anzio all over again. And wholly pointless, given Russian advances in Romania. The terrain beyond Trieste—the so-called Ljubljana Gap—was hazardous in the extreme. The gap was simply impassable if contested in winter, and thus risked more casualties even than the mountains of Italy. Yet the Prime Minister was insistent. Was Churchill then mad, Brooke wondered—or perhaps ill? When the Prime Minister ran a temperature again that evening, Brooke decided he was "very definitely ill," physically and mentally, and it was "doubtful how much longer he will last. The tragedy is that in his present condition he may well do untold harm!"[5]

The next day proved even worse—the Prime Minister determined to speed up planning without any thought to the enemy's likely response. "He knows no details, has only half the picture in his mind, talks absurdities and makes my blood boil to listen to his nonsense," Brooke railed in his diary.[6]

"I find it hard to remain civil," he continued. "And the wonderful thing is that ¾ of the population of the world imagine that Winston Churchill is one of the Strategists of History, a second Marlborough, and the other ¼ have no conception what a public menace he is and has been throughout this war!"[7]

In terms of public support, it was, of course, "far better" the public should remain unaware of this, Brooke judged, "and never suspect the feet of clay of that otherwise superhuman being"—for without the Prime Minister's brave stand in 1940, "England was lost for a certainty"—yet "with him," as Brooke knew best, "England has been on the verge of disaster time and time again."[8]

It was quite an indictment: "Never have I admired and despised a man simultaneously to the same extent," the field marshal concluded.[9]

The truth, as would become clear, was that the two major leaders of the democracies were traveling to a meeting that was unnecessary—moreover was one that neither man was well enough to direct.

Even their arrival was a botched affair.

Mackenzie King, the President's host, was confounded. At the Château de Frontenac, where he spent the night, he'd received a message "from the President that he would like me to go to his [rail] car if there were a few minutes before Mr. Churchill's train arrived," King recorded in his diary on September 11.

Before Mr. Churchill's train?

When King got to the station, he found it was true—"one of the President's staff came to me and said the President was waiting to see me," without even the governor-general having arrived yet!

A British prime minister, on British Empire soil, being deliberately relegated to second place by an American president? Brushing away King's concerns about diplomatic protocol, the President's staff officer "explained that the car was there and the President was anxious to see" him. Swallowing his surprise, Mackenzie King "went in. [President Roosevelt] was seated," the Canadian prime minister recorded in his diary, and to the President's right was "Mrs. Roosevelt"—who had joined the presidential train at Hyde Park—"standing in front of the sofa."[10]

Roosevelt was quite candid. "The President said at once: I wanted to see you first; also to be ahead of Winston, so I gave orders to have the car moved in."[11]

A sort of schoolboy prank?

The President's behavior was most unlike him. As King commented in his diary, "It seemed to me that the President was rather assuming that he was in his own country." More worrying, however, was the sight of the Commander in Chief—for it was clear this was not the man King had welcomed to Quebec the previous year, in August 1943. "It seemed to me, looking at the President, that he had failed very much since I last saw him. He is very much thinner in body and also is much thinner in his face. He looks distinctly older and worn. I confess I was just a little bit shocked by his appearance."[12]

The President asked Mr. King to accompany him in his open automobile, which was being unloaded, to the Citadel—prompting the embarrassed prime minister to protest that the governor-general, as the representative of Canada's monarch, King George VI, was not there yet. He, Earl Athlone, should be the one to accompany the President, King pointed out, and therefore left the *Magellan* to bring in the governor-general for an audience with the President, when he arrived. After that, like a vizier to the President of the United States, King went out again and fetched Prime Minister Churchill, who by then had arrived in his own train, and was completely nonplussed by the strange procedure. All were being brought aboard the *Ferdinand Magellan,* on Canadian soil, just as if Mr. Roosevelt were the Sun King.

Inasmuch as the United States — as Churchill would admit that day in conversation with the President — was by now far and away the "strongest military power today, speaking of air, sea and land"[13] — the scene at Wolfe's Cove was symbolic. Yet it revealed an equally significant development, one far more important than the reversal of diplomatic protocol: namely the stunning change in the President' health since the officers had last seen him in Cairo, following Tehran.

Churchill's military assistant, for example, would later write how "shocked" he was to see "the great change that had taken place in the President's appearance since the Cairo Conference. He seemed to have shrunk: his coat sagged over his broad shoulders, and his collar looked several sizes too big. What a difference from the first time that I had set eyes on him less than two and a half years ago!" General Ismay reflected. It had been in Washington, at the White House. "Seated at the desk of his study, on what I shall always remember as 'Tobruk morning'" — receiving news of the abject British surrender in North Africa to General Rommel — "he had looked the picture of health and vitality. His instinctive and instantaneous reaction to the shattering telegram had won my heart forever. No formal expression of sympathy, no useless regrets: only transparent friendship and an unshakeable determination to stand by his allies. 'Winston, what can we do to help?'"[14]

Now a different Franklin Delano Roosevelt was before the British contingent: a ghost of his former self.

61

Churchill's Imperial Wars

THE COMBINED CHIEFS OF STAFF soon began their military sessions at the Château de Frontenac. But since the first plenary session of Octagon, as the conference was code-named, was not scheduled to take place until September 13, on the third day of the Combined Chiefs of Staff meetings, the President and Prime Minister were, in effect, free to do whatever they wished in the meantime — Mr. Roosevelt and Mr. Churchill staying in the nearby Citadel.

Unlike the year before, in Quebec, there was now an underlying sense of fraud about the conference, beyond obligatory group photographs, newsreels, and dinners. In some respects it promised to be like a bad play in which the players, for the most part, did not believe in the script — the Prime Minister at odds with his own chiefs of staff, while the President seemed strangely distant, even disconnected from his.

To Churchill's surprise the President had brought with him neither Harry Hopkins (who had largely recovered since his major surgery) nor any senior member of his White House team, beyond Admiral Leahy. And his U.S. Joint Chiefs of Staff and their staffs, who would look good in photographs and newsreels. Quebec was, in effect, literal as well as metaphorical window dressing — Churchill afterwards snorting, in disgust: "What is this conference? Two talks with the Chiefs of Staff; the rest was waiting to put in a word with the President."[1]

At lunch on September 11 the Canadian prime minister discussed with the President, in confidence, the forthcoming election — one that King did not propose calling in Canada until the war was over, himself, lest it put in question French Canadian support for the war — and postwar. Mr. Roosevelt, however, had no such electoral leeway — and seemed anxious about those states in America, he said, that were still unwilling to permit soldiers to vote — especially soldiers "favorable to the President." "I can see that he is really concerned; also that he is genuinely tired and weary," King noted again. "He has lost much weight — 30 pounds."

Thirty pounds? The result was alarming; "he looks much thinner in the face and is quite drawn. His eyes," King confided in his diary, seeking to be more specific, were "quite weary," like those of a dying dog.[2] Prime Minister Churchill confided the same to his young assistant private secretary, John Colville, saying the President was now "very frail"[3] — and his eyes "glazed," as Colville himself later noticed when he met him.[4]

Churchill, relishing the chance to "take some Scotch as well as a couple of brandies," seemed physically, by comparison, "as fresh as a baby," King commented, ironically: waxing loquaciously, and in "a distinguished way" that was "not affected," he felt, "but genuine."[5]

As a dutiful diarist, King could not but record the difference between the two men's states of health, even if nothing could be said in public. At dinner on the first night, for example, Mr. Roosevelt responded to the governor-general's extended toast with a charming reference to the presence of the First Lady, Eleanor, his own wife, and of Mr. Churchill and *his* wife, Clementine, as well as mentioning his long friendship with Mackenzie King, who was unmarried. King, who paid the closest attention to people's language as also to the ever-significant hands of the clock at special moments, had listened carefully — noting, with alarm, not only what the President had said, but also what he had not said. Roosevelt "made one reference which makes a good deal of how he is feeling," King afterwards reflected. "He referred to the abuse and attacks on himself as 'a senile old man,'" which clearly pained him. Yet in the course of his speech he had admitted that, if he lost the election, he would not mind. It would offer him the "freedom" to "enjoy friendships."[6] And with that the President had simply ended his toast.

Churchill, recognizing instantly the President's faux pas, had leaned straight across Princess Alice, the governor-general's wife, who was between them, "and reminded him about proposing the King."[7]

King George VI was, after all, not only king of England but of Canada, too. It was an error, King was aware, the President would never have made in earlier days. Attempting to make up for his mistake, the President had then given a long, belabored account of King George VI's visit to Hyde Park in 1939, and how "close their friendship would always be. He then proposed the health of the King. It seemed to me," King reflected, "that [his] language was very loosely used. There was the oldest friendship with everyone — the same for the King, the same for the Athlones, the same for Churchill, etc. Rather a lack of discrimination," the Canadian prime minister thought.[8]

Mackenzie King's observation was all too penetrating. The once-enchanting tenor of the President's voice, and the nuanced, playful, humorous way in which he had always expressed himself, were gone. "I felt as the President spoke," King lamented, "that more than ever he had lost his old hearty self and his laugh." As a devout Christian, however, King allowed as how the President "has suffered a good deal" in the many trials he'd faced in life, and most obviously, his paralysis. Yet the fact remained that he "really is a man who is losing his strength."[9]

From this point, September 11, 1944, King's diary account of the conference was filled with sadness and disappointment, given that the President was still only sixty-two — seven years younger than King himself.

It was Mrs. Churchill, however, who would pen the most perspicacious account of what was happening, behind the public facade. It came in a letter she sent her daughter Mary, who was back in England, but who had stayed with the President at Hyde Park the previous September, after the first Quebec Conference. Clemmie was amazed how Harry Hopkins had "quite dropped out of the picture." Moreover she could not "quite make out whether Harry's old place in the President's confidence is vacant, or whether Admiral Leahy is gradually moulding into it. One must hope that this is so, because the President, with all his genius does not — indeed cannot (partly because of his health and partly because of his make-up) — function round the clock, like your Father. I should not think that his mind was pinpointed on the war," she described the President, "for more than four hours a day, which is not really enough when one is a supreme war lord."[10]

Despite this, the strange, almost surreal, gathering of Western Allies got under its anointed way in the French Canadian city above the St. Lawrence River — Mackenzie King noticing, as Eleanor pushed her husband's wheelchair while he looked at little models Churchill had proudly brought with him of the D-day Mulberry harbors, constructed in Britain, "that out of sheer weakness there was perspiration on his [the President's] forehead. While he looked better today" — September 12 — "than yesterday, he still looks very weak," King confided that night. "I feel great concern for him," he added — touched by the way Eleanor, normally somewhat indifferent to the President's daily agenda, at last seemed to recognize his distress, indeed seemed unusually protective, saying that she was "anxious to get him off to bed for an afternoon rest."[11]

Churchill meantime stayed behind — and without further ado led Mackenzie King off to his portable map room in the Citadel, to rehearse his plenary speech to the meeting of the Combined Chiefs of Staff the next day.

This was something Churchill liked to do, as King knew, before any speech or parliamentary declamation he was to give, spending not only hours but sometimes whole days practicing to get the words and his diction right. In this case the burden of his great speech, however, would be to appeal for presidential approval of his scheme for a new military thrust from the Adriatic to Vienna — which would require not only American but Canadian soldiers.

Instead of welcoming Churchill's latest brainwave, Mackenzie King felt deeply worried.

Like all Canadians, King was immensely proud of what Canadian forces had achieved on D-day and in the titanic battle that had followed in Normandy, as the Canadian Corps expanded into an entire Canadian Army. This had led to the crescendo of the Normandy battle, at Falaise, where the Wehrmacht's Army Group B was eviscerated, and to Patton's extraordinary race with armored divisions around the German southern flank to seize Paris. King was proud, too, of Canadian divisions fighting in the British Eighth Army in Italy — thereby holding a number of Wehrmacht divisions as far south as possible in Europe, while the great battle for France was fought out between armies numbering in the millions. Churchill's alternative schemes, both political and military, had always made King uncomfortable, however. Now, once again, they did so as Churchill envisioned, in his suite at the Canadian Citadel, *yet another* major front in central Europe — and one using Canadian forces!

King, as prime minister of a "lesser" power than Britain, would not be invited to hear the actual speech at the forthcoming plenary session, but listening to Churchill, he recognized with a sinking heart the Vienna scheme would almost certainly require mandatory conscription in Canada — for Canadian casualties in Normandy and Italy had already been severe, and were still mounting. Nor was this the only new theater in which Churchill wished to commit Canadian troops, on behalf of the British Empire. He was, he said, adamant that British Empire naval and air forces should participate with American forces in the fighting in the Southwest and Central Pacific — this on the assumption that Canada would contribute men (who were all volunteers, beyond the regular Canadian military), munitions, and money. In this way, Churchill argued, the British Empire would not be deprived of credit in defeating the Japanese.

For Mackenzie King the very phrase "the British Empire" raised certain hackles, since it was clear that Churchill's vision — his determination to reassert British colonial and imperial authority in Asia and the Far East, as before the war — was inimical to the liberal vision of the postwar world that King had repeatedly discussed and shared with President Roosevelt ever since the war began. There was, in short, too much Churchillian talk of British "honor and glory" from a prime minister who had not himself set foot in India since 1899, had never visited China, had never ventured to the Pacific, moreover who seemed to have too little concern with the military realities — and casualties — of the battlefield when fighting implacable foes such as the Germans and Japanese.

"I held very strongly the view that no government in Canada would send its men to India, Burma and Singapore to fight with any [U.K.] forces and hope to get through a general election successfully," once the European war was over, King warned Churchill. About this he had been completely forthright, not only to Churchill but to members of his own Canadian War Committee. Patriotism was one thing; imperial refurbishing, at the end of another world war, another. "That to permit this," King would maintain the next day, "would be to raise at a general election, a nation-wide cry of Imperial wars versus Canada as a nation."[12]

So worried was King, in fact, that he warned Louis St. Laurent, the deputy for Quebec and a member of his cabinet, that he would resign as prime minister of Canada if pressed by Churchill, or by members of his own government, to send Canadians to fight "what would certainly be construed as Imperial wars."[13]

Churchill, however, was Churchill — listening neither to his own chiefs of staff, nor to the Canadian prime minister. Only the President had ever been able to control him — leaving the question: would the President be well enough to do so at the first plenary meeting on September 13, 1944?

62

A Stab in the Armpit

On the day of the first plenary meeting at 11:45 a.m. the President began by welcoming the participants. He then invited Mr. Churchill to open the discussion.

The Prime Minister's peroration was pure Churchill: a mix of glittering phrases, fine flattery—and bravado. "I would have given a great deal to tell him what I thought of him," Brooke confessed in his diary, having dreaded what was to come.[1]

The Prime Minister might have been speaking to several hundred members of Parliament, instead of the fourteen members of his audience, for all that it mattered. In a quite masterly way he first lauded the "revolutionary turn for the good" that had recently taken place. "Everything we had touched had turned to gold," the minutes of the meeting recorded, "and during the last seven weeks there had been an unbroken run of military successes. The manner in which the situation had developed since the Teheran Conference," he averred, in praise of the officers present, conveyed "the impression of remarkable design and precision of execution"—overlooking his Anzio debacle. He "wished to congratulate the United States Chiefs of Staff," he added, not only on D-day but on Operation Dragoon (Anvil), "which had produced the most gratifying results," he now admitted—omitting his titanic efforts to sink it. General Eisenhower was managing the campaign in France and the Low Countries, he said, "brilliantly." Moreover the British were, he claimed, playing a significant role in the fighting, especially in Italy . . .

As the U.S. chiefs suspected, the Prime Minister's preamble was clearly intended to soften them up for what was to come—for it was in Italy, the Prime Minister declared, that he now saw a great strategic opportunity arising. General Alexander was poised to smash Kesselring's forces, and "if the Germans were run out of Italy we should have to look for fresh fields and pastures new. It would never do for our armies to remain idle. He had always been attracted by a right-handed movement, with the purpose of giving Germany a stab in the armpit. Our objective should be Vienna."[2]

A "*stab in the armpit*"?

The phrase sounded much less inspiring than the Prime Minister's earlier notion of attacking Germany's "soft underbelly" — an underbelly, which had proved far from soft, as all were aware, and was still proving so. The British Eighth Army, comprising two Canadian, one New Zealand, one South African, four Indian, and eight U.K. divisions, had suffered some *eight thousand* casualties in the past month alone. A strike across the Adriatic to Trieste and from there across the mountains to Vienna, if the Wehrmacht forces miraculously collapsed and fled from Italy and Yugoslavia? It might certainly be worth preparing for, in terms of "contingency" planning. Yet there was no sign whatever of such an imminent German withdrawal, nor could mountain conditions — ones that were ideal for the Wehrmacht to defend, especially in autumn or winter weather — ever be seriously likened to an armpit. "Pit" seemed a more appropriate word.

Brooke, at a meeting of the Combined Chiefs of Staff the previous morning, had acknowledged "great advantages" for the Allies *if* the Wehrmacht miraculously retreated to Berlin, and General Wilson sent forces north to advance via the Ljubljana Gap to Vienna. "However," Brooke had cautioned, "if German resistance was strong, he did not visualize the possibility of our forces getting through to Vienna during the winter."[3] As if completely deaf to his own chiefs of staff, here was the Prime Minister, now advancing the scheme as something to be undertaken *immediately*, in order to take advantage of the grand military victory Churchill was certain General Alexander was about to win in Italy . . .

Considering that, as the British military historian Sir Michael Howard would later acknowledge, the "distance from Rome to Vienna is some six hundred miles — about three times the distance from Naples to Rome which it had taken the Allies six months to cover," and that German forces would be "falling back along their own lines of communication,"[4] Brooke had, in the Combined Chiefs of Staff meeting, shown great loyalty to his boss, the Prime Minister, when tabling the idea, though indicating his own skepticism of it. The Ljubljana Gap was, after all, only thirty miles wide, and two thousand feet high in elevation, flanked by higher ground. It led to a range of mountains six thousand feet high. As Brooke later noted in an annotation to his diary, "We had no plans for Vienna, nor did I ever look at this operation as becoming possible."[5] All Brooke had proposed as a possible contingency plan was for Allied staff officers to look into the implications of a limited Allied occupation of the Istrian Peninsula, using landing craft no longer needed for Dragoon, now that Marseilles was in Allied hands. This could, if thought worthwhile, give the Allies a base from which to launch a 1945 campaign, he'd pointed out, or at the very least one through which to funnel Allied occupation forces "in the event of Germany crumbling" — a worthwhile political consideration "in view of the [recent] Russian advances in the Balkans."[6]

The plan certainly had merit in hypothetical terms. As such it was worthy of discussion by the Combined Chiefs, even if not ultimately favored as worth the effort or the inevitable casualties if the operation foundered, as at Anzio. Unfortunately this was *not* what the Prime Minister had traveled all the way across the Atlantic to North America to secure as a new Allied military commitment. As he had once become fixated on capturing Rome, Churchill now seemed fixated on Vienna and Singapore — with no thought to the obstacles and casualties such commitments would entail; nor to their effect on Eisenhower's drive to Berlin; or Nimitz and MacArthur's operations against Japan.

Brooke, on the voyage across the Atlantic, had feared provoking another Anglo-American dispute, threatening the alliance. Yet to his immense relief he had, on arrival in Quebec, found the U.S. chiefs had almost fallen over themselves to be polite, even friendly, and they had politely moved on to other matters.

Unfortunately the Prime Minister was not interested in "contingency" plans, the plenary participants were now told. Mr. Churchill wanted, as he'd famously attached to the papers that landed on his desk as prime minister, "Action This Day." With his trademark afflatus the Prime Minister thus formally repeated his *grande ideé*. "Our object," he declared, "should be Vienna."[7]

Before the assembled officers could react, Churchill then continued his speech. He had, he informed the meeting, given "considerable thought" to such a drive. The landing craft were available, and an "added reason for this right-handed movement was the rapid encroachment of the Russians into the Balkans and the consequent dangerous spread of Russian influence in the area. He preferred to get into Vienna before the Russians," the minutes ran, "as he did not know what Russia's policy would be after she took it."[8]

Brooke cringed — recording that night "a long statement by the PM giving his views as to how the war should be run. According to him we had two main objectives, first an advance on Vienna," for which they had apparently come to Quebec, and "secondly, the capture of Singapore!"[9]

Brooke was ashamed. For a prime minister who had trained at Sandhurst Military College and had studied not only his ancestor's famous campaigns in Europe but even the battles of the American Civil War, Churchill's utter disregard of terrain, of the cost in men's lives, and of the strategic primacy of first defeating the Wehrmacht in northwest Europe was a grave indictment. Anyone could — and should — voice concern about long-term communist Russian ambitions in eastern and central Europe given the four hundred divisions they had at their disposal. A serious military strategist, however, had to consider what was realistic, or feasible, in terms of potential accomplishment — and the costs, especially in human life. Also the matter of priorities — of what should, in the final analysis, be the most important goal of the Western Allies: *defeating Hitler.*

The following day, mortified and disgusted, Churchill's military chief of staff, General Ismay, showed Brooke his letter of resignation as military assistant to the Prime Minister as minister of defense, or de facto British commander in chief. Ismay had had enough. As Brooke later noted, the "fact that dear old patient Pug had at last reached the end of his tether and could stand Winston's moods no longer is some indication of what we had been through."[10]

For the President, however, Churchill's insistent espousal of fresh schemes and strategic plans were merely part and parcel of Churchill's extraordinary personality and imagination. So too was his sheer obstinacy, which had helped save his country after the debacle of Dunkirk in the summer of 1940, when all had seemed over for the Western democracies. Like Brooke, the President had been less impressed by the Prime Minister's performance since then. Churchill's most recent cable, on August 29, 1944, calling for an amphibious landing at Istria "in four or five weeks," had given ample warning of his new Viennese tack.[11]

The President had been unconvinced. He had remained so in September, as Churchill's forecast of a smashing victory by Alexander in Italy to "converge" with two armies "on Bologna" had proven another empty promise by the Meddler. Summoning his dwindling energies at the Citadel on September 13, he thanked "the Prime Minister for his lucid and comprehensive review of the situation." He remarked, too, that with each conference the level of inter-Allied amity had increased — "an ever increasing solidarity of outlook and identity of basic thought," as well as "an atmosphere of cordiality and friendship. Our fortunes had prospered," he agreed with Churchill. Nevertheless, he cautioned, "it was still not quite possible to forecast the date of the end of the war with Germany."[12]

At a moment when so many were predicting the imminent collapse of the Third Reich — two million Allied troops ashore since D-day, fifty thousand Wehrmacht troops having surrendered in France, and ten thousand killed in the Falaise pocket alone — this was a sober warning. Romania had surrendered to Russian forces, and Wehrmacht forces were now pulling out of Greece, yes — but they had done so to shorten their lines, not because they were ready to give up their fight for the Führer. The Germans were thus, the President pointed out, certainly "withdrawing from the Balkans," and it "appeared likely that in Italy they would retire to the line of the Alps," eventually, with the Russians already on the "edge of Hungary" — but this didn't mean they would magically surrender, any more than would the Japanese. "The Germans have shown themselves good at staging withdrawals," he reminded his audience, and had "been able to save large numbers of personnel," despite their losses in "material." Clarity was essential. While he was all for General Alexander maintaining "maximum intensity" of operations in Italy to keep Hitler from reinforcing his other fronts at a critical juncture, the President foresaw that in the north "the Germans would retire behind the Rhine" now. It was a huge river, the President emphasized, that "would present a formidable obstacle." How to vault it would be tough — and "for this purpose our plans must be flexible. The Germans could not yet be counted out and one more big battle would have to be fought."[13]

The Battle of the Bulge, nine weeks later, would demonstrate the President's quiet realism as U.S. commander in chief. At Quebec, however, it was like cold water poured over the long, overoptimistic discourse the Prime Minister had given. Cold, to be sure, but without reminding the Prime Minister of his almost unending failures as a commander in chief and strategist. The President, once again, thus appeared at his very best, despite his frailty: using his natural charm, his spurning of theatrics, his maintaining of Allied unity, his invoking of common purpose. Moreover, what he said was expressed in such a kind, avuncular manner that it did not resemble in any way the put-down he and Stalin had been compelled to administer to the Prime Minister at Tehran.

Quietly the President then went on to address the ramifications of a longer war in Europe than the Prime Minister was assuming: namely, its effect on the eventual defeat of Japan. He did not think it worth spending too much time on Burma, beyond its avenue to China. The current "American plan was to regain the Philippines or Formosa and [attack] from bridgeheads which would be seized in China" — but the burden of American experience in the vast Pacific theater was that it was unwise to get bogged down in major confrontational battles, when Japanese forces could be bypassed — a "bypassing technique which had been employed with considerable success at small cost of life" by General MacArthur, as in the case of Rabaul. "Would it not be equally possible to by-pass Singapore by seizing an area to the north or east of it, for example Bangkok?" he asked. After all, "Singapore," the President reflected sagely, "may be very strong," and for his part he was "opposed to going up against strong positions," unless part of an overarching strategy, as at D-day — and now the Rhine.[14]

Disappointed, the Prime Minister had attempted to defend his dream. "As far as Singapore was concerned he did not favor the by-passing method," he'd responded. Ignoring the President's point about unnecessary casualties he'd declared there "would undoubtedly be a large force of Japanese in the Malay Peninsula and it would help the American operations in the Pacific if we could bring these forces to action and destroy them in addition to the great prize of the recapture of Singapore."[15]

Brooke could only groan at his prime minister's display of casualness in the face of casualties, his repeated talk of "destroying" the enemy, his indifference to geography, and his rank ignorance of modern military combat — for they reflected, as so often, a sort of jejune approach to modern warfare. Or the realities of Hitler's and Hirohito's mind-set. If "Hitler was beaten, say, by January [1945], and Japan was confronted with the three most powerful nations in the world" — given Stalin's promise the Russians would join the war against Japan just as soon as Hitler "was beaten" — then the Japanese, Churchill had airily claimed, "would undoubtedly have cause for reflection as to whether they could continue the fight."[16]

The President had shaken his head at such naiveté. Responding, the President "referred to the almost fanatical Japanese tenacity" in defending their conquests and islands. To prove his point he gave as a recent example that of Saipan, where, he pointed out, some ten thousand Japanese had leaped off the cliffs at Marpi Point on July 12, 1944, to their deaths — "not only the soldiers but also the civilians had committed suicide," he emphasized, "rather than be taken."[17]

The war, in other words, would take longer, and take a much more realistic direction, than the Prime Minister was blandly assuming. The Combined Chiefs should be left to do this, their job; this was the President's gentle but firm opinion, as U.S. commander in chief.

The chiefs, for their part, were delighted — Admiral Leahy, the senior member of the Combined Chiefs of Staff, wrapping up the plenary meeting on the President's behalf by stating that he did not "foresee any insuperable difficulties in reaching agreement on all points at issue."[18]

With that the chiefs had all gone back to the Château de Frontenac for lunch and further meetings.

Field Marshal Brooke was also relieved. "My mind is now much more at rest," Brooke confided to his diary. With the exception of a brief temper tantrum by Admiral King over the possible use of a Royal Navy fleet entering the Pacific, things "have gone well on the whole," the field marshal considered the next day, in spite of Winston's "unbearable moods."[19] He hadn't been able to stop Churchill sending a cable to Field Marshal Wilson, with a copy to General Alexander, claiming that the "Americans talk without any hesitation of our pushing on to Vienna, if the war lasts long enough," and calling on Wilson to prepare operational plans in a "spirit of audacious enterprise."[20] This was, Brooke knew, a bald-faced lie, yet Brooke was sure he could, at least, put the kibosh on any such nonsense the Prime Minister might try to order without American authorization. For the simple fact was, the Americans were now finally in charge of global strategy, and the best the British chiefs could do, he recognized belatedly, was to help them as far as they were able.

"On the whole we have been very successful in getting the agreement which we have achieved, and the Americans have shown a wonderful spirit of co-operation," Brooke thus recorded on September 15—a mood spoiled only by the Prime Minister, who had been "in one of his worst tempers. Now Heaven only knows what will happen tomorrow at our final Plenary meeting . . . The tragedy is that the Americans whilst admiring him as a man have little opinion of him as a strategist."[21]

The war, in other words, was being left to the professionals, now that its strategic direction had been set by the President—and all British diversions had successfully been squashed.

Behind the scenes, unfortunately, a new monster had appeared in the Citadel: one that would overshadow Allied amity and the clear military strategy achieved and displayed by the generals at Quebec in 1944. One presented, moreover, by an American, who had no connection with the military conference, and had not even been invited to it, when it began: Henry Morgenthau Jr., the secretary of the U.S. Treasury.

63

The Morgenthau Plan

THE TREASURY SECRETARY'S PLAN was an extraordinary document; a vision of how to deal with Germany after the war.

The proposal was radically dystopian, namely to eviscerate German heavy industry and mines, and force Germany back to the agricultural economy of the eighteenth century — forgetting the merciless wars of central Europe that predated even the eighteenth century. Once the existence of the plan was leaked in the United States and became known, it would prove manna to Joseph Goebbels in Berlin. It would offer his Propaganda Ministry the very proof needed for asking Germans to continue waging total war. With documented Allied intentions to destroy not only Nazism but the entire country, the Mephisto of Nazi propaganda would be able to persuade German soldiers and civilians they must be prepared to fight to the very death to protect their homeland from deliberate, outright Allied devastation if ever they surrendered.

In every aspect the Morgenthau Plan thus became a crisis that should never have taken place. Nor would it have happened, in all probability, had the President been well, or if Churchill had not brought with him Lord Leathers. After all, there was no pressing need at Quebec to address an issue that was already being examined by the European Advisory Commission in London, set up in conformity with the Moscow Agreements of November 1943, and also in Washington. At the White House, the President had approved on August 25, 1944, a new cabinet-level committee comprising the secretaries of war, state, and Treasury, with Harry Hopkins acting as the President's committee counselor. It was tasked with addressing the question of Germany's future, from an American perspective. Concerned by reports that Eisenhower, the Allied supreme commander, was already being urged to be "soft" on the Germans in a supposedly official U.S. War Department handbook — in fact was being urged to undertake measures similar to the New Deal, such as the Works Progress Administration and Civilian Conservation Corps — the President had been incensed. Such magnanimity, he'd felt, was premature. "The German nation as a whole," he'd complained to Secretary Stimson, "must have it driven home to them that the whole nation" — not just the Gestapo, or SS — "had been engaged in a lawless conspiracy against the decencies of civilization."[1] The Cabinet Committee on Germany had thus been tasked with advising the President on how best to drive this home, once unconditional German surrender was obtained.

For Secretary Stimson, the "treatment of Germany and the Germans" was clearly a military concern, since guidance would have to be given to General Eisenhower in advance of final victory. For Secretary Hull, it was a matter of international relations, since the major Allied powers would be occupying Germany and must decide what kind of postwar country Germany should be — constitutionally, geographically, and politically. For Secretary Morgenthau, however, it was a matter of economics — which he saw as his preserve. Should the German economy be reconstituted as part of a new postwar Europe, or was it vital to smash the German war-making machine for all conceivable time, Germany having provoked two world wars in thirty years?

At the meetings of the Committee on Germany in late August it had become clear the three U.S. cabinet secretaries were at loggerheads — with Hopkins tending to side with the Treasury secretary's simple-sounding solution: namely to remove all heavy industry from the Ruhr, flood the Saar mines, and make Germany a peasant economy once again. This sounded awfully like biblical punishment. Yet to Stimson's horror, Morgenthau's plan had caught the ailing president's attention — and all practical sense had flown out the window.

There was, to be sure, consensus on demilitarizing Germany upon its surrender; dissolving all Nazi institutions; establishing Allied control over the press and education; deferring discussion of partition for a later date; deciding against reparations to be made to the United States; and on prosecuting and — where found guilty — executing all war criminals. Nevertheless Secretary Stimson had objected to Morgenthau's specific, punitive recommendation that the "great industrialized regions of Germany known as the Saar and the Ruhr with their very important deposits of coal and ore should be totally transformed into a non-industrialized area of agricultural land. I cannot conceive of such a proposition being either possible or effective," Stimson had warned in a memo to the President on September 6, "and I can see enormous general evils coming from an attempt to so treat it."[2] Such a destructive program, for one thing, defied the whole economic march of history in Europe. The result — beyond leading millions of Germans to starve, since there was not enough agricultural land to feed seventy million people — would not only slow Europe's postwar economic recovery, but morally it would place the Allies in the invidious position of being as ruthless as the Nazis — and certainly no better than the Russians, who were expected to dismantle and remove all German industrial machinery they could, taking the materials to the Soviet Union as war reparations.

Aware the matter was becoming deeply contentious the President had therefore summoned committee members to the White House on Saturday morning, September 9, for a two-hour meeting before he left for Quebec. Once in the Oval Office Stimson had realized how ill was the President: no longer able to process their arguments, and unable to manage more than a forty-five-minute discussion, even as he prepared to go meet the British.

Forty-five minutes to resolve the destiny of the main enemy of the United States in World War II?

"I have been much troubled by the President's physical condition," Stimson confided in the privacy of his diary. "He was distinctly not himself on Saturday. He had a cold and seemed tired out. I rather fear for the effects of this hard conference upon him. I am particularly troubled," the war secretary had added, "that he is going up there [to Quebec] without any real preparation for the solution of the underlying and fundamental problem of how to treat Germany. So far as he has evidenced it in his talks with us, he has had absolutely no study or training in the very difficult problem which we have to decide, namely of how far we can introduce preventive measures to protect the world from Germany running amuck again and how far we must refrain from measures which will simply provoke the wrong reaction." If the subject was addressed at Quebec, Stimson commented, "I hope the British have brought better trained men with them than we are likely to have to meet them."[3]

Currently aboard the *Queen Mary* the British hadn't even thought to address the matter! Nor, once they arrived in Quebec, was the subject ever once raised in the daily Combined Chiefs of Staff meetings, or mentioned in the minutes or diaries of their deliberations in Quebec. So why, historians wondered afterwards, had Mr. Roosevelt felt it necessary to raise it with Mr. Churchill — especially when he was not taking anyone with him to advise him on the matter, not even Harry Hopkins?

Certainly no historian could later explain the President's sudden focus on the plan of "pastoralizing" postwar Germany, save in terms of growing dementia. In the weeks before Quebec he had been heard objecting to those arguing for generosity in treating the Germans after the war, given the terrible deeds enacted not only by Nazis but many German soldiers and civilians — and which were ongoing. The systematic arrest, deportation, and presumed "liquidation" of hundreds of thousands of Jews in Hungary sickened him. That they were continuing the deliberate extermination of vast numbers of Jews and other defenseless citizens in Nazi concentration camps across Europe without the least concern among "ordinary" Germans was horrifying — and reports that the Wehrmacht was defending every yard of territory it had conquered to the last man standing did not make the President feel magnanimous, understandably, in terms of bringing the people of Germany back into the fold when they were finally forced to surrender. But deliberately and forever laying waste to the Ruhr and the Saar, the very kernel of the European economy, as a punishment?

Stimson's several memoranda on the subject, copied to the President, were among the most articulate the aging lawyer had ever composed, assembled with the help of his large staff at the Pentagon — including General Marshall; Harvey Bundy, special assistant to Stimson; and John McCloy, assistant secretary of war. Such a "draconian" punishment of Germany would amount, Stimson had argued, to cutting off the Western Allies' nose to spite their face — and, if they were seriously to contemplate such a radical measure, would it not require far more serious discussion than Secretary Morgenthau's seemingly vengeful, half-baked proposal?

In this respect the President was undone not only by his own ill health, but by the very concept of a military conference, away from Washington, in which his only role was to take part in two plenary sessions, several days apart. This left Roosevelt more or less alone with Churchill, without agenda or purpose. In short, the President and the Prime Minister were virtually unemployed in the Citadel, with no political discussions supposedly permitted, lest controversy be cast on the President's election campaign. Thus it was — with Churchill seeming "very glum" when shorn of his Viennese and Singapore "diversions" at the first plenary — that on a whim, it seemed, the President had summoned Morgenthau, the Treasury secretary, to Quebec. On his own — that is, without his fellow committee members, namely Stimson and Hull. Or Harry Hopkins.

Such a unilateral invitation by the President completely ruined whatever chance there was for further considered debate or discussion of the issue by experts from different departments. True, the secretary of war had managed to get the President to promise to take "the papers we have written on the subject with him," but he "has not invited any further discussion on the matter with us," or with his longtime White House counselor, Harry Hopkins, Stimson noted in his diary on September 13. "Instead apparently today he has invited Morgenthau up there," Stimson rued, "or Morgenthau has got himself invited." Either way, "I cannot believe that he will follow Morgenthau's views. If he does," he predicted, "it will certainly be a disaster."[4]

It was: Morgenthau triggering a far worse scandal than the mere deindustrialization plan for the Ruhr. For no sooner had he arrived in Quebec than he set about converting Churchill to his cause — by bribing the Prime Minister.

This was, in retrospect, perhaps the most egregious aspect of the crisis. For his part Mackenzie King was neither consulted nor invited to participate.

After escorting Mrs. Roosevelt and Mrs. Churchill to the Château Frontenac for a dinner on September 13—the meal serenaded by the Royal Canadian Mounted Police Band—King had returned to the Citadel. There he had found "Churchill and the President and a small party still at the table" where he'd left them. "They had been seated there when I left at 9, and were still seated at half past eleven," King noted. Churchill was sitting immediately opposite the President, and "both of them seemed to be speaking to the numbers assembled which included Morgenthau, Lord Cherwell, Lord Leathers, Lord Moran, and two or three others. Morgenthau," the U.S. Treasury secretary, had "arrived this afternoon. Anthony Eden is to arrive in the morning."[5]

Looking at her husband's pale, drawn face, Eleanor was alarmed. Telling Mackenzie King the President had not "had any rest this afternoon as he was talking to different persons through the day," she "got the President to come off immediately to bed."[6]

It was too late, however—in both senses. The President had gently but firmly put down Churchill's military dreams concerning Vienna and Singapore that morning, but he was obviously debilitated—and what he had concocted with Winston Churchill, while King escorted their two wives to dinner, would do neither man much historical credit. For when the President had brought up the matter of postwar Germany, the subject had become as contentious as it had been in Washington: the very last thing the President needed with his systolic blood pressure already registering 230, as Dr. Bruenn later noted.

Admiral Leahy, in his diary, recorded that night that the "subjects discussed at dinner were generally international politics, economics and shipping; and the peace terms and punishment that should be imposed upon Germany when that country surrenders to the Allies."[7]

Told by his British table companions that this subject was something that would be addressed in London by his government and civil service colleagues on the tripartite European Advisory Commission, Churchill had sounded furious it was being tabled in London without him. "What are my Cabinet members doing discussing plans for Germany without first discussing them with me?" he'd demanded—prompting the President to suggest that Morgenthau and Lord Cherwell, Churchill's scientific adviser, could talk about the matter the next day. Churchill, however, had always hated to put off to the morrow what could be dealt with today. "Why don't we discuss Germany now?" the Prime Minister had asked the President[8]—and with that, and with the President's approval, Henry Morgenthau had presented his controversial draft plan.

Churchill, facing the President, could not at first believe his ears—in fact the Prime Minister initially "reacted violently against it, even disagreeably so," Morgenthau himself admitted to Stimson when he got back to Washington several days later.[9] He "blew up against it and said he could never accept such a paper"—in fact demanded of the President at one point: "Is this what you asked me to come all the way over here to discuss?"[10]

The plan, as Morgenthau described it, was in Churchill's view "unnatural, unchristian and unnecessary"—words that bespoke the best in Churchill's character. His gift for metaphor was memorably on display, moreover, when he gave vent to "the full flood of his rhetoric, sarcasm and violence"—adding that, far from helping Britain in the postwar years, Morgenthau's deindustrialization of the Ruhr and the flooding of its mines would leave England "chained to a dead body."[11]

"The President said very little in reply to Churchill's views," Harry Dexter White, Morgenthau's deputy, recorded of the "men's dinner" discussion.[12] This had not, however, inhibited others at the table. Churchill's doctor, Lord Moran (as Dr. Wilson now was), recorded in his diary that the Prime Minister "did not seem happy about all this toughness. 'I'm all for disarming Germany,' he said, 'but we ought not to prevent her living decently. There are bonds between the working classes of all countries, and the English people will not stand for the policy you are advocating . . . I agree with Burke. You cannot indict a whole nation.'" In terms of punishment, the Prime Minister "kept saying, 'At any rate, what is to be done should be done quickly. Kill the criminals, but don't carry on the business for years.'"[13]

Moran noted how little the President spoke. "The President mostly listened"—save to say, at one point, that Stalin, at Tehran, had mentioned how even a German factory that "made steel furniture could be turned overnight to war production"[14]—a way of reminding those present that, unless the Western Allies imposed tough terms, the Russians would suspect them of permitting, even financing, the future remilitarization of Germany, as in the 1930s under Hitler.

The discussion might have been considered mere dinner conversation, and of no lasting import. But Morgenthau had not come up to Quebec to converse. He was bearing gifts he thought might interest Mr. Churchill—even persuade him to change his mind on the matter. It was in this way that the next day, September 14, the President and the Prime Minister drew up, literally overnight, a formal plan to punish Germany by reducing its economy to its eighteenth-century agricultural origins. Why?

64

Beyond the Dreams of Avarice

WORKING FIRST THROUGH LORD CHERWELL, Secretary Morgenthau offered the British $6.5 billion in Lend-Lease credit, if his plan for pastoralizing Germany was accepted.

The Prime Minister, who had become alarmed at reports Britain would soon be bankrupt, was electrified — and in a matter of minutes his opposition to Morgenthau's plan evaporated. Moreover, when the President, that afternoon, seemed to be delaying the drawing up of a memorandum of Lend-Lease agreement on such huge financial numbers, Churchill became so "nervous and eager to have the memorandum agreed, he finally burst out: 'What do you want me to do? Get on my hind legs and beg like Fala?'"[1]

This was not a remark the President liked, and indicated how volatile the subject could become. Morgenthau's offer was thus confirmed.

Going to bed that night the Prime Minister seemed over the moon. Britain would be saved from penury; the empire would flourish once more. Boasting to his assistant private secretary of "the financial advantages the Americans had promised us," he seemed almost to levitate as he undressed. "Beyond the dreams of avarice," his secretary commented, as impressed as his boss. It was a phrase that pricked the Prime Minister's sleepy conscience, though. "Beyond," he corrected the young man, "the dreams of justice."

To ensure it was not a mere verbal understanding, the following day, September 15, 1944, the Morgenthau Plan was formally dictated by Churchill himself to a stenographer, to be written up in the manner of a contract between the United States and Britain — to be initialed both by the President and the Prime Minister.[2]

Though the President was too ill to help in drafting the text of the document, he certainly did nothing to question the terms of the plan — in fact he suggested, upon reading the stenographer's first draft, that they widen the scope of the plan to cover *all* potential war-making industries — metallurgical, chemical, and electrical, across the whole of postwar Germany, not simply the Ruhr and the Saar. According to Morgenthau, Churchill became "quite emotional" about the Lend-Lease agreement, and when "the thing was finally signed, he told the President how grateful he was, thanked him most effusively, and said that this was something they were doing for both countries." Britain would not go bankrupt—indeed, with the output of the Ruhr and the Saar mines extinguished, Britain's postwar prosperity would be guaranteed, for Germany would be removed as a postwar economic competitor.[3]

Anthony Eden, arriving posthaste from England, was appalled, hating the very mention of such a deal. "Eden seemed quite shocked at what he heard," Morgenthau later admitted, "and he turned to Churchill and said, 'You can't do this. After all, you and I have said quite the opposite.'"

Churchill had merely batted away his objections.

The President, for his part, had remained silent; he "took no part in it," leaving it to Morgenthau to emphasize the advantages to Britain "in the way of trade," for they would then "get the export trade of Germany."

"How do you know what it is or where it is?" Eden had countered, fully aware of the Prime Minister's ignorance of economics. "Well, we will get it wherever it is," Churchill had responded, "testily"—and had forbidden Eden to return to Britain and inflame the War Cabinet against the plan, before he himself could get back. "After all, the future of my people is at stake," Churchill had declared, "and when I have to choose between my people and the German people, I am going to choose my people."[4]

Secretary Stimson, hearing what the President had done from John McCloy on the long-distance scrambler-phone from Quebec, was aghast. "Apparently he has gone over completely to the Morgenthau proposition," Stimson recorded; moreover had "gotten Churchill and Lord Cherwell"—an individual Stimson considered "an old fool"—"with them."[5] It was essentially, he would later write in hand on his copy of the Morgenthau Plan, a "Bribe to the U.K."[6]

As Stimson lamented in his diary on September 15, "It is a terrible thing to think that the total power of the United States and the United Kingdom in such a critical matter as this is in the hands of two men, both of whom are similar in their impulsiveness and their lack of systematic study." The deindustrializing plan was "Carthaginian"—one that could only shame the United States, and which horrified everyone Stimson spoke to about it.

Unable to live with his conscience if he did not protest, Stimson had penned a rebuke he cabled to the President that night. As he pointed out, he felt he had already objected to the Treasury secretary's proposal, in writing, several times. Nothing he'd learned since then had changed his conviction that the plan "would in the long run certainly defeat what we hope to attain by a complete military victory—that is, the peace of the world, and the assurance of social, economic and political stability in the world."[7]

It was not a matter of "whether we should be soft or tough on the German people, but rather whether the course proposed will in fact best attain our agreed objective, continued peace. If I thought that the Treasury proposals would accomplish that objective, I would not persist in my objections. But I cannot believe that they will make for a lasting peace. In spirit and emphasis they are punitive, not in my judgment, corrective or constructive. They will tend through bitterness and suffering to breed another war, not to make another war undesired by the Germans nor impossible in fact. It is not within the realm of possibility that a whole nation of seventy million people, who have been outstanding for many years in the arts and sciences and who through their efficiency and energy have attained one of the highest industrial levels in Europe, can by force be required to abandon all their previous methods of life, be reduced to a peasant level with virtually complete control of industry and science left to other people.

"The question is not whether we want Germans to suffer for their sins. Many of us would like to see them suffer the tortures they have inflicted on others. The only question is whether over the years a group of seventy million educated, efficient and imaginative people can be kept within bounds on such a low level of subsistence as the Treasury proposals contemplate." It would be "a crime as the Germans themselves hoped to perpetrate on their victims—it would be a crime against civilization itself." It went against the President's own Atlantic Charter, and his Four Freedoms—which sought "freedom of all from want and from fear."[8]

Paragraph after paragraph Stimson panned Morgenthau's plan, not because he thought it unmerited, but because it wouldn't work—and could only rebound on America—and Britain. "The benefit to England by the suppression of German competition is greatly stressed in the Treasury memorandum," he noted. "But this is an argument addressed to a shortsighted cupidity of the victors and the negation of all that Secretary Hull has been trying to accomplish since 1933. I am aware of England's need, but I do not and cannot believe that she wishes this kind of remedy." For, as Stimson remarked, "the total elimination of a competitor (who is also a potential purchaser) is rarely a satisfactory solution of a commercial problem."[9]

Harry Hopkins, reading Stimson's plea to the President, undertook to get it to the President himself. Neither of them, however, was aware Franklin Roosevelt had passed a point of no return: that he was not merely dying of heart disease, but that his once-capacious, genial, and elastic mind was now in severe decline.

65

The President Is Gaga

THE AFTERMATH OF QUEBEC would prove embarrassing. The President rested in Hyde Park, but showed no improvement, in body or spirit. By September 20 the President was predicting Governor Dewey would win the election, as suggested by the latest polls—and seemed actually to welcome the prospect, to Daisy's astonishment. "The Pres. is planning his life after he leaves the W.H. It will be so different, without the many 'services' supplied by the govt. He will 'write'; and can make a lot of money that way—also his corr. [espondence] will be tremendous," necessitating Daisy's assistance, as Grace Tully would want to move closer to her family.[1]

It was at this point that the President's mistake over the Morgenthau Plan caught up with him—for on September 21, 1944, Drew Pearson revealed in the *Washington Post* that trouble was brewing at the War Department. There, senior personnel were "quaking in their boots" after issuing a Handbook on Germany that was ridiculously mild—whereas the President wanted the Germans to receive, Pearson wrote, no more than "three bowls of soup a day."[2]

Three bowls of soup?

Convinced that Morgenthau himself had planted the leak, Stimson's War Department counterleaked: the *New York Times* running an article the next day, written by Arthur Krock, about the Cabinet Committee on Germany: revealing, moreover, that Secretary Morgenthau, but not Stimson or Hull, had traveled to Quebec. The proverbial fat was in the fire—with the President left looking, as even Morgenthau recognized, at the very least, like a "bad administrator."[3]

For a chief executive barely able to function as president or commander in chief, the fiasco was the very disaster Stimson had foretold. Governor Dewey's campaign workers were overjoyed. "After a reporter at the Quebec Conference told Dewey he was convinced that Roosevelt was a dying man, and that the Republican contender had 'an absolute duty' to reveal the true state of the President's health," his biographer recorded, "Dewey and his advisers sweated blood over the issue. 'There wasn't a single night went by we didn't argue that one out,'" Dewey's campaign manager later recalled — with some wanting "'to go all out, stating that he was on his death bed, and getting all the evidence that we could.'"[4] In the meantime, revelation of the Morgenthau Plan would do nicely.

Morgenthau, for his part, was embarrassed and ashamed — especially when Nazi newspapers began quoting the American press about the affair, and General Marshall complained to him in person, over a luncheon, how the publicity could affect the war effort on the Western Front. "We have got loudspeakers on the German lines telling them to surrender," Marshall remonstrated, and "this doesn't help one bit"[5] — Wehrmacht troops being inspired to resist to the bitter end, as even the *Washington Post* warned. Should the Germans believe "nothing but complete destruction" of their country awaited them after surrender, "then they will fight on," the *Post* lectured. "Let's stop helping Dr. Goebbels."[6]

With German radio claiming the Allies would "exterminate forty-three million Germans," and the *Völkischer Beobachter* running a headline, ROOSEVELT AND CHURCHILL AGREE TO JEWISH MURDER PLAN!, Morgenthau was doubly embarrassed — the German press silent on the mass murder, thus far, of over five million Jews in Europe, yet more than willing to quote supposed American intentions to starve many more millions of Germans.

The President did his best to console Morgenthau, whom he'd known as a Dutchess County farming neighbor since 1913. He'd appointed him to the Treasury in 1933, and considered he had done a remarkably successful job — especially in selling war bonds — given that Morgenthau had no training or experience in finance. Knowledge of the ruthless German extermination program had only made the President more compassionate toward Morgenthau — for despite calls for bombing of the concentration camps or the railtracks leading to them, there was nothing the Allies could realistically do to stop the Nazis, save issue warnings of arrest for crimes against humanity and defeat the Wehrmacht as soon as humanly possible.[7]

Secretary Hull was just as disappointed by the scandal as the staff of the *Washington Post* — indeed, he felt particularly sore that Morgenthau, without consultation with the State Department, had gone ahead and promised the British a $6.5 billion get-out-of-debtors-jail card without recompense of any kind, such as imperial trade concessions or even possible military bases. By September 24, 1944, newspapers were widely reporting a "split in the Cabinet"[8] — and the President finally asked to talk with him.

"Now the pack is in full cry," Stimson noted in his diary on September 25.

At a moment when Field Marshal Montgomery's great airborne operation to bounce the Rhine in northern Holland had met with fierce Wehrmacht resistance—an entire British airborne division being rubbed out at Arnhem (the "Bridge Too Far"), and two magnificent American airborne divisions, having secured crossings over the Maas and Waal Rivers, suffering serious casualties—the Quebec fiasco, or "hell of a hubbub," as John McCloy called it, was becoming deeply damaging not only to the President's reelection chances, but to the Allied cause. "The papers have taken it up violently and almost unanimously against Morgenthau and the President himself," the war secretary noted. Morgenthau might cry foul, given leaks to the press, but the President himself "had already evidently reached the conclusion that he had made a false step and was trying to work out of it." As long as the memorandum itself was kept secret, the President would meantime deny all newspaper accounts of his supposed policy. A phone call from the President on September 27 confirmed this—as well as a meeting in person arranged at the White House on October 3.[9]

It was in this way that, behind the scenes, the biggest change in the military direction of World War II since Pearl Harbor was implicitly acknowledged by Secretary Stimson. The President of the United States was no longer really fit to carry out his constitutional role as commander in chief. Word had come from China that the Japanese were attacking Chiang Kai-shek's forces and pushing them back—with huge implications for the course of the war against Japan. Stimson had prepared himself to discuss the situation as a matter of urgency with the President over lunch. "When I got to the White House and saw him," Stimson noted, however, "I saw that it would be impossible. He was ill again with a cold and looked tired and worn"—so much so, in fact, that Stimson was asked to lunch with him "in the main building [mansion] because the mere lunch with me was a violation of his doctor's routine that he should not waste strength at lunches at his office."[10] He even had his daughter Anna there, to see he didn't get stressed or too tired.

Discussing China was clearly beyond the President, so no "business" was conducted during the meal—and then only in the gentlest fashion. When the President denied having signed a paper with Churchill about turning Germany back into an agrarian, preindustrial state, Stimson pulled out his copy of the Morgenthau Plan, with its final sentence calling for "eliminating the war-making industries in the Ruhr and the Saar" and "converting Germany into a country primarily agricultural and pastoral in its character." The intent was plain as a pikestaff. As Stimson recorded that night, the President "was frankly staggered by this and said he had no idea how he could have initialed this; that he had evidently done it without much thought"[11]—openly confessing, "Henry, I have not the faintest recollection of this at all."[12]

No recollection?

The frank admission allowed Stimson to remind the President of the true import of the matter.

"I told him that in my opinion the most serious danger of the situation was the getting abroad of the idea of vengeance instead of preventive punishment and that it was the language of the Treasury paper which had alarmed me on the subject. I told him that, knowing his likeness [*sic*] for brevity and slogans, I had tried to think of a brief crystallization of the way I looked at it. I said I thought that our problem was analogous to the problem of an operation for cancer where it is necessary to cut deeply to get out the malignant tissue even at the expense of much sound tissue in the process, but not to the extent of cutting out any vital organs which by killing the patient would frustrate the benefit of the operation. I said in the same way that what we were after was preventive punishment, even educative punishment, but not vengeance. I told him that I had throughout had in mind his postwar leadership in which he would represent America. I said throughout the war his leadership had been on a high moral plane and he had fought for the highest moral objectives. Now during the postwar readjustment 'You must not poison this position' which he and our country held with anything like mere hatred or vengeance." Stimson was at pains to emphasize the friendship he, Stimson, felt for Morgenthau, and how concerned he'd been by how "misrepresented," in pursuing such a plan, a "man of his race"—i.e., Jewish—would be.[13]

The President was chastened. His excuses that he'd wanted to be nice to the British, who were facing bankruptcy, were lame. Moreover, though it had shown deep and genuine compassion for his friend, and outrage at what the Nazis had done—and were still doing—the plan would not, in the long run, do any good. Nor would it even work, since it would only make seventy million Germans susceptible to a future Hitler. In short, America's moral agenda had to rise above such understandably punitive feelings in order to embrace the future: a future in which the United States would be the new leader of the democracies, whatever the Russians might do.

With that the luncheon visit ended.

Anna, the President's daughter, was clearly grateful for Stimson's kind, even tender manner—Stimson recording in his diary that he'd spoken in front of the two of them "with all the friendliness and tact possible—and after all," he added, "I feel a very real and deep friendship for him."[14]

Their relationship had now changed, though. As King George VI was told in London, the President had "sadly aged. He never ventures to wear his [leg] 'braces,' now—i.e. never walks at all; and in a long sitting" at Quebec, "his jaw would drop, and his thoughts obviously wander"[15]—the king's counselor, Alan Lascelles, being privately told by General Ismay that the President was, effectively and ominously, "gaga."

Secretary Stimson and General Marshall, in other words, would now be running the war — as well as addressing the military aspects of the postwar. On behalf of a president who was now, essentially, a figurehead, with Admiral Leahy caretaker of the White House and linchpin of the chiefs.

With that, Stimson had left the White House — and the unfortunate business of Quebec.

PART TEN

Yalta

66

Outward Bound

ALMOST FOUR MONTHS LATER, at 8:31 a.m. on January 23, 1945, the USS *Quincy*, an American heavy cruiser, cast off from Pier 6 at Newport News, Virginia, bearing the President of the United States.

Sister vessel to the USS *Baltimore*, which had taken the President to Hawaii the previous summer, the two Baltimore-class battle cruisers were almost identical in firepower. The President was not the same man, however, who had sailed to Pearl Harbor in August 1944 — let alone the commander in chief who had journeyed to Oran and back home in late 1943 aboard the USS *Iowa*. Able only to harness his mental faculties for an hour or two a day, and often exhausted by that effort, he was by January 1945 another heart attack, or major stroke, waiting to happen. How, two months before, on November 7, 1944, he had won a fourth term in the White House was something of a miracle — or a monstrous act of deception, depending on how one viewed the presidential election.

From somewhere in his ailing body and spirit the President had drawn upon his final reserves of energy. He had made several campaign speeches, and been driven in an open car through the often rain-swept streets of Philadelphia, Chicago, New York, and Boston to prove he was not only still alive, but kicking. His notorious address to members of the Teamsters union at a dinner in Washington, in which he remarked that he was not bothered by attacks his Republican opponents made on him, but that they did upset Fala, his dog, demonstrated the President still had a sense of humor — a quality his opponent, Governor Dewey, sadly lacked. But with polls showing a swing away from the current sclerotic administration, after twelve years in power, even the President had thought Dewey would win.

And Dewey came close to doing so — in fact, more than was comfortable to the Democrats behind the numbers. In the early hours of November 8, Dewey's campaign cohort was still urging him not to concede, for it had proved the closest election since 1916. With the tide of the country's voters running to the right, the younger Dewey-Bricker ticket had seemed not only menacing to Democrats but almost invincible in the eyes even of their strategists, given the mood of a nation seemingly tired of the same old faces and twelve years of Democratic rule. Only the President's personal popularity could reliably stave off a Republican victory, they had figured—and reports of his worsening health had therefore threatened to make defeat probable, if not certain.

Dewey was done in by his own party, however, which failed to understand the need for idealism. By standing for reelection on the domestic and international platform that he'd chosen—a New Deal for veterans after the war, and an end to isolationism abroad—the President had forced the nation to choose whether, under younger personnel, to go back to Hoover's Great Crash, Depression, and an "America First" retreat from international involvement that had characterized Republican politicking in the 1930s, or to move forward—albeit under increasingly doddery old men.

Compelled by the President to take either the high road in terms of domestic and international policy or the low road of Republican self-interest, Dewey had waffled—Republican voters at his rallies across the country so bored by his initial attempts to articulate a course change that they booed him. What they'd wanted was a fight: an old-fashioned political bar fight. Changing his tune, the young New York governor—a neophyte in terms of international politics—had finally delivered it: railing at the President's New Deal results; at big government; at high taxes; at Jews and communists running the nation's unions. And the sheer senility of the Roosevelt administration.

Roosevelt's stature as the nation's commander in chief in war, and his high-minded appeal to American idealism in terms of postwar security, had thus outweighed the rumors of his ill health—enough, at any rate, to keep the majority of the nation's voters resolute in prosecuting the war and in bracing themselves for high, global responsibilities once the war was won.

This was, in the circumstances, no mean achievement. So irresponsible did Dewey become, in fact, that he was prevented from revealing Magic, the breaking of Japanese codes at the time of Pearl Harbor, only by way of a personal meeting sought by General Marshall. The chief of staff of the U.S. Army had told the young governor that such a disclosure would amount to an act of treason, given that U.S. cryptographers were still using Magic to great effect—moreover that it would be committed solely to embarrass the Roosevelt administration for its failure to warn Hawaii more effectively in December 1941, without concern as to how many American lives would be put in jeopardy at present, should the breaking of the United States' most secret diplomatic and military codes become known to the Japanese.

Tom Dewey had duly backed off—though he subsequently blamed his defeat, in private, on General Marshall, who he claimed had been personally responsible for failing to issue a more emphatic warning to U.S. bases in the Pacific to prepare for a Japanese sneak attack. Other Republicans were less sure, however; a political campaign based on past mistakes, at a time of world war, and without a vision of America's future in the world had probably been a mistake. The ailing president had shown himself to be the more courageous candidate — unafraid, in particular, to address the matter of global security and the need to move beyond isolationism in a postwar world. Thanks to the antiquated electoral college system, a switch of less than three hundred thousand votes in certain states could, in theory, have put Dewey in the White House; instead, the President's ultimate margin of victory had proved to be close to a landslide: 432 college delegates to 99.

In ballots cast across the country, too, the President had trounced his rival, garnering almost three and a half million more votes than his opponent. In a presidential election the Republicans ought on paper to have won, had they been less focused on domestic issues with an eye to their own wealth, the writing was on the wall. No sooner was the election over, therefore, than more thoughtful Republican leaders had begun to rethink their prewar isolationist positions.

How far should the U.S. government go in working with, or appeasing, the Soviets in terms of postwar policy? The question was certainly debatable, but whatever the answer, the Russians could not, as current allies, be ignored. They had provided the essential manpower to defeat the Nazis — and would be a major world power after the war's conclusion: victory, moreover, that was still to be won, whatever casualties it took.

Isolationism would simply not be an option, the former America First senator Arthur Vandenberg had finally acknowledged in the wake of the election. Other Republican stalwarts had begun to follow suit. Even Governor Dewey, following his failed campaign, had spurned his advisers — urging the New York State Legislature to embrace progressive principles in its political and economic programs, from rent control to massive electricity infrastructure projects, antidiscrimination laws, and public education programs, while supporting a new internationalism — earning the sobriquet "a Liberal without blinkers."[1] After the last hurrahs of a bitter general election campaign, idealism was returning, finally, to the Republican agenda.

It was Senator Vandenberg himself who led the new charge. On January 10, 1945, in a dramatic address in the U.S. Senate — one that became known as the "speech heard round the world" — the senator had finally and formally announced his conversion from prewar reclusion to postwar internationalism. Worried that, as victory neared, greed and self-interest would disunite the Allies, Vandenberg — "a big, loud, vain, and self-important man, who could strut sitting down," in the words of one reporter[2] — had appealed for a collective, collaborative international approach to postwar peace—and peacekeeping. Again and again the President's former enemy number one—the very senator who had tried to get MacArthur nominated as the Republican candidate for the presidency in the spring of 1944—had openly admitted he'd been wrong. To the consternation of his peers in the chamber he had repeatedly quoted President Roosevelt's January 6 message to Congress—verbatim.

The President's message had certainly been forceful: delivered, on camera and on radio, in a firm, lilting voice that completely belied his plummeting health. It had been, in truth, an 8,253-word oration crafted by the President's speechwriters, Sam Rosenman and Robert Sherwood, and had addressed, head-on, "the status of the war, the international situation during and after the war, the formation of the peace," as Rosenman later recalled—as well as "the United Nations Organization, and the future of America."[3]

Broadcast in the very midst of the Battle of the Bulge—the massive, initially successful German surprise counterattack by three Wehrmacht armies in the Ardennes—the President had taken pains to declare, as U.S. commander in chief, his "complete confidence" in General Eisenhower. The young general was the nation's "supreme commander in complete control of all the allied armies in France"—a commander who "has faced this period of trial with admirable calm and resolution and with steadily increasing success." Unity among allies in war was, the President had been at pains to emphasize, the ultimate reason no German military counterattack, no matter how lethal (American casualties in the Ardennes and Alsace would amount to more than one hundred thousand killed, wounded, captured, or missing, including nineteen thousand dead), would in the end succeed. Yet it was Roosevelt's appeal for unity and common purpose in creating a United Nations organization *after* the war that had most impressed Senator Vandenberg.

"Again I agree wholeheartedly with President Roosevelt when he says, 'We must not let such differences [between allies] divide us and blind us to our more important common and continuing interests in winning the war and building the peace,'" Vandenberg had declared in Congress, quoting the President. By all accounts Russia was contemplating "a surrounding circle of buffer states, contrary to what we thought we were fighting for in respect to the rights of small nations and a just peace"—including states such as Poland. In terms of Russian national self-interest this was "a perfectly understandable reason," Vandenberg had allowed—but hardly a recipe for postwar peace on a larger, global scale. Isolationism and regional balance-of-power politics would never be enough, he declared, to guard against a third world war—as the rise of Hitler and Japanese militarists had shown, with literally devastating consequences. "The alternative is collective security," Vandenberg had emphasized on the Senate floor, citing again the President's January 6 message to Congress — in fact Vandenberg had gone further: urging, on behalf of his colleagues in the Senate, that the President should enjoy "full power to join our military force with others in a new peace league," or United Nations, in order to keep "Germany and Japan demilitarized." In this way Germany would pose no threat to the Soviet Union after the war. Moreover such "full power" should be accorded the President without his having to "refer any such action back to Congress," to be hobbled once again by isolationists, such as himself in the run-up to Pearl Harbor.

Delighted to read this, the President had ordered fifty copies of Vandenberg's speech to be printed immediately, which he could take with him when he left the Bureau of Engraving station on the *Ferdinand Magellan* two days later, bound for the USS *Quincy*. There were myriad domestic problems to face in Washington — most urgently the fury being unleashed in Congress by the replacement of Commerce Secretary Jesse Jones by the now-former vice president, Henry Wallace — but the President could take satisfaction in having almost single-handedly won the constitutional backing of the American people for his internationalist postwar program, in spite of his debilitated state of health. Not only the support of the voting public, moreover: Republicans and Democrats in Congress, too, pace Senator Vandenberg.

The President's public backing for General Eisenhower was not window dressing.

The British commander of the northern Allied armies, Field Marshal Montgomery, had caused an uproar in the press on January 6, 1945, by posing as the savior of the Allied forces in stopping German panzer armies from reaching the Meuse and splitting the Allies apart, as they'd done in 1940. Like General Patton, who had swiveled his U.S. Third Army to attack the southern flank of the German bulge, in order to smash the side door of the German breakthrough and relieve Bastogne, Montgomery was certainly a field commander of rare distinction in modern warfare. Like Patton, also, he had been endowed, both by nature and his growing fame as a warrior, with a big mouth. Not content with "saving" the American army in the Ardennes, the field marshal had revived his calls to be given back permanently the field command of the main Allied armies he'd held during the Normandy campaign, and which he'd regained once the magnitude of the German offensive in the Ardennes was appreciated. Field Marshal Alan Brooke, moreover, had backed Montgomery in this effort — one in which General Eisenhower's hand would be removed from the Allies' operational tiller.

Though he admired Montgomery's achievements in combat, the President was impressed by General Eisenhower's performance as a coalition supreme commander even more. He had therefore welcomed the chance to state in public, before the nation and the world, his support for the officer he had personally chosen to be supreme commander not only of D-day, but the consequent campaign to defeat the Wehrmacht.

The Germans, Eisenhower had memorably told him at the Villa Dar es Saada in Casablanca before the Battle of Kasserine, were tough, ideologically ruthless, and well trained; they made amazingly good combat soldiers, from senior officers down to corporals and GIs.[4] Where the Italians had become less and less effective soldiers since the time of Caesar's legions, the Germans—who had already been fearsome tribal warriors, north of the Rhine, in Caesar's day—had become steadily more accomplished in military terms. The great technological advances of modern industrialized warfare, from panzers to jet fighters, ballistic missiles to U-boats, suited them: their century-long investment in higher education and science had been rewarded in spades. Moreover the complete separation of technology from morality had permitted the Wehrmacht to conquer democratic Europe in a matter of months—its officers ignoring any consideration outside loyal service to the Führer and the Reich. The secretly executed counteroffensive in the Ardennes—the Battle of the Bulge, as it became known—had shown the Wehrmacht at its most proficient, and morally ruthless—epitomized by the massacre of American prisoners by firing squad at Malmedy, Belgium, on December 17, 1944, committed by units of the First SS Panzer Division.

The President, in disclaiming responsibility for the Morgenthau Plan, had wanted to rise above vengeance as a national policy. The heavy casualties U.S. forces were suffering in the Ardennes, however, had appalled him, and had certainly not inclined him to become more generous in addressing the question of what to do with Germany, once unconditional surrender was achieved: a matter that was now coming to a head. For the question of how to deal with Germany would be a major issue in his second proposed summit with Marshal Stalin and Prime Minister Churchill: a sort of Tehran 2.

Where, exactly, would it take place, though—and when?

67

Light of the President's Fading Life

After much beating around the proverbial bush (with possible venues ranging from Scotland, Athens, Piraeus, Salonika, Jerusalem, Istanbul, Rome, Alexandria, Cyprus, Malta, and the French Riviera)[1] it had been the President himself who had settled on Yalta, in the Crimea: a kind of Russian Riviera on the Black Sea and erstwhile summer resort of the czars. Also the favorite winter resort of the playwright and short story writer Anton Chekhov.

Stalin had insisted the summit be on Russian soil, if it was really urgent—which he doubted. His Russian forces were currently overstretched; some were, however, within striking distance of Berlin, if the Wehrmacht did not launch another Battle of the Bulge on his own front. He needed, he'd therefore claimed, to be in constant supervision of his armies as they fought their way through Poland and across the Oder River to the German border. Why was it necessary that the national leaders meet now?

The urgency was, in the end, the President's health. Roosevelt had already lived longer than many observers, from his doctors in Washington to the *New York Times,* had thought likely or possible the previous summer. In his diary Lord Halifax had noted he'd been told by James Reston, a young reporter, "that before the President went away on his holiday" to Hobcaw, "the *New York Times,* on their information, had made up their mind that it was quite likely he was going to die, and had accordingly prepared a full obituary front page, with all the right photographs."[2]

The President hadn't died, however, and his personal relationship with Stalin, based on their years of correspondence and their meeting in Tehran, still held out strong hope the United Nations authority could at least be set up before war's end, avoiding the likely divisions that would arise between different countries' postwar self-interests, as Senator Vandenberg had also warned. The Dumbarton Oaks Conference had ended with considerable Allied agreement—but without resolving continuing Russian demands for multiple Soviet delegates to the embryonic UN. A personal summit with Stalin could hopefully cut that Gordian knot, before the Russians became too powerful, as well as tying down military-political agreements, such as zones of occupation in a defeated Germany, and Russian entry into the war against Japan. If the President's health held out.

In short, there was not a moment to lose. Russian ground forces were growing stronger by the day, not weaker. Soviet troops were already occupying most of Poland, Romania, and Bulgaria; they would soon be in Hungary — in fact they were close to the Slovakian border in the Carpathian Mountains. By early February they would be only forty miles from Berlin.

Thanks to Hitler's counteroffensive in the Ardennes, the prospects for a major crossing of the Rhine by the Western Allies and a drive on Berlin had been downgraded to somewhere between unlikely and nil, with Eisenhower's broad-front offensive plans, north and south of the Ruhr, put back by months. If the President wished to keep the Allied military and political coalition together and get the United Nations established on a sound preliminary basis, it would have to be done *now,* he was aware — before the sheer tide of national self-interest disunited the prospective member states — before Japan could be defeated.

That the President's medical condition spoke against the strain of such a long trip — especially the air journey from Malta to Yalta — was a serious negative. Using new metal-and-leather braces, and holding on to firm handlebars attached to the lectern on the South Portico of the White House — aided as well by his son James, at his side to assist him if he lost balance — he had managed to give his brief, fifteen-minute fourth inaugural address on January 20, 1945, standing. His cousin Daisy had even persuaded him to try out, in secret, the services of Harry Setaro, a prizefighter turned religious healer who employed massage to revive ailing organs. Setaro had worked on the President's spleen and legs, and the massage had led the President to claim he sensed a tingling in one toe that he hadn't felt since being diagnosed with poliomyelitis in 1921.[3] But in truth it was an illusion: the triumph of hope over experience. The President's doctors had turned a blind eye to it, though, since Setaro's ministrations could do no harm, they reckoned, even if they did no good.

Such obvious quackery, however, had had no relevance, in the doctors' eyes, to the President's ever-worsening congestive heart disease — one that would lead inexorably to failure: a valiant heart still pumping, aided by digitalis, but with no earthly prospect of improvement, let alone recovery. A stroke or heart failure could come at any moment — and in response to any prolonged mental effort or stress, it was noted by both the President's doctors. The President's systolic blood pressure would constantly leap from the 190s to the 230s and 240s. Attending a summit in the Crimea, five thousand miles away, was thus a grave gamble.

Hubris, though, played its part. Rightly or wrongly Roosevelt remained convinced — and was encouraged by all around him to believe — that he was the only national leader who could, by means of his world stature, get the foundation stone of postwar security laid, before the inevitable discords of postwar factionalism. Certainly this was what Lucy Rutherfurd, the light of the President's life at this point, recommended — overcoming not only her concern for his cascading health but any hope of an alternative ending to his life: namely retirement and an arrangement, possibly separation or divorce, with Eleanor, so that she and Franklin could be together in his final months.

Lucy had been among the seven thousand people admitted to the snow-covered South Lawn of the mansion, by special invitation, for the inauguration. What she didn't know was that the President had suffered another attack of abdominal pain, shortly before the event, similar to the one he'd experienced on July 20, 1944, before giving his party-nomination acceptance speech. Ordered by his father to fly from the Philippines to attend the inauguration and assist him, as he stood to give his speech from the flag-draped, pillared portico, Colonel Roosevelt had found himself once again shocked.

"The first moment I saw Father I realized something was wrong. He looked awful, and regardless of what the doctors said, I knew his days were numbered."[4]

Though his father had managed the fifteen-minute speech to the end, he was riven by abdominal pain afterwards, if "somewhat less acute" than in San Diego, James later recalled — shocked as much by his father's reaction to the attack as by its suddenness. "Jimmy, I can't take this unless you get me a stiff drink," he'd said. And had added: "You'd better make it straight." As James later reflected, "In all my years, I had never seen Father take a drink in that manner" — like life-saving medicine. "I was deeply disturbed."[5]

There were other signs to the Marine colonel, moreover, that his father was not going to make it — and was more and more resigned to his "death warrant."[6] He'd already given his secretaries and his cousin Daisy items chosen from his study, which, when he'd pressed them, they'd said they would like to remember him by. To James he said he was going to give him his gold family ring — and discussed his will, since James was to be his executor.

His father's whole demeanor was that of a dying man, James later described, sadly — his days "surely numbered. It showed in the way he talked, the way he looked. There was a drawn, almost ethereal look about him. At times the old zestfulness was there, but often — particularly when he let down his guard — he seemed thousands of miles away."[7] At the reception, indoors, others were similarly shocked, including the new secretary of state, Edward Stettinius — who was to meet up with the President, for the journey to Yalta, once Mr. Roosevelt reached Malta by sea.

"He had seemed to tremble all over" as he gave his inaugural address, Stettinius later recorded the President's condition. "It was not just that his hands shook, but his whole body, as well . . . It seemed to me that some kind of deterioration in the President's health had taken place between the middle of December and the inauguration."[8]

The widow of President Wilson, Edith Bolling Wilson, was also at the reception. She was more fatalistic. "He looks exactly as my husband did when he went into his decline," she told the labor secretary, Frances Perkins. "Don't say that to another soul," she begged, however. "He has a great and terrible job to do, and he's got to do it, even if it kills him."[9]

Given that Mrs. Rutherfurd could hardly be invited to join Eleanor and the VIPs in the White House without a scandal, Daisy Suckley had gone afterwards to Q Street, where Lucy was staying with her sister, to report on the reception — and she did so again, the next day. Thanking Daisy in a letter several weeks later, Lucy referred to "great sorrows and shock" that tended to leave one "with a tremendous incapacity or fear of thinking" — and therefore unable to "think anything through."[10]

Was this a reference to the President's mortal illness, and imminent death? Should the President therefore resign on grounds of ill health? Senator Truman had been sworn in as vice president; he had looked, in contrast to the President, the embodiment of energy and determination behind his broad chest and thick spectacles. Lucy was unsure. "There seems so much to be decided — What is right and what is wrong for so many people & and I feel myself incapable of judging anything. Yes," Lucy had added to her letter to Daisy — "it is difficult when we must speak in riddles but we have spoken to one another very frankly — and it must rest there — One cannot *discuss* something that is sacred — and even simple relationships of friendship and affection are sacred & personal."[11]

Lucy's friendship with the President had certainly become more and more intimate over the preceding months, as his health continued its downward slide. She had joined Franklin at the Little White House, his personal retreat in Warm Springs, Georgia, soon after he'd gone there after Thanksgiving — the President having "lost 10lbs" during the election campaign, which left him looking "very thin," as Daisy herself had noted. Even in Warm Springs, doing almost nothing, the President had "looked pale & thin & tired," she'd described. "I try so hard to make myself think he looks well," she'd mourned, "but he doesn't." He looked positively "grey when tired" — a sign of lack of oxygen in the blood — in fact he "looks ten years older than last year, to me. — Of course I wouldn't confess that to anyone, least of all to him, but he knows it himself."[12]

What had seemed to bring the President back to life had been Lucy's arrival in Warm Springs on December 1, with her daughter Barbara — the President asking both Daisy and Polly Delano, his bosom companions, to move from the Little White House to one of the guesthouses on the property, in order to make way for his former lover. "Mrs. Rutherfurd is perfectly lovely, tall, stately, & with the sweetest expression," Daisy had penned in her diary that night. "She is much worried by the Pres.'[s] looks," in fact she "finds him thin and tired looking," as she'd confessed to Daisy—but the evening had turned quite romantic when "all the lights went out!" and candles were lit. The "Pres. jokes, cocktails are brought, the Pres. mixes them by the light of one candle." He had even compelled Lucy—who didn't really drink—to have "an old fashioned, made half-strength, but is given two, I notice, against her wishes!" Daisy had noted, amused—and touched. "She says she never took one until she was fifty."

The four of them had had dinner together when the lights came back on, as well as "delightful conversation & at a quarter to eleven the Pres. decided to send us all to bed—Polly & I to the guest house,"[13] while Lucy stayed with the President.

The next day they had all gotten into the "big car" and gone for a noonday drive, with the top down and glass windshields on the sides up. "Mrs. Rutherfurd climbs in & sits next to him," Daisy—squeezed in the corner—had described, and together they'd been driven in cold but clear sunshine to "the Knob"—Dowdell's Knob, the highest point in the Pine Mountain range, with its incomparable views. By this point Daisy was wholly won over by the visitor, who "is a perfectly lovely person in every way one can think of," Daisy had felt compelled to admit, "and is a wonderful friend to him."[14]

Lucy had stayed with the President a second night, but had not been deceived by the President's light mood. Before leaving by car on December 3 she'd confided to Daisy she was "worried & does worry terribly, about him, & has felt for years that he has been terribly lonely." Harry Hopkins had been a wonderful companion as well as White House counselor, much like a roommate, but Hopkins's marriage and eventual departure to live in Georgetown with his new wife, Louise, had left the President "entirely alone" after work in the Oval Office, since Eleanor lived pretty much her own life, was constantly traveling, had her own bosom companions, and thus was seldom to be found even in the same house as her husband.

Daisy's occasional companionship, coming over from Wilderstein to the President's home and library, as well as to the White House at times, had been a boon, as had Anna Roosevelt's move to the White House with her husband, John Boettiger. But still . . . "We got to the point of literally weeping on each other's shoulder & we kissed each other, I think just because we each feel thankful that the other understood," Daisy had recorded of her growing bond with Lucy.[15]

After lunch the President had insisted on taking Lucy in the big car all the way to Augusta, almost two hundred miles away, sitting alone with him. There,

Lucy's own chauffeur had caught up with them and had driven Lucy and her daughter on to Aiken, while the President had been brought back to Warm Springs. "We miss them," Daisy had described — adding that "Franklin was 'let down' after the visitors had gone, and I was glad when he went off to bed."[16]

Though the President had resided for three long weeks in Warm Springs, doing almost no work and without even Admiral Leahy, his chief of staff, with him, the stay had done little to help his condition other than to ensure he was spared a medical emergency. Moreover, Roosevelt had continued to miss Mrs. Rutherfurd. On the way back to Washington on December 17 the *Ferdinand Magellan* had stopped in Atlanta, where Lucy met the train. "Lucy has just come on board and I have left her with the Pres.," Daisy had noted in her diary, "so they can have a little talk without an audience."[17]

Christmas had been spent by the President with his family — Eleanor, their children, and their thirteen grandchildren — at Hyde Park, by family custom, and with all the trappings of tradition, from the exchange of gifts to the President's reading of Dickens's *A Christmas Carol* — or part of it, when the grandchildren became too restless and the President's voice too strained. Then, after returning to the White House and giving his State of the Union broadcast in Washington on January 6, 1945, the President had boarded the *Ferdinand Magellan* to travel to Hyde Park on the evening of January 11 — and had taken Mrs. Rutherfurd with him, *alone*. Lucy had even spent the night with him aboard the train, and had remained the following day closeted with him at the Springwood house.

In the afternoon of January 12 the President had then shown Lucy the library, where Daisy worked. "My new cousin," Daisy called Lucy, noting that they had "one very big thing in common: our unselfish devotion to F." "I took myself off as soon as I had finished my coffee," Daisy had added, "so they could talk."[18]

In short, all subterfuge had been dropped. From the President's daughter to his doctors, from the Secret Service to the staff at the White House, Warm Springs, and the Springwood mansion, scores of people had become aware the President had a special friend — and though looking at death's door, seemed to be strangely reconciled to his fate. He had made, his private secretary had noted, practical and legal "preparations" for the end. Yet there had been no sign of sadness, despite the probability that his life would be cut short. And what seemed to best explain his calm acceptance of his approaching death but also the way he was hanging on to life, despite his exhaustion and discomfort, was the strange, late-life romance he'd discovered with the woman he'd loved so passionately at the height of physical well-being, before his poliomyelitis: a still-youthful-looking yet very private, modest woman who had come back to him as a widow, in the evening of his life — and who clearly still loved him as deeply as he loved her.

It was with that special emotional comfort — having prematurely celebrated his sixty-third birthday at the White House on January 21 with his family and the crown prince and princess of Norway as well as the Morgenthaus and others — the four-time elected president of the United States had set forth on what would surely be his last foreign trip: dreading it, as he'd confided to Daisy, since it would undoubtedly be, he forecast, "very wearing." Yet conscious, too, how crucial it would be to the successful ending of the war, and cementing the foundations of postwar international security.

He himself would have to be particularly careful, the President had told Daisy, to be "on the alert, in his conversations with Uncle Joe and W.S.C." His counterparts would assuredly have their own agendas, from Soviet expansion to British reoccupation of lost colonies and revived imperialism. "The conversations will last interminably & involve very complicated questions."[19]

There had been no alternative to going to Yalta, however. No one else possessed his prestige, or stood a chance of getting the three most powerful men in the world to agree and sign off on the establishment and structure of the United Nations, as well as laying the foundations of a new world order.

If all went well.

68

Aboard the USS *Quincy*

WITH HIM, on the journey to Malta and Yalta, the President was taking his daughter Mrs. Anna Roosevelt Boettiger.

Afterwards, the First Lady had expressed disappointment he hadn't taken her. He had, after all, promised to take Eleanor to England in the aftermath of a successful Allied D-day invasion in 1944—yet hadn't done so. In the context of his fourth-term election campaign, the English trip, he'd explained, had become a plan too politically controversial, and would offer an easy target for right-wing attack by Republicans at home. The journey to Yalta, however, was different. The election was over. Eleanor was his spouse. Not, unfortunately, the sort of spouse he could count on to keep down his blood pressure, since she herself felt so strongly about the political issues to be discussed!

Though Eleanor meant well—indeed was memorably said to be her husband's social and political "conscience"—her advocacy frequently exhausted the ailing president. Brittle, opinionated, and energetic, Eleanor seemed to take all too little cognizance of his now-plunging health. Even her young confidant and subsequent biographer, Joseph Lash, would later quote a tart letter Eleanor had posted to Franklin while the President was recuperating in Warm Springs after the election.

Lash had been a sort of idealized son to whom Eleanor was still loyally writing every few days in letters that expressed deep maternal tenderness, thoughtfulness, and sensitivity. Eleanor's correspondence with her own husband, by contrast, had been nowhere near as tender, Lash later observed—especially in the period before the President's trip to Yalta. So hectoring was one letter that Lash felt, as her conscientious biographer, he would have to quote it in full, he explained, "for it shows how tough, relentless, and perhaps unfair, she could be."[1]

The letter had indeed been harsh—in part because Eleanor could see Franklin was not the man he'd once been, even a year before, and thus needed to stand up more aggressively to those she saw as the worst elements in the

Democratic Party and administration: officials who were pro-fascist, ruthlessly capitalist, deeply racist, and often anti-Semitic. "I realize very well," she'd begun her diatribe, "that I do not know the reasons why things may be necessary, nor whether you intend to do them or do not intend to do them." Such reasons would not, nor should not, she wrote, prevent her from speaking her mind. In the State Department, she said, she saw demons who would drastically alter American foreign policy now that Secretary Hull had retired as secretary of state the previous November for health reasons — unlike the President.

Franklin had always maintained that he himself, if truth be told, was the nation's secretary of state — but in his current state of health this no longer held good, Eleanor had pointed out. Men like James Dunn, a senior State Department officer on the U.S. Dumbarton Oaks team, was a man committed to helping restore the British and French colonial empires after the war — and would run rings round him, she warned, if Franklin was not careful. It was, Eleanor chided, "pretty poor administration to have a man in whom you know you cannot put any trust, to carry out the things which you tell him to do. The reason I feel we cannot trust Dunne [*sic*] is that we know he backed Franco and his regime in Spain," she pointed out — using the royal "we." To her horror Dunn was now pressing the U.S. ambassador in London, Guy Winant — "and the War Department," she added — "in favor of using German industrialists to rehabilitate Germany" — and this because, she maintained, "he belongs to the same group" as Will Clayton, the Surplus War Property Administrator under Jimmy Byrnes, and "others, who believe we must have business going in Germany for the sake of business here."[2]

Greedy American capitalism was now in danger of triumphing over higher American morality and idealism, in Eleanor's view — and given her special access to the President, she had felt it her duty to protest. She'd gone on, then, to explain why she'd sent Franklin a long memo on Yugoslavia, since she'd wished to defend Tito's actions there. She had already written Franklin of suspected American skulduggery in Spain, where she saw "the fine Catholic hand" becoming all too "visible in Europe and in our State Department." As she'd summarized, with "Dunne, Clayton and Acheson under Secretary Stettinius," Mr. Hull's successor, "I can hardly see that the set-up will be very much different from what it might have been under Dewey." The fact that Harry Hopkins had supported Clayton made her "even more worried. I hate to irritate you and I won't speak of any of this again but I wouldn't feel honest if I didn't tell you now."[3]

The letter had come with "Much love."

Reading it the President, still feeling "let down" after Lucy Rutherfurd's departure from Warm Springs, had sensed little affection in it. He could only groan — for Eleanor's underlying criticism had been quite fair: he was not showing his former presidential mettle. He simply no longer possessed the energy to direct or appoint, let alone fight, his own new administration. He had, in short, become a virtual passenger aboard the USS *Commander in Chief* after Quebec — and was now becoming a passenger aboard the USS *State Department,* as well as other administration vessels captained by former subordinates. Even his promise to make his outgoing vice president, Henry Wallace, the new secretary of commerce was leading to humiliatingly open opposition from Democrats in Congress: Senate and House of Representatives members who no longer bowed to his authority, despite his brave victory at the polls on behalf of their party on November 7.

To another soldier, who was stationed in China, Eleanor had written that, with regard to race and a raft of other issues, "either we are going to give in to our diehard Southern Congressmen or we are going to be the liberal party." With that prospect in mind Eleanor had felt she could not simply sit idly by and watch the former happen.[4] Chiding Franklin could not make him better, however, when he was clearly dying.

Franklin respected Eleanor's views — especially in relation to racism in America. But taking her to the Crimea, in his ever-deteriorating health, would have been a hiding to hell, the President had felt. She had correctly criticized him for no longer being the steward of his own administration — a sort of fellow traveler in it. But what could he do in the circumstances — i.e., his state of health? He was taking to Yalta his new secretary of state, Edward Stettinius, to be at his side; all he himself could undertake — and was determined to achieve — was the sealing of a formal, signed agreement with Stalin and Churchill to create the United Nations, and an effective UN Security Council. And button down a Russian declaration of war on Japan. Beyond that, the chips would fall where they would in a world in which the interplay of forces, domestically and internationally, he himself could do almost nothing to control.

The President's handsome blond daughter Anna, therefore, not his wife, would accompany him to the Crimea — just as Churchill, the President had learned, was once again taking his daughter, the porcelain-pretty Sarah Oliver.

And so it had been settled. Anna would have her own quarters (and bathroom) on the all-male USS *Quincy,* while the President himself would have two special elevators to take him to different decks. In Malta he would meet up with U.S. chiefs of staff after their deliberations with the British chiefs there, as also with Harry Hopkins and Secretary Stettinius, who would have meantime shared their thoughts and ideas about the coming summit with their opposite British numbers, Alex Cadogan and Anthony Eden.

The long sea voyage, if the January seas weren't too rough, would theoretically allow the President time to read the many documents and reports that had been drawn up for him in preparation for the conference. Yet, as the huge vessel steamed beyond Virginia and American waters (the President telling Anna, on deck, about "the bird life of these shores," then "suddenly and casually" remarking, with an outstretched arm, that "over there is where Lucy grew up"),[5] it was obvious to his small entourage — Admiral Leahy, Justice Byrnes, Steve Early, Democratic political adviser Ed Flynn, his naval and military aides, his two doctors, and three officers from his Map Room — that the President of the United States was not even up to that amount of reading. He tended, they observed, to lose concentration after only a few minutes. He had another cold — with runny nose and cough, which Admiral McIntire treated with "coryza tablets."[6] He seemed desperately, desperately weary.

Day by day the huge American warship zigzagged at speed across the Atlantic to avoid lurking U-boats ("the sea running too high for a sub to keep up with us," Anna noted),[7] protected by a U.S. destroyer escort as well as daylight U.S. air cover. There were movies in Anna's flag cabin in the evening, and after six days at sea, on Sunday, January 28, 1945, the USS *Quincy* entered "the European-African-Middle East Theatre of War" by crossing the 35th meridian — some twenty-five hundred miles from Newport News.

By January 31 — the day after the President's birthday had been celebrated with gifts on the ship's deck, and five candlelit cakes shared with his staff in his cabin — the USS *Quincy* passed through the Straits of Gibraltar in predawn hours, under waning moonlight. The President had explained to his daughter, as she noted in the diary she'd decided to keep, that "the last time he went thru (last year returning from Cairo-Teheran Conf)" three searchlights had illuminated his ship, directed "from somewhere on the Spanish shore. He says no one knows if they were friendly or otherwise. Nothing like that this time."[8]

Conserving his energy, the President hadn't actually risen to see the Rock — in fact he rarely rose before midday. He wasn't feeling at all well, he admitted, though his cold had cleared up. Eleanor's cables wearied him — her signal on February 1, begging him to intercede in a new fight in the Senate over Wallace's nomination, being particularly hard, not least because he'd warned her he would be unable to send radio messages while at sea, lest they thereby give away their location en route to Malta. "A message came this morning from Mrs. Roosevelt urging the President to make some kind of statement in favor of the confirmation of Mr. Wallace as Secretary of Commerce," Admiral Leahy recorded in his diary. "The idealistic attitudes of Mrs. Roosevelt and Mr. Wallace are not very different," the admiral added, amazed that Wallace could have thought, the previous summer, he could win his party's nomination for vice president a second time, when he had great difficulty even getting confirmed in Congress as commerce secretary. The two individuals were "about equally impracticable."[9]

For her part Anna, who was belatedly told their ultimate and still-secret destination, became anxious as to whether her father was really going to be able to manage the subsequent journey by air to the Crimea from Malta, then a long drive to the promised accommodations at Yalta, and after that, tortuous negotiations with Churchill and Stalin. Not to speak of the many people, from aides and advisers to honor guards, whom he would have to meet, inspect, impress.

Time would tell—and soon did, once they passed the snow-covered coast of Spanish Morocco, Oran, Tunis, Cape Bon, and Pantelleria to sail, finally, through the submarine net at 9:30 a.m. on February 2, 1945, and enter the Grand Harbor at Valletta, Malta. The "entrance to the harbor is so small that it seemed impossible for our big ship to get thru, and it looked like a touch and go proposition as we successfully swung our way past the breakwater, then past the USS Memphis, the HMS Cyrius and the HMS Orion."[10]

Anna stood on the flag bridge. "FDR sat by the railing on the deck immediately below," she noted in her diary—able to see, through her binoculars, the British and U.S. chiefs of staff, Ambassador Harriman, Harry Hopkins, Anthony Eden, and other "dignitaries going to the Conf. all lined [by] the rail of the HMS Orion, with the P.M. standing stiffly and trimly at attention."[11]

With ships' crews in dress whites, it was an almost royal welcome to the U.S. president and commander in chief—just as Churchill had specially arranged. From "the very large crowd evident, it appeared that all Malta was out to greet him," the keeper of the President's log, Lieutenant Rigdon, recorded.[12] "Both sides of the channel were lined with people" out to welcome him.[13]

Rigdon, as the President's designated chronicler on the voyage, was not exaggerating. Ed Flynn, accompanying the President, wrote his wife it was "quite an emotional moment," with dignitaries and crowds all seeming to feel the same sense of history being made, after patiently waiting for four days. Finally the President of the United States had arrived—the band on HMS *Orion* playing "The Star-Spangled Banner," while the USS *Quincy*'s musicians played "God Save the King."[14]

Recalling the scene many years later, Eden wrote how "the great warship sailed into the battered harbor"—a harbor in which "every vessel was manned, every roof and vantage point crammed with spectators. While the bands played and amid so much that reeked of war, on the bridge, just discernible to the naked eye, sat one civilian figure"—the President of the United States. "In his sensitive hands lay much of the world's fate. All heads were turned his way and a sudden quietness fell. It was one of those moments when all seems to stand still, and one is conscious of a mark in history."[15]

69

Hardly in This World

THE FIRST TO COME ABOARD the USS *Quincy* were the new U.S. secretary of state, Mr. Stettinius; the U.S. ambassador to Moscow, Averell Harriman; and Harry Hopkins, following his two-week fact-finding mission across Europe.

In cables to the President from London, while the President was *en voyage,* Hopkins had warned how much Mr. Churchill was dreading Yalta. Not for its difficult diplomatic challenges, which he relished, but as a locale. The Prime Minister "says that if we had spent ten years on research we could not have found a worse place in the world than Yalta," Hopkins had signaled — somewhat embarrassed, since holding the conference in the Crimea had been his own inspiration. Once renowned as a holiday resort in the time of the czars, and known for its mild winter climate, it was now only "good for typhus," the Prime Minister had memorably sniffed after reading a bad report, "and deadly lice which thrive in those parts." Lice, moreover, that could only be survived "by bringing an adequate supply of whiskey."[1] A drive over mountain roads — roads that were extremely narrow, steep, winding, and, thanks to snow, often impassable — would be necessary to reach the meeting venue. The decision having been made, however, Churchill — who had already flown to Moscow to see Stalin on his own in October 1944, after the Quebec Conference — had no wish to be left out.

In view of the Prime Minister's misgivings, and at the President's request, Ambassador Harriman had personally checked out the Yalta situation with his daughter Kathleen. Coming aboard the USS *Quincy* he was able to report to the President the good news: the Russians had performed a virtual miracle in recent weeks. The town of Yalta had apparently been razed during the bitter fighting there, but the former czarist homes that would house the conference delegates had mercifully been spared. All were a five- or six-hour drive from the airfield at Saki, where they would land following their overnight flight from Malta.

The Livadia Palace, where the American president and his team would be quartered, had been Czar Nicholas II's summer retreat, looted by the Wehrmacht down to its plumbing pipes before the Germans evacuated the Crimea, Harriman reported. It had been entirely refurbished in recent weeks, however, by a veritable army of Russian workers, prisoners of war, and craftsmen using materials brought from the top former czarist hotels in Moscow. The Livadia Palace, in other words, now offered an excellent setting for the conference. Moreover an American medical team from the USS *Catoctin,* the U.S. Navy's communications vessel for the conference, had thoroughly cleansed and deloused the Livadia's rooms. There were telephones aplenty, and the drive there "would not be too tiring if completed by daylight," Harriman thus assured the President.[2]

Relieved, the President greeted his next visitors to come aboard: the British governor-general, Sir Edmond Schreiber, followed by General Marshall and Admiral King—who in the course of ten minutes told him they had completed three days of meetings with their British opposite numbers, which had gone surprisingly well.

Seeing how frail was the President—"in fact I was quite shocked by his looks," Marshall later confided[3]—the U.S. chiefs had not gone into detail about their discussions with Field Marshal Brooke and his colleagues. In a veritable admission of the futility of Churchill's obsession with Italy the British chiefs had, mercifully, agreed to all the proposals of the American chiefs, including the belated shifting of another six British and Canadian divisions from Italy to reinforce Eisenhower's forces on the German frontier. Later that afternoon, on February 2, they would tell the President that the only point of contention had been whether Eisenhower should be encouraged to cross the Rhine north and south of the Ruhr, i.e., on a broad front. Or, they later explained to the President, to put all Allied efforts into the northern thrust under Field Marshal Montgomery. The field marshal had masterfully blunted Rundstedt's offensive, but his recent January 6 press conference had rubbed salt in General Bradley's wounds (Bradley had been compelled to give up command of his First and Ninth U.S. Armies to Montgomery), as well as managing to infuriate the American reporters and senior officers alike. The U.S. chiefs thus favored General Eisenhower's strategy—as did the President.

In the meantime, for his part, Secretary Stettinius reported to the President that he'd gotten on extremely well with Mr. Eden, and that they were both agreed as to the voting—and vetoing—procedure for the United Nations, which they hoped the Russians would agree to at Yalta. Thus, when Prime Minister Churchill finally arrived on board the USS *Quincy* at 11:48 a.m. for lunch with the President, everything had seemed straightforward. A lighted candle and a cigar had been placed near Churchill's lunch plate, which the Prime Minister took as a touching sign of presidential amity. Throughout the lunch Churchill then "monopolized" the conversation, Leahy noted in his diary, discoursing on "English problems in war time, the high-purpose of the so-called Atlantic Charter, and his complete devotion to the principles enunciated in America's Declaration of Independence."[4]

According to the cabled letter the Prime Minister sent his wife in London that night, the President had arrived "in best of health and spirits." But the Prime Minister had not been paying attention. "No one else who saw the President that day" described Mr. Roosevelt in such positive terms, Churchill's most devoted biographer, Martin Gilbert, later noted — in fact there was consternation over the deterioration in the President's health.[5]

Anthony Eden, present at the luncheon, certainly found himself disturbed. As the British foreign minister confided in his diary, the President "gives the impression of failing powers."[6]

On the American side Harry Hopkins — himself a ghost of his former self[7] — was equally concerned. Charles Bohlen had been asked by the State Department to accompany Hopkins to Europe on his pre-Malta mission, meeting with de Gaulle and the British, so that he could add further memoranda to the President's bulging State Department "briefing books." Like Hopkins, Bohlen was dismayed by the further decline in the President's condition. The young diplomat had watched the USS *Quincy* make its stately entrance into Valletta harbor, had seen the President "acknowledging the salutes from the British men-of-war and the rolling cheers of spectators crowding the quays," like a modern czar. The President had become a global icon, he felt: the embodiment of Allied military success in defeating the Nazis — "very much a historical figure," as he put it. But "when I boarded the *Quincy*," he recalled, "I was shocked by Roosevelt's physical appearance. His condition had deteriorated markedly in the less than two weeks since I had seen him. He was not only frail and desperately tired, he looked ill." Despite "a week's leisurely voyage at sea, where he could rest," Bohlen later recorded, "I never saw Roosevelt look as bad as he did then."[8]

As the President's chosen personal assistant for the trip, Anna knew there was no way her father could manage a whole afternoon session with his British visitors; they were therefore encouraged to leave Mr. Roosevelt's sitting room immediately after luncheon. Instead she took her father for a two-hour drive around the island with Sir Edmond Schreiber — a drive during which he would not have to make more than gentle, polite conversation with the governor-general in the car, while Anna got her "first glimpse of mass destruction in this war," as she scribbled in notes for her diary.[9]

On the President's return, however, the official visits resumed. At 4:30 p.m. Marshall, King, and Major General Laurence Kuter (stand-in for General Arnold, whose heart problems had returned; he would be unable to attend the summit) explained in greater detail the agreements they'd come to with the British. ("President supported the Joint Chiefs of Staff completely," Leahy recorded in his diary that night.[10]) Then at 6:00 p.m. the President received the Prime Minister and his British chiefs of staff—"a plenary meeting on board, at which the Combined Chiefs of Staff were able to report that they had reached agreement on all the points at issue," as General Ismay recorded—though "Roosevelt looked a very sick man."[11]

Churchill finally took note: how much the President was able to comprehend, beyond small talk, seemed unclear. Yet there was still an informal dinner for twelve in the President's cabin to follow, given for the Prime Minister and his daughter Sarah, as well as for Mr. Eden, Harry Hopkins, and Anna Boettiger.

Eden found himself increasingly worried. To the British foreign secretary, at least, it was quite clear that, among the Western Allies, the United States was now the dominant partner. American military and political strategy would henceforth prevail, for good or ill. What was also apparent, however, was that the American team would now be going in to bat with a seriously ill team leader—a president who had deliberately turned down Churchill's request he spend a few days in Malta before continuing to Yalta. And for the simple reason that he was not well enough to spend more time with his coalition partners!

Any hope of flying to Yalta with an agreed Anglo-American approach was rapidly flying out of the cabin window. To Hopkins the foreign secretary thus confided, at the table, his concern they were "going into a decisive conference" with nothing in the bank—having "so far neither agreed what we would discuss nor how to handle matters with a Bear who would certainly know his own mind."[12]

It was too late, however. There was no opportunity to talk after the dinner, either. No sooner was the meal over and the guests departed than the President left the warship and was taken by car to Luqa airfield, where some seven hundred members of the two national Yalta-bound delegations were assembling. The President's new C-54 plane—the latest, custom-equipped Sacred Cow—would be one of the last to take off from Malta. A special elevator had been installed so that the President could board without needing a separate ramp. Aboard the aircraft he was greeted by his old pilot, Major Otis Bryan, but then went straight to his special compartment at the rear, where he tried to rest.

"It is certainly luxurious," wrote Anna, who had not seen the new plane before. "FDR has his own cabin with a nice wide bed. Ross, Leahy, Brown, Watson and I have comfy bunks, and Bruenn and Mike and Arthur Prettyman have to sit up." Unable to rest she "wandered over to Churchill's plane"—a gift from the President—"and someone was nice enough to offer to show me thru it. It's also a C-54—I gather it was presented to the P.M. by this government. Both planes cruise at about 200 miles per hour, which is faster than most C-54s. It is differently appointed than Father's, but very comfortable."[13]

One by one, beginning at 11:30 p.m., the flight of twenty C-54s took off into the night, bound for the Crimea. "We took off at 3.30 and the P.M. at 3.35 a.m.," Anna recorded — both of them accompanied by six fighter-plane escorts, since "there are still some Germans in the Dodecanese Islands" and the Nazis "might have got wind of the trip and have imported some of their own fighters" to the area.[14]

The President seemed unworried about enemy fighters, being more concerned about the looming conference — and his own ability to steer it in the right direction in his dire state of health. Once finally airborne, according to Dr. Bruenn, he "slept rather poorly, because of the noise and vibration."[15]

The omens for success did not look particularly good.

70

"A Pretty Extraordinary Achievement"

THE PRESIDENT WAS NOT the only one having difficulty sleeping. Adolf Hitler had also been on the move. As the President sailed across the Atlantic, the Führer had boarded his special armored train on January 15, 1945, and had moved back to Berlin from the Wolfsschlucht (Wolf Canyon II), from which he'd directed his grand counteroffensive through the Ardennes.

Even to Hitler it had become clear that his surprise attack, though inflicting a major initial defeat on Eisenhower's armies, had failed to reach the Channel and slice Allied forces in two. Moreover that the Soviets, meantime, were massing for a renewed offensive on the Eastern Front. To prevent a complete Soviet breakthrough he would therefore have to transfer troops from the west to the east. To try and keep the Allies on the defensive he had ordered an armored counteroffensive in Alsace on January 16, using Panzer Mark V Panthers, which produced a tactical victory for the Wehrmacht and inflicted a further fourteen thousand casualties on the U.S. Sixth Corps in ten days' heavy fighting.[1] It had not changed the tide of war, however.

As Dr. Goebbels and others noted, victory on anything more than a local scale was now a chimera. The Führer's only hope his Thousand-Year Reich could survive was that the Allies would become un-Allied. Eisenhower had an estimated two million troops assembling in the Netherlands, Belgium, and France to cross the Rhine from the west, and a similar number of Soviet troops were advancing toward the Oder in the east. But if the inherent differences between the Allies' national objectives, as well as their political and social systems, were to erupt, finally, in a clash of goals, then the amazing unity the Führer had sought to inculcate in the *Vaterland* might yet prevail, as Hitler had assured Goebbels on January 22.[2] The structures he had established as Nazi overlord — with himself at the center, as Führer — were proving extraordinarily robust, despite seemingly overwhelming enemy military superiority.[3]

Aware that a Big Three Allied summit was about to take place — though not knowing where — Hitler feared that Stalin would come to genuine agreements with Roosevelt and Churchill. By contrast, Dr. Goebbels clung to the conviction the Western Allies would inevitably break with the Soviets. Faced with daunting Russian successes on the battlefield, surely the peoples of the West would recognize, he argued, how noble had been Nazi intentions in conquering all Europe by military force. "We Germans have never wanted to do anything save to offer Europe peace and social happiness," Goebbels maintained in his diary — "good intentions" that "the Jews with their infernal propaganda" had managed to subvert.[4] But if Roosevelt, with his vision of a "so-called United Nations" based on the "principles of the Atlantic Charter," succeeded in harnessing Stalin to the cause, then the Führer was right: the whole Nazi edifice, ethos, and extermination programs would come down and be obliterated, together with Hitler's attempt to present himself as a great "European." Goebbels thus agreed with Hitler that a Big Three summit, if it succeeded, would be very bad for Germany, permitting Stalin to "integrate Anglo-American war aims with his own." Goebbels could not bring himself to believe this would happen, though; Stalin was "not that kind of character. He's not one of those people who is easily won over." Bolshevism marched to an "inner law," which not "even Stalin, with his huge dictatorial powers," would be able to thwart. "Bolshevism will undoubtedly follow its world-imperialist and revolutionary aims — and it's here we have a great political opportunity. Hopefully the Big Three conference will shortly take place soon, at a moment when Stalin is at the pinnacle of his military triumph. He'll be bound to argue and fall out with Churchill and Roosevelt, which may develop into a real break behind the scenes. If that happens, that would present us with a favorable political situation" — the Allies divided, and Hitler able to parlay his way out of imminent German defeat by negotiating with the Western Allies. "Either way, these will be the most fateful weeks of the war," Goebbels noted in his diary. "The crisis on the Eastern Front is potentially fatal. But the Führer" — who seemed to Goebbels once again "reinvigorated" by the crisis, one that was not tiring him but making him wonderfully mentally "elastic," so that he "radiated an unbelievable confidence and faith"[5] — "is convinced we'll be able to deal with it. He believes in his destiny. There's no doubt, nonetheless, we have to overcome the military crisis, and that we can leave no stone unturned to do so."[6]

In pursuit of "total war," Goebbels had successfully culled, he claimed, a million more German civilians and had put them into combat-ready uniform in German barracks. For his part the Führer was delighted by newsreel footage of his V-1 and also his V-2 weapons, which, if a "thousand of them" could be launched against England over the coming month, might lead to a "fundamental change" in the British political and diplomatic approach to negotiations to end the war, Hitler mused.[7] Fickle press and public opinion in the democratic West could be manipulated to German advantage — Western journalists and commentators having no idea how contemptuous Hitler remained with regard to the "antiquated" bourgeoisie of Europe, which he proposed to "liquidate" in due course.[8] The very idea of smaller countries like Norway or Denmark being given any say in the future of the continent, as recently advanced by Norway's puppet leader, Vidkun Quisling, in a visit to the Führer, was equally foreign to Hitler's thinking—a notion Goebbels dismissed as utter *Quatsch*, or nonsense.[9]

Splitting the Allies by playing on their inherent political contradictions, and the risible innocence of their commentators regarding the nature and intentions of the Nazi leadership, was, in short, the Führer's "elastic" vision: holding the Russians at bay in the east and hammering England and even Paris, now, with V-bomb weapons, Messerschmitt jet fighter-bombers, and Wehrmacht counterattacks that were—according to Allied POWs—disheartening American troops, who had no business being in Europe.

In this respect Dr. Goebbels was amazed at how astutely Roosevelt had managed to lead American public opinion. In terms of Normandy and then the Ardennes he had asked a "sacrifice in blood" that Goebbels had imagined would be unacceptable to most Americans, who had scant interest in Europe or its fate. "Yet he's managed to bring the U.S. right behind him," Goebbels acknowledged in his diary. "That's a pretty extraordinary achievement, for one has to remember Americans only have a secondary interest in the war in Europe and are only inspired by the war against Japan."[10]

Day after day Goebbels had visited the Führer in the Reich Chancellery, and its warren-like bunker downstairs—the two men on tenterhooks about the forthcoming Allied summit. Surely Stalin would cancel or defer it, given that he was now the military master of eastern Europe, able to dictate the postwar setup by simple virtue of Russian bayonets. But to both Hitler's and Goebbels's consternation there seemed to be no sign of a split, yet, in the Allied coalition. Nor—as Hitler hoped—did it look as if Stalin was going to decline Roosevelt's request that the Russians join the war against Japan, Germany's war-partner, once Germany was defeated. *The Allies were holding together*—leaving the Nazis no hope of survival unless the Germans could defeat the Allies on the battlefield. And this—despite their V-bombs, their phenomenal counteroffensive in the Ardennes, and even their recent efforts around Strasbourg—they were unlikely to accomplish, if truth be told. In which case, Goebbels recognized, his and Hitler's lives would be over.

By January 28, when Goebbels again visited Hitler at the Chancellery, the Führer looked "tired and stressed."[11] He was working sixteen- to eighteen-hour days, he told Goebbels, but was still hoping to withdraw forces from Norway, Italy, Hungary, and other fronts to mount a new counteroffensive. "The enemy coalition—and the Führer is quite right in this—is bound to break apart,"

Goebbels recorded, noting that the Führer's radio broadcast the next night, January 30, would be one of huge significance, not only within the Reich but in its likely impact across the world.[12]

The broadcast — Hitler's first since the July bomb plot — duly impressed listeners by its relatively philosophical, reflective tone, surprising even those who despised Hitler, such as Churchill's private secretary John Colville, who thought it "gloomy but more eloquent than of late," as he recorded in his diary.[13]

Certainly the Führer's tone was less belligerent and more philosophical. The "era of unbridled economic liberalism has outlived itself," Hitler insisted, a liberalism that "can only lead to its self-destruction," resulting in world disorder: a world in which "the tasks of our time can be mastered only under an authoritarian co-ordination of national strength," he argued, as well as "duties" — such as the extermination of Jews and others, which went unsaid. After all, the Führer claimed, the "fight against this Jewish Asiatic bolshevism had been raging long before National Socialism came to power." Bolshevism had begun "systematically to undermine our nation from within," with help from a "narrow-minded" German bourgeoisie who refused to recognize that "the era of a bourgeois world is ended and will never again return."[14] Dictatorship was the only way forward.

Continuing to resist the Allied onslaught was, in this case, "the safest guarantor of final victory" for the Third Reich. "God the Almighty," the Führer asserted in unusually religious terms, "has made our nation. By defending its existence we are defending His work. The fact that this defense is fraught with incalculable misery, suffering and hardships makes us even more attached to this nation. But it also gives us the hard will needed to fulfill our duty even in the most critical struggle; that is, not only to fulfill our duty toward the decent, noble Germans," he declared, "but also our duty toward those few infamous ones who turn their backs on their people."[15]

Were there any "decent, noble Germans" left after the barbarism the Nazis had inculcated and demonstrated across Europe, from Poland to Hungary, the Ukraine to the Urals? some listeners in the still-occupied territories wondered. Even Hitler acknowledged the outlook was less promising than when he had attacked Russia in 1941. Mass German civilian and refugee flight was gathering pace in the east, amounting to a veritable "hurricane from Central Asia,"[16] yet the Führer denied he was responsible for starting such a cataclysm, arguing it was the outcome of simple "envy" on the part of Germany's "democratic, impotent neighbors" and a "plutocratic-Bolshevistic conspiracy" — one which must be resisted "until final victory crowns our efforts."[17]

Behind the Führer's sober-sounding assertions, however, Wehrmacht desertions were increasing,[18] leading to kangaroo-court execution of those accused of cowardice or disobedience.[19] The gas chambers at Auschwitz had been ordered shut down by Himmler in November 1944, once word of the Morgenthau Plan and punitive Allied retribution had become public in Germany, yet the concentration camps themselves had continued to be run by the SS. Some seven hundred thousand inmates were forcibly marched, or taken in cattle trucks, to other concentration camps as Soviet forces overran still more death camps in January 1945. An estimated one-third of the defenseless prisoners died in the process.[20]

Goebbels's heart went out to German refugees, he claimed in his diary, but not to German Jews and other concentration camp inmates, whose fate was common knowledge. "The Führer won't let us fall into Russian hands. He'll gas us instead," one German refugee quipped darkly, in East Prussia.[21] As Hitler's biographer Ian Kershaw would later summarize, the "terror which had earlier been 'exported' to the subjugated peoples under the Nazi jackboot was now being directed by the regime, in its death throes, at the German people themselves."[22] The war in Europe was coming to a climax.

In his radio broadcast Hitler repeatedly blamed the Jews, while thanking "Providence" for not having "snuffed" him out on July 20, 1944, with a "bomb that exploded only one and a half meters from me" and could easily have terminated "my life's work. That the Almighty protected me on that day I consider a renewed affirmation of the task entrusted to me," he claimed: namely to fight for National Socialism "with utmost fanaticism" to final German victory.[23]

The Führer's once-omnipotent life was collapsing around him, though, like the Nazi nation. The Reich Chancellery in Berlin had already been destroyed by Allied bombing, and much of the New Chancellery, including his private apartments, was also hit. This now forced him to live two stories below, in the deep Führerbunker: a labyrinthine network resembling "a maze of trenches," Goebbels described. "The enemy coalition would break, and *must* break," the Führer had told Goebbels. The Nazi Party must simply bide its time until the right occasion, before taking the "political initiative"—i.e., negotiations.[24] It was, as Kershaw put it, "a matter of holding out until the moment arrived."[25]

The question was, could Hitler himself hold out? "He was more haggard, aged, and bent than ever," Hitler's biographer later described, "shuffling in an unsteady gait as if dragging his legs. His left hand and arm trembled uncontrollably. His face was drained of colour; his eyes bloodshot, with bags underneath them; occasionally a drop of saliva trickled from the corner of his mouth."[26] He could hardly lift a glass to his mouth without spilling its contents.

Hitler's speech was not considered by his own acolytes to be as inspirational as his earlier broadcasts.[27] It did remind Germans, however, that the Führer was still alive, was still in command—and was still determined that the German nation, collectively, should fight to the bitter end: "that everybody who can fight, fights, and that everybody who can work, works, and that they all sacrifice in common, filled with but one thought: to safeguard freedom and national honor and thus the future of life."[28] It would be a test of national, Nazi-led will, with the underlying hope that the forthcoming Big Three conference — whether in Moscow, Tehran, or Odessa, German intelligence had been unable to decide[29] — would be a disaster for the Allies, forcing them to split apart.[30]

That possibility was something President Roosevelt, however ill he felt en route to the Crimea, was equally determined would not happen.

71

One Ultimate Goal

IN BELIEVING THE ALLIED COALITION was now in danger of dissolution, the Führer and Dr. Goebbels were not entirely wrong. The British contingent, for example, had spent four days in Malta waiting for the President. They were both angry and disappointed "by Roosevelt's refusal to discuss either tactical or substantive questions regarding Yalta" when he arrived, which concerned the diplomat Charles Bohlen as much as it did Harry Hopkins.

The matter continued to puzzle Bohlen in later years, and the distinguished diplomat wondered if the President's health at the time "might" explain it.[1] "Everyone noticed the President's condition," Bohlen acknowledged, "and we in the American delegation began to talk among ourselves about the basic state of his health," which was perplexing. "Our leader was ill," he recalled the feeling, and just at the moment when the American team was making its way to "the most important of the wartime conferences." Yet the President himself seemed almost unconcerned — in fact, almost removed from reality. Perhaps the ailing president wanted to "save his energy for Yalta," Bohlen later posited, rather than get immersed in black books and negotiating tactics. But if so, would this not put the U.S. team at a disadvantage vis-à-vis the well-prepared Soviets?[2]

The President had said little at Malta, in part because he felt so ill, but in part deliberately. For what concerned him was not what his diplomatic team was anxious about. Where they had assembled a small library of black books to back their negotiating positions in the impending conference, the President was neither mentally fit enough to argue on their behalf in the looming meetings, nor did he feel it would be the best use of his fading but still-world-renowned skills as a statesman. He was thus determined, simply, to make sure the Big Three conference would *not* result in a breakdown of the Allied coalition at the very brink of Allied military victory in Europe. Eisenhower's delayed offensive aimed at closing up to, and then crossing, the northern Rhine would kick off the next week, on February 8, 1945, General Marshall had informed him (Operation Veritable, to be followed by Operation Plunder), but those forces were still three hundred miles from Berlin. The Wehrmacht was fighting fanatically, on all fronts. The Nazis must therefore be given no chance of evading their fate by Allied dissension. The establishment of a United Nations organization should be agreed and announced, *jointly,* along with a signed, formal, though necessarily secret undertaking by the Soviets that they would declare war on Japan once Hitler surrendered, and would then aid the United States in completing the swift defeat of Japanese forces in the Pacific — and in China.

Simplistic or not, the Allies, the President felt, must present to the world — and especially to the Germans and Japanese — an inviolable unity. Hitler's latest statement of Nazi resolve in his broadcast to the people of Germany — which had included mention of the Morgenthau Plan, unfortunately — *must* be countered by an even *greater* demonstration of common, higher Allied purpose: one that would inspire the men and women of the many United Nations fighting the Axis powers enough to attain complete surrender of the Third Reich. The symbolism of unity between the Big Three Allies at Yalta, in other words, was as important — perhaps more important — than the details of the negotiated agreements that would be hammered out by the national teams.

Winston Churchill, normally so aware of the symbolism of events and names, seemed initially slow to appreciate this larger aspect of the approaching summit. He had been heard, in particular, to question the notion of unconditional German surrender. He wanted Britain to be given an extra seat in the proposed United Nations organization for India — still a British colony, not a self-governing dominion. In fact the Prime Minister had had, he wrote his wife, Clemmie, from Malta, "for some time a feeling of despair about the British connection with India, and still more about what will happen if it is suddenly broken." He was determined "to go fighting on as long as possible" for the British colonial empire and to "make sure the Flag" was "not let down while I am at the wheel."[3]

Stalin, too, would fight for the resumption of *his* empire, now that his forces had shown such military resilience in facing Hitler's legions, and had demonstrated their growing effectiveness on the battlefield, not simply in policing their communist realm. This was something the President understood — and worried about. If the world was to have confidence in an international postwar order modeled not on Hitler's social Darwinism but on the United Nations organization and its Security Council, then the Yalta summit must inspire faith in the unity and higher purpose of the Allied coalition.

A demonstration to the world of the willingness to differ over political systems of government but agree on confronting aggression and the occupation of other countries by military conquest would at least present a defining contrast to the rampant militarism and inhumanity symbolized by the gas chambers of the Third Reich, as well as the atrocities still being committed by the Japanese across Asia. Winning national benefits in Allied negotiations over details, as evinced in the State Department black books, was thus perfectly understandable in terms of American self-interest, but something the President would best leave to Stettinius, Justice Byrnes, and the U.S. chiefs of staff teams. For himself, with so little strength left in his failing body and mind, he could only focus, ultimately, on one goal — ensuring the Big Three summit did not lead to a breakdown of Allied unity, but enhanced it; ensuring the imminent defeat of the Third Reich, the swift defeat of Japan, and the establishment of a postwar world order that would be more effective than the League of Nations had been in the aftermath of World War I.

Whether, in these circumstances and within his failing powers, the President would be able to hitch his wagon to his star remained to be seen. There would be casualties: issues, from Polish independence to French recolonization, that would doubtless constitute major obstacles to the outcome he hoped to achieve. Yet he was all the more determined, tossing and turning on the Sacred Cow, that he could at least achieve a significant symbolic demonstration of joint Allied purpose.

Admiral Leahy did not sleep much, either. "Air travel in the President's special four-motored transport plane is luxurious in comparison within my previous experience," the admiral noted in his ponderous, somewhat antiquated style in his diary, "but I continue to prefer travel by ship, by railroad, or even on foot if time is available."[4]

Time wasn't. "The President's bed was perpendicular to the plane's axis," the President's cardiologist recalled forty-five years later. "It was a big, wide bed. But he refused to have a safety belt" — raising fears that, if they encountered bad weather or were fired upon, "he would be tossed right out." Mike Reilly, heading the White House Secret Service detail, "and the rest of us" had talked it over "before we took off." They had "decided that when all the lights were out, I would creep in and position myself on one side of the bed so that if he fell out of bed he'd fall on me. We took off without any problems."[5]

Bruenn crept in.

"The next morning, the President said, 'It's lucky I recognized you as you came in.'"[6]

72

In the Land of the Czars

AT 12:10 P.M. on February 3, 1945, the Sacred Cow landed at Saki airfield, near Sevastopol. Stalin had not yet arrived in the Crimea, the President was told, so on his behalf the plane was met by Vyacheslav Molotov, the Soviet foreign minister, and Averell Harriman, the U.S. ambassador to the USSR.

The President waited aboard his plane with Molotov and Harriman until Prime Minister Churchill's own C-54 touched down: "Quite a sight to watch," Anna noted in her diary, "with its six accompanying fighters (American P38's)." This was at 12:30. "Molotov left Father to go to Churchill's plane, and soon they were gathered around our plane. FDR came out," she described, "via the neat little elevator, got into a jeep and with Molotov and Churchill walking beside him, the procession proceeded to an open roadway where a Russian guard of honor and band were most smartly drawn up. There were Russian still and movie camera men by the peck"—something that concerned Anna, who was "a bit worried because FDR did look tired after his hard day yesterday and a short night's sleep on the plane."[1]

Formalities, however, had to be observed. The Russian band "played the Star Spangled Banner first, then God Save the King, then the new Russian anthem—which seems to me a bit sad for this type of song. Then the guard snapped to attention, marched away from us and came back, in review, before us—marching in goose-step, though I'm sure the Russians don't call it that! The soldiers' faces were most interesting to me because they represented so many different races," Anna added. "They were all a fine, strong and healthy looking bunch."[2]

Lord Moran, the Prime Minister's physician, was not so impressed. The seven-hour flight, covering fourteen hundred miles, had finished with a bumpy landing on the snow-swept runway. The commander of the guard had "held his sword straight in front of him like a great icicle," Moran recorded sniffily in his diary. "They were preceded by a crowd of camera-men, walking backwards as they took shots."[3]

Clearly, Moran was unaware of the need for visual symbolism: the President having brought Steve Early all the way to the Crimea with him to ensure the tripartite nature of the historic conference was stressed — without mishap this time. In black and white, and even in color, photographs of the Big Three would subsequently be published in newspapers across the world, and shown on cinema screens. Early's job, too, was to ensure that these images mask, as far as possible, the President's gray, ashen countenance beneath his winter hat.

In truth, close-up, the President "looked old and thin and drawn," Moran noted in his later-revised, heavily edited diary; "he had a cape or shawl over his shoulders and appeared shrunken; he sat looking straight ahead with his mouth open, as if he were not taking things in. Everyone was shocked by his appearance and gabbled about it afterwards."[4]

Moran's most upsetting memory, as a proud Briton, was watching as the "P.M. walked by the side of the President," who sat in his jeep, just as "in her old age an Indian attendant accompanied Queen Victoria's phaeton."[5]

Moran's description of the President might be honest, but his sneer was unfair. Churchill had often expressed hostile feelings about the President and the increasing difference in power they represented, to be sure, but Winston was not without affection for his soi-disant American cousin. Whatever he might write to Clemmie in a cable that other eyes might well see, he was becoming belatedly aware the President was not simply unwell, but failing dramatically. In fact Churchill, having recently turned seventy, felt almost guilty at his own ability to bounce back after bouts of fever, pneumonia, or angina, whereas the President, having only just turned sixty-three, was not so blessed. The Prime Minister thus felt like a protective, still-energetic older brother to the desperately frail-looking president. Walking beside Mr. Roosevelt as he sat in his jeep was, if anything, an act of compassion and unity, however Moran might worry about the implied deference — the Prime Minister ready to march forward and go into battle beside his ailing warrior chieftain.

An hour later, at the inspection's end, "we all got into autos and made our way toward Yalta," Anna noted that night[6] — Churchill driven to the British compound at the Vorontsov Palace, twelve miles from the town, while the President was taken in a Lend-Lease armored Packard to the Livadia Palace, two miles from the city.

The President had survived the ordeal, but Anna was taking no chances. She was now sufficiently anxious over her father's condition that she'd insisted she "ride with FDR" in the Russian-chauffeured limousine, "so that he could sleep as much as he wanted and would not have to 'make' conversation," with staff or advisers who might exhaust him on the five-hour, eighty-mile journey, before the summit even began.

At the Livadia Palace the President's systolic blood pressure was found, thankfully, to be back below 230, despite the flight and the tortuous, twenty-mile-per-hour car journey up the winding mountain roads, and then down. The accommodations seemed duly palatial — with the maître d' addressing the President as "Your Excellency." "I can't understand Winston's concern," the President was heard to remark, relieved. "This place has all the comforts of home."[7]

For herself Anna didn't like the architecture (built in 1910, in Renaissance Italianate style), but admired the setting, overlooking the ocean. Also the surprising warmth inside — certainly warmer than Hyde Park. "You go in the front door to a huge reception hall. To the right of the hall is a ball room — to be used as the Conference room. Straight ahead to the left is a door to the study off the room which FDR is to occupy," she noted, "— and which must have been the Tsar's main bedroom suite. Straight ahead to the right is a corridor leading to the dining room, study, Father's bedroom," and other rooms, which "all have big, handsome fireplaces. When we arrived, fires were blazing merrily in all the downstairs rooms — and were most welcome as we were pretty frozen after our five hour drive."[8] The President thus felt up to having supper in his dining room with Stettinius, Harriman, Leahy, Watson, Brown, and several others, including Anna and Harriman's daughter, Kathleen. As at Tehran, he told Harriman, he'd like a short personal talk with Stalin the next afternoon, before the first plenary meeting, and asked Harriman to extend an invitation to dinner, too, to Stalin, along with Churchill, to be held immediately after the plenary finished.

Harry Hopkins, who had been too ill to come from his first-floor room to the dinner — a five-course affair, with five glasses at each setting — had been ordered straight to bed by Admiral McIntire. Alone in his room he was limited to a diet of cereal and cabbage soup; but he himself was in a "stew," Anna found when visiting him after her own meal — especially when hearing the President would not be seeing Churchill *before* the plenary the next day. "He gave me a long song and dance," saying "that FDR must see Churchill in the morning for a longer meeting." When Anna protested her father would simply not be able to manage such a meeting, then the one with Stalin, and then the long plenary in the afternoon, Hopkins "made a few insulting remarks to the effect that after all FDR had asked for this job, and that now whether he liked it or not, he had to do the work, and that it was imperative that FDR and Churchill have some prearrangements before the Big Conference started," as Anna jotted in her diary. "I had never quite realized how pro-British Harry is."[9]

Anna, in her caregiver role, calmed "the Hop" down. She trusted her father knew what he was doing, she assured Hopkins — and with good reason. Her father had not the energy or the wherewithal now to "do the work," in terms of conducting complex international negotiations. That, however, was not the reason why her father was not communing privately with Churchill; nor did her father feel that an Anglo-British "front" would achieve what he wanted from the Yalta summit. All he hoped to obtain, beyond military cooperation between the U.S., British, and Russian military chiefs of staff, was a signed agreement he could announce to the world: one that would put an end to any speculation in the press, in diplomatic circles, or in governments spanning the globe, including Berlin, about whether the Allies were allied — and relentless in their insistence upon the unconditional surrender of the Third Reich.

Unity of *military* purpose was thus goal number one.

Second, he hoped to get formal agreement as to the Allied military occupation and treatment of Germany.

Third, establishing the United Nations organization.

Fourth, the problem of Poland.

And finally fifth: Russian entry into the war against Japan.

Beyond that, for good or ill, he had neither the strength nor the will; it would be, he figured, up to others, later, to negotiate the details — as well as other issues, such as what would, or should, happen to former colonized territories such as Indochina, currently occupied by the Japanese. In other words, another, later conference would have to be arranged. In the meantime, though, the Yalta summit would show the Germans and the world that the Allied coalition was still watertight, and would proceed, inexorably, not only to defeat Hitler but to defeat Japan — even though, since the Soviets were not at war with Japan and could not marshal the forces to conduct a successful war on both fronts, Stalin would not be able to announce this publicly. A preconference huddle with Churchill would not help in this. In fact the likelihood was that he, the President, would have to act the arbiter, or mediator, in the arguments that promised to arise between the Prime Minister and the Russian dictator, given their conflicting imperialisms.

If this was disappointing to Hopkins and others like General Deane in Moscow,[10] who felt the United States should use more muscle and bang the table, given its global military power — in the air and at sea, especially — as well as its increasing economic predominance, then so be it. With Eisenhower's forces still west of the Rhine, and Soviet forces having crossed the Oder at Frankfurt and battling their way toward Berlin, it was the best he felt he could do. The two northern Soviet fronts under Marshals Konev and Zhukov alone amounted to two and a quarter million men, backed by thirty-three thousand guns, seven thousand tanks, and forty-seven hundred aircraft.[11] Possession would be nine-tenths of the law — and the United States still needed Soviet help to defeat Japan, even after the surrender of the Third Reich.

General Deane had protested that by being so generous towards the Russians, in terms of Lend-Lease, and asking so little of them in return, the picture was "neither dignified nor healthy for U.S. prestige."[12] But the President didn't feel that way, nor did Harriman, for all that he deplored Russian xenophobia. The Soviets had borne the brunt of the casualties and destruction in the war Hitler had declared and waged against the Allies. They had engaged some three-quarters of the Wehrmacht, in contrast to the Western Allies — thus giving the United States time to build up its military forces, and to learn how to conduct modern war successfully on two fronts after a disastrous start. Moreover, the President did not in any way feel he had sacrificed American dignity or prestige. In fact he felt the opposite: that the world now looked towards the United States with a greater respect than ever in its history. He himself had conducted an amazingly effective war strategy since the initial U.S. defeat at Pearl Harbor. He could take justified pride in now having U.S. forces assembling for their own big offensive across the Rhine, as well as the extraordinary advances U.S. forces had already made in the Pacific, and with more to come. Given the intrinsic and inevitable weaknesses of a democracy compared with military dictatorships, it had been a miraculous American recovery, in barely two years. In short, he saw no reason to "refuse assistance," as Deane recommended, unless the Russians ceased being Russians. And besides, he had an American secret weapon.

After Admiral McIntire had given the President some nose drops and Dr. Bruenn had taken his blood pressure, Arthur Prettyman had withdrawn and Roosevelt had turned out his light. He would need a good sleep, for the next day, February 4, 1945, the historic Yalta Conference would begin.

73

The Atom Bomb

THE ENTIRE EIGHTY-MILE ROUTE from Saki to Yalta had been lined by Russian soldiers, many of them female, and civilians. Anna was fascinated by people's "drab" clothing and felt boots. "Others wore what looked like down-at-heel, low or ankle high leather shoes with thick soles. Men's trousers were as drab and unshapely as the women's skirts. Children wore all kinds of clothes — but equally drab. Most were warmly dressed, but fairly often you would see a child with long, heavy cotton stockings held up by garters, then bare skin before coming to a too short skirt or pair of pants"[1] — much like Astrid Lindgren's Pippi Longstocking, whose first appearance would be made that year.

Hitler had described such people as Asiatic *Untermenschen*, but to the President and his daughter they simply looked like Menschen — with the President, commenting on the scrubby land beyond Simferopol, "saying that he was going to tell Marshal Stalin how this part of the country should be reforested," especially with evergreens.[2]

The President was ill, but he was neither intimidated by the challenges of the trip, nor the burgeoning military power of the Soviets. The Western Allies had, it was true, suffered a major reverse in the Ardennes, and had endured heavy fighting in Alsace. But he had no real anxiety about the ultimate outcome, or the subsequent dominance on land of the Soviets. And this for a simple reason most American officers, such as General Deane, and State Department officials knew nothing about. An atom bomb, that would shortly be combat-ready.

Despite the fact that the Soviets were known to be bugging all rooms with listening devices, as well as eavesdropping on all telecommunications, the President thus had good reason to be confident about America's future. The Soviet Union was a main ally in the struggle against Hitler, but he had zero illusions about the Russians or about Soviet communism, as he'd confided to Cardinal Spellman in the fall of 1943.[3] Communism was a godless ideology: an idealized system of human government that could only be maintained by operating a ruthless police state — one that was not substantially different from Nazism. All who spoke against either system became traitors — as did those who were merely suspect: "othered" and arrested for mass deportation, incarceration, and execution — such as Tatars who were being secretly evicted en masse from the Crimea on Stalin's orders, even as the Yalta Conference took place.

But Hitler was the more dangerous and egregious of the two, in terms of his military conquests and "liquidations," and was thus the first enemy of the democracies. As Churchill had said, the self-appointed Führer was worse than the Devil, indeed "if Hitler invaded Hell," Churchill, as prime minister of Great Britain, would "at least make a favourable reference to the Devil!"[4] In short, inasmuch as Stalin was the Devil and the lesser of two evils, the British and the American governments had had to work with him, not against him, in order to defeat Hitler. Moreover, given the Führer's astonishing and enduring hold on the people of the Third Reich and its ruthlessly obedient military forces, Hitler *could* only be brought down with Russian military might — a Red Army that was a double-edged sword.

However much the Russian saber hung heavy over eastern and central Europe, the President — thanks to what he understood of the latest reports of his atom bomb — thus showed no sign whatever of being daunted. Out of almost nothing the United States had built its own military might, with more than nine million men now in uniform and able to operate on a global scale, unlike the Soviet Union. Moreover, its secret weapon was one that not even Hitler's V-2 ballistic missiles could match.

The Manhattan Project to develop an atomic bomb, or "S-1" as it was codenamed, was now nearing critical mass. From the start, when Albert Einstein had convinced him of its feasibility in 1939, the President had seen its development as a race, namely to "see that the Nazis don't blow us up" first, as he'd put it to Alexander Sachs, his young, Harvard-educated, Russian-born Jewish adviser, at a White House meeting on October 11, 1939.[5] It had been a prescient decision. The President had thereafter authorized, arranged secret funding for, and watched over five years of development at Oak Ridge, Los Alamos, and elsewhere; he had also shared development of the bomb with the British, once Japan and Germany had declared war on the United States. The outcome would be potentially war-winning — or war-losing if the experiments failed in the United States but proved successful in German and Japanese weapons laboratories.

They didn't. As the project had grown — employing 130,000 people and costing $2 billion — and the weapon had approached completion, the President had even invited to the White House the distinguished scientist Niels Bohr. On July 5, 1944, Bohr — who had left Copenhagen on word he would soon be arrested as half-Jewish, making his way then via Sweden and Britain to Los Alamos — had spoken to Roosevelt privately. This was before the

President accepted the Democratic nomination for a fourth term, and left on his journey to the Pacific—where, if Emperor Hirohito and his government still declined to surrender following the defeat of the Third Reich, an eventual atomic bomb could be used to save further American and Japanese lives.

Bohr had argued for sharing the knowledge and know-how with scientists in other United Nations, so that there would develop no arms race, since no one could "win" such a war. The nuclear fallout could well destroy the planet.[6]

The President had agreed, both in theory and in practice. But the weapon itself was *not* yet ready, and the war was *not* yet won. The Allied landings in Normandy, four weeks before the President's meeting with Dr. Bohr, had proved a triumph, but fighting in Normandy in July 1944 was still hand-to-hand in places, with almost eight hundred miles more to go to reach Berlin. Operation Bagration had begun, as Stalin had promised at Tehran, but even with the Allies attacking on three fronts—including Italy—there had been no guarantee the war would be over in Europe by Christmas, given the way the Wehrmacht was continuing to fight. War in the Pacific would take much longer, with even more loss of life. Indeed, the current prediction by U.S. planning staffs was that it would require Soviet assistance and another *year and a half's* fighting, after German surrender, to defeat Japan militarily. All manner of things in the meantime could go wrong in U.S.-Soviet relations, given their contrasting ideologies. It had therefore seemed to Secretary Stimson and most of those involved on the American military team unwise to give the S-1 secret to Russian scientists, yet.

Churchill, for his part, had argued even more fiercely for continued secrecy—i.e., maintaining exclusive Anglo-American development of, and information about, the weapon. At Hyde Park, in September 1944, together with Harry Hopkins, the Prime Minister had persuaded the President to sign a secret aide-mémoire, which Churchill himself had written out. "The suggestion that the world should be informed regarding tube alloys [British code word for S-1], with a view to an international agreement regarding its control and use, is not accepted," Churchill had declared on behalf of the President and himself. "The matter should continue to be regarded as of the utmost secrecy; but when a 'bomb' is finally available it might perhaps, after mature consideration, be used against the Japanese, who should be warned that this bombardment will be repeated until they surrender." As for Niels Bohr the idealist, he should now be considered a prime liberal suspect, Churchill had added, and put under American surveillance to ensure there was "no leakage of information to the Russians."[7]

The President, in other words, was far closer to Churchill than anyone—save those at the highest echelons of atomic research—could know. In his role as commander in chief the President had seen to it that, in just five years and in total secrecy, the United States would be the first combatant nation in the war to produce an atomic weapon — thus successfully beating both the Germans and the Japanese. Axis attempts, mercifully, were still way behind that of the United States, General Leslie Groves, the commanding officer of the Los Alamos program, had then been able to confirm to the President in September 1944, after the fall of Paris.[8]

And the Russians? For their part the Soviets, having heard from Soviet spies in America something of the Manhattan Project,[9] had independently set up their *own* covert institute to work on nuclear fission in 1943, without informing the United States. Known as Laboratory No. 2, it was supposed to produce the necessary purity of uranium and graphite for an atomic bomb, as well as separating uranium isotopes,[10] all in the strictest secrecy, but Stalin had not given it priority or much in the way of financing, as the President had. As a result Laboratory No. 2 had numbered only twenty scientists and thirty technicians.

In the race to make a usable atomic bomb the United States was thus still leagues ahead of its enemies — and its Russian ally. As one veteran scientist would put it, the "formidable array of factories and laboratories" comprising the Manhattan Project was "as large as the entire automobile industry of the United States at that date."[11] Even Bohr himself was impressed, later telling Edward Teller of the U.S. Manhattan team: "You see, I told you it couldn't be done without turning the whole country into a factory. You have done just that."[12]

The United States, the President was aware in Yalta, now possessed an important advantage over the Soviets, not simply the Germans and the Japanese. Indeed Mr. Roosevelt had been told by Secretary Stimson, before he left Washington, that "we would not gain anything at the present time by further easy concessions to Russia." Spurred by General Deane's memorandum from Moscow, Stimson had recommended to the President that "we should be more vigorous on insisting upon a quid pro quo. And in this connection I told him of my thoughts as to the future of S-1 in connection with Russia; that I knew they were spying on our work but that they had not yet gotten any real knowledge of it and that, while I was troubled about the possible effect of keeping from them even now that work, I believed that it was essential not to take them into our confidence until we were sure to get a real quid pro quo from our frankness. I said I had no illusions as to the possibility of keeping permanently such a secret but that I did not think it was yet time to share it with Russia. He said he thought he agreed with me."[13]

The Russians, then, would not be informed about the weapon's progress for the moment. Knowing of it, however, would give the new secretary of state, Edward Stettinius, extra, if hidden, leverage, and on Stimson's advice the President had authorized his war secretary to take Stettinius into his confidence about the atomic program. This had been done four days later, on January 3, before Stettinius left for Malta. Thus, before the President and his team left the United States on January 22, 1945, for the Crimea, the President, Admiral Leahy, General Marshall, the U.S. chiefs of staff, and Mr. Stettinius had all been told via General Groves that the first atomic bomb "should probably be ready about 1 August 1945." Moreover, a second one ought to be "ready by the end of the year."[14] If it worked.

At Malta the President and Mr. Stettinius, as well as the chiefs of staff, had agreed to say nothing aloud. Or between themselves — even in the privacy of their rooms at the Livadia Palace. Nevertheless, General Groves's news had been a welcome new arrow in their quiver: something that helped explain why, despite the recent American reverses in the Ardennes, Secretary Stettinius had appeared to all in Malta to be in "tremendous form," as Sir Alexander Cadogan noted.[15]

However ill their President might appear, and however little prior discussion there had been as to an Anglo-American approach to the conference at Yalta, the leaders of the American delegation, in short, gave the impression of being amazingly assured. And with good reason — for the American team had not come four thousand miles to be pushed around, or treated as anything but top global wardog. Even their Russian hosts seemed unusually deferential. It was as if in Yalta, a thousand miles from their common objective, Berlin, an extraordinary international amity, prevailed, rather than the suspicion, tension, and awkwardness that had been present in Tehran — almost as if Stalin were aware of the ace up the American sleeve.

74

Riviera of Hades

STALIN HAD IN REALITY arrived in Yalta the day before the President and Mr. Churchill — installing himself, with his entourage, at the old Yusupov Palace outside Yalta, two miles from the Livadia Palace. He'd got in on the evening of February 2, 1945, following a three-day armored-train journey from Moscow,[1] concerned to have everything ready for his visitors, and that they should feel safe on Russian soil. Four entire NKVD regiments guarded the Soviet, American, and British enclaves, with batteries of antiaircraft guns and 160 fighter aircraft patrolling the skies: a deliberate demonstration of Russian efficiency and power. Looking down at the Black Sea, Churchill would call it the "Riviera of Hades" — ruler of the Greek underworld.[2]

Marshal Stalin seemed on his best behavior. Before driving to greet the President, as arranged, at 4:00 p.m. on February 4, the dictator paid a call on the Prime Minister at the Vorontsov Palace. He even suggested to the Prime Minister and Field Marshal Alexander, when they proudly showed the Marshal their wall maps of the military situation in Churchill's portable map room, that although the primary Soviet and Anglo-American forces should continue to crush the Wehrmacht between them in northern Germany, there was now no real danger of a German counteroffensive in Italy. The bulk of the British-led forces in Italy might therefore safely be transferred to Yugoslavia, if the British wished, and fight their way north to meet up with Red Army forces in Vienna.

Whether Stalin was testing or teasing Churchill by offering *un petit cadeau,* a consolation prize, neither Churchill nor Field Marshal Alexander could be sure (nor were historians later). Given Stalin's ice-cold response at Tehran to any British alternatives to, or diversions from, the mounting of the long-promised Second Front in the spring of 1944, the idea of such an amphibious new invasion and subsidiary campaign through Slovenia seemed positively open-minded. Now that Churchill's Mediterranean obsession was no longer an obstacle to defeating Hitler, Stalin seemed relaxed, even genial. The Wehrmacht was at last truly close to defeat, and Stalin, some officers conjectured, was perhaps more of an opportunist than depicted.

Clearly, though, *Yalta was not Tehran*—the dictator's irritation with Churchill's endless reasons for not mounting D-day in the spring of 1944 a thing of the past. Overlord had proved a phenomenal success: the deciding battle of the war. Blazoned across Churchill's wall maps at the Vorontsov Palace, the results could be seen in stark contrast to the Western Allies' positions in December 1943, when their armies were stuck north of Naples, in southern Italy. Though Alexander's forces were still not much farther north than Pisa, Hitler's control of western Europe was crumbling—just as the President and Stalin had predicted at Tehran. Paris had been liberated in seven weeks, with the Wehrmacht streaming back from Normandy in tatters and Soviet forces smashing all German attempts to hold a defensive line in the Ukraine.

Churchill, aware he'd blundered over D-day's decisive importance, shook his head. Given the huge logistical and organizational requirements of such a seaborne switch of British, American, French, and other forces from Italy to Yugoslavia for a Ljubljana campaign, he thanked the Marshal, but said it was now unnecessary: thanks to its current rate of progress, "the Red Army might not give us time to complete it."[3] Left unsaid was the fact that, following their reverses in the Ardennes, the U.S. military had expressed even greater opposition to yet another Churchillian diversion from their main effort, just at the moment when Eisenhower's forces, stretching all the way from Antwerp to Switzerland, were now seriously short of infantry.

Politeness, in sum, was a more probable reason for Stalin's suggestion—and surprising. The dictator, after all, was not known for his politeness—indicating, possibly, a major change had taken place in the Russians' attitude to their coalition Allies. If true, it presaged the very thing which neither Hitler nor Goebbels, in their conversations at the Reich Chancellery in Berlin, had considered possible: *increasing*, not decreasing Allied unity.

As Stalin departed the Vorontsov Palace to meet the President, the British team was left mulling over the signs. Had the leopard changed its spots?

Even at a local, practical level the omens seemed better than anticipated. For years—certainly since the German invasion of the USSR in 1941—the Russians had displayed an almost paranoid fear of personal contact with Western officials, even when accepting crucial Lend-Lease help. Suddenly things seemed different. At Saki, for example, Soviet, American, and British air force personnel shared control of the large airfield, the Russians responsible for the two big runways, fuel, provisioning, and security; the U.S. team for transport; the British for meteorology, communications, the control tower, and briefing and clearance of aircraft. The airfield had thus received twenty Allied four-engine planes in five hours, escorted by fighters, without mishap and within seconds of their estimated time of arrival, following their fourteen-hundred-mile flights.[4] Similarly, in the three designated palaces and at Allied personnel quarters in Sevastopol, there was found to be a level of hospitality and determination to make the conference a success that could only bode badly for Hitler and Hirohito.

In any event, at 4:15 p.m., with a bevy of armed guards and accompanied by Foreign Minister Molotov, Stalin arrived at the Livadia Palace for a half-hour pre-plenary private session with the President.

To Bohlen, who was to act as the President's sole personal interpreter for the length of the conference, the President, waiting in his palatial, gilded study, did not look good, physically. He was, however, "much better than at Malta."[5]

75

Russian Military Cooperation

THE PRESIDENT HAD BEEN at the height of his physical and mental powers during his last meeting with the Russians at Tehran, fourteen months earlier. Now, they recognized immediately, a terrible change had taken place. They were stunned to see him so shrunk. Yet Mr. Roosevelt acted as if nothing were amiss, displaying something of his old charm as he bade Marshal Stalin sit beside him on the sofa, next to the wooden table in his study with its delicate diamond-design inlay and ashtray, beneath a huge painting from czarist times (one of only two that had not been looted by the Germans).

Stalin certainly allowed no shock to show on his face, which "cracked in one of his rare, if slight, smiles," Bohlen recalled, as he "expressed pleasure at seeing the President again."[1] The small but stocky Georgian wore his new, high-collared military tunic bearing the epaulettes of a marshal of the Soviet Union; the President sat beside him in a light-gray suit, with a flowery silk tie and his trademark pince-nez — his face thinner, and his hair, too, in fact grayer than before, but his blue eyes still alert and friendly.

With sincerity the President thanked Mr. Stalin for Soviet hospitality — especially the decision to hold all plenary meetings at the Livadia Palace, thus saving him from having to travel to other buildings. They both expressed satisfaction with the military situation — the Western Allies massing on the Rhine, the Russians on the Oder. It was clear the Russians were poised to reach Berlin — though whether the Soviets would take the city before the Americans took Manila, in the Philippines, was a moot point. It had been the subject of bets taken aboard the USS *Quincy*, the President explained, but Stalin rightly doubted how straightforward the capture of Berlin would prove, given the intensity with which the Germans were resisting the Russians on the Eastern Front. There was, he confessed in all honesty, "very hard fighting going on for the Oder line."[2] (It was small wonder; Dr. Goebbels noted in his diary how *Volksturm* units were fighting to defend German-occupied Poland, or Posen, with more fanaticism even than Wehrmacht units.[3])

There was very hard fighting going on in the Philippines, too, following MacArthur's invasions of Leyte and Luzon — the Japanese fighting no less hard than Germans to hang on to their conquests, or obliterate them if they could not. Beginning the previous day, the battle for the city of Manila would alone take four weeks for U.S. forces to win. The result was not only almost seventeen thousand Japanese dead and almost six thousand wounded, but the destruction of virtually the entire walled city. (In what became known later as the Manila Massacre, the Japanese local commander had deliberately disobeyed General Yamashita's order to evacuate the city. Instead he led a suicidal campaign, using Filipinos as human shields, before eventually committing ritual seppuku on February 26, 1945.[4])

The war was becoming a deliberate *Götterdämmerung*, or twilight of the gods — both the Japanese and the Germans aware they had committed evil on a scale so vast they would never be forgiven, and would be held accountable once they surrendered, with many (including Yamashita) facing execution for their atrocities.

The sheer destruction the President had seen on his five-hour car journey from Saki to Yalta, however, was nothing, Stalin corrected him, compared to what the Germans had done in the Ukraine, with "method and calculation," as Bohlen's record ran of their conversation. "He said the Germans were savages and seemed to hate with a sadistic hatred" the monuments and buildings their opponents had built over the centuries — while complaining bitterly at Allied bombing of their own cities. The President agreed; the same was the case in western Europe, where German garrisons were still holding out in bypassed French ports on the Atlantic and the English Channel, such as Lorient and Saint-Nazaire, Boulogne, Calais, and Dunkirk, as well as threatening to destroy the Dutch dikes rather than retreat before the Canadian Army as it advanced into the western and northern Netherlands.[5]

What emerged, then, in the first hour in the President's study in the Livadia Palace was that the two most powerful commanders in chief in the world were beginning their historic conference on the same military page: namely their determination the Third Reich should be brought to unconditional surrender by their combined armies. The President raised the matter of the three proposed occupation zones, to follow German surrender, and the possibility of adding a fourth which the French could police. And given German culpability in the recent extermination of more than half a million Hungarian Jews, in addition to Hitler's liquidation program since 1942, the President even suggested that Stalin repeat, at the dinner to follow the plenary later that evening, his Tehran toast regarding the execution of fifty thousand SS officers responsible for running the Nazi concentration camps, summary executions, mass murder without trial or indictment, and other war atrocities.

Some later historians would consider the President's suggestion to be in bad taste. He didn't, in all truth, mean it, given his respect for law, which he'd studied after college. But would law bring the culprits to justice? He had cause to wonder—a concern that would prove, in the event, well founded. The majority of atrocities committed by SS officers and men would, in the war's aftermath, simply go unpunished. In the meantime, though, he had at least made clear as president of the United States that, due to the enormity of German war crimes, he was no advocate of carte blanche forgiveness.

Above all, the President was keen, in his brief preconference meeting with his Russian counterpart, to carry out the promise he'd given his Joint Chiefs of Staff that morning, when they'd held a short meeting in the study. The military teams of the three primary combatant nations would be meeting to hammer out their combined plans to defeat both the Third Reich and—though in secrecy at this stage—the Empire of Japan. General Marshall had therefore asked if the President could help them. There had been commendable communication between Washington and Moscow regarding military operations in Europe, but not much in the way of action. It was not that the Russians were uncooperative so much as they seemed stuck in a kind of bureaucratic pyramid, General Deane had reported from Moscow. No serious decision could be made without authorization from above—that is, Marshal Stalin, as Soviet commander in chief. Could not the President tackle Stalin on the issue?

The President had said he would—and did. Assured the three Allies were sincere in their approach to the surrender of their common enemies, Stalin immediately promised greater Russian military cooperation and decisiveness.

As the President was wheeled into the Grand Ballroom at 5:15 p.m. and took his seat at the huge Arthurian table for the first plenary session of the Yalta Conference, then, he and Stalin seemed remarkably collegial. As did Churchill, who had meantime arrived from the Vorontsov Palace.

76

Making History

ONCE ALL WERE PRESENT in the Livadia Palace ballroom, the three main leaders of the United Nations got down to business, aided by their interpreters and advisers, at the sixteen place settings, and further back, additional advisers seated around the hall. Stalin had brought "General Antonov, deputy chief of staff of the Red Army, Fleet Admiral Kuznetsov, and Air Marshal Khudikov" with him, as Admiral Leahy noted in his diary,[1] as well as Ivan Maisky and Andrei Gromyko, his former Russian ambassador to Britain and his current ambassador to the United States.

Some thirty Soviet and more than a dozen Western photographers had gathered to record the arrival of the participants, but they were only briefly allowed into the chamber, where the proceedings themselves would be secret until official communiqués were drawn up and given out. All, however, were aware that the imagery of that day — and the days to come — would be as important in what the conference represented to the world as the actual decisions made there.

For Steve Early this would become a challenge, for although the President still looked relatively good as he was wheeled into the ballroom after the tête-à-tête with Stalin across the front hall, the deliberations at the big table would, he knew, soon exhaust his boss and the result appear all too visible on camera, unless careful supervision was made of photos before release. He therefore urged the American contingent to agree that Robert Hopkins, a young U.S. Army photographer and son of the presidential counselor Harry Hopkins, should be the sole designated photographer at the palace. He could be relied upon not to abuse his privileged status.[2]

Once the photographers had been barred from the ballroom, Marshal Stalin, as conference host, asked if the President would open the proceedings and act as chairman of this and subsequent plenary sessions, all of which would take place in the Livadia ballroom.

Accepting the task, the President duly thanked Marshal Stalin, for the record. As he'd just shared with the Marshal, they would be tackling further aspects of military strategy across the globe during the coming week. However, he wanted this first session to address only the situation on the Eastern and Western Fronts, together with questions arising, so that the senior military staffs of the three countries could thereafter address them together in military discussions, on their own.

Stalin nodded. On the Marshal's instruction General Antonov gave a detailed report of Soviet operations on the Eastern Front, including the recent capture of the Silesian coal and oil fields — Stalin pointing out not only the Red Army's two-to-one superiority in numbers across the whole front, but its huge preponderance in heavy artillery support. This had led to "300,000 Germans killed, and 1,000,000 prisoners taken" — a fearsome example of the kill-to-mere-casualty ratio. Antonov's account finally climaxed with a main concern he expressed on behalf of the Soviets: that the Western Allies should do everything possible to block more attempts by Hitler to withdraw Wehrmacht divisions from Norway, the Western Front, Italy, and the interior of Germany to reinforce the Eastern Front, where the Wehrmacht was fighting to the literal death.

Understandably impressed by the professionalism, scale, and candor of the Russian account, the Western Allies' military team responded. At Churchill's request and with the President's approval, General Marshall was asked to give the latest picture: namely General Eisenhower's recent regrouping of his forces to close up to and cross the Maas and Lower Rhine in the next few days. Thereupon major American and British armored forces, under command of Field Marshal Montgomery, would be funneled into Germany north of the Ruhr, to be matched by U.S. divisions south of the Ruhr, tasked with completely enveloping Germany's industrial heartland. Significantly, there was no mention now of distant Berlin, which had been their goal in the heady days following the fall of Paris — and which the major German counteroffensive had successfully parried.

Even the diplomats present at Yalta were awed by this first plenary hour of the summit. Although the conference would later become famous — or infamous — for the political discussions held necessarily in secret, it was the military deliberations of the "triumvirate" that would make Yalta unique in the annals of warfare. Never before in this, the most destructive war in history, had the most senior military representatives of the anti-Axis coalition partners all met together. In November 1943 Stalin had taken only one major military adviser to Tehran, on the mistaken assumption the get-together would be a political meeting. But now the situation was very different. The Russian military delegation was in full force, with specific requests to make of its Western partners, and vice versa — from bomb lines to ultimate occupation lines, Schwerpunkts to interdictions, as their teams began trading details of what worked in fighting Hitler's Wehrmacht, and what didn't.

The level of shared Allied military discourse was especially stunning, given Russian paranoia and fear of execution for imparting "secrets" to a coalition ally without higher NKVD authorization. At least and at last, however, it was *happening*, with Stalin's approval.

The Big Three listened in the Livadia ballroom as their generals held forth, and commented as the reports were delivered. Stalin's military acumen, in this respect, astonished even his own team as the dictator, after discussing the ratio of heavy artillery to kilometers in Russian offensive operations, openly debated the difference between Russian and Allied infantry and tank superiority over Wehrmacht forces in battle. Stalin found himself astonished, for example, to learn that where the Russians needed at least a four-to-one ratio to break through at chosen points in German defenses, the Western Allies often had no more than an equivalence in infantry, relying instead largely on their tactical and heavy-bomber air power to penetrate.

Such talk had led Stalin to wax (for him) lyrical on the nature of military cooperation that, as allies, they were now reaching. He instanced, rightly, how in Moscow he had received Air Marshal Tedder, General Eisenhower's deputy supreme commander, during the Battle of the Bulge, and as a result had brought forward the timing of the next Soviet offensive, despite terrible weather.[3] His aim, he claimed, was "to emphasize the spirit of Soviet leaders who not only fulfilled formal obligations but went farther and acted on what they conceived to be their moral duty to their Allies."[4]

Bohlen, as interpreter, would specifically recall the words three decades later, since they had represented, for a brief moment in history, what promised to be a genuine tripartite military alliance, on similar lines to that which existed between the United States and Britain — a first real indication the three nations *could* work together, not only to ensure the unconditional surrender of Nazi Germany and Japan, but to develop a global security system in a United Nations to be set up thereafter.[5]

The President, for his part, was tired but moved — applauding this indication of genuine military cooperation, rather than fighting separately for the same goal. However, it was Churchill himself — ever the historian — who best articulated this new turning point in the war: namely the "highest importance" they were now attaching to the business of cooperation. The "three [military] staffs which were assembling here for the first time" were making history, he declared, as if speaking before Parliament, and were about to "really work out together detailed plans for the coordination of joint blows against Germany." In this way, "if the current Soviet offensive were to come to a halt because of the weather or road conditions," the Western "Allied armies" would still attack. Indeed, the Allies would attack the shrinking German Reich "simultaneously" if at all possible.[6]

Given that Churchill had refused to countenance a meeting of the Combined Chiefs of Staff with their Russian opposite numbers in Cairo, before Tehran, or to include them as observers in Combined Chiefs of Staff meetings, this was a historic turnabout for Churchill — and certainly one that dashed any hope that, by appealing to the Prime Minister, the Führer and his government might still remain in power. Stalin even asked that the meeting instruct the three Allied military teams to prepare, together, joint U.S.-British-Soviet plans for a contingency Allied summer 1945 offensive, "because he was not so sure that the war would be over before summer." Churchill agreed to this, saying they should "take full advantage of this gathering."[7]

The plenary had now been going on for almost two hours. The President was clearly wilting; moreover dinner was ready for the summit leaders, who were invited to attend a small banquet with their foreign ministers in his dining room. Before they did so, however, it was agreed the three military staffs would meet the next morning at Stalin's residence, the Yusupov Palace — former property of the wealthy Oxford-educated prince who had assassinated Rasputin before the Russian revolution — and continue their joint military deliberations there.

The President, in other words, had done exactly as General Marshall had requested: he had broken the proverbial ice, and gotten Stalin to sanction the kind of inter-Allied military discussions, even decisions, that would increasingly be needed as their forces converged — or collided — on German soil. He'd also gotten agreement that the issue of zones of military occupation would be formally addressed the next afternoon, at the Livadia Palace, when there would be a second plenary of the three leaders, this time with their foreign ministers at hand. They would discuss, he said, the "political treatment of Germany" — a euphemism for occupation zones; dismemberment of the Third Reich; reparations Germany would be required to pay for the destruction the Wehrmacht had caused; and the punishment of war criminals.

With that the military teams were dismissed. The President retired to his private quarters with Stalin, the two men "talking together in the study," as Anna Roosevelt noted in her diary, while Churchill "freshened up" in Stettinius's room.

It was at this point that poor Anna was approached by Dr. Bruenn, warning her of an impending disaster. Judge Byrnes was apparently refusing to attend the first formal summit dinner on the grounds that, although he was director of war mobilization, he had not been permitted to attend the plenary dinner for security reasons. "A tantrum was putting it mildly! Fire was shooting from his eyes!" Anna described in her diary — managing, mercifully, to put out the fire with the help of Averell Harriman, who warned that they could not expose the ailing president to such childishness, when the conference had kicked off to such a hopeful start.

77

A Silent President

THE DINNER IN THE PRESIDENT'S DINING ROOM consisted of vodka, five different wines, fresh caviar, bread and butter, consommé, sturgeon with tomatoes, beef and macaroni, sweet cake, tea, coffee, and fruit. It lasted three hours.

The President made a brave effort to keep awake and alert, but the day's meetings had clearly drained him. Apart from friendly chitchat and a multitude of toasts, there was in fact no political discussion, despite the presence of the foreign ministers of the Big Three — the President, as dinner host, simply too weary to discuss anything. Only in the last half hour did the matter of the United Nations organization come up — and without enthusiasm on the part of Stalin, who ridiculed the notion that "small" countries should have the same postwar representation as the Three Great Powers, as he put it. Those small countries were, after all, not the nations who, at great cost in blood and treasure, had "won the war," and whose representatives "were present at this dinner," as Bohlen noted in his minutes.[1] Churchill, from his capacious mind, tried to nip the developing argument in the bud by quoting an old saying, to wit: "the eagle should permit the small birds to sing and care not whereof they sang."[2]

It was clearly a struggle for the President to stay awake, let alone direct or follow the conversation. Word of his condition was already causing concern, indeed dismay, especially at the British team's residence. There, dining with the British chiefs of staff on their return from the plenary session, Churchill's physician listened as the generals told him how pleased they were with the plenary's outcome, yet how worried they were by what they'd seen of the President. "Everyone seemed to agree," Moran wrote in his diary, "that the President had gone to bits physically," and had "intervened very little in the discussions, sitting with his mouth open; they kept asking me what might be the cause."[3]

Moran, despite being a fine doctor, was unsure, having been lied to by Admiral McIntire at the Quebec Conference, six months earlier. There, too, the

President had been largely and unusually silent during discussions, the chiefs said. His deteriorating health hadn't been so apparent in Quebec, since "what he did say was always shrewd," covering up his relative detachment from the details of the proceedings. "Now, they say, the shrewdness has gone, and there is nothing left," Moran recorded. "I doubt, from what I have seen, whether he is fit for his job here."[4]

Moran's comment sounded cruel, yet was all too close to the mark. Brooke, delighted with the day's work, did not remark on the President's performance in his own diary, but Admiral Cunningham complained in his that the "President, who is undoubtedly in bad shape & finding difficulty, did not rise to the occasion." Fortunately, "Stalin was good & clear in his point, the PM also very good but the President does not appear to know what he is talking about and clings to one idea."[5]

That idea would be the making of the conference, however.

It was not only among the British contingent that there was anxiety expressed at the decline in the President's health. Everyone on the American delegation, most of all Admiral McIntire, was aware of it, yet there was nothing that could be done medically. Or politically. There was no one else on the America team, perhaps even on the planet, who could take the President's place at that moment. In fact, there was no one who could really take the place of any of the three Allied commanders in chief at Yalta. They had become the three icons of the Allied coalition — not only symbols of the Grand Alliance, as Churchill called it, but the only leaders whose command of their military forces was incontestable, and irreplaceable if they collapsed or died. As Alexander Cadogan, the permanent undersecretary to the British foreign secretary, would write privately to his wife from Yalta, the "President in particular is very woolly and wobbly. Lord Moran says there's no doubt which of the three will go first."[6]

Somehow, however, with the conference currently slated to last only a further four or five days, the President would have to get through it, or be brought through.

"The dinner broke up early — about midnight," Anna recorded. As the opening act of the summit, despite the President's condition, all had gone amazingly well. "FDR seemed happy about both the Conf. and the dinner," Anna noted — in fact her father told her, before turning in, that "Byrnes had made a fine toast," rescuing him from a faux pas when he'd revealed in his meandering toast that he and Mr. Churchill were wont to call Stalin, between themselves, "Uncle Joe."[7]

The President had meant it as a sign of friendship, not disrespect. Stalin, who had by then drunk a great deal, took it as a slight, however. The Marshal had even made as if to leave the dinner, in pique. Byrnes, who was considered smart as a whippet, had stepped in. Proposing a new toast, he had pointed out that the President's term was one of affection, similar to the way Russians would refer to "Uncle Sam."

The day had thus come to a friendly close, leaving only Churchill and Eden to have a first-class row over voting rights in any putative UN organization, in the ballroom, while Stalin went back to the Yusupov Palace.[8]

78

Kennan's Warning

BOTH STALIN AND CHURCHILL were opposed to a United Nations organization that might tie their communist or colonialist hands. But it was not only Stalin and Churchill who did not care for the idea. A number of American officials, too, opposed the President's proposal. An urgent letter from Moscow had been given to the President's interpreter immediately upon Bohlen's arrival at Yalta. It had come from George Kennan, the top political counselor to Ambassador Harriman at the U.S. Embassy in Moscow: a long, preconference warning in which Kennan confided how much he disagreed with the President's war and postwar policy. The President's idea of a United Nations organization, Kennan advised, should, despite the preliminary Dumbarton Oaks accords, be dumped "as quickly and quietly as possible" at Yalta.[1]

This came as something of a shock to the President's interpreter and rising star in the State Department. As Bohlen understood things, the whole purpose of holding the Yalta summit, from the American point of view, was to show the Nazis and the world that a new postwar world order was being created, under the aegis of a United Nations authority, as discussed at Dumbarton Oaks: one that would authorize military power to be applied via a small Security Council if any nation ever attempted again to do what Hitler or Hirohito had done. Ambassador Harriman had discussed this and the President's vision at length with Stalin, and with Molotov, in the weeks and months since the Dumbarton Oaks Conference — the preliminary agreements of which were only held up by differences of opinion over voting numbers and rights. Kennan now argued, however, that Germany should be broken up and partitioned into separate states, and Europe divided "frankly into spheres of influence." Americans should keep "ourselves out of the Russian sphere," and the Russians stay "out of ours."[2] And no UN created, lest it hamstring future American use of "armed force."[3]

Roosevelt disagreed with this free-for-all — which looked suspiciously like the world before the First World War. The President hoped to announce agreement on the UN to an anxious world in February 1945. Without some kind of international body the future looked dark, and would offer little practical hope or ideals to mankind, after a devastating world war. Whereas, if a UN organization could be established "with teeth" this time, as the President had put it, the prospect of renewed wars of conquest could hopefully be averted—and much positive, constructive postwar work be done meanwhile, in a collective spirit of improvement, from global health initiatives to economic development and education.

Moreover, there was the matter of timing: something which Kennan, as a diplomat, should have been the first to understand. "First things first" had been the President's mantra, reflected in his methodical military strategy since Pearl Harbor: "Germany First," then Japan. How could Germany be broken up and partitioned, as Kennan claimed he wanted it to be, unless its unconditional surrender was first obtained? From all reports of the fighting, as from diplomatic efforts, this would not be possible without military victory by the Allies. And for this to happen, the cooperation of the Soviets was still crucial—cooperation that could be used to promote a more responsible approach to world peace and development, since refusing to join such a United Nations organization would make the Soviet Union into a pariah state.

The Allies were now within a few months of defeating the Wehrmacht, if all went well. How would it possibly help attain the unconditional surrender of the Third Reich if, at a conference with Marshal Stalin and Mr. Churchill in Yalta, the United States President and his secretary of state, Mr. Stettinius, should now announce they were *burying* the Dumbarton Oaks plans, largely agreed with the Russians, for a United Nations organization—and "as quickly as possible," as Kennan suggested?

In an all-out world war that had not yet been won this seemed daft to Bohlen. Those nations next to, or close to, Russia's borders which had contributed military forces to help Hitler attack the Soviet Union—Romania, Finland, Hungary, Slovakia, Romania, and Croatia—faced a future that was hardly rosy. Yet the United States had not entered the war to take responsibility for the democratic future of such enemies—which they had effectively become as German allies, following Hitler's declaration of war on America on December 11, 1941. There was in reality little that the United States could actually do to "save" them from communism; this did not mean, however, abandoning the notion of a United Nations organization. Or that the United States should not do everything in its power to liberate, via its military, as much of western Europe as it currently could. By setting out the principles and ideals of a United Nations organization at this moment, when the Big Three were fighting as allies, the President might yet get the Soviets to back the concept, if Roosevelt could persuade Stalin, upon whom all Russian decisions depended.

Kennan's suggestion the United States should either announce at Yalta it would defend the integrity and independence of democratic eastern and southeastern European states — even to the point of going to war with the Soviets — or simply and openly "write off" the region for Soviet domination and retreat into its own "spheres of influence," thus seemed to Bohlen not only morally wrong at this moment in the war; it did not even reflect what Americans were fighting *for.* Would American voters really subscribe to such a new strategy of open abandonment of international leadership and ideals, before the war against Hitler was even won? In the aftermath of nearly a hundred thousand American casualties suffered in the Ardennes and Alsace — combatting ruthlessly indoctrinated Nazis, who had openly massacred American prisoners taken on the battlefield? Equally, would U.S. troops or voters want their leaders to announce in public, in advance of the end of the war, they were going to "write off" Poland and all those central European Russian borderland states, which had been Hitler's war-partners, and simply withdraw to "spheres of interest"? If so, who then would enforce the breakup of Germany which Kennan supposedly wanted, after the war? Would the U.S. not retreat into isolationism — as had happened, after Versailles?

Bohlen certainly admitted that the prospects of "saving" Poland and Hitler's allies in eastern and central Europe from Soviet communism — i.e., political and social suppression and oppression — were currently slim, as things stood. The vision of a United Nations organization, and a Security Council acting on its behalf, was nonetheless worth supporting, dedicated to principles of democracy and self-determination, surely, even if those principles proved in the short term difficult, if not impossible, to fulfill, given Russian forces in control of eastern and central Europe. The United States would at least have *tried* — which it had *not* done after World War I. As Bohlen later wrote, "I recall feeling quite strongly that to abandon the United Nations would be an error of the first magnitude" — not because he thought the UN would necessarily or automatically "prevent big-power aggression," but because, as a trained diplomat, he felt it would "keep the United States involved in world affairs" — and without necessarily "committing us to use force when we did not want to," assuming the U.S. would have the right of veto against military action in the Security Council. Moreover to openly subscribe, at this point, to a "formal, or even an informal, attempt to give the Soviet Union a sphere of influence in Eastern Europe" would, if it became known in February 1945, be akin to Chamberlain's Munich agreement: indeed would be playing into Hitler's hands, rather than asserting the *principle* of democracy, at least, for post-fascist central and east European territories — especially those who had assisted Hitler in Operation Barbarossa. As Bohlen wrote back in "a hasty reply" to Kennan: "Quarreling" with the Soviets at Yalta "would be so easy," given the contrasting democratic-communist ideological beliefs held by the two military allies, "but we can always come to that."[4]

In his reply to Kennan, Bohlen had confessed that his personal views were "tempered" by his association with the President, who believed in idealistic realism, and by the recent weeks he'd spent with Harry Hopkins, meeting leaders in Europe. The existence of tyrant-led communism, whether in Russia or in other countries, was "a political fact of life" that Americans had every right to deplore, in terms of individual citizens' rights and freedoms. But, after fighting a "long, hard war," the United States surely "deserved at least an attempt" to "get along with the Soviets," whatever it felt about their political system. As the diplomat later put it, the President had but two "major goals" at the impending conference: namely "to pin down Stalin on the timing and extent of entering the war in Asia," once Germany surrendered; and to create a United Nations organization to stop the United States from "slipping back into isolationism," as the Kennans of the world wanted, or were willing to risk.[5]

The fact was, Bohlen felt, the United States had ultimately failed in its recent attempt to remain an isolationist, "America First" nation — forced to watch as the Axis powers overran country after country with impunity. Now, having become the world's most powerful nation economically and militarily, in a bare three years of war, the United States had a duty to itself and to the world to become a world *leader* of partner nations, not simply an isolationist, or limited "sphere of interest," power, Bohlen was convinced. For if not the United States, who else would step up to the challenge, as the United States had done in leading the Western Allies after Pearl Harbor? The British Empire? The French Empire?

Britain, at the end of the day, was close to bankruptcy, economically. Politically, too — more concerned with clasping its far-flung colonial territories (when recovered with American help) than creating a world organization based on the Atlantic Alliance. Which left the problem of possible future German revanchism, à la Hitler, if Germany was broken up and partitioned, pace Kennan. Who, then, in the years to come, would act?

But if neither Russian nationalists nor British imperialists had a genuine interest in establishing a United Nations organization — one in which the Soviet Union would, in the proposed UN Assembly and in its Security Council, be outnumbered by other countries, as would Britain — what chance had the President of persuading them to support his vision? Were "spheres of interest" the inevitable face of the postwar future — British, Russian, French, Chinese — as it had been in the 1930s? Had the President undertaken an impossible challenge at Yalta?

• • •

Looking at the President in his study that morning, greeting Stalin with a warm smile and an outstretched hand from his sofa, Bohlen had been fully aware not only of what the President was up against, but also how slim was the chance that his ailing boss could make it happen — at least in his current state of health. Yet there was, in reality, no one else on the American team who could get the Soviets and the British to back the establishment of a United Nations! Edward Stettinius, the new secretary of state? Justice Byrnes, the director of war mobilization? Harry Hopkins, the White House counselor, or political adviser, to the President? Averell Harriman, the U.S. ambassador to Moscow? Not one of these men had been elected to his present position. And the vice president, Harry Truman, was not even there: unable to leave the country while the President was abroad, and in any case a neophyte with regard to international issues.

Only the President could achieve it, at that moment, Bohlen recognized. Nothing would be lost that would not have been lost anyhow, if he failed. It was, therefore, worth trying, in Bohlen's view. The President, after all, now appeared to have the backing not only of most Democrats but of most Senate Republicans in Congress, swayed by the recent 180-degree turn by Senator Vandenberg. Moreover, unlike Churchill — and certainly unlike Stalin — the President had just won a popular mandate from American voters for such an attempt: one he had laid squarely before the entire American electorate in the November '44 election. With 25.6 million ballots cast and 432 electoral college votes he had won a resounding fourth term as president and commander in chief. It was worth an attempt, surely — unless the effort killed the President, at the very moment Hitler and his regime were on the ropes.

This, sadly, was far from unlikely. Plagued by a "paroxysmal cough" that had kept waking him the previous night (though he "denied dyspnea, orthopnea, or cardiac pain," as Bruenn recalled),[6] the President had dreaded the long, unavoidable dinners, with their endless toasts — just as he'd shuddered at the prospect of the disagreements that would inevitably arise over political issues, especially Poland. But as Hopkins had said, he had signed up for this, however reluctantly. And if his ill health precluded him from following all the arguments in detail: in some ways this was also a blessing. He would stick, he was determined, to his simple, single-minded agenda: close military cooperation; a signed agreement by the Big Three to establish a United Nations organization; and a necessarily undisclosed, but formal Russian agreement to join the United States in the war against Japan once the Third Reich surrendered.

Whether the President would leave Yalta on a plane or in a casket remained to be seen, but he was determined to do his best. For the rest, he could not answer.

79

A World Security Organization

THUS THE SAGA of Yalta began to unfold on February 5, 1945. Over the next several days the generals, admirals, and airmen of the three nations, aided by their staffs, conferred not only about the endgame in Germany — bombing targets and airfield arrangements, in particular — but over detailed plans for American air bases to be secretly set up in the Soviet Union, ready for the day when Russia would declare war on Japan — which Stalin would formally agree to do in secret.

This left, for the President, the final setup and establishment of a United Nations organization to be agreed and announced. Behind the scenes tempers flared, dissension arose, and arguments threatened at times to derail the main proceedings. As Bohlen later put it, the "conference was organized in such a way that there was no orderly discussion and resolution of each problem by the leaders. Instead, issues were brought up, discussed, then shunted off to the Foreign Ministers or military chiefs or just dropped for a few hours."[1] As Sir Alexander Cadogan reported to his wife, the place became, initially, "a mad-house," and took — like all conferences — "days to get on the rails."[2]

Gradually it became clear to all participants, however, that it was too early to make hard and fast agreements about the future political map of the world when the war itself was still not won, and the United Nations organization had not been set up.

Territorial discussions among the various political and diplomatic advisers were the most contentious matters, resembling shooting stars before dawn — leaving Churchill, in the plenary sessions, to display his rhetorical virtuosity and Stalin his more focused acerbity, in equal measure. This left the ailing American chairman often reeling. "The President has certainly aged,"[3] Cadogan noted on February 7 — Roosevelt trying, as chairman, to keep the place, the pace, the peace, and the purpose of the deliberations advancing each day.

Whether the President realized he only had weeks to live no one would ever know, since Lucy Rutherfurd subsequently burned all his letters, and Anna —charged with the daily care of her father—discontinued her diary after the first two days. For their part, though, the President's doctors monitored his condition with increasing concern as the conference reached the fifth plenary session in the Livadia ballroom at 4 p.m. on Thursday, February 8, 1945.

For the President this was to be the climax of the Yalta Conference: getting agreement on what was termed the "World Security Organization": namely those who had signed the "United Nations Declaration" on January 1, 1942, and who would now be "summoned for Wednesday, 25th April, 1945," before the war ended, for an inaugural meeting to "be held in the United States of America."

Stalin expressed concern that some of these states had no diplomatic relations with the Soviet Union, but the President deftly dealt with this by saying the Soviet Union had already "sat down with these states at Bretton Woods and the UNRRA conferences." Moreover Stalin's next grumble that there was a difference between those "nations who had really waged war and had suffered," in the struggle against the Nazis, and "others who had wavered and speculated on being on the winning side" was equally deftly dealt with by the President. It was time not to look back, but for the world to move on—and by inviting to the UN, he proposed, all those nations who now "declared war" on Germany by "the first of March," 1945, they would have the corpus of an organization they could be proud of creating.[4]

Stalin and Churchill had thus reluctantly given their assent—Stalin even agreeing to drop his suggestion that White Russia (Belarus) and the Ukraine qualify separately for invitation by signing retrospectively the 1942 United Nations Declaration, though pressing for this; it would be left up to the nascent UN, Stalin agreed, however, to decide on whether White Russia and the Ukraine should have separate seats, distinct from the USSR.

The President had thus good reason to be proud of what he'd accomplished by February 8. But with the military side of the conference done (the generals slated to hold a final tripartite military meeting the next day and report to the President), and the UN/World Security Organization settled by the early evening, the plenary would have to return to the knotty issue of Poland: its future frontiers and its provisional government. A much more difficult proposition.

80

Poland

AT THE FULCRUM of central Europe, Poland had endured a long, often savage, often contested history. The country had recovered its independence in 1918 — one that was ratified in the Treaty of Versailles, and which Poles had successfully defended against Russian forces in 1919–21 in what was called the "miracle of the Vistula." That independence had been lost yet again when it was invaded and overrun by Hitler's Wehrmacht forces on September 1, 1939. Two weeks later the Soviets had invaded Poland from the east, via the Ukraine, and had annexed eastern Poland in accordance with secret protocols in the Molotov-Ribbentrop Pact regarding "spheres of influence" — the term for occupation or control of puppet regimes.

The spheres had not lasted long, however, thanks to Hitler's agenda, backed by the Wehrmacht. The Soviet-controlled Polish territories east of the Curzon Line had been steamrollered in Operation Barbarossa on June 22, 1941, in which almost four million troops from Germany, Romania, Finland, Italy, Hungary, Slovakia, and Croatia were thrust into full-scale war against the Soviet Union. Of the five million Russians captured during the Barbarossa invasion, almost none returned to Russia alive.

Now, in 1945, Poland had been triumphantly "liberated" — but by Russians, not Poles themselves or the Western Allies. Stalin had been pressing both Churchill and Roosevelt since before the Tehran Summit in 1943 to acknowledge a permanent new eastern Polish frontier in the event of German defeat, which could be done by formally shifting the future border between Poland and the Ukraine. This would follow the so-called Curzon Line — a frontier that had been suggested, the Russians had pointed out, as the ethnic boundary between Poland and Russia by Lord Curzon and the Supreme Allied War Council during World War I. It had also been the line which the Russians had already established in the 1939 partition of the country in the Molotov-Ribbentrop Pact.

What, then, Stalin had kept asking, had changed the minds of the Western Allies about this border since then?

Communism, for one thing. Although Churchill and the British government actually favored such a new, or renewed, Polish border, with a sidestep of the country's western boundary into Prussia to compensate for the territory "lost," the President had cited potential political repercussions in the U.S., where there was a significant Polish-American population (sometimes estimated at six million). He had therefore declined to agree such a new frontier line at Tehran, in spite of Churchill's urging. Now that he had been successfully reelected, Stalin could not see why the President should fear a Polish-American political backlash. In the area east of the Curzon Line, as this became liberated by the Red Army, the Soviets had installed a provisional Polish government, drawn from "Lublin" Poles in exile in Moscow—just as the Allies, liberating France, had permitted one in Paris, under de Gaulle. These "Moscow Poles" accepted both the Curzon Line and Churchill's suggestion of compensation to be given them in East Prussia, once Germany surrendered.

To Churchill's chagrin, however, the Polish government in exile in London, under Stanislaw Mikolajczyk, had adamantly refused to agree either to the Curzon Line boundary or to sidestepping the nation's western frontiers into German East Prussia. Nothing Churchill had done, or could do, seemed sufficient to change their minds. This had set the stage for an international dispute as potentially volatile as that of September 1939.

For the ailing president this was the last problem he wished to deal with at Yalta—yet Stalin, obsessed with his notion of a cordon sanitaire against future German attack, was determined it should be addressed forthwith, *before* the United Nations organization was established. The UN, he figured, might well rule against the boundary change. Since Stalin was willing to accept de Gaulle's French Committee of National Liberation as the provisional government of France, legitimated by the Allies, why should the Polish Committee of National Liberation, legitimated by the Soviet Union, not be accepted by the Allies?

This was a truly Gordian knot.

Principles of self-determination and self-government through popular election, as well as common agreement on borders, were easy enough to urge on paper, in the midst of a world war, the President was aware; enforcing such an outcome against the wishes and security concerns of the nation that had actually liberated Poland (which the Western Allies had been unable to do) was a very different proposition. Moreover even on paper, Western diplomats were on uncertain ground; the Monroe Doctrine as cordon sanitaire had, after all, characterized American policy in the Americas since the early nineteenth century. It remained still the underlying strategic principle of U.S. diplomacy and military power in early 1945—a reality that made it difficult for Americans, whose continent had never been invaded since the British in 1812, to object to the wishes of Russians, who had twice been invaded via Poland by huge armies in the past thirty years alone.

Que faire?

The principle of self-determination mattered to the President, indeed was the bedrock of his pride as an American, whose country had cast off its oppressive British straitjacket in 1776. He had made it the kernel of his Atlantic Charter, and had then made that charter the core principle of the United Nations Declaration, which he'd gotten all the anti-Axis nations to sign in Washington immediately after Pearl Harbor, on January 1, 1942.

At the same time, however, the ailing president was too much a realist to imagine the United States could now *compel* the Soviets to cut their own communist throats by encouraging the creation of a democratic state on their own borders, on the very cusp of their costly victory over the Axis powers. And how, in any case, could such an insistence be safeguarded after the war when there was currently almost zero support in the United States for a continuing American military presence in Europe beyond the unconditional surrender of the Third Reich?

Poland's fate thus hung in the paper balance at Yalta. Certainly no one could have spoken more eloquently than had Churchill on behalf of Poland in the plenary on February 6. The Prime Minister did so once again on February 8: Mr. Churchill citing Poland's long, distressing history, its brief but courageous resistance against Hitler in 1939, and its Polish army in exile that had fought the Nazis with such distinction under British command in Italy, Normandy, and Holland. Yet for all the Prime Minister's grandiloquence, the simple fact had remained: British forces were no more able to defend Poland in 1945 than they were when Hitler invaded the country on September 1, 1939—whereas Italy, ironically, which had been Hitler's main ally, and still had Italian units in the north of the country training and fighting with the Wehrmacht, was set to become a free democratic nation again, protected by the Western Allies and with the pope still on his throne.

The situation was, in short, unjust, even tragic—yet in the circumstances unavoidable. Poland, the central European nation which had *not* been a partner to Hitler but a victim of German aggression, and whose military forces had fought so bravely for the Western Allies, now faced Soviet postwar rule, while perfidious Italy got away scot free . . .

To add to this dark prognosis was the manner in which Poland had already suffered at the hands of Russian communists. Not only had the Russian NKVD deliberately murdered more than fifteen thousand Polish officers in cold blood in the Katyn forest in 1940, but on Stalin's orders the Stavka, or high command of the Red Army, had deliberately decided *not* to provide military assistance during the Warsaw Uprising in the summer of 1944 — the heartbreaking, forlorn attempt by Polish resistance fighters to liberate their capital from German occupation that had resulted in a further fifteen thousand Poles being ruthlessly slaughtered by the Wehrmacht.

Churchill portrayed the issue of Poland's postwar political independence, morally speaking, as the "crucial point of this great conference." Failure to agree on this matter would, he warned, represent a "cleavage" between the Allies — and at the very moment when the entire free world so fervently wanted the representatives at "this conference to separate on a note of agreement."[1]

In reality Churchill knew it was futile. As a lifelong parliamentarian, the Prime Minister was fully aware his appeal was *pour l'histoire,* rather than practical: a valiant *cri de coeur* on behalf of brave Poles. In the melting pot that would comprise central and southern Europe at the end of hostilities, Soviet possession would inevitably be nine-tenths of the law, wherever Russian security strategy was concerned. It was a bitter truth Churchill not only knew but secretly accepted, as per the confidential agreement he'd made in Moscow, closeted alone with Stalin, the previous October, with regard to the fates of Greece, the Balkans, and Yugoslavia. There, in the Russian capital, he had somewhat callously drawn up his famous (some would say infamous) "percentages" agreement on "spheres of influence" in Europe that he'd written on a piece of paper he'd passed across the table to Stalin — and which Stalin had ticked, approvingly.

Greece, however, was not Poland. The Greeks were not located on the Soviet Union's western border; Poland was.

Over the issue of Poland, then, Stalin simply remained intractable and implacable — and the plenary on February 8, 1945, which the President had intended to represent the climax of his efforts at Yalta, in terms of war with Japan and the creation of the United Nations, now turned sour as Churchill proclaimed he could not reconcile himself to accepting Stalin's position on Poland.

As tensions rose in the discussion, so did hypocrisy — especially Russian. The Prime Minister could only roll his eyes when Stalin claimed the Poles, who the Marshal openly admitted had every reason to hate the Soviets after their country had in the past been partitioned three times by Russians, were now filled with "good will" for the Red Army, whose liberation of their country from the Nazis had "completely changed their psychology."[2] But Stalin's point that the Western Allies were opposing the provisional Lublin government whereas he had not objected to de Gaulle forming a provisional French government, ahead of elections to be held at the end of the war — indeed had agreed that France should have an occupation zone when Germany surrendered — put Churchill in a serious bind. It was one the President could only do his best to paper over when wisely suggesting they defer further discussion, and leave it to their foreign ministers.

In the meantime, the President summarized, they were all at least in agreement over the recognition of Poland's borders as the Curzon Line in the east, and to be sidestepped somewhere yet to be determined in German East Prussia in the west.[3] Under pressure from both the President and Mr. Churchill, Marshal Stalin even agreed that "free elections" should be held in Poland, as soon as the war was over, as in France—though how "free" they would be remained open to doubt.

It was a most imperfect, somewhat Munich-like deferment of the issue, but it would have to do. It was all the President could manage at the moment, without the conference being derailed. And with that, at his suggestion, the fifth plenary was thus adjourned to the next day.

81

Pulsus Alternans

Churchill, both as a historian and as a prime witness to events in Europe in the 1930s, had been genuinely tormented, and his passionate plea for Poland's democratic future had been sincere — something Stalin, for his part, respected.

As if to reassure the President he harbored no ill feelings towards the Prime Minister, the Russian dictator was the soul of goodwill at the dinner that night at the Yusupov Palace, or Villa. He toasted the Prime Minister with unmistakable sincerity: lauding Churchill's stand in 1940 as "the bravest governmental figure in the world," according to the minutes which Charles Bohlen wrote up that night.

As Stalin put it, "Due in large measure to Mr. Churchill's courage and staunchness, England, when she stood alone, had divided the might of Hitlerite Germany when the rest of Europe was falling flat on its face before Hitler. He said that Great Britain, under Mr. Churchill's leadership, had carried on the fight alone irrespective of existing or potential allies. He concluded that he knew of few examples in history where the courage of one man had been so important to the future history of the world" — and he'd raised his champagne glass with this accolade to his "fighting friend and a brave man."[1]

After Churchill had responded, Stalin had gone on to propose the health of the President, in similar vein: pointing out that where he and Churchill had been "fighting for their very existence against Hitlerite Germany," the President had had "a broader conception of national interest." Even though his "country had not been seriously threatened with invasion," the President had become "the chief forger of the instruments which had led to the mobilization of the world against Hitler. He mentioned in this connection Lend-Lease as one of the President's most remarkable and vital achievements in the formation of the Anti-Hitler combination and in keeping the Allies in the field against Hitler."[2]

Nothing in such toasts had suggested Stalin was being disingenuous. He seemed genuinely to admire Churchill's character. Moreover he respected the

Prime Minister's rhetorical skills, even when finding them long-winded, or directed against himself. By the same token he genuinely admired the capitalist president of the United States for his global perspective, and was profoundly grateful for American military and economic help — without which, he acknowledged, the Soviet Union could not have held out against the Wehrmacht, or have driven it back. He accepted that what lay ahead of their three countries would not necessarily be easier than what lay behind. As he remarked, it was "not so difficult to keep unity in time of war" when there was "a common enemy." The "difficult task came after the war when diverse interests tended to divide the allies" — Bohlen recording his exact words in another toast the Marshal had made that evening. "He said he was confident that the present alliance would meet this test also and that it was our duty to see that it would, and that our relations in peacetime should be as strong as they had been in war."[3]

Time would tell. Admiral Leahy thought it had been an extraordinary day. "The dinner, starting at nine o'clock, lasted until 1:00 a.m.," he wrote in his diary, "with great quantities of food, 38 standing toasts, and mosquitoes under the tables that worked very successfully on my ankles." Glasses had to be physically clinked, Russian-style, with "the person toasted" — Leahy approached by Molotov, Stalin, and Churchill, no less.

Leahy was not entirely won over, however. "With the amount of important work we have each day," the admiral added, "such dinner celebrations are in my opinion an unwarranted waste of time. We did not succeed in returning to our quarters in Livadia Palace until after one o'clock a.m."[4]

It was at this point that Lieutenant Commander Bruenn was finally able to check how the President was faring, medically. He and Admiral McIntire were shocked. The President's color "looked very poor (gray)," Bruenn revealed twenty-five years later.

Alarmed, Bruenn asked if he might examine the President's lungs and his heart rate before Mr. Roosevelt went to bed. These appeared to be OK. However, "for the first time," Bruenn chronicled in his clinical notes, the President's blood pressure "showed *pulsus alternans.*" Strong, then weak, heartbeats.

It was the classic indication of the onset of serious left-sided heart failure.

82

The Prime Minister Goes Ballistic

IN HIS FAMOUS, deeply autobiographical working-class novel *Sons and Lovers,* D. H. Lawrence describes the last hours of his hero's mother, Mrs. Morel. Looking out the window Paul Morel sees the Nottinghamshire countryside "bleak and pallid under the snow," Lawrence had written. "Then he felt her pulse. There was a strong stroke and a weak one, like a sound and its echo. That was supposed to betoken the end," which came soon after—accelerated by the extra morphine Paul and his sister give their mother in filial compassion, without the doctor's knowledge, since "he says Mrs. Morel cannot last more than a few more days."

Lawrence was recording very much what he himself had witnessed. No such candor was shown by Admiral McIntire, the President's doctor who afterwards maintained that the President was fine at Yalta, merely tired. "As a result of malicious and persistent propaganda," McIntire would write the year after the Crimea Conference, "it has come to be accepted as a fact that the President was not himself at Yalta, either physically or mentally." Retiring as surgeon general of the U.S. Navy in 1946, the admiral declared "these charges" to be "every whit as false and baseless as the whisper about his breakdown in Teheran." The President had reached Yalta "in fine fettle," McIntire lied. And though the Livadia Palace sessions were "long and exhausting, and there were evenings when he confessed to being 'pretty well fagged,'" there was "never" any "loss of vigor and clarity."[1]

By 1946, of course, McIntire was trying to contest growing right-wing Republican claims that the President had been deathly sick and had more or less given away the capitalist store, as McIntire summarized their critique of the dead commander in chief—the President acting merely "as a rubber stamp for Marshal Stalin," and "weakly yielding to his demands at every point."[2]

In this respect, at least, McIntire was right to protest, for the President's performance at Yalta had been no rubber stamp, as McIntire was aware. In his last weeks of life, the President had in fact been able to achieve all his chosen objectives. The USSR being what it was — a police state, under the absolute rule of an often psychopathic dictator — the President had never seriously imagined he could make postwar *political* decisions acceptable to all, especially Polish patriots. What he could do, though, was at least rehearse the issues there, in the open, while the Big Three and their advisers were gathered together, in advance of the establishment of the United Nations organization and Security Council: a UN organization that neither the Soviet Union nor Britain favored, but which the President had been determined to establish — and which, however reluctantly, Stalin and Churchill now declared they were willing to support.

Shocked as his physicians were by the President's heartbeat variations at Yalta on the fourth night of the Yalta Conference, meantime, Bruenn administered more digitalis in the early hours of February 9, 1945, and a sleeping draft. To Anna the doctors were both adamant, however: no one, but no one, should henceforth be allowed to see the President before noonday. Also the President must have at least an hour's sleep or rest before each subsequent plenary began, or he would die in the Crimea, not in the aftermath.

The physicians, of course, could do nothing to lessen the rising tensions over the Polish issue — and on the afternoon of February 9, the oral ructions began all over again.

Fortunately the President had slept well — and thanks to the digitalis his *pulsus alternans* had seemed miraculously to vanish. His Joint Chiefs of Staff arrived to report they'd reached complete agreement with their Soviet counterparts on all matters relating to the war in Europe, and that they had begun detailed planning for the combined defeat of Japan. The President was delighted — and congratulated them as commander in chief.

This left just the Declaration on Liberated Europe, which the President had asked his secretary of state, Ed Stettinius, to draft — and which, he was told, seemed to have found favor with all three foreign ministers of the Big Three. As chairman at the 4:00 p.m. sixth plenary session, the President would only have to get Stalin and Churchill to approve the declaration draft, which they could do just as soon as they ended the previous day's unfinished discussion of Poland.

The sixth plenary began politely enough. On behalf of the three foreign ministers Secretary Stettinius gave his account to date on how they viewed the Polish conundrum — saying they'd decided it was best to defer discussion of "this question to a later date and to report that the three Foreign Ministers thus far had not reached an agreement on this matter."[3]

Had the foreign ministers had their way, the issue of Poland would therefore have been postponed, and dropped from the Yalta agreements. For Churchill, however, this was unacceptable. His motto, in heading the British government since 1940, had been "Action This Day"—and on February 9, 1945, he was determined to live up to it. "It was decided, at Mr. Churchill's request, that the Polish question would be discussed before Mr. Stettinius proceeded with the balance of his report," Charles Bohlen recorded in his minutes.[4]

"The Polish question." This was not, at Yalta, the matter of Poland's proposed new frontiers, but the business of "free" elections. The foreign ministers had agreed, between them, that the provisional government of Poland should be broadened "on a wider democratic basis with the inclusion of democratic leaders from Poland itself and from those living abroad"—i.e., London—"to be called the National Provisional Government." To this American proposal Mr. Molotov had formally assented. Yet Molotov wanted to "eliminate" his colleagues' proposal for a special commission of their nations' future three ambassadors to Warsaw—one that would be asked to "observe and report" to their three governments "on the carrying out of the free elections." And this because, as Molotov put it, it would "be offensive to the Poles and needlessly complicate discussions."[5]

What Molotov meant, of course, was that it would be offensive to Stalin and the Soviets—implying the Russians would otherwise manipulate the results.

The President could only groan—for although "free elections" were the very bedrock of democracy, the Soviet Union was not, by any stretch of the imagination, a democracy. Moreover it was unlikely ever to permit such a state to flower on its very doorstep. Once Churchill had the floor, though, it was too late; the plenary now became a sort of verbal food fight—one that quickly threatened to get out of hand and wreck the summit.

Churchill began by declaring he was glad "an advance had been made" by the foreign secretaries, and to hear Mr. Molotov's report regarding the "urgent, immediate and painful problem of Poland. He said he wished to make some general suggestions that he hoped would not affect" the President's game plan for the conference, but "here, in this general atmosphere of agreement, we should not put our feet in the stirrups and ride off. He said that he felt it would be a great mistake to take hurried decisions on these grave matters. He felt we must study the Polish proposals before giving any opinion."

To this the President could only agree—suggesting that, once Stettinius had finished his report, which would include the Declaration on Liberated Europe, they could adjourn for half an hour to discuss Molotov's proposed amendment. Stettinius thus resumed giving his foreign minister's report regarding reparations and, following this, the "machinery in the World Organization [i.e., UN] for dealing with territorial trusteeships and dependent areas."[6]

At the mention of "trusteeships," however, Churchill went ballistic. "The Prime Minister interrupted with great vigor to say that he did not agree with one single word of this report on trusteeships. He said that he had not been consulted nor had he heard of this subject up to now. He said that under no circumstances would he ever consent to forty or fifty nations thrusting interfering fingers into the life's existence of the British Empire. As long as he was [Prime] Minister, he would never yield one scrap of their heritage. He continued in this vein for some minutes," Bohlen minuted.[7] For his part, Admiral Leahy recorded that night how Churchill had "refused to consider permitting any agency to deal with any territory under the British flag, saying, 'While there is life in my body no transfer of British sovereignty will be permitted.'"[8]

Alger Hiss, on behalf of the U.S. State Department team, also noted Churchill's vehemence: "I will not have 1 scrap of the Brit Empire [lost], after all we have done in the war," the Prime Minister had declared. "I will not consent to a repres[entative] of Brit Em[pire] going to any conference where we will be placed in the dock & asked to defend ourselves. Never, Never, Never . . . Every scrap of terr.[itory] over which Brit flag flies is immune." Churchill's interpreter, Major Birse, wisely omitted the Prime Minister's outburst from his plenary minutes that night, but the damage, in a sense, had been done. Churchill had demonstrated, to Eden's chagrin, that the British were, in a sense, frauds: demanding observation rights in Poland, a country the Soviets had, at great cost, cleared of Nazis, yet denouncing in fury any idea that once liberated, British colonial or mandated possessions should be subjected to any kind of "observation." "I must be able to tell Parliament that elections will be free and fair," in Poland, Churchill had insisted—but not in the British Empire, it appeared.[9]

The Prime Minister's position was now tainted, in front of the entire plenary entourage—Stalin, with a trace of mockery, asking about elections to be held in Greece, where British troops were being used to guard against a communist insurgency. Churchill was compelled to assure him Russian observers would be welcome there, as well as in Italy. He even extolled Egypt as a model of free elections—causing the Russian dictator to say he'd heard "that the very greatest politicians spent their time buying each other" in Egypt, "but this could not be compared with Poland since there was a high degree of literacy in Poland. He inquired as to the literacy in Egypt, and neither the Prime Minister or Mr. Eden had this information at hand."[10]

This was not the direction the ailing president wanted for the summit. The election in Poland, he said—trying to pour oil on roiling waters—was "the crux of the whole matter, and since it was true, as Marshal Stalin had said, that the Poles were quarrelsome people not only at home but also abroad, he would like to have some assurance for the six million Poles in the United States that these elections would be freely held." If such an assurance were given "that elections would be held by the Poles," then "there would be no doubt as to the sincerity of the agreement reached here."[11] In other words, the conference was not being held to lay down specific rules or systems in any one country, only to define at least the principles of future democratic government, which rested upon elections. Agreement over this was not only important to quarrelsome Poles, he urged. Just as important, it would give hope to people in occupied or now liberated countries that the age of tyranny by military conquest was coming to an end. The President hoped therefore that Stalin would accept his proposed Declaration on Liberated Europe.

The Declaration was long — and directed as much to Germans as to those in occupied countries. "The establishment of order in Europe and the rebuilding of national economic life," it ran, "must be achieved by processes which will enable the liberated peoples to destroy the last vestiges of Nazism and Fascism and to create democratic institutions of their own choice. This is the promise of the Atlantic Charter — the right of all peoples to choose the government under which they will live — the restoration of sovereign rights and self-government to those people who have been forcibly deprived of them." And it ended with the summons to a new world order. "By this declaration," it stated on behalf of its signatories, "we reaffirm our faith in the principles of the Atlantic Charter, our pledge in the Declaration by United Nations, and our determination to build in cooperation with other peace-loving nations a world order under law, dedicated to peace, security, freedom, and general well-being of all mankind."[12]

To the President's relief Stalin assented, suggesting (though not insisting) only that they add a sentence about helping especially those nations that had actually "taken an active part in the struggle against the German invaders."[13]

The Prime Minister, however, was appalled. For his part he "said he did not dissent from the President's proposed Declaration as long as it was clearly understood that the reference to the Atlantic Charter did not apply to the British Empire."

Did not apply to the British Empire?

Since Churchill had drawn up the Atlantic Charter with the President in 1941, this sounded ominous. So much so that Churchill, embarrassed, felt obliged to explain that "on my return from Newfoundland" in 1941, he had "read to the H[ouse] of C[ommons] that we were pursuing these aims in Brit Em[pire]. That is part of our [British] interpretation"[14] of the Atlantic Charter. Moreover that he had "given Mr. [Wendell] Willkie a copy of his statement on the subject,"[15] when Willkie was in London on his world tour, back in 1942: to wit, that although the Atlantic Charter might be appropriate for countries to be liberated in Europe, it did not pertain to those countries Britain ruled — that "every scrap of terr.[itory] over which Brit flag flies is immune."[16]

Considering the way Churchill had telegraphed the British viceroy in India to ask "why Gandhi hadn't died yet" in 1944,[17] had only approved Gandhi's release from prison on medical grounds that the Mahatma might otherwise die in British custody,[18] and that Wendell Willkie had been deeply disappointed by Churchill's refusal to countenance Indian self-government, the President could only smile — "inquiring if that was what had killed Mr. Willkie"?[19]

Clearly, a conference in which the deeds and misdeeds of the Big Three were hung out like washing would be a recipe for failure. Sensing the President's attempt at levity, Stalin assured Churchill "he had complete confidence in British policy in Greece,"[20] and would not think of sending a Soviet observer to Athens. By the same token, he did not favor British monitors in Warsaw.

With that — and brief discussion of the treatment of war criminals — the contentious plenary came to a quieter close. The various advisers went their ways: the foreign ministers tasked with drawing up a suitable, face-saving pledge regarding Poland, while the "Big Three" retired to their respective palaces to rest, and have dinner on their own.

The President was once more exhausted, but still alive, at least. If he could manage one more day, he reckoned, they could issue a communiqué before his health gave out completely. He'd gotten all he'd really hoped for from the summit; the rest was only the final wording and signatures. Thus after the seventh plenary the next afternoon, February 10, and a private, confidential meeting with Stalin and Harriman regarding Russian entry into the war against Japan, followed by dinner at Churchill's quarters in the Vorontsov Palace in the evening, the way was open to agree and sign the crucial announcement to the world.

The final dinner held at the Vorontsov Palace was a relatively small and short affair, with just the three heads of government, their foreign secretaries, and their interpreters present. Stalin claimed to be disappointed he had not gotten Mr. Churchill to agree an actual figure for war reparations that he could take back to Moscow — that he "feared to have to go back to the Soviet Union and tell the Soviet people they were not going to get any reparations because the British were opposed to it."

Since Stalin had seldom referred to the Soviet people, the fear sounded somewhat unlikely. Churchill assured him the Russians would get "large quantities" of reparation, to be assessed by the War Reparations Committee they'd agreed to be set up — but, recalling his own bitter experience at Versailles after World War I, he cautioned Stalin that the sum should not be a "figure at more than the capacity of Germany to pay."[21]

With his vast experience of war and politics since the 1890s, Churchill was clearly a match for anything Stalin — or anyone else — could throw at the elderly but combative and astute Prime Minister. But Stalin did insist that a clause about reparations be inserted in the communiqué, in order that the Germans know what lay ahead, even if this made them fight the harder to the bitter end.

The mention of "the people," however, led to discussion of the Prime Minister's chances of success in the general election which would soon have to be held in Britain, once the Labor Party left the current National Government. Marshal Stalin — not an expert on free elections — assured Mr. Churchill he would be a shoo-in, after all he had done for his country in the war.

The President, for his part, said he also hoped so. But he warned "that in his opinion any leader of a people must take care of their primary needs. He said he remembered when he first became President" the United States "was close to revolution because the people lacked food, clothing and shelter," thanks to the Great Depression. He had campaigned on the platform that, "If you elect me President I will give you these things" — and his New Deal intervention, using the powers of the federal government, had worked. Since then, he stated with pride, "there was little problem in regard to social disorder in the United States."[22]

It was a simplistic reflection, but a salutary one. Churchill, sparring with Stalin, consuming glass after glass of wine and Georgian champagne, gave the President's warning no more than cursory note — something he would soon have cause to regret.

For his part, the President had no election to face, only Congress.

And his maker.

83

The Yalta Communiqué

THE YALTA SUMMIT was now over, save for signing the requisite documents. Before leaving the Vorontsov dinner on February 10, the President had proposed that the final plenary at the Livadia Palace begin earlier, at noon the next day, to be followed by a working lunch during which they could all put their signatures to the Yalta communiqué, as well as to the secret protocol on Soviet entry into the war against Japan. After that they would be free to go home.

Every participant in the conference would have his or her memory of the summit — and personal verdict. For his part Charles Bohlen felt it historic that the President had changed position in the seventh plenary on February 10 by supporting Churchill's recommendation that France should have a place on the Allied Control Commission in Germany. "I had translated at his [Stalin's] private meeting with the President on the first day of the conference when Stalin expressed an unfavorable view of de Gaulle and the French. We had been opposed to giving France a seat on the Control Council for Germany when the subject first came up at the conference. When Roosevelt switched to Churchill's position, there was little Stalin could do without engendering French hostility."[1] On the negative side, Poland's democratic future would probably not be secured beyond window-dressing — but democracy across western Europe would at least be bolstered by a stronger France than the President had originally envisaged.

Together with the Russian protocol on war against Japan, and the agreed setup for the creation of the United Nations organization (with the Soviet Union dropping its claim on initial seats to just one by the end of the conference, with two more to be discussed in San Francisco), Bohlen felt the President had done astonishingly well for a leader so direly ill. Even the decision not to inform Chiang Kai-shek with regard to a continued Russian presence in Mongolia, once Russia joined the war against Japan, was, in the context of Mr. Roosevelt's insistence on China being one of the five permanent members of the proposed UN Security Council, along with France, a wise one, given the notorious lack of security in Chungking, and the likelihood the Japanese would otherwise learn of the Soviet commitment to enter the war. Moreover the agreement reached on voting procedure in the UN Security Council — giving its members the right to veto action, but not the right to veto discussion of a major security issue or threat, even where its own country was under scrutiny — was a major American achievement at the conference. As was, in Bohlen's view, persuading Stalin to at least sign the Declaration on Liberated Europe, which gave dignity, objectives, and moral idealism to the Yalta summit — however much Churchill had railed against it, and Stalin first hesitated to commit the Soviet Union to such a formal document.

All in all, then, Bohlen was proud of the President's performance. "Our leader was ill at Yalta, the most important of the wartime conferences, but he was effective. I so believed at the time and still so believe," the interpreter-diplomat wrote almost thirty years later.[2]

Not all were so impressed, however — including Admiral Leahy, the President's chief of staff. The signing of the official communiqué at lunch on February 11, and in particular the secret, formal agreement with Stalin on Soviet entry into the war against Japan, were undoubtedly historic, Leahy granted. With regard to the war against Hitler and the Third Reich, he found himself "deeply impressed by the unanimous and amicable agreement of the President, the Prime Minister of Great Britain and Marshal Stalin of Russia on the action that shall be taken to destroy Germany as a military power," he wrote that night in his diary. But he admitted to being far more worried by the leaders' agreement to "dismember" postwar Germany than their questionable agreement regarding Polish elections. "These three men, who together control the most powerful military force ever assembled, sitting about a round table in the Crimea with their military and political staffs, have agreed to disarm and dismember Germany, to destroy its industry that is capable of manufacturing war material, to transfer territory from Germany to Poland that will necessitate the deportation of the survivors of between seven and ten million [German] inhabitants thereof, and to exact reparations in kind and in forced labor that will practically reduce the present highly industrialized Germany to the status of two or more agricultural states," he noted of what he took away from the conference's decision.[3] In the long term this, though a gross exaggeration of the Declaration, did not bode well, in Leahy's conservative view, despite the welcome cooperation between the triumvirate's militaries.

"While the German nation had in this barbarous war of conquest deserved all the punishment that can be administered," Leahy allowed, "the proposed peace," nevertheless, "seems to me a frightening 'sowing of dragon's teeth' that carries germs of an appalling war of revenge at some time in the distant future," under another Hitler-like tyrant. "I do not know of any other way to punish this nation of highly intelligent, highly reproductive, and basically military minded people for their war crimes, but the prospect of their reaction in desperation at some time in the more or less distant future is frightening," the admiral confessed. And while the prospect of a second, future Hitler worried him, the Soviets did also. "One result of enforcing the peace terms accepted at this conference will be to make Russia the dominant power in Europe, which in itself carries a certainty of future international disagreements and prospects of another war."[4] By actively encouraging Russia to become a responsible global superpower, the civilized nations of the West were taking a huge gamble.

Like Kennan, Leahy had grave concerns about the United Nations organization — the "United Nations Association to Preserve Peace," as he mockingly called the President's nascent body. Far from applauding the President's late-conference switch to supporting France on the forthcoming UN Security Council, Admiral Leahy deplored, moreover, the "fiction that France is a great nation." Though Stalin had agreed to the President's recommendation, it would lead, in Leahy's opinion, to the UN's inevitable "disintegration," since France would then have veto power in its Security Council. And France, under de Gaulle, could be counted upon to be irritating, if not impossible to deal with. It would simply destroy the effectiveness of the UN that Leahy had, in "great hope," initially relied upon.[5]

In short, Admiral Leahy was not sanguine — though Field Marshal Brooke, on the British bench, was surprisingly contented by the conference's results. "A satisfactory feeling that the conference is finished and has on the whole been as satisfactory as could be hoped for, and certainly a most friendly one," Brooke had already jotted in his diary in his slashing, emphatic green pen as chief of the Imperial General Staff on February 9, 1945, before leaving Yalta and the Crimea the next morning, together with all the military chiefs of staff.[6] Photographs and film, including color film, had duly been taken outside; their work was done.

And the President? His *pulsus alternans* had miraculously "subsided," Dr. Bruenn later recalled. His blood pressure was passable (for him), and his facial coloration had gotten somewhat better.[7] He didn't look well in the photographs, in fact looked ten years older than either Stalin or Churchill, but in some respects he looked the most seigneurial of the three, with a commanding presence — his hair having turned white, but his demeanor almost monumental, his eyebrows pronounced and his jaw firm as, with an arm resting on the side of his chair in the Livadia courtyard, he held a cigarette between his elegant fingers, and his other hand, placed upon his thigh, drew attention to his long, though long-paralyzed, legs.

He had gotten pretty much all he'd counted on. When, at the final dinner at the Vorontsov Palace, the Prime Minister had begged for more time — a further day for debate over Poland — the President had simply said no, he could stay no longer, since he had "three Kings" to meet in the Middle East on his way back home — Ibn Saud, king of Saudi Arabia; Haile Selassie, the emperor of Ethiopia; and King Farouk of Egypt.

This was true; but it was not the real reason. The fact was, he did not think he would get more from Stalin over Poland, especially after Stalin's concession regarding France, and Greece; nor did he think, in all honesty, he could summon the energy and concentration necessary to try.

In some ways it had been a marvel the President had survived nine long days of discussion and argument. Thanks to digitalis and his no-visitor regime, he had kept going somehow. "We certainly put the clamps on him by cutting down on his activities for the next 24 hours," Bruenn recalled of the crisis on the night of February 8 — and it had succeeded.[8]

Churchill's doctor later referred in his diary (much doctored before publication) to the President's "decrepitude" at Yalta — but Moran had not been present himself during a single one of the plenary meetings, or the dinners. Bohlen's minutes, by contrast, testified to the way the President had steered the plenaries toward his preferred ends, and the extent to which Stalin had deferred to him — the Soviet dictator determined to deny Hitler the chance of splitting the Allies, and therefore willing to sign the Declaration on Liberated Europe and accommodate the President's idea of a postwar United Nations system. For these concessions the Soviet Union would, the President assured him, continue to get American Lend-Lease assistance in rebuilding its country and economy.

"I'm in the last stretch of the conference," the President had written on the last day at Livadia to his loyal cousin Daisy, "& though the P.M. meetings are long and tiring I'm *really all right,*" he'd claimed, "& it has been I think a real success. I either work or sleep! I am in the palace of the Czar!" he'd added — tickled pink by the unlikeliness of such regal accommodations, in a communist country.[9]

Thus the Yalta Conference duly wrapped up — the President handing to Stalin the version of the joint communiqué which his staff had typed up, and the dictator approving and ticking each of its nine paragraphs.[10] With that done, the three world leaders bade each other goodbye. The President thanked Stalin once again for his Crimean hospitality, and gifts (vodka, caviar, etc.) were exchanged, as well as decorations for the military commanders. Then "Stalin, like some genie, just disappeared," Churchill's daughter Sarah noted in astonishment in a letter home.[11]

The President likewise disappeared — leaving the Livadia Palace by car at 4:00 p.m., bound for Sevastopol, where it had been arranged he would board the USS *Catoctin,* dine, and get a good night's rest before his onward flight aboard the Sacred Cow to Egypt, where the USS *Quincy* was moored in the

Great Bitter Lake. "Three hours after the last handshake," Churchill's daughter noted, "Yalta was deserted."[12]

"I am a bit exhausted but really all right," the President wrote home to Eleanor that night, before turning in.[13] It wasn't true. He was very far from "all right." With the help of Harry Hopkins — who was convinced the conference had been a triumph for the Western Allies — he now intended to sail home very slowly from the Great Bitter Lake. This would give him time to recover from the stress of the meetings, and to prepare with Hopkins, aboard the USS *Quincy,* the presentation he'd have to give to Congress on his return, in hope of securing backing for the April 25 inaugural meeting of the United Nations organization in San Francisco.

Hopkins was by then more ill than President Roosevelt. Unable to withstand the car journey to Sevastopol owing to his dysentery, he was driven to Simferopol, and from there he was conveyed by wooden Pullman train to Saki, more dead than alive, to meet the President at the airfield.[14] General Watson, the President's army aide, became desperately sick during the drive, and though there was grilled steak for all on the *Catoctin,* he suffered a heart attack that, along with a "serious prostate problem," rendered him comatose. He was only kept alive, in fact, Dr. Bruenn recalled, with oxygen.[15]

For Roosevelt things were not much better. "The President had a ghastly night and I think it affected his health," Harriman would later recall, owing to the fact that his cabin was overheated, although Dr. Bruenn recollected no specific medical crisis aboard the *Catoctin,* apart from that of General Watson. In any event it was an exhausted and debilitated American presidential team, despite its summit achievement, that assembled at Saki airfield at 10:30 the next morning, February 12, 1945, to board the Sacred Cow. The President appeared to most people to be completely spent. One army planner who'd seen him during the conference later noted how "gaunt" he'd looked — "his eyes sunken deep in his lined face; he looked very tired and ill, as though he were existing on pure iron determination to see the war to the end."[16]

It was an apt description. "He looked ghastly, sort of dead and dug up," another U.S. diplomat, Carmel Opie, would note.[17] Even the comforts of the USS *Quincy,* which the President boarded after being driven the seventeen miles from RAF Deversoir airfield, northeast of Cairo, on the afternoon of February 12, didn't help — his days now filled with the meetings he'd arranged with King Ibn Saud, Emperor Haile Selassie, King Farouk — and Winston Churchill, who insisted they meet once again, in Alexandria harbor, on February 15.

The President had asked General Eisenhower to join them, but with the final Allied offensive having started in inauspiciously wet weather in Holland on February 8, Ike had had to decline — not only a sign of the supreme commander's anxious focus on the imminent Allied crossing of the Rhine, but an indication that the war, in its final stages, would now be the province of the generals, no longer the Commander in Chief. In a cable drafted by Admiral Leahy, the President thus responded that he was "following your grand offensive with the greatest attention," and "shall always welcome a statement from you of what we can do to help and how you plan the future."[18]

With General Watson belowdecks on life support, and Harry Hopkins resembling death only slightly warmed up, the cruiser had the air of a hospital ship rather than a war vessel. Churchill brought Lord Cherwell with him, for he was anxious to talk about possible use of the atom bomb, as well as Britain's right to develop its own atomic program after the war. The President "made no objection of any kind," Churchill later recalled.[19]

The President had, in truth, now become simply too weak to do more than follow the gist of conversation much of the time. As Admiral Alan Kirk, commander of U.S. naval forces in the Mediterranean, who met the President when the *Quincy* put in at Algiers several days later, remarked to a colleague: it was "really a ship of death and everyone responsible in encouraging that man to go to Yalta has done a disservice to the United States and ought to be shot."[20]

Churchill later described their final meeting more elegiacally. "I felt that he had a slender contact with life. I was not to see him again. We bade affectionate farewells."[21]

84

The End of Hitler's Dreams

IN BERLIN, as the Yalta Conference had continued, there had been only gloom. Day after day the mood had become steadily more anxious — indeed the longer the summit had gone on, the more concerned Dr. Goebbels had become. The conference had not collapsed, as he and the Führer had assumed it would; on the contrary, its very duration spelled doom for the Third Reich.

The Führer at first refused to believe it. "Coalition wars never survive the coalitions with which they start," Hitler airily informed Goebbels; "it's possible that overnight the whole war picture could change completely, according to the political and military situation."[1] He was confident, he assured Goebbels, that with the Wehrmacht forces he was withdrawing from other conquests such as Norway, he would in a matter of "days" be in a position to launch counteroffensives against the Russians on the Eastern Front. On the Western Front, meanwhile, the miserable winter weather had inundated the ground, so that the Allies were effectively stalemated; the Wehrmacht merely had to stand firm there. Shorter lines to defend, he now claimed (in contrast to his military strategy for the past several years), meant the Wehrmacht held the advantage.

Goebbels attempted, as always, to be encouraging. Tentatively he suggested that Mr. Churchill might now be open to negotiation, given growing concerns in certain quarters in England over the future domination of Europe by the Soviets. Hitler, however, shook his head; it was too soon for this, he said. The military situation must first favor Germany, then would be the time to talk. One must be patient; the British, in frustration, might even resort to nerve gas, to hasten Germany's end, he speculated. In that case he was determined that the quarter-million British and American POWs currently in German captivity were, in reprisal, to be slaughtered, "en masse."

Gas?

The world war that the Führer had launched, along with the mass murder of Jewish, political, partisan, and other *Untermenschen,* had now come full circle: "gas warfare" an almost inevitable extension in Hitler's eyes. For him hostilities had now "reached a level of national German suffering," thanks to Allied bombing, the Führer claimed, "where one would be forced" to turn to means such as the execution of POWs — "the only thing left that can impress the British and the Americans."[2]

Much the same was being discussed in Tokyo, where biological warfare had already been tested in China; also, in the Philippines, where General MacArthur had felt he must liberate Allied prisoners in Manila by U.S. paratroop forces before Admiral Iwabuchi could use them as human shields, and perhaps slaughter them.

"The Fuhrer is still convinced that the enemy coalition will fall apart this year," Goebbels had nevertheless noted. "We just have to hold on, defend ourselves and stand." The Führer "believes unwaveringly in our coming victory, albeit without knowing exactly where and how it can be wrought," he recorded on February 11, 1945 — the very moment when the President was leaving the Livadia Palace with the nine signed Allied agreements, plus the still-secret Soviet undertaking to assist the United States in defeating Japan, as soon as the Third Reich surrendered.

Trying not to sound defeatist, Goebbels had suggested to the Führer it might be time at least for Germany to put out a "clear statement of its war aims" — *ein klares Kriegsziel vor Augen stellen* — if the Führer wanted to rally fascist Europe.[3] It was too late, however. The announcement of the Yalta communiqué on February 12, the next day, took the wind out of the propaganda minister's prospective sails — in fact the Allied document left Goebbels utterly deflated, as much by its length and detail as by its unity and firmness. The nine separate sections of the joint agreement covered not only how the Allies proposed to end the war, with specified zones of occupation and the treatment of a defeated Third Reich, but also how they would establish a new security system for the postwar. Having brought such chaos and suffering to the world, Germany would not be allowed to take part in this, at least in the form Hitler had intended; it was to be completely demilitarized as a menace to mankind.

One by one Goebbels listed the communiqué's themes pertaining to Germany, beginning with the Third Reich's unconditional surrender. Four zones of Allied military occupation: Soviet, American, British, and French; total German disarmament and the destruction of its warmaking potential; an Allied commission in control in Berlin; the dissolution of the German high command; the arrest of German war criminals and their punishment; the complete denazification of the country (the Nazi Party weeded out "root and branch"); massive reparations to be paid; "provisional" governments in the liberated countries to be followed by free elections and the wholesale shifting of Poland's new borders westward, taking a part of German East Prussia . . .

Hitler's dream of an inevitable breakdown in the Allied coalition had thus turned to dust. If Germany was defeated in the next months, as seemed likely, his great war to expand the nation's frontiers and establish control of all Europe would result not only in ruin, but in a smaller, defeated, and disarmed Germany. It was a prognosis Hitler could not bear to think about. He instructed Goebbels never to speak of reaching out to either the Russians or the Western Allies again. It could only demoralize German troops and civilians at home, he warned, in their hour of suffering.

The two princes of darkness continued to speak together for a while in the bunker. "He looks to me rather tired and sick," Goebbels admitted frankly in his diary. "He tells me he didn't sleep much last night. It's because he is working so hard. The last fortnight he's carried huge responsibilities that would have felled any ordinary mortal."[4]

Could a German counteroffensive possibly succeed in staving off the inevitable, as the Führer promised?

The Ardennes offensive had initially stunned the Allies — but the massive armored assault by three Wehrmacht armies had required *months* of secret preparation, had involved almost half a million troops, and had been launched in weather that precluded Allied air defense.

As Reich propaganda minister, nevertheless, Goebbels had no option but to maintain his faith in the leader for whom he'd sacrificed his first hopes as a novelist, and for whom he'd dedicated his whole career of unmitigated evil. In four or five days, the Führer had reassured him, German armored attacks would be ready to be launched on the Eastern Front. Meantime there was to be no letup on the Western Front. If the Wehrmacht could just succeed in striking a significant blow on the enemy "on one front or the other," Goebbels summarized, "then we'll be able to talk."[5]

There would be no talk, however. The Thousand-Year Reich, begun with such military fanfare, was almost finished. Yalta had betokened its end. And with it, Goebbels's and Hitler's lives.

PART ELEVEN

Warm Springs

85

King Odysseus

SIX WEEKS LATER, at 4:00 p.m. on March 29, 1945, the President boarded the *Ferdinand Magellan* for what would be his final journey. Destination: the Little White House, Warm Springs, Georgia.

Once again the President was exhausted — indeed by rights he ought to have put down his burden and resigned, now that he had gotten the endgame of the war agreed with his British and Russian allies. On March 7 enterprising U.S. troops had seized the Ludendorff railroad bridge across the Rhine River at Remagen, and had held it against repeated counterattack; then on March 23 three vast Allied armies under Field Marshal Montgomery — First Allied Airborne Army, Ninth U.S. Army, and Second British Army — launched a series of coordinated assault crossings north and south of Wesel, threatening the Ruhr, the industrial engine of the Third Reich. In the Pacific, the crucial island of Iwo Jima had been invaded and cleared of Japanese troops in hand-to-hand combat by March 16, providing fighter air cover for B-29 bombers over Japan. A second, major assault landing on Okinawa was scheduled to take place on April 1, aimed at putting ashore a quarter of a million men.

As U.S. commander in chief the President had done his work; like King Odysseus, he had returned from Yalta and deserved, surely, a peaceful life after his travails. His enemies were reeling and close to defeat — Manila, capital of the Philippines, liberated on March 3, Tokyo firebombed on March 9.

Like Odysseus, however, the President seemed driven to complete his victory by ensuring peace in the aftermath — telling Polly Delano and Daisy Suckley, his two companion-carers, on April 6 he aimed now to retire "by next year," as Daisy noted in her diary, "after he gets the peace organization started."[1]

Daisy was skeptical. Not even the President really believed this, she acknowledged. "I don't believe he thinks he will be *able* to carry on," she recorded, for it seemed a miracle he was still alive. Polly Delano did not believe it, either. On February 28, when Polly had first seen the President again, after his return from Yalta, "she didn't think," in all frankness, "he would live to go to the San Francisco conference," scheduled for April 25.[2] Nor had there been much improvement thereafter. To Daisy herself the President had looked "terribly badly — so tired that every word seems an effort," she'd noted while they were still in Hyde Park on March 25.[3] The slow train journey and his first days spent in Georgia without visitors, away from the stress of Washington, had seen him recover a little; in fact, what had seemed like a deathwatch gradually looked slightly less forbidding.

Not enough, however, to convince Daisy, his loyal and devoted friend. "On thinking further, one realizes that if he cannot, physically, carry on, he will *have* to resign. There is no possible sense in his killing himself by slow degrees, the while not filling his job," she confessed. "Far better," she reflected on April 6, to hand over the reins of office while he could, and avoid the specter of a stroke, like President Wilson, when he wouldn't even "be *able* to function."[4] He was, after all, still managing to sign congressional bills as they were sent from the White House. He could initial cables that Admiral Leahy, remaining at his post in the White House, drafted, on the President's behalf, to be sent from the Map Room — messages to Churchill, Stalin, MacArthur, Eisenhower, and others. Even, in fact, to dictate a few letters, at the Little White House, to his private secretary, Bill Hassett, or his office secretaries, Grace Tully and Dorothy Brady.

Essentially, though, it was a sham — as everyone close to the President knew. He had no military aides or political advisers with him now, only doctors — one of whom later admitted that the President "began to look bad. His color was poor, and he appeared to be very tired," with questions being asked about possible dementia, in the wake of small strokes. It was, however, a tribute to his long and kingly leadership that no one dared tell the President he simply must prepare to step down, and should immediately begin helping his vice president, former senator Harry Truman, to assume the mantle of command.

"He is slipping away from us and no earthly power can keep him here," Bill Hassett confessed to Dr. Bruenn on March 30 when they arrived in Warm Springs.[5] Bruenn had asked what made Hassett so sure, given the President's survival since his fatal diagnosis twelve months before.

Normally so discreet and proper, the stalwart private secretary said he understood Bruenn's Hippocratic oath to keep a patient alive to the bitter end. For himself, he could no longer maintain the fiction. He'd wondered already if it would happen the previous November, after the President's election victory; now, however, he was certain. "To all the staff, to the family, and with the Boss himself I have maintained the bluff; but I am convinced that there is no help for him."[6]

The two men — doctor and secretary — were close to tears. For a year, since the spring of 1944, Hassett had known the President was in trouble —

especially the previous July when Roosevelt hadn't acted "like a man who cared a damn about the election." Only Governor Dewey's campaign barbs had roused him — getting "his Dutch up," as Hassett explained to Bruenn. "That did the trick. That was the turning point to my mind" — the President determined to fight, especially in contesting Dewey's wild lies, such as his claim the President opposed demobilization. Yet "I could not but notice his increasing weariness as I handled his papers with him, particularly at Hyde Park, trip after trip. He was always willing to go through the day's routine, but there was less and less talk about all manner of things — fewer Hyde Park stories, politics, books, pictures. The old zest was going." In his opinion, Hassett told Bruenn on the evening of March 30, 1945, "the Boss" was now "beyond all human resources" — even those of Dr. Bruenn.[7]

The President wasn't quite willing to depart the world yet, Hassett now found. He seemed absolutely determined to attend the inaugural meeting of the United Nations organization in San Francisco on April 25, to which he would take the train, he said, spending a day in the Golden Gate City. The UN was, after all, his creation, which he had envisioned and nurtured since 1942; and, with the Soviet Union now a promised member, irrespective of whether the two extra Soviet republics were granted seats, the organization and its proposed Security Council were crucial to the President's concept of a new world order.

Roosevelt had another reason to keep on living, one which the bookworm Hassett — aware that his diary, kept since shortly after Pearl Harbor, might one day be worth publishing — hesitated to set down in ink: namely that the dying president, despite having lost "twenty-five pounds" in the past few weeks and looking now "worn, weary, exhausted,"[8] was surprisingly happy.

86

In the Well of Congress

THE PRESIDENT'S ENTOURAGE at Warm Springs were not the only ones on deathwatch — or dying. Harry Hopkins had felt so ill he'd refused to return to the United States by ship, insisting on flying instead to Marrakesh — forcing the President to cable for Judge Rosenman to fly out from Washington, if possible, and join him on the USS *Quincy* in Algiers on February 18.

The President had been more than disappointed; Eisenhower too busy to meet him,[1] de Gaulle refusing his invitation (incensed he had not been invited to Yalta), and General Pa Watson in a coma, belowdecks. "All in all it was a sorry ship," Rosenman described.[2]

Interviewing both Hopkins and Bohlen in Hopkins's cabin, the judge — who was charged with helping to draft the President's forthcoming speech to Congress — had done his best to piece together an idea of what had happened at the Yalta summit, using Bohlen's minutes and a special memorandum Bohlen had drawn up for him, too. Thinking he could draft the speech in twenty-four hours, go over it with the President, get it typed, and then disembark at Gibraltar and fly back to Washington directly, the President's loyal speechwriter "soon saw that was going to be impossible, and after the first few hours I gave up the idea and resigned myself to the long trip home."[3]

To Rosenman the situation was worse than sad. "I had never seen him look so tired. He had lost a great deal more weight; he was listless and apparently uninterested in conversation — he was all burnt out."[4] When the ailing Hopkins had come to the President's quarters to say farewell at Algiers, the President had merely muttered "Goodbye," and turned away, as if he did not know who Hopkins was. "All the buoyancy of the campaign, all the excitement of arranging and preparing for the conference, had disappeared," wrote Rosenman; "in their place was gray fatigue — sheer exhaustion." Yalta had been the "climactic project of his life — to arrange for a permanent peace." Yet despite spending the subsequent week with the President aboard the USS *Quincy* —

lunching and dining with Roosevelt every day—Rosenman had been unable to get the President to speak an intelligible word on Yalta—the President either reading, sleeping, or sitting on deck with Anna, his daughter, if weather permitted, "or just smoking and staring at the horizon."[5]

Two days out from Algiers, General Watson died of a cerebral hemorrhage. Even that event did not appear to greatly impact the President, who seemed either vacant or in depression. Asked later if the President had shown any grief, Dr. Bruenn said, frankly, "No. He showed what you would if a good friend of yours passed on. He felt very sorry and reminisced a bit about their past and so forth," but nothing more.[6] In fact, as Rosenman recalled, Pa Watson's death had merely "increased the President's reluctance to go to work."[7] He had simply left Rosenman, Leahy, and Anna to work together on the proposed speech, slated to be delivered the day after he reached the White House, on March 1.

In Rosenman's opinion the result had been a near-catastrophe in terms of presidential leadership. The occasion was to report to Congress and the world the most important achievement of the President's life to date: the creation of the UN and the inevitability of the defeat of Nazi Germany and Japan—his confidence about the latter resting upon the secret agreement with the Russians for their part in the endgame of the war. In the event, the triumphal moment had been reduced, in the well of the House of Representatives at midday on March 1, 1945, to a long, desultory, rambling, often incoherent talk from his wheelchair: the President openly admitting that he had felt too weak to stand upright in his heavy metal braces for the occasion—the first time he'd failed to stand when giving an address to Congress in the Capitol since he became the nation's chief executive, and a shock to many who had never known he was paralyzed from the waist down.

Worse still was the President's mental deterioration. "I was dismayed at the halting, ineffective manner of delivery," Rosenman later recalled. "He adlibbed a great deal—as frequently as I had ever heard him. Some of his extemporaneous remarks were wholly irrelevant, and some of them almost bordered on the ridiculous," the judge noted, sadly.[8] Many of those who'd listened to the President's address to Congress on radio, Dr. Bruenn admitted, noticed "that his speech was hesitant and that he occasionally appeared to be at a loss for words."[9] He claimed afterwards that he'd "spoken at intervals from memory and 'off the record' and then had slight difficulty in finding the proper place when returning to read the printed words of his address,"[10] but few were convinced. In a speech that was broadcast on national radio, "off the record" was a dubious description.

For his part the British ambassador, Lord Halifax, had found it "rather difficult to hear." What he did hear, however, the one-armed ambassador "did not think" was very good, as he noted in his diary. "There were no fireworks and no surprises in it"; the details had long since been printed in the national press, and there were "no philosophical reflections, such as add quality to a factual narrative," only inconsequential memories.[11]

Mercifully Halifax had learned from Senator Vandenberg that the senator was still enthused with the United Nations idea. The upcoming San Francisco Conference was now attracting much excitement in the press and public. The UN would be established, supported wholeheartedly by Congress — a tribute to the President's great statesmanship. Yet the clear signs of dementia were frightening — deeply distressing those who had witnessed him in his prime, a bare year before.

87

Appeasers Become Warmongers

HAD FRANKLIN ROOSEVELT, like General Pa Watson, died of a cerebral hemorrhage aboard the USS *Quincy*, his stature and reputation might not have suffered as they now did, darkened by his last weeks of life: the sight and sound of the once-lyrical but now wandering, incoherent president of the United States, seated in a wheelchair, finally admitting, on radio and film camera, to being disabled by polio and dependent on heavy metal stilts to stand (a fact hitherto known only to a tiny percentage of the electorate)—leaving people stunned and confused.

Was this the six-foot-three-inch giant of a man—a politician who had successfully stood for reelection a fourth time, as recently as three months ago—they had admired, even when disagreeing with his liberal political agenda? As rumors of still-secret Yalta protocols began to leak out in the days of early March, 1945, there had been no commanding president to shoot down critics—only a dying head of state whom his cabinet and colleagues looked to despairingly for leadership.

Churchill had addressed his own parliament on February 27, two days before the President's speech. The Prime Minister had spoken almost nonstop for two long hours, in sparklingly restored health—and parliamentary authority. He was confident the Yalta summit was a harbinger of peace, he'd declared, for the "impression I brought back from the Crimea, and from all my other contacts, is that Marshal Stalin and the Soviet leaders wish to live in honourable friendship and equality with the western democracies. I feel also that their word is their bond."[1] Using his unsurpassed rhetorical skills he had gotten the House of Commons to approve—or agree not to contest—the "declaration of joint policy agreed to by the three Great Powers at the Crimea Conference," as well as the "new world structure" of the United Nations.

Some of Churchill's own Conservative Party colleagues had hated both ideas—in fact they had forced a vote over an amendment which they'd tabled, regretting "the decision to transfer the territory of an ally [Poland]" to Russia.

The vote itself they duly lost, 396 to 25—but Churchill, not only a lifelong student of history, but a witness to and maker of it, had been conscious of the irony. As he noted afterwards, drinking with colleagues in the Smoking Room, "The warmongers of the Munich period have now become the appeasers"—such as himself—"while the appeasers have become the warmongers."[2]

It was an apt aperçu. All too soon the Prime Minister would come under increasing attack for his failure to "save" Poland—indeed for this, and his tone-deaf "failures" of government on the domestic front, he and his Conservative Party colleagues would be voted out of office in the British general election four months later.

Though Churchill would be deposed as prime minister, he would at least survive—and "repackaged himself," as the British historian David Reynolds later put it, "as a fierce Cold Warrior with his 'Iron Curtain' speech in March 1946, whereas Roosevelt, being dead, could not retrieve his reputation."[3]

Professor Reynolds was right, in retrospect. The President would not live to "repackage" himself. By March 1945 night had closed in on the President's career—his press secretary, Steve Early, doing everything possible to keep photographers away from the White House, and the press from reporting that the President of the United States was becoming, to all intents and purposes, a passenger aboard the USS *America*—with the War and Navy Departments pretty much taking responsibility for the war's finale.

Perhaps the best insights into the President's condition in his final weeks of life in Washington were noted in diaries that were only made public decades later—namely those of Daisy Suckley and William Lyon Mackenzie King, the prime minister of Canada.

The President had arrived at Hyde Park on Sunday, March 4, 1945, looking "very tired and sleepy," Daisy Suckley had noted. She herself was unsurprised, since he "has had, & is having, an exhausting time seeing people—'fixing' things which have gotten out of hand during his absence," according to the description he had given her of his nonstop stint at the White House since his return.[4] Daisy's only answer had been to bring back Harry Setaro, the quack doctor or masseur—though both of them knew only real rest could now keep his heart from simply stopping.

The three days at home in Hyde Park had certainly helped. The President returned to Washington on March 8 in good spirits. In the Oval Office he'd seen his vice president, Harry Truman, along with the Senate majority leader, Alben Barkley; Sam Rayburn, the Speaker of the House; and John McCormack, the second-ranking House Democrat: trusting they would finally get the contentious national service bill passed (ensuring enough infantry to continue the war), and Congress lined up to support his United Nations project. He'd also lunched with Admiral Bill Halsey, whom he'd decorated with the

Congressional Medal of Honor. The next day he had felt well enough to give a brief press conference, then had lunch with Admiral Nimitz in the mansion and in the late afternoon, in the Oval Office, discussed with the admiral the latest plans for the eventual seaborne invasion of Japan. And then greeted Mackenzie King, the Canadian prime minister, who had arrived from Ottawa and was going to stay the night.

Mackenzie King was by then already sitting with Eleanor in the mansion —and had been warned by her that he might be shocked by the change in the President's state of health. The President, she said, had been "pretty tired after his journey" to Yalta, and had lost a lot of weight and was looking "pretty thin." Also, she said, there was "the unpleasant side of politics; how ungrateful people were, terrible pressure, etc," on him. "She herself looks a bit older and more worn," King observed in his diary.[5]

Nothing, however, had quite prepared the gentle Canadian premier, who'd known Roosevelt for over forty years, for the man who had now appeared at 5:30 that evening, March 9, 1945.

88

Mackenzie King's Last Visit

WILLIAM LYON MACKENZIE KING had recently turned seventy; Franklin Delano Roosevelt was only sixty-three. As Eleanor's secretary, Malvina Thompson, had left the room, "the President came in in his chair," King described. "When I saw him, I felt deep compassion for him," he admitted, for he was shocked — the face of death upon his longtime liberal friend. "He looked much older; face very much thinner, particularly the lower part. Quite thin. When I went over and shook hands with him, I bent over and kissed him on the cheek. He turned it toward me for the purpose."[1]

It was an affecting moment, almost in the vein of Admiral Nelson and Captain Hardy.

Over the next several days King would talk with Mr. Roosevelt about the President's life; about the Crimea; about recent events; current problems; future hopes. King's nightly dictation to his secretary, Edouard Handy, remains probably the last intimate record of conversations with Franklin Roosevelt by a contemporary of the same political standing.

Mackenzie King had many times been with the President at moments of high decision. It was to King, after all, the President had confided, in December 1943, his decision to pursue the "unconditional surrender" of the Axis powers, a month before he flew to Casablanca — as well as, in confidence, his military strategy, in opposition to his own headstrong Joint Chiefs of Staff, of ensuring American troops and commanders learn how to defeat the Wehrmacht in combat in the Mediterranean *before* launching a cross-Channel invasion in 1944.[2] Also his battles to guard against the threat from Churchill's incessant plans to lose the war or cause it to end in stalemate in 1943, when Churchill had done everything possible to subvert, sabotage, and postpone D-day. King had witnessed in person how, with extraordinary patience and steely consistency, the President had nevertheless prevailed — allowing the United States to take the lead in prosecuting the war against Hitler, and the approach of victory.

For this reason it was all the more galling, in fact tragic, to witness his good friend's spiraling state of health.

The President "got from his chair on to the sofa and asked me to sit beside him," King recorded — with Eleanor on the other side. Roosevelt related the story of Yalta, and of his meeting with Ibn Saud — who had rebuffed all attempts by the President to use his influence to rehouse the million or more Jewish survivors of the Nazi mass-murder program. Rather than put pressure on the people of Palestine, Saud had responded to the President, the Allies should instead give to Jewish survivors land taken from the Germans responsible for the bloodshed in Europe — not from Arabs who had done nothing to deserve more Jewish immigration: millions of refugees whose ancestors had had no connection with Palestine for the past two thousand years. As Harry Hopkins later recalled, the President had been utterly naive to assume he could charm the "born soldier" Ibn Saud to leave Jidda, meet with him on the USS *Quincy,* and change the old warrior's mind. Inevitably, despite all blandishments, including the offer of a private plane and pilot, when the President had pressed the Saudi Arabian king, saying the number of proposed immigrants/survivors of Nazi death camps was "such a small percentage of the total population of the Arab world," he had gotten the bluntest of answers. Ibn Saud had merely stared at him and "without a smile, said, 'No.'"[3]

It was Ibn Saud's repeated no, and with zero indication he would ever change his mind — in fact would happily go to war with the Jews rather than permit another immigrant influx, especially when the United States itself had curtailed further Jewish immigration at thirty thousand — that had caused the President, in an aside during his subsequent address to Congress, to confess that, "on the problem of Arabia, I learned more about that problem — the Muslim problem, the Jewish problem — by talking with Ibn Saud for five minutes than I could have learned in the exchange of two or three dozen letters."[4]

"I asked him about Stalin," Mackenzie King also recorded that night. "He mentioned that Churchill had done about 90% of the talking at the Conference," but that "Stalin had quite a sense of humor. That once when Churchill was making a long speech, Stalin put up his hand to the side of his face, turned to the President and winked one of his eyes as much as to say: there he is talking again." The humor, however, had been a good sign, in the President's view — "Stalin's relations and Churchill's are much friendlier in every way than they were," Roosevelt had reflected. Stalin's humor, in a man known to be merciless in the way he ruled Russia, was beguiling, the President admitted, saying he "liked him. Found him very direct. Later told me he did not think there was anything to fear particularly from Stalin in the future," outside of the Soviet Union and its cordon sanitaire. "He had a big programme himself to deal with": rebuilding the country and its economy following the devastation wreaked by the German invasion and Hitler's *Untermenschen* program. "He also told me privately at night that Stalin would likely break off with the Japanese but wished to be sure to be able to have his divisions up to the front near Manchuria before taking that formal step," and that in the meantime he would authorize preparations to be made, in secret, to "give the Allies bases to operate from."[5]

In terms of his military strategy, then, the President was confident the war against Germany was almost won — and Japan's fate also sealed. "News of the American First Army having crossed the Rhine" at Remagen, released that day, had caused King to think it "the greatest day since the war began." The President clearly had no anxieties on that account. "I asked him about what he thought about the duration of the war. He said he has not ventured to make any statement," in public, "but he, himself, felt that before the end of April, as far as Europe was concerned, it should be over." Japan "would collapse very soon thereafter. Spoke of possibly 3 months."[6]

Such accurate predictions, almost to the week, showed that however ill he was, the President had been *au fait* with the military situation, despite the death of General Watson, his army aide, who'd been replaced by Colonel Park, Watson's assistant; moreover that he was completely realistic. Still, the gentle Canadian hadn't been fooled as to the President's deteriorating condition — not only physically, but mentally.

Upstairs, after dinner, Anna and Eleanor had left the two leaders to talk together in the President's study — Eleanor having begged King to stay for two or three days, since she was leaving the next day for Philadelphia with her companion, Lorena Hickok. The President, she'd admitted, would be alone — implying he would be without company thereafter. King, who had planned on going to Williamsburg to work on the major speech he was slated to give to the Canadian parliament, was torn.

"The President and I then went into the circular room. He asked me to sit beside him on the sofa. I suggested it might be easier for him if I sat opposite so I pulled up a chair. We talked steadily from 8.30 until 20 past 11 when I looked at the clock." Roosevelt, however, "said he was not tired; was enjoying the talk. We talked then until a quarter of an hour of midnight."[7]

The subjects had included S-1. "When I asked about certain weapons that might be used," King dictated coyly to his secretary before finally going to bed, the President "said he thought that would be in shape by August; that the difficulty was knowing just how to have the material used over the country itself" — Japan.[8]

Having been the father of the atom bomb, the President, in other words, was willing, if necessary, to use it.

• • •

Given that the atom bomb, outside of the Los Alamos complex, was a secret known only to a handful of people other than the President and Prime Minister Mackenzie King, and given that the secret had not been shared thus far with the Soviets, King had asked Roosevelt directly about his view on that matter.

In answer, the President "said he thought the Russians had been experimenting and knew something about what was being done."[9] Henry Stimson, the war secretary, thought the time had not come to share progress, though . . .

The President had thought this the wrong way to proceed. *Informing* the Russians privately about advances being made on the bomb would demonstrate coalition solidarity — and ensure the Soviets not underestimate American resolve to stand by the Yalta agreements on the zones of occupation of Germany. It would, in short, exemplify his fifth cousin's famous maxim: "Speak softly and carry a big stick." The President thus "thought the time had come to tell them how far developments had gone. Churchill was opposed to doing this" — for economic reasons. "Churchill is considering the possible commercial use later," the President explained[10] — on the grounds that the development could, if the Soviets were excluded, be kept as an Anglo-American monopoly.

King had found himself disappointed by Winston's cupidity. "I said it seemed to me that if the Russians discovered later that some things had been held back from them, it would be unfortunate" in terms of future international relations. Moreover, withholding the information might even threaten Russia's preparations to fight the major Japanese army and air forces in Manchuria.

Whatever might be decided on the information side, however, Roosevelt was pleased by the latest reports he had from Los Alamos, where "great progress was being made at present" in bringing the bomb to readiness, the President confided to King.[11] With the Russians threatening to engage the Japanese army in Manchuria, the Japanese government might eventually see the light and surrender without the need to drop it.

Which left the United Nations.

If the atom bomb was America's defining contribution to modern warfare — thwarting Nazi Germany (and the Japanese) in the race to develop such a weapon of mass destruction — the United Nations organization was to be the President's greatest potential contribution to postwar world peace, Mackenzie King noted.

"Speaking of the San Francisco Conference, he thought it might last a month. That it would be a mistake if it lasted longer. Work would be done by half a dozen main committees. Some things might be left over" for subsequent decision. "He himself would go to the conference to open it but would not stay. I asked if Churchill was likely to come over. He thought not" — the Prime Minister never having been much of a supporter of the UN idea, as opposed to "spheres of influence."[12]

Cordell Hull, who had worked so hard to promote the concept of the new organization, had been in hospital at Bethesda for almost five months, an invalid; it was unlikely he would be well enough to go to San Francisco. Edward Stettinius, Hull's successor, would therefore take charge — he should probably go to Ottawa now, the President had suggested, to discuss possible subsequent, post–San Francisco UN locations — perhaps the Azores to address European affairs, Canada for North American, Hawaii for Pacific matters.

"I pointed out that there might be considerable difficulty" in using multiple locations, King recorded. The permanent staff would be larger than the current number of officials involved in the Dumbarton Oaks talks and committees, and they would need accommodation, making diverse locales difficult, "particularly in relation to the Assembly." Whatever location was chosen, the President had remarked, "Geneva had not a good name" — seat of the ill-fated League of Nations. Perhaps naming one location subsequent to San Francisco would, for the moment, be worth considering, King had suggested.[13]

It was clear that the two politicians, who'd done so much to infuse the Allies with ideals beyond patriotism — ideals that men and women could fight for, not simply against — were extraordinarily close. Not even King had credited, though, how much the President trusted him as a personal friend — as the next few days would show.

The next morning King had once again spent time with the President at lunch in the mansion and then, privately, in his study — Roosevelt telling him more about his plans to visit not only England, but Holland and France, probably in June 1945. He intended to go "from the ship to Buckingham Palace and stay there, and then drive with the King through the streets of London and at the week-end, spend time with Churchill at Chequers. Also giving an address before the Houses of Parliament and get the freedom of the city of London." Then, he'd said, he would inspect American troops "on the battlefields," followed by "a visit to Queen Wilhelmina in Holland," where he would "stay at the Hague. From there, he would perhaps pay a visit to Paris" — though in view of General de Gaulle's prickly nature he "would not say more about that till the moment came."[14]

The Canadian prime minister was amazed by such ambitious plans for such a clearly sick, in fact dying, president. "It is clear to me from this that he and Churchill have worked out plans quite clearly contemplating that the war will be over before June," King noted — something that would have important political implications both for King himself and for Winston Churchill, in terms of new elections. Meantime the President's trip to Europe would be a "sort of triumphal close to the war itself."

"This in some ways is the most important information he has given me thus far," King dictated to his diary, with respect to his own career.

And with respect to the President's? How much longer could Franklin possibly go on? The President had spoken of a time "three years from now when I am through with here" — for he was, he said, thinking of starting a "newspaper which will be the size of four pages of foolscap." It would be "something like what is used for daily news on board ship. Have no editorials; just give the main news truthfully" — an aim he saw as crucial to the survival of democracy. "By means of radio photography, have it distributed in every city and town of America and sold for one cent."[15]

King had scarcely been able to believe his ears. Three *years* hence? To combat right-wing political agendas, and promote reliable information? Was he dreaming?

Charles Bohlen, now acting as liaison between the President and the State Department, observed that the President's health was becoming dire — Roosevelt's hands "shook so that he had difficulty holding a telegram. His weariness and general lassitude were apparent to all," yet even for Bohlen loss of the President was unthinkable. Thus "the thought did not occur to me that he was near death," Bohlen later recalled. "Although I was in the White House daily, I did not become sufficiently intimate with anyone there to talk to him about the President's health."[16] Admiral Leahy was too remote a figure; Hopkins was in hospital; the engine of U.S. government was working, fitfully but nevertheless working, in Bohlen's view — and that of many others.

In King's eyes, it was a conundrum. For all that the President seemed to have such a tenuous, precarious hold on life, he had, after all, traveled fourteen thousand miles to the Crimea and back in just the last month. This made King's journey by railcar from Ottawa, and need for quiet in which to prepare his report to Parliament in Ottawa, look pretty tame. "I confess he seemed to be in better shape physically than I had thought when I saw him yesterday," King allowed in his diary on March 10 — though he had to admit that much of what the President shared with him was given in broad brushstrokes, the President lacking the mental energy or even the capacity, now, to go into sharp detail, or complexity.

"I find he repeats himself a great deal," King had added. The President had, for example, shared with King his reflections on Jimmy Byrnes, his top civil administrator, currently guiding the national service bill through Congress. In the President's telling, the "renegade" Catholic possessed a brilliant mind, but had married a Presbyterian lady, which in the 1944 election would have lost one or two million votes on that score alone, had he won the Democratic Party vice presidential nomination. The next day the President had repeated the same account. Another repetition was the story he'd heard tell of Winston Churchill bathing naked at Miami, and "defying the waves for rolling him over" — determined nevertheless to go on and swim, before admitting defeat: a paradigm, he felt, of the Prime Minister's obstinacy. "Tonight he repeated the same story without apparently recollecting that he had told it last night," King noted — the President's wife, Eleanor, his daughter Anna, and even his son-in-law John Boettiger, too "embarrassed" to say "anything" to stop him. "There have been several of these occurrences. Indeed some of the stories he is telling he has told me on previous occasions. This of course is a sign of failing memory. I noticed in looking at his eyes very closely, that one eye has a clear, direct look — that is the left eye as he faces one. The right one is not quite on the square with the left one but has a little sort of stigmata appearance in the centre."

The President had seemed to be losing connection with reality, too, at times — claiming personal or sole credit for things in a way he'd never done before — such as the design of the Navy Hospital in Bethesda, and the United Nations.[17]

After dinner, to which Ambassador Robert Murphy was invited together with one of de Gaulle's young Free French officers, Jean Woirin, they had watched newsreel film of the President's trip to Yalta; his address to Congress; discovery of "German massacres in Poland," such as the concentration camps at Auschwitz and Chelmno; and "the landing of marines at Iwo Jima" — the pictures amazingly realistic, "being photographs actually taken during the battle. They surpassed anything I had seen," King recorded — the Japanese determined to kill ten Americans for every Japanese soldier who died, either in combat or by suicide, rather than live with the dishonor of defeat after failing their emperor's orders to conquer all Asia.

The incendiary bombing of Tokyo, the Japanese capital, the night before had shocked even the President, who "told me tonight that the destruction had been terrible, a big part of the city had been burnt out."[18]

King — by nature and religion a pacifist, and thus a reluctant wartime prime minister — had been unsurprised by the news, or its necessity. Earlier that day he'd been warned by the acting secretary of state, Joseph Grew, visiting the President, that "there will be very hard fighting still" to defeat the Japanese: the toughest enemy the United States had ever faced. Grew, who had been ambassador to Japan at the time of Pearl Harbor, had said, however, "he thought it was best not to destroy the Emperor at present, as he was the only person who could issue a rescript which would lead to a conclusion of the war," King noted — a recommendation the President had taken seriously. More U.S. Marines were being killed than Japanese defenders in the ongoing battle for Iwo Jima; "each island would be more difficult to take than the last," Grew had warned[19] — just as, in Europe, fanatical elements of the Wehrmacht seemed still willing to fight to the death rather than admit defeat in Holland, where they were threatening to breach the seawalls and dikes that would ruin the land for generations if the Allies dared come a step closer.

Would the war thus *really* be finished in Europe by the end of April, and in the Pacific by August? King had wondered. The President assured King it would be. He had faith in the supreme commander he'd appointed, and in Eisenhower's plans. Yet he himself had seemed almost disconnected from events—as if watching them unfold rather than directing them, now. And nowhere had this been more evident than in the brouhaha over Operation Sunrise.

89

Operation Sunrise

For some historians the military fracas that arose between the Allies in March 1945 presaged the Cold War. If so, there was little the President, as commander in chief, could do, in his condition, other than watch — and try not to despair. Secretary Stimson and General Marshall, together with Admiral King, were, to all intents and purposes, now running the war — from the culmination of the S-1 Manhattan Project to final efforts, on land, at sea, and in the air, to obtain the unconditional surrender of Germany and Japan by conventional means.

Inevitably, perhaps, the first intimations that the Germans were getting closer to surrendering had come from an SS general in benighted Italy — the country that had for so long been the source of strategic strife, misunderstanding, and contested Allied plans in conducting the war.

General Joe McNarney, commanding U.S. Army Forces in the Mediterranean, had sent a series of messages to the Pentagon regarding secret German offers to surrender, in defiance of Hitler's orders. On March 11, Secretary Stimson had been informed — General Marshall being away from the office, moving house. And when Stimson learned that Winston Churchill had deliberately become involved, in fact had informed the Russians of the offers, the "McNarney Affair" had turned into more than a military surrender offer in Italy. It became a deeply divisive, explosive issue: causing Stimson at 12:30 p.m. on March 11 to go see the President in person.

Unknown to Stimson, sadly, was the fact that the Nazis were playing the Western Allies, including Allen Dulles, the head of the OSS office in Switzerland, for fools — claiming the surrender offer was coming from Field Marshal Kesselring, the Wehrmacht commander of Army Group C in Italy (and Kesselring's replacement, General Vietinghoff, once Kesselring was ordered to take command of the Wehrmacht on the Western Front). In reality the "feeler" was coming from a particularly odious Waffen-SS general, Karl Wolff: a die-hard Nazi seeking (and succeeding) to save his own skin by "offering" to get his former boss, Heinrich Himmler, commander in chief of the SS, to parley with the Allies.[1] With Hitler's secret consent, Himmler was working to sow discord between the Allies, and get the Western nations to join with the Nazis in fighting the Soviets. His operation was called, appropriately, Operation Wool.

It had been agreed since Casablanca that, in pursuing the President's unconditional-surrender policy, commanders of Allied armies — whether Soviet, British, or American — would only negotiate with direct emissaries of an enemy seeking to surrender, under a flag of truce. Field Marshal Alexander, unfortunately, had informed Prime Minister Churchill that an offer had come from General Wolff, commanding Waffen-SS forces in Italy, *though not the Wehrmacht.* And Churchill, overruling his own chiefs of staff, had insisted not only on becoming involved, but on involving Stalin, instead of simply refusing to parley unless the offer was under a white flag from Kesselring or Vietinghoff, the Wehrmacht commanders.

Realizing the matter could easily "backfire," as Stimson noted in his diary, the U.S. war secretary had thereupon decided that, if Prime Minister Churchill was getting involved, so too should the President. "So at about half past twelve" on March 11, "I motored over to the White House and found the President in his study and told him what we were doing. He had heard of the matter through Admiral Leahy but had not heard of the action of the British with Churchill so he was very glad that I told him. He approved the course that we were taking over at the Pentagon so I felt easy in going ahead"[2] — without Stimson realizing it was a Nazi ruse.

Thus the minor tragedy had unfolded. Aware from Russian intelligence the Germans were trying to get the Western Allies to turn against the USSR, the Soviets had protested against such negotiations being conducted by the OSS in Switzerland. What should have been nipped in the bud — as Molotov demanded[3] — became an Allied mess, involving Himmler, Wolff, the OSS, British intelligence, the State Department, Russian intelligence, the British Foreign Office, the Russian Foreign Office, Ambassador Harriman, Ambassador Kerr, General Deane, half the Pentagon, Stalin, Churchill — and the ailing president. As they neared straightforward military victory, in the wake of the triumphantly successful Yalta Conference, the matter had then escalated into a first-class row between the Allies.

By March 13 the Wolff negotiations were spinning out of control.[4] Stettinius had told Stimson that, already at Yalta, Churchill had been "very erratic," as well as unpredictable.[5] Stimson, hearing what Churchill was up to over the Wolff affair, had felt this to be yet another example of "Churchill's erraticness" — moreover one that he himself would have "to settle in regard to what I have called the McNarney negotiations," on behalf of the President. As Stimson summarized in his diary, the "British and American [military] staffs both wanted those negotiations to be on a strictly military level without going into political affairs at all and not being therefore handled by politicians. It was to be purely the surrender of an army. We felt however that we must notify the Russians simply of the existence of the negotiations so that they would not feel they were being intentionally kept in ignorance. Somehow or other the English papers [documents] fell into the hands of Churchill and he overruled his staff and sent a note inviting the Russians to come and did this after our people sent off their letter which simply notified the Russians." Churchill's involvement had aroused immediate suspicion in the Russian camp — "a difficult place to wriggle out of," Stimson noted. With Ambassador Harriman — frustrated by Russian intransigence over Poland and backtracking over repatriation of U.S. POWs — firing off telegrams to the State Department and War Department from Moscow, advising his colleagues to stand firm and not allow Russian interference in military matters on a front (Italy) where Russian forces were not engaged, Stimson had taken advice from his War Department staff, and then "hurried over to the White House" with the cables, "and explained it to the President before the Joint Chiefs of Staff had yet acted." The President "backed me up and the Joint Chiefs of Staff took their position in the afternoon and passed the papers in the way which I had recommended."[6]

It proved a terrible mistake. By the time Prime Minister Mackenzie King had returned to the White House from Williamsburg at 3:30 p.m., to see or stay with the President on his way back to Canada, Stimson had inadvertently locked the U.S. government into a corner, struggling vainly to undo Churchill's unwitting mischief, and unaware the Western Allies were being fooled by the Nazis. As the President said to Mackenzie King, before asking him to sit with him for his weekly press conference, he was "afraid Winston had acted too suddenly and they [the British] had made the situation very difficult. He had cabled Churchill this morning, pointing out there was no objection to the Russian Generals coming as observors [*sic*], but that this should be between the British and the Americans and the Germans in Italy," not the Russians. "He says he is actually waiting further word from Churchill, but feels quite anxious about the situation lest this chance [i.e., army-to-army surrender] may have been lost. Alexander, he said, was agreeable to it, but it was a matter entirely for the military and should be kept as such. This is interesting indeed," King commented.[7]

It was. The negotiations — which Dulles called Operation Sunrise — were disastrous to Allied unity, just as Himmler and Wolff had hoped. They poured poison into the relationship between the Western Allies and the Soviets in the worst possible way — and over a nefarious SS general, who not only had no power to negotiate military surrender of a Wehrmacht army group in Italy, but who was allegedly responsible for the murder of three hundred thousand Jews and Russians. In truth Operation Wool — as recorded in SS Lieutenant

Guido Zimmer's diary—had been approved by Hitler himself in a meeting with Wolff in Berlin on February 4, 1945, the very day the Yalta Conference began. By mid-March its goal of splitting the Allies had proven extraordinarily successful, with Harriman in Moscow completely taken in, even as Russian intelligence learned it was a scam.[8]

The affair was enough to break the President's heart, after all he'd done to improve trust and relations between the Big Three, and to prepare for a clear, decisive, and *united* Allied victory over the Third Reich. Time and again over the following three weeks poor Admiral Leahy was to be tasked with having to draft explanatory, apologetic, and sadly untruthful cables to Stalin, who became incensed at what he saw as Anglo-American double-dealing with the Nazis.

For the President, too ill to find out himself what was the real truth behind the "McNarney negotiations," the subject became a nightmare, as he tried, amid the flurry of accusatory cables from Stalin—cables that vainly warned the Western Allies had been duped by Wolff—to try and rescue their plummeting relationship: hoping against hope it would not affect Stalin's commitment to declare war on Japan, and that he would continue with preparations to send major Russian forces to defeat the Japanese army in Manchuria. Moreover, that it would not compromise Soviet commitment to join the United Nations organization—especially when, in pique, Stalin announced that Mr. Molotov, furious that the Western Allies would not break off negotiations with the nefarious Wolff, would not be going to San Francisco, as he'd initially planned.

It was in the midst of this depressing military finale to the war against Hitler that the President had asked Mackenzie King to dine with him on March 13—and meet the light of his life, who was probably doing more to keep him alive than his doctors: Mrs. Rutherfurd.

90

No More Barbarossas!

THE PRESIDENT HAD YEARNED to get away to Warm Springs, where he'd be able, he thought, to work in peace and quiet on his United Nations opening address in San Francisco. In the meantime he had left Henry Stimson, James Forrestal, Admiral Leahy, and the Joint Chiefs of Staff — as well as Eisenhower, Nimitz, and MacArthur on the battlefields — to run the military side of the war. And had turned to Lucy Rutherfurd for emotional support.

In later years there would be inevitable speculation about the President's relationship with Eleanor in the final period of his life, fueled by assertions that Mrs. Roosevelt tormented her husband with her demands, her lack of patience — and her failure to understand he was at death's door. Moreover that it was Eleanor's remoteness that caused Franklin to turn to Lucy.

This was not the case. By March 1945 Eleanor knew Franklin's days were numbered — just not how much time he had left. She had ceded to her daughter Anna the responsibility for day-to-day care of her husband, and the need to keep down the number of visitors to the White House. It was true, however, that after the election a long phone call she'd made to him while he was at Warm Springs had caused his blood pressure, according to Dr. Bruenn, to rise fifty points — the "veins in his forehead standing out."[1]

Having earlier tested Mrs. Roosevelt for a possible medical condition, Bruenn had discovered Eleanor's metabolism was low. "Just imagine what she would have been like if she had been up to par!" the doctor later remarked, mockingly.[2] Yet Franklin had spent a lifetime with Eleanor; he knew, accepted, and respected her for the woman she was: mother of their five children, as well as a devotee of humanitarian and social causes — not least those of civil rights, refugees, women's inequality, and poverty. "It is very hard to live with someone who is almost a saint," Roosevelt's labor adviser, Anna Rosenberg, later commented[3] — yet Franklin had managed to do so for almost a quarter of a century since they made their pact to stay married. And though both were under great strain in March 1945, there was every indication they were as devoted to each other as they had ever been, indeed possibly more so: writing each other and telephoning whenever apart, sharing news and concern about their four sons in uniform — three in the Pacific — and making preparations for the end of the war and Franklin's resignation on grounds of ill health, whichever came first. "I say a prayer daily that he may be able to carry on till we are at peace & we are set in the right direction," Eleanor wrote her aunt Maude Gray.[4] For it was clear, once Hitler's Germany was defeated — to be followed by the swift surrender of Japan — the President would not continue in the Oval Office. "I am all ready to hand over to others now in all that I do and go home to live in retirement," Eleanor would tell her friend Margaret Fayerweather on March 28 — adding how Franklin (who had for the first time in their marriage asked her to drive him in his beloved Ford Phaeton car, with its special hand controls) had spoken to her of their moving to the Middle East, after his run-in with Ibn Saud; "I believe I'd like to go and live there," he'd said a few days before. Now that he'd met with representatives of both sides in the dispute over a larger Jewish home in Palestine, "I feel quite an expert," he'd said to her. Eleanor had been appalled. "Can't you think of something harder to do?" she'd asked him. At which point he'd suggested India, China, Thailand, and Indochina.[5]

Eleanor had admired his spunk. "*I'm* all ready to sit back. *He's* still looking forward to more work," she'd reported to Margaret.[6]

Franklin was joking, of course — as she knew. The very business of staying alive was about as much as he could manage. Why, then, in the circumstances, did he not summon Harry Truman, his chosen vice president, to come and discuss, in private, the challenges the former senator would soon enough have to face? This was something no biographer or historian would ever be able to comprehend. Exhaustion? A psychological block in taking such a step? Denial? Delusion that he still had enough time to do so?

The President had told Truman on March 1, the day he'd addressed Congress, that he was tired and intended to go down to Warm Springs before addressing the United Nations conference. Yet in the subsequent four weeks before he left the capital he met with Truman only once, for ninety minutes: and that was in the company again of Speaker Rayburn, Rayburn's deputy McCormack, and the Senate majority leader, Barkley, to discuss the domestic political agenda. Moreover Truman had not complained. Highly intelligent, a quick study, and a *bon viveur* when it came to whiskey and cards, Truman had not thought to request a private meeting, contenting himself with the weekly cabinet meeting at the White House.

Thus the President — having decided to leave Washington for Warm Springs on March 28 — had sought to keep going. And to see Lucy as much as he decently could, if she was agreeable. Which she was.

In other circumstances this would qualify as one of the great love stories of the century: of lost love that had been refound, and remained as powerful and romantic for both parties as it had ever been, even if physical love was no longer possible — or perhaps even wanted by the parties. What the President wanted, it seemed, was just to be near Lucy. Whether or not Eleanor facilitated this consciously or unconsciously would never be known; it seems hard to believe that with the President's personal staff, secretaries, White House staff, Warm Springs staff, Hyde Park staff, Secret Service officers, doctors, drivers, and others all "in the know," Eleanor — whose own longtime lover, Lorena Hickok, still lived in the White House, on the top floor — could have remained ignorant of the resumed relationship.

There was certainly very little attempt to hide it. Almost every day the President had arranged to see Lucy. He'd collected her for a drive in his big seven-seat Lincoln V-12 "Sunshine Special" armored convertible on March 12 — avoiding the need to be lifted out of the car and into Lucy's sister Violetta's house at 2238 Q Street, Georgetown — and had then brought Lucy back to the White House to have dinner with him and his daughter, as well as Anna's husband, John Boettiger. Anna and John had then left them together to talk for several hours in the President's study after dinner, when the car duly took her home. The following evening he'd asked her back to dine with him and his closest political friend, Mackenzie King.

"I noticed in talking with the President that he seemed very tired," King had noted in his diary, after the press conference he'd just given with King in the Oval Office, attended by many reporters. But "he seemed to enjoy talking on with me — seemed in no hurry about sending for his letters to be signed or other things. I spoke a couple of times about letting him rest, or not waiting for dinner" — since King was to spend the night back on his train, en route to New York — "but he said to me he would like me to stay. He said for dinner he was having just his daughter and her husband and another relative, Mrs. Rutherfurd. It would be just a quiet family meal which we would have early if I wished, so as to get off to the train early. I felt the President has pretty well lost his spring," King recorded, sadly. "He is a very tired man. He is kindly, gentle and humorous, but I can see is hardly in shape to cope with problems. He wisely lets himself be guided by others and has everything brought carefully before him."[7]

This was probably the most succinct and accurate description of the President of the United States in mid-March, 1945, four weeks before his death. Returning to his room, King had dictated the entry, taken a twenty-minute nap, and then rejoined the President, Anna, and John. And also "Mrs. Rutherfurd of Carolina a relative of the President, a very lovely woman and of great charm."[8]

King — who had dined with Eleanor only two nights before — was evidently dazzled by Lucy, as many men were. A lifelong bachelor who knew and accepted his own limitations in terms of personality, King was always curious about people — their self-importance, reliability, agendas. Mrs. Rutherfurd stunned him, not having heard about her before. "I should think she has an exceptionally fine character," he reflected later that night, aboard his special train — so taken with her, in fact, that "I made an exception in my rule and took a cocktail which the President himself mixed before going into dinner. We dined upstairs in the little hall and the five of us had a very happy talk together."[9]

It was the last time King would see the President. Roosevelt had said they would "soon meet again at San Francisco." By then, King had remarked, the war in Europe would be hopefully over, to which the President had said "there is a very good chance" it might be. And King, who had not quite dared believe the long, dark years of hostilities would soon be done, had been relieved by that at least. "I do not know," King confessed to his stenographer, "that I would have felt sure enough to say so with such assurance until today."[10]

The President might be wrong about meeting King in San Francisco, but he was certainly correct about the war's approaching end. With the Ruhr swiftly surrounded by American forces, and the Ninth U.S. Army advancing to the Elbe River, within striking distance of Berlin, the military situation once again tilted back in favor of the Western armies rather than the Eastern allies. Taking his cue from the Yalta agreements, including the agreed zones of Allied occupation, as well as the predicted casualties that would be suffered in attacking Berlin from the west, Eisenhower decided unilaterally not to authorize an assault on the capital, but to leave it to the Russians to take, house by house — with Hitler refusing to leave the city and remaining in the bunker below the Chancellery, now that the Chancellery itself had been more or less erased by Allied bombing.

Churchill had thought Eisenhower's decision not to attack Berlin from the west a terrible mistake — indeed he found himself aghast at what he saw as a naive deference to the Yalta map. All too soon, in fact, he would order plans to be drawn up for an Anglo-American Barbarossa: an attack from the Dresden area "so as to impose upon Russia the will of the United States and the British Empire" — in part to attain "a square deal" for Poland, though this aim would "not necessarily limit the commitment." It would involve almost fifty — mostly American — divisions, and up to a hundred thousand Wehrmacht troops! It was to be launched, moreover, four days before the British general election: Operation Unthinkable.

The Prime Minister's secret plan, mercifully, was not put into effect, and would never have been authorized by an American commander in chief. It was duly dropped when Churchill lost the election in 1945, though only shelved. Concealed for the next forty years, it was considered by later historians to be appropriately titled, save in defense against a Soviet invasion of western Europe.

The very preparation of plans for Operation Unthinkable would, however, indicate how views on the future of Europe were shifting at senior levels in the spring of 1945 — men like Ambassador Harriman, General Deane, and others indignant that too much was being "given away" to the Russians simply because they had furnished — and were continuing to provide — the essential warrior-manpower for the war. When Eisenhower refused pleas by Field Marshal Montgomery to be allowed to press on with American forces under his command to Berlin — the supreme commander removing the Ninth U.S. Army from Montgomery's Twenty-First Army Group, lest the often insubordinate general turn a Nelson-like telescope to his blind eye and ignore the signal — there had been an outcry at 10 Downing Street, with Churchill sending more exhorting cables to the President to protest the loss of such an opportunity.[11]

The President ignored Churchill. He had ceded his role as U.S. commander in chief to Stimson and Marshall, and for their part they were tired of Churchill's repeated warnings of missed opportunities or even doom if his proposals were not accepted. Mr. Churchill was not president of the United States. Without American forces — and inevitable casualties — his British Empire forces were now junior allies in Europe and even more so in the Pacific; it was a case of the tail perpetually trying to wag the dog — moreover a tail that had, in all truth, a deeply suspect record in terms of strategic or tactical military success. Stimson and Marshall had therefore had no hesitation in backing General Eisenhower, however much the Prime Minister might complain. Besides, Eisenhower had on March 28 already sent his signal to Stalin, explaining his decision.[12] Without starting another McNarney-like arousal of Soviet suspicion that the Western powers were about to switch tack, and combine with German forces to strike at the Soviets, it was too late to revoke.

In subsequent years General Marshall, especially, would be accused of political naiveté by the smear-artist senator Joseph McCarthy. Yet the fact was, there had been remarkable unanimity of American thinking among U.S. military minds at this time — Roosevelt approving every decision about which, via Admiral Leahy, they informed him.

Germany first. Then Japan.

For better or for worse the President had come to see his role as that of underwriter: endorsing the generals who were bringing the war to a successful military end against ongoing, fanatical German and Japanese opposition. And declining to plot a new war, which few if any serving American soldiers at that time would have been prepared to fight. There would be problems enough in getting western Europe back on its feet, once the Nazis surrendered unconditionally. With German officers already plotting to raise a Freikorps if and when defeat came, the President was determined to ensure they would never be able to claim later that German forces had been winning the war but were "stabbed in the back" by their own politicians. Moreover, peaceful postwar coexistence with the Soviets was worth *trying* to achieve, at least—and as history would have it, would prove more or less effective, despite a number of "hot" incidents.

In short: no more Barbarossas!

Thus, as planned, the President of the United States had briefly appeared in the Oval Office on the morning of March 28, 1945, and left the White House that same afternoon for Warm Springs.

At breakfast with Eleanor—the last he would ever take with her—he told her friend Margaret Fayerweather he'd now chosen the spot in the Hyde Park rose garden where he wished to be interred—a spot where "to his certain knowledge, have been buried an old mule, two horses, and a dozen or so of the family dogs," as he joked.[13]

The President clearly still had a vestige of his old sense of humor, but he was "far from well," the British ambassador noted that day[14]—markedly sicker than three weeks before, when Halifax had last seen him. By 3:00 p.m. even his longtime office secretary, Grace Tully, felt the President had "failed dangerously" just in the hours since lunchtime. "His face was ashen, highlighted by the darkening shadows under his eyes, and with his cheeks drawn gauntly," Tully later recalled.[15] His new press secretary, Jonathan Daniels, agreed—as did others.

Alben Barkley, the Senate majority leader, who saw the President shortly after midday in the Oval Office, said to a colleague: "I'm afraid he'll never return alive."[16]

However exhausted he felt, the President was intent upon hanging on to life for a few more days. He'd seen Lucy Rutherfurd every day from March 18 to March 21. As one biographer put it, "A romance which endures for thirty years is not an affair"[17]—and perhaps at some deeper level, the President was simply not willing to go into the dark without seeing Lucy one more time in Warm Springs. They'd discussed Lucy driving down to stay with him, perhaps bringing her friend Elizabeth Shoumatoff, who'd painted a small portrait of FDR in 1943. Once back at Aiken, Lucy asked Elizabeth to paint another portrait of the President; the artist was at first dubious, the photographs of the President at Yalta being, in Elizabeth's word, "ghastly."[18]

Lucy hadn't denied this to Elizabeth. "He is thin and frail," she'd acknowledged, "but there is something about his face that shows more the way he looked when he was young," she'd said. Clearly reminiscing, she'd added: "Having lost so much weight, his features, always handsome, are more definitely chiseled, I think." And then, "in an even lower tone, 'if this portrait is painted, it should not be postponed.'"[19]

Thus was the arrangement made. Lucy would come with Mme Shoumatoff and stay with the President at Warm Springs, after Easter — in approximately a week's time.

Having seen Franklin so often in the previous two weeks, Lucy herself cannot have been under any illusions; she had, after all, watched her husband, Winty Rutherfurd, die the year before, after a long illness.

And the new, life-size portrait, in watercolor? It might well be a last such portrait, Lucy was aware — Douglas Chandler having finished a fine one at the White House on March 16, 1945. Not a death mask, per se, but a farewell-to-life portrayal — painted by a Russian Orthodox Christian whose brother, an expert, had met with the President at Hyde Park and become an unofficial adviser to him on icons of the saints. Beginning in the Late Middle Ages it had been traditional in Europe to take death masks in wax or plaster; in the United States, however, there had been no such tradition. Elizabeth's portrait, then, would have to suffice — accompanied by photographs, which had come to perform a similar role. And since it was Elizabeth Shoumatoff's practice to have a photographer aid her in achieving accuracy in her compositions, she would bring one with her — to which the President had, in principle, agreed.

Whether he would last until they came, however, was another matter.

91

The End

PULLED BY TWO LOCOMOTIVES, the nine-car *Ferdinand Magellan* arrived at the Warm Springs halt at 1:30 p.m. on March 30, 1945, bearing the President of the United States; a handful of his office and personal staff; his communications team; his law partner, Basil O'Connor; and the Canadian ambassador, Leighton McCarthy—but no senior military staff, or even his White House doctor, Admiral McIntire.

In her diary Daisy Suckley—who with Polly Delano was also traveling with the President to provide him with care and company—had noted how both O'Connor and McCarthy, after boarding the train with the President, were "alarmed at his looks." These did not improve.

He "looks really ill—thin & worn—but joking & laughing & carrying the conversation on as usual," Daisy recorded.[1] He even asked the Catholics in his entourage whether they wished to disembark and celebrate Easter Mass in Atlanta, then follow on later. If so they should feel free to do so, he insisted.[2] By the time he himself was transferred to the waiting car at Warm Springs, however, the President looked almost comatose—unable to lift even an arm to help Mike Reilly carry him from his wheelchair. "Just like setting up a dead man," a railwayman commented. Once he'd been wheeled off to bed, his valet, Arthur Prettyman, overheard Leighton McCarthy saying to Basil O'Connor, "Our friend is dying."[3]

It was no wonder. The President's blood pressure was 240 over 130, though dropping at times to 170 over 88. Calling Admiral McIntire in Washington to tell him how worried he was, Bruenn was advised to continue as normal, and not under any circumstances to call on outside medical assistance or advice, which might make the three accompanying press agency reporters suspect it was a final, fatal journey.

Buoyed perhaps by knowing Lucy would be coming to stay on April 9, the President thus clung to life a little longer. In Washington, Stimson, Marshall, and Leahy dealt with the increasingly admonitory (and all too justified) cables from Stalin, culminating in that of April 3, in which Stalin gently commented the President had "not been fully informed" about the true negotiations by General Wolff in Bern, and claiming, rightly, that "my colleagues are close to the truth." If such negotiations were designed to ease the path of the Western Allies into the "heart of Germany," then "why was it necessary to conceal this from the Russians?" he asked. It might give a "momentary advantage" to the Western Allies, but not one of which they could be proud, if they were concerned about "the preservation and strengthening of trust among the Allies."[4]

Leahy was understandably alarmed. "The president today received a disturbing telegram from Marshal Stalin which stated that the Soviet Army had information that the Anglo-American Command had entered into an agreement with the German Command, which arranged for the Allied break-through of the western front in exchange for softer surrender terms than would be accepted by the Soviet Government," he noted in his diary. "This message clearly shows Soviet suspicion and distrust of our motives, and of our promises, a sad prospect of any successful cooperative agreement at the approaching Political conference in San Francisco."[5]

It was too bad—and would probably never have arisen had the President been well, for he had resolutely refused to countenance negotiations other than unconditional surrender, under army-to-army aegis. But he was now not even well enough to draft a reply, Leahy recording with complete frankness in his diary how "I prepared for the President, and sent to Marshal Stalin, a sharp reply to his message that approaches as closely to a rebuke as is permitted in diplomatic exchanges between states."[6]

Leahy was not exaggerating. Sent back to his chief of staff via the radio railway car and then in original by daily pouch, the reply was simply marked APPROVED—the President's message denying the Soviet assertion, saying he had "complete confidence in General Eisenhower and know he would certainly inform me before entering into any agreement with the Germans"; moreover that, as supreme commander on the Western Front, Eisenhower was as determined as ever "to bring about together with you an unconditional surrender of the Nazis." Leahy had then added, "It would be one of the great tragedies of history if at the very moment of victory now within our grasp, such distrust, such lack of faith should prejudice the entire undertaking after the colossal losses of life, materiel and treasure involved. Frankly I cannot avoid a feeling of bitter resentment toward your informers, whoever they are, for such vile misrepresentation of my actions or those of my trusted subordinates."[7]

Crafted in Washington by Leahy, with the support of his colleagues, the message did the trick—causing Stalin, who remained certain the President was being either deliberately or inadvertently misinformed by his subordinates, to back down. The Marshal cabled back that he had "never doubted your honesty and dependability, as well as the honesty and dependability of Mr. Churchill" — though he quietly insisted his informants were telling the truth, and "have no intention of insulting anyone."[8] Accepting the President's assurance there would be no eleventh-hour deal with Hitler or his henchmen such as Himmler and Wolff, he made clear he would stand by the Yalta agreement he'd made with Mr. Roosevelt. And he did.

The Japanese ambassador was summoned by Foreign Minister Molotov to the Kremlin. There Mr. Sato was told that, in view of Japanese hostilities "against the United States and Britain, which are allies of the Soviet Union," the Soviet-Japanese nonaggression pact was finished — the prelude to war.[9] As Leahy noted, this could mean an attack by Japan on the Soviet Union, but "the Soviet Government is well informed in regard to Japanese history in the past century and should be adequately prepared. Hostilities between Japan and the Soviets would be definitely advantageous to our present war effort in the Pacific," he concluded gratefully.[10] His tough language had been effective; Stalin was holding to the secret Yalta agreement.

The President, informed of this in Warm Springs, breathed a sigh of relief. Dulles's misbegotten Operation Sunrise had threatened, but mercifully had not wrecked, the military finale to the war against Hitler's Third Reich — nor the defeat, thereafter, of the Empire of Japan. This, together with the prospect of seeing Lucy again, had heartened him. He'd already stopped Churchill, on March 11, from sending "any message to Uncle Joe at this juncture — especially as I feel that certain parts of your proposed text might produce a reaction quite contrary to your intent"[11] — and Stalin had now borne out the President's trust in the Russian commitment regarding Japan.

Reluctantly, Churchill cabled on April 5 to say that he finally accepted the President's decision not to order Eisenhower to send American forces to take Berlin from the west — particularly in view of the suspicions the Wolff imbroglio had raised of an Anglo-American plot — and regarded "the matter as closed." Moreover to "prove my sincerity, I will use one of my very few Latin quotations, '*Ammantium irae amoris integration est*'" — meaning "Lovers' quarrels always go with true love."[12]

Lucy, the President's actual lover, certainly looked forward to seeing Franklin as much as the President longed to see her. On April 5 she'd written to Warm Springs to thank Grace Tully for making the necessary preparations for her visit. Mindful of how weak the President was, however, she'd added: "If you should change your mind & think it would be better for me not to come — call me up. I really am terribly worried — as I imagine you all are."[13]

For his part Franklin could hardly wait, however — his excitement worrying Daisy Suckley lest it raise his blood pressure still higher. Granted, it would be a "pleasure," Daisy noted in her diary, after confirmation that Lucy would arrive on April 9, but it would be "another interruption in the routine we are trying to keep."[14]

That interruption was precisely what the President yearned for. On the morning of Friday, April 6, he was working on his stamp collection when he hit upon the idea of a new three-cent stamp to be issued on the opening day of the San Francisco Conference, bearing the words "April 25, 1945; Towards United Nations."[15]

Could the stamp be issued in time? He called the postmaster general, Frank Walker, who said it could. Moreover, Walker assured the President designs for his approval could be with him by April 10 or 11. "So," Daisy commented in her diary, it was still possible that "people in high places sometimes get things done in a few minutes!" Later, before going to sleep, the President smoked a cigarette and "talked seriously about the S. Francisco Conference, & his part in World Peace, etc. He says again that he can probably resign some time next year, when the peace organization — the United Nations — is well started."[16]

Some time *next year?*

It seemed risible. Yet each day, as April 9 approached, the President looked better and better — and sunnier. "He sits a little straighter in his chair, his voice is a little clearer and stronger, his face less drawn & he is happier!" Daisy noted on April 8. "We are now looking forward to Lucy Rutherfurd's visit. She comes tomorrow, bringing Madame Shoumatoff, for another portrait of F.D.R., and a photographer." To her relief "Lucy and Mme S. go into the guest house," Daisy penned with a certain satisfaction, for the last time Lucy had come, after the '44 election, Daisy and Polly had been asked to move out of the Little White House to make way for her. On the President's instructions, however, "Polly & I are going to get flowers" for the guesthouse, "to make it look attractive. Lizzie the maid had been cleaning for two days — and it will be very nice," she assured the President — since, given his poor health, "F. has never gone in there to see it."[17]

Unable to wait until Lucy arrived at Warm Springs in Madame Shoumatoff's car, though, the President had told Lucy by phone he would meet her in person at 4:00 p.m., in Macon, Georgia, some eighty miles away, so that she could transfer and travel with him the last part of her journey. After a good afternoon's sleep the President thus set off for Macon on April 9, taking Daisy with him and leaving Polly to arrange the flowers.

"It was a beautiful evening for a drive & we enjoyed it tremendously — on & on," Daisy jotted that night, "away from the sun" and wondering whether, if Lucy was early, she might motor on and they would pass each other — a prospect that left the anxious president "scanning every car that headed towards us, imagining it was slowing up."[18]

If the President was aware how strange this might seem to others — the

Commander in Chief of the United States being driven in his trusty Ford Phaeton, with one Secret Service vehicle in front and one behind, to meet his former lover — he was completely indifferent. They made their way along "old Route 41 and down the old narrow road between the low-slung clay cliffs and fields of wildflowers to level ground, the road full of sudden turns between hummocks of tall trees," Jim Bishop later wrote, reconstructing the drive.[19]

To the President's chagrin, however, there was no sign of Lucy's car in Macon, at 4:00 p.m.

"Finally, after driving 85 miles," Daisy recorded, "we turned around & started toward the setting sun." The President was disappointed. "It got quite chilly" in the open-topped car, "& F put on his cape & I my rain coat which, though not warm, is a good windbreaker. We stopped in front of the drugstore in Manchester, for a 'Coke' & at that moment Lucy & her party also drove up on the curb!"[20]

As it turned out, the map-reading skills of the Russian photographer Nicholas Robbins, né Kotzubinsky, had been insufficient for them to arrive in Macon on time. As Elizabeth later wrote, Robbins had been in love with Mrs. Rutherfurd since meeting her in 1943 and, sitting in the back with the maps, had spent more time gazing at Lucy than at the signposts. "As a result we reached Macon way after four o'clock. The beauty of that town, with its enchanting old houses and Civil War atmosphere, turned us away for a while from the rather annoying feeling that we were late. Lucy powdered her nose and seemed very nervous. Driving out of Macon we began carefully looking for the presidential car. Nothing in sight. We drove for quite a while. 'Nobody loves us, nobody cares for us,' sighed Lucy in a joking fashion, but I felt she was really disappointed. It was a beautiful evening and the sun was beginning to set. As we entered Greenville, a village near Warm Springs, we suddenly noticed, by a corner drugstore, several cars and quite a crowd gathered around them. We drove up and there in an open car was FDR himself, in his Navy cape, drinking Coca-Cola! We pulled to the curb. Lucy and I got out of the car."[21]

It was thus, on this somehow quintessential American 1940s stage, that the two former lovers were reunited. "The expression of joy on FDR's face upon seeing Lucy," Elizabeth Shoumatoff later recorded, was something she would never forget — in part because she found herself, as intended portraitist, so shocked by the change in the President "since I painted him in 1943." He had been so vibrant, then: so humorous, so filled with life and curiosity, both in the White House and at Hyde Park. "My first thought," now, "was how I could make a portrait of such a sick man? His face was gray and he looked to me like President Wilson in his last years."[22]

The President was over the moon, however. "Lucy and Shoumie got into the car, I on the little [jump] seat, & we drove home to this 'Little White House,'"

Daisy recorded contentedly before she went to sleep — the President's blood pressure up, and looking "awfully tired all evening," but, in the dying day, almost deliriously happy.[23]

It is there we shall leave the President. His private secretary, Bill Hassett, was aware "the Boss's" hours were now numbered. The President knew it, too — whatever he might say to Daisy about planning to retire the following year. He'd told his office secretary, Dorothy Brady, that she'd soon enough be able to visit her farm, outside Washington, "more often." Every day, in the meantime, he would go for a drive with Lucy — and Fala, his beloved terrier — to Dowdell's Knob.

Watching the President and Lucy together, Daisy was filled with concern. Eleanor was tough, and very much in control of her own life and emotions. Without Franklin, Daisy asked herself, where would Lucy be, once he died? A "very different future, rather alone," Daisy reflected — especially as "she isn't very well and that makes it more difficult to face life & make decisions." The President, Daisy recorded, was worried about her, and how she would manage — which Daisy rather resented, feeling "that she should face her own life & not put too much of its difficulties on *his* shoulders."[24]

The President didn't mind, though — indeed it was a mark of his generous character that he was always "helping others & making others happy," Daisy acknowledged — excited that, if the President survived the next few weeks, she and Polly could travel with him, he'd told them, to San Francisco on the *Ferdinand Magellan*.[25]

Concern, hope, and sadness thus mixed together. The President was only sixty-three, Lucy fifty-three. Yet there was no way, in all candor, his death could be far off. Admiral Leahy had told him how senior French officials, on behalf of de Gaulle, were pressing for U.S. transport to be arranged for their forces to be shipped to India, ready to move into Indochina, while Dutch officials were pressing for the same, to reestablish colonial rule in the Dutch East Indies,[26] while the British were putting pressure on Admiral King to permit British naval forces into the Pacific for the same reason. In his current condition, all the President felt up to saying was: discuss this with the Joint Chiefs of Staff, as well as Secretaries Henry Stimson and James Forrestal.

During the drive to the Knob on April 11 he had asked Daisy to come with them, and Daisy had watched the couple closely. "Lucy is so sweet with F — No wonder he loves to have her around — Toward the end of the [2-hour] drive, it began to be chilly and she put her sweater over his knees," Daisy noted in her diary. This was still a very private romance, but something beautiful to witness, in terms of Lucy's caring. "I can imagine just how she took care of her husband — She would think of little things which make so much difference to a semi-invalid, or even a person who is just tired, like F."[27]

Henry Morgenthau, the Treasury secretary, who happened to be in the area, had asked to come dine with them, and that night he and the President had shared stories about events and people they both knew, especially Winston Churchill. Morgenthau — who was anxious over his wife, who'd recently had a stroke and heart attack — asked to speak privately with Dr. Bruenn. But his question was not about his spouse. The President was slipping far faster than his own wife — the President's hands "shook so that he started to knock the glasses over," as he'd offered Henry one of his famous cocktails.[28]

The climax came quicker even than expected, the following day, shortly before lunch was served.

To pose for his portrait, at Mme Shoumatoff's suggestion, the President was wearing his double-breasted gray suit with his Harvard Crimson tie, "looking very fine," Daisy noted.[29] Elizabeth even complimented him on his good color that day — to be told later that this might well have presaged what now happened.

Bill Hassett had dried the fifty-odd papers the President had signed, in ink. Once gathered, like laundry, the private secretary had put them neatly into a folder on the card table which the President used as his desk. Lucy and Daisy were sitting on the sofa, watching Mme Shoumatoff at work on the life-size watercolor. She was painting as fast as she could, filling in the sitter's eyes, but became aware suddenly that "his gaze had a faraway look and was completely solemn." He'd just told her about the stamp he'd asked for, to celebrate the upcoming conference — "Wait 'til you see the San Francisco stamps, with the United Nations"[30] — but seemed then to have moved somewhere else in his mind, staring at Lucy, next to him. It was about 1:15 p.m.

To the Filipino butler, Joe Esperancilla, the President had said they needed "fifteen more minutes to work" before taking lunch, which he was looking forward to. "Suddenly," Elizabeth recalled, "he raised his right hand and passed it over his forehead several times in a strange jerky way, without emitting a sound" — at least as far as she could hear.[31]

Daisy Suckley, crocheting on the sofa, recalled the President "looking for something: his head forward, his hands fumbling." Immediately she rose. "I went forward & looked into his face. 'Have you dropped your cigarette?'" she asked him, alarmed. "He looked at me with his forehead furrowed in pain and tried to smile. He put his left hand up to the back of his head & said: 'I have a terrific pain in the back of my head.'"[32]

These would be the President's last words — Daisy quite certain of them, afterwards. "He said it distinctly, but so low that I don't think anyone else heard it — My head was not a foot from his — I told him to [put] his head back on his chair." "The President is sick, call the doctor," Mme Shoumatoff meanwhile yelled.[33]

Dr. Bruenn was at the bottom of the hill, in the big rehabilitation pool with other polio patients. By the time he'd dressed and arrived up at the Little White House, fifteen minutes had passed. Arthur Prettyman and Joe Esperancilla had carried the "dead weight" of the President, with Polly holding his feet, to his bedroom, and there they'd laid him on his bed, where he'd lost all consciousness — loud snoring sounds coming from his throat.

Injections of papaverine, nitroglycerine, and amyl nitrite were administered by Dr. Bruenn — for the President's heart, paradoxically, was still pumping. He had, however, suffered a "massive cerebral hemorrhage," or catastrophic stroke; his blood pressure was 300 over 190, and there was nothing — despite attempted artificial respiration by the President's masseur, Lieutenant Commander Fox — that could be done, except to wait for the end.

Bruenn telephoned Admiral McIntire, the President's White House doctor, who was still in Washington, and warned him that "a long siege" was ahead. Bruenn was forcing himself to show no emotion — having told Daisy only a few days before that he had come to love the President so much that he would "jump out of the window" for him, "without hesitation."[34]

In the event, the siege did not last long. Lucy Rutherfurd, recognizing immediately the end was approaching, told Elizabeth Shoumatoff to pack her easel and bags and summon Nicholas Robbins. In the white Cadillac they set off from the estate before the press could arrive. They would only hear whether or not the President had actually passed away when they stopped to telephone the Little White House, on their journey home. The flag at Macon was already at half-mast. The operator, before putting the call through, asked if they knew what had already become national — in fact global — news at 3:35 p.m., local time, April 12, 1945.

The Commander in Chief was dead.

Acknowledgments

I began FDR at War thinking it would be a single, stand-alone volume, not a trilogy. It was to be about 350 pages: a modest attempt by an Anglo-American author to fill a strange gap in World War II historiography. Over the decades since the war ended we have as readers been the beneficiaries of a cornucopia of books about wartime leaders, generals, and “ordinary” servicemen and women, but no real account of Franklin D. Roosevelt’s role as commander in chief of the Armed Forces of the United States in what was the most violent and destructive war in human history. This had seemed to me, when writing *American Caesars,* my history of postwar American presidents, to be quite wrong — a gap that had puzzled me both as a military historian and as a biographer.

Thus I proposed to my agent such a work, and began my quest, beginning to research it ten years ago. My editor, Bruce Nichols, of Houghton Mifflin Harcourt, welcomed and helped prefinance the project, but neither he nor I initially had any idea it would stretch across a decade — or that my British publisher, Random House, would balk at the multivolume size. I therefore wish here to thank Bruce for his faith and loyalty to the project — a project which has, I hoped, made a valuable contribution both to our understanding of World War II and the singular, hitherto unacknowledged role of Franklin Roosevelt in achieving Allied military victory. Without FDR’s extraordinary military leadership after Pearl Harbor, the course of World War II might well have turned out differently — and I would probably not be here, writing about it. Or at such length. I will always remember Bruce’s pained email to me, on reading the first draft of FDR at War. “I’ve reached page 800 in the manuscript, Nigel, and have enjoyed it immensely. But we’re only in November 1942 — where’s the rest of the war?”

The “rest,” thanks to Bruce’s patience and goodwill, took a bit longer. I wanted not only to document FDR’s performance in the role of commander in chief in World War II, but to help put the reader in the room with him as he made the decisions which, for good or ill, would determine the war's outcome. Marrying history and biography, in other words, I attempted to reconstruct the saga of World War II at the very highest level of decision-making—and on both sides of the drama, once Hitler declared war on the United States on December 11, 1941.

On the Allied side, FDR's primary partner in the great conflict was, of course, Winston Churchill, Prime Minister of Great Britain, and quasi commander in chief of British Empire forces in World War II. Churchill duly recounted his own role in six masterly volumes of memoir and history, *The Second World War,* between 1948 and 1953, which certainly helped him win the Nobel Prize for Literature. To some extent my task therefore became that of countering Winston's version, especially in relation to the dominating role of President Roosevelt as U.S. commander in chief. For behind the scenes Churchill's military leadership, after his valiant stand in 1940, had been deplorable. As the head of the British Army under Churchill, Field Marshal Brooke would confide in his famous diary, after D-day, three quarters of the world imagined the Prime Minister to be a great strategist, whereas in truth he had been a constant "menace," responsible for unending disasters.

Such was the power of Winston's postwar prose, however, and the veneration in which he was held in the United States, especially, that Brooke's candid view (omitted from Arthur Bryant's 1959 version of Brooke's diaries, and only published in 2001) was considered hard cheese by a general who had been passed over for command of D-day.

Addressing that misperception and rechronicling the war from FDR's perspective as, ultimately, the military mastermind of World War II has thus been for me a challenging yet deeply rewarding task as a historian who believes in the power of biography to revise and correct history. The FDR who emerges from this decade-long tapestry is both human and fallible. His greatest virtues are his patience and his inner resolve, on behalf of his country and the free world. How he masterminded a two-ocean war, how he overruled his own Joint Chiefs of Staff, how he "delivered" D-day despite everything Churchill did to sabotage the Allied invasion, from 1943 to the very eve of its mounting, is the core thread of my trilogy—and one of which I have a measure of special, personal connection and pride, inasmuch as my father took part in the triumphant landings as a twenty-five-year-old British infantry battalion commander, later winning the DSO in battle. The landings did not fail; and as Hitler had warned his own generals in December 1943, they did decide the outcome of the war in Europe.

The saddest part of my long, earlier years as official biographer of Field Marshal Montgomery, commander in chief of the D-day invasion armies, had been my father's illness and his death, soon after I finished the Monty trilogy in 1987. This time, reaching the end of the FDR at War trilogy thirty years later, I was beset by similar sadness: recounting the way in which FDR's fatal heart disease was secretly diagnosed before the D-day landings, and — hidden from the public — transformed his role as U.S. commander in chief thereafter. The triumph of his D-day project was subsumed in the tragedy of his ill health, and though the final book became a story of great personal courage and determination, it was strangely hard for me, as author, to narrate. Only when a member of my writers group in New Orleans pointed out that I myself had reached *un certain âge*, namely when death seems so much nearer than in earlier decades, did I realize how deeply invested I was in surviving to set the record of this man's contribution to the history of humanity straight — and how much, in the course of researching and writing the trilogy, I had come to respect, understand, and admire him for it. At a moment when the new world order he created seemed to be fraying, perhaps collapsing, the sheer magnitude of FDR's accomplishment — an accomplishment that had framed my life since birth in 1944 — appeared monumental.

Now that the trilogy — though only half the size of Churchill's! — is complete, I'd like to thank some, at least, of the many people who, over the years, encouraged and assisted its making.

First and foremost, for her patience and faith in the outcome, my wife, Raynel Shepard, who simultaneously launched her new career as a jazz vocalist as I sought to complete mine as a military historian. Also my fellow author-historians Carlo D'Este, Roger Cirillo, James Scott, Mark Schneider, David Kaiser, Douglas Brinkley, David Reynolds, Ron Spector, Fredrik Logevall, Kai Bird, Mark Stoler, Niall Barr, Rick Atkinson, Hans Renders, Doeko Bosscher, Robert Citino, Susan Butler, Lynne Olson, and the many other scholars, friends, and readers who, over the years, have contributed to our better understanding and knowledge of World War II, and of FDR.

At the Franklin Roosevelt Presidential Library Archives I'd like to thank, for their help in preparing this book, archivists Virginia Lewick and Matt Hansen; at the Eisenhower Presidential Library, deputy director Timothy Reeves, and audiovisual archivist Kathy Struss; at the Library of Congress the head of Reference and Reader Services, the indispensable Jeffrey Flannery; at the Marshall Foundation, Director Rob Havers and Jeffrey Kozack; at the Imperial War Museum in London the Keeper of Documents, Anthony Richards; and at the National World War II Museum the president emeritus, Dr. Nick Mueller, and the new president, Stephen Watson, and his staff — especially Jeremy Collins — for enabling me to meet in person so many fellow WWII scholars and history devotees at the annual WWII and Churchill Society conferences. The members of the New York Military Affairs Symposium have been unfailingly encouraging. My especial gratitude also to my colleagues in the Boston Biographers Group, the New Orleans Non-Fiction Writers Group, and the Biographers International Organization (BIO).

At the University of Massachusetts Boston I'd like to thank the dean of the McCormack Graduate School of Policy and Global Studies, Dr. David Cash, as well as Robert Turner and my colleagues there, including the university library staff—as also the staff of the Widener Library at Harvard and Boston College. I was fortunate to have been able to interview the late Lieutenant Commander George Elsey, of FDR's wartime Map Room, and FDR's grandchildren, Mrs. Ellie Seagraves and the late Curtis Seagraves. My fellow members of the Tavern Club, Boston, have been supportive throughout—and I miss the late Tom Halsted there, who shared his memories of his stepmother, Anna Roosevelt Halsted. Most especially I'd like to thank once again my agent Ike Williams, and his colleagues Hope Denekamp and Katherine Flynn at Kneerim & Williams.

Without a publisher an author cuts a sorry figure. At Houghton Mifflin Harcourt I want to mark my great gratitude to my ever-patient and supportive editor and publisher, Bruce Nichols, together with his staff, especially Ivy Givens, Larry Cooper, and publicist Michelle Triant. Without the dedicated, painstaking efforts, moreover, of my copyeditor, Melissa Dobson—who has performed the crucial task, including fact-checking, on all three volumes—the trilogy would have been sadly flawed. It has truly been a team effort. I can only hope *War and Peace*—indeed the FDR at War trilogy—does justice to their great contributions.

NIGEL HAMILTON

John W. McCormack Graduate School of Policy and Global Studies, UMass Boston

Photo Credits

A Trip to Tehran. Arlington Cemetery, November 11, 1943: UPS / INS / FDR Library; on board with Admiral King, Admiral Leahy, and General Marshall: USASC / FDR Library; USS *Iowa:* Naval History and Heritage Command

Interviewing Eisenhower. The specially fitted C-54, Sacred Cow: FDR Library; FDR and Eisenhower on the Sacred Cow: U.S. Army / Eisenhower Library; FDR in Tunis with Generals Eisenhower, Spaatz, Bedell Smith, and others, November 20, 1943: Eisenhower Library

Cairo. FDR with Churchill in the garden of Kirk villa, Cairo: FDR Library; FDR visiting the pyramids: FDR Library; Chiang Kai-shek, FDR, Churchill, and chiefs of staff before departure to Tehran: FDR Library

Tehran. FDR in jeep, reviewing troops, Tehran: FDR Library; FDR with Stalin at birthday dinner for Churchill, British Legation, November 30, 1943: FDR Library

Saving D-day. FDR posing with Stalin and Churchill on the steps of the Soviet Embassy, Tehran: FDR Library; close-up photo at Soviet Embassy, with Hopkins, Molotov, and Eden behind: FDR Library

Who Will Command Overlord? FDR in jeep with Eisenhower, Castelvetrano airfield, Sicily, December 8, 1943, en route home: FDR Library; awarding General Clark the Distinguished Service Cross: FDR Library; FDR riding in jeep with Ike, Patton standing at left: FDR Library

Triumphant Return. Battleship USS *Iowa:* Naval History and Heritage Command; FDR greeted by members of cabinet, Congress, and others at White House, December 17, 1943: FDR Library; FDR greeted by Secretary of State Hull and Judge Byrnes: FDR Library

Christmas 1943. Eleanor, FDR, and their family assembled at Hyde Park: FDR Library; FDR's Christmas Eve broadcast, December 24, 1943: FDR Library

Anzio. Ike warns Churchill against hasty Anzio invasion: Imperial War Museum; Allied landings, Anzio, January 22, 1944: Getty Images; hospital ship evacuates Allied wounded in March 1944: U.S. Navy / FPG / Getty Images; bombed and overcrowded Allied military hospital at Anzio, February 1944: George Silk / *Life* / Getty Images

The Triumph of D-day. Eisenhower and Montgomery inspect U.S. assault troops preparing for D-day: Frank Scherschel / *Life* / Getty Images; FDR's D-day prayer: FDR Library; U.S. troops landing on Omaha Beach, June 6, 1944: Robert Sargent / U.S. Coast Guard / National Archives

The Bomb Plot. FDR attending Marine Corps amphibious assault rehearsal with son Colonel Jimmy Roosevelt and Admiral Davis, Oceanside, California, July 20, 1944: U.S. Navy / FDR Library; Hitler and Mussolini at Wolf's Lair HQ, Rastenburg, East Prussia: Heinrich Hoffmann / Ullstein / Getty Images; Hitler shows Mussolini the operations hut where he was nearly killed, July 20, 1944: Bundesarchiv, Germany; Hitler broadcasts from the Wolf's Lair, July 20, 1944: Heinrich Hoffmann / Getty Images; FDR simultaneously broadcasts from special car on *Ferdinand Magellan,* San Diego, July 20, 1944 (Dr. Bruenn in foreground): George Skadding / *Life* / Getty Images

To Be, or Not to Be. Democratic convention, Chicago, July 1944: Keystone-France / Gamma-Keystone / Getty Images; election poster: FDR Library; the heavy cruiser USS *Baltimore* off California, 1944: Fahey Collection / Naval History and Heritage Command

Hawaii. Aboard USS *Baltimore,* FDR receives General MacArthur and Admiral Nimitz, with Admiral Leahy, July 26, 1944: U.S. Navy / FDR Library; FDR tours installations, Hawaii, with General MacArthur and Admiral Nimitz, July 27, 1944: U.S. Navy / FDR Library; Admiral Nimitz and General MacArthur present their plans to FDR and Admiral Leahy, Holmes mansion, Waikiki, July 28, 1944: U.S. Navy / FDR Library

The Fall of '44. FDR has minor heart attack while giving speech aboard USS *Cummings,* Puget Sound Navy Yard, August 10, 1944: FDR Library; FDR campaigning with Senator Truman: Corbis / Getty Images; Inauguration Day, White House, January 20, 1945: Library of Congress

Yalta. FDR aboard USS *Quincy* arriving in Malta, saluted by Winston Churchill (foreground), February 2, 1945: UPI (Acme) / FDR Library; FDR on arrival at Saki airfield, Crimea, with Churchill and Molotov: U.S. Army Signal Corps (USASC) / FDR Library; FDR and Churchill in ballroom, Livadia Palace, Yalta, February 4, 1945, USASC / FDR Library; FDR and Stalin in FDR's study, Livadia Palace, February 4, 1945: USASC / FDR Library; the Big Three at Livadia Palace, February 9, 1945: USASC / National Archives, courtesy Naval History and Heritage Command

Warm Springs. FDR reports to Congress, March 1, 1945: FDR Library; FDR at his work table, Little White House, Warm Springs, April 1945: Margaret Suckley / FDR Library; the widow Lucy Rutherfurd at the Little White House, April 11, 1945: Nicholas Robbins / FDR Library; FDR posing for Elizabeth Shoumatoff portrait, April 11, 1945: FDR Library; "FDR Dies," front page, *San Francisco Chronicle,* April 13, 1945: *San Francisco Chronicle* / Polaris Images.

Notes

PROLOGUE

1. Second of Churchill's three major speeches to the House of Commons during the Battle of France, given on June 4, 1940: "A Colossal Military Disaster," in Winston Churchill, *The War Speeches of the Rt. Hon. Winston S. Churchill*, comp. Charles Eade, vol. 1 (London: Cassell, 1951).
2. "Evening Situation Report, probably December 20, 1943," section titled "The West, danger of invasion," in Helmut Heiber, ed., *Hitler and His Generals: Military Conferences 1942–1945* (New York: Enigma Books, 2003), 314 and 313.
3. Evan Thomas, "War Comes to America" (review of Nigel Hamilton, *The Mantle of Command: FDR at War, 1941–1942*), *New York Times*, August 1, 2014.
4. Lieutenant Commander George Elsey, interview with the author, September 12, 2011.
5. Entry of September 14, 1944, Diaries of William Lyon Mackenzie King, Library and Archives Canada, Ottawa.
6. Entry of September 20, 1944, in Geoffrey C. Ward, ed., *Closest Companion: The Unknown Story of the Intimate Friendship Between Franklin Roosevelt and Margaret Suckley* (Boston: Houghton Mifflin, 1995), 328.
7. Elsey, interview with the author.
8. See David Reynolds, *Summits: Six Meetings That Shaped the Twentieth Century* (New York: Basic Books, 2007), 161.

1. A TRIP TO THE MEDITERRANEAN

1. John McCrea, "'Iowa'— President and Joint Chiefs of Staff to Africa and Return," manuscript memoir, John L. McCrea Papers, FDR Presidential Library, Hyde Park, NY.
2. Argentia had made history in August 1941, when President Franklin Roosevelt and Prime Minister Winston Churchill had drawn up the famous Atlantic Charter aboard their respective warships. "Tiny villages cluster in the shoulders of her hills," the *Iowa*'s chronicler described, "and mists that shroud the valleys lend an ethereal quality to this corner of Newfoundland. The population is mostly Irish and the

climate is wind-swept, foggy and crisply sombre." Twin peaks on the horizon were likened by the crew to the legendary bosom of Mae West—who, when informed of this, was said to have replied: "Thanks, boys. I hope they are standing up." The ocean ranged "from rolling, roughish grey to lake-water calms of sapphire brilliance. It is a quiet place, a place to drill, to study and prepare": "Iowa" (no author), unpublished TS, John L. McCrea Papers, 1898–1984, Library of Congress, Washington, DC.

3. The mishap took place on July 16, 1943, and led to a court of inquiry. President Roosevelt sent McCrea a box of good Cuban cigars in sympathy, as the fault was entirely McCrea's: John L. McCrea, *Captain McCrea's War: The World War II Memoir of Franklin D. Roosevelt's Naval Aide and USS* Iowa*'s First Commanding Officer,* ed. Julia C. Tobey (New York: Skyhorse, 2016), 171.
4. McCrea, "'Iowa'—President and Joint Chiefs of Staff to Africa and Return."
5. Nigel Hamilton, *Commander in Chief: FDR's Battle with Churchill, 1943* (Boston: Houghton Mifflin Harcourt, 2016), 4–9 and 63–129.
6. Ibid., 55–60.
7. Ibid., 35–38.
8. McCrea, "'Iowa'—President and Joint Chiefs of Staff to Africa and Return."

2. THE MEETING IS ON

1. Hamilton, *Commander in Chief*, 228–34.
2. Ibid., 265–66.
3. Ibid., 360–66.
4. Ibid., 19–23 et seq.
5. Elizabeth Maclean, *Joseph E. Davies: Envoy to the Soviets* (Westport, CT: Praeger, 1992), 108.
6. Entries of October 30, 1943, and November 6, 1943, Ward, *Closest Companion*, 250 and 253.

3. MAXIMUM SECRECY

1. McCrea, "'Iowa'—President and Joint Chiefs of Staff to Africa and Return." See also McCrea, *Captain McCrea's War*, 77 et seq.
2. Ibid.
3. Ibid.
4. Ibid.
5. Ibid.
6. "An Historic Voyage," chapter in "Iowa" (no author), McCrea Papers, Library of Congress.
7. Ibid.
8. See Hamilton, *The Mantle of Command: FDR at War, 1941–1942* (Boston: Houghton Mifflin Harcourt, 2014), 3–18.

9. Cable 146, Stalin to Roosevelt, November 10, 1944, in Susan Butler, ed., *My Dear Mr. Stalin: The Complete Correspondence Between Franklin D. Roosevelt and Joseph V. Stalin* (New Haven, CT: Yale University Press, 2005), 182.
10. "Iowa" (no author), McCrea Papers, Library of Congress.
11. See Hamilton, *Commander in Chief*, 179–91.
12. Foreword to "Log of the President's Trip to Africa and the Middle East, November-December 1943," Franklin D. Roosevelt, Papers as President: Map Room Papers, 1941–1945, Box 24, FDR Presidential Library, Hyde Park, NY.
13. Ibid.
14. Entry of November 5, 1943, Ward, *Closest Companion*, 252.

4. SETTING SAIL

1. Entry of October 30, 1943, Ward, *Closest Companion*, 250.
2. Ibid., entry of November 8, 1943, 254.
3. Entry of November 12, 1943, Leahy Diary, William D. Leahy Papers, Library of Congress, Washington, DC.
4. McCrea, "'Iowa'—President and Joint Chiefs of Staff to Africa and Return."
5. Ibid.
6. Typed FDR Diary, in "War Conference in Cairo, Teheran, Malta, etc., November 11–December 17, 1943," Franklin D. Roosevelt, Papers as President: The President's Official File, Part 1, 1933–1945, 200-3-N, Box 64, FDR Presidential Library, Hyde Park, NY. (This is a typed version of FDR's handwritten original diary, also in file, bearing note: "This typed copy is the one corrected by F.D.R."; it contains FDR's additions, such as extracts from his letters, as dictated to his secretary, Grace Tully, upon his return.)
7. William M. Rigdon, with James Derieux, *White House Sailor* (Garden City, NY: Doubleday, 1962), 61.
8. Entry of November 13, 1943, Typed FDR Diary, "War Conference in Cairo, Teheran, Malta."

5. SHEER MADNESS

1. Entry of November 13, 1943, Typed FDR Diary, "War Conference in Cairo, Teheran, Malta."
2. Entry of November, 4, 1943, Stimson Diary, Henry L. Stimson Papers, Yale University Library, New Haven, CT.
3. Field Marshal Bernard Montgomery to author, multiple occasions, 1963–66.
4. Carlo D'Este, *Warlord: A Life of Winston Churchill at War, 1874–1945* (New York: HarperCollins, 2008), 230–34.
5. Ibid., 251–57.
6. "Prime Minister's Personal Minute, D. 178/3, Most Secret," October 19, 1943, in Martin Gilbert, *Winston S. Churchill*, vol. 7, *Road to Victory* (Toronto: Stoddart, 1986), 533.

7. C-441, October 8, 1943, in Warren F. Kimball, ed., *Churchill and Roosevelt: The Complete Correspondence*, vol. 2, *Alliance Forged, November 1942–February 1944* (Princeton, NJ: Princeton University Press, 1984), 503.
8. Max Hastings, *Winston's War: Churchill, 1940–1945* (New York: Knopf, 2009), 323–40.
9. Kimball, *Churchill and Roosevelt*, 498.
10. Ibid., R-379, October 7, 1943, 379.
11. Ibid., C-441, October 8, 1943, 503.
12. Ibid.
13. Ibid.
14. Ibid.
15. David Reynolds, *In Command of History: Churchill Fighting and Writing the Second World War* (New York: Random House, 2005), 376.
16. Entry of October 8, 1943, in Arthur Bryant, *Triumph in the West* (London: Collins, 1959), 51.
17. Ibid.
18. Ibid.
19. Ibid.
20. Entry of October 19, 1943, Bryant, *Triumph in the West*, 55, and Alan Brooke, *War Diaries, 1939–1945: Field Marshal Lord Alanbrooke*, eds. Alex Danchev and Daniel Todman (Berkeley: University of California Press, 2001), 461.
21. Entry of October 19, 1943, in Brooke, *War Diaries*, 461.
22. C-471, October 23, 1943, Kimball, *Churchill and Roosevelt*, 555.
23. Ibid.
24. Elsey, interview with the author.
25. C-475, October 26, 1943, Kimball, *Churchill and Roosevelt*, 562.
26. Ibid., C-471, October 23, 1943, 556.
27. Ibid.
28. Ibid., 555, and C-472, 558.

6. CHURCHILL'S IMPROPER ACT

1. See Hamilton, *Commander in Chief*, 235–42.
2. Entry of October 24, 1943, in the diaries kept by Sir Charles Wilson, Lord Moran: Moran, *Churchill: The Struggle for Survival, 1940–1965* (Boston: Houghton Mifflin, 1966), 122.
3. Cable W 3325/#1806, October 25, 1943, in Dwight D. Eisenhower, *The Papers of Dwight David Eisenhower: The War Years*, ed. Alfred Chandler, vol. 3 (Baltimore, MD: Johns Hopkins Press, 1970), 1529. To Eden, the Prime Minister cabled that the British "would do our very best for 'Overlord' but it is no use planning for defeat in the field in order to give temporary political satisfaction." Overlord was, in Churchill's view, however, still a nonstarter. Any assurances Eden had given "about May 'Overlord'" had now to be "modified by the exigencies of the battle in Italy." Nothing would "alter my determination not to throw away the battle in Italy at this

juncture," the Prime Minister stated emphatically; "Eisenhower and Alexander must have what they need to win the battle, no matter what effect is produced on subsequent operations. This may certainly affect the date of 'Overlord,'" he cabled (CHAR 20/122/43, Papers of Sir Winston Churchill, Churchill Archives Centre, Cambridge, UK).

4. "A radiogram came from General Deane, our military man with Mr. Hull, and this radiogram said that Churchill had sent to Moscow a copy of Alexander's rather pessimistic summary of the situation in Italy and had directed Eden to read it aloud to Stalin," together with "comments I think by Churchill that this would mean that the Second Front would be delayed or abandoned": Entry of October 28, 1943, Stimson Diary.
5. Ibid.
6. Ibid.
7. "Proposed Draft of Cable, From: The President, To: The Prime Minister," October 27, 1943, Pentagon Office, 1938–1951, Correspondence, Box 81, Papers of George Catlett Marshall, George C. Marshall Foundation Library, Lexington, VA.
8. Entry of October 31, 1943, Stimson Diary.
9. Ibid., entry of November 4, 1943.
10. Sent November 4, received November 6, 1943: *Foreign Relations of the United States: The Conferences at Cairo and Tehran, 1943* (hereinafter *FRUS I*) (Washington, DC: U.S. Government Printing Office, 1961), 65.

7. TORPEDO!

1. McCrea, "'Iowa' — President and Joint Chiefs of Staff to Africa and Return."
2. Ibid.
3. Ibid.
4. Rigdon, *White House Sailor*, 64.
5. Entry of November 14, 1943, in Henry H. Arnold, *American Airpower Comes of Age: General Henry H. "Hap" Arnold's World War II Diaries,* ed. John W. Huston, vol. 2 (Maxwell Air Force Base, AL: Air University Press, 2002), 76.
6. McCrea, "'Iowa' — President and Joint Chiefs of Staff to Africa and Return," and similar in McCrea, *Captain McCrea's War*, 186.
7. Ibid. Harry Hopkins's friend Robert Sherwood expressed his skepticism with this wording later, when writing Hopkins's biography. As an accomplished playwright Sherwood thought the shout would far more likely have been: "This ain't no drill!": Robert E. Sherwood, *Roosevelt and Hopkins: An Intimate History* (New York: Harper, 1948), 768.
8. McCrea, "'Iowa' — President and Joint Chiefs of Staff to Africa and Return."
9. Ibid.
10. Ibid.
11. Ibid.
12. Ibid.

13. Ibid.
14. Sherwood, *Roosevelt and Hopkins,* 768.
15. H. H. Arnold, *Global Mission* (New York: Harper, 1949), 455.
16. Rigdon, *White House Sailor,* 64.
17. Ibid.
18. In his memoirs King wrote that "many people thought the ship had been hit": Ernest J. King and Walter M. Whitehill, *Fleet Admiral King: A Naval Record* (New York: Norton, 1952), 501.
19. McCrea, "'Iowa'—President and Joint Chiefs of Staff to Africa and Return."
20. Entry of November 14, 1943 (Atlantic Ocean), Arnold, *American Air Power Comes of Age,* 76.
21. Arnold, *Global Mission,* 455.
22. McCrea, "'Iowa'—President and Joint Chiefs of Staff to Africa and Return."
23. Entry of November 14, 1943 (Atlantic Ocean), Arnold, *American Air Power Comes of Age,* 76.
24. Arnold, *Global Mission,* 455.
25. "Nov 18. From Letter," in Typed FDR Diary, "War Conference in Cairo, Teheran, Malta."
26. Ibid., entry of November 19, 1943.
27. McCrea, "'Iowa'—President and Joint chiefs of Staff to Africa and Return."
28. Sherwood, *Roosevelt and Hopkins,* 768.
29. Entry of November 14, 1943 (Atlantic Ocean), in Arnold, *American Air Power Comes of Age,* 76.
30. Quoted in Arnold, *Global Mission,* 455.
31. Rigdon, *White House Sailor,* 64. The captain of the USS *William D. Porter* swiftly signaled to explain what had happened. The torpedo was not meant to be actually fired; he had merely been trying to rehearse the procedure, in time-honored rules of such exercises, where a friendly vessel is used as the putative target. Salt spray had, he claimed, "bridged an open electric switch on one of the destroyer's torpedo tubes, thus setting off the ejection mechanism and sending the armed torpedo on its way": ibid. As McCrea later noted, this did not sound at all convincing. In truth the *Porter*'s chief petty officer, after testing "to see if it would fit" properly, had failed—as the President correctly noted afterwards—to remove the firing primer before the voyage began. Once the captain decided to rehearse the firing of a torpedo—but without having been told of the dignitaries aboard the *Iowa*—the aiming mechanism, directed from the bridge, made it almost certain the torpedo would not only be fired but would hit the innocent target. Had the seas been still heavier that day, the lookout on the *Iowa* might not have seen the torpedo racing toward the battleship, twenty feet beneath the waves, and "it no doubt would have hit us," McCrea later reflected soberly. Fate had been kind—though not, McCrea recorded sadly, to the USS *Porter*—which was subsequently posted to the Pacific, only to be hit and sunk there in a Japanese kamikaze plane assault, with the loss of all hands, the following year.

8. A PRETTY SERIOUS SET-TO

1. Entry of November 9, 1943, in Brooke, *War Diaries,* 468.
2. Ibid., entry of November 10, 1943, 468.
3. Entry of November 11, 1943, Bryant, *Triumph in the West,* 67. See also Brooke, *War Diaries,* 469.
4. Recommendation on strategy by the British chiefs of staff, November 11, 1943, in Bryant, *Triumph in the West,* 65.
5. Max Domarus, ed., *Hitler, Speeches and Proclamations 1932–1945: The Chronicle of a Dictatorship,* vol. 4, *The Years 1941 to 1945* (Wauconda, IL: Bolchazy-Carducci, 1977), 2843.
6. Ibid.
7. Ibid., 2837.
8. Ibid., 2838.
9. Ibid.
10. Ibid.
11. Ibid.
12. Ibid., 2840
13. Ibid.
14. Ibid., 2842.
15. Ibid., 2843.
16. Entry of 14.11.1943, in Joseph Goebbels, *Die Tagebücher von Joseph Goebbels* [The Diaries of Joseph Goebbbels], ed. Elke Froehlich (Munich: K. G. Saur, 1993), Band 10 (hereinafter *Tagebücher 10*), 290. Quotes from this source have been translated by the author.

9. MARSHALL: COMMANDER IN CHIEF AGAINST GERMANY

1. "Minutes of Meeting, Between the President and the Chiefs of Staff, held on board ship in The President's Cabin, 15 November 1943, at 1400," 4, Franklin D. Roosevelt, Papers as President: Map Room Papers, 1941–1945, Box 29, FDR Presidential Library, Hyde Park, NY.
2. Reynolds, *In Command of History,* 383.
3. Ibid.
4. Ibid., 381.
5. Ibid., 385.
6. Entry of October 27, 1943, in Alexander Cadogan, *The Diaries of Sir Alexander Cadogan, O.M., 1938–1945,* ed. David Dilks (London: Cassell, 1971), 571.
7. See entry of November 16, 1943, Moran, *Churchill,* 126–27.
8. Reynolds, *In Command of History,* 389.
9. "Minutes of Meeting, Between the President and the Chiefs of Staff, held on board ship in The President's Cabin, 15 November 1943."
10. Ibid. (Author's italics.)
11. Ibid. The word "all" would be important.

10. A WITCHES' BREW

1. Entry of November 16, 1943, in Harold Macmillan, *War Diaries: Politics and War in the Mediterranean, January 1943–May 1945* (London: Macmillan, 1984), 294. (Italics in original.)
2. Ibid.
3. Martin Gilbert, *Churchill and America* (New York: Free Press, 2005), xxiii–xxiv. "No lover ever studied every whim of his mistress as I did those of President Roosevelt," Churchill also reflected: ibid., 386.
4. Entry of November 16, 1943, Macmillan, *War Diaries*, 294.
5. Ibid., 295.
6. CHAR 20/122/43, Churchill Archives.
7. Ibid.
8. Quoted in Churchill, *The Second World War*, vol. 5, *Closing the Ring* (London: Cassell, 1952), 259.
9. CHAR 20/122/80–83, October 29, 1943, Churchill Archives.
10. Entry of November 25, 1943, Macmillan, *War Diaries*, 303.
11. Entry of November 18, 1943, Bryant, *Triumph in the West*, 7.
12. Entry of November 18, 1943, Brooke, *War Diaries*, 472.
13. Ibid.
14. Ibid., 472–73.
15. Entry of November 17, 1943, in Harry Butcher, *My Three Years with Eisenhower: The Personal Diary of Captain Harry C. Butcher, USNR, Naval Aide to General Eisenhower, 1942 to 1945* (New York: Simon & Schuster, 1946), 442.

11. FULLEST GUIDANCE

1. Letter of December 2, 1943, Papers of Sir John Martin, Churchill Archives Centre, Cambridge, UK.
2. "Former Naval Person to Admiral Queen," C-505/1, November 18, 1943, Kimball, *Churchill and Roosevelt*, 602.
3. Maurice Matloff, *Strategic Planning for Coalition Warfare, 1943–1944* (Washington, DC: Office of Chief of Military History, U.S. Government Printing Office, 1959), 338.
4. Forrest Pogue, *George C. Marshall*, vol. 3, *Organizer of Victory, 1943–1949* (New York: Viking, 1973), 301, quoting John Kennedy, *The Business of War*, 305.

12. ON BOARD THE *IOWA*

1. "Minutes of Meeting, Between the President and the Chiefs of Staff, held on board ship in The Admiral's Cabin, on Friday, 19 November 1943, at 1500," Franklin D. Roosevelt, Papers as President: Map Room Papers, 1941–1945, Box 29, FDR Presidential Library.
2. Ibid.
3. Ibid.
4. Ibid.

5. Ibid.
6. Ibid.
7. Ibid.
8. Ibid.
9. Ibid.
10. Entry of November 18, 1943, Bryant, *Triumph in the West,* 70.
11. "Minutes of Meeting, Between the President and the Chiefs of Staff, held on board ship in The Admiral's Cabin, on Friday, 19 November 1943."
12. Ibid.

13. IN THE FOOTSTEPS OF SCIPIO AND HANNIBAL

1. Entry of November 20, 1943, Typed FDR Diary, "War Conference in Cairo, Teheran, Malta."
2. Elliott Roosevelt, *As He Saw It* (New York: Duell, Sloan and Pearce, 1946), 132.
3. Ibid.
4. Entry of November 20, 1943, Typed FDR Diary, "War Conference in Cairo, Teheran, Malta."
5. Roosevelt, *As He Saw It,* 136.
6. Elliott Roosevelt, quoting his brother Franklin Jr., in ibid., 136–38.
7. Dwight D. Eisenhower, *Crusade in Europe* (New York: Doubleday, 1948), 195.
8. "Log of the President's Trip to Africa and the Middle East."
9. Eisenhower, *Crusade in Europe,* 195.
10. Kay Summersby, *Eisenhower Was My Boss* (London: Werner Laurie, 1949), 88.
11. Ibid., 86–87.
12. Entry of November 16, 1943, Moran, *Churchill,* 126–27.

14. TWO PIECES IN A CHESS GAME

1. Eisenhower, *Crusade in Europe,* 194.
2. "Mr. Churchill told me, confidentially, that it had been most embarrassing to him to have to tell Brooke" — to whom the supreme command had hitherto been promised — "of the change because he appreciated and sympathized with the anguished disappointment that Brooke was thus compelled to suffer": Dwight. D. Eisenhower, TS of handwritten draft chapter, unpublished memoirs, 11/17/66, "Churchill-Marshall (1)," 43, Kevin McCann Papers, in Papers, Post-Presidential, 1961–1969, Eisenhower Presidential Library, Abilene, KS.
3. Ibid., 44.
4. Eisenhower, *Crusade in Europe,* 195.
5. "Minutes of Meeting, Between the President and the Chiefs of Staff, on board ship in The Admiral's Cabin, on Friday, 19 November 1943."
6. Ibid.
7. Eisenhower, *Crusade in Europe,* 197.
8. Eisenhower, draft memoir TS, 45.

9. Ibid.
10. Ibid., 47.
11. Ibid., 46.
12. Eisenhower, *Crusade in Europe*, 197.
13. Ibid.
14. "He (admiral Leahy) did not feel we should accept this until we have fought out the matter of a Supreme Allied Commander" for all Europe: "Minutes of Meeting, Between the President and the Chiefs of Staff, on board ship in The Admiral's Cabin, on Friday, 19 November 1943."
15. Eisenhower, draft memoir TS, 46–47.
16. Entry of November 21, 1943, Typed FDR Diary, "War Conference in Cairo, Teheran, Malta."
17. Roosevelt, *As He Saw It*, 137.
18. Ibid.
19. Summersby, *Eisenhower Was My Boss*, 86.
20. The route flown, according to the President's log, was 1,851 miles: Entry of November 22, 1943, "Log of the President's Trip to Africa and the Middle East."
21. Roosevelt, *As He Saw It*, 138.
22. Entry of November 22, 1943, Leahy Diary.

15. AIRY VISIONS

1. Speech at the Lord Mayor's Day Luncheon at Mansion House, London, November 9, 1943: "No Time to Relax," in Winston Churchill, *The War Speeches of the Rt. Hon. Winston S. Churchill*, comp. Charles Eade, vol. 3 (London: Cassell, 1951).
2. Ibid., 65–68.
3. "A Decade of American Foreign Policy 1941–1949: Connally Resolution November 5, 1943," Avalon Project: Documents in Law, History and Diplomacy, http://avalon.law.yale.edu/20th_century/decade10.asp.
4. Entry of 7.11.1943, in Joseph Goebbels, *Die Tagebücher von Joseph Goebbels* [The Diaries of Joseph Goebbbels], ed. Elke Froehlich (Munich: K. G. Saur, 1993), Band 10 (hereinafter *Tagebücher 10*), 244. Quotes from this source have been translated by the author.
5. Ibid., entry of 10.11.1943, 267.
6. Ibid., entry of 20.11.1943, 322.
7. Ibid., entry of 25.11.1943, 356.

16. THE AMERICAN SPHINX

1. Entry of November 22, 1943, Typed FDR Diary, in "War Conference in Cairo, Teheran, Malta, etc., November 11–December 17, 1943," Franklin D. Roosevelt, Papers as President: The President's Official File, Part 1, 1933–1945, 200-3-N, Box 64, FDR Presidential Library, Hyde Park, NY.

2. Michael F. Reilly, as told to William J. Slocum, *Reilly of the White House* (New York: Simon & Schuster, 1947), 170.
3. Entry of November 22, 1943, "Log of the President's Trip to Africa and the Middle East, November–December 1943," Franklin D. Roosevelt, Papers as President: Map Room Papers, 1941–1945, Box 24, FDR Presidential Library, Hyde Park, NY.
4. Entry of November 22, 1943, Typed FDR Diary, "War Conference in Cairo, Teheran, Malta."
5. Entry of November 25, 1943, in Harold Macmillan, *War Diaries: Politics and War in the Mediterranean, January 1943–May 1945* (London: Macmillan, 1984), 203.
6. Entry of November 22, 1943, "Log of the President's Trip to Africa and the Middle East."
7. See ibid., entry of November 25, 1943, on Bryan's and Reilly's return from Tehran.
8. See Nigel Hamilton, *Commander in Chief: FDR's Battle With Churchill, 1943* (Boston: Houghton Mifflin Harcourt, 2016), 204–58.
9. Entry of November 22, 1943, Typed FDR Diary, "War Conference in Cairo, Teheran, Malta." Churchill was heard to say, of their outing: "The two most talkative people in the world meeting the most silent": Letter of December 2, 1943, Papers of Sir John Martin, Churchill Archives Centre, Cambridge, UK.

17. CHURCHILL'S "INDICTMENT"

1. Entry of November 18, 1943, Lord Moran, *Churchill: The Struggle for Survival, 1940–1965* (Boston: Houghton Mifflin, 1966), 129; also Winston Churchill, *The Second World War*, vol. 5, *Closing the Ring* (London: Cassell, 1952), 291.
2. Winston Churchill, quoting his "indictment of our mismanagement of operations in the Mediterranean," given to the British chiefs of staff, quoted in Churchill, *Closing the Ring*, 293.
3. Churchill Minute, "Future Operations in the European and Mediterranean Theatre," November 20, 1943, in Martin Gilbert, *Winston S. Churchill*, vol. 7, *Road to Victory: 1941–1945* (Toronto: Stoddart, 1986), 558–59.
4. Ibid., 558.
5. Ibid.
6. Entry of November 20, 1943, in Arthur Bryant, *Triumph in the West* (London: Collins, 1959), 74.
7. Ibid., entry of November 21, 1943, 74–75.
8. Ibid., entry of November 22, 1943, 75.
9. Ibid., annotation to entry of November 21, 1943, 75.
10. Hastings Lionel Ismay, *The Memoirs of General Lord Ismay* (London: Heinemann, 1960), 337.
11. Annotation to entry of November 21, 1943, in Bryant, *Triumph in the West*, 75.
12. Churchill, *Closing the Ring*, 289–90.
13. Elliott Roosevelt, *As He Saw It* (New York: Duell, Sloan and Pearce, 1946), 144.
14. Ibid.

15. Ibid.
16. Ibid., 142.
17. November 22, 1943, *Foreign Relations of the United States: The Conferences at Cairo and Tehran, 1943* (hereinafter *FRUS I*) (Washington, DC: U.S. Government Printing Office, 1961), 303.
18. Entry of November 23, 1943, in H. H. Arnold, *American Air Power Comes of Age: General "Hap" Arnold's World War II Diaries,* ed. John W. Huston (Maxwell Air Force Base, AL: Air University Press, 2002), 85.
19. Joseph W. Stilwell, *The Stilwell Papers* (New York: Sloane, 1948), 245; Barbara W. Tuchman, *Stilwell and the American Experience in China, 1911–1945* (New York: Macmillan, 1971), 403.

18. SHOWDOWN

1. "Meeting of the Combined Chiefs of Staff with Roosevelt and Churchill, November 24, 1943, 11 a.m., President's Villa," *FRUS I,* 334.
2. Ibid.
3. Entry of November 24, 1943, Bryant, *Triumph in the West,* 82.
4. Entry of November 24, 1943, Leahy Diary, William D. Leahy Papers, Library of Congress, Washington, DC.
5. "Meeting of the Combined Chiefs of Staff with Roosevelt and Churchill," *FRUS I,* 334.
6. Entry of November 24, 1943, in Arnold, *American Air Power Comes of Age,* 86.
7. The President's Log at Cairo, *FRUS I,* 298–99.
8. Entry of November 25, 1943, Leahy Diary.
9. In his written rebuttal, Churchill strenuously objected to the notion of a single supreme commander "to command all United Nations operations against Germany" — especially if it were to be an American officer who could then pronounce "in favour of concentrating on Overlord irrespective of the injury done to our affairs in the Mediterranean": "Memorandum by Prime Minister Churchill," Cairo, November 25, 1943 (Roosevelt Papers), in *FRUS I,* 407.
10. "Note by the British Chiefs of Staff," CCS 409, "Overlord and the Mediterranean," Cairo, November 25, 1943, *FRUS I,* 409.
11. Entry of November 26, 1943, Bryant, *Triumph in the West,* 84.
12. Entry of November 26, 1943, Leahy Diary.
13. Entry of November 26, 1943, Typed FDR Diary, "War Conference in Cairo, Teheran, Malta."

19. A VISION OF THE POSTWAR WORLD

1. See Nigel Hamilton, *Commander in Chief: FDR's Battle with Churchill, 1943* (Boston: Houghton Mifflin Harcourt, 2014), 22–23.
2. Memorandum by the Assistant Secretary of War (McCloy), Attachment 1, November 25, 1943, *Foreign Relations of the United States: The Conferences at Cairo and Tehran, 1943* (hereinafter *FRUS I*) (Washington, DC: U.S. Government Printing Office, 1961), 418.

3. Ibid.
4. See Hamilton, *Commander in Chief*, 370–72.
5. Memorandum by the Assistant Secretary of War (McCloy), *FRUS I*, 418.
6. Ibid.
7. See Hamilton, *Commander in Chief*, 69.
8. Entry of November 27, 1943, "Log of the President's Trip to Africa and the Middle East, November–December 1943," Franklin D. Roosevelt, Papers as President: Map Room Papers, 1941–1945, Box 24, FDR Presidential Library, Hyde Park, NY.

20. IN THE RUSSIAN COMPOUND

1. Susan Butler, *Roosevelt and Stalin: Portrait of a Partnership* (New York: Knopf, 2015), 42.
2. Hastings Lionel Ismay, *The Memoirs of General Lord Ismay* (London: Heinemann, 1960), 337.
3. Martin Gilbert, *Winston S. Churchill*, vol. 7, *Road to Victory: 1941–1945* (Toronto: Stoddart, 1986), quoting J. H. Colegrave, a British staff officer at Tehran, 568.
4. Letter of December 4, 1943, in Sarah Churchill, *Keep On Dancing: An Autobiography* (London: Coward, McCann & Geoghegan, 1981), 70.
5. Entry of November 6, 1943, in Geoffrey C. Ward, ed., *Closest Companion: The Unknown Story of the Intimate Friendship Between Franklin Roosevelt and Margaret Suckley* (Boston: Houghton Mifflin, 1995), 253.
6. Entry of November 28, 1943, "Log of the President's Trip to Africa and the Middle East."
7. Averell Harriman and Elie Abel, *Special Envoy to Churchill and Stalin, 1941–1946* (London: Hutchinson, 1976), 264.
8. "Minutes of Meeting, Between the President and the Joint Chiefs of Staff held in the American Legation, Tehran, Iran, on Sunday, 28 November 1943, at 11:30," 1, Franklin D. Roosevelt, Papers as President: Map Room Papers, Box 29, FDR Presidential Library.
9. Ibid.
10. Ibid., 3.
11. Ibid., 2.
12. Ibid., 3.
13. Ibid., 4.
14. Ibid.
15. Ibid., 3.
16. Ibid., 5.
17. Ibid., 4.

21. THE GRAND DEBATE

1. Keith Eubank, *Summit at Teheran: The Untold Story* (New York: William Morrow, 1985), 190.
2. Charles Bohlen, *Witness to History, 1929–1969* (New York: Norton, 1973), 135–36.

3. Ibid., 136–37.
4. Ibid., 137.
5. Ibid., 139.
6. Ibid.
7. Ibid.
8. Ibid.
9. "Roosevelt-Stalin Meeting, November 28, 1943, 3 p.m., Roosevelt's Quarters, Soviet Embassy," Bohlen Minutes, *FRUS I,* 483.
10. Ibid., 484.
11. Ibid., 484–85.
12. Ibid.
13. H. H. Arnold, *American Airpower Comes of Age: General Henry H. "Hap" Arnold's World War II Diaries,* vol. 2, ed. John W. Huston (Maxwell Air Force Base, AL: Air University Press), 89.
14. Harriman and Abel, *Special Envoy,* 265.
15. Ibid.
16. A. H. Birse, *Memoirs of an Interpreter* (London: Michael Joseph, 1967), 155.
17. Bohlen, *Witness to History,* 142.
18. "First Plenary Meeting, November 28, 1943, 4 p.m., Conference Room, Soviet Embassy," Bohlen Minutes, *FRUS I,* 487.
19. Ibid.
20. Birse, *Memoirs of an Interpreter,* 155.
21. "First Plenary Meeting," Bohlen Minutes, *FRUS I,* 488–89.
22. Ibid., 489.
23. Ibid.
24. Ibid.
25. Ibid., 489–90.
26. Ibid., 490–91.
27. Ibid., 491.
28. Entry of November 28, 1943, in Arthur Bryant, *Triumph in the West* (London: Collins, 1959), 89.
29. Ibid.
30. Entry of November 28, 1943, in Lord Moran, *Churchill: The Struggle for Survival, 1940–1965* (Boston: Houghton Mifflin, 1966), 135.
31. "First Plenary Meeting," Bohlen Minutes, *FRUS I,* 491.
32. Ibid., 495.
33. Ibid.
34. Ibid.
35. Ibid.
36. Ibid., 496.
37. Ibid.
38. Ibid., 497.
39. Annotation to entry of November 28, 1943, Bryant, *Triumph in the West,* 90.
40. Ibid.

41. Ismay, *Memoirs,* 338.
42. Entry of November 28, 1943, Moran, *Churchill,* 134.
43. Ismay, *Memoirs,* 338.
44. Ibid., 339.
45. "Discussion following 'Lecture on Some Aspects of the High Command in World War II by Fleet Admiral Ernest J. King,'" National War College, Washington, DC, April 29, 1947, in Papers of Ernest J. King, 1908–1966, Box 29, Library of Congress.
46. Ibid.
47. "First Plenary Meeting," Bohlen Minutes, *FRUS I,* 497–508.
48. Raymond Clapper, "Confidential" — TS notes on "sixth off-the-record seminar with Admiral King," Alexandria, VA, July 26, 1945, Raymond Clapper Papers, 1908–1962, Library of Congress, Washington, DC.
49. Entry of November 29, 1943, Moran, *Churchill,* 135.

22. A REAL SCARE

1. Bohlen, *Witness to History,* 143.
2. Ibid.
3. "Tripartite Dinner Meeting, November 28, 1943, 8:30 p.m., Roosevelt's Quarters, Soviet Embassy," *FRUS I,* 509.
4. Ibid., 510.
5. Ibid.
6. Bohlen, *Witness to History,* 143–44.
7. Robert E. Sherwood, *Roosevelt and Hopkins: An Intimate History* (New York: Harper, 1948), 781.
8. Birse, *Memoirs of an Interpreter,* 156.
9. Entry of November 28, 1943, Leahy Diary, William D. Leahy Papers, Library of Congress, Washington, DC.
10. William M. Rigdon, with James Derieux, *White House Sailor* (Garden City, NY: Doubleday, 1962), 84.
11. Ibid.
12. Steven Lomazow and Eric Fettmann, *FDR's Deadly Secret* (New York: PublicAffairs, 2009), 11; also Robert H. Ferrell, *The Dying President: Franklin D. Roosevelt, 1944–1945* (Columbia: University of Missouri Press, 1998), 23.
13. Ferrell, *The Dying President,* 17, 23.
14. Harry Goldsmith, *A Conspiracy of Silence: The Health and Death of Franklin D. Roosevelt* (New York: iUniverse, 2007), 172.

23. IMPASSE

1. Entry of November 29, 1943, Leahy Diary.
2. Rigdon, *White House Sailor,* 84.
3. Bohlen, *Witness to History,* 145.
4. Entry of November 29, 1943, Leahy Diary.

5. "Tripartite Military Meeting, November 29, 1943, 10:30 a.m., Conference Room, Soviet Embassy," *FRUS I*, 515–27.
6. Ibid.
7. Ibid., 520.
8. Ibid., 524.

24. PRICKING CHURCHILL'S BUBBLE

1. Ward, *Closest Companion*, 299.
2. Entry of November 29, 1943, Moran, *Churchill*, 136.
3. Ibid.
4. Elliott Roosevelt, *As He Saw It* (New York: Duell, Sloan and Pearce, 1946), 179.
5. "Roosevelt-Stalin Meeting, November 29, 2:45 p.m., Roosevelt's Quarters, Soviet Embassy," Bohlen Minutes, *FRUS I*, 530. The sketch, reproduced in Sherwood, *Roosevelt and Hopkins*, 789, showed three separate bubbles for the three parts of the proposed organization—with forty members of the main "U.N." body.
6. "Roosevelt-Stalin Meeting," Bohlen Minutes, *FRUS I*, 530. Roosevelt's annotation on his sketch read: "ILO—Health, Agric, Food": Sherwood, *Roosevelt and Hopkins*, 789.
7. "Roosevelt-Stalin Meeting," Bohlen Minutes, *FRUS I*, 530.
8. Bohlen, *Witness to History*, 145.
9. "Roosevelt-Stalin Meeting," Bohlen Minutes, *FRUS I*, 531.
10. Ibid., 532.
11. Roosevelt, *As He Saw It*, 180.
12. Entry of November 29, 1943, Moran, *Churchill*, 136.
13. "Roosevelt-Stalin Meeting," Bohlen Minutes, *FRUS I*, 535; entry of November 29, 1943, Moran, *Churchill*, 137.
14. Bohlen, *Witness to History*, 145.
15. Entry of November 29, 1943, Moran, *Churchill*, 137.
16. "Second Plenary Meeting, November 29, 1943, 4 p.m., Conference Room, Soviet Embassy," *FRUS I*, 535.
17. Ibid.
18. Ibid.
19. Entry of November 29, 1943, Bryant, *Triumph in the West*, 92.
20. Ibid.
21. "Second Plenary Meeting," *FRUS I*, 535.
22. Ibid., 536.
23. Ibid., 538.
24. Bryant, *Triumph in the West*, 93. Also further entry in Alan Brooke, *War Diaries, 1939–1945: Field Marshal Lord Alanbrooke*, eds. Alex Danchev and Daniel Todman (Berkeley: University of California Press, 2001), 485.
25. Entry of November 29, 1943, Leahy Diary.

26. "Second Plenary Meeting," *FRUS I*, 546.
27. Ibid.
28. Ibid., 545.
29. Ibid., 546.
30. Ibid., 547.
31. Ibid.
32. Ibid.
33. Ibid.
34. Ibid., 548.
35. Ibid., 549–50.
36. Ibid., 550–51.
37. November 18, 1943, Nigel Hamilton, *Master of the Battlefield: Monty's War Years, 1942–1944* (New York: McGraw Hill, 1984), 455.
38. Ibid.
39. Sherwood, *Roosevelt and Hopkins*, 788–89.

25. WAR AND PEACE

1. Entry of November 29, 1943, Leahy Diary.
2. Entry of November 29, 1943, Moran, *Churchill*, 138.
3. Ibid.
4. Entry of November 30, 1943, "Log of the President's Trip to Africa and the Middle East." See also "Tripartite Dinner Meeting, November 30, 1943, 8:30 p.m., British Legation," Bohlen Minutes, *FRUS I*, 584.
5. "Tripartite Dinner Meeting," Bohlen Minutes, *FRUS I*, 585.
6. Entry of November 29, 1943, Brooke, *War Diaries*, 485.

26. A COMMANDER FOR OVERLORD

1. Appendix D, "Log of the President's Trip to Africa and the Middle East, November–December 1943," Franklin D. Roosevelt, Papers as President: Map Room Papers, 1941–1945, Box 24, FDR Presidential Library, Hyde Park, NY.
2. William D. Leahy, *I Was There* (New York: McGraw-Hill, 1950), 207.
3. Appendix D, "Log of the President's Trip to Africa and the Middle East."
4. Ibid.
5. Entry of 1.12.1943, Joseph Goebbels, *Die Tagebücher von Joseph Goebbels* [The Diaries of Joseph Goebbbels], ed. Elke Froehlich (Munich: K. G. Saur, 1993) Band 10 (hereinafter *Tagebücher 10*), 399. Quotes from this source have been translated by the author.
6. Joseph Lelyveld, *His Final Battle: The Last Months of Franklin Roosevelt* (New York: Knopf, 2016), 40.
7. Entry of December 2, 1943, Leahy Diary, William D. Leahy Papers, Library of Congress, Washington, DC.

27. A MOMENTOUS DECISION

1. Joseph Persico, *Roosevelt's Centurions: FDR and the Commanders He Led to Victory in World War II* (New York: Random House, 2013), 340.
2. Stimson to Hopkins, for the President, November 10, 1943, Stimson Diary, Henry L. Stimson Papers, Yale University Library, New Haven, CT.
3. General Pershing to the President, September 16, 1943, in Forrest Pogue, *George C. Marshall*, vol. 3, *Organizer of Victory, 1943–1949* (New York: Viking, 1973), 272.
4. Roosevelt to Pershing, September 20, 1943, *The Papers of George Catlett Marshall*, vol. 4, 129.
5. Ibid.
6. Ibid., 343.
7. See Nigel Hamilton, *The Mantle of Command: FDR at War, 1941–1942* (Boston: Houghton Mifflin Harcourt, 2014), 330–42, and Hamilton, *Commander in Chief: FDR's Battle with Churchill, 1943* (Boston: Houghton Mifflin Harcourt, 2016), 55–60 and 97–99.
8. George C. Marshall, *Interviews and Reminiscences for Forrest C. Pogue* (Lexington, VA: George C. Marshall Foundation, 1996), 343.
9. Later still General Marshall even wondered whether the President had been pressed to make such an urgent decision because the matter had been "stirred around politically over here in this country (the U.S.A.)," or "in the press, rather": ibid., 343. This was, however, a speculation Marshall made more than a decade after the events, when attempting to recall the exact circumstances, and is not indicated by any contemporary documents.
10. Robert E. Sherwood, *Roosevelt and Hopkins: An Intimate History* (New York: Harper, 1948), 803.
11. Winston Churchill, *The Second World War,* vol. 5, *Closing the Ring* (London: Cassell, 1952), 340 and 370.
12. "Meeting of the Combined Chiefs of Staff with Roosevelt and Churchill, December 4, 1943, 11 a.m., Roosevelt's Villa," *Foreign Relations of the United States: The Conferences at Cairo and Tehran, 1943* (hereinafter *FRUS I*), 675. Also Churchill, *Closing the Ring,* 362.
13. Sherwood, *Roosevelt and Hopkins,* 803.
14. Ibid.
15. Ibid., 802.
16. See Churchill, *Closing the Ring,* 370 inter alia.
17. Sherwood, *Roosevelt and Hopkins,* 803.

28. A BAD TELEGRAM

1. Entry of December 8, 1943, Stimson Diary.
2. Ibid., entry of December 11, 1943.
3. Ibid., Weekly Survey, December 9, 1943.

4. Entry of December 4, 1943, H. H. Arnold, *American Air Power Comes of Age: General "Hap" Arnold's World War II Diaries,* ed. John W. Huston (Maxwell Air Force Base, AL: Air University Press, 2002), 95.
5. Entry of December 4, 1943, in Arthur Bryant, *Triumph in the West* (London: Collins, 1959), 105.
6. Churchill, *Closing the Ring,* 370.

29. PERFIDIOUS ALBION REDUX

1. Entry of December 3, 1943, Arnold, *American Air Power Comes of Age,* 94.
2. Ibid., entry of December 4, 1943, 95.
3. Entry of December 3, 1943, Bryant, *Triumph in the West,* 104.
4. Churchill, *Closing the Ring,* 800–801.
5. Entry of December 4, 1943, Leahy Diary.
6. Ibid., entry of December 5, 1943.
7. See draft FDR cable to Chiang Kai-shek of December 5, 1943, in Sherwood, *Roosevelt and Hopkins,* 801.
8. William D. Leahy, *I Was There* (New York: McGraw-Hill, 1950), 213–14.
9. Ibid.
10. Marshall, *Interviews and Reminiscences,* 622. Concerning Churchill's obsession with Rhodes and the Aegean, as well as his deliberate failure to consult the U.S. Joint Chiefs when ordering the unilateral invasion of the islands in September 1943, see Marshall interview of November 15, 1956 (ibid., 321).
11. Ernest J. King and Walter M. Whitehill, *Fleet Admiral King: A Naval Record* (New York: Norton, 1952), 525.
12. Joseph W. Stilwell, *The Stilwell Papers* (New York: Sloane, 1948), 251–54.
13. King and Whitehill, *Fleet Admiral King,* 525.
14. Facsimile of original in Dwight D. Eisenhower, *Crusade in Europe* (New York: Doubleday, 1948), 208. According to the Map Room files at the White House, declassified three decades later, the historic cable was received in Washington that night "as Black 79 from the President" for "transmittal to Marshal Stalin." It had been rephrased as: "The decision had been made to appoint General Eisenhower immediately to command of cross-Channel operations," and was signed "Roosevelt." It was then released to be dispatched onward to Moscow by the Navy Code Room at 9:15 p.m., December 6, 1943, Eastern Standard Time, or 4:15 a.m. in Cairo, December 7, 1943. A note, in hand, confirms it was received in Moscow at 10:20 a.m.: Franklin D. Roosevelt, Papers as President: Map Room Papers, 1941–1945, Box 8, FDR Presidential Library.

30. IN THE FIELD WITH EISENHOWER

1. Transcript of part of a letter (undoubtedly December 8, 1944) to Daisy Suckley, in Typed FDR Diary, in "War Conference in Cairo, Teheran, Malta, etc., November 11–December 17, 1943," Franklin D. Roosevelt, Papers as President: The

President's Official File, Part 1, 1933–1945, 200-3-N, Box 64, FDR Presidential Library.

2. Entry of December 11, 1944, in Geoffrey C. Ward, ed., *Closest Companion: The Unknown Story of the Intimate Friendship Between Franklin Roosevelt and Margaret Suckley* (Boston: Houghton Mifflin, 1995), 261.
3. Entry of December 7, 1943, Typed FDR Diary, "War Conference in Cairo, Teheran, Malta."
4. Ross T. McIntire, *White House Physician* (New York: Putnam's, 1946), 177.
5. Eisenhower, *Crusade in Europe*, 207.
6. Ibid., 204.
7. Letter of November 20, 1943, Typed FDR Diary, "War Conference in Cairo, Teheran, Malta."

31. A FLAP AT MALTA

1. McIntire, *White House Physician*, 178.
2. Ibid.
3. Ibid.
4. William M. Rigdon, with James Derieux, *White House Sailor* (Garden City, NY: Doubleday, 1962), 93. See also entry of December 8, 1943, in "Log of the President's Trip to Africa and the Middle East."
5. Lord Gort had won the Victoria Cross for gallantry as a battalion commander in World War I, and had then commanded the British Expeditionary Force in the ill-fated advance into Belgium, then retreat to Dunkirk in 1940.
6. Entry of December 8, 1943, "Log of the President's Trip to Africa and the Middle East."
7. Entry of December 8, 1943, Leahy Diary.
8. Entry of December 8, 1943, "Log of the President's Trip to Africa and the Middle East."
9. Ibid.
10. Entry of December 8, 1943, Leahy Diary.
11. Sherwood, *Roosevelt and Hopkins*, 803.
12. Ibid.
13. Keith Eubank, *Summit at Tehran: The Untold Story* (New York: William Morrow, 1985), 287.
14. Eisenhower, *Crusade in Europe*, 194.
15. Eisenhower, draft memoir, tss 45, Eisenhower Library.

32. HOMEWARD BOUND!

1. Entry of December 8, 1943, Typed FDR Diary, "War Conference in Cairo, Teheran, Malta."
2. Entry of December 8, 1943, "Log of the President's Trip to Africa and the Middle East."

3. Leahy, *I Was There,* 215.
4. Rick Atkinson, *The Day of Battle: The War in Sicily and Italy, 1943–1944* (New York: Henry Holt, 2007), 297.
5. Michael F. Reilly, as told to William J. Slocum, *Reilly of the White House* (New York: Simon & Schuster, 1947), 188.
6. Eisenhower, *Crusade in Europe,* 207.
7. Kay Summersby, *Eisenhower Was My Boss* (London: Werner Laurie, 1949), 101.
8. Extract from letter, December 9, 1943, Typed FDR Diary, "War Conference in Cairo, Teheran, Malta."
9. John McCrea, "'Iowa'—President and Joint Chiefs of Staff to Africa and Return," manuscript memoir, John L. McCrea Papers, FDR Presidential Library.
10. Entry of December 10, 1943, in Typed FDR Diary, "War Conference in Cairo, Teheran, Malta."

33. THE ODYSSEY IS OVER

1. Entry of December 16, 1943, Typed FDR Diary, "War Conference in Cairo, Teheran, Malta."
2. Entries of December 9, 11, and 16, 1943, Leahy Diary.
3. Entry of December 16, 1943, Typed FDR Diary, "War Conference in Cairo, Teheran, Malta."
4. Entry of December 15, 1943, "Log of the President's Trip to Africa and the Middle East."
5. Entry of December 16, 1943, Typed FDR Diary, "War Conference in Cairo, Teheran, Malta."
6. Entry of December 16, 1943, Leahy Diary.
7. Appendix K, "The President's Remarks on Leaving the USS *Iowa,* 16 December 1943," in "Log of the President's Trip to Africa and the Middle East."
8. Ibid.
9. Ibid.
10. McCrea, "'Iowa'—President and Joint Chiefs of Staff to Africa and Return."
11. Ibid.
12. Entry of December 16, 1943, Leahy Diary.
13. Entry of December 17, 1943, Stimson Diary.
14. Entry of December 5, 1943, Ickes Diary, Ickes Papers, Library of Congress.
15. Entry of December 17, 1943, Ickes Diary.
16. Ibid.
17. Entry of December 17, 1943, Ickes Diary.
18. Entry of December 17, 1943, Stimson Diary.
19. John H. Crider, Special to the *New York Times,* December 17, 1943.
20. Press and Radio Conference #927, December 17, 1943, 4:17 p.m., Press Conferences of President Franklin D. Roosevelt, 1933–1945, Series 1: Press Conference Transcripts, FDR Presidential Library.
21. Ibid.

22. "Senators Hold Up Patton Promotion," UP Report, *New York Times,* December 16, 1943.
23. Press and Radio Conference #927.
24. Eisenhower, *Crusade in Europe,* 208.
25. Admiral King was a great deal less cautious, or simply more trusting: at a private meeting with five chosen press correspondents, including the Associated Press reporter Lyle Wilson, King shared with them an account of what had happened, including the choice of Eisenhower to command Overlord, though asking them to hold back publication until after the President's Christmas Eve broadcast: "Admiral King, Saturday night, December 18, 1943," in Raymond Clapper Papers, Library of Congress.

34. CHURCHILL'S RESURRECTION

1. Entry of December 17, 1943, in "Secret Diary" of Lord Halifax, Papers of Lord Halifax, Hickleton Papers, Borthwick Institute of Historical Research, University of York, Yorkshire, England.
2. Ibid.
3. Ibid.
4. Letter of December 16, 1943, Martin Gilbert, *Winston S. Churchill,* vol. 7, *Road to Victory: 1941–1945* (Toronto: Stoddart, 1986), 609.
5. Entry of December 18, 1943, Stimson Diary, Henry L. Stimson Papers, Yale University Library, New Haven, CT.
6. Ibid.
7. Ibid.
8. Ibid.
9. Ibid.
10. Ibid., entry of December 10, 1943.
11. Ibid., entry of December 18, 1943.
12. "Evening Situation Report, probably December 20, 1943," section titled "The West, danger of invasion," in Helmut Heiber, ed., *Hitler and His Generals: Military Conferences 1942–1945* (New York: Enigma Books, 2003), 311.

35. IN THE PINK AT HYDE PARK

1. Entry of December 23, 1943, in William D. Hassett, *Off the Record with F.D.R.: 1942–1945* (New Brunswick, NJ: Rutgers University Press, 1958), 222.
2. Ibid.
3. Ross T. McIntire, *White House Physician* (New York: Putnam's, 1946), 180.
4. Entry of December 24, 1943, Hassett, *Off the Record,* 223.
5. Russell D. Buhite and David W. Levy, eds., *FDR's Fireside Chats* (Norman: University of Oklahoma Press, 1992), 272–81.
6. Press and Radio Conference #929, December 28, 1943, 4:07 p.m., EWT, Press

Conferences of President Franklin D. Roosevelt, 1933–1945, Series 1: Press Conference Transcripts, FDR Presidential Library, Hyde Park, NY.

7. Ibid.
8. Entry of December 28, 1943, in Geoffrey C. Ward, ed., *Closest Companion: The Unknown Story of the Intimate Friendship Between Franklin Roosevelt and Margaret Suckley* (Boston: Houghton Mifflin, 1995), 264.
9. Ibid., entry of December 30, 1943, 266.
10. Ibid.

36. SICK

1. McIntire, *White House Physician,* 183.
2. Ibid., 182.
3. Entry of December 19, 1943, Diaries of William Lyon Mackenzie King, Library and Archives Canada, Ottawa (hereinafter Mackenzie King Diary).
4. Washington, DC, Radio Address to Nation, State of the Union Message to Congress, January 11, 1944 (speech file 1501), Franklin D. Roosevelt: Master Speech File, 1898–1945, Box 76, FDR Presidential Library.
5. Samuel L. Rosenman, *Working with Roosevelt* (New York: Harper, 1952), 422.
6. Ibid, 421.
7. Annual Message to Congress — State of the Union (speech file 1501), January 11, 1944, Franklin D. Roosevelt, Master Speech File, 1898–1945, Box 76, FDR Presidential Library, Hyde Park; also in Buhite and Levy, *FDR's Fireside Chats*, 283–93.
8. Ibid.
9. Ibid.
10. Ibid.
11. Rosenman, *Working with Roosevelt,* 427.
12. Ibid.

37. ANZIO

1. Lord Moran, *Churchill: The Struggle for Survival, 1940–1965* (Boston: Houghton Mifflin, 1966), 152.
2. Winston Churchill, *The Second World War,* vol. 5, *Closing the Ring* (London: Cassell, 1952), 292.
3. Ibid., 378.
4. Grand #736, December 22, 1943, in Gilbert, *Road to Victory,* 618.
5. Mark Clark, *Calculated Risk: His Personal Story of the War in North Africa and Italy* (London: Harrap, 1951), 243–45.
6. Entry of December 13, 1943, in Arthur Bryant, *Triumph in the West* (London: Collins, 1959), 120.
7. For Montgomery's predictions regarding Anzio, see Nigel Hamilton, *Master of the Battlefield: Monty's War Years* (New York: McGraw-Hill, 1983), 441 et seq.

8. Dwight D. Eisenhower, *Crusade in Europe* (New York: Doubleday, 1948), 212.
9. "Record of the Conference of 25 December 1943 (General Hollis to War Cabinet Offices)," in Gilbert, *Road to Victory*, 620.
10. Ibid.
11. C-521, Prime Minister to President Roosevelt, December 26, 1943, Warren F. Kimball, ed., *Churchill and Roosevelt: The Complete Correspondence*, vol. 2, *Alliance Forged, November 1942–February 1945* (Princeton, NJ: Princeton University Press, 1984), 633.
12. Ibid., R-427 cable, "Personal and Secret for the Former Naval Person," December 27, 1943, 636.
13. Entry of December 27, 1943, Leahy Diary, William D. Leahy Papers, Library of Congress, Washington, DC.
14. Gilbert, *Road to Victory*, 606.

38. THE PRESIDENT'S UNPLEASANT ATTITUDE

1. Entry of 26.12.1943, in Joseph Goebbels, *Die Tagebücher von Joseph Goebbels* [The Diaries of Joseph Goebbbels], ed. Elke Froehlich (Munich: K. G. Saur, 1993), Band 10 (hereinafter *Tagebücher 10*), 550. Quotes from this source have been translated by the author.
2. Cable 6513 to Bedell Smith, January 5, 1943, in Dwight D. Eisenhower, *The Papers of Dwight David Eisenhower: The War Years*, ed. Alfred Chandler, vol. 3 (Baltimore, MD: Johns Hopkins Press, 1970), 1652.
3. Ibid., note 3, 1651, and note 4, 1653.
4. Entry of January 12, 1944, Stimson Diary.
5. Ibid.
6. Ibid.
7. Ibid.
8. Entry of January 7, 1944, Harold D. Smith Diary, quoted in Robert H. Ferrell, *The Dying President: Franklin D. Roosevelt, 1944–1945* (Columbia: University of Missouri Press, 1998), 30.
9. Francis Bacon, first Viscount St. Alban, Aphorism 46, *Novum Organum*, 1620.
10. Entry of January 19, 1944, in Alan Brooke, *War Diaries, 1939–1945: Field Marshal Lord Alanbrooke*, eds. Alex Danchev and Daniel Todman (Berkeley: University of California Press, 2001), 515.
11. Ibid., entry of January 22, 1944, 515.
12. Entry of January 31, 1944, Bryant, *Triumph in the West*, 142.
13. Ibid., entry of February 3, 1944, 143.
14. Addendum to entry of September 23, 1944, Moran, *Churchill*, 188.
15. Entry of February 29, 1944, Bryant, *Triumph in the West*, 160.
16. Ibid., entry of February 25, 1944, 154.
17. Entry of February 25, 1944, Brooke, *War Diaries*, 525.

18. Ibid., entry of February 28, 1944, 527.
19. Ibid., annotation to entry of February 25, 1944, 525.
20. Ibid., entry of March 3, 1944, 528.
21. Ibid., entries of March 8, 13, and 14, 530–31.

39. CRIMES AGAINST HUMANITY

1. McIntire, *White House Physician*, 182.
2. Entry of March 13, 1944, Stimson Diary.
3. R-506, March 20, 1944, Warren F. Kimball, ed., *Churchill and Roosevelt: The Complete Correspondence*, vol. 3, *Alliance Declining, February 1944–April 1945* (Princeton, NJ: Princeton University Press, 1984), 59.
4. Entry of March 23, 1944, Ward, *Closest Companion*, 286.
5. Entry of March 24, 1944, Hassett, *Off the Record*, 239.
6. Press and Radio Conference #944, March 24, 1944, 11:09 a.m., Press Conferences of President Franklin D. Roosevelt, 1933–1945, Series 1: Press Conference Transcripts, FDR Presidential Library, Hyde Park, NY.
7. Ibid. Italics added.
8. Ibid.
9. Ibid.
10. Ibid.
11. Ibid.
12. Adam Hochschild, *King Leopold's Ghost: A Story of Greed, Terror and Heroism in Colonial Africa* (Boston: Houghton Mifflin, 1999), 96.
13. Press and Radio Conference #944.
14. Ibid.

40. LATE LOVE

1. Jean Edward Smith, *FDR* (New York: Random House, 2007), 156.
2. Elliott Roosevelt, *An Untold Story: The Roosevelts of Hyde Park* (New York: Putnam, 1973), 81.
3. Jonathan Daniels, *Washington Quadrille* (New York: Doubleday, 1968), 145.
4. Resa Willis, *Franklin and Lucy: Lovers and Friends* (New York: Routledge, 2004), 38–40; Smith, *FDR*, 156–57.
5. Joseph P. Lash, *Eleanor and Franklin: The Story of Their Relationship* (New York: Norton, 1971), 220.
6. Willis, *Franklin and Lucy*, 30.
7. James Roosevelt with Bill Libby, *My Parents: A Differing View* (Chicago: Playboy, 1976), 101.
8. Willis, *Franklin and Lucy*, 33.
9. Smith, *FDR*, 214.
10. Eleanor Roosevelt, "If You Ask Me," *McCalls*, January 1952.

11. Willis, *Franklin and Lucy*, 61.
12. Elizabeth Shoumatoff, *FDR's Unfinished Portrait: A Memoir* (Pittsburgh: University of Pittsburgh Press, 1990), 72.

41. IN THE LAST STAGES OF CONSUMPTION

1. Entry of March 25, 1944, Ward, *Closest Companion*, 287.
2. Entry of March 24, 1944, Hassett, *Off the Record*, 240.
3. Entry of March 27, 1944, Ward, *Closest Companion*, 288.
4. Ferrell, *The Dying President*, 37.
5. Howard Bruenn, "Clinical Notes on the Illness and Death of President Franklin D. Roosevelt," *Annals of Internal Medicine* 72, no. 4 (1970): 579–80.
6. Doris Kearns Goodwin, *No Ordinary Time: Franklin and Eleanor Roosevelt: The Home Front in World War II* (New York: Simon & Schuster, 1994), 494.
7. Howard Bruenn, typewritten notes, undated, Harold Bruenn Papers, 1944–1946, FDR Presidential Library.
8. See Ward, *Closest Companion*, 289, and Steven Lomazow and Eric Fettmann, *FDR's Deadly Secret* (New York: PublicAffairs, 2009), 98.
9. Bruenn, typewritten notes, Bruenn Papers.
10. Bruenn, "Clinical Notes on the Illness and Death of President Franklin D. Roosevelt."
11. Jan Herman, Historian, "Oral History with LCDR (ret) Howard Bruenn," January 31, 1990, Office of Medical History, Bureau of Medicine and Surgery, Washington, DC, 3.
12. Ibid.
13. Bruenn, "Clinical Notes on the Illness and Death of President Franklin D. Roosevelt."
14. Lomazow and Fettmann, *FDR's Deadly Secret*, 103.
15. Howard Bruenn, typewritten notes, Bruenn Papers.
16. Ibid.
17. Ibid.
18. Bruenn, "Clinical Notes on the Illness and Death of President Franklin D. Roosevelt."
19. Bruenn, typewritten notes, Bruenn Papers.
20. Entry of March 28, 1944, Ward, *Closest Companion*, 289.
21. Ibid.
22. Bruenn, typewritten notes, Bruenn Papers.
23. Ibid.
24. Bruenn, "Clinical Notes on the Illness and Death of President Franklin D. Roosevelt."
25. Ibid.
26. Lomazov and Fettmann, *FDR's Deadly Secret*, 104.
27. Herman, "Oral History with LCDR (ret) Howard Bruenn," 4.
28. Bruenn, "Clinical Notes on the Illness and Death of President Franklin D. Roosevelt."
29. Quoted in Joseph Lelyveld, *His Final Battle: The Last Months of Franklin Roosevelt* (New York: Knopf, 2016), 104.
30. Herman, "Oral History with LCDR (ret) Howard Bruenn," 3.
31. Bruenn, typewritten notes, Bruenn Papers.

32. Bruenn, "Clinical Notes on the Illness and Death of President Franklin D. Roosevelt."
33. McIntire, *White House Physician*, 187.
34. Quoted in editor's annotation, Ward, *Closest Companion*, 289.
35. Ibid., entry of May 5, 1944, 296.
36. McIntire, *White House Physician*, 185.

42. "THIS ATTACK WILL DECIDE THE WAR"

1. "The West, danger of invasion"; "Evening Situation Report, probably December 20, 1943," in Helmut Heiber, ed., *Hitler and His Generals: Military Conferences 1942–1945* (New York: Enigma Books, 2003), 311.
2. Hitler Rede from the Wolfschanze, or Wolf's Lair, January 30, 1944, Deutsches Nachrichtenbüro, Max Domarus, ed., *Hitler, Speeches and Proclamations 1932–1945: The Chronicle of a Dictatorship*, vol. 4, *The Years 1941 to 1945* (Wauconda, IL: Bolchazy-Carducci, 1977), 2872.
3. Ibid., 2875–76.
4. Statistisches Reichsamt (Hrsg.): *Statistisches Jahrbuch für das Deutsche Reich*, 1919–1941/42.
5. See, among others, Ian Kershaw, *The End: The Defiance and Destruction of Hitler's Germany, 1944–1945* (New York: Penguin, 2011).
6. Entry of 25.2.1944, Joseph Goebbels, *Die Tagebücher von Joseph Goebbels* [The Diaries of Joseph Goebbbels], ed. Elke Froehlich (Munich: K. G. Saur, 1993), Band 11 (hereinafter *Tagebücher 11*), 347–48.
7. Entry of 4.3.1944, Goebbels, *Tagebücher 11*, 399–40.
8. Ibid., 340.
9. Entry of 18.4.1944, Goebbels, *Tagebücher 11*, 131.
10. Detlef Vogel, "German and Allied Conduct of the War in the West," Part 2 of Horst Boog, Gerhard Krebs, and Detlef Vogel, *Germany and the Second World War*, vol. 7 (Oxford: Oxford University Press, 2006), 509.
11. Ian Kershaw, *Hitler, 1936–1945: Nemesis* (London: Allen Lane, 2000), 631.
12. Santi Corvaja, *Hitler and Mussolini: The Secret Meetings* (New York: Enigma Books, 2008), 282.
13. Ibid., 283.

43. SIMPLICITY OF PURPOSE

1. Entry of May 5, 1944, in Geoffrey C. Ward, ed., *Closest Companion: The Unknown Story of the Intimate Friendship Between Franklin Roosevelt and Margaret Suckley* (Boston: Houghton Mifflin, 1995), 296.
2. C-624, Prime Minister to President, March 18, 1944, Warren F. Kimball, *Churchill and Roosevelt: The Complete Correspondence*, vol. 3, *Alliance Declining, February 1944–April 1945* (Princeton, NJ: Princeton University Press, 1984), 54.
3. Ibid., R-506, "From the President to the Former Naval Person," March 24, 1944, 60.

4. Entry of March 16, 1944, Stimson Diary, Henry L. Stimson Papers, Yale University Library, New Haven, CT.
5. Ibid., entry of March 22, 1944.
6. The statistical figure was later recorded as 75 percent: Mark Wells, *Courage and Air Warfare: The Allied Aircrew Experience in the Second World War* (London: Frank Cass, 1995), 46.
7. C-643, Prime Minister to President, April 12, 1944, in Kimball, *Churchill and Roosevelt*, vol. 2, 87.
8. Memorandum for the Secretary of State, April 1, 1944, in Franklin D. Roosevelt, *FDR: His Personal Letters, 1928–1945*, ed. Elliott Roosevelt, vol. 2 (New York: Duell, Sloan and Pearce, 1950), 1504.
9. Ibid., Memorandum for the Secretary of State, April 5, 1944, 1505.
10. Eleanor Roosevelt, letters of March 30 and April 2, 1944, to Joseph Lash, in Joseph P. Lash, *Eleanor and Franklin: The Story of Their Relationship* (New York: Norton, 1971), 697.
11. Ibid.
12. Footnote, quoting Eleanor Roosevelt and Anna, her daughter, in Lash, *Eleanor and Franklin*, 697.
13. Quoted in Joseph Lelyveld, *His Final Battle: The Last Months of Franklin Roosevelt* (New York: Knopf, 2016), 103.
14. Lieutenant Commander George Elsey, interview with the author, September 12, 2011.
15. Quoted in Lelyveld, *His Final Battle*, 103.
16. Jan Herman, Historian, "Oral History with LCDR (ret) Howard Bruenn," January 31, 1990, Office of Medical History, Bureau of Medicine and Surgery, Washington, DC.
17. Ibid.

44. THE HOBCAW BARONY

1. Bernard Baruch, *My Own Story* (New York: Holt, Rinehart, and Winston, 1957), 268.
2. Ibid., 271.
3. Entry of April 4, 1944, Ward, *Closest Companion*, 291.
4. Herman, "Oral History with LCDR (ret) Howard Bruenn," 5.
5. Undated draft letter, 1944, Ward, *Closest Companion*, 295.
6. Herman, "Oral History with LCDR (ret) Howard Bruenn," 8.
7. Ibid., 4.
8. Ross T. McIntire, *White House Physician* (New York: Putnam's, 1946), 188.
9. Entry of April 28, 1944, in "Secret Diary" of Lord Halifax, Papers of Lord Halifax, Hickleton Papers, Borthwick Institute of Historical Research, University of York, Yorkshire, England. No suspected cancer was found.
10. William M. Rigdon with James Derieux, *White House Sailor* (Garden City, NY: Doubleday, 1962), 98.
11. Entry of April 18, 1944, Leahy Diary, William D. Leahy Papers, Library of Congress, Washington, DC.
12. Rigdon, *White House Sailor*, 99.

13. McIntire, *White House Physician,* 187.
14. Howard Bruenn, "Clinical Notes on the Illness and Death of President Franklin D. Roosevelt," *Annals of Internal Medicine* 72, no. 4 (1970): 579–80.
15. Entry of April 18, 1944, Leahy Diary.
16. Clark, Mark Clark, *Calculated Risk: His Personal Story of the War in North Africa and Italy* (London: Harrap, 1951), 318.
17. Entry of April 22, 1944, Leahy Diary.
18. Ibid.
19. Joseph Persico, *Franklin and Lucy,: President Roosevelt, Mrs. Rutherfurd, and the Other Remarkable Women in His Life* (New York: Random House, 2008), 299.
20. Bruenn, "Clinical Notes on the Illness and Death of President Franklin D. Roosevelt."
21. Press and Radio Conference #947, April 28, 1944, at 8:55 p.m., EWT, Press Conferences of President Franklin D. Roosevelt, 1933–1945, Series 1: Press Conference Transcripts, FDR Presidential Library, Hyde Park, NY.
22. Persico, *Franklin and Lucy,* 300; A. Merriman Smith, *Thank You, Mr. President: A White House Notebook* (New York: Harper, 1946), 140–141.
23. Entry of May 4, 1944, Ward, *Closest Companion,* 294.

45. A DUAL-PURPOSE PLAN

1. Entry of May 6, 1944, Leahy Diary.
2. Herman, "Oral History with LCDR (ret) Howard Bruenn," 6.
3. Ibid., 7.
4. Ibid.
5. Ibid., 5.
6. Ibid.
7. Ibid.
8. Lash, *Eleanor and Franklin,* 698, quoting letter to Maude Gray, May 1, 1944.
9. Nigel Hamilton, *Master of the Battlefield: Monty's War Years, 1942–1944* (New York: McGraw-Hill, 1983), 591.
10. Entry of May 15, 1944, in Arthur Bryant, *Triumph in the West* (London: Collins, 1959), 189–90.
11. See Hamilton, *Master of the Battlefield,* 495.
12. Entry of May 15, 1944, Leahy Diary.
13. Ibid.
14. Ibid.

46. D-DAY

1. Entry of May 19, 1944, Ward, *Closest Companion,* 300.
2. Ibid., 301, 305, 308.
3. Ibid., entry of June 4, 1944, 308.
4. Ibid., entry of May 22, 1944, 301.
5. Ibid., 302.

6. Ibid., entry of June 6, 1944, 309.
7. Elizabeth Shoumatoff, *FDR's Unfinished Portrait: A Memoir* (Pittsburgh: University of Pittsburgh Press, 1990), 72.
8. Resa Willis, *FDR and Lucy: Lovers and Friends* (New York: Routledge, 2004), 95; Anna Roosevelt Halsted Papers, 1886–1976, Box 70, FDR Presidential Library.
9. Ibid.
10. Radio Address re "Fall of Rome" (speech file 1518), June 5, 1944, in Franklin D. Roosevelt, Master Speech File, 1898–1945, Box 78, FDR Presidential Library; also "Report on the Capture of Rome," in Russell D. Buhite and David W. Levy, eds., *FDR's Fireside Chats* (Norman: University of Oklahoma Press, 1992), 294–98.
11. Ibid.
12. Entry of June 5, 1944, Ward, *Closest Companion,* 309.
13. Ibid., entry of June 6, 1944, 309.
14. "Communiqué No. One," quoted by John Snagge, BBC Radio, Special Bulletin, 12 Midday, June 6, 1944, BBC Archive, British Library, London.
15. Carlo D'Este, *Eisenhower: A Soldier's Life* (New York: Henry Holt, 2002), 500.
16. Winston Churchill, C-643 to President Roosevelt, April 12, 1944, Kimball, *Churchill and Roosevelt,* vol. 3, 87, and (after Presentation of Plans at Montgomery's headquarters) in Dwight D. Eisenhower, *Crusade in Europe* (New York: Doubleday, 1948), 245.
17. Press and Radio Conference #954, June 6, 1944, 4:10 p.m., Press Conferences of President Franklin D. Roosevelt, 1933–1945, Series 1: Press Conference Transcripts, FDR Presidential Library.
18. Ibid.
19. Ibid.
20. Ibid.
21. Ibid.
22. Ibid.
23. Ibid.
24. Samuel L. Rosenman, *Working with Roosevelt* (New York: Harper, 1952), 433.
25. Address to the Delegates of the American Youth Congress (speech file 1273), February 10, 1940, Master Speech File, Box 50, FDR Presidential Library.
26. Franklin D. Roosevelt, "Letter on Religion in Democracy," December 16, 1940, American Presidency Project, http://www.presidency.ucsb.edu/ws/?pid=15911.
27. Entry of June 4, 1944, in Ward, *Closest Companion,* 308.
28. FDR-61: D-Day Prayer, Original Reading Copy, June 6, 1944, Significant Documents Collection, Box 1, FDR Presidential Library.
29. Ibid.

47. THE DECIDING DICE OF WAR

1. Entry of 6.6.1944, in Joseph Goebbels, *Die Tagebücher von Joseph Goebbels* [The Diaries of Joseph Goebbbels], ed. Elke Froehlich (Munich: K. G. Saur, 1993), Band 12

(hereinafter *Tagebücher 12*), 406. Quotes from this source have been translated by the author.

2. Ibid.
3. Ibid., 405.
4. Ibid.
5. Ibid., 415.
6. Ibid., entry of 7.6.1944, 416.
7. Ibid.
8. According to infantry general Günther Blumentritt, one advanced Kampfgruppe from each of the two armored divisions, Panzer Lehr and Twelfth Panzer, had been ordered to Normandy before dawn on D-day, following reports of Allied paratroop landings: "OB West on D-Day," in David C. Isby, ed., *Fighting the Invasion: The German Army at D-Day* (Barnsley, UK: Frontline Books, 2016), 173. The two divisions were formally committed at 4:00 p.m. — ibid.
9. Entry of 7.6.1944, Goebbels, *Tagebücher 12,* 418.
10. Ibid., 421.
11. Ibid.
12. Ibid.
13. Ibid.
14. Ibid., entry of 8.6.1944, 424.
15. Typed radio message to the President, June 14, 1944, in *The Papers of George Catlett Marshall,* eds. Larry I. Bland and Sharon Ritenour Stevens, vol. 4 (Baltimore, MD: Johns Hopkins University Press, 1996), 479–80.
16. Press and Radio Conference #957, June 13, 1944, 4:19 p.m., Press Conferences of President Franklin D. Roosevelt, 1933–1945, Series 1: Press Conference Transcripts, FDR Presidential Library.
17. Ibid.

48. ARCHITECT OF VICTORY

1. Entry of June 22, 1944, Stimson Diary.
2. Ibid.
3. Nigel Hamilton, *The Mantle of Command: FDR at War, 1941–1942* (Boston: Houghton Mifflin Harcourt, 2014), 295–98.
4. Nigel Hamilton, *Commander in Chief: FDR's Battle with Churchill, 1943* (Boston: Houghton Mifflin Harcourt, 2016), 49–50.
5. Ibid., 343–45.
6. Entry of June 22, 1944, Stimson Diary.
7. Ibid.
8. "From the President for the Former Naval Person," R-541, May 18, 1944, in Kimball, *Churchill and Roosevelt,* vol. 3, 134–35.
9. Ibid., C-680, May 25, 1944, 142–43.
10. Ibid., 149.

11. From the President for the Former Naval Person, R-551, June 6, 1944, in Kimball, *Churchill and Roosevelt,* vol. 3, 167.

49. TO BE, OR NOT TO BE

1. Walter Warlimont, *Inside Hitler's Headquarters* (London: Weidenfeld and Nicolson, 1964), 434.
2. Ibid.
3. See Ian W. Toll, *The Conquering Tide: War in the Pacific Islands, 1942–1944* (New York: Norton, 2015), 477–97. Also, James Hornfisher, *The Fleet at Flood Tide: America at Total War in the Pacific, 1944–1945* (New York: Bantam, 2016), 178–211.
4. C-712, June 23, 1944, in Kimball, *Churchill and Roosevelt,* vol. 3, 203.
5. Entry of June 20, 1944, in Ward, *Closest Companion,* 311.
6. Ibid.

50. A SOLDIER OF MANKIND

1. Entry of July 5, 1944, in Geoffrey C. Ward, ed., *Closest Companion: The Unknown Story of the Intimate Friendship Between Franklin Roosevelt and Margaret Suckley* (Boston: Houghton Mifflin, 1995), 316.
2. Ibid., 317.
3. Harry Goldsmith, *A Conspiracy of Silence: The Health and Death of Franklin D. Roosevelt* (New York: iUniverse, 2007), 171–72.
4. Ibid.
5. Entry of June 28, 1944, in Ward, *Closest Companion,* 301.
6. Joseph Persico, *Franklin and Lucy: President Roosevelt, Mrs. Rutherfurd, and the Other Remarkable Women in His Life* (New York: Random House, 2008), 303.
7. Jim Bishop, *FDR'S Last Year: April 1944–April 1945* (New York: William Morrow, 1974), 71. See also Joseph Lelyveld, *His Final Battle: The Last Months of Franklin Roosevelt* (New York: Knopf, 2016), 94–95 and 148–50.
8. Press and Radio Conference #961, July 11, 1944, 11:07 a.m., Press Conferences of President Franklin D. Roosevelt, 1933–1945, Series 1: Press Conference Transcripts, FDR Presidential Library, Hyde Park, NY.
9. Ibid.
10. Ibid.
11. Entry of July 11, 1944, Ward, *Closest Companion,* 318.

51. MISSOURI COMPROMISE

1. Entry of July 11, 1944, Ward, *Closest Companion,* 318.
2. Ibid.
3. David McCullough, *Truman* (New York: Simon & Schuster, 1992), 308.
4. Ibid., 317–18.

5. Samuel L. Rosenman, *Working with Roosevelt,* (New York: Harper, 1952), 439.
6. Ibid., 445.
7. McCullough, *Truman,* 302.
8. Entry of July 18, 1944, Leahy Diary, William D. Leahy Papers, Library of Congress, Washington, DC.
9. Merle Miller, *Plain Speaking: An Oral Biography of Harry S. Truman* (New York: Berkley, 1974), 181.
10. Ibid., 182; Rosenman, *Working with Roosevelt,* 451.
11. Miller, *Plain Speaking,* 182. In fact Roosevelt's train had left Chicago on July 15, 1944, the day before Senator Truman arrived in Chicago. All accounts testify to Truman's surprise, however: McCullough, *Truman,* 314, and Rosenman, *Working with Roosevelt,* 451.

52. THE JULY PLOT

1. James Roosevelt and Sidney Shalett, *Affectionately, F.D.R.: A Son's Story of a Lonely Man* (New York: Harcourt, Brace, 1959), 351.
2. Ibid.
3. Ibid.
4. Ibid., 352.
5. David Irving, *The Trail of the Fox: Field Marshal Erwin Rommel* (New York: Morrow, 1977), 566–67. At an earlier meeting at Hitler's Wolf's Canyon (*Wolfsschlucht 2*) Western Front headquarters in Margival, near Soissons, on June 17, 1944, Rommel had similarly argued that the "struggle was hopeless," but had been assured that the revenge weapons — V-1s — which had finally been launched en masse against London on June 15, would change the tide: see ibid., 549–52, and Robert M. Citino, *The Wehrmacht's Last Stand: The German Campaigns of 1944–1945* (Lawrence: University Press of Kansas, 2017), 247.
6. Ian Kershaw, *Hitler, 1936–1945: Nemesis* (London: Allen Lane, 2000), 674.
7. Max Domarus ed., *Hitler, Speeches and Proclamations 1932–1945: The Chronicle of a Dictatorship,* vol. 4, *The Years 1941 to 1945* (Wauconda, IL: Bolchazy-Carducci, 1977), 2922.
8. Ibid., 2925.
9. Ibid., 2926.
10. Kershaw, *Hitler: Nemesis,* 688.
11. Roosevelt, *Affectionately, F.D.R.,* 352.
12. Ibid.
13. Trip Log: President's Pacific Inspection Trip, July–August 1944, Franklin D. Roosevelt, Papers as President: Map Room Papers, 1941–1945, Box 24, FDR Presidential Library.
14. Text of the President's Acceptance Speech, *New York Times,* July 21, 1944.
15. Rosenman, *Working with Roosevelt,* 453.
16. Entry of August 6, 1944, Ickes Diary, Harold L. Ickes Papers, 1815–1969, Library of Congress, Washington, DC.

17. Lelyveld, *His Final Battle,* 176.
18. Rosenman, *Working with Roosevelt,* 453.
19. David Irving, *The Secret Diaries of Hitler's Doctor* (New York: Macmillan, 1983), 171.
20. Ibid., 175.
21. Erwin Rommel, *The Rommel Papers,* ed. B. H. Liddell Hart (London: Collins, 1953), 487.

53. WAR IN THE PACIFIC

1. Christopher Thorne, *Allies of a Kind: The United States, Britain and the War Against Japan, 1941–1945* (London: Hamish Hamilton, 1978), 337.
2. E. B. Potter, *Nimitz* (Annapolis, MD: Naval Institute Press, 1976), 280.
3. Ibid.
4. Ian Toll, *The Conquering Tide: War in the Pacific Islands, 1942–1944* (New York: Norton, 2015), 444.
5. Potter, *Nimitz,* 288.
6. Ibid., 289.
7. Arthur Vandenberg, *The Private Papers of Senator Vandenberg* (Boston: Houghton Mifflin, 1952), 85.
8. Ibid., entry of April 30, 1944. See also Hendrik Meijer, *Arthur Vandenberg: The Man in the Middle of the American Century* (Chicago: University of Chicago Press, 2017), 217–21.
9. Vandenberg, *Private Papers,* 86.
10. Ibid., 84.
11. Ibid., letter of March 18, 1944, 82–83.
12. George C. Marshall, *Interviews and Reminiscences for Forrest C. Pogue* (Lexington, VA: George C. Marshall Foundation, 1996), 626.
13. Ibid., quoting Henry Stimson.
14. Ibid., 365.

54. DEUS EX MACHINA

1. Entry of July 26, Trip Log: President's Pacific Inspection Trip, July–August 1944, Franklin D. Roosevelt, Papers as President: Map Room Papers, 1941–1945, Box 24, FDR Presidential Library, Hyde Park, NY.
2. Jan Herman, Historian, "Oral History with LCDR (ret) Howard Bruenn," January 31, 1990, Office of Medical History, Bureau of Medicine and Surgery, Washington, DC, 8.
3. Samuel L. Rosenman, *Working with Roosevelt* (New York: Harper, 1952), 456.
4. Trip Log: President's Pacific Inspection Trip, July–August 1944.
5. Herman, "Oral History with LCDR (ret) Howard Bruenn."
6. Ibid.
7. Rosenman, *Working with Roosevelt,* 457.
8. Potter, *Nimitz,* 316.

55. SLOW TORTURE

1. Entry of July 26, Trip Log: President's Pacific Inspection Trip, July–August 1944.
2. William Manchester, *American Caesar: Douglas MacArthur 1880–1964* (Boston: Little, Brown, 1978), 368.
3. Douglas MacArthur, *Reminiscences* (New York: McGraw-Hill, 1964), 199.
4. Frazier Hunt, *The Untold Story of Douglas MacArthur* (New York: Devin-Adair, 1954), 332.
5. Ibid.
6. Trip Log: President's Pacific Inspection Trip, July–August 1944.
7. Ibid.
8. Ibid.
9. Ibid.
10. D. Clayton James, *The Years of MacArthur,* vol. 2, *1941–1945* (Boston: Houghton Mifflin, 1975), 528.
11. Trip Log: President's Pacific Inspection Trip, July–August 1944.
12. MacArthur, *Reminiscences,* 197.
13. James, *The Years of MacArthur,* vol. 2, 530.

56. IN THE EXAMINATION ROOM

1. Manchester, *American Caesar,* 368.
2. MacArthur, *Reminiscences,* 197.
3. Entry of July 27, 1944, Leahy Diary, William D. Leahy Papers, Library of Congress, Washington, DC.
4. Potter, *Nimitz,* 318.
5. William D. Leahy, *I Was There* (New York: McGraw-Hill, 1950), 251.
6. Ibid.
7. See Nigel Hamilton, *The Mantle of Command: FDR at War, 1941–1942* (Boston: Houghton Mifflin Harcourt, 2014), 177–90.
8. MacArthur, *Reminiscences,* 198.
9. Ibid.
10. James, *The Years of MacArthur,* vol. 2, 533.
11. MacArthur's usual plane, *Bataan I,* was a converted B-17E, but according to the pilot, Weldon "Dusty" Rhoads, a Pan American Airways C-54 was commandeered for the trip, with three rows of seats removed, while *Bataan II,* a C-54 Skymaster, like the President's Sacred Cow, was being readied for MacArthur: Walter R. Borneman, *MacArthur at War* (New York: Little, Brown, 2016), 396
12. Entry of July 28, 1944, Leahy Diary.
13. Press and Radio Conference #962, at Waikiki, Honolulu, July 29, 1944, 4:45 p.m., Press Conferences of President Franklin D. Roosevelt, 1933–1945, Series 1: Press Conference Transcripts, FDR Presidential Library.
14. William M. Rigdon with James Derieux, *White House Sailor* (Garden City, NY: Doubleday, 1962), 118.

15. Rosenman, *Working with Roosevelt,* 458–59.
16. Press and Radio Conference #963, held on the train en route to Washington, DC, August 15, 1944, Press Conferences of President Franklin D. Roosevelt, 1933–1945, Series 1: Press Conference Transcripts, FDR Presidential Library.
17. Franklin D. Roosevelt, *FDR: His Personal Letters, 1928–1945,* ed. Elliott Roosevelt, vol. 2 (New York: Duell, Sloan and Pearce, 1950), 1527.

57. A TERRIBLE MISTAKE

1. Martin Sheridan, "FDR Confers in Hawaii to Speed War on Japs," August 11, 1944, *Boston Globe.*
2. Howard Bruenn, "Clinical Notes on the Illness and Death of President Franklin D. Roosevelt," *Annals of Internal Medicine* 72, no. 4 (1970), 586.
3. Herman, "Oral History with LCDR (ret) Howard Bruenn," 6–7.
4. Leahy, *I Was There,* 254.
5. Rosenman, *Working with Roosevelt,* 461.
6. Ibid., 462.

58. A REDUNDANT CONFERENCE

1. Entry of September 10, 1944, Diaries of William Lyon Mackenzie King, Library and Archives Canada, Ottawa (hereinafter Mackenzie King Diary).
2. Entry of August 29, 1944, in Henry Wallace, *The Price of Vision: The Diaries of Henry A. Wallace, 1942–1946* (Boston: Houghton Mifflin, 1973), 384.

59. THE COMPLETE SETTING FOR A NOVEL

1. Merle Miller, *Plain Speaking: An Oral Biography of Harry S. Truman* (New York: Berkley, 1974), 183.
2. Ibid.
3. Entry of August 22, 1944 (misdated), in Geoffrey C. Ward., ed., *Closest Companion: The Unknown Story of the Intimate Friendship Between Franklin Roosevelt and Margaret Suckley* (Boston: Houghton Mifflin, 1995), 321.
4. Entry of August 16, 1944, in Wallace, *The Price of Vision,* 379–80.
5. Robert E. Sherwood, *Roosevelt and Hopkins: An Intimate History* (New York: Harper, 1948), 820.
6. Ibid.
7. Entry of August 19, 1944, in "Secret Diary" of Lord Halifax, Papers of Lord Halifax, Hickleton Papers, Borthwick Institute of Historical Research, University of York, Yorkshire, England.
8. Resa Willis, *FDR and Lucy: Lovers and Friends* (New York: Routledge, 2004), 131.
9. Two of the death sentences on the eight saboteurs were commuted to thirty years' imprisonment by the President for having betrayed the plot — see John L. McCrea, *Captain McCrea's War: The World War II Memoir of Franklin D. Roosevelt's Naval*

Aide and USS Iowa's *First Commanding Officer,* ed. Julia C. Tobey (New York: Skyhorse, 2016). In 1948 the two men were returned to Germany.

10. Entry of August 29, 1944, Ward, *Closest Companion,* 322.
11. Entry of August 22, 1944, in William D. Hassett, *Off the Record with FDR: 1942–1945* (New Brunswick, NJ: Rutgers University Press, 1958), 267.
12. Entry of September 1, 1944, Ward, *Closest Companion,* 323.
13. Ibid.
14. Ibid., 324.
15. Ibid.
16. Ibid.

60. TWO SICK MEN

1. Annotations, Alan Brooke, *War Diaries, 1939–1945: Field Marshal Lord Alanbrooke,* eds. Alex Danchev and Daniel Todman (Berkeley: University of California Press, 2001), 588.
2. Ibid.
3. Quoted in Andrew Roberts, *Masters and Commanders: How Four Titans Won the War in the West, 1941–1945* (New York: Harper, 2009), 512.
4. Entry of September 8, 1944, Brooke, *War Diaries,* 589.
5. Ibid., entry of September 9, 1944, 590.
6. Ibid., entry of September 10, 1944.
7. Ibid.
8. Ibid.
9. Ibid.
10. Entry of September 11, 1944, Mackenzie King Diary.
11. Ibid.
12. Ibid.
13. Ibid.
14. Hastings Lionel Ismay, *The Memoirs of General Lord Ismay* (London: Heinemann, 1960), 373.

61. CHURCHILL'S IMPERIAL WARS

1. Martin Gilbert, *Winston S. Churchill,* vol. 7, *Road to Victory: 1941–1945* (Toronto: Stoddart, 1986), 971, quoting diary of Sir Andrew Cunningham.
2. Entry of September 11, 1944, Mackenzie King Diary.
3. Entry of September 12, 1944, in John Colville, *The Fringes of Power: 10 Downing Street Diaries, 1939–1955* (London: Hodder & Stoughton, 1985), 513.
4. Annotation, Colville, *The Fringes of Power*, 514.
5. Entry of September 11, 1944, Mackenzie King Diary.
6. Ibid.
7. Ibid.
8. Ibid.

9. Ibid.
10. Gilbert, *Road to Victory*, 969.
11. Entry of September 12, 1944, Mackenzie King Diary.
12. Ibid., entry of September 13, 1944.
13. Ibid.

62. A STAB IN THE ARMPIT

1. Entry of September 13, 1944, Brooke, *War Diaries*, 591.
2. "Meeting of the Combined Chiefs of Staff with Roosevelt and Churchill, September 13, 1944, 11:45 a.m., the Citadel," in *Foreign Relations of the United States: Conference at Quebec, 1944* (hereinafter *FRUS II*) (Washington, DC: Government Printing Office, 1972), 314.
3. "Meeting of the Combined Chiefs of Staff, September 12, 1944, Noon, Main Conference Room, Chateau Frontenac," in *FRUS II*, 303.
4. Michael Howard, *The Mediterranean Strategy in the Second World War* (London: Weidenfeld and Nicolson, 1968), 66.
5. Entry of September 13, 1944, Brooke, *War Diaries*, 591–92.
6. "Meeting of the Combined Chiefs of Staff, September 12, 1944, Noon, Main Conference Room, Chateau Frontenac," *FRUS II*, 303.
7. Ibid.
8. Ibid.
9. Entry of September 13, 1944, Brooke, *War Diaries*, 591–92.
10. Ibid., annotation, 593.
11. "Prime Minister to President Roosevelt, C-772, August 29, 1944, in Warren F. Kimball, ed., *Churchill and Roosevelt: The Complete Correspondence*, vol. 3, *Alliance Declining, February 1944–April 1945* (Princeton, NJ: Princeton University Press 1984), 300.
12. "Meeting of the Combined Chiefs of Staff with Roosevelt and Churchill, September 13, 1944, 11:45 a.m., the Citadel," in *FRUS II*, 316.
13. Ibid.
14. Ibid.
15. Ibid.
16. Ibid.
17. Ibid.
18. Ibid.
19. Entry of September 14, 1944, Brooke, *War Diaries*, 593.
20. Gilbert, *Road to Victory*, 960.
21. Entry of September 15, 1944, Brooke, *War Diaries*, 593.

63. THE MORGENTHAU PLAN

1. Memorandum of August 25, 1944, in Carolyn Eisenberg, *Drawing the Line: The American Decision to Divide Germany, 1944–1949* (Cambridge: Cambridge University Press, 1996), 36.

2. Memo of September 6, 1944, in Stimson Diary, Henry L. Stimson Papers, Yale University Library, New Haven, CT.
3. Ibid., entry of September 11, 1944.
4. Ibid., entry of September 13, 1944.
5. Entry of September 13, Mackenzie King Diary.
6. Ibid.
7. Entry of September 13, 1944, Leahy Diary, William D. Leahy Papers, Library of Congress, Washington, DC.
8. Harry Dexter White Memorandum dated 9/25/44, in *FRUS II,* 326.
9. Entry of September 20, 1944, Stimson Diary.
10. Harry Dexter White Memorandum, in *FRUS II,* 327.
11. Ibid., 325.
12. Ibid., 327. Also Henry Morgenthau Jr., "Our Policy Toward Germany: Morgenthau's Inside Story," *New York Post,* November 28, 1947, 18, in ibid., 326.
13. Entry of September 13, 1944, in Lord Moran, *Churchill: The Struggle for Survival, 1940–1965* (Boston: Houghton Mifflin, 1966), 177.
14. Ibid.

64. BEYOND THE DREAMS OF AVARICE

1. Treasury Files, in *FRUS II,* 348.
2. "He dictated the memorandum, which finally stood just the way he dictated it. He dictates extremely well," Morgenthau noted in his presidential diary, "because he is accustomed to doing it when he is writing his books": *FRUS II,* 361.
3. Ibid.
4. Morgenthau Presidential Diary, *FRUS II,* 362.
5. Entry of September 15, 1944, Stimson Diary.
6. Michael Beschloss, *The Conquerors: Roosevelt, Truman and the Destruction of Hitler's Germany, 1941–1945* (New York: Simon & Schuster, 2002), 118.
7. "Memorandum for the President," included with entry of September 15, 1944, Stimson Diary.
8. Ibid.
9. Ibid.

65. THE PRESIDENT IS GAGA

1. Entry of September 20, 1944, Ward, *Closest Companion,* 328.
2. Beschloss, *The Conquerors,* 139.
3. Ibid., 142.
4. Richard Norton Smith, *Thomas E. Dewey and His Times* (New York: Simon & Schuster, 1982), 432.
5. Eisenberg, *Drawing the Line,* 45.
6. Beschloss, *The Conquerors,* 145.
7. Calls to bomb railway lines to Auschwitz had prompted War Department and U.S.

Army Air Forces feasibility studies, but with the Germans able to repair tracks within days, if not hours, and no way in which the gas chambers could be effectively destroyed without killing large numbers of inmates, they had been turned down as impracticable — see Robert Beir, *Roosevelt and the Holocaust* (Fort Lee, NJ: BarricadeBooks, 2006), 249–52. On November 26, 1944, Himmler, recognizing ultimate defeat was inevitable, "stopped the carnage" at Auschwitz. "He turned his attention to destroying the evidence" — ibid., 252. Auschwitz was only one of many extermination and killing programs, however. It is believed another quarter of a million Jews were deliberately murdered between the cessation at Auschwitz and German surrender — see Richard Breitman and Allan J. Lichtman, *FDR and the Jews* (Cambridge, MA: Harvard University Press, 2013), 287–88.

8. E.g., in the *New York Times* and *Tribune*, as well as the Washington press and syndicated Associated Press: Entry of September 24, 1944, Stimson Diary.
9. Ibid., entry of September 27, 1944.
10. Ibid., entry of October 3, 1944.
11. Ibid.
12. Beschloss, *The Conquerors*, 149.
13. Entry of October 3, 1944, Stimson Diary.
14. Ibid.
15. Entry of November 8, 1944, in Alan Lascelles, *King's Counsellor: Abdication and War; The Diaries of Sir Alan Lascelles* (London: Orion, 2006), 268.

66. OUTWARD BOUND

1. Richard Norton Smith, *Thomas E. Dewey and His Times* (New York: Simon & Schuster, 1982), 444.
2. James Reston, quoted in Classic Senate Speeches, Arthur H. Vandenberg, January 10, 1945, www.senate.gov/artandhistory/history/common/generic/Speeches_Vandenberg.htm.
3. Samuel L. Rosenman, *Working with Roosevelt* (New York: Harper, 1952), 510.
4. See Nigel Hamilton, *Commander in Chief: FDR's Battle with Churchill, 1943* (Boston: Houghton Mifflin Harcourt, 2016), 90–92.

67. LIGHT OF THE PRESIDENT'S FADING LIFE

1. S. M. Plokhy, *Yalta: The Price of Peace* (New York: Viking, 2012), 26.
2. Entry of June 12, 1944, "Secret Diary" of Lord Halifax, Papers of Lord Halifax, Hickleton Papers, Borthwick Institute of Historical Research, University of York, Yorkshire, England, referring to a conversation with Reston "a day or two ago," i.e., circa June 10, 1944.
3. Not even Setaro — nicknamed Lenny — really believed this. "He's all excited. I say, 'Prez, what are you tryin' to do? Kid me?'": Robert H. Ferrell, *The Dying President: Franklin D. Roosevelt, 1944–1945* (Columbia: University of Missouri Press, 1998), 100.

4. James Roosevelt and Sidney Shalett, *Affectionately, F.D.R.: A Son's Story of a Lonely Man* (New York: Harcourt, Brace, 1959), 354.
5. Ibid., 355.
6. Ibid., 347.
7. Ibid., 355.
8. Bert Edward Park, *The Impact of Illness on World Leaders* (Philadelphia: University of Pennsylvania Press, 1986), 258.
9. Frances Perkins, *The Roosevelt I Knew* (New York: Viking, 1946), 391–94, and Ferrell, *The Dying President*, 103.
10. Letter of February 9, 1945, in Geoffrey C. Ward, *Closest Companion: The Unknown Story of the Intimate Friendship Between Franklin Roosevelt and Margaret Suckley* (Boston: Houghton Mifflin, 1995), 394.
11. Ibid.
12. Ibid., entry of November 29, 1944, 348.
13. Ibid., entry of December 1, 1944, 352.
14. Ibid., entry of December 2, 1944, 353.
15. Ibid., entry of December 3, 1944, 353.
16. Ibid.
17. Ibid., entry of December 17, 1944, 365.
18. Ibid., entry of January 12, 1945, 380.
19. Ibid., entry of January 22, 1945, 390.

68. ABOARD THE USS *QUINCY*

1. Joseph P. Lash, *Eleanor and Franklin: The Story of Their Relationship* (New York: Norton, 1971), 918.
2. Ibid., letter of December 4, 1944, 919.
3. Ibid.
4. Ibid., 922.
5. John R. Boettiger, *A Love in the Shadow* (New York: Norton, 1978), 256.
6. Entry of January 23, 1945, in Anna Roosevelt Boettiger, Yalta Diary, Anna Roosevelt Halsted Papers, Notes Folder, Box 84, FDR Presidential Library, Hyde Park, NY.
7. Ibid., entry of November 24.
8. Ibid., entry of February 1, 1945.
9. Entry of February 1, 1945, Leahy Diary, William D. Leahy Papers, Library of Congress, Washington, DC.
10. Entry of February 2, 1945, Anna Roosevelt Boettiger, Yalta Diary, FDR Presidential Library.
11. Ibid.
12. "Log of the Trip," in *Foreign Relations of the United States: Conferences at Malta and Yalta, 1945* (Washington, DC: U.S. Government Printing Office, 1945) (hereinafter *FRUS III*), 459.

13. Plokhy, *Yalta*, 19.
14. Anthony Eden, *The Reckoning* (Boston: Houghton Mifflin, 1965), 511–12.
15. Ibid.

69. HARDLY IN THIS WORLD

1. C-894 of January 24, 1944, in Warren F. Kimball, ed., *Churchill and Roosevelt: The Complete Correspondence*, vol. 3, *Alliance Declining, February 1944–April 1945* (Princeton, NJ: Princeton University Press, 1984), 518.
2. "Log of the Trip," *FRUS III*, 460.
3. George Marshall, *George C. Marshall: Interviews and Reminiscences for Forrest C. Pogue* (Lexington, VA: George C. Marshall Foundation, 1966), 406.
4. William D. Leahy, *I Was There* (New York: McGraw-Hill, 1950), 294.
5. Martin Gilbert, *Winston S. Churchill*, vol. 7, *Road to Victory: 1941–1945*, (Toronto: Stoddart, 1986), 1167.
6. Anthony Eden, Diary, February 2, 1945, in David B. Woolner, *The Last 100 Days: FDR at War and at Peace* (New York: Basic Books, 2017), 60.
7. David Roll, *The Hopkins Touch: Harry Hopkins and the Forging of the Alliance to Defeat Hitler* (Oxford: Oxford University Press, 2013), 361.
8. Charles Bohlen, *Witness to History, 1929–1969* (New York: Norton, 1973), 171.
9. Entry of February 2, 1945, Anna Roosevelt Boettiger, Yalta Diary, FDR Presidential Library.
10. Leahy, *I Was There*, 295.
11. Hastings Lionel Ismay, *The Memoirs of General Lord Ismay* (London: Heinemann, 1960), 385.
12. Eden, *The Reckoning*, 512.
13. Entry of February 3, 1945, Anna Roosevelt Boettiger, Yalta Diary, FDR Presidential Library.
14. Ibid.
15. Howard Bruenn, "Clinical Notes on the Illness and Death of President Franklin D. Roosevelt," *Annals of Internal Medicine* 72, no. 4 (1970), 588.

70. "A PRETTY EXTRAORDINARY ACHIEVEMENT"

1. Entry of 16.1.1945, in Joseph Goebbels, *Die Tagebücher von Joseph Goebbels* [The Diaries of Joseph Goebbbels], ed. Elke Froehlich (Munich: K. G. Saur, 1993), Band 15 (hereinafter *Tagebücher 15*), 135. Quotes from this source have been translated by the author. See also Ian Kershaw, *Hitler, 1936–1945: Nemesis* (London: Allen Lane, 2000), 757; and Robert M. Citino, *The Wehrmacht's Last Stand: The German Campaigns of 1944–1945* (Lawrence: Kansas University Press, 2017), 417–19.
2. Entry of 23.1.1945, Goebbels, *Tagebücher 15*, 193.
3. Kershaw, *Hitler: Nemesis*, 753.
4. Entry of 18.1.1945, Goebbels, *Tagebücher 15*, 144.
5. Ibid., entry of 23.1.1945, 192.

6. Ibid., 196–97.
7. Ibid., 192.
8. Ibid., entry of 25.1.1945, 220.
9. Ibid., 217.
10. Ibid., entry of 18.1.1945, 146.
11. Ibid., entry of 29.1.1945, 262.
12. Ibid., entry of 31.1.1945, 285.
13. Entry of January 31, 1945, in John Colville, *The Fringes of Power: 10 Downing Street Diaries, 1939–1955* (London: Hodder & Stoughton, 1985), 557.
14. Adolf Hitler, last radio address, January 30, 1945,' Internet Archive, https://archive.org/stream/AdolfHitlerLastRadioSpeechJan301945/AdolfHitlerLastRadioSpeechJan301945_djvu.txt.
15. Ibid.
16. Ibid.
17. Ibid.
18. Ian Kershaw, *The End: The Defiance and Destruction of Hitler's Germany, 1944–1945* (New York: Penguin, 2011), 211.
19. Kershaw, *Hitler: Nemesis,* 762.
20. Ibid., 767.
21. Ibid., 762.
22. Ibid., 763–64.
23. Hitler, last radio address.
24. Entry of 26.1.1945, Goebbels, *Tagebücher* 15, 232.
25. Kershaw, *Hitler: Nemesis,* 771.
26. Ibid., 780.
27. Ibid., 773.
28. Hitler, last radio address.
29. Entry of 30.1.1945, Goebbels, *Tagebücher* 15, 273.
30. Kershaw, *Hitler: Nemesis,* 771.

71. ONE ULTIMATE GOAL

1. Bohlen, *Witness to History,* 172.
2. Ibid., 171–72.
3. Martin Gilbert, *Winston S. Churchill,* vol. 7, *Road to Victory: 1941–1945* (Toronto: Stoddart, 1986), 1166.
4. Entry of February 3, 1945, Leahy Diary.
5. Bruenn, Clinical Notes.
6. Ibid.

72. IN THE LAND OF THE CZARS

1. Entry of Febrary 3, 1945 in Anna Roosevelt Boettiger, Yalta Diary, Anna Roosevelt Halsted Papers, Notes Folder, Box 84, FDR Presidential Library.

2. Ibid.
3. Entry of February 3, 1945, Lord Moran, *Churchill: The Struggle for Survival, 1940–1965* (London: Constable, 1966), 218.
4. Ibid.
5. Ibid.
6. Entry of February 3, 1945, in Anna Roosevelt Boettiger, Yalta Diary, FDR Presidential Library.
7. Ibid., and Michael F. Reilly, *Reilly of the White House* (New York: Simon & Schuster, 1947), 210–11.
8. Entry of February 3, 1945, in Anna Roosevelt Boettiger, Yalta Diary, FDR Presidential Library.
9. Ibid.
10. E.g., General Deane Memorandum for General Marshall of December 2, 1944, shown by Secretary Stimson to President Roosevelt, January 3, 1945, in *FRUS III,* 447–49.
11. Citino, *The Wehrmacht's Last Stand,* 424.
12. "Commanding General, U.S. Military Mission in the Soviet Union (Deane) to the Chief of Staff, U.S. Army (Marshall), Moscow, December 2, 1944," forwarded by the Secretary of War, Henry Stimson, to the President, January 3, 1945, *FRUS III,* 447–48.

73. THE ATOM BOMB

1. Entry of February 3, 1945, in Anna Roosevelt Boettiger, Yalta Diary, FDR Presidential Library.
2. Ibid.
3. See Hamilton, *Commander in Chief,* 360–66.
4. Accused by his private secretary of "bowing down in the House of Rimmon" by expressing support for Russia on the eve of Hitler's Operation Barbarossa in 1941, Churchill had responded that "he had only one single purpose — the destruction of Hitler," and making the quoted remark: Diary entry of June 21, 1941, in Colville, *The Fringes of Power,* 404.
5. Richard Rhodes, *The Making of the Atomic Bomb* (New York: Simon & Schuster, 1986), 314.
6. Ibid., 533.
7. Ibid., 537.
8. Ibid., 606–7.
9. Lavrenti Beria, head of Soviet state security, had informed Stalin of the Manhattan Project already in March 1942: see Simon Sebag Montefiore, *Stalin: The Court of the Red Tsar* (New York: Random House, 2003), 497.
10. Rhodes, *The Making of the Atomic Bomb,* 500–502.
11. Ibid., 605.
12. Ibid., 500.
13. Entry of December 31, 1944, Stimson Diary, Henry L. Stimson Papers, Yale University Library, New Haven, CT.

14. "General Groves to the Chiefs of Staff, Subject Fission Bombs, December 30, 1944," *FRUS III*, 383.
15. Letter to Lady Theodosia Cadogan, February 2, 1945, in Alexander Cadogan, *The Diaries of Sir Alexander Cadogan, O.M., 1938–1945*, ed. David Dilks (London: Cassell, 1971) (hereinafter *Cadogan Diaries*), 701.

74. RIVIERA OF HADES

1. Plokhy, *Yalta*, 53.
2. Martin Gilbert: *Churchill: A Life* (New York: Henry Holt, 1991), 809.
3. Gilbert, *The Road to Victory*, 1173.
4. Brian Lavery, *Churchill Goes to War: Winston's Wartime Journeys* (London: Conway, 2007), 331–32.
5. Bohlen, *Witness to History*, 174.

75. RUSSIAN MILITARY COOPERATION

1. Bohlen, *Witness to History*, 180.
2. Roosevelt-Stalin Meeting, February 4, 1945, 4 p.m., Livadia Palace, Bohlen Minutes, *FRUS III*, 570.
3. Entry of 24.1.1945, Goebbels, *Tagebücher* 15, 208.
4. See James Scott, *Rampage: MacArthur, Yamashita, and the Battle of Manila* (New York: Norton, 2018).
5. Roosevelt-Stalin Meeting, February 4, 1945, 4 p.m., Livadia Palace, Bohlen Minutes, *FRUS III*, 571.

76. MAKING HISTORY

1. Leahy Diary.
2. Robert Hopkins, "'How Would You Like to Be Attached to the Red Army?'" *American Heritage*, June/July 2005.
3. See Citino, *The Wehrmacht's Last Stand*, 430, and Plokhy, *Yalta*, 84, inter alia.
4. Bohlen Minutes, *FRUS III*, 579.
5. Bohlen, *Witness to History*, 180.
6. Roosevelt-Stalin Meeting, February 4, 1945, 4 p.m., Livadia Palace, Bohlen Minutes, *FRUS III*, 579.
7. Ibid., 580.

77. A SILENT PRESIDENT

1. "Tripartite Dinner Meeting, February 4, 1945, 8:30 p.m., Livadia Palace," Bohlen Minutes, *FRUS III*, 590.
2. Bohlen, *Witness to History*, 181.
3. Entry of February 4, 1945, Moran, *Churchill*, 223.

4. Ibid.
5. Gilbert, *Road to Victory,* 1175.
6. Letter of February 11, 1945, *Cadogan Diaries,* 709.
7. Entry of February 4, 1945, Anna Roosevelt Boettiger, Yalta Diary, FDR Presidential Library.
8. "Tripartite Dinner Meeting, February 4, 1945, 8:30 p.m., Livadia Palace," Bohlen Minutes, *FRUS III,* 590.

78. KENNAN'S WARNING

1. Bohlen, *Witness to History,* 175.
2. Ibid.
3. Ibid.
4. Ibid., 176.
5. Ibid., 177.
6. Bruenn, "Clinical Notes," 589.

79. A WORLD SECURITY ORGANIZATION

1. Bohlen, *Witness to History,* 179.
2. Letter of February 6, 1945, *Cadogan Diaries,* 704.
3. Ibid., letter of February 7, 1945, 705.
4. "Fifth Plenary Meeting, February 8, 1945, 4 p.m., Livadia Palace," *FRUS III,* 773.

80. POLAND

1. Ibid., Matthews Minutes, *FRUS III,* 788.
2. Ibid., 789.
3. Ibid., 779–80.

81. *PULSUS ALTERNANS*

1. "Tripartite Dinner Meeting, February 8, 1945, 9 p.m., Yusupov Palace," Bohlen Minutes, *FRUS III,* 798.
2. Ibid.
3. Ibid.
4. Entry of February 8, Leahy Diary.

82. THE PRIME MINISTER GOES BALLISTIC

1. Ross T. McIntire, *White House Physician* (New York: Putnam's, 1946), 217.
2. Ibid., 216.
3. "Sixth Plenary Meeting, February 9, 1945, 4 p.m., Livadia Palace," *FRUS III,* 842.

4. Ibid.
5. Ibid., 843.
6. Ibid., 844.
7. Ibid.
8. Entry of February 9, 1945, Leahy Diary.
9. "Sixth Plenary Meeting, February 9, 1945, 4 p.m., Livadia Palace," Hiss Notes, *FRUS III,* 856.
10. Ibid., 847.
11. Ibid., 848.
12. "United States Draft of a Declaration on Liberated Europe," Yalta, February 5, 1945, Hiss Collection, *FRUS III,* 860.
13. "Sixth Plenary Meeting, February 9, 1945, 4 p.m., Livadia Palace," Bohlen Minutes, *FRUS III,* 848.
14. Ibid., Hiss Notes, 856.
15. "Sixth Plenary Meeting, February 9, 1945, 4 p.m., Livadia Palace," *FRUS III,* 848.
16. Ibid., Hiss Notes, 856.
17. Richard Toye, *Churchill's Empire: The World That Made Him and the World He Made* (London: Macmillan, 2010), 255.
18. See Stanley Wolpert, *Shameful Flight: The Last Years of the British Empire in India* (Oxford: Oxford University Press, 2009), 69–72 inter alia.
19. "Sixth Plenary Meeting, February 9, 1945, 4 p.m., Livadia Palace," Bohlen Minutes, *FRUS III,* 849. Willkie had died on October 8, 1944.
20. Entry of July 4, 1944, referring to telegram from Churchill, in *Wavell: The Viceroy's Journal,* ed. Penderel Moon (Delhi, India: Oxford University Press, 1973), 78.
21. "Tripartite Dinner Meeting, February 10, 1945, 9 p.m., Vorontsov Villa," Bohlen Minutes, *FRUS III,* 921–22.
22. Ibid., 923.

83. THE YALTA COMMUNIQUÉ

1. Bohlen, *Witness to History,* 185.
2. Ibid., 172.
3. Entry of February 11, 1945, Leahy Diary.
4. Ibid.
5. Ibid.
6. Entry of February 9, 1945, in Alan Brooke, *War Diaries, 1939–1945: Field Marshal Lord Alanbrooke,* eds. Alex Danchev and Daniel Todman (Berkeley: University of California Press, 2001), 661.
7. Bruenn, "Clinical Notes."
8. Jan Herman, Historian, "Oral History with LCDR (ret) Howard Bruenn," January 31, 1990, office of Medical History, Bureau of Medicine and Surgery, 16.
9. Letter of February 12, 1945, Ward, *Closest Companion,* 395.
10. Trilateral Documents: Communiqué Issued at the End of the Conference, Report of the Crimea Conference, *FRUS III,* 968–75.

11. Sarah Churchill, *Keep On Dancing* (London: Coward, McCann & Geoghegan, 1981), 77–78.
12. Ibid.
13. Franklin D. Roosevelt, *FDR: His Personal Letters, 1928–1945*, ed. Elliott Roosevelt, vol. 2 (New York: Duell, Sloan and Pearce, 1950).
14. Roll, *The Hopkins Touch*, 374.
15. Herman, "Oral History with LCDR (ret) Howard Bruenn," 16.
16. Andrew Roberts, *Masters and Commanders: How Four Titans Won the War in the West, 1941–1945* (New York: Harper, 2009), 549.
17. Woolner, *The Last 100 Days*, 175.
18. "The President's Trip to the Crimea Conference and Great Bitter Lake, Egypt, Janaury 22 to February 28, 1945" in "Logs of the President's Trips," Grace Tully Papers, Box 7, FDR Presidential Library, Hyde Park, NY.
19. Winston Churchill, *The Second World War*, vol. 6, *Triumph and Tragedy* (Boston: Houghton Mifflin, 1953), 397.
20. Woolner, *The Last 100 Days*, 175.
21. Churchill, *Triumph and Tragedy*, 397.

84. THE END OF HITLER'S DREAMS

1. Entry of 12.2.1945, Goebbels, *Tagebücher* 15, 368.
2. Ibid.
3. Ibid., 369.
4. Ibid., entry of 13.2.1945.
5. Ibid., 382.

85. KING ODYSSEUS

1. Entry of April 6, 1945, in Geoffrey C. Ward, ed., *Closest Companion: The Unknown Story of the Intimate Friendship Between Franklin Roosevelt and Margaret Suckley* (Boston: Houghton Mifflin, 1995), 411.
2. Ibid., noted in entry of April 4, 1945, 409.
3. Ibid., entry of March 25, 1945, 401.
4. Ibid., entry of April 6, 1945, 411.
5. Entry of March 30, 1945, in William D. Hassett, *Off the Record with FDR: 1942–1945* (New Brunswick, NJ: Rutgers University Press, 1958), 327.
6. Ibid, 327–28.
7. Ibid., 328.
8. Ibid., entry of March 31, 1945, 329.

86. IN THE WELL OF CONGRESS

1. Lieutenant William Rigdon, accompanying the President, noted later the "surprise of those of us who considered a Presidential invitation the same as a command":

William M. Rigdon with James Derieux, *White House Sailor* (Garden City, NY: Doubleday, 1962), 174.

2. Samuel L. Rosenman, *Working with Roosevelt* (New York: Harper, 1952), 522.
3. Ibid.
4. Ibid.
5. Ibid., 523.
6. Jan Herman, Historian, "Oral History with LCDR (ret) Howard Bruenn," January 31, 1990, Office of Medical History, Bureau of Medicine and Surgery, Washington, DC.
7. Rosenman, *Working with Roosevelt,* 524.
8. Ibid., 527.
9. Howard Bruenn, "Clinical Notes on the Illness and Death of President Franklin D. Roosevelt," *Annals of Internal Medicine* 72, no. 4 (1970), 591.
10. Ibid.
11. Entry of March 1, 1945, in "Secret Diary" of Lord Halifax, Papers of Lord Halifax, Hickleton Papers, Borthwick Institute of Historical Research, University of York, Yorkshire, England.

87. APPEASERS BECOME WARMONGERS

1. Martin Gilbert, *Winston S. Churchill,* vol. 7, *Road to Victory: 1941–1945* (Toronto: Stoddart, 1986), 1234.
2. Entry of February 27, 1945, in Harold Nicolson, *Diaries and Letters, 1939-1945,* ed. Nigel Nicolson (London: Collins, 1967), 437.
3. David Reynolds, *Summits: Six Meetings That Shaped the Twentieth Century* (New York: Basic Books, 2007), 161.
4. Entry of March 3, 1945, Ward, *Closest Companion,* 397.
5. Entry of March 9, 1945, Diaries of William Lyon Mackenzie King, Library and Archives Canada, Ottawa (hereinafter Mackenzie King Diary).

88. MACKENZIE KING'S LAST VISIT

1. Entry of March 9, 1945, Mackenzie King Diary.
2. See Nigel Hamilton, *Commander in Chief: FDR's Battle with Churchill, 1943* (Boston: Houghton Mifflin Harcourt, 2016), 35–38.
3. Robert E. Sherwood, *Roosevelt and Hopkins: An Intimate History* (New York: Harper, 1948), 873.
4. Richard Breitman and Allan J. Lichtman, *FDR and the Jews* (Cambridge, MA: Harvard University Press, 2013), 304; David B. Woolner, *The Last 100 Days: FDR at War and at Peace* (New York: Basic Books, 2017), 155–65; Joseph Lelyveld, *His Final Battle: The Last Months of Franklin Roosevelt* (New York: Knopf, 2016), 291–92.
5. Entry of March 9, 1945, Mackenzie King Diary.
6. Ibid.
7. Ibid.
8. Ibid.

9. Ibid.
10. Ibid.
11. Ibid.
12. Ibid.
13. Ibid.
14. Ibid.
15. Ibid., entry of March 10, 1945.
16. Charles Bohlen, *Witness to History, 1929–1969* (New York: Norton, 1973), 206.
17. Entry of March 10, 1945, Mackenzie King Diary.
18. Ibid.
19. Ibid.

89. OPERATION SUNRISE

1. "It is noteworthy that all the German participants in Sunrise negotiations — Zimmer, Wolff's adjutant Wenner, and Wolff himself — survived relatively unscathed in the immediate postwar period," despite their known crimes against humanity: Professor Richard Breitman, Interagency Working Group (IWG) Director of Historical Research, "RG 263 Detailed Report, Guido Zimmer," National Archives, Washington, DC.
2. Entry of March 11, 1945, Stimson Diary, Henry L. Stimson Papers, Yale University Library, New Haven, CT.
3. S. M. Plokhy, *Yalta: The Price of Peace* (New York: Viking, 2012), 361.
4. Kerstin von Lingen, *Allen Dulles, the OSS, and Nazi War Criminals: The Dynamics of Selective Prosecution* (Cambridge: Cambridge University Press, 2013); Kerstin von Lingen, "Conspiracy of Silence: How the 'Old Boys' of American Intelligence Shielded SS General Karl Wolff from Prosecution," *Holocaust and Genocide Studies* 22, no. 1 (2008): 74–109.
5. Entry of March 13, 1945, Stimson Diary.
6. Ibid.
7. Entry of March 13, 1945, Mackenzie King Diary.
8. Plokhy, *Yalta*, 360, and Woolner, *The Last 100 Days*, 252.

90. NO MORE BARBAROSSAS!

1. Robert Ferrell, *The Dying President: Franklin D. Roosevelt, 1944-1945* (Columbia: University of Missouri Press, 1998), 114.
2. Ibid.
3. Joseph Persico, *Franklin and Lucy* (New York: Random House, 2008), 325.
4. Letter of April 1, 1945, in Joseph P. Lash, *Eleanor and Franklin: The Story of Their Relationship* (New York: Norton, 1971), 925.
5. Ibid.
6. Ibid.
7. Entry of March 13, 1945, Mackenzie King Diary.

8. Ibid.
9. Ibid.
10. Ibid.
11. See Nigel Hamilton, *Monty: Final Years of the Field Marshal, 1944–1976* (New York: McGraw-Hill, 1987), 445–60. Also Churchill cables to Roosevelt, C-931, C-933, C-934, in Warren F. Kimball, ed., *Churchill and Roosevelt: The Complete Correspondence,* vol. 3, *Alliance Declining, February 1944–April 1945* (Princeton, NJ: Princeton University Press, 1984), 602–5, and 612–13.
12. Rick Atkinson, *The Guns at Last Light: The War in Western Europe, 1944–1945* (New York: Henry Holt, 2013), 578.
13. Woolner, *The Last 100 Days,* 234.
14. Entry of March 28, 1945, Halifax Diary.
15. Grace Tully, *F.D.R., My Boss* (Chicago: People's Book Club, 1949), 357.
16. Lelyveld, *His Final Battle,* 310.
17. Jim Bishop, *FDR'S Last Year: April 1944–April 1945* (New York: William Morrow, 1974), 750.
18. Elizabeth Shoumatoff, *FDR's Unfinished Portrait: A Memoir* (Pittsburgh: University of Pittsburgh Press, 1990), 98.
19. Ibid.

91. THE END

1. Entry of March 29, 1945, Ward, *Closest Companion,* 401.
2. Bishop, *FDR's Last Year,* 711.
3. Ibid., 723.
4. Personal and Secret from Marshal J. V. Stalin to President F. D. Roosevelt, April 3, 1945, Cable 300, *My Dear Mr. Stalin,* ed. Susan Butler (New Haven, CT: Yale University Press, 2005), 312–13.
5. Entry of April 4, 1945, Leahy Diary, William D. Leahy Papers, Library of Congress, Washington, DC.
6. Ibid.
7. Ibid., Personal from the President for Marshal Stalin, April 4, 1945, Cable 301, 314–15.
8. Ibid., Personal and Secret from Marshal J. V. Stalin to President F. D. Roosevelt, April 7, 1945, Cable 302, 315.
9. Ibid., 445.
10. Entry of April 6, 1945, Leahy Diary.
11. Personal from the President for the Prime Minister, March 11, 1945, Cable R-714, in Kimball, *Churchill and Roosevelt,* vol. 3, 562.
12. Prime Minister to President Roosevelt, April 5, 1945, Cable C-933, in Kimball, *Churchill and Roosevelt,* vol. 3, 612.
13. Woolner, *The Last 100 Days,* 260.
14. Entry of April 5, 1945, Ward, *Closest Companion,* 410.
15. Ibid., entry of April 6, 1945, 411–12.
16. Ibid.

17. Ibid., entry of April 8, 1945, 412.
18. Ibid., entry of April 9, 1945, 413.
19. Bishop, *FDR's Last Year,* 749.
20. Entry of April 9, 1945, Ward, *Closest Companion,* 413.
21. Shoumatoff, *FDR's Unfinished Portrait,* 101.
22. Ibid.
23. Entry of April 9, 1945, Ward, *Closest Companion,* 413.
24. Ibid., entry of April 11, 1945, 415–16.
25. Ibid., 416.
26. Entries of April 10 and 11, 1945, Leahy Diary.
27. Ibid.
28. Entry of April 11, 1945, in Henry Morgenthau Diary, Morgenthau Papers, FDR Presidential Library, Hyde Park, NY.
29. Entry of April 12, 1945, Ward, *Closest Companion,* 417.
30. Shoumatoff, *FDR's Unfinished Portrait,* 116.
31. Ibid., 117.
32. Entry of April 12, 1945, Ward, *Closest Companion,* 418.
33. Shoumatoff, *FDR's Unfinished Portrait,* 118.
34. Entry of March 31, 1945, Ward, *Closest Companion,* 418.

Index

COMMANDER *in* CHIEF

ALSO BY NIGEL HAMILTON

Royal Greenwich: A Guide and History to London's Most Historic Borough (with Olive Hamilton)

Nigel Hamilton's Guide to Greenwich: A Personal Guide to the Buildings and Walks of One of England's Most Beautiful and Historic Areas

The Brothers Mann: The Lives of Heinrich and Thomas Mann, 1871–1950 and 1875–1955

Monty: The Making of a General, 1887–1942

Master of the Battlefield: Monty's War Years, 1942–1944

Monty: Final Years of the Field-Marshal, 1944–1976

Monty: The Man Behind the Legend

JFK: Reckless Youth

Monty: The Battles of Field Marshal Bernard Law Montgomery

The Full Monty: Montgomery of Alamein, 1887–1942

Bill Clinton, An American Journey: Great Expectations

Montgomery: D-Day Commander

Bill Clinton: Mastering the Presidency

Biography: A Brief History

How to Do Biography: A Primer

American Caesars: Lives of the Presidents from Franklin D. Roosevelt to George W. Bush

The Mantle of Command: FDR at War, 1941–1942

COMMANDER *in* CHIEF

FDR'S BATTLE WITH CHURCHILL, 1943

Nigel Hamilton

HOUGHTON MIFFLIN HARCOURT
BOSTON • NEW YORK

Copyright © 2016 by Nigel Hamilton

All rights reserved

For information about permission to reproduce selections from this book, write to trade.permissions@hmhco.com or to Permissions, Houghton Mifflin Harcourt Publishing Company, 3 Park Avenue, 19th Floor, New York, New York 10016.

www.hmhco.com

Library of Congress Cataloging-in-Publication Data
Names: Hamilton, Nigel.
Title: Commander in chief : FDR's battle with Churchill, 1943 / Nigel Hamilton.
Description: Houghton Mifflin Harcourt : Boston, 2016.
Includes bibliographical references and index.
Identifiers: LCCN 2015037253
ISBN 9780544279117 (hardcover)
ISBN 9780544277441 (ebook)
Subjects: LCSH: World War, 1939–1945 — United States. | Roosevelt, Franklin D. (Franklin Delano), 1882–1945 | Churchill, Winston, 1874–1965. | World War, 1939–1945 — Diplomatic history. | Command of troops — Case studies. | World War, 1939–1945—Campaigns. | Great Britain — Foreign relations — United States. | United States — Foreign relations — Great Britain.
Classification: LCC D753 .H249 2016 | DDC 940.53/2273 — dc23
LC record available at http://lccn.loc.gov/2015037253

Maps by Mapping Specialists, Ltd.

Printed in the United States of America
DOH 10 9 8 7 6
4500773120

The author is grateful for permission to quote from the following: Diary of Lord Halifax, 1941–1942, reprinted by permission of the Borthwick Institute for Archives, University of York. Letters and diaries of Margaret Lynch Suckley, reprinted by permission of the Wilderstein Preservation, Rhinebeck, N.Y.

For Lady Ray

Contents

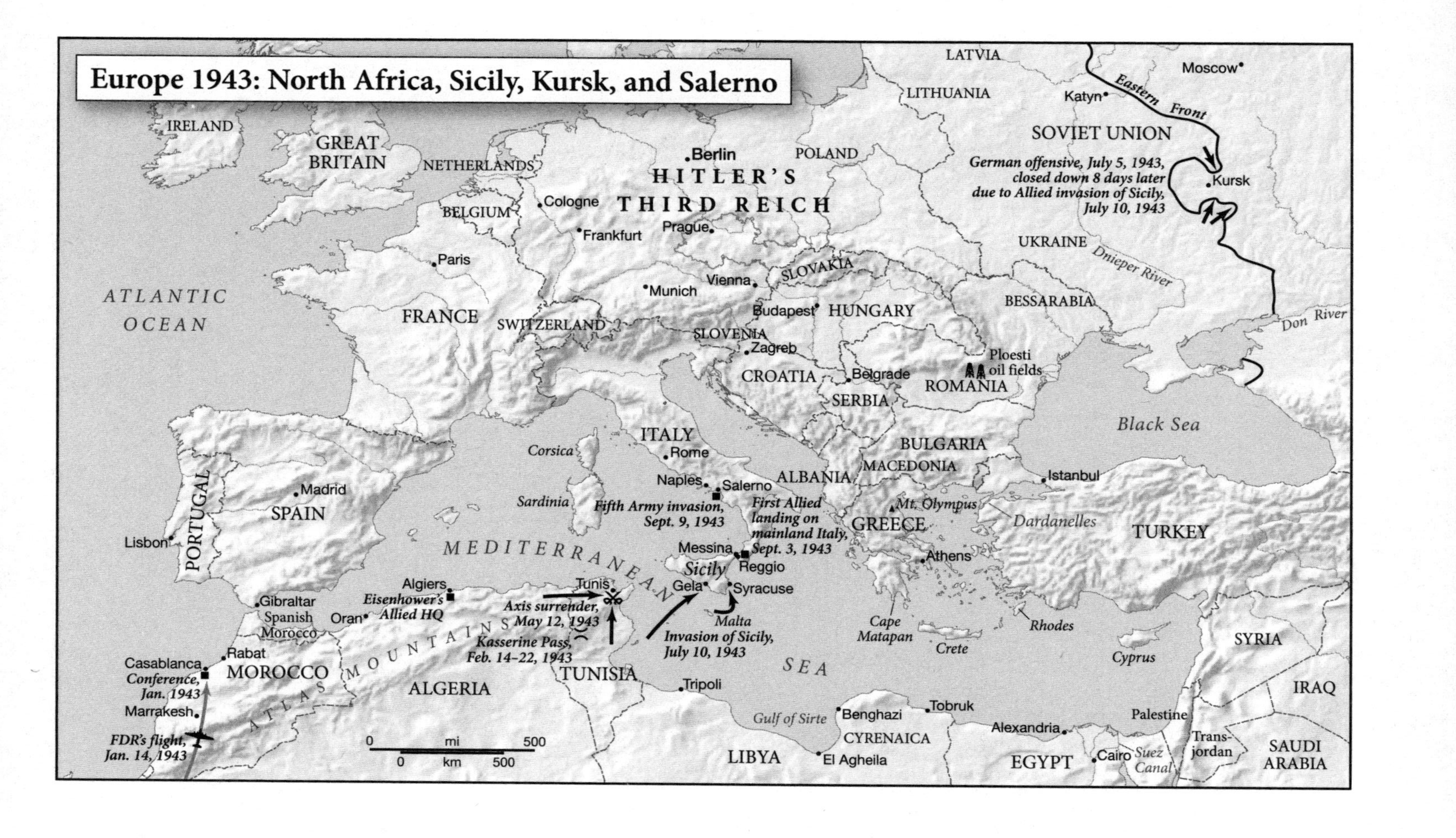
Europe 1943: North Africa, Sicily, Kursk, and Salerno
IRELAND
GREAT BRITAIN
NETHERLANDS
BELGIUM
Cologne
Frankfurt
Paris
Berlin
HITLER'S THIRD REICH
Prague
POLAND
LATVIA
LITHUANIA
Moscow
Katyn
Eastern Front
SOVIET UNION
German offensive, July 5, 1943, closed down 8 days later due to Allied invasion of Sicily, July 10, 1943
Kursk
UKRAINE
Dnieper River
Don River
ATLANTIC OCEAN
FRANCE
SWITZERLAND
Munich
Vienna
SLOVAKIA
Budapest
HUNGARY
BESSARABIA
SLOVENIA
Zagreb
CROATIA
Belgrade
SERBIA
Ploesti oil fields
ROMANIA
Black Sea
ITALY
Rome
Corsica
Sardinia
Naples
Salerno
Fifth Army invasion, Sept. 9, 1943
First Allied landing on mainland Italy, Sept. 3, 1943
ALBANIA
BULGARIA
MACEDONIA
Istanbul
Mt. Olympus
Dardanelles
GREECE
TURKEY
Athens
PORTUGAL
Lisbon
Madrid
SPAIN
MEDITERRANEAN SEA
Messina
Reggio
Sicily
Gela
Syracuse
Malta
Invasion of Sicily, July 10, 1943
Algiers
Eisenhower's Allied HQ
Tunis
Axis surrender, May 12, 1943
Kasserine Pass, Feb. 14–22, 1943
Gibraltar
Spanish Morocco
Oran
ATLAS MOUNTAINS
Rabat
Casablanca
Conference, Jan. 1943
MOROCCO
Marrakesh
FDR's flight, Jan. 14, 1943
ALGERIA
TUNISIA
Tripoli
Gulf of Sirte
Benghazi
CYRENAICA
El Agheila
LIBYA
Tobruk
Alexandria
Cairo
EGYPT
Suez Canal
Cape Matapan
Crete
Rhodes
Cyprus
Palestine
Trans-jordan
SYRIA
IRAQ
SAUDI ARABIA
0 mi 500
0 km 500

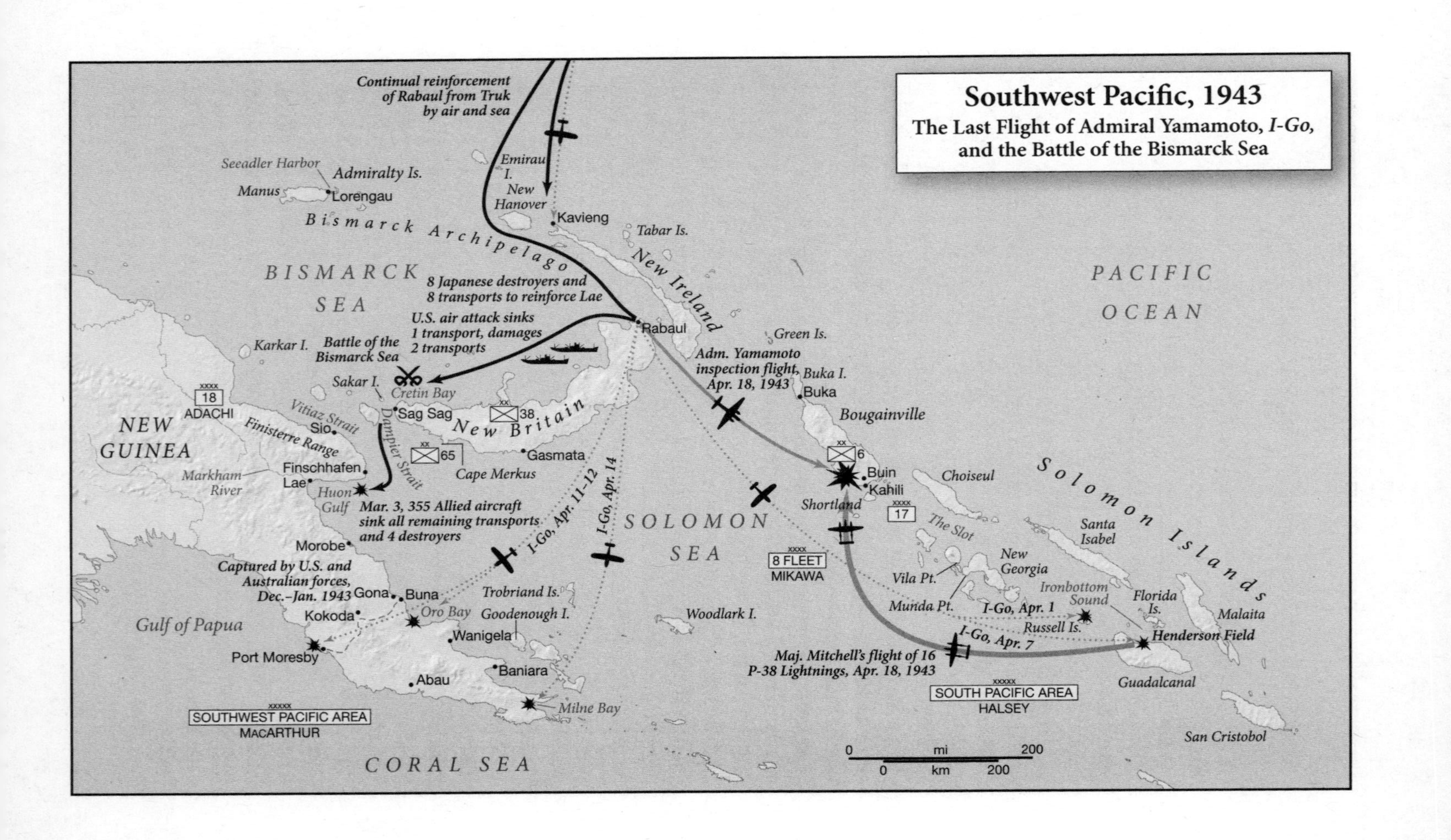

Southwest Pacific, 1943
The Last Flight of Admiral Yamamoto, I-Go, and the Battle of the Bismarck Sea
Continual reinforcement of Rabaul from Truk by air and sea
Seeadler Harbor
Admiralty Is.
Manus
Lorengau
Emirau I.
New Hanover
Kavieng
Tabar Is.
Bismarck Archipelago
New Ireland
BISMARCK SEA
PACIFIC OCEAN
8 Japanese destroyers and 8 transports to reinforce Lae
U.S. air attack sinks 1 transport, damages 2 transports
Rabaul
Green Is.
Karkar I.
Battle of the Bismarck Sea
Adm. Yamamoto inspection flight, Apr. 18, 1943
Buka I.
Buka
Sakar I.
Cretin Bay
18
ADACHI
Vitiaz Strait
Sio
Sag Sag
38
New Britain
Bougainville
NEW GUINEA
Finisterre Range
Dampier Strait
65
Gasmata
6
Solomon Islands
Markham River
Finschhafen
Lae
Huon Gulf
Cape Merkus
Buin
Kahili
Choiseul
Mar. 3, 355 Allied aircraft sink all remaining transports and 4 destroyers
I-Go, Apr. 11–12
I-Go, Apr. 14
SOLOMON SEA
Shortland
17
The Slot
Santa Isabel
Morobe
8 FLEET
MIKAWA
New Georgia
Vila Pt.
Captured by U.S. and Australian forces, Dec.–Jan. 1943
Gona
Buna
Trobriand Is.
Ironbottom Sound
Florida Is.
Oro Bay
Goodenough I.
Woodlark I.
Munda Pt.
I-Go, Apr. 1
Malaita
Kokoda
Gulf of Papua
Wanigela
Russell Is.
I-Go, Apr. 7
Henderson Field
Port Moresby
Maj. Mitchell's flight of 16 P-38 Lightnings, Apr. 18, 1943
Baniara
Abau
Guadalcanal
SOUTH PACIFIC AREA
HALSEY
SOUTHWEST PACIFIC AREA
MACARTHUR
Milne Bay
San Cristobol
0 mi 200
0 km 200
CORAL SEA

Prologue

IN *THE MANTLE OF COMMAND: FDR AT WAR, 1941–1942,* I described how President Franklin Roosevelt first donned the cloak of commander in chief of the Armed Forces of the United States in war — a world war stretching from disaster at Pearl Harbor to his "great pet scheme," Operation Torch: the triumphant Allied invasion of North Africa in November 1942, which stunned Hitler and signified one of the most extraordinary turnabouts in military history.

Commander in Chief: FDR at War, 1943 addresses the next chapter of President Roosevelt's war service: a year in which, moving to the offensive, the President had not only to direct the efforts of his generals but keep Prime Minister Winston Churchill, his "active and ardent lieutenant," in line. Roosevelt's struggle to keep his U.S. subordinates on track toward victory, without incurring the terrible casualties that would have greeted their military plans and timetable, proved mercifully successful that fateful year, but his assumption that Churchill would abide by the strategic agreements they had made proved illusory. Thus, although Roosevelt's patient, step-by-step direction of the war led to historic victories of the Western Allies in Tunisia in the spring of 1943, and again in Sicily in August of that year — results that assured the President a cross-Channel assault would be decisive when launched, in the spring of 1944 — the British prime minister did not agree. The President's resultant "battle royal" with Churchill — who was in essence commander in chief of all British Empire forces — became one of the most contentious strategic debates in the history of warfare.

This dramatic, repeated struggle forms the centerpiece or core of this volume, for it is not too bold to say that upon its outcome rested the outcome of World War II, and thus the future of humanity. The struggle took most of the year—*das verlorene Jahr,* as German military historians would call it. Had Churchill prevailed in his preferred strategy, the war might well have been lost for the Allies, at least in terms of the defeat of Hitler. Even though the President won out over the impetuous, ever-evasive British prime minister, the fallout from Churchill's obstinacy and military mistakes would be profound. Not only was American trust in British sincerity severely damaged, but the need to keep the Prime Minister sweet, and loyal to the agreements he had only reluctantly made for Operation Overlord, led to dangerously naive plans for the Allied invasion of mainland Italy in September 1943—plans involving an airborne landing on Rome, and a gravely compromised amphibious landing in the Gulf of Salerno, south of Naples: Avalanche.

The reality was, Winston Churchill had remained a Victorian not only in his colonial-imperialist mindset, as President Roosevelt often remarked, but in his understanding of modern war—and the Wehrmacht. He grievously underestimated the Wehrmacht's determination to hold fast to the last man at the very extremity of the European mainland, giving rise to fantasies of easy Allied victory, and a possible gateway to central Europe that would make Overlord unnecessary.

Fortunately, the President's absolute determination in 1943 to prepare his armies for modern combat and to then stand by the Overlord assault as the decisive battle of the Western world rendered Churchill's opposition powerless. The Prime Minister's strategic blindness would prove tragically expensive in human life, but mercifully it did not lose the war for the Allies. The President may justly be said to have saved civilization—but it was a near-run thing.

To a large extent the facts of this dark saga are well known to military historians. However, because President Roosevelt began to assemble[1] but did not live to write his own account of the war's military direction, and since others did go on to recount their own parts—sometimes with great literary skill—the President's true role and performance as U.S. commander in chief has often gone unappreciated by general readers. Churchill, who was nothing if not magnanimous in victory, certainly attempted in his memoirs to pay tribute to Roosevelt's leadership, but in his concern to regain the prime ministership he had lost in 1945 he could not always bring himself to tell the truth. Nor was he ashamed of this. As he had boasted after the Casablanca Conference, he fully intended to tell the story of the war from his point of view—and where necessary to suborn history to his own agenda: "to wait until the war is over and then to write his impressions so that, if necessary, he could correct or bury his mistakes."[2] During the war itself he had openly and publicly expressed his loyalty to President Roosevelt as the mastermind directing Allied strategy — a surprise even to Joseph Goebbels — but in private he nevertheless let it be known that he himself was the real directing genius. As King George VI's private secretary, Sir Alan Lascelles, noted in his diary on November 10, 1942, though in his Mansion House speech extolling the successful U.S. invasion of Morocco and Algeria the Prime Minister "gave the credit for its original conception to Roosevelt," Sir Alan believed "it belongs more truly to himself."[3] By the time General Eisenhower took the surrender of all Axis forces in North Africa six months later, this notion of the Prime Minister as sole military architect of Allied strategy and performance had grown to ridiculous levels. Not only was Churchill given credit for having "built up the 8th Army into the wonderful fighting machine that it has become" — despite Churchill's original refusal to appoint General Montgomery to command the army, and his opposition to the new military tactics Montgomery was employing — but Lascelles was convinced, like King George, that "Winston is so essentially the father of the North African baby that he deserves any recognition, royal or otherwise, that can be given to him . . . He has himself publicly given the credit for 'Torch' to Roosevelt, but I have little doubt that W. was really its only begetter."[4]

Aided by his "syndicate" of researchers, civil servants, and historian-aides, Churchill was able to have his day in literary court, in his six-volume opus, *The Second World War,* which helped win him the Nobel Prize for Literature in 1953 — a work that, as Professor David Reynolds has shown,[5] was often economical with the truth. For the memory of President Roosevelt — whose funeral Churchill had not even attended — it was, however, near-devastating, since its magisterial narrative placed Churchill at the center of the war's direction and President Roosevelt very much at the periphery.

In many ways, then, this book and its predecessor are a counternarrative, or corrective: my attempt to tell the story of Roosevelt's exercise of high command from his — not Churchill's — perspective.

In my first volume I selected fourteen episodes, centering on President Roosevelt's "great pet scheme": his Torch invasion of Northwest Africa, and the near mutiny of his generals to stop this and plump instead for a suicidal invasion of northern France in 1942. In this new volume I have selected twelve representative episodes of 1943, beginning with the Casablanca Conference in January and ending with the invasion of Salerno in September. While this has entailed omitting many important events and aspects of Roosevelt's presidency as U.S. commander in chief—some of which, like the development of the atom bomb, progress in the Pacific, and questions of saving the Jews in Europe, will be addressed in a final volume—they continue to give us a clearer picture of how President Roosevelt operated when wearing, so to speak, his military mantle in World War II. By following him closely in his study, in the Oval Office and the Map Room at the White House, at his "camp" at Shangri-la and his family home at Hyde Park; on his historic trip abroad to Africa (the first president ever to fly in office, and the first to inspect troops on the battlefield overseas); and on his long inspection tour of military installations and training camps in the United States (during which he authorized the secret air ambush of Admiral Yamamoto), we are able to see him at last as we have previously been able to see so many of his subordinate military officers and officials of World War II—that is to say, from *his* perspective.

It will be noted that, as hostilities approach their climax in the fall of 1943, the political ramifications take on a more urgent role. Churchill may have been completely wrong in his understanding of the Wehrmacht, and a menace to Allied unity in his Mediterranean mania—one that drove even his own chiefs of staff to the brink of resignation. But Churchill's understanding of the deepening rift and rivalry with the Soviets bespoke his greatness as a leader. Many thousands of miles removed from the continent of Europe, President Roosevelt needed the Prime Minister by his side not as military adviser—given that Churchill's judgment and obstinacy were more millstone than help, as Churchill's doctor himself recognized—but as the President's political partner in leading the Western democracies.

As the final pages of this volume demonstrate, Winston Churchill was thus invited to spend long weeks with the President in Washington and at Hyde Park, in a deeply symbolic act of unity—as much in confronting Stalin as Hitler.

The degree to which President Roosevelt began to rely on Churchill's loyal political support and his political acumen in the summer and fall of 1943—before the Tehran and Yalta Conferences—are thus a testament to the importance of their relationship in world history. Churchill had rattled the unity of the Allies that year to the very brink of collapse by pressing for a military strategy that would arguably have lost the war for the Allies had not President Roosevelt overruled him. In political terms, however, it was to be his steadfast, statesmanlike partnership with the President of the United States that would ensure the democracies, under their combined leadership, had at least a chance of ending World War II with western Europe under safe guardianship in relation to Soviet "Bolshevization."

To better understand FDR's direction of the military is thus to me important not only in terms of a greater appreciation of President Roosevelt's actions, but in understanding the foundations of the world we live in today. From boasting only the world's seventeenth-most-powerful military in 1939, the United States gradually took upon itself the successful leadership of the democratic world under Roosevelt's command—and became the most powerful nation on earth, bar none. How exactly President Roosevelt directed this transformation and the operations of his armed forces across the globe—with what aims, with what challenges, with what lessons—is to me of abiding interest in the world we've inherited. For good or ill, America's military power under a freely elected president remains in large part the basis of the continuing role of the United States in attempting to provide leadership and world security, however imperfect.

This, then, is the record of FDR as U.S. commander in chief in the crucial year 1943—a year in which the United States went on the offensive both in the West and in the East—as seen from the President's point of view. Upon his leadership depended the outcome of the world war: success or failure.

PART ONE

A Secret Journey

1

A Crazy Idea

IT WAS LATE in the evening of Saturday, January 9, 1943, when a locomotive pulling the *Ferdinand Magellan* and four further carriages[1] departed from a special siding beneath the Bureau of Engraving in Washington, D.C. — the federal government's massive printing house for paper money, and thus a sort of Fort Knox of the capital.

Aboard was the President of the United States, his secretary, his White House chief of staff, his naval aide, his White House counselor, and his doctor, all traveling to Hyde Park for the weekend, as usual. Or so it seemed.

The Secret Service had insisted the President use for the first time the massive new railway carriage reconstructed for him — the first such railcar to be made for the nation's chief executive since Lincoln's presidency. Boasting fifteen-millimeter armored steel plate on the sides, roof, and underside, the carriage had three-inch-thick bulletproof glass in all its windows. Best of all, it had a special elevator to raise the President, in his wheelchair, onto the platform of the car — which weighed 142 tons, the heaviest passenger carriage ever used on U.S. rail track.

The car was "arranged with a sitting room, a dining room for ten or twelve persons, a small but well arranged kitchen, and five state rooms," Admiral William Leahy, the President's military chief of staff, recorded in his diary. "Dr. McIntire, Harry Hopkins, Miss Tully, and I occupied the state rooms, and Captain McCrea joined us in the dining room. Other cars accommodated the Secret Service men, the apothecary, the communications personnel, and the President's valet," Chief Petty Officer Arthur Prettyman.[2]

Their luggage had been taken to the baggage car separately, an hour earlier. But was the President really going to Hyde Park? If so, why the thousand pounds of bottled water? Why clothes for two weeks away? Why the four Filipino members of the crew of the USS *Potomac,* the presidential yacht, replacing the normal Pullman staff? Why Eleanor, the First Lady, and Louise Macy, the new wife of Harry Hopkins, bidding them goodbye at the underground siding?

Something was up — something unique. Even historic.

Among the few who did know of the President's real destination, most had counseled against it. Even the President's naval aide, Captain John McCrea, opposed the idea when the President tricked McCrea into supplying information on the geography, history, and significant towns of the region of North Africa. Following the successful Torch landings in Algeria and Morocco on November 8, 1942, the President had explained to McCrea — whose knowledge of the sea exceeded his knowledge of land — U.S. troops would be fighting in battle, and he'd found himself, as U.S. commander in chief, sadly ignorant of the terrain. "See if you can help me correct that deficiency," he'd instructed McCrea, "by means of travel folders, etcetera, put out by travel agencies."

Travel agencies? As the President had quickly assured McCrea, "in the planning and preparatory stages" of Operation Torch, he hadn't wanted to draw attention to "that area." "But now that the troops are there," he'd added, "that restraint is removed."

Innocent of any ulterior motive, McCrea had assembled a raft of informative material. "The President was pleased with it and confided: 'Just the sort of information I want.'"

Some weeks later, though, "late one afternoon, early in December the President sent for me, sat me down at the corner of his desk and this is about the way it went.

"The Pres: 'John, I want to talk to you in great confidence and the matter about which I am talking is to be known to no one except those who need to know.' Since this was the first time the Pres. had ever spoken to me thus, naturally I was greatly curious," McCrea later narrated in his somewhat stilted literary style. The President had then confided, "'Since the landing of our troops in No[rth] Africa, I have been in touch with Winston by letter. I feel we should meet soon and resolve some things and that that meeting should take place in Africa. Winston has suggested Khartoum — I'm not keen on that suggestion. Marrakech and Rabat have been suggested. I'm inclined to rule out those areas, and settle for Casablanca.' And then to my amazement the President said: 'What do you think of the whole idea?'"

McCrea had been stunned.

"As quickly as I could," McCrea recalled, "I gathered my wits and proceeded about as follows. 'Right off the top of my head Mr. Pres. I do not think well of the idea. I think there is too much risk involved for you.'"

The President had been unmoved. "Our men in that area are taking risks, why shouldn't their Commander in Chief share that risk?"

McCrea was a seasoned sailor — an aspect he thought might be a more effective counter. "'The Atlantic can be greatly boisterous in the winter months,'" he had pointed out, "'and a most uncomfortable passage is a good possibility — '

"'Oh — we wouldn't go by ship. We would fly,' said he."

Fly?

McCrea was shocked. No U.S. president had flown while in office — ever. "This was a great surprise to me because I knew he did not regard flying with any degree of enthusiasm," McCrea recounted. Mr. Roosevelt had not flown in a decade, in fact, since traveling to Chicago from New York before the 1932 election. In terms of the President's safety, waging a world war, it seemed a grave and unnecessary risk — especially in terms of distance, and flight into an active war zone. But the President was the president.

McCrea had therefore softened his objection. "I quickly saw that I was being stymied and I tried to withdraw a bit.

"'Mr. Pres.,' said I, 'you have taken me quite by surprise with this proposal. I would like to give it further thought. Right off the top of my head I wouldn't recommend it.'"

When, the next morning, Captain McCrea went upstairs to the President's Oval Study, carrying with him some of the latest reports, secret signals, decoded enemy signals, and top-secret cables from the Map Room — of which he was the director — he'd recognized the futility of opposing the idea. It was a colossal risk, he still thought, but he knew the President well enough to know that, if Mr. Roosevelt had raised the matter, it was because his mind was probably already made up, and he was simply looking for the sort of reaction he would be likely to meet from others.

"He laughed lightly," McCrea recalled — informing him that Prime

Minister Churchill had already responded positively to the suggestion, in fact was gung ho for such a meeting — "'Winston is all for it.'"

McCrea had remained concerned, though. Security would present a problem not only during the broad Atlantic crossing, he warned, but in North Africa itself. "I still think the risk is great and if you are determined to go I will do all possible to manage that risk," he'd assured the President. But the risks were real. "From what I have read in the despatches and the press," he'd said, for example, "affairs in No[rth] Africa are in a state of much confusion." Casablanca itself was a notorious gathering place for spies and expatriates. And worse. "I would suppose that No[rth] Africa is full of people who would take you on for $10 — "

Assassination?

"Why I said that I'll never know," McCrea later reflected. It was almost rude, " — but I did and at the moment, of course, I felt it. He laughed heartily."

McCrea was not being timorous. Several weeks later his concern was validated — Admiral François Darlan, the new French high commissioner under the Allied commander in chief in the Mediterranean, General Eisenhower, was murdered in broad daylight in Algiers.

By then, however, the trip had been prepared in great detail, and the President would hear no more attempts to dissuade him.

Maintaining secrecy for the trip had not been easy, however. There was, for instance, the problem of idle gossip. The British had been making their own travel arrangements for Prime Minister Churchill. By secret cable from his "bunker" beneath Westminster, in London, Mr. Churchill's office had duly informed the British ambassador in Washington, D.C., Viscount Halifax. Halifax had told his wife.

It had been McCrea who had then taken the telephone call from a distraught, elderly Colonel Edmund Starling, who — going back to the days of President Wilson — was chief of the Secret Service detail responsible for the President's safety at the White House. "The Colonel said it was urgent he see me at once," McCrea recalled. "He came to the Map Room and we went out into the corridor, out of earshot of the Map Room personnel. This is about the way it [went]:

"Col — Is anything going on here about the movements of the President of which I should be apprised?"

McCrea had been noncommittal. "I don't understand what you are driving at, Colonel. Could you be more specific?" he'd responded.

"Col — Well, it is this. A taxi cab driver here in Washington called the W.[hite] H.[ouse] today and told the telephone operator that he wanted to talk to someone in authority who had to do with the movements of the President." On being put through to Colonel Starling, he was asked to come straight to the White House. He'd left "just a few minutes ago. His story was that he had answered a call to the British Embassy this forenoon and there he had picked up a couple of ladies and had driven them in town to a Woodward & Lothrop Dept. store. On the way in they had talked at some length and that one lady had said to the other that the President was going soon to North Africa where he would meet with Mr. Churchill. He, the driver, had no way of knowing whether or not it was so, but nevertheless if it was, he thought it was something that shouldn't be talked about."

This was a serious understatement.

Oh, the British. Often so pompous about rank and privilege — and so casual with regard to high-level gossip shared in the presence of the "servant class."

It hadn't boded well, but there was little McCrea had been able to do; an important summit of wartime leaders could hardly be canceled or reconvened because of an ambassador's wife's shopping trip.

The President was more amused by the incident than concerned. What he worried about was his longtime White House military aide, Major General Edwin "Pa" Watson. The general wouldn't be going, the President had told McCrea. "Pa has suffered a heart attack last spring," the President had explained, "and while he is now back on active duty Ron [McIntire, the President's doctor] thinks he is in no condition to stand the stress and strain of a long air trip across the Atlantic and on to Casablanca. I dread telling Pa that I have decided he should not go with us."

McCrea could only marvel at a president more concerned not to upset his loyal military aide than for his own safety. The President had reason to be concerned, however. "I intentionally put off telling Pa as long as possible and when he brought the appointment list to me this morning," Roosevelt told McCrea on January 7, 1943, "I broke the news to him and told him that on Ron's advice because of the considerable flying involved and his recent heart attack that I was not taking him on this trip. Pa was shocked—slumped in his chair and broke into tears—and remarked perhaps his usefulness around the W.H. was about at an end. I comforted him as best I could but to little avail. After a bit he recovered his composure and withdrew. Now John, I told you last evening I would enter the House Chamber [of Congress, for the upcoming State of the Union address] this noon on your arm. If I do that I think it would be a further shock to Pa. Will you please run Pa down at once and tell him that I neglected to tell him this a.m. that as usual I would enter the House Chamber this noon on his arm. That might soften the blow a bit of his not going to No. Africa with us."[3]

Once again Captain McCrea had been amazed at the President's concern for the feelings of others, while directing the administration of his country in a global war. Also the President's innocence, too: for it would be the President's naval aide who would suffer the full force of General Watson's disappointment at being excluded from the North Africa trip, however much the President wished to sugar the pill.

It had not taken long. In General Watson's room next to the Oval Office, where Pa Watson acted as the President's appointments secretary, guarding all access to the Chief Executive, McCrea had endured a tirade from the general. If he himself was forbidden to travel, Watson said, why should the President—who'd had his own heart problems—go? Watson "thought the Pres. was badly advised about making the trip—the risk was too great for him to take. Why hadn't I informed him about the trip? 'I've always taken you in my confidence,' said he, 'and in this important instance you have not taken me into your confidence.'

"I calmed Pa down as best I could," McCrea related. "I told him of the charge given me by the President that no one, absolutely no one, should know about this trip except those who needed to know—and he [the President] laid great emphasis on that point. That he would tell you himself in due course that you could not make the trip and that he would tell me when he had done so."

This did little to solace the Army general—who was, after all, still the President's military aide. The Navy had trumped him. "There was just no comforting Pa," McCrea recalled. "He was deeply disturbed and repeated over and over again that the Pres. was badly advised in the decision to make this hazardous trip. 'I hope you didn't encourage him in that,'" he'd demanded accusingly. "I told Pa that I had done everything I possibly could to dissuade the President — but to no avail. That insofar as I knew the deal had been made with Mr. Churchill and that was it. And then Pa exclaimed with much emphasis: 'There is only one so and so around here who is crazy enough to promote such a thing, and his name is Hopkins'" — the President's White House counselor.[4]

2

Aboard the Magic Carpet

GENERAL WATSON WAS wrong about Harry Hopkins. Recently married, Hopkins had no great wish to go to Casablanca. His wife said goodbye to him "at the rear door" of the *Ferdinand Magellan,* Hopkins jotted in his diary that night. Eleanor had shown no emotion, but Louise had been a bag of nerves — as was Hopkins, who worried about the weeks he'd be away from Washington. A survivor of stomach cancer and major intestinal surgery before the war began, Hopkins required constant medication. Above all, though, he had no wish to leave his new bride. Over Thanksgiving, at a cast party for S. N. Behrman's new play on Broadway, *The Pirate,* he'd been heard to say to a friend, as he introduced his young consort: "Look, Dyke, — I ought to be dead — and here I am married!"[1]

A charming and pretty gadabout, Louise was a socialite who, to her discomfort, had swiftly found herself accused of impropriety after the wedding, thanks to people envious of Hopkins's proximity to the President — people such as the financier Bernard Baruch, who'd failed to obtain a job in the Roosevelt war administration. "I must say that I didn't like the idea of leaving a little bit," Hopkins confided to his diary before going to sleep, for "Louise had been very unhappy all evening because of the political attacks on us."[2]

For his part, Admiral Bill Leahy — the President's chief of staff at the White House, but also now the chairman of the Combined Chiefs of Staff — was equally reluctant to go. The sailor had suffered a bad bout of flu in recent days, and did not relish the long journey by train and then air. Nor did he savor what was awaiting him at the secret destination:

a continuing international political imbroglio that in his view had been pretty much screwed up by people who didn't understand the military difficulties of the situation.

There were others, too, who were anxious. Daisy Suckley, the President's cousin and longtime confidante, had already said her goodbye the day before, and wasn't therefore at the Bureau of Engraving platform. She'd argued strenuously against such "a long trip," she noted in her own diary, one "with definite risks" that included enemy interception, accidents, even assassination. "But one *can't* and *mustn't* think of that."[3] On the plus side there were, she acknowledged, exotic places the President would get to see. And people, too. "W. Churchill first and foremost, of course," she'd added. Others, however, he would not. He'd asked to meet Stalin, "but Stalin answered that he could not possibly leave Russia now—One can understand that," she allowed, given the great winter battle still being fought to the death at Stalingrad.[4]

Fala, the President's beloved Scottish terrier, was not going, either, Daisy noted. The President had asked his wife if she would look after him. Like Stalin, the First Lady had said she was too busy. The President had therefore asked Daisy, who'd originally given him the terrier, as a gift, and she'd agreed to do so.

"I wished him all the best luck on this secret trip," Daisy recorded the next night, after saying goodbye—more devoted to him than ever. "He is leaving as if to go north to Hyde Park," which was near her own baronial home, Wilderstein. "At a certain siding," though, "the train will be picked up by the regular engine & start south for Miami—He goes with all one's prayers."[5]

At Baltimore the locomotive was, indeed, decoupled. Instead of continuing north, a new locomotive bore it south, toward its destination a thousand miles away: through Virginia, North Carolina, South Carolina, Georgia, and Florida, to the former Pan American Clipper terminus.

For his own part, President Roosevelt was glad to get away. Despite the winter cold, the capital was a cauldron of rumor, gossip, political rivalry, and competitive ambitions. Looked after by his valet, Petty Officer Prettyman, and his Filipino crew from the presidential yacht, the USS *Potomac*, he ate and slept well. Rising late on Sunday, January 10, 1943, he lifted the shades of his compartment. The passengers had been instructed that, in order to maintain absolute secrecy, they were to "keep the shades down all day," as he wrote Daisy that night—confiding that he "found myself waving to an engineer & fear he recognized me."[6]

Minor mishaps always amused FDR.

All day, as the heavy, shaded train bore on, the President went through his White House papers, dictating final letters and memoranda to Grace Tully, his secretary, who would be leaving the train in Florida before they reached Miami. He then said goodnight and retired early, knowing they would all have to rise before first light the following morning.

Woken early on January 11, Hopkins donned a robe and made his way to the President's stateroom, where he "found the President alone." Together they "laughed over the fact that this unbelievable trip was about to begin. I shall always feel that the reason the President wanted to meet Churchill," Hopkins surmised, "was because he wanted to make a trip."[7]

Roosevelt had become "tired of having other people, particularly myself, speak for him around the world. For political reasons he could not go to England," Hopkins noted—despite Eleanor having found her husband a nice potential apartment in London, complete with elevator, where he could stay if he chose to meet Churchill there. But the President had balked at the political ramifications. The new, potentially more hostile, isolationist Congress, elected the previous November, would have a field day, he feared. Certain members of Congress and rich, right-wing newspaper owners would accuse the President either of kowtowing to the British or colluding with foreign allies without first telling members of his visit, let alone getting their consent.

London, then, had been out—and the North African battlefield in. Roosevelt would travel as U.S. commander in chief, not as president—thus permitting him to insist upon absolute secrecy, with no press correspondents following him. He "wanted to go to see our troops," Hopkins noted, and "he was sick of people telling him that it was dangerous to ride in airplanes. He liked the drama of it. But above all, he wanted to make a trip."[8]

Whether Hopkins was right was debatable, but the sheer drama of the President's secret getaway from Washington was—like his "escape" from the press to Newfoundland for the Atlantic Charter meeting in 1941—undeniable.

Grace Tully duly disembarked to stay with relatives. Then, at Miami, the party detrained and was driven by car to the former Pan American

Airways terminal by the harbor. Two huge flying boats were waiting, bobbing on the water.

"My God! Why, that's the Pres[ident]. Why didn't they let me know he was to be one of my passengers?" the captain of the first boat, the *Dixie Clipper,* exclaimed. "It's somewhat of a shock to know you are flying the Pres. of the U.S."[9]

With its giant 152-feet cantilevered wingspan, four fifteen-hundred-horsepower Wright Twin Cyclone engines, plus sponsons attached to both sides of its hull to provide extra lift and ease of embarkation, the Clipper — leased by the U.S. Navy, and its crew given Navy rank — duly took off from the predawn waters of the harbor and made first for Trinidad, in the Caribbean, fourteen hundred miles away. "The sun came up at about 7:30," Roosevelt wrote to Daisy, "& I have never seen a more lovely sunrise — just your kind. We were up about a mile — above a level of small pure white clouds so we couldn't even see the Bahamas on our left — but soon we saw Cuba on the right & then Haiti."[10]

The President had known she'd continue to worry on his account, and wanted to reassure her — not only the first president to fly abroad while in office, but the first since Lincoln to visit a battlefield in war. Taking a celestial fix of sun and moon, the captain turned the forty-four-ton behemoth, like a flying carpet, southeastward. "Then out over the Caribbean — high up — I felt the altitude at 8 or 9,000 feet — and so did Harry and Ad. Leahy — The cumulus white clouds were amazingly beautiful but every once in a while we could not go over them & had to go through one —

"At last — 5 p.m. — we saw the N.E. Coast of Venezuela & then the islands of the Dragon's Mouth with Trinidad on the left — The skipper made a beautiful soft landing & Ad. Oldendorf came out & took us ashore to the U.S. Naval Base — one of 'my' eight which we got for the 50 destroyers in 1940. It is not yet finished but operating smoothly."[11]

The U.S. naval base at Trinidad had come with a hotel, situated at Macqueripe on the north coast, "& thither we went for the night," the President related. However, there had then occurred a serious hiccup, unrelated to the dinner he was served. "Ad. Leahy felt quite ill — he had flu ten days ago — Ross McI[ntire] is worried as he is 68 & his temp. is over 100 — we will decide in the a.m."[12]

In the morning, on January 12, the doctor found Admiral Leahy still feverish. "Up at 4 a.m. This is not civilized," the President joked in his letter to Daisy. However, "Leahy seemed no better & we had to leave him behind — He hated to stay but was a good soldier & will go to the Naval Hospital & get good care — I hope he won't get pneumonia — I shall miss him as he is such an old friend & a wise counselor."[13]

If the President was concerned, though, he did not show it, for he never mentioned Leahy again in his letters to Daisy, despite the fact that Leahy was to have chaired daily meetings not only of the U.S. chiefs of staff but the British chiefs of staff, in their role as the Combined Chiefs of Staff, in Casablanca. The President was on top form: confident he could manage the summit quite successfully on his own, even without Leahy's wisdom.

Thus the *Dixie Clipper* flew on a further thousand miles to Brazil, filled its tanks with fuel at Belém, and set off for its great "hop" across the Atlantic, carrying its august passenger and small, slightly diminished entourage — followed closely by the backup Clipper, lest the *Dixie Clipper* experience engine trouble and have need to ditch.

Reflecting their earlier days as transoceanic first-class passenger planes, each Clipper boasted a lounge, a fourteen-seat dining room, changing rooms, and beds normally for thirty-six passengers — with a honeymoon suite at the rear. They required considerable piloting skills, however — takeoffs and landings in choppy, windswept water always an especial concern. The *Dixie Clipper*'s sister plane, *Yankee Clipper,* for example, would snag its wing several weeks later in Lisbon Harbor, with the loss of twenty-four lives.

Meantime, landing smoothly at the old British trading post of Bathurst (later renamed Banjul) on the Gambia River on January 13, after a twenty-eight-hour flight, the *Dixie Clipper* moored offshore. Arrangements had been made for the President to transfer to the light cruiser USS *Memphis,* ordered up from Natal by FDR's chief of naval operations, Admiral King — there to provide the President with a secure overnight stay where he would not be exposed to tsetse fly. As it was still light, however, he took the opportunity to tour the waterfront — the President seated in a whaleboat as the local British naval commander acted as his guide during a forty-minute cruise amid dozens of tenders and oil tankers. Loading and unloading beneath the evening sun, their crews seemed oblivious to the fact that the upright figure seated in the midst of the whaleboat party, in his civilian clothes and hat, together with Hopkins, McIntire, and McCrea, was the President of the United States.

Finally, hoisted aboard the USS *Memphis,* the President was given the flagship admiral's stateroom, "where I've had a good supper & am about to go to bed," he described, delighted to be in African waters.[14]

Given Roosevelt's childhood dream of going to naval college instead of Harvard (a hope dashed by his mother),[15] his long love of naval history, and his nearly eight years as assistant secretary of the Navy, being piped aboard an American warship as commander in chief for the first time in World War II was inspiring for the President. Yet the sentiment paled beside thoughts of what was to come. The next morning would see him embark for a further "1,200 mile hop in an Army plane," this time overland, as he wrote to Daisy — bound for "that well known spot 'Somewhere in North Africa.' *I* don't know just where," he added, in self-censoring mode. "But don't worry — All is well & I'm getting a wonderful rest." He felt positively refreshed. "It's funny about geography — Washington seems the other side of the world but not Another Place — That is way off," he wrote of Hyde Park, "& also very close to — "[16]

There Roosevelt left the sentence, however — unwilling to give hostage to fortune, lest prying eyes open, or see, his letter to the distant cousin whose romantic adoration he'd encouraged, especially after his mother's death two years earlier. "Lots of love — Bless you," he ended.[17]

To his wife, Eleanor, he meanwhile wrote in a similarly informative, if less tender, vein — telling her he'd be seeing their son Elliott when he arrived, and signing off: "Ever so much love and don't do too much — and I'll see you soon. Devotedly, F."[18]

He was almost there: not only the journey of a lifetime, but bringing the agenda of a lifetime. At Casablanca the President wished not only to map the defeat of the Axis powers in World War II, but commence discussions of the world to follow.

PART TWO

Total War

3

The United Nations

EVEN BEFORE THE war began for the United States, the President had been thinking of the postwar world.

Enlisting the help of his protégé, Assistant Secretary of State Sumner Welles, the President had begun drafting ideas immediately after drawing up the Atlantic Charter, in August 1941. What he wanted to create, he'd told Welles, was a postwar organization that the Americans, British, and Russians would embrace as military guardians, and that all sovereign democratic nations could subscribe to. The Japanese attack on Pearl Harbor several months later had made the need for a viable postwar system all the more urgent: a new world order that would make such wars of imperial conquest difficult if not impossible. He'd therefore charged Welles with modeling the project on the twenty-six countries whose representatives he'd assembled over Christmas 1941 in Washington — a group the President had decided, in a moment of inspiration, to announce to the world as "the United Nations."[1]

Properly constituted, the United Nations authority would, the President determined, avoid the disaster of the League of Nations — which neither the United States nor the Soviet Union had joined when it was formed. Building on the "Declaration of the United Nations," which had been signed in Washington on January 1, 1942, the United Nations would, this time, have teeth: the world's "Four Policemen," as the President called them.[2]

First, the Germans and Japanese would have to be defeated — but the military might of the three foremost antifascist fighting nations could then be turned into a global peacekeeping coalition: the United States, the Soviet Union, and Britain. He had then added China, a nation that had been fighting the Japanese since 1937 — thereby forcing the Japanese to keep an army of more than a million men on the Chinese mainland. Once the war was won, the President proposed, this same group of the world's major military powers could be employed not only to disarm the Axis nations for all time, but to police the world thereafter on behalf of the United Nations authority, ensuring that no Hitler or Mussolini or Hirohito would ever again upset global security by force of arms or conquest.

With laudable dedication, Welles—running the U.S. State Department under the sickly secretary of state, Cordell Hull—had thereupon set about the business, leaving the President to focus, meantime, on the best military strategy to defeat the Axis powers.

Under the aegis of the State Department, Welles had quietly set up a host of secret committees and subcommittees, asking members to think ahead on the President's behalf and produce for Mr. Roosevelt at the White House their specific recommendations and alternatives, on a regular basis.[3] "What I expect you to do," Roosevelt had instructed Welles, "is to have prepared for me the necessary number of baskets so that when the time comes all I have to do is to reach into a basket and fish out a number of solutions that I am sure are sound and from which I can make my own choice."[4]

Welles had done as ordered—magnificently, in retrospect. As historians would later note, neither Britain nor the Soviet Union, the other two primary nations conducting the war against Hitler, did anything in 1942 to address the needs or opportunities of the postwar world on an international scale—a "disastrous blockage at the top" in the case of the British.[5] By contrast, bringing together an extraordinary cross section of the nation's foremost minds and political figures in once-weekly meetings in Washington, Welles had single-handedly, in the midst of a global war being fought from Archangel to Australia, gotten his various teams working on the political, military, economic, labor, and even social (health, drug trafficking, refugees, nutrition, etc.) blueprints the President wanted for his vision of the democratic postwar world.

An extraordinary bipartisan group of Democratic and Republican senators and congressmen from the Capitol—including the chairman of the House Foreign Affairs Committee, the ranking minority member of the House Foreign Affairs Committee, a former chairman of the Senate Foreign Relations Committee, the current chairman of the Senate Foreign Relations Committee, and a senior current Republican member of the Senate Foreign Relations Committee—had joined with Welles's handpicked, representative minds from the State Department, the Agriculture Department, and the Board of Economic Warfare, as well as members outside government, including individuals from the press, the Council on Foreign Relations, and the academy, to provide the President with the necessary guideposts and alternatives he wanted at hand.

Though at first Welles had assumed the issues would be handled by the President in a peace conference after the conclusion of the war, as had been the case in the aftermath of World War I, the President had soon changed his mind — reckoning that if the postwar system could be settled before the war's end, it could avoid the unfortunate fate of the Versailles conference of 1919. Instead, the President had asked the committees to report their interim findings, via Welles, as swiftly as possible: concerned that America's allies, too, should help him address the challenge *before*, rather than after, the end of hostilities. By October 1942, therefore, as American troops readied for the Torch invasion of Northwest Africa — a draft outline of the postwar UN organization had begun to take shape.

Welles's teams, the President found, had done a grand job — indeed, Welles suggested that the putative "United Nations authoritative body" could already start functioning during the war itself. It would comprise a General Assembly of United Nations, seating representatives of all eligible countries of the world. It would also have a small Executive Council, incorporating the four major powers to arm and lead the organization with strength and simplicity. By April 1942, in fact, after discussing the matter with Mr. Roosevelt, Welles (who had made himself chairman of the Subcommittee on Political Problems, an international organization) had suggested the way the Executive Council should be set up: the President's Four Policemen — the United States, Britain, the Soviet Union, and China — being given permanent seats on an Executive Council together with a small number of further, rotating seats reserved for members elected by the full United Nations authority, in order to give the council more balance and connection with the main Assembly.[6]

As Welles's committees had advanced their confidential proposals in Washington, British Foreign Office officials in London had become anxious lest Churchill's lack of interest in postwar planning leave Great Britain out on a limb. "His Majesty's Government have not yet defined their views on questions or made any response to Mr. Welles's expression of opinion," the head of the British Economic and Reconstruction Department had complained as late as September 3, 1942, only weeks before Torch.[7]

Little was done to rectify this failure, however, in view of the Prime Minister's full-time preoccupation with Britain's military operations, and his aversion to postwar planning—which would inevitably involve the continuing transformation of the British Empire into a Commonwealth of Nations rather than a colonial enterprise directed by Parliament in London.[8] "I hope these speculative studies will be entrusted mainly to those on whose hands time hangs heavy," Churchill had mocked his foreign minister's attempt to produce a British version of Welles's work, "and that we shall not overlook Mrs. Glass's Cookery Book recipe for Jugged Hare—'First catch your hare.'"[9]

In Moscow, too, there had been a complete lack of interest in planning for a democratic future—Stalin's Soviet government simply refusing to comment on or respond to cables from its Russian ambassador in Washington, imploring the USSR to get involved in international postwar proposals.

For FDR, the failure of Joseph Stalin to participate in discussions about the postwar world was galling if perhaps inevitable, given the history of the Soviet Union since the Russian Revolution: its protracted civil war, Stalin's Great Terror and purge trials, and its ever-darkening development as a communist police state based on intimidation, arrest, torture, imprisonment, deportation, and execution. Nevertheless, as president of the world's biggest and most advanced economy, Roosevelt wanted to give the Russians—who were bearing in blood the brunt of Hitler's war of conquest—at least the chance to be a party to his proposals. And if Stalin, the absolute dictator of the Soviet Union, would not sit down to discuss them, then the President would begin the discussions without him—in Casablanca.

Hitler had declared that democracy was a relic of the past. The President, working with Winston Churchill, would now show him he was wrong: that democracy was, in fact, on the move.

Casablanca, then, was to be much more than a military powwow. Weeks before he left the White House on his secret journey, the President had begun to rehearse his developing vision with other world leaders such as Jan Christiaan Smuts, the prime minister of South Africa, whom he'd known since the summer of 1918.

Then, too, the end of the world war had seemed at hand. Twenty-four years later, Roosevelt was "drawing up plans now for the victorious peace which will surely come" and hoped to discuss them with the former guerrilla leader of the anti-British Boers, if Smuts could see his way to come to Washington. A more durable and effective outcome was necessary than the ill-fated Versailles Treaty. "As you know," the President explained, "I dream dreams but am, at the same time, an intensely practical person, and I am convinced that disarmament of the aggressor nations is an essential first step, followed up for a good many years to come by a day and night inspection of that disarmament and a police power to stop at its source any attempted evasion of the rules."[10]

This time, then, postwar peace would not be guaranteed by treaties that could be broken with impunity, but by irresistible force — on behalf of the community of nations. There were "many other things to be worked out," Roosevelt had added in his letter to Smuts, such as decolonization, effected over time, but with no backsliding by the old European powers, as after World War I — whether by the British, the Dutch, the French, the Belgians, the Spanish, or the Portuguese. "Perhaps Winston has told you of my thought of certain trusteeships to be exercised by the United Nations where stability of government for one reason or another cannot at once be assured. I am inclined to think that the [colonial] mandate system" — instituted in the wake of the Versailles Treaty and the League of Nations — "is no longer the right approach, for the nation which is given the mandate soon comes to believe that it carries sovereignty with it."[11]

Colonialism, in other words, was to be gradually but responsibly phased out in the aftermath of World War II, and a new postcolonial world ushered in.

As prime minister of a former British colony now enjoying self-government and Dominion status within the British Commonwealth, Smuts's reaction was important as Roosevelt sought to picture a viable postwar world and the problems he might encounter in getting international agreement.

Smuts — whose Boer countrymen had, like the Americans, risen up against British colonial rule — understood the President's strong feelings on that score, but was facing a new election and could not travel. The prime minister of Canada, Mackenzie King, could, however — and once more the President had asked if King could come spend a few days with him at the White House, after the Torch invasion, so that he could rehearse his notions of what would, effectively, be the endgame.

Arriving from Ottawa by train, King had thus made his way to 1600

Pennsylvania Avenue on the morning of December 4, 1942, three weeks after Torch. Alert to the dangers of premature leaks, rumor, and outright hostility among Republican politicians and newspaper owners who still hated him for his New Deal program, the President was chary of committing thoughts to paper lest they be used against him. Thankfully, the Canadian prime minister dictated each night a careful record of his day — and it is to this diary we owe our most authentic account of the President's military and political strategy for ending the war, and the peace he hoped to mold thereafter, before leaving for Africa.

The President, King had found on arrival, was "sitting up in his bed" on the second floor of the White House mansion, "wearing a gray sweater," smoking. He'd been "reading newspapers. Gave me a very hearty welcome. Began at once by saying he was having a [hard] time with the new Congress," given the loss of so many Democratic seats in the November midterm elections, "but hoped that would go by."[12]

Senator Robert Taft of Ohio, the leading isolationist opponent of the New Deal in Congress and eldest son of Republican president William Howard Taft, was a particular sore, the President had remarked — quoting to King an account in that day's paper. Taft was reported to be opposing the President's attempt to make a new deal with the Panamanian government over the Panama Canal area.

The President had smiled mischievously. "Asked me," King recorded, "if I knew the U.S. owned the largest red-light district anywhere."[13]

Mackenzie King — a staunch, Bible-obsessed Presbyterian who had forsworn alcohol for the duration of the war — was well aware how much Roosevelt enjoyed teasing him. When King confessed his ignorance, the President had "described how one of his ancestors," William H. Aspinwall, had given up hope of building a transcontinental railway across "the isthmus of Panama, having mortgaged [his] homes in the States." Then suddenly he'd heard gold had been "discovered in California. He knew at once that his railway would be a success and half a dozen offers were immediately made by wealthy men to complete his road. Later, when De Lesseps came to develop the canal, the red-light district developed in that area. The U.S. are now wishing to get control of certain parts and had to purchase this area . . ."[14]

It was typical of Roosevelt to use the irony of a vexing situation to render it less frustrating — U.S. senators "querulous about different things" such as this while the President struggled to win a global war and create the basis for subsequent peace.

Beneath African palms, in complete privacy and in secret, the President would soon, he told Mackenzie King, be able to discuss his vision for the world that would follow war: especially his idea of a United Nations Security Council led by the four powers.

First, however, the war had to be won by the United Nations. Roosevelt had already explained his current strategy to the supreme commander of the Soviet Armies, Joseph Stalin, in a cable he'd dispatched from the White House on November 19, 1942. "American and British Staffs are now studying further moves in the event that we secure the whole south shore of the Mediterranean from Gibraltar to Syria," the President had informed Stalin. "Before any further step is taken, both Churchill and I want to consult with you and your Staff because whatever we do next in the Mediterranean will have a definite bearing on your magnificent campaign and your proposed moves this coming Winter." U.S. and British armies were not only forcing Hitler to keep substantial numbers of troops, artillery, tanks, and planes in Norway, the Netherlands, Belgium, and northern France to defend against the threat of Allied invasion in the West, but were forcing Hitler to do so in the Mediterranean now, in order to keep Italy fighting as a primary Axis partner; this would make it difficult, if not impossible, for Hitler to achieve unilateral military victory against the Russians.[15]

Stalin had not immediately responded to the request, however, as the President confided to King — the Russian dictator's focus having been on the Russian counteroffensive that began that day at Stalingrad. A week later, on November 25, Roosevelt cabled again. The President had congratulated Stalin on the Russian breakthrough west of Stalingrad, which threatened to cut off the German salient stretching as far east as the Volga. In order to remind his Russian counterpart that the United States was fighting a *global,* not simply regional, war, however, he'd informed Stalin of a similar U.S. game changer in the Pacific, where the U.S. Navy had decimated the Japanese fleet attempting to reinforce the Japanese army on Guadalcanal. The Japanese had been compelled to evacuate the island, and U.S. forces were now "sinking far more Jap ships and destroying more planes than they can build."[16]

This time Stalin *had* responded. "As regards operations in the Mediterranean, which are developing so favorably, and may influence the whole military situation in Europe, I share your view that appropriate consultations between the Staffs of the United States, Great Britain and the USSR have become desirable," the Russian premier wired back to the White House on November 27, 1942. But beyond this—and his congratulations on the U.S. Navy's success in the Pacific as well as American-British operations in North Africa—he declined to be specific. In particular he had ignored the idea of a meeting of national leaders, not even according it a mention.[17]

Finally, on December 2, the President had decided to get to the point. In yet another cable sent from the Map Room at the White House, he'd urged Stalin to address "the necessity for reaching early strategic decisions" through "an early meeting." This was not simply because military staffs, conferring on their own, would be unable to reach decisions "without our approval," but because Roosevelt felt "we should come to some tentative understanding about the procedures which should be adopted in the event of a German collapse"—i.e., the postwar.

For this, it would be vital to meet in person, the President had emphasized. "My most compelling reason is that I am very anxious to have a talk with you," he'd written. "My suggestion would be that we meet secretly in some secure place in Africa that is convenient to all three of us. The time, about January fifteenth to twentieth [1943]. We would each of us bring a very small staff of our top Army, Air and Naval commanders." He thought a rendezvous in "southern Algeria or at or near Khartoum," in Egypt, would fit the bill.[18]

Mackenzie King had been awed and delighted by the President's initiative—touched that Roosevelt would share with him both the background and his confidential intentions: his game plan.

Still waiting for Stalin's response, the President had the next day discussed with King the problem of Churchill and Great Britain—to which Canada, as a Dominion of the British Empire, was constitutionally tethered. The President still deplored Churchill's stand over India and unwillingness to abide by the terms of the Atlantic Charter he'd signed up to; also Churchill's dislike of the Beveridge Plan for postwar social security in Britain. Churchill's obstinacy in pursuing the postwar revival of the "British Empire," rather than inspiring and leading a new, postcolonial "British Commonwealth of Nations," came under the President's caustic fire—as well as Churchill's aversion to the notion of postwar United Nations trusteeships. "When I asked him about Churchill's attitude, he said the reply which he [Churchill] had made in discussing these things was a rather sad one," King recorded that night. "It was to the effect that he [Churchill] would not have anything to do with any of these questions. That when the war was over he [Churchill] would be through with public life," and would turn to "writing."[19]

That evening, December 5, 1942, there had been cocktails at 7:30 p.m., mixed by the President himself. ("The President said: we will not ask Mackenzie to take any cocktails tonight. I appreciated," the wartime tee-totaler noted, "his anticipating my refusal.")[20] There was then dinner with Harry Hopkins and his wife, Louise. And several short documentary and newsreel films, in black and white and in color.

Looking at the documentary footage that had been spliced together — some of it "going back to the days of his governorship in the N.Y. state," as well as events like Roosevelt's "flight to Chicago at the time of the [1932] Convention" — "reviews of troops, etcetera" — King had found himself amazed. "It made me marvel how a man had ever stood what he did in dealing with crowds over so many years," the quiet Canadian had noted, given the President's physical disability and the demands of America's almost continuous electoral process, compared to the Canadian parliamentary system. "What was the most interesting was the way in which he, from the outset, had stood for the new deal and the rights of the common man in all his addresses," King dictated. "It was a real recreation and most pleasant," he'd added. "As it was getting on toward 10, I asked the President if he did not think he should retire and let him rest. He said no, we want to have a talk about another matter. He had said earlier in the day: 'I want to speak particularly about Stalin tonight.'"[21]

In the President's Oval Study (or "chart room," as King called it in the nightly diary he dictated), "the President sat on the sofa and told me to come and sit beside him there, to get his good ear, the right ear.

"He then ordered a horse's neck for each of us: ginger ale and the rind of an orange. Harry Hopkins came in and sat down for a few minutes and then retired. The President then started in at once on what he has in mind as a post-war programme. I looked up at the clock at that moment. The hands were exactly at 10 to 10."[22]

A fervent believer in spiritualism as well as Christianity, King was forever watching the hands of the clock for signs of significance. Given the magnitude of the President's global problems — war in the Pacific, war in China and Asia, war in the Soviet Union, war in the Mediterranean, war in the Atlantic, war in the Aleutians, preparations for eventual cross-Channel landings — it had seemed extraordinary to Mackenzie King that the President of the United States could set these concerns aside, in his mind, and share his thinking on the world that would come *after* the war was won.

Before addressing the matter of Stalin, the President had given his own views on the objectives or principles that should guide the victorious nations. "We talked of the 4 freedoms. Two of which," the President remarked, "we cannot do much about." Freedom of religion, Roosevelt had explained, was "something that the people have to work out for themselves. The State cannot impose anything. The freedom of speech: that too is something that will take care of itself" — though the President wished something could be done "to prevent exaggerated and untrue statements" from being broadcast or printed, especially the near-treasonable articles constantly being published by "sensational papers," such as Colonel McCormick's right-wing, isolationist *Chicago Tribune.*

This left "the other two" freedoms, which were perhaps more crucial, at least in planning a postwar universe: "freedom from fear and freedom from want."[23]

Of the two, the President told King, "the first is necessarily the most important, as the second depends on it. As respects freedom from fear," the President had continued, "that can only be brought about when we put an end to arming nations against each other." In the case of Germany and Japan, the arms treaties signed after World War I had proven useless. The German and Japanese capacity for making war must therefore be completely and irrevocably destroyed, once and for all time, he felt. This was something that could not be secured by negotiation, à la Versailles Treaty, but only achieved by a policy of "unconditional surrender" to the Allies, the President explained.

It was the first time King had heard President Roosevelt use the term, and he listened most carefully as the President explained.[24]

"My great hero in all of this today," Roosevelt remarked, "is General Grant. In bringing the civil war to an end, Grant demanded to Lee unconditional surrender. He would make no agreements, no negotiated peace."[25] Repeating the point, the President said: "I think there should be no negotiated peace" at all with the Germans and Japanese. "It should be an unconditional surrender. After Grant had gathered in all the guns, ammunition, etc., there were quantities of horses remaining. Grant turned to Lee and said to let the horses go back to the field. For the people to use them in the cultivation of the soil, get back to the art of peace. That I think at present is what we should do with Germany. Deprive her of all right to make planes, tanks, guns, etc., but not take away any of her territories nor prevent her development in any way."[26]

King had asked if the President was confident of dictating unconditional surrender to the forces of the Third Reich on that basis. "He replied: 'yes'" — with Japan to follow: "that he thought what should be done was to defeat Germany first; demand unconditional surrender and then for the 3 powers: Britain, U.S. and Russia to turn to Japan and say: now we demand the same of you. If you want to save human life, you must surrender unconditionally at once. If not, the 3 of us will bring all our forces to bear, and will fight till we destroy you. Russia would then be persuaded to attack Japan. It would not take a year to bring about her defeat. He was not sure the Japanese would accept any unconditional surrender, and would probably seek to fight on. However that plan of campaign would bring its results" — the world finally and definitively spared the possibility of a renascence of German or Japanese militarism — ever. "If the Japanese did not accept unconditional surrender," he added in a remark that would have immense significance later, "then they should be bombed till they were brought to their knees."

Thinking of the ever-burgeoning size of America's air forces as well as the terrifying new bomb — using vital Canadian minerals such as radium and uranium[27] — that he'd ordered to be developed, the President confided to King how the United States would have to push its forces to within bombing distance of Japan, while the war in Europe went on. "At present, Russia is too busy to attack Japan, and Japan is too busy to attack Russia."[28]

Unconditional surrender it was, then, as the President's war aim — not to placate or encourage the Russians, as some subsequently assumed, nor in punitive revenge against the Germans or the Japanese, as others did. And especially not to mollify liberals in America, who were complaining that the United States was installing former fascists, like Admiral Darlan, to administer liberated territories, instead of getting rid of them, as still others later speculated. Rather, the President saw unconditional surrender quite clearly at that moment as the basis for lasting postwar security — leading to a postwar peace to be overseen by the United Nations, using the four most powerful antifascist nations: the United States, Russia, and Britain, together with China (which the United States was supporting as the most populous and potentially important nation in the Pacific) — acting as the world's "international police" on the UN's behalf: the basis of the UN Security Council.

Each of the policing powers would have its "own air force" to enforce the disarmament of Germany and Japan, the President explained to King — able thereby to cauterize any attempt by such nations to step out of line and make war on others. Colonialist imperialism would, over time, become a thing of the past, with nations reverting to their original boundaries after Hitler's war. "Keep everything as it was" was how the President explained his vision — territorial changes agreed only by democratic plebiscite, not war. "Russia to develop Russia but to make an agreement not to take any territory. Not to try to change system of government of other countries by propaganda."[29]

Was all this a pipe dream?

"The President then said: 'to effect all this, of course, [we] would have to get Stalin to agree,'" King recorded, careful to note Roosevelt's actual words.

Stalin was a dictator — but a dictator more concerned with absolute rule over his own vast territory, stretching from Archangel to Vladivostok, than elsewhere: especially in a world revolution that he would find less easy to control. "He said that he believed he [Stalin] would. That he thought Molotov was an Imperialist but he believed Stalin was less and less on those lines."

However abhorrent Russian communism was, one had to be realistic. Since "it was clear that the U.S., Britain, China could not defeat Russia" by force of arms — something even Hitler was failing to do, with more than two hundred Wehrmacht divisions and his Luftwaffe — it would be futile to try. Better, he thought, to see if the Soviets could be drawn into an international system that guaranteed Russia would never be attacked by the Germans or another Hitler. Quoting a former Republican Senate leader, Jim Watson, the President said to King: "If you cannot beat him, join him. The thing to do was to get them all working on the same lines," under the aegis of a supranational United Nations.

Communist or not, the President had continued, Russia was going to be, after the war, "very powerful. The thing to do now was to get plans definitely made for disarmament" of Germany and Japan, with Russia onboard[30] — hence his attempt to set up a secret summit with Stalin and Churchill.

The Canadian prime minister, in his diary, had acknowledged being thrilled. It was clear that, in sharing his notion of unconditional surrender both of the Third Reich and the Empire of Japan, the President was speaking without vengeance or rancor, but almost as a surgeon might, prior to taking out a tumor.

Postoperatively, in the case of the Third Reich after the war's end, "there should be put into Germany immediately a committee or commission of inspectors," Roosevelt said to him, "say 3 — one to be chosen from Canada; one from South America, and one from China," on behalf of the United Nations. "They should have their staffs, and their business would be to inspect day after day, year in and year out, all the factories of Germany to see that no war material should be manufactured. If any such were discovered, the Germans were to be told that unless that stopped within a week's time, that certain of their cities would be bombed. The cities would be named: Frankfurt, Cologne, and probably the cities where the manufacturing is taking place. If they went ahead, despite this threat, they might then be told that from now on, all imports and exports in and out of Germany would be stopped. That no trains passed out of their countries. Persons would be stopped at the borders."[31] Blockaded, in other words, or "ostracized," as King had reflected.

Yet if Mackenzie King had been impressed by the President's visionary thinking early in December, 1942 — less than twelve months after Pearl Harbor — he'd been equally moved by the depth of Roosevelt's *moral and social* purpose. In Britain, Winston Churchill had dismissed the Beveridge Report, which outlined possible future British social and health policies — as pie in the sky:[32] a dismissive view that was echoed by the British ambassador on Mackenzie King's visit to Washington, when King met with him. Americans were "all much excited about the Beveridge Report," Lord Halifax had confided in his own diary. "I told them all that, just as with the Malvern Conference Report, so with this, they [Americans] always know much more about it than I do!"[33]

The President of the United States certainly knew more than the British ambassador about the postwar social blueprint for Britain. "The President said the Beveridge report has made a real impression in this country," King recorded.[34] "The thought of [medical and employment] insurance from the cradle to the grave. 'That seems to be a line that will appeal,'" Roosevelt had said to King at dinner. "You and I should take that up strongly. It will help us politically as well as being on the right lines in the way of reform" — a remark King correctly interpreted as meaning "the President has in mind a fourth term and that he feels it will come as a result of winning the war, and the social programme to be launched."[35]

As president of the United States, Mr. Roosevelt "did not think the country will stand for socialism," King recorded the President's caution, but he did make clear that improving the condition of America's working people was as much a part of his vision of the postwar world as would be international security achieved through unconditional surrender of the Axis warrior nations, and disarmament closely monitored by the United Nations. "I felt in listening to the President that he was naturally anxious to be responsible for planning the new order," King reflected[36] — a new order that would snatch the wind from the sails of those idly or idealistically espousing communism, since it would guarantee the well-being and security of the majority of ordinary people, without communist barbarity or oppression.

As a deeply devout Christian who read the New Testament first thing in the morning and last thing at night, Mackenzie King had thus listened to the President's *tour d'horizon* with growing "relief," he admitted — the opportunity to discuss with the President of the United States "social questions and reform, instead of these problems of war and destruction. I felt tremendously pleased. It may be that when the war is over, new force and energy will come forward toward the furtherance of these larger social aims. It was midnight when I got to bed . . . From the moment I turned out the light until waking I slept very soundly."[37]

The world, after this war, was clearly going to be very different from the one bequeathed by the victors of World War I.

4

What Next?

WHAT HAD MOST moved Mackenzie King on his stay at the White House early in December, 1942, were the little details that went hand in hand with his discussions with the President.

After lunch in the small dining room upstairs in the White House mansion one day, King noticed "on the President's desk" among the bric-a-brac, "a little bronze of his mother," which touched him deeply.[1] Fresh from her trip to England, the President's wife, the First Lady, had been present — the President proud, King had happily noted, of what Eleanor had accomplished there as a spokesperson, so to speak, of American idealism.

The President's health, though, was another matter. On the afternoon of December 4, 1942, for example, King had been somewhat alarmed by Roosevelt's physical condition. "Had tea alone with the President at 5:20 in his circular chart room [the Oval Study]. The President poured tea himself."[2] The two leaders had spoken of manpower and mobilization — problems common to both countries. However, "I noticed that his hand was very, very shaky," King had dictated — the tea in danger of spilling. The President looked "rather tired," but as they talked he'd "brightened up."

Here again Prime Minister Mackenzie King's testimony, in the detailed diary he was keeping, would offer the most intimate clues to the President's mind in late 1942. No other war leader was exploring a postwar vision such as the President was doing; Churchill could only dream of the past; Hitler, only of the German *Volk* and of ruthless conquest. And who knew what Stalin dreamed of? Would the President be well enough, however, to get his allies to cooperate and carry out his grand vision of the postwar world? Would Congress and the American public embrace it, or go back to isolationism? And what of the war itself?

Turning to confront the President on the sofa, King had therefore asked him, face to face: "What are the immediate plans, supposing you get complete possession of North Africa, what next?"[3]

From the point of view of military strategy in order to achieve political ends, it was a most interesting question.

The President seemed glad that King had raised it. "That of course is the next problem," Roosevelt replied. "I wanted to speak of it."[4] To his great disappointment, despite the success of the President's Torch operation, which they'd opposed almost to the point of mutiny, his U.S. generals and admirals were still out of sync with their commander in chief. In fact his generals and admirals were now out of sync with each other, and the British.

"For some time past," Roosevelt confided to the Canadian prime minister — whose country was supplying a vast amount of war material to the Allied effort, as well as significant numbers of troops, and the crucial materials for development of an atomic bomb in the United States — "we have had the Chiefs of Staff both here and in England working on the strategic side of things." There were "at least 10 different places" where the Allies *could* advance, from northern Norway to the Balkans. "No decision was reached as yet," though, Mackenzie King recorded the President's lament, since "it was very hard," the President said, "to get the different Chiefs of Staff to agree on a plan."[5]

Harry Hopkins had been little help in this respect. Hopkins was by nature and ability a "fixer" — a highly intelligent man, brilliant at absorbing reports, and able to see beyond hurdles. Never having fired a gun or seen war at close quarters, however, Hopkins had erratic military judgment, to say the least. He had urged the President to declare war early in 1941, before the nation was ready to fight a one-ocean war, let alone two.[6] Then — having become convinced the Russians were not going to be defeated by the Germans in 1941 — he had urged throughout 1942 a cross-Channel invasion of France rather than the President's "great pet scheme" of Torch, believing the North African operation might actually fail. Even if successful, it would be a diversion of decisive American effort, he felt.

Hopkins, as a civilian, could at least be forgiven for his ignorance of military realities — especially the lack of American experience in fighting an enemy as battle-hardened, ideologically driven, and professional as German troops marching to Hitler's triumphant tune. However, Hopkins's military innocence had been mirrored by most senior, professional desk generals and admirals in the U.S. War and Navy Departments in Washington. Despite the success of Torch, the U.S. chiefs were *once again* urging, early in December, that a cross-Channel Allied invasion be mounted in the spring of 1943, or latest by the summer of '43 — without American soldiers or their field commanders having seen more than a few days of battle, and that only against conflicted Vichy French forces.

Mackenzie King was as skeptical of the chances of a cross-Channel attack succeeding in 1943 as the President — indeed, more so given the "fiasco" of the Canadian raid on Dieppe three months earlier. Still mourning the loss of so many thousands of Canadian soldiers killed or wounded and captured on the beaches of the little French seaport on August 19, 1942 — almost all of them brave volunteers, sacrificed to no real purpose other than to demonstrate the futility of a premature cross-Channel assault — the Canadian prime minister had been alarmed by Hopkins's views the day he arrived at the White House. The President's counselor had shared with him "the need of a decision being fairly quickly made as to what the campaign for next year [1943] was to be. He said the military heads could not yet make up their mind but he thought that decision would have to be made at once if supplies were to be gotten in to the right place. It would seem to him it would probably have to be from England on Europe" — i.e., for a cross-Channel assault in 1943 — "and that great quantity of supplies would have to be gotten across immediately."[7]

The notion that Allied forces could defeat the Wehrmacht simply by *supplies* had seemed to the Canadian premier unbelievably optimistic. To King's profound relief the President, however, declared he didn't agree with Hopkins — or with his U.S. War Department staff. He reminded King how it was only through his own and Churchill's combined efforts that the Allied war against Hitler had been saved from disaster that year, 1942, by insisting on Operation Gymnast (which was then renamed Torch). "It is a good thing Winston and I kept it out [on the table] as we did," the President had remarked of the invasion of Northwest Africa — for "during early 1941, army and navy were all for direct attack across the Channel in the spring of 1942." The plans for a Second Front invasion "kept taking longer and longer, after spring of 1942. Then it was to be on in the summer." Again, this had proved impossible, at least in sufficient force to assure success. "Could not get ships, etc. Then the next plan was that they would try in the spring of 1943." At that prospect, the President had, in the summer of 1942, finally drawn the line as U.S. commander in chief, convinced that U.S. forces would have to gain actual combat experience fighting the Wehrmacht if a difficult cross-Channel invasion were to have any chance of success. "The President then said that he and Winston [had decided they] would get together in June" of 1942, to work out a new strategy. Tobruk's fall, and the failure of the British to halt Rommel's advance in Libya, had put the kibosh on any hopes of British-American success across the English Channel, where twenty-five German divisions were awaiting their arrival. "The President then said he had told Churchill: 'I go back to my first love, which is to attack via North Africa.'" Such a strategic blow would secure the Atlantic port of Dakar and, in terms of lines of communication and resupply, enable the Allies to use "the short route from Britain to Africa, and short route from U.S. to Africa." Torch would coincide with the British, reinforced with U.S. tanks and air groups, getting "control of North Egypt and with good luck" lead to "control of the Mediterranean."[8]

"I felt the soft place was Southern Europe," Roosevelt had reminded King—who'd been staying with him at the White House the previous spring, when the strategic debate had burned fiercely. Side by side with that southern European/North African strategy, the President had meanwhile wanted "a strong hitting force pointed at Germany from the North" as a permanent threat[9]—forcing Hitler to keep his twenty-five or thirty German divisions stationed along the North Sea and Atlantic coasts of Europe, well away from Russia.

It was in the Mediterranean, however, that U.S. forces could best actually fight and gain crucial command and battle experience, the President had explained to King—at the very extremity of German lines of communication and resupply. The campaign in Northwest Africa was already drawing huge Axis military forces to the Southern Front, across the Mediterranean, forcing Germany and Italy to meet the Allies in combat there—the Germans using vital, battle-hardened and battle-worthy troops, planes, and military resources that could not, as a result, be sent to reinforce their war on the Eastern Front.

The Mediterranean thus offered the U.S. Army, Air Forces, and Navy a priceless opportunity: namely to rehearse and perfect the command and combat skills they would need in fighting ruthless, highly disciplined, strongly motivated German forces in Europe, *before* being expected to undertake anything as daunting as a contested cross-Channel invasion—

an operation of war that had not been successfully attempted, after all, in almost a thousand years, since the time of William the Conqueror.

Battle experience, then, was the crux of the matter: the reason why the President so profoundly disagreed with Hopkins; with Secretary of War Henry Stimson; with General George Marshall, U.S. Army chief of staff; and with all the voices in Washington baying again for an immediate cross-Channel Second Front. As U.S. commander in chief he, President Roosevelt, had a responsibility to ensure the nation did not embark on a course of military action that would fail—especially when there was no need to do so, as he confided to Mackenzie King in another talk on December 6, 1942, as King prepared to return to Canada. He had, "this afternoon, sent word to the Chiefs of Staff in Washington and also to the joint staffs in England to ask exactly what they had thus far decided about the next moves, and what were the points they were still debating. He said when you think it took from January till June before we settled on Africa and definite plans for the campaign, you see it is time we get the next step settled or next move determined." As president and commander in chief, however, he had his own view—which he now shared with King.

"In many ways," the President confided, "he wished for nothing more than let the fighting continue in Africa indefinitely. We are able to get supplies across, so much easier to Africa than to any other place. We can wear down the Germans there by a process of attrition"—just as U.S. forces were doing in the Pacific, in the Solomon Islands, while learning the art of modern combat. "He said: I feel the same about the Japs. As long as we can go on the wearing down in the one place, we are coming nearer to certain victory in the end."[10]

Mackenzie King, as prime minister of Canada, had breathed a sigh of relief. When asked by the President if his Canadian generals were also pressing for an immediate Second Front, King responded that, unlike the generals in Washington, the Canadian generals "felt it was better to keep a strong hitting force pointed at Germany from the North," but *not* to launch such an actual invasion before there was a reasonable chance it would succeed. The President "said he felt that very strongly" too, King recorded.

"It would be a great mistake," Roosevelt had remarked, "to do anything which would take away the German armies that are now concentrated in occupied France and in the North—anything which would make them less fearful of an enemy invasion," in terms of threat. Beyond that potent menace, however, the President had explained to King, he had no actual wish to launch a D-day landing any time soon across the English Channel, with forces and commanders still inexperienced in combat. "He thought that what the Canadians had done at Dieppe" — where almost a thousand men were slaughtered in a matter of a few hours, and two-thirds of their forces were killed, wounded, or captured by the Germans, without even getting off the beaches — "was a very necessary part of the campaign," for it had "made clear how terribly dangerous the whole business of invasion across the Channel was."[11]

These had been the President's own words. They explained why the President was so determined to stop his top military staff from insisting upon a suicidal assault in the wrong place, at the wrong time. The Second Front should be kept as a *threat* — but no actual cross-Channel invasion be launched until 1944, when U.S. mass production could ensure superiority in arms; more important still, it would be a time by which Allied forces in Africa and the Mediterranean would have learned the lessons of modern combat: how to defeat the Germans in battle. Only then would it be fair to ask huge numbers of American sons — perhaps two million — to land across the defended beaches of northern France and fight their way to Berlin.

Mackenzie King thus set off to return to Ottawa that evening, December 6, 1942, deeply relieved: knowing the President would do nothing rash before the military forces of the Western Allies — Canada, the United States, and Britain — had proven themselves in combat and were ready: preferably in 1944, unless by some miracle the Germans collapsed. In the meantime, the President hoped, he confided to King, that Stalin would cease parrying his appeals for a summit, and would help start an international dialogue on the postwar world — with history at stake.

5

Stalin's *Nyet*

FINALLY, ON DECEMBER 6, 1942, shortly after Mackenzie King left Washington, the President heard back from Stalin. Although the Russian dictator "welcomed the idea of a meeting of the leaders of the Governments of the three countries to determine a common line of military strategy," he himself would "not be able to leave the Soviet Union. I must say that we are having now such a strenuous time that I cannot go away even for a day." Around Stalingrad, he explained, "we are keeping encircled a group of German troops and hope to finish them off."[1]

The question of postwar agreements was not even mentioned.

Given the amount of aid — more than 10 percent of Russia's war needs — that the United States was supplying the Russians, the President cabled back that he was "deeply disappointed that you feel you cannot get away for a conference with me in January." Stalin was known never to have gone near the Russian front.

The President urged Stalin to reconsider. The date proposed was still five weeks in the future. "There are many matters of vital importance to be discussed between us. These relate not only to vital strategic decisions but also to things we should talk over in a tentative way in regard to emergency policies we should be ready with if and when conditions in Germany permit. These would include also other matters relating to future policies about North Africa and the Far East which cannot be discussed by our military people alone." If Stalin could not see a way to leave Moscow in January, what about "meeting in North Africa about March first"?[2]

To this plea, however, there was no response from Moscow for a week. When the reply came, it was only to say Stalin regretted "it is impossible for me to leave the Soviet Union either in the near future or even at the beginning of March. Front business absolutely prevents it, demanding my constant presence near our troops."[3]

About this patent untruth the President could only shake his head, knowing Stalin never went anywhere near his brave Russian troops. The rest of Stalin's message — asking what exactly were the "problems which you, Mr. President, and Mr. Churchill intended to discuss at our joint conference," and wondering if these could not be dealt with "by correspondence" — had been similarly disappointing, despite the Russian dictator's assurance "there will be no disagreement between us."[4]

The chances of that, Roosevelt knew, were slim — especially given Stalin's hope that "the promises about the opening of a second front in Europe given by you, Mr. President, and by Mr. Churchill in regard of 1942 and in any case in regard of the spring of 1943, will be fulfilled, and that a second front in Europe will be actually opened by the joint forces of Great Britain and the United States of America in the spring of next year."[5]

That hope — as the President had confided to Prime Minister Mackenzie King — was pie in the sky. Unless the Germans showed signs of collapse in 1943, he was simply not going to approve such a strategy until U.S. forces and commanders were battle-hardened in the Mediterranean that year — just as was taking place in New Guinea and the Solomon Islands in the South Pacific, at the extremity of Japanese lines of communication.

It was Stalin who would inevitably be disappointed, then, once he learned of Roosevelt's implacable decision. Though the President's own generals and admirals, his war secretary, his counselor Hopkins, and even his ambassador to London, John Winant, might echo Stalin's appeals for an immediate cross-Channel assault, the President was simply not going to authorize mass American — and Canadian — suicide. Each day, by contrast, the President was more confident of "certain victory in the end" — if the Allies made no more mistakes.

With Stalin still saying *nyet* to a summit meeting, however — whether in January or in March, 1943 — the President had cabled Churchill on December 14, 1942, to say they should go ahead without him. In Casablanca, as he confided to Captain McCrea.

Before he left, however, the President decided he must do two important things. First, get the nation behind him. And second, his generals.

6

Addressing Congress

As the seventy-eighth Congress prepared to reassemble with a much-diminished Democratic majority, the President decided to use his annual State of the Union address, on January 7, 1943, not only to review the past year but to share something of his vision of the future.

The speech went through no fewer than nine full iterations over "many days," starting before Christmas and extending beyond the New Year, Judge Rosenman (the President's primary speechwriter, together with the playwright Robert Sherwood and Harry Hopkins) later recalled.[1] Finally, at noon on January 7, the President was driven to the Capitol to deliver his "sermon."

"The past year," the President began boldly, "was perhaps the most crucial for modern civilization. The Axis powers knew that they must win the war in 1942—or eventually lose everything. I do not need to tell you," he added to loud cheers, "that our enemies did not win the war in 1942."

Step by step the President reminded members of Congress and those listening on radios at work, or in their homes, of the year's most significant military actions. "In the Pacific area our most important victory in 1942 was the air and naval battle off Midway Island. That action is historically important because it secured for our use communication lines stretching thousands of miles in every direction. In placing this emphasis on the Battle of Midway, I am not unmindful of other successful actions in the Pacific, in the air and on land and afloat, especially those on the Coral Sea and New Guinea and in the Solomon Islands. But these actions were essentially defensive. They were part of the delaying strategy that characterized this phase of the war. During this period we inflicted steady losses upon the enemy—great losses of Japanese planes and naval vessels, transports and cargo ships. As early as one year ago, we set as a primary task in the war of the Pacific a day-by-day and week-by-week and month-by-month destruction of more Japanese war materials than Japanese industry could replace. Most certainly, that task has been and is being performed by our fighting ships and planes. And a large part of this task has been accomplished by the gallant crews of our American submarines who strike on the other side of the Pacific at Japanese ships—right up at the very mouth of the harbor of Yokohama. We know that as each day goes by, Japanese strength in ships and planes is going down and down, and American strength in ships and planes is going up and up. And so I sometimes feel that the eventual outcome can now be put on a mathematical basis. That will become evident to the Japanese people themselves when we strike at their own home islands, and bomb them constantly from the air"—just as Japan had begun the war with aerial bombing.

Japan was not the nation's first priority, however. Nazi Germany was—and would remain so, in terms of global American strategy, as long as Roosevelt remained president. "Turning now to the European theater of war," the President explained, "during this past year it was clear that our first task was to lessen the concentrated pressure on the Russian front by compelling Germany to divert part of her manpower and equipment to another theater of war. After months of secret planning and preparation in the utmost detail, an enormous amphibious expedition was embarked for French North Africa from the United States and the United Kingdom in literally hundreds of ships. It reached its objectives with very small losses, and has already produced an important effect upon the whole situation of the war. It has opened to attack what Mr. Churchill well described as 'the underbelly of the Axis,' and it has removed the always dangerous threat of an Axis attack through West Africa against the South Atlantic Ocean and the continent of South America itself. The well-timed and splendidly executed offensive from Egypt by the British 8th Army was a part of the same major strategy of the United Nations. Great rains and appalling mud and very limited communications have delayed the final battles of Tunisia. The Axis is reinforcing its strong positions. But I am confident that though the fighting will be tough, when the final Allied assault is made, the last vestige of Axis power will be driven from the whole of the south shores of the Mediterranean.

"I cannot prophesy," he added sternly. "I cannot tell you when or where the United Nations are going to strike next in Europe. But we are going to strike — and strike hard. I cannot tell you whether we are going to hit them in Norway, or through the Low Countries, or in France, or through Sardinia or Sicily, or through the Balkans, or through Poland — or at several points simultaneously. But I can tell you that no matter where and when we strike by land, we and the British and the Russians will hit them from the air heavily and relentlessly. Day in and day out we shall heap tons upon tons of high explosives on their war factories and utilities and seaports. Hitler and Mussolini will understand now the enormity of their miscalculations — that the Nazis would always have the advantage of superior air power as they did when they bombed Warsaw, and Rotterdam, and London and Coventry. That superiority has gone forever. Yes," he concluded his strategic survey, "the Nazis and the Fascists have asked for it — and they are going to get it."

To reinforce this message the President announced with pride that, "after only a few years of preparation and only one year of warfare, we are able to engage, spiritually as well as physically, in the total waging of a total war."

The phrase, for the United States, meant complete focus on war production on a scale that dwarfed anything ever done before — exceeding the production figures of America's enemies combined. In the past year the United States had manufactured "48,000 military planes — more than the airplane production of Germany, Italy, and Japan put together," as well as "56,000 combat vehicles, such as tanks and self-propelled artillery" — figures that would double again in 1943. "I think the arsenal of democracy is making good," the President congratulated America. "These facts and figures that I have given will give no great aid and comfort to the enemy," he explained his reason for releasing such numbers. "On the contrary, I can imagine that they will give him considerable discomfort. I suspect that Hitler and Tojo will find it difficult to explain to the German and Japanese people just why it is that 'decadent, inefficient democracy' can produce such phenomenal quantities of weapons and munitions — and fighting men." For, along with the "miracle of production, during the past year our armed forces have grown from a little over 2,000,000 to 7,000,000" men in uniform.

Seven *million?* And that figure rising?

Though the figures were astounding, and though the strategic initiative was now in the President's hands (his personal secretary noting how the "President becomes more and more the central figure in the global war, the source of initiative and authority in action, and, of course, responsibility"[2]), the President was clearly unwilling, it became clear, to leave matters there. "In this war of survival we must keep before our minds not only the evil things we fight against," he asked his audience, "but the good things we are fighting *for.* We fight to retain a great past—and we fight to gain a greater future." With that, he proceeded to outline the terms on which he proposed to end the war. And what to do after the war was won.

"We, and all the United Nations, want a decent peace and a durable peace. In the years between the end of the first World War and the beginning of the second World War, we were not living under a decent or a durable peace. I have reason to know that our boys at the front are concerned with two broad aims beyond the winning of the war; and their thinking and their opinion coincide with what most Americans here back home are mulling over. They know, and we know, that it would be inconceivable—it would, indeed, be sacrilegious—if this nation and the world did not attain some real, lasting good out of all these efforts and sufferings and bloodshed and death."

The good he wanted was, he proceeded to explain, a sort of renewed New Deal:

> The men in our armed forces want a lasting peace, and, equally, they want permanent employment for themselves, their families, and their neighbors when they are mustered out at the end of the war.
>
> Two years ago I spoke in my annual message of four freedoms. The blessings of two of them—freedom of speech and freedom of religion—are an essential part of the very life of this nation; and we hope that these blessings will be granted to all men everywhere.
>
> The people at home, and the people at the front, are wondering a little about the third freedom—freedom from want. To them it means that when they are mustered out, when war production is converted to the economy of peace, they will have the right to expect full employment—full employment for themselves and for all able-bodied men and women in America who want to work.
>
> They expect the opportunity to work, to run their farms, their stores, to earn decent wages. They are eager to face the risks inherent in our system of free enterprise,

the President allowed. On the other hand,

They do not want a postwar America which suffers from undernourishment or slums — or the dole. They want no get-rich-quick era of bogus "prosperity" which will end for them in selling apples on a street corner, as happened after the bursting of the boom in 1929.

When you talk with our young men and our young women, you will find they want to work for themselves and for their families; they consider that they have the right to work; and they know that after the last war their fathers did not gain that right.

When you talk with our young men and women, you will find that with the opportunity for employment they want assurance against the evils of all major economic hazards — assurance that will extend from the cradle to the grave. And this great government can and must provide this assurance.

I have been told that this is no time to speak of a better America after the war. I am told it is a grave error on my part.

I dissent.

And if the security of the individual citizen, or the family, should become a subject of national debate, the country knows where I stand.

I say this now to this 78th Congress, because it is wholly possible that freedom from want — the right of employment, the right of assurance against life's hazards — will loom very large as a task of America during the coming two years.

I trust it will not be regarded as an issue — but rather as a task for all of us to study sympathetically, to work out with a constant regard for the attainment of the objective, with fairness to all and with injustice to none.

These were the fighting words of a president who, quite clearly, was intending to stand for a fourth term, as Mackenzie King had inferred.

Not content with this domestic sally, however, the President then waded into national security on an international scale — national security that would require the end of American isolationism. "We cannot make America an island in either a military or an economic sense," he pointed out. "Hitlerism, like any other form of crime or disease, can grow from the evil seeds of economic as well as military feudalism. Victory in this war is the first and greatest goal before us. Victory in the peace is the next. That means striving toward the enlargement of the security of man here and throughout the world — and, finally, striving for the fourth freedom — freedom from fear." However, to attain freedom from fear meant taking a new role as peacekeeper in a "shrinking" globe, thanks to the "conquest of the air." It was fruitless to imagine the clock could be turned back, once the war was won.

> Undoubtedly a few Americans, even now, think that this nation can end this war comfortably and then climb back into an American hole and pull the hole in after them.
>
> But we have learned that we can never dig a hole so deep that it would be safe against predatory animals. We have also learned that if we do not pull the fangs of the predatory animals of this world, they will multiply and grow in strength — and they will be at our throats again once more in a short generation.
>
> Most Americans realize more clearly than ever before that modern war equipment in the hands of aggressor nations can bring danger overnight to our own national existence or to that of any other nation — or island — or continent.
>
> It is clear to us that if Germany and Italy and Japan — or any one of them — remain armed at the end of this war, or are permitted to rearm, they will again, and inevitably, embark upon an ambitious career of world conquest. They must be disarmed and kept disarmed, and they must abandon the philosophy, and the teaching of that philosophy, which has brought so much suffering to the world.

Step by step the President was leading his audience, and radio listeners, toward his notion of a United Nations authority.

> After the first World War we tried to achieve a formula for permanent peace, based on a magnificent idealism. We failed. But, by our failure, we have learned that we cannot maintain peace at this stage of human development by good intentions alone.
>
> Today the United Nations are the mightiest military coalition in all history. They represent an overwhelming majority of the population of the world. Bound together in solemn agreement that they themselves will not commit acts of aggression or conquest against any of their neighbors, the United Nations can and must remain united for the maintenance of peace by preventing any attempt to rearm in Germany, in Japan, in Italy, or in any other nation which seeks to violate the Tenth Commandment — "Thou shalt not covet."[3]

The President's words, clearly, were not only directed against isolationists in America, but were a preview of what he would announce internationally in the next few weeks. An announcement, to be given from the podium of a global stage, that would make public the fact that the United States was stepping up to the plate; would not this time back off, following victory, but was going to embrace a new, world-historical role as a leader of the democratic nations — if he could get those nations to support his vision.

7

A Fool's Paradise

REACTION TO ROOSEVELT'S ambitious State of the Union address was, somewhat to the President's surprise, decidedly positive.

The British ambassador, certainly, was impressed. Viscount Halifax had spent almost an hour at the dentist before going to the Capitol, where he was "herded on to the floor" with other diplomats "where we had good places. The President's speech was forceful and well-delivered and well-received," he recorded that night in his diary, remarking on the "very warm personal reception both at the beginning and end. The warmth of applause for China as compared with Russia and ourselves was very noticeable," he'd added — with understandable concern. The President, however, had spoken "with great confidence. I thought what he said on the domestic side was pretty strong and likely to be provoking to his domestic critics, as it seemed to be 'Let us have as much unity as we can, but I am going to go ahead with my social policy, and if you don't like it, let the country judge, and I know what their verdict will be,' but the general impression of it seems to have been that it was conciliatory. The informality of all Congress proceedings on these occasions is striking," he'd reflected, "by contrast with our affairs at home" — the ambassador amazed when, in reelecting its Speaker, Sam Rayburn, the day before, "the House [of Representatives] sang 'Happy birthday to you'!"[1]

If the President was delighted by the reception, however, he had little time to bask in it. Following a quick lunch at the Capitol, he returned to the White House — there to face in the Oval Office a smaller but equally critical audience he'd summoned: the U.S. Joint Chiefs of Staff, who would be flying to Casablanca that very evening, ahead of the President, aboard C-54 transport planes.

• • •

Admiral Leahy had warned the President that there had been no breakthrough in the Joint Chiefs' continuing dissension over U.S. global strategy. They were at loggerheads not only about whether to launch a cross-Channel invasion that year, but what to do in the Pacific.

The ringleaders of the continuing argument against a Mediterranean strategy in 1943 were — as had been the case throughout the previous year — the President's Republican secretary of war, Henry Stimson, and General Marshall. Colonel Stimson had openly bet the President that his "great pet scheme" — the Torch invasion — would fail. When it didn't — in fact proved a triumphant success — Stimson had found himself embarrassed. On November 20, 1942, for example, former ambassador William Bullitt, who had been U.S. envoy to Russia and France and who was currently working for Secretary Frank Knox at the Navy Department on Constitution Avenue, had rubbed salt in Stimson's wounded pride. He'd asked Stimson "how I liked to be a mere housekeeper of the War Department now that the President had taken over all relations with the military men."

Stimson had been infuriated by the remark. "I told him that so long as I was constitutional adviser to the President, he would not do it," Stimson had countered. "But Bullitt's remark," he confessed in the privacy of his diary, "irritated and annoyed me."[2]

Fortunately the President was nothing if not sensitive to people's feelings. Some days later he'd spoken with Stimson on the phone. They'd had "a talk on the situation and on my duties as Secretary of War. I told him then of Bill Bullitt's recent fresh visit to me and his remark asking me how I liked being merely a housekeeper for the Army. The President said 'What!' and made it very clear that he was going to use me for a great deal more than that. He said Bullitt was always a problem child." This was "very reassuring and satisfactory and balm to my soul after the troubles and suspicions," Stimson had confessed, "that I had been through for the last two or three days."[3]

The President's solace, however, hadn't stopped the elderly war secretary from working behind the scenes to question the President's military strategy in the Mediterranean — which he still thought utterly misguided.

On December 12, 1942, for example, Stimson had recorded he'd had a "long talk" with General Marshall and also Jack McCloy, the assistant secretary of war, as to "what we are going to do after the North African campaign and what it is going to lead to; and from this talk and other talks that I have had with Marshall and particularly the talk which I had with the President last evening, I am very much more relieved because the trend is now to get back onto the sound line of an attack up in the originally planned route" — namely the possibility of a cross-Channel attack "next summer."[4]

Summer '43?

As the President had confided to Mackenzie King, as president and commander in chief he did not favor a cross-Channel attack until U.S. forces had had ample time, in the Mediterranean, to first learn the arduous business of how to defeat the Wehrmacht in combat — which might well take all of 1943, given that the English Channel became too rough to cross by September.

Clearly the secretary was not listening. Inviting General Stanley Embick, the former deputy U.S. Army chief of staff, to his office at the Pentagon two days later, on December 14, 1942, Stimson had been determined not to be seen as a mere War Department housekeeper. Embick had been requested to report not to General Marshall but directly to Stimson on "the question of what we shall do after the African adventure."

The relationship between the two men went back decades — Embick having attended the Command and General Staff College with Stimson in World War I. At age sixty-four, he was now "head of our elder statesmen in military matters and has been made the head of a board of strategy together with Admiral Wilson of the Navy," Stimson noted. "I knew that he had always been very skeptical about the North African adventure," the war secretary added — anxious to know if the general had changed his mind.

Embick hadn't. "His position was very much the same as mine," Stimson had recorded with satisfaction in his diary, "and I found it was confirmed today. We both feel that the North African adventure has done a great deal of good," he allowed — though only because of luck, he maintained. As a result, "we have thus far gotten through without being knocked out by a great many of the perils that we might very well have had fall upon us and spoil the whole expedition." Among these "perils" was German forces being ordered to invade or granted passage through Spain and shipped across the Mediterranean into Spanish Morocco — there to strike American forces in the flank. "Embick laid great stress on keeping the gate open [to Tunisia] and not impairing the forces that were under George Patton in Casablanca for that [defensive] purpose"[5] — i.e., denying Patton the chance to fight in Tunisia rather than guarding U.S. lines of communication running back from Tunisia to Casablanca. Morocco, both Stimson and Embick felt, should be kept well supplied and protected by large numbers of U.S. troops, to guard against a mythical German riposte through Spain, across the Mediterranean, and then across Spanish Morocco. "We regard that as the sine qua non of the whole adventure," Stimson dictated in his diary.[6]

"Embick was strongly of the opinion that it would be impossible to go any further in adventures in Sardinia or Sicily after we are successful in Tunis," Stimson noted frankly — though making sure to exclude such passages from his later memoirs.[7] "Our shipping absolutely forbids that," he asserted, "and the line of supplies has become so long that it would be intolerable. So his thought is that after Tunis is cleaned out, if it is, if there is any surplus of American troops left over, we should send them up to Great Britain to be ready for the next attack there," Stimson recorded, together with his own approval. In fact the secretary had called in the head of the War Department's Operations Division, General John Hull, to join the discussion; "I found that he was in complete accord with both Embick and myself."[8]

Interrupted by visitors — a senator and governor from Idaho there to discuss the equipping of the National Guard in Idaho — Stimson had then returned to his office, where talk of the "next attack" across the English Channel in the summer of 1943 moved yet deeper into fantasy — in fact, seemed even more reckless than what Stimson had promoted in 1942. "Both Embick and Hull feel that the next step after we get back on the rails again up in Great Britain," he'd confided in his diary, "and are prepared to go forward, is not an attack on one of the peninsulas" — i.e., Brittany or Normandy — "but an attack on the flat coast near Havre and the port of Calais, landing in a large number of places."[9]

The problem of first gaining experience in battle and amphibious operations against the German Wehrmacht — the President's main objection to launching a cross-Channel attack prematurely — was thus simply ignored by Stimson and the senior officers of the War Department. The name Dieppe was simply never mentioned in Stimson's diary. Or the savage lessons of the 1942 disaster, only four months earlier.

Stimson was seventy-five years old; Embick, sixty-five.

In younger men, such ill-considered ardor could perhaps have been forgiven. But for two individuals who had enjoyed distinguished military careers and had themselves served in war, albeit in a different age, to task tens of thousands of inexperienced U.S. servicemen and their field commanders with a perilous invasion across the English Channel, at the most heavily defended area — the Pas-de-Calais — was willful fixation. The tragic slaughter of so many Canadian troops at Dieppe was well known in Washington professional military circles, despite attempts by the British to cover up the appalling number of Canadian casualties.[10] To imagine U.S. forces would, without more experience in amphibious operations, do better than brave Canadians in invading the fortified Pas-de-Calais area of northern France was pure hubris — the secretary and his colleagues at the Pentagon steadfastly refusing to see the Mediterranean as a necessary proving ground for the armed forces of the United States.

The first serious encounter-battles after Torch had, after all, begun to take place already in the last days of November and early in December, 1942, in the Medjerda Valley, outside Tunis. There, Eisenhower's U.S. and British forces, including amphibious units and paratroopers, had gotten a rude shock. The sheer professionalism of German armored and infantry units, backed by Mark IV panzers with long-barreled 75mm guns as well as deadly 88mm antiaircraft artillery used in an antitank role, had stunned the inexperienced Allied forces. Aware it could take months before he could break through to Tunis — especially since General Marshall insisted that a whole U.S. army be kept back, guarding against the improbable threat of a German attack out of Spanish Morocco — General Eisenhower had therefore begun plotting an alternative end-run further south, which in theory could strike through the thinly held German flank in Tunisia. If successful this would, the young commander in chief of Allied forces in the western Mediterranean hoped, reach the Mediterranean coast, east of Tunis, before the Germans could be reinforced by Rommel's Panzerarmee Afrika, retreating from Libya. Eisenhower had wanted to use George Patton for the job. He was overruled by General Marshall, who, with Secretary Stimson, remained obsessed with Patton being kept in the rear, defending the Allied flank.

Quite how, if Stimson, Marshall, and Embick, together with a whole cohort of planners and operations officers at the Pentagon, so feared a German counteroffensive via Spain, they could seriously imagine a cross-Channel invasion by virgin U.S. troops against twenty-five or thirty German divisions in northern France would magically succeed was something the President found hard to comprehend.

For the moment, however, the President had not interfered: trusting that, as American troops met German forces in Tunisia and the penny dropped, they would see sense. The British, after all, had taken three years of war to find a combat commander and the troops who could, at Alamein, defeat the Wehrmacht in battle. How on earth did Marshall, Stimson, Hull, and Embick imagine American forces would do so overnight?

Not even reports of mounting casualties and the need for reinforcements in Tunisia — not Morocco — seemed to dent Stimson's obsession, however. Indeed, by early January, 1943, Stimson seemed to be living in a fool's paradise in his huge new office suite inside the vast 2.3-million-square-foot Pentagon building, on the south side of the Potomac River, completed only a week after the Torch landings.[11]

Stimson had not been invited to the Casablanca meeting, but learning that the President wanted the Joint Chiefs of Staff to assemble for a briefing at the White House after his State of the Union address on January 7, 1943, before they left for Africa by plane, Stimson had decided he must have a "long talk with General Marshall this morning on the subject of the future strategy of the war. There are some conferences impending between the war leaders of America and Britain," he anxiously noted in his diary — wisely withholding the location.

Stimson was relieved to hear that Marshall and the senior officers of the War Department opposed further operations in the Mediterranean that year, once Tunis was reached. "Our people are adhering to their old [cross-Channel] line — the one I have approved throughout — and Marshall said that thus far they had the backing of the President," he noted — erroneously. "In a word, it is that just as soon as the Germans are turned out of Tunisia and the north coast of Africa is safely in the hands of the Allies we shall accumulate our forces in the north and prepare for an attack this year upon the north coast of France — preferably one of the two northwest peninsulas" — Cherbourg and Brittany.

At least the notion of landings in the Pas-de-Calais had been dropped — even the War Department's most gung ho planners conceding that the Pas-de-Calais might be tough. "The one is selected," Stimson noted of Brittany, "but I do not care to mention it yet. We think that we can probably hold such a lodgment but even if we don't, even if our forces should be finally dislodged, it would be at such a terrible cost to Germany as to cripple her resistance for the following year."[12]

An amphibious cross-Channel invasion of France — the Brittany peninsula — in the summer of 1943, on the open acceptance it might fail and require its survivors to be evacuated, like the British at Dunkirk in 1940? An invasion that would nevertheless "cripple" the Germans, in order to facilitate a United States relaunch of the invasion the following year, 1944?

It was small wonder Admiral Leahy had noted that "no agreement could be reached by the opposing elements"[13] on December 28, 1942—Stimson and Marshall's discussions at the Pentagon epitomizing, sadly, the complete lack of realism exhibited in the higher echelons of the U.S. War Department, only hours before the U.S. chiefs were due to leave to meet their British opposite numbers in Casablanca.

The President, however, was not of like mind. And was about to correct them, in the nicest way he knew how.

8

Facing the Joint Chiefs of Staff

AT 3:00 P.M. on January 7, 1943, Admiral Leahy and the U.S. Joint Chiefs of Staff, as well as the secretary of the Joint Chiefs, General John Deane, sat down beside the President in the Oval Office to discuss the strategic impasse.

Mr. Roosevelt proceeded to run the two-hour meeting in his inimitable manner: refusing to follow an agenda but rather, with the greatest friendliness, asking each of the chiefs to present the case as they saw it: once Tunis was secured, where next? "At the conference the British will have a plan, and stick to it,"[1] the President warned. Were they all, he asked innocently, "agreed that we should meet the British united in advocating a cross-Channel operation?"[2]

They were, they said. But *when?* And *where?*

To his credit, General Marshall, on behalf of the chiefs, was too honest to lie. All were not agreed about the timing, he confessed. Somewhat sheepishly, he explained to the Commander in Chief "that there was not a united front on that subject, particularly among our planners"—especially his own chief Army planning officer, Brigadier General Albert Wedemeyer.

"The Chiefs of Staff themselves regarded an operation in the north"—i.e., across the English Channel—"more favorably than one in the Mediterranean" once Tunis was secured, "but the question was still an open one," he admitted. "He said that to him the issue was purely one of logistics; that he was perfectly willing to take some tactical hazards or risks but that he felt we had no right to take logistical hazards. He said that the British were determined to start operations," after Tunisia, "in the Mediterranean"—leaving "Bolero [an early code name for a cross-Channel invasion] for a later date. He said the British pressed the point that we must keep the Germans moving. They lay great stress on accomplishing the collapse of Italy which would result in Germany having to commit divisions not only to Italy but also to replace Italian divisions now in other occupied countries," regions such as southern France, Corsica, the Balkans, and the Eastern Front.

The advantage for the British, Marshall continued, would be a secure Allied sea route to Suez and India, and a base for major operations in southern Europe — not only knocking Italy out of the war but holding out the possibility that Turkey might abandon its neutrality and join the United Nations. In this scenario, were it to be selected, the island of Sicily, Marshall said, was considered by him to be the best target of assault, once the campaign in Tunisia was completed: "a more desirable objective" than Sardinia, he explained, but one that, in terms of amphibious assault, "would be similar to an operation across the Channel," since the "Germans have been in Sicily longer," and "there were many more and much better airfields for them than in Sardinia."[3]

Sicily, then, was the British preference. An amphibious assault on the island dominating the Mediterranean would offer a kind of rehearsal for a future cross-Channel invasion — certainly a better one than Sardinia. But should the United States consent to further operations in the Mediterranean at all? By continuing offensive operations in the northern Mediterranean, whether in assaulting Sardinia or Sicily, Allied forces would be subject to "air attack from Italy, southern France, Corsica, possibly Greece, as well as a concentrated submarine attack," Marshall argued, which could lead to a loss of 20 percent of Allied ships. To this logistical nightmare the general "also pointed out the danger of [neutral] Spain becoming hostile, in which case we would have an enemy in possession of a defile [across the Mediterranean] on our line of communications."[4]

Fear of superior German forces in the Mediterranean and scarcity of Allied shipping thus led General Marshall to "personally favor," instead of further difficult operations in the Mediterranean, "an operation against the Brest peninsula" — i.e., the Brittany coast of northern France, across the widest part of the English Channel. "The losses there will be in troops," Marshall acknowledged, according to the minutes of the meeting, taken by General Deane, the secretary of the committee, "but he said that, to state it cruelly, we could replace troops whereas a heavy loss in shipping" incurred in further operations in the Mediterranean against Sardinia or Sicily, "might completely destroy any opportunity for successful operations against the enemy [across the English Channel] in the near future."[5]

The President was shocked — as historians would be, years later, when the minutes of the meeting were published. Taking vast U.S. casualties in order to hit the ports or beaches of northern France that year, rather than waiting until commanders and men had successful battle experience in the Mediterranean?

What was the hurry? Landing as yet completely inexperienced U.S. forces — commanders and infantry — across the widest part of the English Channel, to be set upon by upwards of twenty-five German divisions? Why invite such a potential disaster when they did not have to? Very politely, the President "then asked General Marshall what he thought the losses would be in an operation against the Brest Peninsula."[6]

Marshall, placed on the spot, had "replied that there would of course be losses but that there were no narrow straits on our lines of communication" like Gibraltar — both in terms of reinforcement or evacuation — "and we could operate with fighter protection from the United Kingdom."[7]

The President could only rub his eyes. No mention of the *two hundred miles* that Allied fighters would have to fly before they, like the assault ships, even reached the heavily defended invasion points, nor the proximity of twenty-five all-German infantry and armored divisions already stationed in western France, waiting and in constant training to repel an assault on its Atlantic coast, as they had done at Dieppe. No mention of the ease with which Germans could reinforce their Wehrmacht troops there, using short lines of communication from the Reich — and further armored forces they could quickly commit to battle. No mention, either, of the Luftwaffe's ability to use French airfields to attack the invading forces. Above all, no mention of the Canadian catastrophe at Dieppe the previous August, only four months ago. Merely a heartless disdain for the U.S. casualties that would be suffered, in comparison with landing craft — and a deeply, deeply questionable assumption that the invasion would, as Marshall had assured Stimson that morning, be at such "terrible cost to Germany as to cripple her resistance for the following year."

Marshall's presentation of the strategy he recommended the United States should best adopt, as chief of staff of the U.S. Army, was thus lamentable — as even Marshall himself seemed aware, once forced to defend his position.

The President, however, was a model of tact — unwilling to humble Marshall before his fellow chiefs. How, exactly, he then questioned Marshall, was such a landing at Brest to be actually mounted by U.S. forces — and how did Marshall expect the Germans to respond?

Marshall twisted in the wind. "The President had questioned the practicability of a landing on the Brest Peninsula," General Deane noted in the minutes of the meeting; "General Marshall replied that he thought the landing could be effected but the difficulties would come later in fighting off attacks from German armored units" — though "U.S. airplanes, flown from the United States, could give the troops help."[8]

Again, the President was amazed. U.S. air power such as U.S. Army Air Forces were giving U.S. and British ground forces in Tunisia, in the battles of Medjez-el-Bab and Longstop Hill — where American casualties were reported as heavy, and the Allies were just beginning to learn how tough it was to defeat the Germans in battle? Tellingly, the President therefore "asked why," if Marshall thought a cross-Channel invasion was the best course, "the British opposed the Brest Peninsula operation?"[9]

Embarrassed, Marshall had to concede "he thought they feared that the German strength would make such an operation impracticable."[10]

To Admiral Leahy's equally direct question as to when Marshall thought such a U.S. invasion of the Brest Peninsula could be "undertaken," Marshall had responded: "some time in August."[11]

August 1943.

It was clear to both President Roosevelt and Admiral Leahy that General Marshall had not done his homework. Above all, the Army chief of staff had no practical idea how a U.S. cross-Channel assault could possibly succeed that very year — in six months' time.

American armed forces currently had only eight weeks' battlefield experience — and most of this fighting ill-armed Vichy French forces, not German troops. How, then, were they to miraculously produce by August of that year the commanders and warriors capable of mounting a successful contested Allied landing in German-occupied Brittany, so close to the German Reich, and then hold out against — let alone defeat — Hitler's concentration of dozens of German infantry and panzer divisions stationed in northern France? And was Marshall really contemplating — as he'd said to Stimson that morning — the possible, even likely, defeat of

U.S. armies on the field of battle, and a Dunkirk-like evacuation from Brest? How would the public at home in America — who in any case favored winning the war against Japan over the difficulties of war in Europe — react to that?

The President had not been impressed. Choosing, by contrast, to back further operations in the Mediterranean, where the Allies had "800,000 or 900,000 men" and were currently in the ascendant, would furnish U.S. forces with a good opportunity to gain tough, amphibious battle experience against retreating German troops, far from the Reich, and in a relatively safe theater of war. U.S. operations in the South Pacific were, after all, providing such experience at the very extremity of Japanese lines of communication and resupply, on the other side of the world. With half a million troops that "might be built up in the United Kingdom for an attack on either Brest or Cherbourg," in Normandy, there was certainly every reason to consider a plan for their commitment to battle, if the Germans showed signs of collapse — but the President saw no reason to rush such a decision. He therefore asked whether "it wouldn't be possible for us to build a large force in England and leave the actual decision" as to its use "in abeyance for a month or two."

General Marshall took the point — saying he "would have a study prepared as to the limiting dates before which a decision must be made."[12]

General Henry "Hap" Arnold, the Army Air Forces commanding officer, did not dare say a word — and Admiral King, embarrassed, very few.

There would, then, be no immediate decision on a U.S. Second Front in France that year — leaving the chiefs ample opportunity to discuss, with the British at Casablanca, the question of whether to assault Sardinia or Sicily if they crossed the Mediterranean after securing Tunis.

This left only the overall politico-military strategy of the war to be addressed. Which, without further ado, the President now rehearsed. "The President said he was going to speak to Mr. Churchill about the advisability of informing Mr. Stalin that the United Nations were to continue on until they reach Berlin," the minutes of the meeting recorded, "and that their only terms would be unconditional surrender."[13]

In the months and years that followed, wild claims would be made that, at Casablanca, the President had thoughtlessly and unilaterally announced a misguided war policy that "naturally increased the enemy's will to resist and forced even Hitler's worst enemies to continue fighting to save their country," as the chief planner on Marshall's team at Casablanca put it.[14]

Moreover, that it was a policy his own staff vainly disagreed with,[15] and that neither Churchill, his staff, nor his government had had any idea of it, prior to the President's announcement.[16]

Like so much popular history, this allegation lacked substance. Not only had the President discussed the matter with Prime Minister Mackenzie King a month prior to the White House meeting with the U.S. Chiefs of Staff, but the President's determination to pursue unconditional surrender of the Axis powers had been widely discussed by Sumner Welles's committees when conceptualizing the United Nations authority and end-of-war requirements — which were in turn shared with senior British government officers. In speaking of it to his generals on January 7, 1943, the President made clear his wish that the chiefs factor this objective into their discussions on military strategy with the British at Casablanca. Thanks to Torch, the war against Germany and Japan was no longer one of defense against Axis attack, but of Allied offense — offense that would not stop until Berlin was reached, and then Tokyo.

No negotiations. No ifs and buts. No concessions, or anything that could later be revoked. Nothing but *complete and unconditional* surrender of the Germans and Japanese, and their "disarmament after the war," as the President put it to his Joint Chiefs of Staff, sharing with them as well his notion of a four-nation postwar policing force on behalf of the United Nations, which they, as the U.S. Joint Chiefs of Staff, would have to lead.[17]

As for the cross-Channel invasion, he would, the President said, follow the Combined Chiefs' advice on the timing "as they thought best." For himself, he was anxious to hammer out with America's allies not only the matter of German and Japanese postwar disarmament but other "political questions" that he would discuss with Mr. Churchill at Casablanca — and hopefully then at another "meeting between Mr. Churchill, the Generalissimo [Chiang Kai-shek], Mr. Stalin and himself some time next summer," perhaps at the port of Nome, in Alaska, which was also the final stop for planes flying Lend-Lease supplies to the Soviet Union.

The Joint Chiefs did not demur. With that — save for a brief discussion of planes for Russia, and French sovereignty versus U.S. military government in North Africa — the meeting ended. The Commander in Chief had spoken, and the chiefs had been given their orders. They would depart that very evening for North Africa, where the President was to join them on January 14, if all went well.

PART THREE

Casablanca

9

The House of Happiness

EARLY ON THE morning of January 14, 1943, the President and his party boarded a four-engine Douglas C-54 Skymaster of U.S. Air Transport Command about twenty miles outside Bathurst. "Normally, the air route from Bathurst to Casablanca would be entirely over land," Captain McCrea later recalled. "On this occasion," however, "a swing to seaward was made in order to afford the President an aerial view of Dakar and St. Louis, Senegal, French West Africa."[1]

The route would allow the President to see the coastline he'd studied for over a year, when thinking about a possible U.S. invasion of Northwest Africa — especially Dakar. First occupied by tribal Africans, then Portuguese, Dutch, British, and finally French slave traders, Dakar was a fabled port. Following the French surrender at Compiègne in 1940, it had posed the danger that, if occupied by the Germans, it could become an impregnable African base for German naval and U-boat operations in the southern Atlantic. Thanks to Torch, however, it was now under American control — the port and fortress having been ceded by its governor-general to General Eisenhower and his French commissioner, Admiral Darlan, on December 7, 1942.

Passing over Dakar, "the French Battleship *Richelieu* was clearly observed alongside a seawall as were several other ships at anchor in the harbor," McCrea noted. There was a special reason, also, that Roosevelt wished to see the battleship, for it symbolized both the challenge and the success of Torch. With its eight fifteen-inch guns, eighty-five-hundred-mile cruising range, and fifteen-hundred-man crew, the *Richelieu* had been the first modern battleship built by the French since the 1922 Naval Treaty. Completed in 1940, too late to defend France against the Nazis, it had nevertheless helped defeat France's former allies and anti-Nazis —

Major General Charles de Gaulle's ill-fated attempt to seize the seaport on behalf of the Free French having been ignominiously repelled that year. For two years the *Richelieu* had then stood sentinel against the Allies, on behalf of the Vichy government and in accordance with the terms of Maréchal Philippe Pétain's capitulation to the Third Reich.

An hour later, two hundred miles further north, at the mouth of the Senegal River, the C-54 flew over "the very old French port" of Saint-Louis as well — giving the Allies two American-controlled ports which, thanks to their strategic importance on the Atlantic seaboard, were to be of inestimable significance to the Allies for the remainder of the war. "Then inland over the desert," the President described his route in a letter he penned that night to Daisy.

On the early-morning drive to the Bathurst airfield, Roosevelt had been upset by the extreme poverty of the people — which said very little for the British and their colonial rule over Gambia, despite having suppressed the slave trade in 1833. Now, over the West African desert, there were no people at all. "Never saw it before — worse than our Western Desert — Not flat at all & not as light as I had thought," he described to Daisy, " — more a brown yellow, with lots of rocks and wind erosion."[2]

The Skymaster, with its wingspan of 117 feet and space for forty-nine troops, was not nearly as luxurious as the Pan American Clipper. For five hours they flew at six thousand feet, until at last they caught sight, inland, of "a great chain of mountains — snowy top," Roosevelt recorded — explaining that the "Atlas run from the Coast in Southern Morocco East and North, then East again till they lose themselves in Tunis": the goal of General Eisenhower's current campaign.

"In approaching the Atlas mountains the cruising altitude was gradually increased from 8,000 feet to 12,000 feet," Captain McCrea remembered the flight — adding his own vivid recollection of how he'd persuaded the President to take oxygen for the first time. "The President was seated amidships, on the starboard side of the plane," he recalled — ever the naval officer. "I was seated directly across the aisle from him, & Ron McIntire was seated immediately in front of me. Harry Hopkins was seated well forward in the plane. Both Ron and I were quickly aware that the pilot was increasing altitude gradually. Ron suggested that I enquire from our pilot as to how much altitude he was going to level off at. This I did." Told that the pilot expected to cruise at about twelve thousand feet, "I squared away in my seat," and the President's doctor, "turning outboard, addressed me in a low tone of voice over his shoulder. 'John,' said he, 'how about putting on your oxygen mask? I want the President to put his on but if I suggest it to him he will no doubt make a fuss. If he sees you put on your mask he no doubt will follow.' In a few seconds I reached for my mask and proceeded to adjust it. Sure enough when the Pres. saw me putting on my mask he started to fumble with his. I promptly moved across the aisle, straightened out his mask harness and adjusted it for him." The doctor then put his own mask on, as did Hopkins — "And thus we were all set when shortly thereafter we reached 12,000 feet — an altitude which [was] maintained while crossing the Atlas Mountains."[3]

"We flew over a pass at 10,000 ft. & I tried a few whiffs of oxygen," the President wrote that night to Daisy. In truth he was more interested in the terrain than the air. "North of the Mts. we suddenly descended over the first oasis of Marrakesh — a great city going back to the Berbers even before the Arabs came — We may go there if Casablanca is bombed."[4]

They were approaching the battlefield.

In Washington the President had done his homework on the Berbers — Lieutenant George Elsey, in the Map Room, managing to get Lieutenant Commander S. E. Morison, a distinguished naval scholar from Harvard, a fifteen-minute interview with the President, "who asked questions I was unable to answer," as Morison subsequently wrote Elsey. Morison had therefore researched a "brief memorandum" on the subject of the Berbers for the President.[5] In this, the historian had pointed out that the Berbers, according to Egyptian inscriptions, dated as far back as 1700 B.C., and were "an entirely distinct race from the Phoenecian [*sic*] Carthaginians, who are comparative newcomers in Africa." The Berbers, by contrast, were "the aborigines of North Africa, with a distinct language and writing," and possibly the original inhabitants of the Iberian Peninsula. "They are a 'white' or 'Nordic' race, brown or hazel eyed, and no darker than the North American Indians in complexion."[6]

Morison's report had only whetted the President's curiosity aboard the C-54 Skymaster as they approached Casablanca — which, despite air raid sirens going off at various times, was not in fact targeted by long-range German bombers from Tunisia. For all their vaunted efficiency, it seemed the Germans had no idea the President was planning to meet Churchill there, let alone intending to stay almost two weeks — Goebbels recording, afterward, his near-disbelief that the *Sicherheitsdienst* of the great Third Reich had actually intercepted enemy phone calls, yet had taken the name Casablanca to be Casa Blanca, or White House, Washington, D.C.[7]

Those "in the know" at the real White House, however, had remained on tenterhooks lest the President, whose leadership of the Allies seemed so crucial to winning the war, fall victim to accident or assassination.

In particular, Mike Reilly, head of the White House Secret Service detail, had furiously objected to the idea of such a well-known venue—fears that had only increased when he arrived in Casablanca in advance of the President. Concerned the city was full of agents, assassins, and former Vichy officials of dubious reliability, Reilly had instantly tried to have the summit moved to Marrakesh, several hours' drive further south. Told that only the President could order this, he'd nevertheless persuaded the U.S. Joint Chiefs of Staff, who'd arrived on January 12, not to go meet the U.S. Commander in Chief in person on his arrival at Medouina airport on the afternoon of January 14, lest they attract unwarranted attention.

Landing at the airfield, the President was not in the least put out. Filming of a new Hollywood movie called *Casablanca* had, by complete coincidence, recently been completed in Los Angeles, and had been flown to Washington and shown to the President at the White House on Christmas Eve. The film—starring Humphrey Bogart, Paul Henreid, and Ingrid Bergman—had charmed him, and the climax at the faux-Medouina airport ("Round up the usual suspects!") had made him much more interested in the fabled city, its kasbah, its émigrés and spies, than in presidential protocol. "At last at 4 p.m. Casablanca & the ocean came in sight—I was landed at a field 22 miles from the town," he recorded in his letter to Daisy. "Who do you suppose was at the airport?" he wrote rhetorically. Not the U.S. chiefs but Lieutenant Colonel Elliott Roosevelt, his second son, standing beside Mike Reilly! And "looking very fit & mighty proud of his D.F.C. [Distinguished Flying Cross]"[8]—awarded for dangerous low-level reconnaissance missions, flown both before and during the Torch invasion.

As Roosevelt proudly told his son when they got into the camouflaged car, this trip marked the first time he'd flown in a decade. And with that the President shared with Elliott his amazement at the progress in air travel in only a few decades. There had been some flights in his early career that had been positively hair-raising, he recounted, when he was assistant secretary from 1913 to 1920. "In naval airplanes. Inspection trips.

The kind of flying," he chuckled, thinking of the open biplane cockpits, "*you'll* never know." By contrast, this, his first transatlantic trip, had given him a dramatic idea of "what so many of our flyers are doing, the sort of thing aviation's going through these days, and developments of flying. Gives me," he'd told Elliott — who knew this far better than his father — "a perspective."[9]

Father and son were then "driven under heavy guard & in a car with soaped windows" not to Rick's Café, the President wrote Daisy, but to "this delightful villa belonging to a Mme. Bessan whose army husband is a prisoner in France — She & her child were ejected as were the other cottage owners & sent to the hotel in town."[10]

He had, in short, arrived.

Selection of the villa, indeed of the city, had only been made a few weeks earlier by Eisenhower's chief of staff, Walter Bedell Smith, and Churchill's military assistant, Brigadier Ian Jacob.

Jacob had judged Casablanca a far better location than Fedala, further north and also on the ocean. The Villa Dar es Saada, in particular, had "the most magnificent drawing room leading out on to a large verandah," plus a dining room at one end, and a "principal bedroom complete with private bathroom" at the other, on the same floor. There were two further rooms upstairs. Along with thirteen other villas it was situated in a "garden suburb" of Casablanca known as Anfa: an area a mile wide, built "on a knoll about a mile inland and 5 miles south-west of the center of town." There was a large forty-room hotel nearby for the Combined Chiefs of Staff and their staffs, with a "view out over the Atlantic, or overland to Casablanca" that was "truly magnificent," as Jacob noted in an account he wrote at the time. "The dazzling blue of the water, the white of the buildings in Casablanca, and the red soil dotted with green palms and bougainvillea and begonia," he recorded, "made a beautiful picture in the sunlight"[11] — bounded, by the time the President arrived, with hastily erected barbed wire, antiaircraft guns, and an entire U.S. infantry battalion restricting all access to a single checkpoint.

Although elaborate steps, Elliott Roosevelt later recalled, had been taken to keep news of the President's impending arrival quiet, the heavily guarded compound could have fooled no one — least of all the "French fascists" left behind by the hastily departing French-German 1940 armistice team. Such individuals were armed, as Elliott caustically put it, "with German money in their pockets."[12] After several air raid alarms — though no German planes—Mike Reilly had certainly had enough, however. Having persuaded the U.S. Chiefs of Staff not to greet the President at the airport, or even at his villa, he now begged them to use their collective military influence, once they did see the President, to get him to change his plans and move south to safer quarters in Marrakesh.

Warned of this, Roosevelt dismissed the very idea. As president he was U.S. commander in chief. He felt on top form—even without his chief of staff, who he'd counted on to keep his Joint Chiefs of Staff in line. Casablanca was the scene of recent battle, and one of the largest artificial ports in the world. Having spent four days and nights getting to the city in a succession of trains, floatplanes, tenders, transport aircraft, and limousines, he was "'agin' it," and "said so, often enough and forcefully enough," Elliott remembered, "to carry the day."[13]

In the meantime he wanted to see where he'd sleep.

"When Father got his first look," Elliott remembered, "he whistled."[14]

The bedroom's décor reminded the President of a French brothel. "Now all we need is the madam of the house," he laughed, throwing back his head. "Plenty of drapes, plenty of frills," Elliott recalled. "And a bed that was—well, perhaps not all wool, but at least three yards wide. And his bathroom featured one of those sunken bathtubs, in black marble."[15]

The plumbing, too, worked fine. Wheeling his father around the house, Elliott found him more at home than he could have imagined possible. Guarded by a battalion of U.S. troops, in an area of Morocco under American rather than French or British military command, the Villa Dar es Saada—meaning "House of Happiness"—was the finest private residence in the suburb. It boasted almost twenty-eight-foot-high ceilings, steel-shuttered windows, and looked out over a beautifully terraced garden with vine-covered trellis. The two rooms upstairs could be used as bedrooms—one for Hopkins, and one for Elliott.

Another of the President's sons, Lieutenant Franklin Roosevelt Jr., would also be coming—unannounced. His destroyer, the USS *Mayrant*, had covered the Torch invasion and was still stationed offshore. Learning of this and having once served as a midshipman with the regional naval commander (the brilliant Rear Admiral John L. Hall), Captain McCrea immediately arranged for FDR Jr. to be brought the next day to Anfa, without being told the reason. "He sighted me and burst out 'My God, Captain, is Pa here?" McCrea recalled humorously. "I told him his suspicion was correct and I took him across the street"—telling him to be "prepared for a surprised parent. The Pres. indeed was surprised, and father and son indulged in fond embrace" — followed by "an invitation to stay for lunch which, of course, Franklin did."[16]

All in all the Villa Dar es Saada was a house of happiness, thanks to Brigadier Jacob: the President's pro tempore White House and his family residence, established in an American realm, guarded by American soldiers — not a British colony or quasi-colony, such as Khartoum or Cairo, the two cities Churchill had recommended.

Once installed, the President asked Harry Hopkins to go fetch Churchill, whose villa, the Mirador, was only "fifty yards away," as Hopkins recorded.[17] It would be the first time they'd seen each other since Churchill's fateful visit to Washington at the time of the British surrender of Tobruk, seven months before. The President could only marvel at how times had changed.

Churchill, for his part, was equally excited — in fact had arrived two days early to prepare for the arrival of the "Boss." In his speech at the Mansion House in London on November 10, 1942, announcing the success of Torch, Churchill had openly revealed that the "President of the United States, who is Commander in Chief of the armed forces of America, is the author of this mighty undertaking and in all of it I have been his active and ardent lieutenant."[18]

Reading the text of the speech, Hitler's Reichsminister für Propaganda had been fascinated. Churchill, Goebbels had noted in his diary, was not only openly ascribing the Allied victories to the huge superiority now enjoyed by American arms, but "he also admits that the whole invasion plan came from Roosevelt's brain, and that he is only a loyal servant to Roosevelt's plans."[19]

Hitherto, Goebbels had assumed from British newspaper articles that Churchill was the brains behind Allied operations in the European theater — something he'd found "comforting," as he'd noted cynically, "since all previous military operations he's been behind have ended up as disasters."[20] Churchill's public acknowledgment that the U.S. president was now in charge heralded something different — indeed, alarming.

"The Americans are now out of the starter's block. Their next target is Tripoli," Goebbels had recorded; in fact, idle armchair strategists in America and England were assuming the Allies would soon clear Axis forces from North Africa entirely. "They already imagine themselves invading Italy and foresee themselves invading Germany via the Brenner

Pass. All this, of course, a very simple and plausible calculation," Goebbels had added sarcastically, " — if it weren't for us being there!"[21]

This was the crux of the matter — for the quality of armed German resistance was something the prognosticators of whom Goebbels spoke, whether in the United States or Britain, seemed incapable of appreciating. Tunisia was to be the key to thwarting Allied strategic ambitions, Goebbels had been told by the Führer — who saw the battle for Tunis becoming a new Verdun. "If we hang on to Tunis, then nothing is lost in North Africa," he'd recorded. And already, as General Eisenhower and his invasion forces attempted to come to terms with the business of real combat with real Germans — as opposed to ill-armed Vichy defenders — Hitler was being proven right, on the field of battle rather than in the print of newspaper columnists.[22]

Winston Churchill had been educated as a soldier at Sandhurst and boasted a lifetime's military experience, from the North-West Frontier to the Sudan and South Africa. Like so many commentators in the press, he had visualized a swift Allied advance — by Montgomery's Eighth Army marching from the east and by Eisenhower's First Army from the west. "I never meant the Anglo-American Army to be stuck in North Africa," the Prime Minister had chided his British chiefs of staff on November 15, 1942, only a week after the Torch invasion — irritated by the celerity with which Hitler had reinforced his meager forces in Tunisia by air, and the slowness of British and American ground troops spearheading Eisenhower's thrust from Algeria. In one of his instantly memorable turns of phrase, the Prime Minister had berated them, saying the Torch invasion was "a springboard, and not a sofa."[23]

It became a classic Churchillian metaphor, oft repeated. In truth, though, it masked a huge difference between Allied and German soldiery. For the simple fact was, whether volunteers or conscripts, Allied soldiers were not like the Germans or the Japanese. As Roosevelt had confided in 1942 to Field Marshal John Dill, the British liaison to the Joint Chiefs of Staff, Allied troops did not have, for the most part, the kind of ruthless, even fanatical obedience to orders and discipline that characterized German and Japanese forces. Only by adopting a careful, step-by-step approach to war, evading ventures that posed unnecessary risks; only by undertaking offensive operations within the capabilities of Allied troops; only by applying the advantages of U.S. mass production; and only by pursuing global military strategies that built upon Allied strengths — fusing air, naval, and ground forces — could the Allies actually defeat the Wehrmacht and the Japanese. Not by prime ministerial exhortation, however inspiring.

Churchill's bon mot reflected an aging yet still wonderfully indefatigable English leader. At heart he remained a dashing young cavalryman, as on the North-West Frontier in his early days of service, or in the Sudan fighting the self-proclaimed Mahdi at the turn of the century, in 1898. Half a century later, his "Action this day" tags — the red stickers he would attach to his brilliantly written memos demanding immediate response by his staff — were a tribute to his abiding energy as he approached seventy: spearing lethargy and electrifying traditional British bureaucratic pen-pushers, sclerotic after centuries of imperial paperwork. However, they masked a profound flaw in the Prime Minister's makeup as his country's quasi–commander in chief in 1943: the irreconcilable difference between his grand strategic ideas and his too-often ill-considered opportunism — a difference affecting tens of thousands of soldiers' lives.

After the war the former prime minister would go to great lengths to cast himself, in his six-volume epic *The Second World War,* not only as a lonely oracle but architect of war. Inasmuch as he saw better than any of his contemporaries the ebb and flow of military history, necessitating that Britain withstand the predations of Hitler's Third Reich until it could be rescued, he was by 1943 being proven right. He had, after all, lost every battle against the Germans since 1940, yet with his U.S. partner in war was helping to force Hitler, thanks to Torch, onto the defensive. Once Tunis fell, the Allies would possess a springboard for eventual victory in Europe, he felt — provided the Russians continued to face the brunt of Hitler's Wehrmacht in the East. But beyond that his military strategy did not go, since he did not believe a cross-Channel attack could possibly succeed. In reality he had no idea how, in fact, the Third Reich could be defeated, beyond constant peripheral pressure and air attack.

As the two Allied leaders met at 6:00 p.m. at the President's villa on January 14, 1943, there was thus, behind the bonhomie and goodwill, a distinct divergence of opinion. The Prime Minister's agenda was how to placate the United States, defer operations against Japan, and by "closing the ring" around the Third Reich — sheering off its allies, such as Italy, as they went, and hoping to get the peoples of occupied Europe to rise up against the Germans — engender Hitler's fall, followed by that of Hirohito. Thence to return the world, as he saw it, to its former European imperialist setup, before the Führer, the Duce, and Tojo's gang of admirals and generals had upset the balance of power.

The President, for his part, had a quite different vision. Not only a vision of the future, but how to achieve that future: the endgame. Their clash of objectives in Casablanca, behind the scenes, thus promised to be historic.

10

Hot Water

KNOWING VIA Field Marshal Dill, in Washington, that the U.S. chiefs of staff did not favor a delay in launching a Second Front that year, Churchill had told his British chiefs they would have to do again what they had done the previous spring: show willing, while stringing the Americans along, in order to pursue a more opportunistic course in the Mediterranean. The chiefs of staff were thus merely to pretend to be agreeable to closing down "the Mediterranean activities by the end of June with a view to 'Round-up'" — an Allied 1943 cross-Channel invasion — "in August." The final decision on a Second Front, however, would be made, he instructed them, "on the highest levels" — i.e., by himself and the President.[1] For Mr. Roosevelt, he was sure, would agree with him it was impossible: the Germans, in northern France, were just too strong in the number of divisions they had there.

With this in mind Churchill had made haste to set off for Casablanca on January 11, together with a huge retinue of staff officers and clerks. Bad weather threatened to vitiate his plan — but had not stopped it.

Serving as Churchill's military assistant, Brigadier Ian Jacob had been wary of the contingent the Prime Minister was taking, instead of the small staff the President had requested. "I was rather horrified at the size of the party which had been gotten together," Jacob wrote in his contemporary account. "The whole added up to a pretty formidable total."[2] Some members of the party could, of course, be concealed and housed onboard the communications ship that was being sent out, HMS *Bulolo,* he recognized. "But knowing as I did from conversations with Beetle [Bedell] Smith that the Americans would bring a very modest team, I was rather afraid that the President or the Chiefs of Staff might take offense at the size of our party, & that the success of the conference might be endangered. I put this point to the P.M. on Sunday morning before I left Chequers [the prime ministerial retreat], & he said the party was to be cut down. However, when we went into the question on Monday with the Chiefs of Staff we found that there were few if any people who could be discarded, & it was decided that the best policy would be to take a full bag of clubs, leaving some of them concealed as it were in the locker — i.e. the ship."[3]

Despite bad weather delaying the takeoff of the main Boeing Clipper, Churchill had insisted the primary team fly still on January 13 using land-based aircraft. The staff were thus farmed out among four RAF American-built Liberator (B-24) planes, each of which could normally take only seven "passengers." As a result the Prime Minister had found himself cramped in a bomb bay bunk, flying without heat, which had not left him in the best of moods. This had not improved when, after asking his manservant, Sawyers, to run him a bath on arrival at the Villa Mirador, he had found it neither hot nor deep enough. "You might have thought the end of the world had come," Jacob described. "Everyone was sent for in turn, all were fools, and finally the P.M. said he wouldn't stay a moment longer, & would move into the hotel [Anfa] or to Marrakesh" — where he'd spent a pleasant month in 1932.[4]

In the event, food and drink — drink especially — had "had its mellowing influence," as the Prime Minister lunched with General Marshall and the Fifth U.S. Army commander, Lieutenant General Mark Clark, and "the excitement died down. Plumbers were assembled from all directions, and somehow or other the water was kept hot in the future."[5]

Bath or no bath, Churchill did, however, take great pains to be amenable to Marshall and Clark — reporting to the British chiefs of staff on the evening of their first day's work in Casablanca, on January 13, that "some kind" of cross-Channel "Sledgehammer" operation in Brittany would have to be undertaken that very year, if only to support U.S. efforts. "Only in this way should we be taking our fair share of the burden of the war," he'd told them at their first meeting with him, at 4.30 p.m.[6]

Brigadier Jacob also noted, however, the Prime Minister's openness to undertaking different operations. The son of a field marshal, Jacob was a first-class administrator, with a crystal-clear mind, fair judgment, loyalty to his superiors, and a talent for lucid exposition, which the Prime Minister particularly valued. A U.S. agenda had been lined up and sent from Washington, which "contained a list of every topic under the sun, but the most important thing," the military assistant noted, "was to get settled in broad outline our combined strategy for 1943, and then to get down to brass tacks and decide how exactly to carry it out. One couldn't decide in detail what to do unless one knew what one's strategic aim was to be. At the same time one could hardly fix one's strategic aims unless one examined in detail what operations we were capable of carrying out and what we were not."[7]

"Not" meant a cross-Channel invasion that year.

Jacob did try to see the problem from a U.S. perspective, however, asking Sir John Dill's view as the British representative on the Combined Chiefs of Staff in Washington—and was not surprised when Dill warned that there was a "general fear of commitments in the Mediterranean, and secondly, a suspicion that we did not understand the Pacific problem and would not put our backs into the work there once Germany had been defeated. Thus although the Americans were honestly of the opinion that Germany was the primary enemy, they did not see how quite to deal with her, especially as they felt there were urgent and great tasks to be done in Burma and the Pacific." These tasks involved logistical and operational struggles between General MacArthur and Admiral Nimitz—the Army versus the Navy in terms of distribution of resources—and the right combination of those forces and campaign strategy. They had already led to much infighting, as well as uneven effort, such as at Guadalcanal, "where the U.S. Marines were thrown ashore, and then it was found that there was no follow-up, no maintenance organization, and no transport."[8]

Along with the British chiefs of staff, the Prime Minister also interrogated Dill, who repeated his assessment of U.S. positions and problems. Delighted that he'd come early to the conference, Churchill was sure he could handle the President. The Prime Minister's "view was clear," Jacob recorded Churchill's approach, expressed now in front of Dill and the British chiefs of staff. "He wanted to take plenty of time. Full discussion, no impatience—the dripping of water on a stone." The big British contingent was to methodically wean the small American team away from its fixation on a major cross-Channel assault that year to more gradual, peripheral operations in the Mediterranean, with a smaller operation in Brittany, perhaps, to get a toehold at least on the continent. "In the meantime," while the chiefs met their opposite numbers in the daily Combined Chiefs of Staff meetings at the Anfa Hotel, "he would be working on the President, and in ten days or a fortnight," Churchill was confident, "everything would fall into place. He also made no secret of the fact that he was out to get agreement on a programme of operations for 1943 which the Military people might well think beyond our powers, but which he felt was the least that could be thought worthy of two great powers."[9]

To the alarm of the British chiefs, then, the Prime Minister was all for action, on multiple fronts. To start with, he "wanted the cleansing of the North African shore to be followed by the capture of Sicily. He wanted the reconquest of Burma, and he wanted the invasion of Northern France, on a moderate scale perhaps. Operations in the Pacific should not be such as to prevent fulfilment of his programme. The Chiefs of Staff were dismissed on this note, and the rest of the evening," before the President's arrival the next day, "was given up to ice-breaking dinner parties."[10]

11

A Wonderful Picture

PRESIDENT ROOSEVELT, for his part, was all for icebreaking. "I marveled at the way the Army just moved in and took charge and ran the whole operation," even Captain McCrea later noted — speaking as a sailor.[1]

Brigadier Jacob, who had reconnoitered and recommended the venue, was equally delighted by the U.S. Army's efforts, especially the catering: "certainly excellent, mostly U.S. Army rations, too. Of course it was supplemented with local produce," including oranges, which — "large and juicy, with the best flavor of any oranges in the world" — formed "a part of every meal." The "cooking too was good, and as again the whole thing was free, a genial warmth spread over our souls," he recorded.[2]

McCrea agreed. "So well was this done that on the first evening of our arrival the President was able to entertain at dinner Prime Minister Churchill and the Combined Chiefs of Staff (both the U.S. Joint Chiefs and British Chiefs of Staff), plus Col. Roosevelt and Averell Harriman — some twelve persons in all."[3]

Churchill had hurried over with Hopkins at around 7:00 p.m. No formal notes were made of what was said between the President and Prime Minister, but the "three of us had a long talk over the military situation," Hopkins wrote in his notes that night.[4] The winter weather in the mountains had slowed the campaign in Tunisia, and Montgomery had yet to take Tripoli and advance toward Tunisia, but the plan of campaign was that the two Allied pincers would eventually trap the German and Italian forces in the Cape Bon Peninsula: forcing them either to attempt a Dunkirk-style evacuation or surrender. It was what would happen thereafter — locally, regionally, and internationally — that was the biggest problem to be resolved in the coming days.

The Combined Chiefs of Staff had been having cocktails at the Anfa Hotel when Hopkins arrived with the presidential summons. Dutifully, the bevy of generals and admirals — Generals Marshall and Arnold, Admiral King, General Alan Brooke, Admiral Dudley Pound, Air Marshal Charles Portal, Admiral Louis Mountbatten, as well as the Lend-Lease administrator, Averell Harriman — trooped over.

The dinner, in the President's villa, went rather well. "People were tired, that first night," Elliott recalled, "but it didn't stop anybody from enjoying himself" — particularly as there was no attempt to limit the consumption of wine or liquor.[5] General Hap Arnold recounted how he had just been down to the harbor to see the damage inflicted on the brand-new French battleship *Jean Bart* during the Torch invasion — the airman delighted to see that American thousand-pound bombs had smashed "holes in bow and stern large enough to take a small bungalow."[6] Others gave their own impressions of the city and its kasbah.

Admiral King "became nicely lit up towards the end of the evening," Brooke scrawled in his diary that night. "As a result" the admiral became "more and more pompous, and with a thick voice and many gesticulations explained to the President the best way to organize the Political French organization for control of North Africa!" — something King would never have dared do when sober. "This led to arguments with PM who failed to appreciate fully the condition King was in. Most amusing to watch."[7]

The dinner was certainly a far cry from life at Hitler's headquarters in East Prussia, where the Führer had ceased to dine with his senior military staff. He'd stopped listening even to music at night, refused to go near the battle front or to make any public pronouncements — and was still demanding that no mention be made in the Nazi media of the increasingly catastrophic situation around Stalingrad.

"I busied myself filling glasses," Elliott Roosevelt later recalled. "After dinner, Father and Churchill sat down on a big, comfortable couch that had been set back to the big windows. The steel shutters were closed. The rest of us pulled up chairs in a semicircle in front of the two on the couch."[8] "Many things discussed," Arnold noted in his diary — including their leaders' safety. "Everyone tried to keep President and Prime Minister from making plans to get too near front," given that both men "seemed determined" to go, and "could see no real danger."[9]

"We have come many miles and must stay long enough to solve very important problems," Arnold finished his nightly jotting—aware how much responsibility the President carried, for good or ill. And he quoted the British prime minister, whose words had the sober ring of history, despite the immense quantity of alcohol the Prime Minister had imbibed. "Churchill: 'This is the most important meeting so far. We must not relinquish initiative now that we have it. You men are the ones who have the facts and who will make plans for the future.'"[10]

An air raid siren then wailed, bringing the postprandial get-together almost to a close. "At about 1:30 a.m. an alarm was received," General Brooke noted in green ink in his own leather-bound diary, "lights were put out, and we sat around the table with faces lit by 6 candles. The PM and President in that light and surroundings would have made," he scribbled, "a wonderful picture."[11]

Rembrandt might have painted it, but sadly, no photographs were taken that evening—though other, iconic photos would be, at the climax of the conference, ten days later. None could fail to be aware, however, just how symbolic was the meeting: the leaders of the two main Western democracies, gathering together with their chiefs of staff on the still-scarred battlefield of a foreign land, there to plan the further strategy and military operations against Hitler's Germany, Mussolini's Italy, and Hirohito's Japan. There would clearly be problems, especially political; but the new, dominating role of the United States was unmistakable—visible not only in the planes, tanks, artillery, and equipment factories were churning out at an ever-increasing rate across North America but their presence now in Northwest Africa, thousands of miles from American shores, barely eight weeks since the huge and successful amphibious U.S. invasion.

Not all was positive in the House of Happiness, however.

"Well after midnight, the P.M. took his leave," Elliott later recalled. The President "was tired but still in a talkative mood"—and talk he did to his son. To Elliott the President confided his continuing distrust not only of the French, in regard to their tottering colonial empire, but of Churchill, too, in that respect. This might well complicate his dream of a United Nations authority committed to the principles of the Atlantic Charter, after the war. "The English mean to maintain their hold on their colonies. They mean to help the French maintain *their* hold on *their* colonies. Winnie is a great man for the status quo," the President said sadly to Elliott. "He even *looks* like the status quo, doesn't he?"[12]

Elliott's version of events was considered suspect by some, but Prime Minister Mackenzie King's contemporary record of his stay at the White House the previous month, as well as the diary kept by Daisy Suckley, the President's cousin, would lend credence to the overall veracity of Elliott's account, published immediately after the war's end. The President had disliked Admiral Darlan — but that did not mean he approved of de Gaulle, who harbored dictatorial ambitions. "Elliott," the President said to his son, "de Gaulle is out to achieve one-man government in France" — and was committed to the revival of its colonial empire. "I can't imagine a man I would distrust more. His whole Free French movement is honeycombed with police spies — he has agents spying on his own people. To him, freedom of speech means freedom from criticism . . . of him."[13]

Which led the President to turn to "the problem of the colonies and the colonial markets, the problem of which he felt was at the core of all chances for future peace" across the globe. "'The thing is,' he remarked thoughtfully, replacing a smoked cigarette in his holder with a fresh one, 'the colonial system means war. Exploit the resources of an India, a Burma, a Java; take all the wealth out of those countries, but never put anything back into them, things like education, decent standards of living, minimum health requirements — all you're doing is storing up the kind of trouble that leads to war. All you're doing is negating the value of any kind of organizational structure for peace before it begins." And with that he'd chortled: "The look that Churchill gets on his face when you mention India!"[14]

To Elliott the President then explained his notion of trusteeships: that "France should be restored as a world power, then to be entrusted with her former colonies as a trustee. As trustee, she was to report each year on the progress of her stewardship, how the literacy rate was improving, how the death rate declining, how disease being stamped out, how . . ."[15]

Phased decolonization, under the aegis of the United Nations, in other words.

"Wait a minute," Elliott had countered, "Who's she going to report all this to?"

And with that his father had set out — to Elliott's amazement — his vision of the "United Nations" postwar "organization." Also his notion of policemen: the "big Four — ourselves, Britain, China, the Soviet Union —

we'll be responsible for the peace of the world" — once the war was won. "It's already high time for us to be thinking of the future, building for it," Roosevelt remarked.[16]

"Three-thirty, Pop," his son pointed out.

"Yes. Now I *am* tired," the President acknowledged. "Get some sleep yourself, Elliott."[17]

And with that the eve of the defining conference of World War II came to a close.

12

In the President's Boudoir

ELLIOTT OVERSLEPT. When he staggered downstairs for breakfast on January 15, it was to find the U.S. Joint Chiefs of Staff already assembled in his father's boudoir.

It was 10:00 a.m. — and the President was listening to what had transpired at the preliminary Combined Chiefs of Staff meeting the previous day, at the Anfa Hotel.

In insisting the conference take place in American-held Morocco, the President had chosen wisely — for it was vital the U.S. chiefs be exposed to the actual Torch battlefield in Northwest Africa. Instead of concocting strategy and operations thousands of miles away, in the safety and comfort of the Pentagon and the Mall, they would have a chance to meet the men and commanders on the ground who were fighting Germans now, not Vichy French troops. It was also crucial that the U.S. chiefs be separated for a time from their dangerously irresponsible planners, who had very little idea of modern hostilities in facing the Wehrmacht — or the fanatical Japanese. In the many documents accumulating in his Map Room at the White House — U.S. Joint Chiefs of Staff minutes, Combined Chiefs of Staff minutes, Joint Strategic Survey minutes, Joint Intelligence Committee minutes, Joint Staff Planners Reports, recommendations and analyses of the differences between British and U.S. strategic views since November 1942 — he had never seen a single mention of the need for American combat experience.

The plethora of paper evinced dutiful, unstinting research and statistical evidence gathered in Washington — but no common sense. That the Western Allies were holding half of the Luftwaffe's entire operational strength on the Western and Mediterranean Fronts was calculated down to the nearest plane; the number of Wehrmacht divisions capable of offensive and defensive operations was tallied and enumerated; the amount of German naval vessels and U-boats estimated. Yet the need for American battle *experience* — and lessons — in matching up to professional German foes had seemed a closed book to such bureaucrats and staffers.

At the time of Casablanca — as well as after the war — there would thus be righteous indignation over the President's decision to allow only a handful of staff officers to accompany the U.S. chiefs on the trip to North Africa. Led by the War Office's chief planning officer, Brigadier General Albert Wedemeyer — who was one of the few permitted to travel with General Marshall to Casablanca — these men would complain they had been thrown to English wolves: a British prime minister taking with him a vast retinue of planners and operations officers and clerks committed to a vague British, rather than an Allied, military strategy for 1943.

Wedemeyer, in particular, would complain they'd been duped; that the President had made a terrible mistake; had through naiveté brought a military team simply too small to confront the host of staff officers accompanying the wily Churchill. Moreover, that the British had tricked the American contingent into abandoning their preferred Second Front invasion that summer, 1943. The British staff officers, Wedemeyer would complain, had been backed by yet *more* staffers aboard HMS *Bulolo,* anchored for their special use in Casablanca Harbor. Using this communications ship, the British planners were able to cable London and put their hands on any fact or figure they needed to support their alternative British plans, and thus defeat American counterproposals; whereas the U.S. contingent, despite being in a U.S. compound in a U.S. military area guarded by U.S. artillery and antiaircraft guns, was virtually captive in terms of British bureaucratic firepower.

"They swarmed upon us all like locusts," Wedemeyer lamented in a letter from North Africa to General Thomas Handy, the assistant chief of staff in the War Department's Operations Division (OPD) in Washington, and "had us on the defensive practically all the time" — backed by "a plentiful supply of planners and various other assistants, with prepared plans to insure that they not only accomplished their purpose but did so in stride and with fair promise of continuing in the role of directing strategy the whole course of the war."[1]

General Wedemeyer was certainly not alone in perceiving a British conspiracy to subvert the swifter course of World War II. General Handy, who received Wedemeyer's letter in Washington and passed it on to other generals at the Pentagon, was of like mind, bewailing afterward that "the British on the planning level just snowed them under."[2]

Yet another U.S. planner later recalled how "we were overwhelmed by the large British staff."

Brigadier General J. E. Hull, heading up the OPD at the Pentagon, was even more embarrassed than Wedemeyer and Handy by the U.S. unpreparedness for paper battle. "The British had come down there in droves," he later recalled, "and every one of them had written a paper about something that was submitted by the British Chiefs of Staff to the American chiefs of Staff for agreement."[3]

In sum, "We came, we listened," Wedemeyer recoined Caesar's famous epigram, "and we were conquered."[4]

All this was true — bearing out Brigadier Jacob's nervousness at the size of the British team Churchill insisted should be flown to Casablanca. Yet in terms of the Allied strategy that President Roosevelt was now to lay down at Casablanca, it completely missed the point. For the reality was: Wedemeyer and his colleagues were still living in a fool's paradise. And the moment of truth — not only the President's truth, but truth on the ground — had arrived.

Inexperienced U.S. planning officers like Brigadier General Wedemeyer were the real problem — not the British.

The U.S. War Department's final planning document, produced by the Joint Strategic Survey Committee and sent over to the White House Map Room for the President to review before he departed from Washington, had said it all: stating baldly that once Tunisia and Libya were secured, the Allies' Mediterranean Front should be closed down. Mussolini's Italy, in the planning committee's view, could be forced to surrender by air bombing alone. All U.S. Army forces should be switched to Britain "for a land offensive against Germany in 1943" — without gaining any further amphibious experience, or campaign lessons in facing and fighting German forces.[5]

Thankfully the President had confided to Mackenzie King, in early December, his unwillingness to tackle a cross-Channel invasion before American commanders and troops were blooded and had the measure of their opponents — which could best be done in the Mediterranean, where this could be achieved, as he said, without risking a major setback. His interrogation of General Marshall at the White House before the chiefs left for Casablanca on January 7 had only convinced him more deeply that a premature invasion of northern France would be a disaster — and he had insisted no decision should therefore be made for several months. It was thus with decided relief that President Roosevelt found, at midmorning in his bedroom on his first full day in Casablanca, January 15, 1943, that the penny had finally dropped, at least among his chiefs of staff.

Making his chiefs fly to Casablanca had, he found, already worked — without British intervention. Once in Morocco, Marshall had finally talked to Allied commanders on the battlefield. As Marshall now confessed in the villa's bordello-like bedroom, he had spoken not only to Admiral Mountbatten — the British chief of Combined Operations, who'd been responsible for the disastrous Dieppe landings — but at great length with General Mark Clark. Clark had been Eisenhower's deputy in the Torch invasion, and had just been promoted to command the U.S. Fifth Army in Morocco, both to defend against mythical German invasion across the Mediterranean and prepare for future offensive operations. General Clark had informed the chief of staff of the U.S. Army that there was no chance of a cross-Channel operation succeeding in the summer of 1943.

No chance whatever.

This was music to the President's ears — for he had half-expected to have to do battle once again against his own team, lest in the interval since their meeting in the Oval Office on January 7 they revert to their insistence on a cross-Channel invasion in 1943.

General Clark's battlefield testimony, however, had applied the necessary dose of cold reality. Clark — a man who certainly did not lack courage, having fetched General Henri Giraud from Vichy-held southern France in person by submarine to assist in the Torch invasion — had been emphatic. To Marshall he'd explained that "there must be a long period of training before any attempt is made to land against determined resistance" — especially Wehrmacht resistance of the kind that would meet a cross-Channel invasion. In particular he'd "pointed out many of the mishaps that occurred in the landing in North Africa which would have been fatal had the resistance been more determined," as Marshall now relayed to the President.[6] In fact, General Clark had himself undergone a Pauline conversion. In London, the previous summer, he'd deplored the idea of landings in Northwest Africa as an unnecessary "sideshow."

Now, however, he felt American amphibious operations in 1943 "could be mounted more efficiently from North Africa"—and certainly with less loss of life—than from the British Isles and the United States, across the English Channel.[7]

Hitler's Atlantic Wall, the fighting general had made clear to Marshall, was no joke. The American military was not up to such a gargantuan task, he'd realized—and was backed by the latest British planning reports. The British, it seemed, had done the numbers that Marshall's team had failed to appreciate in Washington. They looked formidable. As General Brooke had pointed out in his first presentation on January 14, the day before, at the Anfa Hotel, "the rail net in Europe would permit the movement of seven [German] divisions a day from east to west which would enable them to reinforce their defenses of the northern coast of France rapidly."

A day?

By contrast, Marshall now acknowledged to the President, in the Mediterranean theater the Germans "can only move one division from north to south each day, in order to reinforce their defense of southern Europe."[8]

If the U.S. armies were to acquire the combat experience necessary to assault Hitler's Atlantic Wall, then it would best be gained in the Mediterranean, at the extremity of German lines of communication. Northern France was, by contrast, the very closest to the German border. General Clark thus favored a continuation of the war in the Mediterranean, Marshall admitted to the President, where "the lines of communication" for the Allies would be "shorter," and where the "troops in North Africa have had experience in landing operations." Not only did the Allies already possess sufficient American and British forces in the Mediterranean—naval, air, and ground—to knock Italy out of the war that year by invading Sardinia or Sicily "once the Axis had been forced out of Tunisia," but, as Clark had pointed out, the Mediterranean offered the opportunity for U.S. units to gain the battle exposure they needed, in a relatively secure environment where even local setbacks would not be disastrous to Allied strategy. This was something that could not be said for a premature cross-Channel assault.

Lest there be any misunderstanding, the necessary "training" for eventual combat against tough German defenders of the West Wall, Clark had repeated to Marshall, would be infinitely "more effective if undertaken in close contact with the enemy," in current combat. Not in the United States or Britain, Clark had insisted, but in real time, in the Mediterranean.[9]

In the strangest of venues, then, reality had finally set in. The Mediterranean, not northern France, should be the proving ground for as-yet-untested U.S. troops, Marshall now agreed — not only in terms of combat but in developing effective coalition command in 1943 — operations involving British, Canadian, French, and other forces, on a front where the Allies could steadily improve their fighting skills, however much the Russians would, doubtless, complain. Not to mention Marshall's Pentagon team.

Headed by Brigadier General Wedemeyer, Marshall's operations planners would be devastated, the President was aware — as would the secretary of war, Mr. Stimson. But Wedemeyer and Stimson were suffering from delusion — dangerous delusion. However doggedly they urged a cross-Channel attack that year, it was not their lives that would be on the line, but the lives of tens of thousands of Americans — facing a Wehrmacht whose true fighting ability they had not even begun to measure.

It was, as Captain McCrea recalled, a "long conference in the President's bed room" and one that only "broke up well past noon."[10]

This was, in retrospect, the turning point of the war, in terms of the Allied military struggle against the Axis powers — clinching not only the strategy but the timing of America's game plan in conducting World War II. Mass American suicide in a premature Second Front would once again be avoided that year, thanks to the President's military realism. Instead, mercifully, the United States military would back only those operations that promised success: success that would boost morale at home and validate the President's step-by-step strategy for prosecuting the war.

Victory rather than disaster: this would now be the order of the day.

As General Clark had now recommended, U.S. forces would be instructed to learn their deadly trade on the periphery of Europe that year, before meeting the deadliest challenge in 1944: one that even Hitler had balked at attempting in 1940, when Britain was on its knees: a massive cross-Channel invasion. Finally, after thirteen months of war, the Commander in Chief and his chiefs of staff were on the same page.

General Marshall's belated recognition, on January 15, that the President's strategy was probably right would now cement the methodical, stone-by-stone U.S. progress in World War II. The question of "What next?" after Tunisia was, effectively, over—before the first plenary session of the Combined Chiefs of Staff meetings began at the Anfa Hotel that afternoon.

American grand strategy for 1943 was clear: attritional warfare at the extremity of the enemy's lines of communication, enabling U.S. forces to learn how to defeat the Wehrmacht in battle.

And to make sure this policy had a good chance of succeeding, the President said he wanted to see the general commanding the Allies in the Mediterranean from his headquarters in Algiers: young General Dwight D. Eisenhower.

Eisenhower seemed to the President to be a bit "jittery" as, shortly before 4:00 p.m. on January 15, the two men sat down by the picture window in the President's villa, Dar es Saada.

It was for a good reason. General Dwight Eisenhower, or Ike, as he was familiarly known, had undergone a hair-raising flight from Algiers to see the President and the chiefs of staff. Two of his Flying Fortress's engines had conked out, and he'd been told he must get ready to parachute from the aircraft. This he'd begun to do—chiding, as he did so, his naval aide for the time it was taking him to refasten one of his general's shoulder pins, which had been accidentally knocked out. "Haven't you ever fastened a star before?" Ike had barked at the hapless officer, whose hands were shaking uncontrollably. "Yes, sir, but never with a parachute on," the aide had squeaked.[11]

Fortunately the pilot had nursed the surviving two engines long enough to land at Medouina airport, and Eisenhower had immediately been driven to the Anfa Hotel to appear before the Combined Chiefs of Staff.

Dissatisfied by Allied progress—or lack of progress—in Tunisia, the Combined Chiefs of Staff had treated the young general roughly, expressing frank disappointment at his failure to seize Tunis at the very start of the campaign, and also at the German rebuff given to his second attempt, earlier that month.

Eisenhower's logistical excuses had seemed somewhat lame to the chiefs, more than two months after the Torch landings. On paper, after all, he possessed more than three hundred thousand soldiers under his command in Northwest Africa, ranged against "only" sixty-five thousand

German troops in Tunisia. His latest plan for an end run — an armored right-hook thrusting out of the Tunisian mountains toward the sea at Sfax, designed to carve a wedge between von Arnim's army in Tunisia and Rommel's retreating army in Libya — sounded ill-conceived to General Brooke, who'd had actual battle experience against German forces in the spring and summer of 1940. Ultra decrypts that very day had shown Rommel to be dispatching the veteran Twenty-First Panzer Division from Libya to deal with just such an Allied threat. Instead of dividing and conquering the German forces in North Africa, the Anglo-American forces might themselves be split apart. Where was the doctrine of concentration of force rather than dispersion of effort — dispersion that could only encourage the Germans to see their chance to counterattack and defeat the Allies in detail?[12]

"Eisenhower is hopeless!" Brooke had noted in frustration in his diary, in late December, reflecting that the American general "submerges himself in politics and neglects his military duties, partly, I'm afraid, because he knows little if anything about military matters."[13]

Brooke was not alone in his criticism of Eisenhower. While "spying out the land" for the President and Prime Minister's visit, Brigadier Jacob, too, had been appalled by Eisenhower's Allied headquarters. "The chief impression I got was of a general air of restless confusion, with everyone trying their best in unnatural conditions. I was assured on all sides that there was no Anglo-American friction at all. But the simple fact of having a mixed staff is quite enough to reduce the overall efficiency by at least a half . . . The British members of the staff, who occupy many of the key positions, have to work with U.S. officers who are entirely ignorant and inexperienced, and have to operate on a system which is quite different from the one to which they are accustomed. They find their task harassing and irritating in the extreme. Many are inclined to doubt whether a combined Allied Staff is a practical arrangement, and think the experiment should not be repeated, and should be brought to an end as soon as possible."[14]

Emerging from his interrogation by the chiefs at the Anfa Hotel, Eisenhower suspected his number might be up as coalition commander in chief in the theater. "His neck is on the noose, and he knows it," even his naval aide, Lieutenant Commander Harry Butcher, noted in his diary.[15] At his headquarters in town General Patton recorded the same. "He thinks his thread is about to be cut," Patton would scribble, after talking with Ike — urging Eisenhower to "go to the front" instead of returning to the huge Allied headquarters in Algeria, many hundreds of miles behind the fighting.[16]

Roosevelt, however, saw things differently — very differently. From the President's point of view, Eisenhower had done extremely well — indeed, given the friction that would arise between de Gaulle and General Giraud over control of French anti-Vichy forces, Ike had achieved miracles in planting the American flag across Vichy-held Algeria and Morocco in only a few weeks, leaving no real chance the Germans could strike the Allies in the flank, as Secretary Stimson and General Marshall feared.

The fact remained: whether General Brooke liked it or not, only an American supreme commander was going to be able to direct the campaign. For good or ill, a system of workable coalition command, in combat, had still to be developed — and it was a blessing that the President had journeyed, as U.S. commander in chief, to see Eisenhower in person, in the active theater of war, whatever might be the disappointments of the British and American chiefs of staff.

Having first put the young general at ease, the President thus listened with interest as Dwight Eisenhower explained what the Allies were up against in advancing across the same mountain range the President had overflown in Morocco, as well as the atrocious mud and winter weather conditions in Tunisia. Hitler had managed to get sixty-five thousand German troops across the Mediterranean from Italy, together with high-quality equipment, including new Panther and Tiger tanks — the latter armed with 88mm guns — but he had been aided, too, by French pusillanimity, the French forces in Tunisia failing to fire a single shot to delay, let alone stop, the Germans.

Even that French timidity had been outweighed by the political and military leadership problems with which the French had confronted Eisenhower as Allied commander in chief in Northwest Africa. General Giraud had succeeded Admiral Darlan as French commissioner, but was proving a disappointment — a "good Division Commander," possibly, but wholly lacking in "political sense" and with "no idea of administration. He was dictatorial by nature and seemed to suffer from megalomania," Eisenhower had already explained to the chiefs — a view he now repeated to the President. "In addition," Giraud "was very sensitive and always ready to take offense. He did not seem to be a big enough man to carry the burden of civil government in any way. It had been far easier," Eisenhower remarked candidly, "to deal with Admiral Darlan," despite Darlan's record as a Pétainite Nazi appeaser.[17]

The President laughed. If only his many critics in America knew! Feckless French troops were deserting by the hundreds, in the field, rather than risk their lives against the ruthless Wehrmacht. So much for coalition fighting. Getting the French to stop squabbling amongst themselves over currency, supplies, pensions, and administrative aspects of the U.S. occupation had also proven a minefield — permitting Eisenhower, as the general himself acknowledged, too little time to focus properly on the battlefront, where progress had been painfully slow. The Wehrmacht forces facing U.S. troops in the Gafsa and Tébessa sectors were, Eisenhower made clear, first class. The "opposition was tough," Elliott Roosevelt — who was acting as his father's aide-de-camp — recorded, "while we were just beginning to learn about war first hand."[18]

This was exactly the kind of honest appraisal the President wanted to hear, from the lips of the top U.S. commander in the theater — confirming what General Clark had told General Marshall.

"No excuses, I take it," the President commented.

"No, sir. Just hard work." Or fighting.

In which case, the President raised the next question, what was the general's estimate of how long it would take to clear North Africa of Axis forces?

At the White House in late November, 1942, General Bedell Smith, Eisenhower's chief of staff, had personally assured the President that fighting would be over by mid-January 1943. It was now mid-January.

"What about it? What's your guess?" the President pressed Ike. "How long'll it take to finish the job?"

"Can I have one 'if,' sir?"

The President chuckled, Elliott remembered, and bade him give his best estimate.

"With any kind of break in the weather, sir, we'll have 'em all either in the bag or in the sea by late spring."

"What's late spring mean? June?"

"Maybe as early as the middle of May. June at the latest."[19]

Elliott recalled being surprised by the young general's cautious estimate, as this — five months of further campaigning — would make a switch of naval, air, and army forces to England, in order to mount a massive amphibious invasion of France across the English Channel that summer, almost impossible. The notorious fall weather would preclude a late-summer amphibious assault — as Hitler, too, had similarly decided in the summer of 1940, after the Luftwaffe had been rebuffed in the Battle of Britain.

"Father looked satisfied," Elliott clearly remembered[20] — and summoned the Combined Chiefs of Staff, once again, to his villa, at 5:30 p.m.

The President asked Winston Churchill to attend the meeting at the Villa Dar es Saada, too — for the session would be, in effect, a presidential briefing, backed by the President's "active and ardent lieutenant."

One by one the generals and admirals — Marshall, King, Arnold, Brooke, Pound, Dill — entered, together with Air Marshal Arthur Tedder, Admiral Mountbatten, General Hastings Ismay, and Harry Hopkins. Once seated, the President asked General Eisenhower to give yet another presentation of "the situation on his front"[21] — an indication that, as President, he was fully behind his protégé.

The President then briefly reviewed the outlook with the assembled chieftains — and made clear to them his own preference. As Brooke noted in his diary, "we did little except that President expressed views favouring operations in the Mediterranean."[22]

Aware that General Marshall might feel he'd lost face among the Combined Chiefs of Staff after arguing so hard for an end to Mediterranean operations and a switch to the U.K. for a cross-Channel attack that year, the President asked Marshall to stay behind and have dinner with him. He also invited Eisenhower.

Elliott made old-fashioneds (sugar, bitters, and whiskey) for the generals — and, joined by Franklin Jr., the five men sat down in the President's dining room for a first-class Moroccan meal.

Typically, Roosevelt wanted Marshall to feel he was respected, even if his advice had been wrong. He therefore deliberately raised again his wish to inspect troops not only in Morocco but closer to the frontline, near the Tunisian border.

"Out of the question, sir," Marshall stated unequivocally.[23] Even with a fighter escort, the President's slow C-54 could be attacked by Luftwaffe planes — "it would just draw attackers," Eisenhower added frankly, "like flies to honey."[24]

The President reluctantly backed off the idea — allowing Marshall to feel he had won at least a tactical victory.

Satisfied, Marshall and Eisenhower departed the villa after dinner and the President then spent quality time with his sons, talking about the family. "Father got to bed early that night: before midnight," Elliott recalled.[25]

The President had cause to feel the conference was off to a good start. The flight to North Africa had been historic. But so, too, had been the President's first full day in Casablanca. By its end he'd ensured that the great Allied military conference would result in compromise and cooperation, *not contention*—thus injecting not only unity of Allied military purpose but a transfusion of realism into the veins of the U.S. Joint Chiefs of Staff, who were, in all truth, more green regarding modern warfare than Eisenhower.

Instead of insisting upon mass American slaughter on the beaches of northern France that August, the U.S. chiefs could now set about mapping a detailed course of operations that year that would, above all, be *within the capabilities* of the Western Allies—whatever Stalin might plead, when eventually informed.

Besides: if the Russian dictator had wanted to argue for a Second Front in Europe that year, he should have taken the trouble to show up.

With that, having bidden his guests goodnight, the President retired and went to sleep, confident that, though the Allies had much to learn in combat, they would do so in the coming months, and that all would be well—with 1944 the year when the coup de grâce could be given and the Third Reich brought to an ignominious end.

How difficult it would be to steer his coalition partners, however, remained to be seen. If the British were difficult, how much more so were the French. Moreover, how to keep the Russians happy with such a timetable, when they were facing two-thirds of the Wehrmacht on the Eastern Front, would be tougher still.

PART FOUR

Unconditional Surrender

13

Stimson Is Aghast

AT THE PENTAGON, Secretary Stimson was aghast on hearing the "bad news" from General McNarney.

Joseph McNarney was the U.S. Army deputy chief of staff, standing in for General Marshall. His news related to "how the British were forcing us to do some more in the Mediterranean" after Tunis, rather than switching U.S. forces to a cross-Channel invasion, to be launched from England that year.[1] In Washington, D.C., however, the secretary of war could do nothing.

Two days later Stimson's heart sank still further with the "somber news that I had been getting yesterday from the conference in Africa where it seems to be clear that the British are getting away with their own theories," he recorded, "and that the President must be yielding to their views as against those of our own General Staff and the Chief of Staff. So it looks as if we were in for further entanglements in the Mediterranean, and this seems to me a pretty serious situation unless the Germans are very much less strong than I think we should assume."[2]

Stimson's continuing lack of realism was deplorable, given the lack of U.S. experience in mounting an operation as vast and serious as a cross-Channel invasion would be, if undertaken that year. At the same time, the war secretary's fear of "perfidious Albion" was well warranted. Could British assurances they would eventually participate in a Second Front honestly be believed? The answer was clearly no.

For all their criticisms of Eisenhower's tardiness in Tunisia, the British were not actually willing or able to say how the Third Reich could be *defeated*, rather than surrounded. As Admiral King reported to the President when the U.S. chiefs of staff came to the Villa Dar es Saada the following evening, January 16, for a two-hour session with Mr. Roosevelt, "the Joint Chiefs of Staff have been attempting to obtain the British Chiefs of Staff concept of how the war should be *won*"[3] — and had had little luck. It seemed the British had no idea.

In his diary General Sir Alan Brooke, after his own experience in battle against German forces in 1940, remained implacably opposed to a cross-Channel attack unless the Wehrmacht was first weakened and brought to its knees elsewhere. He complained, in his diary, at the "slow tedious process" it was to get the U.S. team to accept his "proposed policy." In a postwar annotation to the diary, he would even pen a diatribe against General Marshall. Among "Marshall's very high qualities he did not possess those of a strategist," Brooke (by then Lord Alanbrooke of Brookeborough) would claim. "It was almost impossible to make him grasp the true concepts of a strategic situation. He was unable to argue out a strategic situation and preferred to hedge and defer decisions until such time as he had to consult his assistants" — assistants who were "not of the required calibre."[4]

Brooke was being disingenuous — for Marshall, like Admiral King, was an excellent strategist; what he lacked was the ability to see how important it was to match U.S. strategy to reality. Neither general properly understood the need to create armies and army commanders who could defeat the Wehrmacht *in battle* — irrespective of wearing down German forces on other fronts.

Hour after hour Marshall thus pressed Brooke and the British to explain how exactly a further campaign in the Mediterranean would, in itself, *defeat the Third Reich* — something neither Brooke nor his colleagues Admiral Pound (who was suffering from an undiagnosed brain tumor) and Air Marshal Portal could answer. Brooke's assertions that the Germans would thereby be "worn down" to a point where they could not send reinforcements to northern France seemed particularly lame, given the likelihood that, if the Allies fought on in Italy, as Brooke envisioned, it would be the Allies who would be worn down rather than the Germans.

Marshall and Brooke thus went at each other hammer and tongs. Almost five hours of discussion at the Anfa Hotel — however irritating to Brooke — did at least permit the American team to challenge and rehearse the different possible military alternatives for 1943 with relentless honesty within the framework of overall war strategy.

The result was a consensus: there were no alternatives. If forced to fight on that year in the relative safety of the Mediterranean theater, the U.S. chiefs accepted, then it would be best to tackle Sicily, once North Africa was cleared — giving the Allies the amphibious-assault-landing experience necessary for a 1944 cross-Channel invasion.

The President had been right, they reluctantly agreed as they went over the requirements for a successful Second Front with their British opposite numbers. General Brooke had pointed to forty German divisions available in or close to France — and a Luftwaffe that was still a potent weapon of war. By contrast, after the expected capture of Tunis in the spring of 1943, the Allies would have but twenty-one to twenty-four divisions ready to assault northern France even by the fall — and as Admiral Pound, the British navy chief, pointed out, "this was too late since the weather was liable to break in the third week of September and it was essential to have a port by then." August 15, 1943, would be the cutoff date, weatherwise, were a cross-Channel invasion to be undertaken that year — moreover, according to the commander of the British amphibious forces, Vice Admiral Mountbatten, it would take all of three months to get the necessary landing craft from the Mediterranean to the United Kingdom. Any hope that the RAF or USAAF could interdict German air forces over Brest were scotched by Air Marshal Portal, the RAF chief, "since it was out of range." Even if the Cherbourg-Normandy area was chosen, "with limited air facilities in the [Cherbourg] Peninsula we should possibly find ourselves pinned down at the neck of the Peninsula by ground forces whose superiority we should be unable to offset by the use of air," Portal pointed out. And once the Germans realized the Allies were not actually going to attack Italy and southern Europe from North Africa, they would "quickly bring up their air forces from the Mediterranean, realizing that we could not undertake amphibious operations on a considerable scale both across the Channel *and* in the Mediterranean."

The simple fact, then, was: "no Continental operations on any scale were in prospect before the spring of 1944," General Arnold concluded.[5]

If the Combined Chiefs were agreed on 1944 for a major cross-Channel assault, at what point should "further operations" in the Mediterranean be halted, though? How exploit Allied strength in the Mediterranean, once achieved, without risking stalemate requiring more and more reinforcements — thus vitiating the success of the cross-Channel campaign planned for 1944? As General Marshall memorably put it, the Mediterranean could become a dangerous "suction pump" on American manpower and arms. What Marshall therefore wanted from Brooke, Pound, and Portal, as a strategist, was an acknowledgment of that danger: an agreement that, if operations in the Mediterranean became stalled or an expensive dead end, the very combat experience the Allies were seeking would thereby be wasted, and a successful cross-Channel invasion in 1944 be rendered impossible.

This danger General Brooke refused to validate, as only an owl-eyed, intelligent, but obstinate Ulsterman could—while paying lip service to the notion of an eventual cross-Channel attack in 1944.[6]

Would Brooke keep his word, though, the U.S. team wondered? Would the British even undertake an offensive to reopen the Burma Road they had lost to the Japanese in 1942, which was vital in order to supply United Nations forces in China?

Marshall had to hope they would. The British, after all, were America's primary allies in the global war. At the Villa Dar es Saada Marshall therefore reported to the President on January 16 his understanding that, after the amphibious invasion of Sicily that summer, "the British were not interested in occupying Italy, inasmuch as this would add to our burdens without commensurate returns."[7]

These were famous last words—or hopes.

The President was as concerned as Marshall over getting bogged down in Italy, and "expressed his agreement with this view."[8] Between them, however, they would have to make the British back off such a potential dead end—the President working on Churchill, Marshall on Brooke. Neither of them had any idea of the nightmares ahead, though, in this regard.

In the meantime, crediting British good faith, the chiefs moved on to other strategic concerns. By Monday, January 18, in fact, Roosevelt had been able to get outline agreement on pretty much all he had wanted at Casablanca. The Combined Chiefs had agreed to his strategy for 1943: further operations in the Mediterranean, after the capture of Tunis, targeted on Sicily, with simultaneous preparations for a cross-Channel assault to be made earliest in late 1943, if there were signs of sudden German collapse; otherwise a full-scale assault early in 1944 on the Cherbourg Peninsula, targeted on Berlin. In Asia there was to be a 1943 British offensive in Burma to open the overland supply route to China. And in the Pacific, further advances that would take the Allies closer to the Japanese mainland — which would be ultimately bombed into submission, or subjected to land assault if required, after the defeat of Nazi Germany.

In this respect the President had invited Churchill to lunch with him privately at the Villa Dar es Saada on the eighteenth, before the afternoon meeting he'd convened with the chiefs of staff — for he wanted something of major importance from Churchill: formal agreement to his "unconditional surrender" policy.

Churchill raised no objection whatever — in fact the Prime Minister found himself positively inspired by the President's proposals for prosecuting the war to the bitter end, gaily promising not only that the British would launch their offensive into Burma (Operation Anakim) under General Wavell that year but would "enter into a treaty," if necessary, to assure him that Britain would fight alongside the United States to ensure the ultimate "defeat of Japan." In reporting the day's deliberations to his cabinet that night, Churchill informed his colleagues in London that the Combined Chiefs of Staff in Casablanca were "now I think unanimous in essentials about the conduct of the war in 1943," and that in respect of the strategy decided upon at the meeting held in the President's villa with the Combined Chiefs, "Admiral Q [FDR] and I were in complete agreement." Moreover, Churchill cabled, he and the President were in agreement that, at the conclusion of the conference, there would be a public "declaration of firm intention of the United States and the British Empire to continue the war relentlessly until we have brought about the 'unconditional surrender' of Germany and Japan."[9]

Historians would later argue over the merits and demerits of such a war policy,[10] but the fact that neither the U.S. chiefs of staff nor the British prime minister and his War Cabinet in London opposed the President's "unconditional surrender" policy gives some idea of how much in control of such war strategies was the President. Time would tell how it would go down once announced to the world, but in the meantime Mr. Roosevelt had come too far to remain closeted in the Anfa camp. He had chosen as his *nom de plume* Admiral Q, in prior secret communications with Churchill — a humorous reference to his Spanish literary hero, Don Quixote. (Hopkins was "Mr. P." for Sancho Panza.) Whether he was tilting at windmills in seeking unconditional surrender of the Axis nations would only become clear in the fullness of time — and war. In the meanwhile he wanted to get out and visit with his commanders and the troops in the field, like Lincoln.

On the evening of January 19, the President went to dine with General Patton at his palatial headquarters in Casablanca — listening with fascination and amusement to the cavalryman as, in his distinctive high-pitched voice, he described his recent landings under French fire, and expounded upon the primacy of the tank in modern warfare.

Two days later, at 9:20 a.m., the President left Casablanca by car with Patton "for an inspection of the United States Army forces stationed in the vicinity of Rabat, some 85 miles to the northeast," as Captain McCrea recorded.[11] U.S. troops lined the entire route as the fifteen cars in the cavalcade made their way north, covered by a U.S. Air Force umbrella.

Recalled General Clark, the President "started asking questions, and I don't think he stopped all day. He transferred to a jeep at Rabat, where Major-General E. N. Harmon, commanding the Second Armored Division, was introduced and joined us for that part of the trip. The President was driven within a few feet of the front rank of the troops, which were lined with their vehicles." Then on to review the men of the Third Infantry Division. "A stiff wind made the flags and banners stand out smartly, and the outfits were polished and alert, so that the President had a fine time, seemed pleased with what he saw, and showed his pride for what they had accomplished."[12] And in the afternoon, the Ninth U.S. Infantry Division, commanded by Major-General Manton Eddy.

"I went 'up the line' this a.m. beyond Rabat," the President wrote Daisy that night, and "reviewed about 30,000 Am[erican]. Troops," followed by a visit to Fort Mehdia — "a very stirring day for me & a complete surprise to the Troops."[13]

Given his paralysis, driving in an army jeep caused the President intense pain, but he bore it with equanimity: pleased as punch to review combat-readying soldiers on the battlefield — the brim of his soft Panama hat turned up as he held onto the jeep's guardrail.

One British staffer, witnessing the inspection, later recalled how "fortunate" he was "in being invited by an American colonel to watch President Roosevelt inspecting an American battalion. I was the only British officer present and I was told it was an historic occasion — the first time a President of the United States had ever inspected an American unit on foreign soil. Instead of the parade receiving the visiting officer with a general salute, being inspected and then marching past, the President arrived first and took up his seat (in his jeep because of course he was paralyzed) at the saluting base. Then the photographers got busy, taking him from all angles, from above and below" — the brigadier disgusted by the photographers "who buzzed round the commanding officer and the leading ranks like flies round a horse's ears. They put down wooden boxes to stand on and photographed the leading ranks from above; they lay on the ground and photographed them from the snake's eye view, rolling out of the way to avoid being kicked. Even I, on the touch line, wanted to kick them. The proceedings were most undignified. Then the battalion formed up in line and the President, with two fierce and heavily armed detectives on his jeep and four others, one looking to each point of the compass, in a following jeep, drove down the line. Finally he decorated a soldier and then drove off."[14]

Brigadier Davy had been a highly decorated commander and then staff officer in Egypt, but like so many British colleagues he had simply no understanding of America: of its immigrant history, or the miracle by which ethnically and socially disparate citizens were being molded into a world power based on democratic principles and the President's four freedoms. No picture was ever taken indicating the President's paralysis, but press photographers were aware the whole nation would respond to images of the U.S. commander in chief out in Africa, inspecting his troops. Moreover, from the point of view of public opinion in America, where the majority of people favored dealing with Japan before Germany, such patriotic images were of inestimable importance.

Telling his son Elliott about the trip that evening, the President certainly brimmed with pride and excitement. "I wish you could have seen the expression on the faces of some of those men in the infantry division. You could hear 'em say, 'Gosh — it's the old man himself!' And Father roared with laughter," Elliott recalled. He'd eaten field rations there with Generals Clark and Patton. And Harry Hopkins. "Harry!" he now called upstairs. "How'd you like that lunch in the field, hunh?"[15]

Hopkins, running a bath, thought for a moment. Then he called back down that, although the food had been somewhat Spartan, he'd loved the music. "'Oh yes,' said Father. '*Chattanooga Choo-Choo, Alexander's Rag-Time Band,* and that one about Texas, where they clap their hands, *you* know . . .'

"'*Deep in the Heart of Texas?*'

"'That's right. And some waltzes.'" The President paused. "'Elliott, tell me,'" he continued: "'Would any army in the world but the American army have a regimental band playing songs like that while the Commander-in-Chief ate ham and sweet potatoes and green beans right near by? Hmmm?" He even showed Elliott the mess kit he'd eaten lunch out of, which he'd brought back with him. When Elliott said he would surely have been able to obtain one in America, if he wanted, the President was appalled. "But I *ate* out of *this* one, at Rabat," he told his son with childlike pride, "the day I saw three divisions of American soldiers, who are fighting a tough war. It's a good souvenir. I'll take it home with me."[16]

The President had also visited Port Lyautey, he told Elliott, and seen the sunken warships. He'd laid a wreath at the American section of the local cemetery—and had looked at the graves of the French who'd opposed them.

In a world at war, the Commander in Chief wanted to do right by those men—and if it was hubris to imagine he could in person get America's allies to combine in effecting his two-part vision of the world war and the postwar, then that was a designation the "Emperor of the West"—as Eisenhower's British political adviser, Harold Macmillan, described him[17]—accepted. Inspecting three entire U.S. divisions in the theater of combat, he felt his vision was at least grounded in America's burgeoning emergence as a world power: a power that would soon become capable, with its allies, of slaying the Nazi monster, *unconditionally*—and the Japanese demon thereafter. For this he would need, however, not only the Prime Minister, but the Emperor of the East: Joseph Stalin. Also, probably, the two rivals for leadership of the French empire: Generals Henri Giraud and Charles de Gaulle—the latter due to arrive the next day.

14

De Gaulle

GETTING Major General de Gaulle to appear in Casablanca had been a trial from the start. "On our arrival at Casablanca at the first military meeting with the Pres.," Captain McCrea later recalled, "the Prime Minister informed the Pres. that General De Gaulle, despite his invitation to the Casablanca conference by the P.M., had decided not to attend."[1]

Since Major General de Gaulle was the leader of the Free French movement in London, it was considered vital to get him and General Giraud, the French high commissioner under Eisenhower, to meld the forces under one authority, if they were to contribute to the liberation not only of North Africa and France but of Europe.

De Gaulle's refusal to come to Casablanca had thus been a nasty shot across the President's bows. Roosevelt was "greatly" annoyed, McCrea recalled. "The Pres. told the P.M. rather sternly, I thought, that it was up to the P.M. to get De Gaulle there. At this the P.M. took off on De Gaulle about as follows: 'I tell you Mr. President, Gen. De Gaulle is most difficult to handle. We house him. We feed him. We pay him and he refuses to raise a finger in support of our war effort. He states vigorously every time he gets a chance to do it that he is entitled to military command. I ask you Mr. President what sort of a military command could either of us give him?'

"The Pres. acknowledged that no doubt De Gaulle was hard to handle and there continued about as follows: 'Winston, this is a shotgun marriage'—referring of course to the hoped-for collaboration between De Gaulle and Giraud—and continued, 'We have our party here, referring to Giraud, and I feel it is up to you to get *your* party here.' I inwardly squirmed a bit," McCrea confessed, "at the bluntness of the Pres. remarks, but he, of course, put a light touch on the proceedings with a hearty laugh. I felt easier."[2]

This was typical FDR. Whether it was wise was another matter. It bespoke, however, Roosevelt's urgency — for it was vital, in his mind, for the Western Allies to retain the cohesion of their military coalition if they were to persuade the Soviets to go on fighting the Wehrmacht on the Eastern Front. Especially once the time came to inform Stalin that the Allies were *not* going to launch a Second Front in 1943 unless the Germans collapsed that summer — which seemed unlikely.

Day after day Churchill had duly attempted to get de Gaulle to fly out to Casablanca. "De Gaulle refused Churchill's invitation to come from London," the President himself wrote Daisy with a mixture of amusement and irritation. "He has declined a second invitation — says he will not be 'duressed' by W.S.C. & especially by the American President — Today I asked W.S.C. who paid De Gaulle's salary — W.S.C. beamed — good idea — no come — no pay!"[3]

The next day Roosevelt heard that de Gaulle had finally consented. "De Gaulle will come! Tomorrow!" the President wrote Daisy in excitement on January 21.[4] But if the President thought that by bringing de Gaulle and Giraud together, he could achieve a genuine marriage, he was to be profoundly mistaken. By contrast Winston Churchill, who had been dealing with the quirky, proud, and imperious Major General de Gaulle for two and a half years, knew exactly what was to be expected.

Quite why the President would take personal charge of negotiations with the senior French leaders and officials was a mystery to the British prime minister — who possessed a far deeper understanding of political realities on the European side of the Atlantic than the President.

American political policy in Northwest Africa seemed disastrously amateur, even the U.S. vice consul at Marrakesh acknowledged. From public relations to economics and intelligence, the various Washington agencies "who came to North Africa were at loggerheads with State Department policy," Kenneth Pendar afterward recorded. "The heads of all the agencies cooperated, but their subordinates left the French feeling that we, as Americans, had no clear policy or ideology of any kind."[5]

This was all too true. It was also inevitable, perhaps, as the United States emerged from its long isolationist slumber and felt its way as the world's foremost military power. Early in the twentieth century the United States had considered, then balked at, becoming an empire; now, however, it had little alternative, whether that empire was to be territorial or post-territorial. And this exposed a major weakness in the American system of government—for though the President might make military decisions as U.S. commander in chief, political decisions were another matter. Not only Congress but the free media of the country were entitled to "weigh in"—making unity of approach virtually impossible. Secretary Hull was even more skeptical of de Gaulle than the President. He was equally opposed to the restoration of France's colonial empire in the postwar world save as trusteeships—for how could American sons be expected to give their lives merely to reestablish a colonial yoke they themselves had thrown off in 1783?

The President—like General Eisenhower—was thus faced with an awkward military task: harnessing British and French forces to the yoke of the Western Allies, without committing the United States to restitution of their colonial empires.

Not even Roosevelt's personal representative at Eisenhower's headquarters, Robert Murphy, had had any idea of the President's long-range political plans when preparing the Torch invasion: namely, that "Roosevelt was planning to encourage extensive reductions in the French empire," as the diplomat delicately put it in his memoirs. Once he met with the President at the Villa Dar es Saada, however, Murphy had been quickly brought up to speed—and recognized the postwar agenda the President was seeking. Having congratulated Murphy on the "Darlan deal" that had brought such quick Vichy surrender, the President had then looked reproachfully at his emissary. "But you overdid things a bit in one of the letters you wrote to Giraud before the landings, pledging the United States Government to guarantee the return to France of every part of her empire. Your letter may make trouble for me after the war."[6] Without further ado, the President had gone on to discuss "with several people, including Eisenhower and me, the transfer of control of Dakar, Indochina, and other French possessions, and he did not seem fully aware how abhorrent his attitude would be to all empire-minded French including De Gaulle and also those with whom I had negotiated agreements."[7]

It was the President's long-term political agenda that set the cat among the pigeons, rather than his modest military expectations. And late on the evening of Friday, January 22, 1943, after a delightful meal with the Sultan of Morocco at the Villa Dar es Saada, the President realized he was playing with fire.

• • •

Captain McCrea remembered the fateful night in Casablanca vividly. He had hand-delivered the President's invitation to the Sultan at his palace near Rabat the day before. "No Hollywood director could have put on a more colorful spectacle," McCrea recalled. "The Court Yard ankle deep in white sand," the cavalry "dressed in colorful costumes, the white horses draped in red blankets"[8] — and the Sultan asking if he might bring with him his young teenage son, the Crown Prince, to meet the President.

At 7:40 p.m. on the twenty-second, the Sultan had duly arrived with his "entourage" — "magnificently attired in white silk robes" and "bearing several presents — a gold-mounted dagger for the President in a beautiful inlaid teakwood case, and two golden bracelets and high golden tiara for Mrs. Roosevelt."[9] In return, the Sultan was given a signed photograph of the President in a heavy silver frame, engraved with the presidential seal.

It was hardly a fair exchange — yet the Sultan of Morocco and his son were delighted, for the evening was historic: it was the first time the Sultan had ever been allowed to meet the head of any foreign state other than France.

Seating the Sultan on his right, the President had proceeded to lay out, verbally over dinner, a magic table of postcolonial dreams for the country. Morocco, after all, had only been colonized by the French early in the twentieth century, becoming a "protectorate" in 1912; it could become a sovereign country once again, in the war's aftermath.

Churchill, seated on the President's left, had grown "more and more disgruntled," Elliott Roosevelt recalled, as the President discussed living standards for the nation's Muslims, better education, and "possible oil deposits" in the country. "The Sultan eagerly pounced on this; declared himself decidedly in favor of developing any such potentialities, retaining the income therefrom; then sadly shook his head as he deplored the lack of trained scientists and engineers among his countrymen, technicians who would be able to develop such fields unaided," Elliott wrote. "Father suggested mildly that Moroccan engineers and scientists could of course be educated and trained under some sort of reciprocal educational program with, for instance, some of our leading universities in the United States."[10]

General Charles Noguès, who as the French resident general had also been invited to the dinner but had been placed further down the table, "had devoted his career to fortifying the French position in Morocco," according to Robert Murphy's account, and "could not conceal his outraged feelings" at Roosevelt's talk of postcolonial development and American investment.[11] At the end of dinner "the Sultan assured Father," Elliott recalled, "he would petition the United States for aid in the development of his country. His face glowed. 'A new future for my country!'"[12]

It was also a new approach to decolonization: discussion, both at table and beyond. As word spread, in the days and weeks afterward, the story of the dinner would become legendary among Moroccans as a "proof of our sincerity in the Atlantic Charter," another American official remembered — almost every Arab in Morocco feeling "he knew the whole story of this *diffa* and everything that was said, just as if he had been there."[13]

"It was a delightful dinner, everybody — with one exception — enjoying himself immensely," Elliott later recalled.

The exception was not General Noguès, however; Elliott meant Mr. Churchill. For his part, Robert Murphy remembered the Prime Minister, thanks to his "rare abstinence," being "unnaturally glum throughout the evening" — as well as uncomfortable at the mention of the end of colonial empires. Captain McCrea, however, recalled Churchill's clever solution to the alcohol problem.

"As to no alcoholic beverage being served [in deference to the Sultan]," the President's naval aide recalled, "the P.M. I think was taken by surprise. At any rate he started to glower, the glower being more pronounced during the small talk which preceded the dinner. The Pres. noted this and I think was rather amused." In the meantime, "directly dinner was announced seats were taken," and shortly after dinner started, an "amusing incident took place. One of our Secret Service men entered the dining area and whispered to me that a Royal Marine, the P.M.'s orderly, wanted to speak to me . . . He informed me that a most important message had been received at the P.M.'s nearby villa which required immediate attention. I indicated where the P.M. was seated and told the Marine to so inform the P.M." This he did. "The P.M. after a word with the Pres, withdrew. In about twenty minutes or thereabouts the P.M. returned. No doubt the message referred to was urgent," McCrea allowed, "but on his return it was evident that the P.M. had taken time out to have a quickie or so while handling the urgent dispatch. After dinner and when the guests had departed the Pres. had a good laugh about it all, remarking 'Winston did not tell me what the message was about. Do you suppose he can have arranged it?'"[14]

It was already 10:00 p.m. "The Sultan obviously wanted to stay and discuss more specifically and with loving emphasis some of the points Father had raised during the dinner," Elliott recounted, "but Father's work for the evening was cut out for him. A signal to Captain McCrea then, to stay and take notes; one to Robert Murphy and Harry Hopkins; one to me to hold myself in readiness to act as Ganymede — and all the others left. The stage was set for Charles de Gaulle."[15]

It had been Theodore Roosevelt's dictum — using a supposed West African proverb — that a successful leader should "speak softly and carry a big stick." Franklin Roosevelt preferred, however, to keep his stick well concealed, relying on the force of his personality, his high intelligence, his self-confidence, and his passionate interest in the future to steer people in what he considered the right direction. Even the generally dismissive General Patton, who despised politicians, had been won over by him.

General Marshall had disappointed Patton when dining with him at Casablanca on arrival — "Never asked a question," Patton had noted in his diary.[16] The President, by contrast, never stopped asking questions. Patton had spent one and a half hours with him on January 16 — the President (whom Patton referred to as A-1) "most affable and interested. We got on fine." The next day Patton had seen the President again, and "we all talked over one and one-quarter hours, then went to see B-1" — Churchill.

Churchill, the general had sniffed in his diary, "speaks the worst French I have ever heard, his eyes run, and he is not at all impressive."[17] On January 18 Patton had again ridden in the President's car for an inspection of the battalion guarding the Anfa enclave. Then on January 19 he'd invited the President to dinner at his headquarters — the President afterward asking Patton to sit and talk with him, alone, "in car while P.M waited, for about 30 minutes. He really appeared as a great statesman," Patton jotted in his diary[18] — and on January 21 the President asked Patton once again to ride with him in his car, together with General Clark, following lunch and his inspection of the three U.S. divisions at Rabat. "Coming back we talked history and armor about which he knows a lot," Patton recorded. "F.D.R. says that in Georgia," in the Soviet Union, "there are Crusaders' Castles intact and that hundreds of suits of armor exist. Then he got on to politics"[19] — with somewhat withering remarks about Vice President Henry Wallace as his potential heir, or even Harry Hopkins; "neither of them had any personality," he claimed, which would rule out any hope of their winning election. Even Churchill drew the President's less-than-complimentary appraisal in terms of empire and future global security. "He also discussed the P.M. to his disadvantage. Says India is lost and that Germany and Japan must be destroyed."[20] Above all, however, the President listened—especially to Patton's military judgment. The general pointed out how green American forces still were, in terms of fighting. "People speak of Germany and Japan as defeated," the general warned sagely, but "we have never even attacked them with more than a division."[21]

Churchill's ill grace at dinner with the Sultan particularly irritated Patton—who claimed the Sultan had "especially asked" to see the President in private, "before Churchill arrived," as he did not seem to like the Prime Minister. Already on arrival the Prime Minister appeared, it had seemed to Patton, "in a very bad temper . . . No wine, only orange juice and water. Churchill was very rude, the President was great, talking volubly in bad French and really doing his stuff," Patton recorded that night. The tanker had personally driven the Sultan home. "On way Sultan said, 'Truly your President is a very great man and a true friend of myself and my people. He shines by comparison with the other one"—the "boor" Churchill.[22]

Patton was being unfair, however—for neither he nor Captain McCrea had any idea of the real cause of Churchill's distemper.

The President did. After the Sultan's departure, the Prime Minister quickly explained. De Gaulle had just visited him, before dinner, at the Villa Mirador—and had scotched any prospect that his arrival would lead to the unification of the Free French movement in London and the French Imperial Council in Algiers, under Giraud.

De Gaulle had been not only intransigent, but rude to the point of insult—"a very stony interview," as Churchill described it to the President. The Prime Minister thus begged the President not to see de Gaulle that night, but to put off the meeting to the next day, when de Gaulle would have had more time to simmer down.

The President, however, insisted de Gaulle be brought straight to him. Thus did the Free French major general arrive at the Villa Dar es Saada, along with two aides, at 10:20 p.m., "with black clouds swirling around his high head and with very poor grace" according to the President's son:[23] there to meet the U.S. commander in chief whose troops had "liberated" Algeria and Morocco.

15

An Acerbic Interview

In a cable to his secretary of state, the President had explained just why he was attempting to accomplish a "shotgun" wedding of the Free French leader from London, where anti-Vichy feeling was high, and the French high commissioner under General Eisenhower from Algiers, where former Vichy administrators and officers still predominated. Though Roosevelt claimed it to be for unity of the French cause, the truth was, the President felt he must give critics of his use of former Vichy personnel in North Africa a sign — a symbol not just of reconciliation but proof that though the United States had acted out of expedience, it was fully resolved to defeat fascism in all its forms.

De Gaulle, unhappily, was loath to oblige — raising serious questions about what kind of "liberation" the Americans were intending to bring to Europe. "It had been my hope that we could avoid political discussions at this time," the President cabled to Hull, in part to explain why he hadn't thought to bring the secretary of state to Casablanca, "but I found on arrival that American and British newspapers had made such a mountain out of a rather small hill that I should not return to Washington without having achieved settlement of this matter."[1]

Knocking de Gaulle's and Giraud's heads together, he imagined with presidential hubris, he would show the free world there was a good, just, fair, and effective alternative to Nazi rule, illustrated by men of goodwill coming together to make democracy work once again, as the Nazis were forced to retreat.

Sitting on the large sofa in the villa's drawing room, the President thus bade de Gaulle sit beside him, and attempted, in his best conversational French, to apply salve to the major general's wounded pride as a Frenchman summoned to appear before an American on what de Gaulle had always thought of as French soil: the President beginning by explaining how he'd come to Casablanca, as U.S. commander in chief, to discuss military operations against the Axis powers in the Mediterranean for the coming year. Mr. Stalin had been invited, but had been unable to leave the Stalingrad front. The purpose of the Casablanca meeting was, therefore, to "get on with the war," and answer the question "Where do we go from here?"

In this context, the President elaborated, he appreciated there were different political views on how North Africa, once liberated from the Nazi yoke, should fare, but the war was not yet won; the "problem of North Africa should be regarded," therefore, "as a military one and that the political situation should be entirely incident to the military situation." How to bring "as much pressure as possible to bear on the enemy at the earliest possible moment" in Tunisia was the order of the day, he claimed;[2] Admiral Darlan, for all his faults, had done his best to make this happen, and General Giraud, his successor, was doing the same. Surely, by moving his London Free French committee to Algiers and fusing it with Giraud's organization, the war could be won more swiftly than if the French war effort were to be hobbled, right at the beginning, by political dissension?

De Gaulle, however, seemed to be a man from a different planet. That American forces had come thousands of miles, and suffered a thousand deaths at the hands of French troops while attempting to roll back the Axis tide and evict the Germans in North Africa as the first step toward the defeat of Hitler was — at least at that moment — a matter of complete indifference to the French general. He'd hoped, rather, for an invitation to come to Washington to meet with the President as the leader of the Free French movement, and for security reasons (Free French headquarters was reputed to leak like a proverbial sieve) had not been told of the Casablanca Conference — just as he had not been told beforehand of the Torch invasion. Feeling insulted, he'd therefore resisted Churchill's invitation to fly out to Casablanca, not only out of pique, but because he foresaw matters of political importance being decided and would have no time to prepare for such discussions, he claimed. Forced nevertheless to present himself, on pain of the Free French movement being stripped of all funds and support in London, he'd reluctantly agreed to travel — promising nothing, however. His arrival at Medouina airport had then given him an indication how low he was on the American totem pole: no band playing

"La Marseillaise"; the windows of the car taking him to Anfa soaped lest he be recognized; American troops and sentries everywhere—and in a country he considered a part of France, not a protectorate.

Interrupting Roosevelt, de Gaulle "made some remark to the President with reference to the sovereignty of French Morocco," Captain McCrea wrote in his notes of the meeting that night—having been asked to stand outside while the President and the general talked. It was, he added, "a relatively poor point of vantage—a crack in a door slightly ajar," and with the Frenchman's voice so low "as to be inaudible to me."[3]

Moroccan sovereignty was not what the President was prepared to discuss with the somewhat mad major general from London, however—especially after spending the evening with Morocco's rightful ruler. Morocco had become a French protectorate only in 1912, barely thirty years ago; it could not by any stretch of the imagination be considered "French" soil, in the President's eyes, and de Gaulle's assumption that the country was to be reestablished as part of the "French Empire," thanks to American blood and courage, aroused Roosevelt's deepest anticolonial feelings.

Reestablishing imperial French sovereignty over colonized peoples promised a hiding to nothing, whereas the opportunity to get "advanced" Western nations to embrace the notion of responsible development in former colonies, encouraging global trade and education, would offer, he felt, mutual benefits. Above all, it would give moral *purpose* to the postwar democracies, especially if the struggle between capitalism and communism worsened. The President therefore dismissed de Gaulle's remarks over French sovereignty over Morocco, "stating that the sovereignty of the occupied territories"—territories occupied now by U.S. forces of liberation—"was not under consideration." Moreover, he stated, it would be up to the occupied countries—like mainland France, once liberated—to elect their own postwar governments to help decide such matters, not jump the gun and be saddled with decisions made by warring factions in exile; in fact, "none of the contenders for power in North Africa had the right to say that he, and only he, represented the sovereignty of France," Roosevelt claimed—neither Giraud nor de Gaulle. "The President pointed out," McCrea recorded, "that the sovereignty of France, as in our country, rested with the people, but that unfortunately the people of France were not now in a position to exercise that sovereignty. It was, therefore, necessary for the military commander in the area [General Eisenhower] to accept the political situation as he found it and to collaborate with those in authority in the country at the time that the occupation took place so long as those in authority chose to be of assistance to the military commander. The President stated that any other course of action would have been indefensible."[4]

Nor did Roosevelt stop there. It was not, he said, simply a matter of temporary accommodation and practicality. With the whole of mainland France now under German occupation[5] and no legitimate or elected French government in exile, it was the task of the Allies — the United Nations — "to resort to the legal analogy of 'trusteeship,'" not committees of self-appointed exiles. It was the President's view "that the Allied Nations fighting in French territory at the moment were fighting for the liberation of France and that they should hold the political situation in 'trusteeship' for the French people. In other words, the President stated that France is in the position of a little child unable to look out and fend for itself and that in such a case, a court would appoint a trustee to do the necessary." He pointed out that General Giraud understood this very well, and wanted only "to get on with the war" — namely, the "urgent task of freeing French territory of the enemy." Only then could questions of sovereignty, empire, and the like be addressed. "The President stated that following the Civil War in our home country, there was conflict of political thought and that while many mistakes were made, nevertheless, the people realized that personal pride and personal prejudices must often be subordinated for the good of the country as a whole, and the contending French leaders could well follow such a program. The only course of action that could save France, said the President, was for all her loyal sons to unite to defeat the enemy, and that when the war was ended, victorious France could once again assert the political sovereignty which was hers over her homeland and her empire. At such time all political considerations would be laid before the sovereign people themselves and that by the use of the democratic processes inherent throughout France and its empire, political differences would be resolved."[6]

De Gaulle looked stunned by such paternalistic American arrogance. He'd endured, he felt, one insult after another that day. "No troops presented honors," he later recalled of his arrival at Medouina, "although American sentries maintained a wide periphery around us." Instead, some American cars had driven up to the plane. "I stepped into the first one," he recorded — as well as his shame when Brigadier General William Wilbur, "before getting in with me, dipped a rag in the mud and smeared all the windows. These precautions were taken in order to conceal the presence of General de Gaulle and his colleagues in Morocco," de Gaulle lamented, using the third person. Once inside the barbed-wire compound, moreover, he'd felt even more insulted. "In short, it was captivity," he remembered feeling—a giraffe incarcerated in an American zoo. "I had no objection to the Anglo-American leaders' imposing it on themselves, but the fact that they were applying it to me, and furthermore on territory under French sovereignty, seemed to me a flagrant insult." Meeting five-star General Giraud, his former commander from 1940, that afternoon, de Gaulle—though a mere major general—blamed Giraud for not feeling similarly aggrieved. "What's this? I ask you for an interview four times over and we have to meet in a barbed-wire encampment among foreign powers? Don't you realize," de Gaulle sneered, "how odious this is from a purely national point of view?"[7]

Giraud didn't. In fact, given that U.S. troops had now liberated Morocco from Nazi control, as laid down under the 1940 armistice agreement, he considered de Gaulle the one who was odious and insulting, especially when de Gaulle had pulled from his pocket a copy of Giraud's letter of loyalty to Marshal Pétain, written the previous spring after his escape from a German prison in Germany and seeking safety from the Nazis in Vichy France.[8] Once Hitler had ordered the occupation of the whole of metropolitan France, Giraud had immediately revoked his letter, and had consented to be brought by Allied submarine to Algeria to take military command of the anti-Axis forces. He'd found it typical of de Gaulle to commence discussions of French unity by producing a copy of such a past document from his pocket; with de Gaulle, you were either subordinate to him or against him. Worst of all, de Gaulle's main opponent seemed neither Hitler nor even Giraud, but the U.S. president.

"Franklin Roosevelt was governed by the loftiest ambitions," de Gaulle allowed later—but not the sort of ambitions of which de Gaulle approved. "His intelligence, his knowledge and his audacity gave him the ability, the powerful state of which he was the leader afforded him the means, and the war offered him the occasion to realize them. If the great nation he directed had long been inclined to isolate itself from distant enterprises and to mistrust a Europe ceaselessly lacerated by wars and revolutions, a kind of messianic impulse now swelled the American spirit and oriented it toward vast undertakings," the major general described in his haughty prose—undertakings, at any rate, that were antithetical to de Gaulle and to the reconstitution of the French Empire under him. Once America had "yielded" to "that taste for intervention in which the instinct for domination cloaked itself," he recorded — ignoring France's capitulation to Hitler, Japan's attack at Pearl Harbor, and Hitler's declaration of war on the United States — "from the moment America entered the war, Roosevelt meant the peace to be an American peace, convinced that he must be the one to dictate its structure, that the states that had been overrun should be subject to his judgment, and that France in particular should recognize him as its savior and arbiter."[9]

This was not far from the truth. The fact that Americans, not Frenchmen, were being asked by their president and commander in chief to die, if necessary, to liberate de Gaulle's country — a country that had put up the most feeble fight against the Germans in 1940, and had submitted to an abject armistice with almost no protest ever since, indeed had attempted to prevent U.S. forces from landing in Morocco and Algeria while not lifting a finger to stop the Germans from occupying Tunisia — was of zero interest to de Gaulle, who deprecated Roosevelt as "a star actor" unwilling to share the limelight de Gaulle craved. "In short, beneath his patrician mask of courtesy," de Gaulle wrote, "Roosevelt regarded me without benevolence."[10]

At 10:55 p.m. the interview came to an end. The "Frenchman unfolded his complete height," Elliott recalled, "and marched with formality and no backward glance to the door."[11]

The President was as put out as was de Gaulle. A seminal political encounter of the war had taken place, pitting American progressive political ideas against recalcitrant French imperialist ideology. There had certainly been no meeting of minds. What it showed was that the President's views on postwar world democracy, as enshrined in the Atlantic Charter, were going to be very, very difficult to apply.

Churchill then came back to the Villa Dar es Saada and, together with Harry Hopkins, Robert Murphy, and Harold Macmillan, they rehashed the evening's discussions and their implications.

Outwardly, the President seemed unconcerned. "Father seemed unperturbed by the mighty sulk to which de Gaulle treated him," Elliott recalled, as well as his father's philosophical attitude. "The past is past, and it's done," the President pronounced, attempting to be positive. "We've nearly solved this thing now. These two:" — meaning Generals Giraud and de Gaulle — "equal rank, equal responsibility in setting up the Provisional Assembly. When that's done, French democracy is reborn. When that Provisional Assembly starts to act, French democracy takes its first steps. Presently French democracy will be in a position to decide for itself what is to become of Giraud, or of de Gaulle. It will no longer be our affair."[12] A democratically elected French government would decide.

In his own mind, however, Roosevelt was far from happy. He was already worried whether his notion of global postwar democracy, based on the Atlantic Charter, would be honored in Europe, once the Russians began pushing back the Germans on the Eastern Front.

The Soviets were a major concern. His brief interview with de Gaulle had indicated all too clearly, though, just how obstinately the old imperial powers would seek to reestablish and then hang on to their colonial possessions — the "unity of her vast Empire," as de Gaulle proudly called it[13] — rather than pursue the ideal of postwar, postimperial commonwealths of sovereign countries bound by history and culture, not the gunboat. And in this respect, Churchill was little different from de Gaulle.

How, though, persuade those dying empires to embrace the *future* rather than the past? How encourage them to join in creating a new world order, not reestablish the tottering colonial empires that had doomed Europe and the Far East after World War I?

When Churchill finally left the House of Happiness at half past midnight, the President went to bed but asked Elliott to sit with him, and in the quiet of his Casablanca villa, he unburdened his soul.

Though he'd said to Churchill they must move on with the prosecution of the war and not permit themselves to be sidetracked by French factionalism, the President was in truth deeply affected by his contretemps with de Gaulle.

"We've talked, the last few days," the President told his son, "about gradually turning the civil control of France over to a joint Giraud–de Gaulle government, to administer as it is liberated. An interim control, to last only until free elections can again be held . . . but how de Gaulle will fight it!" he snorted. Not only did de Gaulle speak of himself as a sort of Joan of Arc, but his dream was the restoration of France on the back of its colonies. "He made it quite clear that he expects the Allies to return all French colonies to French control immediately upon their liberation. You know," Roosevelt confided to his son, "quite apart from the fact that the Allies will have to maintain military control of French colonies here in North Africa for months, maybe years, I'm by no means sure in my own mind that we'd be right to return France to her colonies at all, ever, without first obtaining in the case of each individual colony some sort of pledge, some sort of statement of just exactly what was planned, in terms of each colony's administration"[14] — much as Congress had done with regard to the Philippines in 1932.

Elliott was amazed. "Hey, listen, Pop. I don't quite see this. I know the colonies are important — but after all, they *do* belong to France . . . how come we can talk about not returning them?"[15]

Roosevelt's retort was instant. "*How* do they belong to France?" he countered. "Why does Morocco, inhabited by Moroccans, belong to France? Or take Indo-China. The Japanese control that colony now. Why was it a cinch for the Japanese to conquer that land? The native Indo-Chinese have been so flagrantly downtrodden that they thought to themselves: Anything must be better, than to live under French colonial rule! Should a land belong to France?" he demanded. "By what logic and by what custom and by what historical rule?"[16]

"I'm talking about another war, Elliott," the President told his son, "his voice suddenly sharp," Elliott recalled. "I'm talking about what will happen to our world, if after *this* war we allow millions of people to slide back into the same semi-slavery."

He looked deadly serious. "Don't think for a moment, Elliott, that Americans would be dying in the Pacific tonight, if it hadn't been for the shortsighted greed of the French and the British and the Dutch. Shall we allow them to do it all, all over again?"[17]

It would be hard enough to revive the battered economies of the world and guard against the insidious, antidemocratic ideology of communism, but how much harder it promised to be if Britain, France, and the Netherlands committed themselves to huge military and financial outlays to perpetuate imperialism. They would then be, he predicted, sucked into vain efforts to stop calls for self-government and self-determination in their former colonies — a recipe for postwar disaffection, revolt, and wars.

"One sentence, Elliott. Then I'm going to kick you out of here. I'm tired. This is the sentence: When we've won the war, I will work with all my might and main to see to it that the United States is not wheedled into the position of accepting any plan that will further France's imperialist ambitions, or that will aid or abet the British Empire in *its* imperial ambitions."[18]

And with that the President pointed to the door — and the light switch.

• • •

Before he returned to his photoreconnaissance unit in Algiers, Elliott Roosevelt had one more talk with his father. It was clear de Gaulle's determination to reassert French imperialism still enervated the President. De Gaulle had at least been open about his aims, however — to the point of outright rudeness. Churchill was, by contrast, keeping his own counsel for later. The President therefore interrogated Elliott as to opinion among U.S. troops and airmen — what did they, who were risking their lives, think?

Before Elliott could respond, his father launched into another deeply felt articulation of his views. "You see, what the British have done, down the centuries, historically, is the same thing. They've chosen their allies wisely and well. They've always been able to come out on top, with the same reactionary grip on the peoples of the world and the markets of the world, through every war they've ever been in.

"*This* time," his father continued, "*we're* Britain's ally. And it's right we should be. But . . . first at Argentia, later in Washington, now here at Casablanca," the President reminded Elliott, "I've tried to make it clear to Winston — and the others — that while we're their allies, and in it to victory by their side, they must never get the idea that we're in it just to help them hang on to the archaic, medieval Empire ideas."[19]

Elliott agreed, but his father wasn't done. "I hope they realize they're not senior partner." America was — and would be more and more so, as the war progressed and the postwar world took shape. The United States was "not going to sit by, after we've won, and watch their system stultify the growth of every country in Asia and half the countries in Europe to boot," he warned. Britain had "signed the Atlantic Charter" at Argentia, and "I hope they realize the United States government means to make them live up to it."[20]

These were perhaps the most impassioned words Elliott had ever heard his father say — spoken after inspecting thirty thousand young Americans preparing for imminent combat, and having visited the cemetery of those who had already fallen. They also helped explain his father's determination to insist upon unconditional surrender of the Axis powers, precluding any possibility of negotiated armistice with nations simply too dangerous to be allowed ever to rearm.

Operation Symbol had been the code name given to the Casablanca Conference. The biggest symbol of Roosevelt's intent to end German, Italian, and Japanese military empires and establish a completely new, postimperialist global order would be, the President had planned, his forthcoming announcement to the world, on the field of battle, of his implacable condition for ending the war.

Hour after hour the President had hoped that de Gaulle would make at least a tentative agreement to work with General Giraud — one that could be announced at the President's looming press conference.

De Gaulle refused, however, to make any accommodation with his French rival. In particular he turned down a draft communiqué drawn up by Robert Murphy and Harold Macmillan, Eisenhower's British political adviser — declaring he would not be party to any solution to French political matters "brought about by the intervention of a foreign power, no matter how high and how friendly."[21]

The President, with a kind of bemused amazement, breathed another sigh of vexation. "Finished the staff conferences — all agreed — De Gaulle a headache — said yesterday he was Jeanne d'Arc & today that he is Georges Clemenceau," he scribbled to Daisy — for de Gaulle now insisted on a compact with Giraud in which he, Major General de Gaulle, would be the French political leader in exile, while Giraud would be merely the French military commander in chief — whom de Gaulle could dismiss. Giraud, who had come to hate de Gaulle with Gallic venom over the past few days, refused. There would thus be no unification of the Free French and Algiers committees — and with that, de Gaulle prepared to leave Casablanca.

The President was disappointed, but tried not to be unduly concerned.

De Gaulle, for his part, believed he'd made his point: proving to the President of the United States and the world that he, on behalf of *La France,* was not going to toady to American wishes, or dollars. He certainly seemed to have no idea how rude he'd been, or how small he appeared, in the President's eyes, despite his six-foot-six-inch height. History had given him the chance to lead a great reconciliation of neutral, Vichy, and Free French nationals in the struggle to defeat the Axis powers and to usher in a new world. Instead, he'd pursued the politics of personal ambition and an implacable view of French honor. However laudable the latter, it was sheer obstructionism in terms of the war against the Third Reich — something de Gaulle seemed unable to comprehend. To his aide, Hettier de Boislambert, he confided, the night he'd met the President at the Villa Dar es Saada: "You see, I have met a great statesman today, I think we got along and understood each other well" — but the truth was the very opposite.

For his part, Churchill was dumbfounded. Learning that de Gaulle was refusing to sign the proposed communiqué, prior to the President's press conference, "he was beside himself with rage," the historian of the Churchill–de Gaulle relationship later chronicled. "General de Gaulle's farewell visit to the Prime Minister was therefore uncommonly animated, even by Churchillian standards; the latter chose to omit any reference to it in his memoirs. Not so General de Gaulle," François Kersaudy chronicled.

Kersaudy was not exaggerating. De Gaulle's account, recording how Churchill had threatened to "denounce" him "in the Commons and on the radio" unless he signed the communiqué, pulled no punches. The Prime Minister was "free to dishonor himself," de Gaulle had retorted. "In order to satisfy America at any cost, he was espousing a cause unacceptable to France, disquieting to Europe, and regrettable to England."[22]

Churchill was apoplectic, but in one sense de Gaulle was right. The President was a true statesman, and even if he disliked de Gaulle for making difficulties, he understood him, for all his foibles, as a statesman in the making. Thus at the Villa Dar es Saada shortly before noon on January 24, Roosevelt accepted that de Gaulle would simply not sign an interim communiqué or agreement of a three-man Committee for the Liberation of France — and did not turn away from Charles d'Arc, so to speak. Instead, in his inimitable fashion, the President asked for at least a *symbol* of French purpose in fighting the Nazis. "In human affairs the public must be offered some drama," Roosevelt said to the general. "The news of your meeting with General Giraud in the midst of a conference in which both Churchill and I are taking part, if it were to be accompanied by a joint declaration of the French leaders — even only a theoretical agreement — would produce the dramatic effect required."[23]

The President's almost Olympian approach and charm moved de Gaulle, as Churchill's did not. "'Let me handle it,'" de Gaulle later recalled his response. "'There will be a communiqué, even though it cannot be yours.' Thereupon I presented my [French] colleagues to the President and he introduced me to his."[24]

The press conference was due to take place at midday, but Harry Hopkins, mistaking de Gaulle's sudden graciousness, rushed out and grabbed General Giraud, asking him and Churchill to enter, in the hope that, if he could get "the four of them into a room together," then "we could get an agreement."[25]

This was silly, in view of de Gaulle's stalwart refusal to allow a "foreign power" to dictate French agreements. Moreover, Churchill's renewed "diatribe and his threats against me, with the obvious intention of flattering Roosevelt's disappointed vanity," as de Gaulle put it in his memoirs, only made matters worse. But Roosevelt would not have been Roosevelt, the leader of the United Nations and a man of almost heartbreaking humanity, if he had not attempted a different approach. He therefore made one last request of de Gaulle, "on which he had set his heart," as de Gaulle recalled.

"Would you agree to [at least] being photographed beside me and the British Prime Minister, along with General Giraud?" he asked, in "the kindest manner."

"By all means," de Gaulle responded, "for I have the highest regard for this great soldier."

"Would you go so far as to shake General Giraud's hand in our presence and in front of the camera?"

"My answer, in English, was, 'I shall do that for you.' Whereupon Mr. Roosevelt, delighted, had himself carried into the garden where four chairs had been prepared beforehand, with innumerable cameras trained on them and several rows of reporters lined up with their pens poised."[26]

16

The Unconditional Surrender Meeting

"CASABLANCA, FRENCH MOROCCO," the AP reporter (once his report was cleared for release) described, was "probably the most important gathering of leaders of two great nations in history." It was also, the reporter maintained, even more extraordinary for the setting — the results "disclosed in the most informal press conference ever held."[1]

The picture, in the midst of a global war, was certainly unique — the scene a "garden of a villa on the outskirts of Casablanca," where "the entire area for blocks around was full of troops, anti-aircraft equipment and barbed wire. The correspondents were told they would have a conference at noon. They assembled in the rear garden of the villa, which is a luxurious gleaming white home with many windows overlooking the Atlantic. In the garden were two white leather chairs. A microphone was in front of them for newsreel camera men. Red flowers were in profusion. Inside, reporters could see Harry A. Hopkins and his son, who is now a corporal, rushing around making arrangements. Then Lieut. Colonel Elliott Roosevelt appeared at the rear door carrying two more chairs. The President appeared. He wore a gray business suit and a black tie, and, as usual was smoking a cigarette in a long holder. A minute later Prime Minister Churchill walked out with a cigar in his mouth."[2]

The two giant, giraffe-like French generals in their kepis were also brought out. "Some photographers called out, 'Generals shake hands!'" Captain McCrea recounted — and, as agreed with de Gaulle, when the President said, "Why not? You two Frenchmen are loyal to your country and that warrants a handshake anytime," they did so — not only once but twice, since cameramen complained they'd failed, in their surprise, to get a good photo the first time.

"The four actors put on their smiles," de Gaulle later recorded — in a chapter of his memoirs that he titled "Comedy." "The agreed-upon gestures were made. Everything went off perfectly! America would be satisfied, on such evidence, that the French question had found its *deus ex machina* in the person of the President."[3]

Giraud was as sniffy of the proceedings as de Gaulle, but for the opposite reason: to wit, his profound hostility to de Gaulle, a mere major general who had only minimal support from Frenchmen in Northwest Africa, appearing on the same stage — especially after having been so rude to him ever since he'd arrived. The very suggestion the two Frenchmen might work amicably together seemed to him dishonest, however noble its intention. "Excellent photos that will be transmitted across the world, and be seen as documentary evidence of irrefutable veracity," he recalled sarcastically several years later. "That," he added, "is how public opinion is fashioned."[4]

It was — unabashedly, since Allied unity was as important a weapon in the war as military arms. Even the President's naval aide was amazed. "The pictures went all over the world and I would suppose contributed to French unity in all parts of the globe," McCrea reflected, for even he had not foreseen the power of such simple imagery. "The President literally cajoled the two proud and greatly different persons into making a gesture of friendship — and did it well, indeed. The generals bade farewell to the President and the Prime Minister and then withdrew — forthwith," leaving the President to explain to reporters from across the world the purpose of the summit that had just concluded.

In his business suit and tie, sitting with his long legs crossed, the President "invited the assembled newsmen to seat themselves on the lawn and make themselves comfortable for the discussion which was to follow," the AP reporter described. "It was a beautiful day — brilliant sunshine and with these two heads of state the correspondents heard a complete description of the purpose and the reasons of bringing the British and our own Chiefs of Staff together in North Africa for discussions necessary for further prosecution of the war."[5]

The President certainly looked the picture of confidence and good health. Referring to Torch, he began by reminding his audience how the current campaign in North Africa had begun. "This meeting," he explained, "goes back to the successful landing operations last November, which as you all know were initiated as far back as a year ago, and put into definite shape shortly after the Prime Minister's visit to Washington in June.

"After the operations of last November," the President went on, "it became perfectly clear, with the successes, that the time had come for another review of the situation, and a planning for the next steps, especially steps to be taken in 1943." It was for this reason he'd arranged for Churchill to come to Casablanca, "and our respective staffs came with us, to discuss the practical steps to be taken by the United Nations for prosecution of the war. We have been here about a week."[6]

For the journalists who had been kept in the dark since the President's State of the Union address on January 7 in Washington, D.C., two weeks before, this was something of a bombshell. The very fact that the two leaders of the Western democratic alliance could have spent *an entire week* on the recent field of battle without anyone knowing was a shock—the more so as no American president had ever previously traveled abroad in wartime, or even flown in an airplane while in office. Yet here he was, in bright Moroccan sunlight, addressing them—largely extempore and in person.

> I might add, too, that we began talking about this after the first of December [1942], and at that time we invited Mr. Stalin to join us at a convenient meeting place. Mr. Stalin very greatly desired to come, but he was precluded from leaving Russia because he was conducting the new Russian offensive against the Germans along the whole line. We must remember that he is Commander in Chief [of the Soviet armies], and that he is responsible for the very wonderful detailed plan which has been brought to such a successful conclusion since the beginning of the offensive.

Knowing the Russians had cornered the German Sixth Army at Stalingrad, the President had felt certain the surviving Germans would now be killed or forced to surrender—whatever Hitler might order to the contrary. It was a tremendous Soviet victory in the making, after months of the most lethal, often hand-to-hand, combat of the war, involving vast casualties. Soon the Western Allies would be achieving a similar, momentous victory, however, the President implied. "In spite of the fact that Mr. Stalin was unable to come, the results of the staff meeting have been communicated to him, so that we will continue to keep in very close touch," Roosevelt assured the reporters. Meantime, with regard to the many meetings and discussions between the U.S., British, and French generals, the President expressed his great satisfaction as U.S. commander in chief. What had taken place was different, he said, from, say, Lincoln's visits to his generals in the field, or those of Allied leaders in World War I. This was now *coalition* warfare, on a global scale, but with the leaders and their military staffs working in the closest cooperation and harmony:

> I think it can be said that the studies during the past week or ten days are unprecedented in history. Both the Prime Minister and I think back to the days of the first World War when conferences between the French and British and ourselves very rarely lasted more than a few hours or a couple of days. The [U.S. and British] Chiefs of Staffs have been in intimate touch; they have lived in the same hotel. Each man has become a definite personal friend of his opposite number on the other side.
>
> Furthermore, these conferences have discussed, I think for the first time in history, the whole global picture. It isn't just one front, just one ocean, or one continent — it is literally the whole world; and that is why the Prime Minister and I feel that the conference is unique in the fact that it has this global aspect.
>
> The Combined Staffs, in these conferences and studies during the past week or ten days, have proceeded on the principle of pooling all of the resources of the United Nations. And I think the second point is that they have reaffirmed the determination to maintain the initiative against the Axis powers in every part of the world.

Over the past ten days, the President explained, the talks had examined how the Western Allies were to keep "the initiative during 1943," moreover to keep sending "all possible material aid to the Russian offensive, with the double object of cutting down the manpower of Germany and her satellites, and continuing the very great attrition of German munitions and materials of all kinds which are being destroyed every day in such large quantities by the Russian armies. And, at the same time, the Staffs have agreed on giving all possible aid to the heroic struggle of China — remembering that China is in her sixth year of the war — with the objective, not only in China but in the whole of the Pacific area, of ending any Japanese attempt in the future to dominate the Far East."

It was at this point that the President, looking down at his notes, came to the crux of his outdoor statement — its historic import belied by the lush surroundings. "Another point," he began:

> I think we have all had it in our hearts and our heads before, but I don't think that it has ever been put down on paper by the Prime Minister and

myself, and that is the determination that peace can come to the world only by the total elimination of German and Japanese war power.

Some of you Britishers know the old story—we had a General called U.S. Grant. His name was Ulysses Simpson Grant, but in my, and the Prime Minister's, early days he was called "Unconditional Surrender" Grant.

The elimination of German, Japanese, and Italian war power means *the unconditional surrender* by Germany, Italy, and Japan. That means a reasonable assurance of future world peace. It does not mean the destruction of the population of Germany, Italy, or Japan, but it does mean the destruction of the philosophies in those countries which are based on conquest and the subjugation of other people.

In order to give extra emphasis to the announcement, the President now declared: "This meeting is called the 'unconditional surrender meeting.'"

Unconditional surrender. No negotiation or acceptance of a compromise peace or armistice. And an implacable aim that would be pursued West and East.

While we have not had a meeting of all of the United Nations, I think that there is no question—in fact we both have great confidence that the same purposes and objectives are in the minds of all of the other United Nations—Russia, China, and all the others.

And so the actual meeting—the main work of the Conference—has been ended. Except for a certain amount of resultant paper work, it has come to a successful conclusion. I call it a meeting of the minds in regard to all military operations, and, thereafter, that the war is going to proceed against the Axis powers according to schedule, with every indication that 1943 is going to be an even better year for the United Nations than 1942.[7]

The fifty journalists in the garden of the Villa Dar es Saada were stunned. So, too, was Churchill.

True, the Prime Minister had agreed to the unconditional-surrender policy and even recommended it be part of the President's final pronunciamento, at the conclusion of the conference. Yet he seemed visibly surprised at the emphasis the President had placed upon it, as Captain McCrea vividly recalled. "I was standing nearby and when the President made that remark the P.M. snapped his head toward the Pres., giving the impression, to me at least, that the phrase came as a surprise to him."[8]

Pondering this in later years, McCrea could not quite explain the Prime Minister's body language—"I shall never forget," he wrote, "the quick turn of the head by the P.M. when the Unconditional Surrender of the Axis Forces was announced as to how the war would end."[9]

The fifty journalists, for their part, sat mesmerized. If they found themselves disappointed that the President was not willing to be more specific in terms of actual, forthcoming military operations, the Prime Minister followed up the President's statement by asking them to understand why the enemy should not be told in advance what the Allies would undertake that year—and why the Allies could be grateful for what had already happened, now that the United States was in command. "Tremendous events have happened. This enterprise which the President has organized—and he knows I have been his active Lieutenant since the start—has altered the whole strategic aspect of the war ... We are in full battle, and heavy action will impend." He asked reporters therefore to convey to the world at home "the picture of unity, of thoroughness, and integrity of the political chiefs." The Allies were going to win the war. "Even when there is some delay there is design and purposes," he insisted, "and as the President has said, the unconquerable will to pursue this quality," he sought to find a quotable phrase, "until we have procured the unconditional surrender of the criminal forces who plunged the world into storm and ruin."[10]

Unconditional surrender, then, it was—the news soon flashing across the world, once the two leaders were out of harm's way.

Reports and images of the "unconditional surrender meeting" and the President's trip sent shockwaves across the Third Reich.

The President and Commander in Chief of the Armed Forces of the United States: Inspecting his troops on the battlefield. Ten days of U.S.-British military discussions—and with the French, too. Every battlefront of the globe examined, and its needs factored into the Allies' strategy for the prosecution of a global, offensive war—a war not only to win against the Axis powers, but to permit no compromise, no negotiated armistice, no agreement save unconditional surrender. And the President seated in the sun on a Moroccan lawn, speaking with such naturalness and confidence regarding the inevitable defeat of the Third Reich that those who'd experienced the German victories of the previous summer—the fall of

Tobruk, the second massive German offensive toward the Volga and the Caucasus — could only rub their eyes in wonder. "F.D.R.'s 'unconditional surrender' pronouncement" had swept "practically all other news from today's newspapers . . . It will, no doubt, prove to be," predicted King George VI's private secretary, "one of the most momentous of all such conferences since that of Lucca" — when in 56 B.C. Caesar, Pompey, and Crassus had renewed their triumvirate.[11]

In the wake of Torch the tide had truly turned. In Berlin, Reichsminister Goebbels — who had been busy preparing the final touches for a forthcoming address of his own — was literally speechless.

At first Goebbels could scarcely believe what he read and saw in newsreel film being distributed throughout the neutral countries. In his diary Goebbels expressed utter consternation — especially at the failure of the German intelligence services to learn the whereabouts of a ten-day, top-level enemy conference involving the political leaders of the Western world, together with the chiefs of staff of their air, ground, and naval forces. Even on January 26, 1943, two days after the conclusion of the actual press conference and departure of the principals, Goebbels had been idly noting — alongside secret reports that the terrible battle of attrition at Stalingrad was "reaching its end"[12] — that it seemed "pretty certain that Churchill is in Washington."[13]

The next day Goebbels noted that the rumors of a parley between Roosevelt and Churchill were gaining strength, "only we still don't know where these gangster bosses are meeting."[14]

Goebbels, ever skeptical, had made nothing of the speculation. His own attention was locked on the approaching tenth anniversary of the Nazis' assumption of power, when he would make his own grand announcement at a huge, mass rally of Nazi Party stalwarts in the Sportpalast — urging them with all the declamatory zeal he could summon to devote themselves to their fresh task: to make available to the Führer the men, materiel, and conviction necessary for Germany to embark on a third, this time successful, great offensive on the Eastern Front . . . *totaler Krieg*: total war.

Goebbels was thus floored by the seemingly authentic reports that finally reached Berlin on January 28, 1943. "The sensational event of the day, is the news that Churchill and Roosevelt have met in Casablanca," the Reichsminister dictated in his diary. He made no effort to conceal his amazement. "So the discussions have not, as we assumed, been taking place in Washington but on the hot coals of Africa. Once again our intelligence services have completely failed — unable even to identify the place where the talks were taking place," he fulminated. "They've been held now for almost a fortnight, and they're being heralded by the enemy press as the gateway to victory."[15]

Ever anxious to see signs of Allied dissension, Goebbels had assumed Churchill and Roosevelt, if they were meeting in Washington, might well be sparring over which man should take the reins of the Allied offensive war effort.[16] Reading the transcript of the Roosevelt-Churchill press conference in Casablanca, the Reichsminister became aware, however, that the earth had shifted. "It's worth noting," he reflected in his diary, "that Churchill officially designates himself now as Roosevelt's adjutant; no such humiliation has probably been seen in British history."[17]

Humiliation or not, the threat was becoming daily more real. Not only were the anti-Axis armies targeting Nazi Germany, Goebbels was aware, but so were their political leaders, Roosevelt, Churchill, and Stalin — leaders who, like Hitler, had taken command of their country's armed forces, and were now coordinating those forces against the Third Reich — in complete contrast to the motley democratic forces of the late 1930s.

Barely a year since Pearl Harbor and the Führer's ill-considered declaration of war on the United States, the President's appearance in Casablanca was a startling turnaround — his "unconditional surrender meeting" all the more disturbing to Goebbels, since it made clear there would be no peace feelers or possibility of a negotiated settlement with the leaders of the Third Reich. Along with the imminent extinction of von Paulus's Sixth Army at Stalingrad, Goebbels knew, Hitler's dreams of conquest and declaration of war on the United States now looked not only an unwise gamble, but raised the specter of the Thousand Year Reich — so gloriously proclaimed in the 1930s — being crushed in the nearest future, unless the Nazi Party, under their once victorious Führer, found some way to turn the tables.

Since the Führer still refused to appear in public at such a fateful time — he had not been seen in Berlin since the previous September[18] — his propaganda minister recognized that he, Joseph Goebbels, would have to work the harder to rally the German nation at home.

Accepting that the battle of Stalingrad would now end in utter defeat — the Führer confiding to Goebbels he'd had to sacrifice General Paulus's army lest the whole extended German frontline in Russia be broken — the Reich minister had intended to use the approaching German catastrophe in Russia to new advantage: namely as a wake-up call to the German Volk, once the battle of Stalingrad ended. The fate of so many hundreds of thousands of German soldiers would illustrate, as nothing else could, the mortal threat of Bolshevism — and the need for supreme, self-sacrificing heroism on the part of the loyal German soldier if they were to survive the struggle against Soviet communism.

The news from the Western (in fact, Southern) Front, however, eclipsed even Stalingrad. Coming after what the President had revealed in his State of the Union address on January 7, that the United States was on course to outmanufacture the collective output of the Axis powers several times over, the Casablanca declaration by the leaders of the world's foremost capitalist democracies — democracies working with the Soviet Union — now deprived Goebbels of his "anti-Bolshevist" German master card.

Every day there was more news in neutral countries about Casablanca. Film, photographs, newspaper stories — and discussion of what the conference would now presage for Nazi Germany. It completely turned the world's attention to the *Western,* not Eastern, Front, Goebbels lamented — removing the primary fear of Bolshevism. Overnight, in short, the communist threat had been replaced, thanks to Casablanca, by a dramatically announced determination of the Western democratic powers to destroy all vestige of Nazism as a danger to, and scourge of, mankind — far worse, in effect, than the dangers of communism.

The very words *unconditional surrender* — following the President's use of *total war* in his State of the Union address to Congress — infuriated Goebbels as a master of propaganda. Curt and harsh, they gave no hint of dissension or disunity among the United Nations now lined up against the Third Reich — unsettling Goebbels's ever-maneuvering assumptions, since it showed just "how confident the enemy now feels, or claims to feel, and how much we'll have to do," he noted, "to counter their machinations."[19]

For the second time that month, then, the President of the United States had beaten Goebbels to the punch. Instead of the Reich minister's still-undelivered declaration of total war surprising the world and striking fear in the hearts of Germany's enemies, his *totaler Krieg* speech, if Hitler authorized it, would now be viewed outside Nazi circles as a desperate effort, at best, of an unashamedly totalitarian regime to meet the prospect of defeat; at worst a sort of glorified willingness to countenance the complete destruction of the German nation rather than sparing it by surrender.

To make matters worse for Dr. Goebbels, however, the Führer had declined to make him the sole director of the *totaler Krieg* initiative, lest the Reich minister (and gauleiter of Berlin) become too powerful in Germany. Instead, Hitler had agreed only to a triumvirate of mediocrities to steer the extended mobilization program, enjoying circumscribed powers — with Goebbels granted a "watching brief." Isolated, ill, frustrated, depressed, and blaming others rather than himself for the Wehrmacht's failure on the Eastern Front, the Führer even rejected Goebbels's renewed appeals that Hitler return to the capital and rally the nation at such a time of crisis both on the Eastern and Southern Fronts.

For Goebbels as Reichsminister für Propaganda, this made Roosevelt's dramatic appearance in Morocco especially galling: the U.S. president seen by photographers, cameramen, and reporters so relaxed in the garden of a sunlit villa in Casablanca, while the Führer remained unseen by anybody: hiding out of sight at his freezing headquarters in East Prussia, moaning helplessly as he surveyed on his tabletop maps the sharp arrows of Russian advances, lancing into his besieged remaining forces at Stalingrad . . .

It was in this context that Goebbels had been heard to say — by Albert Speer, the Reich armaments minister, no less — that Germany did not have a leadership crisis, but a "Leader crisis."[20]

Goebbels was not alone in thinking this — though few if any dared say so aloud. Goebbels was especially disturbed by reports from the *Sicherheitsdienst* concerning new anti-Nazi graffiti appearing on the walls of German cities. Some of these openly accused the Führer of mass murder — not of Jews, but of German soldiers, in forcing the Sixth Army to fight to the death at Stalingrad rather than allowing the men to retreat.[21] There were even rumors circulating that Hitler was either dead or suffering mortal sickness in Prussia.

However hard Dr. Goebbels tried, then, it seemed impossible to "counter" the sensational international effect of the Casablanca Conference. In his diary the minister thus cursed the way he and the Führer had been outmaneuvered.

The very lack of military specifics in the President's Casablanca press conference — or even in the final official conference communiqué issued after *weeks* of military discussions held by the most senior Allied generals and admirals — aroused still further concern in Goebbels's suspicious, ever-calculating, yet in many ways brilliant mind. "They're trying to conceal the real decisions they've made at the conference," he dictated in his diary, "clearly to lull us into complacency. But there's no possible doubt in my mind the Anglo-Saxons are planning to invade the mainland of Europe when it suits them. We'll have to prepare for surprises," he noted on January 28. "From week to week," he added, "the war is moving into a bitter, ruthless stage."[22] And two days later, at the stated request of the absent Führer, Goebbels delivered before an audience of invited Nazis in the Berlin Sportpalast — and on German radio — Hitler's tenth-anniversary proclamation, celebrating the Nazis' seizure of power in 1933.

Compared with the President's Casablanca announcement, the proclamation was a dud.

Without new victories to boast of, indeed with the Russians erasing the last pockets of resistance in Stalingrad, Hitler had been reduced in the proclamation to a vague catalog of Nazi "achievements" over the past decade, as well as an assertion that National Socialism would "inspire everybody to fulfill his duty." If not, the Führer warned, woe betide the slacker. The Nazi Party "will destroy whoever attempts to shirk his duty," he'd written — having agreed with Goebbels on the phone that the most savage measures, including execution, were to be taken against any who dared contest the increased mobilization measures that would now be enacted.[23]

Thanks to the Führer and his accomplices, the war — Hitler's war — would indeed move now "into a bitter, ruthless stage."

Three weeks thereafter Dr. Goebbels would, finally, be permitted by the Führer to deliver, in person, his long-awaited *totaler Krieg* speech at the Berlin Sportpalast.

Goebbels was careful, in the days before, to pass word around that he'd be issuing more than a proclamation. One Goebbels biographer later described it as "the most important mass meeting" of Goebbels's egregious life.[24] Ignoring the President's recent reference to Germany's war of conquest and its subjugation of other peoples, Goebbels intended instead to portray Germany's struggle as a noble European battle, waged by the Third Reich and its allies against "international Jewry," and a fight to vanquish the forces of Jewish-sponsored chaos and aggression.

"Behind the Soviet divisions storming toward us we see the Jewish liquidation commandos, and behind them the specter of terror, mass hunger, and complete anarchy," Goebbels described. The goal of Bolshevism, he declared, "is Jewish world revolution. The Jews want to spread chaos across the Reich and Europe, so that in the resultant despair and hopelessness they can establish their international, Bolshevist, concealed-capitalist tyranny." International Jewry, he sneered, was an "evil fermentation of decomposition"—a threat that "finds its cynical pleasure in plunging the world into chaos, and thereby bringing about the fall of thousand-year-old cultures to which it has contributed nothing."[25]

Considering Jews had made German culture and science world famous, and that the Jewish percentage of Germany's population in 1933 had been less than 1 percent, Goebbels's claims were not only preposterous, but malevolent beyond belief—masking, sadly, the real truth: the SS liquidation teams that Hitler and Heinrich Himmler had unleashed when attacking the Soviet Union, as well as the deliberate extermination of innocent Jewish civilians across Europe. Yet before an audience of fifteen thousand Nazi stalwarts, Goebbels's newsreel cameramen "captured extraordinary scenes of emotion," his biographer would describe. "Within minutes the audience was leaping to its feet, saluting, screaming, and chanting"—their cries of "Führer command! We obey!" foreshadowing the shrill madness of Orwell's *Animal Farm.*

"The orgiastic climax was reached by the question: 'Do you want total war? Do you want war more total, if need be, and more radical than we can even begin to conceive of today?' And then, almost casually, 'Do you agree that anybody who injures our war effort should be put to death?'"

"The bellow of assent each time was deafening," Goebbels's biographer would record[26]—the Reich minister's speech interrupted more than two hundred times by literally hysterical applause. Not least would be the climax, when Goebbels reached his frenzied, rhetorical "masterpiece," modeled on Hitler's earlier "masterpieces."

"Nun, Volk, steh auf, und Sturm, brich los!"—"Now, people of Germany, rise up—and storm: *break loose!*"[27]

PART FIVE

Kasserine

17

Kasserine

SHORTLY AFTER THE President's return to Washington, the last pocket of forty thousand starving soldiers of the German Sixth Army at Stalingrad raised the white flag — knowing their chances of survival as prisoners of war were dim.

The SS and Wehrmacht had ruthlessly conquered, murdered, executed, pillaged, and despoiled too much, too mercilessly, since the launch of Operation Barbarossa to expect much mercy. Of the 113,000 German soldiers taken prisoner in the battle for the Russian city of Stalingrad, few would ever return to their Vaterland.[1] "I'm not cowardly, just sad that I can give no greater proof of my bravery," one soldier had written in his last, despairing letter home, "than to die for such pointlessness, not to say crime."[2]

For the Soviet armies, Stalingrad, not Torch, was the turning point of the war. Russian forces had been fighting the Wehrmacht and its Romanian, Hungarian, Finnish, Italian, Dutch, and other Axis assault forces relentlessly since June 1941. In those seventeen months, the Soviets had taken phenomenal casualties before finally learning how to halt and defeat Germans in battle. Americans had been in battle barely a few weeks.

The campaign in Tunisia against predominantly German forces would now evidence the same learning process, if on a considerably smaller scale.

As Allied units began to meet German rather than Vichy French forces in combat, the situation suddenly resembled chaos in Russia at the start of Barbarossa. Even as Roosevelt presided over the conference at Casablanca, in fact, armored German forces struck at the Allied line in the Eastern Dorsal region of the Atlas Mountains, manned by poorly armed

French troops. Some thirty-five hundred troops immediately surrendered; the rest ran for their lives. "The French began showing signs of complete collapse along the front as early as the seventeenth," Eisenhower jotted in his diary on January 19, 1943. "Each day the tactical situation has gotten worse."[3]

The President, who seldom if ever interfered in tactical dispositions, had urged while at Casablanca that another well-armed U.S. division be sent up the line from northern Morocco, but Marshall and Stimson's obsession with a possible German counterinvasion via Spain and Spanish Morocco had tied Eisenhower's hands. Transportation was a further fetter. "We've had our railroad temporarily interrupted twice," Ike lamented. "I'm getting weary of it, but can't move the troops (even if I had enough) to protect the lines."[4] Wisely waiting for better weather and more troops, he wanted to hold fast until Montgomery's British Eighth Army drew closer from Libya, and a proper, integrated Allied offensive could be readied within the capabilities of largely green troops.

The Germans, however, would not oblige. On January 30, 1943, five days after the President's departure from Morocco, the Twenty-First Panzer Division "struck Faid Pass in a three-pronged attack as precise as a pitchfork," campaign historian Rick Atkinson aptly described—killing almost a thousand French defenders in a day.[5] Then, luring counterattacking U.S. armored forces into a trap, the Germans decimated both U.S. infantry and tanks—leaving Wehrmacht forces, backed by Stuka dive-bombers, in control of Faïd Pass and the Eastern Dorsal.

This, however, was just the beginning. On February 14 the Germans launched a Valentine's Day massacre. Warned that German armor was on the move, Eisenhower wanted to withdraw fifty-nine-year-old Lloyd Fredendall's II Corps to safer positions in the Grand Dorsal, but General Fredendall resisted, and Eisenhower felt too much of a tenderfoot to insist, especially since Fredendall was a protégé of General Marshall's. The result would soon be a bloodbath—this time American.

A German officer "could not help wondering whether the officers directing the American effort knew what they were doing."[6] They didn't. Their forces were dispersed and were mutually unsupporting, as well as lacking effective air cover. They were, in short, completely unprepared for the two German armored contingents about to hit them: General von Arnim's *Frühlingswind* assault through the Faïd Pass to Sidi Bou Zid, and Field Marshal Rommel's attack further south: *Morgenluft*. "We are going to go all out for the total destruction of the Americans," Field Marshal Kesselring, the German commander in chief South, declared.[7]

"You're taking too many trips to the front," General Marshall had criticized Eisenhower at Algiers, after flying there from Casablanca. "You ought to depend more on reports," he'd advised — obtusely. Patton had counseled the opposite.

Eisenhower's deference to Marshall's authority pretty much condemned the Allies to defeat — Eisenhower still too young to defy the U.S. Army chief of staff. "Absolute priority" alerts had been sent out, once Ultra intelligence decrypts of German signals recognized something big was up, but it was too late. As German forces smashed their way forward with the latest Tiger tanks, new Nebelwerfer multiple-nozzled mortars, and Stuka ground-attack dive-bombers dovetailing with the Wehrmacht advance, American officers began openly yelling at their men to flee for their lives. In less than twelve hours von Arnim and Rommel's pincers had closed, having seized the high spine of the central Dorsal and threatening to end run the entire Allied line in Tunisia.

Absolute pandemonium characterized the initial U.S. response — followed once again by brave American tankers, ordered to counterattack, being lured into German 88mm mobile-artillery traps: almost a hundred American tanks destroyed with their crews, twenty-nine artillery guns, seven half-tracks, and sixteen hundred casualties suffered at Sidi Bou Zid alone. And this was just the start. Open Allied radio communications allowed the Germans to know American whereabouts and moves without difficulty. Huge Allied gasoline and ammunition dumps were blown up or surrendered, as were three U.S. airfields. The German 88s and Tiger tanks had a field day. The battle became a rout as American troops retreated, pell-mell. Fredendall abandoned his laboriously carved subterranean hideout, far behind the frontlines. By February 17 his corps had been thrown back fifty miles — in three days. On February 19 Rommel then attacked at Kasserine. Panic ensued, with the British First Army commander of the overall Tunisian front ordering "no further withdrawal," and to "fight to the last man." Or last American, wags sneered.

Fredendall even began moving his headquarters back to Constantine — more or less where Torch had begun, in November.

Concerned that he had not sufficient supplies or reserves to fight much beyond Kasserine, Rommel was satisfied with what he'd achieved; he obtained grudging consent to withdraw, sowing forty-three thousand mines as he did, and blowing up all bridges. He had given the Allies a "bloody nose" — inflicting six thousand casualties, destroying almost 220 tanks and over 200 artillery guns, for less than a thousand German casualties, and had set back the Allied timetable for advance by months.

Joseph Goebbels, ecstatic at the reception given to his *totaler Krieg* speech, was further delighted by the news of German victory in Tunisia, which went some way to overcome public despondency when word of the surrender of Stalingrad was finally released.

Once again German troops had proven they were the best soldiers in the world and could not be beaten, even by numerically larger forces. "The Americans have made a really terrible showing," Goebbels noted in his diary — "absolutely awful. Which is reassuring, in the event the Americans try to mount an invasion of continental Europe against German troops. They will probably be so smashed up," he commented, "they won't know what hit them."[8]

The next day, still savoring the news from Tunisia, he reflected: "This U.S. defeat gives us an excellent insight into American fighting ability in case of an American invasion of Europe. I think our soldiers would sooner rip their throats out than let them into Germany. At any rate, the spirit here among the German people is hard to beat."[9]

Hitler was *außerordentlich zufrieden* with Rommel — extraordinarily pleased, Goebbels added, after speaking with the Führer.[10]

18

Arch-Admirals and Arch-Generals

IN WASHINGTON, NEWS of the American defeat at Kasserine was met with disbelief.

The secretary of war and senior officers in the War Department who had urged the President to mount a cross-Channel invasion in the summer of 1943, as soon as Tunisia was cleared of enemy forces — even in tandem with an invasion of Sicily, should the President insist on Operation Husky, as it was code-named, to placate British anxiety to clear the Mediterranean sea route to Suez and India — were chastened. The prospects for a successful cross-Channel assault now looked pretty dire, even to Pentagon fantasists. For a moment, in fact, it looked as if Tunisia might be cleared not of Axis forces, but of American.

Stimson, sadly, took this as a sign the Allies should not have landed in Northwest Africa at all. The President demurred. The lesson, in his view, was the opposite: namely the need for more battle experience against German troops.

Combat, command, and campaign experience: these were crucial — not only at unit level, but in senior command and international-coalition cooperation. It was not only Fredendall who failed in battle. Colonel — later Brigadier General — Paul Robinett would afterward write, "One would have to search all history to find a more jumbled command structure than that of the Allies in this operation."[1] Until the onset of battlefield defeat, however, no one had seemed interested in command structures or battle techniques against a German enemy. In Tripoli, General Montgomery had organized a special "study week" or teach-in to "check up on our battle technique," launching it with a two-hour address that one British general thought "one of the best addresses I have ever heard and that is saying a lot." Thanks to Rommel's attack only a handful of U.S. officers were sent to attend, however, and among those who did go, General Patton was heard boasting, "I may be old, I may be slow, I may be stoopid, and I know I'm deaf, but it just don't mean a thing to me."[2]

On his return to Morocco, once the true extent of American debacle became clear, even Patton began to rethink his supercilious judgment. "The show was very bad — very bad indeed," he confided in a letter to his wife.[3] The matter of how to fight the Germans in battle had come to mean life or death to ninety thousand American soldiers in Tunisia. Even Stimson, in Washington, was shocked. "Heavy fighting is going on," he'd noted in his diary on February 15, "and we have yet to see whether the Americans can recover themselves and stand up to it."[4] That they hadn't, in the days thereafter, was galling.

Two days later Stimson was acknowledging that Rommel had mounted a veritable "coup" in southern Tunisia. "He has attacked our thin line of American troops in that region with a comparatively overwhelming force of tanks and has driven them back some thirty miles. Eisenhower has been expecting it and two or three days ago sent a full appraisal of the situation and of his expectation," Stimson recorded, "and he has withdrawn his force to a new line I hope without suffering irretrievable losses." The secretary worried, nevertheless, that the very distance that reinforcements would have to travel would count against the Allies. "We had such good luck in the beginning but these things were lost sight of" — thanks largely to his and Marshall's obsession with a German flank attack across Spain and the Mediterranean. "Now they will begin to count against us," he lamented. "Nevertheless we must not forget the tremendous and permanent gains which our adventure has brought us — the thus far safe occupation of northwest Africa; the acquisition of Dakar and west Africa; the diversion of Hitler's troops from the eastern front, and the irretrievable losses which he has suffered aided by that fact. All of these gains to us are, I hope, permanent and well worth any local setbacks."[5]

Elderly and obstinate to a fault when it came to the stark, bloody business of fighting real Germans in real battle, Stimson was still thinking of "gains" in strategic terms, however — not in combat and command experience: the blooding of those who had to do America's fighting, and who deserved better of their senior officers. Yet as the hours went by and reports came in of panic, desertions in the field, mass flight, surrenders, and demolitions, Stimson felt it was time to be honest. On February 18 he

Total War

On January 7, 1943, President Roosevelt announces "total war" to Congress, then secretly embarks for North Africa aboard a Boeing clipper. He will be the first U.S. president to fly while in office, and the first to visit the battlefield abroad in time of war.

Via Trinidad and Brazil, the President flies across the Atlantic to Gambia, where he tours the harbor in an American tender and spends the night on the USS *Memphis.* Then, using a special ramp for his wheelchair, Roosevelt (above right, with Captain Bryan) flies in a C-54 transport up the coast of northwest Africa to Casablanca, Morocco.

Casablanca

German intelligence mistakes "Casablanca" for "Casa Blanca," the White House, concluding that FDR and Churchill planned to meet in Washington. Meanwhile, in secret, FDR establishes his headquarters in a Moroccan villa (left, with his sons Elliott and Franklin Jr. and Harry Hopkins). His task: to set the Allies—and the U.S. chiefs of staff—on an implacable course for offensive victory in World War II.

At his villa headquarters, FDR assembles the Combined Chiefs of Staff. He must stop his own generals from committing U.S. forces to mass suicide before they have combat experience. He must also get the British to agree to a 1944 cross-Channel strategy. And get the fractious French to fight the Nazis, not each other.

Visiting Troops on the Battlefield

Generals Eisenhower, Clark, and Patton agree with FDR: U.S. forces need more combat experience before launching a cross-Channel invasion. The presence of the President on the North African battlefield is meanwhile inspiring.

Churchill has mixed feelings, but his British government applauds the policy Roosevelt announces to the press and to the world on behalf of the Allies: no negotiation with tyranny, and "unconditional surrender" of the Axis powers.

What should the Allies fight for? FDR and Churchill do not share the same vision, the President tells his son. They are at loggerheads over colonization: FDR is unwilling to sacrifice American lives just to restore British and French empires.

At Casablanca, FDR invites the Sultan of Morocco to dine, and admires the sunset with Churchill in Marrakesh. Before flying home, he insists on visiting Liberia, which became independent in 1847.

Totaler Krieg

At the Sportpalast in Berlin, Goebbels announces *totaler Krieg* (total war), not only as a battle of ideology, but of will.

Back in the States, the President tours the nation's military training camps where soldiers prepare for combat overseas. In secret, he orders P-38s from Guadalcanal to "get Yamamoto," the man (left) directing Japan's war in the Pacific.

finally gave a press conference that even Joseph Goebbels found "extraordinary in its frankness."[6]

"Today I had a sharp reverse to report to the press at the press conference," Stimson admitted frankly in his diary, having "decided to make no effort to whitewash it but to present it in its sharp outlines and simply in my own language to admit that it was a sharp setback and it would be folly to try to minimize it and it would be still greater folly to exaggerate it . . . I talked it over with Marshall afterwards. The only thing Marshall was worried about is that there are two extra divisions that apparently Rommel hasn't used of armored forces and is wondering where those are. Incidentally he told me that when they were in Casablanca the President wanted to divert another one of the divisions from George Patton's force at the gates of Gibraltar and ship them up into the attack in Tunisia. The Staff, however, had refused to agree to this."[7]

The true lesson—that Tunisia was America's military training ground—still eluded Stimson, though. Despite his own trip to Casablanca and then Algiers, General Marshall seemed similarly blinkered. At the Pentagon, Stimson shared with Marshall, on Marshall's return, his feeling they'd had extraordinary "luck so far and all the excitement of the success of the first attack, but now the length of communications is going to tell and we are going to be under constant pressure from the President, among others, to strip our force at the Gate [in Morocco] and send them out to Tunisia to meet the pressure that is going on there. He agreed with me that this would be disastrous."[8]

Disastrous?

That Marshall and Stimson should have continued to take counsel of such fears of a German invasion of Morocco through Spain and across the Straits of Gibraltar, even in mid-February 1943, was almost risible; certainly it made their continuing urging of a cross-Channel attack that coming summer, in tandem with the plan to invade Sicily, jejune beyond belief.

Fortunately, saner minds saw the situation differently. In the press, at least, the U.S. debacle in Tunisia did at least serve to dampen public ardor for a cross-Channel assault that year.

In Berlin, Dr. Goebbels was derisive. Reading British reports of the battle, expressing ill-concealed contempt for American fighting skills, Goebbels likened the situation to that of German forces having to fight with disappointing allies. "So the English now have their own Italians," he mocked. "We can grant them that. The British have always known how to get others to fight their battles; now they have to acknowledge Americans are even better at it,"[9] he sneered. As if this were not enough, he went on: "The Americans prefer to fight their battles in Hollywood rather than on the rough ground of Tunisia, where instead of facing paper tanks they're up against German panzers."[10] And given the awe of Rommel once again being expressed in London newspapers, his endlessly suspicious mind made him wonder if the British, in the aftermath of the battle, were using the American defeat to quieten calls for a Second Front, which the British wisely knew would fail — or at any rate "delay" the cross-Channel assault Stalin was calling for. "They're seriously doubting if they can really put together a successful Second Front."[11]

Kasserine, then, provided a wake-up call for the Allies. Obtaining authorization from Marshall to dismiss Fredendall and replace him with General George Patton, Eisenhower told Patton to fire the incompetent and "to be perfectly cold-blooded about it."[12]

Dimly — despite the lurid stories that General Wedemeyer and others had spread about how the British had "put one over" the American team at Casablanca, resulting in the outrageous delay of a Second Front — even the senior officers of the War Department began to come to their senses: accepting the President was right. A Second Front would never work until U.S. forces were battle-hardened and had had a chance to rehearse large-scale amphibious landings in Sicily.

Kasserine, moreover, put a temporary damper on the U.S. War and Navy Departments' ridiculous obsession with rank rather than experience.

No sooner had the Torch landings taken place than Admiral King had begun pressing his colleagues to back his bid to be promoted *above* four-star rank. "It seems to me that the time has come to take up the matter of more 'full' Admirals and more 'full' generals," he had written in a special memorandum to Admiral Leahy and General Marshall (though not to Lieutenant General Arnold) soon after the Torch invasion. Theater commanders in chief now needed to have four-star rank, to keep up with the British; this then meant that the Joint Chiefs, though not Arnold, should have even higher rank, he felt. "I therefore suggest that we consider the matter and make appropriate recommendations to the President," he'd urged.

Not satisfied with the idea of merely a fifth star for the chiefs, King wanted wholly new ranking nomenclature in the U.S. Armed Forces — indeed, he had his own pet proposal. "We need also to recognize that there is need to prepare for ranks higher than that of Admiral and General. As to such ranks, I suggest Arch-Admiral and Arch-General," he gave his considered view, "rather than Admiral of the Fleet and Field Marshal."[13]

Arch-Admiral King? Arch-General Marshall?

No one was impressed. King had continued to push, however. In the days leading up to Kasserine, Secretary Stimson had learned from Secretary Knox that Marshall was now slated to become a field marshal. He was appalled.

True, before Kasserine, on February 12, General Eisenhower had been promoted to temporary four-star rank in order to give him further authority as Allied commander in chief in Algiers. But for General Marshall to become an American "field marshal" when he had never actually held a field command as a general?

Stimson had asked Marshall what he thought of the idea. "Marshall was dead against any such promotion," Stimson noted with relief in his diary. "He said it would destroy all his influence both with the Congress and with the people, and he said that it really all came from the lower Admirals of the Navy Department forcing this upon King and Knox and upon the President." Stimson thus immediately wrote to the President, on February 16, in the midst of the battle in Tunisia, to try and scotch the idea — which the President did. Fiddling with more stars and "field marshal" titles — which would require Stimson and Knox putting the proposal in person before the Senate and House Armed Services Committees — seemed to Roosevelt a very poor way of defeating Field Marshal Rommel.

For his part, Dwight Eisenhower had not wanted a fourth star, even. He'd immediately cabled to thank the President for his temporary promotion to full general, in the field — but seven days later he wrote privately to his son John, at college. "It is possible that a necessity might arise for my relief and consequent demotion,"[14] he warned — glad to be able to say that his colonel's silver oak leaf in the regular army couldn't be taken away, whatever happened.

Eisenhower's untrumpeted humility did him proud. True, Ike had placed too much trust in Marshall's protégé, Fredendall. Only the cordite of *Blitzkrieg* combat could have exposed the dire weaknesses in American command and battlefield skills in the end, however. Along with many thousands of platoon, company, battalion, brigade, division, and corps commanding officers, Eisenhower himself would have to learn the "hard" way. As he wrote to his son, "You are quite mistaken in thinking that the work you are now doing is useless in the training of yourself for war." He was there, at college, to train his "mind to think. That is essential. No situation whether general or special, is ever the same in war as it was foreseen or anticipated. You must be able to think as the problem comes up."[15] And he instanced having to use Admiral Darlan to obtain swift surrender of Vichy forces in Morocco and Algeria to save American lives.

It was this very quality that President Roosevelt liked in Eisenhower. The President was certain Ike would mature in theater command. Far from demoting Eisenhower, the President was proud of the way he had handled himself and his relations with the British — authorizing General Alexander, his new field deputy, to take over day-to-day handling of the battlefront on February 19; his treatment of the press (Eisenhower accepting "full responsibility" for the debacle, off the record, with reporters); his quiet removal of General Fredendall; and his patient refusal to advance the launch date for Husky, the invasion of Sicily, by a month, as Winston Churchill pressed him to do, lest it prejudice the conditions for Husky's success.

No: in the President's view young General Eisenhower, at age fifty-two, was doing just fine — and U.S. troops, too. The President had seen them, in person, at Rabat, and was confident they'd learn the crucial lessons soon enough. Rommel was withdrawing from Kasserine, and would shortly be given his own drubbing by Montgomery at Medenine, on the Gulf of Gabès — probably the most perfect defensive one-day battle of the twentieth century.

It would all turn out for the best. It was Stalin who worried the President — for the ramification of the President's patient war strategy was this: that the United States would, by its step-by-step approach, win the global war, yet in delaying a Second Front, might well risk Russian domination of Europe in the war's aftermath.

Military prosecution of the war, in other words, was becoming every day more freighted with political consequences.

19

Between Two Forces of Evil

THE ENIGMA THAT was Russia — its communist purges in the late 1930s; its appeasement of Hitler in its infamous Molotov-Ribbentrop Pact in the summer of 1939; its subsequent occupation of eastern Poland, in the wake of Hitler's *Blitzkrieg* attack on western Poland; its similarly egregious invasion of Finland and the Winter War that had resulted in Russian annexation of 10 percent of Finnish territory, in Karelia; its fearsome NKVD police-state methods to maintain absolute communist control of the entire Soviet Union; its displacement and enforced migration of vast populations to Siberia; its veritable paranoia in terms of capitalistic, foreign influence or sway over its citizens — had not given most Americans much reason to support the Soviets, save as opponents of the even more egregious Germans.

The sheer refusal of Russian soldiers and citizens — often ill armed and ill trained — to cede their country to the German troops who had overrun all of Europe had aroused belated popular admiration in America, and growing confidence that Hitler — despite his control of Europe from Norway to Greece, and the whole of central Europe to the Crimea and Ukraine — could, in truth, be beaten. What would happen then, though? Would the Soviets, obedient on pain of death to absolute communist rule from the Kremlin, permit those countries of Europe liberated by the Soviets to become genuine, capitalist, functioning democracies? Or would they be "Sovietized" by Russia?

The President had been thinking of such matters with increasing concern in the fall of 1942, as he'd confided to the Canadian prime minister, Mackenzie King. The prospect, however, had become all the stronger once it became clear that Hitler had overplayed his hand at Stalingrad, and his Sixth Army was going to get a hammering on the Volga — indeed, that he might lose not only his Sixth Army at Stalingrad but his armies in the Caucasus, by the Black Sea, too. From being on the desperate defensive, the Russians would then begin to pose a mortal threat to the Third Reich — with or without a Second Front.

Certainly this was an eventuality that, in the privacy of his diary, Joseph Goebbels pondered. Far from causing him to question the Nazi ideology that he and Hitler had pursued over the past two decades, it only caused him to dedicate himself the more determinedly to the *Ausrottung* of the people he blamed for Europe's travails: the Jews, as he'd declared in his *totaler Krieg* speech. He now gave orders for the last Jews left in Berlin to be rounded up and sent to be liquidated in SS concentration camps — noting how much more psychologically free this made him feel. He also recorded his determination to stamp out any protest in Germany to his total-war policy in the most ruthless manner — in other words, via execution — as well as ruthless reprisals to be taken against any acts of disrespect or attempted assassination of Nazi officials in the occupied countries of Europe.

Between these two forces of pure evil, it was difficult to say which was the worst. The President had therefore, on November 19, 1942, asked his former ambassador to the Soviet Union, William C. Bullitt, to furnish him with a private report on how he saw the future of Europe, following the successful Torch landings. In particular, Roosevelt wanted to have the former ambassador's reading of Russian intentions.

Bill Bullitt had taken his time. He'd recently been used by the President as an ambassador at large, conducting a presidential mission to West Africa, Egypt, Libya, Palestine, Syria, Iraq, and Iran in the spring of 1942. Independently wealthy thanks to his second marriage, he had then become director of public relations to the secretary of the Navy, Frank Knox. It had taken him two months to comply with the President's new request — too late, unfortunately, for the Casablanca Conference — but when it was ready, it was dynamite.

"Dear Mr. President," Bullitt's covering letter ran, written on January 29, 1943, the day before the President's return to Washington. "The appended will take thirty minutes of your time. It is as serious a document as any I have ever sent you." He warned that "its conclusion is that you should talk with Stalin as soon as possible" — and wished Roosevelt good luck in the attempt.[1]

The President *did* read it — and swiftly invited Bullitt to lunch at the

White House to discuss its implications for Allied political and military strategy.

Bullitt's report pulled no punches. Having served as America's very first envoy to Moscow, he had, after all, an almost unrivaled perspective both on Stalin and the Russians. In addition, for an American he had a keen perspective on Europe, having been U.S. ambassador to France for four years, right up to the German conquest of France in 1940.

Bullitt's portrait of Stalin and his Soviet aims was uncompromising. His memorandum began by trashing former Republican president Herbert Hoover's notion that Stalin had changed his philosophy, wanted no annexations, and was only interested in Russian security. The dictator was reported, in the view of such innocents as Hoover, "to be determined to have the Soviet Union evolve in the direction of liberty and democracy, freedom of speech and freedom of religion. We ought to pray that this is so," Bullitt allowed; "for if it is so, the road to a world of liberty, democracy and peace will be relatively easy." If this was not so, however, "the road will be up-hill all the way." The free world would be tilting in one direction, the oppressive Soviet Union or empire in another.

In these circumstances, Bullitt felt, America must do everything to halt the Russian tilt before it was too late. It was, he wrote, "in our national interest to attempt to draw Stalin into cooperation with the United States and Great Britain, for the establishment of an Atlantic Charter peace," such as the President's teams in Washington were mapping out. "We ought to try to accomplish this feat, however improbable success may seem," for America would then be on the side of right, not merely might. But in dealing with Stalin, Bullitt was adamant, it was imperative to strip away any illusion.

"The reality is that the Soviet Union, up to the present time, has been a totalitarian dictatorship in which there has been no freedom of speech, no freedom of the press, and a travesty of freedom of religion; in which there has been universal fear of the OGPU [secret police] and Freedom from Want has been subordinated always to the policy of guns instead of butter." Stalin might well be persuaded to close down the nefarious Comintern, fomenting world communist revolution — but only because, in the end, Stalin actually *had* no real interest in world revolution by communists; his interest was only in communist-controlled nations serving as "5th column for the Soviet State" or empire. World communist revolution was but "a secondary objective."

As Bullitt pointed out from his intimate, personal knowledge of the dictator, Joseph Stalin had no illusions — or even belief in communism as a motivating faith. He "lets no ideological motives influence his actions," Bullitt warned. Whether the global future lay with the ideal of communism or the president's four freedoms, Stalin was indifferent. His only goal was to maintain and extend the power of the Soviet Union: greater Russia, in effect, as a police state ruled by fear. "He is highly intelligent. He weighs with suspicious realism all factors involved in advancing the interests or boundaries of the Soviet Union. He moves where opposition is weak. He stops where opposition is strong. He puts out pseudopodia" — amoeba-like tentacles — "rather than leaping like a tiger. If the pseudopodia meet no obstacle, the Soviet Union flows on."

The moral, then, was that the United States must do everything in its power to show genuine *desire* for cooperation, as well as to "prove to Stalin that, while we have intense admiration for the Russian people and will collaborate fully with a pacific Soviet State, we will resist a predatory Soviet State just as fiercely as we are now resisting a predatory Nazi State." If not, "we shall have fought a great war not for liberty but for Soviet dictatorship."

This was a sobering eventuality.

"How can we make sure that this will not happen," Bullitt asked rhetorically, "and achieve our own aim in a world of freedom and democracy?"

It would be a case of America Inc. versus Russia Red.

President Roosevelt nodded his head in agreement — for whatever was published in liberal-minded newspapers and journals in the United States, he himself had no illusions about Stalin, or the nature of the Soviet terror state, maintained entirely by patriotism and fear. Moreover, he was pleased to see Bullitt supporting his presidential policy of unconditional surrender of the Nazis and Japanese — whereas there were many, including Third Secretary George Kennan in Moscow, who favored making a deal with the German generals, or non-Nazis, to help fight the Soviet regime. To the President this would be tantamount to condoning German militarism, wars of conquest, and use of terror against its own citizens — whether Jews or gypsies, political prisoners or priests — just as it would be were Japan's example of inhumanity — its savage, genocidal war in China and its atrocity-ridden rampage across the Southwest Pacific — to be condoned. Unconditional surrender and disarmament of the Axis powers was a sine qua non of a permanent postwar peace in the world, beginning on a new page, the President felt strongly—and Bullitt, in his report, did not contest this.

One by one the President ticked Bullitt's points: that unless Russia was pressured into declaring war on Japan, for example, following German capitulation, the United States would be tied down, having to fight its way unaided, island by island, until it could finally bomb and invade the Japanese heartland, which might take years—while in the meantime Russia's amoeba would be left free to spread across Europe, infecting defenseless nations and hitching them to Stalin's Sovietizing wagon.

The answer to that, in Bullitt's view, was to press Stalin, while U.S. Lend-Lease assistance was still critical to Soviet military victory against the Wehrmacht, into agreeing not only to enter the war against Japan in due course, but to sign up to a formal agreement committing the Allies to establish a postwar democratic Europe—not, as was the alternative, a group of communist puppet states subservient to the Soviet Union.

This, as Bullitt articulately put it, could only be done by securing an early meeting with Stalin, since "our bargaining position will be hopeless after the defeat of Germany," when Russian troops would in all likelihood be in occupation of all central Europe up to the Elbe—perhaps even up to the Rhine. Churchill, too, must be harnessed to a European, rather than imperial British, cause, alongside the United States—with everything done, from this moment forth, to prepare the governments in exile and future European leaders to establish strong democratic structures that Stalin's fifth columnists could not successfully subvert.

World disarmament was, in Bullitt's realistic view, impossible—yet he doubted whether U.S. public opinion would willingly support yet another war in Europe to defend defenseless individual states. Ergo, rather than disarm those states, or press such states to disarm, they should be encouraged to arm themselves against Russian interference—forming a U.S.-and-British-supported coalition or alliance. They should be urged to become a European bloc of "Integrated Europe," which Stalin would not dare challenge. "Soviet invasion finds barriers in armed strength," Bullitt emphasized, "not in Soviet promises."

This prediction—an early 1943 version of what became, in 1948, the Western European Union and NATO—was very much the President's thinking. Using Lend-Lease as a lever, it would involve a carrot-and-stick approach to get Stalin, as soon as possible, to dissolve the Comintern as the instigator of world communist revolution; to agree to eventual entry into the war against Japan, once Germany was defeated; and to agree, in a formal document, to sign up to a United Nations world authority guaranteeing the independence and self-government of sovereign states — with the United States and Britain, as two of the world's Four Policemen, ready and willing to use air, naval, and, if necessary, ground forces to counter any attempt, by anyone, to invade such sovereign states.

Would Joseph Stalin, dictator of a police state supposedly wedded to Marxist-Leninist communist ideology, willingly sign up to a democratic concept like this, however — a charter that would be a permanent indictment of the Russian police state?

As Bullitt acknowledged, the Russians would have the "whip hand" at the end of the war. In all frankness, moreover, there seemed little evidence the Soviets, led and ruled by Stalin, were going to undergo a Pauline conversion and become guardians of democracy and freedom, together with the United States, Britain, and China, across the globe — at least not anytime soon. It was therefore imperative that the United States and Britain — since China, for all its millions of people, was in no position to police anyone, indeed would probably have to cede Manchuria to the Soviets — ensure that their own troops reached, as soon as possible, a demarcation line in Europe beyond which Stalin's troops could not march without going to war with the United States, the Soviet Union's great provider.

Where, exactly, as Bullitt surveyed the world in January 1943, would this line be, however — *and how could the Western allies hope to reach it before the Russians?*

This, indeed, was an interesting question.

A colleague of Bullitt's — Bullitt did not name him in his report — had recently posited the end-of-war Sovietization of Europe would include "at least Finland, the Baltic States, Poland, Rumania, Hungary, Czechoslovakia and the entire Balkan peninsula including probably European Turkey" — unless, Bullitt argued, the President beat Stalin to the punch. The United States should therefore "define as Europe the Europe of 1938," he suggested — "minus Bessarabia, which should go to the Soviet Union" — and seek to save that version of Europe from the predatory clutches of the Russian bear. In this respect there was, Bullitt reemphasized, "only one sure guarantee that the Red Army will not cross into Europe — the prior arrival of American and British Armies in the eastern frontiers of Europe."

The eastern frontiers of Europe — when U.S. forces still did not have a single soldier on the European continent?

Anticipating the President's frown, Bullitt had admitted in his report: "To state this is to state what appears to be an absurdity, if the assumption is made that we can reach the eastern frontiers of Europe only by marching through France, Italy and Germany" before the Russians. However Bullitt had a better alternative. "It may . . . be possible to reach this frontier before the Red Army," the former ambassador and now assistant to the U.S. Navy secretary wrote, "if we make our attack on the Axis not by way of France and Italy but by way of Salonika and Constantinople."[2]

Oh dear! the President sighed. Bullitt clearly had less idea of geography as it pertained to military matters than a schoolchild. Had he never heard of the disastrous "Salonika Front" in World War I, or Churchill's fatal Allied assault in the Dardanelles in 1915 — not to speak of the First and Second Balkan Wars of 1912 and 1913?

The President had recently heard similar Balkan proposals being trotted out by Prime Minister Churchill and the French high commissioner, General Giraud, at Casablanca. Given the disaster in the Dardanelles in 1915, it had been utterly amazing to hear Winston recommending such a military strategy. Yet General Giraud was just as unrealistic, the President had found. Both were men of great courage — but in the search for alternatives to "war by attrition," they were given to fantasies that were almost criminal in terms of the loss of human life to which their ill-considered ventures would lead — Churchill's Gallipoli fiasco having cost the Western Allies no fewer than a quarter million casualties.

Ignoring this, Churchill had at Casablanca favored pressing Turkey's president to declare war on Hitler, and revival of the idea of a Dardanelles campaign. He'd asked Giraud whether he agreed — at which the five-star French general had countered with his own equally amazing notion of Allied military strategy.

"*Tout simple,*" Giraud had opined. "First, liberate Africa. Which is being done. This should be finished by spring this year. Then, without wasting a minute, occupy the three big Mediterranean islands: Sicily, Sardinia, and Corsica. Establish a base there, primarily air forces, to assault the mainland of Europe. As soon as the forces are ready, invade the coast of Italy, between Livorno and Genoa. Seize the Po valley. Clean up the rest of the Italian peninsula, and prepare to strike into the heart of Europe on the axis: Udine [northwest of Venice, between the Alps and the Adriatic] and Vienna, backed by air power serviced from bases across the whole of Italy. In one blow Germany can thus be invaded through the Danube valley: we will isolate the Balkans on the right, and have France on the left, and we will beat the Russians to Vienna, which is not to be sniffed at," he'd announced. "After that, following the fall of Germany, the business of Japan will be a piece of cake. QED."[3]

Clearly Giraud—who still pressed to be made Allied commander in chief in the Mediterranean instead of General Eisenhower, rather than have to deal with political matters he abhorred—saw himself as a modern Napoleon, though about a foot taller.

At Casablanca, Churchill had not discouraged this idea—though the President had refused to countenance such craziness. It was therefore nothing short of galling, at lunch at the White House, to have to listen to former ambassador Bullitt, the director of public relations in the Navy Department, now recommending, as an American, such military bêtises.

Bullitt claimed, to the President's concern, he was not alone in Washington in advancing such a war strategy. His discussions at the Navy Department and elsewhere had convinced him, Bullitt maintained, that "there is a large body of military opinion in Washington that favors—on purely military grounds—striking at the Axis by way of Greece, Turkey, Bulgaria and Rumania rather than by way of France and Italy."[4]

Dump the whole idea of a cross-Channel attack?

On paper the notion appeared bold and imaginative—if wars were conducted on paper. Like Giraud, Bullitt seemed convinced the Western Allies could make straight for central Europe, and secure its boundaries before the Russians got there, without problem—irrespective of the terrain. Or the Germans. "This is a question for you and Churchill, and your military advisors to decide," Bullitt allowed[5]—convinced that Churchill, who was nothing if not imaginative, would be of like mind.

The conclusion to Bullitt's twenty-four-page report to the President had climaxed with a three-point politico-military recommendation. Roosevelt should persuade Churchill to subscribe to a "policy of an integrated, democratic Europe." "Conversations between you and Stalin" should then be arranged. But behind the scenes, while negotiating on paper with Stalin, an "immediate study of an attack on the Axis by way of Greece, Turkey, Bulgaria and Rumania" should be ordered by the President.[6]

There was even a fourth recommendation: namely that Bullitt's archrival at the State Department, Sumner Welles, be fired — thus empowering the deeply anti-Soviet secretary, Cordell Hull, to take Bill Bullitt as his deputy, as the Allies raced through the Balkans into central Europe.

QED.

Bullitt's report — which would later be quoted as a kind of Lost Ark that could have changed the course of history, had it been followed — was, in its military naiveté, as senseless as it was callous in respect to the lives that would have been lost in pursuing such a course. Lunching with Bullitt, the President could only shake his head at a man so right about the Soviets and so wrong about military matters.

Winston Churchill, for his part, did not feel the same way. While the President had returned straightaway from North Africa to Washington via Liberia, Bathurst, Natal, and Jamaica (where a recuperated Admiral Leahy was picked up), the Prime Minister had nevertheless flown, against the advice even of his own cabinet, to Turkey, in a vain attempt to get President Ismet Inonu to join the Western Allies — and thus open the way to an invasion of southern Europe via the Balkans and Constantinople: the dream that had consumed him in 1915 and had led to his resignation as First Lord of the Admiralty when it failed.

Roosevelt had been skeptical whether President Inonu would comply with Churchill's request, any more than Portugal or Spain or Sweden could be expected to give up their neutrality in the war. He had, however, authorized Churchill to share with Inonu the President's notions of unconditional surrender and a postwar United Nations authority — not only as a bulwark against future wars of aggression, but as a counter to future Russian expansionism.

The President had cautioned Churchill, however, neither to promise too much military aid, if Inonu did decide to join the war, nor to suggest that the Allies were planning a new invasion of Salonika, as in 1915. The Germans, he warned, would be tougher even than Atatürk's army at Gallipoli — and the mountains beyond Salonika would make an Allied campaign a dead end. His remit, in terms of the U.S.-British coalition, was merely to explore the possibility of airfields and military staging bases being established in Turkey, and if not, to encourage Turkey in its neutrality: dissuading it from any thought of alliance or cooperation with the Third Reich, and encouraging it as a bulwark against communism. This, to his great credit, Churchill had attempted to do as part emissary, part negotiator, flying to Adana and meeting with Inonu onboard their two trains. To the relief of Sir Alexander Cadogan and General Brooke, the Prime Minister had been surprisingly circumspect — relying on his gifts of ratiocination and literary composition. No sooner had he arrived, therefore, than he handed President Inonu a paper he'd written en route to Turkey called "Morning Thoughts: Note on Postwar Security" — a copy of which he was careful to cable to the President in Washington.

Like Bullitt's report, Churchill's Turkish memorandum was to become an important historical document.

In Churchillian prose (Cadogan noting in his diary, "He was awfully proud of it"), the Prime Minister's paper summarized the outcome of the Casablanca Conference and the outlook for the world at the end of the war. As soon as the "unconditional surrender of Germany and Italy" was achieved, the "unconditional surrender" of Japan, too, would be procured — with subsequent "disarmament of the guilty nations" enforced by the victors. ("On the other hand no attempt will be made to destroy their peoples or to prevent them gaining a living and leading a decent life in spite of all the crimes they have committed," Churchill added the rider.) Reparations would not be demanded by the Western countries "as was tried last time," though Russia would have to be helped "in every possible way in her work of restoring the economic life of her people" after suffering "such a horrible devastation" as Hitler had inflicted. This, then, led to the President's plans for a United Nations authority.

The authority was to be "a world organization for the preservation of peace based upon the conceptions of freedom of justice and the revival of prosperity" — one that would not be "subject to the weakness of former League of Nations." It would be held together under the military protection of the victors, who would "continue fully armed, especially in the air." "None can predict with certainty that the victors will never quarrel amongst themselves, or that the United States may not once again retire from Europe, but after the experiences which all have gone through, and their sufferings and the certainty that a third struggle will destroy all that is left of culture, wealth and civilization of mankind and reduce us to the level almost of wild beasts, the most intense effort will be made by the leading Powers," Churchill summarized, "to prolong their honorable association and by sacrifice and self-restraint to win for themselves a glorious name in human annals." Great Britain would "do her utmost to organize a coalition of resistance to any act of aggression committed by any power;" moreover, "it is believed that the United States will cooperate with her and even possibly take the lead of the world, on account of her numbers and strength, in the good work of preventing such tendencies to aggression before they break into open war."[7]

Though it might not be as magically phrased as some of his prose masterpieces and speeches, Churchill's memorandum reflected the extent to which he now understood and agreed with the President's vision of the United Nations and postwar world security at this moment in the war. Given such a future, then, would not Turkey wish to guarantee its own security "by taking her place as a victorious belligerent and ally at the side of Great Britain, the United States and Russia," Churchill had asked President Inonu?

It was a beguiling prospect, but President Inonu, understandably, had declined. The Prime Minister's paper certainly exuded confidence in the inevitable eventual victory of the Allies — but it seemed oblivious to Hitler's likely actions in the meantime. The document made no mention of this, or of the military problems inherent in mounting an invasion of southern and central Europe through northern Italy and/or the Balkans. Or even of Stalin's possible reaction to such a change in Allied military strategy — a change that, if it stalled in the Mediterranean without a Second Front, would give Stalin every reason to scorn the President's plans for unconditional surrender and the establishment of a postwar United Nations authority as idle nonsense.

Both Bullitt's report and Churchill's memorandum were, to be sure, written before the reality of war against the Wehrmacht finally set in. In this respect the American defeat at Kasserine, two weeks after Bullitt's report, had quickly poured cold water on any idea in Washington or London that the Allies could race anywhere, let alone through the Balkans. At his private luncheon at the White House with the President, Bullitt had thus backed off his Balkan idea — for the moment. It was too early to be contemplating ambitious American campaigns in the Mediterranean when for a moment it looked as if U.S. forces would be driven out of Tunisia. Besides, the public would have to be encouraged to support a more interventionist role in American foreign policy if the President was to have any genuine credence in exploiting its current creditor-status with Stalin.

In this respect, at least, the President's vision of the postwar peace seemed to be gaining traction, unaffected by the reverse at Kasserine — in fact it began to become clear, as the weeks went by, that the President's State of the Union address was bearing amazing, anti-isolationist fruit.

Roosevelt had assumed his State of the Union address, with its description of "total war," its call for the disarmament of America's enemies, and his outline of postwar social programs and international security, would be strongly contested in Congress and outside. Far from it. His speech — and press coverage of the Casablanca "unconditional surrender meeting" — seemed to trigger, the President found, a sort of national American awakening to world responsibility that had never really existed before.

Burgeoning pride at the success of the Torch invasion and MacArthur's advances in the Pacific — where, in the battle of the Bismarck Sea, American B-25s carrying five-second five-hundred-pound bombs, carried out "the most devastating air attack on ships in the entire war," in the words of naval historian S. E. Morison[8] (sinking seven of the eight Japanese transports seeking to reinforce Lae following the evacuation of Guadalcanal) — left the noninterventionist voices of Charles Lindbergh, Joseph P. Kennedy, and Senators William Borah, Robert LaFollette, Hiram Johnson, Arthur Vandenberg, and Burton K. Wheeler looking like defeatists. The Casablanca summit — trumpeted in newspaper reports and pictures, as well as in movie-house newsreels shown across the country — lent a moral grandeur to the turnaround in the fortunes of war: photographs of brave, cigar-wielding Winston Churchill sitting beside the President, declaring himself to be his "active lieutenant," French generals shaking hands, the President inspecting and eating with U.S. forces in the field . . . Even Lord Halifax, the British ambassador, was full of congratulations when writing to the Prime Minister, extolling the results of the Casablanca Conference and noting in his diary how, in America, Republicans and Democrats were beginning to talk of the future in a new and wholly different way.

Lord Halifax was learning, himself, to see the world in a different way. Since his appointment to the embassy and America's entry into the war, he had had to meet with people of every stripe and to learn the complexities and nuances of the American political system, with its checks and balances — and vituperative press. As a result Halifax had become a more astute observer of trends in the United States than in his home country, where his aristocratic airs and way of life (hunting, shooting) had inured him, as a notorious appeaser, to the fact that the younger, post–World War I generation would, in fact, fight Hitler — but not for a colonialist, class-riven British Empire they no longer believed in. The President's latest postwar vision, which Roosevelt had shared with him in private talks at the White House, struck Ambassador Halifax not only as positive, but one that even former isolationist Americans seemed more and more willing to embrace. Even the former U.S. president, Herbert Hoover, who lunched with Lord Halifax on January 8, 1943, after the President's address to Congress, had expressed a more "friendly" view of America's association with Britain than before, the ambassador had found. "We discussed a great many post war things," Halifax recorded in his diary — relieved to hear the former president was "absolutely convinced of the necessity of our working together" as nations. "On the whole I was cheered by my talk with him and by his estimate of what American public opinion will accept in the way of international cooperation." Hitherto, public opinion had opposed any American treaty or involvement with other countries that could be "represented as infringement of [American] national sovereignty. This was the rock on which [President] Wilson broke — the idea that some League or conference should dictate United States action." But now — at least in Hoover's opinion — public opinion was changing, as were former president Hoover's own attitudes. "These difficulties would not in his view arise if you had some international organisation that would content itself with expressing moral opinions and leave it to the joint policemen, whom he sees as the United States, ourselves, and, if she will play, Russia, to take action on their own," Halifax noted. The United Nations — or "whatever the international body was" that would be set up at the war's end — would "make a report and recommendation to the policemen," which the policemen could either carry out or not.[9]

This exploratory notion of a United Nations Security Council was a momentous reversal — and when in Washington Lord Halifax addressed assembled British consuls from main U.S. cities, several days later, the ambassador advised them to push the notion of the "British Commonwealth," rather than "Empire," as having "a biggish part to play" in the coming times — yet to exercise "self-restraint, when Americans threw their weight around."

America, henceforth, would be top dog, Halifax made clear. From that meeting the ambassador had then gone to the State Department "to discuss the draft of a scheme for what the Americans call the rehabilitation of the world."[10]

Rehabilitation it certainly would be. The Russian ambassador, Maxim Litvinov, was present at the State Department meeting, too. "We got along fairly well and all did our best to be accommodating to one another. Some difference of opinion as to whether the inner management committee of the thing should be composed of the four Powers," as a security council, "or, as we [British] had suggested, seven" — which would "permit Canada as a great supplier to be on [it], probably a South American, and one of the smaller European allies. Litvinoff made a strong argument about this thing being used as a pattern for the future, and consequently the importance of keeping the four big powers undiluted. I thought there was a good deal in his argument," Halifax noted, approving the Russian's view.[11]

Ten days later, on January 18, 1943, barely a week after the President's State of the Union address, Halifax was noting that Dr. Alan Valentine, president of the University of Rochester — a Democrat who had campaigned for the Republican Wendell Willkie in the 1940 presidential election, organizing "Democrats for Willkie" in opposition to a third term for FDR — "did not think there was much danger of isolation." In fact, Valentine now found Willkie "too emotional and immature."[12] A new "American State [Department] book about American policy in the last ten years" had shown "how paralyzed their Executive was," after World War I, "owing to the prevalence of isolationist thought."[13] Even Willkie himself, when Halifax dined with him on January 27, emphasized the change of Republican mind — now claiming "that historically the Republican party had not been isolationist and had only accidentally been thrown into isolationism after the last war by Wilson's attempt to monopolise the international ticket. He was apprehensive lest something of the kind should happen again, and spoke very earnestly about the necessity of nothing being said in British quarters" of Republicans backing away from an internationalist stance, lest this actually revive isolationist sentiment. "He spoke with great certainty, as did Claire Luce" — a Republican congresswoman from Connecticut and wife of the publisher of *Life* magazine — "about the Republican party in 1944 being victorious."[14]

By the time the President returned from North Africa, therefore, it had been to find his utopian hen had laid its eggs — indeed, the next day Halifax noted a long talk with Henry Luce "about the prospects of the Republican Party being isolationist after the war." Luce dismissed the very idea, just as Willkie had — in fact claimed, like Willkie, that isolationism had been an aberration — the United States having "only accidentally got into that line in 1920," according to Luce. Halifax was then stunned as Luce proceeded to advance Roosevelt's internationalist agenda. "On the post-war business Luce said that he wanted to make a careful examination of just what an international police force might mean," but was not averse to it. "He said that there had been a curious revolution in American feeling in the last few years" — in fact, in the last few weeks. "A short time ago, if you had listened to any argument between the isolationists and internationalists, the isolationists would at once have clinched the argument by saying: 'You want to police the world, do you?' which was generally held to be conclusive against it. Now, he said, American public opinion was completely convinced that an international police force was desirable."[15]

When Halifax went to see the President in person at the White House on February 15, he was told that columnist Walter Lippmann, no less, was talking of "the United States being established in some European base after the war," so that "any infraction of European peace" should at once be addressed: the forerunner of NATO.[16]

20

Health Issues

THOUGH IT WAS too early to crow, the President thus seemed decidedly proud, Halifax found. His step-by-step military strategy for prosecuting the war had been set in stone at the Casablanca Conference — with the target, in writing, of almost a million U.S. troops and their weapons to be conveyed to Britain by December 31, 1943, ready to launch a full-scale invasion across the English Channel in April or May 1944. As Averell Harriman noted, based on his conference notes, a "new joint command (COSSAC, acronym for Chief of Staff to the Supreme Allied Commander) was created to begin immediate planning for this climactic operation known later as Overlord."[1]

The postwar world, too, with luck, might well turn out the way the President envisioned, with the growing support of the American public, the Republican Party, and America's British partners — though the latter would have a difficult row to hoe if they chose to reestablish their colonial empire as Churchill wished.

It was in this context that FDR's health raised some concern, however. Although in the immediate aftermath of his trip to Africa Roosevelt had seemed energized and rejuvenated, in the weeks following the Casablanca Conference it was evident that the journey had taken a physical toll on the President. At least to those in close contact with him. His cousin Daisy, especially.

Daisy Suckley had been relieved to see the President looking so well on his return — yet she remained disappointed by the meager medical attention her hero appeared to be receiving as president of the United States. After ten years in the Oval Office, Mr. Roosevelt still relied on a simple

U.S. Navy doctor as his personal White House physician: Dr. Ross McIntire, who'd been on his staff since 1933 and could be relied upon "to keep a close mouth" about the President's medical condition.

McIntire was an eye, ear, nose, and throat specialist by early training as an intern. He'd had an undistinguished record thereafter, becoming a simple naval dispensary physician, onboard and onshore. Nevertheless, he'd been recommended to the new president by former president Woodrow Wilson's floundering doctor, Cary Grayson — one incompetent recommending another, it would be claimed.[2] Beyond daily treatment of Roosevelt's notorious sinus problems, McIntire appeared to do little for his patient other than keep at bay those doctors who might offer the President more expert medical attention, in view of his fragile health: practitioners who might equally, however, blab inadvertently to reporters employed by Colonel McCormick, owner of the *Chicago Times-Herald*, or Cissy Patterson, owner of the *Washington Times-Herald* — both of them sworn enemies of the Democratic president and determined to oppose his reelection if he stood for a fourth term.

Despite McIntire's mediocre medical talents, the President, then, had been content to continue with a single doctor — in fact, in 1938 Roosevelt had appointed McIntire surgeon general of the U.S. Navy, in addition to his White House duties, and soon had him promoted to the rank of rear admiral — with responsibility for what became a vast naval medical system, involving 175,000 doctors, nurses, and professional medical staff. Such enlarged duties, however, were plainly incompatible with continuing daily care of the paralyzed chief executive.

Such was Roosevelt's authority, however, that by 1943 no one dared question McIntire's solitary supervision of the President, in spite of worrying signs of deterioration in FDR's health, even in the run-up to his historic flight to Africa.

Staying with the President on December 4, 1942, for example, the Canadian prime minister, Mackenzie King, had been alarmed by the President's physical condition. When he'd first gone in to see him, King reflected, "the President was smoking a cigarette in bed while reading the papers. I felt that even at that hour of the morning, he seemed a little tired and breath still a bit short."[3]

Given the President's vast responsibilities and the fact that his mental acuity seemed in no way impaired, King had given no further thought to the matter. The trip to North Africa, meanwhile, had seemingly done wonders for his state of mind and body, the President's staff felt on the

President's return, as did visitors to the White House. "The President was in fine form," the secretary of war recorded in his diary on February 3—"one of the best and most friendly talks I have ever had. He was full of his trip, naturally, and interspersed our whole talk with stories and anecdotes." Though he found them amusing, Stimson was nevertheless discouraged "to see how he clung to the ideal of doing all this sort of work himself."[4]

The war secretary was seventy-four and in excellent health; the President, sixty-two. Stimson noticed nothing amiss in Roosevelt's form other than his messy approach to administration, which Stimson deplored. "He was very friendly but, as I expected, takes a different and thoroughly Rooseveltian view of what historic good administrative procedure has required in such a case as we have in North Africa," the secretary noted. "He wants to do it all himself. He says he did settle all the matters that were troubling Eisenhower when he was over there"—and even claimed Robert Murphy was in North Africa not "as a diplomat to report to Hull but as a special appointee of his own to handle special matters on which he reported to Roosevelt direct. This was a truly Rooseveltian position. I told him frankly over the telephone that it was bad administration and asked him what a Cabinet was for and what Departments were for," he recorded, "but I have small hopes of reforming him. The fault is Rooseveltian and deeply ingrained. Theodore Roosevelt had it to a certain extent but never anywhere nearly as much as this one."[5]

Stimson's criticism of the President was well founded, though his recommendation, namely that the United States should simply administer French North Africa in the same way as the War Department had ruled Cuba, the Philippines, and Puerto Rico, belied the secretary of war's ongoing turf battle with Cordell Hull rather than the tricky realities of the situation. With regard to the President's health, however, neither Stimson nor the majority of the President's visitors seemed alert to any problems.

Only Daisy Suckley paid attention to what was happening—or not happening. After she met with the President, at Hyde Park, she confided her concern to her diary—noting disorienting symptoms of transcontinental air travel that would later be called jet lag. "All his party have been feeling miserable since they got back," she recorded on February 7. "He just hasn't let himself give in until he got here—Then he 'let go' & feels exhausted—the President finding it hard, he said, to rise in the mornings, and sleeping late."[6]

• • •

Was it merely desynchronosis — disruption of circadian sleep rhythms — though, Daisy wondered? After Pearl Harbor the President had stopped swimming daily in the White House pool. He was still smoking several packs of cigarettes a day, but his doctor seemed to pay little or no attention to the President's elevated blood pressure, or to his cardiac condition — despite the fact that there had been worries on that account even before the 1940 election, when he'd been beset by heart problems he couldn't keep from those around him.

"His color was bad; his face was lined and he appeared to be worn out. His jaw was swollen as a result of a tooth infection . . . And I learned there was worry over strain on Roosevelt's heart," the chairman of the Democratic National Committee, James Farley, recalled later.[7] Bill Bullitt was more specific; he would claim that he'd been present at a White House dinner in early 1940 when the President had suffered "a very slight heart attack" and had collapsed.[8]

A *heart attack?* It was little wonder McIntire had been concerned at the height the airplane would have to fly on its journey to Casablanca and then home. On the positive side, however, the President had shown the world he was on the top of his form at Casablanca. Even the dark areas beneath his eyes had vanished, people noticed. "He looks well," Daisy acknowledged in her diary — her anxiety being more over the risk of a flying accident than his health at this stage. The plane carrying Averell Harriman and Brigadier Vivian Dykes — senior British aide to Field Marshal Dill in Washington — back to England had in fact crashed on landing in Wales. Though Harriman survived, Dykes had perished. As Daisy implored the President: "I told him we all thought he *should not* take the risks of such a trip."[9]

"Well — not for some time anyway," the President had responded at Hyde Park, where he reclaimed Fala, his Scottie — hugging the woolly black dog to his breast. When Daisy left to go back to her job at the president's library, which had been created in 1939 and for which she had been working since 1941, "Fala looked at me," she wrote, "but trotted after the P."[10]

Daisy was not convinced, however — and became less so when the President then fell ill again and again in February. The President's physician showed little concern. "Allied successes lessened the nervous strain," was all Dr. McIntire would later comment in a memoir he wrote, "and the President not only picked up weight but lost some of his care lines."[11]

• • •

Was the President really all right, though? Or were underlying, potentially serious health issues not being sufficiently addressed?

To Daisy, his confidante, the President had once remarked that "he caught everything in sight," as he put it. There was nothing new in this, he'd added—"all his life had been that way."[12] His near-fatal bout with virulent flu in 1918; his contracting of poliomyelitis in 1921; his repeated sinusitis; his collapses from possible heart or vascular failure—these were but the more dramatic examples of his proneness to infection and other ailments, he accepted.

Illness was not something Roosevelt dwelt upon or paid much attention to—an attitude Eleanor, his wife, did not discourage, since it absolved her of marital anxieties at a moment when she herself was undertaking such a demanding schedule as First Lady at the White House, spokesperson for the underprivileged, and mother to six children, not to speak of grandchildren.

This left Cousin Daisy, though, to worry all the more on behalf of the President. "The P. looked very tired, but did his usual part of 'Exhibit A,' as he calls it"—entertaining, for example, a party he was hosting at the White House on Valentine's Day, February 14, without Eleanor, who'd flown to Indiana. "At nine, he said he had to go to work & left the guests, calling to me to go with him. He got on the sofa in his study and said he was exhausted—He looked it. He said: 'I'm either Exhibit A, or left completely alone.'"[13]

Daisy was flabbergasted. "It made me feel terrible—I've never heard a word of complaint from him, but it seemed to slip out, unintentionally, & spoke volumes," she had penned in her diary that night. The wife of the President's military aide and appointments secretary, Mrs. Watson, had "said at lunch, on Friday, that 'he is the loneliest man in the world.' I know what she means. He has no real 'home life' in which to relax, & 'recoup' his strength & his peace of mind. If he wasn't such a wonderful character, he would sink under it."[14]

Toward the end of February, 1943, he did, in fact, sink, laid low yet again, this time by what he afterward called "sleeping sickness or Gambia fever or some kindred bug in that hell-hole of yours," as the President complained in a letter to the Prime Minister in London, that "left me feeling like a wet rag. I was no good after 2 p.m. and, after standing it for a week or so, I went to Hyde Park for five days."[15] Daisy looked after him there, recording in her diary on February 27 that it was "the P.'s. 4th day in bed, & he still feels somewhat miserable though his fever has gone. Last Tuesday, without any warning, he felt ill about noon. He lay on his study sofa & slept 'til 4.30, when he found he had a temp. of 102. The Dr. found it was toxic poisoning, but they can't ascribe it to anything they know of . . . The P. doesn't look well but is improving." After having supper with him, eating from trays, she gave him the aspirin Dr. McIntire had prescribed — and almost wept when he said: "Do you know that I have never had anyone just sit around and take care of me like this before." Apart from nurses when he was very unwell, "he is just given his medicine or takes it himself. Everyone else has been too busy to sit with him, doing nothing."[16]

If the President's condition — his tiredness, his fevers, and his everlasting sinus infections — caused him now to draw back a little from the more commanding role he'd taken in directing the U.S. military, this was understandable — in fact, to many in the War Department it was a relief, as planning for the Husky invasion, slated for July that year, went ahead. Even the U.S. setback at Kasserine had not worried him unduly or diminished his confidence in young General Eisenhower; it was, after all, proof of his wisdom in insisting American forces learn the skills of modern combat in a "safe" region of the Mediterranean, where they could swiftly recover.

When Eisenhower's naval aide, Lieutenant Commander Butcher, was brought to the Oval Office to report to the Commander in Chief on March 26, 1943, one of the first questions the President asked him was to give an account of the Kasserine debacle — from Eisenhower's perspective. "He wanted to know how things were going" in North Africa, Butcher recorded. Naturally, he knew them from "official reports," but he wanted to have the story from the horse's mouth. He was "inquisitive about Fredendall and other commanders at the front, the retreat of the Americans naturally being in his mind. I explained to him the reluctance Ike had in relieving Fredendall, and his hope that the change to Patton could be handled in such a way that Fredendall's fine qualities, particularly for training, would not be lost to the army."

To Butcher's surprise the President — who himself hated to have to fire people — seemed more interested in U.S. intelligence. He "wanted to know the circumstances that caused our G-2 [head of military intelligence] to predict that the main thrust of the Germans would come through the Ousseltia Pass rather than at Sidi Bou Zid. I explained to him

[British Brigadier] Mockler-Ferryman's reliance on one source of information, namely the interception of radio communications [Ultra] and that since this source theretofore had proven reliable, not only Mockler-Ferryman but [British General] Anderson, had relied on the 'Mock's' advice in this instance. This reliance had caused General Anderson to hold his reserve in the North when it may have been used to extra advantage to help Americans farther South."[17] It was unfortunate, but a lesson learned in the use and misuse of — or overreliance upon — Ultra.

Certainly the President's faith in Eisenhower was rewarded in March when Rommel, in ill health, was withdrawn to Germany to recuperate. The day of the Desert Fox was over; that of the President's protégé, Dwight D. Eisenhower, had come. He might not have the battlefield prowess of Rommel, Patton, or Montgomery, but he had something far more valuable to the Allies: the ability to get the soldiers, airmen, and sailors of an international coalition to fight together under his leadership. The result was often messy, sometimes contentious, and media-sensitive. As the President told Eisenhower's naval aide, however, such was the price of democracy. The virtue of the Casablanca Conference had been that it enabled the President, as de facto commander in chief of the Western Allies, to make his historic decisions on a 1944 Second Front as well as on unconditional surrender, without the press (let alone the enemy) even knowing he was in Casablanca. "He said for the first time all participants were enabled to explore each others' minds, get all the cards on the table, and reach decisions without distractions. These distractions, he said, are caused by newspaper men gaining small segments of the complete story and printing them under headlines that frequently mislead the public and failed to portray the complete story. 'In most conference[s], particularly where newspaper men have access to the conferees,' the President said, 'almost every participant has a pet newspaper man. By button-holing such friends, newspaper men can get a part of the story and the whole issue becomes tried in the press on the basis of only a small part of all the facts. The result is distortion to the public and disruption to the conferences.'"

No truer words were spoken by an American president — yet this had been the reality of American democracy since George Washington, and would never change. All one could do was, at certain times, employ a certain guile in order that the job got done. At this, the President, by his third term in office, was a past master — in war as well as in peace. "At Casablanca," he told Eisenhower's aide, "we had a secluded spot, well guarded and free from the press. Thus we were able to talk freely without feeling someone would start promoting his point of view in the press by means of contact with his favorite reporter."[18] So pleased was he with the "result of Casablanca" that he had arranged for the administration's looming "food conference," addressing the needs of allies and liberated countries, to take place in Virginia, "guarded by military police," and with "no press permitted . . . I think the press will cry out against this arrangement," but the "public good" was sometimes more important than "public discussion." Moreover, once the decisions were announced, there was freedom enough to debate the matters. "I am planning to make another swing around the country," he told Butcher, taking a group of White House correspondents who would only be allowed to file reports once the tour was over. "The press will yowl again I imagine, but the public seemed to appreciate that trip. In any event," he made clear, "I am going to do it again," yowls or no yowls.

Subtly, the President had been passing on to Eisenhower his advice on how best to deal with "distortions" and "distractions" of a free press — something Butcher was able to convey to Eisenhower as soon as he returned. Along with the President's parting words. "The principal message the President asked me to convey — and he spoke repeatedly of the General as 'Ike' — was: "Tell Ike that not only I but the whole country is proud of the job he has done. We have every confidence in his success."[19]

As the President prepared for his second "swing around the country" aboard the *Ferdinand Magellan,* Eisenhower duly readied his two Allied armies in Tunisia — gathering his twenty divisions like bloodhounds for the final act of the President's North African invasion: a battle the Germans themselves began to call "Tunisgrad."[20] More than a quarter million Axis troops were now hemmed in on the Cape Bon Peninsula, fighting for their lives. Two thousand German troops were being flown into the arena each day from southern Europe; Mussolini was begging the Führer to make peace with Stalin in order to save the Italian Empire; and three hundred thousand Allied troops were massing for the kill.

PART SIX

Get Yamamoto!

21

Inspection Tour Two

BEGINNING ON APRIL 13, 1943, the President set off by train for his latest two-week, seventy-six-hundred-mile inspection tour of U.S. military training bases: from South Carolina to Alabama, Georgia to Arkansas, Oklahoma to Colorado, Missouri to Kentucky. Following his repeated bouts of ill health in February, these inspections would allow the Commander in Chief to see — and be seen by — tens of thousands of young aviators, Marines, tank crews, infantrymen, Women's Army Auxiliary Corps trainees, and Navy crewmen.

Once again the President took Daisy Suckley with him — since Eleanor had her own agenda to fulfill — as well as his other cousin, Polly Delano, who was considered a "law unto herself," but who amused the President: the two women giving him the sense of being looked after. (Eleanor did agree to join the train, in Texas, for a brief three-day detour to Monterrey, to meet the Mexican president, Ávila Camacho.)

The President wanted to judge for himself whether young American servicemen, currently training at home, would fight abroad. Hundreds of young pilots taking off and landing, parades of ten to fifteen thousand men, tanks firing live shells in mock battle, soldiers in hand-to-hand fighting ("the sort of thing they have in the Pacific jungles, with the Japs — It's all horrible when you stop to analyze it, but it's a fight for survival," Daisy noted, amazed[1]).

The tour, the President was pleased to find, belied any German and Japanese assumptions that U.S. troops were too "soft" for the ruthlessness of modern warfare. Above all, however, it was the sheer magnitude of American mobilization for war — in manpower, munitions, organization — that awed the President's party aboard the *Ferdinand Magellan.*

"The impression I have is of vastness, and a miracle of quick construction," Daisy noted in Denver, where they inspected the Remington Rand Ordnance Plant. Propelled by Japan's sneak attack, America had become a new "melting pot," with "50% men and women at Remington, cheerful, well-fed human beings, who, with all their lack of culture, are the backbone of the country, & probably the finest 'mass' of population in the world," she noted proudly. "The women were dressed in pale blue 1-piece overalls (much like Mr. Churchill's air raid zipper suit) and red bandannas tied tightly about their heads . . . People were collected all along the route full of spontaneous enthusiasm. Women & girls jumping, waving, laughing & cheering. The men grinning broadly & waving."[2]

At the President's polio-rehabilitation center at Warm Springs in Georgia, Roosevelt stood tall, kept upright by his heavy steel leg braces, "holding on to his chair," and "made a serious, soft voiced little speech" to the hundred patients assembled in Georgia Hall, then was "wheeled to the door of the dining room where he stayed to shake hands with each patient that filed through."[3] Using "the little car he has had *for years* down here" — a 1938 Ford Roadster with brakes and accelerator he could operate by hand, as well as a license plate reading "F.D.R.-1 — The President" — Roosevelt himself drove his guests around the area.[4]

Daisy — visiting the Warm Springs center for the first time — was deeply moved. "It is certainly a monument to him, his imagination and his faith & his love for his fellow sufferers, and it is very lovely. Peaceful and beautiful. The houses homelike and attractive, mainly white, among trees." For the first time in months the President swam — and insisted Daisy and Polly swim too. He had seemed desperately tired when they left Washington. Now he was "visibly expanding and blossoming."

One night — after a simple, homely dinner which he loved, in contrast to the "pallid" White House food that Eleanor's cook, Mrs. Nesbitt, made and which Roosevelt "detested"[5] — Roosevelt took out "his stamps; the rest of us read. F. complained of a headache" and the women took his temperature — which was fortunately normal.[6] He'd seemed actually happy, though.

With the physical support of his new naval aide, Rear Admiral Wilson Brown — Captain McCrea having been assigned to command of the new U.S. cruiser, *Iowa* — the President was still able to stand and, by swinging his muscular torso, even walk. Visiting Fort Riley, in Kansas, he actually proceeded on foot to the exit of the amphitheater, where fifteen thousand troops had gathered for an Easter service. At the railway station, as the *Ferdinand Magellan* slowly pulled out, officers and men saluted the Commander in Chief. "It was a beautiful sight and the kind of thing that brings a lump in your throat, specially when the commander in chief is a man like F. & crippled besides — Our driver told us he had not the slightest idea that F. couldn't walk, that his brother officers also had never thought of it," Daisy noted. "F. is all the more an inspiration to them —."[7]

At dinner on the train on April 19, they were joined by Sumner Welles, and the Mexican ambassador, Francisco Castilia Nájero. "We stayed up until 10 listening to them talk about the future peace — Very interesting," Daisy recorded in her diary. There was, she recognized, a steeliness in her champion that was never going to allow him to let up until he'd achieved his dream — with little trace of magnanimity toward those responsible for the global holocaust the Nazis and Japanese were so adamantly pursuing. The "perpetrators of the war, like Hitler, Himmler, etc. shall be court-martialed in their own country" and hanged or "liquidated," as Daisy noted, quoting Hitler's sickening word — "not sent to some distant island to turn into heroes and martyrs, with the danger of their trying to come back."[8]

The President might show deep and natural empathy for his fellow polio sufferers, she recognized, and great charm toward visitors of every stripe — but his forgiveness did not extend to the Nazi "Aryans" who were exterminating not only the handicapped but, it was becoming increasingly evident, millions of Jews, homosexuals, gypsies, and political prisoners; sickening atrocities, moreover, that the Japanese were also reportedly committing, not only in the treatment of the populations of the countries Japanese troops had overrun, from China to the Philippines, but American POWs.

In his diary, Secretary Stimson made note of what the Operations Division of the War Department had learned. Colonel Ritchie "gave me a dreadful picture of what is happening to our prisoners of war at the hands of the Japanese in the Philippines. I have been thoroughly churned up over it ever since. They are being killed off and are dying off under mistreatment. The situation is frightful. Yet it is very dangerous for us to make it public because of the reprisals which would be undoubtedly visited upon these," he wrote — aware American prisoners would be tortured and executed for smuggling out news of their mistreatment. Nor could the United States threaten retaliation, "because we have only a few hundred prisoners" thanks to the Japanese code of Bushido, "while they have a good many thousands of our men . . . MacArthur is vowing vengeance and is keeping the score of injuries to our men which he has heard of which some day he hopes to live to avenge."[9]

News of the execution of captured crewmembers from Doolittle's air raid on Tokyo the previous year had aroused similar outrage — Stimson wanting Secretary Hull to issue a warning there would be American "reprisals" for such "an act of barbarism" if it went on.[10]

For such barbarians the President possessed, Daisy recognized, no sympathy. He would not permit MacArthur to carry out reprisals in the Pacific. But when, during his tour of U.S. training camps and manufacturing plants, a decrypt arrived via the communications car of the President's train of a Japanese signal giving the forthcoming flying itinerary of the Japanese commander in chief — the man who had launched the sneak attack on Pearl Harbor in peacetime, killing twenty-four hundred Americans in a single morning — the President, aboard the *Ferdinand Magellan*, had had no hesitation whatever.

22

Get Yamamoto!

SEVERAL WEEKS BEFORE leaving on his inspection tour the President had invited MacArthur's air commander, General George Kenney, to give him a literal bird's-eye view of the campaign in the Southwest Pacific, when Kenney accompanied MacArthur's chief of staff to Washington to ask for more reinforcements.

In Kenney — a World War I pilot almost as highly decorated (Silver Star and Distinguished Flying Cross) as MacArthur himself — MacArthur had recognized the right man to revolutionize the U.S. Army Air Forces in war: not only in combatting Japanese fliers and in bombing ground installations, but in decimating Japanese supply vessels. The result had proved transformative — and the President wanted to know how Kenney had done it.

Whereas carrier-plane pilots of the U.S. Navy had become expert at low-level attacks on Japanese shipping, the U.S. Army Air Forces' pilots had not, Kenney explained. He had therefore hurled himself into the challenge — developing new skip-bombing techniques and modified B-25 mast-height gunship tactics, which he'd ordered to be rehearsed against a partially sunken vessel off Port Moresby. Under his leadership the vessel-attack planes had adopted a new technique: to fly in at 150 feet — with P-38s and 40s providing higher air cover, and B-17s higher still.

At the White House the President had thus been enthralled as Kenney described his new approach. "I talked for some time with President Roosevelt, who wanted to hear the whole story of the war in our theater in detail," Kenney later recalled his first visit to the Oval Office, "as well as a blow-by-blow description of the Bismarck Sea Battle."[1]

Kenney's description of the battle had been especially telling, for the flier had explained how the Magic secret decrypts of Japanese communications that the President was seeing in his Map Room in Washington had enabled Kenney to put into effect his deadly new aerial war tactics in the field.

"In the nose of a light fast bomber," as Kenney also explained to Eisenhower's naval aide — for transmission to Ike in Algiers — the general had installed "eight 50-caliber machine guns. Two planes thus equipped would approach a merchant vessel at low level, one from stern to bow, the other from the side. The one approaching lengthwise the ship would open fire with the eight guns at 1500 yards. No pilot was permitted to go on a mission until he could shoot so accurately that with the first burst at 1500 yards, he could sweep the ship from bow to stern or vice versa. The purpose was to keep the anti-aircraft fire from the ship so the accompanying member of the team could approach the side of the vessel just above the wave tops and drop a bomb that would skip on the water and hit the ship on the side just above or below the water line."[2]

The President had been intrigued to learn not only of such American air force ingenuity and specialized training, but the integration of air and naval tactics.

Code breaking had been the key, though. In the battle of the Bismarck Sea, between March 2 and 4, Admiral Yamamoto's order for Japan's vital troopship-reinforcement convoy, bringing fresh troops up to Rabaul and from there to New Guinea, had been deciphered.[3]

The first Japanese division to arrive in Rabaul, the Fifty-First Division, had hugged the coast of New Britain and then set off by convoy across the Solomon Sea — unaware the Americans knew its route and composition. Flying at high altitude, Kenney's B-17 bombers had sunk two of its transport ships, but the remaining six, escorted by eight destroyers and a hundred planes, had ploughed on. Expecting B-17s at high altitude again over the Dampier Strait, the Japanese fighters giving air cover to the convoy had failed to spot Kenney's one hundred retrained American and Australian B-25s, A-20s, and Beaufighter pilots skimming low across the water — the aircraft so low the Japanese sailors thought they were torpedo planes. In short order all surviving Japanese transports had been sunk, the infantry drowned, and four of the eight destroyers destroyed — the core of the Japanese Fifty-First Division extinguished in a single day.

The President was clearly delighted. His grasp of the intricate mosaic of islands in the Southwest Pacific amazed Kenney, given the President's other responsibilities. "I found the President surprisingly familiar with the geography of the Pacific, which made it quite easy to talk with him about the war out there," the general recalled with admiration and affection after the war. "He wanted to know how I was making out on getting airplanes. I told him that so far my chances didn't look very good. When he asked why, I said that among reasons given me was that he had made so many commitments elsewhere that there were no planes left to give me."

The President had taken this in good spirit; he had "laughed and said he guessed he'd have to look into the matter and see if a few couldn't be found somewhere that might be sent me. He said that if anybody was a winner, he should be given a chance to keep on winning."[4]

Backing winners was important in war, the President recognized — and was a key to Roosevelt's growing style of military command: assessing, encouraging, and supporting those whom he saw as inspirational and effective.

The President did manage to find Kenney more planes, to Kenney's relief. However, if the President was keen to back a winner in Kenney's air force leadership in the Pacific, he was similarly open to depriving the Japanese of *their* outstanding military leadership, if he could.

General Kenney's visit to the White House — a visit repeated on March 25, 1943, when Kenney had attended a Congressional Medal of Honor ceremony — had convinced the President that American fliers were finally proving better than their opponents in the Pacific. Especially when given the advantage of Ultra intelligence.

As the *Ferdinand Magellan* made its way across the American Midwest and South, stopping at military training camp after training camp, the President found himself, as Daisy noted, more and more confident in American professionalism. And though the matter was too secret to share with Daisy or her companion, Polly Delano — moreover, too secret ever to be revealed in his lifetime — he now had an opportunity to show his faith in his American airmen.

Over several days, starting on April 14, an extraordinary series of further decrypts had been brought to the President aboard the *Magellan* by Ship's Clerk William Rigdon, an assistant working for Admiral Brown, the President's naval aide, whose job it was to bring the latest fruits of Ultra to the President's attention twice a day.[5]

"The communications car housed a diesel-powered radio transmitting and receiving station," Rigdon recounted later, "that kept the President in constant touch with the Map Room at the White House. Special codes, held only by the Map Room and the car, were used. This car was just behind the engine. The Magellan was at the rear. Between the two I walked many miles taking messages to the President and picking up those he wished to send."[6]

Some were trivial. Others were more serious. One of them, in particular, related not to security for the President's train schedule — the twenty-six members of the Secret Service traveling with him, as well as military details protecting him at every stage of his 7,668-mile trip — but to the travel plans of another dominant figure in the war, with perhaps even more control over the struggle in the Pacific than the President: Admiral Isoroku Yamamoto.

"From Solomons Defense Force to Air Group #204, Air Flotilla #26," it began. "On 18 April C in C Combined Fleet will visit RXZ [Ballale Island, off Bougainville], R_ [Shortland] and RXP [Buin] in accordance following schedule: 1. Depart RR [Rabaul] at 0600 in a medium attack plane escorted by 6 fighters. Arrive RXZ at 0800. Proceed by minesweeper . . . At each of the above places the Commander-in-Chief will make a short tour of inspection and at _ he will visit the sick and wounded, but current operations should continue."

In case of "bad weather" the preliminary message had ended, "the trip will be postponed."[7]

A trip by the commander in chief of Japan's Combined Fleet, Admiral Yamamoto, to Ballale, Bougainville? An inspection tour by air and sea to a forward area of the Solomon Islands within reach of U.S. Air Force planes? Times and details of his itinerary?

The decrypted signal seemed almost impossible to believe. Admiral Yamamoto usually stayed on his grand battleship, the *Musashi,* at Truk, in the Caroline Islands, eight hundred miles to the north of Rabaul. Moving to a temporary forward command post at Rabaul, however, his strategy after the loss of Guadalcanal had been to pummel the Americans with massive air attacks before they could bring up enough forces to exploit their victory: Operation I-Go. By assigning not only Japanese ground-force pilots operating from airfields in the Solomons but hundreds of well-trained carrier pilots to assist them, Yamamoto had been able to apply massive Japanese air power to the initiative, involving more than 350 planes — the largest Japanese air assault since Pearl Harbor.

Mercifully, the Japanese air armada had been thwarted by Ultra intelligence — allowing Allied naval ships to disperse in good time, and U.S. Army and Naval Air Force units to be ready, off the ground, to meet the approaching aerial fleet each time it flew. A single Allied destroyer, a corvette, several Dutch merchant vessels, and an oiler had been sunk, and twelve Allied aircraft lost, but these were small pickings for such a concentrated and expensive air offensive — a fact that Japanese pilots, despite acknowledging the loss of forty-nine Japanese planes, had misconstrued in their after-action reports. Admiral Yamamoto had, instead, been told the fliers of his Third Fleet and Eleventh Fleet had sunk one American cruiser, two destroyers, and twenty-five transport ships, moreover had shot down 134 U.S. planes, as well as destroying 20 on the ground.[8]

The admiral had been well satisfied — in fact had sent Emperor Hirohito in Tokyo his own version of the triumph, which could be seen as avenging, in part, the recent losses of Guadalcanal and Buna. The Emperor had immediately responded with a congratulatory signal: "Please convey my satisfaction to the Commander in Chief, Combined Fleet, and tell him to enlarge the war result more than ever."[9]

The radio messages giving the itinerary of Admiral Yamamoto's inspection tour raised a number of questions. Was Yamamoto planning an extension of I-Go attacks? Was it a morale booster by the Japanese commander in chief, in person?

The message did not say — but its import was clear to all, from the South Pacific to the *Ferdinand Magellan.* Just as Yamamoto's planes had been able to hit Guadalcanal as part of I-Go, so could U.S. fliers hit Bougainville, on the admiral's itinerary — either attacking the admiral on the minesweeper to which he was slated to transfer, or in the air.

Why, though, had Yamamoto chosen to send such a message by radio?

As it later transpired, the admiral's administrative staff officer had wanted the warning order to be couriered by air, and then hand-delivered to its recipients. He'd been told by the communications officer at Rabaul not to worry, however; the Japanese naval code JN25 had recently been changed and was unbreakable.[10] The signal that first went out was dated April 13, 1943.

"We've hit the jackpot," the U.S. watch officer of Station Hypo, the two-thousand-man decoding unit in Hawaii, declared when the decrypt was handed to him. "This is our chance to get Yamamoto."[11]

If, that was, Admiral Nimitz, the commander in chief in the Central

Pacific, agreed. And if the U.S. commander in chief in Washington signed off on the attempt.

There were important repercussions to be considered. An aerial interception of Japan's most famous — or infamous — admiral might well squander, whether successful or not, America's prize weapon in the struggle against Japan: Ultra. Was it worth such a gamble? And what if it did not, in fact, succeed? Not only would Yamamoto be left in command of all Japanese forces in the Pacific, but the war-winning contribution of Ultra would have been given away, for nothing.

Ironically, Admiral Nimitz worried about something else when first shown the decrypt. Would a successor to Admiral Yamamoto prove a better Japanese commander in chief?

In view of the fact that Yamamoto enjoyed almost godlike status,[12] not only among Japanese forces in the Pacific but at home in Japan, killing him would, without doubt, make a huge dent in Japanese war morale, just as Japanese forces dug in for a do-or-die struggle in the countries they had conquered.

There were other questions, too. In the time-honored ethics of American warmaking, was it even acceptable to assassinate an enemy commander — since assassination was what such an interception, if successful, would be? As at Midway, Nimitz felt it would be a mistake *not* to use such a priceless intelligence breakthrough — but deferred, as before Midway, to Washington's decision.

The matter was thus passed for authorization to Admiral King — who passed it to Navy Secretary Frank Knox.

Knox — who had himself recently carried out an inspection tour of U.S. naval forces in the Pacific — passed the information via the White House Map Room to the *Ferdinand Magellan.*

The President's response was immediate and uncompromising: "Get Yamamoto."[13]

Secretary Knox, at the Navy Department, needed no further prompting. He instructed Admiral Nimitz to go ahead. In turn Nimitz gave the final green light on April 17, 1943, to Admiral Halsey — commanding Allied forces in the Solomons area.

There was now only one day to go. F4F Wildcats and F4U Corsairs had insufficient range for such a mission, but new U.S. Army Air Forces P-38s, flown by Army Air Forces, Marine, and Navy pilots, could do it. Vice Admiral Pete Mitscher, commander air, Solomon Islands, had already begun to explore different proposals with his subordinates. His U.S. Navy fliers recommended they attack the Japanese vessel that Yamamoto was to board at Ballale, but the U.S. Army Air Forces ace, Major John Mitchell, commanding the 339th Fighter Squadron, assigned to carry out the attack, felt it would be easier to spot the admiral's plane — a heavily armed but slow (265 mph) Mitsubishi G4M "Betty" bomber — than it would be to identify an indeterminate Japanese minesweeper. Such a ship would undoubtedly be escorted by other vessels, as well as shielded by extensive Japanese air cover, to judge by the U.S. planes sent up to protect Secretary Knox on his recent visit.

Mitscher wisely yielded the decision on April 17 to the man who would have to carry out the mission. Assembling a fighter group of eighteen Lockheed P-38G Lightnings, Mitchell — who planned a five-leg, low-level end run way out to sea before reaching Bougainville, so as to have the advantage of surprise — asked for special auxiliary-fuel drop tanks flown up from MacArthur's men on New Guinea. He also decided that the best point to intercept Yamamoto's flight would be just as the admiral's Betty bomber — similar to the Betty bombers that had sunk HMS *Prince of Wales,* the battleship on which the President had attended divine service with Churchill off Argentia, in August 1941 — reduced speed to land.

Tension was high. Mitchell was certainly aware just how much hinged on his mission, for Secretary Knox had sent a further signal to Nimitz on April 17, which Admiral Nimitz immediately forwarded to Admiral Halsey. Halsey sent it with a covering note to Admiral Mitscher — who placed it before Mitchell. It read, as the fliers later recalled: "SQUADRON 339. P-38 MUST AT ALL COSTS REACH AND DESTROY. PRESIDENT ATTACHES EXTREME IMPORTANCE TO MISSION."[14]

Almost seventy years later a similar targeted killing of an enemy commander in war — Osama bin Laden — would be revealed to an astonished world within hours. Because of the need to preserve the secret of Magic in 1943, however, the President had accepted that the mission to assassinate Admiral Yamamoto, code-named Operation Vengeance, would not — perhaps ever — be made public. Other than acknowledging Japanese media reports, if the operation was successful, nothing would be said.

Early on the first anniversary of the Doolittle Raid the U.S. Squadron

339 fighter group — reduced to sixteen planes owing to two aborts — thus set off from Fighter Strip Number Two, Henderson Field, Guadalcanal, on its thousand-mile mission for the President.

Resembling huge flying catamarans — their single-pilot cockpits strung between two pontoon-like fuselages, each mounting a massive 1,325-horse-power engine and capable of speeds up to 400 miles per hour — the Lightnings wave-hopped in complete radio-silence for some six hundred miles to the west of the Solomon island chain and across the open Solomon Sea, in order to avoid radar and visual detection. Using a special naval compass, Mitchell then swung to the east, aiming to circle in from the ocean — four of his best pilots designated as killer sharks, while the remainder dealt with the six Japanese Zeros protecting their commander in chief.

The attack — the longest-distance fighter-intercept mission of World War II — went like clockwork. A stickler for punctuality, Admiral Yamamoto had rejected his staff's protests that his inspection tour would be too risky. Precisely on time, his bomber slowed to land at Ballale, Bougainville, at 9:35 a.m. on April 18, 1943.[15]

Mitchell's men were already there, sixty seconds early: surprised only that there were two Bettys, not one. Both would have to be shot down.

The ensuing melee took but a few minutes — the "killer" fighters attacking from below the Japanese bombers and their escort, the rest climbing above the encounter to fend off Zeros that would inevitably begin to take off from Buin airfield. As the lumbering Betty bombers dived to escape, the two lead P-38 pilots closed up on them, using the 20mm cannons and machine guns mounted in the planes' noses. Inside his bomber, Yamamoto, dressed in his green field-combat uniform but wearing white gloves, was killed instantly in his seat — the plane soon plummeting to earth in the jungle, amid smoke and flames. The other bomber was also shot down, crashing into the ocean. One American P-38 was lost, but the rest then turned for home, taking the direct route and encountering no opposition.

Ignoring all prohibition against radio transmissions that might give away the specific target of the operation, one of the P-38 pilots, with one engine already feathering for lack of fuel after a thousand miles of flying, radioed to fighter control at Henderson Field as he came in to land: "That son of a bitch will not be dictating any peace terms in the White House."[16]

23

"He's Dead?"

So secret was the Yamamoto operation — and so worried did Admiral King become, when leaks of the mission to journalists were only censored at the last minute — that the men of 339 Squadron, unlike Doolittle's Tokyo team, could not be decorated for their extraordinary bravery and professionalism. So shocking, however, was the death of Admiral Yamamoto to the Japanese government, that news of his passing was kept from the Japanese public for more than a month[1] — and only confirmed to American code breakers, in the meantime, by the absence of Yamamoto's name or rank in Japanese naval signals decoded in Hawaii and Washington.

Aboard the *Ferdinand Magellan* on April 18, after visiting Camp Gruber, Oklahoma, and messing with the troops training there, the President was informed that the mission had been successful. Pearl Harbor had finally been avenged — the author of the sneak attack dead in the jungle his men had so ruthlessly conquered, and where so many Japanese atrocities had been committed.[2]

Keeping the success of Operation Vengeance a secret among American forces, ironically, proved harder than keeping the Ultra secret from the Japanese. In the days after the mission, more sorties were flown up the Solomons "Slot" to Bougainville, in full view of Japanese radar and spotters, to make the fatal interception seem less extraordinary; reporters, meanwhile, were forbidden to file stories directly linking Guadalcanal's U.S. air aces to the great admiral's death.

In the context of a titanic battle of wills, courage, and morale in the Pacific, it was almost impossible to maintain the fiction of an accidental death of an enemy commander as iconic as Yamamoto, however, either in

American signals or media reports. Admiral Mitscher had immediately reported to Admiral Halsey by telegram, for example, informing him Mitchell's P-38s had "shot down two bombers escorted by 6 Zeros flying close formation . . . April 18 seems to be our day" — a reference to Colonel Doolittle's raid the year before.[3] This had not necessarily given the game away, but Bull Halsey — known for his ebullient, take-no-prisoners personality — had signaled straight back, "Congratulations to you and Major Mitchell and his hunters. Sounds as if one of the ducks in their bag was a peacock."[4]

Inevitably, as the weeks went by, an AP reporter dutifully tracking a story that was common knowledge on Guadalcanal, blew it in Australia in May 1943 — despite being warned it should not be used. Admiral King was incensed — just as he had been the year before, when a similar report had been published ascribing the great victory at Midway to naval codebreaking. Nimitz ordered a full-scale investigation and disciplinary action, with orders that every pilot on Guadalcanal, as well as staff officers and flight mechanics, be questioned. Four citations for Medal of Honor awards were withdrawn, and Halsey, blaming the pilots, declared they should be court-martialed, stripped of their rank, and jailed.[5]

By a miracle, the Japanese, however, proved unable or unwilling to investigate too closely their suspicions. The sheer shock and shame of the admiral's death — the six Zero pilots wishing to commit ritual suicide for their utter failure to protect him — caused a wave of gloom to spread from Bougainville to the Emperor's palace in Japan. Yamamoto's ashes were taken back to Japan aboard his battleship, the *Musashi,* and after lying in state in Tokyo, were interred in a state funeral.[6]

In later years there would be claims that Admiral Yamamoto's death had been a dangerous gamble and counterproductive, given that his earlier objections to war with the United States would have made him, if brought back into the Japanese government, more willing than his successor (who was killed in 1944) or other Japanese admirals to negotiate an armistice.[7]

President Roosevelt certainly had no truck with such hypotheticals. He had laid down a policy of unconditional surrender on the basis that neither the Germans nor the Japanese could ever again be trusted to keep the peace unless forced into complete surrender — and the way Japanese troops were fighting, from the Aleutians to the Solomon islands they had conquered, and the atrocities they were committing, bore out his contention. In Yamamoto's cabin aboard the *Musashi* the admiral had left a poem he'd recently written, lamenting his "dead comrades" in the war—but declaring "with an iron will I will drive deep / Into the camp of the enemy / And will show the true blood of a Japanese man."[8]

Japan's retreat into medievalism, like Germany's, was something America could only end by force of arms—not negotiation. In depriving the Japanese of their greatest admiral in the war, Roosevelt had struck an incalculable blow to the Japanese military machine and national morale at a critical moment in the war.

Certainly, as the President continued his national inspection tour, he was seen to be in great form. The training establishments he'd visited had given him a potent sense of American willingness to fight—and to win. At a press conference he convened onboard the *Ferdinand Magellan* on April 19, he spoke of the "great improvement I have seen since last September in the training of troops of all kinds," and referred to the "cutting down of the age of the higher officers than in the last war." There was, too, higher "morale" to be seen among the troops as they learned the deadlier skills of modern combat—"there is a great eagerness on their part to get into the 'show' and get it over with."[9] Moreover, the strategy he'd settled at Casablanca seemed to be working out. In North Africa, putting Kasserine behind them, American troops were moving in for the kill in Tunisia, close to the time frame Eisenhower had given him at the Villa Dar es Saada. And in the Atlantic—following the President's order to Admiral King to resolve the interservice argument regarding air coverage of the mid-Atlantic and the alarming success of U-boat wolfpacks or face dismissal[10]—King had buckled.

Finally setting aside his childlike struggle with the U.S. Army Air Forces over which service should be responsible for antisubmarine air patrolling, King—in fear of losing his job—had convened a conference of all parties in Washington at the beginning of March. Chastened, he'd belatedly established a special headquarters in the Navy Department, the so-called U.S. Tenth Fleet, to direct the anti-U-boat campaign: a campaign that would use new ASDIC 271M centimetric radar capable of detecting a submerged submarine four miles away; high-frequency radio direction finding (HFRDF) to pinpoint where U-boats were signaling from; "baby flattops" the President had earlier ordered to be constructed from merchant ship hulls as convoy escort carriers; incoming new Bogue-class aircraft carriers; and most important of all: American Liberator long-distance bomber planes.

Based on the Consolidated B-24 USAAF bomber, but equipped with torpedo-like depth charges, radar, and Leigh lights to illuminate U-boats surfacing for night attacks on shipping, the Liberators would now almost instantly turn the tide of war in the Atlantic — completely disproving King's assumption that convoying was the only answer. Within weeks the new combination had worked — the results beginning to show already in April 1943. Sinkings of German U-boats increased — dramatically.

The President was relieved — and Admiral King relieved not to be relieved of his command. By May the demise of the wolfpack menace would become a rout, forcing the commander of the German navy, Admiral Dönitz, to admit defeat that month and recall his entire submarine fleet to safety in Europe, pending the construction of more modern submarines with "snorkels" that could hopefully evade air detection.

And with regard to the death of Admiral Yamamoto? The President was careful to say nothing to anyone until May 21, 1943, when giving his 898th press conference.

"Mr. President," one reporter asked innocently, "would you care to comment on the death of the Japanese admiral (Isoroku Yamamoto), who forecast he would write the peace in the White House?"

"He's dead?" the President asked, as if stunned.

"Q[uestioner]: The Japanese radio announced it. Yamamoto. Killed in action while directing operations in an airplane."

"The President: Gosh! (loud laughter)"

"Q: Can we quote that, sir?"

"The President: Yes. (more laughter)."[11]

To his own staff the President was less deceptive. In truth he could never forgive Yamamoto for his role in attacking Pearl Harbor, causing the deaths of so many thousands of Americans there and in the aftermath, after the many years of hospitality and education Yamamoto had enjoyed in the United States. Two days after the press conference, the President thus had his secretary, Grace Tully, type a letter, headed The White House, dated May 23, 1943, and which he signed.

"Dear Bill," the President scrawled across the top of the letter in his own hand as a memo to Admiral Leahy, "Please see that the Old Girl gets the following:

'Dear Widow Yamamoto,

Time is a great leveler and somehow I never expected to see the old boy at the White House anyway. Sorry I can't attend the funeral because I approve it. Hoping he is where we know he ain't.

Very sincerely yours,

Franklin D. Roosevelt'

"And ask her to visit you at the Wilson House this summer," Roosevelt added in a postscript to Leahy.

It wasn't kind, or gracious; indeed the President never sent the letter.[12] But it reflected something of what, in his heart of hearts, he really felt about Japanese perfidy. And his profound satisfaction that he'd been able to see Admiral Yamamoto get his just deserts.

Ending his long inspection tour at Washington's Union Station on April 29, the President certainly had good reason to be in high spirits.

April had been a bountiful month for the Allies. Admiral Mineichi Koga would be a very poor replacement for Yamamoto — indeed, I-Go was called off, and on Guadalcanal the 339th Squadron's seventy-six pilots did not encounter a single Japanese plane in combat for the rest of April and the whole of May.[13] Staging out of the Ellice Islands — which Marines had captured the previous October — Admiral Nimitz's long-range bombers had been able on April 20 to hit Tarawa, the atoll that was impeding future invasion of the Marshall Islands — some twenty-four hundred miles from Hawaii.[14] General MacArthur had begun to revise his strategy in order to do more with less — persuaded by Washington, in fact, to drop his costly notion of step-by-step advance and merely bypass Japanese "fortresses" in the Pacific, such as Rabaul, if possible. Such strongpoints would thus be allowed to wither on the vine as MacArthur's air forces, ground forces, and naval vessels pursued a leapfrogging, or island-hopping, campaign instead.

All in all, then, the Allies stood fair to succeed in a two-ocean war — if they made no more mistakes, and capitalized on their growing productive and fighting strengths. By the end of the current year, the President had been told by Secretary Stimson, the U.S. Army would have some 8.2 million well-trained men and women in uniform — including more than 2.5 million U.S. Air Forces personnel. With a target of a million U.S. combat troops to be ferried in the coming months to bases in Britain, the President felt the Allies had every prospect of mounting a successful

1944 Second Front, and be on course to win the war that year, or early in 1945—after which the unconditional surrender of the Japanese could be obtained, he was confident, within months.

Such heady confidence, though, rested on a fundamental assumption: that Winston Churchill and the British would stand by the agreements made at Casablanca.

That, however, as Admiral Leahy and General Marshall informed the President at the White House on April 30, was probably misguided. Instead, the Prime Minister, they'd learned, was intent upon coming to Washington with a huge new posse of military advisers and clerks—determined to convince the President his whole strategy was wrong.

PART SEVEN

Beware Greeks Bearing Gifts

24

Saga of the Nibelungs

WITH EVERY NEW day, the news from North Africa had been getting better, the President had felt — under American leadership and arms.

By the end of April, American forces in North Africa outnumbered British, French, and other national contingents 60 percent to 40. From a frontline west of Bizerte that ran first south, then east to Enfidaville on the Tunisian coast, more than three hundred thousand Allied troops were preparing to launch Operation Strike: the final Allied offensive in North Africa to drive the Axis forces into the sea. Italian troops were beginning to desert in increasing numbers, but German troops were paradoxically selling their lives ever more dearly in battles to hold onto djebels and hilltops many thousands of miles from their homes — infused with a blind, arrogant loyalty to their comrades, scorn for their opponents, and a suicidal unwillingness to question either what they were doing in North Africa or why they maintained such slavish faith in their führer.

Certainly the Führer was indifferent to their fate. At his meeting with Mussolini near Salzburg on April 8, he'd dismissed out of hand the notion of a negotiated armistice with Stalin, or revival of the Ribbentrop Pact. As at Stalingrad, he was banking upon his understanding of the unique German psyche: that the members of his chosen Volk would stay loyal to each other, whatever happened; and that, in the manner of the Nibelungen myths, they would only gain greater nourishment for their national pride from stories of heroic valor and self-sacrifice, even death in distant fields. *Nibelungentreue* — whether on the Volga or in the mountains of Tunisia. Not dishonorable retreat or evacuation.

The tenacity and blind courage shown by soldiers of the Wehrmacht to their comrades in battle in North Africa certainly suggested Hitler was right. German casualties had escalated as the end approached, yet in contrast to Italian troops, far from dispiriting the German survivors, the likely outcome appeared to make no discernible dent on their morale in the field. Nor would the Führer countenance plans for Axis flight. Just as he had ordered von Paulus to die rather than surrender his last remaining forces at Stalingrad, so now Allied code breakers read with amazement the decrypted signals in which, from his East Prussian headquarters at the Wolfschanze, or Wolf's Lair, the Führer not only ordered more infantry reinforcements to be flown into the last Axis redoubt — which was now down to only sixty-seven panzers — but declined to permit the word *evacuation* to be spoken.

The Saga of the Nibelungs was thus being enacted — in real life. Allied planners had assumed in early April that Hitler could, if he chose, save as many as thirty-seven thousand men of the Wehrmacht per day by evacuation — the better to defend the shores of mainland Europe. There came, however, no such order. Instead, on April 13, the Führer had dispatched his historic cable to General von Arnim, in command of the quarter million Axis troops in Tunisia. Except for a few "useless mouths" to be airlifted or shipped out of Tunisia, the Axis forces were ordered to fight to the death[1] — killing as many of the Allies as possible before they were themselves felled.

It was a bloody, tragic prospect. Yet thanks to his insistence on Torch as the means by which American forces could first learn how to defeat the vaunted Wehrmacht in battle before embarking on a Second Front, it was also a tribute to the President's patience and determination not to undertake military operations beyond the capabilities of his forces. General Patton — "our greatest fighting general," he called him[2] — had restored morale in II Corps after Kasserine, and was now slated to command all American troops in Husky, the invasion of Sicily, in July — which would allow the Allies to rehearse a major assault landing, this time against Axis defenders, not Vichy French. Meantime, U.S. air forces were beginning to take a huge toll of Axis shipping as well as of the Luftwaffe. Above all, despite the mischief being sewn by the American and English press — delighting in the rivalry between U.S. and British exploits in the field — General Eisenhower was doing a magnificent job in holding together the Allied military coalition in North Africa.

This, more than anything, was what reinforced the President's faith in the outcome of his grand strategy. Hitler and Hirohito might well wish to see their populations obliterated rather than save them, but as long as the Allies held together and continued to build upon their combined strength, they would prevail, he was certain. The timetable General Eisenhower had given him at Casablanca for clearing North Africa of Axis forces, as the final Allied offensive kicked off on May 6, 1943, looked remarkably prescient—indeed, in a brilliant armored coup, British tanks from Montgomery's Eighth Army, stalled beneath the high ground at Enfidaville, performed a magnificent end run, or left hook, which took them into the city of Tunis itself within twenty-four hours, on May 7—where they took the unconditional surrender of all Axis troops there. Infantry and tanks of General Bradley's U.S. II Corps force simultaneously smashed their way down from the mountains in the northwest—including famously bloody combat around Hill 232—into the port city of Bizerte.

General von Arnim's days, perhaps hours, seemed numbered—U.S. and RAF planes swooping on any German or Italian vessel attempting to leave North African shores, while Luftwaffe attempts to fly in final supplies were shot down.

By contrast, the *Queen Mary*—the vessel bearing the British prime minister—was making a mercifully safe passage across the Atlantic—indeed was approaching the East Coast of the United States surrounded by U.S. destroyers and escort vessels, the sky above thick with U.S. planes watching for U-boats as it made its way toward the Statue of Liberty without mishap. Ensconced in the grand staterooms he'd ordered to be reconstructed for his voyage (the transatlantic liner having earlier been converted into an Allied troopship), Mr. Churchill was toasting every new report from London and Algiers: drunk not so much from champagne as sheer excitement over the imminent Allied victory in Tunisia—one that would soon exceed the German Sixth Army surrender at Stalingrad.

After his long years of military failure, the Prime Minister felt wonderfully, arrogantly alive, his staff later recalled: seemingly certain he could, by the force of his ebullient personality and the scores of staff officers and advisers he was bringing with him, reverse the agreements he'd made on behalf of his country at Casablanca.

25

A Scene from *The Arabian Nights*

At the White House, the President, having discussed with Admiral Leahy and General Marshall what the British might be plotting, still found it hard to believe.

The President had hitherto been under the impression that his partnership with his "active and ardent lieutenant" was a firm and happy one. Had the two leaders not motored together, after the Casablanca Conference, to Marrakesh — the fabled Berber city at the foot of the Atlas Mountains? Had they not spent the night there, in the house occupied by the American vice consul, Kenneth Pendar? Had they not settled together into "one of the showplaces of the world," as Pendar afterward described it, a "stylized, modernized version of a south Moroccan *kasbah*"?[1] Had not Churchill asked to be shown up the famous tower, and had he not counted the sixty steps before asking whether Pendar thought it possible "for the President to be brought up here? I am so fond of this superb view that it has been my dream to see it with him"?[2] And had it not so been arranged — the six-foot-three-inch paralyzed president of the United States borne up the sixty-foot tower by his attendants "with his arms around their shoulders, while another went ahead to open doors, and the rest of the entire party followed"? With Churchill humming "Oh, there ain't no war, there ain't no war," had not the President "amidst much laughing on his part and sympathizing with his carriers" been brought up to the open terrace of the tower, and had not a wicker chair been "fetched for him and the Prime Minister," allowing the two leaders to sit and survey the vast Atlas range? "Never have I seen the sun set on those snow-capped peaks with such magnificence," Pendar — who ordered highballs to be fetched and served — certainly recalled. "There had evidently been snow storms recently in the mountains, for they were white almost to their base, and looked more wild and rugged than ever, their sheer walls rising some 12,000 feet before us."[3]

Had not Pendar explained the history of the great twelfth-century Koutoubia Mosque tower that they could see some distance away, dominating the city? As the sun finally disappeared, had not electric lights come on at the top of every mosque, calling the faithful to prayer? "From where we were, we could see the going and coming of the innumerable Arabs on camel- and mule-back, as they made their way in and out of the city gate. Both Mr. Roosevelt and Mr. Churchill were spellbound by the view," Pendar recalled.[4] It had been a far, far cry from Washington, D.C. — followed later by more cocktails in the salon.

Had not the President — after slicing off the top of a huge profiterole representing the Koutoubia tower, in the manner of Alexander the Great — then raised a toast to the English king? Had not Churchill responded by raising a toast to the head of state before them: the President? For his part Pendar, sitting between such exalted modern rulers, had found himself "surprised," he later recorded. Traveling in England, he'd seen Mr. Churchill often in the prewar days, and had felt "sure that no one could eclipse his personality." Now, however, he was "struck by the fact that, though Mr. Churchill spoke much more amusingly than the President" — mesmerizing listeners with his antiquated yet masterly use of language, and his descriptive, imaginative storytelling ability — it was Mr. Roosevelt who "dominated any room they were in." Reflecting on this, Pendar attributed it not to Roosevelt's larger physique when compared to the diminutive prime minister, nor to his rank "merely because he was President of the United States;" no, Pendar had afterward mused, it had resided much more in the radiance of the President's "being": his leonine head and Caesar-like *presence.* Also an intense curiosity about others — others as real people, not simply an audience to entertain or impress. In this respect, despite the Prime Minister's extraordinary mind, the President exhibited, Pendar thought, "a more spiritual quality than Mr. Churchill, and, I could not help but feel, a more profound understanding of human beings," rather than just the course of history.[5] Most surprising to Pendar, perhaps, had been the President's seeming indifference to his own disability — as if employing his abundant interest in people not only to engage those he met in conversation, but to deflect attention from his own paralyzed lower limbs. Nor did he seem to mind being contradicted or corrected, as Churchill did — Pendar recalling how he'd talked "at length about the Morocco and the Arab problem" with the

President at dinner — who was not only well informed, but *listened.* "To my amazement and delight, I found that the President had an extraordinary and profound grasp of Arab problems, of the conflict of Koranic law with our type of modern life and its influence on Mohammedans, and of the Arab character with its combination of materialism and highly developed intuition." In the presence of a diplomat steeped in the history and culture of Morocco, the President had seemed fascinated to hear Pendar's views — Pendar subsequently recalling how, "some six months later, when I was in London talking with Averell Harriman," who had attended the dinner, Harriman "began to laugh and said: 'I will never forget your conversation with the President. I enjoyed hearing you explain to him, in no uncertain terms, that the New Deal simply wouldn't work in Morocco.'"[6]

Had not the two potentates then "set to work," after dinner, writing cables to Stalin and Chiang Kai-shek to tell them, cautiously, of the Casablanca meeting — cables that Churchill's ubiquitous secretaries typed, then retyped to incorporate further corrections and revisions? The President, at one point, had been "wheeled into his room so he could work alone at his dressing table which he used as his desk" — anxious not to dismay Stalin by revealing the Western Allies would not launch a Second Front before 1944, by which time their forces would have sufficient combat and command experience to make such landings in northern France decisive for the outcome of the war . . .

Why, then, three months later, was Winston Spencer Churchill on his way to Washington with an army of staff officers and advisers to argue against a cross-Channel invasion even in 1944? What alternative plan did Mr. Churchill have for continuing the war?

26

The God Neptune

THE PRESIDENT WAS as much in the dark about Churchill's plans as were his Joint Chiefs of Staff. In fact the more so, since he had fondly imagined that he and the Prime Minister were very much in unison with regard to the Allied prosecution and timetable of the war.

Instead, according to the President's best information, Churchill's ever-fertile mind was changing from day to day. According to sources known to the British representative on the Combined Chiefs of Staff Committee in Washington, Field Marshal Dill, the Prime Minister was said to be settling more and more on ditching the notion of a cross-Channel attack, and instead exploiting the Allies' impending victory in North Africa in the Mediterranean.

Churchill's preference, it was reported, was to pursue, instead, an opportunistic strategy of multipronged Allied attacks following the invasion of Sicily: not only on the Italian mainland but in the Aegean and the Balkans in late 1943 and 1944, especially if — President Inonu's unwillingness nothwithstanding — Turkey could be persuaded to enter the war on the Allied side. By this scattershot, indirect method Churchill apparently hoped the Allies would not only draw away from the Eastern Front crucial German forces that Hitler might otherwise employ to hold back the Russian armies, but would provide the Western Allies with the launch pad for a drive into central Europe via the "soft underbelly" of southern Europe: an Allied advance such as the one General Giraud had outlined to him at Casablanca. Or through the Balkans — an avenue of advance that harked back to Churchill's abiding Dardanelles obsession. Either way, such a peripheral strategy would serve to avoid a Second Front bloodbath across the English Channel, which the Prime Minister had always feared.

More disturbingly — again, according to Field Marshal Dill — the British chiefs of staff were now deferring to their prime minister's ideas. The result would be to delay, if not rule out, the agreed Second Front assault across the English Channel to 1945 — two years away — at the earliest.

How the Russians would respond to such delay was predictable. So too would be the response of the American press and public, if they learned of it.

It was small wonder, then, that the U.S. chiefs of staff had grown each day more worried as the *Queen Mary*, which had left port on May 5, drew closer — even as Allied forces moved in for the kill in Tunisia.

"Some of our officers have a fear that Great Britain is desirous of confining allied military effort in Europe to the Mediterranean Area in order that England may exercise control thereof regardless of what the terms of peace may be," Admiral Leahy had noted in his diary on May 2 — his contacts in the State Department fueling his fear that the British were "principally concerned with a post war control of the Mediterranean."[1] Moreover, in view of rumors the Russians were already exploring peace feelers with German representatives, Leahy was doubly concerned lest the Soviets would fight only to liberate Soviet republics, not to defeat the Third Reich. In this potential scenario, Hitler would remain master of western and central Europe, making nonsense of the President's "Germany First" strategy since Pearl Harbor.

For his part, Secretary of War Henry Stimson worried about Churchill's eloquence — and what he saw as the President's unwillingness to put Churchill in his place.

Stimson had not attended the Casablanca Conference, but what he had gleaned of it had been alarming — an account obtained in large part from officers such as Major General Wedemeyer. The British team had run rings around their American "opponents," he'd been told, not only because the U.S. team had been too small, but because the President himself was too accommodating to the British.

By May 7, with U.S. troops entering Bizerte and British troops entering Tunis, Stimson was cock-a-hoop at the "great victory" at hand — one that would "hearten the Russians and discourage the Germans." The Western Allies should therefore be thinking big, not small, in his view: of direct assault, not peripheral piddling.

In Britain and America, where "we are now deliberating over the future conduct of the campaign," the impending triumph in Tunisia "will I hope stiffen the resolution of our British allies for a northern [European] offensive," Stimson wrote in his diary[2] — and he became especially nervous when Marshall told him, the next day, that the President had only agreed "in principle" to what the U.S. chiefs were going to say to the British chiefs when they finally arrived.

Would the President be swayed by Churchill's anti–Second Front rhetoric, once the Prime Minister arrived at the White House for the new conference — code named, ominously, Trident? Was Churchill a new version of the great god Neptune, rising out of the sea to defeat American strategy for winning the war?

As the British arrival-day neared, General Marshall, for his part, "expressed his reservation as to how firmly the President would hold to his acquiescence" to the U.S. chiefs' position. "I fear it will be the same story over again," Stimson despondently recorded in his diary. Repudiation redux: "The man from London will arrive with a program of further expansion in the Mediterranean and will have his way with our Chief, and the careful and deliberate plans of our Staff will be overridden. I feel very troubled by it," Stimson lamented[3] — the British contingent expected to arrive in Washington the next evening, May 11, 1943.

27

A Battle Royal

When hearing the sheer size and composition of the approaching British contingent — 160 officers, with their assistants and chief clerks — the Canadian prime minister thought it a crazy gamble. "I was astonished when I saw the list of names," Mackenzie King noted in his diary, the day of their scheduled arrival in New York. "It is a tremendous risk to have so complete a representation of the military heads, chiefs and their experts and advisers cross the ocean at one and the same time."[1]

Why, though, had they come at all? Had not the overall strategy and timetable for the war in 1943 and 1944 been agreed at Casablanca?

The President had summoned all his chiefs of staff once again to the White House on May 9. There, in the Oval Office at 2:30 p.m., he'd rehearsed with Leahy, Marshall, King, and Arnold "the attitude that should be taken by the U.S. Chiefs of Staff at the conference with the British war officials who will arrive in Washington Tuesday," as the President's chief of staff noted dryly in his diary.[2]

All had been agreed. "The principal contention of the American government will be a cross Channel invasion of Europe at the earliest practicable date and full preparation for such an invasion by the Spring of 1944," Leahy had recorded that night — adding sniffily: "It is expected that the British Chiefs of Staff will not agree to a cross channel invasion until Germany has collapsed under pressure from Russia and from allied air attack."[3]

No cross-Channel Second Front before the Germans *collapsed?*

The likely British proposal seemed to Leahy a pretty awful way to run a war — one that would either leave Hitler in control of mainland Europe, or if not, give the Russians a head start in the overrunning of western

Europe. Though thanks to his fever he had not attended the Casablanca Conference, Leahy had read all the minutes and final agreements, as well as hearing firsthand from the President, Marshall, Arnold, and King the accords the British had made. What on earth were the British up to now, he wondered?

Early on the evening of Tuesday, May 11, the U.S. chiefs of staff congregated for the third time in a week at the White House. The arrival of the *Queen Mary* in New York Harbor had been reported, and the Prime Minister's huge retinue had apparently entrained for the capital. Then at "six forty-five p.m. the American Chiefs of Staff accompanied the President," Leahy recorded in his diary, "to meet a special train bringing to Washington the British Prime Minister and his War Staff."[4]

"Reached Washington at 6:30 pm where we were met by Roosevelt, Marshall, Dill, etc," a tired General Alan Brooke recorded in his own diary that night.

Ambassador Halifax was there to greet them, too. Churchill and his secretaries were immediately whisked off by the President to the White House; Brooke was invited to stay with Field Marshal Dill, his former boss.

It was a "hot and sticky night," Brooke noted before he went to bed at Dill's rented house in Virginia.[5]

He was nervous — embarrassed at the friendliness being shown by his American hosts, given that he was carrying a veritable bombshell. He'd been required first to go to the recently opened Statler, where the rest of the British party would be accommodated, to attend "a cocktail party given in our honour" by his hosts, the American chiefs. "From there," he recorded in his diary, "we did not escape till 8.15 pm." "I must now prepare my opening remarks for tomorrow's Combined Chiefs of Staff conference and muster up all our arguments," he added to his entry. "We have a very heavy week's work in front of us!"[6]

At the White House, the Prime Minister and his closest personal staff were meantime shown to the rooms where they would stay. The First Lady, however, was nowhere to be found. Irritated that the President had seen fit to receive Churchill for an unspecified length of time in the White House, and knowing her husband was having to steel himself for the confrontation he was rather dreading, she had simply decamped—going in the opposite direction, to their house in New York.

For his part, Churchill had begun to show signs of anxiety over his mission—in fact he'd suggested he might stay at the British Embassy on Massachusetts Avenue. Roosevelt had refused to hear of it—figuring it might be better to suborn the recalcitrant prime minister under lock and key, so to speak, in the White House mansion, where he'd have a better chance of countering whatever it was that Winston was harboring or plotting in his brilliant but sometimes dangerously inventive mind.

Instead of cocktails, then, the President wined and dined Churchill on Pennsylvania Avenue, with just his daughter, Anna Boettiger, and Harry and Louise Hopkins, present. Though the Prime Minister's office assistant, Leslie Rowan, and Churchill's aide-de-camp, Commander Tommy Thompson, were asked to eat with them, no invitation was extended to Churchill's military advisers. The dinner ended shortly after 9:00, after which the President invited Churchill to his Oval Study on the second floor. There the two men talked until after midnight.

Secretary Stimson, at his own house in Woodley Park, remained on tenterhooks. As he noted anxiously in his diary the following day: "Churchill arrived last night with a huge military party, evidently equipped for war on us."[7]

"I fear it will be the same story over again," the secretary lamented—furious that Churchill had come with such a huge contingent. He was all the more concerned since General Arnold had suffered "a severe heart attack" immediately after the May 9 Joint Chiefs meeting at the White House. The U.S. chiefs would thus be fielding a man short at the top, and might, Stimson feared, now be overwhelmed by their British colleagues in the talks—talks he had not been asked to attend.

Though the air in Washington remained warm and sticky on the morning of May 12, the atmosphere in the White House seemed somewhat frosty—a far cry from the happy spirit that had invested the Casablanca Conference. The President only went to his office at 11:10 a.m., where he had a succession of appointments—the American Legion, the mayor of Chicago, American labor leaders (regarding the national coal strike—the largest single strike ever called in the United States, involving more than half a million miners demanding more pay). And then lunch in his Oval Study with Hopkins, Churchill, and Lord Beaverbrook—former British minister of munitions, who had come without portfolio, as he was no longer in the British cabinet or government.

If there was open debate at the White House lunch, none recorded it. Indeed, no one recorded the luncheon — reflecting, perhaps, the awkwardness. Given that Beaverbrook was an outspoken advocate of a Second Front to be mounted as soon as possible — at the very latest, he pleaded, in the spring of 1944 — and since Harry Hopkins remained an unrepentant advocate of priority being given to such a direct, cross-Channel strategy, even by inexperienced troops, the Prime Minister was on his own at table. At all events, Churchill's narrative of the trip, written seven years later, jumped straight from joyful arrival in Washington, the night before, to a fictitious account of the discussion that took place that afternoon with the Combined Chiefs of Staff — pretending in his memoirs that he, too, was in favor of a spring 1944 Second Front.

This was, in truth, mendacious — for minutes of the meeting, held in the Oval Office immediately after lunch, were kept by General Deane, secretary of the Combined Chiefs of Staff: minutes that documented, in writing, the rift between the President's and the Prime Minister's views on global strategy.

Anxious to maintain at least a semblance of Allied unity, the President opened the meeting at 2:30 p.m. with a look back across the past year — reminding the generals how far the United Nations had come since their last get-together in Washington. It was, he said, "less than a year ago when they had all met in the White House, and had set on foot the moves leading up to TORCH. It was very appropriate that they should meet again just as that operation was coming to a satisfactory conclusion" — for Allied troops had already "seized Bizerte and British troops had fought their way into Tunis," General Deane noted the President's words. Given complete Allied air and naval control of the southern Mediterranean now exercised by the Allies under General Eisenhower, no Dunkirk-like evacuation of German or Italian forces was possible. It had taken time, but Torch had led, methodically, to a great Allied victory.

What a turnaround the campaign in Tunisia had brought, he remarked. The final surrender of German and Italian troops was expected momentarily, and might possibly number over 150,000 men — perhaps even a quarter million.[8] The invasion and subsequent combat had thus provided the Allies with the safe learning experience they needed. Its sequel had been decided upon at the recent Casablanca Conference, the President recapitulated: namely "operation HUSKY," the invasion of Sicily, which he hoped "would meet with similar good fortune," as the Allies made ready to throw "every resource of men and munitions against the enemy" under July's full moon.[9] The chiefs were assembled now, however, in Washington, to review what should happen after the fall of Sicily: "What next?"

With that, the President asked Mr. Churchill to give his own introductory remarks.

It was a delicate moment.

Churchill's lengthy *tour d'horizon* in the President's study, delivered with his characteristic rhetorical flair, bons mots, cadences, and flattering flourishes, certainly impressed his listeners for its brilliance ("very good opening address," General Brooke noted in his diary).[10] However, it completely failed to dispel the U.S. chiefs' fears of what the British were plotting. With every word, in fact, it became clearer that, whereas the President had seen Torch operations in the Mediterranean in 1943 as a means to gain the vital battle and command experience necessary for a cross-Channel Second Front in 1944, the British were not so confident—indeed, were not seriously interested in crossing the Channel anytime soon, unless the Germans collapsed. Thus the U.S. chiefs were compelled to listen as the Prime Minister lyrically described the triumph of Torch and the imminent conquest of Sicily as the means to a much richer, more byzantine, strategic end: not the defeat of Germany but merely the further clearing of Britain's vital seaway to India, and a staging post for expeditions into the "soft underbelly of Europe," beginning with the knocking of Italy out of the Axis coalition.

Before the assembled generals and admirals, Winston Churchill proceeded to outline how, surely, it ought to be the objective of the Allies, after securing Sicily, to invade Italy, obtain its surrender, then exploit the huge gap this would leave in the Adriatic and the Balkans, where twenty-five Italian divisions were currently helping the Germans in Yugoslavia. Once Italy fell out of the Axis alliance, those Italian forces would be hors de combat—offering an even softer European "underbelly." If, in turn, the Turks saw such a door into southern, mainland Europe opening, they might be persuaded to join the Allies or at least be encouraged to permit the Allies to use Turkish positions and airfields in order to attack the Third Reich from the south and southeast—thus disposing of the need for a cross-Channel operation at all, unless it were to be conducted as a pro forma operation, following the "collapse" of the Germans, similar to 1918, after the "defection" of Bulgaria.

1918?

Bulgaria?

There was a deathly hush in the Oval Office. Admiral Leahy, as chairman of the Combined Chiefs of Staff, was caustic in the entry he made in his office diary that evening. "The prime Minister spoke at length on the advantages that would accrue to the allied cause by a collapse or a surrender of Italy through its effect on the invaded countries of the near East and Turkey. In regard to a cross channel [Second Front] invasion in the near future," the admiral added with ill-concealed disgust, "it is apparently his opinion that adequate preparations cannot be made for such an effort in the Spring of 1944." Such an invasion, Churchill had allowed, "must be made at sometime in the future."[11] Sometime—but not 1944.

Even though Admiral Leahy, Admiral King, General Marshall, and General McNarney (deputizing for the literally heart-stricken General Arnold) had all been told by Field Marshal Dill to expect something on these lines, they still found themselves speechless. That Churchill would openly contradict and defy the strategy laid down by the President of the United States and agreed to at Casablanca, in front of the President and to his face, before his top military advisers, seemed incredible. "There was no indication in his talk of a British intention to undertake a cross channel invasion of Europe either in 1943 or 1944," Leahy repeated in frustration. In order to be quite clear as to the Prime Minister's precise argument, he added that the Prime Minister was recommending that no such invasion take place "unless Germany should collapse as a result of the Russian campaign and our intensified bombing attack."[12]

No cross-Channel invasion, then, even in 1944, unless there was a German collapse.

All eyes thus turned to the President.

To General Brooke's irritation, the President contradicted the Prime Minister. In the nicest yet firmest way possible the President made abundantly clear he did *not* agree with Churchill's new alternative strategy. "The President in a brief following talk," Leahy noted, "advocated a cross channel invasion at the earliest practicable date and not later than 1944."

To the relief of the U.S. chiefs of staff, the President explained that, in order to make certain of success in mounting a spring 1944 cross-Channel assault, U.S. operations in the Mediterranean *must* be curtailed as soon as possible after the fall of Sicily. The Allies would by then have all the command and battle experience they needed from the Mediterranean — in the air, on land, and at sea — for a Second Front invasion from Britain. In combat skills, in field command, in coalition planning and fighting, and in logistics. Mr. Roosevelt therefore categorically "expressed disagreement with any Italian adventure beyond the seizure of Sicily and Sardinia."

The President's tone had now turned from warm politeness to firmness. With regard to the Far East, he made clear he was disappointed by the latest British refusal to carry out the Anakim offensive that had been agreed upon at Casablanca, and stated "that the air transport line to China" — which Chiang Kai-shek was pleading be intensified — must "be placed in full operating condition without any delay, and that China must be kept in the war."[13]

With that, the strange meeting in the President's study came to a close.

Brooke, in his diary, was alarmed, noting the President "showed less grasp of strategy" than the Prime Minister.

The two top military teams then filed out. As the chief of staff to the Prime Minister, General Ismay, later recounted, "there was an unmistakable atmosphere of tension" and "it was clear there was going to be a battle royal."[14]

28

No Major Operations Until 1945 or 1946

EVEN CHURCHILL'S OWN wife, Clementine, worried lest the United States abandon its "Germany First" policy. In fact, Clemmie sent Winston cable after cable, while he was staying at the White House, expressing her abiding fear that, in the aftermath of the massive German surrender in Tunisia—with the numbers of German and Italian prisoners reportedly mounting by the hour—the United States might consider the campaign at an end, and choose to redirect its primary efforts to the Pacific. "I'm so afraid the Americans will think that a Pacific slant is to be given to the next phase of the war," she wrote him on May 13. "*Surely* the liberation of Europe *must* come first," she confided. And in a PS she added that she'd just heard of the "terrific" RAF bomber raid on Duisburg, in the Ruhr. "Do re-assure me that the European front will take 1st place all the time," she begged.[1]

Winston, however, was Winston: endowed with inspirational intellectual energy and romantic imagination yet burdened, too, by an often fatal penchant for peripheral rather than direct, frontal attack. It was a tendency that went back to his justifiable indignation as an infantry battalion commander in the trenches of the Western Front in World War I before the Battle of the Somme, and the bloodbath he witnessed on the plains of France.

Churchill's alternative—his Dardanelles landings—had proven just as futile as Allied offensives on the Western Front in World War I, however. There had simply been no easy military alternative to frontal attacks in World War I in the West—attacks that did, when no diplomatic solution could be found, ultimately decide the outcome once U.S. troops were committed to battle in France in 1918. Certainly the Prime Minister was fully entitled to ask his own chiefs of staff and then the Combined Chiefs to explore other scenarios before confirming the Casablanca decision to pursue a cross-Channel invasion — but that was not how Churchill presented his case at the White House.

Nor was it the case the next morning, when the first so-called Trident meeting of the Combined Chiefs of Staff opened in the Board of Governors Room of the Federal Reserve Building on Constitution Avenue. There, to Admiral Leahy's disgust, it became clear that an extension of the war in the Mediterranean and the Balkans rather than the agreed assault of northern Europe was no mere Churchillian fantasy. General Brooke, the bespectacled, owlish-looking "strongman" on the British team, announced he was even *more* opposed to a cross-Channel Second Front in 1944 than Prime Minister Churchill.

Brooke's apostasy in seeking to overturn the Casablanca agreement on a 1944 Second Front was potentially crippling to the Allied military alliance.

A solitary, self-contained man of incisive mind, Brooke had done his best since succeeding Sir John Dill as British Chief of the Imperial General Staff in 1941 to curb the Prime Minister's penchant for madcap schemes — especially red herrings that detracted from the Allies' primary strategic effort. Now, however, as CIGS, Brooke was supporting Churchill's alternative strategy.

How, though, the U.S. chiefs countered, would an as-yet-unplanned invasion of the mainland of Italy, the Balkans, and Greek islands miraculously lead to the collapse or defeat of the Third Reich?

In hindsight — given German determination to pursue the war to the bitter end — it couldn't. But in truth that was not Brooke's real reason, in May 1943, for backing rather than dissuading the Prime Minister. The fact was, despite the success of the Western Allies in North Africa, he too had lost faith in the essential feasibility of a Second Front in 1944.

Brooke had commanded heavy British artillery in World War I and large numbers of troops in France early in World War II — command that had ended in tearful defeat. The humiliation of British evacuation first at Dunkirk and then Brest, Cherbourg, and Saint-Nazaire in 1940, on top of the complete collapse of the French armies, had cut to his heart. Half-French himself, he simply lacked belief that an Allied cross-Channel invasion could ever succeed. The Wehrmacht drubbing given to Operation Jubilee, Mountbatten's August 1942 mini-rehearsal at Dieppe for an eventual cross-Channel assault landing, had in Brooke's view proved the point. The German massacre of an entire Canadian brigade on the beaches of the little French seaport was clearly a beach too far, given the literally dozens of tough Wehrmacht divisions stationed across northern Europe to repel such an attempted invasion — including panzer divisions.

The result was that in Washington, General Brooke exuded not energy — which at least his Prime Minister did — but a kind of dour, Northern Irish Protestant skepticism amounting to obstructionism. Not only about plans, moreover, but about people.

At Casablanca he had gotten a very poor impression of General Eisenhower as a fighting commander — blind to the way the young Allied commander in chief was not only learning on the job, but inventing a new kind of coalition command that might be messy and might result in many an upset or failure, but which brought together the collective *power* of Western arms — naval, air, and army — in a way that even the most disciplined of German troops could not stand up to, in the end. Even news of the surrender of General von Arnim together with many hundreds of thousands of Axis troops at Cape Bon, when he received it immediately after the meeting in the President's study, failed to change Brooke's mind, or convince him the Allies would ever be ready to fight whole German armies in northern France, unless the German government collapsed, as in 1918.

Wearing his trademark round black spectacles, Brooke sat in his chair at the Combined Chiefs of Staff meeting the next day, May 13, 1943 — his thick black mustache and slightly hooked nose giving him a fierce, intimidating countenance. He listened silently as Admiral Leahy was first acknowledged as the chairman of the proceedings and then read aloud to the meeting the U.S. chiefs' opening paper. This was titled "A Global Strategy, A Memorandum by the United States Chiefs of Staff." Copies of the document, moreover, had been handed to all the chiefs around the table. Looking through the document as Leahy spoke, Brooke hated it.

Word for word the document set down in typed script the strategy the President had outlined the day before at the White House. The "concept of defeating Germany first involves making a determined attack against Germany on the Continent at the earliest practicable date," the U.S. chiefs' document stated, "and we consider that all proposed operations in Europe should be based primarily on the basis of contributions to that end."

Lest there be any misunderstanding on this score, Admiral Leahy spelled it out in the simplest of sentences: "It is the opinion of the United States Chiefs of Staff that a cross-Channel invasion of Europe is necessary to an early conclusion of the war with Germany."[2]

Not to be outdone, General Brooke responded by handing across the table copies of the British chiefs of staff counterpaper — a paper that Brooke then read aloud to the meeting.

Entitled "Conduct of the War in 1943–44," the document was three times as long as Leahy's. In it the British chiefs argued that Italy might *not* surrender after the fall of Sicily, or by the threat of Allied bombing.

In order to achieve Italy's capitulation, the British paper contended, there would probably be need for "amphibious operations against either the Italian islands or the mainland." This "continuance of Mediterranean operations" would, "of course have repercussions elsewhere and will affect BOLERO," the cross-Channel assault, as well as operations in the Pacific, the document allowed. However, the fruits of Italian collapse would, the British chiefs argued, be worth the cost of delaying the cross-Channel invasion for several years, for it would make possible "increasing supplies to the Balkan resistance groups, and by speeding up our aid to Turkey."[3]

Silence again ensued — the two Allies at a strategic stalemate.

After a pause, General Marshall, the chief of staff of the U.S. Army, pointed out that, as the President had said the day before, there was no reason to venture into southern Europe at all. The Ploesti refineries, in Romania, which provided Germany's all-important oil supplies, should certainly be bombed by long-distance B-25 and B-17 bombers, operating from the Mediterranean. In fact, Marshall continued, the use of vastly superior and constantly increasing Allied air power "might enable us to economize in the use of ground forces in the Mediterranean Area," since footling amphibious and ground operations would not achieve more than local advantage — while merely delaying the Allies' main offensive capability. The Allies would then "deeply regret not being ready to make the final blow against Germany, if the opportunity presented itself, by reason of having dissipated ground forces in the Mediterranean Area."[4]

Again, there was silence.

Brooke countered that Allied air power was all very well, as in bombing the Ploesti oil refineries, "but this must be examined in relation to the whole picture of knocking Italy out of the war."

To this Marshall delivered the stunning rejoinder: namely that the aim of the "Europe First" strategy had never been to focus on *Italy*—Germany's junior partner in crime. The objective was to defeat *Nazi Germany,* their real adversary. Thus, rather than dispersing their forces in subsidiary ventures, he rebuked Brooke, "we should direct our attention to knocking Germany out of the war."[5]

The first formal Combined Chiefs of Staff (COS) meeting of the Trident Conference now turned into a free-for-all, as General Brooke, under attack, revealed more and more of his hand—this time claiming that by dumping the Casablanca agreement they would help Stalin—for if Italy fell, the Germans would be compelled to deny reinforcements to their Eastern Front and instead occupy and defend the Italian mainland, as well as defending the Balkans and Aegean Islands, just as they had done when compelled to send German reinforcements to Tunisia. Hitler would thus be able to provide "20 [percent] less on the Russian front," aiding the Soviets.

This aspect might well be so, Leahy, Marshall, and King accepted. But would not the mere *threat* of Allied invasion compel Hitler to station that number of divisions in Italy and the Balkans—much as he had stationed four hundred thousand German troops in Norway, and twenty-five divisions in France? Brooke's other claim, namely that successful Allied amphibious operations to seize yet more Mediterranean islands and occupy the Italian mainland would then provide a springboard from which to mount an attack on southern France, sounded equally irrelevant. Since when had *southern France* been deemed a way of "knocking Germany out of the war"?

Pushed against the ropes, General Brooke was thus driven to confess his deeper fear: that, unless fighting continued in the Mediterranean, "no possibility of an attack into [northern] France would arise"[6]—*for it would surely fail.* Even if Allied troops succeeded in achieving a beachhead across the English Channel, the subsequent battle or campaign in northern France, he believed, would be a disaster—for the Allies. Even after a bridgehead had been established, "we could get no further," he predicted. "The troops employed would be for the most part inexperienced."

With only fifteen to twenty U.S. and British divisions, the Bolero operation would be "too small and could not be regarded in the same category as the vast Continental armies which were counted in 50's and 100's of divisions" in the previous war.[7]

At this defeatist assertion, however, General Marshall really bridled. The discussion was "now getting to the heart of the matter," he acknowledged. The big lesson of Torch — and in planning for the forthcoming invasion of Sicily — was the way such a campaign, inevitably, "sucked in more and more troops." If "further Mediterranean operations were undertaken," Marshall pointed out, "then in 1943 and virtually all of 1944 we should be committed, except to a Mediterranean policy." Not only would this subsidiary campaign detract from the war in the Pacific, in terms of supplies, but it would mean "a prolongation of the war in Europe, and thus a delay in the ultimate defeat of Japan, which the people of the U.S. would not tolerate. We were now at the crossroads — if we were committed to the Mediterranean" rather than northern France in 1944, then "it meant a prolonged struggle and one which was not acceptable to the United States."[8]

Pinned against the ropes, poor Brooke now blamed a paucity of men. He explained that the "British manpower position was weak," and its forces were, in all candor, not up to the challenge of a cross-Channel invasion — neither that year, 1943, when a lodgment area in Brittany might possibly be attained (though one that would be easily cauterized by the Germans, he claimed), nor in 1944, either.

The U.S. chiefs were stunned by Brooke's open confession.

"No major operations," Brooke affirmed, adding insult to injury, "would be possible until 1945 or 1946."[9]

Again, the U.S. chiefs could hardly believe their ears, especially when Brooke explained "that in the previous war there had always been some 80 French Divisions available to our side." Now there would only be a handful, if that. Any advance from the Channel "towards the Ruhr would necessitate clearing up behind the advancing Army and would leave us with long lines of communication," subject to German air and land counterattack. Not only was British manpower "weak," but the RAF lacked mobility, having concentrated on bombing German cities, not supporting land armies; its planes and crews were therefore ill-equipped to support an invasion or subsequent campaign.[10]

Despite the current Allied victory in Tunisia, the picture that Brooke presented was, then, bleak in terms of the defeat of Hitler's Third Reich.

The two Western Allies were at loggerheads.

Without the British as allies, an American invasion of Europe was a nonstarter, and the President's "Germany First" strategy — as well as unconditional surrender of the Axis powers — would be in tatters.

No major cross-Channel operations until 1945, perhaps 1946?

When the President, at the White House, heard what Brooke had, at the Federal Reserve Building, openly declared — an assertion going further even than the Prime Minister had revealed at the meeting at the White House the previous day — he was amazed. So this was the "vast amount of work" the British chiefs had been doing — as Churchill had boasted in a telegram to the President from the *Queen Mary* as it neared New York![11]

The President was disappointed; in fact he was shocked. The British position seemed not only disingenuous but deceitful, in retrospect. Where the President had seen Torch and its Sicilian sequel as a crucial learning curve and rehearsal for a Second Front to be launched across the English Channel in 1944, the British had clearly backed Torch and the impending invasion of Sicily only to secure the Mediterranean as a shorter sea-lane to their occupation troops in India — while doing their best to *avoid* frontal combat with the Nazis in northern Europe.

The President shook his head. That very morning he had been discussing with the president of the Czechoslovak government in exile, Dr. Edvard Benes, the unconditional surrender of Germany, and what might be done to partition or police the country to ensure the Germans could never threaten world peace for a third time. Now, at lunchtime, he was hearing from Air Marshal Charles Portal that the British chiefs had no intention of launching a cross-Channel attack before 1945 or, possibly, 1946, three *years* away. How could this new stance be explained to the majority of Americans who saw Japan, not Germany, as the nation's primary enemy, yet had loyally backed the President's "Germany First" strategy?

The President had been told, in December 1942, that almost two million Jews had in all probability already been "liquidated" by Hitler's SS troops.[12] How many more Jews and others would Hitler exterminate by 1946? And all this so that Britain could sit out the war in Europe, at its periphery — not even willing to open the road to China, but hanging on to India and merely waiting for the United States to win back for Great Britain its lost colonial Empire in the Far East? It seemed a pretty poor performance.

Although disappointed, the President was not defeated. Great leadership demanded positive, not negative, thinking, and Portal, as an airman, did not sound quite as obstinate or defeatist as Brooke.

The British were visitors in a foreign land, and the best way to coax them out of their funk was, the President felt, to encourage them to overcome their understandable fears, not berate them; to help, not shame, their generals into recovering the confidence they would need to partner the U.S. military in mounting a cross-Channel invasion next spring.

The home team must therefore, the President decided, be firm in class, but as nice as possible outside. He'd already planned with Marshall that the Combined Chiefs were all to be taken to Williamsburg, in Virginia, at the weekend — any talk concerning conference matters strictly forbidden. For his own part, while the U.S. chiefs of staff hosted their opposite numbers at the site of an early British settlement in America (a source of cultural pride for the British visitors, but also a reminder of the successful American revolution to wrest independence from the British), the President now decided he wouldn't in fact take Churchill to Hyde Park, as he'd originally planned. Instead he would take him to his little mountaintop camp at Shangri-la. There he would work on him — insisting that Lord Beaverbrook, as an ardent supporter of an immediate Second Front, come too.[13] And Eleanor, who'd returned from New York, would be asked to at least drive with them to the cottage — thus prohibiting Churchill from any attempt to talk alternative Allied military strategy.

Extreme hospitality would thus be the order of the day. By burying the British with kindness, after working hours, the American hosts, in Williamsburg and at Shangri-la, would hopefully encourage their visitors to overcome their fears and confirm the Casablanca commitment to a fully fledged trial-by-combat cross-Channel invasion of northern France next spring: April or May 1944.

Such was the plan. Whether it would work was another matter.

At Shangri-la, once they settled in, the President took Churchill fishing. They settled by a local stream — the wheelchair-bound president "placed with great care by the side of a pool," Churchill recollected, where he "sought to entice the nimble and wily fish. I tried for some time myself at other spots."

It was in vain. "No fish were caught,"[14] Churchill recalled. Nor was Winston's mind changed about a doomed Second Front.

The three days in the Maryland mountains thus became something of a test of wills.

Shangri-la and the President's handling of Churchill on May 14, 15, and 16, 1943, mirrored Casablanca and the president's handling of de Gaulle — *prisonnier,* as de Gaulle had complained, in the President's Anfa camp. Now it was the Prime Minister's turn to feel that way.

Shangri-la was neither the White House nor Hyde Park. Instead it was, as Churchill later put it, "a log cabin, with all modern improvements." He watched "with interest and in silence" as General Pa Watson brought the President not war documents but colorful stamps: "several large albums and a number of envelopes full of specimens he had long desired," after which Roosevelt "stuck them in, each in its proper place, and so forgot the cares of State."[15]

For all the pretty mountain setting, the proximity to nature, and the restful quiet, the Prime Minister would not yield. The more the President and his supporting cast worked on him — both Hopkins and Beaverbrook attacking Churchill's obsessive argument for the invasion of Italy and the Balkans rather than northern France — the more determined Churchill became. So testy, in fact, that he even declined the President's request that he accept an invitation from Madame Chiang Kai-shek to go to New York, where she was staying while receiving medical treatment — risking, as Churchill candidly described his refusal, the "unity of the Grand Alliance," given the importance of the Generalissimo's struggle against the Japanese in China.[16]

Refusing to commit Britain to the 1944 cross-Channel invasion threatened, however, a far greater schism in the unity of the Grand Alliance than Madame Chiang Kai-shek's wrath. As obstinate as de Gaulle, the Prime Minister relentlessly clung to his Mediterranean preference, fearful of a cross-Channel debacle.

Pondering Churchill's behavior at the time, Sir Charles Wilson, the Prime Minister's doctor, wondered if the Prime Minister was suffering some sort of physiological problem. Churchill had, after all, hitherto pursued the "special relationship" with the United States with extraordinary patience, deference, and understanding. Now he was neither patient nor deferential, and certainly unwilling to conceive the strategic problem from an American perspective. His failure to grasp the import of what he was demanding — an extra year, perhaps two, of war in Europe without a Second Front, and a further year after that to defeat Japan — raised serious questions about the Prime Minister's state of mind. He'd come down with pneumonia in February (at the same time the President had fallen ill, after returning from Casablanca), which had been more serious than could be made public at the time — and in its aftermath, Dr. Wilson wondered whether it might have affected Churchill's judgment. Wilson himself had been stricken by fever on the voyage to America aboard the *Queen Mary*, and had had to be hospitalized in New York. When finally he caught up with his patient in Washington on May 17, he was frankly shocked. The Prime Minister had just returned from Shangri-la with the President — and what Wilson heard was amusing, but not encouraging.

The Prime Minister had, according to members of the President's entourage and Lord Beaverbrook, lost nothing of his extraordinary memory. On the return trip to Washington the presidential party had passed several Civil War battlefields, and Harry Hopkins regaled Dr. Wilson with an account of how Winston, hearing Hopkins could recite only two lines of John Greenleaf Whittier's famous Civil War poem, had recited the entire poem. "While we were still asking ourselves how he could do this when he hadn't read the darned thing for thirty years, his eye caught a sign pointing to Gettysburg. That really started him off," Hopkins recounted in awe — Churchill's summary of the battle, with character portraits of the rebel generals Jackson and Lee, being equally amazing. About the current war, however, "Hopkins was a good deal less flattering about the P.M.'s contribution to the discussions which had begun on May 12 in the oval study of the White House," Dr. Wilson recalled. "Indeed, he looked pretty glum as he assured me that I had not missed anything."[17]

The impasse appeared to be the same as the one the year before, when the Prime Minister had journeyed to the White House for the same reason: namely to explain why the British could not agree to a cross-Channel invasion that year. The British surrender at Tobruk, moreover, had made his point: the British were simply not ready for such a challenge in 1942, at a moment when they might even lose control of the Middle East to Rommel's advancing Panzerarmee Afrika.

Now, eleven months later, "damn it all," Churchill was back, with "the old story once more, shamelessly trotted out and brought up to date," Dr. Wilson recalled with concern, recording in his diary the sense of frustration felt by Hopkins: Churchill simply refusing to countenance the Casablanca strategy, unless Italy was swiftly defeated and the Third Reich miraculously fell apart.[18] Hopkins had even imitated Churchill, saying: "Bulgaria's defeatism in 1918 brought about the collapse of Germany; might not Italy's surrender now have similar consequences? It will surely cause a chill of loneliness to settle on the German people and might very well be the beginning of the end."[19]

Loneliness as the beginning of the end — without the Wehrmacht actually being defeated in battle, or even forced back onto German soil? To those who remembered the consequences of the "collapse of Germany" at the end of World War I, this was understandably alarming.

Dr. Wilson had asked Hopkins "what the President made of all this."[20]

"'Not much,'" Hopkins had answered. "'This [idea of] fighting in Italy does not make sense to him,'" he'd explained the President's view. United States naval, air, and ground forces had been sent to the Mediterranean — against the advice of Hopkins, Stimson, Marshall, and the U.S. chiefs of staff, it was true — to learn *how* to defeat German troops in close combat, the President had insisted. As soon as the Sicily invasion and campaign were won, those forces — commanders and troops — were to be switched to England for the invasion of northern France in the spring of 1944, in accordance with the President's strategy. "He wants the twenty divisions, which will be set free when Sicily has been won, to be used in building up the force that is to invade France in 1944," Hopkins made clear.[21]

At the Pentagon and Navy Department, the U.S. chiefs of staff were similarly frustrated.

Brooke's stonewalling, once the chiefs returned from Williamsburg, was especially irritating. "A very decided deadlock has come up," Secretary Stimson noted in his diary on May 17, after speaking with General Marshall. "The British are holding back dead from going on with Bolero. They have done the same thing in regard to Anakim [the campaign to retake Burma] and are trying to divert us off into some more Mediterranean adventure. Fortunately," he added, "the President seems to be holding out."[22] Stimson decided he must call the White House and make sure, though. "I called up the President, told him that I had prepared myself fully by reading all the minutes and was ready to talk with him at any time that he wanted to, although I did not want to intrude myself on him. He told me he was coming to the conclusion that he would have to read the Riot Act to the other side and would have to be stiff."[23]

Stimson, conscious of how the President liked to quote Lincoln, told him how President Lincoln had remarked of General Franz Sigel that, though he couldn't "skin the deer," he "could at least hold a leg." By his intransigence, however, the Prime Minister was in danger of causing the Western Allies — Americans and British — to be Sigels in the war against Hitler: only daring to hold the Nazi leg while the Russians did the skinning. "Stalin," he told the President, "won't have much of an opinion of people who have done that," he warned, "and we will not be able to share much of the post-war world with them."[24]

The President did not need reminding. Yet how *compel* an ally like Britain to conform to American strategy?

The most worrying thing was that Churchill was now threatening to disrupt the Western military alliance just at the moment when the President was becoming more and more anxious to pressure Stalin to sign up to a postwar United Nations authority while the United States — furnishing more than 10 percent of Russia's war needs — still had significant leverage. All in all it was too bad — with no breakthrough in sight.

Whatever Stimson, Hopkins, the U.S. chiefs of staff, and later critics might say about Churchill's sudden intransigence in May 1943, however, it is important to note that Churchill and his British contingent were not the only ones arguing in Washington against a Second Front. The prospect of heavy casualties in head-on combat with the Wehrmacht in northern France was sobering. Outside the War Department more and more people were objecting — especially people in the Navy Department who foresaw a long war with Japan if the "flower of our army and air force" was first expended "in combat with Germany," as Bill Bullitt, assistant to the secretary of the Navy, warned in a renewed memorandum he wrote for the President on May 12.

It was vital the President should, Bullitt argued, put more pressure on Stalin to declare war on Japan at the conclusion of the war against Hitler, lest the United States should have wasted its manpower and resources in a cross-Channel campaign that could get bogged down, as in World War I — thus leaving itself, even after assumed victory, having to fight against Japan "while the Soviet Union is at peace," and Britain contributing only insignificantly to the defeat of the Japanese. In that situation, "we shall have no decisive voice in the settlement in Europe," Bullitt warned. "Europe will be divided into Soviet and British spheres of influence — according to present Soviet and British plans — and further wars in the near future will be rendered inevitable."[25]

Bullitt's recommendation, once again, was the same as Churchill's — to drive swiftly into central Europe through the Dardanelles.

After Roosevelt's death, Bill Bullitt would spend the rest of his own life lancing the memory of the President for having failed to take his recommendation. Only American "boots on the ground" in central Europe would stop Stalin's "Sovietization," Bullitt pointed out again in his memorandum — and the Balkans was the place to plant those boots.

The President could only groan at this extra pressure from his own American side, given the latest British intransigence. Bullitt might have an excellent understanding of Russian communism; his Balkans strategy, however, remained militarily illiterate. Moreover, his latest political recommendation, namely that the President should threaten Stalin with a switch of American forces to the Pacific unless he agreed in writing not to Sovietize central Europe, was, at a time when the Western Allies did not have a single boot on the ground in Europe, less than realistic.

No, the fact was, the President had little option but to stick to his own program: refusing to countenance a quagmire in the Balkans or the northern Italian mountains, and instead holding to the timetable for a U.S.-British Second Front that had been agreed at Casablanca: spring 1944. This strategy, if followed, would at least take U.S. and British forces to Berlin, ending the Third Reich and saving the western part of Europe from Sovietization. He would meantime continue to press Stalin, in order to see if he could get the Soviets to sign up to his postwar plan and to declare war on Japan as soon as Germany was defeated. Without a genuine plan to launch a Second Front by 1944, however, it was unlikely to get very far, as Secretary Stimson had commented.

To produce such a genuine plan, he would have somehow to bring the British back into the fold, or the Second World War might well end in failure.

PART EIGHT

The Riot Act

29

The Davies Mission

ON THE SURFACE, the great victories at Stalingrad and then Tunisgrad boded well for Allied cooperation in eventually defeating the Third Reich.

In truth, however, relations with the Soviet Union were not good—indeed were getting worse. Stalin's rejection of the President's invitation to the summit at Casablanca (or alternative venues the President had offered) had resulted in the sheer scale of the Russian war effort being underappreciated in the West. Even Stalin's own ambassador to Washington, Maxim Litvinov, had warned the Russian Foreign Ministry that such standoffish behavior was counterproductive, indeed would make it harder, not easier, to get the Western Allies to commit to a timely Second Front.

Stalin had paid no heed. This was hard for even the most sympathetic of American observers and reporters to understand. In terms of Allied military cooperation, Russia was, sadly, a write-off—Stalin constantly demanding more U.S.-British convoyed deliveries of war materials to Murmansk, yet refusing to order Russian aircrews to fly out of northern Russia to protect them, lest they leave the borders of the Soviet Union and not come back. This had led to, and continued to result in, terrible British and American shipping losses, not only in Lend-Lease war materials and food but Allied lives as well. Nor would the paranoid dictator allow Allied officers, or representatives, to monitor whether the contents of the convoys were being efficiently unloaded at Murmansk, or were appropriate to actual Russian war needs. The Russians had also refused for months to respond to whether U.S. bomber crews could land in the Soviet Union if they bombed the Ploesti oil fields in Romania—and when they finally did respond, they refused to allow Ploesti raids to be launched from Russian airfields, despite being at war with the Third Reich and its eastern European partners, Romania and Hungary. Whether it was paranoid fear that Russians might become infected by rich capitalist partners, or that Russia's capitalist allies might obtain genuine, accurate, and detailed information—military, political, economic, social—about the Soviet Union, no one really knew. Nor had this changed as the tide of war against Hitler turned. As Western diplomats and journalists—who were forbidden to venture outside Moscow without close supervision—complained, there was virtually not a single Russian who dared question, counter, or ignore Stalin's oppressive policies for fear of arrest, imprisonment, or even execution.

More troubling still had been the sickening revelation, in April 1943, that more than twenty thousand Polish officers, police officers, and members of the intelligentsia had, on Stalin's orders, been murdered in cold blood by Soviet occupation forces in 1940, during the time of the German-Soviet Nonaggression Pact.

That disclosure—the decomposing Polish bodies unearthed by the Germans in the Katyn forest near the Russian city of Smolensk, but the Soviets denying culpability—had given cause for grave trepidation in the West, especially among Polish forces in exile.

Nothing, but nothing, could excuse such mass murder. News of the massacre, at a moment when the tide of war had turned and the forces of the Third Reich seemed to be everywhere on the defensive, had offered the embattled Dr. Goebbels a heaven-sent opportunity to demonstrate to the German Volk,[1] as well as people abroad, just how merciless a Russian victory in the war, and a subsequent Russian-imposed "peace," would be.

Stalin naturally protested it was a Nazi ruse. He denounced the leader of the Polish government in exile for suggesting Russian complicity, loudly claiming the Nazis, not the Soviets, had been responsible for the massacre. Both Roosevelt and Churchill had on good authority been told the bitter truth, however: that it was Stalin himself who had given the orders for the mass execution in 1940.

With Stalin's Soviet Union such an uncooperative, undemocratic, often downright evil partner of the Western democracies—though one that was still taking the brunt of casualties in the war against Hitler—both Roosevelt and Churchill were put in the iniquitous position of publicly accepting, or declining to comment on, Russian lies over the Katyn massacre. Besides, in the balance of atrocities, the Germans were still way ahead of the Soviets, both in SS mass-murder concentration camps and in the treatment of Russian POWs.[2] Continued do-or-die Russian resistance to Hitler on the Eastern Front was crucial—no matter how ungrateful, paranoid, deceitful, and barbarous the Russians, and however chilling the prospect of postwar Sovietization.

How maintain that morally dubious anti-Nazi coalition, though—let alone seek to move the Russian communists from their reign of terror into a more positive postwar world?

It was in this respect that the relationship, or partnership, between the President and the Prime Minister was of the highest importance for the history of humanity. And in Washington, in May 1943, Prime Minister Churchill was coming very close to breaking it.

Hitherto, Churchill had taken the same view as the President—that the enemy of my enemy is my friend, however odious in certain respects. But with Churchill threatening to pull out of the Casablanca accords and refusing to mount a Second Front in 1944, the question arose: would Stalin remain a friend? As Secretary Stimson warned, without a cross-Channel invasion—one that would force Hitler to fight on two fronts—would not the Russians lose military respect or faith in the Western Allies, and be minded to seek an armistice with the Third Reich, even a new Ribbentrop Pact that would leave Hitler master still of all western and central Europe?

Roosevelt didn't think Stalin would stoop to that, after the millions of casualties the German onslaught had already cost the Russians. But it could certainly undermine the President's attempts to get Russian agreement to make air bases available and declare war on Japan, if and when the war with Hitler was successfully concluded, as well as getting Soviet participation in the postwar security system the President had in mind. The Second Front, in other words, was a sine qua non: a test that the Western, democratic Allies *must* meet if they were serious not only about the war but the postwar. Not footling around in the Mediterranean, but a willingness to face up to the war's greatest challenge: D-day, as it would become known.

The President thus changed his mind about a summit with Stalin—feeling it would be better to keep Churchill *out of* any meeting for the moment, if one could be obtained, lest the Prime Minister's opposition to a cross-Channel operation give away their weak hand: namely the fundamental unwillingness of the British to countenance the heavy casualties involved in a Second Front. Somehow, Roosevelt was aware, to defeat the Nazis he must keep the Russians fighting in the East—and get the British to *fight* in the West, not footle about in the South!

This was easier said than done. His cables to Stalin after Casablanca had deliberately, perhaps disingenuously, held out the possibility of a Second Front being mounted in the summer of 1943, after Husky; how then was he to explain to Stalin the Western Allies were not only abandoning any plans to launch a Second Front in 1943, but that the purpose of Churchill's current visit to Washington, together with a military staff of 160 advisers, was to argue against a Second Front *even in 1944?* In fact, according to Churchill's Chief of the Imperial General Staff, that no Second Front should be planned before 1945 or even 1946?

"The Soviet troops have fought strenuously all winter and are continuing to do so," Stalin had assured the President in March. The Führer had lost more than a whole army at Stalingrad, but he had many more at hand—perhaps as many as two hundred divisions, including whole panzer armies. The Germans were preparing for "spring and summer operations against the USSR," Stalin wrote; "it is therefore particularly essential for us that the blow from the West be no longer delayed, that it be delivered this spring or in early summer"—i.e., 1943.[3]

It was in this context that the President had summoned another former ambassador to Moscow, Joseph Davies, to the White House the day after Churchill set sail for Washington. As the President explained to Davies, he'd decided to send Stalin a new letter by hand, to be delivered in such a way that Stalin would be forced to respond to the President's renewed request for a private meeting.

Davies was elderly and had been particularly naive in his acceptance of Russian propaganda regarding their communist show trials, arrests, and deportations in the 1930s. He was sincere in his judgment of Hitler and the barbarity of Nazism, however, and his evaluation of the Soviet will to defend Russia had proven more sophisticated than that of the U.S. military attaché in Moscow—in fact, he'd been the man who correctly reported to the President that Operation Barbarossa, Hitler's invasion of the USSR in June 1941, was going to fail. As an emissary to show goodwill and firmness of American purpose in prosecuting the war against Hitler, the President could not have chosen better. The new, private letter Davies would hand carry would be a direct, personal invitation from the President to meet somewhere that summer and resolve their differences over strategy and timing—one the Russian dictator could not now refuse without giving offense to the President of the one country in the world supplying the Soviet Union with a significant amount of its war needs.

The Prime Minister was not now to be invited to the proposed summit, the President made clear in the letter — though he could not give the true reason, even to Davies, who would doubtless be asked by the dictator, once he reached the Kremlin. Since the President could not reveal Churchill's impending visit to Washington and his reported unwillingness, supported by his chiefs of staff, to launch a timely Second Front, he had merely told Davies he wished to meet Stalin, informally, to discuss the long-term future with him. Not, in other words, to address the matter of impending operations, but rather the conclusion of the war: unconditional Axis surrender, winning the war against Japan, and the establishment of a postwar United Nations authority. It would be, the President explained to Davies, a preliminary discussion, man to man, without risking, Roosevelt told his emissary, any international arguments over British — or French — postwar colonial empires. "Churchill will understand," the President had assured Davies when giving him his instructions in the Oval Office on May 5. "I will take care of that."[4]

As Davies set off for Moscow via the Middle East, Churchill had arrived in Washington — and the Prime Minister's refusal to countenance a Second Front had only reinforced the President's determination to meet Stalin alone. Davies would hopefully convince the Russian dictator that the Western Allies were united and sincere in their commitment to launch a Second Front — the President's willingness to travel halfway across the world to meet in person with the Russian leader surely a gauge of that sincerity.

In the meantime, however, the President was determined to bring Winston Churchill to heel, lest he and his huge military team cause the Grand Alliance, rather than the Third Reich, to collapse.

This, in essence, was the challenge of Trident: suborning Neptune.

Adding to the behind-the-scenes war drama was the fact that the British chiefs of staff now parted company — physically and metaphorically — with their own Prime Minister.

The chiefs' weekend in Williamsburg, Virginia, went well — the officers glad to be out of Washington not only to be able to relax but to get to know their Allied counterparts as human beings. Talks had then resumed at the Federal Reserve Board building at 10:30 a.m. on Tuesday, May 18, 1943.

Admiral Leahy, General Marshall, and Admiral King had feared the worst in terms of British intransigence, once back in uniform, so to speak. So worried, in fact, were the U.S. Joint Chiefs that they came to the table with a compromise whereby they would ask only for a minimum "lodgment area" across the English Channel in 1944, if the British were still so afraid of failure, and would only seek to expand it the following year, 1945.

Once seated in the room, however, it was to find the "battle royal" was already won. To their astonishment, the President's tactic of extreme hospitality appeared to have worked — the weekend away in Williamsburg, with wholesome food and wine and civil conversation having seemingly done the trick. Aided also by Field Marshal Dill — who'd reasoned with his successor as CIGS, General Brooke, that he must give in or risk a breakdown in what was a historic military coalition between the United States and Great Britain. The American people, the field marshal had made clear to his British compatriot, would not stand for the war in the Pacific being deliberately starved of men and resources for years, simply so the British could fiddle around in the Mediterranean — leaving Hitler's legions in almost complete control of continental Europe. A firm date for a cross-Channel invasion *must* be tied down, and the necessary forces assembled to make it work.

A new paper on "The Defeat of Germany" — not Italy — had therefore been ordered from both planning staffs over the weekend, while the Prime Minister was away at Shangri-la, to define exactly how Hitler was to be brought to unconditional surrender — namely by defeating Germany, not simply Mussolini's Italy. By Monday night, May 17, the British version, approved by General Brooke, had been ready. When General Marshall read it through at the meeting on Tuesday, May 18, he was delighted. Though it talked a lot about further interim operations in the Mediterranean, it "appeared that [even] if Mediterranean operations were undertaken in the interval, a target date for April 1944 should be agreed on for cross-Channel operations." In writing.

General Marshall breathed a sigh of relief. Brooke then confirmed this was the case, the date formally recorded in the minutes of the meeting.[5]

April 1944.

Mirabile dictu, Marshall reflected. General Brooke had seemingly dropped his call for a postponement of a Second Front until 1945 or 1946, and was now definitely onboard — if, in the meantime, operations in the Mediterranean were allowed to continue that summer. "The rate of build-up of German forces in western Europe would greatly exceed our own on the Continent unless Mediterranean operations were first undertaken to divert or occupy German reinforcements," Brooke maintained. "If these operations were first undertaken," Brooke conceded, "April 1944 might well be right for a target date, though the actual operation would be more likely to be possible of achievement in May or June."[6]

Genuine, serious military preparations for a massive spring 1944 cross-Channel invasion by the Western Allies could now commence, the generals agreed — with only the thorny question left as to how far to limit interim 1943–44 operations in the Mediterranean so that they did not prejudice preparations for D-day.

Leahy, Marshall, and King were still skeptical. The matter of "interim" operations in the Mediterranean would, they predicted, prove tortuous — but at Marshall's insistence a formal commitment to D-day had been given by the British, in writing. Some seven battle-hardened U.S. and British divisions would be withdrawn that very fall from the Mediterranean theater to the United Kingdom. There they would begin training and rehearsals for the spring 1944 D-day assault. It seemed a reasonable compromise.

For Secretary Stimson, at the Pentagon, the British climbdown was as much a relief as it was to General Marshall.

The Prime Minister, meantime, had been kept well away from the daily Combined Chiefs meetings. Instead, he had been pressed by the President to go address a joint session of Congress again — "a very good speech, noteable for its good, downright eloquence," Stimson recorded, after attending the performance on Capitol Hill, "on the main lines of war history and strategy and also for the adroitness with which he avoided any allusions to the real points of issue which are now being fought over between the staffs of the two countries.[7]

"These points of difference have come out sharply in the two plans and it is taking all Marshall's tact and adroitness to steer the conference through to a result which will not be a surrender but which will not be an open clash. The President seemed to be helping us," Stimson added — Mr. Roosevelt adopting the same approach as his U.S. team, as "indicated by his telephone talk with me the other evening." The President, Marshall had reported to the Secretary, was not only "taking the same line" but "insisting that the planners decide what will be the cost in shipping and men for the 'big point' (as the President called it)": the cross-Channel invasion. Only when these requirements had been met would the planners be permitted, the President had said, to "determine from what is left over what can be done otherwise" in the Mediterranean.[8]

The Second Front, in other words, would now be First Priority for the Western Allies.

As to the sincerity of the British volte face not all were convinced, however. Admiral King, in particular, remained less than happy. Though the British seemed resigned to join the U.S. in launching a Second Front invasion in April or May 1944, they were insisting on so many landing craft, naval forces, air forces, ground forces, and logistics being assigned in the "interim" to the Mediterranean that—in King's eyes—this could well prejudice the success of the primary cross-Channel strategy. More significantly to King—a true believer in prosecuting the war in the Pacific more robustly, now that the Americans and Australians were on the successful offensive there—such an interim policy threatened to slow down Nimitz's and MacArthur's plans, thus allowing the Japanese to "dig in." The result would inevitably be grave American casualties—an aspect that seemed not to register with the British, whose main forces were being held in India as an army of colonial occupation, and were making every excuse not to take the offensive against the Japanese.

There was, moreover, public impatience in America to consider.

"I am very much afraid that, if the British succeed in getting us pulled out any further onto the limb in the Mediterranean," Stimson noted, "we shall face a widespread loss of support for the war among our people." This was serious. "Polls show that the public would be very much more interested in beating Japan than in beating the European Axis [powers]," he acknowledged, thanks to Pearl Harbor—something that could easily translate into "all kinds of personal and party politics" that could damage the bipartisan, "Germany First" war effort.[9] This danger extended, he knew, to his fellow Republicans across the country, who were once again demanding that General MacArthur be recalled from Australia to stand in the 1944 presidential election—a campaign in which MacArthur would doubtless call for a switch of U.S. priority to the Pacific to face not Hitler, but America's "true" enemy, Japan.

What Stimson and Marshall failed to realize, however, was that General Brooke had now parted company with his prime minister—and that Winston Churchill would be the problem, not the British chiefs of staff.

30

A Dozen Dieppes in a Day

Seated at the Federal Reserve Board in the Combined Chiefs of Staff meetings, General Brooke had failed to noticed the Prime Minister's increasingly divergent trajectory. Even the President, living with the Prime Minister each day at the White House, had been unaware of what Churchill was saying behind his back.

Hearing from Admiral Leahy on May 18 that the British chiefs had backed off their opposition to a 1944 or 1945 invasion of France, the President had been delighted by news of his team's success. This would be of inestimable help when and if he met with Stalin, since he would now be able to reveal to the Russian leader, in person and in all honesty, a firm date for the Second Front. It would be a formal U.S.-British military commitment that, even though Stalin had fervently hoped it would take place in 1943, would nevertheless encourage the Soviets to hold out against Hitler's impending summer offensive on the Eastern Front.

The President was crowing too early, however.

The first intimation the Prime Minister was charting his own course in opposition even to his own British team had come on the evening of May 18, 1943 — reported to the President by none other than the Canadian prime minister, Mackenzie King, who had accepted the President's invitation to attend the latest meeting of the Pacific War Council and to stay at the White House.

From his train, Mackenzie King had gone to Pennsylvania Avenue to settle in and have a word with his fellow prime minister. It had been 6:00 p.m., but Churchill was in bed, in the Queen Elizabeth Room on the second floor. He had looked "very frail," and was wearing "a white nightgown of black and white silk," King described in his diary. "He has lost the florid coloring and his face was quite white. Looked soft and flabby. He had a glass of Scotch beside him near his bed," and "looked to be very tired"—as well he might. On a special writing tray Churchill was still, after some seven hours, working on the draft of the address the President had asked him to give to Congress the following day. He was keen for Mackenzie King to read the text—anxious not to say anything impolitic, given that people in Washington were already talking about the 1944 presidential election, still more than a year away. "He indicated that he had not completed his speech and would be taking a little sleep before dinner, which I took to mean that he would not wish the conversation to take up too long."[1]

The two prime ministers had first talked of the recent Allied victory in Tunisia, where General von Arnim had finally surrendered on May 12. It was "really shocking" Churchill claimed, "the way the Germans came in at the end"—"giving themselves up, falling and crawling; some of them waving plumes [white flags], and [he] said that an hour before, when they thought they could win, they were most savage and brutal. He imitated their different attitudes in his own face."[2]

This was vintage Churchill: his vivid imagination running free (since he had obviously not been present at the surrender), yet amazingly astute in his reading of German moral duplicity: able to switch from barbarous hubris toward other humans to shameless appeals for "humanitarian" clemency when they themselves were overpowered.

Once again Mackenzie King found himself entranced by the British prime minister's mind and his colorful use of language. They swiftly moved on to the reason for Churchill's presence in Washington, however. "Churchill began to tell me about the conferences here," King noted in his diary that night. "Said that they were discussing the plans. That he and the Americans were very good in accepting Roosevelt's decisions in the end"—as they had at Casablanca. "He thought that he and Roosevelt were very much of the same view," even if there were "differences of emphasis."[3]

Prime Minister King was baffled. This was not what he'd been told that very afternoon on his visit to the Canadian legation in Washington. There he'd been informed that "the Americans were pressing for a cross-Channel Second Front"—"invasion from the North"—whereas "the British plan was for invasion [of Europe] from the South, either through the Balkans or [southern] France. Views had not yet been reconciled."[4]

How, then, could Roosevelt and Churchill be on the same page? Was Churchill now accepting the President's Second Front strategy? Or was the President accepting Churchill's new strategy—and what was it, in fact? A second Dardanelles? Had he misunderstood? What was Churchill really saying—or not saying?

It was at this point that Churchill made clear "that as far as he was concerned, the plan was to follow on the decisions of the Casablanca conference," which had authorized landings in Sicily in July that year, in Operation Husky—*but had not explicitly gone further than that,* he now claimed. "The thing to do was to get Italy out of the war," Churchill explained. "Altogether he believed this could be done, and said he would not treat them [Italians] too badly if they were to give up and particularly if they were to yield up their fleet. If he could get the fleet, he would be prepared to use it to attack the Japanese." Meantime, however, there was the matter of Europe—and the defeat of Hitler. "The plan was to start the invasion of Europe through Sicily and Sardinia," Churchill now told Mackenzie King, confidentially, "either on through the Balkans or possibly through [southern] France depending on how matters developed." It would be easier than a cross-Channel attack.

"They would be getting footholds all along the way, and Russia might put on a very strong offensive and they [the Allies] would be working toward Russia"—via "southern Europe," the Prime Minister explained. "There was a chance, too, that Turkey might come in," King noted Churchill's words, "though not until she got plenty of equipment. He was not pressing her at present."

King—aware that the Pacific Council would have to wrestle with the implications of Churchill's alternative new strategy, so similar to his notorious failure in World War I—pressed Winston to explain in more detail.

Lest there be any misunderstanding, Churchill privately confided that he remained as implacably opposed to the notion of a cross-Channel Second Front as he had been the year before—indeed *more* so, now, after the catastrophe of Dieppe. "Speaking of invasion [of France] from the North," across the English Channel, "he said that he did not want to see the beaches of Europe covered with slain bodies of Canadians and Americans. That there might be many Dieppes [suffered] in a few days," were such an operation to be launched. "That he, himself, could provide 16 divisions which would include ours [i.e., Canadians] but there was only one American division in England. This was all they had against the numerous divisions Germany could muster; unless Americans were prepared to send a large number of divisions to cross at the same time, he did not see how they could attempt anything of the kind."[5] It would be, King again recalled Winston's actual words, "slaughter" — "a dozen Dieppes in a day." "I thought," King noted, "this was pretty strong language."[6]

Mackenzie King was now doubly dubious as to Churchill's claim that he and the President — let alone the U.S. chiefs of staff — were of the same mind. "I asked if the Americans were likely to make much difficulty over these particular plans," King noted. "He replied that the President and he were very close together; that they could not settle all these things at once. They had to run along for a time" — in order to dupe the U.S. Joint Chiefs. "The President was inclined more his way and he thought that his [U.S] chiefs of staff would accept loyally his decisions in the end."[7]

Mackenzie King said nothing. In truth he was gobsmacked, however.

Yes, the President had indeed insisted, at their last meeting, in December 1942, that further operations should first be carried out in the Mediterranean in 1943, in order to learn the hard, attritional lessons of modern war before attempting anything as hazardous as a cross-Channel invasion. But the President had never said anything to suggest he believed the Allies should attempt to defeat the Third Reich by attacking from the south. Was Churchill, with his "glass of Scotch" on the table beside his bed, making this all up? Was he living — as he tended to do, in the eyes of the abstemious Canadian who had vowed not to drink liquor for the duration of the war — in an alcohol-laced cocoon? Alcohol seemed certainly to fuel Churchill's fertile imagination and brilliant rhetorical skills — but did it equip him to *listen* to what President Roosevelt and the Combined Chiefs were telling him rather than to his own voice?

Dimly, though, Churchill seemed aware the President had been keeping him away from the Combined Chiefs of Staff over the weekend — indeed from anyone who might become alarmed over his Mediterranean ambitions. "He said that the President and he had been off together at Blue Ridge over the week-end," at Shangri-la. The following weekend, however, Churchill "wanted to see a few friends," and was going to insist he be allowed to stay at the British Embassy on Massachusetts Avenue, where he could meet with and telephone anyone he wanted. "Thus far, he had not seen hardly any."[8]

King was somewhat alarmed, but held his tongue, unwilling to disconcert Churchill on the eve of his important appearance before Congress—which, as prime minister of Canada, King had been invited to attend.

Mackenzie King's worst fears indeed materialized the next day when, at the Capitol at midday on May 19—the very day Ambassador Davies arrived in Moscow bearing the President's private letter to Stalin—Churchill followed up his congressional address by talking frankly to senior members of Congress.

"After the luncheon, members of the Senate and representatives of the foreign committee came into the room, and Mr. Churchill was subjected to a quiz," Mackenzie King—who attended this meeting, too—recorded that night in his diary. "He faced squarely the question as to strategy. Told those present that he felt the great objective now was to knock Italy out of the war." This would, he said, "clear the Mediterranean which would mean a through route to India, China; make all the contacts with the Orient much easier. He believed the great offensive was coming against Germany on the part of Russia," and "in the Southern part by allied forces pressing up through the Balkans, and there would be a relief of the pressure on Russia. They might, too, get some of the satellite states of Germany to change their attitude. They would also get additional help from Yugoslavia where some 10 [Italian] divisions were tied up there which could be added to the allied numbers. Thought that all this would be helpful to Stalin. He thought the Germans could be driven entirely out of Italy and would probably leave Italy to look after herself."

King was puzzled. Driven entirely out of Italy? Churchill's forecast of Hitler's likely reaction to an Allied invasion of Italy and the Balkans—especially after the example of German tenacity in reinforcing Tunisia over the past six months—sounded disturbingly naive, even schoolboyish. His prediction, moreover, seemed at odds—very poor odds—with his defeatism concerning the prospects for a Second Front. To the postprandial group of senators and congressmen, Churchill "made it pretty plain," King noted, "he did not favor any immature attack on Europe from the North," across the English Channel. "He spoke of the few divisions they have in Britain—I think 18 altogether including our own, only 1 American division, and that Hitler was able to move many divisions from one part of the continent to the other in a very short time. Referred to the scarcity of ships, etc"—going "pretty far in making clear the plan is to attack across the Mediterranean into Europe either via [southern] France,

Sicily or further East [in the Balkans], without designating what locality would be first."[9]

Even more astonishing to the Canadian prime minister was Churchill's complete lack of shame or caution in opposing the President's strategy in front of U.S. lawmakers, behind the President's back — having "instructed them," Mackenzie King noted, that with regard to questions they were welcome "to try and knock him off his [strategic] perch."[10] He even outlined the idea of a "peace conference," similar to Versailles in 1919, that would take place, perhaps in England, at the end of hostilities — with both Republican as well as Democratic members of Congress "invited" to participate.[11]

Versailles, then, moved to Westminster . . .

That Churchill was playing a dangerous double game became clear later that afternoon when the President invited the Combined Chiefs of Staff to the White House, following their afternoon meeting at the Federal Reserve Building. The President had heard via Admiral Leahy that the chiefs had confirmed their agreement to an April or May 1944 cross-Channel Second Front — but that tempers in the morning's meeting, when addressing remaining "interim" operations in 1943, had become so frayed the secretaries had been asked to leave the room while the chiefs dueled it out.

General Marshall's contention that further operations in the Mediterranean that fall would inevitably suck in the forces needed for a successful cross-Channel attack had hit home — Brooke defending his own strategy by claiming a cross-Channel attack would never succeed unless the Wehrmacht was first forced to fight hard not only on the Russian front but in Italy. Heavy fighting in Italy was thus the prerequisite of a successful invasion in the spring of 1944. "After the capture of a bridgehead" in northern France, "a Cherbourg might be seized, but the provision of the necessary forces to cover this would be difficult unless the Germans were greatly weakened or unable to find reserves," Brooke had warned.[12] A serious military campaign in Italy, in other words, would be the weakening blow: essential in order to make the April or May 1944 operation work.

Marshall had countered that such a strategy might very well achieve the opposite. The British, he'd summarized, were exaggerating the ease of a campaign in Italy, while perilously underestimating the need to throw maximum logistical effort into the real priority: the cross-Channel invasion. It should, Marshall reminded Brooke and the other committee members, "be remembered that in North Africa a relatively small German force had produced a serious factor of delay to our operations," given the mountainous terrain. "A German decision to support [defend] Italy might make intended operations extremely difficult and time consuming."[13]

No truer warning to the British was ever given in World War II — though Brooke would never admit, either then or in retrospect, that Marshall was right. Marshall had, Brooke merely confided to his diary that night, "suggested that the meeting should be cleared for an 'off the record' meeting between Chiefs of Staff alone. We then had a heart to heart and as a result of it at last found a bridge across which we could meet! Not altogether a satisfactory one, but far better than a break up of the conference."[14]

The compromise was certainly vague and open-ended. Rather than halting major offensive operations in the Mediterranean after the successful seizure of Sicily, as the President and Marshall wished, Eisenhower would be authorized to capitalize on any signs of an Italian collapse to seize airfields in southern Italy — but only assigning experienced Allied forces for the remainder of the summer. Then — at the very latest on November 1, 1943 — the best battle-hardened U.S. and British divisions were to be withdrawn from combat and transferred to Britain to prepare for D-day. This, they all agreed, should be mounted either in April or in early May, 1944.

This compromise, confirmed by all, had duly been reported by the Combined Chiefs when summoned to meet with the President in the Oval Office at 6:00 p.m.

They were then joined by the Prime Minister, on his return from the Capitol.

Nine Allied divisions were to be ferried in the assault across the English Channel on D-day itself, with twenty more in the days that followed — a massive rolling offensive backed by Allied air power and naval support. Whatever was left in the Mediterranean could be used by Eisenhower to "eliminate Italy from the war and contain the maximum number of German divisions."

According to the minutes of the Oval Office meeting, "the PRIME MINISTER indicated his pleasure that the Conference was progressing as well as it was and also that a cross-Channel operation had finally been agreed upon. He had always been in favor of such an operation and had to submit to its delay in the past for reasons beyond control of the United Nations."[15]

Given what Churchill had told U.S. congressional representatives *that very afternoon* — namely, that he did not favor what he saw as a "dozen Dieppes in a day" on the beaches of northern France — and given that he favored, instead, an Allied offensive through Italy and the Balkans, this was tantamount to perjury, unless the Prime Minister had truly had a Pauline conversion.

Only time would tell.

31

The Future of the World at Stake

HALF AN HOUR after the Combined Chiefs departed the White House, the President dined upstairs with Mackenzie King, Churchill, and Crown Princess Martha of Norway.

In deference to Princess Martha, the three leaders put aside any discussion of military strategy, and after the meal the President arranged for a Sherlock Holmes film to be shown as light relief. Churchill then "begged off" and went to bed, as did Princess Martha, leaving the President to talk quietly with his Canadian guest.

Gingerly, Mackenzie King sought to find out the President's intentions, in terms of Allied military strategy. "Tonight when I was talking alone with the President and asking how he and Churchill had got on, he said he thought an agreement was practically in final shape by now; that he, himself, would probably want to recast it a little more in the way of bringing up to the beginning some matters that were near the end."

The British had said they couldn't carry out the Anakim offensive to which they'd committed themselves at Casablanca, and there had been initial, heated discussion of this; the primary decision, however, was the Second Front in 1944. The President wanted to ensure the British commitment was not only firm but set down in ink, on paper, and in official accords — which Admiral Leahy, the Combined Chiefs chairman, had assured him would be drawn up formally by the weekend. As the President explained to Mackenzie King, it was vital to tie down and chain the wily British to a solid commitment, not simply rely on the understanding he thought they had come to at Casablanca. "He wanted to emphasize the building up of the forces in Britain so as to be certain of an attack from the North in the spring of 1944. He said he felt that this was the top feature of it all. He did not use that expression but that was the inference. It meant the determining blow in the spring of next year."

Listening to this, King was somewhat perplexed. Given what Churchill had said openly at the Capitol, in King's hearing, it seemed the President and the British prime minister, though sleeping under the same roof, were poles apart. Mackenzie King therefore relayed to the President what Churchill had said at the Capitol — including the Prime Minister's remarks about a Versailles-type conference in London.

President Roosevelt "put his hands to his face and shook his head, a bit as much as to say he wished that part had been left well alone," King recorded the President's pained reaction. "He then said to me that he did not know that there would be any peace conference," given its connotation with Versailles 1919. "As far as he was concerned, there would be total surrender" of Germany and Japan. And certainly nothing "in the nature of a Versailles conference," which Congress would have to ratify.[1]

Hearing of Churchill's behavior at the Capitol, the President had reason to be anxious, however. He liked Winston, in fact he felt enormous affection, bordering on love, for him at times. But he had cause never to quite trust him — and for that reason he preferred to see Stalin alone, without the Prime Minister. Who knew if Churchill would start hedging over the Second Front, if they met *à trois?*

It was going to be difficult enough to explain to Stalin that the Western Allies were not going to launch a Second Front before spring 1944. If Churchill, in a tripartite meeting, were to begin talking in front of Stalin of dumping the invasion of France and concentrating Allied efforts instead in the Mediterranean and the Balkans, the Soviets — preparing at that very moment for the onslaught of *fifty-nine* concentrated German divisions aimed toward Kursk — would rightfully be incensed: vitiating any hope of the Third Reich being defeated any time soon, or of Russian assistance in the war against Japan, or of arriving at a common postwar security agreement. The notion of a United Nations assembly, with a security council of the Four Policemen acting in concert, would thus be out the window.

The President's postwar vision still filled King with awe — as did King's possible role in it. According to the President, the United Nations organization would have a "supreme council representing all the United

Nations," and would need at its head "someone who would fill the position of moderator — someone who would keep his eye on the different countries to see that they were complying with the agreements made in connection with the peace, for example, limitation of armaments, not rebuilding, munitions, etc — not be allowed to build airplanes or any of the paraphernalia of war. It would be the Moderator's duty possibly to warn in advance and, if necessary, to have the council meet to take such action as necessary" — a person who would "have the confidence of all the nations."[2] And, having abjured any idea he himself might take that role, after the presidency, Roosevelt intimated he thought Mackenzie King, at the end of the war, would make an excellent such secretary general.

King was understandably flattered — but in the meantime, like Roosevelt, he remained perplexed by the contradictions in Churchill's character. At the Pacific War Council, Winston had flatly denied in front of the Chinese representative that he'd ever made a formal undertaking to mount Operation Anakim, a British offensive from Indian territory to help China — even though Dr. T. V. Soong had documentary evidence of the commitment.

Like Mackenzie King, the President had shaken his head over such unnecessary falsehoods — "The President said that the trouble with Winston is that he cannot get over thinking of the Chinese as so many pigtails."[3] Similarly, over India, Churchill was as stubborn and indifferent to world opinion as he could get away with — having instructed the viceroy of India to make sure the American minister in Delhi not be permitted to interfere in any way with Mahatma Gandhi's 1943 hunger strike — and cabling Lord Halifax to tell all Americans in Washington that the British government "will not in any circumstances alter the course it is pursuing about Gandhi," even if this resulted in Gandhi's death.[4] He'd insisted, moreover, on speaking in public of "British forces" rather than "British Commonwealth forces"[5] — which was much resented in Canada, and would be even more resented once Canadian forces went into combat in Sicily. Churchill was, in short, a law unto himself — and yet the repository of such underlying humanity, understanding of history, and noble sentiment that it was impossible not to admire him.

The question, then, remained: Would Churchill stand by what he'd told the President and Combined Chiefs of Staff earlier that evening in the Oval Office — or by what he'd told members of Congress that afternoon at the Capitol?

The matter was not academic; the future of the world was literally at stake — and Prime Minister Mackenzie King now watched Churchill's double game with growing concern.

When addressing the Pacific War Council the next day, May 20, at noon, Churchill refrained from discussing strategy in Europe in front of the President. Late in the afternoon, however, the Prime Minister addressed a special meeting of the chiefs of staff of Britain and Canada and representatives of other parts of the British Empire, held in the White House dining room, which the President had kindly made available to him.

Lord Halifax, who attended this "imperial" meeting, dismissed it in his diary. Churchill's "long speech of fifty minutes about the war" had been "very well done but with nothing very fresh in it except two or three things that could have been said in five minutes. I never saw anybody who loves the sound of words, and his own words more."[6] General Brooke, exhausted by the Combined Chiefs of Staff meetings (involving yet another "off the record" battle), dozed off, but Mackenzie King listened very, very carefully.

"After a moment's pause," King recorded that night, Churchill "started in saying he would sharpen and heighten somewhat the points he had made in his address before Congress." This the British prime minister proceeded to do, "following pretty much the sequence" King had heard on the Hill. In this there was "little else that was new." "The most interesting part," King noted that night, however, "was the account he gave as to why it would be advisable to proceed against Europe from Africa as a base." It was, Churchill stated, "advisable to get ahold of a few islands in the Mediterranean, use them as stepping-stones toward Europe. The great effort should be made to get Italy out of the war."

Few could argue with this — or with Churchill's magnanimity. Unconditional surrender was an agreed Allied policy, but one should not be "too particular about the terms on which peace could be made with Italy," Churchill suggested. "Her people had never had their hearts in the war. He was not anxious to see their country destroyed. If he could get the Italian fleet, that would be an immense gain. He would then have more ships to be employed against the Japanese . . ." With regard to the Second Front, whether in 1943 or 1944, he was strangely reticent, however — and King remained uncertain whether Churchill had really changed his view that it "would be slaughter."

Given the loss of so many Canadian lives in the Dieppe assault the previous August, Mackenzie King was understandably sensitive to this, having noted it was "pretty strong language and indicated a feeling that Dieppe had been a real sacrifice, perhaps an unnecessary one." At any rate, the "picture he presented was of the beaches being long and in stratas; in some places, water deeper than others. Very difficult to land troops. He was determined not to have men sacrificed anymore than could be helped."[7]

Whatever the U.S. and British chiefs might agree upon, then, it was still questionable whether Churchill was really willing to commit British and Canadian troops to a Second Front.[8] In fact he "spoke emphatically about not being in too great a hurry to invade Europe even from the South," Mackenzie King noted. "He said opinion was divided as to the best way to win against Germany. Some thought bombing would be sufficient. There was no harm, however, in trying other methods, as well, while trying to do the best they could with bombing."[9]

If bombing was Churchill's only plan to defeat Germany, it did not sound very convincing to King. Moreover, it was certainly not how the President and the Combined Chiefs of Staff, in their long and trying meetings, were approaching the question of how to vanquish the Third Reich and move on to defeat Japan. Churchill seemed unabashed, though. He was not, as King recognized, a strategist in the true sense of the word, but an *opportunist*—opposed down to his entrails to "giving commitments versus tactics," as King noted.[10] And with that the Commonwealth meeting had ended.

Mackenzie King was to spend the night aboard his train, since he would be returning to Canada, via New York, on May 21. Harry Hopkins had asked King to see the President before leaving the White House, however, and this the Canadian did after midday, on the twenty-first, in the Oval Office. Despite the heat the President "looked very fresh and cool. Was seated on his swing chair. I sat to his left looking out of the window toward the garden. A lovely feeling. An ideal office with a little court opening out of the room." The President seemed confident the Allies now had an agreed plan for winning the war—and one he could put to Stalin, confiding again to King his invitation to meet the Russian dictator in the Bering Strait.

What might be Churchill's reaction at being deliberately excluded, the President then asked King, given Churchill's erratic position over a

Second Front? To encourage the President, King assured him that Winston — who had, after all, had his own private meeting with Stalin the previous summer — would get over it. Besides, the main thing was not the Prime Minister's pride or dependability, but the President's hugely important global goal — moral, military, and political — that promised to shape the postwar world.

On that note the two leaders parted company — though King wanted also to say goodbye to Churchill. He therefore walked from the Oval Office through to the White House mansion and up to Churchill's guestroom, just after 1:00 p.m. There he found the Prime Minister still in his underwear, dressed "in his white linen under-garments; little shirt without sleeves and little shorts to his knees, otherwise feet quite bare excepting for a pair of slippers. He really was quite a picture but looked like a boy — cheeks quite pink and very fresh."[11]

King said he wanted "to be perfectly clear in my own mind what is to be done," in terms of military strategy — strategy that involved tens of thousands of Canadian lives — lest there be any misunderstanding.

To this, Churchill responded by saying "there will be no invasion of Europe this year from Britain. I tell you that" — but the Mediterranean was another matter. There, Canadian troops would shortly take part in the invasion of Sicily in July, to gain battle experience. Canadians would, in fact, "be in the forefront of the battle."[12] Moreover, instead of returning direct to London, Churchill himself was going "to Africa from here" — a "dead secret."[13]

What of Allied war strategy *beyond Sicily,* though?

Delicately, King "did say that I thought there was a certain possibility of divergence of view" between the senior Canadian forces' commander and "some of the plans he, Churchill, had in mind; also between some of the plans that our own chiefs of staff or the British chiefs might have." The Canadian War Cabinet was prepared to go along with what would "best serve" the need to win the war — but only a strategy that was feasible "in the opinion of the military advisers who had charge of the strategy of the war."[14] In other words, the Combined Chiefs of Staff.

Churchill, somewhat surprised, reassured King there was no divergence — indeed that King was at liberty to speak with General Brooke, the CIGS, before returning to Canada, if he was in any way unsure or confused.

Still the Canadian prime minister remained skeptical, however. He had another talk with the Canadian minister of national defense that afternoon — who said he had it direct from Brigadier Jacob, Churchill's military assistant, that the strategy agreed by the Combined Chiefs of Staff would now stand. "In the light of this," King noted, "I thought it was just as well not to attempt to see Sir Allen [*sic*] Brooke. It might have looked to the Defense Ministers that I was distrustful of them" — and of Churchill.[15]

That he had every right to be, however, would only become clear after Mackenzie King's departure.

32

The President Loses Patience

EVEN GENERAL BROOKE was disbelieving.

The President had spent the weekend at Shangri-la, while Winston Churchill moved for a few days to more comfortable quarters at the British Embassy on Massachusetts Avenue. Once the two leaders returned to the White House, however, the Combined Chiefs of Staff were asked to come to the Oval Office—and on the afternoon of Monday, May 24, they did: there to present the final terms of the Trident agreement. When they sat down before the President and Prime Minister, however, it was to find Neptune flatly refusing to accept the agreement they had reached.

Brooke had known his prime minister to be an occasionally maddening individual—obstinate, brilliant, sometimes tender, sometimes rude, and with a predilection for chasing red herrings. But to behave like a spoiled adolescent in front of the President of the United States of America—a president who was not only directing a global war but was furnishing the materials and fighting men to win it—seemed to Brooke the height of folly.

As Brooke understood it, the Combined Chiefs had been summoned to be thanked. Instead, Brooke found, "the PM entirely repudiated the paper we had passed, agreed to, and been congratulated on at our last meeting!!" as he recorded with exasperation that night. "He wished to alter all the Mediterranean decisions! He had no idea of the difficulties we had been through," the Ulsterman exploded in the privacy of his diary, "and just crashed in 'where angels fear to tread.' As a result he created [a] situation of suspicion in the American Chiefs that we had been [going] behind their backs, and had made matters far more difficult for us in the future!"[1]

Brooke was riven by shame and embarrassment. "There are times when he drives me to desperation! Now we are threatened by a redraft by him and more difficulties tomorrow!"[2]

General Marshall was equally furious. Admiral King boiled. Admiral Leahy, as chairman of the Combined Chiefs, was simply outraged. "From four-thirty to seven p.m. the British and American Chiefs of Staff presented to the President and the Prime Minister their report of agreements reached during the present conference," he noted in his own diary that night. "The Prime Minister refused to accept the Mediterranean agreement."

The Combined Chiefs' report had made no commitment by the Allies to invade mainland Italy, but instead only to "plan such operations in exploitation of Husky as are best calculated to eliminate Italy from the war, and to contain" — either by threat or by operations — "the maximum number of German divisions" while the cross-Channel invasion of northern France was readied for launching on May 1, 1944.

Mr. Churchill, Leahy noted in exasperation, had other ideas. He "spent an hour advocating an invasion of Italy with a possible extension to Yugoslavia and Greece."[3]

Leahy was as incredulous as Brooke. An "extension" of operations to Yugoslavia, Greece, and the Aegean that risked making a May 1944 cross-Channel Second Front impossible? Churchill was undeterred, however — and adamant.

Since Churchill was not only British prime minister but quasi–commander in chief of all British Commonwealth forces, this was a major stumbling block. "Final decision was by his request postponed until tomorrow," Leahy recorded.

As Brooke feared, this made the U.S. team almost apoplectic. Oh, perfidious Albion! "The Prime Minister's attitude is an exact agreement with the permanent British policy of controlling the Mediterranean Sea, regardless of what may be the result of the war," Leahy noted in disgust in his diary. "It has been consistently opposed by the American Chiefs of Staff," he added, "because of the probability that American troops will be used in the Mediterranean Area" — "at the expense of direct action against Germany." It was a Churchillian demand "which in our opinion [will] prolong the war."[4] If, that was, it did not lose it.

In shock and no little confusion, the British and American chiefs were ushered out of the White House and into their cars.

Churchill went straight to his room. After dinner and a movie there was a meeting in the President's Map Room, with Harry Hopkins and the Prime Minister's chief of staff, General Ismay.

In the narrow, windowless room, its walls hung with giant maps and thousands of the most secret reports, cables, and memoranda locked in filing cabinets in the center, the President pulled no punches. The date for the cross-Channel invasion was now set, he told Churchill, and the forces for it must be withdrawn from the Mediterranean by November 1, 1943 — period.

Churchill was furious. Returning finally to his room at 2:00 a.m., the Prime Minister summoned his doctor.

Sir Charles Wilson found "the P.M. pacing his room" — and blaming the President. "There was no welcoming smile. When I asked him how he had been he did not answer. He had other things to think about besides his health. He stopped and said abruptly, 'Have you noticed that the President is a very tired man? His mind seems closed; he seems to have lost his wonderful elasticity.'"[5]

Dr. Wilson — ignorant of the cause — watched as Churchill "went up and down his room, scowling at the floor." "The President is not willing to put pressure on Marshall," he explained. "He is not in favour of landing in Italy. It is most discouraging. I only crossed the Atlantic for this purpose. I cannot let the matter rest where it is."[6]

Dr. Wilson could prescribe sleeping medication, but he could do nothing to change the situation. Nor could Churchill. The President had said no — and there was little that could be done without seriously undermining, even wrecking, the Western alliance. The Prime Minister would have to accept defeat. The die, after all, was now cast. Even the Canadians were getting ready for a cross-Channel assault in 1944, with no interest in fighting in Italy — let alone Yugoslavia. Once back in Ottawa, Prime Minister Mackenzie King was preparing to tell his War Cabinet that it had been agreed in Washington that "the big battle will come early next year." Moreover, that "the Canadian army will be used along with the American army and the British army to make the final assault on Europe from the North" — not Italy or the Balkans. And "that we may expect the end of the war not before the end of this winter but before the end of another winter (1945)."[7]

Churchill continued to pace. He had not become prime minister of Great Britain and the one leader able to stand up to Hitler, however, by caving in to force majeure — particularly pressure from his own countrymen. Dr. Wilson's medication would permit him to sleep, briefly — but not to alter his convictions.

Before the Combined Chiefs of Staff could reappear before their political and military masters at the White House on the morning of May 25, therefore, Churchill began a new attack on the Trident agreement.

That the Prime Minister meant well was not at issue. Long-term geopolitical British considerations had to be taken into account. But single-handedly to attempt to bend the president of the United States to follow a British agenda was foolhardy — especially in opposition to his own military team.

Dr. Wilson had already been worried lest the Prime Minister, by undertaking so many responsibilities, by refusing to delegate, by drinking so much, and by making so many wild trips abroad, might be approaching a mental breakdown, or "a gradual waning of his powers, brought on by his own improvidence, by his contempt for common sense and by the way he has been doing the work of three men. There is no hour of the night when I can be certain that he is in his bed and asleep. Of course, this cannot go on forever."[8]

It couldn't — and explained in part the Prime Minister's amazing behavior, to the embarrassment of all, especially the President.

The Combined Chiefs assembled again in Roosevelt's office at 11:35 a.m. Once again they were treated to Churchill at his most petulant. "We were therefore exactly as we had started so far as the paper we had submitted to the President and PM was concerned," Brooke recounted in despair in his diary, adding, in his slashing hand, "the PM had done untold harm by rousing suspicions as regards ventures in the Balkans which we had been endeavouring to suppress."[9]

Churchill, Secretary Stimson afterward learned from the President, "acted like a spoiled boy the last morning when he refused to give up on one of the points — Sardinia — that was in issue. He persisted and persisted until Roosevelt told him that he, Roosevelt, wasn't interested in the matter and that he had better shut up."[10]

For that, at least, General Brooke was grateful to the President.

With the President's final loss of patience and his stern word to Churchill, the meeting had mercifully come to an end.

Debate was over — and with that dramatic finale, the Trident Conference done. D-day, to be called Operation Overlord, would take place, come hell or high water, in the spring of 1944.

A grand, celebratory luncheon for the Prime Minister, Combined Chiefs of Staff, and all the senior staff officers involved in the conference was given by the President at the White House at 1:30 p.m. on May 25.

Early the next morning, Roosevelt drove down with Churchill to the special Clipper terminal on the Potomac River. The 160 members of the British military contingent would be sailing home from New York, but the Prime Minister and General Brooke were to board a huge Boeing seaplane that would fly them first to Newfoundland, and from there to North Africa.

The President had been skeptical regarding Churchill's new mission — as the Prime Minister was aware. It had not stopped Churchill, however, and the two leaders had come to a compromise. Churchill had assured the President he had no motive other than to review British and Allied HQ preparations in Algiers for the impending assault on Sicily, Operation Husky. Given the President's chariness, Churchill had felt compelled to suggest that General Marshall accompany him, as a gage of his fealty to the President and the war strategy finally and formally agreed between allies. The President had thought this an excellent idea — General Marshall flying, so to speak, as a U.S. marshal.

Poor Marshall had not been consulted.

At the Pentagon, Secretary Stimson had been furious — on Marshall's behalf. "Marshall told me of it," Stimson recorded in his diary, "and said he rather hated to be traded like a piece of baggage."[11]

The U.S. war secretary remained deeply suspicious, moreover. Churchill was "going to take Marshall along with him" for no other purpose, Stimson protested, than "to work on him to yield on some of the points that Marshall has held out on in regard to the Prime Minister's excursions in the eastern Mediterranean." This was too bad. General Marshall was worn out having to deal with Churchill's two-week visit, along with his vast retinue of military chiefs and advisers seeking to overturn the Casablanca agreement. Of all people, the general now surely deserved a break. In this respect, "to think of picking out the strongest man there is in America, and Marshall is surely that today, the one on whom the fate of the war depends, and then to deprive him in a gamble of a much needed opportunity to recoup his strength by about three days' rest and send him off on a difficult and rather dangerous trip across the Atlantic Ocean where he is not needed except for Churchill's purpose is I think going pretty far," Stimson frothed in his diary, outraged by the iniquity. "But nobody has any say"[12] — the President being the elected president, by far and away the most powerful man in America, and this his will.

For his part, the President found it ironic he was having to send Marshall to North Africa to keep the irrepressible Churchill on the rails. Roosevelt was not sorry, though. Seeing Eisenhower, Clark, Patton, perhaps, and the general lay of the land following General Eisenhower's great victory at Tunis would be no bad thing for his Army chief of staff. The number of Axis troops that had surrendered was now said to have exceeded even those at Stalingrad; the omens were good.

Word had also come from Moscow, moreover, that Stalin had finally agreed to a personal meeting. This would probably now take place in August. The President would pretend to be going to Canada to see Prime Minister King — and secretly fly north across Alaska to the projected rendezvous with the Russian dictator.[13] It was a relief, in these circumstances, that he, the President, would be able to convince Stalin that the Western Allies were united in their resolve to mount the Second Front in the spring of 1944 — and important that Marshall hold the Prime Minister tightly to this agreement. No more reneging, or alternative ventures, or pessimistic doomsaying behind his back!

The cross-Channel invasion would not take place in 1943, to Stalin's likely disappointment, but it would definitely be mounted in overpowering, U.S.-dominated force in May 1944 — and would, the President was confident, lead to the end of the war, either at the end of 1944, or early 1945. Only Churchill, in his unpredictable way, could possibly mess this plan up.

Admiral Leahy remained suspicious. As he noted with scarcely concealed distrust, the "agreements finally reached" were excellent, and would advance the American cause: to defeat Hitler. "This is, of course, based on an assumption that the agreements will be carried out by our allies."[14]

• • •

Ironically, General Brooke was just as skeptical as Leahy—at least with regard to his boss, the Prime Minister. Churchill's great qualities did not include consistency of military strategy. As Brooke had noted in exhaustion on May 24, summarizing his colleagues' contributions to the "Global Statement of Strategy" that the Combined Chiefs had drawn up, Admiral King was still besotted by war in the Pacific theater; General Marshall was too bold, willing to chance a cross-Channel invasion that would risk putting into the cauldron of battle "some 20 to 30 divisions, irrespective of what happens on the Russian front, with which he proposes to clear Europe and win the war"; Air Marshal Portal, by contrast, was imagining the war in Europe would be won by "bombing" alone; Admiral Pound was believing "anti-U-boat warfare" was the key; while Brooke himself favored all-out war in the Mediterranean, not to defeat Germany, per se, but to "force a dispersal of German forces, help Russia, and thus eventually produce a situation where cross-Channel operations are possible."

"And Winston???" Brooke had continued, rhetorically, in the privacy of his diary. "Thinks one thing at one moment and another at another moment. At times the war may be won by bombing and all must be sacrificed to it. At others it becomes essential for us to bleed ourselves dry on the Continent because Russia is doing the same. At others our main effort must be in the Mediterranean, directed against Italy or the Balkans alternatively, with sporadic desires to invade Norway and 'roll up the map in the opposite direction to Hitler'! But more often than all he wants to carry out ALL operations simultaneously irrespective of shortages of shipping!"[15]

To his credit, Churchill was not unaware of or even embarrassed by his own impetuous, pepper-spray, relentlessly demanding/urging nature—"I am arrogant, but not conceited," he told a companion in 1943[16]—but had no idea Brooke was keeping such a candid journal, especially one that might be used to indict him, later, as a volatile commander in chief of lamentably poor and inconsistent judgment. After all he, Winston Spencer Churchill, would ensure his own skills as a writer and historian would make certain he came out smelling of roses—as he openly confided in North Africa some days later. Veracity would not be his objective as an eventual memoirist/historian, he would tell General Eisenhower and a dozen top American and British generals invited to dinner at Eisenhower's headquarters in Algiers. Having imbibed several whiskeys, he announced that "it was foolish to keep a day-by-day diary because it would simply reflect the change of opinion or decision of the writer"—a diary "which, when and if published, makes one appear indecisive and foolish."[17]

To illustrate his dictum Churchill instanced the daily journal of British Field Marshal Sir Henry Wilson. Sir Henry had left copious diaries detailing his role before and during World War I. In one entry he had unwisely forecast: "There will be no war." This was unfortunate because "on the next day war was declared," Churchill told his enthralled listeners.

Since Eisenhower's naval aide was himself keeping a daily diary this was unwise, but Churchill had by then imbibed too much alcohol to care.

The English field marshal, Churchill went on happily, had subsequently been assassinated on his doorstep by Irish republicans, in 1922—leaving the question of what to do with his precious war diaries. "The wife had insisted the diary be published post-humously," Eisenhower's aide recorded, "and, consequently, General Wilson was made to appear foolish. For his part, the Prime Minister said, he would much prefer to wait until the war is over and then to write his impressions so that, if necessary, he could correct or bury his mistakes."[18]

Bury his mistakes. It was a telling phrase.

Flying to North Africa with Winston, however, General Marshall would at least exert adult supervision, the President was satisfied. Marshall could be counted upon not to permit the Prime Minister to veer off into any wild ventures now that the cardinal issue of the Second Front and its timing had been formally resolved.

This still left open, however, the question of command.

Who should be the cross-Channel assault supreme commander—an appointment that, in order to help bolster the somewhat tentative British commitment, the President had at Casablanca suggested should go to a British officer?

In the wake of Trident, however, the President was not so sure. General Marshall's faith in Bolero, now renamed Operation Overlord, had been constant and unremitting. Might not General Marshall, an American, be a surer bet as supreme commander—not only in making certain the assault was actually carried out on time, but in dealing at close quarters with a British prime minister whose penchant for meddling in battles was now notorious?

By spending time not only with General Eisenhower but with English field and staff generals at Ike's headquarters in Algiers, Marshall would get to know potential British colleagues, generals, and subordinates better, the President felt. As well as the British prime minister.

By contrast, Churchill was concerned that, if the Second Front was indeed to be launched at American insistence, his candidate for supreme command, General Alan Brooke, should be on the best of terms with the President. There thus arose, on May 26, an added irony, as the two army chiefs of staff of their respective nations boarded the former British Overseas Airways Boeing 314A seaplane, registration number G-AGBZ, bobbing on the Potomac early that morning. Churchill had duly boarded the Clipper, having made his farewells. Aware that he'd promised Brooke command of Overlord, however, he had told the CIGS to go and sit for a few minutes with the President in his car, in case the President decided to raise the matter.

The President gave nothing away. "He was as usual most charming," Brooke noted in his diary that night, "and said that next time I came over I must come to Hyde Park to see where my father and Douglas [Brooke's brother] had looked for birds."[19]

Roosevelt's invitation was typical of the President — wanting the conference to end on a happy, personal note. Brooke was certainly touched, and the two men shook hands.

The ornithologist and his thorny opposite number, General Marshall, then took their seats inside the body of the huge Boeing seaplane as its engines roared to life, ready for takeoff — the President waiting to watch. Both Marshall and Brooke were now contenders to command the greatest amphibious invasion in human history — one that would undoubtedly, as Hitler himself remarked, "decide the war."[20]

PART NINE

The First Crack in the Axis

33

Sicily—and Kursk

AT THE WHITE HOUSE on the evening of July 9, President Roosevelt was giving a state dinner for General Giraud. He was also waiting patiently for word from General Eisenhower as to how the invasion of Sicily, timed to start soon after midnight in the Mediterranean, was going. Had Allied deception measures worked? Were the Germans waiting for the Allied armies to come ashore in the south? How would Italian forces fight on their home soil?

Finally Admiral Brown, his naval aide, brought him the news.

Taking General Giraud upstairs to his study, Roosevelt met Daisy Suckley, who was staying in the Blue Room, on the landing. The President had told her the dinner would go on until a quarter to eleven, so Daisy was happily sewing a seam on her new nightgown when "the elevator door suddenly opened—I heard the P's voice—I grabbed my diary, my pen, my workbox, & my nightgown—started to flee! The President stopped me, laughing, halfway down the hall already, & followed by the General. My thimble flew to the right, my spool to the left. The General laughed & we shook hands—the P. spoke over his shoulder as he was wheeled into his study: 'The General & I are going to have a heart to heart talk—We have landed in Sicily! The word has just come!"[1]

For his part, Admiral Leahy noted in his own diary: "During the dinner the president announced that British-American-Canadian troops were in process of invading Sicily. Our best information indicates that the enemy force now on the island consists of 4 or 5 Italian divisions and two German divisions, which we should be able to defeat in time if the landing is successful."[2]

The President was pleased, but like Leahy, he was determined not to give way to overexpectations. Failure would delay but by no means wreck the agreed timetable for a cross-Channel assault the next year; victory, however, would give the Allied forces — including French troops fighting under Eisenhower's command — further confidence that they could mount a major amphibious invasion and defeat the Wehrmacht in combat: the prerequisite for a successful Overlord.

And with that quiet confidence the President set off the next day to spend the weekend in Shangri-la with his de facto domestic deputy president, former Justice James Byrnes — his head of the Office of War Mobilization — and Byrnes's wife, as well as Harry Hopkins and his wife, and Daisy Suckley. After watching a movie in the mess hall, "We sat around," Daisy Suckley, "to get news about the invasion of Sicily — During dinner, we had tried also, but static is very bad and reception not good up on this hill, even when the weather is clear . . ."[3]

The President had every reason to be hopeful.

Operation Husky was the largest amphibious invasion ever attempted in war: three thousand Allied vessels, troop planes, and hundreds of gliders setting 160,000 soldiers ashore in Sicily in a single day from across the Mediterranean, departing from ports and airfields in Algeria, Tunisia, Malta, Libya, and Egypt in appalling weather (which caused almost half the gliders from Tunisia to land in the sea) to their rendezvous at dawn on July 10.

General Eisenhower had overruled his own planners and had accepted General Montgomery's preference for a concentrated invasion of the southeastern corner of Sicily, stretching from Gela to the Gulf of Noto and Cassibile. This was just as well, since the German commander in chief, General Kesselring, sent the first of his two panzer divisions (with 160 tanks and 140 field guns) to the west of Sicily — leaving only a single panzer division in the east. However hard they fought, the men of the remaining Hermann Göring Panzer Division were unable to prevail against Allied troops debouching across twenty-six beaches there. Italian defenders, ill armed and ill motivated, for the most part crumpled under the weight and power of the Allied bombardment.

Despite the poor weather — with gale force 7 winds — the invasion thus proved brilliantly successful.

• • •

At the Pentagon in Washington there was an air of near jubilation, especially when the casualty rolls turned out to be less than a seventh of what had been estimated. Once again it was the President, in his capacity as U.S. commander in chief, who had made victory happen. Over the objections of his top generals and secretary of war in January, he'd insisted upon success in the Mediterranean in 1943, rather than sure defeat in France. How wise he'd been proven, all now agreed; only two German divisions in Sicily, instead of more than two dozen in France.

Many things went wrong in the landings, not simply owing to the high wind but also because of friendly fire: trigger-happy naval gunners shooting down dozens of Allied aircraft. Patton's Seventh Army landing at Gela was initially touch-and-go, requiring naval artillery to beat off determined Axis counterattacks — Kesselring having instructed the Hermann Göring tanks and troops to move "at once and with all forces attack and destroy whatever opposes the division. The Führer has ordered all forces to be brought into operation immediately in order to prevent the enemy from establishing itself."[4]

For the Germans, it proved a losing battle, as it had for Vichy defenders in Torch. For the Allies, however, the military lessons provided by Husky would not only be legion but gold — not least in terms of intelligence, deception measures, command experience, army air and naval cooperation, and cohesion. Launched in such overwhelming, concentrated Allied force, there was little the Germans could do to halt it. A U.S. general, Dwight D. Eisenhower, was the Allied supremo, with one American and one British army field commander serving under him. George Patton, who had commanded the invasion forces at Casablanca, now led the U.S. Seventh Army, with excellent U.S. corps and divisional commanders such as Omar Bradley, Geoffrey Keyes, Manton Eddy, and Terry Allen coming to the fore. Montgomery again commanded the British Eighth Army — this time with both veteran and untried troops, including a full Canadian corps determined to obliterate the "fiasco" of Dieppe. Inter-Allied coalition command was rehearsed in real time, as well as interservice cooperation — improving exponentially as the battle for Sicily progressed.

With the Allies achieving complete naval and air superiority over Axis forces in the Mediterranean, moreover, and Patton and Montgomery's ground forces threatening to strike out from the beaches of Sicily, there arose a real prospect that the Italians — who for the most part were refusing to fight to defend their homeland — might overthrow Mussolini and submit to unconditional surrender without the Allies needing to invade Italy.

Hitler's hand was forced, therefore. He would have to call off his latest offensive on the Eastern Front and deal with the Western Allies before they dealt with him.

34

The Führer Flies to Italy

ON JULY 13, three days after the Western Allies landed in Sicily, the Führer summoned his army commanders to his headquarters in East Prussia.

He had changed his mind. Operation Citadel, his massive, long-awaited offensive on the Eastern Front, was doomed. Nervous lest the Western Allies stab him in the back just as the Wehrmacht attacked in Russia, he had already scaled back his objectives for the battle. Instead of seeking to push deeper into the Soviet Union, he had decided to destroy the Russian armies in situ, near the city of Kursk, where their forces formed a salient that could be pinched off by German armies thrusting north and south. In this way, the Soviet armies would be decimated — destroying any chance of a Russian offensive that year, and allowing Hitler to deal decisively with any Allied operation in the west or south.

To their consternation, Hitler now told his generals he was going to call off the Kursk offensive — the biggest tank onslaught yet of the war — in mid-battle. It had been raging for eight days and the Wehrmacht, according to Field Marshal Erich von Manstein, was now on the cusp of victory: ready to close its pincers and destroy Russian forces left in the salient. But as Hitler explained, "the Western Allies had landed in Sicily," and "the situation there had taken an extremely serious turn," Manstein recalled. "The Italians were not even attempting to fight, and the island was likely to be lost. Since the next step might well be a landing in the Balkans or Lower Italy [the heel], it was necessary to form new armies in Italy and the western Balkans. These forces must be found from the Eastern Front, so 'Citadel' would have to be discontinued."[1]

Manstein felt as if the wind had been knocked out of him. The Allied invasion of Sicily, in other words, would now save the Russians from the drubbing the Wehrmacht was poised to administer in the East — Manstein later cursing that "Hitler ruled that 'Citadel' was to be called off on account of the situation in the Mediterranean. And so," the field marshal went on, "the last offensive in the east ended in fiasco."[2]

Even more symbolic for the course of World War II, however, it caused Hitler to fly south, to Berchtesgaden, hopefully to meet with Mussolini in person there.

In the event, Mussolini refused to fly to Berchtesgaden. Ignoring the poor performance of the 230,000 Italian troops he'd stationed in Sicily to defend the island against Allied assault, the Duce blamed the Führer for the success of the U.S., Canadian, and British invasion. The Luftwaffe, he complained, had withheld the necessary planes and equipment with which to defend such a big island, and the responsibility, he claimed, was therefore Hitler's.

Returning to the White House from Shangri-la, by contrast, the President was intensely proud of the Allied performance in Operation Husky. "The news from Sicily is pretty good. Thank Heaven," the President's cousin Daisy noted in her diary on July 14[3] — the President confident Stalin would now see the merit of Allied strategy, which had clearly taken Hitler completely by surprise, and was threatening, overnight, to sever the German-Italian partnership in the Axis Pact.

From all he'd heard, Stalin was nothing if not astute. With Hitler now compelled to send major forces to southern Europe, rather than to the Eastern Front, indeed to move forces away from battle on the Russian front, Stalin would eventually recognize both the political and military ramifications, he was sure. On July 15 the President therefore cabled to congratulate Stalin on the stalwart Russian defense of Kursk — urging him, however, to respond "about that other matter which I *still* feel to be of great importance to *you* and *me*": namely their meeting together to discuss the end of the war — and the postwar.

"The P. is awaiting word from Stalin as to when they can meet — I hate to have the P. take the risk, but he feels it essential for the future," Daisy recorded in the privacy of her diary. "If it occurs now it will be in Alaska; if it occurs late in the Fall, it will be North Africa." Stalin might "not feel able to leave Russia now," she allowed; in fact she actually hoped so, as she considered "the risk of the trip" for the President, "is very great."[4]

There was no response from Stalin, however.

At Hyde Park the following Monday, July 19, Daisy noted the President "looked preoccupied & a little worried." He had "on his mind his possible meeting with Stalin in Alaska — Stalin has set no date & has [still] not committed himself." The President had, however, finally informed Churchill of his invitation to Stalin — but had not extended the invite to include the Prime Minister. "The P. said W.S.C. wanted to go to the meeting, but F.D.R. won't let him," Daisy noted, surprised, but accepting the President's logic. The stakes, in terms of postwar peace and international security, were too high to take the risk of Churchill embarrassing him by his opposition to a cross-Channel assault. "He wants to talk, man to man, with Stalin, & try to establish a constructive relationship. He says that the meeting may result in a complete stalemate, or that Stalin may refuse to work along with the United Nations, or, as he hopes, that Stalin will be willing to work with the U.N.," but it was, surely, worth trying. "How much F.D.R. has on his shoulders! It is always more & more, with the passing months, instead of less & less, as he deserves," she mused — and, she added sagely, as "he gets older."[5]

Hitler rushed two more German divisions to Sicily to stiffen the Axis line, as well as warning his panzer reserve group on the Eastern Front to prepare to head south to Italy. He knew, however, it was hopeless to imagine he could hold on to Sicily itself, given the weight of the Allied assault and the flight of his supposed Italian partners. With Patton racing forces northeast to Palermo, and Bradley and Montgomery pushing the German panzer, paratroop, and infantry defenders back toward Mount Etna, the war seemed to many observers, on all sides, to be, if not won, then winnable in the near future: Sicily the keystone to a possible collapse of the Axis Pact, and even German solicitations for peace . . .

For his part, Hitler agonized over what to do about Mussolini — knowing he would have to breathe fire into the Duce's soul if he was to stop an Italian surrender that would expose his entire southern European flank to Allied invasion. Yet to his chagrin, waiting at the Berghof — the holiday home in the Bavarian Alps he'd bought with the royalties earned from *Mein Kampf* — he simply could not persuade Mussolini to come meet him in Germany.

Every day the situation had become more menacing — for both men. Even Hitler's most loyal supporter, Dr. Goebbels, was forced to acknowledge that, thanks to the Western Allies, Operation Citadel in the East had failed. The Allied forces invading Sicily were simply too massive. "The English and the Americans are expanding their bridgehead on a scale that's really stunning," the Reich minister had already noted in his diary on July 17.[6] "The question keeps coming up, how on earth we will be able to deal with war on two fronts, which we're slipping into. It has always been Germany's misfortune, past and present," he mused.[7] In the circumstances, it would be "almost a miracle were we able to hang on to Sicily."[8]

All Goebbels could think of now was to drive a wedge between the Russians — who were still demanding a Second Front that very year — and the Western Allies. "We haven't really any other alternative than to try to ease the situation through political means," he reflected.[9] Ignoring the millions of Jews and others the Nazi SS had "liquidated" — with more being "exterminated" every day — he wondered how, in view of the evidence of the massacre of Poles at Katyn, the Western Allies could imagine they could seriously do business with Russian barbarians. Could Katyn be the wedge issue?[10]

Political possibilities aside, the Führer had meantime to hold together his military alliance, Goebbels recognized: the Third Reich, the Empire of Italy, and their satellites and puppet regimes, from Norway, Hungary, and Romania to Bulgaria. It was an Axis military coalition that suddenly appeared in grave jeopardy — the once-triumphant Axis forces rocked on their heels both in Russia and the Mediterranean. "What's undeniable is that we find ourselves in a really critical situation," Goebbels admitted in his diary. "In previous summers," he reflected — thinking of 1940, 1941, and 1942 — "that was never the case." Now, however, it was different. "For the first time since the beginning of the war we've not only nothing to show for our summer offensive but we're forced to fight tooth and nail to defend ourselves — something that is casting a dark shadow over world opinion in the neutral countries."[11]

If the mountain would not come to Mohammed, then Mohammed must to the mountain go, Hitler was forced to accept. "The Führer has flown to Italy," Goebbels thus noted on July 19, on hearing the news from his liaison officer. "It's good the Führer is going to have it out with the Duce," the Reich minister added, having learned the meeting would take place north of Venice, "for Mussolini is the heart and soul of Italian resistance, and it's always been noticeable that after he's only been a couple of hours with the Führer, Italian politics and its war effort get a whole new infusion of blood."[12]

• • •

Flown to Treviso airport, the Führer was then taken by train to Feltre, and from there by limousine to the chosen meeting spot: Villa Gaggia. There the two fascist leaders finally conferred.

Despite a two-hour monologue by the Führer there was no infusion of blood or confidence, however; midway through the meeting the Duce was told the Allies were bombing Rome.

The summit proved so disappointing the two dictators decided neither to issue a communiqué nor make the meeting public. Confiding to his diary the inevitable, bitter conclusion behind the false bonhomie, Goebbels recognized that "we will have to move into Italy."[13]

"We" meant the Wehrmacht. And with this decision the war took a new, yet more ruthless and destructive turn.

Mussolini's protestations of loyalty to the Axis Pact, Hitler knew, were sincere, but they were not backed by the Italian people — especially the aristocracy, royal family, and upper middle class. Flying back to his Wolf's Lair headquarters, the Führer ordered Field Marshal Rommel to prepare something akin to Operation Anton, the previous November, when German and Italian troops had secretly readied themselves to occupy the remaining Vichy-administered area of metropolitan France. German troops would now be ordered to occupy the country of their own war partner, Italy, *by force;* it would, cynically, be called Operation Axis.

It was not a moment too soon, from Hitler's perspective. Days later, on July 25, the Italian Grand Council of Fascism convened its first meeting to take place since the early days of the war, in the Palazzo Venezia, in Rome. By a vote of 19 to 7, the members affirmed asking the king to save Italy from destruction, in view of the critical situation in Sicily and the bombing of the Rome rail yards, which President Roosevelt had personally authorized on the very day Hitler met with Mussolini.

Goebbels had assumed the American bombing might stiffen Italian resolve to defend their mother country, as it had in Germany. Instead, however, it caused the Rome police to arrest Mussolini as he left the palace — bundling him into an ambulance and taking him to a destination unknown. Marshal Pietro Badoglio, former chief of the Italian General Staff who had resigned in 1940 after disagreeing with Mussolini's war strategy, was tasked with heading a new government — "our grimmest enemy," as Hitler referred to Badoglio[14]: knowing Badoglio would, inevitably, terminate the Axis Pact.

• • •

It was war in the Mediterranean, then, not in Russia, that had seized the world's headlines and seemed suddenly to bring the global struggle against predatory fascism to a climax. Hitler fired off instructions for the arrest, if possible, of the new Italian government and the members of the royal family, before they could pursue surrender to the United Nations. They were too wily, however, and German forces in Italy still too thin on the ground to effect such a move.

So anxious did the Führer become that he now decided the Mediterranean must take priority over the Eastern Front. He therefore gave final orders to transfer to Italy his top SS armored divisions from Russia — telling Field Marshal von Kluge, who protested at the removal of the Wehrmacht's vital striking reserve, "We are not master here of our own decisions."[15]

The Western Allies were — or seemed to be.

35

Countercrisis

As HITLER CONFRONTED the crisis caused by the overwhelming Allied invasion of Sicily and the imminent defection of Italy from the Axis Pact, there arose a countercrisis or dilemma for the Allies — their biggest, in many ways, since Pearl Harbor.

This would be one of the great ironies of history: that at a moment when victory seemed to many to be within reach that year, the prosecution of the war by the Allies lurched and wobbled — with recriminations, accusations, and blame that have continued among war historians and writers to this day.[1]

The President had pressed Stalin again and again for a one-on-one meeting — determined to assure him, in person, that the United States, as the dominant partner in the Western Alliance, was committed to opening a Second Front at the earliest possible time.

"Referring to the Second Front," former ambassador Davies had told Stalin as the President's personal emissary on May 20, "no-one, I told him, had been more disappointed when, after consideration of all the risks and logistics involved in a cross-channel operation, and also the hazards as affecting the world battleground — the Pacific as well as the Atlantic and the Mediterranean — that for the sake of an assured victory, he [the President] had to agree to postponement of the Second Front cross-channel operation. No one, I said, was more firm in the belief that the quickest and most direct way to defeat Hitler was by a cross-channel invasion, when it could be done after every available means had been exhausted to prevent disaster and assure success."[2]

Overcoming Stalin's leeriness of Churchill with regard to a Second Front had been a tough assignment, Davies had told the President on his return to Washington on June 3. "Stalin said to me expressly that he could accept neither the African invasion [Torch] nor the Air Attack on Germany as the Second Front . . . He was suspicious, not only of the British, but of us, as well," Davies had reported. "They are convinced that Churchill, if he can help it, will consent to a cross-channel crossing only when there is no risk to them" — the British. "They believed that Britain is stalling on a cross-channel operation," both to "save her manpower" and to "divert the attack through the Balkans and Italy" in order to "protect the classic British Foreign policy of walling Russia in, closing the Dardanelles, and building a countervailing balance of power against Russia."

This was a pretty astute reading of British policy — but one that completely ignored the problem of defeating Hitler and the Nazis. As Stalin had pointed out to Davies, Allied operations in the Mediterranean were simply not on the scale of war as on the Eastern Front — where the "Germans had not less than three million" troops "attacking another three million of the Red Army — a total of at least six million — ten times as many as engaged in the African campaign."[3]

Stalin had seemed to Davies to be disappointed in the Western Allies, yet mollified by Davies's sincerity — and the President's firm commitment to mounting a Second Front as soon as feasible.

For his part, the President nevertheless continued to worry lest his "active and ardent lieutenant" become too ardent in terms of Mediterranean operations in the wake of success in Sicily. He'd heard from General Marshall that the Prime Minister was once again seized by excitement, and was plotting a new course in London — one he'd coyly revealed to Secretary Stimson, who was visiting American forces in Britain.

The Prime Minister was still only paying lip service to the Trident agreement, Stimson reported to Washington, after meeting with Churchill — and might well go off on a Mediterranean tangent unless leashed by the President. So worried had Stimson become, in fact, that he'd made a transatlantic telephone call to the Pentagon on July 17, a week after the invasion of Sicily. "The scrambling noise over the wires produced a peculiar effect on Marshall's voice," Stimson noted in his diary that night, "rendering the tones quite unrecognizable," but the secretary found he could "recognize the peculiarities" of Marshall's speech. "I began telling him of my conferences with the P.M., particularly last Monday the 12th. I summed up what I thought was his position, namely, that he was honestly ready to keep the pledge as to 'Roundhammer' [Overlord] but was impulsively likely to branch out into commitments which would make it impossible" — tying up in the Mediterranean the very battle-hardened U.S. and British forces and landing craft needed for a successful cross-Channel invasion early in 1944. Churchill seemed to Stimson to be fixated on seizing the Italian capital — that "he was very set on a march to Rome." More worrying still in terms of the suction-pump effect of the Mediterranean, Anthony Eden, the British foreign secretary, "was dead set on the Balkans and Greece."

The President had winced at the news. Churchill had even claimed that General Eisenhower's "heart," too, was invested in a bold new stroke in the Mediterranean — such as an airborne drop near the Italian capital: Operation Giant. Stimson was concerned that a dangerous overconfidence seemed to be infecting not only Churchill's bunker in London, but possibly Algiers.

The acting commander of American forces in Britain, General Jacob Devers, had assured Stimson, however, that Eisenhower's three service commanders in the Mediterranean — all of whom were British — had poured as much cold water on Eisenhower's idea as did Devers. Not only was this because of "the danger of executing an operation beyond the reach of air cover," but because of the "drain on landing craft" — craft that would be needed for Overlord. Others, too, were putting an oar into the debate — Stimson even told Marshall of a telegram to Churchill from Field Marshal Smuts, supporting Anthony Eden's Balkan aspirations. Marshall responded that he had not seen this — and was worried by the news. "Marshall said that in the light of these circumstances he thought I ought to go as promptly as possible to see Eisenhower where I would be able to round out what I had gotten here in London with the views of the people in Africa."[4]

The U.S. secretary of war having to fly to North Africa to try and head off an abrogation of the Trident agreements?

The situation, from the point of view of clear Allied purpose, was alarming, but it only became worse in the days that followed. On July 19, the day Hitler flew to Italy, the Prime Minister had warned his chiefs of staff, Stimson learned, to prepare plans to dump the Second Front if operations in the Mediterranean prospered and the seven battle-hardened divisions were not sent back to the U.K. In which case, the Prime Minister had said, he favored Allied assault landings in Norway, mounted from

England with whatever forces remained in Britain or could be scraped together.

Norway?

Occupied by some four hundred thousand German troops, Norway was the mountainous country where Churchill's ill-fated Franco-British Expeditionary Force had been completely worsted by a German counter-invasion and its survivors evacuated in the spring of 1940.

It was a disturbing scenario. On Thursday, July 22, Stimson had had it out with Churchill — who was soliciting Stimson's help in getting U.S. restrictions on the sharing of atom-bomb research lifted between the two nations.

The latest reports of heavy fighting around Catania had only reinforced Churchill's continuing skepticism regarding Overlord. He "said that if he had 50,000 men ashore on the French channel coast, he would not have an easy moment because he felt that the Germans could rush up in sufficient force to drive them back. On my direct questioning he admitted that if he was C-in-C, he would not figure the Roundhammer [Overlord] operation [as feasible]; but being as it was, he having made his pledge, then he would go with it loyally. I said to him that was like hitting us in the eye and he said 'Oh, no, if we start anything we will go through with it with utmost effort.'"[5]

In the meantime, Churchill pointed out, there was Italy — a country begging to be invaded by the Allies. The Prime Minister was, as he told Stimson, surely "justified in supporting his faith in the Italian expedition," given the potential rewards. "He spoke of two possibilities; one, going to Rome with the advantages that would come from this, even without capitulation; and second, with an Italian capitulation, it would throw open the whole of Italy as far as the north boundary and would give us opportunities to go and attack southern France. He asserted that he was not in favor of attacking the Balkans with troops, but merely wished to supply them with munitions and supplies."[6]

This was, at least, a mercy. For an hour and a half the two men — one approaching sixty-nine, the other, seventy-six — battled over strategy and tactical operations: Stimson attempting to point out the inevitable suction effect of major operations in the Mediterranean that would "hinder" Overlord, Churchill denying this; Stimson claiming he had the support of the "entire General Staff" in the "Roundhammer [Overlord] proposition,"

Churchill claiming Eisenhower to be "strongly in favor of going as far as he could in Italy."[7]

Stimson had been understandably perturbed — unaware, even as he spoke and exchanged cables with Marshall, that it was not only Churchill who now favored immediate exploitation of seeming Allied success in the Mediterranean. For Churchill's excitement was being replicated among senior U.S. generals in the Pentagon, in Marshall's own War Department.

On July 17, as tanks of Patton's Seventh U.S. Army raced to Palermo in the west of Sicily, the War Department's chief of Operations Division, Lieutenant General John Hull, declared he'd had a change of heart.

Hull's defection from the Trident strategy aroused fierce debate in the Pentagon. From "the very beginning of this war," Hull — who had hitherto been General Marshall's most loyal subordinate — wrote, "I have felt that the logical plan for the defeat of Germany was to strike at her across the channel by the most direct route." He'd now changed his mind, he declared. In a document he drew up for his deputy, General Handy, and his War Department team, Hull pointed out the strategic harvest to be garnered in the Mediterranean. As he put it, "it is a case where you cannot have your cake and eat it."[8] With half a million U.S., British, and Canadian troops in the Mediterranean, and barely 180,000 U.S. troops in England, he'd come to "the belief that we should now reverse our decision and pour our resources into the exploitation of our Mediterranean operations." Summarizing his extraordinary change of mind, he concluded: "As to Germany, in my opinion, the decision should be an all-out effort in the Mediterranean."[9]

Not only did General Hull's renunciation set off furious disputation at the Pentagon, it played straight into the hands of senior admirals in the Navy Department: sailors who had refused to move their offices into the Pentagon and were now in favor of backing out of a cross-Channel "Germany First" strategy, too — a change of objectives that would permit Japan to become America's Enemy Number One. Reexamining the Trident agreement to send seven battle-hardened divisions from the Mediterranean to the United Kingdom by November 1, 1943, the Joint War Plans Committee, representing the three U.S. armed services, now declared it "unsound."

The chief of staff to Admiral King, Admiral "Savvy" Cooke, agreed with the JWP Committee. He'd never been convinced a cross-Channel assault could succeed, and thought that, if the Western Allies simply limited their future operations to the Mediterranean, more U.S. vessels and resources would be available to send to the Pacific.[10]

General Marshall was understandably aghast. Other planners like Brigadier General Wedemeyer, who was actually visiting American headquarters in the Mediterranean, vociferously protested, feeling it folly to divide and disperse impending Allied effort in Europe, when all logistical and fighting focus should be concentrated on an agreed *Schwerpunkt*, or focal point. Analyzing combat reports from Sicily—where German troops were fighting to the death to defend an Italian island that not even Italians were willing to defend—Wedemeyer recognized how tough it was going to be to defeat the German enemy; he felt "our [English] cousins" must somehow be made aware "that this European theater struggle will never be won by dispersing our forces around the Axis citadel," as he responded to General Handy, referring to Churchill's "closing the ring" policy. "Even though HUSKY is successful after a bitter struggle," he'd warned from Algiers the week before the invasion of Sicily, "we could never drive rampant up the boot, as the P.M. so dramatically depicts in his concept of our continued effort over here." Not only would an Italian campaign require "greatly increased resources than those now envisaged or available in the area," but to ensure success—or even security against German counteroffensive measures—the cross-Channel invasion "would be even more remote, in fact, maybe crossed off the books for 1944."[11]

But if not Rome, where next? Even Wedemeyer had to concede the Allies must continue to do *something* in the next nine months, before Overlord was mounted.

This, then, was the strategic conundrum facing the President as U.S. commander in chief in the summer of 1943, even as the war seemed, for the Allies, to be so nearly won.

36

A Fishing Expedition in Ontario

IT HAD BEEN agreed at the Trident Conference in Washington in May that another high-level military parley would probably have to be convened, once the invasion of Sicily was completed. Though Churchill had suggested Washington as the venue, once again the President had demurred. As one of the President's White House Map Room officers, Lieutenant Elsey — who encrypted and decoded almost daily signals between the White House and 10 Downing Street — recorded, "the President recommended to the Prime Minister that this Anglo-American conference be held in Quebec, a happier place in summer than Washington. Quebec offered the advantages of a delightful climate and appropriate and comfortable quarters at the historic Citadel and the Chateau Frontenac."[1]

Before meeting with Churchill and his chiefs of staff, however, Roosevelt still hoped to meet with Stalin. "By mid-July when it seemed unlikely that Marshal Stalin would be able to leave his armies, even briefly, during their first summer offensive, the President suggested to Mr. Churchill that time would be ripe for their conference around the first of September."[2]

The triumph of the Allied landings on July 10 had, however, made even this date seem too distant to Churchill, who now had the bit between his teeth — his wonderfully pugnacious head spinning with romantic excitement as he saw himself entering Rome like a victorious Caesar in the next few weeks. "The very rapid changes on the several fronts and, in particular, the overwhelming success of the Sicilian campaign made it imperative to hold the meeting earlier," Lieutenant Elsey recounted. "The degeneration of Italian resistance and the possibility of complete Italian collapse, greatly increased by the unexpected fall of Mussolini on July 25th, gave birth to new problems only faintly foreseen in the spring. As Mr. Churchill said, 'We shall need to meet together to settle the larger issues which the brilliant victories of our forces have thrust upon us about Italy as a whole.' The Prime Minister pressed for a very early date in August but the President replied that he would be unable to arrive in Quebec earlier than August 17."[3]

The tragedy of late 1943 was now to unfold, almost inexorably—Churchill seemingly blind to Hitler's likely response to the Allied invasion of Italy. As Hitler's war aims crystallized into a ruthless German defend-or-die strategy, without having to rely on weak allies, the Allies' conduct of the war fractured—with grave political as well as military ramifications. If Churchill was right, the Third Reich might, if the Allies put every man into the field in Italy, collapse—with vast political ramifications on top of military, since the Wehrmacht still held a solid front deep inside Russia. But what if Churchill and the generals like Hull in the Pentagon were *wrong*? What if Hitler meant to fight to the bitter end on all fronts—and was backed wholeheartedly by his Volk?

Holding the reins of global political as well as military unity on behalf of the Western Allies, at least, the President decided he must present to the people of America and the world a clear picture of the war's positive progress—and ultimate aims. Calling in Robert Sherwood, Judge Rosenman, and Harry Hopkins, he therefore spent many days at Shangri-la and in the White House working on a major new Fireside Chat.

Broadcast from the White House on the evening of July 28, 1943, the President's radio address certainly seemed a success: the President sounding confident, inspiring, and clear-minded: conveying to listeners a sense of wise direction in prosecuting the war to its appointed end—and beyond.

ROOSEVELT HAILS "FIRST CRACK" IN THE AXIS; OUTLINES POST-WAR AID FOR ALL U.S. FORCES, ran the *New York Times* front-page headline on July 29, the newspaper giving extended coverage to every aspect of his speech—the President's first since February that year, "when he predicted invasion of the Continent of Europe." It was, the *Times* described, "a radio address as varied in its subject matter as the vast pattern of total war," one in which the President had "counseled against complacency, urged much greater efforts if Hitler and Tojo are to be defeated, as he promised, 'on their home grounds,'"—and one which, the *Times* added, "announced the end of coffee rationing due to the improved shipping situation."[4]

Was the war's direction really so clear, though? Was the speech not in truth window-dressing? Was not the "first crack" a split less in the Axis ability to wage war — given that German troops were reported moving ruthlessly into new, former Italian positions across the Mediterranean, and Field Marshal Rommel was reported to be preparing for the German defense of Greece — than in the Allies' *own* situation? Were not the Allies the ones with a problem?

At the State Department there were problems, as well. Former ambassador Bill Bullitt had been circulating throughout top circles in Washington a new paper urging an American invasion of the Balkans, before Soviet forces could reach central Europe, regardless of the military inanity of such a scheme. And to cap this, there were stories that Secretary of State Cordell Hull — at the insistence of his wife — was demanding the head of the President's right-hand man in postwar planning, Sumner Welles, on the grounds he was a homosexual — a story Bill Bullitt, who coveted Welles's job as assistant secretary of state, had been leaking to the press.[5]

And as if all this was not enough, there were indications that the Russians were exploring possible peace-feelers with the Germans — suggestions bruited in "official circles" that Stalin "may have forsaken President Roosevelt and Prime Minister Churchill on unconditional surrender" and was planning "to establish a European order on his own [communist] concepts and under the aegis of Moscow," as the *New York Times* reported.[6]

So much happening, so fast — and so many conflicting voices and calls in the great democracies of America and Britain!

It was small wonder the President felt, once again, exhausted. He longed to get away from Washington, and in cooler climes think things through, so that he could hopefully keep the Allied coalition together and pointing in the same direction.

On July 30, he thus went ahead with his latest plan. It would be another secret trip: this time "to Canada on a fishing and vacation expedition," as Admiral Leahy, his White House chief of staff, noted in his diary. On the beautiful lakes of Ontario, the weather would be less hot — and devoid of journalists, or anyone else. Away from the madding crowd the President could fish in peace, and think for a whole week.[7]

• • •

The *Ferdinand Magellan* duly pulled out from the Bureau of Engraving's special siding in Washington at 9:45 p.m.

The President was far from alone. In addition to Admiral Leahy he was taking his doctor, Vice Admiral McIntire; his naval aide, Rear Admiral Brown; his military aide, Major General Pa Watson; his two secretaries, Grace Tully and Dorothy Brady; twenty Secret Service men; his secret communications personnel; and Filipino crewmembers from the USS *Potomac* and Shangri-la.

The President had still heard nothing from Stalin, and thus knew as much or as little as the *New York Times* correspondent in Moscow concerning the dictator's intentions or thoughts. "The Stalin meeting is 'on,'" his cousin Daisy had noted in her diary on July 28 after speaking to the President on the phone[8] — which was to say the meeting wasn't off, and might still take place at Fairbanks, Alaska, to which he could fly from Ontario, using the nearest air base to his fishing expedition.

Daisy hated the idea. "It is much too dangerous," she recorded her anxiety about such "trips by air." "But he feels he has to, so he has to — His feelings are mixed about them, he told me — He doesn't want to go, but he has to put every possible effort into going because he thinks it will help in planning the future of the world — so — all we can do is wish him Godspeed and pray that all will go well."[9]

The fishing part of the plan, at least, went well — the bulk of it spent in McGregor and Georgian Bays, Ontario. "The days were interesting in providing fresh air and sunburn for all of us, and for me the nights were reasonably busy with messages to and from our British Allies in regard to the Italian campaign, the proposal to make Rome an open city which military authorities do not look with favor upon, and the general war situation," Leahy recorded.[10]

Harry Hopkins joined the party on August 4, in case they were to fly from there to meet Stalin, and though Leahy felt the vacation was a "success in giving all of us a change and exposure to the air and the sun," he did acknowledge "that on a vacation for relaxation we should have gone to bed earlier than midnight which was the usual hour."[11]

"For a week we lived in the train which remained a few yards from a landing from which we embarked each day on our daily fishing expeditions — each member of the party contributing a dollar a day for prizes," Leahy added. "The fish caught were small-mouth bass, wall-eyed pike, and what was either a pickerel or pike that the guides called snakes. In the final settlement of our pool at the weekend only the President and I were the winners."[12]

Air, sun, and pool winners, however, were not enough to solve the growing strategic crisis: one the President knew would be waiting for him once he got back to the White House.

37

The President's Judgment

To a considerable extent, the brewing Allied crisis was inevitable in a coalition, the President accepted, for each ally had its own concerns and war aims.

The President certainly did not take amiss Churchill's excitement over Mediterranean operations, or even the Prime Minister's loyalty to a decaying British empire. Churchill was, he felt, merely misguided—the product of high Victorian imperialism. As the President had discussed with former ambassador Davies, in a conversation that Davies had then related to Stalin, "British imperialism had contributed much to civilization, as well as to their people but, under modern conditions, there were now some aspects of it which did not conform to the American viewpoint. These variances in points of view were not such, however, as would or should be permitted to jeopardize a common effort for victory, and for the preservation of post-war peace. The Statute of Westminster [the Adoption Act of 1942, legally recognizing the independence of the Dominions] had given proof that modern England was conscious of the need for change, and with great courage and nobility had given independence of action in foreign affairs to the colonies and dominions."[1]

Stalin, looking up from his doodling pad, had queried the reason for excluding Churchill from the proposed meeting with the President, demanding to know "Why?"

"I replied," Davies recounted, "that Roosevelt and Churchill respected and admired each other, and although they did not always see eye to eye, they were always loyal. They were 'big' men, and on matters of difference, each could be relied upon. In fact, each would insist on finding common ground to win the war."[2]

This was, however, to tiptoe around the matter of the Second Front—

and the more the Italians caved in, the more brazen had become Churchill's call for exploitation in the Mediterranean—leaving the notion of Overlord as an Allied cross-Channel invasion to wither, the Prime Minister hoped, on the vine.

For the President this was not a surprise. He had gotten to know Churchill, on the Prime Minister's repeated visits to the White House, probably better than any American during the course of the war. The Prime Minister's moods, swinging from gravity to elation, were part and parcel of his colorful character as a leader. Churchill's approach to modern war was, the President accepted, wonderfully exuberant if often flawed. Certainly, with respect to what would now happen in Europe, the Prime Minister seemed to be giving way to a dangerous assumption: namely that Hitler might be toppled in the same way that Mussolini had been brought down.

Unhappily, despite a lifetime spent as a military officer and warrior-politician, Churchill seemed not to understand the nature of the problem confronting the Allies. This was not so much the Führer as the Germans themselves. The Prime Minister's abiding belief was that, once shorn of its allies, the men of the Third Reich would be unable to defend the vast territories they had so rapidly overrun when the Allies had been weak and disjointed. The Allies had now only to "close the ring," as the Prime Minister saw it, and sooner or later Germany would collapse, as it had in 1918—without the Western Allies having to take the great gamble of a cross-Channel assault and campaign in northern France. Churchill thus failed, in the President's view, to fully credit what had happened in Germany under Hitler—and how German forces, like the Japanese, would fight to the bitter end to defend the territory they had seized, *even without the Führer's orders;* that they would make the Allies pay for every meter of advance in blood, whether in the Balkans, Greece, Italy, or France.

Roosevelt's insistence on unconditional surrender had therefore been no aberration, or momentary thoughtlessness, as certain writers—even Churchill himself in a forgetful moment[3]—would later aver. Rather, it went back to the President's childhood sojourns in Germany, and spoke to the President's deep-seated cognizance of the collective German mentality. Roosevelt's unwavering judgment was that, whatever happened with the Italians, the Germans—like the Japanese—would go on fighting until beaten in battle; moreover, that their nations must be completely disarmed after the war's end and the world kept safe from any prospect of their military renascence. "The President believed also," Davies had told

Stalin on May 20, "that, despite differences in ideology and methods of government" between capitalism and communism, "it was entirely possible that our countries could live together in peace, in a decent world, with mutual respect, reciprocal consideration and joint safety, against a possible militant Japan, Germany or any other would-be disturbers of the Peace . . . Together, they could maintain and enforce law and order to preserve a just Peace, or there would be renewed disastrous wars."

Stalin had affirmed his complete agreement. "You can tell your President that so far as Germany is concerned, I will support him to any length he thinks necessary, no matter how soever, to destroy Germany's war potential for the future. Our people and our country have suffered immeasurably because of it. It is vital to us that Germany's war potential be destroyed. As to Japan, he said, the President already knew what their position was and needed no assurance."[4]

Goebbels's April announcement of *totaler Krieg*, in Roosevelt's view, had merely confirmed his judgment of Germany as the world's most dangerous nation, given the size and ruthlessness of its Wehrmacht and the abiding belief that *Macht ist Recht*: might is right. Any notion that the Germans would be easily pried from their conquests across Europe was therefore wishful thinking. The struggle to defeat the Germans and the Japanese would be hard and bloody, as he'd stated in his White House broadcast on July 18 — for there was no alternative to battle. And bloodshed.

What was important was for the Allies therefore to make no mistakes. To proceed methodically, building up command and combat experience, and trained, well-armed forces in order to defeat the Wehrmacht completely and relentlessly in combat, as Grant and his generals had done in the Civil War. Fantasies of victory merely by peripheral operations were seductive in terms of saving lives, but in the end they were idle. Only by relentless concentration of force, in focused application of America's growing output as the arsenal of modern democracy, would the Allies be enabled to win within a reasonable time frame.

The President's judgment of Wehrmacht intentions and abilities, moreover, was reinforced by reports he was receiving from his intelligence services. Access to the extraordinary riches of Ultra posed the danger that one sought in the decrypts for what one wanted to see. Churchill, who read Ultra decrypts every day and often "raw" ones — i.e., uninterpreted by his military staff — had fastened tightly on those indicating that

Hitler only intended his troops to stand in northern Italy, at the foot of the Alps. By contrast, the President was less beholden to that one source, and remained skeptical. On July 10, as the troops of the Western Allies had stormed ashore in Sicily, the latest OSS intelligence bulletin from Brigadier General Bill Donovan, who directed U.S. espionage services abroad, had been couriered to the White House Map Room. Rear Admiral Brown, the President's naval assistant, had himself brought it to the President — a report predicting the Italians would soon betray their Axis partner and sue for surrender. With stark realism, however, Donovan's report had also warned that the Germans "are quite prepared to treat the Italians as they would an enemy."[5] They would thus squash their former partners like cockroaches — and fight the harder once free of coalition allies they largely despised.

The President had agreed with Donovan — who had won the Medal of Honor and the Distinguished Service Medal fighting the Germans in France in World War I. The Allies, the President was sure, must not be complacent, or be swayed by armchair strategists. Italy's collapse would certainly be a political triumph for the Allies, and was certainly worth pursuing. It could also be dangerous, however, if it encouraged Churchill and like-minded peripheralists to jump to conclusions about Italian assistance, or German unwillingness to fight in southern Italy.

As the days of high summer unfolded, Donovan's prediction did indeed become reality. The Germans, it would become clear to even the most starry-eyed, were very, very different from their southern neighbors — neighbors the Germans had always held in suspicion but now began to treat with merciless, murderous contempt.

The profound cultural difference between the two Axis enemies would, in fact, climax in the summer and fall of 1943 — exposing the fault lines of what Churchill called the Grand Alliance, and threatening to sunder what had, in July, appeared to be the approach of Hitler's end.

PART TEN

Conundrum

38

Stalin Lies

As the *Ferdinand Magellan* made its way back from Ontario to Washington, the President finally heard from Stalin. Once decoded, the cable — dated August 8, 1943 — was handed to him. It was a long message agreeing to a meeting. Not, however, the meeting Roosevelt was hoping for.

In surprisingly friendly English, the Russian marshal — who had gotten himself promoted as the first civilian to hold that rank by the Presidium of the Supreme Soviet of the USSR on March 6, 1943, in recognition of his role as supreme commander in chief of the armed forces of the Soviet Union — began by apologizing. His focus as a Russian marshal had had to be on his "primary duty — the direction of action at the front. I have frequently to go to the different parts of the front and to submit all the rest to the interests of the front," he kept repeating — blatantly lying, since he had only once ever gone near the front, and that only for a few hours. "I hope that under such circumstances you will fully understand that at the present time I cannot go on a long journey and shall not be able, unfortunately, during this summer and autumn to keep my promise given to you through Mr. Davi[e]s. I regret it very much, but, as you know, circumstances are sometimes more powerful than people who are compelled to submit to them." He was, however, willing to agree meantime to a later "meeting of the responsible representatives" of the United States and the Soviet Union at Archangel, on the north coast of Russia, or Astrakhan, on the south, Caspian, coast — i.e., on Russian territory, and terms.

If the President was unable to go to such a summit, so distant from Washington, Stalin continued, Mr. Roosevelt could send a "responsible and fully trusted person"; moreover, he was quite happy for Churchill to attend the get-together — thus making it a "meeting of the representatives of the three countries." In the meantime they should raise, in advance, the "questions which are to be discussed," and the "drafts of proposals which are to be accepted at the meeting." He added his belated congratulations to "you and the Anglo-American troops on the occasion of the outstanding successes in Sicily which are resulted [*sic*] in collapse of Mussolini and his gang."[1]

The dictator's excuses for not meeting the President might be specious, but what was clear, now that the battle of Kursk was over and Mussolini toppled, was that Stalin saw no need to travel to America or to Alaska, cap in hand. He could afford to play hard to get — or please.

The President was understandably disappointed, given the phenomenal amount of Lend-Lease equipment, food, chemicals, and metals being shipped to the USSR. Even Marshal Zhukov, Russia's greatest general, would admit after the war that "the Americans shipped over to us *materièl* without which we could not have equipped our armies held in reserve or been able to continue the war." As Zhukov explained, "We did not have enough munitions [and] how would we have been able to turn out all those tanks without the rolled steel sent to us by the Americans?"[2] — let alone the four hundred thousand trucks dispatched.[3]

"Drafts of proposals," meantime, made the President frown. Not only might it be more difficult to get agreement on the President's United Nations authority plan if preconference proposals had to go through the endless (and appropriately colored) red tape of Russian communist bureaucracy, but Churchill's presence might let the cat out of the bag — namely, that Churchill and his generals were once again tilting away from a cross-Channel Second Front in favor of exploitation in the Mediterranean. And dangerous overoptimism in London.

One American chaplain in London, Colonel Maurice Reynolds, had openly forecast that the war might be over in five months — that he would not be "surprised if we all went home for Christmas. The rats are beginning to leave the sinking ship — one [Mussolini] has left already," he'd been quoted in *Stars and Stripes,* the U.S. Army newspaper.

This was an almost tragic assumption, given the tough fighting that lay ahead with the Germans. Not only was Allied strategy in danger of being compromised by naive opportunism, but if the Western Allies pulled out of their commitment to a Second Front, the President recognized, there would be tough problems with America's Russian partner — with grave consequences for the peoples of central and even western Europe.

The disagreement between the U.S. generals at the Pentagon, and the growing continental divide between the Allies, was thus the unhappy scenario that faced the President when he finally entered the White House on the morning of August 9 for a whirlwind round of meetings. He'd agreed to meet Churchill and the British chiefs of staff in Quebec around August 15. This gave him only a few days to get his ducks back in a row.

He saw Secretary Hull for lunch, General Marshall at 2:00 p.m., Lord Halifax, the British ambassador, at 2:30, and dined with Hopkins that evening. He called his cousin Daisy to tell her what a great fishing trip he'd had — "a real success — the place much like the Maine Coast — rocky, wooded, 100s of islands, cool on the whole, very nice — He says he'll take me there, perhaps, next year!"[4] But he also confided to her his latest plan: that he was determined to do his best to head off another Trident-like battle royal in the Canadian capital. He would therefore see Churchill in private at Hyde Park *before* the Quebec meeting of the Combined Chiefs of Staff even began — and twist Churchill's arm there until the Prime Minister backed off.

39

War on Two Western Fronts

IN HIS FIRESIDE CHAT radio broadcast, the President had denied there was any disunity between the Allies. "You have heard some people say that the British and the Americans can never get along well together—you have heard some people say that the Army and the Navy and the Air Forces can never get along well together—that real cooperation between them is impossible." He'd denied the assertions, as U.S. president and commander in chief. "Tunisia and Sicily have given the lie, once and for all, to these narrow-minded prejudices. Ahead of us are much bigger fights. We and our Allies will go into them as we went into Sicily—together. And we shall carry on together," he'd claimed—lauding the achievements of the Soviet Union as America's ally, too.

Behind the façade of unity, however, the conduct of the coalition war was in grave peril. Moreover, with Stalin calling for a meeting of foreign ministers in the fall, before the Allied leaders or their representatives got together, it would become impossible to conceal British pressure to defer or abandon the Second Front in favor of further operations on the Southern, or Mediterranean, Front.

Even the President's postwar vision was in danger of unraveling—from within. Two *New York Times* journalists, John Crider and Arthur Krock, had now openly reported, while the President was fishing in Canada, on the growing rift between the secretary of state, Cordell Hull, and Undersecretary Welles—reports that had been carried in other newspapers, too.[1]

The President had therefore summoned both Welles and Hull to the Oval Office on August 10, the day after his return—a meeting at which

Hull declared he could not work with Welles, and that one of them must resign.

As if this was not enough, the President had read carefully his war secretary's "Brief Report on Certain Features of Overseas Trip," which Henry Stimson had sent over to the White House, following his return from London and North Africa — a report so alarming in terms of Allied strategy that the President had asked Stimson to lunch with him on August 10, immediately after his meeting with Hull and Welles. The lunch would precede the meeting he had convened with the Joint Chiefs of Staff at 2:30 in the Oval Office, to discuss "the attitude to be taken by the U.S. Chiefs of Staff at the coming conference [in Quebec] with our opposite members from London," as Admiral Leahy put it in his diary.[2]

An unfortunate breakdown in Allied war strategy was approaching — at the very moment when, in the Southwest Pacific, American destroyers had sunk all four Japanese destroyers of the "Tokyo Express" seeking to reinforce their troops on Kolombangara in the Solomons; a moment when, in the North Pacific, U.S. and Canadian troops were preparing to land on Kiska Island in the Aleutians; a moment when, in Sicily, the retreating German troops were beginning to evacuate their forces across the Strait of Messina to the Italian mainland; and when, in Russia, the Wehrmacht was being forced to retreat on a three-hundred-mile front, giving up Orel and Belgorod — cities occupied by German troops since October 1941.

Of one thing the President was absolutely certain at this strategic crossroads for the Allies, however: that whatever anyone said or posited, the war might very well *not* be over by Christmas — even by Christmas 1944. He must therefore redouble his efforts to keep the Allied coalition together, marching to the same tune.

And place. Berlin. Then Tokyo.

Reading over the materials the Joint Chiefs of Staff had sent him, prior to their meeting at the White House, the President appreciated their clear strategic reasoning, especially their August 9 memorandum, with its various enclosures.

The President was not, however, amused by the wording of one enclosure: a paper prepared by the Operations Division of the War Office, dated August 8, which stuck in his craw. In it the authors, headed by General Handy, painted the Torch invasion and year of victories since 1942 as wasteful and unnecessary — in fact as having set back the defeat of Germany by a year. A cross-Channel invasion in 1943 "was the one chance to end the war in Europe this year. If this had happened," General Handy claimed, "all that has been gained would be insignificant by comparison."[3]

Clearly the authors had never reflected on Dieppe. Or Kasserine. Or Sicily, for that matter, where the fighting had become remorseless. They had certainly never faced a German soldier in battle. It was, arguably, one of the most egregious underestimations of the enemy ever produced by a senior general of the U.S. military — neither *combat* nor *battle experience* ever appearing in the document. All arguments had merely been laid out in terms of numbers of men furnishable to the front.[4]

The President had shaken his head over that. Would these armchair planners never learn?

Fortunately, the Joint Chiefs of Staff had themselves prepared two far more mature papers, on August 7 and 9, attempting to "develop a strategic concept for the defeat of the Axis in Europe."[5] These papers concluded that, thanks to the Allied invasion of Sicily, the German offensive at Kursk had had to be curtailed; that the Wehrmacht would no longer be able to go on the attack or seek victory on the Eastern Front; and that therefore German strategy would now likely be one of fierce fighting to attain "a satisfactory negotiated peace."[6] As they warned, however, the Axis "still retains strong defensive power. A defensive strategy on the part of the Axis might develop into a protracted struggle and result in a stalemate on the Continent." Therefore, "the rapidly improving position of the United Nations in relation to the Axis in Europe demands an abrogation of opportunistic strategy and requires the adoption and adherence to sound strategic plans which envisage decisive military operations conducted at times and places of our choosing — not the enemy's."

This, at least, was sensible. "We must not jeopardize our sound over-all strategy," the memorandum argued, "simply to exploit local successes in a generally accepted secondary theater, the Mediterranean, where logistical and terrain difficulties preclude decisive and final operations designed to reach the heart of Germany."[7]

This new memorandum, Roosevelt felt, was far better argued than Handy's counterdocument. What it did not do was explain how the Western Allies could simply put major offensive operations against Germany on hold for nine months while they prepared Overlord. Not only would a nine-month hiatus be difficult to excuse to people at home, but it would be harder still to excuse to the Russians, currently facing three-quarters of the German Wehrmacht in combat on the Eastern Front. As the President chided Marshall, who had brought the memorandum to the White House for him to read on August 9, at 2:00 p.m., "the planners were always conservative and saw all the difficulties"; he was sure "more could usually be done than they were willing to admit," as Marshall noted on his return to the Pentagon. By 11:00 a.m. the next day Marshall therefore wanted new planning documents that would meet the President's concerns. As Marshall reported Mr. Roosevelt's wishes:

> That between Overlord and Priceless [further major operations in the Mediterranean] he was insistent on Overlord but felt that we could do more than was now proposed for Priceless. His idea was that the seven battle-experienced divisions should be provided for Overlord but that an equal number of divisions from the U.S. should be routed to Priceless.
>
> He stated that he did not wish to have anything to do with an operation in the Balkans, nor to agree to a British expedition which would cost us ships, landing craft, withdrawals, etc. But he did feel that we should secure a position in Italy to the north of Rome and that we should take over Sardinia and Corsica and thus set up a serious threat to southern France.[8]

Marshall was stunned — able only to protest that "we had strained programmed resources well to the limit in the agreements now standing." Moreover, though Overlord would have priority of resources, a multi-front strategy by the Western Allies, if adopted too heavily in the Mediterranean, would impose grave constraints on Overlord and its chances of success.

The President seemed unimpressed by Marshall's response — as the general was aware. Clearly the President saw Marshall as maintaining a kind of ideological focus on Overlord, which seemed not only bureaucratic but wooden and out of touch with political reality. The American people, furnishing the weapons and the soon-to-be eleven million soldiers for the war — as well as paying the taxes to fund it — could not be expected to condone *nine months* of a virtual cease-fire at this juncture, during which anything might happen — both positive and negative.

Rather, the President sounded determined the Allies should keep the initiative, now that they had the Germans on the run. General Brooke, the British CIGS, was right to see Italy as a major theater of war, where major German forces could be forced to fight, rather than switch divisions back to the Eastern Front. Once in Italy, moreover, the Allies could maintain the offensive initiative: possessing the airfields from which to bomb southern Germany, and bases for the troops from which to mount an invasion of southern France if it were deemed opportune, thus helping Overlord — indeed offering an alternative lodgment if Overlord did not succeed. The President, Marshall penned in the note he made at the Pentagon, therefore wanted to see him "at noon tomorrow" with the logistical implications of a two-front campaign on the European mainland. "Incidentally, he said he did not like my use of the word 'critical' because he wanted assistance in carrying out his conception rather than difficulties placed in the way of it — all of this in a humorous vein," Marshall reported to his staff.[9]

The President had spoken. He was U.S. commander in chief, and it was for Marshall, as U.S. Army chief of staff, to ensure the President's conception be carried out, not keep harping on "critical" insufficiencies, or jeopardy. Period.

Once the Joint Chiefs of Staff were seated in the Oval Office at 2:15 on the afternoon of August 10 — and with the secretary of war, Mr. Stimson, having been invited to witness the meeting, following his lunch with the President — Mr. Roosevelt held forth: explaining the political context behind the next decisions that must be made at Quebec, where they would be conferring with their British opposite numbers.

The "British Foreign Office does not want the Balkans to come under the Russian influence," he told them. Therefore, "Britain wants to get to the Balkans first" — understandably. However, he himself rather doubted the Russians wanted or were in a position to "take over the Balkan states" such as Yugoslavia, Bulgaria, Albania, and Greece. They would, he predicted, prefer rather to "establish kinship," or associative relationship, "with the other Slavic people" in southern Europe.[10] "In any event," he went on, he "thought it unwise to plan military strategy on a gamble as to political results," rather than what was possible or desirable militarily.[11] A major U.S. campaign in the Balkans would have no guarantee of succeeding, indeed might well fail. It would certainly distract from the cross-Channel invasion scheduled for May 1944 — a gamble the United States couldn't take.

General Marshall agreed wholeheartedly. If the shift of the seven battle-hardened Allied divisions designated for Overlord from the Mediterranean did not take place, it would, he pointed out, "simply invite having these extra divisions used for invasion in the Balkans. This would meet the Prime Minister's and Mr. Eden's desires, but would make the Mediterranean operation so extensive as to have a disastrous effect on the main effort from England"[12] — a warning that prompted Admiral King to suggest "to the President that if the British insist upon abandoning Overlord or postponing the operation indefinitely, we should abandon the project as in carrying it on we would simply waste our substance."[13]

Admiral King's disgust with the British prompted the President to reassert the crucial necessity of mounting Overlord as the number one priority. Indeed, to the amazement of his own advisers, the President then "said we can, if necessary, carry out the project ourselves. He was certain that the British would be glad to make the necessary bases in England available to us."[14]

The United States mounting the cross-Channel invasion *without* British participation?

In the long months since Pearl Harbor, there had been threats to switch military priority to the Pacific, but never such a gesture of scorn for British timidity and avoidance of decisive battle — certainly never before by the President.

In part the bleak picture of British cowardice was the result of Secretary Stimson's journey to London and Algiers. On his return he had painted a disturbing portrait of Churchill's intentions, but the President didn't think, in the end, it would come to a breach in the alliance. The British, he was certain, could not afford to let down, before the whole world, the major ally that had saved them from German and Japanese victory. Moreover, the United States was not averse, the President explained, to establishing air bases in southern Italy and opening a fighting front in Italy; it was just a matter of saying no to an advance further north than the capital, lest the Allies be drawn into Hitler's web.

"He was for going no further into Italy than Rome," Stimson noted with satisfaction that night, "and that for the purpose of establishing bases. He was for setting up as rapidly as possible a larger force in Great Britain for the purpose of Roundhammer [Overlord] so that as soon as possible and before the actual time of landing we should have more soldiers in Britain dedicated to that purpose than the British. He said he wanted to have an

American commander and he thought that would make it easier if we had more men in the expedition at the beginning. I could see that the military and naval conferees were astonished and delighted at his definiteness."[15]

They were. "The President stated that, frankly, his reason for desiring American preponderance in force," General Deane wrote in his minutes of the meeting, following discussion of the American divisions that could be shipped to England before D-day (fifteen in number), was "to have the basis for insisting on an American commander. He wished that preponderance of force to be sufficient to make it impossible for the British to disagree with the suggestion."[16]

The new strategy was thus clear. War on two western fronts—but the Italian front limited to a line just north of Rome. And an American supreme commander for Overlord, lest the British try to renege on their commitment to a cross-Channel invasion.

"The President then summed up the discussion by stating that our available means seem to fit in pretty well with our plans. He outlined these as insistence upon the continuation of the present Overlord buildup and carrying out that operation as our main effort," Deane recorded. Moreover, the President wanted to have enough Americans in Britain "in order to justify an American commander" for Overlord, he restated. Together with this, he was in favor of leaving Eisenhower with sufficient forces in the Mediterranean to establish U.S. air power in southern Italy (where weather conditions permitted takeoff and landing almost every day, compared to often prohibitive flying conditions over England). Such forces on the southern European front would give the Allies strategic flexibility if for any reason the Overlord operation was repelled, but the President was emphatically "opposed to operations in the Balkans."[17] Yes, it would be good to have an army able to stop the Russians from overrunning countries in south-central Europe as they advanced—but the Western Allies still did not have a single boot on the mainland of Europe, and the Balkans were in any case a nightmare in terms of terrain. Knowing the Germans, the Wehrmacht would contest every yard. It was, therefore, "unwise to plan military strategy based on a gamble as to political results."[18]

"I came away with a very much lighter heart on the subject of our military policy than I have had in a long time," Stimson dictated at home in his diary, delighted with his commander in chief's stance. "He was more clear and definite than I have ever seen him since we have been in this war."[19]

What pleased Stimson even more was that the President now wanted an American in charge of Overlord—something Stimson, in a letter he'd brought with him to the White House for the President, had also recommended. He'd shown his draft letter to Marshall before leaving the Pentagon that morning, pleading for Marshall to be the man, and Marshall had not demurred (though anxious that Stimson not tell the President he had seen the recommendation, lest he be seen to be pursuing personal ambition).

The loss of Marshall from Washington—were he to be Overlord's supreme commander—would be dire, but it was necessary, Stimson felt, to show the British that America meant business: Overlord the only way that "Germany can be really defeated and the war brought to an end."[20]

Whether the President would select Marshall as supreme commander, however, was quite another matter. As would be the British chiefs' reaction to the President's strategy, once they all reached Quebec.

And with that the President prepared to meet his counterpart, the Right Honorable Winston Churchill and his wife at Hyde Park on August 12, 1943.

The planning for the endgame in World War II in Europe was now coming to a climax.

40

The Führer Is Very Optimistic

THE PRESIDENT'S INSISTENCE that Churchill meet him at Hyde Park before the Quebec Conference was not motivated by politeness or hospitality. For good or ill, the President was aware the meeting with the British prime minister might well determine the course of World War II — and its aftermath.

Strategic flexibility or inflexibility — that was the question in Churchill's eyes. Opportunism or strategic determination — this was the question in Roosevelt's.

The question of who was right and who was wrong would vex political and military historians for the next seventy years. Time was certainly of the essence — the President having received reports of ever-increasing German atrocities in the occupied countries of Europe. The latest of these had come on August 10, the day he met with the Joint Chiefs at the White House. From London, the U.S. ambassador to the Polish government in exile, Tony Biddle, had reported German mass murder — genocidal pogroms — on a scale never seen before in human history.

The President, in his broadcast on July 31, had already warned neutral countries not to give asylum to war criminals, but Biddle felt this would not be enough. As he put it, "apart from the punishment of war criminals for the crimes they have committed, it has become more imperative than ever to restrain the Germans from committing further the mass murder of the Polish population in Poland. This becomes all the more urgent since it may be anticipated that the policy of exterminating the population of entire provinces, as is practiced in Poland, may also be applied by the Germans in the present final stages of the war to the people in other German-occupied territories, like the Czechs, Yugoslavs, French and those in the occupied parts of the U.S.S.R.," his report warned — noting the Germans had already "exterminated" the majority of the Jewish population, and were deporting to concentration camps hundreds of thousands of Poles, while men between the ages of fourteen and fifty were being taken to Germany as slave labor. "Women, children and old people are sent to camps to be killed in gas chambers which previously served to exterminate the Jewish population of Poland," he reported. As if this were not enough, it "may be presumed that the Germans are reckoning in the possibility of a defeat, and have consequently decided to exterminate the largest possible proportion of the Polish population" in a kind of apocalyptic conflagration — quoting Fritz Sauckel, the Reich minister of slave labor, saying as recently as June 19, 1943, in Kraków: "If the Germans lose the war, we shall see that nothing remains either here or elsewhere in Europe."[1] The Germans would, in other words, not only resort to a scorched-earth policy, but torch peoples as well.

It was thus imperative, Roosevelt considered, to end the Nazi nightmare in Europe as soon as could be achieved — something that would never be accomplished by opportunistic operations in the eastern Mediterranean and Aegean, however much Churchill and the British Foreign Office feared an eventual Sovietization in central Europe.

Churchill, too, was all for finishing the war as swiftly as possible; he merely saw the challenge differently. Imaginative, impetuous, and excitable, he was pulled in all directions, as General Brooke noted in his diary — but least of all in the direction of a cross-Channel landing and campaign. When writing his epic, six-volume memoirs of the conflict, Churchill would title his fifth volume *Closing the Ring.* The "Theme of the Volume" (a mantra he liked to insert in the frontispiece to each work) was the story of "How Nazi Germany was Isolated and Assailed on All Sides."

Could Hitler and his regime be swiftly toppled by being "isolated," however? In the excitement of the summer of 1943, Churchill was minded to think so. The Germans would surely cave in once they saw — like the Italians — the game was up, and only destruction faced their nation if they sought to fight on. His July 19 minute to his chiefs of staff maintained the "right strategy for 1944" would be to pursue the Germans "certainly to the Po," after Husky, with the option of attacking westward to the south of France or northeastward toward Vienna, "and meanwhile to procure the expulsion of the enemy from the Balkans and Greece." Moreover, rather than launch a costly cross-Channel assault, to prepare an Allied invasion of Norway, to be mounted "under the cover of 'Overlord.'" Encirclement around the fringes of mainland Europe, he'd considered, would lead to "Hitler and Mussolini" being "disposed of in 1944."[2]

It was this vision — tantamount to fantasy, unfortunately — that had come to obsess the Prime Minister, and which underlay his latest, enlarged mission to North America: some 230 men embarked on the *Queen Mary* to take Churchill's latest strategy to Quebec, which they reached on August 10.

It was a mission, however, that the President was determined to preempt by insisting on meeting Churchill first — in private. The British must be held to the only policy that would actually defeat Nazi Germany, rather than merely ringing it. This meant a spring 1944 cross-Channel invasion, with no holds barred — and no more backtracking by the British. Once the two men got together at Hyde Park, the President had decided, moreover, he would have to use his trump card.

As Roosevelt prepared to meet his recalcitrant ally, Hitler had meantime asked Dr. Goebbels to fly from Berlin to the Wolf's Lair to "discuss the whole situation from every point of view."[3]

The Führer, Goebbels found when he arrived on August 10, had decided to abandon his main Axis ally. The new Badoglio government would "betray" Germany, he was certain, despite its assurances of loyalty to the Axis Pact. A telephone conversation between Churchill and Roosevelt had been intercepted, and both Goebbels and Hitler were scornful. "These plutocrat leaders imagine things in Italy as being much more positive for them than they really are. The Führer knows every trick in the book, though," Goebbels noted with evil satisfaction in his diary. "German troops are now streaming into Italy" — in fact, German flags were already flying over Mantua and Genoa. "There's no danger of anything too terrible happening. The Führer is absolutely determined he's not going to surrender Italy as a battlefield. He has no intention of letting the Americans and the British get to northern Italy. The worthwhile part of the country, at least, will remain in our possession."[4]

Having overcome his initial panic and fury over Mussolini's arrest, Hitler now actually *welcomed,* he said, the impending capitulation of Italy: a nation that had no national will, he felt, beyond the popular fascist speeches of the Duce. Mussolini no longer impressed him, in retrospect. The Duce had failed to declare war in September 1939, when his intercession might have gotten England to back away from world war; moreover, at home in Italy, Mussolini had failed to crush the monarchical, aristocratic conservative elements who considered him a fly-by-nighter.

To maintain his power, a dictator must be perceived by his own people to be ruthless, Hitler understood — and by preparing to crush the forces of Italy and move Wehrmacht troops immediately to occupy the entire Italian peninsula, the Führer would be seen to be asserting his absolute will — Aryan, egalitarian, merciless. Actions that would speak louder than any words.

Was Hitler assuming too much of his *Volksgenossen* — his fellow citizens?

In 1943 Hitler would only appear in public twice — an isolation at his headquarters that filled Dr. Goebbels, as a master of public relations, with disappointment, even anxiety. Yet the Führer seemed to know better than his chief propagandist that he had no need to show himself in public, or even inspect his troops. With drums, banners, swastikas, and film and press fanfare, he had as Führer given the German Volk what they yearned for: order, authority, a new place — a supreme place — in the European sun; a sense of belonging to a dynamic, productive community with rational if draconian goals — and sufficient pride in their country's history and extraordinary military achievements to defend it and its conquests now to the death, literally. He was therefore confident he had the willing, even enthusiastic, obedience of his troops — troops who would fight all the harder and more effectively *without* the hindrance of an always-unreliable Axis ally. Italy's military missteps in Greece and North Africa in 1941 had dragged German forces into a southern theater of war that had distracted from the Führer's main priority, the defeat of Russia — in fact were now affording the Western Allies a possible stepping-stone into Europe. Without the millstone of Italian allies, however, that stepping-stone could be transformed into a *Sumpf*: a bog, where the Western Allies could become enmeshed, ensnared, mired. If, that was, the Allies could be tempted into further fighting in the Mediterranean and Aegean.

While always keeping the Atlantic Wall as strong as possible, ensuring there were enough divisions stationed in France to defeat any attempt at a Second Front, he would now lure the British and Americans to the south of *Festung Europa,* Hitler told Goebbels. He would thereby buy time without taking great losses, or facing a real threat to the Reich itself: the Vaterland. True, in the probable event of Italian surrender to the Allies, vital German divisions and air force groups would have to take over the positions hitherto held by Italian units in the Mediterranean and Aegean, stretching from Sardinia to Samos—forces that could not then be used on the Eastern Front. But with no-nonsense German military control not only of Italy and of the eastern Adriatic—from Slovenia to Albania, Greece and Crete, with all their airfields—the Western Allies would be at a disadvantage. Given the mountainous terrain and the fighting efficiency of Wehrmacht and Luftwaffe units, the Americans and British would be hard put to form a Mediterranean front that had any hope anywhere of reaching Germany.

With every week and every month the Western Allies would be held in the Mediterranean *Sumpf.* Germany would thus have time, Hitler calculated, to finish the development of the Third Reich's secret weapons—long-range rockets and ballistic missiles—and to deploy them, from January 1944 onwards.

Roosevelt had boasted that America was the arsenal of democracy. But as Führer of the Third Reich he would show how, using not only slave labor in Germany but products manufactured in France and the occupied countries for the Third Reich, the Third Reich was the arsenal of Europe. Not even mass RAF night raids and U.S. Air Force daylight bombing could turn the tables. Allied air force losses in conducting mass raids of German cities were unsustainable in the long run. Under the leadership of Albert Speer, the production of German armaments was being dispersed away from big cities, while evacuation of families would deny the Allies the collapse of German civilian morale.

Hitler's fascination, in fact obsession, with the minutiae of weaponry might be mocked by some of his generals, but in a war of numbers, it was the quality of weapons that counted, along with the sheer discipline of German Wehrmacht soldiery in combat. U.S. and British industrial output might statistically outstrip that of the Third Reich, but the technical truth was, he sneered, their weapons were inferior, their soldiery less ruthless, and the demands of their various theaters of war too global. By contrast, now that Italy's ill-fated campaign in North Africa was over, the Third Reich had the advantage: cohesion. A single continent as its battlefield, with Germany at its epicenter—its high command able to furnish reinforcements in any direction, especially if Goebbels, Göring, and other senior officials could squeeze out still more military personnel from the workforce, and more slave labor from the occupied countries—including Italy now.

With the Japanese reaffirming their pact with Berlin, and drawing an

Churchill on the Wrong Warpath

With hundreds of advisers and staff officers, Churchill arrives in New York in May 1943 aboard the *Queen Mary* (here bringing back U.S. troops two years later) to oppose agreed-upon U.S.-British strategy. On Capitol Hill he inveighs against a cross-Channel Allied invasion before 1945 or 1946, citing impossible odds.

Axis Surrender in North Africa

In Tunisia, a quarter million Axis troops in North Africa surrender to Eisenhower on May 12, 1943, the culmination of the President's "great pet scheme." For FDR this proves the Allies are on the road to military victory in Europe; for Churchill it means the Allies should stay in the Mediterranean.

Reading Churchill the Riot Act

Churchill is intransigent; the U.S. secretary of war accuses the British of cowardice. The President takes the Prime Minister to Shangri-la to fish, while the U.S. chiefs of staff work on Churchill's military team. Eventually FDR has to give Churchill a talking-to. Following the invasion of Sicily and southern Italy, U.S. troops will be withdrawn to England for a definite 1944 D-day. Churchill is furious.

Sicily — and Kursk

On July 10, 1943, the Western Allies take the war to Europe, invading Sicily. The magnitude of the landings stuns Hitler — who calls off Operation Citadel, his great summer battle on the Eastern Front to destroy Stalin's Soviet armies at Kursk.

With Italian troops running from the Allies in Sicily, Hitler flies to Italy to encourage Mussolini to fight on rather than surrender. Six days later, the Duce is arrested by his own people. The Germans will have to fight for Italy instead.

Once again Churchill returns to Hyde Park, this time with his daughter Mary, to persuade FDR to abandon U.S. strategy for an Allied cross-Channel invasion in 1944. The Prime Minister is convinced it will be a disaster, and that better results will be obtained from Mediterranean operations.

The President refuses to listen to such defeatism. He threatens to withhold the Manhattan Project's atomic bomb discoveries from Britain. The two men join the Canadian PM and the Combined Chiefs of Staff in Quebec. There the D-day plans for spring 1944 will be cast in stone.

The First Crack in the Axis

In Ottawa, the President announces "the first crack in the Axis." The Allies suffer their own cracks, however. Churchill misunderstands German determination to fight, even in other people's countries, and Stalin — facing two-thirds of the Wehrmacht — loses faith in Allied willingness to defeat the Axis.

At Hyde Park (driving his own car), FDR tries to keep the Allies focused on the defeat of Nazi Germany, rather than become too embroiled in the Mediterranean. For all of Winston's faults, though, FDR needs the Prime Minister to help save western Europe from postwar Soviet domination once the Nazis are beaten.

At Salerno, south of Naples, on September 9, 1943, FDR's worst fears are realized. Churchill's vision of a "soft underbelly" is not soft. Fortunately, as Commander in Chief the President has stood firm, and U.S. preparations for D-day in 1944 will, he hopes, erase Churchill's near-fatal misjudgment.

ever more significant portion of U.S. and British military effort to the Far East and Pacific, the chances of the Western Allies mounting a Western Front — at least a successful Second Front — if they were tied down in the Mediterranean were, in Hitler's mind, distinctly dim.

Goebbels was thus delighted by the führer he'd met at the Wolfschanze. It was no mask of confidence Hitler was putting on for his generals, his headquarters staff, or his visitors in the high and late summer of 1943, the Reich minister judged: it was real.

The South Tyrol, annexed by Italy from the Austro-Hungarian Empire in 1918, would be occupied by German forces and henceforth become part and parcel of the Third Reich. "We just have to keep our nerve and not be distracted by the enemy's panic-machine," Goebbels noted. "Whatever they cook up in Washington and London, they won't find consuming as easy as preparing. It'll take quite a while for the Italian crisis to sort itself out." The Führer was determined to hold the Alps and Italy as far as the Po, but "the rest of Italy is worth nothing," intrinsically, Goebbels recorded — adding, though, that "in private and in the greatest secrecy," Hitler had stated he would not only try and arrest Marshal Badoglio, King Emmanuel, and "the whole baggage" in Rome, but was planning to "defend the Reich as far south in Italy as possible,"[5] as he'd confided to Goebbels already in June.[6]

This latter intention would have the gravest implications for the Allies — who could decode high-grade German communications from Hitler's headquarters, but not read Hitler's mind. "The fundamental principle of our war strategy is to keep the war as far as possible from the borders of the homeland," Goebbels noted on August 10. "It is absolutely the right principle," he reflected. "As long as we can master the war in the air" — especially with better *Flak* and new jet fighters — "the German people can be trusted to stick it out for a pretty long time." The harvest in Germany looked good, with more that could be brought in from occupied countries. Using slave labor, the outlook for the Third Reich was thus far more positive than the way the foreign press was depicting it. If the Western Allies could be lured to commit themselves to all-out war on the Southern Front rather than a Second Front, they could be savaged — perhaps even repelled — by the sheer professionalism and ruthless energy of the Wehrmacht. Certainly the Allies could be held at bay, far from the Reich, and in close combat — with London and the British Isles, meantime, under aerial bombardment by secret weapons: "our planned measures to be taken in the coming months."[7] "With regard to our countermeasures against the British," Goebbels confided, "the Führer thinks they can be launched in great numbers by January or February [1944] at the latest. He's going to set upon London with a fury never witnessed before. He's anticipating great things from our missiles. They've been fully tested; we just have to accelerate production to the level we need. So we have to be patient."[8]

As for the Eastern Front, moreover, Hitler seemed confident the Russians could be held — indeed had been beaten badly, in effect, at Kursk. As such they could be thrashed again that winter, if the Western Allies were kept at bay in the south.[9] "It will take time, and we have to be patient," Goebbels repeated[10] — Hitler interested in why Stalin had recently withdrawn his two ambassadors, Litvinov and Maisky, from Washington and London.

Puzzled, Hitler and Goebbels wondered if there was an opportunity to cleave the Allies apart. Stalin could not defeat the Wehrmacht without the Western Allies mounting a Second Front — something that, if the Allies still balked at such a mission, might well lead to a breakdown in the Allied alliance far more momentous than the looming collapse of the Axis Pact. "We have to let our apples ripen. It would be a real irony of world history if we were to be courted by both the Soviets and the Anglo-Americans in this situation — which is not inconceivable," Goebbels noted. "It sounds absurd, but it is a possibility. In any event we've got to do our best to work on the current difficulties between them. As long as we don't have a disaster on the Eastern Front, our situation will be secure."[11]

If Germany's Eastern and Southern Fronts were held, and the enemy's air offensive was parried, Germany would remain politically and militarily in the ascendant. "The Führer is very optimistic," Goebbels described, "perhaps too optimistic. But it's good to see him in such good form. Either way we're going to put everything we can, to the last breath, into the struggle." He hadn't seen Hitler looking so fresh and on such a high for ages — "he told me that as soon as things get dangerous, all his aches and pains disappear and he feels healthy as never before."[12]

A renewed German peace with Stalin, as in the Ribbentrop Pact of 1939? It didn't seem likely, as things stood. But over time?

Providing the Soviets were willing to leave Germany in control of the Ukraine — with its all-import grain harvest, and the Donets Basin, with its huge reserves of coal — the Führer seemed willing to parley. In the meantime 1943, far from being a *verlorene Jahr,* a lost year, the Third Reich would remain in almost complete military control of the whole of Europe — moreover able to deal with Jewish and Resistance problems more ruthlessly than ever.

This, then, was Hitler's strategy — one that was far more effective in the short term than his enemies or even his own generals admitted, then or later.[13] Mussolini's arrest and the probable defection of the Italians as his Axis ally were removing the biggest burden from Hitler's back — not increasing it, as so many assumed in their excitement. Winston Churchill and his British parliamentary colleagues, especially Anthony Eden, were known to be pressing for exploitation in the eastern Mediterranean. This was all to the good, as Goebbels had joyfully discussed with Hitler. Dissension among the Allies would make the Führer's task the simpler,[14] with no sign of Stalin willing to confer with his Western counterparts — only ever-increasing scolding in the Moscow press at the failure of the Western powers to mount a Second Front.

Just one thing could thus imperil the Führer's warplan: American insistence on the mounting of a massive, all-out cross-Channel invasion of northern France in 1944, in coordination with the Soviets, that would crush Germany between them — Hitler's age-old nightmare.

41

A Cardinal Moment

Mrs. Churchill had felt unwell, following the turbulent sea voyage from Scotland, so Churchill, summoned to stay a few days with the President at Hyde Park, arrived by train from Canada at the small railway halt near the President's home around midday on August 12, 1943, without her. He had, though, his twenty-year-old daughter, Mary, in tow, a bubbly, charming subaltern in the Women's Auxiliary Territorial (equivalent to National Guard) Service.

Contrary to General Brooke's sour mien in Quebec, where he was preparing to meet the U.S. chiefs, the Prime Minister was full of joy at the latest news of secret negotiations for Italian surrender with Marshal Badoglio's representative. Ever the historian in attempting to set current events within the larger picture of the past, he felt the coalition nations had reached "one of the cardinal moments" of the war, as he'd put it in a cable to the President when calling for a tripartite meeting with Stalin, rather than waiting for the President to meet with Stalin one-on-one. In his telegram he'd claimed "our Mediterranean strategy" had already gained all that Stalin had "hoped for from a cross-Channel second front" that year — dooming the German offensive at Kursk — and that a Big Three meeting "would be one of the milestones of history."[1]

The President, for his part, worried that, far from being a milestone of history, it would be a millstone, if the Russians learned Churchill and his 230-man entourage were all for pulling out of Overlord yet again. Worse, in fact, if the Russians — who were still facing some two hundred Wehrmacht divisions on the Eastern Front — lost all faith in the Western Allies. Stalin's unwillingness to meet before the fall had at least given the President time to reassert his role as leader of the Western nations. Roosevelt therefore arrived at Springwood ahead of Churchill on August 12, shortly after breakfast.

Rather than incurring an immediate contretemps, the President had decided to show no outward concern, but to treat Churchill and his daughter with his usual affable hospitality and respect. He'd therefore instructed that Winston's lovely painting of Marrakesh, which the Prime Minister had brought over in person in May, at the time of the Trident Conference, should be hung in the main room of the new Library at Hyde Park before the Prime Minister's arrival, as well as Raymond Perry Neilson's vibrant new canvas of the both of them at the Atlantic Charter meeting, on the deck of the *Prince of Wales,* flanked by their chiefs of staff.

Greeting Churchill and his daughter — who was in uniform — the President drove them in person to Hyde Park. Given the oppressive summer heat, he also arranged, once they were settled in, for swimming at Val-Kill, Eleanor's cottage: the President driving Churchill there in his special Ford, the swim to be followed by fish chowder from an old Delano family recipe, and hot dogs cooked by the First Lady herself.

The Prime Minister certainly showed no disappointment at the simple outdoors fare, indeed he entered into the spirit of the country weekend as to the manor born — even eating the hot dogs he was served. "Mr. C. ate 1 & ½," Daisy Suckley recorded with amusement in her diary, "and had a special little ice-pail for his scotch."

Daisy thought Churchill "a strange little man," though. "Fat & round, his clothes bunched up on him. Practically no hair on his head" — a fact that compelled him to seek shelter from the sun's harsh glare under "a 10-gallon hat."[2] When he undressed, Daisy was even more amused. "In a pair of [swimming] shorts, he looked exactly like a kewpie," she described.

Returning to her own family mansion at Wilderstein that evening, Daisy noted that "Churchill adores the P" — "loves him, as a man, looks up to him, defers to him, leans on him. He is older than the P. but the P. is a bigger person, and Churchill recognizes it. I saw in Churchill, too, an amount of real greatness I did not suspect before. Speaking of South Africa, Ch.[urchill] said General Smuts is one of the really great men of the world — a prophet — a 'seer' — his very words — He wants to get him to London, for his 'mind on post war Europe' . . ."[3]

The President, too, was an admirer of Jan Christiaan Smuts.

Smuts's support for Roosevelt's vision of the United Nations had certainly been encouraging. But in terms of military strategy, Smuts's over-optimism was as worrying as Churchill's, the President felt — seemingly unaware, despite or perhaps because of his great reputation as a guerrilla fighter in the Boer War, just how difficult it would be to fight the Wehrmacht head-on, and thus, like Churchill, now contesting the feasibility of Overlord.

To add to the British Commonwealth preference for peripheral rather than head-to-head combat, also, there was Anthony Eden, the British Balkanist — who would be attending the Quebec Conference and clashing horns with Secretary Hull. This would make the President's task doubly difficult. The President gave no hint of anxiety, however, even to Daisy.

"The P. was relaxed and seemingly cheerful in the midst of the deepest problems," she described. As the President explained to Churchill, the imminent surrender of Italy was a most welcome development — but it would not win the war against Germany. Nor could it be counted upon, in all likelihood, to keep Russia as an ally in the war against the Axis. Germany would arguably be more powerful alone than burdened by an ally like Italy. This could have serious ramifications — not only in the event that Stalin sought a separate peace with Hitler or an alternative German government, but in terms of Russian cooperation in the war against Japan, slated to follow the defeat of Germany.

There was also the question of whether Russia would agree to be a participant in a United Nations security system thereafter, if the Western Allies failed to carry out Overlord — and instead put their energies into a doomed campaign in the Dardanelles, to spite the Russians. As Secretary Stimson had put it in the memorandum he'd brought with him to lunch with the President two days before, the "Prime Minister and his Chief of the Imperial Staff," General Brooke, were still "frankly at variance with" Overlord. "The shadow of Passchendaele and Dunkerque still weigh too heavily over the imagination of these leaders of his government. Though they have rendered lip service to the operation, their hearts are not in it." Nor were their heads — though it was difficult to understand British reasoning that "Germany can be beaten by a series of attritions in northern Italy, in the eastern Mediterranean, in Greece, in the Balkans, Rumania and other satellite countries, and that the only heavy fighting which needs to be done will be done by Russia. To me, in the light of the post-war problems which we shall face, that attitude towards Russia seems terribly dangerous," Stimson had written. "None of these methods of pinprick warfare can be counted on by us to fool Stalin," he'd warned. And he'd pointed to the year 1864, "when the firm unfaltering tactics of the Virginia campaign were endorsed by the people of the United States in spite of the hideous losses in the Wilderness, Spottsylvania [*sic*], and Cold Harbor."[4] Overlord was the only way "Germany can be really defeated and the war brought to an end."[5]

Stimson was certainly right to question the Prime Minister's loyalty to the Trident agreement. The day after his first talk with Churchill in London, on July 13, the Prime Minister had minuted his chiefs of staff with an immortal phrase that would come to personify his irrepressible but often unrealistic spirit. In the minute he had scorned the notion of landing merely on the toe of Italy, across the narrow Sicilian strait at Messina; "why should we," he'd asked his generals, "crawl up the leg like a harvest bug, from the ankle upwards? Let us rather strike at the knee" — an amphibious assault north of Naples, "thus cutting off and leaving behind all Axis forces in Western Sicily and all ditto in the toe, ball, heel and ankle. It would seem that two or three good divisions could take Naples and produce decisive results if not on the political attitude of Italy then upon the capital. Tell the planners to throw their hat over the fence," the Prime Minister had declared in July, adding it was "of the utmost urgency."[6]

Two or three whole divisions, to be transported by sea, put ashore by landing craft, and reinforced more than two hundred miles behind the current German-Italian frontline?

The feasibility of this was something that had not concerned the Prime Minister. He'd seemed on the Mediterranean warpath, delighted with Smuts's supportive cable, and responding to it with excitement. "I believe the President is with me: Eisenhower in his heart is naturally for it. I will in no circumstances allow the powerful British and British-controlled forces in the Mediterranean to stand idle." He would bring a Polish division from Persia, he would use Canadians and Indians — all rushed in to exploit the imminent "collapse" of Italian forces. "Not only must we take Rome and march as far north as possible in Italy but our right hand must give succour to the Balkan Patriots." If the Americans declined to cooperate "we have ample forces to act by ourselves."[7]

Churchill's claim, in retrospect, was as ridiculous as the President's remark to his chiefs that U.S. troops could mount Overlord on their own, if the British reneged on the operation. The two statements were, however, an alarming indication as to how much the two Allies were now separating, not converging, in their war strategy. It was therefore up to the President to stitch them back together — if he could.

In the circumstances, the President felt he had no option but to play his biggest card: the atom bomb.

42

Churchill Is Stunned

Before leaving Washington, the President had rehearsed over lunch with Stimson the latest position over U.S. atomic bomb research — which he'd placed under the war secretary's direction the previous year.

When swift development of research had been in danger of stalling for lack of sufficient funding, early in January, 1943, Roosevelt had found the necessary money. Critical Canadian supplies of the necessary raw materials, moreover, had been contracted with the cooperation of the President's friend, Prime Minister Mackenzie King — leaving the British, essentially, with only a cadre of theoretical physicists and no possibility of producing such a weapon by themselves. For months Churchill had been pressing for a bilateral agreement to pool research and its dividends — the U.S. authorities refusing to cooperate, however, on grounds of American national security. Only the President had the authority to decide.[1]

If Churchill would not adhere to the American Overlord strategy, as per the Trident agreement reached in May, the President thus quietly indicated to the Prime Minister that the United States would have to withhold an agreement to share development of the atomic weapon. If, by contrast, the British were willing to stand by the agreed Anglo-American Overlord strategy, then the President would go ahead and sign an agreement to share its atomic research program with the British — *and not the Russians.* This would, in itself, assure the Western Allies of a reserve weapon that could, if indeed it worked, stop the Soviets from spilling into western, perhaps even central, Europe.

The Prime Minister was shocked by the President's proposed deal. For Churchill personally, it would be a climbdown even more embarrassing than at the climax of the Trident Conference. Before leaving for Hyde

Park on August 10, Churchill had gaily assured Prime Minister Mackenzie King in Quebec that the "president is a fine fellow. Very strong in his views, but he comes around."[2] This had not only been smug but clearly presumptuous, it seemed.

The President's firmness certainly surprised Winston. How would he explain backing off his opposition to Overlord, in Quebec, after bringing 230 staffers to argue his case? An agreement on the atomic bomb project must, of necessity, remain as secret as the research itself; he would thus not be able to reveal, let alone explain, the quid pro quo arrangement, save to a handful of his British team back in Quebec. It would also be politically problematic at home in England. A groundswell of resentment was already forming there against the United States, given that America was so clearly becoming the dominant partner in the Western Alliance. It might well affect the Prime Minister's support in Parliament, and room for maneuver in the War Cabinet.

In his heart of hearts, Churchill therefore continued to hope events on the ground in Europe would make Overlord unnecessary: that if the Allies' fall and winter operations against the Germans prospered in Italy and the Mediterranean, they would find Overlord unnecessary. Or if German defense forces in northern France swelled to an even greater extent, threatening disaster for Overlord, then he could always request the right to cancel the Overlord landings . . .

In any event, after swallowing the bitter pill, Winston Churchill recognized he would have to agree to the President's terms — for the moment. He thus gave his assent.

Overlord would go ahead as the number one Allied operation — the decisive Allied operation — with priority over all other commitments.

Churchill was disappointed, but took his defeat graciously.

There was one further potion, however, Churchill must take before the two men left Hyde Park, the President made clear.

Churchill waited to hear it.

The supreme commander of Overlord must be an American, since the largest contingent in the cross-Channel invasion would ultimately be from the United States. This decision, too, the Prime Minister would have to convey to General Brooke.

Churchill was shocked — the President's insistence an understandable blow to his patriotic British pride.

In the circumstances, though, there was nothing he could say, other than: Yes, Mr. President.

The historic deal, then, was struck.

Churchill was not happy with the outcome — indeed, he woke in the night "unable to sleep and hardly able to breathe." He got up and "went outside to sit on a bluff overlooking the river," where he "watched the dawn," he later recalled.[3]

The worst, at least, was over, however — leaving the Western Allies with a clear, unified timetable and strategy for defeating Hitler's Third Reich. Considering that, at the Oval Office on August 10, Admiral King had suggested switching U.S. priority to the Pacific, Churchill had been skating on very thin ice — with the gravest consequences for world history.

Fortunately the President had gotten the Prime Minister to concur. And with their new accord, the brief Hyde Park summit came to a happy end — the Western Allies on the same page.

Churchill tried to persuade the President they should both now go straight to Quebec to meet with the Combined Chiefs — and thus spare Winston the humiliation of reporting his change of stance alone. The President said no, however.

Mrs. Roosevelt was about to tour American forces, hospitals, and installations in the Pacific theater for six weeks, and the President wanted to see her off. He wished, in particular, to give her a personal letter for General MacArthur in order to facilitate her tour once she arrived in Australia. Though they conducted more or less separate lives, Roosevelt was more proud of Eleanor as First Lady, and guardian of his social conscience, perhaps, than ever. He also wanted to have lunch with Secretary Hull in Washington and concert their approach to Italian government after unconditional surrender, before they both went to Quebec.

Taking Churchill to the station, meantime, the President bade him and his daughter farewell. The following evening, August 15, Roosevelt himself boarded the *Ferdinand Magellan* together with Harry Hopkins, who did not look at all well — "white, blue around the eyes, with red spots on his cheek bones," Daisy Suckley commented[4] — and set off, southwards. Traveling through the summer night the little presidential party made its way back to the White House. It had been quite a weekend.

PART ELEVEN

Quebec 1943

43

The German Will to Fight

IN THE GENERAL narrative of the Second World War, the famous Quebec Conference of August 1943 would be seen as the moment when the Allies — the Western Allies — laid down their D-day strategy and timetable — an Overlord operation scheduled to take place on May 1, 1944.

In reality, however, the decision had already been taken in May 1943, at the Trident Conference — and in writing. Overruling Churchill — and General Hull's brief planning revolt at the Pentagon — the President had thereafter stuck to his guns. There was therefore no reason for Churchill to have brought his 230-man team to the Canadian capital, from a military point of view — or for them to stay. General Eisenhower was handling the secret Italian surrender negotiations with Marshal Badoglio's representative, and the decision to appoint an American, not a British, supreme commander for Overlord had been agreed by Churchill at Hyde Park, in deference to the President's wishes. Had Churchill simply told his British team of the new deal — trading partnership in the atomic bomb's development for British commitment to a clear American D-day strategy — and had they returned to their ship, the *Queen Mary,* the Quebec Conference need not have taken place.

Instead, of course, it did take place — bringing the British and American military teams almost to blows. At one point the noise of a revolver being fired in the conference room — which had been cleared of clerks and junior officers — would be thought to be the start of a gunfight.[1]

The British, in short, acted at Quebec with extraordinary ill grace — loath to accept a policy in the Mediterranean that did not envisage or permit exploitation of what they saw as a unique opportunity, after the toppling of Mussolini, to strike at the outer pillars supporting the Third Reich. In his war memoirs Churchill would title this section of his account "Italy Won." But as the historian of Churchill's *opus magnum* would later point out, Italy was *not* won.[2]

Instead, the Allied campaign in Italy would arguably prove the most ill-conceived Allied offensive of the war thus far: a sad reflection, in all truth, of Churchill's misconception of modern combat. Far from being a victory, it would drag on for almost two years, never putting the Allies anywhere near a breakthrough, and causing the deaths of hundreds of thousands of Italian civilians long after their government had surrendered unconditionally. It would incur almost a third of a million Allied casualties—killed and wounded—for no other gain than could have been made at virtually no cost in September 1943. And this largely because Churchill and his military team completely underestimated the German will to fight, not for their homeland but for every inch of other people's territory as if it was their own: a demonstration of blinkered yet also professional approach to battle that had few parallels in the history of war.

The difference between Germans and Italians in their response to the Allied onslaught would say it all. On July 19, 1943, the largest single bombing raid of the war had taken place in Italy. More than five hundred B-17s and B-24s of Major General James Doolittle's North African Strategic Air Forces had pounded Rome's railway marshaling-yards and nearby airfields.

The raid had destroyed the equivalent of two hundred miles of railway track, and—in spite of millions of warning leaflets dropped the previous day—had resulted in some seven hundred civilian deaths:[3] enough, when rumors of vast casualties spread among the Italian population, not only to frighten the Italian government to end Mussolini's long reign as Duce, but to begin surrender negotiations with the Allies under a different leader, Marshal Badoglio, in order to avoid more destruction of their Italian homeland.

The Italians had not reckoned on the German response to their imminent capitulation, however—not only German forces in Italy, but Germans at home in the Fatherland, where German cities faced the same, indeed far worse, bombing than Rome experienced. Five days after the U.S. bombing of the Rome railway network, there had taken place an even bigger air raid, or series of raids: this time the combined heavy bombers of the RAF and USAAF attacking from airfields in Britain the northern German city of Hamburg—Operation Gomorrah. Employing not hundreds but thousands of bombers in rolling attacks, night and day for an entire week, the Allies created a literal firestorm—with temperatures of 1,000 degrees Celsius, hurricane winds of 150 miles per hour, and melting asphalt in the streets. By its end, Operation Gomorrah had killed some forty-two thousand people—the majority, civilians—injured thirty-seven thousand more, left the center of Hamburg in utter ruin, and had caused a million people to evacuate the burning city. Yet the result was the very opposite of reaction in Italy.

Instead of calls for the arrest of their country's dictator and immediate unconditional surrender to the Allies, as in Italy, there was reported to be an even more relentless determination in Germany to continue to prosecute war to the death. It was as if any hope of conscience—*Gewissen*—had now been incinerated in Germany. Certainly it removed any sense of guilt at having been the first to launch such a war of ruthless conquest by *Blitz* and *Blitzkrieg.* The Allied raid on Hamburg—which would soon be replicated on Berlin—merely reinforced German stoicism: a collective will that was expressed in yet deeper loyalty to the nation's leader and calls for the Führer, in their fury, to exact German revenge. In particular, for him to use, finally, the secret weapons he and Goebbels had publicly alluded to.

At his Wolfschanze headquarters near Rastenburg, far from Hamburg and Berlin, Hitler thus viewed the Allied bombing raids on those cities as more inherently counterproductive than his own earlier raids on London and Coventry in 1940 and 1941. Bombing would not bring Germany to its knees. The fact was, the Allies could not defeat the Third Reich, Hitler reasoned, unless they could defeat his primary weapon, the Wehrmacht. The German armies embodied the highest Teutonic virtues of obedience, courage, group loyalty, and self-sacrifice—and as he studied his maps and daily Abwehr intelligence reports, he rightly saw no signs whatever of an Allied intention to follow up the mass bombing of German cities with a ground offensive on Germany via the beaches of northern France—at least not for another year.

To the extent that, if they did attempt to breach the Atlantic Wall in the late summer of 1943, the Allied invasion forces would be crushed by his German divisions in France, the delay was disappointing to the Führer—and to Goebbels. However, if the Allies didn't dare launch such an assault in 1943, as Hitler pointed out to his panicky generals, then there was no cause for alarm. The German Volk and the German Wehrmacht were too unified and imbued with too much resolve to simply collapse;

rather, they would hold fast at home, despite the bombing, and fight hard and harshly abroad. They would treat every attack on German forces in the occupied countries as if it was an assault on the Vaterland — in fact on the very honor and courage of the German nation. Meantime, German scientists and engineers would make available the new, secret weapons they had devised that would give Germany the wherewithal, if not to win the war, then to negotiate favorable armistices with Germany's enemies.

The war was not over, Hitler thus made clear. There was everything to gain by continuing to fight, implacably and fearlessly, to preserve the Third Reich they had so heroically created out of the ashes of World War I and the stupid Weimar Republic. Once he overcame his fury over Italy's imminent defection, in fact, Hitler was seen to regain his composure — and confidence. Fall weather was approaching; it would soon make conditions for an Allied cross-Channel assault impossible. For all their superiority in the air, at sea, and on land, the Western Allies were, in sum, *in no position to invade Germany* — and without such an invasion there was little chance the Russians could, either. In fact, judging by the American and British performance against modest numbers of German troops defending Sicily — where only sixty thousand Wehrmacht troops had been committed — the Allies might not be able to seize control of much of the Italian mainland, whether or not the post-Mussolini government surrendered unconditionally.

By stamping on the Italians and by using the Italian mainland — with its mountain ranges that would provide good defensive positions — as a Hindenburg Line of the Third Reich, the Germans had nothing to lose, Hitler reckoned. And much to gain scientifically in the meantime.

For all his mistakes — holding on in North Africa, despite Rommel's recommendation of evacuation after Alamein and Torch; holding on at Stalingrad rather than a calculated withdrawal; launching Operation Citadel instead of using his armored forces to entice and then crush a Russian offensive — Hitler was about to show that he had, in fact, a better grasp of the German war machine he'd built up over the past decade than his own generals. He was backed loyally and enthusiastically by the spirit of a whole nation, he felt, and was in a position to fight the war to the bitter, bitter end.

That Hitler was not wrong was certainly the view later taken by German official historians, in a kind of bemused, retrospective awe.

In fanning the flames of *Volksgemeinschaft,* involving a profound sense of national German community, identity, and destiny, Hitler had built upon quite the opposite of what most observers — even the Nazi elite, on occasion — assumed. The "belief that under National Socialism the Germans were, so to speak, subjected to total communicative and ideological brainwashing" by Hitler and his Nazi accomplices was simply not fact, the official historians concluded. "The widespread view that systematic government propaganda kept the population ready and willing for war, or even created a unified 'national' feeling among them, ignores reality," the historians pointed out. "Identification with the nation could not be produced on command, and as a rule propaganda was convincing only to those already converted."[4]

German nationalism, stretching back decades before Hitler, was in truth "the precondition for propaganda being successful, not the other way around." Hitler and Goebbels's propaganda had succeeded so well, in other words, because it hinged upon "established nationalist beliefs." The "spreading of racist, xenophobic, or authoritarian stereotypes" had, as instanced in the conquest of Poland and huge swaths of the Soviet Union, worked so effectively because such propaganda was directed at "soldiers already predisposed to them."[5] In a country like Germany, given the country's warring history since ancient times, Hitler had understood as an Austrian outsider that the very concept of democracy was foreign. German intellectuals had for centuries sneered at it — and had avoided practical politics, preferring philosophy, the arts, and science. With its rich history of land warfare at the epicenter of Europe, and its distaste for thinking through or dealing with the necessary compromises involved in civilized society, Germany's people could therefore, in the wake of deep economic depression and defeat in World War I, be encouraged to focus on a supposedly egalitarian, simplistic expression of nationalist German identity: one that, in order to cohere and remain strong, must see others — whether foreigners or Jews, communists or non-Aryans — as enemies: enemies to be excluded, disrespected, defeated. And where deemed necessary, simply liquidated, without remorse or compunction.

Anyone who objected to the nationalistic program in Germany was "othered," while "in foreign affairs" the "seed was planted for the future offensive war of extermination," the German official historians concluded. "War, established as a permanent component of German politics as an inheritance from the First World War, from then on became the natural means of achieving political ends both at home and abroad."[6] Far from becoming a nation of warrior-serfs obeying a draconian führer, in other words, nationalistic Germans had become loyal and obedient members of a community — proud and arrogant citizens of a revived empire: a third Reich, a *Volksgemeinschaft,* a "master race" of individuals each cognizant at some level and largely supportive of the genocide being directed against Jews in Germany as well as outside Germany on their behalf; supportive, too, of barbarous treatment of enemies such as Russian *Untermenschen,* since the denigration of "others" only increased and inflamed this powerful sense of national German identity.

What Hitler had intuited, then, as Italy's new leaders prepared to defect from the Axis Pact, was what many of his own generals did not: namely that the war would not be won or lost by cleverness or better tactical strategy in the East, the South, or the West, per se — tactics such as the fighting withdrawals that these German generals suggested, or the marshaling of armored reserves using the latest panzers in German counterstrikes. The war could only be won, in the end, by employing Germany's national *spirit:* the amazing solidarity of its people, bonding in a nationalist saga that Hitler saw as mythic in the noblest, Nibelungen sense: a demonstration of national pride and unity by seventy million people at home, but especially so abroad when acting as military overlords — an achievement unmatched, in many eyes, since the Romans.

This national German unity that the Führer had channeled and directed would never be broken by aerial bombing or by peripheral Allied operations, let alone by the cowardly defection of Germany's partners. It was not, in the end, a matter of winning or losing; it was a matter of hunkering down and asserting German moral and military strength, in dark days as well as fair ones. German forces had won, in the shortest time, almost unimaginable victories and territories. All genuine Germans were participants, implicated in its sins and a part of Germany's new trial by fire. No one would be spared. There was thus no talk of the future, the postwar world, because the concept no longer existed — only the current defend-or-die struggle.

By fighting offensively in the Mediterranean in the hopes of German collapse, then, the Allies might well, Hitler recognized, play into German hands. Wehrmacht forces would be operating closer to home, the Allies further away from theirs. Moreover, German forces would have the advantage of mountainous terrain, easily defensible lines, highly disciplined and well-armed troops who would fight *even better* when shorn of their weak, former ally, Italy, once the Italian government capitulated. Moreover, by continuing — in fact expanding — its massed daylight and nighttime bomber attacks that inevitably killed so many German civilians, the Allies could truthfully be portrayed as barbarians — giving Hitler not only the "right" to use new weapons of mass destruction in reply, but impelling the German Volk to *demand* he use them: *Vergeltungswaffen,* as the secret weapons were soon called — weapons of revenge. Winged but pilotless flying bombs, launched from easily constructed concrete ramps and aimed to fall indiscriminately on Allied cities. And also ballistic missiles, with even greater range — and so high and fast they were impossible to shoot down.

As Hitler had assured Mussolini at Feltre, there was no need to fear the Allies — especially the British: their cities would, the Führer forecast, be "razed to the ground," as they deserved. And unless the Allies dared take the risk of attacking Germany proper with ground troops, the Allies could not win. Moreover, if they tried to do so by launching a cross-Channel invasion, they would be easily repelled. Ergo, the Third Reich was bound, the Führer predicted, to prevail.

This, then, was the challenge facing the Allies even at the very moment when they seemed to be winning the war, both in Russia and the Mediterranean in the high summer of 1943. By underestimating German determination to fight on mercilessly in southern Europe, the Allies were heading toward disaster.

Only the President could now steer the Allies through these rapids, and to his great credit Roosevelt tried. Yet in truth he failed — forcing him to paper over the true military debacle, which now, as in a Greek tragedy, unfolded.

44

Near-Homicidal Negotiations

AFTER A HECTIC day at the White House on August 16, 1943, the President prepared to set off by train to Canada — hoping Churchill had done as he'd promised: getting his chiefs of staff to back off proposals for more extensive operations in the Mediterranean, once southern Italy was in Allied hands. And to start putting all British efforts into Overlord under an American commander.

Churchill certainly did the latter — to the consternation of General Brooke, who took the news badly. As Brooke noted in his diary, the Prime Minister "had just returned from being with the President and Harry Hopkins" at the President's home in Hyde Park. "Apparently the latter pressed very hard for the appointment of Marshall as Supreme Commander for the cross Channel operations and as far as I can gather Winston gave in, in spite of having promised me the job!!"[1]

Since Churchill still did not believe, in his heart of hearts, that Overlord would ever really be mounted, he had shown no sympathy when speaking with Brooke on his return to Quebec. Nor did he tell Brooke that it was the President's decision, not Hopkins's. More importantly, however, he did not tell Brooke what had been agreed with the President regarding the prioritization of Overlord over all other opportunistic operations — hoping that the redoubtable Brooke would fight hard in the Combined Chiefs meetings for maximum possible interim American support in the Mediterranean.

As the Prime Minister admitted to General Marshall, when dining with the general in Quebec on the evening of August 15, he'd "changed his mind regarding Overlord," and now agreed "that we should use every opportunity to further that operation." But when Marshall said the first meeting of the Combined Chiefs of Staff that day had been pretty contentious over the issue of Overlord priority, and that the U.S. chiefs were adamantly opposed to prejudicing the success of a spring 1944 Overlord by overambitious "bolstering" operations in Italy in the coming months, the Prime Minister had "finally dropped the subject, saying 'give us time.'"

In relaying Churchill's comment to his fellow American chiefs the next day, General Marshall assumed Churchill meant time for the British chiefs to swallow the inevitable, and put their energies behind Overlord rather than Italy. Marshall was wrong, however. Churchill was not one to give up so easily — and one way or another, the Prime Minister remained bent on pursuing his "soft underbelly" strategy, whether or not it prejudiced the success of Overlord.

Marshall had, after all, agreed to an amphibious American landing south of Naples, at Salerno, in two or three weeks' time, as Churchill knew — in fact the operation, codenamed Avalanche, had filled Churchill with excitement. If all went well, not only would the amphibious assault secure the unconditional surrender of Italy but it would cut off German troops in the foot of Italy and open the road not only to Naples but to Rome. The consequences were irresistible. Once the Italians — who were still occupying positions all across southern Europe, from the south of France to Greece and the Balkans, on behalf of the Axis Pact — came over to the Allies, the soft underbelly of Europe would, Churchill remained certain, become the gateway to central Europe, promising to make a cross-Channel assault either unnecessary or pro forma. And a Russian inundation of central and western Europe impossible.

Churchill's duplicity, in other words, arose not from a perfidious British effort to extend British imperial influence, as some U.S. generals such as Admiral Leahy posited at the time, but from a genuinely held belief that Overlord would fail. And, conversely, out of a genuine belief that opportunistic Allied operations in the Mediterranean — especially if Turkey could be persuaded to join the Allies — would succeed.

In both matters Churchill would be proved utterly wrong. As the historian of Churchill's memoirs, David Reynolds, would write, Churchill was profoundly if understandably deceitful in writing his fabled account of that fateful summer and autumn[2] — but the Prime Minister was not insincere in his faith in a Mediterranean rather than a doomed Normandy strategy. His was a faith based not only upon fear of failure in northern France, but also a deep and abiding fear of Russian motives and intentions — and in this respect the President was just as concerned. It was certainly something that he was taking very, very seriously as on August

16, 1943, Roosevelt set forth from the secret siding near the White House at 8:20 p.m. to join the Prime Minister in Quebec.

By the time the President's train arrived in Quebec, via Montreal — where Fala's presence on the platform banished any attempts by the Secret Service to maintain secrecy[3] — on the evening of August 17, 1943, the Combined Chiefs of Staff of the United States and Britain had been at loggerheads for three days, and were getting close, he was informed, to homicide.

General Brooke, chairing the Combined Chiefs of Staff meetings (since they were being held on "British," or non-U.S., soil), felt he was being driven almost out of his mind by American unwillingness to see the connection between operations in the Mediterranean and Overlord. "I entirely failed to get Marshall to realize the relation between cross Channel and Italian operations, and the repercussions which the one exercises on the other," Brooke noted in exasperation on August 15. "It is quite impossible to argue with him as he does not even begin to understand a strategic problem."[4]

This was the pot calling the kettle black. If anything, the reality was the reverse. Brooke's obstinate insistence, along with that of his irrepressible prime minister, upon overambitious Allied operations in Italy would, just as Marshall had feared, become a near-fatal drag on trained Allied manpower and logistical support for Overlord, as well as incurring a far higher Allied death toll than was necessary. Had Brooke devoted himself to how best to achieve the maximum German commitment of troops and reserves in the Mediterranean by the minimum of effective Allied operations, the course of World War II for the Allies would have been far better served. Far from later acknowledging his mistake, Brooke — who was promoted to the rank of field marshal in 1943 and then raised to the peerage as Lord Alanbrooke in 1946 — would go to his grave in 1963, arguing he'd always been right: that the German defense of Italy and the casualties the Werhmacht suffered by the summer of 1944 had contributed mightily to Overlord's success.

This was ridiculous. The Western Allies were to suffer 312,000 troops killed, wounded, and missing — including 60,000 Allied deaths — in the eighteen-month Italian campaign, without ever getting much further than the Po. The Wehrmacht would suffer 434,000 casualties, including 48,000 men killed in Italy by May 1945[5] — but though it did keep German divisions from the Eastern Front, it had little or no effect upon Overlord, since the Germans would have been forced to keep troops stationed across southern Europe (as they did in Norway) in fear of invasion, whatever happened in Normandy. It would become a heavy price in blood, destruction, and civilian misery to have paid for British strategy — a strategy based on a fatal illusion, or delusion: that the Allies would be able to achieve great things in a country that was ideally suited to defense, not offense.

The fact was, as Churchill's military biographer, Carlo D'Este, would write, Churchill and Brooke had utterly failed to predict "the casualties that would be incurred" by their obsession with warmaking in Italy. "During their twenty months in Italy the Allies fought one bloody battle after another, for reasons no one ever understood," D'Este would lament. "Allied strategy in Italy seemed to be not to win, but rather to drag out the war for as long as possible," he would write in retrospective frustration, a tragedy that "simply distracted the Allies from their real task: crossing the English Channel and opening the endlessly delayed second front."[6]

Nor was this hindsight. Marshall's understanding of the "strategic problem," far from being ignorant, as Brooke described it, was prophetic — and Marshall's unrelenting argument with the British chiefs of staff was greatly to his credit in counseling caution before sending tens, even hundreds of thousands of men — American, British, Canadian, French, Polish, and others — to their deaths in Italy and southern Europe. In this respect Brooke's diary gave but a glimpse of the fierce altercations and traded accusations coloring their meetings.

Brooke was implacable. "Dined by myself as I wanted to be with myself!" he noted on August 15, after hearing he was no longer to command Overlord, and having learned from Field Marshal Dill that General Marshall, now the presumed supreme commander-to-be, was "threatening to resign if we pressed our point" on overambitious Mediterranean operations. The next day Brooke himself was near resignation — "Marshall has no strategic outlook of any kind, and [Admiral] King has only one thought and that is based on the Pacific," he penned in his special green ink — the traditional color reserved for chiefs of the Imperial General Staff in Britain. The Combined Chiefs had had to ask all secretaries, stenographers, and planners to leave the room, and had argued for three hours without agreement. "This is the sixth of these meetings with the American chiefs that I have run," Brooke noted, "and I do not feel that I can possibly stand any more!"[7]

Admiral Leahy was certainly stunned by the extreme acerbity. "The

British and U.S. Staffs today got into a very frank discussion of a difference of opinion as to the value of the Italian campaign to our common war effort against Germany," Leahy recorded in his diary that night. He felt Marshall's willingness to go ahead with occupation of southern Italy to secure the Foggia airfields, from which the U.S. Army Air Forces could bomb southern Germany and the Ploesti oil fields in Romania, was "very positive in his attitude toward the Mediterranean committment [*sic*]," but Brooke seemed ungrateful, and dissatisfied. When Brooke suggested the Combined Chiefs divert Allied forces on their way out to the Far East and the Pacific to mount a bigger campaign in Italy, King's language lit the borealis lights. "Admiral King was very undiplomatic to use a mild term for his attitude," Leahy confided.[8]

Admiral King was once again reaching the end of his tether. If the British devoted too much effort to Italy, then the "build up in England would be reduced to that of a small Corps" for Overlord, as Brooke mocked King's approach — in which case King would favor "the whole war [being] reoriented towards Japan."[9]

It was small wonder. Ignoring King, Brooke had argued for an immediate, major Allied campaign in Italy to reach as far north as Turin and Milan. Not content with those objectives, Brooke had even pressed to "retain" in the Mediterranean three of the seven battle-hardened divisions earmarked for Overlord, and perhaps all seven, if German resistance in Italy was fierce . . .[10]

Marshall was almost apoplectic at this, causing Brooke to grudgingly admit, under pressure, that "'battle experienced' troops were required for Overlord" if it was to succeed.[11] Brooke remained furious, however, noting that night, "It is not a cheerful thought to feel that I have a continuous week of such days ahead of me!"[12]

The American chiefs felt the same.

As the Combined Chiefs of Staff discussions became ever more strident, Brooke had descended into the foulest of moods. He later confessed that "it took me several months to recover" from what he called the "blow" at being passed over for the cross-Channel supreme command[13] — something doubly disappointing since he had begun to yearn to get away from the Prime Minister, he confessed, and be able to command troops in battle once again.

The imminent arrival of the President had made it imperative that the chiefs come to an accord, however. Though the British team pushed the struggle over strategy to the very brink on August 16, 1943, they finally and reluctantly gave in. The decisions made at the Trident Conference in May would stand. Overlord would, they confirmed, be top Allied priority—and the seven battle-hardened divisions the President wanted would be transferred from the Mediterranean to Britain by November. Whatever could be achieved in Italy in the interim would be undertaken jointly by the Allies with remaining forces in the Mediterranean, on an ad hoc basis, to keep as many German forces away from France and Russia as possible—but under no circumstances were operations to be considered in the Balkans or elsewhere in the eastern Mediterranean.

There were, besides, equally important decisions still to be reached concerning Southeast Asia and the Pacific. The difference of opinion over strategy in Europe was therefore papered over at the conference. It was not a perfect result, but better than an outright split.

Meeting the President on his arrival in Quebec and bringing him up to speed regarding the recommendations of the Combined Chiefs of Staff, Admiral Leahy told him of the long days of contention—and the result.

As Lieutenant Elsey later recalled, Leahy was very much the President's lynchpin. "He was already at Quebec, and Roosevelt looked to him, in the summer residence of the governor general, the Citadel, as the top dog. Roosevelt looked to him rather than reaching out to King, Arnold, and Marshall. Leahy was the channel of communication from the chiefs to FDR. He, Leahy, really *was* the chief of staff to the President, and was dealt with as such, and Roosevelt saw relatively little of the Joint Chiefs during the Quebec Conference. Things came to him from Leahy, their views."[14]

The President, after all, had not come to Quebec to do their job. In truth he'd come for a very different reason.

45

A Longing in the Air

PRIOR TO THE President's arrival in Canada, a team from his White House Map Room had traveled to the Citadel in Quebec to set up a map room there, as Lieutenant Elsey afterward explained. They were "standing by the President's map room on his arrival at the Citadel to acquaint him with all the latest developments of the war. War reports had been radioed to the train during our trip up from Washington, but a more complete picture was available here for the President. The Prime Minister had his own map room in another part of the Citadel."[1] Special telephone communications with Washington and the White House had also been set up, "so that the President was never out of instantaneous communication with Washington." "Direct telegraph wire service," also, "was available between the Citadel and the White House."

Once the President was established in the Citadel, the wires grew hot with new cables — for the President was found to be batting drafts of a big speech back and forth with Judge Rosenman and Robert Sherwood, his speechwriters in Washington. On August 14, before leaving, he'd told them he wanted something he could broadcast to the whole world from Ottawa, on the eve of what looked like imminent Italian collapse and surrender. As he'd explained, he'd set the strategy and the timing of Overlord in stone. Though there would be much fighting still to be done, the war was moving into a new gear, political as well as military — and it was time to speak to the people of the United Nations: to make sure the moral aims and objectives of the Western Allies were clear and noble, before their first soldiers set foot on the mainland of Europe, early in September.

Before the President could give his speech, however, his Map Room received a very different kind of cable — this time from Moscow.

• • •

The telegram was from Marshal Stalin, dated Kremlin, August 22, 1943. It was not friendly.

Having refused every invitation to meet with the President for the past ten months, the Soviet dictator now declared he was fed up with the Soviet Union being treated as "a passive third observer" of agreements made by the United States and Britain with liberated countries, as well as with others "dissociating themselves" from Hitler. "I have to tell you that it is impossible to tolerate such a situation any longer," the quasi-emperor cabled. "I propose to establish," he declared, a three-power military-political "Commission" to handle such matters, immediately, "and to assign Sicily as the place of residence of the Commission."[2]

Stalin's arrogant new signal from Moscow made the President "mad," Harriman recalled, as it did Mackenzie King.[3] When, two days later, Stalin sent *another* cable, yet again turning down the President's invitation to meet at Fairbanks, Alaska, but demanding that a "Soviet Representative" be part of Eisenhower's secret negotiations with the Badoglio government for unconditional Italian surrender, the President became doubly incensed.

Except for his epistolary relationship with Stalin, the President had come to feel proud of the way his war strategy since Pearl Harbor had played out — thus far. He'd even treated Churchill with extraordinary patience and good humor when the Prime Minister had gotten into an interminable argument over Sumatra, after the President's arrival.

Sumatra?

"Mr. Churchill strongly advocated the establishment of an allied aviation base on the north end of Sumatra instead of the west coast of Burma," Leahy had protested, amazed at the Prime Minister's chutzpah. Instead of helping reestablish Burmese road communications with China — which Chiang Kai-shek considered vital for U.S. supplies — Sumatra would offer the prospect of air cover for a British invasion of Singapore, Churchill had argued: an objective that had never hitherto been raised before the Combined Chiefs. Even General Brooke had cringed. The Prime Minister's latest obsession had led to distracting arguments that continued for three long days — leaving Brooke furious and ashamed of his boss. An assault on Sumatra had never been seriously examined by the British chiefs — in fact the idea had only come to Churchill on the transatlantic voyage to Canada, Brooke railed in his diary, "in a few idle moments," yet here was the Prime Minister "married to the idea that success against Japan can only be secured through the capture of the north tip of Sumatra" — and "wants us to press the Americans for its execution!" The Prime Minister was acting like "a peevish temperamental prima donna," and proving "more unreasonable and trying than ever this time."[4]

Churchill would not give up his bone of contention, however — as if in lockjaw. Not even the President had been able to silence him on the subject. When the two leaders went on a quick fishing trip for the day in Laurentides Park, forty miles from Quebec, on August 20, and in the governor-general's cabin were eating the small trout they'd caught, Averell Harriman witnessed the sight of Churchill *still* going at the subject hammer and tongs with the President — who responded with glass and silverware.

There was simply insufficient shipping for such a venture, the President patiently pointed out to Churchill, even if they wanted such a strategy — which they didn't. Reopening the supply route to China was the real priority. The President "used most of the glasses and salt-cellars on the table making a 'V'-shaped diagram to describe the Japanese position" from western China to the South Pacific, "indicating the advantages of striking [Japan] from either side." Instead of laboriously trying to "remove the outer ones," such as Singapore and Sumatra, "one by one," the Allies should, the President said, simply go for the enemy's jugular — "thereby capturing the sustaining glasses" behind the outliers — Roosevelt corralling the glassware with a sweep of his hand.[5]

Churchill had remained unpersuaded, though — the argument mirroring, Harriman later reflected, their earlier "disputations over striking across the Channel or in the Mediterranean. Roosevelt once again favored the straight-line approach," Churchill the peripheral.[6] As the President shared with Mackenzie King, however, Winston's military misjudgments might be truly appalling, but they were vastly outweighed by Churchill's profound *political* wisdom: wisdom that would be crucial in the next phase of the war — especially when both men saw the tone of Stalin's cable. The war against Hitler, and then Hirohito, was set — but avoiding future war with Russia was not.

Despite the war of words traded by the chiefs of staff at the Château de Frontenac, then, the irony was this: that an extraordinary measure of harmony seemed to persist between the President and the Prime Minister — both of them staying in the Citadel, where they lunched and dined together every day.

What the President, in contrast to his chiefs, recognized was that the very unity of the Allies was being tested — not merely by the challenge of defeating the forces of Hitler and Hirohito in battle, but by the need to deal with Stalin. *And that the Western Allies must not fail this test.*

To the world, the Allied summit at Quebec in the summer of 1943 thus held a symbolic importance far outweighing any recommendations the Combined Chiefs of Staff might make: an alliance that must be seen by the world as growing closer and closer, not further apart. Though it could not be revealed to the public, possession of an atomic bomb, if nuclear fission worked, would give the Western Allies huge authority in ensuring a world free of German- or Japanese-style militarism and aggression — or Russian. The President had even gotten British acceptance of the draft Joint Four-Power Declaration he'd asked Sumner Welles to draw up in writing before his meeting with Churchill at Hyde Park, together with a suggested United Nations Protocol document.[7] All in all, this was a tremendous achievement for such an alliance between the Old World and the New: an achievement the President was determined to emphasize in the speech he intended to deliver in Ottawa, the capital of Canada. And Stalin's rude new cables only made him the more determined to make it strong.

To outsiders, the President thus appeared in an even more confident frame of mind than usual on August 25, as Mr. Roosevelt and the governor-general, the Earl of Athlone, were driven to the seat of Canadian government, having traveled to Ottawa by train from Quebec. "It was estimated that there were a crowd of approximately 30,000 people on hand at Parliament Hill and its vicinity to welcome President Roosevelt and to hear his address," the official chronicler of the President's trip noted. "This was said to be the largest crowd ever to welcome a distinguished visitor to Ottawa, even exceeding the welcome accorded to King George VI and Queen Elizabeth" in 1939.[8]

"The setting of the ceremony was one of the finest if not the finest ever provided for a Presidential speech," the *New York Times* agreed the next day. "Through streets packed with people and lined with sailors, soldiers, airmen and uniformed women in all services, the President drove from the railroad station into the parliamentary grounds in his open car." As people swarmed over the lawn, the President, "looking down on the waving crowd, turned to Prime Minister King and said, 'I shall never forget this sight.' The crowd in turn looked up at the magnificent Gothic building with its tower stretching up into the blue sky, as imposing a monument to parliamentary government as exists in the world. Carved over the portal are the lines by a Canadian poet descriptive of this great country: 'The wholesome sea is at her gates, her gates both east and west.'"[9]

The President was determined to give no hint of weakness — physical or moral. He "stood throughout the ceremony," Mackenzie King recorded in his diary. "Quite an effort as one could see, and shaking a good deal as he held on a chair and the stand" — kept upright by his steel leg braces.[10]

Robert Sherwood and Rosenman — "the firm of Sherwood and Rosenman, astrologers," as they signed their first draft — had been tasked with helping the President compose an address that would encompass his political philosophy, as well as his vision of the war's purpose — and the future. The two speechwriters had thus done their best and had arranged for the first finished draft to be flown to Canada, along with a plea that they be invited to Quebec to incorporate the new military decisions being made there. "It was signed with a drawing (by Sherwood) of a tall thin man — Sherwood; and a short fat man," Rosenman recalled humorously.[11]

The President had turned down their request, however — for he had no intention of announcing the military decisions he'd made, either to the Germans or to the Japanese. He liked the "astrologers'" initial draft, however, which he then worked on "very carefully, making many changes in language here and there, which strengthened it," Rosenman later recalled. Most significantly he'd added a whole new section relating to the postwar. "The Ottawa speech was not a major policy speech in any sense of the word," Rosenman explained its tenor. "It was, however, important," for in it the President declined to mention the Russians — at all.[12]

What Roosevelt had decided to do, instead of talking about the things his military advisers were discussing with their counterparts, was to raise instead, at a critical moment in the prosecution of the war, a rallying cry for the democracies. A call for the United States and the United Nations to put isolationism finally and forever behind them, and embrace his larger, moral vision of the future. He therefore "discarded the last few pages of our draft," Rosenman recalled, "and wrote a new conclusion with an optimistic note."[13] A note that would follow his grimmer picture of the turmoil that Hitler and the Japanese had brought to mankind. "We did not choose this war," the President reminded his audience — "and that 'we' includes each and every one of the United Nations. War was violently forced upon us by criminal aggressors who measure their standards of morality by the extent of the death and destruction they can inflict upon their neighbors."

With war forced upon them, the United Nations were now pulling harder and harder *together,* the President emphasized. He mocked the panickers who, after Pearl Harbor, had "made a great 'to-do' about the invasion of the continent of North America"—especially the Aleutian Islands. "I regret to say that some Americans and some Canadians wished our Governments to withdraw from the Atlantic and the Mediterranean campaigns and divert all our vast supplies and strength to the removal of the Japs from a few rocky specks in the North Pacific"—from which the Japanese had now wisely retreated, he pointed out. America was, he made clear, taking upon itself a much, much larger challenge. "Today, our wiser councils have maintained our efforts in the Atlantic area, and the Mediterranean, and the China Seas, and the Southwest Pacific with ever-growing contributions." It was in this context he himself had come to Canada—"Great councils are being held here on the free and honored soil of Canada—councils which look to the future conduct of this war and to the years of building a new progress for mankind."

> During the past few days in Quebec, the Combined Staffs have been sitting around a table—*which is a good custom,*

the President explained,

> —talking things over, discussing ways and means, in the manner of friends, in the manner of partners, *and* may *I* even say in the manner of members of the same family. (applause)
>
> We have talked constructively of our common purposes in this war—of our determination to achieve victory in the shortest possible time—of our essential cooperation with our great and brave fighting allies.
>
> And we have arrived, harmoniously, at certain definite conclusions. Of course, I am not at liberty to disclose just what these conclusions are. But, in due time, we shall communicate the secret information of the Quebec Conference to Germany, Italy, and Japan. (applause) We *will* (shall) communicate this information to our enemies in the only language their twisted minds seem capable of understanding. (laughter and applause).[14]

As the *New York Times* reporter described, "Thirty thousand persons had gathered on the lawns in front of the building to welcome the President and to hear him speak and their cheers rolled up in a storm when he uttered that warning."[15]

> Sometimes I wish that that great master of intuition, the Nazi leader, could have been present in spirit at the Quebec Conference — I am thoroughly glad that he wasn't there in person. (laughter) If he and his generals had known our plans they would have realized that discretion is still the better part of valor and that surrender would pay them better now than later.

Hitler and his Volk were, however, unlikely to surrender without a great deal more bloodshed.

> The evil characteristic that makes a Nazi a Nazi is his utter inability to understand and therefore to respect the qualities or the rights of his fellowmen. His only method of dealing with his neighbor is first to delude him with lies, then to attack him treacherously, then beat him down and step on him, and then either kill him or enslave him. *And* the same thing is true of the fanatical militarists of Japan.
>
> Because their own instincts and impulses are essentially inhuman, our enemies simply cannot comprehend how it is that decent, sensible individual human beings manage to get along together and live together as (good) neighbors.
>
> That is why our enemies are doing their desperate best to misrepresent the purposes and the results of this Quebec Conference. They still seek to divide and conquer allies who refuse to be divided just as cheerfully as they refuse to be conquered. (applause)
>
> We spend our energies and our resources and the very lives of our sons and daughters because a band of gangsters in the community of Nations declines to recognize the fundamentals of decent, human conduct . . .
>
> We are making sure — absolutely, irrevocably sure — that this time the lesson is driven home to them once and for all. *Yes,* we are going to be rid of outlaws *this time.* (applause)

Under the heading "Much Post-War Discussion," the *Times* reporter noted the President's speech then addressed a much bigger challenge than merely winning the war. "There was much talk" in the speech, he added, "of the post-war world."

> Every one of the United Nations believes that only a real and lasting peace can justify the sacrifices we are making,

the President claimed,

> and our unanimity gives us confidence in seeking that goal.
>
> It is no secret that at Quebec there was much talk of the postwar world. That discussion was doubtless duplicated simultaneously in dozens of nations and hundreds of cities and among millions of people.
>
> There is a longing in the air. It is not a longing to go back to what they call 'the good old days.' I have distinct reservations as to how good 'the good old days' were. (laughter) I would rather believe that we can achieve new and better days.
>
> Absolute victory in this war will give greater opportunities to the world, because the winning of the war in itself is certainly proving to all of us up here that concerted action can accomplish things. Surely we can make strides toward a greater freedom from want than the world has yet enjoyed. Surely by unanimous action in driving out the outlaws and keeping them under heel forever, we can attain a freedom from fear of violence.
>
> I am everlastingly angry only at those who assert vociferously that the four freedoms and the Atlantic Charter are nonsense because they are unattainable. If those people had lived a century and a half ago they would have sneered and said that the Declaration of Independence was utter piffle. If they had lived nearly a thousand years ago they would have laughed uproariously at the ideals of Magna Carta. And if they had lived several thousand years ago they would have derided Moses when he came from the Mountain with the Ten Commandments.
>
> We concede that these great teachings are not perfectly lived up to today, but I would rather be a builder than a wrecker, hoping always that the structure of life is growing — not dying.
>
> May the destroyers who still persist in our midst decrease. They, like some of our enemies, have a long road to travel before they accept the ethics of humanity.
>
> Some day, in the distant future perhaps — but some day, it is certain — all of them will remember with the Master, "Thou shalt love thy neighbor as thyself."[16]

Mackenzie King, standing behind the President, was deeply moved. "I noticed that he had the speech in a ring binder so as to prevent the leaves slipping away. He followed what he was saying by running his little finger along the lines as he spoke. He was given a most attentive hearing and a fine ovation at the close."[17]

It was small wonder. Rosenman was both right and wrong in writing that the Ottawa address was not a "policy speech." In its deeply personal way, using the simplest of language, it was perhaps the most heartfelt moral speech the President would ever give, cutting to the essence of what he believed: spoken to an audience in the open sunshine and through microphones and radio to the world, on behalf of a country that was rapidly becoming the most powerful nation on the earth: a nation that, with enough determination, would be able with its democratic allies to safeguard at war's end the future of humanity.

46

The President Is Upset — with the Russians

FOLLOWING THE PRESIDENT'S speech, there was lunch at Government House, following which Mr. Mackenzie King took FDR on a drive through the city, and showed him his two homes, Kingsmere and Laurier House, where they had tea.

On the drive Roosevelt confided to King how glad he was to have gotten from Churchill the now firm, formal British commitment to launch "an attack from Britain to the North of France," and that "he believed he could get a million men across [to England] during the remainder of the summer and on in the autumn," that very fall, 1943, ready for D-day on May 1, 1944. He felt, in retrospect, that the Quebec discussions had been, despite the fierce arguments, ultimately satisfactory and boded well for the successful prosecution of the war. Was that all, though?

"The most important of all he told me," King dictated that night, "was in answer to the question which I asked him: how satisfied he was with the conclusions of the conference."

Expecting the President, from what he'd shared, to say that he was — especially in view of his rapturously received speech that day — the Canadian prime minister was stunned. "He replied instantly that everything was most satisfactory until last night — just after 6. A telegram came from Stalin at that time which was most disconcerting; very rude and wholly uncalled for. It was the reply to the invitation that had been sent him to meet Winston and himself somewhere, the suggestion having been made in particular of Nome, Alaska. Stalin had replied that he, himself would arrange a conference, and it would be in Sicily" — but not between leaders, only "on a lower level." He, Stalin, had other things to attend to, of "greater importance."[1]

"I asked the President what it meant. He said there were only two interpretations. The most charitable one is that [just] like the Russians, they are one day with you, and the next day, they are prepared to take a very opposite course and be against you. You never feel sure of them. They may one day be very cheerful on your side; later, very down against you."[2]

And the other?

"The other interpretation — which is a very serious one and which is quite possible — is that Stalin is trying to work up a record against us" — in order to have an excuse "to make a separate peace with Germany. In this way, get us out of the war [with Hitler], leaving it to us to bring the war to a conclusion. He said that would be a very serious matter as the German armies would then be quite free of the Russian attack from the rear and," as King noted with alarm, "could devote all their energies to fighting against our forces."[3]

Certainly, the two leaders of North America agreed, the very threat of making a separate peace with Germany would give Stalin more leverage in making further demands for American aid and for more Allied operations in the West — as well as more concessions in terms of the end-of-war/postwar. Demands, amounting to blackmail, that Stalin "could not hope to get out of a peace conference," as King noted.

The President was clearly upset — as was Churchill, who had received the same cables. Roosevelt "then said: Winston was terribly annoyed. Was all for sending him [Stalin] a sharp answer. That he, the President, had strongly urged him to do nothing of the kind but to wait a day or two." Instead of responding, he himself would arrange for a message to be sent to Stalin by his secretary that he was traveling back to Hyde Park and would be in touch again in a few days. "'In the meantime, Winston and I will have a chance to think over what is best to say in reply.'"

"We talked a good deal of the conference," King summarized; "of what had been achieved. He was greatly pleased that all had been so harmonious" with the British, in the end. But the President "made no bones about telling me how deeply concerned he was" — about Russia.[4]

The President's concern was palpable — and understandable. Not only was Stalin an unreliable ally, constantly refusing to get together to discuss the prosecution of the war, but more worryingly still, refusing to get together to discuss either the endgame or postgame.

Roosevelt found himself amazed not only by the arrogance of the Soviet leader, after refusing to attend the summit meetings, but at his shameless hypocrisy. Cocooned in secrecy and almost pathological security in the Kremlin, Stalin was still forbidding any but the barest information about Russian forces and operations to be shared with the United States or Britain, his allies, and had declined to meet with the President and Prime Minister—yet was now fiercely decrying them for not including him in their deliberations. Given that the Soviets would permit virtually no Allied access to Russian cities, organizations, or individuals, his sudden demand that Soviet politico-military representation be set up in Sicily, an island thousands of miles from Moscow and which the Western Allies had only captured a few days previously, was significant. Stalin, clearly, was flexing his muscles: dictator of a power or quasi-empire now boasting two hundred army divisions in the field, on the Eastern Front—and the Western Allies still without a single division on the mainland of Europe.

It was, in other words, Stalin taking stock of the resolve of the Western Allies—a test that would best be met by a demonstration of Allied unity, not irritation.

Mackenzie King and Harriman—who had few illusions about the nature of the Soviet police state—seemed nevertheless surprised by how offended the President seemed to be over Stalin's two cables, once they heard their content. As Harriman put it, at the time, surely "one can't be annoyed with Stalin for being aloof and then be dismayed with him because he rudely joins the party."[5]

The President *was* dismayed, however—and not merely at Stalin's gatecrashing with regard to imminent Italian surrender negotiations. There was the question of how Stalin would behave, once he arrived at the bigger "party," when German surrender was in sight: a dictator running an impenetrable police state at home, yet announcing he wished to be treated as a controlling presence in Western councils: even dictating where the political-military surrender commission should be located.

However dismaying, it was clear that the balance of power within the United Nations was changing. In the spring of 1943, nervous about Hitler's impending Kursk offensive, Stalin had felt compelled to make certain concessions to his capitalist allies, such as closing down the egregious Comintern—which he'd finally done in May 1943. As the dictator explained to a Reuters correspondent, "the dissolution of the Communist International" as the purveyor of world communist revolution since 1919 would, once effected, increase the "pressure by all peace-loving nations against the common foe, Hitlerism, and expose the lie of the Hitlerites that Moscow allegedly intends to interfere in the life of other states and to 'bolshevize' them."[6]

Now, in late August, 1943, Stalin sounded quite different. With the great German offensive at Kursk called off thanks to the Allied invasion of Sicily, and with Russian armies pushing the Wehrmacht out of Orel and Kharkov — moreover with a huge Russian battle in the offing to move forward their forces from the Dnieper River in the south — Stalin clearly felt he could bang the Allied drum without having to leave Russia to meet with the President or Prime Minister.

Far from intimidating the President, however, let alone hammering a wedge between the President and British prime minister, Stalin's outburst now served to bring the leaders of the Western alliance closer to each other than Stalin could ever have imagined.

Behind the scenes at Quebec and Ottawa a deep and consequential *political* shift began to take place — a "sort of changed attitude," as Roosevelt put it to Mackenzie King on the way back to Quebec.[7] The President and his Joint Chiefs of Staff might hold Churchill's fantasies of defeating the Wehrmacht via the Mediterranean — whether through Italy, Yugoslavia or Greece, Turkey or the Balkans — to be just that: fantasy. But in terms of Soviet intentions, the President was very much on the same page as the Prime Minister. Stalin might make outward concessions to the Western Allies, such as closing down the Comintern, and even easing Soviet restrictions on religion — which the Soviet leader now also did. But the dictator himself remained a godless Russian psychopath — "Ivan the Terrible," directing two hundred divisions on the field of battle.

It was at this point that the President — who still nursed serious qualms not only about Churchill's military judgment but his backward, Victorian views on colonial empire and postwar social reconstruction — paused to reconsider his approach to the Grand Alliance. He still hoped he could come to a military understanding with Stalin, since neither he nor Stalin could defeat Hitler without the other. Moreover he still hoped he could come to a political understanding with the Russians, where the two powers — who clearly would be the dominant world powers at the war's end — could agree to disagree in terms of their own ideologies. But he needed, he recognized, a Plan B if Stalin did not cooperate in the postwar world the President envisioned — or even failed to cooperate in the end-of-war scenario that would come either in 1944 or 1945.

Winston Churchill might be the most infuriating partner in terms of his military obsessions, his impetuous whims, and his failure to follow a consistent strategy. He was, nevertheless, a political partner of huge and possibly historic importance in the world that was fast approaching: a democratic partner more important, in terms of dealing with Russia, than Harriman, or Hopkins, or former ambassador Davies — all of whom had been to Moscow and had firsthand knowledge of Stalin — perhaps realized.

What, though, was Stalin's real plan — if indeed he had one? Stalin was refusing to meet the President, either one-on-one or with the Prime Minister. Clearly the dictator wanted to conceal and safeguard Russian intentions behind a wall of paranoid secrecy, using a front of apparatchiks and spokesmen who never dared speak with authority, but referred everything to Stalin, on pain of dismissal or death.

For his part, Churchill didn't necessarily believe that Stalin would conclude a separate peace with Germany, despite the 1939–1941 Ribbentrop Pact, since "the hatreds between the two races" — the millions killed — "have now become a sanitary cordon in themselves," as Churchill told the President, and cabled his deputy prime minister in London that night. To Mackenzie King Churchill said the same: that the Russians and the Germans had "come to hate each other with an animal hate." So conscious were the Germans of their crimes against humanity in Russia and the likely repercussions, in fact, that they would probably "prefer to open their Western front to British and American armies and have them conquer Germany rather than Stalin," Churchill thought (and hoped), if "Stalin went on winning."[8] But the tone of Stalin's message boded ill for agreement between the Allies themselves in prosecuting the war. Which raised the question: what *was* Stalin's version of the endgame?

Without a single American or British boot on the mainland of Europe, the Western Allies were in a weak position, still, to inhibit Stalin. Would he use the new power of his many hundreds of Russian divisions — four hundred in total, it was calculated, stationed across the entire Soviet Union — to dictate the territorial and political outcome of the war in Europe?

To both Harriman and Mr. Roosevelt, the Prime Minister said he "foresaw 'bloody consequences in the future'" — "using 'bloody' in the literal sense," as Harriman noted. "'Stalin is an unnatural man. There will be grave troubles,'" Churchill declared — and openly rebuked Eden, who, like Harriman, considered the cable from Stalin "not so bad." "There is no need for you to attempt to smooth it over," he snapped, "in the Foreign Office manner."[9]

The President felt the same as Churchill — the Soviets a strange yet brave people, in the service of another psychopath.

Some historians would later resent and question Churchill's anti-Soviet stance,[10] but there can be little doubt in retrospect that, though Churchill would be proven completely wrong about Hitler, the Wehrmacht, and the progress of the war in Italy, he was extraordinarily prescient about Stalin and the Russians — and that the President was of like mind. Where Roosevelt and Churchill differed, however, was in how to deal with the Russian threat to freedom and democracy, as the Western Allies understood those ideals.

With no American or British forces yet on the mainland of Europe, the Western Allies were hamstrung. By the same token, however, without an Allied Second Front the Russians could not defeat the forces of the Third Reich. Ergo, if the Third Reich was to be defeated and the Nazi nightmare brought to an end, there would *have* to be a military agreement between the three countries, irrespective of political considerations. It would be up to the President and the Prime Minister, if possible, to turn that military agreement into a political accord, setting out a road map for postwar Europe and the world that both sides could live with. It might not prove possible to reach, but it would be worth trying to. The example of the disastrous Versailles Peace Conference in 1919, which had been tasked with solving end-of-war issues that had not been discussed or agreed in advance, was too awful to contemplate.

Encouraging Russia, then — a country or empire that had not been a party to the 1919 peace talks, but which had been the elephant in the room there — to take a responsible role in the postwar world was now the biggest challenge for Roosevelt and Churchill.

Both men had been present at Versailles, and both knew the task of an enduring postwar security settlement would be no easier. Also that Russia would be, together with the United States, the key player. In spite of the terrible losses the Soviets had suffered since Hitler's invasion in 1941, the USSR still comprised more than 170 million people: the largest nation, or union of so-called republics, in Europe. It was more than twice the size of Nazi Germany in population — and many times its size in territory. Its wartime economy might be a disaster, its industry dependent largely on slave labor, its military dependent on American Lend-Lease aid, and its society ruled by fear, incarceration, deportation, and execution; nevertheless, it now boasted the largest number of troops in the world — in excess of thirteen million men in 1943. After utter disarray and retreat in the summer of 1941, Soviet forces had finally turned the tables on the mighty Wehrmacht: by numbers, willingness to take casualties, determination to fight for the homeland, fear of what further atrocities the dreaded Nazis would commit if they were not repelled; and younger, better, nonpolitical professional military leadership on the field of battle. Patriotic pride, moreover, had swelled and grown as the Soviet armies successfully defended Mother Russia — the Communist Party taking a secondary, background role. Stalin's Great Terror and his Purges of the 1930s had been set aside — for the moment at least. As supreme commander in chief of the Soviet Armed Forces, Joseph Stalin was not only the effective, single ruler and dictator of the USSR, but he was in a position to begin moving Russia toward a less repressive future, *if he so chose:* something his ambassador to Washington, Maxim Litvinov, had begged him to do, before Stalin had recalled him from America.

Would Stalin dare — or want — to take that course, though?

It seemed unlikely, as Churchill intuited — especially after reading the latest, more detailed investigation of the Katyn massacre.[11] Behind Soviet propaganda, directed and controlled from Moscow, Marshal Stalin remained an arguably certifiable psychopath: a mass murderer living with his own terror, namely that of being assassinated. And of flying. His refusal to meet with the President, as well as his recent decision to withdraw his highly experienced Soviet ambassadors from Washington and London and replace them with apparatchiks, offered little hope of an open, democratic future for the Russian-dominated world — at least one based on the four freedoms to which the President referred in his Ottawa speech.

Stalin's latest cables to Quebec were thus dispiriting, at a moment of Allied joy and hope, on the eve of Italian surrender. The telegrams convinced both Churchill and the President that the defeat of the Third

Reich and Japan — which would still entail a vast military effort — would be but the first act in a new struggle for control of those occupied nations: nations such as Poland, currently ruled by the Germans, whose people innocently hoped for independence, self-determination, free elections, and freedom from fear.

This was, in actuality, the saddest of prospects, even as the unconditional surrender of Italy loomed.

Churchill, too, saw the moment as a watershed. He had earlier favored Sumner Welles's notion of grand regional or hemispheric councils, representing "spheres of influence" across the postwar world; now, suddenly, it became clear to him — as to the President — that there would, essentially, be but two such spheres: Anglo-American versus Russian. A rivalry, moreover, that would not necessarily be confined to central Europe, if Stalin's talk of southern Europe — of Sicily and Italy — was anything to go by.

In many ways it was a tragedy: a road not taken. Had Stalin been a different leader, a statesman willing to rise to the challenge of advancing and protecting a postwar world based upon the four freedoms, the challenge of the future could, in the aftermath of Hitler and Hirohito's demise, have been that of a secure, spirited, economic, social, and cultural opportunity for the progress of all nations. Instead, a very different prospect arose: a darker world of communist dictatorships and puppet states, modeled on the Soviets, answering to Moscow.

Unless Stalin were assassinated, or the Allies could somehow prevail upon the Russians to abandon the notion of tyrannous rule by fear, the postwar future thus suddenly looked bleak to Churchill and the President — despite the grandeur of a United Nations coalition that had successfully turned the tide against two empires, German and Japanese, still committing crimes against humanity on a scale of mass murder not seen for centuries, if ever.

All therefore now seemed to depend on a Russian dictator: a Soviet supreme commander in chief who was, as Averell Harriman later remarked, "the most inscrutable, enigmatic and contradictory person I have ever known."[12] It was a sobering prospect.

The Canadian prime minister accompanied the President to the station at 7:00 p.m. on August 25. "As the last word," King recorded in his diary, "I reached over to the President and said quietly God bless and help you."[13]

Roosevelt's talks with the Canadian prime minister had left Mackenzie King at once awed and anxious. In his library at Laurier House, the Canadian premier had shown the President and Grace Tully, the President's secretary, not only his private library but a "photograph of Hitler." The President had "instantly reacted to it with a shudder at the appearance of the man." King had also "pointed out the handbill of the time of Lincoln's assassination" — a reminder how seldom violence was separable from politics, their chosen profession.

"I was unfortunately pretty tired and unable to take in or contribute to the conversation as much as I would have liked, but I felt throughout how real was the affection the President had for myself and felt drawn more closely to him than ever," King recorded that night. "I confess, too, one came to feel he had a much more profound grasp of the situation than I had, at times, believed him to have. By that, I mean not a knowledge of the facts but the understanding of history and places and the like which are so essential to the understanding of great movements. The kind of thing that Churchill possesses in so great a degree."[14]

In his pedantic, cautious way the Canadian had come to see, increasingly, just how blessed was the free world in having such titans of humanity as their two great leaders — and how vital it was to create a durable system of international security and development *while they were still in office*. Moreover one that would survive them — since, "unhappily, we could not rely on having the President and himself at the head of affairs for all time," as King remarked to Churchill. "That any post-war order would have to take account of the persons who might take their places, and that each nation would want its say."[15]

Winston Churchill, Mackenzie King reflected, was "not so democratic at heart as the President. He still remains a monarchist and a Conservative," whereas "Roosevelt is clearly for the people and they know it." To be sure, "Churchill is for his country and its institutions" — including its "great Empire," King allowed. Thinking especially of India, though, King deplored continuing colonialist complacency at high levels in England, where "less believe in the abilities of people to govern themselves" than was the case in the United States and Canada.[16]

The future shape and peace of the world, however, was at stake: leading inexorably to the question of whether the President would stand for an unprecedented fourth term. "We talked," King had already noted, "of the next elections," which would take place the following year, in November. Health was a factor, the President had acknowledged. "He quite clearly has it in mind to run again but says he will not travel about; will not do any speaking over the radio and not make many speeches. He dislikes Willkie"—his Republican opponent in the 1940 presidential election—"but says he has been encouraging Willkie's renomination in order to get more or less a split in the Opposition [Republican] party which he believes will come if Willkie is nominated. He said that Willkie was all right on foreign policy which was important, but it would be dangerous if a Republican isolationist were to get the nomination."[17]

American participation, even leadership, in the new world order was quite clearly the President's goal—thus redeeming the failure of President Woodrow Wilson to get Senate ratification of U.S. membership in the League of Nations in 1920. As Churchill pointed out, it was not the League of Nations that had failed; rather, it was the nations who had failed the League of Nations—something the President was determined would *not* be the case this time.

Thus arose, at Quebec in the summer of 1943, the greatest irony of the war: that the United States and its Western Allies were, in effect, faced with two potentially competing struggles. The first, to pursue the fight against the odious, genocidal Axis powers to obtain their unconditional surrender; the second, to achieve a global postwar democratic system that would not be prejudiced or sundered by the emerging power of a Russian-directed Soviet Union—a communist quasi-empire ruled by a psychopath scarcely less dangerous to humanity than Adolf Hitler.

PART TWELVE

The Endgame

47

Close to Disaster

THE PRESIDENT HAD laid down the strategy and timetable of the war to defeat the Third Reich and then Japan, on behalf of the Western Allies; Churchill, for all that he feared a bloodbath in northern France, had had to comply. Yet to achieve the political results of the war that he wanted — a new world order — the President had need of Stalin. And in dealing with Stalin, he also had need of Winston Churchill, as a demonstration of unity between the U.S. and British governments. Roosevelt therefore asked the Prime Minister to come stay with him in Washington after Quebec. They would be together when the Italians, as seemed likely, surrendered. Above all, though, they would be together in showing Stalin there was no rift in the Western Allies: that the U.S.-British coalition was inviolable, and would remain so.

Early on August 26, 1943, the *Ferdinand Magellan* pulled into the little halt at Highland, north of New York, and the President was driven up to his family home. "The P. came from Ottawa, looking well," Daisy Suckley noted in her diary, "but tired. He said he would try to get rested before Churchill comes to Wash.[ington] next Wednesday. The Quebec Conference was a success but Russia is a worry — the P. said a message had come from Stalin which was 'rude — stupidly rude.' Churchill wanted to send back an answer — even ruder!"[1]

Stalin was not the only problem. There was the question of how Hitler would react, once the Italians surrendered and U.S. and British armies landed on the mainland of Europe, as they planned to do in the coming days. Though in a sense it was only a diversion in order to keep the Germans from beating the Russians and away from the eventual beaches of Normandy, the Allied assault on Italy would reveal whether Churchill was right, or the President: whether the Germans would collapse, or whether Italy would turn out to be a hornet's nest.

All too soon they would find out.

In the domestic comfort of his Hyde Park home, Roosevelt meantime took things easy—with Admiral Leahy at his side. "Today ends a three day restful visit with the President at Hyde Park, where there were no demands on any of us at any time," Leahy noted in his diary on August 29, 1943, "and where we were completely relaxed after our strenuous Staff Conference in Canada."[2] "The P. was very cheerful & seemed relaxed," Daisy recorded. Taking the sun at his cottage he "sent for some eggs & bread & butter—He toasted the bread on the electric toaster, sitting by the fire on the sofa," and to unwind "talked about a good many phases of the present situation."[3]

The first landings in mainland Italy would begin on September 3, 1943: a crossing of the Strait of Messina by troops of Montgomery's British Eighth Army, which would hopefully draw Axis forces into close combat—and away from the primary invasion site: Salerno, where the major Allied assault would take place a week later. The Salerno landings would be a massive three-division invasion in the Gulf of Salerno, 270 miles north of Montgomery's army; it would plant major Allied forces close to Naples under U.S. general Mark Clark, and hopefully cut off German forces facing the British. Not content with this sweeping plan, Marshall had urged Eisenhower to use his Eighty-Second Airborne Division—not to ensure the success of the Salerno landings but to mount a yet more ambitious landing, 200 miles further north still. On Rome, from the air.

This operation would be called Giant II—perhaps the most misguided military undertaking of the war thus far. Churchill thought it a masterstroke, which would enable the Allies to seize Rome, the Italian capital, by coup de main. Were Rome to fall to the Allies, and the Italians turn against their former partners, who knew what might then transpire? The Third Reich might collapse like a house of cards.

The President remained doubtful; taking Rome would be nice in terms of morale and publicity, but it led nowhere, strategically, given the terrain in northern Italy. And Hitler, he was sure, would not fold his hand that easily. He thus left the campaign to Marshall and the chiefs of staff, confident that General Eisenhower would not be pressed into doing anything too foolish.

As it turned out, however, Eisenhower *was* so pressed—and the Allies, in the days that followed, came very close to disaster in Italy.

48

A Darwinian Struggle

HITLER, FOR HIS part, was contemptuous of Roosevelt's call for Allied unity in Quebec, as was Goebbels. The President's speech in Ottawa was dismissed as rhetoric. "It consisted of dull, stupid scolding and lacked any political substance," Goebbels sniffed in his diary. "It's not worth bothering about. One can see from the speech, and the terrific reception it was given by Canadian members of Parliament, though, just how half-witted the public is over there. Roosevelt threatens military operations, but refrains from being specific, because he probably can't be. He ends by quoting Jesus, which is all-of-a-piece with his bizarre and misbegotten character."[1]

For both the Reich minister and the Führer, the world was now entering the vortex of a great Darwinian struggle of survival—a struggle in which only the strongest would emerge. Italy was exhibiting weakness, would most likely crumble, and would have to be sacrificed on an altar of blood; Germany, by contrast, would only grow stronger, more savage—and more ruthless once unencumbered by allies.

Those among the Allies who hoped Germany's generals or Wehrmacht soldiers would lose heart in their leader were to be disappointed. Hitler and Goebbels's *Weltanschauung,* including their *Nibelungentreue,* was to be largely replicated among German civilians across the Third Reich—and among German troops across the occupied countries. Goebbels noted, for example, how contented were German soldiers, returning on leave from the frontline to their relatives in the German homeland—yet how angry, stunned, and surprised they were at the effects of RAF and U.S. Air Force bombing on the civilian population.

The question, then, arose for Goebbels as the propaganda genius of the Third Reich: could the war be prolonged for another six months or a year by stubborn German defense of the nations the Wehrmacht had conquered in western, southern, and eastern Europe — keeping the enemy as far as possible from Germany until the Führer's *Vergeltungswaffen,* or V-bombs, were ready to be launched?

On his last visit to the Wolf's Lair, Goebbels had found the Führer disinclined to think Kharkov as being in danger — at least, if it was in danger, there should be no mention of it in public. "Wir kämpfen an allen Fronten, im Süden wie im Osten, möglichst weit vom heimatlichen Boden entfernt, um den Krieg vom Reichsgebiet fernzuhalten" — "We are fighting on all fronts — in the South as well as the East, as far from home ground as possible — in order to keep the war as far as we can from the Reich," Hitler had declared — while doing everything in his power as führer to counter Allied air power, from expediting German antiaircraft guns to greater priority for jet-engined Messerschmitt fighters, better radar, and new interception tactics.[2]

As this was ordered, the question of a political solution had meantime become more and more tempting. How could Goebbels and the German Foreign Ministry exploit the widely suspected split between the Western Allies and the Soviets? Could they persuade either Churchill or Stalin to negotiate an armistice with the Third Reich, and thus avoid war on two fronts?

"I ask the Führer whether he thinks we might be able to make an accommodation with Stalin, over time," Goebbels noted on his next visit to the Wolf's Lair. "For the moment, however, the Führer thinks not," he recorded, disappointed. Moreover, the Führer was unwilling to surrender, if negotiations could be started, the Ukraine: the breadbasket of Europe and crucial for Germany's food needs. "In general," Goebbels noted, "he thinks it more likely we would have more success in doing something with the British rather than the Soviets." As the British came to realize that fighting the Wehrmacht on European soil was very different from war in faraway North Africa, they would surely "come to their senses" — especially once German V-bombs began to rain down on London.

"It's true Churchill is an absolute anti-Bolshevist," Goebbels agreed with Hitler — a Churchillian stance that might be manipulated to get him to abandon his antifascist rhetoric and agenda in favor of anticommunism. Given Churchill's Mansion House speech, in November 1942, warning that he would never allow the dissolution of the British Empire, the Prime Minister might well be open to new peace feelers, Hitler intimated, if convinced Britain could not win the war militarily. Churchill was "naturally pursuing British imperial objectives in this war, as in the last. Now that he has Sicily in his pocket he's in a good position," Goebbels recorded their conversation. "The Italians will never get Sicily back, for with Calabria and Sicily in British hands Churchill will control the whole Mediterranean as an English ocean, for all time . . . So the Führer thinks the English rather than the Russians will be more willing to come to an arrangement in the end."[3]

Knowing Churchill, Goebbels was skeptical, however. "I don't see any sign of this happening," he admitted in the privacy of his diary, whatever Hitler might think — though he did not dare say so to the Führer. Besides, the matter of an armistice either with the British or the Russians was academic, since the split between the Allies had not reached the point where they could be prised apart — yet. Nevertheless, the "controversy between the Soviets and the Anglo-Americans is really serious," Goebbels noted with satisfaction. "Our information from Quebec is quite clear about that." However, "the Führer doesn't think the crisis in the enemy camp is ripe enough to exploit at the moment. So we have to wait, and make sure we get both our fronts back under control. That is a sine qua non: that we have to stand firm where we are. A faltering military power can't be looking for an arrangement."[4]

September 1943, then, would reveal whether the Allied coalition was going to hold together, or could be brought to stalemate on the battlefield and either the Western Allies or Russia be persuaded to sue for an armistice with Germany.

49

A Talk with Archbishop Spellman

"I LEFT QUEBEC by train, and arrived at the White House on September 1," Churchill recorded in his memoirs. "I deliberately prolonged my stay in the United States in order to be in close contact with our American friends at this critical moment in Italian affairs."[1]

News had come from General Eisenhower that the post-Mussolini government of Italy, under Marshal Badoglio, had secretly agreed to surrender, once American and British troops were established on the mainland of Italy—and two days later, on September 3, Montgomery's troops crossed the Strait of Messina to Reggio, where they encountered negligible opposition. Italian forces simply abandoned their posts, in anticipation of imminent surrender, while Wehrmacht forces laid mines, detonated bridges, and staged a fighting withdrawal from Calabria.

Staying in Washington with the President, the Prime Minister seemed dangerously overconfident about impending victory in Italy. "Churchill does not think," Mackenzie King had already noted in his diary on August 31, as the British prime minister set off from Quebec, that "the further fighting in Italy will occasion anything like the loss of life that the fighting in Sicily has occasioned."[2]

The President certainly wished Winston to be by his side when the Italian surrender took place. But in truth, there was a more important, underlying reason for Churchill to stay at the White House—a purpose both men had agreed was vital not only to the winning of the war, but the postwar. For whatever happened on the ground in Italy, it was understood by the two leaders, the unity of the *Western* Allies must be further symbolized, beyond the conference in Canada, and an incontrovertible message of common purpose be sent not just to Hitler and the world, but to Stalin in Moscow.

The Western Allies, this message went, would hold together in pursuing the defeat of Germany—*and beyond.*

Though he could barely contain his excitement over Montgomery's crossing at Messina and Clark's impending invasion at Salerno, Churchill made every effort to be patient and good company to the President. Mrs. Roosevelt was still on her tour of the Pacific, requiring Daisy Suckley to stand in for her as White House hostess, and at Hyde Park. In her diary she noted the "intensely interesting" conversations at table—the President "full of charm, always tactful, even when he has to be 'painfully' truthful & perhaps harsh. He is harsh, but with a smile which tells you you are wrong, but there is no ill-feeling toward you because of the wrong—It's more that you are mistaken—in all probability because you don't know the facts. I've never known a person who so consistently tries not to hurt people."[3]

Churchill, by contrast, "snaps out disapproval. They say he fights with everyone"—not just Hitler; "jumps all over them. One person alone he doesn't jump on," Daisy added, however, "& that is the P.! The P. laughs about it: he says that if the P.M. ever did jump on him, he would just laugh at him! As I have said before, the P.M. loves F.D.R." Moreover, Daisy had had this confirmed, from the highest authority, for "Mrs C[hurchill]. told me that, too, out of a clear sky."

"The P.M. recognizes in the P. a man with a greater soul & a broader outlook than his own—It is very evident to a person who has had such wonderful opportunities to see them as I have. I consider W.S.C. a 'great man,' also, but he has not yet achieved the spiritual freedom of F.D.R. . . . They get along beautifully, and understand each other. The P. is all for the Democratic ideal because he loves it & believes in it. The P.M. is working for it because he thinks it is inevitable . . ."[4]

Daisy was naturally biased, but she was also perspicacious—and one of the only people, other than the President's White House doctor, who was watching Roosevelt's health. Churchill's daughter Mary, traveling now with her mother, Clementine, found "the pres magnetic & full of charm" as she wrote in her own diary; "his sweetness to me is something I shall always remember—But he is a raconteur," she noted, and in all honesty, aged only twenty, she found his stories "tedious" at times, though "at other times it is interesting & fun"—a "cute, cunning old-bird—if ever there was one. But I still know who gets my vote," she added loyally—her father probably the most eloquent raconteur alive. "Every evening FDR makes extremely violent cocktails in his study. Fala attends — & it is all very agreeable & warm. At dinner Mummie is on his right, & several nights no other guests being there I've been on his left. I am devoted to him & admire him tremendously — He seems to have fearless courage & an art of selecting the warmest moment of the iron."[5]

Still so young, Mary thought both her father and the President indestructible. She did, however, find herself intrigued, as was Daisy Suckley, by the "contrast" between their two characters. "To me," she noted in her diary, Roosevelt "seems at once idealistic — cynical — warm hearted & generous — worldly-wise — naïve — courageous — tough — thoughtful — charming — tedious — vain — sophisticated — civilised — all these and more for 'by their works ye shall know them' — And what a stout hearted champion he has been for the unfortunate & the battling — and what a monument he will always have in the minds of men. And yet while I admire him intensely and could not but be devoted to him after his great personal kindness to me — yet, I must confess [he] makes me laugh & he rather bores me."[6]

The truth was, the President had had other things on his mind, despite doing his best to keep the Churchills and their daughter entertained. He'd dined on September 2 with Winston and Averell Harriman — who was to be his new ambassador to the Soviet Union — to discuss Russia. Also present at the meal was Francis Spellman, the archbishop of New York, who was returning after a long inspection tour of American units overseas, as the vicar military responsible for all Catholic chaplains in the armed forces.

The next morning — the day Montgomery's troops crossed onto the mainland of Italy — Spellman came to see the President privately at the White House for another hour. The Archbishop was concerned about the Allied bombing of Rome — where he'd spent the greater part of his adult life. Spellman had shown as little concern about the alleged extermination of European Jews as his mentor who had promoted him to the top American see in 1939: the pope, Pius XII. Now that Rome itself was threatened with heavy bombing, however, Spellman was deeply worried. Moreover, he was becoming concerned over Russian designs on European countries yet to be liberated — especially Catholic Poland.

As Spellman had found on his tour overseas, American officials in Iran (where the majority of U.S. Lend-Lease supplies were now being delivered) were disgusted at the way the Russians behaved. It was as if every

Russian lived in terror of being accused of cooperation with their allies, or worse: sharing secrets with a quasi-enemy. Spellman's "information is that two of the four freedoms as we understand them, — freedom of expression and freedom of religion, — do not exist in Russia."[7]

The President was all too aware of this. What to do, though? Spellman had hitherto raised no protest over the President's conduct of the war, and at the White House he now found the President extraordinarily frank about the chances of American forces being able to stop two hundred Russian divisions from doing whatever Stalin pleased, at a time when the United States did not yet have a single boot on mainland Europe. The President certainly hoped to "get from Stalin a pledge not to extend Russian territory beyond a certain line," but there was little the United States could do when Stalin "had the power to get them anyway" — "them" being Finland (which had been a Russian duchy from 1809 through 1917), the Baltic States (a part of the Russian Empire from the eighteenth century until 1917), the eastern half of Poland (partitioned by Russia, Prussia, and Austria in the wake of the Russo-Polish war of 1792, and much of it a czardom until 1918), and Bessarabia (a czarist governorate from 1812 until 1917). Such countries might *want* to retain their recent independence, but the President was sanguine. "There is no point to oppose these desires of Stalin, because he has the power to get them anyhow. So better give them gracefully."[8]

"Give" them?

In later years — especially once the United States became a nuclear superpower, with global military reach — Roosevelt's acceptance of the inevitable would be seen as shocking, even immoral, especially for a president who was so idealistic.[9] Right-wing American critics of Roosevelt such as Senators Robert A. Taft and Arthur H. Vandenberg, fired by American exceptionalism, would deplore such a "giveaway," but the criticism reflected their historical ignorance and lack of realism.[10] No one at that time had any idea how the United States could have approached the matter differently, given U.S. military weakness, with no soldiers yet in mainland Europe — and little idea how effective those soldiers would be, once they reengaged with Wehrmacht forces on European soil.

At a moment when the Third Reich still extended from the shores of France to the Ukraine, and when Hitler, Goebbels, and Ribbentrop seriously hoped to split the Allied alliance and compel the British to negotiate an armistice in the manner of Munich in 1938, or get Stalin to renew the Ribbentrop Pact, the President saw his main priority in avoiding a premature collapse of the Grand Alliance before the Western Allies even landed in force on the European mainland. As he made clear to Spellman, one had to be realistic. Over time, he was sure, the Russians would become more civilized — especially when having to interact and compete with Western economies. Unless they somehow remained a closed society, under lock and key, they would eventually be forced to adapt to Western cultural influences.

Such a long-term view left open the question of the imminent fate of western European nations, however — nations the United States *could*, realistically, hope to save, as long as the President could get Churchill and the British to throw themselves wholeheartedly into the cross-Channel assault the following spring. Once these nations were liberated, however, would the American public support tough American peacekeeping, in countering Russian influence, after the war? Taking soundings nationwide, Judge Rosenman had warned the President that, politically, he would have to be more careful in his speechmaking with regard to postwar security, if he or any Democratic nominee wished to prevail in the 1944 election. "People are almost twice as much interested in domestic affairs as international affairs," Rosenman passed on to the President the conclusion of a recent opinion poll. Two-thirds of those polled did not wish even to provide "aid to foreign countries after the war," let alone have to keep the peace in Europe.[11]

Such findings did not stop the President from pursuing his vision of a United Nations authority, with the Four Policemen acting on the UN's behalf. It did, however, cause him to wonder how far he could single-handedly change or guide American public opinion to back such a vision. What would be the fallout, the President and the Archbishop wondered, if the United States did *not* take the leading role? Would Britain — virtually bankrupt and, pace Churchill, far more concerned with avoiding the dissolution of its colonial empire than maintaining European peace — be able to marshal sufficient will and force of arms to do the job: namely holding the Soviets at bay, if and when they began to "Bolshevize" the continent after the fall of Hitler?

From reports of communist governments in exile in Moscow it was evident Stalin intended, if possible, to install communist puppet regimes beyond Russia's borders: in Germany, Austria, and probably other bordering states. This would make it unnecessary for the Soviets to keep their forces there, beyond establishing bases. Roosevelt "agreed this is to be expected. Asked further, whether the Allies would not do something from their side which might offset this move in giving encouragement to the better elements, just as Russia encourages the Communists, he declared that no such move was contemplated [by the United States]. It is therefore probable that Communist Regimes would expand, but what could we do about it?"[12]

Archbishop Spellman was disappointed—but could see the problem: namely the American electorate. Although, in the wake of the President's State of the Union speech and his Casablanca summit, there had been a growing acceptance in Republican circles of the idea of American involvement in international decision-making once the war was won, there was still a deep core of the American public wedded to isolationism.

To push through American membership in a United Nations organization, given President Wilson's failure in 1920 with respect to the League of Nations, would already be a tremendous challenge. To achieve this, Roosevelt was ready to stand for a fourth term in 1944. But offering a platform of American intercession in European politics, with the possibility of yet another war to be fought there—this time with the Soviet Union, which not even Hitler's two hundred divisions had been able to defeat—was unlikely to fly.

The President sounded, for once, almost defeatist. "France might possibly escape" such a puppet fate, if its people elected a sufficiently socialist government, so that "eventually the Communists might eventually accept it. On the direct question whether the odds were that Austria, Hungary or Croatia would fall under some sort of Russian protectorate, the answer was clearly yes." Hopefully, with the Soviets industrializing their economy, the outlook would not necessarily be so terrible in terms of European people's standard of living. "It is natural that the European countries will have to undergo tremendous changes in order to adapt to Russia, but he hopes that in ten or twenty years the European influences would bring the Russians to become less barbarian."[13]

The President's hopes on this score would, in the end, take more than forty years to be met—not ten or twenty. Spellman, however, did not contest the President's crystal ball after his own foreign trip, for it seemed too grounded. The archbishop was only disappointed that Mr. Roosevelt, normally such a figure of moral as well as physical courage, should be so laissez-faire. As the President put it: "The European people will have to endure the Russian domination, in the hope that in ten or twenty years they will be able to live well with the Russians" — the Russians gradually becoming more civilized, while the Europeans became more egalitarian. "Finally he hopes, the Russians will get 40% of the Capitalist regime, the capitalists will retain only 60% of their system, and so an understanding will be possible. This is the opinion of Litvinoff," too, the recent Soviet ambassador to Washington, the President averred.[14]

Litvinov had been recalled to Moscow, however, not simply for talks, but to be replaced in October by a "barbarian" apparatchik: Andrei Gromyko.

Spellman, who had spent so many years at the Vatican earlier in his career, wondered at the almost dispirited view of the President regarding the future of Europe: the very cradle of civilization, and the home of so many Christians. Roosevelt had always been against "spheres of influence" in the world, but was now talking of "an agreement among the Big Four. Accordingly the world will be divided into spheres of influence: China get the Far East; the U.S. the Pacific; Britain and Russia, Europe and Africa. But as Britain has predominantly colonial interests it might be assumed that Russia will predominate."[15]

It was an unenviable scenario for Europe. "Although Chiang Kai-shek will be called in on the great decisions concerning Europe, it is understood that he will have no influence on them," the President explained. "The same thing might become true — although to a lesser degree — for the U.S.," in terms of meager American "influence on decisions concerning Europe." The President "hoped, 'although it might be wishful thinking,' that the Russian intervention in Europe might not be too harsh."[16]

Stalin "not too harsh"?

Was the President serious? American knowledge about the Soviet regime, thanks to Russian secrecy, was admittedly minimal, but from the head of U.S. foreign intelligence, General Donovan, the President had received an all-too-real picture of Stalin's system of mass deportation, arrests, executions, and rule of fear. The President's realism concerning Russia, in fact, went way back to his instructions when sending Ambassador Bullitt to Moscow to establish the first U.S. embassy in Soviet Russia: "You will be more or less in the position of Commander Byrd — cut off from civilization."[17]

The President's view of communist Russians as "barbarians" had not changed since then. Tragic though it might be, Stalin the Barbarian had survived as dictator of the USSR. As had communist Russia itself, despite facing the greatest war-assault ever mounted in military history.

Was Russian nationalist barbarianism reason enough, though, for the United States to hold back and watch while the struggle for Europe was—as in the 1930s—left to others? Could a near-bankrupt Britain be expected to master events in western Europe any better than it had in 1940, let alone in central or eastern Europe? Its military forces had been evacuated from the continent in Norway and at Dunkirk in 1940, been trounced in North Africa and rebuffed with ease at Dieppe in 1942, and even in 1943 its prime minister was really only backing a 1944 cross-Channel attack in deference to the President's will—Churchill concerned, still, that it could be a disaster if indeed it took place . . .

Britain, in short, could not be depended upon as a military power in Europe in its own right.

If Soviet domination of Europe was to be the ultimate price of defeating the Nazis, then, should American sons be sent to Europe at all? Here the President sounded more positive, for he was by no means defeatist about the larger, global picture. The League of Nations had been "no success, because the small states were allowed to intervene," he said—leading to a state of anarchy that Hitler had exploited, allowing him to conquer most of Europe by force. The lesson was therefore simple. Once Hitler was defeated, it had to be assumed postwar peace could only be guaranteed by "the four big powers (U.S., Britain, Russia, China)." The United States would be supreme in its own hemisphere, and across the Pacific. But did that mean that the security of the heartland of modern civilization—a civilization built on the foundations laid by the Greeks and the Romans—should be handed over to the British, who were weak, and the Russians, "because they are big, strong and simply impose themselves"? the vicar military asked.[18]

The President shrugged—unsure how the future would play out, and whether American voters would support a permanent American presence in Europe. All that could be said with certainty at this juncture was that, after waging two vast and destructive wars in Europe in the space of thirty years, Germany was clearly too powerful a nation to be allowed to threaten world peace again. It should, he thought, be divided up into numerous states—"Bavaria, Rhineland, Saxony, Hesse, Prussia," and "disarmed for forty years," he asserted. "No air force, no civilian aviation, no German would be authorized to learn flying." Austria, though

Catholic, could not be saved from a "Russian dominated Communist Regime." Hungary, by contrast, might be saved—"He likes the Hungarians. He wants them to come over," Spellman quoted the President's view. "He would be ready to accept them on the Allied side as they are, if they come over." The only states where self-determination would actually be guaranteed—presumably by Britain—would be in western Europe: "Plebiscites would be held in the following countries: France, Italy, Netherlands, Belgium, Norway, Greece"—but not even Czechoslovakia, which he doubted could be saved in time.[19]

Western Poland, on September 3, 1943, went unmentioned.

The Allied "side."

Hungary coming "over" . . .

It was clear the President foresaw a division of Europe into an Allied West and a Russian-dominated East in the not-too-distant future, now that British Eighth Army troops, having crossed onto mainland Italy, were beginning to fight their way north.

Spellman, who had supported Roosevelt against the bitter denunciations of Father Charles Coughlin in 1936, as well as in confronting the Axis powers after Pearl Harbor, was made acutely aware by the President of the domestic political challenge: how to get the American public to endorse, after the fall of the Third Reich, even the remotest possibility of another war to "save" specific nations in central Europe, and push back the Soviets, once they established themselves there.

Leaving the conundrum up to the President, the Archbishop focused, for his part, on pressing for Rome—as well as its environs in a twenty-mile safe zone—to be considered an "open city" in order to protect the Vatican and Rome's historic churches: "what to do for Holy See," as he put it.[20]

Nevertheless, the President's somewhat dispirited "realism" worried him. Was the President ailing?

50

The Empires of the Future

THOUGH NOT PRIVY to the President's discussions with the British prime minister, young Mary Churchill was aware that, though almost a decade younger than her father, Mr. Roosevelt was not one hundred percent well.

Decades older than Mary, Daisy Suckley was noticing the same — and was concerned. The President had returned from Canada in apparent good health, yet sported "dark rings under his eyes" — and was finding it more difficult to exhibit the abiding confidence and humor that were his trademark as a leader.

For his part Harry Hopkins seemed ill — but that, at least, showed in public. In Roosevelt's case, the President refused to show weakness, let alone signs of illness. "This is one noticeable way in which the P. is so outstanding," Daisy noted. Others seemed positively "shell-shocked" by the pace and demands of government and command in war, whereas the President "is completely normal mentally & spiritually, although he has in a way, more responsibility than anyone," she described.[1] Roosevelt would not even permit Ross McIntire, his doctor, to accompany him to Hyde Park — nor would he allow McIntire to bring in a medical consultant to assess his cardiac and circulatory health, lest word leak out he might not be up to the trials of a fourth presidential election, were he to stand.

Daisy thus worried that Churchill's extended stay at the White House, with his wife, daughter, and immediate staff — military, clerical, private — to boot, was simply too demanding at a time when the First Lady was still away: leaving the President to have to take care of even the most basic aspects of hospitality.

What she did not quite understand was that Churchill was now the only man in the world who could help the President not only shoulder his great burdens, but stop the "barbarians" from occupying too much of eastern, central, and western Europe as the war progressed.

The responsibilities of being a national leader, and on top of that commander in chief in a world war, were almost literally crushing—and Daisy was certainly right to be anxious.

Hitler, for his part, was unwell, living in isolation, intimate only with his mistress, Eva Braun, and his dog, Blondi; Stalin associated only with those in literal terror of him—even instructing the NKVD to "investigate" his son and daughter by his second wife, Nadezhda Alliluyeva, who had allegedly committed suicide in 1932.[2] (Of his first wife, Ekatarina, who had died of tuberculosis in 1907, a year after their marriage, Stalin had reportedly said: "With her died my last warm feelings for humanity."[3])

Given that the Quebec Conference was over and that its military decisions would henceforth be carried out by the Combined Chiefs of Staff, twelve days of entertaining the Churchills did seem rather long, however, to Daisy. It appeared so even more to the press, who wondered why Churchill needed to spend so much time in Washington with the President. What was Churchill busy plotting now, if the big decisions had supposedly been made at Quebec? skeptics wondered.

It was in this respect that Dr. Goebbels was more insightful than Allied journalists. The fact that Churchill was spending so much time with the President in America spoke volumes to him. The Reichsminister and Führer might still hope for signs of a split in the Allied coalition, one that might help preserve the Third Reich and its armies. The President, however, seemed still master of world opinion. Hitler had spent only a few hours with Mussolini at Feltre, before the Duce's arrest. By contrast, hosting Churchill for almost two weeks, the President was demonstrating to the world the *solidarity* of the Western Allies—an even more symbolic demonstration, in fact, than the Quebec Conference.

Taking his cue, Churchill had settled in and talked to the President at length about the Russian menace, in front of Cardinal Spellman and others. As a result of those conversations, in fact, a new idea of Western unity began to emerge in the Prime Minister's fertile brain.

At 10:00 p.m. on September 5, Winston Churchill left the President and with his wife, Clemmie, and his daughter Mary, departed the White House and took the train to Boston.

The Prime Minister had cabled Field Marshal Smuts, the South African prime minister, writing: "I think it inevitable that Russia will be the greatest land power in the world after this war which will have rid her of the two military powers, Japan and Germany, who in our lifetime have inflicted upon her such heavy defeats. I hope, however, that the 'fraternal association' of the British Commonwealth and the United States together with sea and air power, may put us on good terms and in a friendly balance with Russia at least for the period of re-building. Further than that I cannot see with mortal eye, and I am not as yet fully informed about celestial telescopes."[4]

Unrecognized by most historians, however, this was in fact a new turning point in the war, as Churchill now sought not only to wed Britain to the United States in terms of defeating Hitler, but beyond that in dealing with the Soviet Union.

The truth was, without the help of the United States there was little hope Britain could, on its own, do much of anything to halt the advance of Soviet forces in Europe, or even combat Soviet communist "influence" there. *In partnership with the United States,* however, it could — possibly. It would require girding up the people of the United States to the challenge, but it was perhaps for this reason, rather than to perpetuate British colonialism, that he had been put on this earth. Churchill had earned huge respect for his moral courage in confronting Hitler, when Britain stood alone; as Prime Minister he now felt he must, as far as possible, use that continuing respect and public support to buck up the President; to help Americans, not simply Britons, embrace a new, quasi-imperial global role as the guarantors, as far as possible, of democracy and the four freedoms.

It was a tragedy the present war could not end as the triumph of democracy over fascism and tyranny, but as the President said, it could take a generation or more before the Russians cast off communist dictatorship and embraced anything like the four freedoms.

Distinct from Western norms of civilization, the Soviet Union remained a tyranny based on fear, paranoid secrecy, incarceration, deportation, mock justice, xenophobia, and ruthlessly Russian — as opposed to international — self-interest. How much better would history have been served had Stalin never been born! Stalin had, however — and his tough, dictatorial leadership had at least ensured the Soviet armies succeeded in halting Hitler's mad invasion, just as Napoleon's invasion army had been destroyed in the heart of Russia. Somehow, Churchill mused, it would be for the United States not only to create a United Nations authority that would help preserve the peace after the defeat of Hitler and the Japanese but — in partnership with the British — face up to the Russians . . .

Fortunately the Prime Minister liked to work long and late. No sooner had the train pulled out than "he started to compose his speech," his secretary, Elizabeth Layton, wrote home,[5] and together with Churchill's shorthand stenographer, Patrick Kinna, she took down his words over four hours of nighttime railroad dictation: his speech to be given at Harvard University on September 6, on the acceptance there of an honorary degree.

Churchill had already given some of the most memorable, indeed historic, speeches in the annals of rhetoric — rich in metaphor and in the sheer magnitude of his historical perspective. His Harvard University address, however, was to be special in that, three years before his famous "Iron Curtain" speech, Churchill now made an open appeal to the youth of America to assume responsibility not only to help win the war against current tyranny but to continue to do so thereafter: safeguarding democracy on behalf of those who could not, by virtue of their weakness, do so on their own.

To the "youth of America, as to the youth of Britain, I say 'You cannot stop,'" Churchill declared the next day in Harvard's famous Yard. "There is no halting-place at this point. We have now reached a stage in the journey where there can be no pause. We must go on. It must be world anarchy or world order." As he put it, "We do not war primarily with race as such. Tyranny is our foe, whatever trappings or disguise it wears, whatever language it speaks, be it external or internal, we must forever be on our guard, ever mobilized, ever vigilant, always ready to spring at its throat. In all this," he emphasized as a British prime minister speaking in America, "we march together. Not only do we march and strive shoulder to shoulder at this moment under the fire of the enemy on the fields of war or in the air, but also in those realms of thought which are consecrated to the rights and dignity of man." The British Commonwealth and the United States were now joined at the hip — not only in their common language, but in their willingness to fight alongside, even subordinate to, one another.

The Combined Chiefs of Staff was the clearest manifestation of this development: acting not only as the transatlantic advisory body to the two elected leaders, but as the de facto strategic command center of the forces of the United Nations. Churchill was therefore unapologetic in claiming it would be a "most foolish and improvident act on the part of our two Governments, or either of them, to break up this smooth-running and immensely powerful machinery the moment the war is over. For our own safety, as well as for the security of the rest of the world, we are bound to keep it working and in running order after the war — probably for a good many years, not only until we have set up some world arrangement to keep the peace but until we know that it is an arrangement which will really give us that protection we must have from danger and aggression" — a protection "we have," as Britons, "already had to seek across two vast world wars."

In all but name this was a warning not only to Hitler, but to Stalin: that the English-speaking democracies of the world should — and would — hold together to confront and defeat tyranny and the evils of a police state. "Various schemes of achieving world security while yet preserving national rights, tradition and customs are being studied and probed," the Prime Minister acknowledged: a search to develop a system more durable and effective than the League of Nations. "I am here to tell you," he declared, though, "that whatever form your system of world security may take, however the nations are grouped and ranged, whatever derogations are made from national sovereignty for the sake of the large synthesis, nothing will work soundly or for long without the united effort of the British and American peoples."

"If we are together," Churchill declared, "nothing is impossible. If we are divided all will fail," he warned. How proud Americans and Britons could be, then, "young and old alike" to live at a time in the story of man when "these great trials came upon it" — and had found, he declared, "a generation that terror could not conquer and brutal violence could not enslave."[6]

The speech even contained the most stunning suggestion: that not only should Britons and Americans continue their military alliance after the war, but even resume a "common citizenship."[7]

While there was little enthusiasm expressed in America for common citizenship — the United States, after all, having waged a revolutionary war to achieve independence from the British Empire — Churchill's remarks, at the very moment when cables were being exchanged between the President's Map Room and the Kremlin regarding the need for high-level U.S., British, and Soviet meetings, were welcomed by newspapers in

America, England, and United Nations countries. An initial meeting of the Big Three's foreign ministers was in the works for October; then a Big Three summit to be held hopefully in November or December, 1943 . . .

Beyond the imminent amphibious Allied assault at Salerno and airdrop on Rome, then, there were larger issues at stake.

The United States was at last entering upon its manifest destiny not only as a world power, but as *the* leading power of the free world, Churchill accepted—and while no one knew which way France and other occupied countries would eventually turn, there was no doubt as to where he, the President's "ardent lieutenant" and "representative of the British War Cabinet," stood.

It would not be easy. "The price of greatness is responsibility," Churchill solemnly warned at Harvard. "Let us go forward in malice to none and good will to all. Such plans offer far better prizes than taking away other people's provinces or lands or grinding them down in exploitation. The empires of the future," the once-implacable British imperialist maintained, "are the empires of the mind."[8]

Such a bold assertion of Anglo-American solidarity would not stop Stalin from controlling those eastern and central European countries the Soviet armies might well overrun, as they combined with the United States and Britain to defeat the forces of the Third Reich. It left no doubt, however—whether in Hitler's mind, Goebbels's, or Stalin's—that the Western Allies, led by the United States and Britain, would not rest until the evils of the Third Reich were ended, and in the aftermath that they would remain united: intent upon blocking any attempt by Stalin to expand into western Europe a Soviet empire of gulag and fear.

51

A Tragicomedy of Errors

WHILE CHURCHILL GAVE his support to the notion of a new, internationalist America, General Eisenhower faced the problem of the military and political prosecution of the current war in the Mediterranean.

It did not go quite as planned. Indeed, blame for the near-catastrophe that befell the Allies in Italy ultimately rested with the two commanders in chief in Washington, historians would rightly aver[1] — for the failure to make clear to Eisenhower that his task was merely to occupy southern Italy while the Overlord invasion of northern France was prepared permitted the most dangerous optimism and false hopes to spread among the senior ranks of U.S. and British forces in the Mediterranean.

Thus the tragedy unfolded.

Churchill, so magnificent in his appreciation of the larger forces of history and tyranny, once again demonstrated an impetuous military opportunism — an aspect of his character he had never been able to control. Without General Brooke at his side in Washington to restrain him, he yearned for the Allies to swiftly seize Rome, as in the days of the Caesars — rightly seeing in it a prize whose capture would electrify both the free and the occupied countries of the world. The image across the world evoked by Italian unconditional surrender and the Allied occupation of Rome would be the second "crack in the Axis" that the President had spoken of in Ottawa.

These were understandable political and moral ambitions for the Allies — achievements that would impress the Soviets (who were still nowhere near evicting the German armies from the USSR).

Unfortunately, neither agenda took account of the Wehrmacht's likely response. Nor did it account for the invidious dilemma into which it placed Badoglio's Italian government: whether the country was to be destroyed alongside the Germans — or by the Germans.

As the days of early September passed, then, the various headquarters in the Mediterranean suffered a fatal lack of clear strategic direction from the President, the Prime Minister, and the Combined Chiefs of Staff. The President favored only a limited Allied military campaign, but was less than clear where it should end — whether in the south of Italy, in order to secure the important all-weather Foggia airfields, or as far as Rome. In fact, in a moment of levity, having summoned the chiefs to the White House to discuss the "strategic situation in light of Italian collapse," he suggested that "a new slogan should be adopted" for the campaign in Italy: "Save the Pope"![2] He was not anxious to go further, however, lest the buildup for Overlord be compromised.

By contrast the Prime Minister wanted to drive right up to the mountains of Tuscany, and there "establish a fortified line to seal off the north of Italy; a line prepared in depth which Italian divisions should help us man and so strong that it would make it very costly for the Germans to do anything effective against us." In the meantime, he urged, the Allies should do everything in their power to seize the Dodecanese islands such as Rhodes and put pressure on Turkey to enter the war.[3] The Allies would then possess a huge staging post in southern Europe to strike, in the event of a German collapse, toward southern France, the Balkans, Greece, or even northern Italy through the so-called Ljubljana Gap and Austria.

Behind the rejoicing over the recent conquest of Sicily and the first Allied boots on the mainland of Europe, across the Messina Strait — where Italian forces simply fled, and British Eighth Army troops had only to follow retreating Wehrmacht survivors of the Sicilian campaign — the real situation for the Allies began, in all truth, to border on the farcical.

"He is host & hostess & housekeeper all in one," Daisy reflected of her hero, the President — for it seemed really amazing with what ease Roosevelt had switched from a meeting of the Joint Chiefs of Staff at the White House to arranging trips of his English guests to Williamsburg, Virginia, or from reviewing British Eighth Army progress with General Ismay, at Hyde Park, to showing his guests his library before they finally left. Major military forces — land, sea, and air — were being committed to battle in the Mediterranean, but without clear and realistic strategic objectives passed down the Allied chain of command, the situation in the Mediterranean became daily more complicated.

Tasked with obtaining, if possible, the unconditional surrender of all Italian forces in Italy, southern France, the Balkans, and Greece, General Eisenhower had begun parleys with the emissaries of Marshal Badoglio, while having to decide what to do about General Patton's latest scandal (a report by the U.S. chief medical officer in the Mediterranean claiming Patton was psychologically and behaviorally unfit to command U.S. forces after striking battle-traumatized soldiers);[4] planning and commanding an invasion of Italy with limited resources (since Overlord was now to have logistical priority) and unclear strategic objectives; and having to meld as supreme commander in the Mediterranean the international ground, navy, and air force contributions to that uncertain challenge.

In the Torch invasion and campaign, the Allies had made a plethora of errors — errors that had taken place in an area occupied only by Vichy troops. This had permitted the U.S. and supporting British troops to establish themselves in overwhelming force before Hitler could react. In Husky, again, only two German divisions were on hand to repel boarders — even Hitler conceding it would be impossible to hold Sicily for more than a few weeks. But now, as the Allies prepared to invade the mainland of southern Europe in considerable force and from two different directions, the challenge changed. Montgomery had already complained on August 19 that his "Baytown" landing across the Messina Strait had no strategic objective; when pressed, Eisenhower's land forces commander, General Alexander, could only say Montgomery was to "engage enemy forces in the southern tip of Italy," and thus give "more assistance" to "Avalanche" — the four-division assault on Salerno, three hundred miles away on Italy's west coast, near Naples.

Three hundred miles, Monty had whistled! "If Avalanche is a success, then we should reinforce that front for there is little point in laboriously fighting our way up Southern Italy," his headquarters staff had protested — vainly. For his part, Montgomery, having faced the cream of the Afrika Corps since the battle of Alamein, was deeply skeptical whether Avalanche, south of Naples, would be the sort of walkover that Eisenhower and Alexander's headquarters assumed. Or Mark Clark — the as-yet-untested commander of the U.S. Fifth Army, tasked with the amphibious assault there. "The Germans had some 15 Divisions in Italy and at least four could be concentrated fairly quickly against the 5 American Army," Montgomery wrote in his diary after listening to Clark's presentation of the Avalanche plan.[5] He vigorously disputed, as the Allies' most professional if slow field commander, any idea of an easy run. So did the swifter Patton, when shown the task given to Clark. Given the hills surrounding the beautiful beaches, the "avalanche" might well come to a halt on the shore without chance of reaching Naples — let alone Rome.

As if this was not all, the plan — pressed by General Marshall — to land U.S. airborne troops on Rome was even less prudent; indeed Giant II, as it was code-named, was arguably one of the most ill-conceived near-blunders of the entire war.

General Eisenhower later confided that he "wanted very much to make the air drop on Rome," and was so "anxious to get in there," at Marshall's urging, that he removed the Eighty-Second Airborne Division from Mark Clark's Avalanche invasion force for the purpose. Somewhat surprised, General Matt Ridgway was thus ordered by the Allied commander in chief Mediterranean to drop his Eighty-Second Airborne Division on the Italian capital instead.[6]

Eisenhower's chief of staff, General Bedell Smith — a brilliant staff officer, but wholly ignorant of combat — proved equally naive, not only then but even after the event. He considered it would have been a "bold move, and it would have caught the Germans off balance" — causing Field Marshal Kesselring to retreat "immediately . . . Caught by the surprise of the American airborne landing in Rome and with his communications cut, Kesselring would have been compelled to retire to the North, and to abandon all southern and central Italy," Smith later asserted.[7]

At West Point, such boldness might have been lauded — in theory. Would the Italians, even if they were ordered to surrender to the Allies by Marshall Badoglio, actually lift a finger, however, to challenge let alone fight the Germans, who had two armored divisions surrounding Rome, and more approaching? Although General Alexander had browbeaten General Castellano, Badoglio's secret representative, into promising four divisions of Italian troops to aid Ridgway's assault from the sky, Montgomery certainly remained deeply skeptical. The "Italians won't do anything" he predicted — and Ridgway and his artillery director, General Max Taylor, feared the same. Indeed — though trashed after the war by both Bedell Smith and general Eisenhower's intelligence chief, Brigadier Kenneth Strong — Ridgway and Taylor refused to commit thousands of their paratroopers' lives to a wild plan, concocted in an "all night session" in a tent in Sicily, without further research.[8] General Taylor and a companion — Colonel William Gardiner — were therefore authorized to go behind the enemy lines, in advance of the airborne drop, to interview the commander of all Italian forces in Rome.

Transported in disheveled uniforms — posing as POWs being taken from the coast at Gaeta to the outskirts of the capital, then in an ambulance with frosted windows to the Italian War Office in Rome — Taylor and Gardiner only got to see General Carbonari at 9:30 p.m. on September 7, roughly twenty-four hours before the 150 C-54 paratroopers' planes of the Eighty-Second Airborne Division were due to take off from Sicily. A draft Instrument of Surrender — approved by President Roosevelt and by Prime Minister Churchill — had been signed on behalf of Badoglio on September 3, but had been held back in order to give the Germans no chance of preempting the Eighty-Second Airborne's drop on Rome, or Clark's invasion at Salerno, at dawn on September 9.

The paratroopers, however, did not go in — mercifully. Once Taylor reached Rome, General Carbonari explained that there were twelve thousand German paratroopers and twenty-four thousand men of the German Third Panzer Grenadier Division, with tanks, encircling the city. The American landing area was *twenty miles* from Rome; only two U.S. battalions could be airlifted in the first wave, and the Italian divisions had ammunition for only a few hours fighting — if that.

Taylor and Gardiner were agog. General Alexander, a Brit, had predicted Clark's land forces would reach Rome from Salerno, hundreds of miles away, in only three — five, at maximum — days, to relieve them.[9] It was a prediction near-criminal in its credulity — and cavalierness. Without genuine Italian assistance from the four Italian divisions, Taylor foresaw, the Eighty-Second's airdrop would be a bloodbath: an American one. He rightly demanded to see Marshal Badoglio — who, when roused from his bed, was even more defeatist, Taylor found.[10]

Badoglio had seen no fighting since 1940, and now disavowed the very Instrument of Surrender he had authorized by cable — saying he had not signed it, and had only given way to temporary telephone agreement when General Alexander threatened his emissary to destroy Rome by bombing. His representative in the negotiation "did not know all the facts," he told Taylor; "Italian troops cannot possibly defend Rome." In fact he predicted grave "difficulties" for the Allies if they landed at Salerno, given the number of German troops in the area and those streaming down with more tanks from the north. When Taylor tried the same tactic as General Alexander — threatening to bomb the Eternal City, unless the Italians carried out the proposed surrender — Badoglio merely looked at him. "Why would you want to bomb the city of the people who are trying to aid you?"[11]

Trying — but not very hard. Certainly not hard enough to save the Eighty-Second Airborne Division from extinction.

There followed a veritable tragicomedy of errors as Taylor's secret wireless signals to Eisenhower's headquarters and to General Ridgway, in Sicily, failed to get through. By the time Eisenhower called off the operation — sending Alexander's American deputy, General Lemnitzer, in person to Sicily to stop it — more than fifty C-47s with their paratroop companies were already in the air, circling the departure airfield. Firing an emergency warning flare, Lemnitzer — crammed behind the pilot in a British Beaufighter — managed to land with the cancel order, and the planes were instructed by radio to return to base.

It was a near-run thing.

But for Mark Clark's Fifth Army there was no cancellation or reprieve as, like the cavalry in the famous Charge of the Light Brigade, they were convoyed through the night toward the beaches of Salerno, and a most unwelcome welcome.

52

Meeting Reality

TOUCHING DOWN AT dawn on September 9, 1943, the Western Allies finally met reality. It was to be the most venomously contested amphibious invasion since Dieppe — contested by major Wehrmacht forces.

With the assent of the President and the Prime Minister, the "unconditional surrender" of all Italian forces had finally been announced on Allied radio in Algiers by General Eisenhower at 6:30 p.m. on September 8, in order to give the Germans the least possible time to man the beaches at Salerno. There was no confirming announcement on Rome radio by Marshal Badoglio, however — and for ten minutes it looked as if the Allies would have egg as well as blood on their faces.

All Eisenhower could do was continue to bluff — by reading aloud on Allied radio in Algiers the text of Marshal Badoglio's supposed surrender proclamation — which the Marshal was still refusing to confirm. This proclamation ordered the Italian military on the mainland and abroad to "cease all acts of hostility against the Anglo-American forces wherever they may be."[1]

Badoglio's hand was thus forced. After much handwringing, the seventy-one-year-old marshal — fearing arrest, even execution, by stalwart Italian fascists — felt he had no option but to confirm the surrender on Rome radio and seek to save himself. At 7:20 p.m. he did so — and immediately made himself scarce. Together with the royal family in the capital he fled the city on the only still-open road, in a convoy of carabinieri-protected vehicles, and bearing boxes of lire to bribe loyal fascists at roadblocks.

It was an ignominious end to the Pact of Steel: the final act of Italy's venal participation in the war — first as Hitler's partner in world crime, then as partner to the approaching Allies, which Badoglio now offered, on behalf of the Italian government, to become.

Others were skeptical. "The House of Savoy never finished a war on the same side it started, unless the war lasted long enough to change sides twice," a Free French newspaper commented sarcastically.[2] The inheritors of Rome's great empire in ancient times, the Italians now merely blew with the wind. "If you analyze the matter in cold blood there is no doubt the Italians have carried out a really good double-cross; they change sides on one day!!!" Montgomery wrote the next day to friends in England. "I wouldn't trust them a yard, and in any case they are quite useless when it comes to fighting."[3]

This was the real issue — for the Germans, by contrast, were very good when it came to fighting. And merciless. As Field Marshal Kesselring remarked of the Italians, "I loved these people. Now I can only hate them"[4] — hate that was now authorized to be channeled into vengeance on an unsparing scale, not only against Italian military units, but women, children, and the elderly. "No mercy must be shown to the traitors," Kesselring instructed General von Vietinghoff, his Tenth Army commander. Nor was it: the Italian general commanding the Salerno coastal division was executed in his headquarters even before the Germans turned on the approaching Allies — and the same fate befell tens of thousands of Italian troops across the country, as well as partisans, indeed anyone who challenged German military occupation or was seen to be aiding the Allies.[5]

General Sir Harold Alexander had willfully overestimated Italian assistance while utterly underestimating German resistance — fatally misleading the Allied commander in chief, General Eisenhower, as well as the Fifth Army commander, General Mark Clark.[6]

As the ground-forces commander of the assault, Clark had meantime hourly become more anxious. He'd thought the removal of his airborne division a terrible mistake, and had not been amused by Eisenhower's offhand dismissal of his doubts. "'Well, Wayne' — he always called me Wayne," Clark recalled Eisenhower's words, "When it drops [on Rome], it passes to your command!' And I said, 'Thanks, Ike, that's five hundred miles away!'"[7]

With the belated decision to call off the airdrop on Rome, Clark had no airborne division to worry about in Rome — indeed, no airborne division at all.[8] "As dusk came I was on the bridge. I could see the silhouettes of a hundred ships with my men in them. And I had never had such a forlorn feeling in my life," he later recounted. Shorn of the Eighty-Second Airborne, the fifty-five thousand men (British and American) of his Fifth Army thus sailed into a trap—"spitting right into the lion's mouth."[9]

Alerted that an Allied armada of almost a hundred ships was anchoring twelve miles offshore in the Gulf of Salerno, Kesselring had sent out his orders. The enemy "must be completely annihilated and in addition thrown into the sea. The British and Americans must realize that they are hopelessly lost against the concentrated German might."[10] Facing a barrage of Luftwaffe planes, lethal 88mm guns, and dense machine-gun fire, the Allied invasion force went into battle—the scene soon resembling something out of Dante's *Inferno* as both the clear Italian water and the sandy beaches ran with blood. As the veteran AP reporter Don Whitehead heard someone remark, "Maybe it would be better for us to fight without an [Italian] armistice."[11]

With operations now in the hands of General Eisenhower, there was nothing President Roosevelt, as the U.S. commander in chief in Washington, could do but leave the battle to the men in combat.

Late in the evening of September 9, once his meeting with the Joint Chiefs of Staff was over, the President therefore set off in the *Ferdinand Magellan* for Hyde Park with his British guests, the Prime Minister and his family. He had delivered another Fireside Chat the previous night, from the Diplomatic Reception Room, to announce the armistice with Italy—and to warn his listeners against complacency or idle assumptions. He welcomed the Italian people, who were "at last coming to the day of liberation from their real enemies, the Nazis." But "let us not delude ourselves that this armistice means the end of the war in the Mediterranean. We still have to drive the Germans out of Italy as we have driven them out of Tunisia and Sicily; we must drive them out of France and all other captive countries; and we must strike them on their own soil from all directions. Our ultimate objectives in this war will continue to be Berlin and Tokyo," he made clear.

"I ask you to bear these objectives constantly in mind—and do not forget that we still have a long way to go before we attain them," he'd warned. "The great news that you have heard today from General Eisenhower does not give you license to settle back in your rocking chairs and say, 'Well, that does it. We've got 'em on the run. Now we can start the celebration.' The time for celebration is not yet. And I have a suspicion that when this war does end, we shall not be in a very celebrating frame of mind. I think that our main emotion will be one of grim determination that this shall not happen again.

"During the past weeks," he continued, "Mr. Churchill and I have been in constant conference with the leaders of our combined fighting forces. We have been in constant communication with our fighting allies, Russian and Chinese, who are prosecuting the war with relentless determination and with conspicuous success on far distant fronts. And Mr. Churchill and I are here together in Washington at this crucial moment. We have seen the satisfactory fulfillment of plans that were made in Casablanca last January and here in Washington last May. And lately we have made new, extensive plans for the future," he added — a coded reference to Overlord. "But throughout these conferences we have never lost sight of the fact that this war will become bigger and tougher, rather than easier, during the long months that are to come.

"This war does not and must not stop for one single instant. Your fighting men know that. Those of them who are moving forward through jungles against lurking Japs — those who are landing at this moment, in barges moving through the dawn up to strange enemy coasts — those who are diving their bombers down on the targets at roof-top level — every one of these men knows that this war is a full-time job and that it will continue to be that until total victory is won."[12]

At Hyde Park, once the party arrived, the Prime Minister found himself on tenterhooks. Though the President tried as far as possible to keep the Churchills, including young Mary, entertained, Winston remained anxious. Giant II had been canceled; fearing savage Wehrmacht reprisals, Badoglio had reportedly attempted to renege on the Italian surrender.

The news that did come through was not good — indeed, with more German troops racing toward the battle zone at Salerno in succeeding days, Clark not only asked Eisenhower's authority to use troops of the Eighty-Second Airborne Division, but to drop them on the very beaches of Salerno, to bolster the infantry — and even ordered contingency plans be made for possible evacuation, á la Dunkirk.[13]

Daisy Suckley, watching the President, was amazed at his sang-froid. "Sunday, September 12, 1943," she wrote in her diary, three days into the invasion. "Sitting on his wheelchair, with all the Churchill party standing around, he sent for Jennings, and, in two minutes arranged for the visit, next week-end," of his son John Roosevelt and John's wife, Anne, "with two children & a nurse, and 6 Norwegians with a maid."[14]

The President had spent the morning driving his visitors about the estate, "at the wheel, his dog Fala beside him," and had arranged for lunch to be served for them all "at his own cottage (higher up the hillside than Mrs R's Val-Kill)." Following this they'd lain on the veranda — Churchill telling his daughter Mary the colors he would use, were he painting the scene, and commenting with a smile on "the wisdom of God in having made the sky *blue* & the trees *green*. 'It wouldn't have been nearly so good the other way round.'"

"To me these moments with Papa are the golden peaks of my life," Mary noted in her diary — aware that, between them, the President and the Prime Minister had it in them to protect and preserve civilization as they knew it. Then, after dinner, where the President had proposed the health of his guests Winston and Clemmie, who were celebrating their thirty-fifth wedding anniversary, "FDR drove us down to the train," the subaltern jotted in her diary.[15]

"God Bless You," Daisy heard Churchill say, leaning into the President's car. "I'll be over with you, next spring."

"Next spring" had meant before D-day. There was a long way to go before Overlord, however. Behind the bonhomie, the war in Europe was now entering a critical time for the Western Allies.

Churchill's moral and political sturdiness had certainly bucked up the President, but his military judgment, once again, was of a different order. The campaign in Italy upon which he'd so set his heart would now, inexorably, prove the very quagmire that General Marshall and the U.S. chiefs had foretold.

Even Churchill's doctor recorded how anxious, at the White House the week before, Winston had become: his thoughts "wandering to the coming landing at Salerno. That is where his heart is. As the appointed day draws near, the P.M. can think of nothing else. On this landing he has been building all his hopes. There are no doubts in his mind; anyway he admits none. It must succeed, and then Naples will fall into our hands. Last night, when the stream of his conversation was in spate, he talked of meeting Alex [General Alexander] in Rome before long — the capture of Rome has fired his imagination; more than once he has spoken about Napoleon's Italian campaign."[16]

At Hyde Park, three days after Clark's landings, Dr. Wilson had then noted the effect on Winston when the troops landed on the Salerno beaches and "it did not prove to be a walk-over. On the contrary, the news that filtered into Hyde Park, where we had followed the President, was disquieting: the Germans had launched a strong counter-attack and the situation was very uncertain."[17]

This was the reverse of what Churchill had so confidently forecast. There would be heavy casualties and loss of life, it seemed—American as well as British. "These things always seem to happen when I am with the President," the Prime Minister confided to Wilson, thinking of Tobruk the previous year—Sir Charles noting: "Poor Winston, he had been so anxious to convince Roosevelt that the invasion of Italy would yield a bountiful harvest at no great cost." Now that the first bill had come in, it was proving almost prohibitively expensive—both in human life and in the very vessels and logistical backup the U.S. chiefs wanted transferred to Britain for D-day. "When we left Hyde Park tonight, on the long journey to Halifax," Wilson recorded in his diary, "the situation was still very obscure."[18]

Churchill was embarrassed—and as his train bore him back to Canada, where he was to embark for Britain, the news from Italy only became more forbidding. Instead of seemingly effortless victory initially—the Italian fleet having sailed south from Taranto to join the Allies, pursued by German U-boats and Luftwaffe that sank many of them—Eisenhower's ground campaign turned sour. By September 16, Eisenhower was admitting to his naval aide that, if the Salerno battle ended in disaster," he himself would "probably be out."[19]

For his part, Churchill saw his once-glorious predictions for an Allied campaign in southern Europe exposed as wishful thinking. He'd earlier called upon his British chiefs of staff to be much bolder in their plans, and to "use all our strength against Italy," even without American help. He'd even recommended making plans for British assault landings as far north along the coast of Italy as possible, in order to "cut off" as many Germans as they could. Far from throwing their proverbial hats further over the fence, as Churchill had urged his military team, the Prime Minister was now faced with having to eat his own. Though from his train he cabled directly Eisenhower's field deputy, General Alexander, urging him to go ashore in person at Salerno and avoid another Dardanelles fiasco, it could not alter the bitter, bitter truth: namely that the Allied campaign in Italy, as planned, had been based upon a false premise: not only that the Italians would help, but that the Germans would fail to offer a serious defense of Italy south of the Po.

The next day, as Churchill's train bore him to Halifax, where HMS *Renown* would take him across the Atlantic, things sounded "no better,"

Sir Charles noted. "I have never seen him more on edge during a battle. Three 'bloodys' bespattered his conversation, and twice, while I was with him, he lost his temper with his servant, shouting at him in a painful way. He got up and walked down the train." Without information he seemed bereft. "'Has any news come in?' he kept demanding. In truth, "the reports that are reaching him only leave him more anxious," his doctor noted. "There is a dreadful hint, though it is carefully covered up, that we might be driven into the sea. It appears, as far as I can tell, that the P.M. is largely responsible for this operation; if anyone is to blame, he is the man; and, from the way things seem to be going, I suppose he is beginning to think that there might be a good deal to explain."[20]

Without the President to calm him, Churchill was metaphorically at sea — and soon was in reality, where he remained "immured in his cabin"[21] the whole voyage home, firing off telegrams to General Alexander to do more, and other wild cables, too, such as to General Maitland Wilson to accelerate a British seizure of the Dodecanese islands in the Aegean — without the agreement of the President — and be ready for potentially war-altering operations in the Balkans, where the Germans might, following the Italian surrender, be forced to withdraw to the Danube . . .

To Sir Charles Wilson this was all of a piece: the Prime Minister a bundle of nerves when things did not go in the way he had optimistically and impetuously planned.

At Hyde Park, however, the President neither blamed Winston nor worried unduly. He'd gotten to talk at length with young General Mark Clark during his stay in Casablanca, and was confident the U.S. troops — many of them in their first battle — would acquit themselves well. Moreover that General Eisenhower would recognize the gravity of the crisis and commit all he could to rectify the situation.

Neither Rome nor even Naples was the point, after all. Even if Clark failed to make much headway, the Italians had surrendered — unconditionally. All the Allies had to do, now, was secure the vital Foggia airfields, and bring the Germans to battle in Italy over the next months, until D-day was launched.

If Clark was forced to evacuate, after all, Allied troops could be sent to reinforce Montgomery's Eighth Army in Calabria. All would be well. Most importantly, the United States had demonstrated to the Soviets an absolute determination to fight on the mainland of Europe — first in southern Italy, then across the Channel. He and Churchill would show Stalin they meant business, and would follow through on their promises — moreover, that it would be best for the Russians to maintain civil discourse with the Western Allies in the fight to defeat the Third Reich.

The President thus slept a full ten hours after Churchill's train left the Hudson railway halt. He would spend only three days out of the next two weeks in Washington.

The fact was, he had bigger things on his mind than Salerno: his meeting with the Russian dictator, who in a flurry of new cables had finally agreed to a meeting of the Big Three — though not outside Russia. His tone had been, however, more "civil," as the principal private secretary to King George VI had noted in his diary at Buckingham Palace in London; "he re-iterates his wish to have a three-party meeting," Sir Alan Lascelles aptly put it, "but he won't go outside Russia, and I don't see how the President is to be got inside it."[22]

What had changed the Russian dictator's attitude?

There was much speculation — though few were sure. Certainly, in terms of public attention, the Allied conference at Quebec, coming on top of the summit at Casablanca, had monopolized the attention of the free world. Russia was losing the very respect it was looking for, internationally — Stalin conspicuous by his absence at such conferences, a fact that, in view of the many invitations to take part, began to suggest an ominous Russian agenda rather than genuine commitment to the anti-Axis cause and the Atlantic Charter/Declaration of the United Nations.

Above all, though, the war had moved into a new phase: the endgame. American, British, and Canadian troops were now on the mainland of Europe, only eight hundred miles from Berlin — while Russians were still fighting deep in the Soviet Union, more than a thousand miles from the German capital. As a result, Russian media calling for an immediate Second Front, instead, sounded silly — however strategically necessary a cross-Channel assault might be in terms of the military defeat of the Third Reich. The President and Mr. Churchill, in short, appeared to be in control of the moral and political dimensions of the war, even the military — leaving the Russians out on a limb, despite the almost obscene casualties they were suffering in their struggle to evict the Germans from their country.

In the new cables, Stalin still speciously claimed his presence was needed on a daily basis to control the battles raging on the Eastern Front ("where more than 500 divisions are engaged in the fighting in all"), but he now went out of his way not only to compliment the President on the "new brilliant success in Italy" but to acknowledge, for the first time, something even more significant. As Stalin put it, in his telegram to the President on September 11, "the successful landing at Naples and break between Italy and Germany will deal one more blow upon Hitlerite Germany and will considerably facilitate the actions of the Soviet armies at the Soviet-German front."[23]

This latter acknowledgment was, for Roosevelt, especially gratifying. Not only was the President relieved that his long, patient striving to convene a Big Three summit seemed about to pay off, but for the first time since Torch, Stalin had conceded that the President's strategy of landing in Sicily and then the mainland of Italy was having a major military impact on the war on Stalin's own Eastern Front. First at Kursk—where the Germans had called off the battle early—and now in the helter-skelter Russian advance in the Ukraine, where the Wehrmacht was being defeated in battle largely because Hitler simply had insufficient reserves to put into the line. The initiatives taken by the Western Allies had effectively spoiled any chance of the Wehrmacht defeating the Soviets that year.

"Everything turns on Italy at the moment," Goebbels had himself acknowledged on September 11, while staying with Hitler. Granted, "the enemy hasn't the faintest idea of the real situation in Northern and Central Italy. They are still imagining we'll pull back our divisions over the Brenner to the homeland, and they'll be able to unleash a huge aerial attack on Berlin from airfields in southern Italy." Clearly the Allies hadn't reckoned on the German genius for combat. The ruthless German occupation of Rome and other Italian cities was being greeted with applause in the Third Reich, evoking shades of 1940 and the German occupation of Paris: the German Volk expressing "rage against the Italians," who had nefariously betrayed them[24]—a people who would now be treated with the same remorseless cruelty that the Wehrmacht had shown their former Ribbentrop Pact partners when launching Barbarossa in 1941.

At the Wolf's Lair, the Reichsminister for Propaganda had even gotten Hitler to deliver the speech he'd desperately wanted the Führer to give, in order to bolster morale in Germany. It had been recorded at the OKW headquarters and relayed by radio in Berlin to the nation on September 10: a speech given in measured tones without the usual Hitler histrionics. Instead, it had soberly denounced those Italians who had failed their

Duce and who were now giving an example of cowardice and treachery that would go down in the annals of dishonor. Germany had consistently been compelled to bail out its ailing partner, in the Balkans and in North Africa — "the name of Field Marshal Rommel will forever be attached to this German effort" — but the Reich had now been "betrayed" by reactionary elements in Italy. "Italy's defection will have little military impact," the Führer claimed, "since the battle in that country has primarily been carried and conducted by German forces. We'll now be freed of all restrictions and constraints."[25]

Hitler's calm, measured tone would be balm to those in Germany wondering at the massive Allied air raids, the arrest of Mussolini, and then the unconditional surrender of the Italian government — his speech worth "seven divisions," as Goebbels put it. Hitler even made open mention of his secret weapons program. With Germany's enemies a thousand kilometers from the Reich, only their bombers could seek to "terrorize" the German population — and in that connection "there are," the Führer announced, "technical and organizational measures now being developed not simply to completely stop the terror bombing attacks, but to repay them with other, more effective measures" — his *Vergeltungswaffen,* or V-1 and V-2 weapons.[26]

The Führer's speech worked "like a refreshing thunderstorm," Goebbels noted — "one of the best," he reflected, "he has delivered in the whole war."[27]

Still and all, Goebbels acknowledged, the Allied invasion of the European continent was now a game changer. Though the Führer was confident the Wehrmacht could hold back the Allied armies south of Rome if they were fortunate, the divisions required for such a campaign would make it impossible to restrain the gathering Soviet tide on the Eastern Front: thus starving the German line of the reserves they desperately needed, especially on the line of the Dnieper, where little had been done to prepare solid defensive positions.

As Goebbels dictated in his diary, not only was the Allied invasion of Italy a lance in the German flank, but the situation in the East was "absolutely critical,"[28] with German troops pulling further and further back. "We see here what the unexampled betrayal of the Italians has caused us. If we'd had at hand the divisions we've had to send to Italy since the fall of Mussolini available to go into action on the Eastern Front, the current crisis would never have arisen. The superiority the Russians have over us is not that big — you can see that in the way we've slowed their advance."[29]

It was, for the Reich, a tragedy, he wrote. "We have about eight divisions in northern Italy and in southern Italy another eight, so about sixteen divisions, fitted out with first-class personnel and equipment. The Führer is convinced that with these sixteen divisions we'll be able to deal with the crisis in Italy," with a further fifty thousand troops in Sardinia and four thousand in Corsica who could be switched to Italy — battle terrain that would be "tabula rasa," with no concern about civilian casualties or destruction.[30] Yet the absence of those very divisions from the Eastern Front was now galling. "If we only had fifteen or twenty intact first-class divisions to put into battle, we'd be able to throw back the Soviets without any doubt whatsoever. But we're having to send those fifteen, twenty divisions south to the Italian theater," Goebbels wailed.[31]

For German leaders accustomed to seize whatever they wished from weakly defended European neighbors, the arrest of Mussolini, the defection of Italy as an ally, and the arrival of the Western Allies on the mainland of Europe now threatened, in other words, to stretch and bring down the whole Axis edifice — with the Allies possessing the upper hand.

If only the Western Allies would have launched a Second Front that year, Goebbels mused. A cross-Channel invasion by the Anglo-Americans would have given the Wehrmacht and Luftwaffe a real chance to defeat the Allies using the forces the Germans had in France, while retaining sufficient first-class divisions to deal with Stalin's forces on the Eastern Front. "But that would be too good to be true," Goebbels lamented.[32] Instead, the Western Allies had brought the war to the mainland of *Italy:* not only obtaining the unconditional surrender of its government at very little cost to themselves, but opening the floodgates, in the east, to Stalin's armies, against whom the Wehrmacht would now have insufficient reserves. And with the Führer too anxious about a fall invasion of France or even the Netherlands by the Allies to dare withdraw German divisions from the Atlantic Wall.

It would be up to valiant German troops both on the Eastern Front and the Southern Front, then, to show the Soviets and the Anglo-Americans the true mettle of the Wehrmacht. In Italy, Goebbels noted with a kind of sneering satisfaction, Allied troops would now face a ruthless German military machine unencumbered by Italians — and with more German divisions streaming down from the north, they would demonstrate their prowess in killing, without question or remorse. Italians who did not lay down their arms, or who sought to impede the German military occupation of Italy in any way, would simply be shot or slaughtered—as would civilians who aided partisans. Ruthlessness had gotten the Third Reich to its hitherto unimaginable string of imperial conquests—and would now be applied as mercilessly as in Russia. *Totaler Krieg.*

"The main purpose of my visit to the High Command Headquarters is fulfilled," Goebbels thus noted with satisfaction, on September 12. "I think Göring was right when he said to me that we have thereby won a battle. The Führer's speech will be worth whole divisions on the Eastern Front and in Italy. I spend a little time chatting with the Führer. He himself seems pleased he's gotten the speech out of the way. He wishes me all the best in my work and for my health . . . He promises to give another speech in the Sportpalast [in Berlin], to start the Winter Assistance Program. I'll make sure he'll get to taste once more just what it's like to be in touch with the Volk. Our farewell is very warm. I wish the Führer all the best."[33]

At 8:00 p.m. the Reich minister heard the latest news on the radio of "our operations in Italy, which are going very well," at Salerno, on top of the Führer's recorded speech—which the Russians had failed to jam. "A little more work, a little more talk. Then I fall into bed, dead tired. There'll be a mountain of work waiting for me in Berlin."[34]

53

A Message to Congress

GOEBBELS WAS, HOWEVER, misinformed about the Italian campaign. Though the Allied assault landings at Salerno came close to the very "brink of obliteration," as the American campaign historian Rick Atkinson recorded six decades later, the line held.[1]

It was touch-and-go, however. According to Rommel's son, Manfred, Hitler — buoyed by early reports of Wehrmacht victory at Salerno — "discussed with my father the possibility of launching a counter-offensive to retake southern Italy and possibly Sicily."[2] Ordered in panic by General Alexander to cease dawdling and save Clark, however, General Montgomery — who feared just such a Rommel riposte, as the Desert Fox had attempted at Medenine — finally renewed his advance. As Clark himself related, years later, "we had a hard time . . . Monty was sending me messages: 'Hang on, we're coming!' And I'd send back: 'Hurry up — I'm not proud, come and get me.' So it was really something."[3]

In truth it was Clark himself who saved Fifth Army, since his U.S. VI Corps commander, Major General Ernest Dawley — another protégé of General Marshall's — proved a broken reed. Clark himself went ashore to take personal command, in a magnificent display of courage and leadership in battle. He persuaded General Ridgway to drop airborne troops on the beaches, and with U.S. and Royal Navy vessels firing almost as many shells as at Iwo Jima and Okinawa, later,[4] Clark's men fought off the German counterassault.

Eisenhower had meantime warned the chiefs of staff of the possible need to evacuate the landings — a message passed on to the President and to Churchill — but the crisis eventually passed. Necessity — the need to fight harder or die at Salerno — had proven the ultimate mother of virtue.

Hitler was furious, as was Goebbels, but the President was relieved.

"You know from the news of the past few days," the President began his Message to Congress on September 17, 1943, as the Allied situation at Salerno seemed to stabilize, "that every military operation entails a legitimate military risk and that occasionally we have checks to our military plans — checks which necessarily involve severe losses of men and materials.

"The Allied forces are now engaged in a very hard battle south of Naples," he admitted. "Casualties are heavy. The desperation with which the Germans are fighting reveals that they are well aware of the consequences to them of our occupation of Italy. The Congress and the American people can rest assured that the landing on Italy is not the only landing we have in mind. That landing was planned at Casablanca," he claimed — bending the truth somewhat, since post-Husky operations had not actually been discussed in more than principle. Still and all, such planning had certainly taken place during the early summer. At Quebec, he explained, "the leaders and the military staffs of Great Britain and the United States made specific and precise plans to bring to bear further blows of equal or greater importance against Germany and Japan — with definite times and places for other landings on the continent of Europe and elsewhere."

Congress should be aware, then, that even though reverses lay ahead, the story of the Allied prosecution of the war was proceeding according to a genuine timetable and a larger, overall strategy — a strategy that was American, not British, but one calculated to succeed on behalf of the United Nations.

The President also pointed to the difference between Allied liberation and Nazi occupation — the "food, clothing, cattle, medicines, and household goods" systematically stolen by the Germans in "satellite and occupied Nations," whereas the Allies had a "carefully planned organization, trained and equipped to give physical care to the local population — food, clothing, medicine." He lauded the advance of the Allied armies from Sicily to the mainland of Italy on September 3 — stating: "History will always remember this day as the beginning of the answer to the prayer of the millions of liberty-loving human beings not only in these conquered lands but all over the world." However, "there is one thing I want to make perfectly clear. When Hitler and the Nazis go out, the Prussian military clique must go with them."

Unconditional surrender — without negotiation. "The war-breeding gangs of militarists must be rooted out of Germany — and Japan — if we are to have any real assurance of future peace," he asserted. Surrender negotiations with the Italian government, of necessity, had had to be conducted in secret, in order that the Nazis not be able to seize Marshal Badoglio or his associates in Rome, but he wanted Americans and Congress to know "that the policy which we follow is an expression of the basic democratic tradition and ideals of this Republic. We shall not be able to claim that we have gained total victory in this war if any vestige of Fascism in any of its malignant forms is permitted to survive anywhere in the world."

Bearing a banner of American democracy, the United States was, in other words, on the move—producing planes, tanks, and matériel on a scale that beggared description: fifty-two thousand airplanes, twenty-three thousand tanks, forty thousand artillery guns in the first six months of 1943 alone, he reported. American shipyards were launching "almost five ships a day."

The war had become "essentially a great war of production. The best way to avoid heavy casualty lists is to provide our troops with the best equipment possible—and plenty of it," the President asserted. Although the nation had come a long way since his State of the Union address just before the Casablanca Conference, he now cautioned that "we are still a long, long way from ultimate victory in any major theater of the war." It would entail "a hard and costly fight up through Italy—and a major job of organizing our positions before we can take advantage of them.

"Likewise," in the British Isles "we must be sure we have assembled the strength to strike not just in one direction but in many directions—by land and sea and in the air—with overwhelming forces and equipment." Moreover, to "break through" the Japanese defensive ring stretching from the "mandated islands to the Solomons and through the Netherlands East Indies to Malaysia and China" would be a challenge. "In all of history, there has never been a task so tremendous as that which we now face," he stated candidly. And warned: "Nothing we can do will be more costly in lives than to adopt the attitude that the war has been won—or nearly won. That would mean a letdown in the great tempo of production which we have reached, and would mean that our men who are now fighting all over the world will not have that overwhelming superiority of power which has dealt so much death and destruction to the enemy and at the same time has saved so many lives."[5]

"Overwhelming superiority of power"—directed at the right time and at the right place—to produce the necessary outcome: the unconditional surrender of the Third Reich and the Empire of Japan. Their total disarmament. And a "national cooperation with other Nations" in order that "world aggression be ended and that fair international relationships be established on a permanent basis"[6]: these were the military and political objectives the President was pursuing on behalf of the United States — on a global scale.

Aware there were those who resented him in his role as U.S. commander in chief as much as they had resented him as president in tackling the Great Depression and New Deal, Roosevelt dismissed the narrow-minded critics who, "when a doughnut is placed in front of them, claim they can only see the hole in it" — people who lacked "war-winning ideals." "Obviously," he added, "we could not have produced and shipped as much as we have, we could not now be in the position we now occupy in the Mediterranean, in Italy, or in the Southwest Pacific or on the Atlantic convoy routes or in the air over Germany and France, if conditions in Washington and throughout the Nation were as confused and chaotic as some people try to paint them" — paintings "eagerly sought by Axis propagandists in their evil work." For himself he remained proud of the "amazing" job that "the American people and their Government" were doing "in carrying out a vast program which two years ago was said to be impossible of fulfillment."[7]

Nothing, the President claimed, could now halt the Allies, whatever the Germans and Japanese threw at them.

54

Achieving Wonders

IN BERLIN, THE master of Axis propaganda read the text of the President's latest Message to Congress carefully.

"The American struggle isn't just against Nazism," Dr. Goebbels noted, "it's also against militarism. We know these words. The British and the Americans have always used them to try to carve the Reich into little pieces," he sneered. "More significant was what he says about American output. The numbers are way behind their needs; nevertheless," he confessed, "as far as airplane production goes, the U.S. has achieved wonders."

In proof of what he saw as his own analytic intelligence, however, the Mephistopheles of public relations and propaganda thought he could discern Roosevelt's deeper motive behind his Message to Congress — and the free world. "It's quite clear," Goebbels noted, "that the whole enemy press is being brought to bear to distract attention from Soviet successes on the Eastern Front — and make sure their own public isn't made uneasy."[1]

At a time of unease in Washington political circles over ultimate Soviet intentions, there was considerable truth in this. The President was certainly banging a proud American drum to remind the American public of the war's *global* dimensions — the manner in which control of the Mediterranean would release naval vessels for the Far East, closing the gap between Northwest Australia and Ceylon, thereby forcing "General Tojo and his murderous gang" to "look to the north, to the south, to the east, or to the west," where he would only see "closing in on them, from all directions, the forces of retribution under Generalissimo Chiang-Kai-shek, General MacArthur, Admiral Nimitz, and Admiral Lord Mountbatten,"[2] the new supreme commander of Allied forces in the Indian Ocean. But the President's target audience went beyond American or even English shores. Published not only in nearly every newspaper in America and abroad, the President's long congressional address was once again directed at Moscow.

Whether at Hyde Park, Quebec, Ottawa, or the White House, the President had taken great pains to demonstrate over recent weeks just how solid was the U.S.-British alliance. Now he wished to back that image, in writing, with quotable numbers: statistical proof of new, global American power that would not be content with Hitler's fall, but was to be harnessed to a postwar democratic agenda.

His conversation with Archbishop Spellman had reflected his unusually despondent mood; two weeks later, though, with American and British troops having established a hard-won lodgment on the mainland of Italy, and U.S. air forces already beginning to bomb factories in southern Germany, he seemed to have recovered his confidence: a confidence he would certainly need if he was to bring the American electorate, via his own efforts and the Congress, to ditch isolationism for good and take responsibility for the survival and development of a democratic postwar world.

Churchill's stay, in other words, had acted as a tonic, despite the crisis at Salerno — and the President was fired up.

Ranked seventeenth in the table of world military strengths in 1939, the United States was now primus inter pares, with an all-American military, economic, and political agenda, based on the clear goals of the four freedoms, that the President was determined to fulfill come hell or high water — with or without Soviet participation.

Exactly what would happen if the Soviets did *not* participate in an endgame agreement — indeed, what exactly such a political agreement should comprise — was still to be decided.

As Churchill had said to Mackenzie King before leaving Quebec to join the President in Washington, it was impossible to predict whether the "Germans will give up this autumn." It was, as Winston put it with characteristic wit, "like trying to bet on the Derby. No one could tell exactly what would happen. He spoke of the small numbers the British and Americans have in armies compared with the Germans" — and with the Russians.[3]

The President had been "very mad" at Stalin at the time, and had deliberately refrained from responding to the Russian's rude cables — ignoring him and working with Churchill to show the strength of the Western, Anglo-American coalition. Somehow, though, in the interests of postwar peace, agreement would nevertheless *have* to be obtained on the "post-war order," however powerful the Russian land armies. Compromise would be necessary, involving sad concessions — ones that would hopefully preserve, at least, the western nations of Europe within an Allied, democratic embrace: the "Allied side." While meantime encouraging the Soviets to join an international security system, not stand outside it.

Could such a system of postwar security be negotiated with the inscrutable Russians? Would it be effective? Would the American public even support security guarantees of foreign countries on another continent, in another hemisphere, at the risk of a further war? It was small wonder that, in relation to the "post-war order," Churchill had given to Mackenzie King a "desultory sort of account of the scheme that he, himself, had in mind, and what the President had talked of, but there was nothing very definite about it. Nothing is to be published at present as coming out of the [Quebec] Conference. Some months will be needed to consider the matter."[4]

With Stalin's sudden willingness to tackle "the matter" — first through a preliminary meeting of foreign ministers, then a summit of the Allied leaders, perhaps — planning for the future of the world was, however, becoming hot: red hot.

In the hopes therefore of obtaining formal Russian participation — participation in a four-power postwar security structure; participation in a global United Nations authority; and formal international agreements to be made on the future of Germany and the countries that had aided Hitler militarily, from Austria to Bulgaria and Rumania — the President thus authorized Secretary Hull to attend the preliminary Moscow Conference of foreign ministers of the Big Three, plus a representative from China.

The President had wanted Sumner Welles to attend, despite his recent resignation as undersecretary of state, but Hull was insistent that *he* should represent the United States as secretary of state, and the President — needing congressional support for the mission and its outcome — had acquiesced. The war was moving toward a climax, as even the Russians were aware. After almost a year of pressure to get together and get with a formally agreed Allied program, the Russian dictator had, it appeared, finally seen the light. The foreign ministers' conference would begin in only four weeks' time — on October 11, 1943 — and might last as long as a month.

There in Moscow, the President hoped, the secretary would pave the way for the leaders of the Four Policemen to sit down together and discuss postwar security—and how to avoid the fate of the League of Nations. The President, Churchill, Stalin, and Chiang Kai-shek would also have to decide the war's endgame: how, exactly, Nazi Germany was finally to be defeated and its military disarmed. Following which, the Empire of Japan.

With that vast challenge looming, the President wheeled himself from the Oval Office to the mansion, stopping by the Map Room to check on messages from London, Moscow, and Chungking. Where the national leaders would meet, when exactly, how he would travel—by air or sea—and how they would get along, Roosevelt had little idea, but he was suddenly supremely confident.

"Of course he knew better than anyone else what was good for the United States," Lieutenant George Elsey remembered the spirit the President conveyed. "That was the attitude at that point. He was supreme in every respect!"—the Map Room off-limits to all but five people in the world, not only to safeguard the most sensitive and secret military information, including Ultra, but because it enabled the President to be the only person with a complete picture of the war's progress—and perils. "'I'm in control; this is the way it's going to be—it's going to be the way I want it'—this was the sense I had of his perception of himself as the war went on," Elsey recalled.[5]

Franklin Roosevelt had every reason to feel supreme. As president he had not only brought America out of the Great Depression without resorting to the kind of tyranny that had been occasioned in Germany and elsewhere, but he had subsequently become—in the least dictatorial yet most dominating manner—a most successful U.S. commander in chief in war. A global "war for civilization," as he rightly called it.

The President's generals and admirals had "no reason to challenge or contradict his leadership,"[6] Elsey pointed out, since in setting the ongoing strategy of the war—at times against their dissenting voices—he had so ably brought the United States now within sight of eventual victory.

Many great battles still lay ahead, as well as further disagreements with Churchill and the British over military operations and policy. Churchill's obsession with war in the Mediterranean would continue, despite disastrous expeditions in the Greek islands that would drive his own generals as mad as it did the American military in the next weeks[7]—Churchill resisting to the bitter end the British commitment to the mounting of Overlord the following spring.[8]

For all his faults as quasi–commander in chief of British Empire forces, however, Winston Churchill's loyalty to the President, as the de facto commander in chief of the forces of the Western Allies, had never snapped; nor had Churchill's acumen in terms of Stalin and the Russians, and his moral courage. This would be of inestimable value in the coming months.

There would be dire problems of agreement with the Soviet Union, the President was all too aware — Russians who would have no gratitude to the United States for having helped save them, nor genuine interest in the Four Freedoms in a postwar world. There was also the matter of a fourth presidential election, and the President's always-precarious health. Moreover, what exactly should be done with Nazi Germany in the aftermath of victory — disarmament or dismemberment. How best to help the Chinese, and plan the defeat of Japan. And how best to then turn Japan from aggression to peaceful coexistence . . .

These were but some of the politico-military challenges remaining, as the President began planning his second trip to North Africa later that fall — hopefully there to meet with Chiang Kai-shek, Churchill, and Stalin.

The road from Torch had certainly been rocky, over the past year, but what a year of achievement it had been!

His secretary of war and Joint Chiefs of Staff were now finally on the same page — *his* page. So was Churchill — if he could be kept there. From faltering first offensive combat in Tunisia, the United States had in less than one year moved to the brink of what would become the greatest global military performance in its history: a massive American-led invasion and campaign in 1944 that would hopefully win the Second World War in Europe. And after that, Japan.

With that, the President left the Map Room and went up to bed.

Acknowledgments

Readers of *The Mantle of Command: FDR at War, 1941–1942* will know how indebted I am to those who have helped me in recounting, afresh, one of the most important stories of the twentieth century.

My task was to challenge the widely held perception of Franklin Delano Roosevelt as a hands-off U.S. commander in chief in World War II, and to demonstrate, rather, just how important was his role in directing U.S. and Allied strategy, even though he did not live to record it. Now, with publication of *Commander in Chief: FDR's Battle with Churchill, 1943,* I am in debt again to a number of people. First off, I'd like to thank my commissioning editor, Bruce Nichols, without whose guidance, editing, and support this book could not have been published. And his assistant, Ben Hyman; my copyeditor, Melissa Dobson; and the indefatigable manuscript supremo, Larry Cooper—indeed the whole team at Houghton Mifflin Harcourt. At a time when my longtime London publisher had rejected the manuscript of *The Mantle of Command* and pulled out of the project completely, claiming there was insufficient interest in Franklin Delano Roosevelt in Britain or in the British publishing territories (Australia, New Zealand, South Africa, India, the West Indies, etc.) for a multivolume work on FDR as commander in chief in World War II, Bruce's encouragement meant the world to me—and something, I think, to the many readers who have enjoyed *The Mantle of Command.* I hope this sequel will again repay Bruce's faith in my project.

Second, I'd like to thank Dr. Hans Renders, Professor of Biography at Groningen University in the Netherlands, who has not only been a stalwart supporter and colleague of my work in biographical studies, but who encouraged me to include *Commander in Chief* as part of my submission for a doctorate at the Biografie Instituut, Research Faculty of Arts, Groningen University. For his and Professor Dr. Doeko Bosscher's suggestions, corrections, and support I am deeply grateful.

Writing history and historical biography is a process of research and constant iteration—factual, interpretive, selective, and architectural—before a book is finished and goes to press. Many fellow historians have assisted me; in particular I'd like to thank Carlo D'Este, Roger Cirillo, James Scott, Mark Schneider, David Kaiser, Douglas Brinkley, David Reynolds, and Ron Spector. I'd like to acknowledge my debt as a historian also to Gerhard Weinberg, Rick Atkinson, Mark Stoler, Warren Kimball, Michael Schaller, H. W. Brands, Andrew Roberts, David Woolner, Evan Thomas, Douglas Porch, Michael Howard, and the late Martin Gilbert, Stephen Ambrose, and Forrest Pogue, for their many works on World War II and FDR's role. Over the years as a military and presidential historian I have learned not only from them but from many hundreds of veterans, from generals to GIs, as well as other students of World War II, ranging from professors to archivists in the United States, Britain, and Germany. Without those years of grounding, going back to the decade I spent as official biographer of Field Marshal Bernard Montgomery, I could not have undertaken such an ambitious project, and I will always be grateful to them.

At the FDR Presidential Library I'd like to thank the Deputy Director, Bob Clark, as well as the staff of the Research Room and Photographic Records. At the Warm Springs Presidential Museum I'd like to thank the Manager, Robin Glass, and his staff. At the Eisenhower Presidential Library in Abilene I'd also like to thank the Deputy Director, Timothy Reeves, and the Research Library staff. In New Orleans, where I winter, I'd like to thank Nick Mueller, the President and CEO of the National World War II Museum, as well as his Director of Research, Keith Huxen, and Conference Director, Jeremy Collins. At the Churchill Society of New Orleans I wish to thank the President, Gregg Collins, and his colleagues. In Washington, D.C., I'd like to thank especially Jeff Flannery, Head of Reference in the Manuscript Division at the Library of Congress, and his staff, as well as the staff of the National Archives in College Park, Maryland. Also John Greco and his colleagues at the Operational Archives of the U.S. Naval History and Command, Washington Navy Yard.

At the Imperial War Museum in London I'd like to thank particularly the Keeper of Documents, Anthony Richards. At the Churchill Centre in Chicago I'd like to thank Lee Pollock and David Freeman. At the Churchill Archives Centre in Cambridge, England, the Director, Allen Packwood.

Closer to home, at the University of Massachusetts Boston I'd like to thank Ira Jackson, the former Dean of the McCormack Graduate School of Policy and Global Studies, where I am Senior Fellow; the new Dean, David Cash; and my many colleagues at the university, especially Provost Winston Langley. My thanks also to the staffs of the Widener Library, Harvard University, and its Microfilm Department, and the staff of Boston College Library.

In Boston I'm indebted, also, to my fellow members of the Boston Biographers Group, whose monthly meetings have offered the kind of support that only fellow biographers can, in the end, offer: collegial understanding, sympathy, advice, and reassurance in our necessarily often lonely biographical profession. I'd like also to acknowledge the friendship and intellectual fraternity of my fellow members of the Tavern Club—as well as my fellow members of Biographers International Organization (BIO), whose annual conference is both an inspiration to biographers and a chance to share a common passion for the study of real lives.

My literary agent, Ike Williams, has been once again my stalwart champion, together with his colleagues Katherine Flynn and Hope Denekamp. To my brother Michael and to my children, my thanks; and to my wife, Raynel, much more than thanks can ever repay.

As in the writing of *The Mantle of Command,* I have kept a portrait of my father, Lieutenant Colonel Sir Denis Hamilton, DSO, above my desk during the writing of *Commander in Chief*—for it is the memory of his service as a fighting infantryman, first at Dunkirk and then at D-day and the grim Battle of Normandy, that cautions me never to forget the men who had to carry out, in combat, the strategies laid down by the "brass hats" in World War II—and pay the price of their decisions, for good or ill.

Finally, to those readers of *The Mantle of Command* who wrote me with corrections as well as expressions of gratitude, my great appreciation.

NIGEL HAMILTON

John W. McCormack Graduate School of Policy and Global Studies, UMass Boston

Photo Credits

Total War. FDR addresses Congress, Jan. 7, 1943: FDR Library; boarding the *Dixie Clipper* at U.S. naval base, Trinidad, Jan. 12, 1943: National Archives

En Route to Casablanca. Stopover at Bathhurst, Gambia: National Archives; aboard the USS *Memphis* with Captain John McCrea: National Archives; FDR aboard C-54 with pilot, Captain Otis F. Bryan, Jan. 11, 1943: FDR Library; President's C-54 and ramp, North Africa, Jan. 1943: FDR Library

Casablanca. The President's villa, Dar es Saada: FDR Library; FDR dines with sons Elliott and Franklin Jr., and Harry Hopkins, Jan. 16, 1943: FDR Library; with Joint Chiefs of Staff, Jan. 20, 1943: FDR Library

Directing World Strategy. At the President's villa, hosting Winston Churchill and the Combined Chiefs of Staff, Jan. 1943: FDR Library; FDR with Admiral Ernest King: National Archives; with Generals Henri Giraud and Charles de Gaulle: National Archives

Visiting Troops on the Battlefield. FDR with General Dwight Eisenhower (seen flying together in December 1943): National Archives; with Generals Mark Clark and George Patton, Jan. 21, 1943: FDR Library; reviewing a U.S. armored division, Jan. 21, 1943: FDR Library; reviewing U.S. troops, Jan. 21, 1943: FDR Library

Unconditional Surrender. FDR and Churchill with Combined Chiefs, Jan. 1943: FDR Library; with Churchill at press conference, Jan. 24, 1943: FDR Library; press conference, Jan. 24, 1943: FDR Library

End of Empires. FDR with son Elliott, Jan. 1943: FDR Library; dining with the Sultan of Morocco, Jan. 22, 1943: National Archives; with Churchill at top of tower of Villa Taylor, U.S. vice consulate, Marrakesh, Jan. 24, 1943: FDR Library; with President Edwin Barclay of Liberia, Jan. 27, 1943: FDR Library

Totaler Krieg. Goebbels at the *Sportpalast:* Bundesarchiv, Federal Archives of Germany; FDR reviews military training camp: National Archives; Admiral Yamamoto saluting Japanese plane, ca. 1942: Corbis; group of U.S. Army Air Forces P-38 Lightning fighter planes, June 1, 1943: Associated Press

Churchill on the Wrong Warpath. The RMS *Queen Mary* in New York harbor, June 20, 1945: Interim Archives / Getty Images; Churchill addresses Congress, May 19, 1943: Corbis

Axis Surrender in North Africa. Roundup of German and Italian soldiers in Tunisia, June 11, 1943: Associated Press; soldiers in a prison camp: The Print Collector / Getty Images

Reading Churchill the Riot Act. FDR fetching Churchill from train, Washington, May 11, 1943: Bettmann / Corbis; fishing with Churchill at Shangri-la: FDR Library; U.S. Chiefs of Staff (Generals Arnold and Marshall, Captain Royal [deputy secretary], Admirals Leahy and King) at a Combined Chiefs Conference, facing British counterparts, 1943: Three Lions / Hulton Archive / Getty Images

Sicily — and Kursk. Allied landing in Sicily, July 10, 1943: IWM © Sgt. Frederick Wackett / Getty Images; retreat from Kursk: Bundesarchiv, Federal Archives of Germany

The Fall of Mussolini. Mussolini greets Hitler upon arrival in Italy, July 19, 1943: ullstein bild / Getty Images; on their way to Feltre, Italy: ullstein bild / Getty Images; before the meeting at Feltre: ullstein bild / Getty Images

Churchill Returns — Yet Again. Churchill with daughter Mary, Aug. 16, 1943: Toronto Star Archives / Getty Images; FDR with Churchill, Canadian Prime Minister McKenzie King, and Combined Chiefs in Quebec, Aug. 18, 1943: FDR Library

The First Crack in the Axis. FDR addresses crowd outside Canadian parliament, Ottawa, Aug. 25, 1943: FDR Library; Stalin with Foreign Minister Molotov (seen later, Feb. 1, 1945): Keystone / Getty Images

The Reckoning. FDR drives Churchill at Hyde Park, Sept. 12, 1943: Associated Press; Allied invasion of Italy, Salerno, Sept. 9, 1943: SeM / UIG / Getty Images

Notes

PROLOGUE

1. Lieutenant Commander George Elsey, interview with author, September 12, 2011.
2. See this volume, chapter 32, 257.
3. Entry of Tuesday, November 10, 1943, in Alan Lascelles, *King's Counsellor: Abdication and War: The Diaries of Sir Alan Lascelles*, ed. Duff Hart-Davis (London: Weidenfeld & Nicolson, 2006), 75.
4. Ibid., entry of Thursday, May 13, 1943, 129.
5. David Reynolds, *In Command of History: Churchill Fighting and Writing the Second World War* (New York: Random House, 2005).

1. A CRAZY IDEA

1. "Exclusive of the President's own car, the train comprised one compartment car, one Pullman sleeper, one combination club-baggage car, and the special Army radio car": "Log of the Trip of the President to the Casablanca Conference, 9–31 January, 1943," Papers of George M. Elsey, Franklin D. Roosevelt Presidential Library, Hyde Park, NY.
2. Entry of January 9, 1943, Leahy Diary, William D. Leahy Papers, Library of Congress.
3. "Because of his infirmity he [the President] could walk only briefly on two canes. It was much easier for him to use one cane and the right arm of an escort" for balance and support while swinging the fourteen-pound steel irons encasing his legs, from his thighs to his shoes, McCrea explained: John McCrea, "Handwritten Memoirs/Recollections," McCrea Papers, FDR Library.
4. Ibid.

2. ABOARD THE MAGIC CARPET

1. "Marriage of Hopkins to Louise Macy," miscellaneous files for "Roosevelt & Hopkins," Robert E. Sherwood Papers, Houghton Library, Harvard University.

2. Robert Sherwood, *Roosevelt and Hopkins: An Intimate History* (New York: Harper, 1948), 669.
3. Entry of January 8, 1943, in Geoffrey C. Ward, ed., *Closest Companion: The Unknown Story of the Intimate Friendship Between Franklin Roosevelt and Margaret Suckley* (Boston: Houghton Mifflin, 1995), 194.
4. Ibid.
5. Entry of January 9, 1943, in Ward, *Closest Companion,* 194.
6. Ibid., entry of January 10, 1943, 196.
7. Sherwood, *Roosevelt and Hopkins,* 669.
8. Ibid.
9. McCrea, "Handwritten Memoirs/Recollections."
10. Letter of January 11, 1943, in Ward, *Closest Companion,* 196.
11. Ibid.
12. Ibid.
13. Ibid., letter of January 12, 1943, 197.
14. Ibid., letter of January 13, 1943.
15. Related to John McCrea by the President: McCrea, "Handwritten Memoirs/Recollections."
16. Letter of January 13, 1943, in Ward, *Closest Companion,* 197.
17. Ibid.
18. Letter to Eleanor Roosevelt, January 13, 1943, in Elliott Roosevelt, ed., *F.D.R.: His Personal Letters, 1928–1945* (New York: Duell, Sloane, and Pearce, 1950), vol. 2, 1393.

3. THE UNITED NATIONS

1. Nigel Hamilton, *The Mantle of Command: FDR at War, 1941–1942* (Boston: Houghton Mifflin Harcourt, 2014), 138 et seq.
2. Ibid.
3. Christopher D. O'Sullivan, *Sumner Welles, Postwar Planning, and the Quest for a New World Order, 1937–1943* (New York: Columbia University Press, 2008), 65 and 72.
4. Sumner Welles, *Seven Decisions That Shaped History* (New York: Harper, 1951), 182–83.
5. Christopher Thorne, *Allies of a Kind: The United States, Britain, and the War Against Japan, 1941–1945* (London: Hamish Hamilton, 1978), 221. Christopher O'Sullivan argued that such foot-dragging was deliberate. "The imperial powers faced growing demands for independence" among colonial and mandated countries — exposing "a paradox at the heart of empires: progress in the political and economic sphere would encourage self-rule, whereas a lack of progress justified continued European rule": *FDR and the End of Empire* (New York: Palgrave Macmillan, 2012), 4.
6. O'Sullivan, *Sumner Welles,* 67.

7. Ibid., 74, quoting Gladwyn Jebb.
8. It was not only the British Empire that was having to face the prospect of dismantlement; Queen Wilhelmina of the Netherlands gave a broadcast on the first anniversary of Pearl Harbor, December 7, 1942, promising that the Dutch colonies in Southeast Asia would be given home rule — see Thorne, *Allies of a Kind*, 218.
9. David Reynolds, *In Command of History: Churchill Fighting and Writing the Second World War* (New York: Random House, 2005), 334.
10. Letter of November 24, 1942, in Elliott Roosevelt, ed., *F.D.R.: His Personal Letters, 1928–1945* (New York: Duell, Sloane, and Pearce, 1950), vol. 2, 1371–72.
11. Ibid.
12. Entry of December 4, 1942, Diaries of William Lyon Mackenzie King, Library and Archives Canada, Ottawa, ON (hereinafter Mackenzie King Diary). The Democrats lost the election in the House of Representatives by over a million ballots in the popular vote. To Roosevelt's relief they nevertheless retained a majority of 222 seats to 209. In the Senate, Democrats lost 8 seats and 1 independent — as well as the popular vote, in ballot numbers cast. Again, however, they held on to their majority, 58 seats to 37.
13. Ibid.
14. Ibid.
15. Susan Butler, ed., *My Dear Mr. Stalin: The Complete Correspondence Between Franklin D. Roosevelt and Joseph V. Stalin* (New Haven, CT: Yale University Press, 2005), 98.
16. Ibid., 99.
17. Ibid., 99–100.
18. Ibid., 101.
19. Entry of December 5, 1942, Mackenzie King Diary.
20. Ibid.
21. Ibid.
22. Ibid.
23. Ibid.
24. Ibid.
25. Roosevelt's speechwriter and later editor of his presidential papers, Sam Rosenman, later pointed out that "it was not at Appomattox but at Fort Donelson that Grant demanded unconditional surrender; it was not of Robert E. Lee but of S. B. Buckner — in 1862": Samuel I. Rosenman, *Working with Roosevelt* (New York: Harper, 1952), 372.
26. Entry of December 5, 1942, Mackenzie King Diary.
27. In June 1941, prior to the war, Roosevelt had set up an Office of Scientific Research and Development to direct science for military purposes. The OSRD was soon tasked with atomic research. The notion that an atomic weapon would have to be launched by ship, given the necessary volume, quickly gave way to the idea of a small, bomb-size weapon that could be flown and dropped on a target. A so-

called S-1 Committee was therefore set up to report directly to the President at the White House, chaired by the president of Harvard University, James Conant. Robert Oppenheimer was made director of fast-neutron research, and by the summer of 1942 the need for substantially more fissionable material was reported to FDR. To accelerate development, General Brehon Somervell, the U.S. Army's senior officer for logistics, appointed Lieutenant Colonel Leslie Groves to take charge of the Manhattan Project — which was moved to Los Alamos, in New Mexico. The 60 tons of Canadian uranium ore, already ordered in March 1942, was judged insufficient. After consultation with the U.S. government, Mackenzie King's Canadian government nationalized the Eldorado Mining and Refining company in June 1942, and another 350 tons of uranium was immediately ordered by the U.S. government — followed by another 500 tons later that year, and 1,200 tons of stored Congolese ore to be refined by the Canadians.

28. Entry of December 5, 1942, Mackenzie King Diary.
29. Ibid.
30. Ibid.
31. Ibid.
32. Churchill referred to the plan as "airy visions of Utopia and El Dorado" (Martin Gilbert, *Winston S. Churchill,* vol. 7, *Road to Victory: 1941–1945* [London: Heinemann, 1986], 292), while Harold Laski, professor of political science at the London School of Economics, wrote to President Roosevelt, hoping he would "teach our Prime Minister that it is the hope of the future and not the achievement of the past from which he must draw his inspiration" (Thorne, *Allies of a Kind,* 144).
33. Entry of Saturday, January 9, 1943, "Secret Diary" of Lord Halifax, Papers of Lord Halifax, Hickleton Papers, Borthwick Institute of Historical Research, University of York, Yorkshire, England.
34. Entry of December 5, 1942, Mackenzie King Diary.
35. Ibid.
36. Ibid.
37. Ibid.

4. WHAT NEXT?

1. Ibid., entry of December 6, 1942.
2. Ibid., entry of December 4, 1942.
3. Ibid., entry of December 5, 1942.
4. Ibid.
5. Ibid.
6. David Kaiser, *No End Save Victory: How FDR Led the Nation into War* (New York: Basic Books, 2014).
7. Entry of December 5, 1942, Mackenzie King Diary.
8. Ibid., entry of December 4, 1942.

9. Ibid., entry of December 6, 1942.
10. Ibid.
11. Ibid.

5. STALIN'S *NYET*

1. Cable of December 6, 1942, in Butler, *My Dear Mr. Stalin,* 102.
2. Ibid., cable of December 8, 1942, 103.
3. Ibid., cable of December 14, 1942, 103.
4. Ibid.
5. Ibid., 103–4.

6. ADDRESSING CONGRESS

1. Rosenman, *Working with Roosevelt,* 366–68.
2. Entry of November 29, 1942, in William D. Hassett, *Off the Record with F.D.R., 1942–1945* (New Brunswick, NJ: Rutgers University Press, 1958), 145.
3. "The Spirit of This Nation Is Strong"—Address to the Congress on the State of the Union, January 7, 1943, in Franklin D. Roosevelt, *The Public Papers and Addresses of Franklin D. Roosevelt,* comp. Samuel I. Rosenman, 1943 vol., *The Tide Turns* (New York: Russell and Russell, 1969), 21–34.

7. A FOOL'S PARADISE

1. Entry of January 7, 1943, Halifax Diary.
2. Entry of November 20, 1942, Stimson Diary, Henry L. Stimson Papers, Yale University Library, New Haven, CT.
3. Ibid, entry of December 11, 1942.
4. Ibid., entry of December 12, 1942.
5. Ibid.
6. Ibid.
7. Henry Stimson and McGeorge Bundy, *On Active Service In Peace and War* (New York: Harper & Brothers, 1947).
8. Entry of December 14, 1942, Stimson Diary.
9. Ibid.
10. Nigel Hamilton, *The Full Monty,* vol. 1, *Montgomery of Alamein, 1887–1942* (London: Allen Lane, 2001), 467–68.
11. Because of its five fortress-like side walls, President Roosevelt referred to the Pentagon as the "Pentateuchal Building," after the first five books of the Bible: Steve Vogel, *The Pentagon: A History* (New York: Random House, 2007), 297. The U.S. Navy was offered space to ensure a combined-services headquarters; "He was very much pleased," Stimson had noted the President's satisfaction with the offer

in his diary, "and told us to go ahead": Ibid., 281. The Navy bureaus declined to integrate their activities, however—as they did racial integration in the seagoing Navy, other than small numbers of black sailors as messmen or glorified bellhops. Even onshore, the Navy insisted their installations be strictly segregated, with no black officers—causing the National Urban League's journal *Opportunity* to declare Japan was not wrong in claiming "the so-called Four Freedoms in the great 'Atlantic Charter' were for white men only": Morris J. MacGregor, *Integration of the Armed Forces, 1940–1965* (Washington, D.C.: Government Printing Office, 1985). The Pentagon saga would epitomize the President's difficulties as commander in chief in a democracy—the U.S. Navy refusing, in fact, to move in with the Army and U.S. Air Force until 1948, under protest, while the Marine Corps held out another four decades, until 1996.

12. Entry of January 7, 1943, Stimson Diary.
13. Entry of December 28, 1942, Leahy Diary, William D. Leahy Papers, Library of Congress.

8. FACING THE JOINT CHIEFS OF STAFF

1. "Joint Chiefs of Staff Minutes of a Meeting at the White House," Washington, January 7, 1943, *Foreign Relations of the United States: The Conferences at Washington, 1941–1942, and Casablanca, 1943* (hereinafter *FRUS I*) (Washington, D.C.: Government Printing Office, 1968), 511.
2. Ibid., 509.
3. Ibid.
4. Ibid.
5. Ibid., 509–10.
6. Ibid., 510.
7. Ibid.
8. Ibid.
9. Ibid.
10. Ibid.
11. Ibid.
12. Ibid.
13. Ibid.
14. E.g., Albert Wedemeyer, *Wedemeyer Reports!* (New York: Henry Holt, 1958), 95, and John McLaughlin, *General Albert C. Wedemeyer: America's Unsung Strategist in World War II* (Philadelphia: Casemate, 2012), 31, referring to "the utter failure of 'Unconditional Surrender.'" For the view that it was a sop to the Soviets, see Frank Costigliola, *Roosevelt's Lost Alliances: How Personal Politics Helped Start the Cold War* (Princeton, NJ: Princeton University Press, 2012), 179–81. The classic condemnatory statement on unconditional surrender was, however, by Hanson Baldwin: "Unconditional surrender was an open invitation to unconditional

resistance: it discouraged opposition to Hitler, probably lengthened the war, cost us lives, and helped to lead to the present abortive peace": Hanson Baldwin, *Great Mistakes of the War* (New York: Harper, 1950), 13.

15. Wedemeyer, *Wedemeyer Reports!*, 186–87.
16. In the House of Commons on July 21, 1949, Labor Minister Ernest Bevin claimed the British War Cabinet had not been consulted over "unconditional surrender," prompting Churchill to claim he "had never heard the phrase until the President suddenly uttered it at the Casablanca press conference": David Reynolds, *In Command of History: Churchill Fighting and Writing the Second World War* (New York: Random House, 2005), 323.
17. Ibid., 506.

9. THE HOUSE OF HAPPINESS

1. John McCrea, "Handwritten Memoirs/Recollections," McCrea Papers, FDR Library.
2. Letter of January 14, 1943, in Geoffrey C. Ward, ed., *Closest Companion: The Unknown Story of the Intimate Friendship Between Franklin Roosevelt and Margaret Suckley* (Boston: Houghton Mifflin, 1995), 198.
3. McCrea, "Handwritten Memoirs/Recollections."
4. Letter of January 14, 1943, in Ward, *Closest Companion*, 198.
5. Letter of December 17, 1942, Elsey Papers, FDR Library.
6. S. E. Morison, "Memorandum For the President," December 18, 1942, Elsey Papers, FDR Library.
7. David Stafford, *Roosevelt and Churchill: Men of Secrets* (Woodstock, NY: Overlook, 1999), 197–98.
8. Letter of January 14, 1943, in Ward, *Closest Companion*, 198.
9. Elliott Roosevelt, *As He Saw It* (New York: Duell, Sloan, and Pearce, 1946), 65.
10. Letter of January 14, 1943, in Ward, *Closest Companion*, 198.
11. Ian Jacob, unpublished Casablanca account, Churchill College Archives, Cambridge, UK.
12. Elliott Roosevelt, *As He Saw It*, 62.
13. Ibid., 67.
14. Ibid., 66.
15. Ibid., 66.
16. McCrea, "Handwritten Memoirs/Recollections," McCrea Papers, FDR Library.
17. Robert Sherwood, *Roosevelt and Hopkins: An Intimate History* (New York: Harper, 1948), 673.
18. "A Gleam of Victory: A Speech at the Lord Mayor's Luncheon at the Mansion House, November 10, 1942," in *The War Speeches of the Rt. Hon. Winston S. Churchill*, ed. Charles Eade (London: Cassell, 1952), vol. 2, 342–45.
19. Entry of 10.11.1942, Joseph Goebbels, *Die Tagebücher von Joseph Goebbels* [The

diaries of Joseph Goebbels], ed. Elke Froehlich (Munich: K. G. Saur, 1995), Teil II, Band 6, 273 (hereinafter *Die Tagebücher 6*). Quotes from this source have been translated by the author.

20. Ibid., entry of 10.11.1942, 265.
21. Ibid., entry of 11.11.1942, 273.
22. Ibid., entry of 15.11.1942, 294.
23. Quoted in Winston S. Churchill, *The Second World War,* vol. 4, *The Hinge of Fate* (London: Cassell & Company, 1951), 583.

10. HOT WATER

1. Martin Gilbert, *Winston S. Churchill,* vol. 7, *Road to Victory: 1941–1945* (London: Heinemann, 1986), 269.
2. Ian Jacob, unpublished Casablanca account, Churchill College Archives, Cambridge, UK.
3. Ibid. The ship, which had been the headquarters ship for the landing at Algiers, had "a complete set of wireless instruments." It could be "placed in Casablanca harbor, & our cipher staff could live aboard and all our telegram traffic with London could thus be handled without the necessity for any elaborate machinery ashore. All that was necessary was a constant carrier service between ship & hotel, and a Defense Registry organization in the latter" — "exactly as" if they were operating out of the War Rooms in London. The *Bulolo* sailed for Casablanca on January 5, arriving on January 10, 1943. Jacob, unpublished Casablanca account.
4. Ibid. See also Brian Lavery, *Churchill Goes to War: Winston's Wartime Journeys* (London: Conway, 2007), 160–64.
5. Jacob, unpublished Casablanca account.
6. Gilbert, *Road to Victory: 1941–1945,* 294.
7. Jacob, unpublished Casablanca account.
8. Ibid.
9. Ibid.
10. Ibid.

11. A WONDERFUL PICTURE

1. McCrea, "Handwritten Memoirs/Recollections," McCrea Papers, FDR Library.
2. Jacob, unpublished Casablanca account.
3. McCrea, "Handwritten Memoirs/Recollections."
4. Sherwood, *Roosevelt and Hopkins,* 673.
5. Elliott Roosevelt, *As He Saw It,* 66.
6. Entry of January 14, 1943, John W. Huston, *American Airpower Comes of Age: General Henry H. 'Hap' Arnold's World War II Diaries* (Maxwell AFB, AL: Air University Press, 2002), vol. 1, 464.

7. Entry of January 14, 1943, Arthur Bryant, *The Turn of the Tide: A History of the War Years, Based on the Diaries of Field Marshal Lord Alanbrooke, Chief of the Imperial General Staff* (New York: Doubleday, 1957), 446.
8. Elliott Roosevelt, *As He Saw It,* 67.
9. Entry of January 14, 1943, Huston, ed., *American Airpower Comes of Age,* 464.
10. Ibid.
11. Entry of January 14, 1943, Bryant, *The Turn of the Tide,* 446.
12. Elliott Roosevelt, *As He Saw It,* 71.
13. Ibid., 73.
14. Ibid., 74.
15. Ibid., 76.
16. Ibid..
17. Ibid., 77.

12. IN THE PRESIDENT'S BOUDOIR

1. Albert Wedemeyer, *Wedemeyer Reports!* (New York: Henry Holt, 1958), 192.
2. Andrew Roberts, *Masters and Commanders: How Four Titans Won the War in the West, 1941–1945* (New York: Harper, 2009), 337.
3. Wedemeyer, *Wedemeyer Reports!*, 337, quoting Brigadier General J. E. Hull of the War Department's Operations Division.
4. Wedemeyer, *Wedemeyer Reports!*, 191–92.
5. Joint Strategic Survey Committee, January 8, 1943, Map Room Files, FDR Library.
6. "Meeting of Roosevelt with the Joint Chiefs of Staff, January 15, 1943, 10 a.m., President's Villa," *Foreign Relations of the United States: The Conferences at Washington, 1941–1942, and Casablanca, 1943* (hereinafter *FRUS I*) (Washington, D.C.: Government Printing Office, 1968), 559.
7. Ibid.
8. Ibid.
9. Ibid.
10. McCrea, "Handwritten Memoirs/Recollections."
11. Harry C. Butcher, *My Three Years with Eisenhower: The Personal Diary of Captain Harry C. Butcher, USNR, Naval Aide to General Eisenhower, 1942 to 1945* (New York: Simon and Schuster, 1946), 237.
12. Entry of January 15, 1943, Brooke, *War Diaries,* 351. Also "Meeting of the Combined Chiefs of Staff, January 15, 1943, 2:30 pm., Anfa Camp," *FRUS I,* 567.
13. Entry of December 28, 1942, Brooke, *War Diaries,* 351.
14. Jacob, Casablanca Diary, Churchill Archives.
15. Butcher, *My Three Years with Eisenhower,* 243.
16. Entry of January 14, 1943, Martin Blumenson, ed, *The Patton Papers, 1940–1945* (Boston: Houghton Mifflin, 1974), 154.
17. "Meeting of the Combined Chiefs of Staff, January 15, 1943, 2:30 pm., Anfa Camp," *FRUS I,* 568–69.

18. Elliott Roosevelt, *As He Saw It*, 79.
19. Ibid.
20. Ibid.
21. "Meeting of the Combined Chiefs of Staff with Roosevelt and Churchill, January 15, 1943, 5:30 p.m., President's Villa," *FRUS I*, 573.
22. Entry of January 15, 1943, Brooke, *War Diaries*, 359.
23. Elliott Roosevelt, *As He Saw It*, 83.
24. Ibid.
25. Ibid., 84.

13. STIMSON IS AGHAST

1. Entry of January 19, 1943, Stimson Diary, Henry L. Stimson Papers, Yale University Library, New Haven, CT.
2. Ibid., entry of January 21, 1943.
3. "Meeting of Roosevelt with the Joint Chiefs of Staff, January 16, 1943," *Foreign Relations of the United States: The Conferences at Washington, 1941–1942, and Casablanca, 1943* (hereinafter *FRUS I*) (Washington, D.C.: Government Printing Office, 1968), 594.
4. Entry of 16 January 1943 and annotation, *War Diaries, 1939–1945: Field Marshal Lord Alanbrooke, War Diaries, 1939–1945*, ed. Alex Danchev and Daniel Todman (Berkeley: University of California Press, 2001), 360.
5. "Meeting of the Combined Chiefs of Staff, January 16, 1943," *FRUS I*, 591.
6. Brooke: "We should definitely count on reentering the Continent on a large scale" — "Meeting of the Combined Chiefs of Staff, January 16, 1943," *FRUS I*, 591.
7. "Meeting of Roosevelt with the Joint Chiefs of Staff, January 16, 1943," *FRUS I*, 597.
8. Ibid.
9. Martin Gilbert, *Road to Victory: Winston S. Churchill, 1941–1945* (London: Heinemann, 1986), 299–300.
10. See, for example, Mark Stoler and Melanie Gustafson, "Creating a Global Allied Strategy," in their *Major Problems in the History of World War II: Documents and Essays* (Boston: Houghton Mifflin, 2003), 74–108.
11. John McCrea, "Handwritten Memoirs/Recollections," McCrea Papers, FDR Library.
12. Mark Clark, *Calculated Risk* (London: Harrap, 1951), 148–49.
13. Letter of Thursday, January 21, 1943, to Daisy Suckley, in Ward, ed., *Closest Companion: The Unknown Story of the Intimate Friendship Between Franklin Roosevelt and Margaret Suckley* (Boston: Houghton Mifflin, 1995), 199.
14. Unpublished autobiography, chapters 12 through 22, Private Papers of Brigadier G.M.O. Davy, PP/MCR/143, courtesy of Documents Department, Imperial War Museum, London.
15. Elliott Roosevelt, *As He Saw It* (New York: Duell, Sloan, and Pearce, 1946), 106.
16. Ibid., 106–7. General Mark Clark, later recalling the episode, confessed that by

the time the President had asked for the mess kit, it had already been washed and mixed with others. Clark had demanded "any mess kit" from the kitchen staff, "And make it fast." The President had been delighted, and had said, "I'll have them put it in the Smithsonian Institution": Mark Clark, *Calculated Risk*, 149.

17. Harold Macmillan, *War Diaries: Politics and War in the Mediterranean, January 1943–May 1945* (London: Macmillan, 1984), 8.

14. DE GAULLE

1. McCrea, "Handwritten Memoirs/Recollections," McCrea Papers, FDR Library.
2. Ibid.
3. Letter to Daisy Suckley, January 20, 1943, in Ward, *Closest Companion*, 199.
4. Ibid., January 21, 1943.
5. Kenneth Pendar, *Adventure in Diplomacy* (New York: Dodd, Mead & Co, 1945), 161.
6. Robert Murphy, *Diplomat Among Warriors* (New York: Doubleday, 1964), 168.
7. Ibid.
8. McCrea, "Handwritten Memoirs/Recollections."
9. Friday, January 22, 1943, "The President's Log, January 14–25, 1943," *FRUS I*, 531.
10. Elliott Roosevelt, *As He Saw It*, 111.
11. Murphy, *Diplomat Among Warriors*, 173.
12. Elliott Roosevelt, *As He Saw It*, 112.
13. Pendar, *Adventure in Diplomacy*, 145.
14. McCrea, "Handwritten Memoirs/Recollections."
15. Elliott Roosevelt, *As He Saw It*, 112.
16. Entry of January 14, 1943, *The Patton Diaries II*, ed. Martin Blumenson (Boston: Houghton Mifflin, 1974), 154.
17. Ibid., entry of January 17, 155.
18. Ibid., entry of January 19, 157.
19. Ibid., entry of January 21, 158–59.
20. Ibid., 158.
21. Ibid.
22. Ibid., entry of January 22, 158.
23. Elliott Roosevelt, *As He Saw It*, 112.

15. AN ACERBIC INTERVIEW

1. Robert Sherwood, *Roosevelt and Hopkins: An Intimate History* (New York: Harper, 1948), 678–79.
2. McCrea, "Handwritten Memoirs/Recollections."
3. McCrea Notes: "Roosevelt–De Gaulle Conversation, January 22, 1943," *FRUS I*, 694.
4. Ibid., 695.

5. Immediately following the Allied Torch invasion, Hitler had ordered Operation Anton, the German and Italian occupation of all Vichy-administered France and Corsica.
6. "Roosevelt De-Gaulle Conversation, January 22, 1943," *FRUS I,* 696.
7. Charles de Gaulle, *The Complete War Memoirs of Charles de Gaulle,* vol. 2, *Unity* (New York: Simon and Schuster, 1964), 388–89.
8. Henri Giraud, *Un seul but, la victoire: Alger, 1942–1944* (Paris: R. Julliard, 1949).
9. De Gaulle, *The Complete War Memoirs,* vol. 2, 392–93.
10. Ibid., 384.
11. Elliott Roosevelt, *As He Saw It,* 113.
12. Ibid., 113–14.
13. De Gaulle, *The Complete War Memoirs,* vol. 2, 384.
14. Elliott Roosevelt, *As He Saw It,* 114.
15. Ibid., 114–15.
16. Ibid., 115.
17. Ibid.
18. Ibid., 115–16.
19. Ibid., 121.
20. Ibid., 122.
21. François Kersaudy, *Churchill and De Gaulle* (London: Collins, 1981), 252.
22. Ibid., 253.
23. Ibid., 254.
24. Ibid., 255.
25. Sherwood, *Roosevelt & Hopkins,* 693.
26. Kersaudy, *Churchill and De Gaulle,* 255.

16. THE UNCONDITIONAL SURRENDER MEETING

1. "Historic Meeting Informal in Tone: Reporters Sit on Garden Grass at Leaders' Feet to Hear of Momentous Talks," *New York Times,* January 27, 1943.
2. Ibid.
3. De Gaulle, *The Complete War Memoirs,* vol. 2, 399.
4. Giraud, *Un seul but,* 96.
5. "Historic Meeting Informal in Tone," *New York Times.*
6. Ibid.
7. "875th Press Conference. Joint Conference by the President and Prime Minister Churchill at Casablanca, January 24, 1943," in Franklin D. Roosevelt, *The Public Papers of Franklin D. Roosevelt,* comp. Samuel Rosenman, 1943 vol., *The Tide Turns* (New York: Russell and Russell, 1969), 37–44.
8. McCrea, "Handwritten Memoirs/Recollections."
9. Ibid.
10. *Roosevelt Presidential Press Conferences,* No. 875, 90–91.
11. Entry of Wednesday, January 27, 1943, in Alan Lascelles, *King's Counsellor:*

Abdication and War: The Diaries of Sir Alan Lascelles, ed. Duff Hart-Davis (London: Weidenfeld & Nicolson, 2006), 93.

12. Entry of 27.1.1943, Joseph Goebbels, *Die Tagebücher von Joseph Goebbels* [The diaries of Joseph Goebbels], ed. Elke Froehlich (Munich: K. G. Saur, 1993), Teil II, Band 7 (hereinafter *Die Tagebücher 7*), 203. Quotes from this source have been translated by the author.
13. Ibid., entry of 26.1.1943, 197.
14. Ibid., entry of 27.2.1943, 203.
15. Ibid., entry of 28.1.1943, 208.
16. Ibid., entry of 26.11943, 197.
17. Ibid., entry of 28.1.1943, 209.
18. Max Domarus, ed., *Hitler, Speeches and Proclamations 1932–1945: The Chronicle of a Dictatorship,* vol. 4, *The Years 1941 to 1945* (Wauconda, IL: Bolchazy-Carducci, 1997), 2671–85.
19. Entry of 28.1.1943, Goebbels, *Die Tagebücher 7,* 209.
20. Albert Speer, *Inside the Third Reich* (New York: Macmillan, 1970), 258.
21. Ian Kershaw, *Hitler, 1936–1945: Nemesis* (London: Allen Lane, 2000), 552.
22. Entry of 29.1.1943, Goebbels, *Die Tagebücher 7,* 216.
23. Domarus, ed., *Hitler, Speeches and Proclamations 1932–1945,* 2749.
24. David Irving, *Goebbels: Mastermind of the Third Reich* (London: Focal Point, 1996), 421.
25. "*Nun, Volk steh auf, und Sturm brich los! Rede im Berliner Sportpalast,*" *Der steile Aufstieg* (Munich: Zentralverlag der NSDAP, 1944), 167–204. Translated by the author.
26. Irving, *Goebbels,* 422–23.
27. Ralf Georg Reuth, *Goebbels* (Munich: Piper Verlag, 1990), 563.

17. KASSERINE

1. Ian Kershaw, *Hitler, 1936–1945: Nemesis* (London: Allen Lane, 2000), 550.
2. Ibid., 548.
3. Entry of January 19, 1943, Robert Ferrell, ed., *The Eisenhower Diaries* (New York: Norton, 1981), 86.
4. Ibid.
5. Rick Atkinson, *An Army at Dawn* (New York: Henry Holt, 2002), 308.
6. Ibid., 317.
7. Ibid., 322.
8. Entry of 18.2.1943, Joseph Goebbels, *Die Tagebücher von Joseph Goebbels* [The diaries of Joseph Goebbels], ed. Elke Froehlich (Munich: K. G. Saur, 1993), Teil II, Band 7 (hereinafter *Die Tagebücher 7*), 366.
9. Ibid., entry of 19.2.1943, 370.
10. Ibid., entry of 21.2.1943, 389.

18. ARCH-ADMIRALS AND ARCH-GENERALS

1. Atkinson, *An Army at Dawn,* 390–91.
2. Nigel Hamilton, *Master of the Battlefield: Monty's War Years, 1942–1944* (New York: McGraw Hill, 1984), 142.
3. Letter of February 23, 1943, in Martin Blumenson, ed, *The Patton Papers II, 1940–1945* (Boston: Houghton Mifflin, 1974), 175.
4. Entry of February 15, 1943, Stimson Diary, Henry L. Stimson Papers, Yale University Library, New Haven, CT.
5. Ibid., entry of February 17, 1943.
6. Entry of 20.2.1943, *Die Tagebücher 7,* 377.
7. Entry of February 18, 1943, Stimson Diary.
8. Ibid.
9. Entry of 20.2.1943, *Die Tagebücher 7,* 377.
10. Ibid., entry of 23.2.1943, 398–99.
11. Ibid., 398.
12. Blumenson, *The Patton Papers,* vol. 2, 183.
13. "'Memorandum' to Admiral Leahy and General Marshall, Copy to the Secretary of the Navy," November 17, 1942, King Papers, Naval Historical Archives.
14. Letter of November 19, 1943, in Alfred Chandler, ed., *The Papers of Dwight David Eisenhower,* vol. 2, *The War Years* (Baltimore: Johns Hopkins, 1970), 964–65.
15. Ibid., 965.

19. BETWEEN TWO FORCES OF EVIL

1. Report to the President, January 10, 1943, Bullitt File, Safe and Confidential Files, FDR Library.
2. Ibid.
3. Henri Giraud, *Un seul but la Victoire: Alger 1942–1944* (Paris: R. Julliard, 1949), 93–94.
4. Report to the President, January 10, 1943, Bullitt File, Safe and Confidential Files, FDR Library. See also Michael Cassella-Blackburn, *The Donkey, The Carrot, and the Club: William C. Bullitt and Soviet-American Relations, 1917–1948* (Westport, CT: Praeger, 2004), 213–14.
5. Report to the President, January 10, 1943, Bullitt File, Safe and Confidential Files, FDR Library.
6. Ibid.
7. C-259-A/1, From the Prime Minister to the President, February 2, 1943, in Warren Kimball, ed., *Churchill & Roosevelt: The Complete Correspondence,* vol. 2, *Alliance Forged* (Princeton, NJ: Princeton University Press, 1984), 129–30.
8. S. E. Morison, *The Two-Ocean War* (Boston: Little Brown, 1963), 272.
9. Entry of January 8, 1943, in "Secret Diary" of Lord Halifax, Papers of Lord Halifax,

Hickleton Papers, Borthwick Institute of Historical Research, University of York, Yorkshire, England.

10. Ibid., entry of January 11, 1943.
11. Ibid.
12. Ibid., entry of January 18, 1943.
13. Ibid., entry of January 26, 1943.
14. Ibid., entry of January 28, 1943.
15. Ibid., entry of February 2, 1943.
16. Ibid., entry of February 15, 1943.

20. HEALTH ISSUES

1. Elie Abel and Averell Harriman, *Special Envoy to Churchill and Stalin, 1941–1946* (New York: Random House, 1975), 183.
2. R. H. Ferrell, *The Dying President* (Columbia: University of Missouri Press, 1998), 10.
3. Entry of December 4, 1943, Diaries of William Lyon Mackenzie King, Library and Archives Canada, Ottawa, ON (hereinafter Mackenzie King Diary).
4. Entry of February 3, 1943, Stimson Diary.
5. Entry of February 1, 1943, Stimson Diary.
6. Diary entry of February 7, 1943, in Geoffrey Ward, ed., *Closest Companion: The Unknown Story of the Intimate Friendship Between Franklin Roosevelt and Margaret Suckley* (Boston: Houghton Mifflin, 1995), 201.
7. James A. Farley, *Jim Farley's Story* (New York: Whittlesey House, 1948), 108–9.
8. Jean Edward Smith, *FDR* (New York: Random House, 2007), 442, quoting Orville Bullitt, ed., *For the President, Personal and Secret: The Correspondence Between Franklin D. Roosevelt and William C. Bullitt* (Boston: Houghton Mifflin, 1988), 398. Also Will Brownell and Richard Billings, *So Close to Greatness: A Biography of William C. Bullitt* (New York: Macmillan, 1987).
9. Diary entry of February 7, 1943, in Ward, *Closest Companion,* 201.
10. Ibid.
11. Ross McIntire, *White House Physician* (New York: Putnam's Sons, 1946), 159.
12. Ferrell, *The Dying President,* 29.
13. Diary entry of February 14, 1943. Ward, *Closest Companion,* 201.
14. Ibid.
15. R-262/1, letter of March 17, 1943, in Kimball, ed., *Churchill & Roosevelt,* vol. 2, 156.
16. Diary entry of February 27, 1943. Ward, *Closest Companion,* 203.
17. Entry of April 8, 1943 (relating March 26, 1943, visit to White House), Butcher Diary, A-292–3, Eisenhower Library; also Harry Butcher, *My Three Years with Eisenhower,* 278–79.
18. Ibid., A-293.
19. Ibid., A-294.

20. Michael Burleigh, *The Third Reich: A New History* (New York: Hill & Wang, 2000), 740.

21. INSPECTION TOUR TWO

1. Entry of April 16, 1943 (regarding Fort Benning), in Ward, *Closest Companion,* 210.
2. Ibid., entry of April 24, 1943, 219.
3. Ibid., entry of April 16, 1943, 211.
4. Ibid., entry of April 17, 1943, 211.
5. Ibid., entry of April 18, 1943, 214.
6. Ibid., entry of April 24, 1943, 219.
7. Ibid., entry of April 25, 1943, 220–21.
8. Ibid., entry of April 19, 1943, 214.
9. Entry of February 24, 1943, Stimson Diary.
10. Ibid., entry of March 30, 1943.

22. GET YAMAMOTO!

1. George Kenney, *George C. Kenney Reports* (New York: Duell, Sloan, and Pearce, 1949), 52–53.
2. Entry of April 8, 1943 (relating March 26 visit to War office), Butcher Diary, A-287, Eisenhower Library.
3. Samuel Morison, *The Two Ocean War: A Short History of the United States Navy in the Second World War* (Boston: Little, Brown, 1963), 272–73.
4. Kenney, *George C. Kenney Reports,* 215.
5. See Christopher Andrew, *For the President's Eyes Only: Secret Intelligence and the American Presidency from Washington to Bush* (New York: HarperCollins, 1995), 123–24.
6. William Rigdon, *White House Sailor* (Garden City, NY: Doubleday, 1962), 19. For an account of how Ultra/Magic messages were relayed to the President, see David Stafford, *Roosevelt and Churchill* (Woodstock, NY: Overlook, 1999), 118–19; Andrew, *For the President's Eyes Only*, 103–11; and David Kahn, "Roosevelt, Magic, and Ultra," *Cryptologia* 16, no. 4 (October 1992).
7. Signal NTF131755, in Japanese Naval Cipher JN-25D, decoded by the U.S. Fleet Radio Unit Pacific in Hawaii: Ronald Lewin, *The American Magic: Codes, Ciphers and the Defeat of Japan* (New York: Farrar Straus Giroux, 1982), 182–83.
8. John Prados, *Combined Fleet Decoded: The Secret History of American Intelligence and the Japanese Navy in World War II* (New York: Random House, 1995), 453–58; Edward J. Drea, *MacArthur's ULTRA: Codebreaking and the War Against Japan, 1942–1945* (Lawrence: University Press of Kansas, 1992), 73.
9. Donald A. Davis, *Lightning Strike* (New York: St. Martin's, 2005), 220.

10. Ibid., 222.
11. Ibid., 227.
12. Rear Admiral Matome Ugaki, chief of staff to Admiral Yamamoto, in Burke Davis, *Get Yamamoto* (New York: Random House, 1969), 207.
13. Lewin, *The American Magic,* 185.
14. Burke Davis, *Get Yamamoto,* 128; Carroll V. Glines, *Attack on Yamamoto* (London: Orion Books, 1990), 9; Thomas Lanphier, "I Shot Down Yamamoto," *Reader's Digest,* December 1966, 48.
15. According to subsequent Japanese accounts, U.S. code breakers may have misinterpreted Admiral Yamamoto's flight schedule, which had the airfield of Buin, not Ballale, as its destination. "In the end it didn't matter," given the proximity of the two: Prados, *Combined Fleet Decoded,* 462.
16. Donald Davis, *Lightning Strike,* 273. Admiral Yamamoto was widely thought to have said, before Pearl Harbor, that he would take the surrender of America riding down Pennsylvania Avenue on a white charger; in truth he had pointed out that the United States would never surrender to Japan unless Japanese forces reached Washington, D.C., and the White House — which, having earlier studied at Harvard and having twice served as naval attaché in Washington, Yamamoto thought unlikely to eventuate.

23. "HE'S DEAD?"

1. Donald Davis, *Lightning Strike,* 304–8.
2. The airfield at Ballale was constructed in November 1942 by the Japanese, using forced labor of British artillery officers and men who had surrendered at Singapore. All 517 men were murdered by the Japanese on completion of the air base, in March 1943. See Don Wall, *Kill the Prisoners!* (Cambridge, UK: Peter Moore, 1996). Also Australian War Memorial Archives.
3. Burke Davis, *Get Yamamoto,* 196.
4. Ibid.
5. Donald Davis, *Lightning Strike,* 306–8.
6. Ibid., 309.
7. Ibid., 289–90. See also Andrew, *For the President's Eyes Only,* 138, and Walter Borneman, *The Admirals: Nimitz, Halsey, Leahy, and King — the Five-Star Admirals Who Won the War at Sea* (New York: Little, Brown, 2012), 315.
8. Burke Davis, *Get Yamamoto,* 210.
9. Presidential Press Conference No. 891, April 19, 1943, FDR Library.
10. In the first three weeks of March, 1943, more than three-quarters of a million tons of Allied shipping were still being sunk in the North Atlantic gap "not yet covered by air search," War Secretary Stimson complained to the President: Henry Stimson and McGeorge Bundy, *On Active Service in Peace and War* (New York: Harper and Brothers, 1948). For the best summary of the interservice controversy

see Samuel Eliot Morison, *The Battle of the Atlantic: September 1939–May 1943* (Boston: Little, Brown, 1947), 237–47.

11. Presidential Press Conference No. 898, May 21, 1943, FDR Library.
12. Grace Tully Archive, Franklin D. Roosevelt Papers, Box 11, Yamamoto (joke letter to), May 23, 1943, FDR Library.
13. Donald Davis, *Lightning Strike*, 315.
14. John C. Fredriksen, *The United States Air Force: A Chronology* (Santa Barbara, CA: ABC-CLIO, 2011), 104.

24. SAGA OF THE NIBELUNGS

1. Rick Atkinson, *An Army at Dawn* (New York: Henry Holt, 2002), 489–90.
2. Ibid., 484.

25. A SCENE FROM *THE ARABIAN NIGHTS*

1. Kenneth Pendar, *Adventure in Diplomacy* (New York: Dodd, Mead & Co, 1945), 43.
2. Ibid., 147.
3. Ibid., 148.
4. Ibid., 149.
5. Ibid., 152.
6. Ibid., 150.

26. THE GOD NEPTUNE

1. Entry of May 2, 1943, Leahy Diary, William D. Leahy Papers, Library of Congress.
2. Entry of May 7, 1943, Stimson Diary, Henry L. Stimson Papers, Yale University Library, New Haven, CT.
3. Ibid., entry of May 10, 1943.

27. A BATTLE ROYAL

1. Entry of May 11, 1943, Diaries of William Lyon Mackenzie King, Library and Archives Canada, Ottawa, ON.
2. Entry of May 9, 1943, Leahy Diary, William D. Leahy Papers, Library of Congress.
3. Ibid., entry of May 9, 1943.
4. Ibid., entry of May 11, 1943.
5. Entry of May 11, 1943 in Alan Brooke, *Field Marshal Lord Alanbrooke, War Diaries, 1939–1945*, ed. Alex Danchev and Daniel Todman (Berkeley: University of California Press, 2001), 402.
6. Ibid.

7. Entry of May 12, 1943, Stimson Diary.
8. The figure of 150,000 Axis troops who had already surrendered by May 12 was announced to the press the next morning, May 13, 1943, by the secretary of war: see entry for May 13, 1943, Stimson Diary.
9. "Meeting of the Combined Chiefs of Staff, May 13, 1943, 10:30 A.M.," in *Foreign Relations of the United States: The Conferences at Washington and Quebec, 1943* (hereinafter *FRUS II*) (Washington, D.C.: Government Printing Office, 1970), 24–25.
10. Entry of May 12, 1943, Brooke, *War Diaries*, 402.
11. Entry of May 13, 1943, Leahy Diary.
12. Ibid.
13. Ibid.
14. Lord Ismay, *The Memoirs of General Lord Ismay* (London: Heinemann, 1960), 296.

28. NO MAJOR OPERATIONS UNTIL 1945 OR 1946

1. Letter of May 13, 1943, in Mary Soames, *Speaking For Themselves: The Personal Letters of Winston and Clementine Churchill, Edited by Their Daughter* (New York: Doubleday, 1998), 479–80.
2. "A Global Strategy: Memorandum by the United States Chiefs of Staff," in *FRUS II*, 222–23.
3. "Conduct of the War in 1943–44, Memorandum by the British Chiefs of Staff," in *FRUS II*, 223–27.
4. "Meeting of the Combined Chiefs of Staff, May 13, 1943, 10:30 A.M.", *FRUS II*, 39–40.
5. Ibid., 41.
6. Ibid., 43.
7. Ibid., 44.
8. Ibid.
9. Ibid.
10. Ibid.
11. C-294, Churchill cable to Roosevelt, May 10, 1943, in Warren Kimball, ed., *Churchill & Roosevelt: The Complete Correspondence*, vol. 2, *Alliance Forged* (Princeton, NJ: Princeton University Press), 212.
12. Richard Breitman and Allan J. Lichtman, *FDR and the Jews* (Cambridge, MA: Harvard University Press, 2013), 206–10.
13. Robert E. Sherwood, *Roosevelt and Hopkins: An Intimate History* (New York: Harper, 1948), 728.
14. Winston S. Churchill, *The Second World War*, vol. 4, *The Hinge of Fate* (London: Cassel & Company, 1951), 713.
15. Ibid.
16. Ibid.

17. Lord Moran, *Winston Churchill: The Struggle for Survival, 1940–1965* (Boston: Houghton Mifflin, 1966), 95.
18. Ibid., 95–96.
19. Ibid., 96.
20. Ibid.
21. Ibid.
22. Entry of May 17, 1943, Stimson Diary.
23. Ibid.
24. Ibid.
25. Letter of May 12 (on "Office of the Secretary, Department of the Navy" notepaper), Bullitt Files, FDR Library.

29. THE DAVIES MISSION

1. "This is a situation full of ugly possibilities, and engendering it is a triumph for Goebbels": entry of Monday, April 26, 1943, in Alan Lascelles, *King's Counsellor: Abdication and War: The Diaries of Sir Alan Lascelles,* ed. Duff Hart-Davis (London: Weidenfeld & Nicolson, 2006), 126.
2. Walter Reich, former director of the Holocaust Museum in Washington, "Remember the Women," *New York Times Book Review,* April 12, 2015, 23.
3. Roosevelt to Stalin, Document 88, March 16, 1943, in Susan Butler, ed., *My Dear Mr. Stalin: The Complete Correspondence of Franklin D. Roosevelt and Joseph V. Stalin* (New Haven, CT: Yale University Press, 2005), 121.
4. Elizabeth MacLean, *Joseph E. Davies: Envoy to the Soviets* (Westport, CT: Praeger, 1992), 100.
5. "Meeting of the Combined Chiefs of Staff, May 18, 1943, 10:30 A.M.," *Foreign Relations of the United States: The Conferences at Washington and Quebec, 1943* (hereinafter *FRUS II*) (Washington, D.C.: Government Printing Office, 1970), 101.
6. Sir Alan Brooke, Proceedings of the Conference, "Defeat of the Axis Powers in Europe: discussion, Combined Chiefs of Staff," *FRUS II,* 101.
7. Entry of May 19, 1943, Stimson Diary.
8. Ibid.
9. Ibid.

30. A DOZEN DIEPPES IN A DAY

1. Entry of May 18, 1943, Diaries of William Lyon Mackenzie King, Library and Archives Canada, Ottawa, ON (hereinafter Mackenzie King Diary).
2. Ibid.
3. Ibid.
4. Ibid., "Conversation with Hon. L. McCarthy, at Canadian Legation after luncheon, Washington."

5. Ibid., "Conversation Mr. Mackenzie King had with Mr. Winston Churchill, Tuesday, May 18, 1943 — White House, Washington, 6.00 p.m."
6. Ibid., "Meeting of the Joint Chiefs of Staffs, May 20, 1943."
7. Ibid., "Conversation Mr. Mackenzie King had with Mr. Winston Churchill, Tuesday, May 18, 1943 — White House, Washington, 6.00 p.m."
8. Ibid.
9. Ibid., "Memorandum re questions asked Mr. Churchill by members of the Senate of the U.S. and representatives of the Foreign Committee and answers given by Mr Churchill, Washington, May 19, 1943."
10. Ibid., entry of May 19, 1943, "Quotations and answers, members of Senate of the U.S. — 19.v.43."
11. Ibid.
12. Meeting of the Combined Chiefs of Staff, May 19, 1943, 10:30 A.M., in *FRUS II,* 113.
13. Ibid., 114.
14. Entry of May 19, 1943, in Arthur Bryant, *The Turn of the Tide: A History of the War Years, Based on the Diaries of Field Marshal Lord Alanbrooke, Chief of the Imperial General Staff* (New York: Doubleday, 1957), 509.
15. Meeting of the Combined Chiefs of Staff with Roosevelt and Churchill, May 19, 1943, 6 P.M.," in *FRUS II,* 122–23.

31. THE FUTURE OF THE WORLD AT STAKE

1. Entry of May 19, 1943, Mackenzie King Diary.
2. Ibid.
3. Ibid., "Meeting of the Joint Staffs — May 20, 1943."
4. Prime Minister's Personal Telegram, 21 February 1943, in Martin Gilbert, *Road to Victory: Winston S. Churchill, 1941–1945* (London: Heinemann, 1986), 343.
5. E.g., entry of Wednesday, May 26, 1943, Mackenzie King Diary.
6. Entry of May 20, 1943, in "Secret Diary" of Lord Halifax, Papers of Lord Halifax, Hickleton Papers, Borthwick Institute of Historical Research, University of York, Yorkshire, England.
7. "Meeting of the Joint Staffs, May 20, 1943," Mackenzie King Diary, Library and Archives Canada.
8. Under the Canadian constitution, command of Canada's all-volunteer forces to serve overseas (conscription was confined to service in Canada only) was vested in the British monarch, and exercised by the Canadian federal Cabinet, who deferred largely to the authority of Winston Churchill in his role as minister of defense and prime minister of Great Britain.
9. "Meeting of the Joint Staffs, May 20, 1943," Mackenzie King Diary.
10. Ibid.
11. Ibid., entry of Friday, May 21, 1943.

12. Ibid., "Conversation with Mr. Churchill, White House — May 21, 1943."
13. Ibid.
14. Ibid.
15. Ibid.

32. THE PRESIDENT LOSES PATIENCE

1. Entry of May 24, 1943, in Bryant, *The Turn of the Tide,* 513.
2. Ibid.
3. Entry of May 24, 1943, Leahy Diary, William D. Leahy Papers, Library of Congress.
4. Ibid.
5. Entry of May 25, 1943 in Lord Moran, *Winston Churchill: The Struggle for Survival, 1940–1965* (Boston: Houghton Mifflin, 1966), 97.
6. Entry of May 24, 1943 Moran, *Winston Churchill,* 97–98.
7. Entry of May 25, 1943 (400c), Mackenzie King Diary.
8. Entry of May 28, 1943 Moran, *Winston Churchill,* 99.
9. Entry of 25 May, 1943, Bryant, *The Turn of the Tide,* 514.
10. Entry of May 27, 1943, Stimson Diary.
11. Ibid., entry of May 25, 1943.
12. Ibid.
13. "He thought the time might be in August . . . He then said: if, by any chance, something should prevent Stalin making the trip, what I would like to do is to come to Ottawa just the same though perhaps this might be in July" — "Conversation with Pres. Roosevelt, White House — May 21, 1943," Mackenzie King Diary.
14. Entry of May 25, 1943, Leahy Diary.
15. Entry of May 24, 1943, Bryant, *The Turn of the Tide,* 512–13.
16. Entry of Tuesday, January 26, 1943, in Lascelles, *King's Counsellor,* 93.
17. Entry of May 31, 1943, Diary of Harry C. Butcher, Eisenhower Presidential Library, Abilene.
18. Ibid.
19. Entry of 26 May, 1943, Bryant, *The Turn of the Tide,* 517.
20. Kershaw, *Hitler 1936–1945,* 606.

33. SICILY — AND KURSK

1. Entry of July 9, 1943, in Geoffrey C. Ward, ed., *Closest Companion, The Unknown Story of the Intimate Friendship Between Franklin Roosevelt and Margaret Suckley* (Boston: Houghton Mifflin, 1995), 225.
2. Entry of July 9, 1943, Leahy Diary, William D. Leahy Papers, Library of Congress.

3. Entry of July 9, in Ward, *Closest Companion,* 226.
4. Ben Macintyre, *Operation Mincemeat: The True Story That Changed the Course of World War II* (New York: Random House, 2010), 294.

34. THE FÜHRER FLIES TO ITALY

1. Erich von Manstein, *Lost Victories* (London: Methuen, 1958), 448.
2. Ibid., 449. Manstein's view has been much contested, especially by Russian military historians anxious to honor the Soviet defense of Kursk and the start of a major counteroffensive by Russian forces at Orel: see, inter alia, Chris Bellamy, *Absolute War: Soviet Russia in the Second World War* (New York: Palgrave, 2007), 586–87. However, it is clear from Joseph Goebbels's private conversations with Hitler at Berchtesgaden before the battle that Hitler was far more worried by the next moves of the Western Allies in the Mediterranean than by what would happen at Kursk — essentially a "show" offensive to write down Soviet armies using the latest German firepower. "The Führer has decided to stay where we are," on the Eastern Front, Goebbels recorded. "We have to keep our reserves up our sleeves. His old plan of seizing the Caucasus and fighting in the Middle East is redundant, thanks to last winter's crisis . . . Under no circumstances is he prepared to give up the Italian mainland — he has no intention of pulling back to the Po, even if the Italians abandon the front. We will simply take over the running of the war in Italy. That is the overriding principle of German strategy: to keep the war as far from the German homeland as possible": entry of 25.6.43, in Goebbels, *Die Tagebücher von Joseph Goebbels,* ed. Elke Froehlich (Munich:K. G. Saur, 1993), Teil II, Band 8, 531–34.
3. Entry of July 14, 1943, Ward, *Closest Companion,* 226.
4. Ibid., entry of July 13, 1943, 226.
5. Ibid., entry of July 19, 1943, 227.
6. Entry of 17.7.43, in Joseph Goebbels, *Die Tagebücher von Joseph Goebbels* [The diaries of Joseph Goebbels], ed. Elke Froehlich (Munich: K. G. Saur, 1993), Teil II, Band 9 (hereinafter *Die Tagebücher* 9), 116.
7. Ibid., 114.
8. Ibid., 116.
9. Ibid., 114.
10. Despite being made aware of Soviet rather than Nazi responsibility for the massacre back in April 1943, both Roosevelt and Churchill had been unwilling to raise the issue in public — or even encourage others to do so, when continued Soviet resistance on the Eastern Front was crucial. General Sikorski, the commander in chief of all Polish forces in the West, on April 15, 1943, was thus begged not to make Katyn, however awful, a matter of contention, just before Hitler opened his expected summer offensive: Martin Gilbert, *Road to Victory: Winston Churchill, 1941–1945* (London: Heinemann, 1986), 385.
11. Entry of 19.7.43, Goebbels, *Die Tagebücher* 9, 126.

12. Ibid.
13. Ibid., entry of 20.7.43, 132.
14. Ian Kershaw, *Hitler 1936–1945: Nemesis* (New York: Norton, 2000), 594.
15. Ibid., 597.

35. COUNTERCRISIS

1. See inter alia Trumbull Higgins, *Soft Underbelly: The Anglo-American Controversy over the Italian Campaign, 1939–1945* (New York: Macmillan, 1968), 91–124; Douglas Porch, *The Path to Victory in World War II: The Mediterranean Theater in World War II* (New York: Farrar, Straus and Giroux, 2004), 459–76; Mark Stoler, *The Politics of the Second Front: American Military Planning and Diplomacy in Coalition Warfare, 1941–1943* (Westport, CT: Greenwood Press, 1977), 97–129; and Mark Stoler, *Allies in War: Britain and America Against the Axis Powers, 1940–1945* (New York: Hodder Arnold, 2005), 123–28.
2. Davies Papers, mss for May 20, 1943, 9, Library of Congress.
3. Entry of June 3, 1943, "Arrival in Washington and Report to the President," Davies Papers, Library of Congress.
4. Davies Papers, mss for May 20, 1943, 9.
5. Entry of July 22, 1943, Stimson Diary, Henry L. Stimson Papers, Yale University Library, New Haven, CT.
6. Ibid.
7. Ibid.
8. Maurice Matloff, *Strategic Planning for Coalition Warfare* (Washington, D.C.: Center of Military History, 1959), 165.
9. Ibid.
10. Ibid., 167.
11. Ibid., Letter to Handy, July 4, 1943, 164.

36. A FISHING EXPEDITION IN ONTARIO

1. George M. Elsey, Introduction to "The Log of the President's Visit to Canada, 16 August 1943 to 26 August 1943," p. 3, FDR Library.
2. Ibid.
3. Ibid.
4. "Will Punish Duce; President in His War Report Demands Total Surrender," *New York Times*, July 29, 1943.
5. FDR finally told Daisy Suckley "the whole story, which is unsavory," later that summer, including Bullitt's part. "The P. never wants to speak to Bullitt again": entry of September 22 and 29, 1943, in Ward, ed., *Closest Companion*, 244.
6. "Warning by Stalin to Allies Is Seen; U.S. Observers in Moscow Said to View German Manifesto as Russian Declaration," *New York Times*, July 29, 1943.
7. Entry of July 29, 1943, Leahy Diary.

8. Entry of July 28, 1943, Ward, *Closest Companion,* 227.
9. Ibid.
10. Entry of August 9, 1943, Leahy Diary.
11. Ibid.
12. Ibid.

37. THE PRESIDENT'S JUDGMENT

1. Davies Papers, mss for May 20, 1943, 9, Library of Congress.
2. Ibid.
3. David Reynolds, *In Command of History: Churchill Fighting and Writing the Second World War* (New York: Random House, 2005), 322–23.
4. Davies Papers, mss for May 20, 1943, 10, Library of Congress.
5. OSS Numbered Intelligence Bulletins, No. 39, 10 July 43, Roosevelt Map Room, Military Subject Files, Box 72, Section 2, MR 203 (12), FDR Library.

38. STALIN LIES

1. From Premier J. V. Stalin to President Franklin D. Roosevelt, August 8, 1943, in Susan Butler, ed., *My Dear Mr. Stalin: The Complete Correspondence of Franklin D. Roosevelt and Joseph V. Stalin* (New Haven, CT: Yale, 2005), 151.
2. Albert Weeks, *Russia's Life-Saver: Lend Lease Aid to the United States* (Lanham, MD: Lexington Books, 204), 1.
3. Ibid, 146–47. By the end of the war, over 30 percent of Russian wheeled vehicles had come from the United States, as also aircraft; almost 60 percent of aviation fuel, and more than 50 percent of Russian ordnance (ammunition): Ibid., 8–9.
4. Entry of August 9, 1943, in Geoffrey C. Ward, ed., *Closest Companion, The Unknown Story of the Intimate Friendship Between Franklin Roosevelt and Margaret Suckley* (Boston: Houghton Mifflin, 1995), 228.

39. WAR ON TWO WESTERN FRONTS

1. Entry of September 29, 1943, in Ward, *Closest Companion,* 244.
2. Entry of August 10, 1943, Leahy Diary, William D. Leahy Papers, Library of Congress.
3. "Memorandum: Subject: Conduct of the War in Europe, 8 August, 1943," in *Foreign Relations of the United States: The Conferences at Washington and Quebec, 1943* (hereinafter *FRUS II*) (Washington, D.C.: Government Printing Office, 1970), 467–72; also Maurice Matloff, *Strategic Planning for Coalition Warfare* (Washington, D.C.: Center of Military History, 1959), 176.
4. Even in the President's meeting with the Joint Chiefs of Staff at the White House, General Marshall was more concerned with logistical waste than vital combat

experience, chiding the President that it was "impossible to calculate the wastage that has accrued to the United Nations war effort from changes made to basic decisions" — i.e., the cross-Channel invasion, planned in 1942. "The first instance was carrying out TORCH which involved moving troops set up from the United States to England and thence to Africa" — "Minutes of Meeting Held at the White House Between the President and the Chiefs of Staff on 10 August at 1415," in *FRUS II*, 503.

5. "Memorandum by the Joint Chiefs of Staff: Strategic Concept for the Defeat of the Axis in Europe, 9 August, 1943," *FRUS II*, 472–81.
6. Ibid., 473.
7. Ibid.
8. "Memorandum for General Handy," August 9, 1943, in *The Papers of General George Catlett Marshall*, vol. 4 (Baltimore, MD: Johns Hopkins University Press, 1996), 85–86.
9. Ibid.
10. "Minutes of Meeting Held at the White House Between the President and the Chiefs of Staff on 10 August 1943 at 1415," *FRUS II*, 499.
11. Ibid.
12. Ibid., 500.
13. Ibid., 500–501.
14. Ibid., 501.
15. Entry of August 10, 1943, Stimson Diary, Henry L. Stimson Papers, Yale University Library, New Haven, CT.
16. "Minutes of Meeting Held at the White House Between the President and the Chiefs of Staff on 10 August 1943 at 1415," *FRUS II*, 501.
17. Ibid., 502
18. "Minutes of meeting held at the White House at 1415 between the President and the JCS, 10 Aug 43, with JCS Memo 97 in ABC 337 (25 May 43)," in Matloff, *Strategic Planning for Coalition Warfare*, 215.
19. Entry of August 10, 1943, Stimson Diary.
20. "Dear Mr. President" letter, August 10, 1943, attached to Stimson Diary.

40. THE FÜHRER IS VERY OPTIMISTIC

1. "The Polish Ministry for Foreign Affairs to the American Embassy Near the Polish Government in Exile," in *FRUS II*, 410.
2. "Prime Minister's Personal Minute," July 19, 1943, in Gilbert, *Road to Victory*, 445.
3. Entry of 10.8.1943 in Joseph Goebbels, *Die Tagebücher von Joseph Goebbels* [The diaries of Joseph Goebbels], ed. Elke Froehlich (Munich: K. G. Saur, 1993), Teil II, Band 9 (hereinafter *Die Tagebücher* 9), 250.
4. Ibid.
5. Ibid., 254.

6. "Er denkt nicht daran, bis zum Po zurückzuziehen" ["He has no intention of retreating to the Po"], entry of June 25, 1943, Goebbels, *Die Tagebücher von Joseph Goebbels* [The diaries of Joseph Goebbels], ed. Elke Froehlich (Munich: K. G. Saur, 1993), Teil II, Band 8, 532.
7. Entry of 10.8.1943, Goebbels, *Die Tagebücher* 9, 255.
8. Ibid.
9. Ibid., 260.
10. Ibid.
11. Ibid.
12. Ibid., 261.
13. See Karl-Heinz Friezer et al., *Das Deutsche Reich und der Zweite Weltkrieg* (Stuttgart: Deutsche Verlags-Anstalt, 2007) Band 8, 1192–1209.
14. "die schon erwähnte Spekulation auf wachsende und letztlich bündnisprengde Divergenzen innerhalb der Feindkoalition": Ibid., 1194.

41. A CARDINAL MOMENT

1. Cable of June 25, 1943, in *Foreign Relations of the United States: The Conferences at Cairo and Tehran* 1943 (hereinafter *FRUS III*), 10.
2. Entry of August 14, 1943, in Ward, *Closest Companion*, 229.
3. Ibid.
4. "Dear Mr. President" letter, August 10, 1943, attachment to entry of August 10, 1943, Stimson Diary.
5. Ibid.
6. Personal Minute of July 13, 1943, in Martin Gilbert, *Road to Victory: Winston S. Churchill, 1941–1945* (London: Heinemann, 1986), 442.
7. Ibid., Cable T.1043/3, July 16, 1943, 443.

42. CHURCHILL IS STUNNED

1. In London, Secretary Stimson had told Churchill that with regard to the sharing of atom bomb development (code-named S-1), "I could only promise to report the matter to the President for the final decision": "Brief Report on Certain Features of Overseas Trip," August 4, 1943, Stimson Diary.
2. Entry of August 10, 1943, Mackenzie King Diary, Library and Archives Canada, Ottawa, ON.
3. Winston Churchill, *The Second World War,* vol. 5, *Closing the Ring* (London: Cassell, 1952), 73.
4. Entry of August 14, 1943, in Ward, *Closest Companion,* 228.

43. THE GERMAN WILL TO FIGHT

1. Vice Admiral Mountbatten used his pistol to demonstrate the toughness of ice

floes — his latest brainwave for floating harbors in the invasion of Normandy: Andrew Roberts, *Masters and Commanders: How Four Titans Won the War in the West, 1941–1945* (New York: Harper, 2009), 405.

2. David Reynolds, *In Command of History: Churchill Fighting and Writing The Second World War* (New York: Random House, 2005), 363.
3. Philip A. Smith, *Bombing to Surrender: The Contribution of Air Power to the Collapse of Italy, 1943* (Maxwell Air Force Base, Alabama: School of Advanced Airpower Studies, 1997), 63.
4. Sven Oliver Mueller, "Nationalism in German War Society 1939–1945" in *Germany and the Second World War*, ed. Jörg Echternkamp (Oxford: Clarendon Press, 2014), vol. 9, no. 2, p. 32.
5. Ibid., 34.
6. Ibid., 30.

44. NEAR-HOMICIDAL NEGOTIATIONS

1. Entry of August 15, 1943, *Field Marshal Lord Alanbrooke: War Diaries, 1939–1945*, ed. Alex Danchev and Daniel Todman (Berkeley: University of California Press, 2001), 441.
2. Reynolds, *In Command of History*, 374–82.
3. "The Log of the President's Trip to Canada, August 16–August 26, 1943," 2, FDR Library.
4. Entry of August 15, 1943, Arthur Bryant, *The Turn of the Tide: A History of the War Years, Based on the Diaries of Field Marshal Lord Alanbrooke, Chief of the Imperial General Staff* (New York: Doubleday, 1957), 578.
5. Carlo D'Este, *World War II in the Mediterranean, 1942–1945* (Chapel Hill, NC: Algonquin, 1990), 196. Since over 200,000 Germans were reported "missing," these may include many who surrendered at the war's end.
6. Carlo D'Este, *Warlord: A Life of Winston Churchill at War, 1874–1945* (New York: Harper, 2008), 626.
7. Entry of August 16, 1943, Brooke, *War Diaries*, 443.
8. Entry of August 15, 1943, Leahy Diary, William D. Leahy Papers, Library of Congress.
9. Entry of August 15, 1943, Brooke, *War Diaries*, 442.
10. *Foreign Relations of the United States: The Conferences at Washington and Quebec, 1943* (hereinafter *FRUS II*) (Washington, D.C.: Government Printing Office, 1970), 865.
11. Ibid., 866.
12. Entry of August 16, 1943, Brooke, *War Diaries*, 443.
13. Annotation to entry of August 15, 1943, in Brooke, *War Diaries*, 442.
14. Commander George Elsey, interview with the author, September 11, 2011.

45. A LONGING IN THE AIR

1. "The Log of the President's Trip to Canada, August 16–August 26, 1943," compiled by Chief Ship's Clerk William Rigdon, 4, FDR Library.
2. Cable of August 22, 1942, in Susan Butler, ed., *My Dear Mr. Stalin: The Complete Correspondence of Franklin D. Roosevelt and Joseph V. Stalin* (New Haven, CT: Yale University Press, 2005), 155.
3. Averell Harriman and Elie Abel, *Special Envoy to Churchill and Stalin, 1941–1946* (New York: Random House, 1975), 225; and entry of August 31, 1943, Mackenzie King Diary, Library and Archives Canada, Ottawa, ON.
4. Entry of August 23, 1943, Brooke, *War Diaries,* 447. Lieutenant General Henry Pownall, who became one of Churchill's many assistants in writing his memoirs, claimed in his 1943–1944 diary that the Sumatra idea, code-named Operation Culverin, was "a typically Winstonian project, advanced with his usual fatuous obstinacy": David Reynolds, *In Command of History,* 404.
5. Harriman and Abel, *Special Envoy,* 224.
6. Ibid.
7. *FRUS II,* 691ff.
8. "The Log of the President's Trip to Canada, August 16–August 26, 1943," compiled by Chief Ship's Clerk William Rigdon, 15, FDR Library.
9. P. J. Philips, "President Is Grim: Only Long Peace Could Justify Sacrifices," *New York Times,* August 26, 1943.
10. Entry of August 25, 1943, Mackenzie King Diary.
11. Samuel I. Rosenman, *Working with Roosevelt* (New York: Harper, 1952), 387.
12. Ibid.
13. Ibid.
14. Text in "The Log of the President's Trip to Canada." (President's own copy. The alternative wording gave the President a choice, for extra emphasis, as he spoke.) Also as "Address at Ottawa, Canada, August 25, 1943," in *The Public Papers and Addresses of Franklin D. Roosevelt,* vol. 12, *The Tide Turns,* comp. Samuel Rosenman (New York: Russell & Russell, 1950; reissued 1969), 365–69.
15. Philips, "President Is Grim," *New York Times.*
16. "Address at Ottawa, Canada, August 25, 1943," in *The Public Papers and Addresses of Franklin D. Roosevelt.*
17. Entry of August 25, 1943, Mackenzie King Diary.

46. THE PRESIDENT IS UPSET — WITH THE RUSSIANS

1. "Memorandum of conversation Mr. Mackenzie King had with President Franklin D. Roosevelt — Ottawa, Wednesday — August 25, 1943," Mackenzie King Diary.
2. Ibid.
3. Ibid.

4. Ibid., "Conversation with Mr. Roosevelt, Ottawa — August 25, 1943."
5. Martin Gilbert, *Road to Victory: Winston S. Churchill, 1941–1945* (London: Heinemann, 1986), 482.
6. Dmitri Volkogonov, *Stalin: Triumph and Tragedy* (Rocklin, CA: Prima, 1991), 486.
7. "Conversation with Mr. Roosevelt, Ottawa, August 25, 1943," Mackenzie King Diary.
8. Ibid., entry of August 22, 1943.
9. Gilbert, *Road to Victory*, 482.
10. E.g. Susan Butler, *Roosevelt and Stalin: Portrait of a Partnership* (New York: Knopf, 2015).
11. Gilbert, *Road to Victory*, 484–85.
12. Harriman and Abel, *Special Envoy*, 536.
13. Entry of Wednesday, August 25, 1942, Mackenzie King Diary.
14. Ibid., "Conversation with Mr. Roosevelt. Ottawa — August 25, 1943."
15. Ibid., entry of August 31, 1943.
16. Ibid., "Conversation with Mr. Roosevelt. Ottawa — August 25, 1943."
17. Ibid.

47. CLOSE TO DISASTER

1. Entry of August 26, 1943, in Geoffrey C. Ward, ed., *Closest Companion: The Unknown Story of the Intimate Friendship Between Franklin Roosevelt and Margaret Suckley* (Boston: Houghton Mifflin, 1995), 231.
2. Entry of August 29, 1943, Leahy Diary, William D. Leahy Papers, Library of Congress.
3. Entry of August 28, 1943, Ward, *Closest Companion*, 231–32.

48. A DARWINIAN STRUGGLE

1. Entry of 27.8.1943, Joseph Goebbels, *Die Tagebücher von Joseph Goebbels* [The diaries of Joseph Goebbels], ed. Elke Fröhlich (Munich: K. G. Saur, 1993), Teil II, Band 9 (hereinafter *Die Tagebücher* 9), 369. Quotes from this source have been translated by the author.
2. Ibid., entry of 10.8.1943, 260.
3. Ibid., entry of 10.9.1943, 464. Interestingly, addressing reporters' questions in Washington, "Churchill said Britain wants no more territory: such as Sicily, Pantelleria, etc," but that "islands of chiefly strategic value probably should be held by the Allies." However, he also made clear the "British did not propose to give up any territory" they considered theirs — "this in answer to a question about Hong Kong": "Churchill Luncheon with Correspondents, September 3, 1943," in Raymond Clapper Papers, Personal File, 1942–43, Box 23, Library of Congress.
4. Ibid.

49. A TALK WITH ARCHBISHOP SPELLMAN

1. Winston Churchill, *The Second World War,* vol. 5, *Closing the Ring* (London: Cassell, 1952), 109.
2. Entry of August 31, 1943, Mackenzie King Diary.
3. Entry of September 6, 1943, Ward, *Closest Companion,* 236–37.
4. Ibid.
5. Mary Soames, *A Daughter's Tale* (New York: Random House, 2011), 275–76.
6. Ibid, 276–77.
7. Robert I. Gannon, *The Cardinal Spellman Story* (Garden City, NY: Doubleday, 1962), 218.
8. Ibid., 223.
9. "To many historians, especially but far from exclusively those writing in the first years of the Soviet-American Cold War that followed World War II, Roosevelt was exceptionally naive and foolish to believe he could collaborate with Stalin": Mark Stoler and Melanie Gustafson, eds., *Major Problems in the History of World War II: Documents and Essays* (Boston: Houghton Mifflin, 2003), 378.
10. John Morton Blum, *V Was for Victory: Politics and American Culture During World War II* (New York: Harcourt Brace Jovanovich, 1976), 271–73.
11. Ibid., 255.
12. Ibid.
13. Gannon, *The Cardinal Spellman Story,* 223.
14. Ibid.
15. Ibid.
16. Ibid.
17. Susan Butler, *Roosevelt and Stalin: Portrait of a Partnership* (New York: Knopf, 2015), 153.
18. Gannon, *The Cardinal Spellman Story,* 223.
19. Ibid.
20. Ibid., 227.

50. THE EMPIRES OF THE FUTURE

1. Entry of September 2, 1943, Ward, *Closest Companion,* 234.
2. Simon Sebag-Montefiore, *Young Stalin* (New York: Knopf, 2007), 193.
3. Ibid., 193.
4. Martin Gilbert, *Road to Victory: Winston S. Churchill, 1941–1945* (London: Heinemann, 1986), 492.
5. Ibid.
6. "Anglo-American Unity: A Speech on Receiving an Honorary Degree at Harvard University, September 6, 1943," in *The War Speeches of Winston Churchill,* ed. Charles Eade, vol. 2 (London: Cassell, 1952), 510–15.

7. Ibid.
8. Ibid.

51. A TRAGICOMEDY OF ERRORS

1. E.g., Douglas Porch, *The Path to Victory: The Mediterranean Theater in World War II* (New York: Farrar, Straus and Giroux, 2004), 459–61.
2. "Review of the Situation in the Light of Italian Collapse," in RG 218: Records of the U.S. Joint Chiefs of Staff, Box 307, National Archives.
3. Ibid.
4. Carl D'Este, *Patton: A Genius for War* (New York: HarperCollins, 1995), 533–55; Rick Atkinson, *The Day of Battle: The War in Sicily and Italy, 1943–1944* (New York: Holt, 2007), 147–49.
5. Nigel Hamilton, *Monty: Master of the Battlefield, 1942–1944* (New York: McGraw-Hill, 1986), 390.
6. Ibid., 398–402.
7. Ibid., 399.
8. Ibid., 388.
9. Atkinson, *The Day of Battle,* 190.
10. General Mark Clark to author, interview of October 26, 1981, in Hamilton, *Monty: Master of the Battlefield,* 414.
11. Atkinson, *The Day of Battle,* 192.

52. MEETING REALITY

1. Atkinson, *The Day of Battle,* 195.
2. Ibid., 196.
3. Hamilton, *Monty: Master of the Battlefield, 1942–1944,* 393.
4. Atkinson, *The Day of Battle,* 197.
5. On the island of Cephalonia, for example, more than five thousand Italian troops were massacred by invading German forces, who were told to take no prisoners: Alexander Mikaberidze, ed., *Atrocities, Massacres, and War Crimes: An Encyclopedia* (Santa Barbara, CA: ABC-CLIO, 2013), 326 and 750.
6. See Hamilton, *Monty: Master of the Battlefield,* 404; and Atkinson, *The Day of Battle,* 190.
7. General Mark Clark to author, interview of October 26, 1981, in Hamilton, *Monty: Master of the Battlefield,* 414.
8. Ibid.
9. Atkinson, *The Day of Battle,* 199.
10. Ibid., 203.
11. Ibid., 205.
12. "Fireside Chat Opening Third War Loan Drive, September 8, 1943," *The Public Papers and Addresses of Franklin D. Roosevelt,* 377–80.

13. See Hamilton, *Monty: Master of the Battlefield,* 405–6 and footnote 403.
14. Entry of Monday, September 13, 1943, Ward, *Closest Companion,* 237.
15. Soames, *A Daughter's Tale,* 278.
16. Entry of September 7, Moran, *Winston Churchill: The Struggle for Survival, 1940–1965* (Boston: Houghton Mifflin, 1966), 118.
17. Ibid., entry of September 12, 1943, 119.
18. Ibid.
19. Entry of September 16, 1943, Diary of Harry C. Butcher, Eisenhower Library.
20. Entry of September 13, 1943, Moran, *Winston Churchill,* 119–20.
21. Ibid., 120.
22. Entry of Friday, August 27, 1943, in Alan Lascelles, *King's Counsellor: Abdication and War: The Diaries of Sir Alan Lascelles,* ed. Duff Hart-Davis (London: Weidenfeld & Nicolson, 2006), 156.
23. Cable of September 11, 1943, in Susan Butler, ed., *My Dear Mr. Stalin: The Complete Correspondence of Franklin D. Roosevelt and Joseph Stalin* (New Haven, CT: Yale University Press, 2005), 164.
24. Entry of 11.9.1943, Goebbels, *Die Tagebücher 9,* 479.
25. Rede des Führers über den Zusammenbruch Italiens am 10. September 1943, http://de.metapedia.org/wiki/Quelle/Rede_vom_10._September_1943_(Adolf_Hitler). Translated by the author.
26. Ibid.
27. Entry of 12.9.1943, Goebbels, *Die Tagebücher 9,* 492.
28. Ibid., entry of 10.9.1943, 463.
29. Ibid., 464.
30. Ibid., 460.
31. Ibid., 464.
32. Ibid., entry of 7.9.1943, 438.
33. Ibid., entry of 12.9.1943, 486–87.
34. Ibid.

53. A MESSAGE TO CONGRESS

1. Atkinson, *The Day of Battle,* 212.
2. Hamilton, *Monty: Master of the Battlefield, 1942–1944,* 413.
3. Mark Clark to author, interview of October 10, 1981, in Hamilton, *Monty: Master of the Battlefield,* 405.
4. Atkinson, *The Day of Battle,* 207.
5. "Message to the Congress on the Progress of the War, September 17, 1943," in *The Public Papers and Addresses of Franklin D. Roosevelt,* comp. Samuel I. Rosenman, vol. 12, 388–406.
6. Ibid.
7. Ibid.

54. ACHIEVING WONDERS

1. Entry of 19.9.43, in Goebbels, *Die Tagebücher 9*, 533.
2. "Message to the Congress on the Progress of the War, September 17, 1943."
3. Entry of August 31, 1943, Mackenzie King Diary.
4. Ibid.
5. George Elsey, interview with the author, September 12, 2011.
6. Ibid.
7. Andrew Roberts, *Masters and Commanders: How Four Titans Won the War in the West, 1941–1945* (New York: Harper, 2009), 412–13.
8. Brooke was equally to blame, plotting with Churchill to postpone Overlord yet again beyond its planned spring 1944 target date, and to demand "another full-scale Combined Chiefs of Staff conference in early November," 1943, "to try to sell" the alternative Mediterranean-exploitation strategy to Roosevelt and Marshall: Roberts, *Masters and Commanders*, 418. "We should have been in a position to force the Dardanelles by the capture of Crete and Rhodes, we should have the whole Balkans ablaze by now, and the war might have been finished in 1943!!" Brooke lamented in one of his wildest diary entries of the war: November 1, 1943, in *War Diaries, 1939–1945: Field Marshal Lord Alanbrooke*, ed. Alex Danchev and Daniel Todman (Berkeley: University of California Press, 2001), 465. Even Roberts, who admired Brooke, was moved to harsh judgment regarding Overlord. "It had probably been the correct decision not to appoint him as its supreme commander after all," he considered — Roberts, *Masters and Commanders*, 419.

Index

The MANTLE *of* COMMAND

ALSO BY NIGEL HAMILTON

Royal Greenwich: A Guide and History to London's Most Historic Borough (with Olive Hamilton)

Nigel Hamilton's Guide to Greenwich: A Personal Guide to the Buildings and Walks of One of England's Most Beautiful and Historic Areas

The Brothers Mann: The Lives of Heinrich and Thomas Mann, 1871–1950 and 1875–1955

Monty: The Making of a General, 1887–1942

Master of the Battlefield: Monty's War Years, 1942–1944

Monty: Final Years of the Field-Marshal, 1944–1976

Monty: The Man Behind the Legend

JFK: Reckless Youth

Monty: The Battles of Field Marshal Bernard Law Montgomery

The Full Monty: Montgomery of Alamein, 1887–1942

Bill Clinton, An American Journey: Great Expectations

Montgomery: D-Day Commander

Bill Clinton: Mastering the Presidency

Biography: A Brief History

How to Do Biography: A Primer

American Caesars: Lives of the Presidents from Franklin D. Roosevelt to George W. Bush

The MANTLE *of* COMMAND

FDR AT WAR 1941–1942

Nigel Hamilton

HOUGHTON MIFFLIN HARCOURT
BOSTON · NEW YORK

This one is for my grandchildren, spread across the world:
Sophie, Oskari, Toby, and Matthew

Copyright © 2014 by Nigel Hamilton

All rights reserved

For information about permission to reproduce selections from this book, write to Permissions, Houghton Mifflin Harcourt Publishing Company, 215 Park Avenue South, New York, New York 10003.

www.hmhco.com

Library of Congress Cataloging-in-Publication Data
Hamilton, Nigel.
The mantle of command : FDR at war, 1941–1942 / Nigel Hamilton.
pages cm
Includes bibliographical references and index.
ISBN 978-0-547-77524-1 (hardcover)
1. World War, 1939–1945 — United States. 2. Roosevelt, Franklin D. (Franklin Delano), 1882–1945 3. World War, 1939–1945 — United States — Biography. 4. World War, 1939–1945 — Diplomatic history. 5. Command of troops — United States — Case studies. 6. World War, 1939–1945 — Campaigns. 7. Great Britain — Foreign relations — United States. 8. United States — Foreign relations — Great Britain. I. Title.
D753.H25 2014
940.54'1273 — dc23 2013045586

Typeset in Minion
Maps by Mapping Specialists, Ltd.

Printed in the United States of America
DOH 10 9 8 7 6 5 4
4500773122

The author is grateful for permission to quote from the following: *War Diaries, 1939–1945: Field Marshal Lord Alanbrooke,* edited by Alex Danchev and Daniel Todman, reprinted by permission of David Higham Associates. Diary of Lord Halifax, 1941–1942, reprinted by permission of the Borthwick Institute for Archives, University of York. Diary of Thomas C. Hart, reprinted by permission of the Operational Archives Branch, Naval Historical Center, Washington, D.C. Letters and diaries of Margaret Lynch Suckley, reprinted by permission of the Wilderstein Preservation, Rhinebeck, N.Y.

Contents

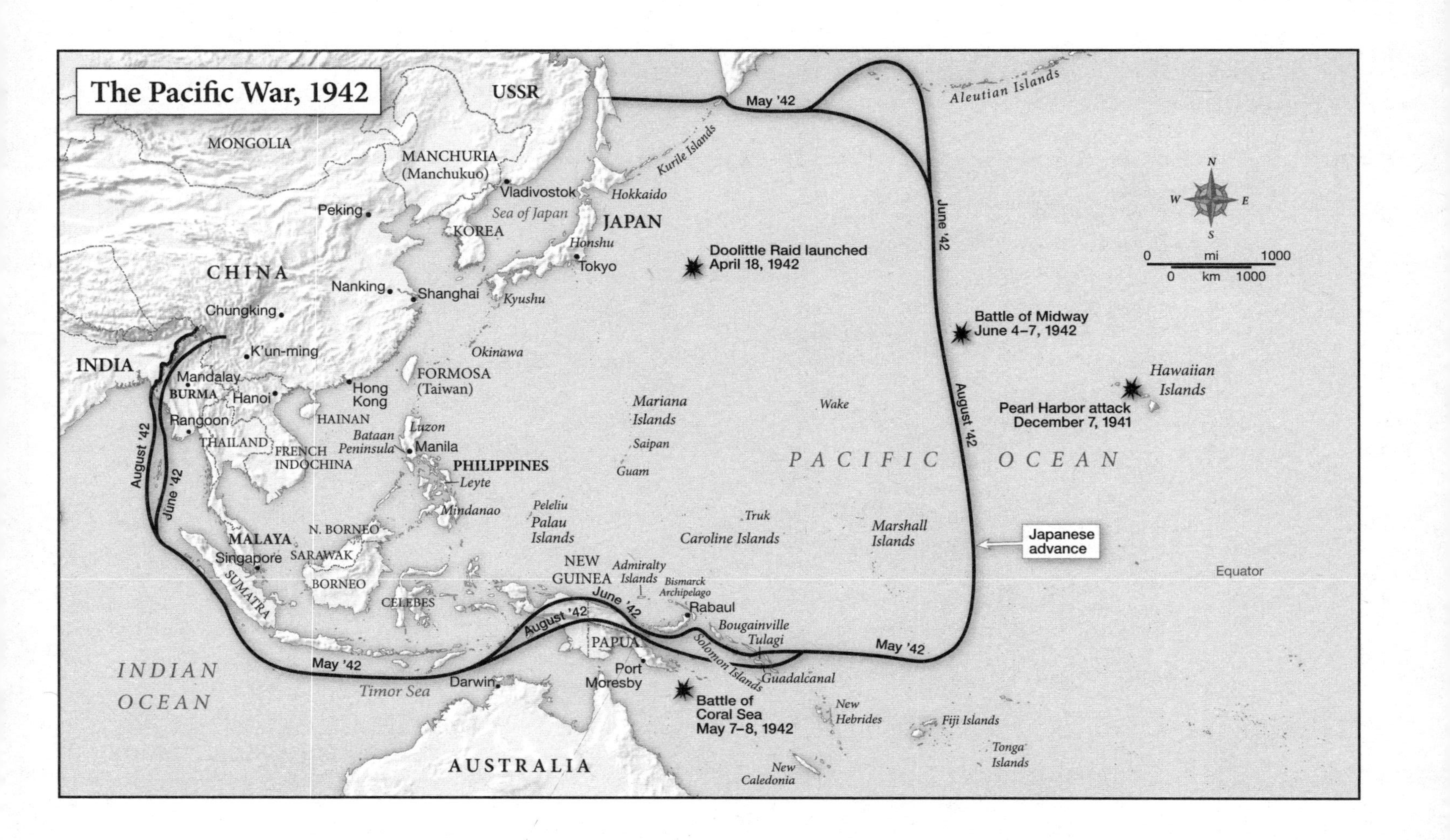
The Pacific War, 1942
USSR
MONGOLIA
MANCHURIA
(Manchukuo)
Vladivostok
Hokkaido
Kurile Islands
Aleutian Islands
May '42
June '42
Peking
Sea of Japan
KOREA
JAPAN
Honshu
Tokyo
Doolittle Raid launched
April 18, 1942
CHINA
Nanking
Shanghai
Kyushu
Chungking
K'un-ming
Battle of Midway
June 4–7, 1942
Okinawa
INDIA
Mandalay
BURMA
Hanoi
Hong
Kong
FORMOSA
(Taiwan)
Hawaiian
Islands
Pearl Harbor attack
December 7, 1941
Rangoon
THAILAND
HAINAN
Bataan
Peninsula
Luzon
Manila
FRENCH
INDOCHINA
PHILIPPINES
Leyte
Mariana
Islands
Saipan
Guam
Wake
August '42
PACIFIC
OCEAN
Mindanao
Peleliu
Palau
Islands
Truk
Caroline Islands
Marshall
Islands
Japanese
advance
MALAYA
N. BORNEO
SARAWAK
Singapore
BORNEO
SUMATRA
CELEBES
NEW
GUINEA
Admiralty
Islands
Bismarck
Archipelago
Rabaul
Bougainville
Tulagi
PAPUA
Solomon Islands
Guadalcanal
Port
Moresby
Equator
INDIAN
OCEAN
Timor Sea
Darwin
Battle of
Coral Sea
May 7–8, 1942
New
Hebrides
Fiji Islands
Tonga
Islands
New
Caledonia
AUSTRALIA
N
S
E
W
0 mi 1000
0 km 1000

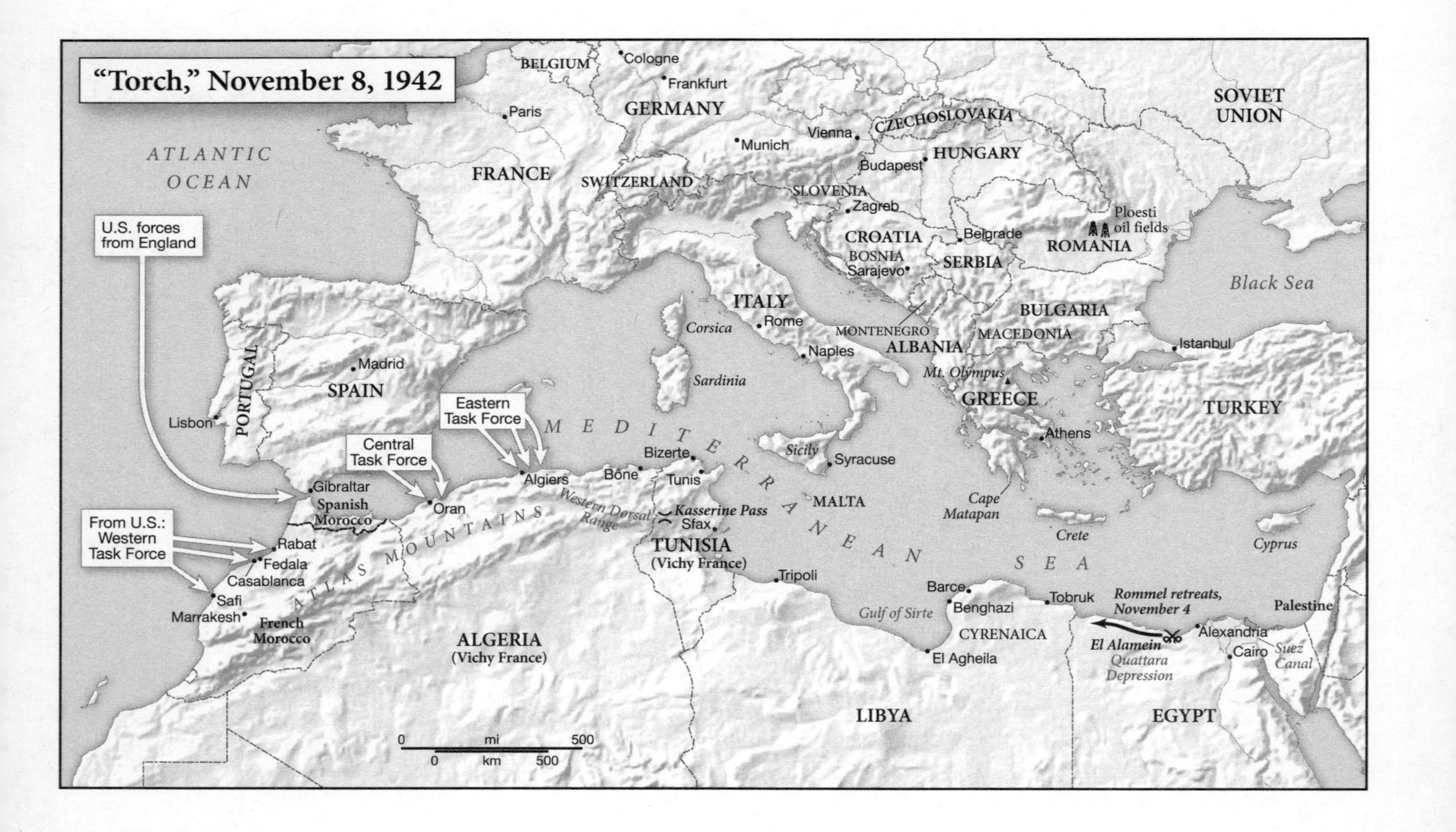
"Torch," November 8, 1942
ATLANTIC OCEAN
U.S. forces from England
From U.S.: Western Task Force
Central Task Force
Eastern Task Force
BELGIUM
Cologne
Frankfurt
GERMANY
Paris
FRANCE
SWITZERLAND
Munich
Vienna
CZECHOSLOVAKIA
Budapest
HUNGARY
SLOVENIA
Zagreb
CROATIA
BOSNIA
Sarajevo
Belgrade
SERBIA
Ploesti oil fields
ROMANIA
SOVIET UNION
Black Sea
BULGARIA
MACEDONIA
MONTENEGRO
ALBANIA
Istanbul
TURKEY
Mt. Olympus
GREECE
Athens
Cape Matapan
Crete
Cyprus
ITALY
Rome
Corsica
Naples
Sardinia
Sicily
Syracuse
MALTA
PORTUGAL
Lisbon
Madrid
SPAIN
Gibraltar
Spanish Morocco
Rabat
Fedala
Casablanca
Safi
Marrakesh
French Morocco
ATLAS MOUNTAINS
Oran
Algiers
Bône
Bizerte
Tunis
Western Dorsal Range
Kasserine Pass
Sfax
TUNISIA (Vichy France)
MEDITERRANEAN SEA
Tripoli
Gulf of Sirte
Barce
Benghazi
CYRENAICA
El Agheila
Tobruk
Rommel retreats, November 4
El Alamein
Quattara Depression
Alexandria
Cairo
Suez Canal
Palestine
ALGERIA (Vichy France)
LIBYA
EGYPT
0 mi 500
0 km 500

Prologue

WE CAN VIEW World War II from many angles, military to moral. Many fine books have been written about the struggle — perhaps the most famous being Winston Churchill's *The Second World War,* in six volumes, which helped the former British prime minister to win the Nobel Prize in Literature.

The Mantle of Command: FDR at War is my attempt to retell the story of the military direction of the Second World War from a different perspective: that of President Franklin Delano Roosevelt, in his role as U.S. commander in chief.

Following Pearl Harbor there were many calls for Roosevelt to hand over direction of America's world war to a military man: a professional like General Douglas MacArthur, the former U.S. Army chief of staff, who was serving in the Philippines. FDR rejected such calls — arguing that, as U.S. president, he was the U.S. commander in chief, and the Constitution made him so. As Alexander Hamilton had written in *Federalist* No. 74, the President of the United States was to have "the supreme command and direction of the military and naval forces, as first general and admiral" of the nation. This Roosevelt was, whether people liked it or not. "What is clearer than that the framers meant the President to be the *chief executive* in peace," he said to his doctor, Ross McIntire, "and in war the *commander in chief*?"[1]

Nevertheless, the military challenges facing Roosevelt as commander in chief were greater than any that had confronted his predecessors: America assailed by a coalition of three twentieth-century military empires — Hitler's Third Reich, Mussolini's Italian Empire, and Hirohito's Empire of Japan — seeking, in a Tripartite Pact, to remake the modern world in their own image. To this end they had revolutionized warfare: Nazi *Blitzkrieg* in Europe, and dazzling, ruthless amphibious invasions in the Far East by the Japanese.

How Roosevelt responded to those challenges as his nation's military commander is thus the burden of my new account. It is a story that, astonishingly, has never really been chronicled. Roosevelt himself did not live to tell it, as he had hoped he would, in retirement;[2] Churchill, surviving the war, did, in incomparable prose—but very much from his own point of view.

Succeeding generations of writers and historians have certainly addressed Roosevelt's career, but primarily as statesman and politician rather than as commander in chief. As far as the military direction of the war was concerned, such writers tended to ignore or downplay the President's role, focusing instead on Allied global strategy or on Roosevelt's subordinates and field commanders: General Marshall, Admiral King, General Arnold, Admiral Nimitz, General MacArthur, General Eisenhower, General Patton, General Bradley, and other World War II warriors.[3] As a result, the popular image of President Roosevelt has become one of a great and august moral leader of his nation: an inspiring figure on a world stage, but one who largely delegated the "business of war" to others—including Winston Churchill.

General George C. Marshall, for example, once remarked to the chief of staff of the British Army, General Alan Brooke, that Brooke was lucky to see the Prime Minister almost every day in London; in Washington, by contrast, Marshall—who was chief of staff of the U.S. Army—often did not see the President "for a month or six weeks."[4]

Marshall was exaggerating; moreover, he was expressing a very different frustration from the one the majority of writers have taken him to mean. Marshall was, in reality, complaining that President Roosevelt was making *all* the major military decisions at the White House, rather than allowing Marshall to make them at the War Department—and worse still, not allowing his U.S. Army chief of staff to contest them, or give advice, unless by appointment with the President.

This was a deliberate stratagem, as I hope *The Mantle of Command* will demonstrate. Deference to the military by political leaders in World War I had permitted the senseless battles of attrition on the Western Front. For this reason the President was unwilling to delegate something as important as world war to "professionals." Keeping General Marshall and Admiral Ernest King as separate though equal supplicants, the President intentionally sought to assert his ultimate authority as commander in chief: a power he kept strictly within the parameters of the U.S. Constitution, but which brooked no real opposition to his wishes or decisions — until the fateful day in 1942 when his military officials attempted a quasi mutiny, which is the centerpiece of this book.

The story of how America's commander in chief conducted World War II in the aftermath of Pearl Harbor, then, is almost the polar opposite of what we have been led, for the most part, to believe.[5] It is also more freighted, since the stakes for America and the free world in 1942 were perhaps the most serious in global history.

Tracing afresh how Roosevelt dealt with the military challenges he faced as commander in chief following the Japanese attack on Pearl Harbor allows us to see him in perhaps his greatest hour — setting and maintaining the moral agenda of the United Nations (as he christened the Allied powers), while slowly but surely turning defeat into relentless victory. His handling of General MacArthur and the manner in which he kept the Filipino forces fighting as allies of an embattled America, rather than giving in to the Japanese, was but one of his extraordinary achievements of the succeeding months as, swatting the persistent machinations and rumblings of near treason in the U.S. War Department, Roosevelt finally overruled his subordinates and, ordering into battle the largest American amphibious invasion force in the nation's history, his legions set out from shores three thousand miles apart to turn the tide of war against Hitler — astonishing the world, as they did so, and giving rise to the slogan that would hearten millions across Europe: "The Americans are coming!"

Side by side with this perspective, *The Mantle of Command* seeks to tell another story that has been largely downplayed or obscured in the decades since World War II: namely the collapse of the British Empire in the weeks and months after Pearl Harbor.

As prime minister of Great Britain, Winston S. Churchill had become an emblem of his island country's noble resistance to Nazi tyranny in 1940 and 1941 — so much so that writers and historians, following in the literary footsteps of Churchill's own multivolume account, have tended to overlook his often suspect leadership thereafter. In particular, Churchill's imperialist obsession over India, and the crisis this led to in his military relations with President Roosevelt in the spring of 1942, have been largely ignored in terms of their significance.

A third perspective that I feel has been neglected or underappreciated in relation to Franklin Delano Roosevelt was his modus operandi in the White House—and the consequences this has had for the writing of history. Paralyzed from the waist down after contracting polio in 1921, the President led a very different life from that of the British war leader. Winston Churchill was from childhood a romantic historian and journalist who loved to travel and put everything he thought or witnessed on paper—indeed, he made his living, his entire life, primarily by his writing. He also loved speechifying, holding forth with inimitable turns of phrase and perception to gatherings small and large. As his own doctor observed, he was not a good listener—and many of his worst mistakes as his nation's war leader stemmed from this.

Franklin Roosevelt, by contrast, was a very good listener. Though he could, as his mother's only child, be perfectly content on his own, reading or pasting items into his beloved stamp albums, Roosevelt also loved getting to know people, and enjoyed true conversation. He had earlier edited his university's newspaper; as a politician in a democracy made vibrant by an unfettered press and deeply partisan Congress, however, he came to distrust paper save as annotated records to be kept locked in his "'Safe' and Confidential Files" in his eventual presidential library at Hyde Park on the Hudson. These were the documents he thought he would eventually employ to reconstruct, once the war was over, the greatest drama of his life: his struggle to impose a moral, postimperial vision on his coalition wartime partners, and how he had been compelled by circumstances to supplant the United Kingdom as guardian of the world's democracies.

The President did not live to write that work. Reassembling from surviving documents his role as commander in chief seventy years later is thus considerably harder than it has been for writers seeking to portray and chronicle Churchill as wartime British prime minister. Piecing together the evidence not only from archival records but authentic wartime diaries, as well as the testimony of President Roosevelt's last surviving Map Room officer, I hope nevertheless that I've been able to restore for the reader something of the drama, the issues, and the confrontations Roosevelt faced, as well as the historic decisions he had to make as commander in chief in the aftermath of Pearl Harbor.

• • •

From the vantage point of Roosevelt's Oval Study, Oval Office, and his ground-floor Map Room at the White House, as well as his mansion at Hyde Park and Shangri-la (his presidential retreat, or camp, in the Maryland hills), the true story of FDR's conduct of the war bears little semblance to the picture Winston Churchill was at pains to chart in later years. Nor was it always perceived by outsiders, who found themselves charmed by Franklin Roosevelt's easy manner, and were not witnesses to the Commander in Chief's iron glove. General "Vinegar Joe" Stilwell, for example, had "retained the family's Republicanism and joined naturally in the exhilarating exercise of Roosevelt-hating" for his New Deal policies, in the words of his biographer; the general's contempt for the President got no better once war began. As Stilwell sniffed in his diary, Roosevelt was a "rank amateur in all military matters,"[6] and "completely hypnotized by the British," who had "his ear, while we have the hind tit."[7] Churchill's right-hand military man, Field Marshal Sir Alan Brooke, reflected in later years that, in contrast to his own master, the "President had no military knowledge and was aware of this fact and consequently relied on Marshall and listened to Marshall's advice."[8]

How little Sir Alan Brooke, as a British officer, knew! Churchill did, however — especially once his beloved British Empire began to collapse. It was not for nothing that the Prime Minister, waving goodbye to the President's plane some weeks after the successful American landings in Northwest Africa which turned the tide of World War II, remarked to the U.S. vice consul in Marrakesh: "If anything happened to that man, I couldn't stand it. He is the truest friend; he has the furthest vision; he is the greatest man I have ever known."[9]

The Mantle of Command, then, focuses for the first time on Roosevelt's *military* odyssey in the aftermath of Pearl Harbor — the personal, strategic, staffing, and command decisions he was called upon to make, in the context of the challenges he faced. In the interests of brevity I have focused on fourteen episodes, beginning with the President's historic meeting with Prime Minister Churchill on August 9, 1941, aboard their battleships in Placentia Bay, off the coast of Newfoundland, and ending with the first major landing of American troops on the threshold of Europe, in North Africa in late 1942. This time frame reveals Roosevelt's evolution from noncombatant supporter of Churchill, to become the master of the Allied effort, a commander in chief who took control of the war not only from his ally but from his own generals.

How accurately President Roosevelt read the demented mind of the Nazi führer; how, after ensuring U.S. naval victory in the Pacific, he turned his attention back toward Europe; how he overruled his generals and insisted upon American landings in French Northwest Africa in 1942, rather than a suicidal "Second Front" assault on the coast of mainland France — these marked a remarkable reversal of fortune for the Allies, and testify to Roosevelt's extraordinary military leadership: the saga reaching its climax as he sent into battle the massive American air, army, and naval forces that, on November 8, 1942, stunned Hitler and changed the course of World War II.

The tough challenges that came thereafter are the subject of another book. In the meantime, though, I hope these fourteen episodes will allow us to better understand the global test that Franklin Roosevelt faced as his country's military leader in the months following America's terrible defeat in Hawaii — and perhaps better appreciate the wisdom of Churchill's valedictory remark, seven decades ago.

PART ONE

Placentia Bay

1

Before the Storm

THE "PLAN OF ESCAPE," as Roosevelt called it, was simple. It was also deceitful — the sort of adventure that the President, confined to the White House by the burden of his responsibilities as well as his wheelchair, loved. He would pretend to go on a fishing trip on his 165-foot presidential yacht, the USS *Potomac,* similar to the vacation he had taken earlier in the spring. In reality he would secretly transfer from the "floating White House" to an American battleship or cruiser lying off the New England coast, then race up to Canadian waters to meet with the embattled British prime minister, Mr. Winston Churchill: the man who for more than a year had been leading his country in a lonely struggle against the Third Reich, following the fall of France and most of Europe.

FDR had suggested such a meeting several times since January 1941, when his emissary, Harry Hopkins, first put the idea to Churchill on a visit to London. The purpose was, according to Roosevelt's own account (which he dictated for the historical record and a magazine article — one that, sadly, he never completed), "to talk over the problem of the defeat of Germany."[1]

The proposed date "mentioned at that time," the President stated in his narrative, was to be "March or April" of that spring. However, the tortuous passage of the vast Lend-Lease bill through Congress and other important legislation made it impossible for him to leave Washington before the early summer, "and by that time the war in Greece — and later the war in Crete — prevented Churchill," Roosevelt explained. "The trip was mentioned again in May and June," the President narrated — but talk of such a meeting was overshadowed by a more momentous event than Hitler's predations in the Mediterranean. For on June 22, 1941, the German invasion of Russia began — Hitler launching several million mobilized German troops in a do-or-die effort to smash the Soviet Union before turning back to the problematic invasion of England.[2]

And — the U.S. secretary of war, Colonel Henry Stimson, feared — eventual war with the United States.[3]

Three weeks later a date for the Anglo-American summit "was finally decided"; it would take place, the President and Prime Minister concurred, in a mutually agreed upon location between August 8 and 10, 1941.[4] The initial site chosen was the British island of Bermuda. Canada, though, considered by the President to be safer, met with final approval by both leaders.

An official U.S. presidential visit to the capital city of Ottawa was mooted as cover for the meeting, with a secret side trip to the coast allowing Roosevelt to meet Churchill on his arrival. A problem was foreseen, however, in other British Dominion premiers asking to join the powwow. Such a gathering would have raised all sorts of political questions back in Washington, where a suspicious, isolationist Congress would have had to be informed — and involved.

It had thus been decided that only the President and Mr. Churchill would meet, aboard their anchored battleships — preferably in a protected sixty-mile-wide gulf off the Newfoundland coast called Placentia Bay, named for a French naval station that had existed there before the British conquest. Though the waters were Canadian, the naval station at Argentia had been ceded to the United States for ninety-nine years as a quid pro quo in the previous year's "destroyers for bases" deal to help Britain fight the Nazis. Following its acquisition it had been expanded to provide U.S. Army Air Force protection, and was already handling U.S. Navy minesweepers. Shore-based communications could also be provided, if required. From the President's perspective, however, it would above all be an *American* venue — putting the Prime Minister at a disadvantage, in the same way that visitors to Louis XIV were made to climb a thousand steps at Versailles before meeting the French monarch.

"Escape," for the President, meant, of course, *from* something, namely the American press: mainstay of the nation's vigilant democracy, but also a millstone in terms of executive privacy and confidentiality — and security. If word of the prospective meeting leaked, it would endanger not only the President's life but the Prime Minister's as well, drawing German U-boats in the North Atlantic to the area.

More threatening to Roosevelt's presidential authority in a time of continuing isolationism, though, would be the fierce debate aroused across America about the purpose of such a meeting. The majority of the American public (as expressed in opinion polls, which Roosevelt watched carefully)[5] remained resolutely opposed to being drawn into the war raging in Europe. At the height of the previous year's election campaign, the President had given his "most solemn assurance" that there was "no secret treaty, no secret obligation, no secret commitment, no secret understanding in any shape or form, direct or indirect, with any other Government, to involve this nation in any war or for any other purpose."[6] In the summer of 1941 there were still isolationists aplenty — encouraged by Hitler's turn to the east — watching to see that the President kept his word. Only Congress could declare war — or alter the terms of the November 1939 Neutrality Act.[7] For the President to end or breach American neutrality without congressional backing would risk his impeachment.

It was for personal reasons as well that the President was anxious to keep away the press and other voyeurs. He wanted the meeting to be intimate: an opportunity to finally get to know in person the British prime minister, with whom he had begun secretly corresponding in 1939, when Hitler invaded Poland and Churchill was made First Lord of the Admiralty. Once Churchill had become prime minister in May 1940, the President had continued to bypass his own U.S. ambassador to London, the nefarious appeaser and isolationist Joseph P. Kennedy, and the communications between Roosevelt and Churchill had become more and more grave, as the President first agreed to provide American mothballed warships to the British, then brokered through Congress the vast Lend-Lease deal to provide munitions, aircraft, and weapons on credit. Instead of being grateful, however, the Prime Minister kept asking for more — indeed, to the President's irritation, Churchill had recently told Roosevelt's emissary in London, Harry Hopkins, that he would be bringing all his military chiefs with him to the Placentia Bay summit. The President therefore had no option but, on Hopkins's advice, to take with him his *own* service chiefs: stern General George Marshall, chief of staff of the U.S. Army; bluff but more junior Major General Henry "Hap" Arnold, chief of the U.S. Army Air Forces (the Army Air Corps and GHQ Air Force, which was still a division of the U.S. Army);[8] and Admiral Harold

R. "Betty" Stark, quiet, bespectacled chief of naval operations. As commander of the Atlantic Fleet, responsible for the small presidential contingent's safe naval passage to Newfoundland, the irascible and somewhat anti-British Admiral Ernest "Ernie" King would be a party to the summit, too.

On a very hot August 3, 1941, Roosevelt left Washington by train. That evening he embarked on the presidential yacht (which had a crew of fifty-four) at New London's submarine base, in Connecticut — unaccompanied by the three Associated Press journalists who typically followed Roosevelt in a separate vessel on other such "fishing trips."

The USS *Potomac* motored north, anchoring that night off Martha's Vineyard, across the water from Cape Cod. At dawn the next morning Roosevelt secretly sped away by launch from the two-deck, 376-ton motor yacht, leaving a group of U.S. Secret Service stand-ins to impersonate him and his private guests when it continued its stately way up the Cape Cod Canal. From the shore the white vessel would be (and was) seen and waved to by peacetime summer holidaymakers. In truth the U.S. commander in chief was by then aboard the flagship of Admiral King's Atlantic Fleet, the USS *Augusta:* a ninety-two-hundred-ton, six-hundred-feet-long Northampton Class heavy cruiser, manned by more than a thousand sailors and armed with nine eight-inch guns, eight five-inch guns, and six torpedo tubes, lurking off Martha's Vineyard.

Admiral Stark and General Marshall were already onboard when the President arrived. A handful of other members of the presidential party, including General Arnold, had embarked on an accompanying heavy cruiser, the New Orleans–class USS *Tuscaloosa.* Escorted by four new American destroyers, the VIPs then sailed north toward a summit that, the officers finally became aware, promised to make history.

Speeding at times at thirty-two knots, the American presidential party raced through patchy fog to reach the Newfoundland rendezvous ahead of time. Roosevelt had not even told his secretary of war, Colonel Stimson, about the conference, nor his secretary of the navy, Mr. Frank Knox — nor even his secretary of state, Mr. Cordell Hull, who was on medical leave. The President had not even told his secretary, Grace Tully! He had only informed General Marshall and Admiral Stark three days before departure — with orders that General Arnold, the air force commander, be invited to attend but not informed of the purpose or destination of the voyage before embarking on the USS *Tuscaloosa.* There was to be no fraternization, or planning, before the meeting: nothing that could later be denounced as preparatory to a secret agreement or alliance.

By contrast, Prime Minister Winston Churchill had great plans for just such an affiliation.

Buoyed by excitement and hope, Winston Spencer Churchill — son, after all, of an American mother — had ordered his chiefs of staff to draw up a "Future Strategy Paper," setting out how Britain could win the war if the United States became an ally. He had also proudly sent secret signals to the prime ministers of all the Dominions of the British Empire to let them know of the impending conference — stating that, although none of them had been invited, he "hoped that from the meeting some momentous agreement might be reached."[9] Setting off "with a retinue which Cardinal Wolsey might have envied" (as his private secretary sarcastically noted in his diary),[10] the Prime Minister had even written in excitement to Queen Elizabeth, consort of the monarch, to tell her of his great expectations.

"I must say, I do not think our friend would have asked me to go so far, for what must be a meeting of world notice, unless he had in mind some further forward step," Churchill confided, explaining why he was leaving his country at such a critical time.[11] He had ordered grouse and rare turtle soup as among the provisions he would take, as well as a full military band. He would travel aboard his latest radar-equipped forty-three-thousand-ton battleship, HMS *Prince of Wales*. As he had presumptuously signaled to the President from Scapa Flow, Scotland, on August 4, 1941: "We are just off. It is twenty-seven years ago today that Huns began their last war. We must make a good job of it this time. Twice ought to be good enough."[12]

Churchill's hope that the President was about to declare war on Nazi Germany, or was going to promise to solicit the backing of the U.S. Congress for such a declaration, or was perhaps willing to engineer a casus belli (as Churchill himself had been accused of doing in 1915, over munitions he had ordered to be taken aboard the ill-fated non-military passenger liner, the SS *Lusitania*), was understandable, but completely erroneous. Roosevelt had no intention whatsoever of entering hostilities in Europe to save the British Empire — especially its colonial empire. Instead, he wished merely to get the measure of the British arch-imperialist — and see if he might bend him to a different purpose.

Poor Churchill, who rested up on the voyage and barely interacted with his own chiefs of staff, had no idea what was coming. Nor, ironically, did the U.S. chiefs of staff, who were not told the object of the meeting, or their roles, beyond that of advising the President.

Roosevelt genuinely respected his chiefs of staff as spokesmen of the armed services they directed, but the truth was, he had as yet little or no faith in their military, let alone their political, judgment. Almost everything they and their war departments had forecast or recommended to him as commander in chief since May 1941 had turned out wrong.[13] The "preparations" that the War Department had reported for a German drive through Spain and Northwest Africa to Dakar, prior to an anticipated assault on South America,[14] had proven but a ruse. Instead, the Führer had invaded the Soviet Union, on June 22, with more than three million troops, thirty-six hundred tanks, and six hundred thousand vehicles, supported by twenty-five hundred aircraft.

Far from conquering Russia in a matter of weeks, however, as forecast both by the secretary of war, the U.S. War Department,[15] and the U.S. military attaché in Moscow,[16] the vast 180-division German Wehrmacht and Luftwaffe forces sent into battle looked by early August as if they were meeting stiff resistance in the Soviet Union.[17] Moreover, far from abandoning their hold on the Middle East—as American military observers were still advising the British to do, but the President was not[18]—the British were holding General Erwin Rommel at bay in North Africa. British forces, in fact, had successfully driven into Iraq and Syria to deter Vichy French assistance to Hitler. As a result, neither Turkey nor Portugal, nor Spain, had moved a finger to help Hitler.[19] Even Marshal Pétain's egregious puppet government in Vichy had refused to alter the terms of its 1940 surrender to Hitler and permit French military cooperation with the Nazis. Hitler, the President was convinced, was not going to have things his own way.

It was not only the predictions of the U.S. War Department that were wrong, the President felt. The advice given by the U.S. Navy Department had seemed to him, as a former assistant secretary of the navy, to be strategically unsound—as well as psychologically naïve. The chief of naval operations and his director of plans favored a one-ocean navy operating solely in the Atlantic, with U.S. forces in the Philippines and Pacific Islands left to defend themselves against potential Japanese attack.[20] By contrast, the President had been determined to bluff both Hitler and the Japanese emperor, Hirohito, by keeping one U.S. fleet in the Pacific to deter Japan — still embroiled in a vicious land war on the Chinese mainland — from new conquests, while using the other U.S. fleet in the Atlantic to assert its naval authority over the waters of the Western Hemisphere. Hitler, the President was certain, had his hands full in Russia, and would not dare declare war on the United States as long as the U.S. was seen to be strong. In the meantime, moreover, the President would do his best to cajole Congress into expansion of the U.S. Armed Forces, to be ready for war once it came — as, inevitably, he was sure it must.

Roosevelt's firm belief about Russian resistance — backed by advice from old Russia hands like Joseph Davies, the former U.S. ambassador to Moscow — had been mocked by senior officers in the War Department. However, Harry Hopkins's latest signals from Moscow, where he had met personally with Marshal Stalin, had given Roosevelt renewed confidence in his own judgment. The President had therefore decided, in his own mind, America's best course of action — but it was not yet something he was willing to share with his chiefs of staff or his cabinet. Nor with Winston Churchill, the political leader of a foreign, fading colonial power.

To the surprise of his military advisers, then, the Commander in Chief had ordered all plans for U.S. military operations against the Azores, the defense of Brazil, as well as possible U.S. occupation of Vichy French territories in the Caribbean (which could be used as German naval bases) to be shelved. Instead, on July 9, 1941, he had formally instructed the secretary of war to draw up, in concert with the navy, a secret military plan or estimate: what exactly it would ultimately require of Congress and the U.S. military in terms of men, money, and machines to win the war against Hitler — and possibly the Japanese — if war came the following year, 1942, or, preferably, in 1943. That report was not due to be completed until September 1941. Thus, as the President set off for his sea summit with Churchill, his purpose was not to declare war on Germany, as the Prime Minister evidently hoped, but to see how war could be *avoided* for the moment.

With the press fooled as to Roosevelt's whereabouts, and his own staff, the cabinet, and even the U.S. chiefs of staff kept deliberately in the dark about the purpose of the powwow with Churchill, the most extraordinary drama now took place. Only the President — who seemed almost absurdly confident — appeared to have any idea what was going on, or of what was intended or likely to happen.

"There is I fear little chance of my getting to Campobello," the President had apologized in a letter to his elderly mother, who was summering at the Roosevelt camping estate off the coast of New Brunswick — but "I am feeling really well & the war is now encouraging to my peace of mind — in spite of the deceits & wiles of the Japs."[21]

To his distant cousin Daisy (an outwardly prim spinster with whom he had formed a quite intimate friendship over the past decade), Roosevelt was more jokey. "Strange thing happened this morning," he wrote her *en voyage* on August 5 — for he, his doctor, and his personal staff, even his beloved little Scottie, Fala, had "suddenly found ourselves transferred with all our baggage & mess crew from the little 'Potomac' to the Great Big Cruiser 'Augusta'! And then, the island of Martha's Vineyard disappeared in the distance, and as we head out into the Atlantic all we can see is our protecting escort, a heavy cruiser and four destroyers. Curiously enough the Potomac still flies my flag & tonight will be seen by thousands as she passes quietly through the Cape Cod Canal, guarded on shore by Secret Service and State Troopers while in fact the Pres. will be about 250 miles away. Even at my ripe old age I feel a thrill in making a get-away — especially from the American press. It is a smooth sea & a lovely day."[22]

The President was, in short, enjoying himself, hugely. Having bypassed Secretary Hull — who did not learn of the trip until Roosevelt was aboard the USS *Augusta*[23] — the President had secretly summoned Hull's undersecretary, Sumner Welles: the handsome professional diplomat, six feet three inches tall, who had attended the same school and college as FDR and had been a page boy at Roosevelt's wedding. Welles was, the President instructed, to travel separately, joining the U.S. team in Newfoundland. In the meantime Welles was ordered to start drafting a declaration of the President's postwar peace aims.

Postwar peace aims?

It was this document, not a putative agreement to enter the war, that the President had determined would make history. Although Roosevelt had, at the last minute, decided to take extra people to the meeting, they would still amount to less than half the number the British were bringing. Churchill had signaled that his party would include twenty-eight officers, military planners, and backup clerks, as well as (unbeknown to the President) two journalists and five photographers. By contrast, the President had limited himself to General Marshall, Admiral Stark, General Arnold, Admiral King, and only a handful of their staff; also his White House doctor, his appointments secretary, and his secret-intelligence officer[24] —

advisers to the President who would, as Roosevelt made clear, be under strict orders to say nothing that would in any way commit the U.S. military, beyond its current Western Hemisphere patrol and military-supply duties under the congressionally authorized Lend-Lease.

In short, any talk of operational military cooperation with the British, let alone an alliance, was *streng verboten,* the President told his military contingent when they finally assembled in his cabin onboard the USS *Augusta,* shortly after their arrival in Placentia Bay on August 7, 1941. They were merely to *listen* to the British.

Marshall, Stark, and Arnold were stunned.

As General Arnold noted in his diary on August 4 aboard the *Tuscaloosa* as it steamed north from New York, where he had boarded, he hadn't even brought enough clothes for the trip. "Thank God there is a laundry aboard. Where are we going? And why? Certainly the crew and the ship's officers do not know. Twice I was about to be informed and twice someone came up and I heard nothing."[25]

When the pioneering airman—the first to fly over the U.S. Capitol, and two-time winner of the Mackay Trophy—was brought by launch to the USS *Augusta,* off Martha's Vineyard, Admiral King, commander of the Atlantic Fleet and the man in charge of the secret expedition, had thrown a fit. "King quite mad because we came aboard his ship and he knew nothing about it. He gave us a look, got mad and went out to cool off prior to our getting in his office," Arnold noted.[26] They did not like each other.

Once King had finally cooled down, however, "Marshall and Stark came in. Marshall told us of our 'Brenner Pass' conference ahead"—a mocking reference to Hitler's earlier meeting with Mussolini to concert Axis strategy, and then his recent meeting at the beginning of June 1941, prior to Operation Barbarossa, the invasion of Russia—an offensive that the Führer had somehow failed to mention in advance to his main ally, lest the Italians leak the date and details! The Anglo-American version, General Arnold now learned from Marshall and Stark, would take place off Newfoundland and begin, General Marshall explained, on Saturday, August 9—with Arnold ordered to remain on the *Tuscaloosa,* away from Marshall and Stark—lest they form a military triumvirate, as in ancient Rome, and spoil the President's plan.

It seemed a strange way to prepare for modern war, let alone fight one.

But then, unknown to the somewhat "unsophisticated" air force general,[27] or even to Stark and Marshall, that was precisely the President's point.

Churchill's plan, concocted with his chiefs of staff before setting out from Scotland, was very different.

Acting as both British prime minister and minister of defense (a position he had created for himself, thus making himself military as well as political supremo), Churchill had decided in advance that he should first present to the American team his own strategic overview of the current war—and his plans for winning it. This would be followed by carefully drawn-up military proposals in the "Future Strategy Paper," a formal document his chiefs of staff would present to the American team as to how to achieve military victory—with American help.

Such an agenda for the summit had by no means been agreed to by the President, however. In fact, the scheme had not even been communicated to him—leaving the British "war party" somewhat anxious as they rehearsed in advance the ceremony of piping the President aboard the HMS *Prince of Wales* (a ceremony, given the President's disability, requiring a reversal of normal naval procedure: British officers would have to file past and salute the President, rather than vice versa).

"The programme is quite unknown at present," the Prime Minister's military assistant noted in his diary on August 8, 1941. "All that is certain is that the Prime Minister will call on the President and the President will call on the Prime Minister, but whether they will be accompanied by their Chiefs of Staff or whether the Chiefs of Staff will go separately will not be known till we reach harbour and there is an opportunity to consult the wishes of the Americans. . . . The Chiefs of Staff met once during the day, at noon. There is little more they can do now until the meetings start."[28]

Onboard the USS *Augusta*, things were not much clearer.

On August 6, steaming through fog and with its radar malfunctioning, the huge cruiser had put out its antimine paravanes, which "made a lot of noise," the President noted in his diary-style letter to his cousin that afternoon, revealing they were "off Halifax and in the submarine area—Tho' there have been no reports of them in these waters recently." Visibility was good, but Roosevelt had gotten word that morning of a "leak" in London regarding the meeting—though "it seems to be pure guesswork," he told Daisy, unworried. "I went up to the deck above — alone in the bow & the spray came over as it has before."[29]

The President seemed entirely in his element, "smiling and cheerful," as Admiral Stark described him[30] — the former assistant secretary of the U.S. Navy in a previous world war now the nation's commander in chief; commander in chief, moreover, not only of the country's navy, but its army and burgeoning air force too.

Emerging on deck at 11:00 a.m. the next day, August 7, the President was glad he'd overruled Churchill's suggestion that they meet at an alternative British location. Under U.S. management, Argentia was bustling with American activity. "[F]ound several destroyers & patrol planes at this new base of ours," Roosevelt boasted to Daisy, " — one of the eight [bases] I got last August in exchange for the 50 destroyers. It is a really beautiful harbor, high mountains, deep water & fjord-like arms of the sea. Soon after we anchored, in came one of our old battleships accompanied by two destroyers — & on one of the latter F[ranklin] Jr. is asst. navigator — so I have ordered him to act as my Junior Naval Aide while I am here," he confided proudly, referring to his son. The "old battleship" was the World War I–era thirty-thousand-ton dreadnought USS *Arkansas:* three times the size of the *Augusta,* mounting a dozen twelve-inch guns and carrying three floatplanes.

"It was a complete surprise to him & to me to meet thus," the President told Daisy. In fact, loath to show favoritism, the President, who hoped to spend the afternoon fishing and to see how the naval station was progressing, soon summoned his chief of the Army Air Forces, General Arnold, and ordered that his other son, Elliott, an Army Air Corps navigator currently stationed at Gander, "80 miles from here," should "join me as Junior Military Aide. Again, pure luck, but very nice."[31]

The President was fortunate in his fishing sally, too — catching "toad fish, dog fish and halibut," General Arnold noted in his diary.[32] Arnold had earlier upset the President by his reluctance to recommend selling, let alone giving, warplanes to Britain, concerned that it would slow deliveries to his own U.S. Air Corps; in fact, "I felt I was about to lose my job," Arnold later recalled as, "looking directly at me," the President had said "there were places to which officers who did not 'play ball' might be sent — such as Guam."[33] In the end it was only on General Marshall's recommendation that the President had relented, and finally, a few weeks before, had forwarded Arnold's name to Congress for promotion from mere colonel to the rank of permanent major general. Relieved to be back in presidential favor, Arnold dutifully congratulated the President on his angling success.

Arnold possessed one advantage over his colleagues, however: he was the only one to have met — indeed stayed — with Prime Minister Churchill in England, that spring. His personal report to the Commander in Chief at the White House — advocating more airplane production and assistance to the British in countering the continuing German bombing of London and other British cities — had saved his career, which was slated to end that fall, after the usual two-year stint. But though he had genuinely admired the courage of Londoners enduring the Blitz — pounded by upwards of five hundred German bombers each night — the experience of being bombed had only increased Arnold's determination to build up America's own heavy-bomber air force, not dissipate its strength by giving most of U.S. airplane production to the Brits. To his boss, General Marshall, Arnold had therefore said that morning: "We must be prepared to put a [U.S.] force into the war if and when we enter. The people will want action and not excuses. We will be holding the sack. Time then will be just as important to us as it is to the British now."[34]

This was a new, more assertive Hap Arnold, aviator and spokesman for air power. As commander in chief, however, the President was determined not to allow the military to decide American policy, which he was intent on holding strictly in his own hands. The airman was thus summoned a second time that afternoon, to the *Augusta,* at 4:30 p.m. "Sort of heavy seas, almost fell into sea when little boat went down," Arnold recorded, "and gangway to big ship went up." Having spoken to General Marshall, Arnold then filed into Roosevelt's cabin, along with Admiral King, Admiral Stark, General Marshall, General James Burns, Colonel Harvey Bundy, and General Edwin "Pa" Watson, the President's elderly appointments secretary and longtime military aide.

Welcoming the officers for the first time on the trip as a group, the Commander in Chief then made clear that the meeting with Mr. Churchill and his staff was to be informal and informational — i.e., neither strategic nor political. The United States was not, repeat not, at war with Germany, and had no congressional mandate to go to war. Nevertheless, it was the policy of the U.S. government to aid both Britain and Russia in their struggle to deal with the Axis menace in Europe, as it was to aid China in its struggle with the expansionist Empire of Japan. Making sure that military aid was manufactured and successfully delivered to Great Britain, Russia, and

China was the point at issue — *without incurring war.* Indeed, it was the President's purpose to dissuade the Axis powers and Japan from risking war with the United States as the U.S. ramped up military production, by *deterrence:* i.e., showing strength rather than weakness. There was to be no collective summit of the U.S. chiefs of staff with the British chiefs of staff; rather, they would simply meet one-on-one with their counterparts, to find out what the British needed in the way of weaponry and help.

The officers got the message. No politics. And absolutely *no* mention of U.S. military strategy, let alone U.S. entry into the war.

"Discussed: convoys," Arnold noted in his diary, and "defense of convoys: US responsibility for getting [Lend-Lease] cargoes safely delivered . . . [L]ine of [U.S.] responsibility extends east of the Azores and east of Iceland; duties and responsibilities of Navy; what British may want from [U.S.] Navy, ships from Maritime Commission; tanks from Army, airplanes; troops in Iceland, Marines, relief by soldiers; airplanes to Russia; aid to Philippines, B-17s, P-40s, tanks, AA guns." The only nod to future strategy related to the question of Japan, whose government's most secret war plans had been revealed by "Magic," the U.S. Army's Signals Intelligence decryption of the supposedly unbreakable Japanese "Purple" diplomatic code. The United States would, the President stated, "turn deaf ear if Japan goes into Thailand but not if it goes into Dutch East Indies."[35]

In later years, General Marshall would look back at the lack of preparation for the Placentia Bay conference with disbelief. Claiming he "had no knowledge" of the impending discussions with the British "until we were well up the coast on the cruiser *Augusta,*" Marshall had had no time to assemble papers or even files in advance. At the President's firm insistence, he'd found, the rendezvous was to be "largely a get-together for the first time, an opportunity to meet the British chiefs of staff, and to come to some understanding with them as to how they worked and what their principal problems were."[36]

Having given his pep talk, the President meanwhile sent his lieutenants back to their quarters — with no instructions even to meet again the next day.

General Arnold was not the only one to be amazed. With nothing to do on August 8, since Churchill's battleship was delayed by heavy weather in mid-Atlantic, Admiral Stark and Admiral King commandeered a Catalina navy patrol plane and flew up to the Avalon Peninsula, while General Marshall suggested to Arnold that they inspect the growing U.S.-

Canadian air base at Gander Lake, the final staging post for U.S. aircraft being delivered by air to the United Kingdom.[37] As they circled Placentia Bay in their twin-engine Grumman Goose seaplane on their return, they saw that even *more* U.S. vessels and floatplanes had arrived in the harbor. "We now have corvettes, destroyers, destroyer leaders, cruisers, one battleship, two tankers, one aircraft tender, about 18 [four-engined] PBYs and PBYMs," Arnold noted. Moreover, as they disembarked and transferred back to their warships "we saw a large 4-engine flying boat arrive. Where from? The U.S.? What for? Carrying two distinguished passengers? Who?" he recorded the questions running through his and Marshall's minds.[38]

One passenger, they learned, was the undersecretary of state, Sumner Welles. Was the President preparing a diplomatic surprise, then, despite his assurances the previous day? Was he contemplating a more formal alliance with the Prime Minister, who was due to arrive first thing the next morning—even American entry into the war?

It was a measure of General Arnold's naïveté—and the success of the President's insistence on keeping his military team lodged on different vessels, with no orders but to listen to the British war needs once Churchill's party arrived—that the primary U.S. air force general had absolutely no idea what was going on. "I can't make up mind as yet whether most of us are window dressing for the main actors," he would write several days later.[39] For the moment, however, finding "everyone taking a nap" onboard the *Tuscaloosa,* he was completely in the dark.[40]

Sumner Welles, for his part, experienced no such puzzlement. A consummate professional of the "striped pants brigade," the assistant secretary of state was both counselor and confidant to the President—who trusted him more than the secretary of state, Cordell Hull, who, a former congressman and senator, was very much a distinguished political appointee.

To Welles the President had stated, before leaving Washington, that he wanted "some kind of public statement of objectives."[41] It should be, he explained, a draft declaration that would "hold out hope to the enslaved peoples of the world,"[42] based upon his famous "four freedoms" address (of speech and worship; from want and fear).[43] That would be all the President wanted of a concrete, or formal, nature from the conference. Such a peace communiqué would quieten the isolationists at home, and give the Prime Minister something positive to take back to Britain.

It would also serve to mask, the President intended, America's complete military unpreparedness for war.

The fact was, for all the outward show of U.S. naval and air strength to impress the British visitors on their arrival at Placentia Bay, the United States had no army to speak of—at least no army capable of mounting anything other than a minor operation overseas; no air force with the capacity to deter a determined enemy, let alone support its own ground troops; and no navy able to operate effectively in one ocean, let alone in two.[44] As the official historians of the U.S. Army later put it, "the United States Army's offensive combat strength was still close to zero."[45]

Worse still, according to General Marshall, the U.S. Army was now in a "desperate plight" unless the Selective Service Bill, or draft, was extended for a further six months. Its belated preparations for possible war were in imminent danger of being put back "a year and a half or two years," if its current eight hundred thousand draftees were sent home, once the draft lapsed. Letting these trainees go home would result in "the complete destruction of the fabric of the army that we had built up," Marshall told the President—who had meanwhile heard from the Speaker that there were insufficient Democratic votes in the House of Representatives to pass the extension bill; in fact, at the very moment when the President was secretly steaming to Placentia Bay on August 6, the majority leader had reported to the White House that he simply had not enough votes to pass the new bill.[46]

Yet to Welles and to Averell Harriman—the U.S. Lend-Lease administrator who had accompanied the undersecretary of state in the flying boat from the capital—the President looked and sounded refreshed, indeed positively ebullient, during their three-hour talk.[47] "Father looked well, and was obviously enjoying his break in routine," Roosevelt's son Elliott also found when, along with his brother Franklin Jr., he was ushered into the presence of the nation's commander in chief.[48]

Captain Elliott Roosevelt had recently been scouting potential bases for air ferry and delivery routes across the Northern Hemisphere. Like General Arnold, he'd stayed with Churchill at Chequers, the British premier's official country residence, on a visit to England. In Elliott's account, published five years later, the President now rehearsed over lunch with his sons the next day's meeting with the Prime Minister: a meeting that he saw primarily as morale-boosting. "*You* were there," the President said to Elliott. "You saw the people. You've even told me how they look—gray and thin and strained. A meeting like this one will do a world of good for British morale," his father asserted — adding that the British would be concerned over "Lend-Lease schedules" now that Russia, too, would be receiving American military aid. "They'll be worried about how much of our production we're going to divert to the Russians," the President predicted — the British still convinced Hitler was going to win on the Eastern Front. "I know already how much faith the P.M. has in Russia's ability to stay in the war," Roosevelt remarked — snapping his fingers to indicate zilch.[49]

"I take it you have more faith than that?" his son queried.

Roosevelt did — his confidence buoyed after receiving Hopkins's recent cables from Moscow. Although the war on the Eastern Front would help England, it wouldn't save Britain in the long run, the President told his son.

"'The P.M. is coming here tomorrow because — although I doubt that he'll show it — he knows that without America, England can't stay in the war. . . . Of course,' my father went on, 'Churchill's greatest concern is how soon we will be in the war. He knows very well that so long as American effort is confined to production, it will do no more than keep England in. He knows that to mount an offensive, he needs American troops. . . . Watch and see if the P.M. doesn't start off by demanding that we immediately declare war against the Nazis.'"[50]

Elliott, who had been the first of Roosevelt's sons to join the U.S. Armed Forces, would become increasingly ambivalent in the ensuing years about Britain's national interests, and may have been dramatizing the conversation he recalled with his father. However, the gist of it was probably correct, judging by contemporary accounts — especially the President's next assertion: namely that the "British Empire is at stake here."[51]

To his sons, FDR portrayed the British and the Germans as having been engaged in a struggle over trade for decades: a struggle that had turned into a new war between the revived German Empire and the ailing British Empire: a war the United States could not simply exploit out of greed — "what will profit us most greatly," as isolationists such as his former ambassador to London, Joseph P. Kennedy, advocated — since its outcome would affect the very future of the world. This did not mean that the U.S. should favor, let alone save, Britain as a *colonial* empire, however.

The United States had a noble Constitution, deriving from its Declaration of Independence from Britain, which the President was proud to uphold, and which as president he felt bound to embody, as far as was possible, in his foreign policy: that "all men are created equal, that they are endowed by their Creator with certain unalienable Rights, that among these are Life, Liberty and the pursuit of Happiness." This fundamental striving for "Liberty" made the U.S. a natural enemy of Nazism. "Leaving to one side for the moment that Nazism is hateful," he told Elliott, "and that our natural interests, our *hearts,* are with the British," there was, he confided to his son, "another angle. We've got to make clear to the British from the very outset that we don't intend to be simply a good-time Charlie who can be used to help the British Empire out of a tight spot, and then be forgotten forever." Taken aback, Elliott had feigned incomprehension.

"I think I speak as America's President when I say that America won't help England in this war," Roosevelt made clear to his son, "simply so that she will be able to ride roughshod over colonial peoples."[52]

Elliott, five years later, claimed to have been astonished at this revelation. "I think," he recalled telling his father, "I can see there will be a little fur flying here and there in the next few days."[53]

Early next morning, Saturday, August 9, 1941, the grand bout began—heralded by the arrival of the Prime Minister's battleship.

Normally, Churchill rose late, liking to work in bed, dictating to a secretary. This time, however, the Prime Minister was up soon after dawn, standing on the admiral's bridge aboard HMS *Prince of Wales*—"eager and restless as a boy, longing for the first sight of the Stars and Stripes," as one of the two journalists he'd unwisely brought with him recorded. "Just out of bed, his sandy hair still ruffled by the pillow, he stood watching the sea that stretched to the New World. In a few hours ceremony and anthems would begin, but in that quiet opening of the day, like a warrior awakened from his tent, he stood unarmed at dawn, surveying the scene, wondering maybe what the day would bring forth."[54]

Things soon went wrong. The battle-scarred *Prince of Wales* (which had narrowly avoided being sunk by the German battleship *Bismarck* in May) was due to anchor in Placentia Bay at 9:00 a.m. When, preceded by an American destroyer and shadowed by two U.S. flying boats circling above, the ship's company fell in at 8:30 a.m.—marines with fixed bayonets, Mr. Churchill standing in his dark-blue uniform as Lord Warden of the Cinque Ports, and a marine band ready to play—the huge thirty-five-thousand-ton battleship began to tilt and "started turning to starboard," Churchill's military assistant, Colonel Jacob, recorded in his diary that night. To Jacob's surprise, "we found ourselves heading out again."[55]

The two nations were, it appeared, observing different times — the U.S. following Eastern Standard Time, the British observing Newfoundland Time.

"We kicked our heels for an hour and a half," Jacob noted, "and then went through the whole process again," steaming slowly past the anchored vessels of the American armada: the men called to attention as they passed each vessel, until they reached a central body of clear water and the USS *Augusta.*[56]

The British band played "The Star-Spangled Banner," while across the water they heard "God Save the King." "The Prime Minister stood with the Chiefs of Staff and others at the after end of the Quarter Deck," Colonel Jacob noted, "and through our glasses we could see President Roosevelt under an awning just below the Bridge of the 'Augusta.'"[57] Dropping anchor some three hundred yards away, the formalities continued with the piping aboard of Admiral King's chief of staff, stepping up the gangway from a launch. There followed, for Churchill, another wait of one and a half hours before he was invited to board the President's gleaming cruiser.

For his part, the President had slept well, and was almost as excited as Churchill — though for a different reason. "All set for the big day tomorrow," he'd written his cousin the night before. "I wish you could see this scene. By the way," he cautioned, "don't ever give any times or places or names or numbers of ships!"[58]

Anxious not to be blindsided by any misunderstanding, Roosevelt had ordered Harry Hopkins to transfer immediately from the *Prince of Wales* — on which he had sailed with Churchill, following his dramatic air journey to the Kremlin — to the USS *Augusta* after the British battleship anchored. General Arnold, too, was summoned from his quarters on the *Tuscaloosa.* The airman was "received on deck by President, Stark, Marshall, King, Watson, Elliott Roosevelt, and F. Roosevelt, Jr.," as Arnold noted in his diary. "First to appear from below Sumner Welles then A. Harriman; soon a boat from the *Prince of Wales,* Harry Hopkins came aboard."[59]

Hopkins had earlier cabled the President a long report of his tête-à-tête with Stalin in Moscow — telling the President that the Russians were not about to cave in. Now, in person, he was anxious to confirm to the President and the chiefs of staff that he had not been whitewashing the Russian situation, as the U.S. Army chief of staff feared. Contrary to General Marshall's military intelligence reports from Europe — most especially those of the U.S. military attaché in Moscow — there was absolutely no doubt in Hopkins's mind that Hitler's invasion of Russia had by then nowhere near succeeded. "The Russians are confident, claim 2,500 plane output a month without counting 15 training planes a day," Arnold wrote in his diary, impressed. "Stalin claims the Russians have 24,000 tanks."[60]

Twenty-four *thousand?* Marshall was skeptical, since Hopkins was no military expert, yet the President chose to trust his emissary's judgment, at that moment, more than Marshall's — perhaps because it was what he wanted to hear. Hopkins also reported that Stalin had begged him — as had Churchill, in May — to ask the President to enter the conflict and declare war on Germany.

For President Roosevelt, Hopkins's verbal summary on the USS *Augusta* that morning became the keystone he needed in putting into effect the latest plan he'd concocted the day before with Welles and Harriman. If the Soviet Union was to hold out until the following year, when America would be fully armed and ready for combat in Europe, it was in America's best interests — as the President had drummed into his staff at the White House throughout July — to provide as much weaponry and aid as possible to the Russians, rather than giving all foreign aid to Britain, let alone enter the war on Britain's behalf, with all the military responsibilities and commitments this would require. Neither Stalin nor Churchill, ironically, seemed to have any idea how puny were current U.S. armed forces, at least with regard to offensive capacity, outside the continental United States. Moreover, the Selective Service extension bill was hanging by a thread in Washington, and the vast majority of the nation (between 75 and 80 percent, according to polls)[61] remained unwilling to go to war to save Britain's imperial possessions — and even less willing to go to war to save Russia's Communist empire, however much they might distrust Hitler.

Swearing his chiefs of staff once again to silence in terms of U.S. military strategy, the President made clear that he alone, as president and commander in chief, would be in charge of the two-day meeting. There would be no U.S. military "team": only a commander in chief backed by his various army, navy, and air officers as advisers.

• • •

Having cleared the air, the President was helped to his feet by his sons, acting as his equerries, ready to receive the British prime minister and his entourage.

Promptly at 11:00 a.m. on August 10, 1941, the admiral's barge of the HMS *Prince of Wales* approached the USS *Augusta*. As the bullheaded, chubby-faced Prime Minister in his peaked cap mounted the gangway, stepped onto the deck, and walked forward to shake hands with the waiting President, standing upright by the guardrail beside his son Elliott—Mr. Roosevelt a head taller than his British counterpart and dressed in a light-gray Palm Beach suit and hat—the introductory ceremonies came to a climax: the Prime Minister handing over a letter of introduction from his sovereign, King George VI, who had met and stayed with President Roosevelt at Hyde Park two years before.

Ironically, tall Lord Halifax, the man King George VI would have preferred to see as prime minister on the resignation of Neville Chamberlain the year before, was now in Washington, demoted from foreign secretary to British ambassador to the United States. Instead, as the Prime Minister of Great Britain, there stood little Winston Spencer Churchill—short, pudgy, menacing, and pugnacious—who had yearned all his life for the post, and had finally got it. Like most of the world, the King, however, had quickly responded to Churchill's rhetoric, if not his style of decision-making. The King's letter, when the President read it through, was brief but nicely phrased. He was glad, George VI wrote, "that you have an opportunity at last of getting to know my Prime Minister. I am sure you will agree that he is a very remarkable man, and I have no doubt that your meeting will prove of great benefit to our two countries in pursuit of our common goal."[62]

"Our common goal" was delicately put—neither mincing nor presumptuous. "We all met on the top deck and were duly photographed & then Churchill stayed on board & lunched with me alone," the President confided to his cousin afterwards.[63]

The Lion of England was, if not on American soil, then under American custodial protection.

At their private luncheon the President was polite, but noncommittal.

As Roosevelt's aide and speechwriter Robert Sherwood would describe, "If either of them could be called a student of Machiavelli, it was Roosevelt; if either was a bull in a china shop, it was Churchill."[64]

Certainly Churchill was bullish. He began by stating how privileged he felt to meet Mr. Roosevelt at last in person. The President corrected him, however, pointing out that they *had* already met. Did Churchill not remember the occasion? It had been twenty-three years earlier, at a Gray's Inn dinner during World War I, when Churchill was a British cabinet minister and minister of munitions. Roosevelt had been in London on an official visit as assistant U.S. secretary of the navy. Churchill had ignored him.

Strike one to the President.

When Churchill then admitted that, in addition to the twenty-eight members of his military, diplomatic, and scientific staff, he had also brought along two British journalists and a five-man camera crew, in direct contravention of Roosevelt's instructions regarding "no media," the President was understandably irritated. Given his own determined efforts to escape the American press and preserve privacy as well as secrecy, the Prime Minister's faux pas seemed extraordinarily gauche. Churchill quickly assured the President that the journalists would not be permitted to board any U.S. vessel, or to interview the President or any American officers, or to publish any account of the meeting when they returned, at least not until the following year.

Strike two to the President.

This agreed, the President, as host, ran over the agenda for the two days of meetings — making it clear that the get-together was not, repeat not, to be seen as a formal conference of political and military leaders gathering to make war. The President and Prime Minister were meeting merely to *discuss* matters as leaders of their respective countries: the one neutral, the other at war with Germany. Each U.S. chief of staff would be permitted to meet with his British opposite number to learn more of British needs — but with no roundtable discussion or semblance of a formal conference that could in any way be construed by people at home as an alliance. Most important of all, the President announced, he wanted to issue with the Prime Minister a joint declaration of principles, or war aims, in order to inspire the peoples of the "enslaved" countries and others.

For Winston Churchill, the President's easy charm — he began straightway calling him Winston — belied a steely American assurance that was close to arrogance: a projection of intelligent confidence in his own judgment that was hard to dent. The President was like a player holding all the cards — at least the cards that mattered — with little indication that he had any intention of declaring war on anyone.

Understandably, Churchill's heart sank. Hopkins had seen his own role as that of catalyst between "two prima donnas," but was unable to do much to relieve the tension. The summit thus lurched into second gear — Hopkins hoping food and libation might ease the encounter.

To Elliott Roosevelt, who was invited for coffee after the "tête-à-tête" luncheon, the atmosphere seemed little better. He found the two world leaders "politely sparring," as they sat facing each other. "My information, Franklin, is that the temper of the American people is strongly in our favor," Churchill claimed. "That in fact they are ready to join the issue."

"If you are interested in American opinion, I recommend you read the *Congressional Record* every day, Winston," the President retaliated tartly.

"Two ideas were clashing head-on," Elliott recalled: "the P.M. clearly was motivated by one governing thought, that we should declare war on Nazi Germany straightaway; the President was thinking of public opinion, American politics, all the intangibles that lead to action and at once betray it."

Finally, "after draining his glass, the P.M. heaved himself to his feet. It was close to two-thirty." The quasi papal audience was clearly over. "Father mentioned he was sending, on behalf of our Navy, gifts to the officers and men of the *Prince of Wales* and her three escorting destroyers," Elliott recalled. "The P.M. acknowledged this information with a nod and a short word, and left" — leaving Sir Alexander Cadogan, his chief diplomatic civil servant as undersecretary of the Foreign Office, to meet with Undersecretary Welles. They could discuss the President's proposed declaration of principles and other matters — such as the threat from Japan.

Given the prohibition against alcohol being served on U.S. naval vessels, it had been, the diminutive Cadogan wrote in his diary, "a very unsatisfactory, dry, déjeuner à la fourchette"[65] — the Prime Minister so disappointed he had simply gone to bed with a stiff drink, once back onboard the *Prince of Wales.*

The Prime Minister could be forgiven for feeling disappointed, even humiliated.

Surely, he reasoned in his bunk, he had not sailed more than two thousand miles through heavy seas, often without air or sea escort (the ocean had turned too rough for the British destroyers to keep up), simply to be given American *food parcels* for his crew: an orange, two apples, half a pound of cheese, and two hundred cigarettes in each seaman's package? He therefore hoped his second meeting with the President, when dining aboard the *Augusta* that evening, would prove more productive.

Harry Hopkins had sent a personal message over to the *Prince of Wales,* informing the Prime Minister that he had "just talked to the President." Mr. Roosevelt was, Hopkins wrote, "very anxious, after dinner tonight, to invite in the balance of the [U.S.] staff and wants to ask you to talk very informally to them about your general appreciation of the war. . . . I imagine there will be twenty-five people altogether. The President, of course, does not want anything formal about it."[66]

The President had also asked whether, as per their lunch conversation, the Prime Minister would be the one willing to try his hand at a first draft of the declaration of principles, in his inimitable English, so that it could become a true Anglo-American document: not a declaration of war, but a declaration of peace — at least, the peace they were seeking in confronting the Nazi menace, and after.[67]

To his cousin later that day the President described Churchill as "a tremendously vital person & in many ways is an English Mayor La Guardia!" — likening Churchill to the diminutive mayor of New York, an authoritarian, excitable, liberal Republican. "Don't say I said so!" Roosevelt enjoined Daisy, since the comparison was in some ways unflattering.

The President did not mean the description maliciously, however. He had, after all, recently made La Guardia his first director of civilian defense. "I like him," he confided to Daisy his feelings about Churchill, " — and lunching alone broke the ice both ways."[68]

The idea of getting Churchill to write a first draft of the President's declaration of principles was certainly brilliant, though; it would force Churchill to own the project as much as the President did. The stratagem may well have issued from Hopkins's fertile brain and his psychological understanding of the Prime Minister's ego — flattering him by the request for a draft couched in high, stirring English prose, as well as a peroration before the American chiefs of staff that would impress them.

Hopkins, often on the point of death because of the stomach cancer he had suffered,[69] had, after all, heard the Prime Minister give a number of spellbinding *tour d'horizon* talks on his two visits to Britain that year: Churchill's rhetoric full of memorable metaphors and demonstrating a command of history and language, with an Olympian perspective that raised him head and shoulders above any English-speaking contemporary. Moreover, in asking Churchill to produce the first draft of a joint declaration of principles, the President would be putting the Prime Minister on the spot, since Churchill, as supplicant at the American court, could scarcely refuse. Vanity, Hopkins assured the President, would do the rest.

Roosevelt left nothing to chance, however. Dinner aboard the USS *Augusta* comprised hors d'oeuvres, broiled chicken, buttered sweet peas, spinach omelet, candied sweet potatoes, mushroom sauce, current jelly and hot rolls, with tomato salad, then cheese and crackers to follow. After dinner there was conversation — and once again, in talking to Churchill, the President emphasized the need for an articulation of common peace aims, or "joint Anglo-American declaration of principles," as Sir Alec Cadogan noted of the evening.[70]

Churchill was, in truth, incensed by the repeated request; he wanted an American declaration of *war,* not a declaration of principles. As a guest of the President of the United States, however, he could only plead — and plead he was determined to do, with all the words at his command.

The President, however, was as well known for his mastery of defensive as for aggressive tactics. In asking Churchill to speak to the whole American contingent, Roosevelt had felt it better to let the Prime Minister show his hand openly, rather than keep it cached, lest there be even a hint of behind-the-scenes transaction.

To remind the gathering that war in Europe involved more than just a commitment to Great Britain, Roosevelt insisted that Hopkins, who had given the President a typed report of his trip to Moscow that afternoon, first entertain the assembled dignitaries with his eyewitness account of his stay in the Kremlin and his one-on-one interviews with the Russian dictator, Joseph Stalin — the "ghost" at the table, so to speak.

It was the Prime Minister of Great Britain whom the assembled brass really wanted to hear, Roosevelt knew, however — and once Hopkins had spoken and the tables were cleared, Winston S. Churchill, the King's First Minister of Great Britain and Northern Ireland, rose to deliver his "strategic overview."

The longer the evening had progressed, the more Churchill — who had been given special dispensation to drink — had imbibed, and the more loquacious, even lyrical, he had become. The President's military aide, General Watson, afterward admitted, for example, that he had been "curious as to whether he [Churchill] was a drunk." As Churchill finally stood before the roomful of generals, he certainly assumed "a broader stance" than before — whether to steady himself or to marshal his thoughts, General Watson was unsure. Drunk or sober, the effect was remarkable, once he began his speech. "He held the floor that evening and he talked," Elliott described. "Nor were the rest of us silent because we were bored. He held us enthralled even when we were inclined to disagree with him."[71]

Even the President was impressed, according to Elliott. "My experience of him in the past," the younger Roosevelt observed, "had been that he dominated every gathering he was part of; not because he insisted on it so much as that it always seemed his natural due. But not tonight. Tonight Father listened."[72]

Watching Laurence Olivier and Vivian Leigh at the screening of the film *Lady Hamilton* the night before, onboard the *Prince of Wales,* one of the two banished British journalists had seen Churchill actually weep—the sight of which deeply moved the newsman. Olivier was a consummate British actor, but watching Winston Churchill, the journalist had found himself even more affected than by the screen icon. "I thought that in some extraordinary way he belongs definitely to an older England, to the England of the Tudors, a violent swashbuckling England perhaps, but a warm and emotional England too, an England as yet untouched by the hardness of an age of steel," H. V. Morton afterward recalled. Why, Morton asked himself, did both ordinary and extraordinary people find themselves "so firmly held," when Churchill spoke—"so silent until the last word?" He wondered if the enchantment might not lie in the fact that Churchill's voice was "not of an industrialist, but of one who has, so to speak, missed the Industrial Revolution and speaks to us as if from the deck of the *Golden Hind.* Churchill's voice is also classless. . . . Like the Elizabethans, he speaks not as an Etonian but as an Englishman."[73]

Sir Alec Cadogan—who *was* an Etonian, and the son of an earl—was disappointed, having heard his master speak in public so often before, but Morton's was an apt insight, and one that chimed with Roosevelt's growing respect for the Prime Minister. The President was skeptical of Churchill's judgment in terms of military operations against the Germans, which thus far in the war had not produced a single victory on the battlefield. Roosevelt was far from impressed, moreover, by the Prime Minister's choice of subordinates, since the British chiefs on first acquaintance appeared that evening to be a polite, characterless, minion-like group of yes men. Rather, the President's admiration belied a sort of compassion: a recognition by Roosevelt of Churchill's *courage:* his tenacity, in a sea of mediocrity, in trying to make the best of the impossible situation he'd inherited from his mealy-mouthed, appeasement-minded predecessor, Neville Chamberlain—who had definitely *not* impressed the President. And admiration, too, for the Prime Minister's remarkable intellect, amounting almost to genius: his insistent, valiant efforts to place the problems of the world within a wider, historical and moral, framework.

Not only was Churchill's knowledge of history formidable, laced with a seemingly photographic memory for lines of poetry and idiosyncratic detail, but alongside his romantic exaggerations the Prime Minister could be disarmingly honest. Churchill thus admitted, freely, to his military audience — both his own countrymen and his American listeners — that his island nation, the previous summer, had been wholly unprepared for German invasion. "Hitler and his generals were too stupid," the Prime Minister asserted. "They never knew. Or else they never dared." According to Elliott, the subtext of this confession was an appeal for the United States, with all its military power, manpower, and industrial potential, to enter the war — Churchill's underlying message being: "It's your only chance! You've got to come in beside us! If you don't declare war, I say, without waiting for them to strike the first blow, they'll strike it after we've gone under, and the first blow will be their last as well!" Though his American listeners "could detect the underlying appeal," Elliott noted, they could not fail to be moved by the Prime Minister's personal courage and determination never to give in. Churchill's "whole bearing," Elliott Roosevelt recalled, "gave the impression of an indomitable force that would do all right, thank you, even if we didn't heed his warning."[74]

The President, sitting at the head of the table, with the Prime Minister on his right, impressed Alec Cadogan, meanwhile, who was placed on Mr. Roosevelt's left. In his diary that night he noted the President's "great, and natural, charm."[75] Listening to Churchill's studded rhetoric, the President seemed content to remain quiet, save when interrupting to ask about Russia, and how long the Prime Minister thought it would hold out (not long, Churchill answered, once Hitler took Moscow and the Germans reached the Urals, perhaps striking even beyond).

The President chose not to argue. If war came to the United States — and as president he was determined to hold off that evil day — it was as well his chiefs of staff see for themselves who was at the helm of the fading British Empire: not only as England's political leader but as a military strategist and commander in chief.

Britain had now been at war for two long years — and had learned many lessons, Churchill confessed. In contrast to World War I he characterized the struggle as "a mobile war, in the air, on the land, and at sea," a war in which science and mechanical science were playing a crucial role.[76] The British could not, and would not, give up their position in the Middle East; for by fighting the Germans at the farthest point from their bases in Germany, the British Commonwealth forces had the best chance of meeting their adversaries "on even terms."[77] Meanwhile, in the air, with enough bombers manufactured in Britain as well as those purchased or leased from the United States, they could "bring home to the Germans the horrors of war, just as the Germans had brought it home to the British." If the United States would take over full convoy protection across the whole North Atlantic, this would enable the Royal Navy to send destroyers to the South Atlantic; if the U.S. would join Britain in "sending an ultimatum to Japan," it could halt Japanese expansion. And if, in the successful aftermath, a new League of Nations could be set up, then the world could perhaps learn the lessons of Versailles, and start afresh, on a new page of history . . .[78]

"[N]ot his best," Sir Alec Cadogan noted in his diary,[79] but to the President it was exactly what he had hoped for. No businessman would want to enter a partnership, Roosevelt felt, with such a diffuse, fading imperial power; but by investing in the company, so to speak, he might well help it stave off bankruptcy. Lend-Lease was, in fact, doing that; the next step would be military. But not yet. Not when the United States was, in all truth, still a military mouse, despite its roar.

As Mr. Churchill rehearsed Britain's strategy — to hang on to its collapsing colonial empire in the Middle East, India, Burma, Singapore, Malaya, and Hong Kong, and to harass Hitler's Third Reich from its margins, in the hope that something might turn up, as in the parlance of Mr. Micawber — it was impossible for President Roosevelt and his chiefs of staff not to shake their heads at the Prime Minister's mix of sentiment, myopia, and imagination. And luck! For not only had Hitler *not* invaded Britain when the country was at its most vulnerable, after the Dunkirk evacuation in May 1940, but something else *had* come up. Not an American entry into the war, as Churchill had so ardently wanted, but something in some ways even more fortunate for Britain and for America: Hitler's crazed decision, after failing to bomb the English into submission in the Battle of Britain, to attack Russia.

No one but a Nazi madman could have undertaken such a gamble. Assuming — as the President did, especially after hearing Hopkins's full report — that the Soviets would hold out, even if they had to retreat to the Urals, Hitler's mistake would inevitably mean the survival of Great Britain. Though not necessarily the survival of the British colonial empire, which was, in President Roosevelt's eyes, a different matter.

As he listened and occasionally prompted the Prime Minister to comment on how he saw certain issues — the Russian campaign, the threat of an expanding Japanese Empire in the Far East — the President felt more and more strongly that America's moment of destiny was approaching. Despite the fact that American isolationists were currently fanning public fears and dictating congressional attitudes, the United States was going to have to fight eventually, the President was certain. Secret decrypts made it quite clear that the Japanese were hell-bent on war, just as the Germans had been — and no amount of Chamberlain-style diplomacy, appeasement, or negotiating would placate them.[80] Yet the war that Winston Churchill was seeking — a war that preserved Britain's colonial empire, while smashing Hitler and Hirohito's empires — was not what President Roosevelt saw in America's tea leaves. If and when war came, America must fight for its *own* role in the sun, as leader of a postimperial, democratic world. America would thus become not just the arsenal of democracy, but — as the world's most prosperous nation by far — the senior partner in a new world order, with open borders and open markets.

To Churchill's consternation, then, once the Prime Minister sat down, the President announced that the dinner was over. It was 11:30 — and the President was going to bed. The British visitors were promptly seen off the ship at 11:45, and reaching his cabin onboard the *Prince of Wales*, the Prime Minister, too, retired. He was exhausted by his own peroration, the disappointments of the day — and the President's request that he begin the drafting of a joint declaration of peace aims.

"Considering all the tales of my reactionary, Old World outlook, and the pain this is said to have caused the President," Churchill later wrote, he was proud to say it was he, not the President, who now produced the "first draft" of the declaration of principles that the President had requested, and that it was "a British production cast in my own words."[81]

As David Reynolds, the British historian, later revealed, this was not, strictly speaking, the case.[82] In truth, the next morning, as Sir Alec Cadogan was enjoying his breakfast of "bacon and eggs" in the admiral's cabin, he was summoned by the Prime Minister, who was already up and on deck. "He wanted an immediate draft of the 'joint declaration,' which he outlined verbally." Cadogan then "worked up a text, about which Churchill 'expressed general but not very enthusiastic approval,' but it was typed up virtually unchanged for the Prime Minister to give to the President."[83]

The first draft was, then, a Cadogan production, rather than the Prime Minister's. Moreover, Churchill later misrepresented his own feelings about the very idea of such a document.[84] For what Churchill could not bring himself to admit, when penning his epic account of his war service, in 1949, was that this declaration was emphatically *not* what he had sailed all the way from Scapa Flow to Placentia Bay to obtain. Nor was it what Churchill wanted to subscribe to, as the prime minister of Great Britain and a servant of the British colonial empire. Biting his tongue, however, he approved Cadogan's first iteration of the joint declaration — and turned to Plan B, set for 11:00 a.m. that Sunday morning, August 10, 1941: the arrival of the President of the United States on the *Prince of Wales*, and a rousing church service.

At first the program went without hitch. As the crews of the two great warships prepared for the difficult maneuver, the clouds above Placentia Bay parted, the sun shone, and the shoreline reminded some who were present of the spare beauty of the Western Isles of Scotland.

An American destroyer wedged lengthways between the main deck of the USS *Augusta* and the stern of HMS *Prince of Wales* allowed the President, his small staff, and three hundred American sailors to be piped aboard the British battleship for divine services without transferring to barges. There then followed a ceremony that Churchill had planned and rehearsed in detail with the crew of the *Prince of Wales* and his own staff, right down to the choice of hymns, even before their arrival in Canadian waters.

The President, an Episcopalian, was delighted to participate in the religious ceremony — even sending a presidential invitation to each member of his own staff to attend. Loath to allow Churchill to control the media rendering of the event, however, Roosevelt had wisely sent for his own camera crew — a group of American army film cameramen and still photographers working in Gander, who had been ordered to fly immediately to Argentia by Grumman floatplane. Churchill might want to give the appearance of an alliance, but President Roosevelt was determined that the imagery reflect his joint declaration of principles of peace — and how better than by showing men worshiping God together?

With the Royal Marine Band playing in the background, the President of the United States was "received with 'honors,'" as he wrote his cousin that night, then "inspected the guard and walked aft to the quarter deck" — where, behind desks draped with the Stars and Stripes and Union Jack, he and the Prime Minister faced almost a thousand sailors grouped under the menacing fourteen-inch guns of the after-turret. The Prime Minister wore the dark blue uniform of the Royal Yacht Squadron, the President a blue double-breasted suit, "without a hat. It is a very great effort for the President to walk, and it took him a long time to get from the gangway to his chair, leaning on a stick and linking his arm with that of one of his sons who is acting as his A.D.C.," Churchill's military assistant recorded in his diary that night. "We heard that this was the longest walk that the President had ever taken since his illness many years ago."[85] As the President "slowly approached the assembled company," wrote another British officer, "it was obvious to everybody that he was making a tremendous effort and that he was determined to walk along that deck even if it killed him."[86]

Recorded on film, the service was profoundly affecting to those who took part — as it was to those who saw it on newsreels across America and the free world in the days and weeks afterward. Urged by their captain to "raise steam in an extra boiler so as to give the hymns extra value,"[87] the British sailors — intermingled with their American guests and sharing with them their hymnals — sang "O God, Our Help in Ages Past," "Onward, Christian Soldiers," and — at the President's urging — "Eternal Father, Strong to Save," better known as "For Those in Peril on the Sea." Six months later the huge warship would be attacked and sunk by Japanese planes in the South China Sea, its captain and many hundreds of the crew drowned.

H. V. Morton noticed once again how the Prime Minister's "handkerchief stole from its pocket" — for it was almost impossible not to be touched by emotion. "A British & an American chaplain did the prayers," the President himself described to Daisy.[88] With their caps off, "it was difficult," Morton later recalled, "to say who was American and who was British; and the sound of their voices rising together in the hymn was carried far out over the sea. In the long, frightful panorama of this War, a panorama of guns and tanks crushing the life out of men, of women and children weeping and of homes blasted into rubble by bombs, there had been no scene like this." It was, he wrote, "a scene, it seemed, from another world, conceived on lines different from anything known to the pageant-masters of the Axis."[89]

Aboard the *Prince of Wales*, as the divine service ended, the President seemed wonderfully confident. "Captain Leach read the lesson — and then we were all photographed — front, sides & rear!" he described to Daisy. "Next I inspected the P. of W. in my [wheel]chair, then sherry in the Ward Room & then a 'beautiful' lunch of about 40 — Toasts followed by two speeches."[90]

Churchill had certainly ordered nothing but the best for his guests, given the dire situation in an England suffering grave food shortages and universal rationing. The menu for the President and his entourage featured smoked salmon, caviar, turtle soup, freshly shot roast Scottish grouse, dessert, coffee, wines and liqueurs, with mood music played by the Royal Marine Band. In addition to the formal toasts to the King and to the President, there was even a risqué joke by Hopkins and welcome news that the German battleship *Tirpitz* had been espied in the dockyard at Kiel, meaning it would not have time to put to sea and prey upon the Prime Minister on his return voyage.

Afterward the President was introduced to the Prime Minister's junior staff. To the chagrin of the British chiefs of staff, however, that was it. As Churchill's military aide lamented, "it had been the intention that the Chiefs of Staff should have a short meeting with the American Chiefs of Staff at which to hand over the Future Strategy Paper. However, this went by the board as Admiral Stark and General Marshall decided to go back to their ships with the President."[91]

The luncheon, intended as the prelude to joint military discussions, had been for naught — the President determined not to be snared by Churchill into a position that isolationists back in the U.S. could interpret as having even the semblance of an alliance.

Once aboard the USS *Augusta* the President then held "a military & naval conference in my cabin"[92] — adamant to ensure that no hint of a military alliance was being suggested, or any whisper of U.S. "war plans" being given to emissaries of a foreign country. His chiefs of staff still seemed bewildered by his tactics, but were too loyal to protest.

Despite the President's ban on joint discussions among the chiefs, however, it was impossible to stop junior staff officers from confiding in one another. At a junior meeting with Colonel Harvey Bundy, General Marshall's director of plans, for example, Colonel Jacob, Churchill's military assistant, learned to his consternation just how different were British and American ideas for conducting the war against Germany.

Some weeks previously, it appeared, the President had ordered a secret new review to be drawn up — later called the Victory Plan — of what the U.S. Army and Navy Departments would deem necessary in a war to defeat the Third Reich. A preliminary report had been presented to the President before he'd left Washington, and Colonel Bundy now unwisely shared the gist of it with his counterpart — who was both amazed and disbelieving. "The Americans are busy trying to draw up a scheme of the forces which they would ultimately raise, and the possible theatres in which they might be utilised. They are tentatively aiming at an Army of 4 million men."

Colonel Jacob was shocked. *Four million men?* "We did our best to point out to Bundy that this was possibly a wasteful use of manpower and manufacturing capacity; it hardly seemed conceivable that large scale land fighting could take place on the Continent of America, and shipping limitations would make it quite impossible for large forces to be transferred quickly to other theatres."[93]

Colonel Charles Lindbergh, one of the leaders of the America First isolationist movement, would have been appalled to know that such secret discussions regarding possible American "intervention" in "other theaters" were being aired; he would have been even more appalled to discover the sheer magnitude of the army the U.S. military was proposing in order to win the war against Germany. Colonel Jacob certainly was — for the American notion of defeating Hitler was almost diametrically opposed to that of the British.

"The day has been almost entirely wasted from the point of view of joint discussion," Jacob lamented in his diary. "We have been here two days and have not yet succeeded in getting the opposite sets of Chiefs of Staff together round a table," he recorded in frustration — unaware that this was happening on the President's specific orders. "We have thus given away the strength of our position, which lies in the fact that our three Chiefs of Staff present a unified front of the strategical questions, while it is quite clear that theirs do not. We have played into their hands by allowing the discussions to proceed in separate compartments."[94]

The President's stratagem worked magnificently — the British were unable to present their "Future Strategy Paper," while their U.S. hosts remained wholly uncommitted either to enter the war, or to follow "unified" British strategic military policy: a policy that assumed there would be no major ground forces landed on the continent of Europe, and that

Hitler could merely be forced into submission by peripheral harassment and aerial bombing — if only the United States provided enough bombers. (The RAF's request, Arnold learned from his opposite number, Air Vice Marshal Freeman, was for ten thousand heavy bombers — the entire output of the U.S.)[95]

As Colonel Jacob rued immediately after the conference, "neither the American Navy nor the Army go much on the heavy bomber" — the mainstay of Britain's only plan to defeat Hitler. Neither the U.S. Navy nor Army "seems to realize the value of a really heavy and sustained aerial offensive on Germany."[96]

Given that England had itself successfully survived the world's most sustained bomber offensive in human history — the Blitz — for an entire year, the notion that Nazi Germany could be defeated by the same tactics in reverse seemed nonsensical to General Marshall and Admiral Stark, the U.S. Army and Navy chiefs of staff — indeed, even to General Arnold, who *did* believe in an important role for heavy bombers in modern war.[97] Yes, heavy bombers could savage an enemy's manufacturing and supply chain to its armies. Even Arnold could not visualize, however, the war against Hitler being *won* by bombers . . .

Thanks to the President's injunction against a formal meeting between the two nations' joint chiefs of staff, however, the difference between British and American military strategy in conducting a full-scale war against Nazi Germany could, at least, safely be deferred.

Thus, when finally the British chiefs of staff managed to present their hollow-sounding "Future Strategy Paper" on Monday, August 11, the day the *Prince of Wales* was supposed to depart, all the U.S. chiefs of staff would say was that they would study the paper "with interest," and respond later.[98] For the only matter the President was determined to nail down was his declaration of principles.

Once the Cadogan draft of the declaration was handed over to the American team, Sumner Welles and the President took over.

Together with Harry Hopkins, Welles was perhaps the President's most trusted senior adviser. He "looks exactly as if he had stepped out of a film," Jacob felt — the sort of film in which Welles would probably be playing "a business lawyer."[99]

The comparison was apposite, for Welles was indeed acting the business lawyer in world politics. In the quiet of the President's cabin, he and

Roosevelt got down to the business of war and peace: pens poised as they went over the preliminary draft of what would become the Atlantic Charter.

"I am very doubtful about the utility of attempts to plan the peace before we have won the war," the Prime Minister had confided his fundamental unwillingness to his foreign secretary, Anthony Eden, in May 1941[100] — but given that it was what the President requested, Churchill had not dared refuse. In its contorted language the first draft reflected the Prime Minister's reluctance to draw up a charter at all, and his attempt to twist it, if he could, into a declaration of war. The preamble had thus opened by claiming that the U.S. president and the British prime minister were meeting at Placentia "to concert and resolve the means of providing for the safety of their respective countries in face of Nazi and German aggression . . ."

This, clearly, suggested an alliance — which was the last thing the President wanted American isolationists to infer. With Sumner Welles at his side, Roosevelt went through the document with the utmost care, taking out anything that could provide free ammunition to his opponents in America — isolationists waiting all too keenly to pounce on the President for any sign he had entered into an unconstitutional agreement with Great Britain and its colonial empire. By evening the American draft two was ready to be given to Churchill — who had gone ashore for a couple of hours to clear his head, but was due to come over to the USS *Augusta* at 7:00 p.m.

The previous night "Churchill had talked without interruption, except for questions," Elliott Roosevelt recalled[101] — the Prime Minister "talking, talking, talking," as irascible Admiral King put it.[102] "Tonight," however, as the Prime Minister arrived for his informal, private dinner with the President in the admiral's cabin of the *Augusta,* "there were other men's thoughts being tossed into the kettle, and the kettle correspondingly began to bubble up and — once or twice — nearly over. You sensed that two men accustomed to leadership had sparred, had felt each other out, and were now readying themselves for outright challenge, each of the other."[103]

The first bone of contention was the British Empire: its restrictive trade agreements, and its colonialism. "Of course," the President opened his attack, "of course, after the war, one of the preconditions of any lasting peace will have to be the greatest possible freedom of trade." Churchill countered by pointing out Britain had long-established trade agreements with its Dominions and colonies. "Yes. Those Empire trade agreements are a case in point," the President agreed.

"It's because of them that the people of India and Africa, of all the colonial Near East and Far East, are still as backward as they are."

Churchill's face, according to Elliott Roosevelt's account, went red with fury. "Mr. President, England does not propose for a moment to lose its favored position among the British Dominions. The trade that has made England great shall continue, and under conditions prescribed by England's ministers."[104]

The President's challenge had been met by Churchill — Roosevelt acknowledging "there is likely to be some disagreement between you, Winston, and me. I am firmly of the belief that if we are to arrive at a stable peace it must involve the development of backward countries. Backward peoples. How can this be done? It can't be done, obviously, by eighteenth-century methods."

Churchill, according to Elliott, became even more furious. "Who's talking eighteenth-century methods?" he snapped.

"Whichever of your ministers recommends a policy which takes wealth in raw materials out of a colonial country, but which returns nothing to the people of that country in consideration," the President explained patiently. "*Twentieth*-century methods involve bringing industry to these colonies. *Twentieth*-century methods include increasing the wealth of a people by increasing their standard of living, by educating them, by bringing them sanitation — by making sure that they get a return for the raw wealth of their community."[105]

At the mention of India, Churchill became, Elliott described, "apoplectic." "Yes," the President had added blithely, ignoring Churchill's rage as he piled accusation upon accusation. "I can't believe that we can fight a war against fascist slavery, and at the same time not work to free people all over the world from a backward colonial policy."[106]

Backward colonial policy? This then was the battle royal the Placentia Bay meeting had built up to, and though Churchill — "a real old Tory, isn't he? A real old Tory, of the old school," Roosevelt described his opponent to Elliott afterward[107] — would give no quarter, the President knew he'd made his point, and had the upper hand, "& now I'm ready for bed after dining Winston Churchill, his civilian aides & mine," the President finished his account of the day's doings for Daisy.[108]

Churchill, by contrast, went to bed with the aching realization that, despite all his oratory, the President of the United States was even *less* likely to enter into an alliance with Great Britain than when the Prime Minister had set sail from Scapa Flow.

The third day, Monday, August 11, 1941, the sparring continued.

"A day of very poor weather but good talks," the President wrote. "My staff came at 12, lunched, & we worked over joint statement. They went and Churchill returned at 6:30 & we had a delightful dinner of five: H. Hopkins, Elliott, F. Jr., Churchill & myself."[109]

Roosevelt was clearly delighted — getting what he wanted by his usual mixture of presidential charm, dogged insistence, occasional compromise, and sincere American hospitality. All day, fresh versions of the declaration of principles sped between the two warships — the President cutting out any implications of military or political alliance, the Prime Minister refusing to relent on "imperial preferences" in postwar trade agreements, and threatening to delay the declaration a further week while he submitted it to the governments of the British Dominions if the President insisted upon that article.

Feeling magnanimous, Roosevelt gave way. The core of his demand had gone unchallenged, after all: that the British government and the U.S. president sought "no territorial" or other "aggrandizement," in fact no "territorial changes that do not accord with the freely expressed wishes of the people concerned." The text of the declaration specifically committed the signatories to "respect the right of all peoples to choose the form of government under which they will live": a postwar peace aim that guaranteed an eventual end to the British colonial system — indeed caused British colonial administrators across the globe to shudder when they read the terms of the Atlantic Charter in the weeks that followed.

In the meantime, however, the President was ecstatic. He had got what he wanted — and not given Churchill what he so dearly hoped for.

For his part, the Prime Minister was resigned to defeat. It was disappointing, but the summit had at least brought the two men together. At dinner on Monday evening harmony finally reigned. "We talked about everything except the war!" the President related to his cousin, "& Churchill said it was the nicest evening he had had!"[110]

At some deeper level, Churchill was well aware the British Empire was doomed as a colonial enterprise, though he prayed the crumbling of the once-proud imperial edifice that had controlled a quarter of the world would not happen on his watch. Elliott Roosevelt afterward claimed that, at the summit, he saw "very gradually, and very quietly, the mantle of leadership was slipping from British shoulders to American." He had seen it vividly when, the night before, Churchill got to his feet and "brandished a stubby forefinger under Father's nose. 'Mr. President,'" he had cried, "'I believe you are trying to do away with the British Empire. Every idea you entertain about the structure of the postwar world demonstrates it. But in spite of that'—and his forefinger waved—'in spite of that, we know that you constitute our only hope. And'—his voice sank dramatically—'*you* know that *we* know it. *You* know that *we* know that without America, the Empire won't stand.'"[111]

In sending the final, revised draft of the declaration of principles to the war cabinet in London by secret cipher on the evening of August 11, 1941, Churchill felt ashamed of what he had been forced to concede. With trepidation he therefore went to bed, wondering what the cabinet's response would be.

"Am I going to like it?" he asked his private secretary, "rather like a small boy about to take medicine," Colonel Jacob noted in his diary, when the response eventually came in.[112]

The British cabinet *did*—mercifully for the President.

In actuality, Roosevelt's hand was much weaker than Winston Churchill had realized. Not only was the United States in no position to wage war on anyone, at that time, but the President's political position was a great deal less powerful than the Prime Minister knew. Reports of a secret meeting with the leader of a belligerent nation were bound to arouse isolationist ire across America—making American intervention in the war even less likely. News of a joint declaration of principles, by contrast, would not.[113] In this sense the President had played a masterly hand.

"W.S.C. to lunch," the President wrote to Daisy the next day, "with Lord Beaverbrook [Churchill's minister of munitions], who landed by plane this A.M. at Gander Lake from Scotland." Churchill had brought with him "approval of statement by his cabinet & King—& after a few minor changes we gave final OKs & drew up the letter to Stalin, & arranged for release dates," the President chronicled.[114]

The Atlantic Charter, as it was swiftly called, was a historic document: a declaration in the great tradition of the American Bill of Rights, guaranteeing the rights of all nations—*including British colonies*—to self-determination, not conquest by rule of force. If the United States were to go to war, it would this time be for a noble cause.

"They left at 3:30, their whole staff having come to say goodbye — It was a very moving scene as they received full honors going over the side," back to their battle-scarred battleship, the President described to Daisy.[115] Then, at 5:00 p.m. that evening, August 12, 1941, after "great activity" getting the battleship ready to weigh anchor, the ill-fated HMS *Prince of Wales* steamed out of Placentia Bay with salutes given as it passed the ships of the U.S. flotilla — strains of the Royal Marine Band still playing as it headed across the still water.[116]

In Placentia Bay, meanwhile, the President breathed a sigh of relief. "At 5 p.m. sharp the P. of W. passed out of the harbor, past all our ships," he described the scene to his cousin. On his desk was the Atlantic Charter. "Ten minutes later we too stood out of the harbor with our escort, homeward bound. So end these four days that I feel have contributed to things we hold dear."[117]

PART TWO

Pearl Harbor

2

The U.S. Is Attacked!

PEARL HARBOR DAY BEGAN quietly. We were expecting quite a large party for luncheon," Eleanor Roosevelt later recalled, "and I was disappointed but not surprised when Franklin sent word a short time before lunch that he did not see how he could possibly join us."[1]

This was not unusual; the President and First Lady led somewhat separate lives. They lived together upstairs at the White House, but on opposite sides of the Central Hall. Their marriage, since FDR's affair with Eleanor's secretary and then his affliction with polio in 1921, had become one of duty, parenthood, and convenience — though they did respect one another, in the manner of English aristocrats. Eleanor acknowledged that in the White House the President "had been increasingly worried for some time and frequently at the last moment would tell me that he could not come to some large gathering that had been arranged. People naturally wanted to listen to what he had to say," she allowed, "but the fact that he carried so many secrets in his head made it necessary for him to watch everything he said, which in itself was exhausting."[2]

Mrs. Roosevelt's explanation seemed, in retrospect, a trifle jejune — yet was closer to the mark on December 7, 1941, than even she, as First Lady, recognized. A veritable army of conspiracy theorists in subsequent decades would come to suspect the Commander in Chief of having received secret warning of the Japanese attack on Pearl Harbor via American, or even British, intelligence — and of having withheld it in order to embroil the United States in war, against the will of the American people.[3] These were grave, posthumous charges — and they rested on undeniable truths. Had not the President received, late on the evening of December 6, 1941, decrypts of a top-secret Japanese signal from Tokyo to its imperial ambassador in Washington, suggesting that "peace" negotiations over U.S.-

Japanese problems in Southeast Asia — where Japan had seized control of southern China and also Indochina — were coming to an end? Had not Roosevelt remarked to his White House assistant, Harry Hopkins, within the hearing of the young officer delivering the decrypt to the President, that "this means war"? Had not the President immediately sought to telephone the chief of naval operations (CNO) of the United States, to discuss that secret intelligence? And had not the President said that "it certainly looked as though the Japanese were terminating negotiations"?

More tellingly still, had not the final fourteenth paragraph of the secret Japanese signal, decoded by American cryptographers early the next morning, been delivered to the President at 9:00 a.m. on Sunday, December 7, 1941, as he lay in bed having his breakfast, in the room next to his study? Had not the secretary of state, Mr. Cordell Hull, received the very same decrypt that morning, along with Mr. Henry Stimson and Mr. Frank Knox, the U.S. secretaries of war and of the navy, meeting together at the Munitions Building on the Mall? And had not Mr. Hull said to his colleagues he was "very certain that the Japs are planning some deviltry"[4] — for even as they read over the fourteenth paragraph of the formal diplomatic "note" that the Japanese ambassadors were being ordered to deliver, had not *further* decrypts been delivered by special messenger? Had not this Japanese cable from Tokyo instructed the ambassadors to present the formal government message, ending all efforts at diplomacy, to the "United States government (if possible to the secretary of state) at 1 p.m. on the 7th, your time"? Why that specific day and hour? Was not mention of an exact moment — lunchtime on a Sunday in Washington, D.C., but dawn of December 8 in the Philippines, and 6:30 a.m. in Hawaii — enough to ring a very loud bell in the minds of the top Roosevelt administration officials? And had not the very last decrypted postscript added a final instruction, to immediately destroy all secret documents and codebooks at the Japanese Embassy, which was then to shut down?

What more warning did the government of the United States *require*, for heaven's sakes? Why had decrypts of Purple communications by the U.S. Magic team (so called for their almost miraculous monitoring and deciphering of Japanese diplomatic radio signals) been denied to the naval and army-air commanders in chief in Hawaii? Why, in sum, had the President and his staff not *warned* the many thousands of brave U.S. servicemen — in the navy, the army, the air corps — who were to lose their lives a few hours later?

It *had,* surely, to be a conspiracy, or so the army of conspiracy theorists would say. After all, how could the Japanese Imperial Navy's First Air Fleet (a veritable armada of modern aircraft carriers — no less than six in number — and two battleships, three cruisers, nine destroyers, as well as eight fueling tankers and twenty-three submarines, totaling almost three dozen vessels) leave Hitokappu Bay in Japan and make its way across thirty-five hundred miles of the Pacific without U.S. detection? Surely the defense forces of the United States could not *all* have been asleep — especially when there was ample intelligence warning beforehand?

The reality of "Pearl Harbor day" — the longest and worst day of President Roosevelt's life — was somewhat different.

On receiving a Magic decrypt of the "pilot" message and the first thirteen parts of the alarming but mysterious Japanese Purple signal being sent from the Japanese foreign minister, Mr. Shigenori Togo, to Ambassador Kichisaburo Nomura in Washington, D.C., at 10:00 p.m. on the night of December 6, 1941, the President had indeed tried to call Admiral Harold "Betty" Stark, his CNO, at Stark's residence at the Naval Observatory in Georgetown.

The admiral, the President was told, was out at the theater with his wife, attending a performance of Romberg's popular operetta *The Student Prince* — famed for its rousing "Drinking Song." It was thus only at 11:30 p.m. that the President had finally spoken to Betty, once the admiral had returned home and had had time himself to read the still-incomplete message. They had agreed, on the telephone, that the news the Japanese were ending peace negotiations looked bleak for America — the two men speculating on what would be contained in the final part of the communiqué, yet to come. A declaration of war with Britain and the Netherlands, whose oil fields the Japanese military were eyeing with impatient, predatory interest, now that the United States had cut off American oil exports following the Japanese invasion of Indochina earlier that summer? Or war, even, with the United States — beginning with an invasion of the Philippines?

Over the past months the President had, as commander in chief, overridden the advice of Admiral Stark and deliberately augmented the U.S. fleet based at Pearl Harbor, at the very center of the Pacific. He had stationed another U.S. fleet at Manila, and — against the reluctance of General Marshall, the U.S. Army chief of staff — had ordered reinforcements of the most modern U.S. warplanes, ammunition, and troops to be sent out urgently to the Philippines as a deterrent against further Japanese predations in Southeast Asia.

Clearly, as the latest decrypts of Japanese diplomatic signals and cumulative American secret intelligence reports were indicating, the President's policy of deterrence had not worked. In fact the very opposite seemed to be the consequence. Like belated British and French rearmament in 1939, America's end of appeasement and its more muscular approach toward Japanese military conquest in Southeast Asia appeared to be producing the contrary effect to the one intended: convincing the leaders of the Japanese militocracy that further "peace" negotiations with the Americans, posing as the guardians of tranquility in the Far East in order to get their way, were pointless. Only a preemptive Japanese attack, similar to Hitler's assault on the West on May 10, 1940, could hope to defeat the United States before it reinforced its Far Eastern bases even further.

Japanese militarists were not mistaken in fearing belated American rearmament. The simple fact was: given the output of the U.S. economy—which was estimated to be more than five times that of Japan—America could only get more powerful. A flight of thirteen of the latest long-range, almost indestructible B-17 Flying Fortress bombers and reconnaissance airplanes, for example, was that day taking off from San Francisco, bound for the Philippines, on the President's orders. And a ship convoy carrying some twenty thousand U.S. troops and military equipment for the Philippines was due to leave San Francisco on December 8, on the President's instructions. The United States was waking up after its long slumber in the Orient.

Rather than continue diplomatic negotiations and allow a more powerful U.S. presence to be built in the Pacific, the Japanese were going to go to war—this was the incontrovertible conclusion of the initial decrypt. The first thirteen paragraphs of the intended Japanese note were dark and disappointing to the President and to his chief of naval operations, Admiral Stark. How much more potent would U.S. forces in the Pacific and Far East become as a deterrent, if only negotiations between the Japanese and U.S. governments could be dragged out still longer, they agreed. Yet the Japanese were not fools. They had done the sums—indeed, for weeks now American intelligence had tracked a vast fleet of warships and military troop transports assembling in Shanghai, then putting to sea. Clearly they were readying to invade somewhere in Southeast Asia, once they terminated negotiations.

Increasingly fatalistic, the President had composed a final appeal for "peace" to Emperor Hirohito. Despite the objection of his secretary of war, Roosevelt had dispatched it in a special personal telegram that was to be delivered by the U.S. ambassador in Tokyo on the evening of December 6 (December 7 in Japan)—a message the President had phrased with great care, so that if leaked or afterward published, it could be appreciated by all as a plea for peace, not war.[5]

In the cable, the President of the United States had assured the Emperor of Japan that the U.S. had no thought of "invading Indo-China" if the Japanese, as the U.S. requested, withdrew its occupation troops. Nervousness about Japanese intentions was understandably rife across Southeast Asia, Roosevelt had pointed out. "None of the peoples" of the Philippines, the East Indies, Malaya, and Thailand could be expected to "sit indefinitely or permanently on a keg of dynamite," he'd written. "I address myself to your Majesty . . . so that Your Majesty may, as I am doing, give thought in this indefinite emergency to ways of dispelling the dark clouds. I am confident that both of us, for the sake of the peoples not only of our own great countries but for the sake of humanity in neighboring territories, have a sacred duty to restore traditional amity and prevent further death and destruction."[6]

It was futile, of course. Unknown to the President, the Japanese foreign minister, Mr. Togo, did not even allow the U.S. ambassador, Joseph Grew, to take the cablegram to the Emperor's palace, lest it upset Japanese war operations already in train.

Togo's reasoning was straightforward: as commander in chief of Japan's Imperial Armed Forces, His Highness Emperor Hirohito had already been informed of Japanese invasion plans, and on December 3 had not only signed off on multipronged amphibious landings all across Southeast Asia—assaulting Malaya, Singapore, the Dutch East Indies, and the Philippines—but a top-secret sneak attack on the main military base of the United States at Pearl Harbor in Hawaii.[7] As one Pearl Harbor historian would later note, "attempting to stop Operation Hawaii at this point would have been rather like commanding Niagara Falls to flow uphill."[8] All was set; once the senior Japanese naval commander, Admiral Yamamoto, obtained the signed imperial order to proceed, his chief of staff noted smugly that, at the very moment when the Japanese ambassador would be performing his appointed tasks in Washington, pretending to be continuing negotiations as part of the Japanese plot or charade, "the biggest hand will be at their throat in four days to come."[9]

Admiral Yamamoto's chief of staff was not exaggerating. The surprise left hook at the American jugular at Pearl Harbor was designed not to win the war overnight, but to administer a savage first shock, ensuring that the Americans could not interfere with massive Japanese invasion forces about to strike across the whole of Southeast Asia, far to the north. The presence of *those* assault troopships could not be — and were not intended to be — concealed. Eight Japanese cruisers, thirty-five transport ships, and twenty destroyers had been observed and reported by the British Admiralty moving toward Kra on December 6, indicating an impending invasion of Singapore, Malaya, or Indonesia (Dutch East Indies) — or all three. And possibly the Philippines, too.

The President had thus gone to bed at the White House, after midnight on December 6, with foreboding. He slept fitfully. Then, at 9:00 a.m. on the morning of December 7, he received from a U.S. naval intelligence courier, Lieutenant Schulz, the top-secret Magic decrypt containing the missing fourteenth paragraph of the Japanese government's official "response" to Secretary Hull's American message of November 26. Hull's message had urged the Empire of Japan to unequivocally cease and desist in its military occupation of southern China and Indochina, in order that U.S.-Japanese relations could be put back upon a peaceful course. The final Japanese paragraph ended, the President noted, with bleak and ominous words. Since "efforts towards the establishment of peace through the creation of a New [Japanese] Order in Asia" had failed over the preceding weeks, the ambassadors were to inform the U.S. government, "it is impossible to reach an agreement through further negotiations."

So this was it. The decrypt was pretty much what the President had expected — it "looked as though the Japs are going to sever negotiations, break off negotiations" he remarked to his naval aide, Captain Beardall[10] — but it did not specifically indicate hostilities would result, or when or where they would take place. Hostilities, nevertheless, were clearly coming — the fleet of Japanese warships and transports openly poised to strike at British and Dutch territories in Southeast Asia, and perhaps the Philippines. Reading the first part of the Japanese message the night before, the President's assistant, Harry Hopkins, had remarked: "since war was undoubtedly going to come at the convenience of the Japanese it was too bad that we could not strike the first blow and prevent any sort of surprise."[11]

"No, we can't do that," the President had retorted. "We are a democracy and a peaceful people"[12] — even if his cabinet, disappointed by continuous American appeasement of the Japanese military government, and sickened by reports of Japanese atrocities in the countries they had overrun, favored preemptive war. Roosevelt had overridden their advice—repeating the well-known story of President Lincoln, when he polled his own cabinet members on whether to go ahead with the Emancipation Proclamation. As the cabinet members all said no, Lincoln had summarized: "Seven nays and one aye, the ayes have it"!

Asking, in the same vein, if they thought the country would back him if the United States were to attack the Japanese Navy preemptively, Roosevelt's own cabinet members had voted unanimously yes. "The Nays have it," Roosevelt had concluded the cabinet meeting—refusing to go down the preemptive route.[13]

Despite mounting evidence of further Japanese aggression being prepared, then, the President had simply refused to budge. Not only did isolationists hold the whip hand in Congress and across the nation, he had reminded Hopkins, but preemptive military attack was not in America's historical, moral vocabulary. Raising his voice, he'd claimed that the United States' policy of nonaggression—of only responding if and when itself attacked—had over the centuries been to America's advantage; "we have a good record," he'd summed up[14]—despite his anxiety.

The latest decrypt, however, was not the end of communications from Tokyo to its embassy in Washington, the President soon learned. A few minutes after 10:00 a.m. a courier arrived with *more* decrypts. These included instructions to deliver the entire fourteen-part message to the State Department that day at one o'clock, Washington time, *precisely.* The concluding part of the message thanked the dual Japanese ambassadors, Admiral Nomura and Mr. Kurusu, for their patient and devoted service to the Emperor—and ordered them to destroy, after reading the message, the cipher machine, all codes, and all secret documents remaining at the embassy.

It was clear to President Roosevelt, as he read this, that Japan was going to war not simply with Britain and the Dutch, but with the United States also—and that war could start any time after 1:00 p.m.

Understandably, then, the President forswore lunch with his wife and her thirty guests downstairs at the White House. After speaking on the telephone with his civilian war council—Secretaries Hull, Knox, and Stimson, who had seen the same decrypts, messengered to them at their meeting in the Munitions Building on the Mall—Roosevelt resigned himself to his doctor's painful treatment of his sinus problem. "The damp weather, though mild that day, had made his sinus bad, which necessitated daily treatment of his nose," Eleanor recalled. "I always worried about this constant treatment for I felt that while it might help temporarily, in the long run it must cause irritation."[15]

On the Magic decrypt delivery list, beginning with the President and the civilian secretaries of state, war, and the navy, there followed the names of the chiefs of staff of the U.S. Army and Navy—though not the Air Corps (recently renamed the United States Army Air Forces), as befitted its still-lowly status in the nation's armed forces. The response of the chiefs of staff to the latest information, however, proved, in the light of history, as poor as that of their civilian masters. General Marshall, for reasons that remain unclear, later claimed not even to have received the decrypt of the first part of the Japanese note the night before; thus, when the missing fourteenth paragraph, followed by the instructions to the Japanese ambassadors about the 1:00 p.m. presentation and subsequent shutdown of the embassy, was hand-delivered to Marshall at his official residence at Fort Myer, the general was out riding and couldn't be contacted. Only at 11:15 a.m. did General Marshall, once alerted, reach his office at the Munitions Building on the Mall. Since he had not seen the earlier, thirteen-part decrypt, it took him a further twenty-five minutes to read and digest the whole message—leading up to its ominous climax regarding destruction of codes, and time of delivery of the Japanese government note.

Finally, at 11:40 a.m., the penny dropped. When General Marshall called Admiral Stark in his office at the Navy Department building next door, the admiral—who was under the mistaken impression there was a Magic decrypting office in Hawaii[16]—did not feel more Washington alerts than had already been sent would help naval commanders in the Philippines, Panama, Hawaii, and the West Coast.[17]

Stark's deputy must have questioned the wisdom of this, however, for Admiral Stark soon called back, and on second thought agreed to add his imprimatur to a cable General Marshall had drafted[18]—even though it threatened to reveal, if the Japanese intercepted the signal, Magic's breaking of the Japanese secret code. Certainly neither officer dared use the "scrambler" telephone to contact the overseas commanders.[19]

"The Japanese are presenting at 1 P.M. Eastern Standard Time, today, what amounts to an ultimatum," Marshall's cable disclosed to the recipient field commanders at midday—a presentation that was now only an hour away. "Also they are under orders to destroy their code machine immediately. Just what significance the hour set may have we do not know, but be on the alert accordingly."[20]

In the meantime, the Japanese ambassador, one of his aides at the embassy later recalled, "peeked into the office where the typing was being done, hurrying the men."[21] It was no use, however; they simply could not get the fourteen-part official note ready in time for 1:00 p.m. presentation to the U.S. State Department. The Japanese government's whole scheme — to hand over the official note only twenty minutes before the arrival of their warplanes over America's moored Pacific Fleet in Battleship Row in Pearl Harbor, Hawaii, thus leaving the Americans virtually no time to defend themselves — would be ruined.

Flustered, Ambassador Nomura telephoned Secretary Hull's office a few minutes after the 1:00 p.m. deadline, asking for a brief extension of the audience he had requested, to 1:45 p.m.[22] This the secretary of state, knowing what was in the note but not knowing what specific "deviltry" it portended, granted, after calling the President.

A guest at an official dinner the previous night, Mrs. Charles Hamlin, had noticed that Mr. Roosevelt "looked very worn . . . and after the meat course he was excused and wheeled away. He had an extremely stern expression."[23] Given the incoming decrypts and approaching winds of war, this was understandable. But the following morning the President's physician, Admiral McIntire, did not think the President unduly stressed, given the circumstances — indeed it was one of Franklin Roosevelt's most attractive traits, he reflected, that the President seldom showed irritation, though his humor could be sharp. He certainly took his medicine — the clearing of his nostrils and sinus ducts — like a man. He seemed more resigned than tense as he waited for news from Secretary Hull of Ambassador Nomura's visit.

Roosevelt had reason to be resigned, rather than nervous. He had spent the last several years as president trying to preserve America's stature as a neutral nation in a world of dictators and competing military empires — while holding off a phalanx of isolationists at home, headed by Senator Burton K. Wheeler and Colonel Lindbergh, opposed to anything but defense of the homeland. Like an expert juggler, Roosevelt had managed this feat without committing the United States to war — a destructive social behavior he had come to despise since his experience in World

War I, when he was still young and at the height of his physical and mental energies.

Roosevelt's character in the intervening years had certainly changed; as president he had become more opaque and manipulative, yet more compassionate, too. He rather liked Admiral Nomura, whom he'd known during the last war; despite the Magic decrypts he was reading, the President remained certain the ambassador, like the Emperor, was now but a pawn of the militarists in the Japanese government, headed by General Hideki Tojo. Fortunately for the United States, the President felt sure, America was not only more economically powerful than Japan, but cleverer, too. The Magic decrypts were giving the U.S. a huge advantage, making it possible not only to read the mind of the Japanese militocracy, but their diplomatic instructions, several hours before they were read by the intended recipients. By steadying the hand of gung-ho American interventionists at home — including most members of his cabinet — and sending strict instructions to commanders in the Pacific and Far East not to provoke any kind of incident that could lead the Japanese to declare war as a response ("If hostilities cannot repeat cannot be avoided, the United States desires that Japan commit the first act," Admiral Stark had signaled in a war-warning to his fleet commanders on November 28),[24] the President had, it seemed, forced the Japanese government to make the first military move. In that way — the same way in which he had forced Winston Churchill to agree to the Atlantic Charter in August, before there could be any question of a U.S. alliance with Britain — America would be in the right, morally speaking, if war came.

So confident was the President in holding to this position of *moral* superiority that when the Chinese ambassador came to see him, as scheduled, in his study at 12:30 that day, Roosevelt had shown him the text of his personal appeal to Emperor Hirohito, dispatched the previous evening. "I got him there; that was a fine, telling phrase," FDR congratulated himself on his language. (The Chinese ambassador, Hu Shih, had a PhD in philosophy, and was an expert on linguistics, being credited with developing a Chinese vernacular.) "That will be fine for the record," he'd added, knowing he'd done everything possible to avoid war, short of appeasement.[25]

For the record? Responding to Dr. Shih's curious look, the President had explained: "If I do not hear from the Mikado by Monday evening, that is, Tuesday morning in Tokyo, I plan to publish my letter to the Mikado with my own comments. There is only one thing that can save the situation and avoid war, and that is for the Mikado to exercise his prerogative" — and cancel Japan's war preparations. "If he does not," the President went on, "there is no averting war. I think that something nasty will develop in Burma, or the Dutch East Indies, or possibly even in the Philippines." Referring to the impending visit of the Japanese ambassadors to the State Department, he remarked: "Now these fellows are rushing to get an answer to Secretary Hull's most recent notes; in fact, I have just been told that those fellows have asked for an appointment to see Secretary Hull this noon. They have something very nasty under way."[26]

The President was thus still thinking as a president — not as a commander in chief. His meeting with the Chinese ambassador ended after forty minutes, at 1:10 p.m.

Hearing from Secretary Hull that the Japanese ambassadorial visit had been delayed for almost an hour, the President then summoned Harry Hopkins, his adviser, and the two men ate a sandwich at his desk in the Oval Study on the second floor of the private residence of the White House, looking out over the National Mall, the Washington Monument, and the Lincoln Memorial.

According to Hopkins's account, written later that evening, the two men talked of "things far removed from war."[27]

Half an hour passed. The telephone rang. The President picked it up himself, expecting it to be Mr. Hull at the State Department, announcing the delayed arrival of the Japanese ambassadors.

It wasn't.

It was Mr. Frank Knox, the secretary of the navy, phoning from his office in the Navy Department on the Mall.[28] He was there with the chief of naval operations, Admiral Stark, and Stark's chief of war plans, Rear Admiral Turner. They had urgent news. The Navy Department in California had just monitored an emergency radio message being broadcast in Honolulu: "Air raid Pearl Harbor. This is no drill."[29]

Pearl Harbor?

Harry Hopkins questioned the veracity of the telephoned report. Not because he disbelieved news of a Japanese attack, but because he found it difficult to believe — despite the decoding of the Japanese note — that the Japanese would be so stupid as to target a primary American territory, instead of starting with the invasion of British and Dutch colonial possessions in Southeast Asia, as Washington expected.

That scenario — attacks by the Japanese on Burma, Singapore, Siam,

Malaya, the Dutch East Indies, without going to war with the United States—had been the President's constant dilemma over the past four months. For the plain political reality in America was: if the Japanese were clever, and attacked only British and Dutch territories in the Far East, Congress could not be counted on to declare war on Japan—leaving the Japanese to "pick off" any country it wished in Southeast Asia, much as Hitler had done in Europe.

It now seemed clear that it had been Japan's intention to attack the United States straight away, judging by the Magic decrypts of their communications to their ambassadors that morning. But Pearl Harbor, thousands of miles from Malaya and the Philippines? Would the Japanese dare strike at America's primary military and naval base in the Pacific, with ample U.S. fighter planes covering the islands, and bombers able to attack approaching Japanese warships from multiple Hawaiian airfields? Surely, Hopkins argued, it must be a mistake.

Pearl Harbor, Hopkins pointed out to the President, was *six thousand miles* away from the Japanese invasion fleet that the U.S. military was currently tracking off Cambodia Point. Not only was Hawaii an American island archipelago, but the impregnable headquarters—army, naval, and air—of the powerful U.S. Pacific Fleet: a fleet of aircraft carriers, battleships, cruisers, destroyers, and submarines, dominating the entire Central Pacific, together with U.S. bombers, patrol planes, fighter aircraft, radar! It seemed ridiculous. And yet . . .

Hopkins noted Roosevelt's pained expression.

Roosevelt had sailed almost since he could walk; had been assistant secretary of the United States Navy for nearly eight years, from 1913 to 1920. He had devoured the works of America's greatest naval strategist, Rear Admiral Alfred Thayer Mahan. Though he had never served as a commissioned officer, Roosevelt knew the navy forwards and backwards—in fact, he felt so committed to it that he had even asked to leave his post as assistant secretary and be permitted to join the service as an ordinary seaman in 1918, as the United States turned the tide of World War I in Europe. To his chagrin the U.S. Navy had by then—thanks in part to him—half a million sailors in its ranks, and had not needed a father of five, aged thirty-six—however able a seaman.

Assistant Navy Secretary Franklin D. Roosevelt's aggressive spirit and sheer energy had nevertheless become legendary in President Wilson's administration—his driving enthusiasm leading, among other achievements, to the laying of an innovative two-hundred-mile-long mine barrier in 1918, all the way from Scotland to Norway across the North Sea, to inhibit German submarines from getting into the Atlantic (or back if they succeeded).[30] Moreover, he was still so much a navy man that his army chief of staff once asked if, as president, he would mind not referring to the navy as "us" and the army as "they."[31]

More to the point: President Roosevelt had himself visited Hawaii in the summer of 1934, in his second year in the White House, aboard the new heavy cruiser USS *Houston.* There he'd been greeted by a crowd of some sixty thousand residents. He'd toured both Big Island and Oahu—the first U.S. president ever to do so. He'd witnessed a military review by some fifteen thousand U.S. troops—the largest ever on the islands. No less than a hundred army and navy planes had performed a fly-past—forming the letters "FR" in the sky. Joseph Poindexter (since 1934 the governor of the territory) had wined and dined Roosevelt at Washington Place, in Honolulu. "Concerning Hawaii as the American outpost of the Pacific," a reporter at the *Honolulu Star Bulletin* had written, "the president is anxious to confer with the heads of the military units first hand to determine for himself the defense needs here. His visit may later lead to an increase in the size of the army and navy posts. As the time approaches for the release [independence] of the Philippines, the president desires full preparedness information regarding the bulwark in the Pacific."[32]

Departing from the Hawaiian Islands, the President had congratulated "the efficiency and fine spirit of the Army and Navy forces of which I am Commander-in-Chief." These American forces constituted "an integral part of our national defense, and I stress the word 'defense.' They must ever be considered an instrument of continuing peace," he'd emphasized, "for our Nation's policy seeks peace and does not look to imperialistic aims."[33]

Little had changed, from President Roosevelt's point of view, in the intervening seven years. Except that America's "bulwark" in the Pacific had increasingly become a thorn in Japan's side—not least because the British and Dutch, as well as Australia and New Zealand, so utterly depended on American naval and air power in the Pacific as a deterrent to Japanese expansion. Should the Japanese seek to redefine their war in Asia, Pearl Harbor would doubtless offer a tempting target—much as Russia's Port Arthur had done in the run-up to the Russo-Japanese War of 1904–5. Then, too, the Japanese had launched a preemptive, sneak attack to trap the enemy's most powerful battleships in a seemingly inviolable harbor, before war was even declared, and to destroy it in situ. Ironically, the name of the defending admiral, in 1904, had been Stark . . .

Looking at Hopkins, the President now shook his head. Harry "the Hop" Hopkins might be skeptical, but the President felt in his very bones the news was right; it was Hopkins who was wrong. An air raid on Pearl Harbor it could well be — the Fort Sumter of World War II.

Belatedly, all *too* belatedly, in the early afternoon of Sunday, December 7, 1941, the pieces of the puzzle began to come together in Roosevelt's mind: in particular, the whereabouts of the fleet of Japanese aircraft carriers that had left port in Japan on November 26, and which had thereafter kept radio silence — a fleet whose whereabouts were currently unknown to American military and naval intelligence.

The President's immediate intuition, as well as his naval experience, thus told him what Hopkins, his closest civilian adviser, could not credit.

No, the President contradicted Hopkins. This was "just the kind of unexpected thing the Japanese would do."[34] A warrior nation defined by its history and culture, the Japanese had no qualms or reticence on moral grounds. If they were to embrace world war rather than a negotiated settlement, would it not make sense for them to strike preemptively at the very heart of America's Pacific defense?

What exactly President Franklin Roosevelt should *do* was unclear, however — just as Joseph Stalin had been unsure, on the morning of June 22, 1941, how to react when unconfirmed reports reached the Kremlin that Hitler's vast 180-division army had attacked Russia across the German frontier, aiming for Leningrad and Moscow.

Stalin had done nothing. In fact, in his suspicious wisdom, the Russian leader had rejected President Roosevelt's repeated warnings of impending German invasion in the long weeks prior to the Nazi invasion, dismissing them as an attempt by the capitalist Western nations to sow discord between the two signatories to the Hitler-Stalin Pact, who'd promised not to attack one another. Stalin had refused to credit first reports of the invasion. (It was even said he'd ordered his men to literally shoot the messenger, a German deserter.) And this barely five months before . . .

President Roosevelt's own case was different — yet bore uncomfortable similarities. It was true that the President, in contrast to Stalin, had heeded all intelligence warnings he'd gotten via Magic decrypts; it was true that, on his specific authority, all U.S. forces in the Pacific, the Philippines, and the Far East, including Hawaii, had been on war alert since November 27. But there were other aspects of the story that were more tellingly similar — such as the matter of supposed deterrence.

Fearing Nazi attack, despite the nonaggression pact he'd signed with Hitler in 1939, Stalin had accelerated Soviet arms production as he became more concerned about the Führer's intentions. In fact Stalin had even invited German officers to inspect Russian Ilyushin aircraft and manufacturing plants in Moscow, Rybinsk, Perm, and other cities as far as the Urals, to convince the Germans that they would be making a big mistake in attacking the Soviet Union — Artem Mikoyan (brother of the foreign minister) warning the Germans they had been "shown everything we have and are capable of. Anyone attacking us will be smashed by us."[35] Stalin had, meanwhile, bent over backwards not to give Hitler a casus belli — instructing Russian forces on the border with the Third Reich not to do anything provocative. Like Roosevelt, Stalin had even rejected the notion of a preemptive attack on Hitler's massing armies in Poland.[36] Worse still, four days before the German attack, Stalin had turned down General Georgy Zhukov's request for an official order that would put Russian forces on the border on full alert. "Do you want a war as you are not sufficiently decorated or your rank is not high enough?" Stalin had ridiculed the request. "It's all Timoshenko's work. He ought to be shot," Stalin had remarked to his Politburo colleagues — dismissing the Russian defense minister's protest that the Soviet Union's forces, in the pursuit of deterrence, were now neither prepared to attack nor to defend. "Timoshenko is a fine man, with a big head," Stalin had mocked his adviser, "but apparently a small brain" — illustrating how small by showing his own thumb. "Germany on her own will never fight Russia," he'd declared in one of the greatest mispredictions of the twentieth century. After he'd walked out of the meeting, he "stuck his pock-marked face" back around the door and in a loud voice sneered: "If you're going to provoke the Germans on the frontier by moving troops there without our permission, then heads will roll, mark my words."[37]

President Roosevelt was nowhere near as coarse as his Russian counterpart — yet the fact was, his "deterrent" posturing had proved just as vain as Stalin's. By compelling the Japanese to fire the first shot, yet refusing to allow U.S. forces to take preemptive action — indeed almost *inviting* a "midnight" Japanese attack by forbidding anything that could be construed as a hostile or threatening act in the Pacific and North Pacific —

had not the President, like Russia's leader, doomed his forces to receive the first blow?

How big a blow was it, though? More reports of the air attack on Pearl Harbor came in over succeeding minutes. They seemed genuine — indicating the start of hostilities, rather than a feint to draw American attention away from the South China Sea.

Still the Japanese ambassadors failed to arrive at the State Department, however. What did *that* mean?

Mr. Hull, the secretary of state, wished to cancel the meeting with Admiral Nomura altogether, he told the President by phone at 2:00 p.m., but at 2:05 the President ordered Hull to "receive their reply formally and coolly and bow them out"[38] — i.e., pretend he did not already know the details of the message they were bringing him, or that Pearl Harbor had already been bombed. That way, Japanese perfidy — continuing the ritual of negotiation, while embarking on a sneak attack — would be all the more unmistakable, once the President announced it.

The secretary of state assured the President he would do so. Yet far from quietly receiving the Japanese ambassadors, Hull lambasted them when they finally entered his office, at 2:20 p.m. "In all my fifty years of public life," he told Admiral Nomura and his assistant ambassador, Saburo Kurusu — who were made to remain standing while the secretary read through their fourteen-point note, in English — "I have never seen a document that was more crowded with infamous falsehoods and distortions — infamous falsehoods and distortions on a scale so huge that I never imagined until today that any Government on this planet was capable of uttering them."[39]

If the secretary of state hoped thereby to shame Admiral Nomura and Mr. Kurusu in their oriental treachery, however, it was to prove short-lived schadenfreude — for Pearl Harbor was under devastating Japanese air and undersea attack as they spoke. In fact, at 2:00 p.m. the President had anxiously called his secretary of war, Henry Stimson, who'd gone home to his mansion outside Washington for lunch. "Have you heard the news?" Roosevelt had asked. Incredibly, the secretary had still heard nothing. "They have attacked Hawaii," the President told him. "They are now bombing Hawaii!"

"Well that was an excitement indeed," Stimson jotted in his diary.[40] Excitement soon turned to horror, however. Less than half an hour later, according to Hopkins's memorandum that night, Admiral Stark, the CNO, called the President from his office at the Navy Department on the Mall. Not only could he officially confirm the aerial assault on Pearl Harbor, Hopkins recorded, but he had grave news. He stated, in Hopkins's words, "that it was a very severe attack and that some damage had already been done to the fleet and that there was some loss of life."[41]

"Some" damage to the fleet? "Some" loss of life?

Giving permission to Admiral Stark to execute War Plan 46 — effectively authorizing U.S. forces to begin unrestricted submarine and naval war in the Far East and in the Pacific — the President immediately called his press secretary, Steve Early, at his home. "The Japanese have attacked Pearl Harbor from the air," the President told him, "and all naval and military activities on the island of Oahu, the principal American base in the Hawaiian Islands. You had better tell the press right away." He added (erroneously at this point) that "a second air-attack is reported on Manila air and naval bases." Then he asked, almost innocently: "Have *you* any news?"

Early's response was almost comical: "None to compare with what you have just given me, sir."[42] The press secretary immediately called the three main U.S. press agencies (AP, UP, and INS) via the White House telephone switchboard, and gave them, at 2:22 p.m., the President's first statement: "The Japs have attacked Pearl Harbor, all military activities on Oahu Island. A second air-attack is reported on Manila air and naval bases."[43]

In actuality, Manila itself had still not been attacked — but a second wave of bombers had descended on Pearl Harbor, and the island's Hickam and Wheeler airfields, a full hour after the first assault. Newly developed shallow-water torpedoes had been used, with devastating results against almost no defensive action. As Admiral Stark reported, it was like a massacre of the innocents.

The conference of advisers that had been planned for 3:00 p.m. at the White House to discuss the President's dilemma — whether, and how, to appeal to Congress for action, if Japan went ahead with an invasion of British or Dutch territories — was now redundant, as the President clarified, once his war council — or quasi war-cabinet, now — assembled in his office: Secretaries Hull, Knox, and Stimson, as well as General Marshall. (Admiral Stark remained at the Navy Department, communicating with Hawaii. General Arnold was at the time in California — unreachable, it was explained, as he was out shooting grouse!)

America was now effectively at war, even if war had not been declared—as the President acknowledged shortly after 3:00 p.m., when taking a call from the U.S. ambassador in England, John Winant. Winant was staying with Winston Churchill at the British prime minister's official country residence, Chequers, together with Lend-Lease administrator Averell Harriman; they had just heard an announcement of the Japanese attack at the end of the nine o'clock BBC evening news.

Like Hopkins, Winston Churchill, suffering one of his periodic bouts of depression, had first failed to believe the news. "He didn't have much to say throughout dinner and was immersed in his thoughts, with his head in his hands part of the time," Harriman afterward recalled—the Prime Minister showing no sign of having understood when the BBC newscaster, Alvar Lidell, referred to reports of an air raid on Pearl Harbor.[44] Churchill's security chief, Commander Tommy Thompson, was equally at a loss, imagining the announcer had said "Pearl River"—wherever that was! Shocked to the core, Harriman and Winant, as Americans, had understood exactly the name of the location the BBC announcer had mentioned—yet as guests of the Prime Minister, they were unwilling to contradict Commander Thompson. It was only when the butler, Sawyers, entered, as in an Edwardian play, that the news was taken seriously. "It's quite true," the butler confirmed, "we heard it ourselves outside [i.e., in the servants' quarters]."[45]

Winant later recalled how the dinner guests looked at each other "incredulously. Then Churchill jumped to his feet and started for the door with the announcement, 'We shall declare war on Japan.' There is nothing half-hearted or unpositive about Churchill," Winant wrote, "—certainly not when he is on the move. Without ceremony I too left the table and followed him out of the room. 'Good God,' I said, 'you can't declare war on a radio announcement.'"[46]

Instead, Winant suggested that he himself should telephone the President to seek confirmation, which he promptly did—offering his sympathies, when told by the President that there had been significant loss of life, and ships sunk. He then passed the phone to Mr. Churchill, telling the President he would recognize the speaker by his voice.

"Mr. President, what's this about Japan?" asked Churchill.

"It's quite true," the President confirmed. "They have attacked us at Pearl Harbor. We are all in the same boat now."[47]

To the President, Mr. Churchill now announced that he wished to declare war immediately on Japan, as he'd told Winant. The President demurred.

They must take things calmly, Roosevelt declared, step by step. He himself would ask Congress for a declaration of war against Japan, the next morning. Churchill promised he would then ask Parliament for a similar declaration, which would "follow the President's within the hour."

Aware in part that he should not burden the President with more at this moment, Churchill rang off. His own mood, however, had shifted from depression to exultation. More than two years into the war, Britain was no longer alone! The United States would now protect British interests in the Far East, as well as safeguarding the sea-lanes to Australia and New Zealand!

All week the President had squirmed and struggled to avoid the Prime Minister's appeals for a mutual commitment to go to war with Japan, if Japan attacked only British and Dutch territories. Now, however, the United States would be at war, and of its own volition: the world's largest economy, untouched as yet by hostilities, bombing, or even blackouts. Even Churchill's two American guests, not knowing the extent of the disaster suffered at Pearl Harbor, seemed "exalted," Churchill later wrote, "—in fact they almost danced for joy," he claimed.[48]

In Washington, the atmosphere among the war council members in the President's Oval Study was very different, however. It was, as yet, "not too tense," Hopkins summarized later that night,[49] since the true extent of the destruction and casualties suffered at Pearl Harbor was, at 3:00 p.m., still unknown. All present agreed that hostilities had been bound to come to the United States sooner or later, and that this way the President, given his unwillingness to "fire the first shot," would be exonerated in the court of public opinion, as well as history.

As the minutes ticked by, however—with fresh reports arriving of the damage inflicted by the second Japanese air attack on Pearl Harbor—the mood in the White House became less confident. It was clear this was not going to turn into an American victory, or even a brave performance in defense of the nation's main Pacific base. Calls "kept coming in, indicating more and more damage to the fleet. The President handled the calls personally," Hopkins noted, "on the telephone with whoever was giving the dispatches. Most of them came through the Navy."[50]

The meeting soon broke up, as General Marshall wanted to return to his headquarters in the Munitions Building—saying he had already ordered General MacArthur to execute "all the necessary movement required in event of an outbreak of hostilities with Japan," including a U.S. air attack on Japanese installations on Taiwan.[51]

Grace Tully, the President's secretary, had meantime arrived from her apartment on Connecticut Avenue; she recalled there was such "noise and confusion" and so many "calls on a telephone in the second floor hall" that she herself moved into the President's bedroom, next to the Oval Study, and took them down in shorthand, then typed them for the President in a tiny office room next door. "The news was shattering," she recalled. "I hope I shall never again experience the anguish and near hysteria of that afternoon" — "each report more terrible than the last, and I could hear the shocked unbelief in Admiral Stark's voice as he talked to me. . . . The Boss [Roosevelt] maintained greater outward calm than anybody else but there was rage in his very calmness. With each new message he shook his head grimly and tightened the expression of his mouth."[52]

There was good reason. Word from the U.S. Army Air Corps in Hawaii was just as terrible as from the Navy — its planes blitzed on the ground, before they had even been able to take off.

It was clear the U.S. Armed Forces at Pearl Harbor — forces the President had himself inspected seven years before as commander in chief, and which he had much reinforced since then — had been caught with their pants down.

As ever more humiliating news came through from Hawaii, President Roosevelt felt something of the same disbelief, even guilt, that had paralyzed Stalin the previous June, following the German invasion of Russia. With shock and near panic gripping those around him in the White House, however, the President did not dare show his feelings.

At the White House there was certainly embarrassment, even shame, at the ever-bleaker news coming from Hawaii — with inevitable questions arising as to how far the President was himself responsible for the catastrophe. Had he not insisted, against the advice of his war council, upon pursuing a meandering course of initial appeasement of Japan, followed by belated military posturing in pursuit of supposed deterrence and moral high grounding — all carried out in spite of intelligence decrypts pointing to hardening Japanese attitudes and, finally, ominous signs of imminent Japanese hostilities? Overruling his team, had not the President refused to order a preemptive American attack on Japanese forces clearly massing for a new invasion in Southeast Asia? Moreover, had he not discouraged the British from carrying out such a preemptive attack on the Japanese fleet approaching Singapore, when they were in a good position to do so?

Given Roosevelt's character, and his absolute authority over the members of his cabinet, there was, however, no call for the President's resignation—something that had never taken place in American history. Nor, to judge by FDR's demeanor, did the President feel he should resign. He looked grave, but far from despair. In any event, there was simply no one who could take his place as chief executive—certainly not his vice president, Henry Wallace, the former secretary of agriculture.

Given the President's role as commander in chief, though, how was it possible that the President's *military* team had not foreseen such a sneak attack, over the months of increasing tension between Japan and the United States? How had the eventuality of an attack on Pearl Harbor not been taken seriously by Generals Marshall and Arnold and Admiral Stark?

Roosevelt had personally chosen and appointed them, as his professional chiefs of the U.S. Army, Army Air Forces, and Navy. Ironically, each one *had* considered the possibility, in the past, of such a sneak attack, and each had attempted, in his own feeble way, to guard against it. "Thinking out loud, should not Hawaii have some big bombers?" General Marshall had asked his operations and planning officers in the summer of 1940, more than a year before the Pearl Harbor debacle. "It is possible that opponents in the Pacific would be four fifths of the way to Hawaii before we knew they had moved. Would five or ten flying fortresses at Hawaii alter this picture?"[53] For his part, Admiral Stark had written to Admiral Husband Kimmel, commander in chief of the Pacific Fleet, as recently as November 25, 1941, to say the Japanese naval forces were at sea and capable of a "surprise aggressive movement in any direction."[54] And General Arnold, on an inspection of air defenses in Hawaii as far back as September 1939, had noted the lack of a supreme commander to ensure integration of air, navy, and army forces on the islands, as well as the vulnerability of battleships moored in Pearl Harbor to aerial attack—as he'd openly remarked in a press conference on the West Coast.[55] Moreover, on his visit to Britain during the Blitz in the spring of 1941 Arnold had been expressly shown (as he'd noted carefully in his diary) the way British aircraft were always kept *dispersed* on their airfields, to minimize damage from surprise attacks. Yet none of the U.S. chiefs had actually visited Pearl Harbor since then. Nor had they asked to see integrated plans or evidence of *rehearsals* for the defense of Hawaii if attacked by naval aircraft, launched from enemy flattops or carriers.

Pearl Harbor had, in sum, been considered inviolable: a vital way station in ferrying aircraft to defend the Philippines, and a platform to service the Pacific Fleet, but too far from Japan to be attacked itself, save by submarines, which could not get past the harbor boom, and would not be able to launch their torpedoes in such shallow water unseen, if they did—a fact that explained Admiral Kimmel's failure to reverse his predecessor's decision not to have antitorpedo nets lowered around his battleships. Thus, in their loyal concern to carry out the President's policy of deterrence in the Far East by building up forces in the Philippines primarily for show, backed by the might of the Pacific Fleet at Hawaii, but with no intention of using such forces aggressively, the chiefs of staff had arguably failed their commander in chief and their country.

Whoever was to blame for Pearl Harbor's unpreparedness for a sneak attack, the question now was what, as U.S. commander in chief, President Roosevelt could *do* about it.

Tragically, the Hawaiian Air Force—as the USAAF group in the islands was called—appeared to have been caught unarmed, literally: the planes' gun breaches empty, and the aircraft standing huddled on the nearby airfields, wingtip to wingtip, to guard against possible sabotage by some of the 150,000 Japanese immigrants and Japanese American citizens living in the islands—a third of the entire population. As a result, the U.S. airplanes were effectively wiped out by Japanese attack planes and bombers, while not a single significant American warship survived the attack in Battleship Row without major damage—four of the eight battleships moored there being sunk. Moreover, there was still the possibility that the news from Hawaii would get even worse, if the Japanese mystery fleet were to catch the rest of Admiral Kimmel's remaining Pacific Fleet warships out at sea, where his two aircraft carriers were returning from weekend maneuvers and nearing home. "Within the first hour," Grace Tully later confessed, "it was evident that the Navy was dangerously crippled, that the Army and Air Force were not fully prepared to guarantee safety from further shattering setbacks in the Pacific. It was easy to speculate that a Jap invasion force might be following their air strike at Hawaii—or that the West Coast might be marked for similar assault."[56]

Speculation mixed fact and fantasy. A telephone call by the President to Governor Poindexter in Honolulu was interrupted by the sounds of planes and antiaircraft fire in the background, suggesting a third Japanese air raid and causing the President to bark aloud to the people in his office: "My God, there's another wave of Jap planes over Hawaii right this minute."[57] American gunners were, unfortunately, mistakenly firing at the few surviving U.S. planes that attempted to take off, either to seek combat or search for the invaders.

However exaggerated such alarums and panic, the stream of incoming reports, taken collectively, gave a growing indication that a veritable catastrophe had taken place in Hawaii: the destruction of virtually the entire American fleet moored at Pearl Harbor, as well as most of the U.S. Army Air Forces' planes.

At times the President felt such disappointment with his chiefs of staff he would readily have fired them. (In contrast to the chief of the decrypting department at the Navy Security Section, Commander Laurance Safford, who wanted to get his gun and "shoot Stark" for the admiral's failure to warn Manila and Hawaii more urgently, given the early decrypting of the Japanese diplomatic note.)[58]

With whom could the President replace his chiefs of staff, though? Almost everything General Marshall, Admiral Stark, and General Arnold had done, or not done, to prepare America for combat in the Pacific and Far East had failed — miserably. But Secretaries Stimson and Knox — both of them Republicans — had not done much better. "Knox, whose Navy had suffered the worst damage, and Stimson were cross-examined closely on what had happened," Grace Tully recalled, "on what might happen next and on what they could do to repair to some degree the disaster."[59] To which they responded: nothing.

Finally, in the "hysteria" at the White House, the President managed to collect his thoughts and focus on the address he would have to give to Congress the next day, requesting an official U.S. declaration of war on Japan. "Shortly before 5:00 o'clock the Boss called me to his study. He was alone, seated before his desk on which were two or three neat piles of information of the past two hours. The telephone was close by his hand," Ms. Tully later described. "He was wearing a gray sack jacket and was lighting a cigarette as I entered the room. He took a deep drag and addressed me calmly."[60]

Hundreds were already gathering in the dusk beyond the White House gates — incredulous at the news being put out by radio stations. Some were singing patriotic songs. Others held candles, in prayer.

"Sit down, Grace. I'm going before Congress tomorrow. I'd like to dictate my message," the President said. "It will be short."[61]

It was short: barely 390 words. "Yesterday December 7, 1941," he began, "a day which will live in world history, the United States was simultaneously and deliberately attacked by naval and air forces of the Empire of Japan . . ."[62]

"As soon as I transcribed it, the President called Hull back to the White House," and the two men "went over the draft," Ms. Tully remembered.[63]

Hull was now seventy years old, and easily unnerved or irritated. With his white hair and courtly demeanor, the handsome former senator from Tennessee had been secretary of state since 1933, but the news that, after so many months of mounting tension and decrypted warnings of Japanese perfidy, the army, navy, and air forces of the United States had all been caught completely unawares, infuriated him. Magic decrypts had indicated throughout the year not only that Japan was preparing for war with the United States, but was doing everything possible to determine the "total strength of the U.S." and train fifth-columnists in America to work on anti-Semites, labor union members, blacks, Communists, and "all persons or organizations which either openly or secretly oppose the war."[64] To his staff, in his office, Mr. Hull had therefore expressed "with great emphasis his disappointment that the armed forces in Hawaii had been taken so completely by surprise," as well as his "bitter feelings" over the invidious way Ambassadors Nomura and Kurusu had behaved. The secretary had thus already prepared his own statement to the press, which duly went out at 6:00 p.m., denouncing the Japanese for their "infamously false and fraudulent" professions of desire for peace, while preparing for "new aggressions upon nations and peoples with which Japan was professedly at peace, including the United States."[65]

With the latest information coming in to the State Department that Japan had formally announced it was at war with the United States, Hull now begged the President to give the American people the whole history of Japan's treachery, not merely a 390-word request to Congress to declare war. As Grace Tully recalled, "The Secretary brought with him an alternative message drafted by Sumner Welles, longer and more comprehensive in its review of the circumstances leading to the state of war."[66] As Hopkins noted that night, Hull's draft was certainly a "a strong document," but one "that might take half an hour to read."[67]

The President didn't like it. What *more* justification did Congress need in order to declare war? Would such a review not simply lead people to question the administration's past efforts? Japan had openly announced that hostilities existed with the United States. It would be enough for the President to say, on record and before Congress, that the nation had been attacked, without warning, at the very moment Japanese diplomats were bringing their response to the latest American peace proposal. In other words, a briefer address would give no member of Congress the excuse to criticize the President or his administration, including Secretary Hull, for not having done yet more to appease or dissuade the Japanese government from going to war.

Hull was unconvinced, but the President was sure in his own mind. The fact of the Japanese sneak attack spoke for itself. The President's own longwinded speech before the governing board of the Pan American Union earlier that year, May 27, 1941, was a case in point. Intended as a refutation of the claims of Colonel Lindbergh and other America First isolationists, it had been a methodical account of the growing threat facing the United States from the Third Reich. It had failed completely. Isolationist sentiment in America had actually increased in the aftermath of the President's exhaustive state of emergency proclamation that day, not diminished. Accusations of warmongering had refused to die down, making it even harder for the President to draw up his contingency plans for war. Only three days before the Japanese attack, Colonel Robert McCormick, the virulently right-wing opponent of Roosevelt's New Deal policy, had published the guts of the President's top-secret "Victory Program," under the headline "FDR'S WAR PLANS," in his newspaper, the *Chicago Tribune*. The article had called the President's plan "a blueprint for total war on a scale unprecedented in at least two oceans and three continents, Europe, Africa and Asia" — a revelation so inflammatory that the President, while denying the existence of such a plan, had called J. Edgar Hoover to ask him to instigate an immediate FBI investigation into the source of the leak.

No, Roosevelt felt, better to let the plain fact of Japanese aggression speak for itself: thus ending the reign of isolationist loudmouths in the country forever. Especially as reports of the sheer scale of the American military disaster multiplied.

That evening, more than a hundred journalists, photographers, radio reporters, and technicians crowded into the small White House press room — the normal capacity of which was twelve — desperate for more information.

By 5:58 p.m. Steve Early, the White House press secretary, confirmed "the report of heavy damages and loss of life" in the sneak attack.[68] "The telegraph boys fairly came out of the cracks, the floor was tangled with a black spaghetti of wires, the motion picture lights were on, cameras were busy, men were telephoning, a radio receiver blared," wrote one White House correspondent. "Men with chattering hand motion-picture machines climbed over and under desks . . . and they were followed by others carrying glaring lamps on black cords."[69]

The Treasury secretary, Henry Morgenthau, had already doubled the size of the small Secret Service detail at the White House. When he sought permission for half a battalion of troops, along with tanks, to be stationed around the Executive Mansion, however, the President told Morgenthau to drop the idea. The Japanese might possibly invade Hawaii, 2,676 miles away, but it would take them longer to get to Pennsylvania Avenue, he argued.[70] Instead, the President ordered that U.S. troops be sent to protect the Japanese Embassy on Massachusetts Avenue, and gave instructions for all the White House lights to be turned *on* as darkness fell.

The White House should be seen, still, as a beacon of democratic hope, not as a military barracks, the President explained—and he asked that the members of his full cabinet assemble with him in his study at 8:30 p.m., to be followed then by the leaders of Congress at 9:00.

As the terrible afternoon and early evening of December 7, 1941, had worn on, Eleanor Roosevelt had "stayed in my sitting room and did my mail and wrote letters." Nevertheless, she recalled, "one ear was alert to the people coming and going to and from my husband's study. He went down in the late afternoon to have his nose treated, and at seven o'clock [Solicitor General] Charles Fahy came to see him for a short time," to discuss the legal terms required of the declaration of war, and the status of Japanese diplomats in the interim. "Again in the evening he had supper in his study, with James [Roosevelt], who was then a captain in the Marines, Harry Hopkins, and Grace Tully."[71]

Hopkins, Grace Tully recalled later, was in a state of sustained shock, indeed "looked just like a walking cadaver, just skin and bones."[72] The "phone was ringing constantly," Hopkins noted, for his part. Admiral Stark "continued to get further and always more dismal news about the attack on Hawaii. We went over the speech again briefly and the President made a few corrections"—including the change from "world history" to "infamy."[73]

"The Cabinet met promptly at 8.30," in Roosevelt's study, or Blue Room, over the South Portico—the members summoned from across the Washington area and the nation. "All members were present. They formed a ring completely around the President, who sat at his desk. The President was in a very solemn mood and told the group this was the most serious Cabinet session since Lincoln met with the Cabinet at the outbreak of the Civil War."[74]

It was an apt analogy. The members of the war council already knew the worst. "The news coming from Hawaii is very bad," Secretary Stimson had by then noted in his diary. "It has been staggering to see our people there, who have been warned long ago and were standing on the alert, should have been so caught by surprise."[75] Secretary Hull, by contrast, blamed Stimson and Knox. He had told his staff at the State Department that he had, "time after time" in recent months, "warned our military and naval men," with all the vigor at his command, "that there was constant danger of attack by Japan"—and how "deeply" he regretted his "warnings had not been taken more seriously." As the note-taker recorded, the initial reaction at the State Department had been that the Japanese attack had been "exceedingly stupid," for it would "instantaneously and completely" unite the American people. "However, after it became evident that our armed forces had suffered tremendous damage in Hawaii, there was less feeling that the Japanese had been stupid."[76] Worse still, reports were coming in of a massive Japanese air raid on the Philippines—with the destruction of pretty much all of General MacArthur's air force at Clark Field, despite *nine hours* of prior warning.

Other cabinet members, however, were still in the literal as well as proverbial dark. "I'm just off the plane from Cleveland. For God's sake, what happened?" asked the attorney general, Francis Biddle. "Mr. President, several of us have just arrived by plane. We don't know anything except a scare headline 'Japs Attack Pearl Harbor.' Could you tell us?" asked another.[77]

As best he currently knew, the President brought the full cabinet—including Vice President Wallace—up to date, recounting the final hours of negotiation. "And finally while we were alert—at eight o'clock," he recounted, "a great fleet of Japanese bombers bombed our ships in Pearl Harbor, and bombed all our airfields." He confided that thc "casualties, I am sorry to say, were extremely heavy. . . . It looks as if out of eight battleships, three have been sunk, and possibly a fourth. Two destroyers were blown up while they were in drydock. Two of the battleships are badly damaged. Several other smaller vessels have been sunk or destroyed. The drydock itself has been damaged. Other portions of the fleet are at sea, moving towards what is believed to be two plane carriers, with adequate naval escort."[78]

The President's summary of U.S. naval losses was all too accurate. His belief that Admiral Kimmel's remaining naval forces — his carriers — were moving toward battle with the Japanese Navy, however, was overly optimistic.

In truth, neither the Japanese nor the American fleet commanders were anxious to join battle at sea. Enough, for the moment, was enough.

The same held true for the President's plans for his appearance before Congress the following day. Rocked by the disaster at Pearl Harbor, the President was reluctant to ask Congress for a declaration of war on Nazi Germany in addition to Japan. Thus when, in his study, the President read out to the cabinet members the draft of his proposed speech to Congress, Secretary Stimson objected that the declaration only covered war with Japan. It was, Stimson wrote in his diary that night, too simple, "based wholly upon the treachery of the present attack." Although in that respect it was "very effective," Stimson allowed, it did not "attempt to cover the long standing indictment of Japan's lawless conduct in the past. Neither did it connect her in any way with Germany," as he and Secretary Hull felt it should — in fact Stimson claimed "we know from the interceptions and other evidence that Germany had pushed Japan into this."[79]

Hitler as Hirohito's éminence grise? The President was unimpressed, and "stuck to his guns," in Hopkins's words, that night[80] — as if more determined than ever to avoid the moniker "warmonger." There was no evidence of collusion between Germany and Japan, Roosevelt countered — despite the suspicions voiced by his military team, such as Admiral Stark's remark to Rear Admiral Bloch in Hawaii, asking about an enemy submarine reported to have been sunk in the harbor: "is it German?"[81] As President he would continue to take one step at a time.

No sooner was his meeting with the cabinet over than the ten invited leaders of Congress — interventionists and former isolationists alike — now herded into the Blue Room.

It was 9:00 p.m. — with yet more bad news streaming in from the Far East. Word had come from Britain that Malaya had been invaded. Hong Kong, Bangkok, and Singapore had also been bombed. The American Pacific islands of Guam and Midway were under attack. Japanese carrier planes were confirmed as having bombed the Philippines — in fact, they seemed to have annihilated General MacArthur's air force at Clark airfield. If anything, the picture was worsening.

Labor Secretary Frances Perkins later recalled how, when she arrived in haste from the airport, the President did not even look up. "He was living off in another area. He wasn't noticing what went on on the other side of the desk. . . . His face and lips were pulled down, looking quite gray. . . . It was obvious to me that Roosevelt was having a dreadful time just accepting the idea that the Navy could be caught off guard. His pride in the Navy was so terrific that he was having actual physical difficulty in getting out the words that put him on record as knowing that the Navy was caught unawares, that bombs dropped on ships that were not in fighting shape and not prepared to move, but were just tied up."[82] And on top of that, the destruction of Hawaii's army air forces.

The mood began with collective shock, but soon gave way to congressional fury, as the President repeated the account that he had already given the cabinet, and then took questions.

Asked about losses suffered by the Japanese, the President was evasive. "It's a little difficult. We think we got some of their submarines, but we don't know," he responded lamely but truthfully. "We know some Japanese planes were shot down." Quoting his own experience in World War I, he cautioned against premature assumptions, or wishful thinking. "One fellow says he got fifteen of their planes and somebody else says five. . . . I should say that by far the greater loss has been sustained by us, although we have accounted for some Japanese." About the rumor that a Japanese carrier had been sunk off the Panama Canal Zone, he was dismissive — "Don't believe it," he warned; the U.S. forces there were "on the alert, but very quiet."[83] It had been, in short, an unmitigated naval and air disaster for the United States in the Pacific.

Unconfirmed reports had come in that the Japanese government had already proclaimed a state of hostilities with America, the President went on. With this in mind, he wished to ask the members of the Senate and House for authority to address Congress the next day, at 12:30 p.m. — though he did not read out his proposed speech, mindful that it would only spur more discussion, and leak within minutes.[84] Assured he would be invited to speak to a joint assembly of Congress, he now had to field more questions from the senators and congressmen about how the fleet and garrison at Pearl Harbor had been so unprepared.

"Hell's fire, we didn't do anything!" asserted Senator Tom Connally, a member of the Foreign Relations Committee from Texas, banging his fist on FDR's desk.

"That's about it," responded the President glumly.

"Well, what did we *do?*" Connally asked the navy secretary, Frank Knox, directly. "Didn't you say last month that we could lick the Japs in two weeks? Didn't you say that our navy was so well prepared and located that the Japanese couldn't hope to hurt us at all? When you made those public statements, weren't you just trying to say what an efficient secretary of the navy you were?"

Poor Knox knew not how to answer. Nor did the President help him out—he merely listened to the verbal attack with "a blank expression on his face."

Connally kept up his assault on the navy secretary—asking why "all the ships at Pearl Harbor" were so "crowded" together, and wanting to know about the log chain he'd heard had been pulled across the harbor entrance, so they could not get out.

"To protect us against Japanese submarines," Knox explained.

"Then you weren't thinking of an air attack?"

"No," the secretary admitted.

Connally was almost apoplectic by this time. "I am amazed by the attack by Japan, but I am still more astounded at what happened to our Navy. They were all asleep. Where were our patrols? They knew these negotiations were going on."

Knox fell silent. Attempting vainly to calm the temper of his meeting, the Chief Executive confided it was "a terrible disappointment to be President" in such "circumstances," in the aftermath of an attack that had "come most unexpectedly."[85]

The meeting went on for two long hours: accusations and disbelief at U.S. military incompetence leaving the President not only weary, but concerned lest the nation now descend into a witch-hunt as to whom to blame.

Fortunately Roosevelt was rescued by one of the legislators. "Well, Mr. President, this nation has a job ahead of it," a member of the delegation summed up, "and what we have to do is roll up our sleeves and win this war."[86]

Most, however, left the White House with unresolved feelings of guilt, anger, disappointment. And anxiety over the country's next steps.

After walking back to the Treasury, next door, at 11:25 p.m., Secretary

Morgenthau railed, like Secretary Hull, against the ineptitude of the professional armed forces — epitomized by the security at the White House, where he still counted only three men guarding the side of the mansion! He had then gone straight back to the Blue Room, and told the President in person that the "whole back of the White House — only three men. Anybody could take a five ton truck with 20 men and they could take the White House without any trouble."[87] It seemed endemic — Pearl Harbor merely the symbol of America's wider complacency. Inside the Treasury, Secretary Morgenthau was accosted by one of his senior staff. "Has there been negligence," the staffer asked, "or is it just the fortunes of war?"[88]

Morgenthau was unsure. The disaster was "just unexplainable," he answered, tormented. The Japanese "walked in just as easily as they [the Germans] did in Norway." At least "they didn't do it in the Philippines," where it was expected. "Let Stimson take the credit for that," Morgenthau remarked, not comprehending how devastating had been the destruction of MacArthur's air force planes, on the ground, as in Hawaii — but with nine hours' warning. He shook his white head; "all the explanations I have heard," he puzzled aloud, "just don't make sense." Hawaii was supposed to be "impregnable. I mean that has been sold to us," the Treasury secretary lamented. "They haven't learned anything here. They have the whole Fleet in one place — the whole fleet in this little Pearl Harbor base. The whole Fleet was there." And if, at the cabinet and congressional leaders' meeting, Secretary Knox had been distraught over the damage to the Pacific fleet, so had Secretary Stimson been over his precious army air forces. "He kept mumbling that all the planes were in one place," Morgenthau recounted to his staff.[89] How an entire Japanese fleet could sneak in, approach within a hundred or two hundred miles, as the President had described, and then make off, without being seen — let alone caught — was beyond the seventy-four-year-old. "Was it a terrible shock to the President?" asked his wife, who had come to take him home. To which Morgenthau could only sigh: "Must be — must be."[90]

The assistant secretary of state, Adolf Berle, noted in his diary: "It was a bad day all around; and if there is anyone I would not like to be, it is Chief of Naval Intelligence."[91]

Another candidate was, however, the devastated commander of the U.S. Pacific Fleet, Admiral Kimmel — who earlier that day had been hit by a spent bullet that crashed through the window of his operations room overlooking Pearl Harbor and ended on his chest, tarnishing his spotless white uniform. As Admiral Kimmel confided to his communications director, Commander Maurice Curts: "It would have been merciful, had it killed me."[92]

"His reaction to any event was always to be calm," the First Lady later described the President's temperament. Instead of getting agitated, he would batten down his hatches, emotionally. "If it was something that was bad, he just became almost like an iceberg."[93]

It had always been so, but now, late on the night of the Pearl Harbor attack, the President had trouble repressing his emotions. After the members of Congress left the White House, only Cordell Hull, the secretary of state, remained. They were joined by Sumner Welles — who arrived with yet *another* draft declaration of war "which the President did not like," Hopkins noted that night, "although Hull pressed very strongly that he use it."[94]

It was Hull's third attempt. "Hull's message," Hopkins noted, "was a long-winded dissertation on the history of Japanese relations leading up to the blow this morning. The President was very patient with them and I think in order to get them out of the room perhaps led them to believe he would give serious consideration to their draft. Waiters brought in beer and sandwiches, and at 12.30 the President cleared everybody out and said he was going to bed."[95]

Whatever his intentions, the President's living nightmare was not quite over. He had asked his son James, a liaison officer on the staff of Colonel William Donovan, to bring his boss to the White House. Known as "Wild Bill," Donovan was Roosevelt's recently handpicked chief of foreign intelligence, under the cover title "coordinator of information" (COI). For his part, however, Donovan had not been listening to his information; like half of America that Sunday, it seemed, he'd been watching a ball game — in Donovan's case in New York City's Polo Grounds (capacity fifty-four thousand) — when summoned to the White House.

The President had also decided to ask CBS reporter Edward R. Murrow, who with his wife had dined with Eleanor that night (a meal of scrambled eggs and pudding, served by the First Lady), to stay and speak with him privately, too.

When the two visitors thus finally went into the President's study, shortly after midnight, the extra chairs for the members of Congress were still out, and the President was still eating his sandwich along with a bottle of beer, alone at his desk.

"Gray with fatigue," the President gave his visitors a frank account of the past weeks — and past hours. His chiefs of staff had sent multiple warnings to all U.S. bases in the Pacific and Far East, yet, as he put it, "They caught our ships like lame ducks! Lame ducks, Bill. We told them, at Pearl Harbor and everywhere else, to have the lookouts manned," Donovan later recalled the President's words. "But they still took us by surprise."[96] Murrow, for his part, remembered how appalled the President was by the destruction of U.S. airplanes. "Several times the President pounded his fist on the table, as he told of the American planes that had been destroyed 'on the ground, by God, on the ground!'" As Murrow remembered, the very "idea seemed to hurt him."[97]

Given both Murrow's and Donovan's work in London, following the German invasion of the West, the President wanted to know from them, firsthand, whether they thought the people of the United States would rally in the same way the British had during the Blitz. Both men stated that they thought Americans would.

With that assurance, at 1:00 a.m. the President finally called it a day — the longest day of his life.

3

Hitler's Gamble

THE CAPITAL OF the United States buzzed with rumor, dread, and disbelief.

"The news of the shocking extent of the casualties and the damage to capital ships spread rapidly through Washington," Robert Sherwood, the President's speechwriter, later chronicled — even though the press were encouraged not to print the numbers of casualties, nor the extent of the destruction at Pearl Harbor. "The jittery conduct of some of our most eminent Government officials was downright disgraceful. They were telephoning the White House, shouting that the President must tell the people the full extent of this unmitigated disaster — that our nation had gone back to Valley Forge — that our West Coast was now indefensible and we must prepare to establish our battle lines in the Rocky Mountains or on the left bank of the Mississippi or God knows where." Sherwood wondered for a while whether Hitler might even be right: "that our democracy had become decadent and soft, that we could talk big, but there were too many of us who simply did not know how to stand up under punishment."[1]

Ignoring such hysteria and riding in an open car, as if to his inauguration, the President was driven to Capitol Hill late in the morning of December 8, 1941, and insisted he walk rather than be wheeled to the podium for the joint session. Congressmen and senators rose to their feet, giving him a standing ovation. Holding the lectern, facing a battery of microphones broadcasting his words to the world, the President delivered his address, beginning with the words: "Yesterday, December 7, 1941 — a date which will live in infamy — the United States of America was suddenly and deliberately attacked by naval and air forces of the Empire of Japan."

The speech — which included mention of the further Japanese attacks that had taken place in Malaya, Hong Kong, Guam, and the Philippine Islands — was a tour de force. Even Sherwood, who had not had a hand in the address, was amazed. There was, he later wrote, none of Winston Churchill's "eloquent defiance in this speech. There was certainly no trace of Hitler's hysterical bombast. And there was no doubt in the minds of the American people of Roosevelt's confidence. I do not think there was another occasion in his life when he was so completely representative of the whole people."[2]

The speech lasted only six minutes: six minutes that, in a way no one could ever have quite predicted, changed the world. "No matter how long it may take us to overcome the premeditated invasion, the American people in their righteous might will win through to absolute victory," the President closed. "I believe that I interpret the will of the Congress and of the people when I assert that we will not only defend ourselves to the uttermost but will make it very certain that this form of treachery shall never again endanger us.

"Hostilities exist. There is no blinking at the fact that our people, our territory, and our interests are in grave danger.

"With confidence in our armed forces — with the unbounding determination of our people — we will gain the inevitable triumph — so help us God." And with that, the President asked Congress to declare that, "since the unprovoked and dastardly attack by Japan on Sunday, December 7, 1941, a state of war has existed between the United States and the Japanese Empire."[3]

War, then, had come to the United States — war with Japan, not with Germany.

Ignoring this, within hours of hearing of the Pearl Harbor disaster, Winston S. Churchill decided he should travel to Washington to see President Roosevelt — Churchill informing the head of state in England, King George VI, as well as his own staff and colleagues, that he would meet with the President of the United States again, in person, so they could coordinate the "whole plan of Anglo-American defence and attack."[4] When cautioned by Admiral Pound, the First Sea Lord, that he must be careful not to be too assertive, given that the United States was not yet at war with Germany, the Prime Minister reacted "with a wicked leer in his eye," as Pound recalled.

"Oh! That was the way we talked to her while we were wooing her,"

Churchill quipped; "now that she is in the harem we talk to her quite differently!"[5]

In actuality Churchill was much more diplomatic. "Would it not be wise for us to have another conference?" he suggested cautiously in a cable to the President. "We could review the whole war plan in the light of reality and new facts, as well as the problems of production and distribution," he explained. "I could if desired start from here in a day or two, and come by warship to Baltimore or Annapolis. Voyage would take about eight days and I would arrange to stay a week so that everything important could be settled between us." He would bring, he added ominously, however, his three chiefs of staff and their staffs, just as he had done at the Newfoundland meeting. "Please let me know at earliest what you feel about this."[6]

Given the date of his cable — December 9, 1941 — and the fact that Hitler still appeared not to have made up his mind whether to declare war on the United States, the President felt this was jumping the gun, literally as well as metaphorically. President Roosevelt "was pretty sure that Germany and Italy would declare war almost immediately," the British ambassador, Lord Halifax, noted, however, in his secret diary, having called on the President before lunch.[7] But a visit planned by Churchill, when no German declaration of war against America had yet been made? Lord Halifax was not a little embarrassed by his prime minister's importuning.

The President, Halifax reported back to London, "was genuinely pleased at the idea of another meeting, and very grateful, I think for the suggestion at this particular moment." However, "publicity could not be avoided" — which would endanger Churchill's safety — and the President "thought it far too big a risk to take unless there is no other alternative."[8]

"I had a slight feeling," Halifax added — employing convoluted English that bespoke his discomfort in having to warn his own prime minister — "that with all these quite genuine anxieties went a certain feeling, the strength of which I could not exactly assess, that he was not quite sure if your coming here might not be rather too strong medicine in the immediate future for some of his public opinion that he still feels he has to educate up to the complete conviction of the oneness of the struggle against both Germany and Japan. I wouldn't overstate it," he apologized, "but I think it was definitely in his mind from something he said at the beginning before he switched on to laying the main weight of his argument on to security."[9]

The majority of Americans, the diplomat meant, still did not see the attack on Pearl Harbor as a casus belli against Hitler—however much the Prime Minister yearned for U.S. help in that struggle. "I seem to be conscious," Halifax added, voicing his own concern as ambassador in Washington, "of a still lingering distinction in some quarters of the public mind between war with Japan and war with Germany."[10]

This was putting the matter mildly. All eyes in America, and public fury, were directed toward Japan—not Germany.

Halifax's cautionary tone was confirmed the next day, December 10, 1941, when in the U.S. Senate the outspoken isolationist Hiram Johnson "stopped another hearing upon another AEF"—the American Expeditionary Force that the President wished to prepare for service overseas. Senator Johnson's adamant and unrepentant feeling was, as he explained to his son at the time, "we ought not to prepare an expeditionary force for Europe."[11]

Nothing the President could say would stop Churchill from coming, however. As to the date for such get-together with the Prime Minister, Lord Halifax reported that President Roosevelt "did pretty well satisfy me that it was almost physically impossible for him to make it earlier" than after the New Year. "He feels he cannot go away immediately with a possible crystallisation of the position vis-à-vis Germany very close."[12]

Besides, the British ambassador pointed out, the President had a raft of legislative and executive matters to attend to in the wake of war with Japan; "he is preparing large appropriation demands on Congress. On all this side of it he really is the only person that can pull all the strands together." Given Churchill's dual role as prime minister and minister of defense, making him Britain's quasi commander-in-chief, Churchill would, Halifax was sure, be understanding, and tame his impetuosity. "You will easily judge his difficulties arising from the immediate position on the defence side. Then he has to prepare in the last two weeks of this month the next annual budget, and also his annual message to Congress, which he will deliver on either the third or the fifth of January." Nevertheless, Halifax didn't want the Prime Minister to feel that he, as ambassador, had been remiss in personally communicating to the President Churchill's urgent request for a conference. "I pressed him as hard as I could about the importance of your meeting as quickly as it could be managed, and I don't think that he was other than perfectly genuine about the reasons that made an earlier date than he gave impracticable." Finally, in mitigation of his failure, Halifax added: "They are terribly shaken here, as you can well suppose, and fully realize that they have been caught napping. I think they realise too what it means."[13]

Churchill was furious to be balked. Halifax was, as he later put it, "a man compounded of charm. He is no coward," he allowed, "no gentleman is, but there is something," he noted, "that runs through him like a yellow streak; grovel, grovel, grovel. Grovel to the Indians [Halifax had been viceroy of India in 1926–31], grovel to the Germans, grovel to the Americans."[14]

Churchill's remark was nasty, but not unmerited. Halifax's role as Neville Chamberlain's foreign secretary had been execrable, and now he was thwarting the wishes of his own prime minister, who wished to come immediately to Washington to direct the next phase of the war against Hitler — who had, however, still not declared war on the United States.

Churchill groaned. He'd sent Halifax to Washington in January 1941 in large part to get rid of a rival — Lord Halifax having been King George VI's preferred choice when Chamberlain resigned in May 1940. He was, moreover, still Churchill's "heir apparent," were Churchill to be forced to resign by continuing British failures on the field of battle. Of this, Lord Halifax — an aristocrat and snob from his bald head and withered arm to his toes — was well aware. "I have never liked Americans," Halifax had confided before leaving London. "In the mass I have always found them dreadful."[15] "In the end we had to go," his wife, Lady Halifax, later lamented, "and I don't think I have ever felt more miserable."[16]

In the months since he'd arrived, however, Halifax had found himself entranced by Roosevelt's graciousness toward him. In comparison with Churchill, the President seemed most charming — yet with a steely underlay, the hidden hand of a strong presidential will. It was clear that, beneath all the politeness with which Roosevelt received Halifax at the White House to discuss Churchill's request, he had no wish to see the Prime Minister at this point, or to allow such a visit to become publicly known before the New Year, lest remaining isolationists cry foul and claim the President was conspiring to go to war with Germany, just at the moment when all attention should be paid to the war that Japan had begun against America in the Pacific. The response that the President drafted to be sent to the Prime Minister on December 10, 1941, was thus negative.

"In August," the President pointed out, thinking back to their Atlantic Charter meeting off Newfoundland, "it was easy to agree on obvious main items — Russian aid, Near East aid and new form Atlantic convoy — but I question whether situation in Pacific area is yet clear enough to make determination of that decisive character. Delay of even a couple of weeks might be advantageous" — a point he reiterated, given the time necessary in order to get "a clearer picture" of the situation in the Pacific. "I suggest we defer decision on your visit for one week. Situation ought to be much clearer then."[17]

Since the message sounded so circular — repetitive and somewhat negative in tone for a new ally — Roosevelt withheld it for several hours.

It was just as well. A report now came through that rocked the President at the White House, when his naval aide, Captain Beardall, brought it to him in his study. And stunned the Prime Minister, at his annex apartment next to 10 Downing Street in London.

"We got the bad news about the Prince of Wales and the Repulse," Lord Halifax jotted in his diary. "This is very bad, especially following after Hawaii."[18]

Britain's only real naval fleet in Southeast Asia, the battleships HMS *Repulse* and *Prince of Wales* — the very ship on whose deck the President and his military advisers had sung such rousing Christian hymns in August — were reported sunk, with great loss of life.

Churchill himself was devastated, since it was he who had sent the two latest battleships out to Singapore without their accompanying aircraft carrier, HMS *Illustrious,* which had put in for repairs at Ceylon. Field Marshal Smuts, the former South African premier, had warned Churchill on November 18, 1941, that the ships would be vulnerable to concentrated Japanese attack, in the case of war — "If Japanese are really nippy there is here an opening for a first-class disaster." The night before the ships were sunk Churchill had finally discussed with his advisers whether the battleships should "go to sea and vanish among the innumerable islands" or even seek safety in Hawaii.[19] No decision had been made, however, and when Churchill answered the phone next morning, Admiral Pound, the First Sea Lord, was on the line. "His voice sounded odd," Churchill later wrote. "He gave a sort of cough and gulp, and at first I could not hear quite clearly. 'Prime Minister, I have to report to you that the *Prince of Wales* and the *Repulse* have both been sunk by the Japanese — we think by aircraft. [Fleet Admiral] Tom Phillips is drowned.' . . .

"I was thankful to be alone," Churchill recalled. "In all the war I never received a more direct shock. . . . As I turned over and twisted in bed the full horror of the news sank in upon me. There were no British or American capital ships in the Indian Ocean or the Pacific except the American survivors of Pearl Harbor, who were hastening back to California. Over all this vast expanse of waters Japan was supreme, and we everywhere were weak and naked."[20]

The President's heart went out to Churchill and the British, who now had no hope of defending their British territories and Dominions in the Far East from invasion without American help—help that, thanks to America's own disaster at Pearl Harbor, could not be given.

It is possible the two men spoke by scrambler telephone; at any event the Prime Minister signaled the President at 6:00 p.m., December 10, London time, that he wasn't worried about his own security in making the voyage to America. There was, however, "great danger in our not having a full discussion on the highest level about the extreme gravity of the naval position"—and he offered again to meet the President either in Bermuda, or to fly on to Washington from Bermuda. It was no longer a matter of whether or not Germany would declare war. It was a question of whether British territories in the Far East—including Australia—could be defended, now that Churchill's fleet was sunk. "I feel it would be disastrous to wait for another month before we settled common action in face of new adverse situation particularly in Pacific." He admitted he'd "hoped to start tomorrow night," even without having been invited, "but will postpone my sailing till I have received rendezvous from you."[21]

Given the magnitude of Britain's new naval disaster, on top of Pearl Harbor, there was little the President could say, other than to repeat that there was still no way he himself could leave Washington before his State of the Union address to Congress, set for January 5, 1942. Once again he therefore sought to postpone Churchill's visit.

"I wholly agree about the gravity of naval position especially in Pacific," he allowed in his second draft reply, attempting to be courteous. "We are both of us reduced to defensive fighting in Pacific Islands and Malaya. At this moment you cannot help us there and we cannot help you except with very small naval forces now retiring southward from Philippines. Only small reinforcements on both sides can be sent to that area immediately. My first impression," the President stated, getting to the point, "is that full discussion would be more useful in a few weeks hence than immediately."[22]

Even this wording the President thought too negative, though, at a time of such fresh disaster for the British—and in a surge of compassion and goodwill, he assured Churchill that, were the Prime Minister to venture across the Atlantic, he would be "Overjoyed to have you here at the White House," adding, "If you come, give consideration to Canadian route, Bermuda route with plane from there, or sea route all the way."[23] Having decided that this signal, too, was too longwinded, he simply sent word: "Delighted to have you here at White House."[24]

Sheikh Churchill, his pasha-like mood crushed by the loss of Britain's two latest battleships and the death of so many brave sailors, gratefully accepted. Moreover, his instinct was right. The next day, December 11, 1941, before the Prime Minister set off from the River Clyde to see the President, the Führer made the second greatest blunder of his life—a mistake that would, in due course, end his odious life.

The Nazi leader had read translations of President Roosevelt's previous speeches, provided to him by his minister of propaganda, Dr. Joseph Goebbels. Whether Hitler read a transcript of President Roosevelt's first wartime Fireside Chat, broadcast from the White House at 10:00 p.m. on December 9, 1941, is unclear, but doubtful. Had he done so, though, it would have caused him to question his own assumption—for the Führer had become convinced that the President was going to have his hands full dealing with war in the Pacific and Southeast Asia, now, and would be forced to withdraw his naval forces from the Atlantic. In this case, Hitler convinced himself, Germany had nothing to fear from declaring war on America.

Roosevelt's broadcast from the White House, recorded in the Diplomatic Reception Room next to a fake fireplace, had made it abundantly clear, however, that as president he saw Nazi Germany, not Japan, as the number one threat to civilization.

"In 1931, ten years ago," the President began, "Japan invaded Manchukuo—*without warning.* In 1935, Italy invaded Ethiopia—*without warning.* In 1938, Hitler invaded Austria—*without warning.* In 1939, Hitler invaded Czechoslovakia—*without warning.* Later in '39, Hitler invaded Poland—*without warning.* In 1940, Hitler invaded Norway, Denmark, the Netherlands, Belgium and Luxembourg—*without warning.* In 1940, Italy attacked France and later Greece—*without warning.* And this year, in 1941, the Axis Powers attacked Yugoslavia and Greece and they dominated the Balkans—*without warning*. In 1941, also, Hitler invaded Russia—*without warning*. And now Japan has attacked Malaya and Thailand—and the United States—*without warning*."

It was a pattern of deceit that illustrated the difference between democracy and totalitarian government; between good and evil. "I can say with utmost confidence that no Americans today or a thousand years hence, need feel anything but pride in our patience and in our efforts through all the years towards achieving a peace in the Pacific which would be fair and honorable to every nation, large or small. And no honest person, today or a thousand years hence, will be able to suppress a sense of indignation and horror at the treachery committed by the military dictators of Japan, under the very shadow of the flag of peace borne by their special envoys in our midst."

"We are now in this war," the President concluded. "We are all in it—all the way. Every single man, woman and child is a partner in the most tremendous undertaking in our American history. We must share together the bad news and the good news, the defeats and the victories—the changing fortunes of war.

"So far, the news has been all bad. We have suffered a serious setback in Hawaii. Our forces in the Philippines, which include the brave people of that Commonwealth, are taking punishment, but are defending themselves vigorously. The reports from Guam and Wake and Midway Islands are still confused, but we must be prepared for the announcement that all these three outposts have been seized . . ."

The President's honesty was as shocking as was his familial, homey tone, which was warm, even intimate. He warned against "rumors" and "ugly little hints of complete disaster" that "fly thick and fast in wartime." News would necessarily be delayed, lest it give valuable information to the enemy, but it would in time be released, and would not be doctored, he promised. He quoted, for example, a statement made on the night of Pearl Harbor that "a Japanese carrier had been located and sunk off the [Panama] Canal Zone," attributed to "an authoritative source." As the President pointed out, "you can be reasonably sure from now on that under these war circumstances the 'authoritative source' is not any person in authority."

If the news from the Pacific was bad, his prognosis for the war was somewhat better, the President maintained. "Precious months were gained by sending vast quantities of our war material to the nations of the world still able to resist Axis aggression. Our policy rested on the fundamental truth that the defense of any country resisting Hitler or Japan was in the long run the defense of our own country. That policy has been justified," he claimed. "It has given us time, invaluable time, to build our American assembly lines of production."

Those "assembly lines are now in operation," he stated. "Others are being rushed to completion. A steady stream of tanks and planes, of guns and ships and shells and equipment — that is what these eighteen months have given us. . . .

"It will not only be a long war," the President had warned, however, "it will be a hard war. That is the basis on which we now lay all our plans. That is the yardstick by which we measure what we shall need and demand; money, materials, doubled and quadrupled production — ever-increasing. The production must be not only for our own Army and Navy and air forces fighting the Nazis and the war lords of Japan throughout the Americas and throughout the world. I have been working today on the subject of production," he said, explaining his agenda: "to speed up all existing production by working on a seven-day-week basis in every war industry, including the production of essential raw materials," and the building of "more new plants, by adding to old plants, and by using the many smaller plants for war needs." The days of labor strife, "obstacles and difficulties, divisions and disputes, indifference and callousness" were "now all past — and, I am sure, forgotten."

The President had one final thing to add, however, which presaged not only American determination to avenge Pearl Harbor, but a far more historic resolve. "In my message to the Congress yesterday I said that we 'will make very certain that this form of treachery shall never endanger us again.' In order to achieve that certainty, we must begin the great task that is before us by abandoning once and for all the illusion that we can ever again isolate ourselves from the rest of humanity.

"In these past few years — and most violently, in the past three days — we have learned a terrible lesson.

"It is our obligation to our dead — it is our sacred obligation to their children and to our children — that we must never forget what we have learned.

"And what we have learned is this," he continued. "There is no such thing as security for any nation — or any individual — in a world ruled by the principles of gangsterism. . . . We have learned that our ocean-girt hemisphere is not immune from severe attack — that we cannot measure our safety in terms of miles on any map any more. We may acknowledge that our enemies have performed a brilliant feat of deception, perfectly timed and executed with great skill. It was a thoroughly dishonorable deed, but we must face the fact that modern warfare as conducted in the Nazi manner is a dirty business. We don't like it — we didn't want to get in it — but we are in it and we're going to fight it with everything we've got." He pointed again to the connection between the aggressions of Hitler's Third Reich and Hirohito's Japanese Empire, codified in the Tripartite Pact of September 1940, by whose terms the world would be divided between Axis and Japanese spoils of conquest and subjugation. It was a global strategy of evil that the United States could no longer tolerate. "I repeat that the United States can accept no result save victory, final and complete. Not only must the shame of Japanese treachery be wiped out, but the sources of international brutality, wherever they exist, must be absolutely and finally broken."

"The true goal we seek is far above and beyond the ugly field of battle," the President emphasized. "When we resort to force, as now we must, we are determined that this force shall be directed toward ultimate good as against immediate evil. We Americans are not destroyers — we are builders. We are now in the midst of a war, not for conquest, not for vengeance, but for a world in which this nation, and all that this nation represents, will be safe for our children." The United States was economically the most powerful nation on earth — and the time had come for America to exert that power, for good. "We expect to eliminate the danger from Japan, but it would serve us ill if we accomplished that and found that the rest of the world was dominated by Hitler and Mussolini."

"So we are going to win the war," the President declared — his broadcast heard by an estimated 92.4 percent of American families[25] — "and we are going to win the peace that follows."

The President's words were those of a new leader on the world stage: one whose rhetoric was backed by a vast military potential that was being unlocked, and could be unleashed against Nazi Germany, not just Japan, unless the Führer was careful not to provoke war with the United States. Without a German declaration of war, the President could give all the Fireside Chats he wished; the Constitution still forbade him to wage war against any nation without the assent of the Capitol.

Why did Hitler then court war with such an opponent, when only Congress could declare war — and would in all likelihood not do so against Germany, given the public fury that was currently directed against Japan?

Certainly Germany was not obliged to go to war with the United States, according to the Tripartite Pact of 1940, any more than Japan was obliged to go to war with Russia.

Neither then nor later could observers and historians of World War II quite explain Hitler's fatal decision. Aware in advance of Japanese intentions with respect to the United States — though not informed of the specific target or the date the Japanese military had decided upon — the Führer had been delighted by the prospect of a sneak attack. He had therefore telephoned from the Wolfsschanze, or Wolf's Lair, his field headquarters near Rastenburg, in East Prussia, to the Reichskanzlei in Berlin several days before Pearl Harbor, instructing his foreign minister, Joachim Ribbentrop, to begin redrafting the existing Tripartite Pact with Italy and Japan. On December 3, 1941, Count Ciano, the Italian foreign minister, had thus noted with alarm in his diary how Ribbentrop was demanding not only that Italy "sign a pact with Japan agreeing not to make a separate peace," but "that Italy declare war on the United States as soon as the conflict begins."[26]

Declare war on the *United States?* The interpreter who was taking down these requests, in Rome, was "shaking like a leaf," Ciano had recorded, as the poor man translated Ribbentrop's "requests." The Duce claimed to welcome the looming global struggle — "So now we come to the war between continents, which I have predicted since 1939," Mussolini boasted — but Count Ciano, despite being Mussolini's son-in-law, was less convinced of its outcome. "Who will have the most stamina?" he'd asked himself — a question he wished the Duce and others would address *before* declaring war, rather than preening themselves over their current prowess in battle.[27]

Days later, when news of the Pearl Harbor attack came through, Ciano's heart had sunk. "One thing is now certain: America will enter the conflict, and the conflict itself will last long enough to allow all her potential strength to come into play," he'd noted — a prediction even the king of Italy, when Ciano discussed it with him that day, admitted "could be right."[28]

Hitler, however, had had other concerns. As leader of a Third Reich bogged down in a winter war with the Soviet Union that it had not envisaged, he saw the Japanese coup de main quite differently from the Italians. The latest attempt by the Wehrmacht to reach Moscow had gotten within a few miles of the city, causing the Russian capital to be largely evacuated.

Winter temperatures had then plummeted, however — and the much-vaunted Wehrmacht, which was meant to have swept over Russia "like a hailstorm,"[29] had failed, in the end, to reach its spectacular goal.

Worse still was what had followed. The Russians not only held the German armies before Moscow, but on December 5, 1941, a *hundred* Russian divisions suddenly appeared out of seeming nowhere, and launched a counteroffensive to throw the enemy back.

Hitler had been stunned as messages had flooded into his headquarters in East Prussia, begging permission to allow the German Army Group Center to retreat to more defensible lines — requests the Führer denied, eventually firing General Guderian, his top panzer commander.[30]

Japan's miraculous achievement in sinking the vaunted U.S. fleet at Pearl Harbor in a single morning, two days after the start of the massive Russian counteroffensive, had therefore given the embattled Führer new hope that, acting in concert with the Japanese, he might yet realize his dream. According to his staff, the Führer "slapped his thighs with delight as news of the report was brought to him. It was as if a heavy burden had been lifted, as with the greatest excitement he explained the new world situation to everyone around him."[31] His change of mood infected his whole headquarters at Rastenburg, which was "caught up in an ecstasy of rejoicing."[32] "We simply can't lose the war now," Hitler declared — "We have a partner who has never been beaten in three thousand years!"[33] Ribbentrop, calling Ciano from Berlin, was "jumping with joy about the Japanese attack on the United States," Ciano noted in Rome. "He is so happy that I can only congratulate him, even though I am not so sure about the advantage."[34]

Hitler was, however. He left the Wolfsschanze on December 8, 1941, and flew back to Berlin, where for several weeks he had been planning to give his own version of a State of the Union address: a speech he would deliver to the assembled deputies of the Reichstag on the progress of his war.

Meeting with Ribbentrop, the Führer there learned that the Japanese — exulting over their successful attack on Pearl Harbor — were pressing that the Third Reich, too, should declare war on the United States, as per Hitler's recent assurances and the gist, if not the letter, of the existing Tripartite Act.

Ribbentrop explained to Hitler that he had deflected the Japanese request. There was surely no need for Germany to engage the United States in open war; American naval forces would inevitably be sent to the Pacific to defend U.S. territories there and in the Far East, from the Philippines to Guam and Wake—thus making Britain even more isolated, and more vulnerable to invasion, once the Russian front had stabilized, or the war with Russia was won. In fact Ribbentrop had pointed out to the Japanese ambassador, as he reported to the Führer, that the Tripartite Pact in its extant form still did not commit signatories, unless themselves attacked, to wage war upon each other's enemies. The Japanese had not been attacked by the Soviet Union, so the Japanese had not felt obliged to declare war on Russia as a third party—to the disappointment of the German government. By the same token, Germany had not been attacked by the United States; Germany was thus not bound to fight America—at least, not now, when it had its hands full in Russia. According to Ribbentrop, Germany was required by the original pact only to supply aid—though a new draft addition was being drawn up, by which each signatory bound himself not to make peace with an enemy without the consent of all.

Hitler did not wait for Ribbentrop to finish. The Führer, Ribbentrop later recalled, cut him off midsentence. Hitler wanted, he said, an updated pact that he could announce in his forthcoming address to the Reichstag—a pact in which Germany would declare its solidarity with Japan by declaring war on the United States, with or without a Japanese declaration of war on Russia. "If we don't stand on the side of Japan, the pact is politically dead," Hitler stated, tellingly.[35]

He was the führer: the leader of a great movement in the world, a New Order. He wanted to appropriate the Japanese triumph in the Pacific as part of that New Order, to give the functionaries and population of the Third Reich, and its allies in Europe, a political message: that the Führer knew what he was doing, despite the reverses in Russia; that his world war was on track. Standing tall with Japan and Italy in a Three Musketeers trio would achieve that. Moreover, from a military standpoint, a German declaration of war on the United States would not impose a new burden on the Reich, for the United States would be locked in do-or-die combat in the Pacific and Southeast Asia. In other words, war with the U.S. would be *kostenlos*—free.

Herr Ribbentrop did not dare argue. He then left Hitler's four-hundred-square-meter marbled office, the Reichskanzlei, promising to get Italy and Japan to sign the latest protocols to the Tripartite Pact, committing them to the new agreement—one for all, and all for one—so that the Führer could include such an announcement in his forthcoming speech.

In his excitement, Hitler seemed completely oblivious to the possible repercussions. At a perilous moment in the Nazi attempt to destroy the Soviet Union—Russian Jews and commissars to be "liquidated" as the armies moved forward, "the intelligentsia" to be "exterminated," cities razed, the survivors returned to serfdom, but under German rulers[36]—the notion of a pact with military partners had taken on a new psychological importance for him, given the latest confidential report on public disaffection in the Reich: namely, warnings of a "1918 mentality" or weariness prevalent among German civilians. Pearl Harbor had galvanized the Axis at a critical moment—Germany no longer alone, with Mussolini's Italy somewhat by its side and a few minor central European satellites. Instead, the Third Reich would henceforth be fighting in a global military alliance with the great Empire of Japan: Japanese forces stretching Britain's diminishing resources to the breaking point in the Far East, and also diverting American naval and mercantile attention from the Atlantic to the defense of U.S. territories in the Pacific.

The die, then, was cast.

No one around the Führer could be in any doubt about his exultant mood. The "East-Asia conflict drops like a gift into our lap," Hitler assured his propaganda chief in Berlin. The revised pact would, he told Dr. Goebbels, fulfill the strategic war directive he'd issued some eight months earlier. In it he'd laid out, in advance, his global strategy for the victory of the Third Reich. "The aim of cooperation based on the Tripartite Pact," he'd explained, "has to be to bring Japan to active operations in the Far East as soon as possible. Strong English forces will be tied up as a result, and the main interest of the United States of America will be diverted to the Pacific."[37] Japan's attack at Pearl Harbor had now conformed to that strategic calculation to a T. As Hitler's adjutant recalled the Führer boasting to his entourage in Berlin on December 9, 1941: "Compelled by the conflict with Japan," America would "not be able to intervene in the European theater of war."[38]

But if America would not be able to intervene, by virtue of the Japanese triumph at Pearl Harbor, why the need for Germany to go to *war* with the United States? individuals like Ciano wondered.

Goebbels, however, did not dare question the Führer any more than Ribbentrop had. Instead, Dr. Goebbels attempted to see the new developments through the Führer's infallible eyes. "Thanks to the outbreak of war between Japan and the USA," he articulated in his diary that day, after speaking with Hitler, "a complete shift in the general world picture has taken place"—a fact that could soon be trumpeted by his propaganda ministry. Great benefits would accrue, aiding Germany's battles in the Atlantic, the Middle East, and Russia. "The United States will scarcely now be in a position to transport worthwhile material to England let alone the Soviet Union," Goebbels summarized the Führer's thinking.[39]

So certain was Hitler that America's attention would now "switch" to the Pacific, in fact, that he ordered unrestricted submarine warfare against American-flagged vessels in the Atlantic to begin that very day, December 9, 1941, even before he officially declared war on the United States. The "American flag will no longer be respected," Goebbels noted in his diary. "Anyone caught on the way to England must reckon with being torpedoed by our U-boats."[40]

Hitler's intention had been to address the Reichstag the next day, December 10, 1941—but he didn't. Did he have second thoughts? Was he still hoping that the Japanese would sign up to declare war on Russia, simultaneously? Or was he simply psyching himself up to deliver a historic speech for the occasion, which he insisted on writing himself?

No one knew. As the clock ticked in the White House, considerable apprehension arose—as it did in London. Suppose the Führer thought better of his decision? Suppose he was advised there was no real need to go to war with the United States: that only Congress could approve war, and was loath to do so against Germany when U.S. defense forces were now desperately needed to fight Japan's predations in the Far East?

It was only the next day, at 8:00 a.m. on December 11, 1941, that the German chargé d'affaires in Washington appeared at the State Department, next to the White House, to deliver an important message. Since Secretary Hull would not receive him, Herr Thomsen was told to wait.

Eventually, at 9:30 a.m., Thomsen was able to hand over to the head of the European Division of the U.S. State Department his note, explaining that Germany's patience with the United States was at an end, and that a state of hostilities, of war, now existed between their two countries.

Simultaneously in Berlin, the U.S. chargé d'affaires was given the same message. Then, at 3:00 p.m. that day, Berlin time, after receiving confirmation that the Japanese had signed the new tripartite agreement—but without the Japanese having agreed, reciprocally, to declare war on Russia—the Führer went before his assembled, fawning "party comrades" in Berlin and ended speculation across the globe. His speech was an hour and a half long — and was soon in the President's hands. Reading it, Roosevelt was stunned by how personal it was: a speech primarily directed at him.

"Providence," the Führer announced, to the Reichstag's amazement — since Hitler had never shown signs of religious belief — had personally entrusted him "with the waging of a historic struggle which will decisively fashion not only our German history for the next thousand years, but also the history of Europe — even the history of the entire world."

Following a review of earlier world history, the Führer brought Reichstag deputies up to date on the progress of the war in Russia. He claimed that Stalin had fully intended to attack the Third Reich and conquer Europe, and thus Operation Barbarossa had been but a preemptive spoiling attack: a necessary act of German self-defense that had, however, succeeded in capturing no fewer than 3,806,865 imminent Russian invaders, for the loss of only 158,773 German lives.

Before the deputies could digest the sheer enormity of these numbers — 158,733 Germans *killed* (which understated German dead by four-fifths), and half a million German casualties since June — the Führer moved on to the president of the United States.

"The course of these two lives!" the Führer said, comparing their careers — and worldviews. Acknowledging that "the philosophy of life and the attitude of President Roosevelt and my own are worlds apart," Hitler proceeded to defame the U.S. president as the dupe of "members of the same people that we once fought in Germany as a parasitic phenomenon of mankind, and which we had begun to remove from public life": the Jews. Sneering at the New Deal, the Führer claimed that Roosevelt had "increased the national debt of his country to enormous proportions, devalued the dollar, continued to ruin the economy, and maintained unemployment" — thanks to "those elements which, as Jews, have always had an interest in ruin and never in order."

The Führer then pointed out that there were many isolationists in America who held similar racist views to his own. "Many distinguished Americans agree with this assessment, or rather, realization. A threatening opposition hangs over the head of this man," he claimed — and accused the President of "sabotage" of "all possibilities for a policy of European pacification," as he put it: possibilities of peace destroyed by Roosevelt's insistence on helping governments in exile, and supplying American weapons to those who were holding out against Nazi "pacification" efforts.

The December 7 attack on America was thus fully deserved, Hitler declared, getting to the point. "I think that all of you felt relieved that now finally one state has protested, as the first, against this historically unique and brazen abuse of truth and law," the Führer said, congratulating Germany's ally, Japan. "It fills all of us — the German Volk and, I think, all decent people of the world — with profound satisfaction that the Japanese government, after negotiating with this falsifier for years, has finally had enough of being derided in so dishonorable a manner. We know what force stands behind Roosevelt. It is the eternal Jew," a race whose international tribe had "destroyed people and property in the Soviet Union," where "millions of German soldiers" had witnessed for themselves what Jewry had wrought: Bolshevism. "Perhaps the president of the United States himself has failed to understand this," Hitler speculated. "This speaks for his mental limitations," he sneered. And with that, the Führer poured scorn on the Atlantic Charter principles for postwar peace — "a new social order" that was, in Hitler's view, "tantamount to a bald hairdresser recommending his unfailing hair restorer."

Reichstag deputies tittered at the simile. In contrast to the homilies of the Atlantic Charter, the Third Reich and its New Order represented the wave of the future, the Führer proudly claimed. "Thanks to the National Socialist movement," he declared, Germany had "never been as united and unified as it is today and as it will be in the future. Perhaps never before has it been so clear-sighted and rarely so aware of its honor."[41]

The Führer came then, at last, to the climax of his speech. "I have therefore had passports sent to the American Chargé d'Affaires," he confirmed — together with a copy of the four new articles of the Tripartite Pact, "signed today in Berlin," which he proceeded proudly to read out: "Article 1: Germany, Italy, and Japan will together fight this war, a war that was forced upon them by the United States of America and England, and bring it to a victorious end by employing all instruments of power at their disposal . . ."[42]

It was, then, global war — a war that Germany, Italy, and their noble ally, Japan, would win. "Today," the Führer boasted, "I head the strongest army in the world, the mightiest air force, and a proud navy." Anyone in the Third Reich or elsewhere who criticized the "front's sacrifices" or sought to "weaken the authority of this regime" would be executed, he warned — without mercy. "The Lord of the Worlds," he ended, "has done so many great things for us in the last years that we bow in gratitude before Providence, which has permitted us to be members of such a great

Volk. We thank Him that, in view of past and future generations of the German Volk, we were also allowed to enter our names honorably in the undying book of German history."[43]

Surprised by Hitler's affirmation of an Almighty, but puffed up by his references to the German Volk, and anxious not to be accused of criticizing the Führer, or "weakening" his authority, the Reichstag deputies gave him a great ovation.

Hitler's remarks about "many distinguished Americans" or isolationists were not entirely wrong.

"I was somewhat surprised at Germany and Italy declaring war upon us," admitted Hiram Johnson to his son in California — blaming his own country, rather than Hitler or Mussolini, since "we had been guilty of many breaches of peace, and have given the greatest causes for war that, under international law, can be given." Still and all, the U.S. senator confessed, "I did not think they would declare war" — in fact, "the day before, when I had made my objections" to a new American Expeditionary Force or AEF, he was confident he had up to "ten votes with me in the Senate," whereas the "day after when war was declared [by Hitler], I did not have a single damned vote."[44]

This, in truth, was the measure of the Führer's historic miscalculation. Without a German declaration of war, Congress would not have authorized the President to declare war on Germany, given the disaster at Hawaii and the worsening military situation in the Philippines and Pacific Islands. Hitler could thus have gotten America off his back, for free. Instead, by declaring war on the United States he now silenced America's isolationists like Senator Johnson — and just as importantly, provided the President with the power to act not simply as president, but as the world's most powerful commander in chief, sending American forces into combat on a global scale.

Shortly before 3:00 p.m. Eastern Standard Time on December 11, 1941, the President therefore sent over to Congress, in response to Hitler's declaration of war, his formal, written request that Congress "recognize a state of war between the United States and Germany," as well as "between the United States and Italy." Not he, but the dictators — the Japanese, and now the Germans and Mussolini's Italians — had chosen to wage war on the United States. In a unanimous voice-vote in the House of Representatives and a unanimous recorded vote in the Senate, the U.S. legislature gave its approval to the President's request. Senator Johnson even withdrew his opposition to an American Expeditionary Force — which would be under the President's sole direction.

"Those who know," the senator confided to his son, "claim that this will be a long war. . . . I doubt this. I think it will be fast and furious for a time, and then it will begin to crumble. . . . We may be certain of one thing, however," he added in one of the most celebrated mispredictions of a member of the august U.S. Senate. "It will last long enough to demolish our internal economy; and we'll find at its conclusion little value to our money and less to our properties."[45]

PART THREE

Churchill in the White House

4

The Victory Plan

UNTIL EARLY ON December 22, 1941, Eleanor Roosevelt had not even been told that Prime Minister Winston Churchill would be staying at the White House. The President had, however, recently spoken with Malvina Thomson, his wife's secretary, asking idly whom the First Lady had invited to stay over Christmas, Eleanor recalled, "as well as the people invited to dinner."

Who to dinner? "In all the years that we had been in the White House he had never paid much attention to such details," Mrs. Roosevelt reflected, "and this was the first time he had made such a request of Miss Thomson." Since the President "gave no explanation and no hint that anything unusual was going to happen," the First Lady and Miss Thomson could only conclude that the President "felt a sudden curiosity."[1]

Given Mrs. Roosevelt's work for civilian defense, the many causes she supported, as well as her efforts to keep the members of the large Roosevelt family connected with each other and with their paterfamilias, her lack of particular concern was understandable.[2] The nation, after all, was now at war. She'd ordered blackout curtains to be made for the White House. She had watched while antiaircraft guns were placed on the roof, gas masks were distributed, three-inch bulletproof glass was installed, and a tunnel was constructed to an air raid shelter beneath the building next door, on the orders of the Treasury secretary, Mr. Morgenthau, who was responsible for the President's physical security. (The President had not been amused. "Henry, I will not go down into the shelter," Roosevelt warned Morgenthau when refusing to have anything to do with such a scheme. Then, smiling, he added: "unless you allow me to play poker with all the gold in your vaults.")[3] But about a visitor from England, not only to dine but to stay with them, barely two days before Christmas, she had had no idea.

The British ambassador to the United States, by contrast, had known about Churchill's upcoming visit ever since the President issued the invitation. "After lunch I had twenty minutes with the President about plans for the talks this week," Lord Halifax had noted in his secret diary on December 21, the day before the Prime Minister's arrival. "The arrangements are all a bit fluid," he'd added, "depending on what time the people concerned can get here, and I foresee a good many last minute changes."[4]

There were. In the course of December 22, "we got news that the arrangements to meet Winston and co. had to be revised owing to a change of time of their arrival, and accordingly aeroplanes and special trains were improvised," the ambassador recorded. "Rather surprisingly, this all worked out pretty well, and they duly arrived at the airport about half past six, the President meeting them."[5]

Emerging from the U.S. Navy Lockheed Lodestar that had flown him from Norfolk, Virginia, once the Prime Minister's battleship, HMS *Duke of York*, had moored there, Winston Churchill espied the familiar figure of the President of the United States leaning against his car. He had come — in person! In the summer they had gotten together only as *potential* partners, meeting aboard their respective battleships. Now they were in the same, if metaphorical, boat.

The ten-day voyage across the North Atlantic had been brutal, the Prime Minister recounted to his host — storms so severe the vessel was reduced at times to six knots out of concern for its destroyer escort. In the end it had been compelled to shed its ocean antisubmarine lifeguards completely.

"Being in a ship in such weather as this is like being in prison, with the extra chance of being drowned,"[6] the Prime Minister paraphrased Dr. Johnson's famous quip in a letter he posted that night to his wife. For his part, Lord Beaverbrook, the Canadian-born British minister of supply who had accompanied Churchill on the voyage, declared he would have preferred traveling by submarine to being tossed about like balsa wood on the surface.

"They didn't enjoy their journey much," Lord Halifax confided to his diary. "Max Beaverbrook told me he wasn't sure whether their journey had been seven days or seven weeks at sea, but Winston seemed in good form, and we had quite a pleasant dinner at the White House — ourselves, Cordell Hull, Sumner Welles, Harry Hopkins, Winston and Max and one or two of their personal staff."[7]

Steaming alone and as fast as possible into the shallow waters of Chesapeake Bay to make up time, they had created so much swell that the Prime Minister's luggage had been inundated in the hold in the ship's stern. Fortunately, his clothes had stayed dry. After dinner at the White House the small group went upstairs to the Blue Room, "where we talked in the President's study about Vichy and North Africa, and various other matters on the Atlantic side of the world," as the British ambassador noted. "Complete unanimity of view and the atmosphere very good, Winston and the President getting on very well indeed together."[8]

In later years, following publication of Churchill's memoirs and the diaries of Churchill's doctor, Charles Wilson, a myth would spawn that the Prime Minister had somehow exercised strange magic on his Christmas visit to Washington: that he had somehow persuaded the legendary sorcerer of the White House into following a Churchillian rather than Rooseveltian strategy for winning the world war.

It is an absurd myth. Dr. Wilson, who became Lord Moran, was not even present at the White House dinner, nor at the long discussion in the President's study afterward. He had, it was true, been summoned from his hotel across the road by the Prime Minister. However, when he got to the upstairs bedroom where Churchill was staying, opposite that of Harry Hopkins, the Prime Minister was not there, he recalled. The White House Rose Room "smelt of cigar smoke and I tried to open the window. The crumpled bed-clothes were thrown back, and the floor was strewn with newspapers, English and American, just as the P.M. had thrown them away when he had glanced at the headlines," for "he always wants to know what the papers are saying about him."[9]

An hour and a half later, "the P.M. came out of the President's room. He looked at me blankly; he had forgotten he had sent for me."[10]

Churchill certainly seemed fired up, though. "I could see he was bottling up his excitement," Dr. Wilson described. Seeing the agitation in Churchill's whole being, he allowed the Prime Minister two "Reds," or barbiturate pills, to ensure he got a good night's sleep. In the elevator with Beaverbrook, thereafter, Churchill's minister of supply remarked to Wilson that he had never "seen that fellow [Churchill] in better form. He conducted the conversation for two hours with great skill"—claiming the "P.M. had been able to interest the President in a landing in North Africa."[11]

The PM had interested the President in a landing in North Africa, rather than the other way around? Either Beaverbrook expressed himself badly in front of Dr. Wilson, or Dr. Wilson (who reconstructed many of his wartime conversations later, under the guise of contemporary diary entries) misheard him. Wilson, on his first trip abroad with Churchill, was a sensitive man and a cautious doctor; he had no idea what he was writing about militarily, beyond what Churchill purportedly said to him. The Prime Minister's excitement that December night had certainly been palpable, but in reality it was owing to profound relief that, despite the disaster at Pearl Harbor and worsening situation in the Pacific and Far East—where Guam and Wake had been surrendered to the Japanese, Hong Kong was besieged, British forces were retreating to Singapore, and the Philippines looked next in line for Japanese conquest—the President was still determined upon a "Germany First" strategy.[12] And, moreover, that the two leaders were of like mind about how that offensive strategy should start: in North Africa.

The question in Churchill's mind the entire time he was onboard the *Duke of York*, crossing the Atlantic, had been: would Pearl Harbor force President Roosevelt to change his mind, and to direct American efforts to the Pacific, as Hitler was hoping? On finding that this was *not* the case, and that the President was still bent on following a "Germany First" policy, beginning with his "great pet scheme" (as Secretary Stimson dubbed it) for American landings in Northwest Africa, Churchill's excitement threatened to run wild.

As Churchill cabled the next day to the war cabinet and rump Chiefs of Staff Committee (whose members had accompanied the Prime Minister to Washington, with the exception of General Sir Alan Brooke, the new chief of the Imperial General Staff, or CIGS), there had been "general agreement" in his late-night discussion with the President about preempting a German occupation of Vichy-controlled Northwest Africa. Also about seizure of French battleships rather than allowing them to fall into German hands; and about invading, if necessary, the Atlantic islands of the Azores and Cape Verde. With regard to the North Africa enterprise, Churchill proudly reported, "it would be desirable to have all plans made for going into North Africa with or without invitation. I emphasised immense psychological effect likely to be produced in France and among French troops in North Africa by association of United States with the undertaking."[13]

Association with the undertaking? It was an odd yet characteristic circumlocution, typical of Churchill's often deliberately antique style of reportage. What he meant was, he was convinced the President was right to make landings in French Northwest Africa — Morocco and Algeria — an *American,* rather than British or Allied, enterprise: embarrassed to admit in writing that, given the unpopularity of the British among Vichy officials following his cruel orders for the sinking of the French fleet at Mers el-Kébir on the coast of French Algeria the year before, any hope of getting Vichy French commanders to welcome British troops alongside American soldiers might well doom the project.

What made Churchill's blood run so much faster, however, were the strategic advantages to the Allies of such an operation. As Churchill explained in detail in his cable to his war cabinet, the President "favored a plan to move into North Africa being prepared for either event, i.e. with or without [Vichy French] invitation. It was agreed to remit the study of the project to Staffs on assumption that it was vital to forestall the Germans in that area." The real meat of the plan was, however, its ramification: the assumption that the British Eighth Army, currently advancing from Egypt into Libya in Operation Crusader, "achieved complete success. I gave an account of the progress of the fighting in Libya, by which the President and other Americans were clearly much impressed and cheered."[14]

Between the two American and British armies, advancing from either end of the Mediterranean in a vast pincer movement, General Rommel's German-Italian Panzerarmee Afrika could thus be squeezed to extinction — if General Claude Auchinleck, commander in chief in the Middle East, fulfilled Churchill's proud expectations.

Given that this North African strategy would, ten months later, prove the very strategy that turned the tide against the Third Reich, leading inexorably to Hitler's downfall, it is extraordinary that its genesis and authorship would become so misunderstood, so misconstrued, and so fought over by historians in subsequent years[15] — especially those who mistrusted Winston Churchill and his tendency to bend the truth with "glittering phrases."

Among those who distrusted Churchill right from the time of his arrival at the White House on December 22, 1941, was General Joseph Stilwell, a fearless, feisty American commander slated to go to China as chief of staff to Generalissimo Chiang Kai-shek. "Vinegar Joe" Stilwell not only resented what he saw as Roosevelt's preference for the U.S. Navy over the U.S. Army, but — after conversing with Secretary Stimson and War Department planners who were convinced the President's French North African landings would fail, and serve no strategic purpose — he proceeded to trash the President in his diary as a "rank amateur in all things military," subject to "whims, fancy and sudden childish notions," a U.S. commander in chief who had been "completely hypnotized by the British" into supporting a "cockeyed" and "crazy gamble" in North Africa, when the better strategy was so obvious. "We should," he wrote in his diary, "clean the Pacific first and then face East" — exactly as Hitler had hoped when declaring war on the United States.[16]

Stilwell, unfortunately, was not alone in holding such views at the time. Colonel Edwin Schwien, of the U.S. Army's G-2 Intelligence Division, had already ridiculed the President's insistence on American landings in French Northwest Africa as "a patent absurdity" in comparison with the opportunity to embrace cross-Channel landings in northern France.[17] Colonel Albert Wedemeyer, a senior planning officer in the War Department who had been tasked with drawing up estimates of what would be required for whatever strategy was chosen, was also opposed to the idea of American landings in Morocco and Algeria — especially when Churchill arrived in the U.S. capital. As Wedemeyer warned his colleagues, Churchill was weaving an imperial "plot" that was against America's interests.

Whispers of such dissension within the War Department, moreover, were quick to circulate in rumor-ridden wartime Washington. Senator Hiram Johnson, the eighty-one-year-old isolationist, continued to believe America could avoid hostilities with the Third Reich, even after Hitler's declaration of war. "The war is getting worse all the time, and we're feeling it more and more. The President is trying to keep all matters within his hands," he would write to his son several weeks later, after Churchill's departure from Washington. Roosevelt "has been made a fool of by Winston Churchill, and in his innermost heart, I think he begins to realize it. We go merrily on spending money until we are pretty nearly over our debt limit, and it will be but a short time that we are. Another billion now for Russia. What a travesty this is! We are going to come out, everyone of moderate means, absolutely broke."[18]

• • •

Such criticism of the President, paraded at the time and in subsequent years, was not helped by Churchill himself.

After being voted out of office in 1945, Churchill was doubly anxious to paint himself, in his six-volume memoirs, *The Second World War,* as a successful wartime prime minister. Against a background of the British losing every single battle against the Nazis in the two years after he became prime minister, he naturally wished to portray himself as the coauthor, at least, of the eventual strategy that reversed this sorry saga. Describing his visit to Washington shortly before Christmas, 1941, he thus introduced the reader to the only extant, documented account of his first discussion with President Roosevelt, on the first night of his trip. This was his cabled message to the British war cabinet, a document Churchill was making public for the first time, eight years after the event — Churchill explaining in his memoirs how, straightway after his arrival at the White House, he had "immediately broached with the President, and those he had invited to join us, the scheme of Anglo-American intervention in North Africa. The President had not, of course at this time read the papers I had written on board ship, which I could not give him till the next day. But he had evidently thought much about my letter of October 20. Thus we all found ourselves pretty well on the same spot. My report home shows that we cut deeply into business on the night of our arrival."[19]

This was, unfortunately, to mislead readers and historians, since it implied that Churchill had suggested U.S. landings in Morocco and Algeria on October 20, and the President had "evidently thought much" about the idea.

The reverse, however, was the truth.

President Roosevelt's preferred strategy for U.S. landings in French Northwest Africa went way back to midsummer 1941, long before he even met Churchill off the coast of Newfoundland.

Recognizing, in the wake of the Nazi invasion of Russia on June 22, 1941, that Hitler could never be ultimately constrained save by force, Roosevelt had instructed the Joint Chiefs of Staff to research and then draw up an estimate of military production requirements — a program that would, of necessity, "make assumptions as to our probable friends and enemies and to the conceivable theatres of operation" in order "to defeat our potential enemies."[20] This secret Victory Plan — as it became known — was not to be too "detailed," the President had ordered in July;

rather, he'd urged, it was to be "general in scope," in order to ensure "the efficient utilization of our productive facilities." To meet the President's directive, the secretaries of war and navy and the chiefs of staff had then taken two months to carefully set out, in consultation with the President, a top-secret, twenty-three-page gauge of what it would require for America to win a probable war against Germany and Japan, both in strategy and production numbers. Their report, or "Joint Board Estimate," had been readied for the President on September 11, 1941, several weeks after he returned from Newfoundland.[21]

The Victory Plan's assumptions and predictions remain astonishing for their clarity and accuracy even seventy years after they were prepared for the President.

"Assumed enemies" were "Germany, and all German-occupied countries whose military forces cooperate with Germany; Japan and Manchuko; Italy; Vichy France; and possibly Spain and Portugal.

"Countries considered as friends or potential associates in warfare are the British Commonwealth, the Netherlands East Indies, China, Russia, Free France, people in German-occupied territory who may oppose Germany, and the countries of the Western Hemisphere."

The "national objectives of the United States" were given as "the preservation of the Western Hemisphere; prevention of the disruption of the British Empire; prevention of the further extension of Japanese territorial dominion; eventual establishment in Europe and Asia of balances of power which will most nearly ensure political stability in those regions and the future security of the United States; and, so far as practicable, the establishment of regimes favorable to economic freedom and individual liberty."

The U.S. document had then lain down the requirements necessary even if "the British Commonwealth collapsed" — necessitating a five-year struggle for the United States, which would last until 1946. With relentless realism, the report had considered "the overthrow of the Nazi regime by action of the people of Germany" to be unlikely — certainly not "until Germany is on the point of military defeat." Even were a new regime to be established in Germany, "it is not at all certain that such a regime would agree to peace terms acceptable to the United States" — necessitating, therefore, unconditional surrender. Since Hitler's Germany "can not be defeated by the European Powers now fighting against her," the report had thus concluded, and "if our European enemies are to be defeated, it will be necessary for the United States to enter the war and to employ a part of its armed forces offensively in the Eastern Atlantic and in Europe or Africa."

Or Africa.

It was this phrase that had, in truth, raised the hackles of the secretary of war and others working in the War Department. Nazi Germany, Secretary Stimson accepted, would be America's primary foe in the event of war. Even so, a simultaneous two-ocean war would probably have to be waged, to hold the Japanese forces at bay until Hitler's Germany surrendered, for the report did not envisage the British or the Free Dutch (whose government was in exile in London) would be able to "successfully withstand" a Japanese assault "against the British in Malaya and against the Dutch in the Dutch East Indies." Therefore it would be up to the United States to furnish the munitions and the troops, not only to deal with Hitler in Europe, but to hold off the Japanese, and eventually turn defeat in the Pacific into victory in Southeast Asia.

The report for the President had then addressed with great realism the thorniest question: how could such a simultaneous, two-ocean war be *fought,* with *Germany* as the primary foe? In accordance with the President's wishes, the report had laid down initial American offensive operations to be directed on those areas where Germany's "lines of communication" were most extended — Morocco, French West Africa, Senegal, and the Azores, if the Germans attempted to occupy those territories. In the case of Japan, likewise, it would mean striking where Japanese "lines of communication" were most extended: a strategy that would allow the United States to stretch the Japanese until Japan's pips squeaked "owing to a lack of adequate resources and industrial facilities." Meanwhile, by building up its own arsenal of trained combat troops and munitions, the U.S. would, over time, work its way into a position to win decisive victory, first over Germany, then over Japan.

With the exception of Russia, the great strength of America's assumed allies in such a war would be in naval and air forces — forces that "may prevent wars from being lost, and by weakening enemy strength may greatly contribute to victory," the Victory Plan acknowledged. "By themselves, however, naval and air forces seldom, if ever, win important wars," the report continued. "It should be recognized as an almost invariable rule that only land armies can finally win wars."

Those land armies would, with the exception of Russia, have to be predominantly American.

The President's Victory Plan strategy, then, had already been clearly established in September 1941, months before the U.S. entered the war. In particular, Roosevelt's strong doubts about the feasibility of successfully landing American forces across the English Channel anytime soon had, at his insistence, been clearly addressed in the document. After all, had not Hitler, even at the height of his military conquests in 1940, balked at attempting a crossing of the Channel to defeat the remnants of British forces after Dunkirk? It was, in the words of the Victory Plan, "out of the question to expect the United States and its Associates to undertake in the near future a sustained and successful land offensive against the center of the German power" — something that was beyond even Soviet Russia, with its *millions* of troops fighting in the field.

How then was Germany to be ultimately defeated? "It being obvious that the Associated Powers can not defeat Germany by defensive operations, effective strategic offensive methods *other than an early land offensive in Europe* must be employed," the report had summarized.[22] A two-part strategy would have to be adopted — defensive at first. The Associated Powers, or Allies, would have initially to concentrate on cauterizing the ambitions of the dictators. This should be done by "a continuation of the economic blockade," and by "the prosecution of land offensives in distant regions where German troops can exert only a fraction of their total strength." In the case of Japan, a "strong defense of Malaysia," would be necessary, as well as an "economic offensive through blockade; a reduction of Japanese military power by raids; and Chinese offensive against the Japanese forces of occupation."

Cauterization would not bring victory, however. In the second stage of American strategy, American armies would have to go on the offensive: aiming first at Rome and Berlin, then at Tokyo.

The route of such offensive action in Europe was here the issue — since the planning and especially the production targets for eventual offensive action would be contingent upon that decision. In setting out the "major strategic objectives" in the ultimate defeat of Germany, the report had emphasized — at the President's specific direction — that an initial, indirect step in an American march on Rome and Berlin be rehearsed: *namely via Northwest Africa* — currently not occupied by German forces — which the President saw as the vital strategic first act in defeating Germany.

American occupation of French Northwest Africa—with or without Vichy help—would not only deny German access to the Atlantic Islands (a possible steppingstone to South America), the President was certain, but would provide the United States with "a potential base for a future land offensive," as the Victory Plan concluded. "In French North and West Africa, French troops exist which are potential enemies of Germany, provided they are re-equipped and satisfactory political conditions are established by the United States. Because the British Commonwealth has but few troops available and because of the unfriendly relations between the British and Weygand [Vichy] regime, it seems clear that a large proportion of the troops of the Associated Powers employed in this region must be United States troops."[23]

This, then, had been the President's blueprint, prepared over the summer of 1941, for prosecuting a future global war, both in the short term and the long term—once hostilities came. The typed report, with numerous appendices, had been dated September 11, 1941, and had borne the President's imprint and imprimatur on every page. Moreover, in the weeks after its preparation, the President's predilection for landings of U.S. forces in Vichy-controlled French Northwest Africa had only grown stronger.

On October 3, 1941, for example, the U.S. secretary of the navy, Frank Knox, had confided to Lord Halifax that "what he wanted to see happen"—as Ambassador Halifax immediately reported to Prime Minister Churchill—"was for the Americans to send 150,000 men to Casablanca and join hands through an assenting Weygand with us [the British] in North Africa. The President, according to him, was much interested in this idea. I should suppose we are some way off that yet but the fact that they should be thinking about it at all is interesting."[24]

Interesting it was! A giant pincer movement, using American naval, air, and ground forces that would, with British forces pressing from Libya, crush the German-Italian Panzer Army in Africa and provide a base and springboard on the very threshold of Europe—forcing Hitler to defend the continent from Norway to the Mediterranean! As Halifax had also learned, however, the President's enthusiasm for such an indirect strategy was by no means shared by the aging Republican secretary of war—or even by General Marshall, who saw North African landings as too dangerous to undertake, and in any event a distraction from the most direct route to Berlin. They had thus favored U.S. troops being sent directly to Britain in the event of war, thence to be put into combat on a more straightforward path to the capital of Hitler's Third Reich, beginning with cross-Channel landings, mounted from England's southern coast against the coast of mainland France.

A week later, on October 10, 1941, over lunch at his desk at the White House, the President had confided his Northwest Africa plan in person to the British ambassador,[25] informing Halifax he'd "told Stimson and Marshall to make a study of the proposal to send an American Expeditionary Force to West Africa. This had greatly excited Stimson and Marshall, who thought he was 'going off the deep end' and embarking on a dispersal of effort that they thought unwise," Halifax had secretly reported to Churchill in London the next day, October 11, 1941. "He had explained to them, however, that he did not contemplate anything immediate, but none the less wanted the question studied," Halifax had cabled the Prime Minister. The titular Vichy leader, Maréchal Pétain, "might die," and his understudy, General Weygand, "might feel himself released from his personal pledge of loyalty" to Pétain and to Hitler, "and things might move. I don't suppose that all this is to be taken very seriously at present, but it is a pointer," the ambassador had added in his letter to Churchill.[26] The long term, in other words, might well become the short term.

Given that America was not at war with Germany at the time, such a difference of opinion between the President and his War Department officials had remained academic — but potentially problematic, too, Halifax had wisely recognized. It would be unfortunate, he felt, if the President's coalition pincer-strategy were to be defeated, not by the Germans, but by the U.S. War Department.

To check out for himself, Halifax had therefore wisely gone to the Munitions Building to sound out the secretary of war in person that very evening — October 10, 1941, as he informed Churchill. Stimson was dead against U.S. landings in North Africa. "Stimson told me that he was inclined to hold the President off schemes that would dissipate United States effort, the possibilities of which were severely limited" — for Stimson, like General Stilwell (who was initially chosen to lead the Casablanca attack), feared such an operation of war could never succeed — indeed, as the war secretary told the British ambassador in somewhat defeatist language, he was now far from confident the United States could do more than send American troops to help defend Britain against a possible Nazi invasion if, as Stimson suspected, the Russians were defeated in Operation Barbarossa, or sued for peace.

Churchill, in contrast to Colonel Stimson, had been *delighted* to hear of the President's "great pet scheme" for North African landings — and had made it his personal task to help the President override Stimson's objections to it.

Nine days after receiving Halifax's cable, Churchill thus wired to tell the President that, if the United States chose to remain out of the war against Germany, and in the event of German pressure on the Vichy government to grant the Nazis "facilities in French North Africa," he himself was "holding a force equivalent to one armoured and three field divisions ready with shipping," that could "either enter Morocco by Casablanca upon French invitation or otherwise help to exploit in the Mediterranean a victory in Libya."[27]

The President had smiled at such well-meant Churchillian bravado. How the British, whose performance in amphibious landings since Norway in April 1940 had been uniformly disastrous, imagined that a unilateral British invasion of French Northwest Africa would succeed against hostile Vichy forces without U.S. help did not say much for the Prime Minister's realism. But the President had said nothing — admiring Churchill's offensive spirit, and his support for Roosevelt's strategy.

There the matter of potential American military operations in Europe had rested, up until December 4, 1941. For it was on that day, in the most egregious act of treachery, that Colonel McCormick had published many of the details of the President's Victory Plan — the entire report to the President having been deliberately leaked by an individual in Stimson's War Department. Printed by McCormick, it had inevitably found its way straight into Hitler's declaration of war on the United States, announced before the Reichstag on December 11, seven days later.

By then, however, events had swamped the scandal — the Japanese attack on Pearl Harbor making McCormick's isolationism risibly naïve. Indeed, in the days after the terrible news from the Far East, there was many a patriot in Washington who, reflecting on the *Chicago Tribune*'s recent revelations, thanked the Almighty that, despite the disaster in Hawaii, the United States at least *had* a plan for fighting the war — and in Britain, an ally willing to fight with the U.S. to win it.

• • •

In days of preparation for Churchill's arrival and "the strategic problems which are confronting us in the coming conferences,"[28] the President had convened meeting after meeting at the White House. The first had been on December 18, 1941, at 3:00 p.m., a conference at which Stimson, Knox, General Marshall, Admiral Stark, Admiral King, Harry Hopkins, and Admiral Nimitz — the new commander the President had appointed to replace Admiral Kimmel in the Pacific — were present.

"The President then told this conference exactly the nature of the conference which is coming next Monday [December 23, 1941] and who will be there, and he said that he desired us to attend it as his advisers. A paper was produced containing [a] suggested agenda which had been drawn up by the British," Stimson noted in his diary.[29] The President, however, had wanted an American version — and had therefore told his advisers to draw up a counteragenda. America, not Great Britain, was to run the war, the President made clear — not only because the United States would be providing the bulk of the necessary forces to defeat Nazi Germany and Japan, but because the performance of the British so far had been, in all frankness, miserable.

The Japanese invasion of Malaya looked increasingly ominous — aiming toward Singapore, Britain's primary naval base in the Far East. The Australians and New Zealanders were panicking — with very few British naval, air, or army forces to defend them. Stimson, for his part, was all for "safety first" — and therefore opposed any discussion of offensive action at this point. Defending the American West and East Coasts, helping to defend the British Isles, providing arms to help the British defend the Middle East and India, using American forces to defend Australia and New Zealand — these were the first priorities to be tabled, Stimson thus insisted. Beyond that, he forecast, there were "a number of things on which there will be sharp divergence" between the U.S. Army and the Navy — and between the United States and Great Britain.[30]

The President was disappointed in his war secretary. He felt it politic, however, not to risk dissension at such a critical moment. *Unity of purpose* was the foremost consideration, since the whole world was now looking to the United States to take the reins in a global war. The President's plan was therefore to focus not on the things that might divide his advisers, the armed forces, and the U.S. and British, but on those that bound them together. He therefore laid down his first priority: namely, the moral basis for a coalition war, involving as many allies as possible in the fight against Germany and Japan. His Atlantic Charter should, he told Secretary Hull, now be enshrined in a new declaration of principles that all of America's allies could sign up to.

Simultaneously, a good working relationship with America's closest new coalition partner, Great Britain, should be fostered. Allied military strategy should be concerted, he explained — but with Washington, not London, as the central headquarters of the Allies. If, in the next days over Christmas, 1941, he got agreement on these two priorities, the President felt, he would be doing very well, no matter how bad the immediate news "from the front" might be. All would fall into place.

Winston Churchill, for his part, had a very different modus operandi from the President. Where Roosevelt sought to achieve agreement with his wishes by charm and goodwill, Churchill simply relied on his commanding personality and sense of privilege — fueled by an intake of alcohol that made him relatively impervious to criticism or rebuke. Immediately upon his arrival in the White House, the Prime Minister had made himself at home. "We had to remember to have imported brandy after dinner," Eleanor complained — deeply aware of her own family's history of alcoholism. "This was something Franklin did not have as a rule."[31] "Never had the staid butlers, ushers, maids and other Executive Mansion workers seen anything like Winston before," recalled Roosevelt's Secret Service detail chief — the Prime Minister consuming "brandy and scotch with a grace and enthusiasm that left us all openmouthed in awe."[32]

Churchill's commandeering of the First Lady's office, the Monroe Room, for his portable maps and files did not endear the Prime Minister to Mrs. Roosevelt, either.

The President did not share his wife's shock, nor her barely concealed disapproval. He liked eccentricity, which made people the more interesting, he found. Observing the diminutive politician setting up shop in his, the President's, own residence, Roosevelt was amused — indeed, he wondered how he might turn the Prime Minister's premature arrival in America into a public relations winner — thus helping him override the concerns of his secretary of war and senior generals in the War and Navy Departments. Hitler's military success, after all, could not solely be ascribed to German tanks and air power. The Führer had spent years perfecting the art of propaganda, using the dark genius of his associate, Dr. Goebbels — and the sense of national unity they had created in the Third Reich was a fundamental component of German military morale.

Well then, the President recognized, the Associated Powers must do even better! In this respect, Churchill's presence in the U.S. capital might prove a perfect foil in concealing American disarray — as well as internal dissension over future operations. The Prime Minister's distinctive, pugnacious personality could be used to advantage — a perception the President put straight into action by asking, over lunch at his cluttered desk in the Oval Study, whether Winston would appear with him at his weekly press conference in the West Wing at 4:00 p.m. on December 23, 1941.

The Prime Minister said he would be delighted to do so. And thus, at one of the darkest moments of the war, before a barrage of cameras, microphones, and journalist's notepads, the first image of a truly Grand Alliance was created, at the White House, in Washington, D.C.

Gazing at the extraordinary scrimmage of a hundred White House reporters jamming the small room — each one having been screened for security purposes — Roosevelt began with an apology. "I am sorry to have taken so long for all of you to get in, but apparently — I was telling the Prime Minister — the object was to prevent a wolf from coming in here in sheep's clothing. (Laughter) But I was thereby mixing my metaphors, because I had suggested to him this morning that if he came to this conference he would have to be prepared to meet the American press, who, compared with the British press — as was my experience in the old days — are 'wolves' compared with the British press 'lambs.' However, he is quite willing to take on a conference, because we have one characteristic in common. We like new experiences in life."[33]

The President's genuine affection for the Prime Minister was instantly clear to all. "Mr. Roosevelt liked Churchill a great deal," wrote the AP reporter A. Merriman Smith, later; "disagreed with many of his ideas and suggestions, but nevertheless found his presence stimulating, often to the point of fatigue."[34]

Before Churchill stood to speak, however, the President said he wanted "to make it clear that this is a preliminary British-American conference, but that thereby no other Nations are excluded from the general objective of defeating Hitlerism in the world. Just for example, I think the Prime Minister this morning has been consulting with the Dominions. That is especially important, of course, in view of the fact that Australia and New Zealand are very definitely in the danger zone; and we are working out a complete unity of action in regard to the Southwest Pacific. In addition to that, there are a good many Nations besides our own that are at war. . . . I think it is all right to say that Mr. Mackenzie King [prime minister of Canada] will be here later on. In regard to the other Nations, such as the Russians, the Chinese, the Dutch, and a number of other Nations which are — shall I say — overrun by Germany, but which still maintain governments which are operating in the common cause, they also will be on the inside in what we are doing."

Churchill's country, then, was but one combatant in a great coalition of nations the President was assembling to stand against Hitler, Mussolini, and the Japanese. "In addition to that, there are various other Nations, for example a number of [South] American Republics which are actually in the war, and another number of American Republics which although not acting under a declaration of war are giving us very definite and much-needed assistance. It might be called on their part 'active non-belligerency.'

"At five o'clock we are having a staff meeting. We have already had a meeting with the State Department officials, and during the next few days decisions will materialize. We can't give you any more news about them at this time, except to say that the whole matter is progressing very satisfactorily.

"Steve [Early] and I first thought that I would introduce the Prime Minister, and let him say a few words to you good people, by banning questions. However, the Prime Minister did not go along with that idea, and I don't blame him. He said that he is perfectly willing to answer any reasonable questions for a reasonably short time, if you want to ask him. . . . And so I am going to introduce him, and you to him, and tell you that we are very, very happy to have him here. . . . And so I will introduce the Prime Minister.

"(To the Prime Minister) I wish you would just stand up for one minute and let them see you. They can't see you. (Applause greeted the Prime Minister when he stood up, but when he climbed onto his chair so that they could see him better, loud and spontaneous cheers and applause rang through the room.)

"THE PRESIDENT: (to the press) Go ahead and shoot."[35]

The reporters shot — though the first question ("What about Singapore, Mr. Prime Minister? The people of Australia are terribly anxious about it. Would you say to be of good cheer?") went so straight to the heart of the war crisis that the President almost fell off his wheelchair. "The President laughed so hard that he nearly choked on his cigarette holder," Smith recalled.[36]

Faced with "wolves" the like of whom he'd never encountered before, Churchill acquitted himself extremely well. Chubby-cheeked, wielding his trademark cigar, and wearing a dark jacket with a polka-dot bow tie and striped pants, he parried calls for predictions and delivered memorable phrases of moral uplift. However, there was never any question but that he was now an admired but junior partner in a new coalition—utterly dependent on the United States for the successful prosecution of the war.

"My feeling is that the military power and munitions power of the United States are going to develop on such a great scale that the problem will not so much be whether to choose between this and that," Churchill responded to questions of strategy, "but how to get what is available to all the theaters in which we have to wage this World War."

Churchill was telling the truth. However defeatist Secretary Stimson and other generals might be, the numbers spoke for themselves. From the President himself and from Harry Hopkins, Churchill had heard preliminary numbers of estimated U.S. war production that had made the blood in the Prime Minister's veins run faster—and did so even more when, at the first formal Anglo-American staff conference that followed, it was made clear the war would indeed be directed from Washington.

The writing was on the wall—and it read quite clearly: "United States of America First."

Convincing the press and public that the democracies were united in confronting the Axis powers was one thing. Convincing the British visitors that the American military was, under its commander in chief, ready for the big leagues was another.

Fresh from bombed-out London, Churchill's huge retinue of staff officers and clerks had expected a land of plenty, yet they found themselves awed by life in America. From thirty-six-page newspapers "an inch thick" to the huge cars in which they were taken to their hotels, Churchill's almost eighty attendants were mesmerized by the very scale of things—much like Gulliver's adventure in the "Voyage to Brobdingnag."

The traffic in Washington, for example, amazed Churchill's military assistant, Colonel Ian Jacob: "a flood of American saloon cars, all looking new, all almost alike, and most of them with only one or two people in them. There is a car to every 2½ people in the city. No-one walks a yard. Cars lie parked everywhere, and no-one bothers about having a garage. You simply leave the car on the side of the nearest street, if you can find a spare hole. The whole effect is as of an ant-heap, or a swarm of beetles."[37]

As at the Newfoundland summit in August, there appeared to him to be a vast difference between the American approach to high command, however, and the British version. "At 5.30," Jacob noted in his diary immediately after the President's press conference, "there took place the first meeting between the Prime Minister and the president and their Chiefs of Staff," held in the Cabinet Room, which was situated "in the West block, on the garden level," next to the Mansion. "It is a pleasant room, about the size of the Cabinet Room at No. 10 [Downing Street] but a little wider. It has four French windows each in an archway, and is devoid of furniture except for a table and a number of chairs including arm chairs disposed around the walls. Two or three pictures of former Presidents hang on the walls which are white." The long, almost coffin-shaped table was unexpectedly Arthurian, widest about halfway along one side, where Mr. Roosevelt presided. The table seated "16 in comfort, all of whom can see each other well." To the right of the President sat Prime Minister Churchill; to the President's left, the secretary of the navy, Mr. Knox, flanked by Admirals Stark, King, and Turner, the chief of naval planning; and to Churchill's right, Secretary Stimson and the other U.S. and British chiefs of staff, occupying the rest of the table, together with Harry Hopkins and Max Beaverbrook.[38]

"The President led off with a statement as to the talks which he already had during the day with Churchill," Secretary Stimson noted in his own diary—expressing surprise and delight as the President pulled out Stimson's somewhat feeble "Memorandum of Decisions Made at the White House on December 21," 1941, "and made it the basis of the entire conference."[39]

"He went over it point by point," Secretary Stimson noted, "telling the conference of their views on each point and then asked Churchill to follow and comment on it, which he did. There was then a little general discussion participated in by the American military and naval members and the British military and naval members, and it became very evident that there was a pretty general agreement upon the views of the grand strategy which we had held in the War Department and which were outlined in my paper. Churchill commented feelingly on the sentence of my summary where I described our first main principle as 'the preservation of our communications across the North Atlantic with our fortress in the British Isles covering the British fleet.'"[40]

In truth Roosevelt was concealing from Stimson his own determination to proceed, as soon as practicable, with an American counterstroke: the invasion of Casablanca, Morocco, and Algeria with 150,000 troops. For the moment, however, the President expressed himself in complete agreement with the U.S. war secretary as to the importance of sticking to a "Germany First" strategy.

Roosevelt's charm and confident air certainly deceived Colonel Jacob. "The President is a most impressive man and seems on the best of terms with his advisers," Jacob described in his diary — evincing no idea of what was going on, in reality, namely an awkward dance between the Democratic president and his elderly Republican secretary of war, whose tone was one of fear and anxiety over the current situation in the Far East, and America's limited ability to do anything offensively. By the side of the Prime Minister, the President appeared to the innocent Jacob to be "a child in Military affairs, and evidently has little realization of what can and can't be done. He doesn't seem to grasp how backward his country is in its war preparations, and how ill-prepared his army is to get involved in large scale operations."[41]

The very opposite was, in verity, the case — the President all too aware of American disarray in the wake of Pearl Harbor and the time it would take, behind the scenes, to ready American forces of any size to be put into battle, yet determined to look beyond the current trials in the Pacific. "To our eyes," Jacob wrote, "the American machine of government seems hopelessly disorganized. The President, to start with, has no private office. He has no real Private Secretary, and no Secretariat for Cabinet or Military business. The Cabinet is of little account anyway, as the President is Commander-in-Chief. But he has no proper machinery through which to exercise command. . . . We found this complete lack of system extended throughout. . . . Their ideas of organization and ours are wide apart, and they have first to close the gap between their Army and Navy before they can work as a real team with us."

As a result, Jacob blamed American mismanagement and disorganization for what he saw as a disappointing military summit. As he put it, sailing back to England, "the Chiefs of Staff meetings which now took place almost daily for the next three weeks never really achieved anything. There was never any settled Agenda, and every kind of red herring was pursued. We thought we had achieved a considerable triumph when we got our general strategy paper agreed to," he lamented, "without amendment by the U.S. Chiefs of Staff. I am pretty sure however that it is regarded as an agreeable essay which all can pay lip service to, while each American Service follows its nose and does the job which seems to stick out at the moment."[42]

Poor Jacob — whose father had been a British field marshal in World War I — was completely bamboozled, unaware that the president's modus operandi was the product of decades of experience in marshaling American talent to serve his purpose and strategies. In this respect, Churchill's modus operandi and vivendi were the polar opposite. Churchill's methods, however, could boast only of having produced in two years the longest series of military disasters in British history. By contrast the President's methods, as U.S. commander in chief, had yet to be tested in war.

All too soon Jacob's ignorance of the American system, and his dedicated bureaucratic approach to high command in the service of the British prime minister/minister of defense, were to embroil Roosevelt in the near-resignation of one of the senior members of the President's own administration, his secretary of war. For what Jacob could not understand was that this was America, not India.

Jacob's faux pas would be telling. British imperial bureaucracy had incontestably served its empire well. Over the centuries, following conquest or annexation, the British had learned to administer vast indigenous populations by imposing a hierarchy of paper wallahs, or colonial bureaucrats, much as the Chinese had done over millennia in their own territories: a governor and, below him, a multitiered hierarchy of civil servants, with small but highly disciplined naval and army forces on hand to impose civil order when and where required.

For generations this British system of colonial administration had proven almost unimaginably successful — but it had also led, in such a huge and scattered global jurisdiction, to a white elitism disguising an almost fatal aversion to manual labor. It was simply assumed that Indians, or coolies, or foreign servants and mercenaries, would provide the necessary muscle to maintain an imperial system that guaranteed order, was generally not corrupt, preserved freedom of religion — and ensured British profits. Thus a tiny cadre of British civil servants, educated at elitist British "public" (actually private) schools, had managed the necessary administrative and clerical paperwork of an empire with admirable diligence — one that elicited even Hitler's admiration.[43]

In times of war, moreover, British Empire administrators and clerics had simply continued to do the same as they'd done in peace: giving orders in writing to those responsible for carrying them out. In recent decades, in the aftermath of World War I, with its terrible loss of life, there had been fewer and fewer competent British administrators—especially soldiers—willing or able to see these orders were executed. And once the fires of World War II had been kindled by Hitler in Europe and the French Republic collapsed, the heart had seemed to go out of the elderly British imperial motif. Against more disciplined, ideologically inspired, and well-armed Nazi troops in Norway, Belgium, France, Greece, Crete, Libya, and Egypt, Britain's valiant forces—its Royal Navy, RAF, and Army—had simply failed to operate together as a unified, cohesive, modern military force on the field of battle, even though the performance of individuals and individual units—as in the Battle of Britain—was often meritorious.

The results, when pitted against Hitler's Nazi legions in France, the Mediterranean, and North Africa, had shamed the island nation: campaigns studded by retreat, evacuation, and surrender. More worrying still, the current British performance in the Far East, following the Japanese declaration of war, held out little hope of being more fortunate. Hong Kong was expected, Churchill had assured the President, to hold out for several months, until reinforcements could be assembled and sent out; in fact, outnumbered four to one, the crown colony would surrender after seventeen days, on December 25, 1941. The same was to hold true of Singapore, the "Pearl Harbor" of the Far East—which Churchill had hoped would hold out for many months. Moreover, despite Churchill's assurances to the President, the case of Libya—where General Auchinleck had already found it necessary to fire the Eighth Army field commander he'd appointed, General Alan Cunningham—was not promising to be any better in terms of British military competence on the field of battle, whatever the Prime Minister might claim.

In a word, the British had perfected the bureaucratic arts, but had let slip the art of fighting—at least in terms of modern combat. And the President, better than anyone, knew it: not from his "advisers," but from his special sources.

However much his cabinet colleagues and subordinates deplored the habit, President Roosevelt liked to send for, see, and hear personal reports on what was going on around the world from those he trusted.

These reports to the President—both verbal and written—were necessarily anecdotal, but they ensured that a highly intelligent and above all curious U.S. commander in chief was able to gauge the reality of what was going on, rather than relying on the sanitized version from his various ambassadors and government officials, as Churchill did. The President's emissaries thus kept him well informed, at least anecdotally, about reality on the ground, acting as the President's "eyes and ears" outside the White House—at home and abroad. Where Churchill relied on his reading of history and his abiding, romanticized Victorian vision of British arms, the President liked to question every visitor, and every report. Moreover, the President's emissaries and informants reflected every area of government and society, high and low. His wife, Eleanor, might—and did—attract opprobrium for her dedication to progressive social causes, for example, but the President, restricted by his wheelchair and the discomfort of traveling, admired her for her openheartedness and willingness to journey forth to see things firsthand. "You know, my Missus gets around a lot," Roosevelt told Churchill's physician, Dr. Wilson. "She's got a great talent with *people.*"[44]

As Dr. Wilson reflected later, "when the President was so immersed in the war that he was in danger of forgetting the hopes and aspirations of the ordinary people, Eleanor was at his elbow to jog his memory." This was in marked contrast to the doctor's formal patient, the Prime Minister of Great Britain, and *his* wife. "Mrs. Churchill played no such part in her husband's political life. They were a devoted pair," Dr. Wilson acknowledged, "but he paid little attention to her advice, and did not take it very seriously."[45]

Wilson may have overstated the case, but his perception was revealing. Winston Churchill might paint his trip to Washington, both at the time and later, as an example of his own great leadership at a moment of world crisis. In many respects it was, in terms of Allied unity and propaganda—certainly when contrasted with the visit by his deputy prime minister, Major Clement Attlee, leader of the Labour Party, only a month before. ("All quite useful for the Americans to hear," Lord Halifax had written in his diary on November 10, 1941, after listening to Mr. Attlee making a speech at the National Press Club in Washington, "but I don't think he is a very impressive personality.")[46] Churchill, by contrast, was out to impress—from the start.

Attempting to save the British Empire as Neville Chamberlain's successor, Churchill had sought to infuse Britain's military hierarchy with new energy from the top, even if this meant riding roughshod over his colleagues and staff. Like a natural, aggressive chess player, Churchill had a wonderful, instinctive grasp of grand strategy, backed by the ability to withdraw his mind from the current fray in order to achieve a transcending perspective across time and globe. This made him an inspiring speaker, and courageous leader in a great cause—giving heart to his staff and to his nation at a time of ever-worsening military defeats. It could not, unfortunately, substitute for good judgment.

Of all Wilson's perceptions about Churchill, it was his recognition of Churchill's poor judgment, in the context of a man so otherwise exceptional, that most testified to the doctor's honesty. Churchill's courage was genuine and exemplary, Dr. Wilson observed. On the other hand, in terms of people and decisions, Churchill was also endowed with desperately poor discernment. As Churchill's own wife put it, Winston had surrounded himself all his life with "charlatans and imposters," because his very genius—his ability to cast his mind across broad horizons, to play with ideas and invent "glittering phrases"—demanded a fawning rather than critical audience. Once asked by King George VI to become prime minister, in May 1940, Churchill had made himself minister of defense, thus licensing himself to become a quasi commander-in-chief. As the Prime Minister of Canada himself told Dr. Wilson, "Winston cowed his colleagues. He stifled discussion when it was critical and did not agree with his views."[47] Lord Hankey, who had for many years been secretary to the British cabinet and was the paymaster general, deplored Churchill's moody behavior and impulsive decisions, referring to the Prime Minister in his diaries and letters as "the dictator," a "Rogue Elephant" whose military disasters would have long since ended his career, had there been an alternative leader in Britain of any real timbre. "It was he who forced us into the Norwegian affair which failed; the Greek affair which failed; and the Cretan affair which is failing," Hankey had lamented in his diary earlier that summer—adding, in October 1941, that the British war cabinet had become a "crowd of silent men," enduring "the usual monologue by Churchill"—a British government of "utter incompetence,"[48] while another senior civil servant, Sir Norman Brook, described the Prime Minister's greatest failure as being "not much interested to hear what others had to say."[49] Leo Amery, the British secretary of state for India, declared that Churchill had reduced his country to "a one-man government, so far as the war is concerned."[50]

Had Churchill's military decisions been sounder, his dictatorial approach might have been forgivable; as it was, even the former British prime minister Arthur Balfour called him a "a genius without judgment."[51]

By comparison, the President — who appeared to lack a foolproof administrative machine to convey dictatorial orders, as Churchill possessed — was blessed by something far more important: the gift of good judgment. Lacking an imperial-style apparatus to assemble information on paper and then impart his decisions, and, moreover, a believer in the American system of checks and balances in making decisions that affected the nation, Roosevelt relied on his own sunny personality, not obedience, Dr. Wilson noted. Churchill "beat down opposition and struck men dumb who had come to the Cabinet to expostulate." By contrast, the President "was not a strong administrator," but "he got the work done by picking the right man for a particular job — he was a good judge of men — and trusting him to do it; he encouraged and inspired his man to get on with his job. If he didn't he got rid of him. And of course the man he trusted gave of his best. The plan worked."[52]

As he later looked back across the course of the war, Dr. Wilson — who knew Churchill probably better than anyone — found himself lost in admiration of Roosevelt. "To lead a nation in this fashion calls for unusual qualities," he wrote. "Roosevelt had them. Men came away from the White House feeling better." The President had not always solved their problems, "he had not even given them directions, but he sent them away determined to carry out their task . . . Roosevelt's detachment was always taking me by surprise; he kept his head above the sea of administrative problems; his task was not to straighten them out, it was to harness the nation to its work."[53]

Confident in his choice of subordinates, even when they failed to match his expectations, the President was thus not only a master politician, but a master of happy, confident delegation — whereas Churchill, who "unlike the President was not a good judge of men" and had "so often been let down when he entrusted to others the solution of anything," attempted to run everything, down to the last detail. "Winston tried to do the work of three men, he had his finger in every pie," Dr. Wilson reflected, recalling Foreign Secretary Eden's heartfelt protest, "I do wish he'd leave me to do my own job!"[54]

"Certainly it is not in that mood that men do their best work," the doctor chronicled. "Roosevelt knew this. The P.M. never did. Once I said to him, 'Hitler seems to tackle not only the strategy of his campaigns, but also the details.' Winston looked up at me with a mischievous smile spread over his face. 'That's exactly what I do,' he said."[55]

Churchill had said this with cheeky pride. Dr. Wilson — like so many of those serving the Prime Minister — found himself disappointed by this aspect of his genius-patient: a man who could see the larger picture so brilliantly, yet would interfere with, chide, overrule, and bring to ruin even his most professional subordinates. "He got caught up in a web of detail," Wilson wrote, "like a fly in a spider's web."[56]

How, then, was it possible for a national leader of such poor military judgment to get on with a U.S. commander in chief blessed with fine judgment, but a completely different approach to man management?

Dr. Wilson, watching the combination over the ensuing days, noted something strange — a key perhaps to how the miracle was effected. For the first time in his life, as Dr. Wilson observed, Churchill *listened.* And accepted his new position, not as the President's equal but as his honored vizier.

For his part, Colonel Jacob also watched Churchill's transformation or submission — though with incredulity. On Christmas Eve, 1941, at a meeting in the Monroe Room, temporarily serving as Churchill's Map Room, "at which the Prime Minister and our [British] Chiefs of Staff were discussing matters of domestic concern," a "peculiar incident" occurred, Jacob recorded in his diary.[57]

"In the middle of the meeting, the door opened and in came the President in the wheeled chair and joined us. He asked various questions, and then said he feared the news from the Philippines was not good, and that it looked as it would very soon be impossible to get any more air reinforcements into Manila as the aerodromes would be in the enemy's hands. He felt it would be for the Joint Chiefs of Staff to consider where, in the circumstances, the reinforcements could go. The Prime Minister agreed, and so did our [British] Chiefs of Staff."[58]

Once the President and Prime Minister had left to have dinner, Colonel Jacob, as a good staff officer, asked the British chiefs of staff — Field Marshal Dill, Admiral Pound, Air Marshal Portal — if he should draw up a record of the meeting for the U.S. chiefs of staff, so they could duly discuss the President and Prime Minister's haunting question about the reinforcements at their conference meeting the next day, Christmas Day, at 3:00 p.m. The British chiefs said yes.

The next day, Christmas Day, there was an explosion that rocked the new Grand Alliance.

"At 3'oclock the meeting was just starting when an urgent call came for [Brigadier] Jo [Hollis] to go to the White House. Off he went," Jacob recorded. "Not long after that, I was called out to answer the telephone, and found Jo on the line. He said there was a regular flutter in the dovecote, and that we had dropped a brick, but he couldn't quite make out what we had done wrong. The Prime Minister had said that we had issued some Minutes containing statements, or a Directive, by the President, and that this had given offence. We must realize that the President had to be treated with ultra respect, and we had been guilty of some kind of lèse-majesté."[59]

"Lèse majesté" was an apt phrase for the realization, among dutiful British staff officers serving Mr. Churchill, that there was a new military monarch whose wishes and concerns must be treated as senior to those of their own, hitherto dictatorial, prime minister/minister of defense. Above all, they must never, ever embarrass the new monarch in terms of his U.S. war council and cabinet—which the President ran in a quite different way to that of Mr. Churchill.

The "dovecote" was, in fact, the U.S. War Department—where Jacob's neatly typed minutes of the British Map Room meeting had caused consternation.

In his own diary, a seething Henry Stimson recorded how, upon returning from a Christmas horseback ride that morning, he'd been bombarded by irate "Generals Arnold, Eisenhower [the new head of the army's Far Eastern Section of the War Plans Division, with a staff of one hundred], and Marshall" who "brought me a rather astonishing memorandum which they had received from the White House concerning a meeting between Churchill and the President and recorded by one of Churchill's assistants"—i.e., Jacob. "It reported the President as proposing to discuss the turning over to the British of our proposed reenforcements [destined] for MacArthur. This astonishing paper made me very angry," Stimson recorded, "and, as I went home for [Christmas] lunch and thought it over again, my anger grew until I finally called up Hopkins, told him of the paper and of my anger at it, and I said that if that was persisted in, the President would have to take my resignation; that I thought it was very improper to discuss such matters while the fighting was going on and to do it with another nation."[60]

Stimson's furious reaction went to the very heart of the new alliance being forged in the prosecution of a modern, global war. The resignation of the Republican secretary of war, in the wake of Pearl Harbor and the Japanese invasion of the Philippines, would certainly have done more than upset the dovecote. "This incident shows the danger of talking too freely in international matters of such keen importance without the President carefully having his military and naval advisers present," Stimson dictated on his tape recorder that night. "This paper, which was a record made by one of Churchill's assistants, would have raised any amount of trouble if it had gotten into the hands of an unfriendly press"[61] — as the Victory Plan had done. Stimson felt, in fact, that he had personally saved the President from personal disaster at the hands of his own leaky War Department staff.

Hopkins, sensing a dual threat — of resignation, and of a deliberate leak — made as if aghast. "He was naturally very surprised and shocked by what I said and very soon called me back telling me that he had recited what I had said to the President in the presence of Churchill," Stimson noted, "and they had denied that any such proposition had been made."[62]

"I think he felt he had pretty nearly burned his fingers," Stimson speculated about the President — causing Roosevelt, in fact, to summon Stimson, Knox, Marshall, Arnold, King, Stark, and Hopkins to a new meeting at the White House on Christmas Day at 5:30 p.m. The President "went over with us the reports up to date of the various matters and we discussed various things which were happening and the ways and means of carrying out the campaign in the Far East," the secretary of war noted. "Incidentally and as if by aside, he flung out the remark that a paper had been going around which was nonsense and which entirely misrepresented a conference between him and Churchill. I made no reply of course as he had given up, if he had ever entertained, the idea of discussing the surrender of MacArthur's reenforcements [to the British]."[63]

Stimson, who felt acutely the impotence of the U.S. Army as well as the Navy in confronting Japanese moves in the Far East, added that the episode had "pretty well mashed up" his Christmas.[64] The episode was, however, a painful illustration of how careful the President, in a democracy such as the United States, was obliged to be, if he was to avoid impeachment or press lynching.

Churchill, whose mother had been born an American and who had himself often traveled to the United States before the war, understood the danger only too well. Wisely he impressed upon his small army of note takers and imperial paper-wallahs that Britain would now have to accept a subordinate role in waging war against the Axis powers, and how important it would be for his staff to mind their p's and q's.

Behind Secretary Stimson's explosion at the War Department and his threat of resignation, of course, lay the awful truth of which both the President and his secretary of war were deeply aware: namely that, thanks largely to MacArthur's poor generalship, the Philippine Islands were now doomed, whatever reinforcements were sent out from U.S. shores.

MacArthur had lost his air force on the first day of war, and although he had insisted to the War Department since the summer that he would be able to hold the islands for six months against a Japanese invasion, there was never any real chance of resupplying him, let alone reinforcing him, once the Japanese sank the American fleet at Pearl Harbor. The distances were simply too immense, as Stimson, Eisenhower, and his team found when they marched their calipers across the Pacific Ocean: more than fifty-three hundred miles from Pearl Harbor to Manila, thirty-six hundred miles from Brisbane, Australia.

"I've been insisting Far East is critical and no other sideshows should be undertaken," Eisenhower would note in his diary a few days later. "Ships! Ships! All we need is ships!" Ike mourned some days after that. "Also ammunition, anti-aircraft guns, tanks, airplanes, what a headache!" The major general even briefly felt the President should reverse course—namely abandon the "Germany First" strategy that Stimson and the chiefs had laid down on December 23, 1941, and give in to Hitler's calculation. The United States should "drop everything else," Eisenhower penned, "and make the British retire in Libya," if necessary abandoning the Middle East. "Then scrape up everything everywhere and get it to the Dutch East Indies, Singapore, and Burma for a new Alamo."[65]

Christmas Eve and Christmas Day at the White House were, in the circumstances, somewhat mournful compared with the prewar Roosevelt family celebrations—the Roosevelts' four sons all in uniform now, and away from home.

Eleanor did her best as First Lady and hostess. "How hollow the words ["Merry Christmas!"] sounded that year!" she recalled. "On this visit of Mr. Churchill's, as on all his subsequent visits, my husband worked long hours every day." While the Prime Minister took a long nap each afternoon, the President caught up on his paperwork. Once Churchill was awake the President had again to play host to his guest — who, refreshed by his afternoon slumber, then kept the President up till all hours of the night. "Even after Franklin finally retired," Eleanor explained, "if important dispatches or messages came in, he was awakened no matter what hour, and nearly every meal he was called on the telephone for some urgent matter."[66]

The President's tradition of "mixing cocktails" at 6:00 p.m., or "children's hour," was one moment of Roosevelt's day that remained sacrosanct no matter who the visitor might be. Confined to his wheelchair, the President had the opportunity to carry out an involved task himself, rather than instruct others to do it. "He would be wheeled in and then spin around to be at the drinks table, where he could reach everything," one visitor recorded. "There were the bottles, there was the shaker, there was the ice. . . . And you knew you were supposed to just hand him your glass, and not reach for anything else," lest this draw attention to the President's disability.[67]

The President's signature creation was an FDR martini — although imbibers were aware that his mix of gin, vermouth, and fruit juice followed no hard-and-fast ratio, and included on occasion rum from the Virgin Islands.

Churchill, however, hated the President's cocktail hour. Moreover, he found the President's favorite concoction foul, as another visitor noted — leading the Prime Minister to pretend to drink the martini, but in fact to take it with him to the bathroom after asking to be excused, and pour it down the sink — replacing it with water from the faucet![68]

The President, however, was the president. Told he should not light the traditional Christmas tree — which had been erected on the south side of the White House, rather than in Lafayette Square — for security reasons, Roosevelt dismissed such advice with a snort of ridicule. Instead he took the Prime Minister with him to witness the lighting, and participate in the prayers and carols on Christmas Eve; then on Christmas Day took his visitor to the Foundry United Methodist Church for an interfaith service, followed by lunch for the Joint Chiefs of Staff and their staffs, and the "biggest Christmas dinner we ever had — sixty people sat down at the table," Mrs. Roosevelt chronicled proudly. After dinner a movie — *Oliver Twist* — more Christmas carols, and "the men worked until well after one o'clock in the morning."[69]

Secretary Morgenthau had sat opposite the Prime Minister at the huge dinner. The next day the secretary told his Treasury staff how puzzled he'd been by the legendary British leader. "You know, he has a speech impediment," Morgenthau recounted (Churchill unable to pronounce the hard *r* sound). The Prime Minister had said very little, because "he just wasn't having a good time," Morgenthau surmised. The brilliant Treasury chief observed the faces of the British team closely. Max Beaverbrook's countenance, creased and lined like that of a lizard, presented "a map of his life." Churchill's skin, by contrast, seemed completely smooth, as if untroubled and unscarred. To Morgenthau, Churchill appeared — erroneously, as it was to turn out, only a day later, when the Prime Minister experienced heart trouble — "literally in the pink of health." Still and all, Churchill seemed preoccupied, Morgenthau was aware. He asked "three times to be excused after dinner so, he says, 'I can prepare these impromptu remarks for tomorrow.'" Sitting next to Morgenthau for the movie, nevertheless, followed by a documentary film on the war so far, the Prime Minister had seemed at least cheered by newsreel shots of the campaign in Libya, remarking: "Oh, that is good. We have got to show the people that we can win."[70]

Unlike the great financier, Churchill understood the huge political import of what the President had next asked him to do: to address not only Congress, but the people of America and the free world, from the rostrum of the U.S. Capitol.

For twenty-four hours the Prime Minister drafted and revised versions of his speech, even reading passages to the President, including the quotation from the 112th Psalm, "He shall not be afraid of evil tidings: his heart is fixed, trusting in the Lord."

Given that there were stirrings of revolt among a number of members of Parliament back home in London, there was good reason for Churchill to be anxious. "I saw Winston for a quarter of an hour before luncheon in the Map Room at the White House, complete in grey romper suit," the British ambassador had noted in his diary on Christmas Eve, "after which Stimson, whom I met as he came out from talking to him, must have reflected that he had never seen anything quite like it before."[71]

Stimson had not. Churchill "was still in dishabille, wearing a sort of zipper pajama suit and slippers," the secretary of war described the Prime Minister, whom he had come to brief on the worrying situation in the Philippines. "This has been a strange and distressful Christmas," Stimson noted the next day, after the "reenforcements" contretemps. "The news around us is pretty gloomy. Hong Kong has fallen; the Japanese have succeeded in making landings not only at Lingayen Gulf but two places south of Manila, and MacArthur has cabled that he was greatly outnumbered, would make the best fight he could, and retreat slowly down the Batan [*sic*] Peninsula and Corregidor," while evacuating the Philippine government and declaring Manila an open city.[72]

With bleak reports such as these, Churchill realized, his speech to Congress would have even more significance in terms of Allied morale. When Halifax visited him late on Christmas morning, he "found him in his coloured dressing gown in bed, preparing his speech for the Senate tomorrow, surrounded by cigars, whiskies and sodas and secretaries!"[73]

The President had tried his best to lighten the atmosphere, ribbing the Prime Minister and saying it had been good for him "to sing hymns with the Methodies" that day, at church. Churchill had attempted to smile, admitting it had been "the first time my mind has been at rest for a long time."[74] Yet in reality he could not relax, and the President eventually decided it would be best if he didn't accompany Churchill to the Capitol on the morning of December 26, lest he distract from the Prime Minister's reception on the Hill. Or worse still, in an atmosphere where all too many legislators still distrusted Churchill and the British for inveigling the U.S. into war with Germany, suggest unconstitutional collusion.

Where the President seemed to have no visible nerves, Churchill suddenly had too many. He worried about support in Parliament, back at home; he worried about the situation in the Far East; he worried about the situation in Libya — unable to sleep if he did not receive his nightly signal of good tidings from General Auchinleck, however bogus "the Auk's" claims.[75] Moreover, he worried about the relationship between Britain and the United States. For it was clear to him, if only haltingly to his staff, that the U.S., whose entry into the war he had so long prayed for, would now be *primus inter pares:* not simply, at last, a world power, but *the* world power. It would be important, Churchill recognized, to do everything he could to maintain good relations with the new imperium: supporting, aiding, advising, and, where necessary, flattering its emperor or pharaoh. At one point, he even wheeled the President into dinner at the White House — likening his act to that of Queen Elizabeth I's famous courtier, Sir Walter Raleigh, spreading his cloak before the sovereign, lest she get her shoes dirtied.[76]

Speaking before Congress was, however, a quite different challenge. "Do you realize we are making history?" Churchill remarked to Dr. Wilson shortly after midday, as he paced across the antechamber to the U.S. Senate chamber, still rehearsing his speech.[77]

For years the Führer had been the one to make history; even at this moment he saw himself being accompanied toward Valhalla by his accomplice, Emperor Hirohito, and his medieval warriors. Now, however, it would be the turn of the Associated Powers.

The "galleries were crowded, the Diplomatic Corps, the Supreme Court and others being accommodated on the floor," the British ambassador noted in his diary that night.[78] One listener, the son of Jewish immigrants, spoke for many Americans hearing the Prime Minister for the first time, describing it in his diary as a triumph — "the first sound of blood lust I have yet heard in the war." Churchill's rhetoric amazed him: "the color and the imagery of his style, the wonderful use of balance and alliteration and the way he used his voice to put emotions into his words. Why at one point he made a growling sound that sounded like the British lion!"[79]

"Winston spoke for about 35 minutes, and was much cheered," Lord Halifax recorded, for his part. "Everybody thought it very good, and he produced a great impression. Personally I did not think it so very good, but naturally I kept my opinion within a narrow circle."[80]

Churchill's doctor also had his doubts about the Prime Minister's speech. As rhetoric it was beautiful — beginning in humility and ending in dignity — but its length and content revealed both Churchill's brilliance and his flaws.

The Prime Minister could not admit to personal error, so was unable to resist lecturing senators and congressmen on their failure to stop Hitler "five or six years ago" when it "would have been easy" to do so. To make matters worse, however, he found himself unable to resist romanticizing British feats of arms to come, under his own military leadership. Buoyed by the newsreel film of fighting in Libya, he unwisely predicted imminent British victory in the desert, fighting a German-Italian army of 150,000 men.

"General Auchinleck set out to destroy totally that force," the Prime Minister announced to the august assembly. "I have every reason to believe that his aim will be fully accomplished. I am glad to be able to place before you, members of the Senate and of the House of Representatives, at this moment when you are entering the war, proof that with the proper weapons and proper organization we are able to beat the life out of the savage Nazi. What Hitler is suffering in Libya is only a sample of foretaste of what we must give him and his accomplices, wherever this war shall lead us, in every quarter of the globe."[81]

Given the drubbing Rommel was now about to administer to the British Eighth Army — a drubbing that Japanese forces were already administering to the British across the globe — this was simply asking for trouble. Yet the Prime Minister could not refrain from more boasting. Answering his own question "why is it" that Britain did not have "ample equipment of modern aircraft and Army weapons of all kinds in Malaya and the East Indies" to defend against the Japanese, he answered: "I can only point to the victories General Auchinleck has gained in the Libyan campaign. Had we diverted and dispersed our gradually growing resources between Libya and Malaya, we should have been found wanting in *both* theatres," he claimed. American generosity in arms shipped, under Lend-Lease, to British forces in North Africa, thus not only explained the weakness of Britain's Far Eastern defenses, he claimed, but "to no small extent" explained why American forces in the Pacific had been "found at a disadvantage" at Pearl Harbor and in the Philippines.[82]

Such exaggerated claims for current British prowess in Libya while salting the American wounds suffered in the Pacific in an effort to affirm his mastery as a British global strategist and commander in chief, were unfortunate, Halifax and others felt.

Such passages would, of course, be excluded from Churchill's memoirs, after the war — for by then Churchill had no wish to be reminded how he had crowed over Auchinleck's accomplishments against Rommel, only weeks prior to the longest British military retreat in the history of its empire.

Who, though, could penetrate the Prime Minister's wall of self-regard? Churchill was the sum of his strengths and weaknesses — and no one who ever met him could doubt the former, while even the latter could be tragicomic: eliciting compassion in a man so gifted, moreover so resolute in defending the values of decency and goodwill.

It was the Prime Minister's "aggressive" quality that most drew the President to him — even if he thought Churchill a figure from England's past rather than its future. "Winston is not Mid-Victorian — he is completely Victorian," the President was heard to remark,[83] while Dr. Wilson quoted the President as saying Churchill was not only "quite Victorian in his outlook," but a "real blimp."[84] Nevertheless, Churchill's *courage* moved Roosevelt — who knew a great deal about the quality, enabling him to differentiate between straightforward courage and principled, moral courage.

To the President the distinction was not academic. The legendary courage of a man like Charles Lindbergh, the pioneering aviator, was of a deeply tainted order, in the President's view. Lindbergh was brave but simple-minded — and more self-serving than was realized by most people. In this respect the President's instinct — as was the case with his perception of Hitler's demonic character — was instinctively sagacious. Lindbergh's acceptance of a Nazi medal, and his gullible, exaggerated reports on German Luftwaffe superiority before the shooting even began, had amounted to rank defeatism, tricked out in isolationist rhetoric, the President judged. Therefore when at Christmas the aviator applied to General Arnold, chief of staff of the U.S. Army Air Forces, to take up his former commission as a colonel in the U.S. Air Force, the President found himself on the spot.

The President asked not only the advice of the secretary of war, whose domain this was, but of the secretary of the interior, Harold Ickes — and of the director of the FBI, J. Edgar Hoover.

Ickes, after analyzing Lindbergh's speeches and articles, concluded that Lindbergh was "a ruthless and conscious fascist, motivated by a hatred for you personally and a contempt for democracy in general. His speeches show an astonishing identity with those of Berlin, and the similarity is not accidental." To achieve political power in the United States, Lindbergh would, Ickes reflected, require "a military service record" — which Secretary Ickes hoped the President would deny him.[85]

The President, as a political tactician of considerable renown, remained unsure. When he received a confirmed report, however, of what Lindbergh had said at a private meeting of America First members in New York ten days after Pearl Harbor, he felt his basic instinct about Lindbergh had been right. The former colonel was a blackguard.

"There is only one danger in the world," Lindbergh had reportedly said to the gathering in New York — that being "the yellow danger. China and Japan are really bound together against the white race. There could only have been one efficient weapon against this alliance, underneath the surface, Germany itself could have been this weapon," Lindbergh explained. "The ideal set-up would have been to have had Germany take over Poland and Russia, in collaboration with the British, as a bloc against the yellow people and bolshevism. But instead, the British and the fools in

Washington had to interfere. The British envied the Germans and wanted to rule the world forever. Britain is the real cause of all the trouble in the world today. Of course, America First cannot be active right now. But it should keep on the alert and when the large missing lists and losses are published the American people will realize how much they have been betrayed by the British and the Administration. Then America First can be a political force again. We must be quiet a while and await the time for active functioning. There may be a time soon when we can advocate a negotiated peace."[86]

Hitler would have ordered the execution of a purveyor of such treason against the Reich. The President, however, simply decided to deny Lindbergh's request to serve in American uniform. (About Lindbergh's self-serving lack of fundamental moral principle, the President would, moreover, be proven right — Lindbergh, a passionate eugenicist, conducted secret adulterous affairs with three German women two decades younger than himself, two of them disabled; he sired seven secret children by them in Germany and Switzerland, after the war.)[87]

In the meantime, however, the President had to decide how best to win the war — which was still going disastrously in the Far East, in the wake of Pearl Harbor. And would soon fare as badly in the Mediterranean, despite the Prime Minister's boasts.

Fueling Lindbergh's fire was the fact that the war in the Far East was going so badly. "Germany First" thus seemed to many, at this time, to be a mistake.

Certainly, by focusing America's primary war effort on the Pacific, the President would, had he been so willing, have been able to play to public opinion all across America — the majority of the nation clamoring to avenge the Japanese sneak attack, to save the Philippines, and to defend Australia and New Zealand if they were invaded by the Japanese.

In reality, however, the Philippine Islands could no longer be saved, Roosevelt knew. Their distance from the United States, the destruction of MacArthur's air force, and the emasculation of the American fleet made for a bleak outlook in the Southwest Pacific. At the same time, however, from his intelligence services and his emissaries, the President was convinced the Japanese had no intentions of invading Australia or New Zealand, despite the continuing panic consuming the Australian capital, Canberra. With sufficient U.S. naval, army, and air force reinforcement he had no doubt the Antipodes could easily be held — and would be, once

American forces were sent out. What was far more important, therefore, Roosevelt felt, was to deal with Hitler before the Führer could defeat the Soviet Union — following which he would undoubtedly turn back to Britain. In that case, the United States would truly become a second-rate nation on the world stage — with calls from Lindbergh-style "patriots" to abandon democracy altogether.

It was the recognition of this danger, as 1941 came to an end, that caused the President to review the question of an Allied Supreme War Council, which the secretary of state was recommending, and face up to the matter of how best to go forward. As commander in chief Roosevelt had a junior partner, vested with poor military judgment but supreme courage: Winston Churchill. With him, the President was certain, he could work. With others — especially a committee of others — he was less certain. Thus, where Secretary Hull had pressed for a politico-military Supreme War Council to represent all the Associated Powers, or Allies, as in World War I,[88] the idea of such a body now died. Instead, Hull was told by the President to concentrate exclusively on the new version of the Atlantic Charter, or war aims that *all* of America's partners in the conflict against Hitler and the Axis powers could sign. With regard to the waging of the war itself, however, Roosevelt made clear, he would take over this responsibility at the White House. In effect he would, with Winston Churchill as his chief lieutenant in London, become the supreme commander of the Associated Powers, or Allies, in Washington. There would naturally be differences of opinion and strategy, but he was now sure he had the skills to keep the Allied high command focused on eventual victory.

The Joint Chiefs of Staff in Washington would form a new military committee, with an officer present to represent the British chiefs of staff, that would henceforth be tasked with carrying out Roosevelt's orders and decisions — decisions he would make with Churchill as his junior partner. In that way the Prime Minister's capacity for poor judgment would for the most part be disabled, while his terrific moral strength could be applied to winning, not losing, the war.

It was an inspired intention. Whether it would work as 1941 came to a close was yet to be seen.

5

Supreme Command

ON DECEMBER 26, 1941, Winston Churchill suffered a minor heart attack while in the White House and was sent to Florida by the President to recuperate for a week. Roosevelt, whose patience had at times been sorely tried by the Prime Minister's visit, was relieved. "It always took him several days to catch up on sleep after Mr. Churchill left," Eleanor Roosevelt later wrote, concerned for her husband's rest.[1]

Churchill's stay in the White House certainly proved exhausting for the President, the First Lady, and for the White House staff. Plans for offensive action—especially the President's "great pet scheme" for U.S. landings in Northwest Africa—had had to be put on the back burner while the Japanese rampage in the Pacific dominated all military planning and operations. Many ideas were nevertheless advanced—perhaps the most important of which was the President's notion of a new declaration of principles by the Associated Powers.

The document's final maturation, indeed the President's whole method of bringing a project to fruition, amazed the Prime Minister's military assistant, Colonel Jacob—symbolized in the "mess" he saw in the President's study on the second floor of the White House. The President "leads a most simple life," Jacob described in his diary. "He moves about the White House in a wheeled chair. His study is a delightful oval room, looking South, and is one of the most untidy rooms I have ever seen. It is full of junk. Half-opened parcels, souvenirs, books, papers, knick-knacks, and all kinds of miscellaneous articles lie about everywhere, on tables, on chairs, and on the floor. His desk is piled with papers and alongside his chair he has a sort of bookcase also filled with books, papers and junk of all sorts piled just anyhow. It would drive an orderly-minded man, or a woman, mad. The pictures on the walls are fine, mostly prints or paintings of ships. There are also good bookcases round the walls, and the furniture is not bad. But the effect is ruined by the rubbish piled everywhere. It is rather typical of the general lack of organization in the American Government." As a proud English bureaucrat, Jacob found the "British Governmental machine" to be, by contrast, "like a motor car or even a train. Provided a reasonably efficient driver is in charge it will go. The American Government is not a machine at all. The various parts are not assembled into a working whole. The President is in the position of a patriarch, with a rather unruly flock, and much depends on the actual men who actuate or influence the various sections of that flock. The patriarch also relies to a great extent on sheep dogs, who are his stand-by, but are regarded with fear and suspicion by the sheep."[2]

One of the sheep dogs was Bill Donovan, a lawyer whom the President had put in charge of a "kind of super intelligence organization," the Office of the Coordinator of Information, forerunner of the OSS (and later the CIA). Another was Harry Hopkins, "a frail anaemic man of great honesty and courage, who lives permanently in the White House and is the President's constant companion. . . . Hopkins is usually to be seen in a magenta dressing gown and pyjamas. Other examples of the President's peculiar method of working are the personal representatives he sends about the place, such as [former ambassador William C.] Bullitt in the Middle East. These report to him direct, and to our way of thinking are irresponsible meddlers."[3]

When Lord Halifax went to the White House to discuss a draft of the revised Atlantic Charter with President Roosevelt and Harry Hopkins, he too was bewildered. "They are the most amazing people," he noted in his own diary, "in the way of what seems to us most disorderly and unbusinesslike methods of working. But somehow the result comes out not too badly and they seem quite happy working like that. It would drive me to drink," the ascetic Catholic reflected. "While the draft [charter] was being retyped Harry Hopkins took me to wait in his bedroom while he dressed, his bedroom serving as bedroom and office. It is the oddest menage I have ever seen."[4]

Equipped with his own silk Chinese-dragon dressing gown and beloved romper suit, Winston Churchill had fitted almost seamlessly into this strange ménage, however. He had certainly made himself at home. "Now, Fields," he had instructed the President's head butler, who at six feet three towered over the Prime Minister. "We want to leave here as friends, right?

So I need you to listen. One, I don't like talking outside my quarters; two, I hate whistling in the corridors; and three, I must have a tumbler of sherry in my room before breakfast, a couple glasses of scotch and soda before lunch and French champagne and 90 year old brandy before I go to sleep at night."[5]

In the end, Churchill stayed at the White House almost a month rather than his planned week. The First Lady might resent Churchill's heavy, all-day drinking and late hours, as well as his egoism, but the President seemed glad of his company, and never once complained — even when he and Churchill clashed over Roosevelt's desire to include India as a British Dominion in the proud list of nations subscribing to his declaration of the Associated Powers. "Being convinced that complete victory over their enemies is essential to defend life, liberty, independence and religious freedom," the draft text began, "and to preserve human rights and justice in their own lands as well as in other lands . . ."

Churchill would not, however agree to India as a signatory beside the Dominions of Canada, South Africa, Australia, and New Zealand — in fact, the Prime Minister refused even to agree to the inclusion of India as a colony. To the President's relief, though, the text of his joint declaration was accepted by the governments of some twenty-six nations in all, including the Soviet Union — despite, to the President's delight, his insistence upon the inclusion of religion as a freedom for which the signatories were fighting. "Let's get it out on Jan 1," Roosevelt penned with relief on December 30, 1941, in a note to Harry Hopkins. "That means speed. FDR."[6]

Awakening on the morning of New Year's Day, prior to a church service in Alexandria to which he was taking Churchill and the British ambassador, then to the laying of a symbolic wreath on President Washington's tomb at Mount Vernon, the President had a final brainwave, however.

Dressing quickly with the help of his valet, the President shifted to his wheelchair and rushed to the Rose Room, where Churchill was staying, to tell him: the revised Atlantic Charter, which currently had no formal name, should be called "A Declaration by the United Nations."[7]

Outside Churchill's door, however, stood W. R. Jones, an assistant to Colonel Jacob. Though an admirable clerk, Mr. Jones had, as Colonel Jacob noted with amusement in his diary, "a most peculiarly pompous and over-correct way of speaking. He never can get a perfectly straightforward sentence out. If you ask him where Brigadier Hollis [Jacob's immediate boss] has gone, instead of saying 'I'm afraid I don't know,' he will say 'I fear it is not within my knowledge where the Brigadier may be at the moment.'" Conveying a message from Colonel Jacob to the Prime Minister, Jones had been told that the Prime Minister was in the bath. Jones had therefore waited in "the central passage" on the second floor, "and stood looking about for a few moments, when what should he see coming towards him but the President in his wheeled chair, unaccompanied by anyone. Jones stood rooted to the spot, and the President addressed him saying:

"'Good morning. Is your Prime Minister up yet?'

"'Well, Sir,' said Jones, 'it is within my knowledge that the Prime Minister is at the present moment in his bath.'

"'Good,' said the President, 'then open the door.'

"Jones accordingly flung open the bathroom door to admit the President, and there was the Prime Minister standing completely naked on the bath mat.

"'Don't mind me,' said the President, as the Prime Minister grabbed a towel.

"Jones' day was made," Jacob recorded. "Not only had he seen the inside of the White House, but he had spoken to the President and seen a meeting between him and the Prime Minister in quite unique circumstances."[8]

Jones certainly had — the two leaders working in what Churchill would correctly call "closest intimacy" as they discussed the new name for the signatories to the upgraded Atlantic Charter: the United Nations.[9]

President Roosevelt's next great achievement during Churchill's time at the White House was his success in junking Secretary Hull's idea of a Supreme War Council and creating, instead, an official (yet never formally instituted, in writing or in law) military body to carry out his directions as commander in chief: the so-called Combined Chiefs of Staff. With its headquarters in Washington, the new body would translate the President's military policies into combined action by the forces of the United States and Britain, as well as subsidiary contingents. Moreover, to make the new system work, the President decided on General Marshall's advice that it would be best to appoint supreme commanders in each theater of combat fighting the Axis powers, to command the Allied air, naval, and ground forces.

This notion of supreme theater commanders had arisen at the meeting between the U.S. and British military chiefs of staff on Christmas Day. General Marshall had proposed the idea—a single commander directing not only the "air, ground and ships"[10] of his own nation in the region, but *all* the combatant forces of the nations contributing to the campaign. The President, still scarred by the dysfunctional, fragmented performance of the three armed services at Pearl Harbor, then leaped at the idea of such "unity of command" at a meeting at the White House on the morning of December 27, as Secretary Stimson recorded in his diary[11]—indeed, he had driven it through against the opposition not only of Admirals King, Stark, and Turner, but of Prime Minister Churchill, as Colonel Jacob noted in his diary.

"When the idea was first put forward there was almost universal opposition," Colonel Jacob wrote when he returned to England, "and the Prime Minister expressed his doubts about the wisdom of such a system as General Marshall proposed to set up. The U.S. Navy were also against it. General Marshall, however, had backing from the President, whom he had convinced that unity of command would be the only solution of the Far East troubles, and that nothing could be worse than having several independent commanders of different nationality, especially in a theater where interests are divergent."[12]

Churchill's opposition had initially threatened to derail the idea. General Marshall then put forward a suggestion the President thought inspired, politically: namely to appoint, as the first such supreme commander in the Far East, a *British* general! The President had therefore sent Marshall in person to convince Churchill—which Marshall did, proposing that the supreme commander should be General Sir Archibald Wavell, the current commander in chief in India. "This very naturally put a different complexion on the affair," Colonel Jacob noted, "as it was hard for the Prime Minister to refuse to back a principle which was undoubtedly attractive in theory and which was to be applied in a way which recognized the pre-eminence in the field of choice of a British general."[13]

The die was thus cast—even Colonel Jacob being amazed at his deeply conservative Prime Minister's apostasy. Churchill's acquiescence was, however, followed by an even more revolutionary innovation: namely that the Combined Chiefs of Staff would be headquartered not in London but in Washington, D.C.—a decision that the Prime Minister meekly accepted, but which Jacob, on his return to England, found had raised the hackles of every English patriot in the War Office. "Special body in Washington to control [military] operations under PM and USA president," General Sir Alan Brooke, the owl-faced new chief of staff of the British Army, sniffed in his diary on December 29, 1941.[14] The "special body" would doubtless comprise General Marshall, Admiral Stark, Admiral King, and General Arnold—leaving the British chiefs of staff with only "representation" at the Combined Headquarters in Washington, invested in a single British officer, yet to be appointed. "The whole scheme wild and half-baked," Brooke had snorted.[15] To Brooke's further chagrin the Prime Minister had then cabled Acting Prime Minister Attlee from the White House, saying it was, in effect, a done deal—with Allied supreme commanders for different theaters of war. Churchill, moreover, had made no attempt to conceal the origin of the idea. "Last night President urged upon me," Churchill explained, "appointment of a single officer to command Army, Navy, and Air Force of Britain, America and Dutch."[16] Since the "President has obtained the agreement of the American War and Navy Departments" to the scheme,[17] and since the first Allied supreme commander was to be General Sir Archibald Wavell, the British cabinet could do little else than wring their hands.

General Brooke had been mortified, having been promoted to be his nation's CIGS, only to discover his war-making powers would be entirely dependent on decisions made in Washington. "The more we looked at our task the less we liked it," he noted not only of the supreme commander business but of the very idea of a global command headquarters in the U.S. capital—recalling later that he "could see no reason why at this stage, with American forces totally unprepared to play a major part, we should agree to a central control in Washington."[18] There was nothing Brooke could say, at least aloud, however—the British cabinet "forced to accept PM's new scheme," as Brooke had lamented, "owing to the fact that it was almost a fait accompli!"[19]

It was—the more so, moreover, when reports had come in of the President's State of the Union address to Congress on January 5, 1942—a speech in which the President had announced arms production goals that had made it clear the United States would win the war by industrial output alone.

"Plans have been laid here and in the other capitals for coordinated and cooperative action by all the United Nations—military action and economic action," the President had declared. "Already we have established, as you know, unified command of land, sea, and air forces in the southwestern Pacific theater of war. There will be a continuation of conferences and consultations among military staffs, so that the plans and operations of each will fit into the general strategy designed to crush the enemy. We shall not fight isolated wars — each Nation going its own way. These 26 Nations are united — not in spirit and determination alone, but in the broad conduct of the war in all its phases.

"For the first time since the Japanese and the Fascists and the Nazis started along their blood-stained course of conquest they now face the fact that superior forces are assembling against them. Gone forever are the days when the aggressors could attack and destroy their victims one by one without unity of resistance. We of the United Nations will so dispose our forces that we can strike at the common enemy wherever the greatest damage can be done him," the President had warned. "The militarists of Berlin and Tokyo started this war. But the massed, angered forces of common humanity will finish it."

"Victory," however, "requires the actual weapons of war and the means of transporting them to a dozen points of combat. It will not be sufficient for us and the other United Nations to produce a slightly superior supply of munitions to that of Germany, Japan, Italy, and the stolen industries in the countries which they have overrun," Roosevelt had pointed out. "The superiority of the United Nations in munitions and ships must be *overwhelming* — so overwhelming that the Axis Nations can never hope to catch up with it. And so, in order to attain this overwhelming superiority the United States must build planes and tanks and guns and ships to the utmost limit of our national capacity. We have the ability and capacity to produce arms not only for our own forces, but also for the armies, navies, and air forces fighting on our side."

Thereupon the President had openly announced astronomical figures for U.S. military output. "I have just sent a letter of directive to the appropriate departments and agencies of our Government, ordering that immediate steps be taken:

"First, to increase our production rate of airplanes so rapidly that in this year, 1942, we shall produce 60,000 planes, 10,000 more than the goal that we set a year and a half ago. This includes 45,000 combat planes — bombers, dive bombers, pursuit planes. The rate of increase will be maintained and continued so that next year, 1943, we shall produce 125,000 airplanes, including 100,000 combat planes.

"Second, to increase our production rate of tanks so rapidly that in this year, 1942, we shall produce 45,000 tanks; and to continue that increase so that next year, 1943, we shall produce 75,000 tanks.

"Third, to increase our production rate of anti-aircraft guns so rapidly that in this year, 1942, we shall produce 20,000 of them; and to continue that increase so that next year, 1943, we shall produce 35,000 anti-aircraft guns.

"And fourth, to increase our production rate of merchant ships so rapidly that in this year, 1942, we shall build 6,000,000 deadweight tons as compared with a 1941 completed production of 1,100,000. And finally, we shall continue that increase so that next year, 1943, we shall build 10,000,000 tons of shipping.

"These figures and similar figures for a multitude of other implements of war will give the Japanese and the Nazis a little idea of just what they accomplished in the attack at Pearl Harbor."[20]

Listening, the British ambassador, Lord Halifax, rubbed his eyes. Could the President be serious? "[C]ertainly if they can make the figures to which they have hitched their wagon on the supply side come out," he noted in his diary with a mix of incredulity and new confidence, "it will be prodigious."[21]

In London, however, General Brooke could only pray the British Empire would hold fast long enough for the United States, with its prodigious output of men and materiel, to save it.

As the veritable new Allied commander in chief in Washington, then, the President successfully imposed his will in the first days of January 1942, not only on his own staff but upon his new primary ally, and in an almost magical way: overcoming his dissenters by dint of his seemingly effortless goodwill, common sense, charm, and positive spirit.

With Churchill's departure on January 14, however, the President suddenly found himself alone in the White House — a relief, but the Prime Minister's absence also created a distinct vacuum. The President realized, in fact, that he missed his British counterpart. Short, squat, bald, chubby-cheeked, Churchill had exuded not only cigar smoke but a fierce, indomitable energy, whatever setbacks he faced. And with his retinue of personal and military staff, he had left, too, an unforgettable image of a traveling chieftain — especially the sight of his staff unrolling the Prime Minister's world maps and charts, marked up with the latest information on British and enemy forces.

In honor of his departing guest, the President, his aides learned, decided to set up in the White House his own Map Room, modeled on the Prime Minister's portable headquarters. It would not, however, be in the Monroe Room — to Eleanor's understandable relief. And definitely not in the underground headquarters or bunker that the President's advisers were urging be constructed in the White House, modeled on Churchill's War Rooms in London.

6

The President's Map Room

THE PRIME MINISTER'S War Rooms in London dated back to 1938, when the British government ordered the construction of a vast underground complex for the war cabinet, complete with military headquarters and communications. This was secretly installed below the "New Public Offices" on Great George Street, close to 10 Downing Street, the prime minister's residence and office. It was on a visit to the underground War Rooms that Churchill, on becoming prime minister in May 1940, had declared: "This is the room from which I'll direct the war."[1]

Known as "the Bunker" or "the Hole,"[2] the underground complex had been designed to allow the closest working (and living) proximity between the members of the war cabinet and the British chiefs of staff, as well as their ancillary civil and military staffs and clerks: a veritable rabbit warren of rooms to house the joint military planning, operations, intelligence, and civil defense departments of the government, complete with bedrooms for all. A further five-foot-thick concrete slab had wisely been inserted over the ever-expanding series of rooms during the London Blitz.

At the heart of the complex, however, lay the Map Room, staffed twenty-four hours a day by officers and ratings from the three services: men and women who not only maintained up-to-date wall maps of the war's progress, with the latest positions of British and enemy forces at sea and on land, but produced daily intelligence summaries for the chiefs of staff, the Prime Minister/Defense Minister, and the King.[3]

Such a secure bunker from which to conduct the war was eminently sensible, given the original RAF forecasts of potential bombing casualties of up to two hundred thousand victims per night. Actual casualties had been considerably lower — yet still daunting. During the Blitz, on October 10, 1940, a German bomb had in fact fallen only yards from 10 Downing Street, destroying the Prime Minister's kitchen, pantry, and offices, and forcing Churchill to move into an apartment immediately above the secret War Rooms, called the No. 10 Annex. Treasury Secretary Morgenthau had heard of the headquarters, and perhaps understandably — given that he was responsible for the Secret Service and President Roosevelt's safety — had been the first to push for something similar to be constructed in Washington, D.C., following the surprise Japanese attack on Pearl Harbor. Moreover, Morgenthau believed that he had, in the somewhat panic-stricken days that followed December 7, gotten the President's approval. But he'd been mistaken.

An entire half-block between Pennsylvania Avenue and H Street had already been acquired as a proposed extension for the State Department, west of Lafayette Park — with more that could be added later, Morgenthau was assured.

Was an underground bunker necessarily the best idea, though? Morgenthau had queried. Hitler's Berlin bunker (Vorbunker) had been constructed beneath the garden of the Reichskanzlei in 1936, and another at Wolf's Lair, his forward headquarters in the Masurian forest near Rastenburg in East Prussia in the winter of 1940–41. But new information from Britain indicated to Morgenthau — an inveterate believer in research — that bunkers below ground were even more susceptible to bomb damage than above-ground fortified premises, owing to the transmission of subterranean shock waves.

"The suggestion," General Fleming told the Treasury secretary as he unrolled a map of Pennsylvania Avenue, was therefore "to build about a five story building on that site, a complete blackout building. No windows at all. It can be made — it would have about two hundred seventy-five thousand square feet of floor space. We can make it so that there is living quarters and everything else there. It can go ahead and just stand a long siege, that building. It has a very heavy reinforced roof"[4] — in fact, at twelve feet thick it would be only eight feet less thick than the latest windowless Admiralty building in London, a veritable fortress near Churchill's War Rooms.

In actuality Churchill had declined to use the Admiralty Citadel, referring to it later as "a vast monstrosity which weighs upon the Horse Guards Parade."[5] Such a windowless, above-ground fortress had definite advantages over a below-ground complex, General Fleming assured the Treasury secretary, however. Within the building, the "farther down the safer it is, so the ground floor and basement are completely safe," he claimed, and extolled its virtues: an above-ground reinforced-concrete building, like a medieval castle, capable of housing "fifteen to twenty thousand people." The building would be "about a hundred and forty-four feet deep," and would accommodate "Treasury, State, and the Executive Offices," as well as the senior military personnel. "I think he [the President] probably wants some of the higher staff officers of the Army and Navy in there," General Fleming had explained to Morgenthau. The President would have his own access via a secret tunnel from the White House that would branch off the current zig-zag tunnel being dug to the basement shelter in the Treasury's gold vaults—avoiding trees.

"That will please the President. He and I both like trees," Morgenthau had commented. Claiming that "the President asked me to get together and have ready a building which would house the White House staff, State, and Treasury," Morgenthau had then telephoned General "Pa" Watson, the President's appointments secretary, on December 15, 1941, to say they had an architect's plan for the bunker, had already purchased virtually the complete site, and were "ready to go ahead. I'd like to show it to the President, with General Fleming, and it would take about five minutes." He even had the money—although, as Morgenthau wisely cautioned, "I don't want to go ahead and order a seven million dollar building without the President seeing it."

The monster complex was not the only fortification Morgenthau wanted in order to protect "the Boss," as the President was called. As in Alice's Wonderland, Morgenthau and his team of security advisers also planned a thirty-feet-deep interim bunker in the grounds of the White House itself, with eight-feet-thick walls and ceiling, inside of which was a second, interior box-like room with walls two feet thick, which "gives you a pretty good protection"—though not against a "four thousand pound bomb." It was to be dug in front of the White House, "for the President and for our communications center," and would protect against . . .

Here the Treasury secretary was unclear. General Fleming had explained that the U.S. Air Corps had told him to expect only "token raids," for the moment—planes that would "come in and terrify the population, show what they can do. They would have to come in from a carrier some place. That therefore limits the size of the bomb that can be carried to about two thousand pounds."

German or Japanese air raids were not the only danger envisaged, it appeared; there might also be enemy ground forces parachuted in and deployed against the President and even Congress, General Fleming had added.

The mention of Congress complicated the matter still further. While the bunker plans for the President were being prepared, there were now other, equally zealous proposals also put forward to protect the capital—and Capitol. There were to be machine-gun nests protected by sandbags everywhere, including on the roof of the White House; bulletproof glass in the President's office and study windows. Moreover, the White House itself would no longer be white. It would be repainted, Fleming had explained, in camouflage colors, with a fifteen-feet-high "sand-bag barricade" that would go "completely around the White House building and Executive Offices" or West Wing.[6]

Even Fleming expressed skepticism as to whether the President, who loved history and had been for some years planning a museum in the new East Wing in order to house the many artifacts and documents he and his predecessors had been given, would tolerate such draconian changes—especially given his love of trees and landscaping. "It is believed the President would not permit it," Fleming had warned Morgenthau about the great earthwork/barricade proposal—or indeed the other plans.

Such wariness as to the President's response was well founded. General Watson called Morgenthau back on the afternoon of December 15, 1941, to say that not only would the President not see the Treasury secretary, but he was not pleased. "He said you were crazy as hell. He's not going to build that building."[7]

"He asked me to!" Morgenthau had vainly protested.

General Watson, however, repudiated this claim, quoting the exact words the President had used and which he wanted Watson to convey to Secretary Morgenthau. "'Why,' he said, 'tell him he's crazy, what is he talking about?'"[8]

Putting the kibosh on Morgenthau's bunker idea had turned out to be but the tip of the President's derision. To Morgenthau's chagrin, Roosevelt had been equally disparaging about Morgenthau's idea to cancel the traditional lighting of the Christmas tree, in front of the White House, on Christmas Eve. General Albert Cox had warned how dangerous such a public illumination would be—a veritable invitation to the enemy—saying, "you might just as well put up an airplane beacon right in front of the White House." The President had remained adamant, however. Morgenthau was thus forced to drop this and any idea of a fortified government control center or bunker. "We'll have to let it rest there until the President changes his mind," Morgenthau had acknowledged wearily on December 22.

The President hadn't changed his mind, however—even after the Prime Minister's long stay over Christmas. The notion of a "token raid," launched from "a carrier some place," seemed too remote a possibility to take seriously—though the idea itself lodged in the Commander in Chief's capacious mind as something the U.S. might well carry out against the enemy.

Maps were a different matter. The President loved maps, just as he loved stamps—a hobby that had become the more passionate the less he himself could travel, owing to his disability. It was no surprise, then, when the President ordered that something similar to Churchill's portable map-and-filing system be installed in the White House.

"I can't think how I'm going to get on when you take your Map Room away," the President had said to Churchill before his departure. "I shall feel quite lost."

"But Franklin, you must have a Map Room of your own. That shall be my parting present," Churchill had responded. "And you shall have my lieutenant to help make it and to run it for you. Lieutenant Cox, how would you like to work for the President?"[9]

Sublieutenant Cox had been thrilled. When first introduced to Mr. Roosevelt, he'd "found my hand taken in a warm, strong grasp and saw two piercing eyes looking into mine with a kindly twinkle, and the wide mouth was curved up in an understanding grin."[10] It had been clear to Cox from the start how deeply Churchill admired the President—Cox writing to his mother in England how the Prime Minister had said to him, "What a wonderful man that is. It is a mercy for mankind he is where he is at this moment."[11]

The Prime Minister, Cox recalled, was like a "miniature whirlwind," his mind forever racing, calculating, preparing. The President seemed the opposite—as if no burden, however great, would ever dent his smile. "He is an amazingly great man," Cox described his feelings to his mother, "though not as fiery as the P.M. . . . He is a great admirer of Churchill's, but the P.M.'s energy seems to have worn him out a bit during the past few days, and I can well understand it!"[12] "Your man has worn me out so I am taking the day in bed," the President had stated in a message to Cox, when Churchill briefly flew to Canada. "I guess he's worn you out too, so I suggest you take the day off," he ordered — an order "I obeyed," noted the sublieutenant, gratefully.[13]

Cox's "temporary" commission was to train a U.S. officer in the mechanics of setting up and running a map room for the Commander in Chief. "Roosevelt wanted something comparable to the Map Room," explained Commander George Elsey seventy years later, and "asked his naval aide [Captain Beardall] to establish some form of communications center at the White House." However, the idea was never, Elsey was at pains to point out, intended to be a version of Churchill's London War Rooms — which soon led to yet another round of Alice-in-Washington misunderstanding when Captain Beardall "looked around and found this young Reserve officer," Lieutenant Robert Montgomery, a famous former Hollywood actor.[14]

"Montgomery had been on duty in London, as an aide to the U.S. naval attaché, and had become acquainted with Churchill's War Room," Elsey (who had been a naval ensign working for Lieutenant William C. Mott, who was transferred to the White House as an aide to Captain Beardall) explained. It was from Captain Mott that Elsey had obtained a firsthand account of how the President had put his *über*zealous military aides in their place.

Roosevelt had "responded positively right away, because Montgomery had quite a reputation as a movie actor." The movie actor, however, completely misunderstood the President's instructions. Whether Montgomery was "got at" by Secretary Morgenthau or other conniving figures such as General Fleming remains unclear, but the suggestions he began to make for a fortified bunker seemed eerily similar to Morgenthau's. "Robert Montgomery had the same sort of grandiose plans," Elsey explained with a laugh more than a half century later. "He had been on duty in London, he was familiar with the catacombs there, and he prepared the same sort of thing to be built on Constitution Avenue, across from where the War and Navy Departments were."

The President, when he heard Montgomery's latest scheme, was as appalled as he had been by Morgenthau's plans the month before. "All Roosevelt wanted," Elsey pointed out, "was simply one secure spot in the White House itself which he and only his immediate associates would have access to — not something for *the whole military* to use!" In particular the President had no wish to set up a "control room." Montgomery's proposition was "beyond anything FDR wanted," Elsey made clear, given the accretion of myths and misconceptions that had built up in subsequent years. "This was *in no sense* to be a command center, as Churchill's War Room was," the President's former aide emphasized, for that was simply not how the President operated — nor wished, instinctively, to operate in directing the war.

"*And that was it!*" Elsey remarked with finality. The idea of a presidential control room on the lines of Churchill's London bunker, or Hitler's bunkers in Berlin and Rastenburg, was irremediably nixed — indeed, its progenitor was soon fired. "[Robert] Montgomery was brought down to earth" and removed from the White House, Elsey recalled. Instead of a grand above- or below-ground bunker, the President merely wanted a small, secure room to house his secret signals to and from Allied leaders and his own military advisers. And with that in mind it was temporarily installed "right across a narrow corridor from the Oval Office in the West Wing," in the Fish Room.

"It was called the Fish Room because that was where Herbert Hoover had mounted his fish trophies," Elsey recounted, amused by the irony. "Roosevelt had replaced Hoover's fish with his own fish, and it became the Fish Room." Instead of fish stretching across its walls, however, there were soon global and campaign maps, just as Churchill had recently mounted in the Monroe Room. Even this setup proved unsatisfactory to the President, however — in fact within a few days Roosevelt decided it was no good. "It was too public, there was too much access to it — the room had been used by the President and senior White House staff members for all kinds of meetings, and there just wasn't adequate security," Elsey explained of the Map Room's demise in the West Wing. "Too many newspaper people and others pushing around! So it was moved."

The new location that the President chose was in the presidential mansion itself — installed, by even greater irony, in "the Museum," a small ladies' cloakroom, next to the Diplomatic Reception Room, where guests had traditionally hung their coats and freshened up. This room was on the ground floor of the White House, below the formal entrance vestibule of the mansion. Barely twenty-four feet by nineteen, the new Map Room would be closer to the President's elevator, so he could visit the secret communications sanctum in private, from his bedroom or study, without needing to go to the West Wing at all.

In due course, the cloakroom was easily converted. The door to the beautiful Diplomatic Reception Room was blocked off, the elegant wood-paneled walls were faced with soft wallboard, and maps of the world and of the various battlefronts were mounted at eye level. "The ground situation was marked with grease-coated pens on plastic sheets over the maps," recalled Elsey, who began working under Lieutenant Mott at the new Map Room in April 1942, after Montgomery's departure. In the middle of the room, filing cabinets were installed for the signals that came in and went out, brought by army and navy couriers in locked leather pouches, with the latest information and intelligence. Most important, Mott explained to Elsey, were the President's secret, direct communications with Churchill, Stalin, and Chiang Kai-shek, as well as top-secret messages and reports from the secretaries of war and navy and chiefs of staff for the President. But what struck Elsey later was not only Roosevelt's deliberate decision not to replicate Churchill's bunker or War Rooms, but the President's reasons for doing so.

"During the early days the President visited the growing room with keen interest," Cox explained, "remarking on the progress of construction and suggesting modifications about the fittings and the placing of the furniture. When the room was at last working, Mr. Roosevelt's visits tended to occur towards the end of his working day, usually just before 7 o'clock, but as the collection, evaluation and display of information increased in efficiency, he came to pay a regular routine visit at 10.45 a.m., immediately before his conference with his Chiefs of Staff."[15]

As time went on, however, the President's visits to the Map Room grew less frequent — the President possessing, as Cox recalled, an almost photographic memory for geography. By the time Elsey joined the staff in April, "the President rarely came into the Map Room," Elsey later recalled. Instead, Roosevelt asked that important communications from his global counterparts — Churchill, Stalin, Chiang Kai-shek — as well as other secret signals be brought to *him,* wherever he was. Even more intriguingly to the young ensign, the President insisted that outgoing presidential messages continue to be enciphered and sent by the Navy Department, while incoming messages be deciphered and sent over by the Army Department. In this way neither department had more than half the story. It was "characteristic of Roosevelt," reflected Elsey with a chuckle — "all *too* characteristic of Roosevelt! Wanting to be the only person who knew *everything.*"

The simple truth remained, however, that although manned eventually by a six-officer staff and guarded twenty-four hours a day, the President's new Map Room was only ever intended to be his own secret store of information or reference.[16] It was never meant to be a control center like Churchill's War Rooms — for by the time Ensign Elsey was posted to the Map Room in April 1942, Roosevelt clearly had developed his own distinctive vision of how he would direct the war as commander in chief.

First off, the President seemed to have no intention of rubbing shoulders continually with his military chiefs, as Churchill did; indeed, the army, air, and navy chiefs would not even have access to his new Map Room, unless the President or one of his immediate staff personally accompanied them into it. The Commander in Chief's "control center" would remain, by contrast, the same as it had always been, upstairs: his bedroom, or his beloved, cluttered, but welcoming Oval Study, next to his bedroom; or his larger, equally cluttered oval-shaped room, the Oval Office in the recently rebuilt West Wing, connected to the White House via a colonnade.

From those rooms, the Commander in Chief would exercise his unique approach to military command — hoping it would prove more effective than had Churchill's, thus far.

PART FOUR

Trouble with MacArthur

7

The Fighting General

The president woke, as usual, around 8:30 a.m., and was served breakfast in bed.

"The President always had a tray in his room," recalled Alonzo Fields, the White House chief butler. "The coffee for the President was a deep black French roast, prepared in the kitchen. We roasted the green coffee beans to any degree we wanted. The President's coffee, however, was a much deeper roast than we used for the family, and it was freshly ground. A coffeemaker was placed on the tray so the President could control the brewing."[1]

As he drank his coffee Roosevelt read through the morning's newspapers, looking to see what new inanities were being published about his commanding general in the Far East, Douglas MacArthur, whose brave Army of the Philippines was fighting a doomed, rearguard battle against the Japanese. Among the absurdities: growing calls to have MacArthur brought back to Washington in order to make him U.S. commander in chief.

The President could only wince at such media madness, trumpeted by an increasing number of Republicans. Among them was the defeated contender for the 1940 presidential election, Wendell Willkie—who had the backing of newspaper magnates such as Ogden Reid of the *New York Herald Tribune,* Roy Howard of the Scripps-Howard newspaper chain, and John and Gardner Cowles, publishers of *Look* magazine, the *Minneapolis Star* and *Tribune,* and the *Des Moines Register.*

Some of the adulation being showered on the Far Eastern general took the President's breath away. The *Baltimore Sun,* for example, had recently proclaimed MacArthur a "military genius"—a general whose skills rose high above the "single field" of battle. "He has some conception of that high romance which lifts the soldiers' calling to a level where on occasions ethereal lights play upon it," the newspaper waxed lyrical.[2] The *New York Herald Tribune,* meanwhile, had run fully half a page of photographs of the general,[3] while towns across the United States were considering renaming their roads, even themselves, in his honor. The TVA's Douglas Dam should be called "Douglas MacArthur Dam," it was proposed in Congress; another congressman had called for MacArthur to be awarded the Congressional Medal of Honor, the nation's highest award for bravery in battle.[4] The U.S. Senate was equally, if not more, adulatory than the House of Representatives. Senator Elbert D. Thomas of Utah, a Democrat and former professor of history, had declared: "Seldom in all history has a military leader faced such insuperable odds. Never has a commander of his troops met such a situation with greater and cooler courage, never with more resourcefulness of brilliant action."[5]

The *Washington Post,* for its part, had declared that MacArthur, by his "last-ditch fight in the bamboo jungles of Bataan," had now shamed the ignorant "prophets of disaster" who had written off the Philippines as a hopeless cause.[6] The *Philadelphia Record* considered Bataan had proved "anew" that MacArthur "is one of the greatest fighting generals of this war or other war. This is the kind of history which your children will tell your grandchildren." Thanks to General MacArthur, Bataan "will go down in the schoolbooks alongside Valley Forge," the newspaper predicted, "Yorktown, Gettysburg and Chateau Thierry."[7]

The President could but shake his head. If the press only knew what a mess General MacArthur had made of the war thus far!

The President had known Douglas MacArthur since before the First World War—a war in which MacArthur had been awarded an unparalleled seven Silver Stars for courage and exemplary combat leadership, becoming the youngest brigadier general in the United States Army.

The relationship between the two men had been cordial, but never easy. Both came from somewhat "aristocratic" backgrounds: Roosevelt's "Dutch" lineage stretching back to the first settlers in America, while MacArthur's father had won fame and the Medal of Honor in the Civil War at age nineteen, and later, as a distinguished major general.

Both men were only-surviving sons born of strong, domineering mothers—mothers who had moved into nearby accommodation, for example, when their sons went to college. Both men were tall, handsome—and charismatic. Where Douglas MacArthur was a traditional Republican, however, Franklin Roosevelt was a compassionate Democrat—a difference that had come to a head in 1932, during the Great Depression.

As chief of staff of the U.S. Army in Washington at the time, General MacArthur had been charged by the president, Herbert Hoover, with the eviction of veterans who had marched on the capital to demand early payment of their promised war bonuses that summer. Despite President Hoover's express order to halt his thousand troops at the Anacostia River, MacArthur had insisted on taking personal charge of the brutal operation, involving tanks, cavalry, gas, and infantry with bayonets. The general had claimed the war veterans had no cause to claim their promised bonus early, indeed that the protest had been planned by the Communist Party, hoping to incite "revolutionary action" in America.[8] Casualties had reached three figures, and there were a number of deaths.

"You saw how he strutted down Pennsylvania Avenue," Governor Roosevelt had commented to Dr. Rexford Tugwell, a member of his famous Brain Trust. "You saw that picture of him in the *Times* after the troops chased all those vets out with tear gas and burned their shelters. Did you ever see anyone more self-satisfied? There's a potential Mussolini for you. Right here at home." And the presidential candidate had gone on to say: "I've known Doug for years. You've never heard him talk, but I have. He has the most pretentious style of anyone I know. He talks in a voice that might come from an oracle's cave. He never doubts and never argues or suggests; he makes pronouncements. What he thinks is final."[9]

Once Roosevelt was inaugurated as the thirty-second president of the United States, in March 1933, a confrontation between two such ambitious men had become inevitable. At a time of deteriorating international relations, MacArthur objected—with good reason—to proposed budget cuts involving more than 50 percent of the U.S. Army's budget appropriation for 1934. Summoning Major General MacArthur to the White House, the President—facing the worst economic crisis in American history—had "turned the full vials of his sarcasm" on the army chief of staff, who in "emotional exhaustion" had retorted "recklessly," as MacArthur himself later admitted, with "something to the effect that when we lost the next war, and an American boy, dying in the mud with an enemy bayonet through his belly and an enemy foot on his dying throat, spat out his last curse, I wanted the name not to be MacArthur, but Roosevelt."[10]

"You must not talk that way," Roosevelt had responded, "to the President."

"He was of course, right, and I knew it almost before the words had left my mouth," MacArthur recalled. The President was clearly furious. "I said that I was sorry and apologized. But I felt my Army career was at an end. I told him that he had my resignation as Chief of Staff. As I reached the door his voice came with that cool detachment which so reflected his extraordinary self-control, 'Don't be foolish, Douglas; you and the budget must get together on this.'"[11]

The bitter confrontation over army funding had in fact cleared the air between the two men: MacArthur respecting the President's amazing way with people, while the President respected MacArthur as a "brilliant soldier," as well as for his "intelligence" and leadership.

"We must tame these fellows and make them useful to us," Roosevelt had said of MacArthur and other prominent right-wing individuals—and he had.[12] Extending MacArthur's term of duty by a year—the first time ever in the history of the U.S. Army chief of staff's position—he had asked General MacArthur to implement his new Conservation Corps, which duly trained over a quarter million recruits, veterans and foresters, for civilian duties, putting them to work in some forty-seven states, at nominal federal expense.[13] When MacArthur's term was finally coming to a close in 1935, however, the President and MacArthur had had yet another falling-out—this time over the Philippines.

MacArthur had a long connection with the Philippine Islands, which in 1902 had become an American "insular area," or territory, at the conclusion of the Spanish-American and Philippine-American Wars. MacArthur's father, Brigadier General Arthur MacArthur, had fought as a brigade commander in the Philippine-American War; he had then become the first American military governor of the Philippines.

First as an army engineer, then in command of the Philippine Military District, then of the Philippine Division, and finally of the Philippine Department of the U.S. Army in the 1920s, Douglas MacArthur had followed in his father's footsteps. He had gotten to know not only the islands and their political leaders intimately, but a number of women—including a certain Isabel Rosario Cooper, half-Scottish, half-Filipino, who became his mistress in Washington, and unfortunate pawn in a failed high-profile libel lawsuit brought by General MacArthur for criticism of his egregious Bonus March operation.[14]

In spite of being humiliated by the Isabel Cooper scandal, Lieutenant General MacArthur had hoped that, on his mandatory retirement as U.S. Army chief of staff in 1935, President Roosevelt would appoint him U.S.

high commissioner to the Philippines — the islands having been granted interim semi-independence in 1935, and by an act of Congress assured full independence, to take place in 1946.

Roosevelt, however, had failed to give MacArthur the political appointment.

"Douglas, I think you are our best general," the President had said to the distraught soldier, "but I believe you would be our worst politician."[15] Instead, MacArthur had had to settle for a reduction in rank to brigadier general, and a posting to the Philippines as head of the small U.S. military mission in Manila.

MacArthur's exile had certainly been to President Roosevelt's political advantage.

Roosevelt had been well aware of the much-decorated general's appeal in the eyes of Republican political kingmakers at home. Although as an orator he required a prepared text, the general shone on paper and in one-on-one conversation, where he conveyed passion as well as incisive analysis. His World War I bravery was legendary; his reforms as commandant of West Point in the 1920s had demonstrated great military and educational vision. Like Roosevelt himself he had an astonishing ability to absorb complex information and pick out essentials. When MacArthur yet again applied for the job of U.S. high commissioner to the Philippines in 1937, the President had been torn. He did not trust, nor could he quite forgive, MacArthur, who was widely known to have backed the candidacy of Republican nominee Alf Landon in the 1936 presidential election, telling all who would listen — including President Manuel Quezon of the Philippines — that Landon would win by a landslide. Landon had lost by a landslide, however.

The President thus turned down MacArthur for a second time for the post of high commissioner. MacArthur's career seemed over.

In retrospect the President wondered if he had made the right decision. But at a time when the Japanese had invaded Manchukuo and were waging a major war of conquest in southern China, elevating Brigadier General MacArthur to the political post of high commissioner would have offered his potential Republican rival a free steppingstone to the Republican presidential nomination in 1940. On the other hand, however, blocking the career of such an outstanding American leadership-figure was, Roosevelt knew in his heart of hearts, unworthy. The political and diplomatic experience MacArthur would have gained as a U.S. high commissioner might well have tempered the general's somewhat lonely, introverted personality and broadened his mind.

Instead, the President had allowed MacArthur to "rot" in the Far East — permitting him to retire from the active list of the U.S. Army in 1937 and to become (at MacArthur's own quirky request) a Filipino "field marshal," replete with his own special uniform and gold braid–splattered hat, taking on the role of "civilian adviser" to the Philippine president on military matters in Manila. There, for four years, MacArthur — who remarried in 1937 — had drawn his U.S. military pension and his Philippine government salary, to become the highest-earning military officer in the world, with a generous expense account and magnificent penthouse apartment in the Manila Hotel.[16] If the war had not come, the President reflected, MacArthur might simply have remained there, in luxurious semiretirement.

But the war had come. In the spring of 1941 President Roosevelt had turned down the general for a *third* time as possible U.S. high commissioner — in spite of a fawning letter from MacArthur to the President's press secretary, lauding the President as "not only our greatest statesman," but "our greatest military strategist."[17] This time the President's rejection was no longer out of pique or political rivalry. As the war clouds over the Pacific darkened, Roosevelt had indicated via his military aide, General "Pa" Watson, that he wanted to use MacArthur in a "military capacity rather than any other." And sure enough, on July 28, 1941, having federalized the Army of the Philippines, Roosevelt had restored MacArthur to the U.S. Army's active list as a brigadier general and then lieutenant general. MacArthur had thus become commanding general of the United States Army Forces in the Far East — USAFFE.

Given their vast distance from the United States (ten thousand miles from Washington), the Philippine Islands could never be successfully defended against a Japanese invasion, the President knew. Nonetheless, the Army of the Philippines could be used as a lever: an interim threat in a last-ditch attempt to dissuade the Japanese from going to war with the colonial powers in the Far East. It was with this strategy in mind that the President embarked on a crash program of reinforcement of all U.S. bases in the Pacific. Assured of major shipments of weapons and airplanes, Lieutenant General MacArthur had mocked his naval counterpart, Admiral Tommy Hart. "Get yourself a *real* Fleet, Tommy, then you will belong!" he'd sneered — boasting that the War Department would be

The Plan of Escape

Worn down by cares at the White House, following Barbarossa, the German invasion of Russia, FDR claims he is leaving Washington on August 3, 1941, for a private fishing trip aboard the presidential yacht, the USS *Potomac*.

Instead of fishing, the President secretly transfers to the USS *Augusta,* flagship of the Atlantic Fleet. He then speeds to Argentia, the new U.S. military base in Placentia Bay, Newfoundland. There, on August 9, 1941, the former assistant secretary of the U.S. Navy invites the former First Lord of the British Admiralty, Winston Churchill, to board — and to dine with him and his "advisers."

Using the U.S. destroyer *McDougal* as a floating bridge, the President boards Churchill's battleship, HMS *Prince of Wales,* where he walks the length of the vessel, supported by his son Major Elliott Roosevelt for a binational Sunday worship.

Parrying Churchill's hopes of a promise to enter the war against Hitler, the President insists first on a declaration of anti-imperialist principles, or the Atlantic Charter.

Pearl Harbor

In his Oval Study at the White House at lunchtime, December 7, 1941, FDR and his assistant Harry Hopkins await the termination of Japanese diplomatic negotiations — unaware that five thousand miles away Japanese carrier-borne bombers are swooping over Pearl Harbor and will destroy the entire Pacific Fleet moored in Battleship Row.

As crowds anxiously gather, the world waits for a reaction from the White House. On December 8, 1941, the President asks Congress for a declaration of war against Japan. Believing that the United States will be preoccupied by war in the Pacific, Hitler declares war on America on December 11, 1942.

At Christmas 1941, Churchill arrives at the White House to help concert direction of the war. In the Oval Office he is thrown to the wolves — the U.S. press. As the British Empire in the Far East collapses, FDR takes supreme command, broadcasting to the nation on February 23, 1942.

Modeled on Churchill's portable map and filing system, FDR's White House Map Room allows him to cable directly to commanders across the globe, including MacArthur and President Quezon in the Philippines — instructing them not to negotiate with the Japanese.

The Raid on Tokyo

On April 19, 1942, Colonel James Doolittle leads a flight of B-25 bombers off the deck of the USS *Hornet* to attack Tokyo, six hundred miles away — the first time in history such a carrier takeoff had been effected — with no possibility of return. For his valor, President Roosevelt personally awards him the Congressional Medal of Honor on his return from China.

sending the bulk of its latest B-17 bombers to the Philippines under MacArthur's army command.[18] Moreover, MacArthur had unwisely assured President Quezon: "I don't *think* that the Philippines can defend themselves, I *know* they can."[19]

Roosevelt's diplomatic gamble—backed by a show of belated but growing American air power—had failed. Japan's military government, or junta, had simply made a careful assessment of the production, supply, and installation rate of proposed American reinforcements, and concluded that the United States would reach naval and air force parity by the spring of 1942, after which it would steadily surpass Japan's military production capacity. It was now or never, if the Empire of the Sun wished to expand its stalled war of conquest in the Far East, while the Western powers were so weak. Japan, their admirals reasoned, had but one sole chance of success if they wished to achieve their aims by force. And on December 7, 1941, they had taken it.

MacArthur's performance in the Philippines, beginning that same day, had been execrable, as President Roosevelt knew better than anyone in America, the decorated general having lost virtually his entire air force *on the ground*—despite nine hours of warning, both from Hawaii and from General Marshall himself in Washington.[20]

"MacArthur seems to have forgotten his losses in the Japanese surprise attack on Manila," the President would later tell his private secretary, Bill Hassett—"despite the fact that Admiral Kimmel and General Short face court-martial on charges of laxity at Pearl Harbor."[21]

Even after losing his air force on day one, MacArthur had performed miserably. As he had earlier told Admiral Hart, the commander of the U.S. Far Eastern Fleet, he refused "to follow, or be in any way bound by whatever war plans have been evolved, agreed upon and approved" by Washington[22]—and he didn't, simply failing to put into effect the plan that the War Department had laid down in the event of war. Instead, he'd ordered his Army of the Philippines to carry out an ill-rehearsed plan of "Beach Defense" without naval or air support—a scheme that failed to stop any of the Japanese landings that began at Lingayen (without Japanese air cover)[23] on December 22, 1941. All too quickly MacArthur's hastily assembled, ill-trained, and poorly armed Philippine troops had run, and within hours Manila, the capital, was threatened.

Belatedly recognizing his beach-defense scheme was a shambles and that the capital could not be defended, MacArthur had reported to the

President and War Department in Washington his decision to declare Manila an open city, occupied only by civilians, while belatedly pulling back his military forces, in conformity with War Department plan WPO-3, to the thirty-mile-long Bataan Peninsula. His own headquarters would move farther back still, to the island of Corregidor, a four-mile-long fortress isle, replete with deep tunnels, guarding the entrance to Manila Bay. In so doing, however — despite *months* to make contingency preparations, as the War Department had instructed him — MacArthur had failed to ensure sufficient provisions were sent back to the Bataan Peninsula and Corregidor. One depot, for instance, at Cabanatuan, on the central plain of Luzon, had held enough rice to "feed U.S. and Filipino troops for over four years,"[24] but was not relocated. Instead, MacArthur had airily assured the War Department — and President Quezon, whom he took with him by boat to Fort Mills, Corregidor, on Christmas Eve, 1941 — that his Army of the Philippines could hold out against the Japanese for six months in the difficult, jungle Bataan Peninsula territory, until reinforcements could be dispatched from Hawaii or the United States.

Not the lack of men or weapons but the lack of food had thereafter become the single most critical factor in the defense of Bataan — MacArthur having completely underestimated the number of mouths he must feed. In his cables to the President and War Department in Washington, he had claimed his army numbered only forty thousand men, while the enemy numbered eighty thousand.[25] In actuality it was the other way around. With the Army of the Philippines bottled up on the Bataan Peninsula, the Japanese could, by blockading it, simply starve out its opponents.

Moving their air force and one of their two divisions to prepare the invasion of the Dutch East Indies in preparation for their next major campaign, the Japanese had done exactly that. However hard they tried thereafter, neither the U.S. Army nor the Navy was able to break the blockade. And without enough food, thanks to MacArthur's error, the garrison was doomed.

None of this was known by the public in America, thanks largely to MacArthur's publicity machine on Corregidor. At his underground headquarters in the Malinta Tunnel beneath the Corregidor "Rock," two miles across the water from the Bataan Peninsula, MacArthur had reserved to himself the sole right to issue press communiqués and press releases, telling each day more stories of heroic combat against the Japanese under his sterling generalship — even though MacArthur only once ever crossed the water to visit his army in the field.[26] He was writing pure propaganda. "General MacArthur personally checks all publicity reports, and writes many of them himself," his chief of staff afterward explained, "always with an eye on their effect on the MacArthur legend."[27]

MacArthur's subsequent air force commander, Brigadier General George Kenney, mockingly described the communiqués as having "painted the General with a halo and seated him on the highest pedestal in the universe."[28] Of almost 150 communiqués put out by the headquarters of the USAFFE in the weeks following Pearl Harbor, 109 mentioned only one individual: MacArthur.[29]

As a battlefield commander MacArthur was, Roosevelt reflected, a fraud — his January 15, 1942, message, "to be read out to all units," declaring that "help is on the way," being "criminal" in its mendacity and the raising "of false hopes," the President told his personal secretary, " — hopes that MacArthur knew could not be fulfilled."[30]

But what of the general's credentials to replace the President as U.S. commander in chief — a replacement that had never taken place in American history, given the Constitution's express condition that only the President of the United States should hold that title, rank, and responsibility?

Here the President was even more disappointed by MacArthur's histrionics — the only term that could describe the "flood of communications" (as Eisenhower called it in his diary)[31] the general had transmitted to Washington by wireless since Pearl Harbor. For in their miscalculations, wild exaggerations, grandiose recommendations, and doomsday warnings, MacArthur's cables had given cause for the President to question MacArthur's mental health.

Among many others, Admiral Tommy Hart had long despaired of MacArthur's contact with reality. "The truth of the matter is that Douglas is, I think, no longer altogether sane," Hart had confided to his wife even before the Japanese assault; in fact, Hart added thoughtfully, "he may not have been for a long time."[32]

In terms of army-navy cooperation, MacArthur had evinced a fatal lack of interest — adamantly refusing Hart's request to let the navy call upon the army's long-range B-17s for reconnaissance or protection purposes over the sea. Instead, the "field marshal" had simply gloried in his refusal to take orders from Washington, telling Hart it was his aim to create a "200,000-man army" and fight "a glorious land war" in defending the Philippines *without naval support.* The "Navy had its plans, the Army had its plans," MacArthur had declared with finality, "and we each had our own fields" — waving away Hart's proposals for combined defense.[33]

The loss of his air force on December 7, 1941 (Washington time), had rendered the U.S. fleet in the Far East sitting ducks. Such a disaster might have chastened a lesser man than MacArthur. Given his monumental ego it had not, at least to judge by his signals to Washington, which Marshall arranged to be messengered to the Commander in Chief in the White House immediately on receipt, given the crisis in the Pacific. "I do not know the present grand strategy," MacArthur had shamelessly cabled, for example, on December 13, 1941, a week after his air force had been destroyed — "but I do know what will follow here unless an immediate effort, conceived on a grand scale, is made to break the Jap blockade. If Japan ever seizes these islands the difficulty of recapture is impossible of conception. If the Western Pacific is to be saved, it will have to be saved here and now. If the Philippines and the Netherlands East Indies go, so will Singapore and the entire Asiatic continent," he warned. "The Philippines theatre of operations is the locus of victory or defeat," he claimed, "and I urge a strategic review of the entire situation lest a fatal mistake be made. The immediate necessity is to delay the hostile advance. This can be effectively accomplished by providing air support" for U.S. ground forces, as well as "bombardment to operate against [enemy] air bases, communications and installations" — the very things he had conspicuously failed to order on the day of Pearl Harbor. "The presence of air forces here would delay the enemy advance," the general summarized, and would moreover serve to protect the "Netherlands East Indies and Singapore, thus insuring the rapid defeat of the enemy. It justifies the diversion here of the entire output of air and other resources. Please advise me on the broadest lines possible. End. MacArthur."[34]

A week later, when the Japanese invasion of the Philippines began in earnest, MacArthur had sent his inexplicable estimate of his forces and those of the enemy he faced. He commanded only "about forty thousand men in units partially equipped," he had signaled on December 22, 1941, confronting "eighty to one hundred thousand" Japanese[35] — when in truth he had no fewer than eighty-five thousand of his own armed troops, facing less than half that number of Japanese. On January 2, 1942, he had then revised his numbers, claiming to have "only seven thousand American combat troops here, the balance of force being Filipino" — and had begged not only for U.S. planes to be urgently delivered, but "the landing of an expeditionary force"—emphasizing his firm belief "that the loss of the Philippines will mark the end of white prestige," and that U.S. forces must "move strongly and promptly," as he put the choice, "or withdraw in shame from Orient. Stop."[36]

Five days later, on January 7, 1942, the general had urged that, as in some pageant of miracles, "an Army corps should be landed in Mindanao"—the largest southern island of the Philippines—"at the earliest date possible." He begged also for "more aggressive and resourceful handling of naval forces in this area"—while reporting that, despite the War Department's long-laid plans for a staged withdrawal of U.S. forces to the Bataan Peninsula, he now had insufficient food to feed them, and had been compelled to place them on "half rations." He therefore demanded "steps must be taken immediately to get in supply ships no matter at what loss."[37]

How such steps could be taken ten thousand miles away was unclear, though the President had ordered everything possible to be done to get supplies through. Troop reinforcements—especially an entire "Army corps"—were not only impossible to prepare and transport to the battlefield, but strategically absurd. In his diary at his headquarters in Java, toward which Japanese forces were now steaming, Admiral Hart painted on January 11, 1942, a mocking image of the field marshal on Corregidor: "Douglas sitting in his tunnel dreaming up suggestions of how the Navy could help him win the war that he actually had lost in the first 24 hours."[38]

Undeterred, a week later, on January 15, 1942, MacArthur had issued his message to be "read and explained to all troops" on Bataan and Corregidor: "Help is on the way from the United States. Thousands of troops and hundreds of planes are being dispatched," the commanding general had assured them. "The exact time of arrival of reinforcements is unknown as they will have to fight through Japanese attempts against them. It is imperative that our troops hold until these reinforcements arrive. . . . If we fight we will win; if we retreat we will be destroyed."[39]

Two days after that, on January 17, 1942, however, MacArthur had appeared at his wit's end. The rations he needed, "measured in ships capacity," were "small indeed," he begged Washington. "Many medium sized or small ships should be loaded with rations and dispatched along various routes. Stop. The enemy bomber formations are no longer here but have moved south. Stop. Unquestionably ships can get through but no attempt yet seems to have been made along this line. Stop. This seems incredible to me and I am having increasing difficulty in appeasing Philippine thought along this line. Stop. They cannot understand the apparent lack of effort to bring something in. Stop. I cannot overemphasize the psychological reaction that will take place here," he warned, "and unless something tangible is done in this direction a revulsion of feeling of tremendous proportions against America can be expected. Stop. They can understand failures but cannot understand why no attempt is being made at relief through the forwarding of supplies. Stop. The repeated statements from the United States that Hitler is to be destroyed before an effort is made here is causing dismay. Stop. The Japanese forces — air, land and ground — are much overextended. Stop. His success to date does not measure his own strength," MacArthur claimed of the Japanese invasion, but rather "the weakness of his opposition" — despite even three-to-one superiority in troops on Bataan.[40] "A blow or even a threatened blow against him will almost certainly be attended with some success. Stop. I am professionally certain that his so-called blockade can easily be pierced. Stop. The only thing that can make it really effective is our own passive acceptance of it as a fact. Stop. If something is not done to meet the general situation which is developing the disastrous results will be monumental. Stop. The problems involved cannot be measured or solved by mere army and navy strategic formulas they involve comprehensiveness of the entire oriental problem. End. MacArthur."[41]

MacArthur's doomsday language had been ridiculed at the War Department — which nevertheless tried again and again to get blockade runners through to Mindanao, on the President's specific orders. Yet there was no one in the War Department willing to put the distinguished general in his place, let alone criticize him for losing his own air force on the first day of battle. Every message had therefore been replied to with courteous War Department assurances, signed by General Marshall himself, explaining that everything possible was being done to get supplies to him, as well as reinforcements to Australia and the Far East in order to assemble an eventual coordinated counterattack.[42] But with Malaya being overrun, Singapore within Japanese sights, the Dutch East Indies vulnerable, and Australians themselves rattled over imminent invasion, there was, undoubtedly, a tendency to see MacArthur and his eight or more thousand beleaguered American troops as "expendable." The view in the Munitions Building on Constitution Avenue — and even more so in the Navy Department next door — was that, having lost his air force and failed to provision his army on the Bataan Peninsula as he had been instructed to do in advance of Pearl Harbor — instructions he had labeled "defeatist" — MacArthur had made his own bed and must lie in it.

MacArthur's cables to Washington had understandably only grown more desperate in the days that followed — indeed MacArthur's wild pleas and warnings seemed to Eisenhower and others at the War Department to "indicate a refusal on his part to look facts in the face, an old trait of his." Highly emotional in their language, envisaging a "fatal" scenario for the Allies if his wishes were not met, his cables showed the hero, by January 29, 1942, to be "jittery!" as Eisenhower jotted in his diary.[43] "Looks like MacArthur is losing his nerve," Eisenhower wrote on February 3, 1942.[44]

Eisenhower — who had come to despise MacArthur, for all his "brilliance," after years working for him in Washington and the Philippines — had simply lost patience with the braggart, as had others in the War Department, despite all the sympathy they felt for their fellow servicemen, beleaguered on Bataan.

MacArthur's pleas were filed in the new ground-floor Map Room at the White House, once the President had read them. "The President has seen all of your messages," Marshall had assured MacArthur in late December 1941,[45] and the President continued to read them over the following days and weeks. "I welcome and appreciate your strategical views," Marshall signaled back to MacArthur in February, "and invariably submit them to the President."[46]

Was the Far Eastern general a "strategical" genius, as the press and Republicans increasingly seemed to believe? Or was he a charlatan — a Mad Hatter, ensconced in a rabbit hole of a tunnel — driven literally to distraction by his unenviable situation in the Philippines?

The President had his own views, knowing MacArthur over so many years. Yet it was part of Roosevelt's genius as a leader that he was able, for the most part, to take a more Olympian view than others, especially in times of stress and ill success. His long political battles and his struggle against polio had given him a stature among his colleagues and contemporaries unrivaled by any other figure in America. And at the heart of this robustness of character, tempered by so many reverses across the years, was his abiding optimism. MacArthur might well be mad — but as King George II had famously said when listening to his advisers' objections to the appointment of young Brigadier General James Wolfe to command an amphibious operation to seize Quebec: "Mad, is he? Then I hope he will bite some of my other generals!" MacArthur was a potentially great leader, whose flaws were as large as his strengths. He was *symbolically* important, as an inspiration to so many.

As Commander in Chief of the Armed Forces of the United States and *de facto* leader of the United Nations, the President thus had a wider perspective than those in the War or Navy Departments who derided MacArthur's histrionics, and who deplored MacArthur's failure to mind his own business in his endless "strategical" outpourings, instead of actually commanding his troops in battle on Bataan. At least MacArthur was determined to strike back against the Japanese!

It was all very well, the President reflected, to point to MacArthur's mistakes from day one of the war — indeed even before war began — but who in the Far East, or the West, was any better as a military commander? As the President noted in a cable to Manuel Quezon that he sent via MacArthur, "The deficiency which now exists in our offensive weapons are the natural results of the policies of peaceful nations such as the Philippines and the United States" — nations "who without warning are attacked by despotic nations which have spent years in preparing for such action. Early reverses, hardships and pain are the price that democracy must pay under such conditions." Roosevelt had assured Quezon that "every dollar and every material sinew of this nation" were being thrown into the fight, a fight whose objective was in part the restoration of "tranquility and peace to the Philippines and its return to such government as its people may themselves choose."[47] Yet patient realism of this kind, sincere as it was, would have no value, the President recognized, without military commanders able to convey confidence and a transcending belief in their own leadership.

"This war can't be won with men who are thinking only about retiring to farms somewhere and who won't take great and bold risks," the President had told Harry Hopkins at dinner on January 24, 1942 — Hopkins noting that Roosevelt "has got a whole hatful of them in the Army and Navy that will have to be liquidated before we really get on with our fighting."[48]

What a contrast MacArthur was, for example, to Admiral Tommy Hart, the four-star commander of the U.S. Navy's Far Eastern Fleet, as the President explained to Hopkins. The President had known Hart for many years. Like Admiral Stark, Hart was a highly professional officer — but one who exuded defeatism, however much he called it realism. From Admiral King and others in the Navy Department, the President was hearing how Hart was spreading little but weariness and resignation wherever he went—prompting the chief of naval operations to send messages from the Navy Department warning Hart of what was being said about him in Washington,[49] and Harry Hopkins to write in his diary, before he himself went into hospital for another prolonged bout of treatment, that the President was "going to have many of the same problems that Lincoln had with generals and admirals whose records look awfully good but who may turn out to be the McClellans of this war.[50] The only difference," Hopkins wrote, "is that I think Roosevelt will act much faster in replacing these fellows."[51] *Faute de mieux*, Hart had recently been appointed commander in chief of all American, British, Dutch, and Australian naval forces in the Southwest Pacific (the so-called ABDA area), under General Archibald Wavell, the new supreme commander in area—but was letting down the American flag. As Hopkins noted, the President felt Admiral Hart "is too old adequately to carry out the responsibilities that were given him and I fancy before long there will be a change in our naval command in the Far East."[52]

It was to ponder over Hart's command, among other matters, that on January 31, 1942, the President set off in his special train from Washington's University Station to spend a few days at his home in Hyde Park, north of New York.

Winston Churchill had urged him to get out of Washington whenever he could, in order to "think about" the war in quiet—Churchill often spending his weekends at Ditchley Park, an American-owned manor house outside London, where he could obtain respite from the German bombing and the cares of Parliament. Despite the inclement weather, which presaged snow, the President had finally taken Churchill's advice. He was "in rare form, full of wisecracks" on the journey, his secretary, Bill Hassett noted in his diary, the "perfect host" to the small staff he took with him. "He seemed a trifle tired to me," Hassett allowed, "but he was in excellent spirits."[53]

In truth the President was worn out. He had even taken, briefly, to wearing "quite a snappy gray-zippered siren suit," like Churchill's,[54] Hassett described, but seemed to enjoy the change of scenery—knowing he would have to return to Washington to face the press on February 6, and a battery of questions about his prosecution of the war.

Four inches of snow had indeed fallen by the time they arrived. The President had developed such a cold that his doctor, Ross McIntyre, had been summoned from New York. Although he seemed "in good shape and good spirits," according to Hassett, the President appeared "reluctant to go back" to Washington.[55] He therefore stayed put.

The days at his family home — so big and quiet and lonely without the presence of his mother, Sarah, who had passed away the previous fall (causing the President to wear a black armband for a full year to mark his grief) — now enabled the President to see what his own military team were missing — even those like Eisenhower and Colonel Handy, who were now urging that the Philippines be written off.[56]

Ike, as he was universally known in the War Department, could only scoff at MacArthur's melodramatic cables, especially the general's recent signal stating that "in the event of my death" he wanted his chief of staff, Richard Sutherland, to assume command in the Philippines — a man Eisenhower considered one of MacArthur's biggest "boot lickers."[57]

Only a handful of officers or men of the Army of the Philippines had seen MacArthur, their commanding general, on his one-and-only visit to the Bataan Peninsula, on January 10, 1942. They had become even less enchanted with MacArthur's generalship when, on January 24, he had given orders reducing the limited rations on Bataan by half — while ordering a doubling of food stocks for the eleven thousand men on Corregidor, who — apart from antiaircraft and long-range artillery units — were not even fighting.[58] It was this, primarily, that had led to the famous ballad, sung by the soldiers to the tune of "The Battle Hymn of the Republic," deriding their commanding general as "Dugout Doug":

Dugout Doug MacArthur lies ashaking on the Rock
Safe from all the bombers and from any sudden shock
Dugout Doug is eating of the best food on Bataan
And his troops go starving on.

Chorus

Dugout Doug, come out from hiding
Dugout Doug, come out from hiding
Send to Franklin the glad tidings
That his troops go starving on![59]

Whether or not the general was "hiding out" in the Malinta Tunnel while his valiant army was continuing the Battle of Bataan, FDR understood, however, a key fact: MacArthur had gotten the Filipinos to fight the Japanese, not welcome them. *This,* the President recognized, represented a potentially war-changing phenomenon in terms of Japan's moral basis for hostilities—and America's destiny in countering the Japanese rampage. Though the American press and Republican politicians might be wildly overstating his claims to fame—mythologizing MacArthur's prowess as a commander in the field and his fitness to be U.S. commander in chief in Washington in place of Roosevelt himself—MacArthur's *spirit,* like that of Churchill's in 1940, might turn out to be more important to American and Allied morale in prosecuting the war than all his errors as a general combined.

The parallel with Churchill fascinated the President. How alike were the two in many ways! Both men loved symbols to mark their individuality—Churchill his cigars, General MacArthur his corncob pipe and cane. Even their distinctive hats were designed to be memorable: Churchill's bowler, MacArthur's special "field marshal's" cap, with its intricate, spaghetti-splash of gold braid. Both men were positively dangerous in terms of their lack of realism, their tactical missteps, their mood swings. Churchill's performance during the fall of France—his wild orders for counterattack that bore no relation to what was possible after Guderian and Rommel's breakthrough, and his instructions to Lord Gort to surrender the British Expeditionary Force rearguard at Dunkirk when it was not necessary—had been as pitiful as MacArthur's defense of Lingayen and ill-provisioned retreat to Bataan. Even so, the performance of Churchill's ally, France, had been even *more* deplorable! What Churchill and MacArthur both exuded, the President acknowledged, were a pluck and defiance that made others seem small and minion. MacArthur's pie-in-the-sky calls for counteroffensive action took no account of what was actually possible in the Southwest Pacific at such an early stage in the war, and the general's own responsibility in failing to prepare for a prolonged siege, as the War Department had instructed him. For all his faults, however, MacArthur was at least expressing the spirit of a fighters, determined to strike back at the Japanese, if possible—not run from them.

Hopkins—who remained in hospital for ten days—thus soon proved correct. When the President returned to Washington, it was, as U.S. commander in chief, to order that Admiral Hart be fired—Hart having confided to his supreme commander, General Wavell, that at age sixty-four he was simply too weary for the stress of the job, had already passed his retirement age, and could not be expected to run the Japanese blockade encircling the Philippines, where his ships would be subjected to heavy attack by hostile land-based aircraft.[60] Instead, Hart had advised, they should concentrate only on trying to attack Japanese naval ships and transports as they approached the Netherlands East Indies and Java, and let MacArthur wither on his vine.

By contrast, the President remained rather proud of his Far Eastern general.

The President was aware, however, that MacArthur's nerves were increasingly on edge — as well they might be. In his almost daily cables of woe — as well as suggestions on how to vanquish the enemy — the general was, in a sense, letting off steam, the President knew: his language reflecting the torment he was going through.

On February 4, 1942, for example, the War Department had flown up to Hyde Park a new signal from MacArthur in which the general had referred to what was, in his view, "a fatal blunder on the part of the democratic allies. The Japanese are sweeping southward in a great offensive and the allies are attempting merely to stop them by building up forces in their front. This method," MacArthur had signaled, "as has always been the case in war, will fail. Such movements can only be negated by thrusts not at the enemy's strength but at his weakness. The lines of weakness from time immemorial have been the lines of communication. In this case they are stretched over two thousand miles of sea, with the whole line subject to American sea thrust. This line is not defended by enemy bombers but is held by scattered naval elements. A sea threat would immediately relieve the pressure on the south and is the only way that pressure can be relieved. A great naval victory on our part is not necessary to accomplish this mission; the threat alone would go far toward the desired end. The enemy would probably not engage his entire fleet in actual combat. If he did and lost, the war would be over. If he did and won, the losses he would sustain would still cripple his advance and take from him the initiative. You must be prepared to take heavy losses," MacArthur warned, "just as heavy losses are inflicted in return. I wish to reiterate that his bomber strength is entirely engaged on his southern front and represents no menace at all to such a naval thrust. With only minor menace from the fleets of Germany and Italy the American and British navies can assemble without serious jeopardy the force to make this thrust. I unhesitatingly predict that if this is not done the plan upon which we are now working, based upon the building up of air supremacy in the Southwest Pacific, will fail, the war will be indefinitely prolonged and its final outcome will be jeopardized. Counsels of timidity based upon theories of safety first will not win against such an aggressive and audacious adversary as Japan. No building program no matter of what proportions will be able to overtake the initial advantages the enemy with every chance of success is trying to gain. The only way to fight him is to fight him immediately. . . . From my present point of vantage I can see the whole strategy of the Pacific perhaps clearer than anyone else. End. MacArthur."[61]

It was easy enough, as Brigadier General Eisenhower did,[62] to ridicule such *pronunciamientos.* MacArthur was correct, strategically. The problem was simply a practical one: his advice could not be followed. But at least MacArthur was thinking *aggressively* as a commander — in fact, he was by the very force of his personality holding together a symbolic partnership between American and Filipino forces that had incalculable importance for the Western alliance. This, the President recognized, was a pearl beyond price — something that the President's advisers (and many later critics, also) simply failed to understand.

On New Year's Eve, already, the President had suggested that President Quezon be rescued from the Philippines and brought to Washington to lead a government in exile, like that of the Dutch government.[63] MacArthur had instinctively demurred, claiming Quezon — who was suffering from tuberculosis — would not survive the flight, and that his presence on Corregidor was vital for Filipino troops fighting alongside U.S. soldiers. Anxious lest Quezon be captured by the Japanese and made into a quisling, President Roosevelt had felt MacArthur's intransigence to be a mistake, but had concurred, leaving the decision as to the best moment for Quezon's evacuation up to MacArthur.

As the days had gone by, however, and it had become more and more obvious the Allies would not be able to mount a relief of the Philippines, MacArthur's insistence on keeping Quezon at Corregidor to fire up his Filipino troops had begun to backfire. It had become obvious to Quezon that, whatever promises President Roosevelt made and MacArthur assured him of, the United States was not going to be able to save the Philippines from Japanese conquest.

Quezon had already sent the President a despairing message on January 13, 1942,[64] which Roosevelt had not answered, but which had certainly disturbed him. Then on January 29, Quezon had written out a formal letter to MacArthur, intended for transmission to the President, pointing out again that the "war is not of our making. Those that dictated the policies of the United States could not have failed to see that this is the weakest point in American territory. . . . We decided to fight by your side and we have done the best we could and we are still doing as much as could be expected from us under the circumstances. But how long are we going to be left alone? I want to decide in my own mind whether there is any justification in allowing all these men to be killed, when for the final outcome of the war the shedding of their blood may be wholly unnecessary."[65]

The President had relied on MacArthur to keep up the spirits of the Philippine president, but Quezon's latest appeal he had had to answer—assuring Quezon that he recognized "the depth and sincerity of your sentiments" with respect to "your own people." He wanted Quezon to know that he himself would be "the last to demand of you and them any sacrifice which I considered hopeless in the furtherance of the cause for which we are all striving. I want, however, to state with all possible emphasis that the magnificent resistance of the defenders of Bataan is contributing definitely toward assuring the completeness of our final victory in the Far East. While I cannot now indicate the time at which succor and assistance can reach the Philippines, I do know that every ship at our disposal is bringing to the southwest Pacific the forces that will ultimately smash the invader and free your country. . . . I have no words in which to express to you my admiration and gratitude for the complete demonstration of loyalty, courage and readiness to sacrifice that your people, under your inspired leadership, have displayed. They are upholding the most magnificent traditions of a free democracy."[66]

MacArthur had been delighted by the response. As he had cabled back to General Marshall in Washington: "The President's message to Quezon was most effective. Stop. Quezon sends following reply for President Roosevelt. Colon. Quote. Your letter has moved me deeply. Stop. I wish to assure you that we shall do our part to the end. Signed Quezon. Unquote. MacArthur."[67] Quezon had even broadcast, on Voice of Freedom radio from Corregidor, that night, urging "every Filipino to be of good cheer, to have faith in the patriotism of valor of our soldiers in the field, but above all, to trust America. The United States will win this war, America is too great and powerful to be vanquished in this conflict. I know she will not fail us."[68]

• • •

What, then, went wrong? Hardly had Quezon's cable arrived in Washington than, two days after the President's return from Hyde Park, a new cable from Quezon arrived, with a very different message. In transmitting President Quezon's text, General MacArthur prefaced it by saying that he himself would be sending a second part to the cable, in which he would give his own thoughts on the matter. That part, however, had not yet been deciphered.

It was the first part, however—the text from Quezon—that filled the President with trepidation as he went to sleep on February 8, 1942. As Secretary Stimson noted in his diary—having spent an hour and a half discussing the ramifications of the partial message with General Marshall that afternoon—the war in the Far East was coming to an inevitable crisis.

Quezon's cable had been blunt. "I feel at this moment that our military resistance here can no longer hold the enemy when he sees fit to launch a serious attack," Quezon had written. "I deem it my duty to propose my solution:

"That the United States immediately grant the Philippines complete and absolute independence;

"That the Philippines be at once neutralized;

"That all occupying troops, both American and Japanese, be withdrawn by mutual agreement with the Philippine government within a reasonable length of time;

"That neither country maintain bases in the Philippines . . .

"It is my proposal to make these suggestions publicly to you and to the Japanese authorities without delay and upon its acceptance in general principle by those two countries that an immediate armistice be entered into here pending the withdrawal of their respective garrisons."[69]

Reading this, Colonel Stimson could only shake his head. It was, in short, yet "another appeal from Quezon who has evidently made up his mind to make a surrender for his people in order to avoid useless sacrifice," as the secretary summarized in his diary. Which now raised the question "not only as to Quezon and the Philippines' future," he mused, "but what we should do with the devoted little garrison that has been holding out"—and "what we should do with MacArthur."[70]

The President, as commander in chief, would finally have to make a decision "ghastly in its responsibility and significance"[71]—but first they must wait for the second part of the cable to be decoded.

• • •

Henry Stimson, aged seventy-four, had good reason to be anxious, since the crisis over the Philippines put his own post as secretary of war on the line.

The U.S. War Plans Division had already produced on January 3, 1942, a "very gloomy study," stating "that it would be impossible for us to relieve MacArthur and we might as well make up our minds about it," as Stimson had noted in his diary.[72] Such a relief would require an unachievable task force to be assembled, overnight: not only 750 more warplanes but between six and nine more battleships or heavy cruisers, five to seven aircraft carriers, almost fifty destroyers, sixty submarines as well as their auxiliary vessels, and "several hundred thousand" troops.[73] As the secretary had noted in his diary two nights later, the truth was written on the wall, but no one was allowed to speak it. "Everybody knows the chances are against getting relief to him, but there is no use saying so beforehand," Stimson had confessed — having ordered the report to be kept secret.[74]

The simple truth was, the Philippines was doomed. The remnants of the U.S. Pacific Fleet could not be risked, so soon after Pearl Harbor, in such a wild adventure — leaving Pearl Harbor undefended. The surviving vessels of Admiral Hart's Far Eastern Fleet had wisely sailed south to Java. And with the U.S. Army Air Forces lacking sufficient available planes to get beyond Mindanao, the southernmost Philippine island, there was no hope of staging a relief mission — indeed, it was questionable whether the Allies could hold on to the Netherlands East Indies, with its crucial oil resources. MacArthur would have to fight it out to the bitter end, with the forces he had. "None of us is likely to make the mistake," Stimson had admitted, "of taking too much risk."[75]

As the inevitable surrender of the Bataan and Corregidor garrisons approached, however, cruel words were spoken in the press and in Washington about Secretary Stimson's performance — indeed, Stimson would be urged to retire by a member of Congress speaking on the floor of the House.[76] It was small wonder that, as the concluding part of General MacArthur's cable was received and successfully decoded early on the morning of February 9, 1942, the secretary of war blanched.

"When I reached the War Department," he recorded that night, "the telegram which had begun coming in yesterday had been finished." First, however, Stimson reread President Quezon's "somber" evaluation of the situation in the Philippines — "arraigning the United States for delinquency in helping the Philippines in many matters which were entirely false," and therefore proposing "that we should declare the independence of the Philippines and retire and that the Japanese should be appealed to on the basis of a recent speech by the Prime Minister of Japan" — a speech to the Japanese Diet in which Prime Minister Tojo had promised eventual independence to the Filipinos if they would stop fighting side-by-side with the Americans.

If Quezon's recommendation — namely a pact with the Japanese to withdraw all forces, Japanese and American, from the Philippine Islands in the midst of a critical battle — was not bad enough, MacArthur's accompanying telegram was even worse. "This telegram was most disappointing," Stimson dictated in his diary, for MacArthur "went more than half way towards supporting Quezon's position."[77]

Having talked over the telegram with Marshall and Eisenhower, "I then called up the President, told him of the [second part of the] message, and said I was on the point of sending it to him by messenger. I gave him an outline of it to break the news and he at once suggested that Marshall and I come over at half past ten."[78]

Stimson and Marshall were duly driven to the White House. In the President's study on the second floor they found not only the Commander in Chief but Assistant Secretary of State Sumner Welles — "Cordell Hull being sick."[79] Quezon's message might be disappointing, but it was MacArthur's telegram recommending "neutralization" of the Philippines in the middle of a battle that most amazed the President.

Japanese forces had already invaded Borneo and Celebes, prior to the seizure of Java and Timor; meanwhile, they had other forces preparing for the conquest of Sumatra. What earthly reason would the Japanese have to accept "neutralization" of the Philippines — the closest major islands to Japan — and the evacuation of American forces, when they already had the islands within their grasp, indeed had withdrawn one fighting division, with the intention of dealing with the Bataan business later, once they had reached Borneo?

It made no sense — yet MacArthur, in his cable, had supported President Quezon's appeal, and had quoted the high commissioner, Francis Sayre, as supporting it too. "I took the liberty of presenting this message to High Commissioner Sayre for a general expression of his views," MacArthur reported in the second part of his cable. "States as follows: 'If the premise of President Quezon is correct that American help cannot or will not arrive here in time to be availing I believe his proposal for immediate independence and neutralization of Philippines is sound course to follow.'"[80] To which MacArthur had appended his own estimate of the military situation.

MacArthur's report was as "gloomy," in Stimson's view, as that of Quezon — claiming "we are near done. . . . Since I have no air or sea protection you must be prepared at any time to figure on the complete destruction of this command. You must determine," MacArthur had addressed himself to General Marshall, "whether the mission of delay would be better furthered by the temporizing plan of Quezon or by my continued battle effort. The temper of the Filipinos is one of almost violent resentment against the United States. Every one of them expected help and when it has not been forthcoming they believe they have been betrayed in favor of others. It must be remembered they are hostile to Great Britain on account of the latter's colonial policy. In spite of my great prestige with them, I have had the utmost difficulty in keeping them in line. If help does not arrive shortly nothing, in my opinion, can prevent their utter collapse and their complete absorption by the enemy. The Japanese," he admitted, "made a powerful impression upon Philippine public imagination in promising independence." In the general's view, then, "the problem presents itself as to whether the plan of President Quezon might offer the best possible solution of what is about to be a disastrous debacle. . . . Please instruct me."[81]

If MacArthur's recommendation was anathema to the war secretary, it was even more so to Roosevelt.

It was, after all, a betrayal of MacArthur's aggressive spirit. On January 31, the President had asked Stimson to look into whether MacArthur could be awarded the Medal of Honor, the nation's highest medal for bravery in the field, in recognition of the general's fighting ardor. Now the general was proposing to parley with the enemy!

The President asked Marshall for his opinion, but it was Secretary Stimson who first responded, allowing that General Marshall "said that I could state our views better than he could." Stimson thus now "gave my views in full and as carefully as I could," as he subsequently explained his action — aware that this was perhaps the most critical moment of the war since Pearl Harbor.

"I arose from my seat and gave my views standing as if before the court"[82] — for Stimson was not only a highly successful prosecuting attorney, but had himself been governor-general of the Philippines, and felt that the notion of asking the Japanese to stop fighting in midconquest and vacate the Philippine Islands, halfway through their campaign to seize the entire Malay Barrier, was ridiculous.

However eloquent the secretary of war, and however much General Marshall agreed with him, the plain fact of the matter was that they were both afraid of MacArthur. In the circumstances, only the President had the moral as well as constitutional authority to respond to what was effectively "surrender."

All now looked at the President—who shook his head, negating any such idea.

"Roosevelt said we won't neutralize," Marshall later recalled the President's emphatic words—and his own relief. At that moment of world crisis, "I decided," he added, "he was a great man."[83]

The President having made his decision, Marshall and Stimson were instructed to go draft replies on the lines they then agreed—a cable over which both Stimson and Brigadier General Eisenhower labored "the entire day," as Eisenhower duly noted in his diary.[84]

The President was dissatisfied when the drafters returned around three o'clock that afternoon, however—Stimson's prose being mealy-mouthed and apologetic.[85] Recognizing the importance of the decision, the President had meantime called in Admiral King and Admiral Stark, as well as Sumner Welles, to assist in the business. The time had come, the President recognized, when he must not only take action as commander in chief, but be seen by his military staff to do so.

"My reply must emphatically deny," Roosevelt warned MacArthur straight off the bat, "the possibility of this Government's agreement to the political aspects of President Quezon's proposal." A full presidential response, to be handed to Quezon, would be contained in the "second section of this message."[86] First, however, the Commander in Chief had personal and confidential instructions for General MacArthur.

"I authorize you to arrange for the capitulation of the Filipino elements of the defending forces," the President's cable began, "when and if in your opinion that course appears necessary and always having in mind that the Filipino troops are in the service of the United States. For this purpose the Filipino troops could be placed by you under the command of a Filipino officer who would conduct actual negotiations with the enemy. Such negotiations must involve military matters exclusively. Details of all necessary arrangements will be left in your hands, including plans for segregation of forces and the withdrawal, if your judgment so dictates, of American elements to Fort Mills [Corregidor]. The timing also will be left to you," the President laid down.[87] With regard to American troops, however, the Commander in Chief minced no words.

"American forces will continue to keep our flag flying in the Philippines so long as there remains any possibility of resistance. I have made these decisions in complete understanding of your military estimate that accompanied President Quezon's message to me. The duty and necessity of resisting Japanese aggression to the last transcends in importance any other obligation now facing us in the Philippines."[88]

For almost thirteen weeks the Commander in Chief had tolerated MacArthur's increasingly melodramatic appeals for an American imaginative counteroffensive strategy in the Pacific — naval, air, and army operations that would have been hard for a nation at the very apex of its military power, let alone one whose forces were depleted in the first, halting weeks of a war it had not sought. Now it was Roosevelt's turn to educate his commander in the field — and he proceeded to do so, uncompromisingly.

"There has been gradually welded into a common front," he reminded MacArthur, a global coalition confronting "the predatory powers that are seeking the destruction of individual liberty and freedom of government. We cannot afford to have this line broken in any particular theater. As the most powerful member of this coalition we cannot display weakness in fact or in spirit anywhere. It is mandatory that there be established once and for all in the minds of all peoples complete evidence that the American determination and indomitable will to win carries on down to the last unit.

"I therefore give you this most difficult mission in full understanding of the desperate situation to which you may shortly be reduced. The service that you and the American members of your command can render to your country in the titanic struggle now developing is beyond all possibility of appraisement."[89]

It was, in short, time for General MacArthur to cease sending home schemes of grand strategy from his tunnel below Fort Mills, and to concentrate on his troops in Bataan and Corregidor. "I particularly request that you proceed rapidly to the organization of your forces and your defenses so as to make your resistance as effective as circumstances will permit," the Commander in Chief ordered, "and as prolonged as humanly possible."[90]

Perhaps no signal in history from a United States president, in his role as commander in chief, to his commanding general in the field had ever been as candid or coldly imperative. It was a directive calculated to pierce MacArthur's *amour propre:* to rouse him out of his temporary mental collapse, and to sting.

It did — transforming MacArthur from a near-wreck into his old self: a great commander.

The President's signal to President Quezon would, too, go down in history — though for another, perhaps even more significant, reason: articulating, at a critical juncture in the unfolding drama of World War II, the goals of an undaunted United States emerging from its long isolationist slumber and beginning its new role as the unchallenged leader of the democracies — not only in the Western world, but the East.

"I have just received your message sent through General MacArthur," Roosevelt began his reply to President Quezon. "From my message to you of January thirty, you must realize that I am not lacking in understanding of or sympathy with the situation of yourself and the Commonwealth Government today. The immediate crisis certainly seems desperate," he allowed, "but such cris[e]s and their treatment must be judged by a more accurate measure than the anxieties and sufferings of the present, however acute. For over forty years," he pointed out — his language carrying shades of Sumner Welles's contributions, reminiscent of the assistant secretary's work on the Atlantic Charter — "the American Government has been carrying out to the people of the Philippines a pledge to help them successfully, however long it might take, in their aspirations to become a self governing and independent people with the individual freedom and economic strength which that lofty aim makes requisite. You yourself have participated in and are familiar with the carefully planned steps by which that pledge of self-government has been carried out and also the steps by which the economic dependence of your islands is to be made effective.

"May I remind you now that in the loftiness of its aim and the fidelity with which it has been executed, this program of the United States towards another people has been unique in the history of the family of nations," the President pointed out. "In the Tydings McDuffy [*sic*] Act of 1934, to which you refer, the Congress of the United States finally fixed the year 1946 as the date in which the Commonwealth of the Philippine Islands established by that Act should finally reach the goal of its hopes for political and economic independence.

"By a malign conspiracy of a few depraved but powerful governments this hope is now being frustrated and delayed," the President went on — but only delayed. Moreover, the Commonwealth of the Philippines was not alone among the United Nations in its suffering. "An organized attack upon individual freedom and governmental independence throughout the entire world, beginning in Europe,[91] has now spread and been carried to the Southwestern Pacific by Japan. The basic principles which have guided the United States in its conduct towards the Philippines have been violated in the rape of Czechoslovakia, Poland, Holland, Belgium, Luxembourg, Denmark, Norway, Albania, Greece, Yugoslavia, Manchukuo, China, Thailand, and finally the Philippines. Could the people of any of these nations honestly look forward to a true restoration of their independent sovereignty under the dominance of Germany, Italy or Japan? You refer in your telegram to the announcement by the Japanese Prime Minister of Japan's willingness to grant to the Philippines her independence. I only have to refer you to the present condition of Korea, Manchukuo, North China, Indo-China, and all other countries which have fallen under the brutal sway of the Japanese government, to point out the hollow duplicity of such an announcement. The present sufferings of the Filipino people, cruel as they may be, are infinitely less than the sufferings and permanent enslavement which will inevitably follow acceptance of Japanese promises. In any event is it longer possible for any reasonable person to rely upon Japanese offer or promise?"

With this the President came to the crux of the matter. "The United States is engaged with all its resources and in company with the governments of 26 other nations in an effort to defeat the aggression of Japan and its Axis partners. This effort will never be abandoned until the complete and thorough overthrow of the entire Axis system and the governments which maintain it. We are engaged now in laying the foundations in the Southwest Pacific of a development in air, naval, and military power which shall become sufficient to meet and overthrow the widely extended and arrogant attempts of the Japanese. . . . By the terms of our pledge to the Philippines implicit in our forty years of conduct towards your people and expressly recognized in the terms of the Tydings McDuffie Act, we have undertaken to protect you to the uttermost of our power until the time of your ultimate independence had arrived. Our soldiers in the Philippines are now engaged in fulfilling that purpose. The honor of the United States is pledged to its fulfillment. We propose that it be carried out regardless of its cost. Those Americans who are fighting now will continue to fight until the bitter end," the Commander in Chief of the

United States made clear. In the meantime the Philippine president could be proud in the knowledge that "Filipino soldiers" were not mercenaries, but "have been rendering voluntary and gallant service in defense of their own homeland."

In sum, the President concluded, the war in the Southwest Pacific was only beginning; Japan had no idea what it had taken on. "So long as the flag of the United States flies on Filipino soil as a pledge of our duty to your people, it will be defended by our own men to the death," the President promised. "Whatever happens to the present American garrison we shall not relax our efforts until the forces which we are now marshaling outside the Philippine Islands return to the Philippines and drive the last remnant of the invaders from your soil. Signed Franklin D. Roosevelt."[92]

At 6:45 p.m. Brigadier General Eisenhower brought the final texts to the President for signature, prior to transmission. It had been a "Long, difficult, and irritating" day. "But now," Eisenhower noted in his diary, "we'll see what happens."[93]

Washington waited — the business of encryption and decryption seeming to take an eternity. At 9:51 a.m. on February 10, however, General MacArthur — who had still not received the President's cable — transmitted *yet another* message from Quezon to President Roosevelt, enclosing the text of the "letter I propose to address to you," Quezon stated, "and to the Emperor of Japan if my recent proposal meets with your approval."[94]

It didn't — indeed the President's response was tarter than the day before.

"1037. From the President to General MacArthur. Transmit the following message from me to President Quezon: 'Your message of February 10 evidently crossed mine to you of February 9. Under our constitutional authority the President of the United States is not empowered to cede or alienate any territory to another nation. Furthermore, the United States ha[s] just bound itself in agreement with 25 other nations to united action in dealing with the Axis powers and has specifically engaged itself not to enter into any negotiations for a separate peace. You have no authority to communicate with the Japanese Government without the express permission of the United States Government. I will make no further comments regarding your last message dated February 10 pending your acknowledgement of mine to you of February 9 through General MacArthur. Franklin D. Roosevelt."[95]

To MacArthur himself General Marshall sent his own, equally tart, message from the White House: "The President desires that no public statement of any kind bearing directly or indirectly on this subject be permitted to go out from any station or source under your control unless and until the President of the United States ha[s] had prior and specific notification and you have received his reply and further instructions." General Marshall added that, as the President had already made clear, Quezon was to be evacuated, if willing, as soon as possible "by submarine" for "a safe and speedy trip" via Australia "to the United States" — in which country he would, "because of the gallant struggle the Filipino soldiers have made under your command," receive "an extraordinary welcome and honors" in leading a Philippine government in exile.[96]

By the evening of February 10, 1942, then, there could no longer be any debate about who was the commander in chief of the United States Armed Forces. MacArthur's self-serving, self-lauding press communiqués were now to be submitted first to the President, if they related in any way to the subject of neutralization, surrender, independence, or relief. President Quezon was to be evacuated from the Philippines as soon as possible, lest he take any steps to parley with the enemy; MacArthur was to concentrate on defending Bataan and Corregidor — not grand strategy.

MacArthur's pride was more deeply dented than at any other time in his life — more so even than in his contretemps with the President in 1934, since his honor as a soldier had now been called into question in recommending parleying with the enemy.

Even MacArthur's secretary and stenographer, a young private first class who had damned "all politicians including roosevelt,"[97] now recorded in his secret, shorthand diary the reality that MacArthur had been struggling to avoid for eight weeks. "All help that can be given at the present time has been given," he summarized, having presented to the commanding general the decoded messages from the President. "If the present American force is destroyed," he noted, "another will be sent" to liberate the Philippines "in the future" — not the present. As for the Filipino leader, "Quezon reminded of America's faithfulness in the past and the deception of the Japanese promises of independence for the Philippines."[98]

So there it was. General MacArthur, in the cordoned-off portion of the Malinta Tunnel where he spent all day with his wife, Jean, and four-year-old son, Arthur, was understandably chastened. By "First Priority" on February 11, 1942, he wired back to assure the President he had "delivered your message to President Quezon," as well as to High Commissioner Sayre, who was also to be evacuated with his family by submarine. MacArthur's own plans, he claimed, "have been outlined in previous radios; they consist in fighting my present battle position in Bataan to destruction and then holding Corregidor in a similar manner."[99]

There was little doubt that the general had spoken with President Quezon, though, for he continued: "I have not the slightest intention in the world of surrendering or capitulating the Filipino elements of my command. Apparently my message gave a false impression or was garbled with reference to Filipinos. My statements regarding collapse applied only to the civilian population including Commonwealth officials the puppet government and the general populace. There has never been the slightest wavering of the troops. I count upon them equally with the Americans to hold steadfast to the end. End. MacArthur."[100]

To this response MacArthur then added a second telegram, assuring General Marshall that "President Quezon's suggested proposal was entirely contingent upon prior approval by President Roosevelt. Replying your 1031 he has no intention whatsoever so far as I know to do anything which does not meet with President Roosevelt's complete acquiescence. I will however take every precaution that nothing of this nature goes out."[101]

Swallowing his wounded self-esteem, MacArthur thus accepted Roosevelt's orders without demur — in fact refused the President's suggestion that his wife and his son should be evacuated by submarine with the Sayre and Quezon families. As he responded, somewhat bathetically, he was "deeply appreciative of the inclusion of my family in this list but they and I have decided that they will share the fate of the garrison."[102]

General MacArthur might be penitent — even resigned to die on Corregidor, as he told two war correspondents[103] — but for his part, President Quezon was not.

Quezon was incensed — at once ashamed of his naiveté and furious with President Roosevelt for pointing it out. Pride, humiliation, ill health, and horrible conditions in the cramped 835-foot-long tunnel at Corregidor, with its twenty-four 160-foot-long "laterals" where the MacArthur, Sayre, and Quezon families lived and slept, made President Quezon see red. Would Americans ultimately triumph over the Japanese, he asked himself? Did the United States really have the determination and the power?

According to James Eyre, a later adviser to Philippines vice president Sergio Osmeña, Quezon "flew into a violent rage" on receiving the copy of the President's signal that MacArthur gave to him. "His anger giving him new-born strength, Quezon got up from his wheelchair and walked back and forth within the narrow confines of the tent in which he was resting near the mouth of the Malinta tunnel. As if addressing an invisible audience he bitterly attacked Roosevelt's direction of policy with reference to the war." Calling for his secretary, he thereupon "resigned as President of the Commonwealth," saying he "would return to Manila. Repeating his verbal attacks upon Roosevelt and American policy, he stated that he wanted no further responsibility for the continued participation of the Filipinos in the war."[104]

Vice President Osmeña urged Quezon to reconsider. Even as a resigned president, Quezon might "bring permanent dishonor to himself and his country" by returning to Manila — the duty of Filipinos surely being to "defend their homeland against the invading armies" of Japan, the vice president argued.[105] Manuel Roxas, the senior liaison officer between the Commonwealth and MacArthur's headquarters, supported Osmeña[106] — as did Lieutenant Colonel Carlos Romulo, a Filipino newspaperman and broadcaster at MacArthur's headquarters.[107]

As Romulo later recalled, Roosevelt's message had, in essence, allowed Quezon the right to surrender his Filipino forces, but had stated that the U.S. Army, for its part, would go on fighting. "Since then I have thought often of this struggle between Roosevelt and Quezon," Romulo wrote — "of Quezon's willingness to yield and Roosevelt's telling him to go ahead, American soldiers would fight on. Would the Filipino soldiers," Romulo wondered, "have stayed with the American fighters or would they have given up?"[108]

According to Osmeña's chronicler, Quezon "remained adamant" all day and night that he would return to the Philippine capital, now under Japanese military occupation. "Because of Roosevelt's insulting message, it is no longer possible for me to serve as President of the Commonwealth," he told his entourage. "I have been deceived and I intend to return to Manila."[109]

A further day passed, and a small boat was actually readied for the President's journey across the bay to Manila.

Early on the morning of February 12, 1942, however, Vice President Osmeña spoke quietly with the Filipino president, pointing out that, if

Quezon returned to Manila, "history might record him as a gross coward and a traitor."[110]

Osmeña himself would have nothing to do with such a move; he would go to Washington as vice president, he made clear, and lead a government in exile on the part of proud Filipinos — not cowards.[111]

MacArthur waited patiently for Quezon to make his decision. So did the White House.

At last it came: MacArthur transmitting on February 12, 1942, the "following message from President Quezon." It was addressed to "The President of the United States: I wish to thank you for your prompt answer to the proposal which I submitted to you with the unanimous approval of my cabinet. We fully appreciate the reasons upon which your decision is based and we are abiding by it."[112]

The U.S. secretary of war and all the senior officers of the War Department, as well as the State Department, breathed a sigh of relief. President Roosevelt had certainly won a great moral battle — despite a pending military defeat. And he had gained the grudging obedience and respect of his theater commander, as well as the support of the elected president of the Philippines.

Quezon, in retrospect at least, was proud of Roosevelt, too. "I first knew President Roosevelt when he was Under-Secretary of the Navy," Quezon afterwards wrote, in exile, before his death in 1944. "From the first time that I had met him, his irresistibly winning smile had attracted me to him. I gave him from the beginning my personal affection. From my official dealings with him, I had come to the conclusion that he was a great statesman — with broad human sympathies and a world-wide knowledge of affairs; a leader of men, with physical and moral courage rarely seen in a human being."[113]

At the point of choice between throwing in his lot with the Japanese or with Washington, Quezon had chosen Washington — because of Roosevelt. "I had become convinced of his extreme regard for the welfare of the Filipino people and his abiding faith in liberty and freedom for the human race," Quezon explained. "When I realized that he was big enough to assume and place the burden of the defense of my country upon the sacrifice and heroism of his own people alone, I swore to God and to the God of my ancestors that as long as I lived I would stand by America regardless of the consequences to my people and to myself. We could not in decency be less generous or less determined than President Roosevelt. Without further discussion with anybody I called my Cabinet and read them my answer to President Roosevelt . . ."[114]

In truth, the process of digestion — or indigestion — had been fraught. There were other things that went missing from the record, too — one of them so egregious it would be covered up for forty years. For at the very moment when the President had shown his mettle as commander in chief, General MacArthur — who had been given a fourth star on December 20, 1941 — did something that would stun and disappoint even his most ardent admirers.

Forty years later, MacArthur's former wartime office secretary at Corregidor could only scratch his head in retrospect as to why MacArthur would do something so stupid — for on February 13, 1942, the day after President Quezon had agreed to turn down the Japanese offer of independence for the Philippines and for the Philippine government to stand by the United States, General MacArthur persuaded Quezon to award him a backdated bonus or bribe of half a million dollars — the sum to be wired into MacArthur's personal bank account in New York.[115] Not only that, but until the general received a cable from Chase National Bank in New York confirming that the money had been credited to his account, Quezon was to give MacArthur half a million dollars in cash (or bonds) as a surety. "God, would I like to be a General!" Private Rogers noted in amazement in his shorthand diary that night — so embarrassed by the transaction and the negotiations for it, in the midst of a significant battle in a world war, that he did not dare set down the true sum — changing it to $50,000.[116]

That MacArthur would take such a huge sum ($4.7 million in today's currency) for supposed "past services" to the president of the Philippines was not only illegal for a serving U.S. officer, but a tremendous risk for General MacArthur in terms of his stature as an officer and a gentleman. Such an urgent, cabled request, from a commanding general in a combat zone in the Far East, would obviously be seen by others — indeed, it would require the authorization of senior U.S. Army and cabinet officers, as well as directors of Chase National Bank. It could not be (and was not) hidden from the President (who kept a copy of MacArthur's secret wire in his files), the secretary of war, the secretary of the interior, the chief of staff of the U.S. Army, the adjutant general, or Brigadier General Eisenhower — who became head of the Far East section of the War Department on February 16, 1942, just as the strange money-transfer cable request was going through.[117]

The President could be forgiven for wondering why MacArthur and his closest staff (who also were offered and accepted lesser sums) should take such fortunes from the president of an American territory, at a critical juncture of the war, when tens of thousands of men were fighting for their lives—and would necessarily be abandoned—on Bataan. Was MacArthur, as Admiral Hart claimed, completely "mad"?

Under Article 94 of the Articles of War of 1920, "Frauds Against the Government" as well as "conduct unbecoming of an officer and a gentleman" could be punished "by fine or imprisonment, or by such other punishment as a court-martial may adjudge."

Rogers was ashamed of the risk MacArthur was taking, but took some comfort in the fact that the bribe, though kept secret, was known to the highest authorities in America. "If Roosevelt had not approved the transfer," as Rogers later wrote, "the entire affair would have been annulled."[118]

In the event, however, the question of corruption or insanity was set aside by the President, and by Secretary Stimson—both of whom had law degrees. For on February 15, 1942, the war in the Pacific turned a new and darker page. Singapore, the "Gibraltar of the East," was surrendered by the British without more than a token fight. More than forty thousand of their Indian troops went over to the enemy, offering to fight *with* the Japanese, against the British.

Like a house of cards, Britain's empire in the Far East collapsed, overnight. And General Douglas MacArthur, the man who had helped President Quezon and the Filipinos to continue fighting with the democracies, instead of becoming a felon became the U.S. Commander in Chief's "indispensable man" in the Pacific.

PART FIVE

End of an Empire

8

Singapore

BRING DOUGLAS MACARTHUR HOME!" shouted Wendell Willkie, Roosevelt's Republican rival for the presidency in the 1940 election, to a big audience in Boston, Massachusetts, on February 12, 1942. "Place him at the very top. Keep bureaucratic and political hands off him. Give him the responsibility and the power of coordinating all the armed forces of the nation to their most effective use. Put him in supreme command of our armed forces under the President."[1]

"Ordinarily it might be hard, it might be impossible to find such a man," Willkie allowed. "But as the last two months have proved, we have the man," he declared "—the one man in all our forces who has learned from first hand, contemporary experience the value and the proper use of Army, Navy and air forces fighting together towards one end; the man who on Bataan Peninsula has accomplished what was regarded as impossible by his brilliant tactical sense; the man who almost alone has given his fellow countrymen confidence and hope in the conduct of this war—General Douglas MacArthur."[2]

Henry Stimson was appalled—having found the same suggestion circulating in his War Department. Secretary Stimson had, in fact, "just come from a series of discussions" with a group of men charged with the "reorganization of the [War] Department," in which the same question of a commander in chief of all, or at least all U.S. Army forces, had been raised. Stimson had been at pains to argue "just the opposite position—that in the United States there should be nobody between the President and the commanders of different task forces, and that the Chief of Staff should be a Chief of Staff," he noted in his diary. In other words, General Marshall's job was to run the War Department under the commander in chief, not take his place *as* commander in chief.[3]

When, the next day, Stimson "picked up the newspapers and saw that Willkie had come out for making MacArthur General-in-Chief of all forces of the United States under the President," he was appalled by such "Republican talk."[4]

The President, for his part, had heard much the same. In the aftermath of his contretemps with General MacArthur and Quezon, however, he was confident he could bat away such nonsense. "No," Roosevelt responded to the White House correspondents gathered at the 806th press conference of his presidency, he would not "comment on the agitation to have MacArthur ordered out of the Philippines and given over-all command." He had little time for such silly talk. "I think that is just one of 'them' things that people talk about without very much knowledge of the situation."[5]

And there the matter of Roosevelt's commander-in-chief-ship of the Armed Forces of the United States was laid to rest for the rest of the war.

Winston Churchill, in London, was also under fire as quasi commander-in-chief of British imperial armed forces. In his case this stemmed from the British Eighth Army's lamentable showing against General Rommel's forces in Libya, and even worse performance in the Far East. He was also being pilloried as a poor protector of Britain's preferential prewar trading rights, in view of President Roosevelt's latest Declaration of the United Nations, on January 1, 1942.

A month later, on February 2, 1942, a fierce new argument had erupted over the looming Master Lend-Lease Agreement. Article 7 stipulated there was to be an eventual termination, after the war's end, to trade tariffs, known as "imperial preference," which unfairly benefited Britain.[6] By February 7, 1942, Churchill had felt compelled to beg the President not to insist upon the article's inclusion in Lend-Lease, lest the United States be accused, he said, of "breaking up the British Empire and reducing us to the level of [a] territory of the Union."[7]

Economic concerns, with the prospect of postwar bankruptcy, were only the tip of the looming iceberg for Great Britain. The once-majestic British Empire, with its king-emperor as lord of a quarter of the world's population and landmass, was approaching a crisis of existence — something that now became vividly, tragically, symbolically, and militarily clear.

With a population of only 50 million in 1939, the notion of the British

Isles continuing to command the fortunes of such a vast Victorian imperial construct was inherently unlikely. Even as far back as 1883, the British historian Sir John Seeley had forecast a day when the game would be up. "Russia and the United States will surpass in power" Great Britain, he had written, just as the emerging nation-states in the sixteenth century had "surpassed Florence."[8] Compared with 140 million Americans and 180 million Russians, what chance did the little United Kingdom have in administering and defending an ill-assorted global empire of 490 million people, from the Faroe Isles to Hong Kong?

"Imperial tariffs" could not, Churchill realized, keep a failing economic network of colonies and Dominions alive unless, as in the nineteenth century, there was at its core a crusading *moral* zeal: exploratory, exploitive, even extortionate, but benevolent, too. As had become obvious even to Churchill, as a stalwart champion of colonial empire during the rise of the dictators in the 1930s, the moral zeal had gone — a reality epitomized in the appeasement policies of Neville Chamberlain and of the "Clivedon set" of mainly aristocratic Britons.

For his part, Churchill would embrace no accord with Hitler merely to preserve the fruits of the British Empire. By his own individual energy, imagination, and military leadership, he was certain he could reverse Britain's moral decline — but he needed American help. By declaring war on the United States, the Axis powers had, to his abiding relief, brought the United States into the war not only on Britain's side, but *at* Britain's side. The British Empire would thus be saved by Professor Seeley's emerging empires: the United States on the one hand, and the Soviet Union on the other. The British Empire would then act as a sort of middleman: rich still in military bases, in mineral resources, and in colonial and Dominion management-manpower.

The heart of this network of military bases in the Far East was Britain's supposedly impregnable naval and military base at Singapore. Holding Singapore was, to Churchill, crucial: a symbol of British willingness to stand up for its imperial assets, in a war that would determine the shape of the rest of the twentieth century and beyond. "The honour of the British Empire and of the British Army is at stake," he had cabled the new Allied supreme commander, General Sir Archibald Wavell, on February 10, 1942[9] — and he meant it.

To preserve that honor, Churchill had, in his own role as generalissimo, employed U.S. transport ships to convoy the British Eighteenth Division to Malaya via the Cape of Good Hope, as he'd informed the President with gratitude on December 12, 1941;[10] then, in his *tour d'horizon* of grand strategy on his way to Washington on December 17, the Prime Minister had assured the President that the island of Singapore and its "fortress will stand an attack for at least six months," since a "large Japanese army with its siege train and ample supplies of ammunition and engineering stores" would be required to take it.[11]

As the relatively small Japanese invasion army, comprising 33,000 troops, made its way down the Malayan peninsula in January 1942, however, Churchill's assurances of stout resistance by the 130,000 British, Australian, and Indian troops had become less convincing. On February 7, 1942, the Prime Minister modified his predictions. "Seventy per cent of our forces which fought in Malaya got back to the [Singapore] Island," he admitted, glossing over the helter-skelter British retreat toward Johore, in a new cable to the President. "Eleven convoys of stores and reinforcements including the whole 18th Division and other strong good A.A. and A/T [antiaircraft and antitank] units are now deployed making the equivalent of four divisions, a force very well proportioned to the area they have to defend. I look forward to severe battles on this front, where the Japanese have to cross a broad moat before attacking a strong fortified and still mobile force."[12]

Churchill prided himself, as a former cavalry officer, on his command of military detail. Decades had passed, however, since he had served in the line in World War I—and even more decades since he had served in Asia. In his underground London bunker, not having traveled to India or the Far East since 1889, he seemed completely unaware of what was going on in the farther reaches of the empire he so doggedly represented. What had begun as a phased withdrawal by British imperial troops from the frontier with Indochina became, in reality, a helpless rout. Now, as Churchill looked forward to a glorious, medieval-style defense at Singapore, even those around the Prime Minister failed to share his optimism.

"News bad on all sides," Churchill's new chief of the general staff in London, General Alan Brooke, had already noted in his diary on January 30, 1942—for in Libya the German panzer commander, General Rommel, had replenished his forces and had struck back at General Auchinleck's British Eighth Army with venomous daring. "Benghazi has been lost again and Singapore is in a bad way," Brooke lamented. Churchill had told the President Singapore would hold out for six months; Brooke was dubious. "I doubt whether the island holds out very long," he scribbled in his slashing green handwriting.[13]

Brooke was right. Three days later he was mauled by members of the cabinet "in connection with defeat of our forces!" he recorded that night. "As we had retreated into Singapore Island and lost [control?], besides being pushed back in Libya, I had a good deal to account for,"[14] he noted — unaided by the head of the British Navy, Admiral Sir Dudley Pound, who — suffering from an undiagnosed brain tumor — was constantly falling asleep, in fact "looked like an old parrot on his perch!" the amateur ornithologist described.[15]

The British Chiefs of Staff Committee, chaired by Admiral Pound, certainly evinced little confidence among British cabinet members. Aware of this, but refusing outwardly to show signs of anxiety, the Prime Minister — having survived a vote of no confidence in the House of Commons by 464 votes to 1 — was becoming hysterical behind the scenes. "He came out continually with remarks such as: 'Have you not got a single general in that army who can win battles, have none of them any ideas, must we continually lose battles in this way?'" Brooke later recounted.[16] Singapore was to be "defended to the death," and the "Commander, Staffs and principal officers are expected to perish at their posts," Churchill had laid down on January 19, 1942: complete with a ten-point plan of defense.[17] It was crucial, he felt, to show the Japanese — and the world — that British imperial troops believed in the British Empire, and would die for it.

To his chagrin, however, they refused to do so.

For his part, Brooke blamed the United States. By agreeing to the Combined Chiefs of Staff system based in the U.S. capital, he felt that the British chiefs of staff — harried, bullied, disparaged, and constantly overruled by their own prime minister — had lost imperial prestige and authority over their own forces.

"Ever since [Air Marshal] Portal and [Admiral] Pound came back from the USA I have told them that they have 'sold the birthright for a plate of porridge' while in Washington," Brooke lamented on February 9, 1942. "They have, up to now, denied it flatly. However this morning they were at last beginning to realize that the Americans are rapidly snatching more and more power with the ultimate intention of running the war in Washington! I now have them on my side," he congratulated himself.

Brooke's satisfaction was short-lived. Several hours later he was summoned to face the wrath of his political masters. "An unpleasant Cabinet meeting," he jotted in his diary—for the news from the Far East was terrible. Churchill's "moat"—the Johore Strait, separating Malaya from Singapore—had been breached by the Japanese in a few hours.[18]

"The news had just arrived that the Japs had got onto Singapore Island," Brooke recorded that night. "As a result nothing but abuse for the army. The Auk's retreat in Cyrenaica is also making matters more sour! Finally this evening, at 10.45, I was sent for by PM to assist him in drafting a telegram to Wavell about the defense of Singapore, and the need," once again, "for Staffs and Commanders to perish at their posts."[19]

Churchill's assurances to President Roosevelt of British stoutheartedness had so far proven empty, despite Brooke's derision of the notion of a Combined Chiefs of Staff operating in Washington rather than London. "The news from Singapore goes from bad to worse," Brooke confided to his diary on February 11. "PM sent for me this evening to discuss with him last wire from Wavell about Singapore from where he [Wavell] had just returned" with somber tidings. The British commander in chief, Lieutenant General Percival, had failed to erect defenses along the "moat"—even barbed wire—lest he be accused of being defeatist. This had allowed the Japanese to cross virtually at will. Some of Wavell's subordinate commanders were already talking of surrender.

"It was a very gloomy wire and a depressing wire as regards the fighting efficiency of the troops on Singapore Island. It is hard to see why a better defence is not being put up, but I presume there must be some good reason," Brooke recorded, hopefully. "The losses on the island will be vast, not only in men but in material. I have during the last 10 years had an unpleasant feeling that the British Empire was decaying and that we were on a slippery decline! I wonder if I was right? I certainly never expected we should fall to pieces as fast as we are, and to see Hong Kong and Singapore go in less than three months plus failure in the Western Desert is far from reassuring!"[20]

Next day, as if the gods were out to further humiliate the Prime Minister and his military chiefs, there was a new disgrace. For early on February 12, 1942, the German Kriegsmarine astonished the world. The 32,100-ton battleships *Gneisenau* and *Scharnhorst*—which had sunk the aircraft carrier HMS *Glorious*—accompanied by the 18,750-ton heavy cruiser *Prinz Eugen*—which had helped sink HMS *Hood*—succeeded "in running the gauntlet of the Channel yesterday without being destroyed," Brooke noted in his diary:[21] the three huge warships racing at almost thirty knots through the entirety of the English Channel from Brest to their naval base in Kiel without serious damage to any of them, despite hapless, even suicidal efforts by Royal Navy destroyers and torpedo boats, RAF bombers, and Swordfish torpedo planes (all of which were shot down), and even radar-directed coastal artillery, capable of covering the width of the Channel at Dover, unable even to "knock any paint" off them, as Sir Alexander Cadogan, the permanent secretary to the British foreign secretary, Anthony Eden, phrased the misfortune.[22] Even the solitary British submarine detailed to stand guard outside the harbor at Brest, it appeared, had left station to recharge its batteries, with no replacement! "We are nothing but failure and inefficiency everywhere," Cadogan noted in his diary on February 12, "and the Japs are murdering our men and raping our women in Hong Kong. . . . I am running out of whiskey and can get no more drink of any kind. But if things go on as they're going, that won't matter."[23]

"These are black days!" General Brooke noted.[24]

The blackest of all, however, came on February 15, 1942.

Hoping to circumvent the eyes of the Combined Chiefs of Staff in Washington, Churchill had sent more and more hectoring, do-or-die cablegrams direct to General Wavell. Thanks to the new Allied command system, however, the supreme commander was, for his own part, duty bound to report to the Combined Chiefs of Staff in Washington. This made it impossible for the Prime Minister to disguise from the President or the U.S. Joint Chiefs of Staff the impending British disaster at Singapore.

Wavell's reports to Washington had been all too candid. One of the supreme commander's staff officers later recorded how, on the way to the front, they had "passed groups of Australian troops streaming towards the harbor, shouting that the fighting was over and that they were clearing out."[25]

Clearing out? Churchill's orders to officers and men to fight and die where they stood seemed to fall on deaf ears. On February 14, 1942, Japanese forces, despite being heavily outnumbered and running out of ammunition, advanced across Singapore Island toward the city and harbor. The 130,000 defenders for the most part refused to fight; desertions were pandemic among Australian troops — even their commanding officer commandeering a boat and sailing away, leaving his men.

Among Indian troops, moreover, there was something more ominous still: a mass refusal to risk their lives for a British Empire that denied them self-government, independence, or even Dominion status, such as the Australians enjoyed.

Hushed up for decades and for the most part ignored by historians of the Malaya campaign and its ultimate debacle, the majority of Indian troops — some forty thousand out of forty-five thousand — captured by the Japanese at Singapore thus volunteered to join the Indian National Army (INA) and fight the British.[26] And still more would do so over the following year.

Churchill, hearing at midday on February 15, 1942, that General Percival had surrendered an army of over a hundred thousand armed men to the Japanese, without fighting, went into shock.

To the White House the Prime Minister had, on February 11, cabled: "We have one hundred six thousand men in Singapore Island, of which nearly sixty thousand are British or Australian," while forty thousand were Indian. "The battle must be fought to the bitter end," he'd assured the President, adding, "Regardless of consequences to the city or its inhabitants."[27]

Instead, four days later, they had surrendered to no more than a few thousand Japanese infantry troops. It was not only humiliating. It was a disgrace.

That night, the Prime Minister gave the saddest broadcast of his life, delivered from his official country residence, Chequers. In it the great orator of empire reached for every metaphor that might, by its brilliance, distract from his dismal sense of shame. Borrowing from the President's State of the Union address, he titled his broadcast "On the State of the War": beginning by spinning out the story of the entire conflict since 1939. How did matters now stand on February 15, 1942, the Prime Minister therefore asked his listeners? "Taking it all in all, are our chances of survival better or worse than in August, 1941? How is it with the British Empire, or Commonwealth of Nations — are we up or down?"[28]

"Commonwealth of Nations" sounded, to the President, as he listened to the speech in Washington, a great deal better than the "British Empire." The President was therefore pleased to hear the Prime Minister's alternative title for Britain's imperial domains — especially when Churchill pointed to Britain's alliance with the United States and repeated it, without mentioning "British Empire" at all, a minute later. "When I survey and compute the power of the United States, and its vast resources, and feel that they are now in it with us, the British Commonwealth of Nations, all together, however long it lasts, till death or victory, I cannot believe there is any other fact in the whole world which can compare with that," Churchill declared. "That is what I have dreamed of, aimed at, and worked for, and now it has come to pass."[29]

At this remark, sitting by the radiogram with Harry Hopkins in his study at the White House, the President shuddered—knowing immediately how it would be parsed and interpreted by former isolationists, anti-British citizens, and reluctant interventionists in America. Why, Hopkins wondered, did the Prime Minister have to mention his almost two-year campaign to entreat, and if necessary inveigle, the United States into entry into the war?

Worse still was to come, however. After a long explanation of how Britain's resources had been stretched to breaking point in holding its "Commonwealth" lines of communication open, the Prime Minister began his announcement of the fall of Singapore by blaming America. By an "act of sudden violent surprise, long calculated, balanced and prepared, and delivered under the crafty cloak of negotiation," the Japanese had "smashed the shield of sea power which protected the fair lands and islands of the Pacific Ocean," in fact "dashed" it "to the ground"—at Pearl Harbor. "Into the gap thus opened rushed the invading armies of Japan. We were exposed to the assault of a warrior race of nearly eighty millions with a large outfit of modern weapons, whose war-lords had been planning and scheming for this day, and looking forward to it perhaps for twenty years—while all the time our good people on both sides of the Atlantic were prating about perpetual peace and cutting down each other's navies in order to set a good example. The overthrow, for a while, of British and United States sea power in the Pacific was like the breaking of some mighty dam," the Prime Minister narrated in his bewitching, metaphorical style. "The long-gathered pent-up waters rushed down the peaceful valley," he continued, "carrying ruin and devastation on their foam and spreading their inundations far and wide."[30]

Only at the very end of his extended broadcast did the Prime Minister finally get to the point. Britain's great outpost in the Far East, its equivalent of Hawaii and Pearl Harbor, he at last confessed to the millions listening, had surrendered.

Churchill's voice dropped a whole register, as he continued his broadcast on a new note: one of sadness and misery. "Tonight I speak to you at home. I speak to you in Australia and New Zealand, for whose safety we will strain every nerve, to our loyal friends in India and Burma, to our gallant allies, the Dutch and the Chinese, and to our kith and kin in the United States. I speak to you all under the shadow of a heavy and far-reaching military defeat. It is a British and Imperial defeat," the Prime Minister acknowledged. "Singapore has fallen. All the Malay Peninsula has been overrun."

It was, Churchill nevertheless urged his British imperial listeners, "one of those moments" when the "British race and nation" could "show their quality and their genius": a moment when "they can draw from the heart of misfortune the vital impulses of victory."[31]

The President had good reason to take a deep breath. Blaming the fall of Singapore and the British Empire in the Far East on America's defeat at Pearl Harbor was decidedly impolitic in terms of American public opinion. As the President remarked to Hopkins, however, "Winston had to say *something.*"[32]

The President's patient magnanimity was an endearing trait—certainly one that contrasted with the characters of the dictators wreaking mayhem and genocidal violence on the innocent. Churchill had called them, in his broadcast, "barbarous antagonists"[33]—and they were. At Singapore the Japanese promptly began murdering tens of thousands of Chinese civilians;[34] and in the Philippines MacArthur had already passed back to Washington reports of Japanese atrocities and mistreatment of prisoners in Manila so disturbing that he recommended the President take a number of Japanese immigrants in America hostage, as a surety against further barbarity[35]—a suggestion that in part persuaded Roosevelt to authorize the removal and internment of over one hundred thousand members of Japanese immigrant families from the California area. It would be one of the most controversial decisions the President ever made—licensing paranoia and xenophobia over the very virtues the President claimed as the moral basis of the democracies.[36]

Wars, in any event, could not be won by hostage taking. In ordering General MacArthur to stand and fight rather than negotiate with the Japanese, the President had rightly been concerned to put down a marker of American intent, both military and moral. The failure of the British imperial troops to fight at Singapore, however, would now compel the President to question his whole concept of the global struggle.

Britain's imperial forces were proving everywhere a broken reed, as even the British Army's CIGS—chief of the Imperial General Staff—recognized. Rommel was once again trouncing British imperial troops in Libya. And in the Far East it was now questionable whether British Empire forces would even fight for Burma: the vital causeway the President was counting on for American supplies to Chiang Kai-shek's army, currently fighting the Japanese in China.

Sitting with the President over dinner at the White House, after listening to Churchill's grim broadcast, Harry Hopkins took notes on how the President proposed to fight the war.

It was, in its way, a turning point, though largely ignored or papered over after the war. The British had failed to perform as competent coalition partners in the global conflict. Having embraced the President's insistence on a "Germany First" policy they had proceeded to lose their main pivot in the Far East, Singapore, without a fight—while blaming the United States for its failure to protect them, after Pearl Harbor. The British could not be depended upon, at least for now. This was the sad conclusion.

The war, as Hopkins wrote down the President's new "list of priorities" that night, would have to become an American undertaking, for the British Empire was proving a figment of Churchill's imagination. As an empire it was over. The fact was, the British Dominions, the still-free nations of the world, and even the occupied and threatened nations, now looked to the United States, not to Great Britain, to liberate and protect them. "The United States to take primary responsibility for reinforcing the Netherlands East Indies, Australia and New Zealand," the President's list began.[37]

It was in this respect that the example of the Philippines, at that precise moment in the war, was in the President's view a key factor. Not because the Philippines could be saved from Japanese occupation, but because the Filipinos were continuing to fight alongside American troops, rather than joining the Japanese—as Indian troops were doing in huge numbers in Malaya and Singapore. More than any other instance, the example of the Philippines would be the core of President Roosevelt's confidence in assuming the mantle of leadership of the United Nations: the Philippine troops fighting alongside U.S. soldiers and thereby presenting a symbolic demonstration of the trust that free governments had in the United States' direction of the war.

There would be consequences, the President knew—with difficult pills for the British to swallow in the days ahead. At a critical juncture of the war, with the Japanese rampage in the Pacific seeming to be irreversible, it was vital, Roosevelt felt, that the Allies should affirm their faith in the Atlantic Charter not as a "Magna Carta" for only the white nations of downtrodden Europe, as Churchill was interpreting it,[38] but as a document for humanity across the globe.

Under the terms of the Atlantic Charter, as attached to the Declaration of the United Nations, the British government, the President felt, *must* be persuaded to accept decolonization, both as an aim of the war and the postwar: an approach that would, at a stroke, delegitimize the propaganda of Japanese forces in their brutal campaign of conquest across Southeast Asia and the Pacific. Whether it was already too late remained to be seen, but Roosevelt would work on Churchill to that effect—much as Churchill had worked on him to wage war on Britain's side.

The ignominious fall of Singapore on February 15, 1942, thus became a pivotal moment in World War II, as the British Empire fell apart and the United States was forced to take over.

9

The Mockery of the World

SINGAPORE HAS WEAKENED his position enormously," the Führer remarked of Churchill — in fact, talking with his propaganda minister, Joseph Goebbels, Hitler thought Churchill's days as prime minister might now be numbered: the sorry chain of British failure in the field of battle inevitably leading "one day," he postulated, "to catastrophe."[1]

Such thoughts led, sickeningly, to ever-greater hubris on the part of the Führer. Satisfied that the British would never be able to interfere with his plans for the extermination of German and European Jewry, an emboldened Hitler was now encouraged by Britain's pusillanimous showing on the field of battle to be more murderous than ever toward the millions of "Hebrews" his henchmen were rounding up and incarcerating across the Third Reich and its occupied territories.

"Once again the Führer expressed his determination to cleanse Europe of its Jews, ruthlessly," Goebbels noted in his diary on the day of Churchill's broadcast announcing the fall of Singapore. "No room for squeamish sentimentality. The Jews have earned the catastrophe they are suffering today. They will be destroyed, at the same time as we destroy our enemies. We need to accelerate the process with utter ruthlessness, and will be doing humanity, for thousands of years tormented by the Jews, a priceless service. We have to make sure we spread this virulent anti-Semitic attitude throughout our own people, whatever the objections. The Führer puts great stress on this, repeating it afterwards to a group of officers, to get it into their thick heads. He is fully aware of the great opportunities the war is providing us. The Führer recognizes he is waging war on a vast canvass and that the fate of humanity depends on its outcome . . ."[2]

The two men then gleefully discussed how the pitiful performance of

British Empire forces could be used to insert a wedge between the trans-atlantic partners: Roosevelt and Churchill.

The "widespread political gloom and plummeting morale" that Dr. Goebbels perceived in reports from London were, in the early months of 1942, all too real — a British Empire without clothes, dangerously increasing the chance that Hitler would get away with mass murder in Europe and Russia on a scale so immense it would eventually be called "the Holocaust."

In Berlin — many hundreds of miles from the fighting — Goebbels began to describe the war against the Allies, in his own metaphor, as a boxing match. The Axis had, unfortunately, not quite "KO'd" its adversaries in the first round, he acknowledged eight months after the launch of Operation Barbarossa; it would require many more punches. The winter, in particular, had been tough going on the Eastern Front, but for the rest he was once again confident the Third Reich would prevail in battle. "As long as we go on fighting as we are, the day will come when England and America are flattened," he noted. "Once again," he prided himself after seeing the latest newsreels on the German Navy's daring seamanship in the English Channel, "we're on our feet, and the German people welcome the successes of the Axis powers with great pride and satisfaction."[3]

Alerted by his U.S. ambassador in London that the Prime Minister was lapsing into depression over recent events,[4] Roosevelt sympathized with his new partner — even though, in the President's eyes, Churchill seemed to have invited the latest battlefield disasters by his inability to pick effective subordinates, and his insistence on meddling in operational matters. Added to this, however, was Churchill's refusal to credit why British soldiers were refusing to *fight*.

The Prime Minister's cables to General Percival in Singapore had been meticulous in their detailed military instructions for defense against siege — instructions entirely worthy of Churchill's childhood fascination with soldiering, his love of military history, his training at Sandhurst military college, and the many savage wars in which he had personally fought, as subaltern on India's North-West Frontier and lieutenant colonel in the trenches in World War I.[5] But although the Prime Minister had every idea of what he was talking *about,* he seemed to have no idea *to whom* he was speaking. The nation he had dreamed his entire life of leading was no longer the Victorian society in which he had grown up as the grandson of a duke and son of a lord — indeed, the people of Britain were now more isolationist than those in the United States. Poorly led by "toffs" and

"Blimps," British soldiers were voting with their feet—for the most part simply no longer willing to lay down their lives in foreign fields on behalf of a colonialist empire in which they no longer believed.

Churchill's myopia in this regard never ceased to amaze President Roosevelt. Reeling from the vituperative criticism in the British press, the Prime Minister neither blamed himself as British quasi commander-in-chief nor even attributed the surrender of Singapore to the fortunes of war, but instead derided the troops—expressing to a friend "a dreadful fear that our soldiers were not as good fighters as their fathers were. 'In 1915 our men fought on when they had only one shell left and were under fierce barrage,'" Churchill confided to Violet Bonham-Carter. "'Now they cannot resist dive-bombers. We have so many men in Singapore, so many men—they should have done better.'"[6]

Churchill was far from alone in such reflections.

The Prime Minister's blindness reflected an older English generation, born in Queen Victoria's reign, unwilling to surrender the privileges of their class. "It is the same in Libya. Our men cannot stand up to punishment. And yet they are the same men as man the merchant ships and who won the Battle of Britain," Harold Nicolson—a member of Parliament, governor on the board of the BBC, and married to the daughter of the third Baron Sackville-West, living in Sissinghurst Castle—noted with equal puzzlement in his diary. "There is something deeply wrong with the whole morale of our Army."[7]

Sir Alec Cadogan, who had helped Sumner Welles draft the Atlantic Charter, felt the same concern—though he at least could see that without commanders able to inspire their troops, Britain was a spent power. "Our generals are no use," Cadogan wrote in his diary, as news came in that the Japanese were within striking distance of Singapore. "[D]o our men fight? We always seem to have 'Indian Brigades' or Colonials in the front line. . . . Our army is the mockery of the world."[8] After the battle was lost, the Prime Minister's Singapore speech certainly did nothing to reassure his senior civil servant in the Foreign Office. "His broadcast not very good—rather apologetic," Cadogan noted, "and I think Parliament will take it as an attempt to appeal over their heads to the country—to avoid parliamentary criticism."[9]

Woven through the anger and distress in the debate in the House of Commons on February 24 was a common, underlying thread: "the implication that our Army has not fought well," Nicolson recorded frankly—

and in shame. "How comes it that we were turned out of Malaya by only two Japanese divisions? How comes it that our casualties were so few and our surrenders so great? This is the most disturbing of all thoughts," he confessed in the quiet of the night. He added that he had wakened from a dream in which a hand on his shoulder had not been that of his wife, as he had thought. It was the hand, he recorded, of "Defeat." "This Singapore surrender has been a terrific blow to all of us. It is not merely the immediate dangers which threaten in the Indian Ocean and the menace to our communications with the Middle East. It is the dread that we are only half-hearted in fighting the whole-hearted. It is even more than that. We intellectuals must feel that in all these years we have derided the principles of force upon which our Empire is built. We undermined confidence in our own formula. The intellectuals of 1780 did the same."[10]

Nicolson, like Churchill, was only half-right. For sure, grand ideals of universal disarmament at the conclusion of the "war to end all wars" in 1918, as well as economic hardship in the Great Depression a decade later, had set the Western powers at a tragic disadvantage in relation to rising military dictatorships. Yet imagining that "principles of force," rather than a postcolonial *moral* vision, could have kept an ailing colonial empire alive did little credit to Nicolson as a historian of the Versailles Treaty. It was this tragic nearsightedness — reflecting an empire that had lost its way, and could find no alternative vision of the future — that threatened to doom Britain to defeat in its struggle against the Axis powers.

Ironically, it was Dr. Joseph Goebbels, analyzing Churchill's broadcast acknowledging the fall of Singapore, who recognized how much the Prime Minister was throwing himself at the mercy of the United States — utterly dependent as Churchill now was on the might of America, as the British will to fight abroad appeared effectively broken. "It had been his great and remorseless effort to get the U.S. into the war, and now he had achieved it,"[11] Goebbels paraphrased the Prime Minister's broadcast. All Churchill could offer Britons were more tears, the propaganda minister mocked — determined, for his own part in Berlin, to let Britain's continuing military disasters "water the plant" of conflict between the two transatlantic allies rather than devote extra Nazi propaganda to the mission.[12]

Buoyed like his führer by Japan's astonishing successes in the Far East, Rommel's rebound in North Africa, and preparations in Russia for a renewed German offensive as soon as the snow melted, Goebbels was once more convinced that German force of arms and military professionalism would prevail. Wherever an Englishman looked, he was faced by "retreat, humiliation, capitulation and white flags," the propaganda minister jeered. "To what depths has Churchill led his great English empire!"[13]

By February 19, 1942, Goebbels was noting in his diary that the fall of Singapore and the successful passage of the German battleships through the English Channel had given rise to a sense of victory among the public in Germany that "actually worries me more than the reverse."[14]

Anxious about reports of Churchill's sinking mood, the President telephoned from the Oval Office to speak with his U.S. ambassador in London, John Winant. He then cabled direct to Churchill at his War Rooms.

"I realize how the fall of Singapore has affected you and the British people," Roosevelt wrote, but it was, he urged, important that the British not lose heart. "It gives the well-known back seat drivers a field day but no matter how serious our setbacks have been, and I do not for a moment underrate them, we must constantly look forward to the next moves that need to be made to hit the enemy."[15]

From there the President went on to tell the despairing British leader — who admitted he was finding it difficult to "keep my eye on the ball"[16] — what the new strategy of the United Nations in the Far East should best be.

It was time, the President made clear, to face facts. America and the United Nations would have to begin again, pivoting not on their farthest outposts in the Orient, but on their backstops, in North and South Asia — i.e., at the extremities of Japan's rampage. In this way, the United States could stretch and disperse Japan's naval forces, while securing bases that could be supplied, expanded, and used as launching pads for later Allied counteroffensive action, as per his original Victory Program, once American war production ramped up.

The prospect was unappetizing in the short term, he allowed. General Wavell's supreme command in the South Asia region would have to be dissolved. Britain's great Dominions in the southern Pacific — Australia and New Zealand — would have to come under American, not British, military protection and control. The Pacific, in fact, would have to become an American theater of war, from Chiang Kai-shek's U.S.-funded and -supplied Chinese forces operating in the north (with an American general, Joseph Stilwell, acting as Chiang's chief of staff) down to the vast Australian continent in the South Pacific — where an American commander in chief would be installed.

In terms of the latter, the President already had in mind the man he wanted — however controversial the half-million dollars the general had just extracted from the president of the Philippines. For what would *win* the war for the United Nations, Roosevelt recognized, was not only the moral purpose and industrial might of America, but the determination of Americans to use that might offensively, under aggressive leaders. And General Douglas MacArthur, for all his faults and fantasies, was nothing if not aggressive.

MacArthur's military blunders in the Philippines had been appalling — even Dr. Goebbels noting that MacArthur's "successes" only existed "on paper"[17] — but in the United States they had been hushed up. In the battle for high morale — especially when compared to General Percival's performance in Singapore — General MacArthur had proven himself a potentially great wartime leader, whose Filipino troops were fighting as hard as his Americans.

Already on February 4, 1942, General Marshall had, on behalf of the President, warned MacArthur he would be needed for a higher command. "The most important question concerns your possible movements," Marshall had warned the general — once Bataan could no longer be held, and dissolved into guerilla warfare conducted behind the Japanese lines. "Under these conditions the need for your services [there] will be less pressing than [at] other points in the Far East."[18] Once spirited out of the Philippine Islands, MacArthur would, the President was certain, quickly put an end to the veritable panic sweeping through Australia — where the port and town of Darwin, on Australia's north coast, now suffered its own Pearl Harbor: an unprotected American convoy devastated in the harbor, as was the town itself, and its airfields heavily bombed, on February 19, 1942, despite ample advance warning.[19]

The air attack on Darwin would at least wake up the sleepyheads in Australia, the President consoled himself. No landings had taken place, at least. In taking full responsibility for the protection of Australia and New Zealand, however, the United States would now expect a quid pro quo. Roosevelt would insist that the British use the forces released by this new southern Pacific strategy to defend Burma, in the north — which offered the only overland supply route for U.S. munitions to Chiang Kai-shek.

Would the British prove any more robust in fighting for Burma than they had in Malaya, though? Roosevelt had cause to wonder. And beyond Burma, could Britain's forces stop the Japanese from invading the Indian

Subcontinent? If not, Japan would control too much of the Far East to be evicted.

In the President's eyes it was thus a thousand pities that Churchill had so set his heart and mind — year after year, month after month, week after week — against Indian self-government: self-government that could ensure the willing cooperation of India's own political leaders and population — as it had in the Philippines with respect to Quezon and his forces.

Somehow, the President reasoned, Churchill must be bucked up — and urged to embrace the future, not the past.

From London Averell Harriman, Roosevelt's Lend-Lease administrator and special envoy, reported confidentially that the "surrender of their troops at Singapore has shattered confidence to the core" in the English capital — so much so, in fact, that the British were losing confidence "even in themselves but, more particularly, in their leaders."

Churchill, according to Harriman, looked as if he had lost the thread. "Unfortunately Singapore shook the Prime Minister to such an extent that he has not been able to stand up to this adversity with his old vigor" — leading "astute people, both friends and opponents" to say it was "a question of only a few months before his Government falls."[20]

For President Roosevelt, attempting to lead the United Nations at a critical moment in the history of humanity, the collapse of British imperial armed forces on the field of battle presented a new and potentially fatal threat to the Allied effort.

Rarely in the history of war had the connection between moral and military goals been more vividly demonstrated than in the cascading collapse of the British Empire in the spring and summer of 1942. The British seemed finished as a martial race. They had no genuine map for the future, only the past. Yet somehow, with a centuries-old network of foreign bases and its island fortress off the coast of Northwest Europe, as well as its not inconsiderable industrial capacity, Britain *had* to be saved — not only from its enemies but from itself.

Churchill had looked to American military industrial production as a panacea that would rescue Britain. In reality it was American moral leadership that would ensure the survival of the democracies.

It was to this challenge, the President recognized in late February 1942, that America must now rise.

10

The Battleground for Civilization

It is a small room, as such rooms go — say, about the length of an ordinary Pullman car and four times as wide," Judge Samuel Rosenman recalled. A portrait of President Woodrow Wilson, Roosevelt's former boss, hung over the fireplace, and paintings of Presidents Jefferson and Jackson stared down from the otherwise bare white walls. Red damask drapes framed the French windows overlooking the Rose Garden and South Lawn. Andrew Jackson's huge but leafless magnolia tree still rose above the pillared, curving balcony of the presidential residence itself.

Sitting in the Cabinet Room in the West Wing of the White House, Judge Rosenman watched as the President suddenly came into view, careening "along the covered walk at a speed which made you fear he could not possibly continue to hold on to his armless little wheel chair. In his hand there was always some document he had been reading. In his mouth was his cigarette holder tilted at the usual jaunty angle. Fala, his Scottie dog, ran along his side. A Secret Service agent raced alongside also; and a messenger followed, holding a large basket containing the mail and memoranda on which the President had worked in his bedroom the night before or that morning. A bell had rung three times to alert the White House police that the President was on his way; and officers were stationed at several different points along the path."[1]

Clearly, the President meant business. Starting with his own thirteen-page copy of the address, they were on the fifth draft by the morning of February 21, 1942. For his part, Harry Hopkins had "stopped in the map room on the way over," and had reported that the news from the Pacific was "going to get worse instead of better," Rosenman recalled.[2]

The President had not spoken to the nation since Pearl Harbor. For several weeks he'd wanted to deliver another Fireside Chat, but the press of work had made it impossible—preparations for such an important broadcast requiring days of research and rhetorical iteration.

"I'm going to ask the American people to take out their maps," the President had told Rosenman when summoning him to help draft the talk. "I'm going to speak about strange places that many of them have never heard of—places that are now the battleground for civilization. I'm going to ask the newspapers to print maps of the whole world. I want to explain to the people something about geography—what our problem is and what the over-all strategy of this war has to be. I want to tell it to them in simple terms of A B C so that they will understand what is going on and how each battle fits into the picture. I want to explain this war in laymen's language; if they understand the problem and what we are driving at, I am sure that they can take any kind of bad news right on the chin."[3]

The cascade of reverses in the Far East, climaxing with the fall of Singapore, had produced an increasing "atmosphere of defeat and despair," as Rosenman called it[4]—an atmosphere that Churchill's oratory, full of pain and suffering, had done little to dispel; in fact, other than appealing for a new display of British "genius," Churchill had made it clear in his February 15 broadcast that he was bereft of ideas beyond passing the buck for winning the war to the United States—financially, productively, and militarily.

The President's new Fireside Chat would thus be of critical import—not only his first since December 9, 1941, but his most significant since "the dark days of 1933 during the banking crisis," when as incoming president, Roosevelt had had to "inform the people of the complicated facts of finance and to reassure them that their government was taking any action," no matter how radical, "necessary to protect them."[5]

It would be a talk "encouraging Americans—and people all over the world—to the belief," as the President explained to Judge Rosenman, "that victory and liberation could be won."[6]

Churchill, in his Singapore speech, had lambasted the men who, in the 1930s, had failed to nip the Nazi menace in the bud, but Hopkins thought the President wrong to trash, in similar fashion, American isolationists like Senator Wheeler and Colonel Lindbergh. "That kind of vindictiveness about the old isolationists is out of place now," Hopkins warned[7]—and Roosevelt heeded his advice.

Out went the President's "I told you so" reference.

The President had also wanted to say something magnanimous about the British misfortune in the English Channel, hoping thereby to "stiffen the morale of the British people," he made clear.

Both Rosenman and his fellow speechwriter, Robert Sherwood, were dead set against the inclusion, however — not because they were in any way unsympathetic to the British, but because they felt it veered off message. As "we read what he had written, it seemed to us too apologetic; none of us liked it," Judge Rosenman recalled. "We spent some time that night rounding up our arguments — and our nerve — to go after him on it, although he had already told us several times that he wanted it in."[8]

Out, in the course of the next draft, went that reference, too.

The fact was, as the speechwriters spread their papers and drafts across the great Cabinet Room mahogany table and the old clock chimed every quarter of the hour, all were aware that this talk, given to the people of the free world, would make history, on Washington's Birthday.[9]

So concerned were the Japanese by President Roosevelt's widely announced forthcoming broadcast — which led to huge sales of maps and atlases across America — that a Japanese submarine was ordered to approach the coast of California and fire some shells ashore, in the hope of stealing the headlines in what was, after all, a war of public relations as well as military prowess.

The submarine's salvo did, indeed, garner attention — if not the kind the Japanese were hoping for.[10] More than a hundred thousand Japanese and Japanese American citizens had been taken into custody in California, in order to shield the huge aircraft and shipbuilding plants on the West Coast from potential sabotage or spying; these citizens were now moved out of state to special internment camps inland, and what remained of American sympathy toward individual Japanese Americans in their midst eroded still further.

The Japanese military had good cause to fear Roosevelt's oratory on February 23. Drawing an analogy with General Washington's winter of survival at Valley Forge, the President proceeded to sketch the lines of communication across the world by which the United States would take the war to the enemy — an enemy that had already passed its maximum war production, whereas the United States was only truly beginning to unveil its manufacturing potential. Axis hopes of isolating the constituent countries of the United Nations were thus doomed, the President maintained, not only because of America's war-making arsenal, but because the enemy's aims were nihilistic and despotic. "Conquered Nations in Europe know what the yoke of the Nazis is like. And the people of Korea and Manchuria know in their flesh the harsh despotism of Japan. All of the people of Asia know that if there is to be an honorable and decent future for any of them or any of us, that future depends on victory by the United Nations over the forces of Axis enslavement. If a just and durable peace is to be attained, or even if all of us are merely to save our own skins, there is one thought for us here at home to keep uppermost — the fulfillment of our special task of production." Existing plants were being extended, and new ones created. "We know that if we lose this war it will be generations or even centuries before our conception of democracy can live again. And we can lose this war only if we slow up our effort, or if we waste our ammunition sniping at each other."

"This generation of Americans has come to realize," the President declared, "that there is something larger and more important than the life of any individual or of any individual group — something for which a man will sacrifice, and gladly sacrifice, not only his pleasures, not only his goods, not only his associations with those he loves, but his life itself. In time of crisis when the future is in the balance, we come to understand, with full recognition and devotion, what this Nation is, and what we owe to it.

"The Axis propagandists have tried in various evil ways to destroy our determination and our morale. Failing in that, they are now trying to destroy our confidence in our own allies. They say that the British are finished — that the Russians and the Chinese are about to quit. Patriotic and sensible Americans will reject these absurdities" — as well as those directed at the United States. "From Berlin, Rome, and Tokyo we have been described as a Nation of weaklings — 'playboys' — who would hire British soldiers, or Russian soldiers, or Chinese soldiers to do our fighting for us.

"Let them repeat that now!

"Let them tell that to General MacArthur and his men.

"Let them tell that to sailors who today are hitting hard in the far waters of the Pacific.

"Let them tell that to the boys in the Flying Fortresses.

"Let them tell that to the Marines!"[11]

Simple, homey, and inspiring, the President took listeners on a tour of the world and the "battlefield for civilization" — a historic example of patient but firm presidential exposition. By the end nobody could doubt the President's confidence in the eventual outcome, or his mastery of the situation, however temporarily bleak. Pointing to the difference between the United Nations and the Axis powers, he again emphasized how the coalition or alliance of twenty-six constituent countries would inevitably prevail. "We have unified command and cooperation and comradeship," he reminded listeners. "We Americans will contribute unified production and unified acceptance of sacrifice and effort. That means a national unity that can know no limitations of race or creed or selfish politics. The American people expect that much from themselves. And the American people will find ways and means of expressing their determination to their enemies, including the Japanese Admiral who has said that he will dictate the terms of peace here in the White House.

"We of the United Nations are agreed on certain broad principles in the kind of peace we seek," he went on. "The Atlantic Charter applies not only to the parts of the world that border the Atlantic but to the whole world; disarmament of aggressors, self-determination of Nations and peoples, and the four freedoms — freedom of speech, freedom of religions, freedom from want, and freedom from fear." As a final flourish, the President quoted Thomas Paine's words that General Washington had ordered "to be read to the men of every regiment in the Continental Army," distinguishing between the "summer patriot" and the true patriot. "'Tyranny, like hell, is not easily conquered; yet we have this consolation with us, that the harder the sacrifice, the more glorious the triumph.'

"So spoke Americans in the year 1776.

"So speak Americans today!"

The President's 10:00 p.m. "Fireside Chat on Progress of the War" was listened to by sixty-one million Americans. The *New York Times* dubbed it "one of the greatest of Roosevelt's career."[12]

Ironically, of course, the homily General Washington had ordered to be read to his troops, two centuries before, concerned British tyranny. Nevertheless the broadcast elicited admiration even in Italy, where Mussolini's foreign minister, Count Ciano, noted the difference between his father-in-law, the Duce, and the American president. Mussolini had told him, sententiously, "Wars are necessary in order to see and appraise the true internal composition of a people, because during a war the various classes are revealed: the heroes, the profiteers, the indolent."[13]

Ciano worried that Italian heroes were now dead — leaving only the profiteers and the indolent. Roosevelt's speech, by contrast, impressed him. "A calm, measured, but nonetheless determined speech. It doesn't sound like the speech of a man who is thinking of suing for peace soon," despite popular Italian beliefs and Japanese claims to that effect. Given the failure of Barbarossa to defeat the Russians in 1941, President Roosevelt's predicted outcome of the war sounded, on reflection, all too likely. Certainly in Germany — in private at least — "they all believe that another winter of war would be unbearable. Everyone is convinced of this, from the supreme heads of the army to the men close to Hitler."[14]

Italy, Ciano confided to his diary (which would eventually lead to his execution by his former fellow Fascists several years later), ought really to begin to sue for peace, or at least offer to act as "peacemaker" between the combatant nations, before the conflagration got out of hand.

"But no one dares tell Hitler," Ciano added.[15] Moreover, it was too late — Italy a puppet partner of the Führer, who was hell-bent on a German fight to the death, if necessary; the Japanese likewise.

PART SIX

India

11

No Hand on the Wheel

PLEASE TREAT IT AS SOMETHING I would say to you if you and I were alone," the President wrote to Winston Churchill, two days after his widely praised Fireside Chat. He was struggling with the first draft of an urgent but rather delicate message he wanted to send the Prime Minister. Before encoding and dispatching it, however, he wanted Ambassador Winant and Averell Harriman in London to offer advice on his draft. "As you may guess," he signaled to Winant at 11:40 p.m., on February 25, 1942, "I am somewhat concerned over the situation in India, especially in view of the possibility of the necessity of a slow retirement through Burma itself."[1]

"From all I can gather the British defense will not have sufficiently enthusiastic support from the people of India themselves," the President confided. "In the greatest confidence could you or Harriman or both let me have a slant on what the Prime Minister thinks about new relationships between Britain and India? I hesitate to send him a direct message because, in a strict sense, it is not our business. It is, however, of great interest to us from the point of view of the war."[2]

The two American envoys did their best to press Churchill about Indian self-government, but the response was not encouraging. Still smarting over the abject surrender at Singapore and the prospects of another disaster in Burma, the Prime Minister was so livid at the mention of India, he would scarcely speak to them when they arrived in his office. For some days, therefore, there was little or no signals traffic between the White House and Churchill's War Rooms — the very alliance darkened by disagreement over Britain's colonial rights.

If Burma fell, Roosevelt reflected, the northern bastion of his whole

Pacific defense strategy would be compromised. Thus, as reports of the deteriorating situation in Burma continued to come in to the White House Map Room, the President became more and more concerned lest the defense of India, too, be at risk — Roosevelt certain that Indians, if only they were given self-government by the British, would fight for their country against the Japanese, as the Filipinos were doing.

Churchill continued to dig in his heels, however, and by March 2, 1942, five days later, the undersecretary at the Foreign Office, Sir Alexander Cadogan, was recording in his diary that his boss, Anthony Eden, the foreign secretary, "feels — as I do — that for the last fortnight there has been no direction of the war. War Cabinet doesn't function — there hasn't been a meeting of Defence Committee. There's no hand on the wheel. (Probably due to P.M.'s health). . . . News from everywhere — except Russia — bad. There's something wrong with *us*, I fear."[3]

Two days after that, Cadogan noted: "Poor old P.M. in a sour mood and a bad way. I don't think he's well and I fear he's played out."[4]

It was small wonder the Prime Minister was cast down. Not only were British and British Empire troops not fighting as stoutly as he hoped, but Churchill was being asked by Stalin to accept, as the Soviet Union's price for shouldering the burden of fighting the Nazis, a new treaty by which the Soviets would keep a slice of Poland as well as all the Baltic States after the war. And now, from across the Atlantic, he was being cajoled by the President of the United States: the head of state of a foreign country telling the British Prime Minister that he ought to grant self-government to India, leading to Dominion status or even complete independence after the war — the President quoting not only the current model of the U.S. relationship with the Philippines, but that of the American Revolution in 1776!

Seen in retrospect, Russia and the United States were thus laying down their own markers for a postwar world, at the expense of the dying British Empire — even as the world war itself reached its most critical point for the survival of the United Nations. Military intelligence revealed to Stalin that the German high command was preparing a vast new mechanized assault into the southern regions of Russia, giving access to the oil fields Hitler needed to fuel, literally, his Nazi rampage. Meanwhile, German-Italian forces in Libya, under the command of General Erwin Rommel, were pushing back the British Eighth Army through Libya — with the possibility that, if successful, Rommel could sweep the British out of the Middle East entirely, and threaten the region's oil resources from the south.

In the Far East the situation was even more menacing. Rangoon, the capital of British Burma, was in its death throes — the British beginning to set fire to the oil depots, and getting ready to abandon the city to the Japanese.

In these circumstances, under pressure from President Roosevelt, Churchill finally gave in. He who pays the piper, the Prime Minister reluctantly acknowledged, plays the tune.

On March 4, 1942, the Prime Minister finally signaled to the President that he and his cabinet were "earnestly considering whether a declaration of Dominion status after the war, carrying with it, if desired, the right to secede, should be made at this critical juncture."[5] Moreover, to try and effect this turnabout, Churchill was seeking enough support from the Conservative members of his cabinet to appoint Sir Stafford Cripps, the leader of the House of Commons, as the British government's express emissary to Delhi, bypassing the viceroy of India, Lord Linlithgow.

Cripps would be empowered to seek an immediate, if provisional, accommodation with Indian leaders, including Mahatma Gandhi and Jawaharlal Nehru, in order to win majority Indian support for what promised to be a last-ditch defense of the subcontinent against Japanese invasion from the north, once Burma was completely overrun.

There was also the possibility of an amphibious invasion by Japanese troops landing from their assault vessels in the Indian Ocean — an ocean that the Royal Navy was currently ordering its ships to abandon, in fear of approaching Japanese fleet carriers and their deadly attack-fighters and bombers. It was time — high time — to act.

Churchill's problem, however, lay in the very quality that had enabled him to stand up to Hitler in 1940, when France collapsed — his pride. And more than pride: namely, his complete unwillingness to embrace a vision of Britain as a postcolonial leader of a commonwealth of English-speaking nations.

"Trouble in Cabinet," Sir Alec Cadogan noted in his diary on March 5, 1942, lamenting the Prime Minister's attempts to sabotage the very policy he'd assured the President he was advancing. "Winston having agreed in War Cabinet to [President Roosevelt's] Indian plan, puts it to other Ministers with a strong bias *against,* and finds them unanimously of that way of thinking! Talk—only talk—of [Conservative] resignations from War Cabinet—who met again at 6. Poor old Winston, feeling deeply the present situation and the attacks on him, is losing his grip, I fear. The outlook is pretty bloody."[6]

Sir Stafford Cripps, the leader of the House of Commons, was one of those threatening to leave the coalition government in London. In Delhi, the viceroy, Lord Linlithgow, threatened, however, to resign his post over any usurpation of his imperial role. With Churchill's mishandling of the war as minister of defense—invoking renewed calls for him to relinquish the post, or at least take a deputy such as Anthony Eden in that position—the question now arose as to whether Winston could survive as coalition prime minister.

Since Churchill's doctor claimed the Prime Minister's very health was being adversely affected by the Anglo-American dispute over India, Ambassador Winant decided to return to Washington to brief the President in person. Averell Harriman, meanwhile, penned a long letter to Mr. Roosevelt on March 6, telling of his own worries about the Prime Minister—"both his political status and his own spirits." Churchill had, it was true, reshuffled his war cabinet, making Clement Attlee his formal deputy prime minister. Though "the British are keeping a stiff upper lip," he reported, "the surrender of their troops at Singapore" had been a terrible shock. As things stood, "they can't see an end to their defeats"—and in all frankness, nor could the Prime Minister, in his view.[7]

Harriman, at least, felt Churchill would survive, if only because there was no other figure on the English political scene who could supplant him. Sir Stafford Cripps vainly imagined he could lead the country—but without popular appeal or support. "Eden you know all about"—a lightweight. "[Sir John] Anderson [Lord President of Council] is an uninspired, competent technician. Bevin has never really risen above labor union politics. And then we have Max!"—Lord Beaverbrook, who had resigned as minister of production in a huff, hoping he could wait in the wings and then be called to lead Britain by acclamation, once Churchill collapsed, from his seat in the House of Lords. "There is no one else on the horizon," Harriman pointed out candidly.[8]

The prospect, for the United States, was discouraging.

In the circumstances, Roosevelt decided, he would have to make do: pressing the Prime Minister, yet not to the point of causing him to fall—

and in the abiding hope that, in terms of the current British panic in Burma, it was not too late.

The President's patient pressure seemed to pay off. Swallowing his pride, the Prime Minister at last did as the President bade him. Recognizing that his own history of fierce objection to Indian independence meant that he himself would never be accepted by Indian leaders as a credible broker if he traveled there in person to negotiate, Churchill now formally asked Sir Stafford Cripps, the former British ambassador to the Soviet Union, on March 10, 1942, to set off for India as the cabinet's chosen representative as soon as possible. "We have resigned ourselves to fighting our utmost to defend India in order, if successful," Churchill wrote mournfully to Mackenzie King, his fellow prime minister, "to be turned out."[9] Yet that was what American troops were doing in the Philippines, proudly.

How genuinely, though, did the Prime Minister intend the British to bow out after the war, Indian leaders asked themselves. And how much fighting would the British really *do* to defend India in the meantime, judging by their performance in Malaya and Burma—where dreadful cases of cowardice, military incompetence, and racial as well as imperial misconduct were reported? Was it really worth Indian Congress Party leaders throwing in their lot with "perfidious Albion," in such circumstances?

For President Roosevelt, there now unfolded the most difficult test of his career as commander in chief, not only of the United States Armed Forces, but of the United Nations in the struggle against the Axis powers.

12

Lessons from the Pacific

THE PRESIDENT DOESN'T KNOW me and besides, I'm no New Dealer," Captain John McCrea had protested his appointment to be Mr. Roosevelt's new naval aide. The secretary of the navy, Frank Knox, had laughed, telling McCrea that "FDR needs some of our kind" — Republicans — "to give him support!"[1]

The President's ability to draw people into his orbit was legendary — and effective. "Most cordial — offered me a cigarette and remarked: 'Up in Dutchess County I have a friend by the name of John McCrea Livingstone. By any chance are you related to him?'"

McCrea was "astounded that the Pres. of the U.S. has time to look up my name in register. Flattered a bit too," the captain confessed. "Invites me to his birthday party. I am completely charmed by him."[2]

The Commander in Chief appeared relaxed and confident, despite the worrying reports McCrea brought him each morning at breakfast — his coffee served in a "very large" cup that "must hold as much as four ordinary cups." He seemed to know "more naval officers by their nicknames," from his earlier days as assistant secretary of the navy and later as president, than McCrea knew as a serving officer.[3]

McCrea was also stunned by how astute the President was, behind his mask of easy affability. "Had luncheon with President at his desk," he recorded — "cream spinach soup, veal on toast, mushrooms, potatoes, asparagus tips, double ice cream and fresh raspberries. Told me some remarkable things about MacArthur — I feel flattered at the confidence!"[4]

Roosevelt's huge, handsome head seemed to McCrea to be like an intelligence-gathering machine — putting people at ease, then encouraging them to be frank with him. One example was the way he asked to see the recently fired Admiral Tommy Hart, MacArthur's former naval counterpart in the Philippines, on his return from Batavia, following his brief but tragic stint as Allied naval commander in chief in the Southwest Pacific, serving under the supreme commander, General Sir Archibald Wavell.

At Wavell's request, Hart had been relieved of his command on the grounds he was "a defeatist" — a view with which Hart had not demurred. "They don't like to hear anything which is not optimistic," Hart had noted in his diary. "I think their idea is that frank statements which openly express something which is unpalatable smacks too much of defeatism — and in that they may well be much nearer right than I."[5] His "blunt fashion" realism[6] had, however, proved all too accurate — his successor, the senior Dutch officer, Admiral Conrad Helfrich, subsequently lost virtually the entire Allied fleet in the South Pacific.[7]

On March 10, 1942, the "defeatist" was "convoyed" at the President's request into the Oval Office by Captain McCrea, accompanied by Navy Secretary Knox and Admiral Ernest King, commander in chief of the U.S. Navy. As Hart wrote in his diary that night, "F.D.R. greeted me as 'Tommy,' turned on all his charm, etc. He asked some searching questions, particularly as regards MacArthur's affairs and therein indicated that he was not sold on Douglas' 'masterful defense' to the extent the public is."[8]

"In the public eye," Hart jocularly noted, "MacA[rthur] now stands as the best soldier, say, since Napoleon!!" but at the White House, Roosevelt was having none of it — in fact, the President "simply astounded me by saying that Gen. Marshall had assured him" — thanks to MacArthur's overoptimistic signals from Manila to the War Department — "that the Army was ready in the Philippines *on 1 Dec* [1941]!! That otherwise he, F.D.R. could have held the Japs off, say, another three months" by spinning out negotiations. Hart was incredulous. "That is difficult to swallow," he noted, "but the President is the man who said it."[9]

In part the President was trying to console the admiral. Yet what most interested the Commander in Chief now, as he talked to Hart, were the lessons to be learned about the war in the Far East.

It was in this respect that the difference between Roosevelt and Churchill — indeed between Roosevelt and Hitler — was revealing, as the President *listened* to the admiral's analysis of the war in the East thus far.

Hart's report from the Pacific was sobering. The admiral had spent the several weeks it had taken him to return by sea and air to the United States pondering and analyzing recent military events.

"The man-in-the-street must have been tremendously surprised when the Japs attacked—probably could not possibly imagine that such a small and poor country would have the temerity to attack BIG US," he'd reflected in his diary, "to say nothing of taking on all of our Allies in the same breath. Moreover I wonder if our *Rulers,* in general, had anything like a true estimate of the danger of a war with Japan." For his part Hart had assumed that war with Japan "was entirely evident to responsible people in Washington. I wonder if it was? Well, what must not have been evident was that if said war did come, in the Pacific, we Allies would find ourselves in a war with a First Rate Power; 1st class in a military way, at least. I guess some of us realized *that;* we didn't guess that, even with the enormous advantage which an all-out surprise attack gives, we should find Mr. Jap anywhere near as high grade as he has been thus far. We Americans in general may have realized the imminence of this war," he'd summed up, "but thought that if it came it would be with a second or even third rate enemy."[10]

Listening to Hart's blunt American appraisal, given before Secretary Knox and Admiral King as well, the President could have taken offence—but he didn't. It was clear Hart was a professional and a realist. The war in the East was currently being lost because Allied military leaders—and their commanders in the field—had been unable to match the professionalism of a "First Rate" enemy. And here, Hart's analysis proved spellbinding to its witness.

"And now: lacking almost everything in the way of natural resources, hampered by so very many other disadvantages, after four years of an exhausting war in China, how *is* it that the Japs have set us so back on our heels in this theater of the war?" Hart had asked himself—"all in less than three months? The advantage of surprise of course but the fact remains that the Japs have done everything very well indeed and have repeatedly accomplished what we of the white race said it was impossible for them to do. How have they done it? Certain wise men have long been saying that the worth, strength and power of a country lies in the quality of its people. Well, in a military sense, the Japs are a strong people. Whatever the basis for it, the Jap's patriotism is first class. No race excels them in willingness to get killed or mutilated in war. Moreover, during all the grind of training for war, they have not had to be hired by high pay, good feeding, 'aids to morale,' etc. They have gone on, in peace and war, with a minimum of everything that we have to have to make life endurable. And they come to the 'push of pike' full of fight, hardy, tough, enduring, almost fanatical in courage and, thus far at least, equipped with the material they need for the task, with sufficient skill in its use and — seemingly — under adequate leadership. Now it has happened!"

"This war, in the western Pacific: it is an offensive war," the admiral reflected, "being the long talked about Southern Advance, or the solution of their national difficulties, which the Jap Navy has advocated for some years. It is the variety of war known as amphibious war — named thus I think by [the] British. My recent experience with the British Army didn't indicate that their generals ever use the term very much or that they understand that kind of war, as much as one would expect. I know that in our own Army there has been no understanding of it, and I have known a general or two who had never heard the term used. One, even though it was immediately confronting him," Hart added with sarcasm — a reference to General MacArthur in the first days of the Japanese invasion of Luzon.

"It is a risky variety — Amphibious War — but the advantage of surprise goes with it. And this war began as a surprise."

Hart correctly assumed that the "Jap Navy is in Supreme Command of this war," since it was "what nationalism would dictate — and the Southern Advance idea is theirs." But this was no old-fashioned independent Japanese navy, he noted — it was the very acme of *combined services* operating in the field, not simply at a distant headquarters. And the Japanese combination of those services began with air.

"We *know* that the control of the air, over the war theater, has been gained and exercised by the Jap Navy air. And that control is what has defeated all the defensive power which the Allies could get into the fight. The Japs know the value of the Ships-Plus-Planes combination, handled under one controlling command and without any of all those restrictive limitations which ham-strung the British Navy and so badly hampered ours. Moreover the Japs were prepared for an Amphibious War to the Enth [degree] because they had available a large population habituated to sea-going on all kinds of vessels. And a considerable part of those people knew the waters of the ABDA [American, British, Dutch, Australian] area, including all the harbors and surroundings, because of earning their living therein, as fishermen, small traders, etc." They had studied not only war, then, but the terrain and waters in which they would fight it. "Their expeditions could always be supplied with adequate pilots and guides, and their air service could always tell the Jap leaders what their enemies were doing."

Aspect by aspect, from advanced weaponry ("The Japs are copyists . . . they codify and simplify") to war supplies, Hart had pondered the extraordinary way the Japanese had developed their war machine — and how they had applied it to the business of amphibious invasion. "It is to be noted that the Japs have landed expedition after expedition from ships directly on to open beaches and that these landings have included all the equipment, munitions and supplies needed to win campaigns in Luzon and Malaya at will," as well as "to seize many other points well scattered over the ABDA area." Once ashore, Japanese Army troops could count on naval supply and air support — but showed their own resourcefulness, too. "In the Malaya campaign the expedition fought southward over 400 miles of rough country, during the rainy season. From what I have seen of our own forces, I would estimate that expeditions having similar tasks would require loading in improved harbors — and need a vast amount of 'Transport' to maintain supply lines as they advanced. So why the difference?" he asked pointedly.

Hart's answer was revealing. For one thing, the Japanese soldier was able and willing to fight with only minimal resources: "he goes out, into the field with his weapons and 5lbs of cooked rice, (he doesn't worry about where the next meal after that is coming from — and no one worries for him), treks, and then fights bravely. He also fights skillfully, whereas we are to suppose that the troops which the Japs have been defeating were really poor troops." For another, he was backed by applied air power of amazing efficiency. The Japanese, he noted, had "gained quite full control of the air, and have taken it by defeating some hundreds in all of British, American, and Dutch planes. Said defeat has been more or less in detail, with the Japs producing superiority of numbers at different points of contact but with all due allowances for everything — including the fact that the Allied Air has comprised no less than six different organizations — it must be regretfully admitted that the white airmen and their planes have not demonstrated superiority over the Japs. I repeat — most evidence indicates that it is mostly Jap *Navy* air which we have been contending against. The Jap observation, including photography, must have been very good. Many of us have seen that their bombing has been very high grade indeed; and their strafing has likewise been very good. As for fighters: the 'Flying Fortresses' and the 'Catalina' have done very well in carrying on in the presence of the Jap fighters. But the results gained by the British and American fighters — of which there must have been 200–300 in the area — have not demonstrated any superiority," he recorded candidly, noting that "relatively few of the Jap pilots, or other airmen, are officers. Neither are they found to be of any particular education," being pilots largely selected "from the enlisted ranks. So much for one specialty as regards skill" — and he duly noted exactly "the same in Japanese land and seagoing forces."[11]

Hart's conclusion was thus uncompromisingly tart. The Allies had, as yet, "not defeated" or even turned the enemy back "at any point. At present," he'd noted, "it does not promise that we can prevent the Japs from taking the rest of the N.E.I. [Netherlands East Indies] or relieve Corregidor in time. Current danger is that they will also take Rangoon and cut the communication to Free China. Yes we got into a war with an eastern Asiatic power that is First Class in a military sense. It has now in its control, (or nearly so), riches sufficient to make it enduringly first class in an economic way," unless "interfered with and driven out from recent gains. That means a long war. Not a cheerful prospect but we must not forget that the enemy won't look so good when he in his turn is surprised, loses the initiative, and gets set back on his heels."[12]

Hart's prediction had proven all too accurate — in fact the Japanese "liberated" Rangoon the very day he met with the President.

Listening to Tommy Hart, it was easy for Roosevelt to see why Wavell had asked for him to be relieved — Hart himself expecting he would be demoted from a four-star admiral to a two-star rear admiral for his alleged defeatism. Even Admiral Stark, the outgoing chief of naval operations, had told Hart on arrival in New York "that I was to go on up home and rest up," which "fitted my desires perfectly as I'm decidedly travel-worn," Hart had noted in his diary.

It was Hart's wife, Caroline, who had disagreed. "Not at all," she had declared. "She says the world then will think that I'm sick and senile. That whatever I've brought back with me is hot right now and that I should get to Washington with it forthwith" — in fact, the next day. "She shows me that I can well give head to the subject — and I shall," Hart had written.[13] Scenting a story, the *Washington Post* published a headline: "Let Hart Speak!" and there arose some concern in the administration that Hart, a die-hard anti–New Dealer and Republican with strong opinions on America's march to war, would prove an embarrassment to the President.

The reverse proved, however, to be the case. Roosevelt's insistence on hearing Hart's side of the story at the White House, unadorned and in person, became a turning point in Roosevelt's conception of the war.

The President's natural charm, his use of Hart's first name, Tommy, and his penetrating questions about the Pacific and about MacArthur in particular, not only won over Hart — who became one of the President's most loyal Republican supporters — but gave the President what he most needed at a critical juncture of the war: the truth.

Following this interview, the President ordered that Admiral Hart keep his four stars; arranged that the admiral appear the next day at the President's own press conference; and insisted Secretary Knox use the admiral to crisscross America, speaking to newspapers and professional organizations, in order to tell the American people the unvarnished verity.[14] It was not enough, the President recognized, for the United States simply to ramp up its output as the arsenal of democracy. Just as Hart had predicted in his diary, Rangoon *had* fallen, as would the Netherlands East Indies — Hart's successor, the bombastic Dutch naval commander Admiral Helfrich, wholly unable to compete with Japanese naval control of the air and Japan's seagoing skills.

Lack of cohesion between the U.S. Air Force, the U.S. Navy, and the U.S. Army, as well as the inevitable problems of inter-Allied coalition warfare, suggested that Hart was not only right about a long war in the Pacific, but that America *must* address the issues that the admiral had raised; issues that went to the core of modern warfare itself. The U.S. high command would have to examine, study, and learn the lessons of modern war against an indoctrinated, cohesive, professional, and skilled opponent. Those advocating that the U.S. Air Force be split apart from the U.S. Army, as had been the case with the RAF and the British Army, must be stopped, at least during the war, the President was adamant — for it was vital that growing potential U.S. air power be used to support U.S. ground forces effectively. Naval ones, too. And with better planes.

Admiral Harold Stark, the President had already decided the day before, would be sent to London to concert inter-Allied naval relations there, and would not be replaced as chief of naval operations, or CNO. Instead Admiral King would now take over Stark's responsibilities, as well as remaining commander in chief of the U.S. Navy. After lunch with the navy secretary, Hart gave a talk to a "jam-packed" meeting of the General Board of the U.S. Navy in New York — including Admirals King and Stark, and "all Bureau Chiefs and an Asst. Sec. or two. I talked about fifteen minutes along narrative lines and then twice as long on what I called the lessons which I had learned. In that I pulled no punches, was plenty critical and, as I said in the body of it, I set forth some quite revolutionary ideas" — ideas that would back the "general changes which King and his entourage would like to make, and may in general have helped toward some realism in certain respects wherein we have long been much too theoretical."

With Stark removed from the helm, it would be up to King's legendary "blowtorch" leadership, the President had decided, to kick the U.S. Navy into the mid-twentieth century. For the first time in American history, the head of the U.S. Navy would be answerable directly and only to his commander in chief, the President. He would be urged not only to ramp up naval aviation and order better interservice cooperation with U.S. Army and U.S. Army Air Forces, but to expand, develop, and employ the U.S. Marine Corps, a division of the U.S. Navy, in the same way as the Japanese were doing: as the spearhead of modern amphibious invasion forces.

It was not for nothing that the President had spent seven years as assistant secretary of the navy — however much Hitler, who had been a messenger in the trenches of the Western Front in World War I, derided him. With Admiral King at his side, the President was determined as U.S. commander in chief to refashion the U.S. Navy into a force the Japanese would learn to fear.

13

Churchill Threatens to Resign

As another ninety-six thousand Allied troops surrendered to the Japanese in the Netherlands East Indies, Assistant Secretary of State Breckinridge Long wrote in his diary that he was now becoming "apprehensive lest we be left mostly alone to carry on this fight." If Russia made peace and the British lost the Middle East, America would be left to face a conflict "in two oceans against a combined navy superior to ours."[1]

Assistant Secretary Long had reason to feel anxious. Not only did British forces still appear unable to fight effectively overseas—whether on land, ocean, or in the air—but their political leaders seemed incapable of embracing a postwar vision that would give British soldiers, as well as soldiers of the British Empire, a reason to do so.

The consequences, for the United States, were thus serious. If Britain refused or failed to set out a postwar vision for the peoples of its former imperium, would Congress permit American sons to continue to give their lives for a crumbling colonial empire no longer capable of fighting, or willing to fight, for itself?

Breckinridge Long had recently reported "a serious undercurrent of anti-British feeling" among the members of the Foreign Relations Committee on Capitol Hill: senators expressing the concern that, unless given some form of self-government, Indians "would not have the desire to fight," once the Japanese reached the Burmese-Indian border, "just in order to prolong England's mastery over them."[2] Like the Burmese, Indians might well aid and abet a Japanese invasion, unless Prime Minister Churchill addressed the problem.

Long, who was advising President Roosevelt and the acting secretary of state, Sumner Welles, had agreed with the senators' view. "Concerning India," he reported from Senate hearings on the Hill on February 25, 1942, "the argument was that we are participating on such a large scale and had done so much for England in Lend-Lease that we had now arrived at a position of importance to justify our participation in Empire councils and such as to authorize us to require England to make adjustments of a political nature within the framework of her Empire."[3]

The senators were, in other words, losing patience with the British. "We should demand that India be given a status of autonomy. The only way to get the people of India to fight," they had concluded, "was to get them to fight for India."[4]

Bowing to appeals from the President, the Prime Minister had felt compelled to give in to pressure directly "from Roosevelt," as Leopold Amery, the British secretary of state for India, explained in a confidential letter to the viceroy in Delhi — the Prime Minister finally seeing the "American red light" that, together with the urgent prodding of Clement Attlee's Labour Party colleagues, had "opened the sluice gates" to Indian self-government. Churchill himself had cabled the viceroy that, thanks to "general American outlook," it would "be impossible to stand on a purely negative outlook"[5] — hence the decision to send out Sir Stafford Cripps to assure Indian leaders of postwar independence, and negotiate meanwhile Indian self-government.

Roosevelt had been delighted by Churchill's climb-down — as had been senators in Congress and newspaper editorial writers across America, who mistakenly welcomed the Prime Minister's decision to send out Cripps as a significant new demonstration, however reluctant, of the sincerity of the Atlantic Charter: putting into practice the moral aims of the United Nations.

None had quite reckoned, however, on the continuing obstinacy of Prime Minister Winston S. Churchill, even at the nadir of British military misfortunes in the Far East.

No sooner did Cripps arrive in India on March 23, 1942, than he met a duo of doubting Thomases: the British viceroy, Lord Linlithgow, and General Sir Archibald Wavell, the British commander in chief in India. Moreover, Churchill and the secretary of state for India, Leo Amery — whose son was an open Nazi sympathizer who hated Roosevelt and would later be hanged for treason — now proceeded to do their best, from London, to wreck the negotiations, in the subsequent view of President Roosevelt's personal emissaries. "Colonel Johnson and Colonel Herrington both reported, without using the word, that in their opinion the British Government had deliberately sabotaged the Cripps Mission and indicated that likewise in their opinion the Government in London had never desired that the mission be other than a failure."[6]

Roosevelt was at first disbelieving. Why, he wondered, were the British deliberately ignoring the Atlantic Charter? In his cable of March 10, the President had recommended "the setting up of what might be called a temporary government in India," a "group that might be recognized as a temporary Dominion Government," with executive and administrative responsibility for the civil government of India, "until a year after the end of the war," when a formal constitution could be settled. "Perhaps the analogy of the United States from 1783 to 1789 might give a new slant in India itself, and it might cause the people there to forget hard feelings, to become more loyal to the British Empire, and to stress the danger of Japanese domination, together with the advantage of peaceful evolution as against chaotic revolution. Such a move is strictly in line with the world changes of the past half century," the President had pointed out, "and the democratic processes of all who are fighting Nazism." Moreover, he had specifically warned against allowing the British colonial authorities in India — the viceroy and his acolytes — to kibosh the mission. "I hope that whatever you do the move will be made from London," he cabled Churchill — urging that the British viceroy of India, the pigheaded Marquess of Linlithgow, be discouraged from claiming that Indian self-government was being forced on him "by compulsion."[7]

Churchill, tragically, did the opposite — blaming American pressure. As his military assistant, Colonel Jacob, later reflected, Churchill was a thorough Victorian — his worldview "greatly coloured by his experiences in India, South Africa, and Egypt as a young man, and by his connection with the central direction of the First World War as a Minister. All these experiences tended to give him a great feeling for the British Empire as something, though diverse and growing, which could be directed from London, the great Imperial centre." Unfortunately, Churchill had "never been further East than India." Moreover, India itself was a country he had not seen since the end of the nineteenth century, four decades in the past. "By training and historical connection he was a European first, and then an American," thanks to his mother, Jennie, Jacob attempted to explain. "He did not seem to understand the Far East, nor was his feeling for Australia and New Zealand deep or discerning" — his assumption being that, once the Japanese forced the United States into the war, America would win the war for the British Empire and that American "power would in the end be decisive."[8]

For Roosevelt, this casual British "assumption" was galling; indeed, the saga over Indian self-government was doubly vexing, coming on top of Churchill's concurrent "duplicity" in dealing with Stalin: the Prime Minister agreeing to a draft treaty with the Russians that would, unless President Roosevelt stopped it, *also* vitiate the principles of the Atlantic Charter, by according Stalin the legal right to seize and rule the Baltic States and a large part of eastern Poland at the war's end. So much for the self-determination and self-government the Prime Minister had signed up to on the USS *Augusta.* Even Sir Alec Cadogan, the permanent undersecretary at the British Foreign Office who had helped Sumner Welles draft the final wording of the Atlantic Charter in the summer of 1941, was appalled — railing in his diary at the perfidy of Churchill's foreign secretary, Anthony Eden, who was "quite prepared to throw to the winds all principles (Atlantic Charter). . . . We shall make a mistake if we press the Americans to depart from principles, and a howler if we do it without them."[9]

In the event, President Roosevelt was able to stamp on such British appeasement of the Russians — for the moment at least. But getting the Prime Minister to back off from sabotaging the very Cripps mission he had authorized to negotiate Indian self-government proved a more difficult battle — threatening to ruin the President's entire strategy.

To ensure that Churchill held to the Cripps mission plan, the President had decided to upgrade his munitions envoy to Delhi, Colonel Louis Johnson. Johnson was now told to fly to India as quasi American ambassador to the Indian government, in the rank of minister, bearing the title "Personal Representative of the President of the United States."

Johnson duly arrived in Delhi on April 3, 1942. To his chagrin he found that the Prime Minister had pretty much destroyed any possibility of the Cripps mission succeeding — for Churchill had not only refused to recall the viceroy to London, as Deputy Prime Minister Attlee had urged him to do, but had deliberately encouraged the Marquess of Linlithgow to thwart Cripps's negotiations with the Indian leaders, once Cripps arrived in Delhi on March 23 — and even as Japanese forces drew every day closer to the Indian border.

So effective was the Prime Minister's sabotage in this respect that, on

April 3, 1942, Cripps had wired London, through the viceroy's office, to give up. His "mission," had failed, he cabled: the Indian Congress Party having decided it could not accept the pathetically emasculated version of self-government ("collaboration," as Nehru called it) that was all Churchill, Amery, and Linlithgow would offer. Every mention Cripps made of a "National Government," or "Indian Cabinet," or "Indian Minister of Defense" to work with the British Commander in Chief in defending India had been immediately denied by the viceroy, Lord Linlithgow.[10]

The President was incredulous, given the deteriorating military situation — of which Churchill was either oblivious or willfully blind. On April 1, the Prime Minister had sent him a cable, claiming he was not in the least disturbed by the Japanese Army's rout of British forces in Burma, since he thought the Japanese would "press on through Burma northwards into China and try to make a job of that. They may disturb India, but I doubt its serious invasion," Churchill had added, complacently. "We are sending forty to fifty thousand men each month to the East. As they round the Cape we can divert them to Suez, Basra, Bombay, Ceylon or Australia."[11]

In view of the fact that similar British troops had failed to fight for Singapore, and were now failing to fight in Burma, Churchill's assurance seemed to the President to be optimistic at best. "In his military thinking," the Prime Minister's own military aide noted, "Churchill was a curious blend of old and new. He tended to think of 'sabres and bayonets,' the terms used by historians to measure the strength of the two forces engaged in battle in years gone by. Thus, when he considered Singapore," Colonel Jacob observed, "his mind seemed to picture an old fortification manned by many thousands of men who, because they possessed a rifle each, or could be issued with one, were capable of selling their lives dearly, if necessary in hand-to-hand fighting."[12]

Had Churchill interviewed a single veteran of fighting against the Japanese, he might have recognized how disastrously he was underestimating the enemy, and the sheer unwillingness of British troops to "sell their lives dearly" for a form of colonialism that was doomed. As a result, no one was more shocked than Churchill when the situation in Burma and in the Indian Ocean now spiraled out of control.

Ominously, U.S. and British intelligence had already reported on March 31, the night before Churchill's cable to the President, that the largest carrier fleet ever sent into combat by the Japanese — indeed the same force that had attacked Pearl Harbor—seemed to be heading through the Malay Barrier into the Indian Ocean.

In Churchill's underground headquarters in London, the Map Room became a scene of high alarm. Was the Japanese fleet moving to support the Japanese conquest of Burma? Or was it out to destroy British maritime shipping in the Indian Ocean, and annihilate remaining Royal Navy warships there? Did the Japanese intend to invade Ceylon as the steppingstone to an amphibious assault on southern India?

In the Map Room on the ground floor of the White House, there was equal concern. In the circumstances, it seemed incomprehensible to the President that the British government would seriously allow the Cripps mission to fail.

As the six Japanese aircraft carriers, five battleships, and seven cruisers were identified steaming into the Indian Ocean, shock turned to dismay. Virtually unmolested, Admiral Jisaburo Ozawa proceeded in the ensuing days to decimate Churchill's naval forces, sinking twenty-three British ships in the Bay of Bengal, while Japanese submarines sunk five more off the Indian coast, and Admiral Chuichi Nagumo attacked the British naval base at Colombo, Ceylon.

Then on April 5, 1942, Nagumo's forces not only found two British cruisers, HMS *Dorsetshire* and HMS *Cornwall,* and sent them to the bottom of the Indian Ocean, but went on to attack and sink the British aircraft carrier HMS *Hermes.*

It was a devastating blow.

"Poor American boys!" Radio Tokyo had recently broadcast a sneering challenge to U.S. troops and sailors in the Java area. "Why die to defend foreign soil which never belonged to the Dutch or British in the first place?"[13]

The President, proud of the way the U.S.-Filipino Army had held out against the Japanese in the Bataan Peninsula for so long, despite virtually no supplies or reinforcement, was deeply disappointed by Churchill's sabotage of self-government for Indians. It seemed to Roosevelt impossible that the British would cling to their colonial "rights of conquest" in India, rather than welcoming Indian participation in the war on the British side, when Gandhi himself had withdrawn from the Indian Congress Working Party to enable his "legal heir," Jawaharlal Nehru, to negotiate a deal with Sir Stafford Cripps and the British government. Nehru had promised in writing that Indians, if given self-government, would defend

India to the last hamlet; and Mohammed Jinnah, the Muslim League leader (and future founding governor-general of Pakistan), had also told Cripps he would go along with an Indian cabinet—yet *still* the viceroy and Churchill resisted.

In American eyes, Churchill's refusal to allow Cripps to make concessions amounted to fiddling while Rome burned. In a cable on April 4, Colonel Johnson begged Roosevelt to intervene personally, since unless the President "can intercede with Churchill, it would seem that Cripps' efforts are doomed to failure."[14] This prompted Welles to respond that he had "personally discussed with the President your telegram no. 145, April 4, 8p.m.," but that he did not "consider it desirable or expedient for him, at least at this juncture, to undertake any further personal participation in the discussion. You know how earnestly the President has already tried to be of help.... In view of the already increasingly critical military situation do you not believe that there is increasing likelihood of the responsible leaders adopting a more constructive attitude?"[15]

For his part, Colonel Johnson, in Delhi, did not blame the Indians. Like Cripps, the colonel thought it was the British who, in such a critical military situation, should have been more reasonable, if they wanted Indians to fight for their own territory.

Instead of offering the post of minister of defense to an Indian with genuine military responsibilities under a British war minister, however, Churchill and Amery would only agree to offer, on April 7, the possibility of appointing an Indian member on the viceroy's Executive Council with responsibility for "storage of petroleum products; welfare of troops; canteen organizations; stationery and printed forms for the Army..."[16]

How the British could be so stupid, at such a menacing time, seemed downright incomprehensible to the President. "I suppose this Empire has never been in such a precarious position in its history!" even Churchill's own army chief of staff, General Alan Brooke, acknowledged in his diary on April 7, 1942.[17]

As was inevitable, Churchill cabled the President in growing desperation that evening to ask if the United States Pacific Fleet in Hawaii could be ordered into action in order to "compel" the Japanese to "return to the Pacific."

The American fleet must save India and the British Empire—forcing the Japanese to retreat, "thus relinquishing or leaving unsupported any invasion enterprise which they have in mind or to which they are committed. I cannot too urgently impress the importance of this."[18]

Churchill's *cri de coeur,* when it arrived, caused consternation. Instead of negotiating with Nehru, Churchill bombarded the President with disingenuous claims about the exclusive fighting skills of Muslim rather than Hindu soldiers — despite the fact that neither Muslim nor Hindu nor Sikh soldiers were fighting the Japanese with anything but halfheartedness, as long as their home country was denied self-government.[19]

Despite Welles's formal response to his cable begging the President to intervene personally, Colonel Johnson was, therefore, quietly encouraged to act on the President's behalf. In a series of accelerando meetings on April 8 and 9, the colonel — who ran a top legal practice in America — knocked heads together and, after a meeting with General Wavell, got a "Cripps-Johnson Plan" unofficially accepted by the Indian Congress Party. "CRIPPS SAID TO HAVE ACCORD ON NATIONAL REGIME IN INDIA," the *New York Times* reported triumphantly on April 9, 1942.

Cripps was delighted — as was Colonel Johnson, who reported proudly to the President that "the magic name here is Roosevelt," and that "the land, the people would follow and love, [is] America."[20]

Colonel Johnson was speaking too soon. Churchill was still determined to counter the success of the Cripps mission, however dire the situation.[21]

Secretary Hull was still away from Washington, recuperating, but when he heard what had been done at this critical juncture, he too was disgusted — indeed, he deliberately titled the chapter of his memoirs covering the episode "Independence for India" — furious that Churchill had, after the signing of the original Atlantic Charter, "excluded India and Burma" from the principles and, as he put it, had already declared, in an address to Parliament in September 1941, that "Article 3 applied only to European nations under Nazi occupation and had no effect on British policy."[22]

For Roosevelt, Churchill's panic-stricken plea for the United States to rescue the Royal Navy marked the end of their military honeymoon. The President had authorized the surrender of all Filipino and American troops remaining on the Bataan Peninsula, after doggedly fighting the Japanese since December 10, 1941. Though Corregidor might hold out a further month, failure of the British to fight the Japanese, and their assumption that their colonial empire would merely be rescued by Americans who did fight, was unacceptable.

Roosevelt would not have been Roosevelt had he allowed his personal feelings of disappointment to affect his judgment, however. Coalition war, the President knew, meant allying oneself with partners who did not necessarily share the same political or moral principles, or vision—as, for example, America's military partnership with Stalin's Soviet Union. Collaborationist, imperialist Vichy France was another possible ally, or at least non-hostile government—a government whose military compliance might be of profound significance in launching a Second Front against the Nazis, either in North Africa or mainland France. Leading a coalition of United Nations, in other words, was bound to involve associations that were at best necessary, and at worst cynical.

Deliberately trivialized by Churchill in his memoirs,[23] and ignored by most historians in the decades following the war, the saga of the Cripps mission—and the President's role in it—marked in truth the end of Britain's colonial empire, as Churchill willfully surrendered Britain's moral leadership of the democracies in World War II.

Sir Ian Jacob, reflecting on the Anglo-American alliance, would later date the change that came over U.S.-British relations as taking place in 1943: the "change which came about when the Americans felt that they had developed enough power to conduct their own line of policy."[24] In reality, however, the change had taken place much, much earlier—as the White House records and diaries of those visiting with the President would show.

The President had absolutely no intention of risking the gathering strength of the U.S. fleet in the Pacific in an unprepared battle with the Japanese Navy. Instead, ignoring what he'd told Welles to say officially to Colonel Johnson, he wired Prime Minister Churchill on April 11, asking him please *not* to recall Cripps, but to make one last effort at accommodating Indian aspirations for a national government.

Why Churchill closed off this possibility will be debated by historians and biographers to the end of time—"one of the most disputed episodes in Britain's imperial ending in India," as Oxford historian Judith Brown would later call it.[25]

Did Churchill fear Sir Stafford Cripps returning to Britain in triumph—given broad British public support for a settlement—and supplanting him as prime minister? Or could Churchill simply not accept the idea of Indian leaders running their own country of four hundred million people, after all the years Churchill had fought to deny India self-government, let alone independence?

In any event, deliberately rejecting the President's advice as well as widespread British and American hopes for an act of statesmanship, Churchill now did the opposite — withdrawing all previous assurances he'd given Cripps, which had been the basis of the tentative agreement with Nehru. "I feel absolutely satisfied we have done our utmost," Churchill cabled the President with finality on April 11 — and instructed Cripps to return to London without any agreement.[26]

The President was dumbfounded. In a cable early that same day, Colonel Johnson had, by contrast, assured the President that, in regard to the impasse with Churchill, Sir Stafford Cripps and Nehru "could solve it in 5 minutes if Cripps had any authority" from Churchill. Johnson was at a loss to understand why the Prime Minister had become so intransigent. The Indian Ocean, after all, "is controlled by the enemy," the Japanese, he pointed out to the President. "British shipping from India has been suspended; according to plan determined many days ago, British are retiring from Burma going north while fighting Chinese go south; Wavell is worn out and defeated." In Colonel Johnson's view, the British were finished not only as an imperial, colonial power, but as a first-rate nation. "The hour has come when we should consider a replotting of our policy in this section of the world," he recommended. "Association with the British here is bound to adversely affect the morale of our own officers. . . . Nehru has been magnificent in his cooperation with me. The President would like him and on most things they agree. . . . I shall have his complete help; he is our best hope here. I trust him."[27] India, in other words, could become a great democratic partner to the United States.

Sir Stafford Cripps had complained that a patronizing speech by Lord Halifax in New York on April 7, broadcast on CBS, had "done the greatest harm at a most critical moment,"[28] and Johnson was of like mind, telling the President that the address had "added the finishing touches to the sabotaging of Cripps. It is believed here it was so intended and timed and I am told pleased Wavell and the Viceroy greatly" — with Churchill breaking into a victory dance in the Cabinet Room below Whitehall "on news the talks had failed," jubilantly declaring: "No tea with treason, no truck with American or British Labour sentimentality, but back to the solemn — and exciting — business of war."[29]

• • •

Johnson was certainly right about the viceroy, the Marquess of Linlithgow, who was if anything more intransigent—and racist—than the Prime Minister. Linlithgow had earlier described the Japanese, before Pearl Harbor, as "Yellow Bellies" who would probably enter the war on Germany's side: "I don't see how they can help it, the silly little things!" he'd mocked them in a letter to Lord Halifax. Burma might be threatened, but the British would prevail, he'd been sure. American weaponry nevertheless remained important. In fact, in terms of the Indian Army, Linlithgow was "greatly dependent upon your constituents in North America," he'd admitted to Halifax, "for heavy gear. So don't tell them what I think of them," he warned.

What he thought was, in truth, mean, despicable, and almost unbelievably snooty. "What a country," he derided America, "and what savages who inhabit it! My wonder is that anyone with the money to pay for the fare to somewhere else condescends to stay in the country, even for a moment! What a nuisance they will be over this Lease-Lend sham before they have finished with it. I shan't be a bit surprised if we have to return some of their shells at them, through their own guns! I love some clever person's quip about Americans being the only people in recorded times who have passed from savagery to decadence without experiencing the intervening state of civilization!" Halifax, he was quite sure, shared his views, but would just have to be stoic and go on with his work in Washington to obtain more free military assistance, while "toadying to your pack of pole-squatting parvenus!"[30]

Why Churchill tolerated such an anti-American, bigoted buffoon as viceroy of India was hard both for Colonel Johnson and President Roosevelt to understand—especially since Linlithgow had begged Churchill for months to be allowed to retire, after almost six years in the post. Churchill had, however, insisted that, despite his unpopularity, Linlithgow should stay—with orders to abort Cripps's mission.

This the Scottish aristocrat did with enthusiasm. Colonel Johnson he described as "Franklin D.'s boy friend," whom Cripps had brought into "close sensual touch with Nehru, for whom J. has fallen." It was too bad—though Linlithgow hoped others could correct "in the President's mind, the distorted notion which I feel sure Johnson is now busy injecting into that very important organ."[31]

Distorted or not, the news from India mystified the President, who became more and more concerned lest the British use their crucial, American-provided weaponry not to combat the Japanese, but to put down the inhabitants of India, who might well revolt if denied at least a semblance of self-government.

In the circumstances the President decided he must try one more time to pressure Churchill into seeing sense. Harry Hopkins had recently set off for London with General Marshall to discuss prospects for a Second Front in Europe, and was staying with the Prime Minister at Chequers, his country residence. To Hopkins the President now sent an urgent cable, asking him to "give immediately the following message to the former naval person." As he added, "We must make every effort to prevent a breakdown."[32]

The President's signal—which Churchill later derided, after the President's death, as "an act of madness"[33]—was a simple request: namely for the Prime Minister to "postpone Cripps' departure from India until one more final effort has been made to prevent a breakdown in the negotiations."[34]

Roosevelt was, as he explained, "sorry to say that I cannot agree with the point of view set forth in your message to me that public opinion in the United States believes that the negotiations have failed on broad general issues. The general impression here is quite the contrary," the President corrected the Prime Minister. "The feeling is almost universally held that the deadlock has been caused by the unwillingness of the British to concede to the Indians the right of self-government, notwithstanding the willingness of the Indians to entrust technical, military and naval defense control to the competent British authorities"—as he had heard from Colonel Johnson himself. "American public opinion cannot understand why, if the British Government is willing to permit the component parts of India to secede from the British Empire after the war," as per the original Cripps mission's declaration, set out by the British government, "it is not willing to permit them to enjoy what is tantamount to self-government during the war."[35]

The President's cable might have been simple, but it was not, sadly, the sort of language the Prime Minister was prepared to tolerate; indeed, the more that the walls of Britain's empire appeared to be tumbling down in the Far East, the more the Prime Minister now dug in his heels. Reading the first lines of the message Hopkins handed him, Churchill's blood rose. Worse followed, however.

The President warned in the cable that if Churchill did not relent and India were invaded by the Japanese "with attendant serious military or naval defeats, the prejudicial reaction of American public opinion can hardly be over-estimated." He asked Churchill therefore to reconsider the Cripps mission, and "to have Cripps postpone his departure on the ground that you personally have sent him instructions to make a final effort to find a common ground of understanding." The President — who had been remarkably polite and helpful till now — had clearly given up pretending. "I read that an agreement seemed very near last Thursday night," he complained. Why had it been allowed to fail? If Cripps "could be authorized by you personally to resume negotiations at that point" — i.e., the position before Churchill had withdrawn the cabinet's approval of the terms Cripps had gotten — "it seems to me that an agreement might yet be found.

"I still feel, as I expressed to you in an earlier message," the President finished his cable, "that if the component groups in India could now be given the opportunity to set up a national government similar in essence to our own form of government under the Articles of Confederation" — on the "understanding that upon determination of a period of trial and error they would then be enabled to determine upon their own form of constitution and, as you have already promised them, to determine their future relationship with the British Empire" — then he was sure "a solution could probably be found."[36]

Poor Harry Hopkins, having handed over the rest of the President's cable, now witnessed Churchill's meltdown.

Hopkins, sickly but willing to do anything for his revered president, had traveled to London with General Marshall to ensure that U.S. planning to aid the Russians was not stymied by British bureaucracy and timidity. The spat over India, however, now banished European war plans to a back seat — Hopkins later telling Robert Sherwood that no "suggestions from the President to the Prime Minister in the entire war were so wrathfully received as those relating to solution of the Indian problem."[37] To the secretary of war, Colonel Stimson, Hopkins even confided, on his return to Washington, "how a string of cuss words lasted for two hours in the middle of the night" in London[38] — with Churchill adamant he would rather resign than permit an American president to dictate British imperial conduct. It was no idle threat.

The fact was, Churchill seemed exhausted, as all around him had noticed. He was drinking more, sleeping less, and busying himself in the minutiae of military operations across the world that he seemed unable or unwilling to delegate. The Australian prime minister had made it clear he had lost confidence in Churchill's leadership, and the Pacific War Council in London was entering its "death throes" — soon to be entirely replaced by the Pacific War Council in Washington. Japanese naval forces were roaming at will like sea monsters off the coast of India — and British forces were in helter-skelter retreat in northern Burma, abandoning their Indian units to be killed or captured. General Rommel was once again forcing British Empire troops to retreat in Libya.

However, as one of Churchill's own "closest and most affectionate associates" later confided to Hopkins's biographer, "the President might have known that India was one subject on which Winston would never move a yard."[39] Certainly Indian self-government, as Robert Sherwood recalled Hopkins telling him, was "one subject on which the normal, broad-minded, good-humored, give-and-take attitude which prevailed between the two statesmen was stopped cold" — indeed Churchill, rounding on the hapless Hopkins, told him he "would see the Empire in ruins and himself buried under them before he would concede the right of any American, however great and illustrious a friend, to make any suggestion as to what he should do about India."[40] Calling in his stenographer, the Prime Minister was determined to put his feelings in writing. To the President he therefore dictated a nasty rebuke.

"A Nationalist Government such as you indicate would almost certainly demand," he deceitfully claimed, "first, the recall of all Indian troops from the Middle East, and secondly, they might in my opinion make an armistice with Japan on the basis of free transit through India to Karachi of Japanese forces and supplies."[41]

For Churchill to send such unqualified "opinions" was sailing close to dishonesty. Both claims were specious, as Churchill and the viceroy (with whom Churchill was in secret correspondence, bypassing Sir Stafford Cripps) well knew — contradicting the assurances Nehru had given Cripps and the President's "Special Emissary," Colonel Johnson.[42] Nevertheless, Churchill argued in his draft response, "From their point of view this would be the easiest course, and the one entirely in accord with Gandhi's non-violence doctrines"[43] — despite Gandhi's express withdrawal from the matter, and public statement that Pandit Nehru would decide the Congress Party's conduct.

In Churchill's lurid forecast, the "Japanese would in return no doubt give the Hindus the military support necessary to impose their will upon the Moslems, the Native States and the Depressed classes." In conclusion, the Prime Minister made clear he would resign rather than permit this to happen — indeed, that he had "no objection at all to retiring into private life, and I have explained this to Harry just now" — a threat he larded with the prospect of a British parliamentary revolt by Conservatives in his favor. "Far from helping the defense of India, it would make our task impossible," he warned the President. And though as prime minister he would "do everything in my power to preserve our most sympathetic co-operation," he wanted the President to be aware that the U.S.-British alliance was now at stake. "Any serious public divergence between the British and United States Governments at this time might involve both of our countries in ruin."[44]

It was, in effect, blackmail.

In a subsequent telephone call to the President some hours later, Harry Hopkins relayed to Mr. Roosevelt the gist of Churchill's draft cable. It was a document that, in fear of the whole Western alliance now collapsing, Hopkins had begged the Prime Minister *not* to encode and dispatch[45] — in fact, so alarmed was Hopkins that he begged Churchill not even to raise the subject with the British cabinet, which was due to meet the following day, lest this become a test of the whole Atlantic coalition.

Roosevelt should simply back off, Hopkins therefore advised the President on the phone to the White House. He, Hopkins, would do his best to calm the Prime Minister down.

Mid-April 1942 now came to resemble, in terms of the Atlantic alliance, something of a French farce. Hopkins was begging Roosevelt, in the interests of Anglo-American cooperation, not to press for an Indian national government, but at the same time Colonel Johnson was cabling the President with the *opposite* plea: forwarding a personal appeal by Pandit Nehru, the Indian Congress Party leader, that contradicted Churchill's dire predictions. Nehru had absolutely no intention of negotiating with the Japanese, if they did invade India, he made clear in his letter — being himself all too aware of the likely consequences of Japanese invasion, if it happened, and the "horrors" that would follow, "as they have followed Japanese aggression in China."[46]

"To your great country, of which you are the honored head, we send greetings and good wishes for success," Nehru's message ended. "And to you, Mr. President, on whom so many all over the world look for leadership in the cause of freedom we would add our assurances of our high regard and esteem."[47]

Hearing the alarm in Hopkins's voice, the President, once again, could only sigh at Churchill's negative attitude—which seemed all the more racially demeaning and self-serving at a moment when the Japanese controlled the Indian Ocean and were nearing the Indian border in Assam.

Churchill's negative behavior at this juncture of the war seemed indefensible—moreover, shameful, given the mess into which the British had gotten themselves in the Far East. With the Japanese fleet causing mayhem off the Indian coast and threatening not only Ceylon and Calcutta with impunity (nine hundred thousand people had evacuated Calcutta, in fear of Japanese bombing and possible invasion), the Prime Minister could no longer fulfill his role in holding the northern flank or cornerstone of the President's two-point military strategy in the Far East. Not only had Churchill been forced to appeal for U.S. air forces to be sent urgently to India to protect the subcontinent, but for U.S. naval forces to be sent to the Indian Ocean to save the Royal Navy.

For Churchill the situation was deeply humiliating—a fact that, in part, explained his psychological resistance to reason. Twice already he had declined to show an act of statesmanship over Indian aspirations. Backed into a corner, he was declining for a third time to do so, threatening resignation as prime minister if the President insisted upon Cripps being told to continue negotiations with Nehru.

It was at this juncture that Harry Hopkins, in the interests of Allied unity, hit upon a solution.

Hopkins might have little understanding of military strategy or tactics or combat, but he was an indefatigable fixer. He was devoted to his president—and in thrall to Winston Churchill. These were the two greatest men of their time, at least in the West—and he, Harry Hopkins, had the privilege of serving them as intermediary. It was crucial, he felt, to find some way of defusing the mine threatening the Atlantic alliance. Hopkins had never been to India, and dumping Indian aspirations for self-government, even so that Indians would defend the subcontinent from the approaching Japanese, seemed a small price to pay for unity between the United States and Britain. Yet how could he bring the two leaders of the Western world back on course?

Rather than supporting President Roosevelt's doomed pressure on the Prime Minister to come to terms with Nehru, Hopkins hit upon an alternative: a new stratagem to persuade the President to rescue Britain's collapsing empire in India and its forces in the Indian Ocean without having to grant self-government to India. It would involve a gigantic pretense, by making the U.S. an offer it could not refuse. In the ensuing months it would bring not Churchill, but the President's military advisers, to the point of resignation, once they discovered how insincere it had all been. But at a critical moment in the war, when Allied unity seemed vital to the eventual victory of the democracies, it was all Hopkins could think of.

Churchill should, Hopkins suggested, simply ignore the President's plea regarding India completely. Instead he should, Hopkins advised, ditch the draft of his resignation letter and commence a new message: beginning on a positive note in terms of the Western alliance, by promising wholehearted British military cooperation in carrying out General Marshall's top-priority plan for a cross-Channel Second Front that very year.

Churchill's new cable—encoded and sent at 3:50 p.m. on April 12, 1942—thus sidestepped the whole issue of India. It began, instead, by congratulating President Roosevelt on the truly "masterly document" that General Marshall had brought with him to London regarding U.S.-British strategy in Europe—adding that as prime minister and minister of defense Churchill was "in entire agreement in principle with all you propose, and so are the Chiefs of Staff."[48]

This was, in actuality, complete moonshine, as even Churchill's own senior military assistant, General Hastings Ismay, later admitted. Perhaps, Ismay confessed in shame, "it would have obviated future misunderstandings if the British had expressed their views more frankly"[49]—for the British chiefs of staff were *not* in "entire agreement" with General Marshall's "masterly document." In fact they were, from the very start, utterly *opposed* to American Second Front plans that would result in untold numbers of British deaths for no purpose. Even Churchill's most slavish chronicler would state that Churchill was being "at best disingenuous."[50]

At the time, however, the chiefs of staff were willing to practice such a deceit on behalf of their prime minister/minister of defense—Churchill calculating that, over time, the Americans would recognize the impracticability of a cross-Channel landing that year. And in the meantime, the charade would be enough to get the American president off the Prime Minister's back with regard to Indian self-government.

Hopkins's suggestion worked. Churchill's "masterly" signal did succeed in getting Roosevelt to back off. However, the problem of India would not go away so easily — indeed, the saga came to a head several days later, when Churchill was compelled to send a second, this time panic-stricken, plea to the President for help.

14

The Worst Case of Jitters

IN HIS MEMOIRS, Churchill would omit his desperate plea to the President on April 15, 1942, for it was simply too embarrassing to quote.

If the Prime Minister hoped there had been no witnesses to its reception in Washington, however, he was mistaken. For on that very day, the Canadian leader, Prime Minister William Lyon Mackenzie King, happened to be staying at the White House.

King, who had been invited to attend his first meeting of the new Pacific War Council in Washington, was suffering from bronchitis and a bad cold, but he found the President remarkably confident when he arrived. "Was conducted from the White House by the garden corridors to the President's Secretary's office, and from there into the Oval Office" in the West Wing, the Canadian prime minister noted in the detailed diary he kept. "The President was sitting working in his shirt sleeves, white shirt, no vest or coat. Gave me a very warm welcome. Laughed a little about his attire" — given King's own, rather formal clothing — "and we went in together into the Cabinet room."[1]

The President's naval aide, Captain McCrea, was there too — having been asked by the President to act as his liaison officer to the council. As McCrea noted in his own diary, Roosevelt told him privately: "Don't keep any minutes, but write out a memorandum afterwards."[2]

King, by contrast, had been in the habit of keeping copious, careful notes of all his meetings in a special account he dictated each day — and his description of Roosevelt's torment over India would provide historians with their most intimate glimpse of the President's reluctant assumption of overall Allied strategic command, as the British Empire collapsed in all but name.

• • •

The President was, in Mackenzie King's admiring eyes, a patrician—the Canadian prime minister proud to be asked "to be seated to his right." At 3:00 p.m. the President of the United States then "opened the proceedings," saying "that what was most upon the minds of all present was the news from France and the situation in India, and in the Indian Ocean."[3]

It looked, the President confided, as if the nefarious fascist Pierre Laval would soon become prime minister of the Vichy government, leading to yet "closer collaboration with Germany." Not only did this mean the Vichy French would now give voluntary aid to Nazi Germany, but "conditions" in France, the President noted, would continue to "become more serious through Laval's ascendancy."[4] It was an ominous prospect. Laval—who was executed after the war—did indeed become Vichy prime minister a few days later and immediately agreed to dispatch three hundred thousand Frenchmen to work in German munitions factories. He also arranged to have all non-French Jews and their children rounded up and transported to German concentration camps, where they would be exterminated.

"With regard to India" the President showed even more concern—incredulous the British could be so obstinate over Indian self-government, yet not wishing to embarrass Sir Ronald Campbell, the representative from the British Embassy in the absence of Lord Halifax, as the council considered "the immense perils which confront us" in India and the Indian Ocean (in Churchill's phrase)—with no representative of India at the table.

The latest British Navy's "loss of the two ships 'Dorsetshire' and 'Cornwall,'" especially, aroused the President's incomprehension as U.S. commander in chief. "He mentioned," King recorded, "that this emphasized the need which the United States had asserted to the British of not allowing ships to get too far away from the coast, and the protection which land-based planes could afford."[5]

The more the President spoke, the more King saw how Roosevelt was quietly but confidently assuming the mantle of the war's overall direction, not just the provision of its weaponry. As the Canadian premier pointed out to the council, "since Japan's entry" it was no longer a European war, with "all plans" made "largely in consultation and co-operation with London"—a former time when "Britain was viewed as the centre of the Empire and the British Isles as the most important of the possible theatres of war." The global conflagration created a new "political" as well as "strategic" reality. Henceforth, in order to avoid "alienation of feeling between different parts of the British Commonwealth and any of the free countries," the old Dominions of the British Empire would look to leadership by the United States of America, not Britain — for the United States, not Britain, was now the glue holding together the antifascist alliance, the Canadian prime minister felt. "I pointed out how Australia's problem had created a problem in Canada such as had scarcely been dreamt of before. Just as the feeling had grown up suddenly in Australia which was causing Australians to look more to the United States than to Britain, so to the amusement of some of us, British Columbians" — on the Canadian Pacific seaboard — "were beginning to adopt a similar attitude toward the Government of Canada," with some of them "saying they would have to look to the United States rather than to Ottawa for an understanding of their problems."[6]

Mackenzie King was not alarmed by that. What struck him was how graciously and yet firmly Roosevelt was handling his enlarged role. Where Churchill had recently become somehow smaller, in both spirit as well as power, the President had seemed to grow larger. The President explained how he had secretly sent Marshall and Hopkins to London "to urge the necessity for offensive action which would help to relieve pressure on the Russians by creating another front" — but that this certainly did not mean he was willing to give in to the "Russian request regarding guaranteeing of [postwar] frontiers. He said any consideration of this meant an ignoring of the Atlantic Charter. The main difficulty was that beginning with one concession would only lead to concessions regarding boundaries of other countries. The Russians," Roosevelt said, "would keep pressing for all they were worth," but as president of the United States he would not alter his stance — leading King to consider "it inadvisable" to contest the matter at the meeting, no matter how much Churchill was imploring him to do so, in pursuit of a British treaty with the Soviet Union. The dispute was, as King noted, "a matter of rather delicate discussion at the moment between the United Kingdom and the United States." Canada, he said, would stay right out of it.

The Chinese, Netherlands, and Australian and New Zealand representatives on the council "all seemed to approve cordially of the President's action," King noted — Roosevelt promising at the meeting to dispatch not only U.S. planes to give backbone to the British in defending Ceylon, but American crews and even troops.[7]

All in all, however, it seemed a veritable tragedy that, with so many hundreds of thousands of British troops and personnel in India and

Burma, the British had so ignominiously surrendered the Allies' overland route via Burma to China — and now looked like they were being pushed back across the Burmese border into India itself.

The meeting of the Pacific War Council ended at 5:30 p.m. Later that evening the President gave a small, intimate dinner for Mackenzie King, who was staying in what was called Queen Elizabeth's Bedroom, or the Rose Room — beginning with cocktails in the President's Oval Study, next to his bedroom, at 7:10 p.m.

"The President himself mixed up the cocktails before going down to the small dining room, and we had a very happy little dinner party during which time the President recounted some of the events in connection with Churchill's visit and his stay in Florida" during his convalescence in January 1942[8] — including the British prime minister's embarrassment when telephoning Wendell Willkie, the 1940 Republican contender for the U.S. presidency, and getting, instead, President Roosevelt.[9]

Churchill, in other words, was admired as a great character, but not quite to be trusted. "We had a little talk with regard to some aspects of the war, but mostly a pleasant social evening during which the President and I talked a good deal across the table to each other and the younger people joined in with their observations," King wrote in his diary that night. "I confess I felt how much it adds to one's life to be surrounded by young people."[10]

The President seemed relaxed — and sanguine. Over dinner he had amused his guests by saying, in the event of an air attack, he would use the underground tunnel to the Treasury Building vaults, where he hoped "they would arrange to have some card tables and poker chips" set up, "so they could appropriately pass the time while concealed there." He had from the Army Air Forces the latest information on U.S. long-distance bombers — and expected that, over time, both the Japanese and the Germans would surely build similar ones that could reach the U.S. eastern and western coasts, and even bomb "the capital," Washington. "The news had just come before dinner of the very successful attack of MacArthur's men from Australia upon the Japs at the Philippines by bomber plane, taking a trip of 4,000 miles and return. Naturally every one was relieved and rejoiced at what had been done in that way," King noted.[11]

After dinner, "the young people withdrew and the President and I went to his circular library" — the President's upstairs Oval Study. There "the President seated himself in the corner of a large leather sofa to the left," King recorded, "and told me to sit on the sofa beside him."[12]

King at first declined, thinking it would be easier to converse facing his American counterpart, and took a chair instead. He quickly noticed, however, that this would cause the President "to be seated at a lower level," which seemed wrong. Moving to the place on the sofa that the President indicated, King was once again drawn into Roosevelt's affectionate orbit: his easy, intelligent charm and goodwill, despite his affliction—indeed the President's disability, as they sat there, seemed almost the opposite, investing Roosevelt, who was six feet, two inches tall, with a strangely powerful aura, at once humble and magnetic.

The President asked what King thought of the afternoon meeting. "I thought from the way he referred to the Pacific Council that he was a little uncertain himself as to its value"—given that it was originally to have had its location in London—"and really wanted to know what I thought of it. He then said to me he wanted to tell me about India, and made the significant remark that this would be of historic interest" in the future. "He then repeated what he had said at dinner, that he believed that the plan he had proposed" for an Indian national government "might have met the situation satisfactorily, and would have been accepted had the British Government been agreeable to it. He said: 'the idea was not my own but I communicated it to Churchill. He [Churchill] had this material before Cripps' interviews in India.'"

Churchill, clearly, had not been amused, the President confided. "'He, Churchill, sent me' (I am not using his exact words but tone) 'long accounts of the situation respecting India, going back to the days of [Governor-General] Warren Hastings and [General] Clive," in the eighteenth century. By contrast, "'What I proposed was along the lines of the way America proceeded at the time of the Revolution'" when American independence was declared. "'It was arranged that the delegates should be sent to an Assembly and an ad hoc provisional government set up which would carry on the government of the different states on matters of general concern, and later allowing a Constitution to be formed which would give full powers of government. What I proposed for India,'" the President explained, "'was to give the Indians complete right of government themselves at once with regard to such matters as tariffs, trade and commerce, post office, communications and external affairs or foreign policy, and also defense with the understanding, however, that for the actual military operations, General Wavell would have control of the strategy and direction of forces, etc. I pointed out that it was a great mistake in the British proposal to allow any part [of India] to secede, and spoke particularly of the civil war in our country which had been the result of an attempt at secession. All should have been prepared to work together for a time admitting there would be problems to be overcome, but that these should be worked out between themselves.'"[13]

King found himself astonished at how closely the President had been involved in the Cripps mission — and its reception in India. "The President went on to say," the Canadian premier stated, "that he had reason to know that his proposals would have been accepted by all of the different groups in India, and that they would have been satisfactory to Cripps."[14] Roosevelt had not, however, been able to get Churchill "to arrange for a plan on those lines" — this account verified by the Chinese foreign minister, Dr. Soon, the following day. (Dr. Soon confided to Mackenzie King "for my strictly personal information," that he had it "on the best of authority that a settlement could have been reached by Cripps, had the British Govt. allowed him to make the settlement on a basis which the Indians and he were prepared to agree on, but that the British Govt. would not give that extent of authority.")[15]

Clearly, the President was frustrated that the British would simply fiddle while Calcutta burned. It was, Roosevelt lamented, too bad. Yet not as bad as the next morning, when the President told King of the latest cable he had received from Churchill.

"It was midnight when I turned out the light. Slept soundly," King recorded in his diary.[16] When he awoke at 7:30 on April 16, 1942, in the Rose Room's four-poster bed, he had two visions, which deeply affected him. They both concerned, he thought, the President's plans for prosecuting the war against the Nazis in Europe — but before he could discuss them personally with Mr. Roosevelt, he met Admiral King (no relation) in the upstairs hallway. The admiral had been urgently summoned, together with General Joseph McNarney, the acting chief of staff of the U.S. Army in the absence of Generals Marshall and Arnold in London, to an early meeting with the President. Admitted into the Oval Study while the officers were asked to wait outside, Mackenzie King listened in amazement as the President told him the latest news.

The night before, Roosevelt had said he thought the British were falling apart — "that they had the worst case of jitters in Britain that he thought they had ever had," as King recorded in his diary. "That they were terribly concerned and fearful of the whole situation" in India and the Indian Ocean.[17] King had agreed — saying it was with good reason, given the British failure to handle the situation in India sensibly. Looking at the President on the morning of April 16, however, King was aware that the President's worst fears had now been realized, as Roosevelt "put his hand to his forehead." On the desk in front of him was a pile of telegrams. "I had a bad night last night," the President confided. "At 11:30, I received a war message from Winston. It is the worst message that I have received." Pointing to the cables, he said: "They are the most depressing of anything I have read."[18]

Calling in Admiral King and General McNarney, the President read aloud Churchill's tale of woe — a long cable "to the effect that he, Churchill, was greatly concerned about his position in the Indian Ocean. That he feared the Japanese were assembling a powerful fleet which might succeed in taking Ceylon and later Calcutta and lead to landing Japanese forces in India with internal situation arising there which might lead to any kind of consequences. That if Ceylon was taken, the getting of assistance to the Middle East, to Egypt, etc, might be cut off with consequent demoralizations of British position there. He [Churchill] did not think the British could hold the situation without some of the American fleet coming to their assistance and asked for a couple of battleships, mentioned one or two additional battleships which the British might be able to send."[19]

Churchill's panic was evident in every line. As the Canadian premier noted afterwards: "the passage that impressed me most in what came from Churchill was a statement to the effect that Madras and Ceylon might both be taken; also the steel industries of Calcutta. The possibility of internal trouble in India which might lead to anything there; also the Japanese might sweep on to the Persian Gulf, and that the whole Middle East might become demoralized. The message was a plea for urgent assistance by the American fleet."[20]

It was understandable that Churchill would later excise this entire saga from his six-volume war memoirs. It was far from his finest hour.

The "situation in India, and in the Indian ocean" marked the turning point in World War II, in the President's eyes: the moment when the collapse of the British as a primary global power became manifest.

"China and the U.S. together would have to settle the affairs in the Far East," Mackenzie King had noted Roosevelt's view the night before. "He did not see how Britain could be expected to do much in that area."[21] If the British fell apart, the United States would have to take over responsibility for the defense of the hemisphere. With "about 100,000" American troops already stationed in Australia and New Zealand[22] — men who would actually fight, rather than running away — the President had no real concern that this transfer of power in the Orient could be achieved — anchoring U.S. power in Hawaii and the Antipodes. Nevertheless, Roosevelt clearly felt the British had made a historic mess of their empire in the Far East.

"He could not understand that there was no reference," in Churchill's latest entreaty, to Roosevelt's "previous communication by which he had offered to supply large quantities of bombing planes to Ceylon which could be sent from Montreal and which were ready to start, just awaiting the press of a button. . . . He then said that he had repeated to Churchill that they should not let their ships get beyond air protection. The loss of the last two ships was like the loss of the *Repulse* and the *Prince of Wales,* getting far enough from land to be unable to take care of themselves. He felt the British were thinking too much of a fleet battle in the Indian Ocean"[23] — for, in his latest panic-stricken telegram, Churchill had informed the President that he was not only immediately dispatching his First Sea Lord, Admiral Sir Dudley Pound, to Washington to "discuss with you and Admiral King the whole position and make long term plans" but expressed the hope that, if the President agreed to use the U.S. Pacific Fleet as he, the Prime Minister, had suggested, in the Indian Ocean, "you will be able to have the necessary orders given without waiting for his arrival. We cannot afford to lose any time."[24]

A "fleet battle" by the U.S. Navy in the Indian Ocean, *at the British prime minister's suggestion?* Roosevelt was, once again, incredulous. "Until we are able to fight a fleet action," Churchill's cable ran, "there is no reason why the Japanese should not become the dominating factor in the Western Indian Ocean. This would result in the collapse of our whole position in the Middle East, not only because of the interruption to our convoys to the Middle East and India, but also because of the interruptions to the oil supplies from Abadan, without which we cannot maintain our position at sea or on land in the Indian Ocean Area. Supplies to Russia via the Persian Gulf would also be cut . . ."[25]

It was an absurd request. Not only had Roosevelt, as U.S. commander in chief, no intention whatsoever of sending an American fleet into the Indian Ocean, he had no intention of allowing the British tail to wag the

American dog. Coming on top of the Prime Minister's earlier threat to resign, Churchill appeared to have lost his mind.

To Admiral King's relief, the President explained he had no intention, nor had he had at any time any intention, of fighting a "fleet action" in the Indian Ocean — particularly in conjunction with a Royal Navy that had no idea how to cooperate with modern air force units. Even the British Army's chief of staff, General Brooke, was in despair at the refusal of the RAF to see its role as supporting British Army or Navy forces — dooming the British to defeat in battle. Why, then, would the President of the United States replicate the antiquated approach of the "Former Naval Person" — as Churchill called himself in his cables to the White House — to modern warfare?

Moreover, to imagine that the dimwitted, brave but ailing First Sea Lord would be able on arrival to sway the President and his chiefs of staff, especially Admiral King, to accept strategic direction from the Prime Minister, was, in this regard, yet another demonstration of Churchill's almost infallible instinct for choosing the wrong commanders of his military forces.

In the presence of the Canadian prime minister, the President told Admiral King and General McNarney to draft a blanket no to Churchill's cable — but to add a brief but guarded mention of American plans that were already in hand in the Pacific.

"Luncheon was served at his desk" in the downstairs Oval Office, Mackenzie King recorded — "which like the one in his library upstairs was literally covered with a lot of political souvenirs. Cloth dunkeys and other fantastic figures" — bric-a-brac from past political campaigns depicting Democrats as donkeys — seemed to the ascetic Canadian to be "incongruous. The President clearly enjoys nothing more than political campaigning and its associations. The game of politics is a great stimulus to him" — affording him "as many personal contacts as possible," despite his disability.[26]

The Canadian premier had been amazed, earlier that morning, to have been invited into the President's study while the chiefs of staff were kept waiting: a deliberate gesture designed, he realized, to remind the President's military advisers of their lower place in his sun — and a warning not to exceed it. Similarly, he would never allow a cable to go out over his signature that did not reflect his genuine views or feelings. "After lunch

General McNarney and Admiral King came in with a telegram prepared for Churchill which the President read over and"—to Mackenzie King's fascination—"revised and softened a bit here and there," as King noted.[27] Roosevelt was clear that, whatever Admiral King or General McNarney might draft, he was not going to humiliate his great ally in the war, despite Churchill's missteps.

The Prime Minister's stature as the embodiment of defiance was, Roosevelt thereby indicated to Mackenzie King, an essential part of the Allied cause. Using the USS *Ranger* aircraft carrier as a fast plane-transport, the President advised Churchill he would send planes immediately to reinforce British forces in India—not only to ensure the defense of Ceylon, Madras, and Calcutta from Japanese naval attack, but to "compel you to keep your fleet under their coverage," as the President added, pointedly. As for sending an American fleet to the Indian Ocean, though, Roosevelt was polite, but firm: "I hope you will agree with me that because of operational differences between the two services there is a grave question as to whether a main fleet concentration should be made in Ceylon with mixed forces." To follow such a course would be to play the Japanese game. Instead, the President had in mind a quite different strategem: to make the Japanese Navy withdraw from the Indian Ocean without the President having to compromise his ever-growing U.S. Pacific Fleet in Hawaii. This plan was something he was not, however, willing to confide to Churchill, as his second paragraph made clear.

"Measures now in hand by Pacific Fleet have not been conveyed to you in detail because of secrecy requirements," the President explained, "but we hope you will find them effective when they can be made known to you."[28]

More, on that score, the President would not say.

Listening to Roosevelt as he read aloud the telegram he was about to dispatch, Mackenzie King found himself intrigued, as he also imagined Churchill would be—"some venture by the Americans" that must be kept secret, the Canadian premier noted in his diary, for reasons that "would be apparent later and which could not be mentioned even to Churchill at this time. It seemed to me," he reflected, "this had reference to some attack the American fleet intended to make, or action to keep the Japanese away from the Indian Ocean."[29]

In a sign of how the balance of power among the Western allies was now changing, the Canadian prime minister arranged with President

Roosevelt to hold a military conference on allocation of Lend-Lease planes to take place in Ottawa the next month — without even bothering to first obtain Churchill's agreement. His work done, King bade farewell to the President and spent the night of April 17, 1942, in the special sleeping car of his train, traveling to New York and arriving in the early hours of April 18.

There, in the afternoon, the Canadian premier visited a spiritualist friend, Mrs. Coumbe — who had her own visions. "She spoke of not being alarmed about either France or India. She had a vision about large fleets and little fleets." Quite what it meant was unclear. "It seemed to her to signify unrest in France, which would, ultimately, be all to the good," he noted.

Back in Car 100 on his special train, King had dinner with a friend. "After dinner, I read the papers," he recorded, "including" — to his utter surprise — "an account of the bombing of Tokyo."[30]

The first of the President's secret "measures," it was clear, had begun. America was on the offensive.

PART SEVEN

Midway

15

Doolittle's Raid

AT 12:45 P.M. ON MAY 19, 1942, Brigadier General James Doolittle — recently promoted from the rank of colonel — was ushered into the Oval Office to meet President Roosevelt.

Doolittle was there to receive the coveted Medal of Honor, escorted by Generals Marshall and Arnold — and pretty Mrs. Doolittle, whom the officer had not seen for more than a month, when he set off from California on his epic mission: to bomb the capital of Japan for the first time in World War II.

Doolittle had stated in the car on the way to the White House that he didn't feel he deserved the Medal of Honor. "General, that award should be awarded for those who risk their lives trying to save someone else," he'd protested.

Marshall had silenced him in six words: "I happen to think you do."[1]

So did the President, as the originator of the April 18 "Doolittle Raid."

Ever since the fateful day of the Pearl Harbor attack, and in the succeeding weeks as Japanese forces cut their swath down the Malay Barrier toward Australia, the President had urged his chiefs of staff to find a way to retaliate; some operation that would shake Japan's sense of its own invincibility and show the rest of the world that the United States would retaliate against aggression — with a vengeance.

The Tokyo raid was "a pet project of the President's," in Secretary Stimson's words.[2] Since Stimson himself had opposed it from the start — considering it a dispersion of effort in the Pacific, and a beacon for "sharp reprisals" by the Japanese — he was not invited to the little Oval Office ceremony, held before newsmen and press photographers. Admiral King was late and missed the photo opportunity, yet in truth he had been as intimately involved in the undertaking as Doolittle, Marshall, and Arnold — and would be significantly more affected by its consequences.

Fleet Admiral Ernie King was the only senior admiral in the U.S. Navy with a pilot's license, as well as command experience in submarines, aircraft carriers, battleships, and cruisers. It was King who had, early in January 1942, first suggested U.S. Navy carriers could be used to launch U.S. Army bombers in invasion operations — the planes then able to land on airfields seized or established by amphibious troops. On January 10, 1942, King had proceeded to give the go-ahead, with General Arnold, for a medium bomber B-25 group of the USAAF to start training, under U.S. Navy supervision, for abbreviated carrier-takeoff. In this instance it was not to support an invasion but to fulfill the President's call for a bombing raid on the capital of the Japanese Empire, launched from the sea.

General Arnold had pleaded for pressure, rather, to be put on the Soviet Union to permit a Russian airbase to be used for the takeoff and landing of the planes. The President had ruled that out as naïve. With German forces massing for a repeat of Operation Barbarossa — code-named Blue, this time — Stalin would not dare incite a Japanese declaration of war that would then force the Russians to fight not just on one but two fronts. Equally, there were no U.S. bombers currently in China for the task — indeed, those that had been sent to India on their way to support China were inevitably reassigned to defend India, once British forces fell apart in Burma and the Indian Ocean.

Only by using U.S. carriers as mobile airfields could the President's directive be carried out — just as the Japanese were doing in the Indian Ocean, spreading mayhem and panic. But with this difference: that American carriers, wisely, would not risk being attacked by enemy land-based aircraft. They would launch their B-25s secretly, from outside the range of Japanese land-based planes — and it had been this operation to which the President, in his cable to Winston Churchill on April 16, 1942, had mysteriously referred.

At the very moment Prime Minister Mackenzie King was staying at the White House, a secret U.S. Navy carrier task force comprising sixteen ships, submarines, and ten thousand sailors had been sailing toward the Japanese mainland in the strictest secrecy. Once airborne, their sixteen long-distance, heavily loaded B-25 army bombers would blitz military targets in the Tokyo area, then fly on and land in western China, and there become, if all went well, the first contingent in Chiang Kai-shek's air force, counterattacking the Japanese.

The President had been thrilled with reports of the plan's progress since inception. Using painted outlines on airfields to resemble mock carrier decks, the pioneering speed-aviator Colonel James Doolittle had not only adapted B-25s for the task, but had trained his crews to take off in a matter of four hundred yards, then find their targets six hundred miles away across the Pacific, avoid anticipated antiaircraft and enemy fighter fire by flying absurdly close to the ground, aim their bombs on specified, strictly military targets — and then fly another thousand miles to reach relative safety in China — the first time land-based air force bombers had ever been so launched in human history.

Bombing their capital, Tokyo, would serve as an unmistakable warning to the Japanese public at home as well as in the front lines, the President calculated, even as their forces shelled and bombed the U.S.-Filipino defenders in the Philippines. More importantly, an air raid on Tokyo would, if successful, draw Japanese naval attention back from the Indian Ocean by demonstrating to the Japanese their failure to defend their own capital and homeland. At the very least it would force the Japanese high command to hold the major part of its navy in the Pacific, instead of sending it back to the Indian Ocean, after refueling.

Above all it was in the Pacific, not the Indian Ocean, that the President believed the U.S. Navy could best deal with the Japanese. Under the command of Admiral Chester W. Nimitz — who had been appointed the supreme Allied commander for all waters of the Pacific east of the Solomon Islands — the U.S. Pacific Fleet could confront the Japanese juggernaut in American-controlled waters. Flying from land bases in the Hawaiian Islands and American atolls such as Midway, U.S. Army Air Force bombers and fighters could, in intimate new cooperation with U.S. Navy aircraft, then deal with the Japanese *Kido Butai* (literally, "mobile force") carrier fleet on its own, American, terms.

Strict radio silence kept by the American task force, until it had withdrawn safely from the reach of Japanese land-based aircraft, meant that no report of the success or failure of the Doolittle Raid could be sent to Washington on April 18, 1942.

Roosevelt had been spending the weekend at Hyde Park, in any case. When Captain McCrea, his naval aide, finally called the President from the White House to tell him that hysterical announcements had been monitored on Tokyo radio about an enemy air raid on the city, Roosevelt was coy.

"My telephone conversation with him (about 10 a.m.) went something like this," Captain McCrea later recalled:

> *McCrea:* The most important item of the morning's report, Mr. President, is that Tokyo has experienced an air raid from U.S. planes.
>
> *The President:* Really (with a laugh) — How? Where do you suppose those planes came from?
>
> *McCrea:* That, Mr. President, is what the Japanese want to know. Our intelligence sources say that is the question in Tokyo everyone wishes an answer to — where did those planes come from?

Roosevelt had said no more until later that afternoon, when he called McCrea, "remarking about as follows: 'I think I can answer the question.'" The President continued, "Ask 'Ernie' King if he doesn't think it a good idea to say that the raid came from Shangri-la. If so, when the word reaches Japan, every Japanese will be busy looking at his or her equivalent of a Rand McNally atlas!"[3]

Captain McCrea had done as instructed. "I called Admiral King and told him what the President had said. Admiral King laughed softly, and thought well of the idea."

"Shangri-la" — the fictional valley in James Hilton's 1933 novel *Lost Horizon,* where a group of downed airplane passengers landed — thus became the President's response when questioned by the press, asking for details, on his return to Washington on April 21, 1942.

The Japanese military, by contrast, were incensed, once they figured out the mystery. An entire U.S. naval task force had approached Japan without being identified! Japanese air defenses had been either nonexistent or deplorable where they did exist — not a single B-25 had been shot down or even hit.

The triumph of the Japanese sneak attack on Pearl Harbor had thus been avenged, leading the Japanese military to embark on reprisals against the populace in China, reprisals of almost unimaginable atrocity: a fury of executions, butchery, and slaughter of civilians accused of harboring the American fliers responsible for bombing their capital city — Chiang Kai-shek cabling in a panic to say the Japanese were murdering as many as *a quarter million* Chinese civilians for harboring and aiding Doolittle's fliers, after they crash-landed or bailed out of their planes at night over western China. "The Japanese troops slaughtered every man, woman, and child in those areas — let me repeat — these Japanese troops slaughtered every man, woman, and child in those areas," the generalissimo wired.[4]

Murdering tens — even hundreds — of thousands of Chinese civilians was not going to win the war against the United States for Japan, however. In the headquarters of the Japanese high command there was consternation that the protective screen around Tokyo had proven as assailable as the American defenses at Pearl Harbor. Thus, failing to catch a task force under Admiral William Halsey's command before it could withdraw, the Japanese Combined Fleet staff argued over how best to respond to what amounted to an embarrassing, even shaming, defeat: one that was seen as a direct insult to their revered Emperor, since the American planes had flown right over the imperial palace, as if mocking its vulnerability. A distraught woman had broken in to Radio Tokyo's broadcast, shouting: "Your lives are in danger. Your country is in danger. Tomorrow — even tonight — your children may be blown to bits. Give your blood. Save them. Save yourselves. Save Japan."[5]

Even Secretary Stimson was astonished by the hysterical Japanese reaction to the Tokyo raid.

The Japanese, Stimson noted in his diary, "have been taken wholly by surprise and were very much agitated by it, and it is quite interesting to see their conduct under such conditions. It has not been at all well self-controlled. I have always been a little doubtful about this project, which has been a pet project of the President's, because I fear that it will only result in sharp reprisals from the Japanese without doing them very much harm. But I will say that it has had a very good psychological effect on the country, both here and abroad and it has had also a very wholesome effect on Japan's public sentiment."[6]

Stimson was right. The raid on Tokyo was taken by many in Japan's homeland as an evil omen: a harbinger of how the nation would one day be punished for its devotion to the god of total war. A Samurai nation that had not itself been invaded for thousands of years had embarked on a rash war of conquest across the entire Pacific and Southeast Asia — and only four months after its triumph at Pearl Harbor was rudely reminded what the cost might be. As one woman wrote her cousin, a Japanese pilot in the South Pacific, the "knowledge that the enemy was strong enough to smash our homeland, even in what might be a punitive raid, was cause for serious apprehension of future and heavier attacks."[7]

The effect was to be far more consequential than public sentiment, however. An April editorial in Japan's *New Order in Greater East Asia* had forecast that the United States and Britain would soon decline into "second-rate or third-rate powers, or even to total disintegration and collapse."[8] The Tokyo raid seemed to contradict that prediction.

Such an unanticipated, successful American air attack demonstrated to the Japanese high command the need to finish what the Japanese attack on Pearl Harbor had started: the destruction of America's remaining naval power in the central Pacific.

For some time Admiral Yamamoto, the commander in chief of the Japanese Imperial Fleet, had been arguing that the U.S. Pacific Fleet should be lured from its base at Pearl Harbor and dealt with, once and for all. His recommendation had been ignored, since Japanese invasion forces had been advancing farther and farther in the South Pacific, virtually without opposition. Why risk that success — which required constant naval carrier and warship support — to embark on an old-fashioned fleet naval battle? Japanese forces were, after all, almost at the southern end of the Malay Barrier, in New Guinea. An amphibious invasion of the south coast of New Guinea around Port Moresby, only three hundred miles north of Australia, would, if backed by a Japanese naval carrier force, enable Japan to cauterize Australia as an Allied military base, and allow Japanese forces then to take New Caledonia, Fiji, and the Samoan Islands — thus severing Australia's tenuous maritime communications with America. This seemed a far more effective strategy than risking a fleet battle in open seas.

The Doolittle Raid, however, had upset such calculations. Admiral Yamamoto, it now appeared, was right: the U.S. Pacific Fleet was too powerful to be left untouched. The humiliating American air attack on Japan's capital now gave Yamamoto the ammunition he needed — claiming that he must, at all costs, cauterize Pearl Harbor as the base of the carrier fleet from which Doolittle's bombers had sprung.

Given the number of bombers and fighter planes that the U.S. Army had since Pearl Harbor sent out to the Hawaiian Islands, another Japanese air attack on Pearl Harbor would now be suicidal, as would an attempted invasion, Yamamoto accepted. But seizure of the atoll of Midway, an

American military and air base 1,325 miles west of Hawaii, would give the Japanese an airfield for land-based planes with which to keep the American fleet locked up at Hawaii, while the Japanese *Kido Butai* carried out its operations in the South Pacific — or Indian Ocean. If the U.S. Pacific Fleet allowed itself to be lured out to contest the Midway landings, so much the better! The carriers that escaped destruction at Pearl Harbor could in that case finally be attacked by superior Japanese fliers' skills, and sunk.

Thus on April 20, 1942, the day after Doolittle's raid, the Japanese plan to seize New Caledonia, Samoa, and Fiji in order to cut America's sea communications with Australia was put on hold. Instead, a massive Japanese air, naval, and amphibious assault on Midway would be undertaken by the Japanese Imperial Navy, with ten thousand troops loaded on transports, just as soon as naval forces assigned to the imminent capture of Port Moresby, the Australian base in New Guinea, had completed their task and could be withdrawn.

And with that fateful decision, in the wake of the President's latest "pet project," the great Japanese march of victories in the Pacific came to its peak.

16

The Battle of Midway

IF THE PRESIDENT WAS NERVOUS about a major Japanese naval reaction to the Doolittle Raid, he refused to let his concerns show. A leader must exhibit confidence, he felt, since the slightest hint of anxiety or dejection would spread like ripples in a pond. It was for others — Stimson, Knox, Marshall, Arnold, King — to worry and second-guess; for him to solicit information, views, and observations before making his ultimate decisions as commander in chief.

Despite being (or in part because of being) surrounded by men and women who revered him, the President was lonely for the kind of relationships in which he could let down his guard — as Daisy Suckley, FDR's cousin and close friend, noted in her diary. His longtime secretary, office manager, and daily companion, Marguerite "Missy" LeHand, had been paralyzed by a stroke in June 1941. Sara Roosevelt, the President's mother, had died in September of that year. He missed their company. Eleanor had her own cottage on the Hyde Park estate, Val-kill (valley stream) — two miles away from the "Big House," which stood empty most of the time now. (In 1938 Roosevelt had built his own cottage, "Top Cottage," on the hill above the Big House, but Sara had forbidden him to spend the night away from the mansion — and her — while she was alive.) "The house has no hostess most of the time," Daisy recorded matter-of-factly, "as his wife is here so rarely — always off on a speaking tour, etc."[1]

Instead of retreating into wheelchair-bound isolation, however, the President had sought to allay it by leading a life of constant activity, meetings, and personal interaction when staying the weekend at Hyde Park. Despite his exalted status, he deliberately treated each person he encountered as an existing or potential personal friend. With each one — whether cabinet member or valet — he would make a swift determination as to their character and interests, and use it to interpret the information he elicited. With his personal secretary Bill Hassett he discussed genealogy and rare books; with Harry Hopkins he rehearsed past and possible power plays of senators and congressmen; with Daisy Suckley, their shared family history, New York society, and filing systems; with Crown Princess Martha of Norway — who had three small children — the business of parenting; with ornithologist and author Ludlow Griscom, the identification of North American birds — even rising at 2:00 a.m. on May 10, 1942, to go birding with him, Daisy, and others at Thompson's Pond, to hear the dawn chorus.

The President was elated — "Total for day 108 species," he noted with satisfaction, signing the checklist "Franklin D. Roosevelt."[2] "It is the kind of thing he has probably given up any idea of ever doing again," Daisy reflected, "so it did him lots of good. In that far-off place, with myriads of birds waking up, it was quite impossible to think much of the horrors of war."[3]

To keep such widely different relationships fresh, genuine, and vital in the complex clockwork of his daily presidency — at least without resorting to cold discipline and intimidation — the President employed both his charm and his sense of humor: at once ironical and teasing, balancing the deadly seriousness of his responsibilities with levity, even occasional mischief. Also: refusing to let anyone in his presence act pompously, or imagine he or she possessed more power than that which the President permitted in his presence.

Such had been the nature of Roosevelt's leadership style for many years — and in assuming the reins of global military leadership on behalf of the United Nations, Roosevelt appeared, as noted by his aides, to apply the selfsame principles now as war leader: dominating his colleagues and subordinates through his status, his high intelligence, and his abiding confidence in himself and his own authority. It was a domination he also maintained, as he had since 1933, by a process of often maddening divide and rule: appointing gifted people to key positions, delegating day-to-day responsibility to them, but ensuring they remained ultimately subordinate to him as chief executive of the nation.

Such delegation, at this time of Roosevelt's life and health, never allowed any question but that he, and he alone, was "the Boss." Around him kings abounded: Admiral King; Prime Minister King; General King; King Peter of Yugoslavia; Queen Wilhelmina of the Netherlands; Crown

Prince Olaf of Norway; the former King Edward VIII of England, now the Duke of Windsor. The President, however, remained the ultimate global monarch, as all around him were aware: charming, well-mannered, and above all, naturally seigneurial.

This was, Daisy Suckley noticed, the difference between the President and his adviser, Harry Hopkins, who was fawning in the presence of Winston Churchill, and was "almost familiar with the Cr. Princess Martha. That's the difference between him and the P.[resident], who doesn't have to show anyone he is not impressed by greatness of any kind. He is one of them," she noted perceptively[4] — a self-assurance, at an otherwise dark moment in the war, that struck those around him as much as it did monarchs in exile.

"He had been President for several years before the war came along," recalled George Elsey, who joined the Map Room staff in April 1942. "He had been Boss for several years. Congress had generally — not always, but generally — done what he wanted. The American public had supported him. He simply felt that he knew what the country needed, what it ought to have, and that he could get his way with what he wanted. Had he not been in his third term, his attitude would have been very different. Had he been a first-termer, or second-termer . . . But here he was: an unprecedented third-term President — and of course he knew better than anyone else what was good for the United States. *That was the attitude at that point!* He was supreme in every respect. And he was in a better position than was Churchill, who had to explain himself to Parliament; he was better than Chiang Kai-shek, who couldn't control his own country, that was riven with civil war," as well as war with a Japanese invader. "'I'm in control; this is the way it's going to be — it's going to be the way *I* want it!'" was the President's frame of mind.

"Now, that may be a false reading, but this is the sense I had of his perception of himself as the war went on," Elsey recalled. "It was accepted by *everyone* that he was the Boss — Stimson, Knox, Marshall, King, Arnold — everyone. *Absolutely!* After all, think how many years he had been President! All those officers had been relatively junior officers, and they had risen with his support and under his command to their present 'lofty' — relatively speaking — positions. And they had no reason to doubt or question his authority. What was Marshall? Marshall was a colonel, I guess, when FDR was elected president. What was King? Probably a captain in the Navy. And they'd all risen to where they were under FDR. They had no reason to challenge or contradict or believe otherwise than to accept his leadership. Except MacArthur — who of course was God, and superior to everybody — as taught him by his mother!"[5]

Elsey laughed at the memory. "In that sense," the commander reflected some seventy years later, "MacArthur and FDR were very much alike: they were 'Mama's boys,' who'd been raised by their mothers to believe they were supreme, superior to the ordinary human being. And by God, their life was going to prove it!"[6]

As, now, they proceeded to do: each in his own way.

Aware of MacArthur's failings as a commander in the field of modern battle, especially in terms of his relations with air and navy colleagues, Roosevelt had wisely turned down Marshall's insistent recommendation that MacArthur, operating in Australia, be made supreme commander in the entire Pacific, with authority over all U.S. and Allied forces operating in the theater. Certainly the campaign to oust the Japanese from their ill-gotten gains would be tough and unrelenting: something MacArthur would have to undertake from secure positions in Australia once sufficient forces were sent his way. In the meantime, however, the Japanese Imperial Navy, with its all-conquering, high-speed carriers, had to be dealt with — and for that, the President decided, Admiral Chester W. Nimitz, whom he had personally appointed to command the U.S. Pacific Fleet, was named supreme commander in the Pacific — commander in chief of all naval, army, and air forces in the north, central, and southern Pacific east of the 159th meridian: his operations monitored and checked on a daily basis by the President. MacArthur, by contrast, would command only the Southwest Pacific Area.

Just how much the President was in personal control of the war astonished even the secretary of war, Colonel Stimson — who had only been invited to go inside the President's Map Room on April 12, 1942, for the first time since the war began. "I had an interesting time with him in his map room which I had not seen before. Captain McCrea was there and acted as expositor," the war secretary had confided to his diary. "We went over all the areas of the globe where we are interested, and I was interested by the fact that the President had here what we are unable to have at the War Department — the naval movements."[7]

The President was clearly way ahead of Stimson, even the Joint Chiefs of Staff — men he deliberately kept firmly under his thumb, charging them to develop and carry out his overall policies as a military high command headquarters, yet making sure each of the committee members reported also to him *individually*—thus depriving them of any hint of collective power. It was in this way that his decision to appoint Admiral Nimitz to command the whole of the Pacific, despite Marshall's opposition, was final—but was now to be tested, as the Japanese Imperial Navy went into silent mode, permitting virtually no communications signals that American cryptographers could decipher. All eyes turned to their impending but surely devastatingly concentrated punitive response to Doolittle's daring raid on Tokyo.

Reading the ambiguous reports of his Washington cryptographic team, Admiral King cautioned that the Japanese might well attack Alaska, the West Coast, even South America.

Admiral Nimitz, fortunately, now had his own Magic unit in Hawaii. Nimitz's team correctly deciphered Japanese signals indicating an imminent amphibious invasion of Port Moresby by a mixed fleet of Japanese troop ships, supply vessels, carriers, cruisers, destroyers, and submarines. Six days after Doolittle's raid, on April 24, 1942, Nimitz flew to San Francisco to meet in person with Admiral King—who belatedly agreed not only to Nimitz's reading of Japanese intentions, but to Nimitz's plan: namely to engage in a sea battle with the Japanese invasion fleet in the New Guinea area with the only two U.S. aircraft carriers he could bring to bear in the region, given that USS *Hornet* and *Enterprise* were still returning to Hawaii from Japanese waters, following the Tokyo raid.

To the President's joy, in the final days of April 1942, Nimitz's task force commander, Admiral Jack Fletcher, duly set upon the Japanese armada—and in the first carrier-to-carrier battle of World War II, as well as first naval battle ever conducted in which the combatant ships were out of sight of each other for the entire engagement, Fletcher's planes fought the Japanese invasion to a standstill: the Battle of the Coral Sea, fought between May 1 and 8, 1942.

A huge American carrier, the venerable thirty-six-thousand-ton USS *Lexington* (known as "Lady Lex") was sunk, and the USS *Yorktown* badly damaged, while only the small Japanese carrier *Shohu* was blown apart on May 7, 1942, by American torpedo and dive-bomber planes. Yet that was the first major Japanese warship to be sunk in World War II—and more importantly, its loss persuaded Admiral Inoue, at his headquarters at Rabaul, to abandon the entire Japanese Coral Sea operation.

Back in Japan, Admiral Yamamoto, the fleet commander, countermanded the pull-back order, but it was too late: the invasion vessels had turned back to Truk—and by nightfall, though both big Japanese carriers, the *Shokaku* and *Zuikaku,* were still afloat, they were down to only thirty-nine planes, and running low on fuel.[8] Besides, Yamamoto was reminded, the *Zuikaku* would be urgently needed for the impending Midway operation. Instead of another amphibious attack on Port Moresby, it was decided, an overland campaign by Japanese Army forces would be conducted across the Owen Stanley Ridge—an impenetrable jungle that was, in due course, to prove beyond even their fabled abilities.

Round one had therefore gone to Roosevelt: ensuring the security of Australia from invasion—indeed permitting it to become the springboard for a counteroffensive that could, in time, take back all the territories the Japanese had stormed since December 1941.

"Delighted to hear your good news," Churchill cabled the White House[9]—aware how wise the President had been to ignore his request for help in the Indian Ocean, only days before.

However, as the *Zuikaku* withdrew to Japanese waters to rendezvous with the gathering Imperial Fleet, there was a far bigger naval battle to be fought—over Midway.

Watching Japanese reactions to the Doolittle Raid, the President held his breath, scarcely able to believe the Japanese Navy would risk the virtually complete ascendancy they had hitherto gained in the Pacific and Indian Oceans in a do-or-die sea battle—especially if its object was little Midway, an atoll in the middle of the Pacific with a simple landing strip. Yet this was what American intelligence in Hawaii was predicting.

Secretary Stimson, ashamed he had not installed a similar Map Room to that of the President, had set about rectifying the War Department's myopia. It was, he realized belatedly, small wonder the President had refused to put Nimitz under MacArthur's supreme command, for the President was able to see not only the global picture, but naval and army perspectives of the picture that were hitherto a closed book to Stimson's team. "It was a great help to go over them tonight with him," Stimson had recorded, noting the movements on the President's maps of U.S. naval forces in the Pacific as well as the Atlantic, after leaving him. "Every task force, every convoy, virtually every ship is traced and followed in its course on this map as well as the position of the enemy ships and the enemy submarines so far as they can be located."[10]

The President, moreover, had "talked very frankly" to Stimson about the inability of the British to marry their naval and RAF forces in India.

However hard Prime Minister Churchill begged for U.S. naval rescue in the Indian Ocean, he'd told Stimson, the U.S. Navy would only engage the Japanese Imperial Navy *on its own terms,* and in its own waters — where U.S. air power, army air force as well as naval, could be husbanded and put into battle *together,* to maximum effect.

Admiral King's intelligence team had initially been unable to fathom where the secret Japanese invasion fleet was heading, following the Doolittle Raid — causing Colonel Stimson to ask General Marshall to fly out to California to make sure, in person, that U.S. coastal forces were prepared for an onslaught, if Yamamoto's forces were heading for America.

The President had been unconvinced, though happy to see California's air and coastal defenses put on a more warlike footing. What would a raid against the West Coast achieve? Why would the Japanese *Kido Butai* fleet take the risk?

Nimitz's team was convinced Midway was the designated target — and in order to overcome the skepticism of Admiral King's cryptographers in Washington, they came up with a simple but brilliant idea. Operating under the foremost cryptanalyst of his time, Commander Joe Rochefort, Nimitz's intelligence analysts ordered a hoax American radio transmission to be sent out from U.S. forces on Midway, requesting water. The Japanese intercepted the signal — and in dutifully forwarding a copy of the message to their headquarters in their own secret naval cypher, they inadvertently gave away the code name they had been using for target Midway, since before their fleet set out. Thenceforth Nimitz had, at his headquarters, an advance copy of Yamamoto's plan, complete with dates, times — and its true destination.[11]

By May 20, 1942, with more than 85 percent of Japanese orders successfully deciphered, it was agreed between King and Nimitz that an intended Japanese feint or diversion toward the far-north Aleutian Islands would be left to local U.S. Army, Army Air Force, and naval units to repel. Instead, the United States would focus on repelling the attack on Midway. The airfield defenses and U.S. Air Force contingent would be substantially reinforced from Hawaii, together with submarines. Meanwhile, the three remaining carriers of Nimitz's Pacific fleet (the fourth having been sunk in the Coral Sea) would be sent out ahead of time from Hawaii, under strict radio silence and hopefully undetected by Japanese long-range patrol aircraft. They were ordered to lie concealed in waters several hundred miles from the atoll, out of Japanese view: an ambush to ambush the ambushers. Only after the Japanese invasion of the island began would they be put into battle, when the four huge Japanese carriers would be awaiting the return of their bomber planes.

Thus was the Midway counterstrike prepared, less than six months after Pearl Harbor.

Like Admirals Nimitz in Hawaii and King at the Navy Department in Washington, the President followed the build-up to the great sea battle at the end of May with mounting anticipation.

"Called me up last night from Wash[ington] about the weekend," Daisy Suckley noted, disappointed, in her diary on May 26, 1942, when Roosevelt canceled his weekend plans to come to Hyde Park. "He can't get away until next Monday or Tuesday."[12] He was, he explained, expecting the Russian foreign minister, Vyacheslav Molotov, who would be pleading for an immediate American invasion of France to draw off German divisions in Russia. He also had a bevy of foreign royals visiting, including the Duke and Duchess of Windsor — "who are in Washington on their own invitation and about as welcome as a pair of pickpockets," the President's secretary, Bill Hassett, noted caustically in his own diary.[13] "He said he was very tired & all alone & was going to bed at 10," Daisy meanwhile recorded. "He said Fala," his faithful Scottie, "was very sweet — jumped into the front seat of the car beside him & went to sleep."[14]

As former assistant secretary of the navy in World War I, and now U.S. commander in chief in World War II, the President was all too aware of the critical importance of Nimitz's plan for the impending naval battle — details of which he did not even share with Churchill. Against Japan's eight aircraft carriers, ten battleships, and twenty-four heavy and light cruisers, seventy destroyers, and fifteen submarines, as the Imperial armada set out from Palau and Japanese waters on May 27, 1942, Nimitz could only furnish three carriers, seven heavy cruisers and one light cruiser, fourteen destroyers, and twenty-five submarines.

On April 9, MacArthur's former U.S-Filipino forces on Bataan had finally been forced to surrender,[15] and a month later, on May 6, 1942, after full-scale Japanese landings across the strait, General Wainwright was forced to surrender his thirteen thousand remaining forces on Corregidor Island to the seventy-five-thousand-strong Fourteenth Japanese Imperial Army under General Masaharu Homma (who was subsequently executed for his role in the infamous Bataan Death March). For weeks thereafter, General MacArthur had been bombarding the President and Combined Chiefs of Staff from his headquarters in Melbourne with new warnings of "catastrophe" unless the President sent him aircraft carriers and more aircraft. "The Atlantic and the Indian Ocean should temporarily be stripped" to provide security for Australia, he cabled to instruct General Marshall. "If this is not done, much more than the fate of Australia will be jeopardized. The United States itself will face a series of such disasters and a crisis of such proportions as she never faced in her whole existence."[16]

The President—aware that crying wolf was par for MacArthur's course—patiently explained to the general that his fears were unfounded. A huge Japanese fleet—the largest amphibious invasion fleet it had ever sent into combat—was currently at sea, invisible thanks to dense fog, but "it looks, at this moment," he cabled MacArthur on June 2, "as if the Japanese fleet is heading toward the Aleutian Islands or Midway and Hawaii, with a remote possibility it may attack Southern California or Seattle by air"—i.e., *not* Australia.[17]

The President's message shut MacArthur up—the general victim of his inveterate localitis. Admiral Yamamoto, as forecast by Nimitz's code breakers, meanwhile split his armada; directing one-third to the Aleutians and Dutch Harbor, the rest to Midway to support his ten-thousand-man invasion of the island, backed by four aircraft carriers and a formidable amount of naval artillery. Assuming the Japanese proved successful, there was no end to what they could next undertake—indeed, in the absence of General Arnold, who had flown to England, General Marshall had sent all currently available Lend-Lease bombers still in the United States to the West Coast rather than to Britain.

As the President politely dealt first with the Russian foreign minister in Washington, then with the visiting royals at Hyde Park, anticipation became torment.

Before lunch on June 3, the President took his guest Princess Martha, and her companion Madame Ostgaard, to the Library, where his personal and presidential papers were to be housed. The midday meal was taken early, however, as his personal secretary recorded that day—for the President had agreed to seek alternative summer accommodation for still other "royal exiles."[18]

That the President of the United States, directing critical military efforts across the entire globe, should be tasked with finding a summer residence for the Dutch royal family in exile struck Hassett as extraordinary. "Roosevelt, Inc, gentlemen's estates and summer homes," he added sarcastically in his diary. "Needs of royalty carefully attended to."[19]

Roosevelt enjoyed driving, though, given his immobility, and was happy to check out locations for the royals. Besides, there was nothing better the President could do while waiting for news from Midway. He had the utmost confidence in Admiral Nimitz, commander in chief of the Pacific Fleet, whom he liked and admired for his cool judgment. On the basis of Nimitz's brilliant intelligence breakthrough, the President had authorized the admiral's ambush. It would be the greatest sea battle of the war thus far — employing America's entire Pacific force of aircraft carriers. It was, in sum, all or nothing — and unlike Winston Churchill, he had no intention of interfering in the operations of his commanders in the field, or ocean.

"Some day," Nimitz had already written to his wife on May 31, 1942, "the story of our activities will be written and it will be interesting."[20]

Interesting it certainly was. On the morning of June 3, 1942, word came through both to Nimitz at his headquarters in Hawaii, and to the President in Washington, that two Japanese light carriers had begun their feint attack on Dutch Harbor. A thousand miles farther south, however, Yamamoto's primary fleet was approaching Midway.

USAAF bombers, operating from the Midway airfield, did their best to hit the approaching vessels of the Japanese landing force from eight to twelve thousand feet. Their aim was appalling — some bombs missing by more than half a mile. The great Japanese armada merely continued toward the atoll — and early on the morning of June 4, 1942, launched their expected attack — using some 108 Japanese naval bombers.[21]

There followed a day of intense drama — the airwaves crackling with claims and counterclaims.

That evening, near Hyde Park, the President's train was made ready, and at 11:00 p.m. it left Highland Station for Washington, complete with royals, Harry Hopkins, "and the rest of us," Bill Hassett recorded.

Waking the next morning, the President left the train at Arlington at 9:00 a.m. and made his way by car to the White House. There, in the Map Room and the Oval Office, tantalizing fresh news began to come through. First three, then *all four* Japanese aircraft carriers in the attack on Midway had been successfully ambushed. They were burning fiercely — indeed, in the course of the day they were ordered to be torpedoed and sunk by their own colleagues, lest they fall into American hands and be towed back to Pearl Harbor in quasi-Roman triumph. The entire carrier contingent of the Japanese *Kido Butai* fleet, under Admiral Nagumo — commander of the attack on Pearl Harbor — had been obliterated.

As Admiral Nimitz proudly declared in the official communiqué he issued on June 6, 1942: "Pearl Harbor has now been partially avenged. Vengeance will not be complete until Japanese sea power is reduced to impotence. . . . Perhaps we will be forgiven if we claim that we are about midway to that objective."[22]

As it transpired, the American ambushers had been extraordinarily lucky. The discipline, professionalism, and bravery of the Japanese aviators and seamen were beyond compare — indeed, at 10:00 a.m. on June 4, 1942, the great naval battle could well have gone the other way. Not a single American plane operating from the U.S. Army airbase on Midway, despite heroic attacks in which most were shot down, even scratched one of Admiral Nagumo's carriers. Nor did a single U.S. torpedo plane, operating from Admiral Raymond Spruance's Task Force 16, under Fletcher's overall tactical command, succeed in hitting the Japanese carriers, despite equal heroism — in fact, the Japanese carriers survived no less than *eight* waves of American attacks without suffering the slightest damage. U.S. Army Air Force bombers simply proved incapable of hitting the nimble Japanese carriers from high altitude, while the heavily laden, trundling TBD Devastator torpedo bombers, flying at fifteen hundred feet from their carriers with their mostly ineffective torpedoes, were sacrificial lambs when pounced upon by Japanese Zero fighters, as the President was informed by three Midway fliers invited afterward to visit him at the White House and tell him in person their version of the battle. What was needed, they said, was "something that will go upstairs faster."[23]

Yet the sacrifice of so many American fliers had not been in vain. With the Japanese Zeros rendered helpless at mounting a defense while flying at low altitude to deal with the Devastator torpedo bombers, a veritable storm of American SBD Dauntless dive-bombers suddenly swooped on the Japanese carriers from nineteen thousand feet. With their thousand-pound bombs they had hit each one of the three Japanese flattops in the main Japanese formation on its pale yellow flight deck, marked with a red "Rising Sun" disk — penetrating the vessel's wood and steel topskin and setting the warship ablaze.

One accompanying destroyer captain, Tameichi Hara, could not at first credit news that all three Japanese carriers were ablaze. "Was I dreaming? I shook my head. No, I was wide awake! . . . The horrifying reports continued until there was no room for doubting their accuracy."[24] Aboard the largest battleship in the world, the Japanese flagship *Yamoto,* hundreds of miles in the rear, Admiral Yamamoto "groaned" as he read the latest dispatch from the commander of the screening force at Midway: "Fires raging aboard *Kaga, Soryu,* and *Akagi* resulting from attacks by enemy carrier and land-based planes. . . . We are temporarily withdrawing to the north to assemble our forces."[25] Admiral Nagumo, victor of the Pearl Harbor sneak attack, was compelled to evacuate his flagship, the *Akagi,* as it burned — only dissuaded from suicide by a subordinate. Then the fourth and last Japanese carrier, the *Hiryu,* was attacked by still *more* SBD Dauntless planes — and though its pilots managed to reach and disable the USS *Yorktown* carrier, it, too, became an inferno, and had to be sunk by its own compatriots.

Four Japanese carriers sunk in a single day?

It was a victory on a scale not even Nimitz had dreamed possible. Moreover, though Admiral Yamamoto attempted to snatch a consolation victory by summoning his remaining cruisers and lighter vessels from support of his Aleutian expedition, he was out of luck. Admiral Fletcher prudently kept his remaining ships out of nighttime range. Instead he called upon U.S. Air Force B-17s and Marine Corps aircraft from Midway to attack and severely damage Yamamoto's luckless, retreating Japanese heavy cruisers in daylight the next day — one of which was finished off by more dive-bombers from Spruance's Task Force 16 on June 6, 1942.

The greatest sea battle of the Second World War was then finally over — its conclusion marked in the early hours of that morning when Admiral Yamamoto reluctantly issued his signal of defeat: Order #161, beginning with the words, "The Midway Operation is canceled."[26]

For Yamamoto, the naval confrontation in the central Pacific had turned into a calamity. Even his hopes that Japanese planes from the *Hiryu,* his last, burning carrier, could instead land on Midway's runway had gone up in smoke, literally.[27] Every single Japanese plane that had set out on the fateful, punitive expedition was therefore lost, together with more than a quarter of the Imperial Navy's best pilots.[28]

Admiral Yamamoto's chief of staff, Matome Ugaki, noted in his diary that those responsible "ought to have known the absurdity of attacking a fortress with a fleet!"[29] Yet it was Ugaki's own commander who had persuaded the Japanese Combined Headquarters to back the plan — dooming Japan to defeat in the Pacific if it failed.

At the White House, by contrast, there was jubilation — followed by alarm when a reporter from the *Chicago Tribune,* Robert McCormick's isolationist newspaper that had earlier leaked the President's Victory Program before Pearl Harbor, wrote a new dispatch detailing exactly what American intelligence had known of the Japanese plan, in advance of the battle. "NAVY HAD WORD OF JAP PLAN TO STRIKE AT SEA," ran the nefarious *Tribune* banner. "UNITED STATES NAVY KNEW IN ADVANCE OF JAP FLEET: Guessed There Would Be Feint at One Base, Real Attack at Another," was the headline as McCormick's exclusive was then repeated in the *Washington Times-Herald* on June 7, 1942 — together with exact details of the Japanese order of battle and three echelons.

The President was so angry, he first considered sending Marines to close down McCormick's building in Chicago, and having McCormick tried for treason — which carried the death penalty. A few days later, however, Secretary Knox talked him out of that — substituting a grand jury investigation of McCormick for violation of the Espionage Act.[30]

Fortunately, the Japanese were too shocked to take note of McCormick's publication of the leak. They had, in any case, already changed their JN25 code a week before the battle, which henceforth proved formidably difficult to break. Besides, nothing good would come from drawing further attention to McCormick's story, any more than it would have after his Victory Program revelation. McCormick would only hate the President more deeply, given the subsequent national rejoicing over the triumphant feat of American arms. Even General MacArthur was reluctantly compelled to signal his personal congratulations to Admiral Nimitz, his fellow supreme commander, on his "splendid" naval victory.[31]

In Washington, the President breathed a sigh of relief. And apprehension — knowing the United States must now turn its full attention to Europe and the defeat of Adolf Hitler, who was hell-bent on conquering Soviet Russia in a second, titanic version of Barbarossa that year.

PART EIGHT

Tobruk

17

Churchill's Second Coming

SHORTLY AFTER THE TRIUMPH of Midway, while on the floor of the House of Commons in Ottawa, the Canadian prime minister, Mr. Mackenzie King, received word that "President Roosevelt wished to speak to me over the 'phone. I left at once," he noted in his diary on June 11, 1942.

After exchanging greetings, King asked the President: "How are you? To which he, the President, replied: I am terribly 'upshot.'"[1]

Upset?

King waited to hear what about.

The President "then asked me if I had any information about a certain lady who was crossing the Atlantic to pay a visit to this side."

If enemy agents were eavesdropping — and subsequent evidence would prove that they were — they would have been puzzled. The President was unable to refer to anyone by name, it being an open line, but King "said that I thought I knew who he meant." In fact the premier had, "this morning, received word that the person mentioned would be leaving almost immediately."

The President corrected him. The "certain lady's" flight from England had been postponed by a day, Roosevelt informed his Canadian counterpart — and "then asked: 'Do you know where she is to land?'

"I replied: I do not know. He said: 'I do not know either.'"

The President tried, elliptically, to explain. He was referring, he said, to "the lady [who] was coming to stay with her daughter at a house which had been engaged for the summer at Stockbridge, Mass[achusetts]." The house wasn't ready for the lady in question — yet the President had found himself simply unable to get her to postpone her trip. As the President complained, he had personally been charged with "the job of getting the house," which had entailed quite some searching, "and in addition, he had done pretty well everything else including obtaining the servants," in fact everything "short of supplying the silver." "He said to me," King recorded the President's expostulation, "Don't you love it?"

"Well," responded the premier with a chuckle, "that comes from your making such a favourable impression" on the daughter of the "certain lady."

And at that the two political leaders burst into laughter.[2]

It was true. The President had a soft spot for Princess Juliana of the Netherlands, her husband, and their children.

To help the President out, the Canadian premier duly agreed to house the princess's mother, Queen Wilhelmina, and her family on their arrival in North America, at least until the Massachusetts summerhouse was ready—the two men shaking their heads at such responsibilities in the very midst of a world war for civilization. "The President was quite amusing about the whole affair," Mackenzie King summarized in his diary that night, "saying that it was pretty much the limit in the way of imposition."[3]

Worse would surely follow, the President added. Once Queen Wilhelmina—a notoriously demanding lady—arrived in the United States, he would, as President, have to invite her over to Hyde Park, for the summer home he'd found for her was not far from his own home. He was "afraid the lady might be shocked by his informality around Hyde Park in the summer," he confided to King. Fortunately, "the Legation had told him she did not wish anything in the nature of ceremonies," he added with relief: "no salute or guard of honour or anything of the kind. . . ."[4]

Again, the two men chuckled at the irony: Mackenzie King hauled out of a crucial debate on military conscription—a knife-edge issue in Canada, given the country's French-speaking population with mixed cultural loyalties to Canada, Vichy France, and General Charles de Gaulle's Free French movement—to speak with the President. Reflecting on the mix of wartime crisis and the accommodation expectations of the royals, King noted that the situation was approaching "opera bouffe."[5]

It seemed unlikely, however, that the President of the United States would be calling the Canadian premier only about a matter so trivial. Aware still of the security issue of an open telephone line, King went on to congratulate his interlocutor on news of the American victory at Midway.

"Was it not splendid?" he recorded Roosevelt's response. "It made a complete difference in regard to the whole situation in the Far East and throughout the Pacific as well." The President didn't have "all the information yet," but was able to tell King cautiously that the situation was "very good."[6]

Only then, finally, did the President get to the point.

Prime Minister King should be ready for a phone call summoning him to Washington "on pretty short notice some day soon."[7]

The President was expecting a yet more important VIP than "a certain lady."

Two days later the anticipated cable came from London.

"In view of the impossibility of dealing by correspondence with all the many difficult points outstanding," Winston Churchill signaled the President, "I feel it is my duty to come to see you."[8]

The President knew what this meant. Since he had personally promised Mr. Molotov, the Russian foreign minister, that the Allies would launch an Allied landing in France that year in order to draw German divisions away from the Eastern Front—German divisions that would otherwise be thrown into Hitler's renewed offensive in Russia—and since the British would have to furnish the majority of troops for such landings in France, the President was pretty sure what the Prime Minister was coming to say.

Whatever he had promised in the spring, when India looked vulnerable to Japanese invasion and he desperately needed American help, he was going to go back on his word. The British wouldn't do it.

Churchill's second visit to Washington would certainly put the cat among the pigeons—and given the feelings of the U.S. War Department on the subject, would require even more finessing than dealing with the Dutch battle-ax, Queen Wilhelmina.

In preparation for the inevitable uproar that the Prime Minister's arrival would cause, the President called a meeting of his old war cabinet at the White House at 2:00 p.m. on June 17, 1942, before heading to Hyde Park for a last weekend of peace and quiet.

It was in the war cabinet meeting that the President "sprung on us a proposition which worried me very much," as Secretary Stimson recorded in his diary that night—full of foreboding.

The President's "proposition" was the one Stimson had feared for over a year, since Roosevelt first ordered the Victory Program to be prepared:

namely, Roosevelt's personal preference for American landings in French Northwest Africa, where there were currently no German forces.

Stimson had always hated the Northwest Africa plan — preferring the idea of a single Anglo-American thrust across the English Channel. So too did General Marshall — who had hoped that his recent appointment of young Major General Dwight Eisenhower to head up U.S. military forces in Britain would bring a sense of urgency to Allied plans for landings on the coast of mainland France: plans that, on a prior trip as an "observer," Eisenhower had thought lamentably unfocused.[9] "I'm going to command the whole shebang," Eisenhower had proudly told his wife, Mamie, as he got ready for his departure to London, slated for June 24, 1942, to stiffen British resolve.[10]

Thanks to Churchill, the "shebang" suddenly seemed in grave jeopardy. "It looked as if he [the President] was going to jump the traces over all that we have been doing in regard to BOLERO [code name for the build-up of U.S. forces in Britain leading to an invasion of France] and to imperil really our strategy of the whole situation," Stimson recorded in his diary. "He wants to take up the case of GYMNAST [the U.S. invasion of French Northwest Africa] again, thinking that he can [thereby] bring additional pressure to save Russia. The only hope I have about it at all is that I think he may be doing it in his foxy way to forestall trouble that is now on the ocean coming towards us in the shape of a new British visitor."[11]

The "British visitor" was not on the ocean, however, but in the air — and already almost over Washington.

Forewarned was, at least, forearmed, Stimson comforted himself. General Marshall, at the meeting with the President, thus had "a paper already prepared against" Gymnast, "for he had a premonition of what was coming," Stimson noted with satisfaction. "I spoke very vigorously against it [Gymnast]. [Admiral] King wobbled around in a way that made me rather sick with him. He is firm and brave outside of the White House," Stimson derided the "blowtorch" sailor, "but as soon as he gets in the presence of the President he crumples up. . . . Altogether it was a disappointing afternoon."[12]

The President insisted — insisted — he wanted a renewed study of what it would require to mount landings in French Northwest Africa.

There was little the secretary of war or the chiefs of staff could therefore do, other than to comply. "The President asked us to get to work on this proposition," Stimson noted in anguish, "and see whether it could be done."[13]

In their collective view, it couldn't.

At the British Embassy on Massachusetts Avenue, meantime, there had been "every sort of minor turmoil" over Churchill's imminent arrival—and how to explain to the American press the reason for the Prime Minister's Second Coming.

For his part, Winston Churchill could not wait to see the President, not the press—and having landed in a Boeing seaplane on the Potomac River on June 18, 1942, after flying direct from Scotland in twenty-six and a half hours, he declared he was ready to fly straight up to meet with Roosevelt at Hyde Park.

The President, to his chagrin, was not ready to see him.

"Winston arrived at 8 p.m.," Lord Halifax noted that night in his diary, "and I brought him to dine and sleep at the Embassy. He was rather put out at the President being away, and inclined to be annoyed that he hadn't been diverted to New York, from where he could have flown on more easily" to Hyde Park. "He got into a better temper when he had had some champagne, and we sat and talked, he doing as usual most of it, until 1.30 a.m."[14] Even General Sir Alan Brooke, and Major General Ismay, Churchill's chief of staff, "took advantage of the darkness on the porch to snatch bits of sleep," Halifax described, "while Winston talked."[15]

Halifax—who had good reason to resent that Churchill, not he, Edward Halifax, had become prime minister in May 1940—was now strangely passive, recognizing he himself would never have had the stamina or the determination to lead Britain in war. "He certainly is an extraordinary man," the ambassador described Churchill in his diary; "immensely great qualities, with some of the defects that sometimes attach to them. I couldn't live that life for long!"[16]

The Prime Minister seemed full of beans. The Battle of Midway had removed his fears of a Japanese invasion of India, as well as any possible Japanese threat to the Middle East from the Orient—so much so that Churchill appeared utterly indifferent to the mounting negative reaction to his refusal to grant self-government to the British colony. In a moment of vexation a few days before, he had even cabled to instruct the viceroy of India to arrest Mahatma Gandhi if he "tries to start a really hostile movement against us in this crisis"[17] — claiming that "both British and United States opinion would support such a step. If he starves himself to death we cannot help that."[18]

Even Halifax found that difficult to swallow. But what of German moves in the opposite direction — the threat of German armies blitzing their way through southern Russia and the Caucasus, and Rommel's Panzerarmee Afrika, aiming toward Egypt and the Suez Canal?

The "news from Libya doesn't look good," Halifax confided in his diary, with the British Eighth Army being forced to retreat toward the Egyptian frontier. This had left the big garrison at the port of Tobruk to be surrounded and to have to hold out, as it had the previous year, on Churchill's express, if meddling, orders. Surely, Halifax mused, "Rommel must be getting a bit strung out himself." He'd quizzed the Prime Minister on the subject. "Winston on the whole, though disappointed with Libya, was in good heart about the general situation."[19]

With British complacency born of American naval victory, the real Allied opera began.

Meeting briefly with Mr. Churchill at the British Embassy before the Prime Minister's departure by air to Hyde Park on the morning of June 19, 1942, General Marshall, for his part, became deeply worried. Not only did Rommel's panzer forces seem dangerously effective in North Africa, but the situation in Russia looked potentially worse than the previous year — with vast numbers of German tank and motorized troops assembling for a breakthrough.

In April that year, Churchill had sworn in person to him that the British government was one hundred percent behind American plans for a joint U.S.-British cross-Channel operation that summer or fall, if at all possible, to help the Russians, and begin at least the march on Berlin. Now, at the embassy on Massachusetts Avenue two months later, the Prime Minister sounded quite different — thanks to Midway.

The U.S. Navy had routed the Japanese — and thereby saved the British position in India. With India externally secure, Churchill's primary concern now seemed to be the defense of the Middle East as the lifeline of the British Empire — not the question of how best to mount an offensive in Europe that would help keep the Russians in the war, and bring down Hitler's odious Nazi regime.

Stimson, when he met with General Marshall later that morning, was as furious as Marshall over the Prime Minister's broken promises. Marshall explained he'd "seen Churchill this morning up at the Embassy," as Stimson confided to his diary, and had described Churchill as being "full of discouragement" about a Second Front, while sounding off with "new proposals for diversions. Therefore the importance of a firm and united stand on our part is very important," Stimson concluded.[20]

Frantic lest the Prime Minister, by flying up to Hyde Park to stay with the President, now ruin the plans which the War Department had been making for a cross-Channel invasion of France, Stimson and Marshall discussed their options. They were adamant they wanted no "diversions" — especially ones that would simply bolster the British position or empire in the Middle East. Sitting down, Stimson thereafter drew up a new strategy paper, or Memorandum to the President. General Marshall thought it brilliant. "Mr. Secretary," Marshall declared, "I want to tell you that I have read your proposed letter to the President [aloud] to these officers" — indicating Generals Arnold, McNarney, Eisenhower, Clark, Hull, and one or two others, who were present at the War Department meeting — "and they unanimously think it is a masterpiece and should go to the President at once."[21]

Stimson thereupon sent his Memorandum to the President by personal messenger, together with a letter saying it also represented the views of Marshall "and other generals" in the War Department. As if this were not enough, Marshall and King drew one up, too, which they also dispatched by courier. Marshall even telephoned the President at Hyde Park personally, asking Mr. Roosevelt to please, please read it, once it arrived — begging him to make no decisions with Mr. Churchill until there had been time for the President to meet with his military advisers in Washington, on Mr. Roosevelt's return to Pennsylvania Avenue.

Thus began Act 2.

Secretary Stimson's letter to the President certainly pulled no punches.

Not only was Bolero, the plan to build up massive American forces in the British Isles, a definitive way of dissuading Hitler from any thoughts of invading Britain, Stimson lectured the Commander in Chief in writing, it promised to be a potent launch pad for the U.S.-British invasion of France — an invasion that would "shake" Hitler's renewed invasion of Russia that year and eventually lead, in 1943, to "the ultimate defeat of his armies and the victorious determination of the war" for the United Nations.

"Geographically and historically," Stimson noted in his diary that night, "Bolero was the easiest road to the center of our chief enemy's heart. The base," in the British Isles, "was sure. The water barrier of the Channel under the support of Britain-based air power is far easier than the Mediterranean or the Atlantic. The subsequent over-land route into Germany is easier than any alternate. Over the Low Countries has run the historic path of armies between Germany and France."[22]

The recent "victory in mid-Pacific" at Midway had now ruled out any possibility of Japanese raids affecting U.S. aircraft manufacturing on the West Coast, Stimson was glad to record. "Our rear in the west is now at least temporarily safe. The psychological pressure of our preparation for Bolero is already becoming manifest. There are unmistakable signs of uneasiness in Germany as well as increasing unrest in the subject population of France, Holland, Czechoslovakia, Yugoslavia, Poland and Norway," he asserted—with rather less evidence. "This restlessness," he claimed, "patently is encouraged by the growing American threat to Germany. Under these circumstances an immense burden of proof rests upon any proposition which may impose the slightest risk of weakening Bolero," he added—fearing that the Prime Minister was going to suggest alternative schemes: schemes that would do nothing to expedite victory beyond frittering away America's huge but overstretched arsenal of men and materiel.

"When one is engaged in a tug of war," the secretary of war warned the President in his letter, "it is highly risky to spit on one's hands even for the purpose of getting a better grip." No "new plan should be whispered to a friend or enemy unless it was so sure of immediate success and so manifestly helpful to Bolero," he noted, "that it could not possibly be taken as evidence of doubt or vacillation in the prosecution of Bolero."

The problem, in Stimson's view, was Winston Churchill—and his predilection for fatuous diversions. The Prime Minister had even told Marshall at the British Embassy that he favored an Allied landing in *Norway*, instead of France, that year—a proposition Stimson could only shake his head over. Yet it was not Churchill's Scandinavian fantasy (Churchill having been responsible as First Lord of the Admiralty for utter disaster when ordering an Anglo-French invasion of Norway in the spring of 1940) that made the U.S. war secretary anxious. Rather, it was the President's own bright idea that *really* worried him: the "President's great secret baby," as Stimson would sneeringly call it: "Gymnast."[23]

Against the President's preference for a U.S. invasion of French Northwest Africa the secretary inveighed, in his letter to the President, with a kind of elderly man's fervor. Gymnast, Stimson claimed, would tie up "a large proportion of allied commercial shipping," thus making the American "reinforcement of Britain" in the Bolero build-up to an eventual cross-Channel landing "impossible."[24]

The Allies were coming, as Stimson saw it, to a real "crisis" in the direction of the global war[25] — with Prime Minister Winston Churchill more interested in saving British interests in Egypt and the Middle East, just as he had done in India, than in attacking Nazi Germany.

Stimson's talks with Marshall "over the crisis which has arisen owing to Churchill's visit" had in fact continued all morning on June 20, 1942, even after dispatching his "masterpiece" letter to the President.[26]

To add insult to injury, Stimson learned, a frank letter to Marshall had been received from Vice Admiral Lord Louis Mountbatten, in London. This indicated that the British chiefs of staff (in whose meetings Mountbatten, as chief of British Combined Operations, participated as an "observer") were not only unanimously opposed to a cross-Channel Second Front in 1942, but hoped to persuade the President of the United States, via Mr. Churchill, to abandon a further Bolero build-up of U.S. troops and eventual assault landing in France. Instead, they hoped the President could be persuaded to send American reinforcements to help the British in holding Egypt.[27]

Stimson was furious — seeing in all this a British conspiracy to turn the war against Hitler into a war to preserve, or even expand, the British Empire. All-out "American support of the Mideast"[28] — Britain's shortest lifeline to its empire — would thus be the British chiefs' game plan in accompanying the Prime Minister, according to Mountbatten's somewhat duplicitous warning — Mountbatten claiming, as chief of Combined Operations, that he personally was all in favor of an Allied assault on the French coast, in which he might be given a starring role.

All thus rested now upon the Commander in Chief of the United States, in Hyde Park — and how Mr. Roosevelt would respond to Churchill's artful arguments *against* a Second Front landing across the English Channel that year.

Stimson feared the worst. "I can't help feeling a little bit uneasy about the influence of the Prime Minister on him at this time," the war secretary noted in his diary that evening. "The trouble is Churchill and Roosevelt are too much alike in their strong points and in their weak points. They are both brilliant. They are both penetrating in their thoughts," the secretary dictated, "but they lack the steadiness of balance that has got to go along with warfare."[29]

Roosevelt was certainly bombarded by arguments that weekend.

Responding to the President's latest June 17 request for an immediate, up-to-date study of U.S. landings in Morocco and Algeria, General Marshall's own couriered letter was emphatic. It informed the President that Marshall, Admiral King, and their army, air, and naval staffs had duly reexamined the "Gymnast project as a possible plan for the employment of U.S. forces against the Axis powers in the summer and fall of 1942, following our conversation with you Wednesday." Their conclusion was devastatingly negative. "The advantages and disadvantages of implementing the Gymnast plan as compared to other operations, particularly 1942 emergency Bolero operations, lead to the conclusion that the occupation of Northwest Africa this summer should not be attempted," they bluntly reported.[30]

To explain their opposition to the President's plan, the U.S. Army chief and U.S. Navy commander in chief enclosed, with their letter, a three-page official army and navy analysis for the President. Gymnast would not work, they claimed, because the Luftwaffe would be able, they asserted, to move aircraft into "Spanish and North African bases, from which they could operate against Casablanca" — air operations that would threaten the U.S. aircraft carriers vital to provide the necessary air support for the invasion. If Gymnast was seriously being advanced as a major undertaking by the President, they argued, necessitating an inevitable diversion of U.S. forces from other projects, why on earth not invade Brittany or the Brest Peninsula — Operation "Sledgehammer" — instead? These would at least bring U.S. forces closer to an eventual Allied path to Berlin.

Gymnast's naval requirements, moreover, promised "disaster in the North Atlantic," the chiefs claimed, owing to the "thinning out" necessary to provide naval support to a Northwest African campaign as it unfolded. Moreover, Vichy French cooperation was essential, yet unlikely — the American vice consul at Casablanca having reported that the Nazis "have already made plans to meet a U.S. invasion of Northwest Africa." The Germans, according to the vice consul, had available three armored divisions, boasting three hundred tanks, on top of the "700 tanks there" and

"200–300 Stukas" and "58 Messerschmitts." There were even, the chiefs claimed, "250 to 300 fast launches collected on the coast of Spain" that Hitler might use . . .[31]

Polite as always, the President duly met Churchill's U.S. Navy plane when it landed at New Hackensack airfield on the evening of June 19, 1942. He was annoyed, however, at Churchill's presumption — the Prime Minister emerging from the aircraft with no fewer than five other people, including Major General Ismay, his military assistant or chief of staff. Faced with the unexpected retinue the normally stoic President said testily to his private secretary, "Haven't room for them" — and gave instructions some would have to be housed nearby.[32]

The President was yet more irritated when Churchill began to use the President's exclusive, personal telephone line to the White House as if it were his own. Churchill had, the President's secretary Bill Hassett recorded, "seated himself in the President's study and had entered upon an extended conversation with the British Embassy in Washington," until it was terminated by the President's chief switchboard operator, Louise Hachmeister.[33]

Churchill, though, was Churchill.

"You cannot judge the P.M. by ordinary standards," General Ismay had written to General Claude Auchinleck, commanding British forces in the Middle East, earlier that year. "He is not in the least like anyone that you or I have ever met. He is a mass of contradictions. He is either on the crest of the wave, or in the trough, either highly laudatory or bitterly condemnatory; either in an angelic temper, or a hell of a rage. When he isn't fast asleep he's a volcano. There are no half-measures in his make-up. He is a child of nature with moods as variable as an April day."[34]

The President could be forgiven for comparing the situation to that of August 1941, when meeting Churchill aboard their battleships off the coast of Newfoundland. Then, too, the President had known Churchill was approaching with a purpose — to get the United States to declare war on Germany — and had his own counterstrategy: to make the British sign up first to a statement of principles, before any idea of a wartime alliance could be discussed. But then, at least, the President had had his own chiefs of staff by his side. Now he was on his own, without even Harry Hopkins.[35] His chiefs were in Washington, imploring him by courier and phone not to make any premature "deal."

In the circumstances, the President resorted to his usual strategy in such matters: charm. Taking Churchill in his specially converted convertible, equipped with only hand controls, he drove the Prime Minister "all over the estate, showing me its splendid views," as Churchill later related—but with some close calls. "In this drive I had some thoughtful moments. Mr. Roosevelt's infirmity prevented him from using his feet on the break, clutch or accelerator. An ingenious arrangement"—designed by the President—"enabled him to do everything with his arms, which were amazingly strong and muscular. He invited me to feel his biceps, saying that a famous prize-fighter had envied them. This was reassuring; but I confess that when on several occasions the car poised and backed on the grass verges of the precipices over the Hudson I hoped the mechanical devices and brakes would show no defects."[36]

The President's performance was clearly intended to keep the Prime Minister quiet, rather than allow him to "talk business" without their advisers present—indeed, the picture of the two men was all too symbolic, careering around Hyde Park in the President's open-topped, dark-blue 1936 Ford Phaeton with Churchill attempting to tell the President that the Allies should invade either Norway, in the far north of Europe, or French Northwest Africa, to the south of Europe, or indeed anywhere but mainland France.

More ominously, Winston Churchill was bearing with him his own memorandum for the President, which he'd dictated to his secretary at the British Embassy that very morning—a document he had in his pocket and was intent on handing in person to Roosevelt, if he could, without any of Roosevelt's military advisers being present. And Churchill was no mean writer, the President knew.

The President promised to read it—side by side with General Marshall and Admiral King's three-page memorandum that night.

Once he'd read the two competing memoranda carefully, the President recognized there would be a tough session on his return to Washington.

"The President wanted to see secretaries of War and Navy, Admiral King, and General Marshall tomorrow afternoon," Hassett noted in his diary. "Said later he might not see them till evening and would notify them after reaching the White House. Does, however, want to see General Marshall at 11 o'clock tomorrow morning—all appointments off the record," Hassett added.[37]

• • •

Had Stimson written a better memorandum, he too might have been invited to attend the "crisis" meeting at the White House the next day. But the war secretary's prose was far from masterful, it was dour.

Stimson's argument for an immediate and exclusive Bolero build-up and cross-Channel invasion of France seemed not only doctrinaire, but his justifications for building up forces in Britain rather than sending them into battle in French Northwest Africa were jejune: among others he now asserted that Britain faced a possible, even imminently "probable," invasion by German paratroopers, "producing a confusion in Britain which would be immediately followed by an invasion by sea."[38]

The President shook his head over that.

Churchill's memorandum, by contrast, addressed squarely and without fear the real question the President had in mind: if a Second Front could not be mounted in 1942, where could the Allies actually *strike?* "Arrangements are being made for a landing of six or eight Divisions on the coast of Northern France early in September," Churchill began his critique. However, he declared, "the British Government would not favour an operation that was certain to lead to disaster," since this would "not help the Russians whatever in their plight, would compromise and expose to Nazi vengeance the French population involved and would gravely delay the main operation in 1943. We hold strongly to the view," he summed up, "that there should be no substantial landing in France this year unless we are going to stay."[39]

The fact was, neither Churchill nor his advisers could see any hope of such a successful "substantial" operation in France in 1942—whatever the U.S. chiefs of staff might argue. "No responsible British military authority has so far been able to make a plan for September 1942 which had any chance of success unless the Germans become utterly demoralized, of which there is no likelihood," Churchill stated categorically. "Have the American Staffs a plan? If so, what is it? What forces would be employed? At what points would they strike? What landing-craft and shipping are available? Who is the officer prepared to command the enterprise? What British forces and assistance are required?"[40]

Churchill was famous for his Macaulayan eloquence, rich in metaphor and tart phrasing; this time, however, he simply ended with what he knew would resonate with the President. If an immediate Second Front landing was impossible in September 1942, "what else are we going to do?" he asked. "Can we afford to stand idle in the Atlantic theatre during the whole of 1942? Ought we not to be preparing within the general structure of BOLERO some other operation by which we may gain positions of advantage and also directly or indirectly to take some weight off Russia? It is in this setting and on this background that the Operation GYMNAST should be studied"[41] — unaware that General Marshall and Admiral King had just studied it yet again, on the President's orders, and had concluded it "should not be undertaken"!

Shortly before 11:00 p.m. on Saturday, June 20, 1942, the President's party left Hyde Park and drove to the local railway station. Churchill got out first, in his black topcoat, and walked up the ramp toward the train. In his diary Bill Hassett recorded the sight of the venerable little Englishman standing at the top, "at just a sufficient height to accentuate his high-water pants — typically English — Magna Charta, Tom Jones, Doctor Johnson, hawthorn, the Sussex Downs, and roast beef all rolled into one. Nothing that's American in this brilliant son of an American mother. The President went at once into his [railway] car and Winnie followed."[42]

The train then moved off, traveling slowly in order not to shake the President, who slept well.

"We all went to the White House together," Hassett recorded, once they had arrived at Arlington Cantonment, just before 9:00 a.m. on June 21, 1942, for what was to prove an historic, calamitous day.[43]

18

The Fall of Tobruk

LORD HALIFAX, a Roman Catholic, had gone to church at St. Thomas — "where everybody was provided with fans" to combat the heat — having been assured that "the party had got back all right" from Hyde Park.[1] He was relieved to know Churchill wouldn't be returning to the embassy — the Prime Minister having been invited by the President to stay at the White House for the remainder of his visit.

Once Churchill was reestablished in his old bedroom on the second floor, he and Major General "Pug" Ismay joined the President and General Marshall in the Oval Study to debate the pros and cons of Bolero and Gymnast. They had barely begun, however, when the whole issue of a Second Front in 1942 was exploded by a bombshell. It came in the form of a piece of pink paper: the copy of an urgent telegram, brought up from the Map Room and handed to the President.

The President read it, then handed it to Churchill.

It came from the war cabinet in London.

"Tobruk has surrendered," the message ran — "with twenty-five thousand men taken prisoners."[2]

Twenty-five thousand British troops? Without fighting?

"The year before," recalled the President's speechwriter, Robert Sherwood, "Tobruk had withstood siege for thirty-three weeks. Now it had crumpled within a day before the first assault. This was a body blow for Churchill. It was another Singapore. It might well be far worse even than that catastrophe in its total effect — for with Tobruk gone, there was little left with which to stop Rommel from pushing on to Alexandria, Cairo — and beyond."[3]

In his diary, Breckenridge Long, the assistant secretary of state, noted that the British "lost 1009 tanks"—largely to Rommel's secret weapon, his dreaded "88" (millimeter) antiaircraft gun, used as a long-range anti-tank weapon—"and just smashed the British. They have had six months to prepare—and are now licked. It is serious now. They have no real fortifications between Rommel and Cairo or Suez—and a broken army. It may easily mean the loss of Egypt—unless we can stop it."[4]

As Churchill himself later wrote, it was "a bitter moment. Defeat is one thing; disgrace is another."[5]

Lord Halifax heard the same news. Meeting Churchill's doctor, Sir Charles Wilson, the ambassador "discussed probable reactions in England" to the catastrophe, "and how it is likely to affect the Prime Minister's plans here." The Prime Minister had "never meant to stay long this time," Halifax reflected, "and he certainly will not make the mistake he made in January of overlooking feeling at home"—critical outcries that might well lead now to a vote of censure in the House of Commons.[6] He would surely have to fly home, instanter.

Back in England there was, indeed, consternation and calls for Churchill to stand down as prime minister. At the Foreign Office, Sir Alec Cadogan had just acknowledged in his diary that "Libya is a complete disaster"[7] when he learned "that Tobruk had fallen." It seemed impossible to believe. The heavily fortified port, with ample munitions, water, and troops, had "held out for 8 months last time, and for about as many hours this. I wonder what is most wrong with our army. Without any knowledge, I should say our Generals. *Most* depressing."[8]

Averell Harriman, who had returned to the United States with Churchill for talks about munitions and Lend-Lease consignments, called it "a staggering blow" to the Prime Minister, which Churchill at first refused to believe. "But when it was confirmed, by telephone from London, he made no attempt to hide his pain from Roosevelt."[9]

General Ismay recalled the same. Churchill had "scarcely entered" the President's study when the news was delivered. "This was a hideous and totally unexpected shock, and for the first time in my life I saw the Prime Minister wince."[10] Before Ismay could get official confirmation, on Churchill's disbelieving orders, he met Churchill's secretary, John Martin, in the corridor bearing a new telegram. This one was from the British naval commander in chief in the Mediterranean, Admiral Harwood. It confirmed not only that Tobruk had surrendered but went on to explain to the Prime Minister that, in the circumstances, Harwood was sending the Royal Navy's Eastern Mediterranean Fleet south of the Suez Canal, toward the Indian Ocean.

General Alan Brooke, the next day, recorded the timing differently—noting news of the disaster had come through only in the afternoon of June 21, after he'd lunched with the President. "Harry Hopkins and Marshall also turned up," and it was only in the midst of a "long conference" that "the tragic news of Tobruk came in!" he scribbled. "Churchill and I were standing beside the President's desk talking to him when Marshall walked in with a pink piece of paper containing a message of the fall of Tobruk!"[11]

Whatever the actual timing, the British team seemed distraught, even lost. As Brooke subsequently admitted, in his annotations to his diary, "Neither Winston nor I had contemplated such an eventuality and it was a staggering blow"[12]—with neither man having any idea what to do.

Certainly, when Sir Charles Wilson was finally summoned and went over to the White House to see the Prime Minister, later in the afternoon of June 21, 1942, around 3:00 p.m., he found Churchill "pacing his room. He turned on me," Moran recorded in his diary notes. "Tobruk has fallen."[13]

"He said this as if I were responsible. With that, he began again striding up and down the room, glowering at the carpet." "What matters is that it should happen when I am here," Churchill confessed—stung by the humiliation in front of his American hosts.[14]

Churchill went to the window. "I am ashamed. I cannot understand why Tobruk gave in. More than 30,000 of our men put their hands up. If they won't fight—"[15]

Churchill paused, midsentence—bitterly aware of the effect abroad of such an abject British surrender, following the disastrous showing of British imperial forces in Malaya, Burma, and the Indian Ocean. Not only might people in the occupied nations lose heart, but the surrender could even drive neutral countries like Turkey, Portugal, or Spain to parley with Hitler . . .

Directing all Third Reich propaganda, Joseph Goebbels reveled in the news from North Africa.[16] The Spanish press agency EFE, he noted, was describing "an atmosphere of catastrophe in Washington."[17]

Even Goebbels had not expected Rommel to be so successful, given the general's skimpy reports to Berlin since the beginning of Operation Theseus, his plan to drive the British out of Libya and Egypt, on May 26, 1942.

Not only had the great port of Tobruk now been captured by Rommel, but it contained enough food, oil, and weapons to keep the Panzerarmee Afrika going for three months. Churchill must have knowingly flown the coop in order to be out of London, Goebbels conjectured cynically — the Prime Minister knowing he'd be blamed for misleading the public into thinking the British were winning the desert war. The British press were now in an uproar, making mincemeat of Churchill's lamentable military leadership and personal "responsibility for the catastrophe."[18]

Goebbels thus gloried in Britain's shame. The British did not possess a "single general who has shown himself a real commander, on any current field of battle. After Wavell, Auchinleck; after Auchinleck, Ritchie" — the Eighth Army commander, appointed by Auchinleck after he'd had to fire General Alan Cunningham, his first commander. "And every one a failure," Goebbels sneered.[19] The latest German 88mm antitank guns and their armor-piercing shells had made a killing of British armored vehicles in the desert — even decimating the American M-3 "Grant tanks" the British had been given.[20] Rommel, who had invented a new tactic — luring the British armor onto his concealed 88mm gun positions, to be "shot like hares" — was "the hero of the hour"[21] and was instantly promoted by the Führer to the rank of field marshal, on the field of battle.

In such circumstances "talk in the U.S. of a Second Front can only be considered a joke," Goebbels sniffed.[22] "Churchill is no longer the leader of the most powerful empire on earth," the propaganda minister dictated in his office diary. "On the contrary, he has had to go to Washington like a pilgrim, seeking help, and Roosevelt has now taken over many of the functions that were once the Prime Minister's. Doubtless the President intends to inherit Britain's empire one day. But that is of no significance to us. We aren't interested in that. We'll be quite content to be allowed a free hand in Europe" — a free hand that would exterminate all Jews and reduce all non-Germans to accomplices, servants, slaves — or ashes. "That," he concluded, "is what we have to achieve in this war."[23]

How much Hitler's mind was focused on the territories to the east of Germany, rather than to the west, had again been made clear to Goebbels the previous month, as Hitler gave the orders for his legions to recommence their armored drive deeper into Russia. The Führer had already gotten a formal resolution from the Reichstag granting him absolute judicial powers as dictator, "without being bound by existing legal precepts." Thus empowered, Hitler had briefly returned from his East Prussian headquarters to Berlin again in May, certain he would overwhelm Russian forces by the end of the summer; if not, his armies would be better prepared for winter, he promised the country's assembled Nazi Party Gauleiters in a two-hour peroration.[24] To Goebbels's surprise, the Führer then confided to the assembled administrators just how critical the previous winter had been — admitting that by the time Japan attacked Pearl Harbor the Third Reich had been facing retreat and possible defeat on the field of battle. The Japanese had saved the Third Reich.[25] Now, thankfully, the situation was reversed — with the Soviet armies facing collapse before German might. Nothing could now stand in his way.[26]

The Führer had respect neither for Roosevelt nor untested American troops.[27] He had blindly forecast — thirteen days before Midway — that the Japanese Navy would destroy the U.S.-British fleet if it came to a major sea battle.[28] American preparations for a Second Front in the West were not worth taking seriously, either, he had mocked, given the current success of Germany's U-boat war. He'd positioned enough German troops in Norway to repel any invasion attempt, as was rumored to be planned by the British; moreover, with some twenty-five German divisions on call in the West, and no less than four first-class German panzer divisions and even a parachute division ready to meet an Allied landing,[29] he predicted a drubbing if the Americans and British attempted to create a Second Front anywhere on the French coast.

"Of course the Führer doesn't believe for a moment that, were there to be a British invasion even lasting ten or fourteen days, it would end with anything but an absolute catastrophe for the British. That might alter the whole course of the war, perhaps even end it," Goebbels recorded the Führer's words.[30] The Allies would hardly be so stupid, surely. In the meantime, German's destiny lay in the East, the Führer had emphasized. It was there that Germany needed room to expand — Hitler's perennial mantra since the 1920s — establishing as it did a sort of Chinese wall that would separate the West from Asia.[31]

"Never," Goebbels had recorded the Führer's strategic imperative in May, "must Germany allow itself to be sandwiched between two military powers, for then the Reich would always be threatened." Thankfully, the Führer had claimed, Germany had "succeeded in destroying the military power on its western front. Over the summer it would now proceed to destroy the military power on its eastern front. Then we'll be able to begin the process of reconstruction," he noted the gist of the Führer's address. "In the East we want, above all, to use our soldiers as frontier settlers. In that way German resettlement will proceed as it had in the greatest days of Germany's first empire. The German diaspora should be brought back from foreign countries, even from America, its men applied to the Reich's skills in colonization. We won't need to be cultural fertilizers for foreign countries any more, we'll be able to develop our own territories culturally, intellectually, and spiritually. . . . *This* was the point of the war," he sketched the Führer's *tour d'horizon*, "for the spilling of so much blood will only be justified by future generations on the basis of swaying cornfields. Of course it would be nice to inherit a few colonies, where we could plant coffee or rubber trees. But our colonial future lies in the East. That is where rich black earth and iron lie, the foundation of our national wealth. By smart demographic measures, above all using Germans returning from abroad, it would be easy to increase the German population to 250 million . . ."[32]

The future — "the next stages straightforward. Our punitive air raids on English cities ["Baedeker" raids on Exeter, Bath, Norwich, and York, following massive RAF bomber raids on Lübeck and Rostock] have already taught them a lesson. Once we have established our eastern frontiers, the British will reflect on whether to pursue such air attacks on German territory, because our Luftwaffe will once again be free for action . . . Once matters are settled in the East — and we all hope that will be done this summer — then Europe can, as the Führer says, get stuffed. For the war will be won for us. We'll be able to indulge in piracy on the high seas against the Anglo-Saxon powers, who won't be able to withstand it. The United States will lose all enthusiasm for the war, once they see the British empire plundered and disemboweled . . ."[33]

On and on in this vein the Führer had shared his apocalyptic vision with his Nazi functionaries — none of whom dissented. "He doesn't take American declarations of intent too seriously. Against their boast of their 120 million people at war with us, we can counter with about 600 million on our side. For now that Japan has entered the war, we are not talking just of a few continental powers, but we shall soon be able to turn it into a global struggle, spreading it across all continents. The United States are still thinking in terms of world war; but world war terms don't begin to describe this war."[34] All that was required to fulfill his demonic dream, Hitler repeated, were Nazis with nerves of steel; also Hitler's own personal survival "to the end of the war," in order that all actually happened as he willed, despite the inevitable trials and setbacks that would occur.

Spring had come, thankfully; the Führer seemed to be in the "best of spirits," Goebbels had noted. "*In glänzendster Form.*"[35]

Four weeks later, as spring had turned to summer and the German armies had driven deeper into southern Russia, and in Libya Rommel turned the tables on General Ritchie's British Eighth Army, Hitler's dream had seemed eminently achievable to Goebbels. In late June 1942, it was simply marvelous, he felt, to be alive.

Gloom had meanwhile descended on Washington in the aftermath of the British surrender of Tobruk.

In the sticky heat, sixteen of the most senior British and American military staff officers met on June 21, 1942. They had intended to discuss offensive strategy, but they now switched to defense: addressing the ramifications of the collapse of the British in North Africa, and the possible fall of Churchill as prime minister.

The President's response, however, was different — indeed, would go down in world history. At the British Empire's nadir of shame, with British Empire soldiers refusing to fight — the number of those surrendering at Tobruk increasing to thirty-three thousand in subsequent hours — the President turned to Churchill and said: "What can we do to help?"[36]

General Marshall was consulted, and to the consternation of Secretary Stimson — who was not summoned — the President offered, with Marshall's approval, to take the Second U.S. Armored Division, currently being equipped with the latest M-4 Sherman tanks with swiveling turrets, and dispatch it immediately to Egypt, with its men and artillery, to defend Alexandria, Cairo, and the Suez Canal.

Churchill, shocked, chastened, and grateful, accepted. Americans and Britons would thus fight side by side.

It was in this way that the strategic Second Front "crisis" of June 1942 was temporarily averted — not by argument but by British disaster.

What earthly hope, after all, could there be of a successful Second Front that fall, at a moment when the British were collapsing in the Middle East? "Nobody seriously believes in the feasibility of a Second Front," even Goebbels noted in his diary on June 25.[37] After all, the majority of the forces for such an invasion would have to be British, given the time it would still take to ship significant numbers of U.S. troops to England. And in the wake of Tobruk's disgraceful surrender, what possible victory could be won on the fields of Europe, if the British lacked generals who could win offensive battles, or soldiers willing even to fight them?

19

No Second Dunquerque

ON TUESDAY, JUNE 23, 1942, three days after the fall of Tobruk — and the day that Major General Eisenhower set off to London to take command of the still-meager U.S. forces rehearsing for the Second Front[1] — the Canadian prime minister received the long-awaited call from Washington.

"[T]he President said: Hello Mackenzie. How are you? I expressed the hope that he was well. He said, Yes, very well. He then said: Winston and I are sitting together here. We want you to come down to Washington for a meeting of the Pacific Council on Thursday."[2]

"We drove to the White House through the private entrance to the grounds, at the rear," Mackenzie King noted in his diary entry for June 25. "We were shown into what I imagine judging from pictures, etc. would be Mrs. Roosevelt's sitting room" — for there was to be held, prior to the Pacific Council meeting, a conference of representatives of all the British Dominions, addressed in person by Mr. Churchill.

As he waited, King found a chance to speak with Field Marshal Sir John Dill, the British representative on the Combined Chiefs of Staff Committee in Washington.

Dill — who had been chief of the Imperial General Staff (CIGS) before General Alan Brooke, but had been fired by the Prime Minister for constantly contesting Churchill's interference in the operations conducted in the field — was gloomy. "He said to me that he regarded the reverse at Tobruk as very serious; from the quiet impressive way he spoke to me, it was clear that he felt this equally. He spoke of the difficulty of a campaign in the desert and of Egypt, and gave me the impression that he felt the whole situation was very grave. While he did not say it, I felt he realized it might be impossible to save the situation so far as the Suez was concerned."[3]

Captain McCrea, Roosevelt's naval aide, then ushered them into the next-door conference room, where the Dominion representatives of the British Empire took their places. Standing to address them in secret session, Churchill was, if anything, more pessimistic than his listeners, according to King: "When he came to speak on the Middle East, Churchill said that as far as he was concerned, he was prepared to lose the Middle East rather than sacrifice Australia's position if it had to come to that."

This was bleak — strangely contradicted, however, by Churchill's appearance. The Prime Minister looked "remarkably fresh," King noted, "almost like a cherub, scarcely a line in his face, and completely rested though up to one or two the night before," as he gave the assembled representatives a "review of the whole situation."

"He started at once with Libya and the fall of Tobruk," King recounted. "Said that we must not conceal the fact that it was a very serious reverse to him; it had been quite unexpected. The reports he had received from [General] Auchinleck had led him to feel that the British forces would be able to hold situations [positions] successfully and to win out, but there it was, and now the next concern was over Egypt. He said he had no doubt in his own mind about the British being able to hold Egypt. The enemy would find fighting over the desert, many miles [distant] of each other, no water, a very arduous business . . ."

The Prime Minister's mixture of brutal frankness and yet confident hope was bewitching to the Canadian prime minister, seated among senior fellow Dominion leaders and representatives — especially when Churchill went on to admit "that the present situation, bad as it appeared, was nothing to what it had been in April last at the time the Japanese fleet were assembled in the Indian Ocean. . . . Pointed out that it looked at one time as though India might readily have fallen to the enemy. There was very little in the way of protection in Ceylon or in India." The United States, however, had saved England's imperial bacon. "Happily since then, the Japanese fleet had encountered the attacks it had" — first in the Tokyo raid, then the Coral Sea and Midway — "and he now felt that India was in a better position than she had been in at any time from [point of view of] the number of soldiers there. She was better protected" — especially by Americans — "than she had ever been. As to the internal situation" — where Gandhi was putting together what would become his historic "Quit India" protest movement — Churchill was indifferent: "the British had made the best offer they could," he declared, claiming that "India was not a country," as King quoted him. "It was a continent, full of different races, etc. and had to be so regarded."

This was not a view that the Canadian premier shared. "I felt, however, that Churchill did not really appreciate the position in India." The Prime Minister spoke about never having changed his views — but claimed he had not allowed them "to interfere with the utmost effort being made at this time to meet the situation through Cripps."[4]

Given Churchill's "sabotage" of Cripps's mission, this was untruthful; yet there was something almost hypnotic about Churchill's oratory even to the Indian representative at the meeting, Sir Girja S. Bajpai. Churchill spoke of Russia, China, Australia. "Explaining the difficult situation generally," King recounted, "Churchill said it was the wide space that had to be covered with only limited numbers of men and supplies. He said it was like a man in bed trying to cover himself with a blanket which is not large enough. When his right shoulder was cold and he pulled it over to cover it, the left became uncovered and cold. When he pulled it back, the situation was reversed. Similarly when he hauled the blanket up to put around his neck and chest, his feet became cold and exposed and got cold. When he went to cover them up, his chest became exposed and he got pneumonia or something of the kind."

The Dominion representatives, spellbound, waited for the British prime minister to tell them how he proposed to deal with this "grave" situation.

Churchill was, however, aware he had his audience in thrall — and deferred the capstone of his talk. "He then suddenly stopped," King wrote, "and in a dramatic way, began to go over the situation compared as it was at the beginning of the war." Russia and now the United States were in the fight, he reminded them, on Britain's side, "and he referred to the heroism of the Russians and the magnificent work which the Americans were doing on their production etc."

Finally, however, the Prime Minister came to the climax of his peroration.

"He then said: but we have an ally which is greater than Russia, greater than the U.S."

The assembled Dominion representatives were agog.

Churchill was nothing if not an actor, when faced by an attentive audience. The Prime Minister, the Canadian premier recorded, "paused for a moment, and said: it is air power."[5]

• • •

Air power?

The Dominion representatives were at first disbelieving.

Troops of the British Empire had in the past two years lost Norway, France, Greece, Hong Kong, Malaya, Burma, and only four days before had surrendered Tobruk without a fight. They now looked to be losing Egypt, the gateway to the Levant. Did the Prime Minister really believe Britain's new "ally," far from any war front, would magically reverse the tide of war and defeat Germany and Japan any time soon?

It seemed ludicrous, given the failure of Germany's vaunted air power, the Luftwaffe, to bring Britain to its knees in 1940 and 1941.

It was all that Mr. Churchill could produce for his Dominion listeners at that moment, however — the Prime Minister extolling the recent, highly controversial RAF thousand-plane "bombing of Cologne and Essen," which had produced an uproar in the House of Commons, as the death toll among German civilians threatened to vitiate the moral principles upon which the Allies were defending "human civilization." "Told of the destruction there," Mackenzie King noted the Prime Minister's response: "He said our objective would continue to be military targets, though, some times, the airmen might go a little wide of the target. That the destruction of Cologne" — one of the glories of European medieval architecture — "had given the Germans great trouble in moving populations, taking care of those moved, trying to rebuild roads, etc., and intimated that there would be more of it, and it would be most demoralizing."[6]

Demoralizing? This was debatable — indeed, the deliberate killing of so many civilians might be counterproductive, *strengthening* rather than diminishing the resolve of ordinary Germans, as the Canadian prime minister — who read the Bible before rising each day — was uncomfortably aware. That morning King had read a passage that had given him guidance "throughout the day": "Chapter VII of Jeremiah with its words: 'Obey my voice and I will be your God and He shall be my keeper. Walk ye in all the ways that I have commanded you that it may be well unto you.'"[7]

"Terror bombing," as the Germans called it, did not sit well with such walking. Even Lord Halifax, a deeply religious man, had blanched when recently told by a British pilot what he had done — "a red-headed young Scottish sergeant-pilot from Motherwell who had been on nearly every trip to Germany including Rostock, but had missed Cologne as he was getting ready to come over here. About Rostock he said the fires were tremendous; 'I'm afraid we not only killed them but cremated them.'"[8]

The Prime Minister's much debatable paean to his new "ally," however, now led him to his ultimate revelation — a confidential admission in the White House that was far, far more welcome to the Dominion representatives.

As Mackenzie King recorded that night, "Churchill reserved to the last part of his talk the reference to the European front; here he spoke very positively" — or negatively.[9] A Second Front, he confided in absolute candor, was simply not on the cards for that year.

"He said, using the expression 'By God,' nothing would ever induce him to have an attack made upon Europe without sufficient strength and being positively certain that they could win. He said that to go there without a sufficient force would be to incur another Dunkirk, and what would be worse than that, they would have of course to supply the French with arms and cause them to rise when any invasion was made, and that to have to leave them to the Huns [in a subsequent evacuation] would be to have the whole of the French massacred, and none of them left."[10]

Churchill's only answer, beyond RAF terror raids on German cities, was therefore more minor raids on the French coast to force the Germans at least to keep their defense forces stationed there, rather than in Russia; but "he did not think they could afford to contemplate the invasion of the continent before the spring of 1943, despite the number of troops that the Americans might be able to send across . . . Perhaps in the spring, shipping facilities will be better and the attack could take place then."[11]

Stopping there, Churchill asked for questions or comments, "and turned to me."[12]

Mackenzie King's testimony would be important to historians, because the Canadian premier's relationship to the President was a sort of marker in the war's changing dynamic. Canada was producing huge amounts of war materiel, food, and shipping; it was also providing a considerable number of volunteer troops to the global struggle against the Axis powers. Canada, in fact, was now as important to the war effort as Great Britain — and King, as Canadian prime minister, was as opposed to a cross-Channel Second Front as was Winston Churchill — a fact the President had been aware of ever since King had stayed with him at the White House, earlier that year, when King had experienced a strange vision relating to Mr. Roosevelt.

"Had a very distinct vision during the night," the Premier had noted in his diary in April. "It seemed as though some being was seeking rest; alongside were forces in the nature of flames, not of a fire but of passion or an animal instinct like fighting, etc. were continually banging at the side of this individual and in a way seeking to compel a yielding to its influence." The person "who came to mind" with respect to this dream "was the President and the influence of those" of his senior military staff "who were forcing him into a line of warfare without sufficiently surveying the whole field. The more I think of the vision," King recorded, "the more I feel it was to let me see that there was a spiritual significance behind the attitude which I took yesterday at the Pacific Council and again with the President last night in discussing the plan of campaign in Europe for this year."[13]

At that time — April 16, 1942 — it had become clear to the Canadian prime minister that the President had no idea how few divisions there were, in reality, in Britain, with Roosevelt imagining there were a hundred. Nor had the U.S. president quite recognized, in King's view, the *magnitude* of military effort required to mount a successful cross-Channel assault. King had therefore warned the President at the April Pacific War Council that a failed assault would make Britain itself vulnerable to attack. There had been "no dissent" from the council members — the Australian representative supporting King, and speaking "very emphatically about the necessity of avoiding the possibility of a second Dunquerque, and how disastrous anything of the kind would be. He said he wished to support very strongly what I had said." The New Zealand representative had spoken "in an equal strain, saying that while everyone believed in a second front, and the need for offensive action, up to the present there had been no one who could say how it could be done."[14]

That was two months ago; now, in the wake of the disaster being suffered by the British Eighth Army in Libya and Egypt, a cross-Channel invasion seemed *even less* mountable that year — especially, Mackenzie King reflected, when the lives of volunteer Canadian troops, training in Britain, were at risk.

It was, in this respect, a tragic moment: Prime Minister King as straightforward as Churchill in responding to the British prime minister's appeal for questions or comments. Addressing the Dominion assembly on June 25, 1942, King warned "that the subject he had dealt with was one about which we [Canadians] were most concerned. That we felt very strongly he was right in what he had said about the necessity of overwhelming forces, and not taking unnecessary risks."[15]

It was abundantly clear, then, that in late June 1942 it was not simply the British who were balking at the implications of a cross-Channel Second Front, but America's crucial ally, the Canadians — whose premier felt no shame or inhibition in sharing with the President his concerns. None of the Dominions, in fact, were willing to risk another military debacle, whatever the U.S. chiefs might favor — especially after the surrender of Tobruk and the flight of the British Empire forces of the British Eighth Army toward Suez and Cairo.

20

Avoiding Utter Catastrophe

IRONICALLY, CHURCHILL AND HITLER agreed on one thing: a cross-Channel attack in 1942 would fail. Hitler and Goebbels naturally prayed that it would be attempted.

After reading secret intelligence reports from Moscow, Goebbels recorded on July 4 that the Kremlin was calling for a Second Front, "but Churchill and Roosevelt are in no position to comply." The situation was giving rise, to Goebbels's delight, to idle talk of the Soviets taking the "most extreme measures unless the English and Americans actually mount a Second Front. But how can Churchill and Roosevelt launch a Second Front in reality," he asked himself, "when they don't have the necessary shipping and are having such difficulties on existing fronts and suffering such fatal defeats?"[1]

Two weeks later Goebbels was noting the British "have no intention of rushing into any invasion adventure. Even they will have learned that we've now stationed, in the West, the quality of first-class German troops, well-trained and hardened by battle on the Eastern Front, who would *welcome* such a confrontation with pure joy."[2]

Scanning the British and American press, as well as reports from German agents, Goebbels noted the continuing pressure on Roosevelt and Churchill to launch just such an invasion. "The man in the street screams for a Second Front, and it remains an open question whether Churchill and Roosevelt might, in certain circumstances, together with the deteriorating military situation for the Russians, have to give in to such pressure. It's one of the disadvantages of democracy that it can't conduct politics or war according to logic and intelligence, but has to respond to the up-and-down swings of public opinion. In such circumstances," he mused, "a concession to public pressure for a Second Front could lead Churchill and Roosevelt to utter catastrophe."[3]

Ironically, it was not only the proverbial man in the street who was driving the President toward "catastrophe," but his most senior military officers in Washington.

General Marshall was becoming the most evangelical of Second Fronters — determined to avoid unnecessary dispersion of effort. Even the President's offer to send the Second U.S. Armored Division to reinforce and revitalize the British Eighth Army in Egypt had been revised on Marshall's instructions, since it would take, it was calculated by his staff at the War Department, three to four *months* to get such a fully armed division to Suez — by which time the war in Egypt would have been decided. Only the tanks themselves, and a cadre of technicians, were thus sent, by fast convoy. On limiting the forces to that, the general would not bend — in fact he actually walked out of the President's study rather than contemplate dispatching yet more American forces to the Middle East beyond the Second Armored Division's tanks — forces that would then not be available for the planned build-up in Britain prior to a Second Front attack.

The President was surprised, and disappointed. But then, as quasi Allied commander in chief, he saw the situation differently from his U.S. Army chief of staff and war secretary. The collapse of the British at Tobruk was disheartening, but might yet offer the Allies a great strategic opportunity for a counterstroke. Rommel had lured the tanks and vehicles of the British Eighth Army into the range of his deadly 88mm antitank guns, winning a historic victory in the desert at Gazala and causing the British to flee eastward across the deserts of Libya for their very lives. But driving his Panzerarmee Afrika onward toward Cairo, Rommel was leaving himself open to a possible strategic counterstrike — a huge American army landing in his rear, in French Northwest Africa, after which the new *Feldmarschall* could, if all went well, be crushed inexorably between the two Allied pincers.

Why Marshall, King, Stimson, and others held so firmly to their idea of a cross-Channel Second Front in 1942, and seemed so naïve about the reception they would be given by Hitler's waiting forces, never ceased to amaze the President. For almost an entire year, ever since July 1941 — long before Pearl Harbor — the President had favored as an alternative a U.S.

landing in French Northwest Africa, in the event of war, to forestall the Germans — yet every time he had pressed for serious operational study of the scheme he had met with opposition from the War Department.

At the very least, Operation Gymnast promised to allow the United States to control the whole Atlantic seaboard of French West Africa, including the vital port of Dakar — thus ensuring German forces would be denied the closest jumping-off point to a potential attack or invasion of South America. Better than that, though, it would give the Allies the chance to jam and crush German and Italian forces in North Africa between the west and eastern ends of the Mediterranean.[4]

Hitler would not be able to ignore such a threat to his Italian Axis partners and to his southern flank. He would *have* to join battle — but on the periphery of Europe, at the very farthest point from his own bases and supplies — while all the time having to hold significant German forces in France to meet a possible cross-Channel invasion from Britain. This would aid the hard-pressed Russians substantially, without risking a devastating reverse in a cross-Channel landing that year. Against nominal Vichy French resistance in North Africa, moreover, U.S. forces would get to rehearse the business of amphibious warfare. Then, in the relative "safety" of an American-occupied Northwest Africa, they would be able to put into practice in real time the command and combat techniques they would later need for an ultimate "Bolero" cross-Channel invasion, using preponderant force.

By contrast, the challenge of Bolero — facing at least twenty-five German divisions charged by Hitler with defending against Allied landings on mainland France — would be a far harder proposition. Allied formations would have to land on hostile shores that were within easy air, road, and rail reinforcement from Germany. And face a far, far tougher prospective enemy there than Vichy French troops in Morocco or Algeria.

In short, Churchill and the British chiefs of staff were right in being disinclined to carry out Bolero. Breaching Hitler's so-called Atlantic Wall and smashing down the gateway to Hitler's Third Reich was, in mid-1942, an almost impossible task — as Hitler and Goebbels knew better than anyone. Whereas U.S. landings in French Northwest Africa would give the Allies the initiative they needed in steadily pursuing a "Germany First" strategy. Shorn of its North African territories, and with Sicily and southern Italy within striking distance, Mussolini's Italy would very likely capitulate — leaving Hitler's Germany alone, battered from all sides.

In the President's eyes the United Nations were looking a gift horse in the mouth — for the Germans had *still* not occupied French Northwest Africa, even two years after their conquest of Western Europe! Morocco and Algeria were there for the taking — with Hitler still obsessively focused on defending the Atlantic Wall.

Yet how to get Stimson and the U.S. chiefs of staff to fall in line with such reasoning, without turning them against him as commander in chief?

Quietly the President began to cast around for a way to bring his reluctant sheep back into the fold, and support — instead of sabotage — Gymnast. It was at this point that he called on Bill Leahy.

The tall, balding, stern-looking Admiral William Leahy was well known to the President — Leahy having preceded Admirals Stark and King as chief of naval operations. Before his mandatory retirement as CNO, Admiral Leahy had even served on the newly minted Joint Chiefs of Staff board that the President had set up in 1939.

The President had come to admire Leahy's calmness of judgment — respect that had prompted him to make the admiral the U.S. ambassador to the French government in Vichy, after the 1940 French capitulation, and to press Leahy to gain and keep the trust of Vichy's military leaders.

Leahy had done well. Almost two years later the United States and Vichy French governments were still at peace with each other; indeed, even in the face of Premier Laval's active cooperation with the Nazis, Roosevelt had insisted that U.S. diplomatic relations with the Vichy government *not* be broken off, disregarding the pleas of the Free French leader, General Charles de Gaulle — who had almost zero influence or authority in Northwest Africa.

As Leahy had reported to the President, the commander in chief of all French Vichy forces, Admiral François Darlan, had told him confidentially that if the Americans landed "with sufficient force in North Africa to be successful against the Nazis, he would not oppose us."[5] It was this prospect, not specious public calls for a Second Front, that had remained the abiding lure of Gymnast for the President. As an extra incentive to French goodwill, Roosevelt had even insisted American food aid must continue to be sent to Vichy France and North Africa.

These had proved wise decisions, despite the political outcry that arose in the United States over appeasement of appeasers. Yet for all that Gymnast could achieve for the Allies in 1942, it had received only derisory responses from General Marshall and Admiral King — trained and experienced officers who claimed to see no point in such an invasion, and all too many military reasons why it was too daunting, and would founder.

The President disagreed. In fact he found their reasoning for the most part fallacious, and in some respects nonsensical, especially the claims in their memorandum of vast numbers of forces the Germans could bring to bear to smash a U.S. assault on Northwest Africa. Far from being in a numerically superior position to defeat such landings, the Germans were in no position to interfere with a U.S. invasion. In fact Admiral Leahy had assured Roosevelt, on June 5, 1942, when he returned to Washington from France for "consultations" with the President and Secretary Hull, that there were not 180 Germans in the whole of French Morocco.

Fewer than 180 Germans in Morocco?

It was in the context of Ambassador Leahy's revelation that, two weeks later, when the British surrendered at Tobruk and Churchill could only clutch at RAF bombing straws, the President's mind was largely made up. Bolero in 1942 was a pipe dream. The British disaster at Tobruk was simply the final straw. The British could no longer be depended upon to fight for their empire in the Far East, perhaps even the Middle East. How, then, could they be seriously expected to fight for the immediate liberation of mainland France, their ancient enemy in Europe, let alone for Europe?

Chances of Allied success in a cross-Channel invasion in 1942 were nil; a Second Front there would be suicidal. Instead, the President was now adamant, if the Western Allies were truly committed to a "Germany First" strategy, then the United States must land forces in French Northwest Africa *before the Germans did*—if he could overcome the skepticism of his own War and Navy Departments. It was they, after all, who would have to assemble and launch such an American invasion force.

As Churchill departed to face the music in the House of Commons (including a vote of no confidence in his government),[6] the stage was set in Washington for a monumental confrontation between the Commander in Chief and his own military staff, which had not taken place since Lincoln and the Civil War.

In a sea ambush at Midway, the U.S. fleet avenges Pearl Harbor, sinking all four Japanese carriers. Thus ends Japan's brief domination of the Pacific. At the White House, FDR is overjoyed — and proud of his supreme commander, Admiral Nimitz. Here Marine General Thomas Holcombe shows FDR a Japanese flag his son Marine Major James Roosevelt helped capture during a raid on Makin Island.

The Fall of Tobruk

On June 19, 1942, Churchill flies to Hyde Park to warn the President that the U.S. generals' plan for a Second Front in France that year is too difficult. Two days later, more than thirty thousand British troops surrender the vital port of Tobruk to Rommel without a fight. It becomes clear to FDR at the Pacific War Council in Washington that the United States will have to intercede in North Africa to save the British, while the Russians must save themselves.

The secretary of state for war, Henry Stimson, and General Marshall, the army chief of staff, do not agree. They threaten to switch to the Pacific unless the British go along with a Second Front invasion in 1942. The result, when Churchill feels compelled to mount a miniversion, is the British "fiasco" of Dieppe on August 19, 1942—where more than three thousand brave Canadians are killed, wounded, or captured in a few hours without getting off the beaches.

The President is mortified. He appoints retired Admiral Bill Leahy as his military chief of staff at the White House to put down the Stimson-Marshall insurrection and help enact his "great pet scheme": U.S. landings in French Northwest Africa, which the Germans have failed to occupy.

FDR Inspects the Nation

Boarding his presidential train, FDR sets off on September 17, 1942, on a fourteen-day inspection tour of U.S. production plants and military training facilities, prior to American landings in North Africa.

Gearing Up for Victory

From East Coast to West Coast, from the Gulf to South Carolina, the President inspects and inspires the "arsenal of democracy" he has fathered. The miracle of mass production — of ships, planes, tanks, guns, and munitions — is stunning: a new ship in ten days, the promise of a plane an hour . . . Together with rigorous training and rehearsal in amphibious operations, the tour confirms the President's faith in Allied victory — under American arms and leadership.

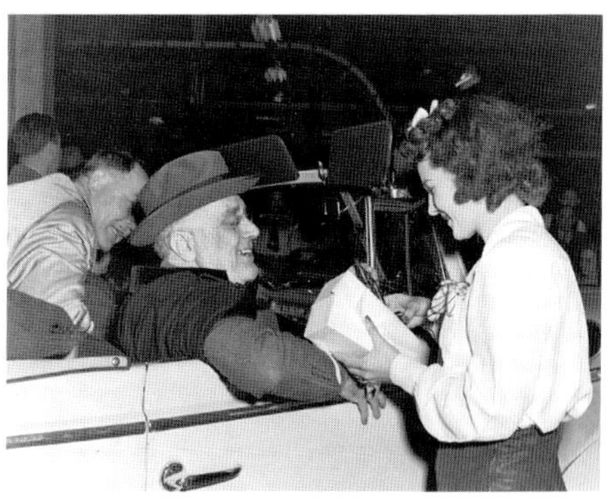

At Shangri-la, the President's secret camp in the Maryland mountains, FDR awaits news of the Torch invasion of Morocco and Algeria. The war secretary has bet him the invasion will fail, but FDR remains optimistic his "great pet scheme" will succeed.

Generals Eisenhower and Patton are FDR's favorite protégés — the one appointed by him to be supreme commander of the Torch invasion, the other his star performer as a saber-rattling armored corps commander at Casablanca.

With the surrender of all Vichy French forces in Northwest Africa to Eisenhower on November 11, 1942, the huge U.S. invasion is heralded around the world as the turning of the tide of World War II. In Washington that day, the President, accompanied by General Pershing of World War I fame, gives thanks at Arlington National Cemetery. "May He keep us strong in the courage that will win the war," he prays, "and may He impart to us the wisdom and the vision that we shall need for true victory in the peace which is to come."

PART NINE

Japan First

21

Citizen Warriors

BY JULY, WASHINGTON, D.C., WAS sweltering in all respects.

If only he could get away, the President sighed to his staff. Fortunately, he did not have long to wait. A "getaway" had been selected several months before, as a presidential retreat — and it was almost ready.

"Early in the spring of 1942, possibly late March or early April while having his sinuses packed one evening in [Dr.] Ross McIntire's office, President Roosevelt remarked about as follows," Roosevelt's naval aide, Captain McCrea, later related — attempting in his somewhat stilted English to recall the President's request.

"'Both of you' — referring to Ross McIntire and me — 'know how very much I like to go to Hyde Park for weekend breaks. With the war on, I am conscious of the fact that I cannot go to Hyde Park as often as I have in the past'" — for the overnight journey from Washington to the Hudson Valley and then back was too time-consuming. Nevertheless, the President had said, "'I would like very much to dodge, as far as possible, the heat and humidity of the Washington summer: additional air conditioning is not for me, as you well know. As I have often told you, Ross, I never had sinus trouble until I became shipmates with air conditioning. The two may not be related but nevertheless I associate this condition'" — at which he tapped his sinus area with a forefinger — "'with air conditioning. Now, cannot we locate an area within easy access of Washington where it would be possible to set up a modest rustic camp, to which I could go from time to time on weekends or even overnight and thus escape for a few hours at least the oppressiveness of the Washington summer? I suggest this as an alternative to the Potomac'" — referring to the USS *Potomac*, the presidential yacht — "'since the Secret Service people are adamant against my using it, except on selected occasions.'

"'Now I know that President Hoover had a camp on the Rapidan in the Catoctin mountains,'" the President had gone on to explain his idea. "'I know nothing about it but that might be a good area to investigate, anyway. Ross, I want John and you and Steve Early [FDR's press secretary] to find some place which will fit not alone my needs, but provide for the housing of the clerical staff which usually accompany me to Hyde Park. Remember now: nothing elaborate—something most modest, functional and within easy distance of the White House. This last requirement is important, so no doubt the choice of location will be limited to nearby Virginia and Maryland. Since the summer is approaching you should get after this as soon as possible.'"[1]

McCrea, McIntire, and Early had diligently begun their search.

"The Hoover camp was quickly eliminated," McCrea later recollected. The thirty-first president's camp had been built "alongside a nearby stream and had very little view of the surrounding countryside. President Hoover was interested in stream fishing and the camp, while ideal for that, did not fill the needs of President Roosevelt."

The general area, however, seemed ideal. "Nearby, atop the Catoctin Mountains at some 1800 feet elevation, was located a modest model recreation camp which had been built by the Civilian Conservation Corps and Works Progress Administration," as part of the New Deal economic-stimulus program in the early days of the Depression. "The exact number of the buildings involved escapes me at the moment but it could not have been more than eight or ten. The buildings were small save for one somewhat larger building which had a mess hall and kitchen. The larger of the cottages could, we thought, with a few alterations be adapted for the President's needs and the needs of his staff."[2]

The chief of the Bureau of Yards and Docks had then been brought in—a man who was no "stranger to action," as McCrea neatly put it. "In a few short days this building was altered by the Seabees to provide for a combination dining and living room, the President's bedroom, and three small bedrooms. All the small bedrooms had a common bath[room], which had no key! The living room area was, by usual standards, small, no larger in size than a modest living room. At one end was a stone fireplace which contributed greatly to the comfort of the place. The President's bedroom was the largest of the four bedrooms, but it too could be said to be of modest size. One side of the President's bedroom was equipped with a large hinged panel which, when tripped, would fall outward, and serve in an emergency as a ramp and escape route for the President's wheelchair. A combination kitchen and pantry adjoined the dining room. A screened in porch — entrance to which was via the dining room, providing a spectacular view of the surrounding countryside — completed the structure."

The man responsible for the camp, the President had decided, should be his naval aide. "'John,' said he, 'since I shan't be using the *Potomac* except on rare occasions you, in addition to your other duties, are hereby appointed the proprietor and landlord of the camp. And by the way, all camps should have a name. Let's see. I think it quite appropriate that we call this camp Shangri-la' — referring of course to the mythical area. . . . 'What do you say to this?'"[3]

McCrea had said he would be honored — the camp accommodations of "USS Shangri-la" somewhat primitive, but the views stunning, and the air mercifully cool compared with Washington. On Sunday, July 5, 1942, the President drove with Harry Hopkins, McCrea, and a party of friends including Daisy Suckley to "what the papers are calling Shangri-La, 'a cottage in the country'!" as Daisy noted. "No one is supposed to know, in order to give the President some privacy, and also for safety. But I am sure alien spies can find it out somehow without the slightest trouble."[4]

Walking with Hopkins to see the swimming pool, Daisy now found the President "cheerful & delighted & rested" — and reading *Jane's Fighting Ships,* the famous encyclopedia of the world's warships.[5]

Back in Washington the next day, refreshed by his trip to Shangri-la, the President took the first step in bending the chiefs of staff to his will over Gymnast. With the War Department still parrying his wishes, the President needed a new stratagem.

In all confidence, the President had admitted to Daisy, he was more "depressed by the situation" than he was letting on. "If Egypt is taken, it means Arabia, Afghanistan, etc., i.e. the Japs & Germans control everything from the Atlantic to the Pacific — that means all the oil wells, etc. of those regions — a bleak prospect for the United Nations."

"I asked where the blame lies for the present situation in Egypt," the President's confidante noted in her diary after speaking with him. The President thought about her question, then answered with surprising candor. "He said partly Churchill, mostly the bad generals."[6]

Was he, the President, any better than Churchill, though? Were American generals any better than their British counterparts?

Roosevelt found it difficult to understand why Marshall remained so intransigent in pressing for a cross-Channel attack, while objecting to an American invasion of French Northwest Africa. Yes, Russia needed help — but so, too, did the British in Egypt. And urgently. The British Eighth Army seemed to be in its death throes in North Africa, as Rommel drove its remnant forces back almost a thousand miles to Alamein, the last defensive position before Cairo and Alexandria. Was this, then, the best moment to launch a supremely risky cross-Channel Second Front assault that had almost no chance of success, given the number of German divisions defending the French coast and interior? Would not an American landing in Northwest Africa — in Rommel's rear — be the saving of the British in the Middle East, as well as a safe area in which U.S. forces could learn the art of modern war?

Field Marshal Dill, the British representative on the Combined Chiefs of Staff Committee, had been even more pessimistic than Churchill — prompting the President to say to him "that the trouble with the British is that they think they can beat the Germans if they have an equal number of men, tanks, etc." As the President pointed out to Dill, this was a serious error. It was simply "not so — the Germans are better trained, better generaled."[7]

It was, the President reflected, a fact of life — "You can never discipline an Englishman or an American as you can a German," he told Daisy what he'd shared with Dill.[8]

Daisy was charmed by the observation, noting it in her diary. Yet its significance in the President's growing realism went deeper than even she realized. For in that casual, consoling remark to Dill, the President of the United States had put his finger on the problem that his own U.S. generals were still not confronting.

Americans might scoff at the British failure to fight the Germans effectively, despite two years' experience of German tactics and interservice skills in action. *Would Americans fare better in battle, though, straight off the mound?* And was it fair to put them into battle in the supposedly right place but at the wrong time — when they would only get slaughtered? American troops, like most English soldiers, were for the most part *citizen* warriors, the President reflected — not professionals. They lacked the sort of self-sacrificing discipline that seemed second nature to German and Japanese troops.

American and British individualism was, in effect, their undoing against such an enemy. Yet it was also, Roosevelt felt, their ultimate strength — *if* they could be encouraged to work together toward a realistic, common cause in which they believed, and were put into battle in operations that had a reasonable chance of success.

This, then, was the insight that came to the President after Tobruk—and marked a profound shift in his thinking as his nation's commander in chief. Millions of American troops were being called to serve their country. They would do fine, he was sure, if they could be given the chance to learn the arts of modern warfare against German or Japanese troops, on ground of their own choosing, *not the enemy's*. The hostile beaches of mainland France, the President emphatically recognized, were not the place to do it, not yet—whatever General Marshall and his cohort of War Department staffers maintained.

Emboldened by his insight, the President formally asked Bill Leahy to resign on July 6, 1942, as ambassador to Vichy France. He was, the President requested, to leave the State Department, go back on the "active" list as a four-star admiral, and become the President's first-ever "military assistant."

The President had, it seemed, hit upon a solution to his Second Front problem. As Roosevelt explained over lunch with the admiral, at his desk in the Oval Study, Leahy would have his own office at the White House, once reconstruction of the East Wing was completed. As to the responsibilities of his new job, Roosevelt was deliberately vague—but Leahy knew the President well enough to know what he was plotting.

Besides, the admiral—whose wife had died unexpectedly after surgery at a hospital in France that spring—was lonely in Washington. He was thus happy to accept the position: bracing himself for what was, undoubtedly, in store.

22

A Staggering Crisis

TWO DAYS AFTER APPOINTING Admiral Leahy to be his new military assistant came the cable Roosevelt had been expecting. Captain McCrea brought it from the Map Room to the President in his bedroom: a personal signal from the Prime Minister of Great Britain, in uncompromising language.

"No responsible British General, Admiral or Air Marshal is prepared to recommend SLEDGEHAMMER"—code name for a cross-Channel invasion of France in the Brittany or Cotentin area that year—"as a practicable operation,"[1] Mr. Churchill commenced his broadside, explaining that the British chiefs of staff would be sending their collective, formal decision to the U.S. Combined Chiefs of Staff that very evening.

The cross-Channel idea was plainly madness in 1942—as it had always been.

"In the event of a lodgement being effected and maintained it would have to be nourished and the bomber effort on Germany would have to be greatly curtailed," Churchill explained the view of his British chiefs of staff. "All our energies would be involved in defending the Bridgehead. The possibility of mounting a large scale operation in 1943 would be marred if not ruined. All our resources would be absorbed piecemeal on the very narrow front which alone is open. It may therefore be said that premature action in 1942 while probably ending in disaster would decisively injure the prospect of well organized large scale action in 1943."[2] The Prime Minister therefore turned to the alternative.

"I am sure myself that GYMNAST"—the American invasion of French Northwest Africa—"is by far the best chance for effective relief to the Russian front in 1942. This has all along been in harmony with your ideas," Churchill acknowledged. "In fact it is your commanding idea."[3]

The relief the President felt was palpable. The Prime Minister might be an exhausting companion — a meddling commander in chief in dealing with his field officers, and a very, very poor selector of army commanders. His chiseled English prose, however, put most of the paperwork the President received to shame. The telegram was not only splendidly worded in its refusal to carry out a currently impossible military undertaking, it brought the evidence the President had been praying for: that Churchill had overridden his own generals in London in order to back the President's "great secret baby." Since his return to England, the Prime Minister had clearly gotten the British chiefs and the rest of his government to support Gymnast — abandoning their preference for large numbers of U.S. troops to be dispatched to Egypt and the Middle East to give more backbone to existing British imperial forces. "I have consulted cabinet and defence committee and we all agree," Churchill's telegram read. Gymnast it was. "Here is the safest and most fruitful stroke that can be delivered this autumn"[4] — with several summer months now to prepare and launch the invasion of French Northwest Africa, before the Germans could stop it.

But if Churchill could override his generals, could he, the President, override *his* — American generals who were still clamoring for a cross-Channel attack that year, more and more loudly, while opposing Gymnast ever more venomously?

There now arose, in Washington, a veritable uprising or quasi mutiny, as General Marshall declared open hostilities on the British.

Marshall had never liked the President's "great pet scheme," and had done his best over the past year to wean the Commander in Chief from it — culminating in his and Admiral King's uncompromisingly negative memorandum on June 19, 1942. Though the President had been unimpressed by their argument — especially their assertion as to the forces Hitler could deploy against a U.S. invasion of French Northwest Africa — the chiefs had held to their insistence on Bolero, the build-up to a cross-Channel attack, and to Sledgehammer, in particular, to be mounted if possible in 1942.

British deceit in the spring — pretending to be in support of a 1942 cross-Channel assault but in truth opposing it from the start — had not helped. To Marshall — and to Secretary Stimson — the fight against Mr. Roosevelt's Gymnast idea had thus been acerbic and exhausting, filling them with suspicion not only of Churchill but of their own revered President. Refusing to countenance the President's Gymnast plan—which Marshall rightly feared would make a cross-Channel invasion impossible even in 1943, owing to the subsequent demands on U.S. reinforcement and supply, especially in shipping—Marshall had returned from his visit to London in April under the illusion that the British supported, albeit reluctantly, his 1942 cross-Channel plan. Churchill's appearance in Washington had disabused him of that notion. Despite the President's caustic remarks about his and Admiral King's memorandum, however, Marshall had thought he had gotten both the President's and Churchill's consent to hold off any decision on such a 1942 cross-Channel invasion until September 1942. Now, only two weeks after Churchill's departure from Washington, the Prime Minister's new cable on the night of July 8, followed by a similar one from the British chiefs of staff to their U.S. colleagues on the Combined Chiefs of Staff, left no room for misunderstanding. The British were *formally* refusing to carry out a cross-Channel operation that year, despite the generosity of the President in sending so much military help to Egypt.

Marshall, as well as his colleagues and staff, and the secretary of war, were incensed.

For his own part, General Marshall felt doubly betrayed. That the British had lied in pretending they were prepared to mount a cross-Channel attack he was willing to swallow. But why were they now rolling over and supporting the President's preferred course, Gymnast? The British chiefs of staff—especially General Brooke—had, after all, assured Marshall in June, only two weeks before, that they were just as opposed to Gymnast as he and Admiral King were, since they wanted all American help to go to Cairo, not to French Northwest Africa.

In disgust Marshall now exploded. Typically, Winston Churchill, in Marshall's view, wanted to use an American assault on French Northwest Africa as a distraction: a clever way of avoiding the challenge to Britain of a Second Front even in 1943—but risking only American lives, since Gymnast would be a U.S. undertaking. And, in his view, a very dangerous one.

Ergo, the American response should be, Marshall decided, a switch of American military effort in the war. From "Europe First" to "Japan First."

Japan First?

George Catlett Marshall was admired by all as the very soul of integrity and loyalty — an officer unimpressed by flimflam, always deeply serious, and a first-class administrator. Tall, trim, calm, and direct, he had served as General Pershing's chief of staff in World War I, and was credited with excellent judgment, no matter what pressure he was under. He had never commanded in combat, but had a good eye for talented younger commanders and potential commanders. The President had come to rely on his administrative ability, working together with Secretary Stimson to transform a tiny professional army — seventeenth in world rankings in 1939, behind Romania — into the world's most powerful potential army-air force in 1942: its numbers slated to reach 7,500,000 by the end of the year.

Marshall was not joking, however. He was convinced that America's expanding army should not be frittered away in diversions from the country's main effort in World War II — and the fact that a British prime minister was refusing to assist in mounting a Second Front in 1942 but was now backing the President's deplorable plan reduced the otherwise wise, loyal, and imperturbable chief of staff of the U.S. Army to apparent apoplexy.

At the weekly meeting of the Joint Chiefs of Staff held on July 10, 1942, General Marshall once again excoriated the President's plan for landings in French Northwest Africa as "expensive and ineffectual." The British veto on a cross-Channel attack that year was, he declared, a mark of British pusillanimity, even cowardice. "If the British position must be accepted," he formally proposed to his colleagues, "the U.S. should turn to the Pacific for decisive action against Japan."[5]

Admiral King agreed — knowing the switch would please Admiral Nimitz in his preparations to contest the Japanese operations at Guadalcanal in the Solomon Islands, which the President had authorized.

General MacArthur, too, would be ecstatic. MacArthur had foretold only doom in Europe, whereas in the Pacific there was a chance for the United States to assert its burgeoning air, sea, and military power to good purpose.[6] Switching America's main effort to the war in the Pacific, Marshall thus asserted to his fellow chiefs, "would tend to concentrate rather than scatter U.S. forces." Moreover, such a move would "be highly popular on the West Coast," where there was still great concern about Japanese raids. The general even claimed that "the Pacific War Council, the Chinese, and the personnel of the Pacific Fleet would all be in hearty accord"; and from a strategic point of view in conducting the global war, switching to the Pacific was "second only to BOLERO" — in its potential. Going flat out in the Pacific, in other words, "would be the operation which would have the greatest effect towards relieving the pressure on Russia."[7]

Admiral King heartily agreed. He had never liked Gymnast. The transfer of aircraft carriers from the Pacific to support the invasion of French Northwest Africa would, he claimed, wreck U.S. naval domination of the central and southern Pacific—indeed, he even stated, "in his opinion, the British had never been in wholehearted accord with operations on the continent as proposed by the U.S. He said that, in the European theater, we must fight the Germans effectively to win, and that any departure from full BOLERO plans would result in failure to accomplish this purpose."[8] General Arnold, still only a lieutenant general and subordinate to General Marshall, kept silent.

The chiefs thus concurred, and in the first outright confrontation with their own commander in chief in World War II, they agreed that afternoon to send a new chiefs of staff "Memorandum for the President" drawn up by General Marshall, and signed by both King and Marshall.

Not content with this formal new July 10 memorandum by the Joint Chiefs of Staff, Marshall felt compelled to send an even more personal *third* memorandum, on his own individual account, explaining the Joint Chiefs' second memorandum: again decrying the President's Northwest Africa alternative as "indecisive" and leading to an "ineffective" Second Front in the spring of 1943, if such an invasion were actually mounted. The United States, he maintained, would "nowhere be pressing decisively against the enemy." Therefore, "it is our opinion that we should turn to the Pacific, and use all existing and available dispositions and installations, strike decisively against Japan."[9]

Marshall's advice to the Commander in Chief, then, was simple: that the President should give the British an ultimatum. Full, exclusive cooperation over a cross-Channel Second Front that fall, or latest by the spring of 1943. Otherwise the United States would switch its forces to the Pacific, and leave only token forces in Britain for defense, together with a few U.S. air missions. "Admiral King and I have signed a joint memorandum to you regarding the foregoing," Marshall ended, ominously, enclosing the somewhat sensational document and giving it to a dispatch rider.[10]

The President had already left for Hyde Park the night before. He'd reluctantly undertaken to spend the weekend entertaining Queen Wilhelmina of the Netherlands—the head of state of a main ally in the war against

Hitler and Japan. "The President rather dreads the coming of Queen Wilhelmina because of the stories of her stiff and stern ways that have preceded her," the President's personal secretary noted in his diary—one story being "that she is a teetotaler and once left the room when drinks were brought in."[11]

The Queen's visit proved not nearly as bad as feared—but the arrival of General Marshall and Admiral King's new memorandum was. Delivered by courier on the evening of July 10, it ruined any semblance of peace or relaxation for the President. The new memorandum, he was told, was once again supported by the secretary of war.

It was—Secretary Stimson having become deeply involved. "In the afternoon," the secretary noted in his diary on July 10, "Marshall told me of a new and rather staggering crisis that is coming up in our war strategy. A telegram has come from Great Britain indicating that the British war cabinet are weakening and going back on Bolero and are seeking to revive Gymnast—in other words, they are seeking now to reverse the decision that was so laboriously accomplished when Mr. Churchill was here a short time ago. This would simply be another way of diverting our strength into a channel in which we cannot effectively use it, namely the Middle East. I found Marshall very stirred up and emphatic over it. He is naturally tired of these constant decisions which do not stay made. This is the third time this question will have been brought up by the persistent British and he proposed a solution which I cordially endorsed. As the British won't go through with what they have agreed to, we will turn our backs on them and take up the war with Japan. That was the substance of a memorandum which he wrote and sent to the President this afternoon. It was fully concurred in by the Navy and secretly concurred in by Sir John Dill and the British staff here. I hope it will be successful in preventing a new series of painful negotiations. But there is no use in trying to go ahead with Bolero unless the British are willing to back up their agreements. I rather think this will serve as an effective block."[12]

Stimson, in other words, saw the ultimatum as a bargaining ploy—which, in effect, it was. The secretary and General Marshall had had enough of the President's niceness to the British—indeed, they were so confident their "showdown" would stun the President and compel him as commander in chief to back them that Marshall took the next day off "for rest and recreation at Leesburg," while Stimson went riding with a friend at Woodley, his estate outside Washington.

• • •

Receiving Marshall and King's memorandum at Hyde Park, the President shook his huge head. Not only was the document a renewed repudiation of his "great secret baby," but the document was filled with unsupported assertions and dictatorial absolutes unworthy of the chiefs of staff of the U.S. Army and U.S. Navy.

The memorandum, once again, ridiculed the President's Gymnast plan as "both indecisive and a heavy drain on our resources." It claimed, moreover, "that if we undertake it, we would nowhere be acting decisively against the enemy and would definitely jeopardize our naval position in the Pacific." The two chiefs acknowledged that the United States could not mount a Second Front on its own, however. For a Second Front to succeed, it needed "full and whole-hearted British support," since the British "must of necessity furnish a large part of the forces." Giving up all possibility of an immediate cross-Channel attack in 1942 "not only voids our commitments to Russia," but neither one of the proposed alternatives or "diversions" for that year — Churchill's idea of an invasion of northern Norway or the President's Gymnast plan — would achieve anything, in their unhumble view. Instead, those diversions "will definitely operate to delay and weaken readiness for Roundup [the actual liberation of France] in 1943." The chiefs' recommendation of a switch to operations in the Pacific would, they claimed, not only "be definite and decisive against one of our principal enemies, but would bring concrete aid to the Russians in case Japan attacks them."[13]

It was perhaps the worst-argued strategic document ever produced by America's highest military officers — as studded with ill-defined "definitely," "decisive," and "definitive" claims as MacArthur's most outspoken missives from the Pacific. The President was disappointed in them.

On Sunday, July 12, 1942, having pondered the best tactics to employ, Roosevelt telephoned the secretary of the General Staff, Brigadier General Walter Bedell Smith.[14] To Smith's consternation it was not, however, to ask for a meeting, but for something far more ominous.

23

A Rough Day

INSTEAD OF RESPONDING with anger or invective to his chiefs' defiance, as Churchill would have done, Roosevelt simply turned the tables on General Marshall and Admiral King. They had recommended switching American priorities to the Pacific. Well, then, what exactly *was* their plan for a "decisive and definitive" campaign in that hemisphere?

As Stimson found when he reached the War Department, a "telephone request had come in from the President for a memorandum in detail outlining the steps that would be necessary to make the alternative change over to the Pacific which General Marshall suggested in his memorandum on Friday that he and King would recommend if the British insisted on sabotaging Bolero," the secretary noted in his diary.

Roosevelt was calling their bluff. "This was important and required very immediate and prompt action, for the President wished to have such a memorandum flown up to him this afternoon."[1]

That afternoon? Marshall was not even in Washington that day. Moreover, as became instantly clear to Stimson as a first-class attorney, neither Marshall nor King had actually considered how a "decisive and definitive campaign" switch to the Pacific could be mounted.

As Stimson confided to his diary in embarrassment that night, it was clearly "impossible to have a careful study" of a new Pacific campaign manufactured in a few hours, without prior preparation; in fact it was unwise, even stupid, of Marshall and King to have proposed such a "decisive" switch without first bothering to rehearse its ramifications — and then disappearing for the weekend!

Panic ensued. General Marshall's senior planner, General Handy, was summoned urgently, and "he and his fellows in the General Staff plunged into work on it and I went down to the Department at three o'clock to be ready for consultation on the subject," Stimson noted — the trial lawyer in him reeling at the situation in which Marshall and King's empty threat had placed him and the whole War Office team. "Marshall was called back from his rest at Leesburg and came into the office at the same moment that I came in. By that time General Handy had a rough memorandum ready and Marshall went over to the Navy Department to consult over it with King."[2]

Had the moment in World War II been less serious — Hitler having moved his advance headquarters from East Prussia on July 6 to the Ukraine, near Vinnitsa, to be closer to his armored forces as they raced beyond Voronezh to the Caucasus — the scramble in Washington might have been comical. At the War Department on Constitution Avenue, Secretary Stimson read over Marshall's proposed reply, together with his assistant secretary of war, John McCloy. McCloy had been a captain in World War I like Stimson, but could hardly be expected to come up with a detailed "definitive" military strategy for the Pacific on his own, let alone in five minutes.

McCloy was quiet — sensing his bosses had been exposed. "He has always been somewhat of a 'Middle-East-ner' but I think my arguments before Marshall came in had pretty well knocked that out of him," Stimson claimed[3] — erroneously.

"Marshall came back and told us that he and King had revised it [the Handy memorandum] rather drastically and tried to put a little more punch into it. As soon as it was written out he brought me the revised draft, a copy of which I attach," Stimson wrote in his diary[4] — the document signed by Marshall, King, and Arnold.

"I told him and Marshall that I approved of the [new] memorandum as the only thing to do in such a crisis. I hope that the threat to the British will work and that Bolero will be revived. If it is not revived, if they persist in their fatuous defeatist position as to it, the Pacific operation while not so good as Bolero will be a great deal better and have a much stronger chance of ultimate effective victory than a tepidly operated Bolero in which the British do not put their whole heart."[5]

Still, George Marshall remained uneasy.

Marshall hated to be challenged in such a way — a general "as cold as a fish," in the words of the secretary of the Joint Chiefs of Staff.[6] As chief of staff of the U.S. Army, Marshall was the first to point out holes in subordinates' logic and to question unsubstantiated claims. He also hated to be contradicted over his decisions. When Major General George S. Patton protested in June over Marshall's decision to send only one U.S. division to Egypt to help fortify the British line at Alamein, rather than the two that Patton had said would be essential for such a mercy mission to be effective, Marshall had responded like a viper. Patton, he ordered, was to be put on the first plane from Washington to California "that morning"[7] for insubordination. "You see, McNarney, that's the way to handle Patton," Marshall had boasted[8] — even claiming later he had "scared him half to death."[9]

Now it was Marshall who was sweating for his career — yet unwilling to admit he might be wrong. He was for the most part an imperturbable officer, but he could be obstinate where his pride was concerned. His dander was up — unwilling to accept he had made a stupid mistake in proposing a course of action he had not thought through: a mistake he would have lashed a subordinate for making, but which he only made worse as, challenged by the President, he now faced the secretary of war — and argued for the seriousness of his overnight Pacific ultimatum.

"Marshall was eloquent and forceful in his advocacy of the plan," Stimson dictated that night. "He is a little more optimistic as to the speed with which he thinks Japan can be knocked out in the Pacific than I am. But he has thought it out more carefully than I and has more facts at his disposal. He told me that Sir John Dill, who is very loyal to Bolero and has been very helpful to us, has sent a telegram to Churchill warning him that, if the British government persisted in their defeatism as to Bolero, we would turn our backs and go to the Pacific. Such a telegram ought to have great force with the British government."[10]

This was to dig themselves only deeper — the U.S. Joint Chiefs of Staff, via General Dill, going behind the President's back to the British government.

As Marshall, King, and Arnold admitted in the new memorandum they now sent by air to the President, "There is no completed detailed plan for major offensive operations in the Pacific. Such plans are in process of being developed," they claimed — untruthfully. "Our current strategy contemplates the strategic defensive in the Pacific and offensive in the Atlantic," they acknowledged; therefore to switch to offensive in the

Pacific and withdraw from Europe, or merely remain on the defensive in the British Isles and Iceland, would, they confessed, be a mammoth planning task. "A change therein would require a great deal of detailed planning which will take considerable time."[11]

The lameness of this excuse was embarrassing even to Stimson. If the heads of the U.S. Army, Navy, and Air Force had no idea how such a switch could be done, why had the chiefs of staff gone out on a limb to propose such a supposedly "decisive" campaign in the Pacific?

Aware the issue was dynamite, Stimson held a war council with all senior army and air officers at the War Department on July 13, 1942, and "cautioned all present that this matter must not be spoken of to anyone whatever."[12] If word should get out to the press that the chiefs of staff were in revolt against their own commander in chief, the entire direction of the war could be compromised. Moreover, if word of Marshall's recommendation of a switch to the Pacific reached MacArthur, there would be no end of pressure from that source to make good on the threat.

The fact was, Colonel Stimson knew, to switch all U.S. military attention to the Pacific was something that had never been seriously studied. ("This shows how little we had really thought of the Pacific," Stimson admitted when annotating the account in his diary, after the war.)[13]

Hastily contrived, Marshall's arguments for such a major change in strategy defied common sense, Stimson recognized. Their argument that such a diversion would dissuade the Japanese from declaring war on the Soviet Union was utterly without merit, since there was no evidence whatever that the Japanese were currently intending to attack the Soviet Union — Japan far too heavily invested in China and across half of Asia and the Pacific. American operations to clear the Japanese from their conquests in the Malay Barrier would take *years,* even if such a Marshall-proposed switch to the Pacific took place — without having any appreciable effect on Hitler's war, or helping the Russians. Moreover, in terms of a reinforced Pacific counteroffensive, American soldiers would be fighting and dying to restore the Netherlands East Indies to a colonialist European power; Malaya and Burma to another colonialist power; and the Philippines to its mandated independence as a sovereign country. What was the urgency for this? And where was the "decisiveness," that such a campaign would offer?

While Marshall oversaw the dispatch to the President of his explanation of what he and Admiral King proposed to do in the Pacific, Stimson went to dinner at the Chevy Chase Club, somewhat perturbed.

It was now the President's move.

At Hyde Park, the President continued with his duties of hospitality toward Queen Wilhelmina — showing her his new presidential library and his estate, holding a picnic, then inviting her to dinner at the Big House, without ever intimating to her that one of the great crises of the war was now taking place: a crisis that would affect her country, the Netherlands, more directly than any other, since a cross-Channel Second Front promised to liberate Holland, while a switch to the war in the Pacific promised to oust the Japanese from the Netherlands East Indies — neither of which he was intending to do that year!

Receiving the new memorandum from Marshall and King, the President chose at first to ignore it — not even summoning his new military assistant, Admiral Leahy, who was in hospital at the Naval Medical Center at Bethesda, outside Washington, having emergency treatment for an abscessed tooth. Messrs. Stimson, Marshall, and King could wait, the President decided, until he returned to the White House to discuss the matter in detail. In the meantime, however, he let them know he was utterly unimpressed by their paper on Pacific strategy.

"My first impression," he notified them in a warning telegram July 14, "is that it is exactly what Germany hoped the United States would do following Pearl Harbor."

As if this was not enough, he added another paragraph.

"Secondly, it does not in fact provide use of American troops in fighting," as he pointed out tartly, "except in a lot of islands whose occupation will not affect the world situation this year or next.

"Third: it does not help Russia or the Near East. Therefore it is disapproved of at present."

He signed himself "Roosevelt C-in-C."[14]

In the entire war President Roosevelt would never express his contempt so forcefully. In addition to this rejoinder the President then added a further message, saying he wanted to see General Marshall at the White House first thing upon his arrival in Washington the next morning; he would then see all three Joint Chiefs in the afternoon. He had meanwhile "definitely" decided, he alerted them in the cable, to send General Marshall with Admiral King along with Harry Hopkins to London "immediately." There they could thrash out the matter of the next steps to be taken in Europe, North Africa, or the Middle East—but they were not to make mention of any empty Pacific threats; they should arrange to fly "if possible on Thursday, 16 July."[15]

The President's devastating response to the Marshall-King memorandum was perhaps the tersest rejection Marshall had ever experienced—or would experience—in his life. He read it aloud to his fellow chiefs of staff at a Joint Chiefs of Staff meeting on the afternoon of Tuesday, July 14. Colonel Albert Wedemeyer of the War Department—the officer reputed (though without proof) to have leaked the Victory Program before Pearl Harbor to the McCormick press—kept notes.

However humiliating, the chiefs would have to take the President's telegram seriously, Wedemeyer recorded, since "unquestionably the President would require military operations in Africa."[16]

The question was, therefore: *where in Africa?*

Once again the "relative merits of operations in Africa and in the Middle East" were discussed—none of the chiefs happy about the President's pressure on them. Despite the tart language of the President's telegram, none showed embarrassment or willingness to rethink their stance over switching to a "Japan First" strategy.

"All agreed to the many arguments previously advanced among the military men in the Army and Navy that operations in the Pacific would be the alternative if Sledgehammer or Bolero were not accepted wholeheartedly by the British. However, there was an acceptance that apparently our political system would require major operations this year in Africa."[17]

"Our political system" meant the United States Constitution, which stipulated that the President of the United States be commander in chief of the nation's armed forces.

The refusal of the U.S. chiefs of staff to consider the folly of a premature cross-Channel invasion was breathtaking in its lack of professional military realism—yet their response, when challenged, had only been to blame the democratic "political system." This did not reflect well on them.

For his part, Secretary Stimson preferred to blame "the other" politician, Winston Churchill. The President had asked to see Marshall, Arnold, and King on his return to Washington, before they left for England, but Stimson decided he should perhaps see the President first, and attempt a plea bargain: explaining perhaps that the chiefs were not *really* pressing for a turn to the Pacific, just trying to put further pressure on the British to be serious about a Second Front that year.

Landing at Bolling Field at 9:15 a.m. on July 15, 1942, Stimson thus went "directly to the White House where the President had just returned from his absence at Hyde Park. I had no appointment but he very kindly saw me and I had a long talk with him about the crisis which is happening in regard to Bolero."[18]

For Stimson, his July 15 interview with the Commander in Chief in the Oval Office was painful. The secretary brought with him a book he had recently been rereading, first at Fort Devens, then at his one-hundred-acre estate on Long Island: Field Marshal William Robertson's *Soldiers and Statesmen,* with its vivid account of Churchill's Dardanelles fiasco in World War I.

The decision "to go half-baked to the Dardanelles is being repeated now as to the proposed expeditions to North Africa and the Middle East which Churchill twenty-five years afterwards is trying to entangle us into," Stimson had noted in his diary two days before. "The trouble is neither he nor the President has a methodical and careful mind. They do not implement their proposal with any careful study of the supporting facts upon which the success of such expeditions must ultimately rest."[19]

Handing over the Robertson book, Stimson "begged" the President "to read the chapter on the Dardanelles in which I had carefully marked important passages," the secretary noted that night. "The President asserted that he himself was absolutely sound on Bolero which must go ahead unremittingly, but he did not like the manner of the memorandum in regard to the Pacific, saying it was a little like 'taking up your dishes and going away.'"[20]

Stimson, as a lawyer, was cut to the quick by the accusation of childish pique. "I told him I appreciated the truth in that but it was absolutely essential to use it as a threat of our sincerity in regard to Bolero if we expected to get through the hides of the British and he agreed to that. He said he was going to send Marshall and King abroad to thrash the matter out in London. I don't know how much effect I had on him although he was very clear in his support of Bolero. I think he has lingering thoughts of doing something in the Middle East in spite of my thumping assertions of the geographical impossibilities of doing anything effective."[21]

Clearly, Stimson was embarrassed by his own threat of a switch to the Pacific — indeed, he later noted that "the Pacific argument from me was mainly a bluff."[22]

A bluff that had been called.

Shaming the secretary of war and the Joint Chiefs of Staff into dropping their recommendation of a diversion to the Pacific was one thing. Getting Marshall and Stimson to back Gymnast, Roosevelt's "great secret baby," was another.

It became even harder once General Marshall, following Stimson's visit, arrived in the Oval Office.

Harsh words were exchanged.

Stimson noted the outcome later that night. "I had a talk with Marshall over our respective conferences in the White House, he having seen the President immediately after my early morning interview. He evidently had a thumping argument with the President and thought he had knocked out the President's lingering affection for Gymnast," the American invasion of Northwest Africa, "and then Middle East. Between us the President must have had a rough day on those subjects. I had told him that when you are trying to hold a wild horse the way to do it was to get him by the head and not by the heels, and that was the trouble with the British method of trying to hold Hitler in the Mediterranean and the Middle East. The better way would be to get a grip on his head."[23]

Marshall had, he said, "pointed out to the President that by going into the Middle East we lost Sledgehammer and Bolero '43 and got nowhere, being everywhere on the defensive, for the Middle Eastern operation at best was a defensive operation even if successful. At the same time by so doing, we put ourselves in great peril on the Pacific. Such a situation therefore cost us a year's delay in which Germany would recuperate herself while we simply imperiled ourselves."

For his part, Marshall was clearly not surrendering his notion of a switch to the Pacific lightly — whether out of pride or genuine strategic belief was unclear. The President had insisted on a campaign to hold Guadalcanal, in the Solomon Islands, as the ultimate stop line in confronting the Japanese. Well, then, switching to the Pacific, Marshall maintained, would permit the United States to build upon that, take the offensive, and win a victory there that "would have a tremendous beneficial effect on the general fortunes of the war. It would clear our Pacific area," halt Japanese thrust in the Indian Ocean, and "thereby make it impossible for Germany and Japan to clasp hands."[24]

The sheer silliness of this argument had, however, left the President speechless.

As Stimson noted, the President had had a rough day—he and the War Department clearly still at loggerheads.

Marshall's "thumping argument" with the Commander in Chief did his cause no good. Nor did it improve Marshall's chances of being chosen to command the Bolero landings, once they took place—indeed, Marshall's obstinacy and his outburst in the White House would ultimately wreck his chances of military fame.

Even in retrospect Marshall could not admit his error—blaming the "politicians" for a decision he deplored. "Churchill was rabid for Africa. Roosevelt was for Africa," Marshall recalled. "Both men were aware of the political necessities. It is something we [in the military] fail to take into consideration," he later said. "We failed to see that the leader in a democracy must keep the people entertained. That may sound like the wrong word, but it conveys the thought. . . . People demand action."[25]

This remark was unworthy of Marshall, who in other respects was an entirely honorable man. The "people" of America and the free world certainly demanded action, as newspapers in the United States and Britain trumpeted in 1942—but in terms of political pressure, the action they wanted, by an overwhelming majority, was Marshall's Second Front.

Far from responding to political, and popular, pressure, the President was, however, doing the opposite: patiently preferring, as U.S. commander in chief, a military operation that had a reasonable chance of success. Moreover, one that might change the course of World War II if it succeeded.

PART TEN

The Mutiny

24

Stimson's Bet

NEITHER THEN NOR LATER did Marshall concede that the President, in his role as U.S. commander in chief, was demonstrating a greater military realism in devising Allied strategy in 1942 than his U.S. Army chief of staff.

In the meantime, however, the President wished to make sure there would be no misunderstanding — or ill will. Gymnast would not succeed unless the War Department got behind the plan wholeheartedly. He therefore wanted Marshall and King to see for themselves, in person, how impossible an imminent Bolero operation was — not because the British were cowards, but because Hitler's forces were waiting, and the Allies could not, that year, assemble preponderant force to ensure its success. Drawing up in his own handwriting General Marshall's and Admiral King's instructions for their mission to London on July 16, Roosevelt gave the document to Harry Hopkins, who was to fly with them — and make certain they stayed to the script.

Paragraph nine was direct and to the point. "I am opposed to an American all-out effort in the Pacific against Japan with the view to her defeat as quickly as possible," the President made clear. Yet some form of Second Front was desirable that year, since Hitler — whose troops were smashing their way deep into the Caucasus, as well as toward the gates of Cairo, at Alamein — would otherwise be given time to achieve total control of Europe. "It is of the utmost importance that we appreciate that the defeat of Japan does not defeat Germany and that American concentration against Japan this year or in 1943 increases the chance of complete [Nazi] domination of Europe and Africa." On the other hand, "Defeat of Germany means the defeat of Japan, probably without firing a shot or losing a life."[1]

There was to be no switch to the Pacific.

Successive military historians would extol General Marshall as the great architect and "organizer" of American military operations in World War II: a "titan"[2] whose strategic grasp and patient handling of his commander in chief would, like Marshall's opposite number in London, General Alan Brooke, entitle him to the highest pantheon in military history.

Such accolades were understandable with regard to a man of noble character — especially in countering the excessive admiration, even adulation, garnered by World War II field generals such as Eisenhower, Patton, Montgomery, and MacArthur. Certainly with regard to Marshall's administrative achievement there would be every reason to laud his record in World War II. But as to his strategic and tactical ability, such tributes were way off the mark.

As commander in chief, the challenge for Roosevelt was thus how to marshal Marshall: how to direct, encourage, and support his work at the War Department, while stopping him from losing the war for America. While Marshall and King journeyed to London on their presidential mission, therefore, the Commander in Chief decided now to put his coup de main into action — in their absence.

Ambassador Leahy, emerging from dental hospital, was summoned to see the President on July 18, 1942. The admiral would not only be his military assistant, the Commander in Chief announced, but his new chief of staff, or deputy. As the sole senior military officer supporting the American invasion of French Northwest Africa, Leahy was critical to the President's success in avoiding American defeat on the beaches of mainland France that year, and instead adopting the President's preferred course: U.S. landings in Vichy-held Northwest Africa, before the Germans could occupy the area. Leahy was therefore instructed by the President to become a member of the Joint Chiefs of Staff Committee, and the Combined Chiefs of Staff. Not only a member, in fact. He was, the President laid down, to be the chairman of the Combined Chiefs of Staff — speaking for the Commander in Chief.

Before submitting his formal resignation as U.S. ambassador to Vichy France, Leahy was asked first, however, to do everything in his power at the State Department, where Leahy still had an office, to ensure the

United States did not side with General de Gaulle, leader of the Free French, lest this lead to a hostile Vichy response to an American landing in Morocco and Algeria, as it had done when de Gaulle's Free French forces had attempted to assault Dakar, in 1940. This, in a meeting with the secretary of state, Cordell Hull, Leahy promptly did. ("Conferred with Secretary of State Hull regarding the advisability of maintaining diplomatic relations with the French Government in Vichy," Leahy noted in his diary).[3]

Four days later, on July 22, Leahy then transferred his papers from the State Department to the Combined Chiefs of Staff Building, at 1901 Constitution Avenue, and "took up my duties as Chief of Staff to the Commander in Chief of the Army and Navy of the United States," as Leahy proudly wrote in his daily diary, "which duties included presiding over the Joint Chiefs of Staff and the Combined Chiefs of Staff."[4]

By the time Marshall and King returned from England, the coup would be complete: the Commander in Chief's own man in control of the Combined Chiefs.

In the meantime, the Marshall-King mission fared badly at the hands of the British. The officers were forbidden by the President to use bluff or blackmail by threatening an American switch to a "Pacific First" strategy. Without the threat, they found, as the President knew would happen, that the British were wholly opposed to a cross-Channel invasion that year for the soundest of military reasons: namely, that it would fail.

In Washington, Secretary Stimson, on July 23, was stunned. Initial cables from Marshall had seemed as if the American chiefs were making headway, he had thought.[5] "A very bad jolt came this morning at nine thirty in the shape of a telegram from Marshall," he recorded in his diary, however — a cable "saying the British War Cabinet had definitely refused to go on with Sledgehammer and that perforce negotiations were going on along other lines."[6]

Stimson knew exactly what "other lines" signified.

"I went at once over to the White House," Stimson recorded, "and got into the President's room before he was up" — only to find he was too late. The President "had received his telegrams to the same effect last night and had replied to them."[7]

There was little that the war secretary could say or do. "Apparently

Marshall tried hard to carry his point," dropping all pretense of switching U.S. priority to the Pacific, while "offering to give up an attempt on the Pas de Calais," which the British said would be suicidal, "and to take instead another place" on the French coast, such as Brittany, using it as a quasi-permanent cross-Channel bridgehead through the winter — Sledgehammer. "But the British were obdurate and Marshall had informed the President that we would be unable to go with any Sledgehammer attack without their cordial cooperation."

This was exactly as President Roosevelt had anticipated. "The President had telegraphed expressing his regrets but saying American troops must get into action somewhere in 1942. He then suggested in order of their priority a number of places to the south, each of which seemed to me to be a dangerous diversion," Stimson lamented, "impossible of execution within the time we have."[8]

Stimson was now hopelessly outfoxed. He had been living, he began to realize, under a delusion in thinking Marshall was making headway with the British over a cross-Channel assault — an initial landing to be made in the fall of 1942, then reinforced in 1943, followed by a drive on Berlin.

Stimson had even met up with Frank Knox, to see if he could obtain *his* support and extend the strategic struggle to members of the cabinet. With the navy secretary by his side he had then approached the U.S. secretary of state, the most senior member of the administration, and had given Mr. Hull a grand *tour d'horizon militaire,* telling him the United States had enough forces to launch a cross-Channel attack *and* continue fighting in the Pacific, but not enough to chase their own President's red herrings. A "diversion of strength to an African expeditionary force would be fatal to both," he told the secretary of state, who seemed to have little or no idea what Stimson was talking about. Nor did Secretary Knox — who looked bemused by Stimson's "rather long-winded explanation," of current army and navy plans — Bolero, Sledgehammer, Roundup, et al. — which the navy secretary "has thus far been unable to assimilate," Stimson noted with irritation.

Stimson had been "amazed again at how little he [Secretary Knox] knows about the plans of his own people. This time I hope I got it across," he added, "and, when I parted with Knox on the street after we had left Hull's office, he expressed his warm appreciation of the entire situation and said he would back us up. We shall need him for we never can tell what is going to happen in the White House, although I hope that the President will this time stick to his confession of faith as to Bolero."[9]

With the arrival of Marshall and King's cable reporting the British war cabinet's latest rejection of a cross-Channel invasion that year, however, the bottom fell out of Stimson's strategic world. Given his open attempts to turn members of the cabinet against the President, he was deeply embarrassed — in fact, after several hours trying to calm down, on his return to the War Department, Stimson dictated a formal letter to the President, deploring the "fatigued and defeatist mental outlook of the British government," and had it couriered to the White House.[10]

The letter was as ill conceived as it was jejune. In a last-ditch effort to save Marshall's preferred strategy, he now urged the President to authorize Marshall and King to insist the Allies put all their eggs in one basket: discard the idea of a cross-Channel attack in 1942, but concentrate all efforts on preparing at least for a 1943 invasion of France — with no question of any "diversions" to Africa. American forces — "young vigorous, forward-looking Americans" — would, he claimed, have "a revolutionary effect" on the British. He even cabled General Marshall to tell him what he had written the President.

Stimson's appeal fared as badly as the chiefs' argument in London, however — as Stimson recognized when there was a knock on his door the next morning, July 24, 1942.

In came four-star Admiral Bill Leahy, the new military chief of staff to the Commander in Chief of the Army and Navy of the United States, who had arrived to talk to him — on the instructions of the President.

Stimson, a first-class prosecutor, did not propose to go down without a fight. He attempted to give Leahy, before the admiral could tell the secretary to get in line with the President's wishes, a brief history of the cross-Channel project since the previous December, hoping to turn Leahy against the Commander in Chief.

As Stimson claimed to Leahy, Marshall had gotten apparent British acceptance of the scheme on his last trip to London, in April, but then "Churchill had come over and tried to break it in June; and how we had rounded him up and again gotten his acceptance; and how he had jumped the boundary for the third time."[11]

Stimson's ranch metaphors had little effect on the new chief of staff, however — a full admiral, a former U.S. ambassador to Vichy France, and a supporter of Roosevelt's plan to invade Vichy French Northwest Africa, *not* German-occupied France. "I also gave him a copy of my last letter to the President about continuing the influx of men and munitions into Britain and he read it. But he dropped remarks which confirmed my fears that the President was only giving lip service to Bolero," Stimson confessed in his diary that night, "and that he really was thinking of Gymnast."

Stimson was at last correct — Leahy noting in his own diary that he returned from Stimson's office to the White House for an interview with the President "in which we discussed the practicability of 'Gymnast' in 1942."[12]

Matters were now moving fast. As Stimson dined quietly with his wife at his Woodley Mansion home, there came the "long awaited message from London giving the arrangement arrived at by the conference": Gymnast.

To me "it was most disappointing, not to say appalling," Stimson confided, "for Marshall and Hopkins had apparently been compelled, in order to get an agreement, to agree to a most serious diversion of American troops."[13] Not only were United States forces to embark on landings in French Northwest Africa, but the plan was to be enlarged! Instead of being just the President's plan for U.S. landings in Morocco and Algeria, the British — the very people who would not land troops across the English Channel — would contribute twenty thousand troops to back up the American landing at Algiers.

Stimson's heart sank. With a sickening sense of doom, he went over the cable with Marshall's deputy, General McNarney, who was with him, and "analyzed it, he agreeing with my analysis, and then after he had gone I got the President on the telephone and gave him my views."

It was fruitless to object. The President was at least compassionate in victory. "He said that he was strongly opposed to the giving up of Bolero but I could see," Stimson noted, "that nevertheless he was anxious to go on with Gymnast. And I felt in my soul that the going on with Gymnast would necessarily destroy Bolero even in 1943 and throw us on the defensive."[14]

The President, by contrast, was clearly delighted. Far from fearing it would put the United States on the "defensive," he thought his plan would put America on the *offensive* — a shot that would be heard round the world in the next few weeks.

"The President asked me to come to the White House bringing Arnold and McNarney to meet him and Admiral Leahy at 11:30 tomorrow Saturday," Stimson recorded.[15]

The President, it appeared, had truly taken over as commander in chief.

Henry Louis Stimson was nothing if not obstinate — a trait that had made him a fortune as a trial lawyer, but something of a millstone as secretary of war. He still thought he could, at the last hour, deflect the President from his preferred course, and therefore now "hurried down to the Department" early on July 25, where he "dictated an analysis of my views," as he called it[16] — driving with Generals McNarney and Arnold to the White House and handing his latest memorandum to the President, who received him with his new chief of staff, Admiral Leahy, standing behind him.

The President did not even look at it. He had, he announced, "decided that the going on with Sledgehammer this autumn was definitely out of the question," Stimson recorded.[17] The President had already telegraphed to Marshall, King, and Hopkins "that he accepted the terms of the agreement they had negotiated with Churchill except that he wished a landing made in Gymnast not later than October 30" — i.e., before the November congressional elections.[18] Hopkins was instructed to tell the British prime minister it was now "full speed ahead" on Gymnast — which was renamed "Torch" that day.[19]

It appeared to be a done deal — yet *still* Stimson objected to the idea, as did the senior air and army officers who had accompanied him to the President's office. "McNarney, Arnold, and I pointed out to him the dangers of the situation produced by this operation as contrasted with an operation in the Pacific."[20]

The Pacific? This was, given the weakness of King and Marshall's amateurish paper on the merits of a "Pacific First" strategy, a mistake. In any event it was of no avail. "I cross-examined him as to his realization that his decision on Gymnast would certainly curtail and hold up Bolero," Stimson added — and with complete frankness the President "admitted that it would," thus delaying, therefore, an eventual cross-Channel operation until 1944.[21]

Stimson was mortified. Pointing to his memorandum, he said he wanted it to be placed on record that he, the secretary of war of the United States, completely opposed the U.S. landings in Northwest Africa — indeed, in perhaps the single most dramatic gesture of the war's direction since Pearl Harbor, Stimson took up the President's offer to wager on the outcome of the landings: Stimson betting, in effect, against the success of his fellow Americans.

"I told him," Stimson recorded in his diary, "I wanted this paper read at the time when the bets were decided"—i.e., when the landings were made, that fall. "The decision," Stimson recorded, "marks what I feel to be a very serious parting of the ways." The secretary of war was distraught. "We have turned our back on the path of what I consider sound and correct strategy," he lamented, "and are taking a course which I feel will lead to a dangerous diversion and a possible disaster."[22]

Stimson's bleak "prophecy," as he referred to it afterward, was dire: that Russia would likely be conquered by the Germans that very year. As a result, if the United States went ahead with the invasion of Northwest Africa, a "large portion" of American troops would be left "isolated in Great Britain, Africa and Australia"—leaving "a Germany victorious over Russia" and "free to turn its forces on us."[23]

In the light of history, Stimson's prediction said little for his acumen. The war secretary's own preferred strategy for America was, if anything, even more fanciful. If the British would not mount a Sledgehammer version of a Second Front that year, he now felt the United States should switch all its forces to the West Coast of America, leaving only enough U.S. forces "consolidated" in Britain to be able to launch an "overwhelming attack on Germany if and when that time finally arrives"—while in the Pacific, Indian Ocean, and Far East the United States should seek "check-mating Japan's attack against Iran and India; falling on Japan's back in Siberia; and opening access to China through Burma."[24]

Japan's attack against Iran and India? Like the President, Admiral Leahy could only rub his eyes in disbelief when he read the war secretary's memorandum.

The Japanese, the admiral knew, had withdrawn their navy to home waters after their devastating defeat at Midway in June 1942; they would not be able to replace their sunken aircraft carriers for *years.* A successful invasion of India from the Burmese border or in the Indian Ocean was now unlikely, given the amount of air power the United States had diverted to protect the British. Moreover, how Stimson hoped to get Stalin to permit U.S. forces to move into *Siberia* and thereby risk Russian forces having to divert their efforts into a war with Japan, at a time when Soviet forces were only holding the German armies by the skin of their teeth, or how they might miraculously reopen the road to China through Japanese-held Burma, was beyond Leahy's comprehension. As Leahy noted laconically in his own diary that evening, the two-hour meeting that was held "with regard to a second front in 1942" had been lamentable. Secretary Stimson and his War Department team had been utterly and wholly negative. The President demanded, as Leahy noted, not pique but action: "an effort in the 'Gymnast' plan this Fall." But "the Army was not favorably disposed."[25]

Adamantly *opposed* would have been a better description. Marshall and Stimson seemed to have infected the entire senior staff of the War Department. General Eisenhower, who had arrived ahead of Marshall in London to take command of U.S. troops in Britain in preparation for a cross-Channel Bolero attack, even went as far as to call the July 22 cancellation of Sledgehammer "the blackest day in history."[26]

Stymied in their efforts at the White House, Secretary Stimson and Generals McNarney and Arnold returned to the War Department on July 25 to lick their wounds.

The next day, July 26, found the war secretary "very depressed." To his shame, he would never admit to the President that he, Roosevelt, had been right. Nor would he apologize, or make good on his bet. In his memoirs, written after the President's death, he glided over the saga,[27] unwilling to be reminded of his great protest in July 1942, in which he had thought of himself as a sort of Revolutionary orator, adopting the language and nobility of the Declaration of Independence. "They have been deaf to the voice of justice and consanguinity," the famous text had described the British under King George III. "We must, therefore, acquiesce in the necessity, which denounces our Separation, and hold them, as we hold the rest of mankind, Enemies in War, in Peace Friends."

Stimson had certainly felt the same way.

Even when reading over his diary, in private, after the war's end, Stimson remained surprisingly churlish. "As I look back on this paper," he wrote of his infamous bet and memorandum, "it seems clear that the one thing which saved us from the disaster" that he'd forecast "was:

1. the unexpected victory of Russia at Stalingrad
2. enormous luck in landing in Africa
3. success over submarines."[28]

Given that the Russian success at Stalingrad took place months *after* Operation Torch, as did Allied success against the U-boat menace, this was ungracious.

At Woodley Mansion, meantime, Stimson confided to his diary his sense of foreboding at "the evil of the President's decision. It may not ripen into immediate disaster. What I foresee is difficult and hazardous and very likely successful attempts made to attempt a landing in northwestern Africa" — to be followed, however, by an American failure, since "even when obtained, it will be a lodgement more or less like that of the British at Gallipoli in 1915 — troops suffering constant attacks from the German air force and possibly German and Spanish land troops."

Stimson's mix of bravado, pessimism, and abject fear had reduced him to a wreck. He now worried for the safety of General Marshall, who was returning by air to Washington in poor flying weather. "On the whole this is written on a blue Monday morning," he confessed on July 27, 1942 — and his mood only got bluer the following afternoon, when Marshall's Stratoliner arrived from London and the secretary of war was given the "full story of what happened" — "a somber tale and I see very little light in it," the war secretary noted.[29]

Knowing that Stimson had spoken not only for General Arnold but for General Marshall, General McNarney, and most senior staff officers at the War Department, the President was acutely aware, in the White House, that he needed to change not only their minds but their hearts: to rally them to his Northwest Africa cause if the landings were to succeed.

25

A Definite Decision

THE PRESIDENT WAS INCREASINGLY disappointed in his aging war secretary — and furious when leaks began to filter into the American press that the Commander in Chief was ignoring the advice of his senior military advisers.[1]

Fortunately, one of the President's qualities as a leader was to focus on the positive aspects of an individual's character. Stimson's behavior seemed uncharacteristic of the secretary, who despite his party affiliation had hitherto been a loyal colleague in cabinet, and an effective administrator of the War Department alongside General Marshall. He had handled the awkward business of internment of Japanese Americans on the West Coast, and the noninternment of German Americans and Italian Americans on the East Coast, with tact and skill, even if the Japanese "relocation" was, in many ways, a tragedy. As reports of Japanese atrocities multiplied and anti-Japanese feeling grew, any hope of being able to release the internees early was dashed. It was, Stimson had found, simply easier to leave them in camps well away from the West Coast, where they would be out of public sight and out of mind.[2] Stimson had also taken quiet responsibility for development of the top-secret "diabolical weapon"[3] that the President had ordered to be developed before the Germans could do so, and was handling that efficiently, too. Above all, though, Stimson was a Republican, and thus represented crucial bipartisanship in the Roosevelt administration in its conduct of the war — especially with midterm elections approaching in November that could affect the passage of all future legislation.

If Stimson's leaks were in part designed to stop Roosevelt from firing him, they worked. They did not, however, change the President's mind — Roosevelt simply bottling up his irritation rather than seeking a further fight with his war secretary. Having initially ignored Stimson's memorandum when the secretary brought it to the White House in person, he now refused to respond to Stimson in writing. He did, however, set down his written response for posterity. Ever the antiquarian and history buff, Roosevelt wanted historians, at least, to have the truth. "Memorandum to go with Memorandum from the Secretary of War dated July 25, 1942," the President therefore dictated a response to his secretary, which he ordered should be filed with the rest of his secret papers at his presidential library in Hyde Park.[4]

"This memorandum from the Secretary of War is not worth replying to in detail," the President wrote bluntly, "because it is contradictory in terms and fails to meet the objective as of the Summer of 1942." The threat of further expansion of Japanese conquest in the Pacific was now almost nil — "They seem to be making little progress westward or southward," the President noted, thanks to the Battle of Midway. His decision to contest Japanese occupation of Guadalcanal in the Solomon Islands would, he was certain, seal off the Japanese rampage in the South Pacific. To mount an American amphibious offensive in the Pacific would, however, take "one to two years — and the total lack of effect on Germany of such a major offensive" would be unconscionable, in his view. It would not "win the war," particularly if "Germany puts Russia completely out of action, occupies the Near East and the Persian Gulf and starts down the west coast of Africa."

Europe or its doorstep, by contrast, was different. "On the other hand," as he pointed out, "helping Russia and Britain to contain Germany this Autumn and undertake an offensive in 1943 has a good chance of forcing Germany out of the war, in which case Japan could not conduct war in the Pacific alone for more than a few months."

In that, at least, the President and Colonel Stimson were as one. "The Secretary of War fails to realize," however, the Commander in Chief went on, "the situation which prompted me to send Hopkins, Marshall and King [to London] to urge 'Sledgehammer' or, failing that, some definite offensive, using American ground troops in 1942." The result had been a foregone — and in his view — correct conclusion. "They find 'Sledgehammer' is impracticable and, therefore, make the other proposal," the President's plan for U.S. landings in French Northwest Africa — a plan "with which the British," having initially objected to it, now "agree."

In the President's eyes the strategic situation was as clear as crystal —

and Stimson's wild alternatives quite unworthy of an American secretary of war. "The Secretary of War says in effect:

(a) Sledgehammer should not be abandoned.
(b) He offers no alternative for 1942.
(c) He agrees to further preparations for 'Bolero' which is, however, one year off.
(d) He speaks vaguely of some kind of major operation in the Pacific area."[5]

With devastating logic the President thereby punctured Stimson's warnings of "disaster" unless his strategy was followed; indeed, he found himself incredulous at how foolish his secretary of war was being — much as Lincoln, as commander in chief, had despaired of some of his colleagues during the Civil War.

In the following days the President did his best, along with his new chief of staff, to get his team to pull together and get behind Torch, as the Northwest Africa invasion was now code-named. On July 28 — having instructed Stimson to leave Washington and take a week's holiday in Maine[6] — he had Leahy, Marshall, King, and Hopkins come to his Oval Study and sign up with good hearts to Torch, wisely telling Leahy to get still "more relief supplies to French North West Africa and infant relief to unoccupied [Vichy] France" to help soften the blow that would be coming — for the Germans would be sure to overrun metropolitan Vichy France, once American troops invaded French North Africa.[7]

Then, after lunch that very day, he ordered Admiral Leahy, his own chief of staff, not only to attend but to chair the weekly conference of the U.S. Joint Chiefs of Staff as the senior officer present.

Leahy was delighted to do so. "At 2:30 p.m. presided for the first time at a meeting of the Joint Chiefs of Staff committee," the admiral noted in his diary.[8]

Still, however, Marshall and King resisted — refusing to postpone cross-Channel attack preparations in Britain, arguing that the "agreement" made during Churchill's trip to Washington in June to leave open any decisions on Bolero was still in force until September! If Torch was to go ahead, it had to be ordered by the Commander in Chief himself, they said.

Admiral Leahy sighed, and scratched his balding widower head. He was, once again, disbelieving — incredulous that the heads of America's army, air, and navy could seek to delay the Commander in Chief's directive with such specious reasoning. As the minutes recorded, however, "he would now tell the President that a definite decision was yet to be made" by his colleagues.[9]

Roosevelt was now losing patience. At 8:30 p.m. on July 28, 1942, he decided to convene a conference at the White House, in his Oval Study. It was time, he felt, to issue a direct order, not hope for collegiate resolution.

"The PRESIDENT stated very definitely that he, as Commander-in-Chief, had made the decision that TORCH would be undertaken at the earliest possible date," the secretary of the Combined Chiefs of Staff, Brigadier General Bedell Smith, recorded. "He considered that this operation was now our principal objective and the assembling of means to carry it out should take precedence over other operations as, for instance, BOLERO."[10]

There was no time to lose. "He mentioned the desirability of sending a message immediately to the Prime Minister advising him that he (the President), as Commander-in-Chief, had made this decision and requested his agreement."[11]

The bickering was over; the doubters were, in the President's view as he said goodnight to his generals, vanquished. The race to mount a successful invasion of the threshold of Europe, before winter came, was on.

26

A Failed Mutiny

TORCH "WAS ONE OF THE VERY few major military decisions of the war which Roosevelt made entirely on his own and over the protests of his highest-ranking advisers," the President's speechwriter, Robert Sherwood, later wrote. The President "insisted that the decision had been made and must be carried through with expedition and vigor."[1]

Sadly, it was not — for though the Commander in Chief might insist, the War Department could desist.

"On no other issue of the war," reflected Forrest Pogue, the doyen of American World War II military historians, "did the Secretary of State and the Chief of Staff so completely differ with the Commander-in-Chief. Their distrust of his military judgment, their doubts about the Prime Minister's advice, and their deep conviction that the TORCH operation was fundamentally unsound persisted," Dr. Pogue admitted candidly, "throughout August" of 1942.[2]

Harry Hopkins's biographer noted the same. "It is evident," Sherwood chronicled after the war, "that even after Hopkins, Marshall and King returned from London on July 27, there were further attempts to change the President's mind about the North African operation."

That Secretary Henry Stimson, General George Marshall, and so many of the "top brass" in Washington would continue to press for the potential fiasco of a cross-Channel venture in 1942, while at the same time exaggerating the danger of failure in terms of the President's French Northwest Africa project, did not say a great deal for their military judgment at this stage of the war. Instead of knuckling down to the Commander in Chief's "full steam ahead"[3] order at the end of July, Secretary Stimson and General Marshall continued, as Marshall's biographer recorded, "by a fine splitting of hairs to insist that the final decision had yet to be made and that preparations for SLEDGEHAMMER," an immediate cross-Channel attack, "must be continued."[4]

This was, in the circumstances, disgraceful. In London the Prime Minister even attempted a ploy to buy General Marshall off, at long distance; in a cable to the President, Churchill suggested that General Marshall be appointed to command the eventual cross-Channel attack, while Lieutenant General Eisenhower — who seemed more amenable — should take command of Torch, the impending North Africa operation.

The President, having confided to Stimson that Bolero would probably not be mounted in 1943, given the paucity of naval vessels, did not even raise the matter with Marshall — who was far too important to him as U.S. Army chief of staff at home.

Roosevelt's dismissal of Churchill's suggestion masked, however, a deeper concern. In truth the Commander in Chief had lost faith in Marshall's judgment and objectivity as a military commander, however much he admired him as an individual and administrator. Appointing him to command, even from his perch in Washington, the planning for the Second Front landings might only encourage Marshall to pursue a course which, in 1942, the President had come to view as unrealistic. With rumors rife in Washington of a staff revolt at the War Department, and reports of a grave disagreement between the President and his military advisers leaking into the press, Roosevelt began to have real concerns about Marshall's loyalty and willingness to subordinate himself to civilian leadership. Relations between the two men became frosty, despite the summer heat — the President worried that Marshall, like Secretary Stimson, was turning the entire War Department against him and thus sabotaging the success of the Torch undertaking. In truth, Marshall declined for two entire weeks in August to give Lieutenant General Eisenhower, in London, an official directive to plan and carry out the Torch invasion of French Northwest Africa.

Time was wasting — the second wave of Hitler's renewed panzer attack in Russia, *Fall Blau,* having reached the Don on July 5, and Operation Edelweiss, to seize the all-important oil wells of the Caucasus, had begun on July 23 with 167,00 men, a thousand aircraft, over a thousand tanks — and 15,000 oil workers in tow.

Somehow the President had to show his armed forces and the world that he was not only in command but confident of American victory. A suicidal Second Front mission was not going to achieve it, whatever Marshall or Stimson maintained. Firing either or both of them might send a message of presidential determination. Given the "distrust" of his Torch plan throughout the War Department, however, this would not have aided preparations for the North Africa venture; nor, to be sure, was it the President's preferred modus operandi.

Once again, Roosevelt swallowed deeply and ignored the rumblings of discontent on the Mall. By remaining absolutely and irrevocably intent upon seeing through Torch, Roosevelt felt certain the gathering *momentum* of operational planning, preparations, and necessary training would gradually overcome his doubting Thomases, steering them toward his goal: American combat in or on the threshold of Europe, in a secure area, one where green U.S. troops and their commanders could be blooded and learn the business of modern combat without inviting catastrophe or risking a major setback to the expectations of freedom-loving people across the globe.

If General Marshall was less than supportive of his Commander in Chief in August, Secretary Stimson, for his part, was reaching a point of despair. Returning from his vacation on August 7 he found the War Department "hard at work on plans for Gymnast [as he still called it] and, as they go into them more and more, the preparations which we have been so carefully making for Bolero and Sledgehammer are being cut and delayed, the shipping reduced and the shipments [to Britain] of men put off or diminished. In fact, if Gymnast goes through, Bolero is out of the window at least until 1944," he noted in his diary, "and that seems to me a dreadful thing."[5]

To the Commander in Chief it did not. At a cabinet meeting in the White House the President attempted to convey a sense of unity in the administration. The United States had taken over direction of the war and was preparing to deliver a first great blow upon the enemy that would reinvigorate the free—and occupied—world. "During the meeting," however, "the President said that he was much troubled by charges which had appeared in some of the papers that he and Churchill were running the war plans of the war without regard to the advice of their military advisers."[6]

This was not far short of the truth—but the President was not to going to admit to such an assertion in front of his full cabinet. Rather, he intended his next remark to be a shot across Stimson's—and Knox's—

bows. "It was a matter evidently on his mind and he put it up before [Navy Secretary] Knox and myself apparently to silence us," Stimson recorded in his diary that night — the President claiming, as Stimson recorded, that he was in complete accord with his war and navy people, and that he never, ever intruded as commander in chief, except to arbitrate "when the Army and Navy differed and it was necessary for someone to decide between them."[7]

There was little the secretary could do about such lies — amazed at the President's ability to say such things with a straight face. Yet it worked: the President able to manipulate the members of his administration into doing what he wanted, even against their will or better judgment, and then sweep everyone along on a tide of goodwill and common purpose. Franklin D. Roosevelt had, in Stimson's thoughtful account that night, "the happy faculty of feeling [at one with] himself and this was one of the most extreme cases of it that I have ever seen because he must know that we are all against him on Gymnast," the secretary sighed, "and yet now that is going to be the first thing probably which is done, and we are all very blue about it."[8]

Far from being rested by his vacation, Stimson began to panic.

Sinkings of Allied merchant ships by German U-boats were reaching record numbers. News from the Pacific was no better: a battle royal taking place at Guadalcanal, in the southern Solomon Islands, where American Marines had seized the newly cleared Japanese airfield, daring the Japanese to retaliate in force — which they did.

Turning seventy-five that fall, Henry Stimson, despite being a first-rate lawyer, had not the flexibility and energy of a younger man. Once again, in his diary, he recounted for his own edification the sorry history of Torch, the "President's great secret baby" — charting yet again the way the President had "hankered and hunkered" to revive the "evil" plan each time General Marshall and the War Department had knocked it down. "Today the whole thing came into my conversations with [Generals] Marshall, Lovett and Handy and all of them feel strongly against Gymnast," Stimson noted with a sort of perverse satisfaction on August 7.[9]

Given that Averell Harriman, the President's personal emissary, was already on his way with Prime Minister Churchill to brief Joseph Stalin personally on the impending Torch operation, Stimson's efforts to find a way to stop the operation were becoming seriously dysfunctional. And detrimental to America's war effort.

The scene was, in fact, little short of tragic at such a critical moment of the war. In Stimson's tortured mind, however, the fact that the decision to mount Torch had "decisively" been taken on July 28 and that the British prime minister — approaching sixty-eight years of age and having already suffered a mild heart attack — was flying halfway across the world to explain the Torch operation to Stalin, was of no consequence. As Stimson saw it, Churchill had become the éminence grise behind America's president, and thus America's Public Enemy Number One.

The President's silencing words in cabinet also rankled. On his return from the cabinet meeting in the White House Stimson thus began, in the quiet of his office on Constitution Avenue, to mull over *yet another* official letter of protest to Roosevelt: one that would either stop the President in his tracks, or at least prove to the world, later, that the President had refused to take his chief military advisers' advice.

If the war secretary was to come out and openly oppose the President, he wanted "to be sure I am on solid ground." Taking the army chief of staff aside, Stimson had a "careful talk with Marshall over the strategic situation, getting our teeth right into it and into each other — a very frank talk and a rather useful one," the secretary confided to his diary that night. As Churchill prepared to fly on from Cairo to Moscow to tell Stalin in confidence about Torch, Stimson thus asked General Marshall point blank if he, Marshall, "was President or Dictator" of America, "whether he would go on with Gymnast and he told me frankly no."[10]

Marshall as president or dictator? Such language, exchanged between the country's most senior military officials on August 9, 1942, was sailing close to sedition — as even Marshall began to recognize the following day, when the war secretary showed him exactly what he had in mind.

Colonel Stimson's draft "Letter to the President" as U.S. secretary of war on August 10, 1942, took Marshall's breath away — for Stimson, having for months insisted on the most perilous American invasion of mainland France, across the English Channel, was now suggesting that Torch, the U.S. invasion of French Northwest Africa, was even *more* perilous.

Beginning "Dear Mr. President," Stimson acknowledged that, thanks to "the refusal of the British to join us in going forward" with a cross-

Channel attack in 1942, the chiefs of staff of the U.S. Army had agreed to substitute an invasion of French Northwest Africa. However, "intensive studies of the conditions and effects involved in the Torch proposal" by the War Department had led the secretary to take the same stance toward Torch as the British had done to the idea of a cross-Channel attack: to say no.

"I am now credibly informed that, in the light of these studies and of the rapidly unrolling world situation now before us, both the Chief of Staff and the General Staff believe that this operation should not be undertaken. I believe it to be now their opinion that under present conditions the Torch undertaking would not only involve serious danger of our troops meeting an initial defeat, but that it could not be carried out without emasculating any air attack this autumn on Germany from the British Isles and would postpone the operation known as Roundup [the invasion and conquest of France] until 1944. Furthermore, being an essentially defensive operation by the Allied Force, it [Torch] would not in any material way assist Russia."

These were bold assertions. Again Stimson urged the President to cancel the Torch landings — indeed any landings in Europe or in Africa. Instead he called for an application of all U.S. energies on concentrated *air* attacks on Germany, "with sufficient effectiveness to affect the morale of Germany more effectively than any of the other proposals which could be carried out this year. I earnestly recommend that before an irrevocable decision is made upon the Torch operation you should make yourself familiar with the present views of these your military advisers and the facts and reasons that underlie them."[11]

Throughout June and July Stimson had decried the British for being "defeatist" and lily-livered for not daring to mount a cross-Channel Second Front that year, relying merely on their RAF thousand-bomber "terror" raids — which shocked the world but showed no sign of denting German civilian or military morale, in fact only seemed to make the Germans more determined to support the Führer. Now Stimson was not only talking up the sole strategy of Allied bombing efforts, but recoiling at the prospect of "initial defeat" if U.S. troops attacked a region on the threshold of Europe where there were virtually no German troops!

A more defeatist protest to the President, two weeks after the "definitive" decision over Torch had been made, could not have been drafted.

General Marshall was embarrassed by it—especially given the manner in which Secretary Stimson was intending to speak for Marshall and his colleagues. "He takes an even severer attitude towards the President than I do," Stimson noted in his diary, "but he pleaded with me not to send it," the secretary admitted, as "he thought it would put him (Marshall) in the position of not being manly enough to do it himself."[12]

In the whole of World War II the United States would never come closer to a military mutiny—which had certainly not been General Marshall's intention. The President was the U.S. commander in chief. Marshall was a serving officer—a soldier. He had been given an order by the Commander in Chief, and it behooved him to carry it out, however much he might disagree with it, as the President well knew. Or resign. Marshall therefore begged the secretary not to send the letter. Stimson retorted by accusing him of "welching" on their opposition to the President's "evil" plan. Marshall was incensed. To Stimson's chagrin, Marshall now made clear he had no intention of making such a protest in his role as U.S. Army chief of staff—neither in writing nor in person. He would have no truck with talk of being president or dictator of the United States.

Seeing Stimson's crestfallen face, however, Marshall took pity on the secretary—who meant well, and had been devoted to the best interests of the U.S. Army since his appointment in July 1940. Marshall therefore assured the secretary "that I could rest confident that he and the Staff would not permit Gymnast to become actually effective if it seemed clearly headed to a disaster."[13]

General Marshall was not alone in stepping away from Stimson's revolt. To Stimson's added chagrin, Secretary Knox, the following day, also withdrew any presumed support for an official protest, mutiny, or further machinations against the President. The doughty Colonel Knox, who had served with courage in Cuba in 1898 with Theodore Roosevelt's Rough Riders, and again as an artillery major in World War I in Europe, was "less worried about Torch than I was," Stimson noted—admitting that, as secretary of the navy, Frank Knox was, by contrast, "more worried about Sledgehammer than I was."[14]

With this evaporation of support, Stimson decided he had best shelve his "Letter to the President."

The possible mutiny was over, for the moment at least. In high dudgeon the secretary went away for another two weeks' vacation.

Given what was awaiting the five thousand brave but inexperienced Canadians who had been assigned to Operation Jubilee, the mini-version of Sledgehammer that Churchill had felt compelled to authorize nine days later in order to silence men like Secretary Stimson, General Marshall, and the senior officers of the War Department, Colonel Knox had had every reason, however, to be worried.

PART ELEVEN

Reaction in Moscow

27

Stalin's Prayer

PERHAPS THE GREATEST IRONY of the 1942 Second Front/Torch imbroglio was Stalin's reaction.

Secretary Stimson, the U.S. chiefs of staff, and the senior generals in the War Department had all claimed that Torch would not aid the Russians. Winston Churchill, flying to Moscow to tell the Russian leader the news that no Second Front would be mounted in France that year, but that instead, U.S. landings would be substituted in Northwest Africa, was understandably apprehensive. Bravely he ventured, on August 12, 1942, to the heart of the Soviet Union — a country whose Communist forces he had himself tried to destroy in 1920, after World War I. With Russian backs to the wall in the Caucasus — Hitler's legions having crossed the Don and now aiming to take Stalingrad — it felt as if he was "carrying a large lump of ice to the North Pole," in the Prime Minister's immortal later phrase.[1] "We were going into the lion's den," one general recalled, "and we weren't going to feed him."[2] Fortunately, however, Churchill was traveling to the Kremlin with the personal representative of the president of the United States, Averell Harriman.

Reading the cables Harriman sent him from Russia, the President considered Churchill's mission to have been nothing short of heroic. As soon as Harriman returned to the United States, the President said he wanted to see him and hear his firsthand account, in person.

Two weeks later, on August 30, 1942, Averell Harriman duly drove to lunch with the President, together with the President's two speechwriters, Robert Sherwood and Sam Rosenman. The President was at his new summerhouse retreat: the USS *Shangri-la,* as Roosevelt called it.

The rustic mountain camp consisted of "a number of rudely constructed, small pine cabins, each of two or three rooms," which did not impress Judge Rosenman—especially the President's hut. "It was furnished with the most rudimentary kind of secondhand furniture, most of which had come from a navy storehouse where unwanted and well-used furniture had been accumulated over the years,"[3] he sniffed.

The President didn't mind; the cabin was as Spartan as the rooms of his beloved presidential yacht, the USS *Potomac,* but possessed something forbidden on the steamer (a converted coast guard cutter): rugs. "The rugs," Rosenman recorded, however, "had come from the same place and were in a bad state of repair."

The view, Judge Rosenman allowed, was magnificent. "The President occupied a bedroom looking out through the woods over a beautiful valley. To it was attached one of the two bathrooms. The other three bedrooms were double bedrooms but none of them had space for more than two simple metal beds, a dresser, and a chair. These three bedrooms were all served by one bathroom. The door to the bathroom never quite closed quite securely, and the President laughingly used to warn each of his guests of that fact; but the door was never repaired."[4]

Most of all Rosenman was amazed at the President's buoyant mood, when the war seemed to be going so badly. The Germans, Rosenman had been informed, had by now advanced more than five hundred miles on the Eastern Front, capturing half a million Russian troops. They had already reached the peaks of the Caucasus: poised, it seemed, to race to the Caspian Sea, seize the crucial Caucasus oil wells, and threaten northern Iran. In North Africa Erwin Rommel, promoted to field marshal by the Führer following his capture of Tobruk, was bringing up hundreds of new and improved long-barreled Mark IV panzers as well as lethal 88mm antitank guns for his final assault on Alexandria, Cairo, the Suez Canal—and then on, if successful, to Palestine.

In the Pacific, Americans were fighting fiercely for a toehold on Guadalcanal, where on August 8–9 they had suffered a naval defeat so great it had had to be kept from the public: no fewer than four Allied cruisers—three American and one Australian—being sunk in the Savo Sea by the Japanese, who suffered no losses at all . . .

Questioned at Shangri-la by the President, Harriman assured him the Germans had still a huge task on their hands—contradicting reports of the War Department's head of intelligence, who was currently predicting the imminent fall of Stalingrad to the Germans.[5] The Russians *would* hold, Harriman assured the President—whatever Secretary Stimson and the men in the War Department might say to the contrary. "Averell gave a lucid analysis of the situation," Sherwood recalled Harriman's verbal report, "and then firmly predicted that Stalingrad would not fall, and that the battle could conceivably end in a major military disaster for the Germans." As far as the Ural Mountains were concerned, "He thought the Russians could prevent the breakthrough which would have cut them off from the Caucasian oil fields and given the Germans a clear road into Iran and the Middle East"—for Stalin had assured Harriman and Churchill he could hold both Baku on the Caspian Sea, and Batum on the Black Sea, for the next few months, when the approaching winter snow would "greatly improve their position."[6] Better still, Mr. Stalin had even confided to Harriman that he was planning a huge counteroffensive that would stun the Germans.

This was greatly encouraging to the President, who listened to his personal emissary's blow-by-blow account of the three-day series of summit meetings in the Kremlin with intense fascination—and sly amusement. Churchill had first off "announced the decision to give up Sledgehammer without mincing words." Stalin, in response, had been rude to the point of deliberate insult, Harriman related—"Stalin gave him hell," and "without mincing words" derided "the timidity of the democracies in comparison to Russia's sacrifices."[7] A tyrant by nature and struggle, Stalin could not resist denigrating the pathetic British military performance in the war so far—sneering at the Royal Navy's failure to protect its convoys to Murmansk, scorning the failure of the British Army to beat the Wehrmacht in open battle. War was war, he had grimly pronounced; to win a battle, one must be willing to accept huge casualties.

It was then that Churchill, according to Harriman, had delivered his tersest riposte.

"War is war," the Prime Minister acknowledged, "but not folly."[8]

The President, at Shangri-la, was utterly delighted by the phrase—and by Winston's refusal to be bowed by the Russian's ill-mannered rebuke. Or to be tempted to pack his bags, once Harriman put his mind at rest by passing him a note, in which he pointed out that this was merely par for the psychopath's bullying course. The next day, Harriman had promised the Prime Minister, the Russian monster would be all sweetness and roses.

The President "appeared to enjoy hearing about Churchill's discomfiture in those long reproach-filled sessions," Harriman later told his ghostwriter, Elie Abel[9]—for there, but for the grace of God, the President might well have been: attempting to explain in person to the Soviet dictator why neither the United States nor Great Britain were willing to make good on their promise of a Second Front in France that fall.

About the President's substitute invasion plan — Torch — however, the Russian leader had been, to even Churchill's astonishment, almost ecstatic.

Churchill, according to Harriman, had gone on to explain, after delivering the bad news, that there was good news, too. The Americans were, indeed, coming — within weeks!

At this the dictator had changed his tune entirely. By way of metaphor, the Prime Minister had described the President's substitute strategy as akin to dealing with a crocodile: instead of hitting the critter on its hard snout, it was best to cut into its "soft underbelly." Moreover, this would be far more than a mere slash: for the invasion that would be mounted by American and British forces would number a quarter of a million men: more troops than Hitler had sent into the Caucasus — dispatched separately from the British Isles and the United States with massive air and naval cover.

Secretary Stimson, General Marshall, General Arnold, Admiral King, General McNarney, General Handy, General Wedemeyer — all had sought to persuade the U.S. commander in chief that Torch was a mistake and would not help the Russians in any way in their hour of need. Instead, according to Harriman, the Russian dictator had *instantly* grasped how the U.S.-led invasion could change the whole dynamic of the war against Hitler.

As Prime Minister Mackenzie King subsequently heard the story, via the Canadian defense minister who'd met with Winston Churchill in London, "Stalin had approved strongly" of Torch. "He had thought for 10 minutes after Churchill had proposed it, and then was greatly pleased. This of course is to be the second front that will be opened this year."[10]

The description of Stalin's response given to Mackenzie King, though given to the President weeks later, was certainly in line with the detailed minutes of the summit meeting, taken down at the time by a stenographer and used by Churchill in his own cables from Moscow to the President in Washington. According to the typed minutes of the meeting, which Harriman then showed the President at Shangri-la, "Mr. Stalin appeared suddenly to grasp the strategic advantages of 'Torch.' He saw four outstanding advantages" of the operation:

1. It would take the enemy in the rear.
2. It would make the Germans and the French fight each other.
3. It would put Italy out of action.
4. It would keep the Spaniards neutral.[11]

Reflecting on this extraordinarily positive response, the President had shaken his head at the irony. Stalin had required but *ten minutes* to recognize the way Torch would turn the tide of World War II, whereas it had taken the War Department more than a year! Moreover, the most senior U.S. generals were reputedly *still* trying to sabotage the operation, by ordering preparations for Bolero to continue in England without interruption, even at the risk of compromising the success of the Torch operation.

However, Stalin's next remark had been, if anything, even more astonishing.

Turning to the President's personal representative, Stalin had said to Churchill and Harriman — as Harriman now told Roosevelt — "May God help this enterprise to succeed."[12]

PART TWELVE

An Industrial Miracle

28

A Trip Across America

RETURNING FROM SHANGRI-LA on August 30, 1942, the President summoned General Marshall to dinner at the White House. He wanted to see whether, with Hopkins and Harriman present, he could use Stalin's positive reaction to the news of Torch to reinvigorate the general, and get him to now put his whole authority at the War Department behind preparations for the U.S. landings. To the President's dismay, however, General Marshall presented him, instead, with a draft cable to Churchill, reducing Eisenhower's plans for a three-pronged invasion of French Northwest Africa to two: one outside the Mediterranean, one within. At Casablanca and Oran only.

"This matter has been most carefully considered by me and by my naval and military advisers," Marshall's draft cable ran, for the President to sign. "I feel strongly that my conception of the operation as outlined herein must be accepted and that such a solution promises the greatest chance of success in this particular theatre."[1]

The President could only laugh — suspecting that Secretary Stimson and Admiral King were behind the maneuver: the War Department still hoping that by limiting the landings to the Atlantic seaboard of Morocco and the westernmost part of Algeria, the United States could continue to pursue plans for a cross-Channel landing in the spring of 1943 — or even a switch to the Pacific, if the battle for Guadalcanal and the Solomon Islands became more and more menacing.

Shaking his head, the President told Marshall the cable was unacceptable. Landings only in Morocco and at Oran would not persuade the Vichy French that America was serious — indeed, facing such meager landings, Vichy French forces in the rest of Morocco, Algeria, and Tunisia would be encouraged to resist invasion by America as an outside power. Worse still, recognizing the weakness of such a force, Hitler would undoubtedly seize the chance to ship troops across the Mediterranean and order them to occupy the Vichy territories — threatening to create the very scenario Secretary Stimson had always feared: a sort of Custer's Last Stand by American troops, or second Gallipoli. Instead of a mighty American operation on the threshold of Europe that would give heart to all those praying for Hitler's defeat, Torch would be a flickering candle.

The President was deeply disappointed in Marshall. Somehow, Roosevelt insisted, enough naval forces must be found for *all three* landing areas to be simultaneously assaulted, however hard this might be. Redrafting Marshall's proposed cable, he turned it on its head.

Torch must succeed, by its very preponderance of men and munitions, convoyed and landed in overwhelming strength. "To this end I think we should re-examine our resources and strip everything down to the bone to make a third landing possible," Roosevelt reworded the cable to Churchill and the British Admiralty.[2] All three initial landings *must* be made by purely American forces, he also laid down in the telegram he eventually sent to Churchill that night, lest the Vichy French be inspired to defend their colonial territories the more determinately, given their hatred of the British. "I would go so far as to say," he wrote, "I am reasonably sure a simultaneous landing by British and Americans would result in full resistance by all French in Africa whereas an initial landing without British ground forces offers a real chance that there would be no French resistance or only token resistance."[3]

Poor Eisenhower, the designated supreme commander for Torch, now found himself torn between instructions from the Commander in Chief of the United States and his U.S. Army chief of staff to whom he owed his meteoric promotion since 1939. "I feel like the lady in the circus that has to ride three horses with no very good idea of exactly where any one of the three is going," he laughingly told General Patton, who'd been chosen to command the Western Task Force's assault landings in the Casablanca area, setting out directly from the United States.[4]

Receiving the President's cable, Churchill was understandably disappointed that no British troops would be landing in the first wave of Torch. He bravely accepted Roosevelt's logic, however — and urgency.

Time was running out, if the invasion was to be mounted that fall — the leaves at Shangri-la already beginning to turn. Over the following days the three-pronged Torch operation, despite Marshall and Stimson's objections, was finally set in stone. The Western Task Force's Casablanca operation was trimmed, the Eastern Task Force's Algiers landings increased. "We are getting very close together," the President cabled Churchill from Washington on September 4, 1942 — adding: "I am directing all preparations to proceed" — meaning that General Marshall and the War Department would now be told to obey, or resign. "We should settle this whole thing with finality at once."[5]

Trying to put together the largest Allied amphibious operation of the war from a headquarters in London, over three thousand miles from the troops that would be embarked in America, Lieutenant General Eisenhower was understandably nervous. For his own part he remained unconvinced that Torch was a better option than Sledgehammer, or Bolero staged in 1943, and gave it only a fifty-fifty chance of success.[6] He was, however, relieved that a final decision had been made — indeed, he was already proving a remarkably patient and intelligent coalition commander. The "Transatlantic essay contest," as he put it, was at least over.[7]

The next day, Churchill cabled his agreement. "It is imperative now to drive straight ahead and save every hour. In this way alone shall we realize your strategic design," he telegraphed the President in cipher, "and the only hope of doing anything that really counts this year."[8]

The President's simple comment was one word: "Hurrah!"[9]

On September 3, 1942, meanwhile, the President had agreed to give, in the White House, a talk to representatives of the International Student Assembly — a speech that was sent out across the world and gave perhaps a better idea of his growing sense of America's destiny in the modern world than any he had previously broadcast.

The talk was, the President explained to listeners, "being heard by several million American soldiers, sailors, and marines, not only within the continental limits of the United States, but in far distant points — in Central and South America, in the islands of the Atlantic, in Britain and Ireland, on the coasts of Africa, in Egypt, in Iraq and Iran, in Russia, in India, in China, in Australia, in New Zealand, in many parts of the Pacific, and on all the seas of the world. There — in those distant places — are our fighting men. And to them," Roosevelt declared in a voice not only of authority but of absolute conviction and confidence, "I should like to deliver a special message, from their Commander in Chief, and from the very hearts of their countrymen."

The speech touched on familiar themes. "Victory is essential," Roosevelt stated, "but victory is not enough for you—or for us. We must be sure that when you have won victory, you will not have to tell your children that you fought in vain—that you were betrayed. We must be sure that in your homes there will not be want—that in your schools only the living truth will be taught—that in your churches there may be preached without fear a faith in which men may deeply believe.

"The better world for which you fight—and for which some of you give your lives—will not come merely because we shall have won the war. It will not come merely because we wish very hard that it would come. It will be made possible only by bold vision, intelligent planning, and hard work. It cannot be brought about overnight; but only by years of effort and perseverance and unfaltering faith.

"You young soldiers and sailors, farmers and factory workers, artists and scholars, who are fighting our way to victory now, all of you will have to take part in shaping that world. You will earn it by what you do now; but you will not attain it if you leave the job for others to do alone. When you lay aside your gun at the end of the war, you cannot at the same time lay aside your duty to the future.

"What I have said to our American soldiers and sailors applies to all the young men and women of the United Nations who are facing our common enemies. There is a complete unanimity of spirit among all the youth of all kinds and kindreds who fight to preserve or gain their freedom."

"This," the President declared, "is a development of historic importance. It means the old term, 'Western civilization,' no longer applies. *World* events and the common needs of all humanity are joining the culture of Asia with the culture of Europe and the culture of the Americas to form, for the first time, a real world civilization. In the concept of the four freedoms, in the basic principles of the Atlantic Charter, we have set for ourselves high goals, unlimited objectives. These concepts, and these principles, are designed to form a world in which men, women, and children can live in freedom and in equity and, above all, without fear of the horrors of war.

"For no soldiers or sailors, in any of our forces today, would so willingly endure the rigors of battle if they thought that in another twenty years their own sons would be fighting still another war on distant deserts or seas or in faraway jungles or in the skies.

"We have profited by our past mistakes. This time we shall know how to make full use of victory. This time the achievements of our fighting forces will not be thrown away by political cynicism and timidity and incompetence."

It would not be straight sailing. "We are deeply aware that we cannot achieve our goals easily. We cannot attain the fullness of all of our ideals overnight. We know that this is to be a long and hard and bitter fight—and that there will still be an enormous job for us to do long after the last German, Japanese, and Italian bombing planes have been shot to earth.

"But we do believe that, with divine guidance, we can make in this dark world of today, and in the new postwar world of tomorrow—a steady progress toward the highest goals that men have ever imagined.

"We of the United Nations have the technical means, the physical resources, and, most of all, the adventurous courage and the vision and the will that are needed to build and sustain the kind of world order which alone can justify the tremendous sacrifices now being made by our youth.

"But we must keep at it—we must never relax, never falter, never fear—and we must keep at it together.

"We must maintain the offensive against evil in all its forms. We must work, and we must fight to insure that our children shall have and shall enjoy in peace their inalienable rights to freedom of speech, freedom of religion, freedom from want, and freedom from fear.

"Only on those bold terms can this total war result in total victory."[10]

Roosevelt was seeking not only to raise national morale in the weeks before the Torch invasion, but to prepare Americans for a far greater challenge: asking young Americans, especially, to step up to the plate in embracing America's moral role in a postwar world.

Along with more proselytizing, though, the President was anxious to make sure American industrial output and its expansion of the military were matching his high expectations. In his State of the Union address in January 1942, he had announced production targets that were ridiculed by Hitler and Goebbels. Not only were the majority of them being reached, however, but many of them were being exceeded, the President was informed, by dint of mass production on a scale never seen before in human history. To check on this, and to spread something of the gospel of inspiration that his personal presence would engender, Roosevelt now set off on what was for him an epic, 8,754-mile train journey across America — and was amazed.

Mrs. Roosevelt accompanied the President only as far as Milwaukee, but FDR's daughter Anna, his secretary, Grace Tully, and his stenographer, Dorothy Brady, as well as his first cousin, Laura Delano, continued on with him. Daisy Suckley was also a member of the party, together with the President's former law partner, Harry Hooker, as her escort, in order that there be no gossip. Steve Early, the White House press secretary, went along to ensure no word of the trip be reported in the press before the President's return;[11] also Ross McIntire, the President's doctor, and Captain McCrea manned the communications car.

"I can't quite believe it even," Daisy scribbled in her diary, ensconced in Stateroom B on Car No. 3, as it left Silver Spring, Maryland, "yet, here I am — on board — to tour the country with the P. of the U.S.!" Donald Nelson, head of the War Production Board, had briefed the President on "munition plants, the new airplanes, etc." that would be viewed on the President's tour.[12] It was only at the Chrysler Tank Arsenal in Detroit, however, that the true magnitude of what was being achieved industrially hit home.

At the tank plant in Detroit, "a boy with a yard-long Polish name plowed through water & mud" in his new M-4 Sherman, "straight up to the President's car, stopped and pushed his head through the hole with a smile. People standing around looked rather alarmed as the tank plowed forward, but the P. had a good laugh."[13]

"Good drive!" the President shouted.[14] "It was a monster performing his tricks & lacked only the final bowing of the front legs, like the elephant in the circus!" Daisy noted on September 18. "30 tanks a day — "[15]

A day? Two hundred a week? Sure enough, by year's end tank production had increased from under four thousand in 1941 to twenty-five thousand in 1942 — hoisted "like ducks on a spit" by thirty-ton jigs as they were assembled.[16]

It was the same story at Ford's new plant at Willow Run on September 18, 1942. Where in March that year there had been but trees there now stood a new aircraft factory a mile and a half long, containing the world's first mass-production assembly line for airplanes, which that month produced its first B-24 Liberator bomber for the President to see. Over succeeding months it would churn out planes at a phenomenal rate that would eventually top one B-24 *every sixty-three minutes* — the plant's contribution to some forty-nine thousand U.S. planes produced that year.[17]

At North Chicago some sixty-eight thousand naval officers and men were training; at Milwaukee a huge turbine manufacturing plant was visited. At Lake Pend Oreille in Idaho there had been nothing in March 1942. A naval training station opened only five days before the President's arrival. It would train almost three hundred thousand sailors over the following thirty months, becoming the second-largest training center in the world.

In Seattle the President drove beneath the wing and fuselage of a B-17F Flying Fortress at the new Boeing Plant 2 — the factory buildings camouflaged with burlap and fake trees to resemble a quiet American suburb. Produced both by men and women ("Rosie the Riveters"), production of the four-engine bomber rose from 60 that fall to almost 100 per week, or 362 per month, as the war progressed.

Factory workers "evidently knew nothing of the P.'s coming & looked up vaguely from their work as we drove between the machines," Daisy recorded. "It seemed incredibly crowded, though in perfect order. When we came out, the word had spread & workers were running to get a view of the P., clapping & smiling. We see many women in these plants, & they tell us more & more are being taken in. Here they are even welding, with masks on."[18]

From there they traveled to the aluminum smelting works at Vancouver, Washington — "Long sheds filled with electrolytic cell furnaces burning 24 hours a day and 7 days a week — No stop is possible."[19] Later that morning they inspected the Kaiser shipbuilding yard on the Willamette River, in Oregon. The country's biggest housing project had been undertaken to provide labor, and it was there that the President witnessed a true miracle of mass production: the launching of a ten-thousand-ton freighter, the USS *Joseph N. Teal,* built in only ten days.

"The work men were all lined up," Daisy noted as the President's car drove up a special ramp to face the bow of the ship. Roosevelt's daughter Anna was given a "bouquet" of Defense Stamps, tied with a "red white & blue ribbon. The ceremony began with a prayer by a priest." Then, as the last of the eight rivets holding the vessel were knocked out, the champagne bottle swung against the bow "showering Anna to the skin, & down the ways went the Ship — it is a most moving scene, specially now, when you realize that that ship may be sunk by a submarine on her very first trip."[20]

How was it possible to build and launch a vessel of that size in mere *days*—an almost biblical achievement? As the President was driven "around the shops where they are making the various parts & assembling them as far as possible," the genesis of Roosevelt's confidence in American industry became clear to Daisy. "Large portions of the ships were loaded on huge trucks with rubber tires ready to be taken to the 'ways,'" Daisy noted. "This is Mr. Kaiser's 'secret' for getting a ship built in 14 days!! The P. likes both Mr K. & his son Edgar who was there with him, said Mr. K is a 'dynamo.'"[21]

Henry Kaiser was. Notable too was the morale of the workforce, and the managerial and engineering masterminding that went into a process in which, like the assembling of a model from a kit, the constituent parts of a vessel were first manufactured, then merged at the appropriate moment with the nascent vessel, from the keel up.

This miracle of mass production was awe-inspiring for the President to witness with his own eyes, barely two years after he himself had secretly begun assembling the team that would cause it to happen: a marvel that had begun on May 28, 1940, when Roosevelt had put in a phone call to the CEO of General Motors, a Danish immigrant by the name of William Knudsen. "Knudsen? I want to see you in Washington. I want you to work on some production matters. When can you come down?"[22]

Knudsen had taken leave from General Motors and become a "dollar-a-year man" in Washington—first as leader of the Advisory Commission to the Council of National Defense, then as director general of the Office of Production Management. When Donald Nelson was made chairman of the War Production Board in January 1942, Knudson became a lieutenant general in the U.S. Army and director of production in the Office of the Under Secretary of War: Knudson and Nelson the equivalent of Hitler's production tsars, Fritz Todt and Albert Speer.

*Pre*fabrication, then, was the key to American military mass production, as the President explained to Daisy Suckley and his other guests—whole sections of a ship, such as deckhouses, built elsewhere, then transported and welded into place on the slipway—the shipyard becoming a literal as well as metaphorical assembly line. To achieve such output, Kaiser's Oregon complex would employ thirty thousand people—30 percent of whom were women.

America's transformation from potential into actual industrial superpower dwarfed in swiftness, scale, and quality anything comparable in the world. It was small wonder the President felt proud; by the end of the year the United States would be producing more war material than all three Axis powers, Germany, Italy, and Japan, put together.[23]

From the Boeing plant the President traveled the next day to Mare Island Naval Shipyard, where fifty thousand workers tested, built, and repaired submarines. From there to Oakland Naval Station, where dozens more submarines and subchaser vessels were under construction and repair. At Long Beach, Los Angeles, he visited the huge Douglas Aircraft plant, which would manufacture upwards of thirty-one thousand aircraft during the war; at Camp Pendleton he inspected more naval training units and visited the San Diego naval hospital — caring for wounded men from the fighting in the Pacific, as well as some still recuperating from Pearl Harbor. Then to the Consolidated Vultee plant at Fort Worth, Texas, where not hundreds but thousands of B-24 Liberators were being mass-produced. And on to Louisiana — where the President visited the Higgins boatbuilding yard.

Employing twenty thousand people, Andrew Higgins had overcome initial U.S. Navy hostility and revolutionized landing-craft production. His accumulated knowledge of shallow-draft vessels required for the marshes and bayous of Louisiana had given him a fierce faith in his own product — enabling him, once contracted, to begin building landing craft for the U.S. Navy on a bewitching scale: more than twenty thousand craft being produced in the months after Pearl Harbor. "He is the same type as Mr. Kaiser, a genius at getting things done," Daisy Suckley wrote in her diary, "constantly inventing new gadgets. His trouble is that he is too blunt & fights with everyone, so that the maritime commission hates him and won't play ball with him — But, he turns out the goods!"[24]

And so the journey had continued: everywhere the same story, that of a nation not only at war, but operating at almost manic speed to provide itself with the means to win it. At Camp Jackson, in South Carolina, thousands of soldiers "marched before the P. and disappeared over the hill, raising a mist of dust, their guns and helmets showing against the sky," Daisy described. "It is our last evening on board. But the P. said the trip had worked so well with us four that he will take us on another! No complexes — no quarrels — etc.!"[25]

29

The President's Loyal Lieutenant

IF ANYONE QUESTIONED, LATER, just how it was that Roosevelt remained so deeply confident of victory in the fall of 1942, after such a summer of reverses for the United Nations, they need only have looked to the President's trip across America that September.

"So I think he has had a real mental rest, & is now ready to go back & 'talk turkey' to a good many people — He can talk from what he has seen with his own eyes," Daisy Suckley noted[1] — for the President knew now, beyond all doubt, that the United States was ready to win the war, whatever it took. He did not want Torch to be "delayed by a single day," as he wrote in a cable to Churchill from his train — and certainly not by diverting troops to a landing in Norway, as Churchill was suggesting once again, to mollify Stalin over the cancellation of further convoys to Murmansk, in view of the heavy casualties. Nor did he see a need to tell Stalin in advance that the latest convoy, PQ 19, would not sail. "I can see nothing to be gained by notifying Stalin sooner than is necessary and, indeed, much to be lost," he cabled. Torch, he was more and more certain, would change the whole dynamic of the war: would give the United States and United Nations the global initiative. "We are going to put everything in that enterprise and I have great hopes for it. . . . I am having a great trip. The training of our forces is far advanced and their morale excellent. Production is good but must be better. Roosevelt."[2]

Dimly, even Winston Churchill began to face up to what was obvious to the rest of the world: that the United States would not only win the war for the Western Allies, but was set to become the dominant world power thereafter.

No sooner had the President returned from his inspection trip than

Churchill cabled Roosevelt to question the incredible numbers that the American Production and Resources Board had given the British for U.S. military production — output amounting to some seventy-six thousand tanks by 1943, enough to equip two hundred U.S. divisions.

"This appears to me to be a provision on a scale out of all proportion to anything that might be brought to bear on the enemy in 1943," Churchill telegraphed in alarm. It was clear to him that if Great Britain was intending merely to cauterize Hitler's Europe by more ad hoc raids — operations like Mountbatten's "Operation Plough" mini-landings in snowbound Norway, or Rumania, or Northern Italy[3] — the President was not. Nazi Germany, Roosevelt was certain, would not be felled by pinpricks but by a sequence of ever-greater amphibious landings that would unroll the true military potential of the United States.

Churchill, having predicated his whole strategy for British survival in World War II upon his alliance with the United States, could hardly complain — and to his credit, beyond his mild protest over American über-production and Roosevelt's insistence on Torch being, in its initial phases, a completely U.S. operation of war, he didn't. When General Brooke, the British Army CIGS, protested against Averell Harriman's recommendation that U.S. teams take over the Persian port and railroad system, which would result in British forces in Iran becoming wholly dependent on America, Churchill had rounded on Brooke with the words: "In whose hands could we be better dependent?"[4]

The fall of 1942 thus marked, in effect, the turning point in the evolution of the modern world, as the British Empire wound down. Though Churchill might shortly declare in public his refusal to preside over its liquidation, the fact was, Great Britain was now to become, to all intents and purposes, the staging post of American power in Europe.

Reluctantly but with dignity, the Prime Minister — who had shown no mercy in putting down protest riots in India, where it was estimated that some 2,500 Indians were killed, 958 were recorded as having been flogged, and 750 government buildings destroyed[5] — accepted his new role. When the President instructed Harriman to return to London in mid-September to make clear to Churchill he wanted no further changes to Torch, and to insist it remain an American, not binational, operation, the Prime Minister gave way with little more than a murmur of protest.

"I am the President's loyal lieutenant," Churchill said to Harriman in person[6] — and in a cable direct to the President on September 14, 1942, the Prime Minister repeated his expression of fealty in writing. "In the whole of TORCH, military and political, I consider myself your Lieutenant," he wrote, "asking only to put my viewpoint plainly before you. . . . We British will come in only as and when you judge expedient. This is an American enterprise in which we are your help mates."[7]

As Harriman cabled the President that same day, the Prime Minister "understands fully that he is to play second fiddle in all scores and then only as you direct."[8]

Great Britain, which had once ruled more than half the earth, was now fated to play a subordinate role to the United States — a momentous comedown, but better than becoming junior partner, puppet, or quisling of Hitler's Third Reich, as the Vichy French had done. Besides, there was the very thrill of imminent battle, which could not fail to excite the warrior in Winston Churchill.

Ultra intelligence — decrypts of top-secret German military signals that Churchill loved to see raw and uninterpreted by his staff — was revealing the Nazis had no conception of what was about to hit them. Exultant, Churchill cabled Roosevelt on September 14, 1942, saying he was counting "the days" to "Torch"[9] — the more so since his own recent British landing, Operation Jubilee, had proven yet again an utter and bloody disaster.

PART THIRTEEN

The Tragedy of Dieppe

30

A Canadian Bloodbath

PRESSED BY GENERAL MARSHALL and Admiral King on their visit to London in July to mount Sledgehammer that year, Churchill had wisely refused. Yet he had also bristled at the accusations of British faint-heartedness — an accusation that in part explained why he allowed himself to be persuaded by his chief of Combined Operations, Vice Admiral Mountbatten, to go ahead with an operation that had already been canceled once evidence revealed the Germans were aware of it: a landing by an entire Canadian infantry brigade, with tanks, on the beaches of Dieppe, a small French fishing port south of the Pas-de-Calais, where Winston had once courted his wife, Clementine.

Churchill had hoped the "reconnaissance in force," as it was termed, would help convince not only Marshall and King but Stalin, too, that the British — which was to say, Canadians — were not lacking in courage. In a few days, Churchill had told the Russian leader, the operation — a sort of exploratory, miniature version of the full-scale Second Front landings planned for 1943 — would be mounted across the Channel on a selected target with "8,000 men with 50 tanks." They would "stay a night and a day, kill as many Germans as possible and take prisoners." The landing, as Churchill had described, could "be compared to a bath which you feel with your hand to see if the water is hot."[1]

Stalin had shaken his head at such military and political naiveté. Whatever happened on the day — whether successful or not — once the troops were withdrawn the Nazis would simply trumpet their withdrawal as "the failure of a British attempt at an invasion" or retreat — which would help no one, he sniffed.[2]

Launched on the early morning of August 19, 1942, the "Dieppe Raid" had proven Stalin's prediction tragically correct. The water had been scalding, the raid a "fiasco," as even Churchill acknowledged.[3] The Germans, whose troops occupied the entire Atlantic coastline of Europe as far as the Pyrenees, were not only waiting, but had even been conducting an exercise the day before to rehearse repelling just such an assault.[4]

As Stalin had predicted, the master of Nazi propaganda was over the moon when hearing of the operation. Goebbels had just landed in the Ukraine and been driven to the Führer's "idyllically concealed" new advance headquarters at Vinnitsa on August 19 when the news was given to him that at "6.05 in the morning a major invasion attempt had been made at Dieppe."[5] The Allies had landed "more than a division, and had established in one place a small bridgehead. The RAF had thrown large forces into the battle. The English had brought 20 panzers"; moreover, a huge number of vessels were reported to be waiting at Portsmouth to "reinforce the landings if successful." In other words, as Goebbels dictated for his diary, "under pressure from Stalin the British have clearly undertaken the attempt to establish a Second Front."

The Reich minister of propaganda had been contemptuous. "Not for a single second does anyone in the Führer's headquarters doubt that the British will be given a resounding whack and sent home."[6]

Goebbels was proved right. In an eight-hour interview with the Führer, the propaganda minister recorded Hitler's complete unconcern about Dieppe. In March that year the Führer had already stationed a top panzer division in the Pas-de-Calais area, with two further motorized divisions in reserve. They were not even needed — for by 2:00 p.m. on August 19 the invasion attempt had been "liquidated."[7] Sepp Dietrich, commanding the Führer's SS Life Guard motorized division, would surely be swearing blue murder, the Führer chuckled, that he hadn't even had the chance "to enter the fray."[8] Churchill must have ordered the landing as a sop to Stalin — the Russian leader a veritable giant in comparison to little Churchill, who could only boast a "few books he'd written, and speeches in Parliament,"[9] while Stalin had re-created a nation of 170 million and prepared it for a huge military challenge, as Hitler conceded. In fact, if ever Stalin fell into German hands, Hitler told his propaganda genius, as Führer he would out of respect spare the Russian premier, perhaps banishing him to some beach resort. Churchill and Roosevelt, by contrast, would be hanged for having started the war "without showing the least statesmanship or military ability."[10]

Flying back to Berlin to direct the Nazi propaganda response to the Dieppe invasion, Goebbels could only mock at how Churchill then sought to cover up the "true catastrophe," censoring and concealing in the press the huge casualties the Canadians had suffered. The Prime Minister had tried to parlay the attack into an "experiment"—but if it was such, it had achieved the opposite effect, Goebbels crowed. Not only had it shown how devastatingly effective were German defenses in the Pas-de-Calais and nearby region, but it had made the Führer decide to *further* fortify the entire Atlantic coast against invasion: a "full-blown defensive line in the same manner as the Atlantic Wall." "If the British mount a real invasion next spring, where they're planning, they are going to be battering against reinforced castle gates," the Führer had assured him. "They'll never set foot again on European soil. The Atlantic coast and the Norwegian coast will then be one hundred percent in our possession, and we will no longer be threatened by invasion, even if mounted on the most massive scale."[11]

The Führer had then turned to other, more important matters: his decision to seize Leningrad that very year, but to spare Moscow until the next year—though both cities were in due course to be completely "erased"[12] as part of the complete destruction of any kind of Russian national heritage or pride. Plus the thorny problem of the German churches, which were to be threatened with the same solution as was being meted out to the Jews, given their Christian leanings toward Bolshevism and their failure to support Nazism wholeheartedly . . .[13]

Churchill might ask the British and Allied press not to reveal the true extent of the Dieppe fiasco, but it proved impossible to conceal it from the Canadian prime minister, a thousand of whose soldiers had been killed in cold blood on the beaches of the harbor town, with further thousands wounded and taken into German captivity for the duration of the war—their feet even manacled, after an Allied operational order was intercepted and translated, detailing how manacling of captured German troops was to be carried out by the assault troops.

Mackenzie King had opposed the idea of a major cross-Channel landing that year, as long as the Allies lacked preponderant naval and air forces, as well as experienced soldiers. Nevertheless, his defense minister had gone along with the revived operation—and was the first to hear reports that night of the catastrophe. "While [War] Council was sitting," King recorded, "the first authentic word of its extent and probable extent of our losses"—completely contradicting a mendacious press release put out by Lord Mountbatten. In truth, the Canadian premier noted, "casualties were heavy. Number of Canadians taken prisoners but also many killed and wounded. One felt inclined to question," he added, "the wisdom of the raid unless it were part of the agreement reached when Churchill was with Stalin."[14]

Stalin, to be sure, was blameless — having argued against such an operation. Well over half of the sixty-one hundred troops who had taken part in the fiasco had been killed, wounded, or captured.

Two days later King's heart sank still further, as more news of the fatalities came in. "Reports received of raid make one very sad at heart for losses, which have been considerable," he noted again in his diary — German newsreel footage, bruited across neutral countries by Goebbels's propaganda team, making it impossible to maintain Mountbatten's fiction. How much better, Prime Minister King reflected, would it have been "to conserve that especially trained life for the decisive moment. . . . It makes me sad at heart."[15]

And on August 24, 1942, King lamented: "I keep asking myself was this venture justified, just at this time?"[16]

In Washington, the President felt deeply for his Canadian ally: aware that, had Stimson, Marshall, and King gotten their way and launched Bolero that year, it would have been Americans who perished at the hands of the waiting Germans.

PART FOURTEEN

The Torch Is Lit

31

Something in West Africa

WITH HUGE NUMBERS OF AMERICAN troops preparing to embark from ports in Britain and the United States for the invasion of Northwest Africa, how was it possible that neither Hitler, the commander in chief of the forces of the Third Reich, nor his Oberkommando der Wehrmacht (OKW), or German high command, saw the American invasion coming?

Historians could never quite decide. It was not, after all, as if there were no indications of an American surprise attack. On Columbus Day, October 12, 1942, for example, the President had given a special Fireside Chat from the White House. Encouraged by what he had seen of American industrial output, he explained to listeners across the nation—and abroad—that the worst times were now over for the United Nations. The Axis powers had already reached their full strength, the President described; "their steadily mounting losses in men and material cannot be fully replaced. Germany and Japan are already realizing what the inevitable result will be when the total strength of the United Nations hits them"—moreover hits them "at additional places on the earth's surface."[1]

Where, though? Reading British and American newspapers, Goebbels was less sure than the Führer that the Allies would now risk a Second Front, especially after the pasting the Canadians had received at Dieppe. "The Second Front seems to be definitively shelved," the propaganda minister recorded in his diary on September 8, 1942—noting how even Russians were now contenting themselves with calls for the RAF to do heavier bombing rather than harping on an impossible cross-Channel invasion.[2] But if not a cross-Channel Second Front, might the Americans invade elsewhere? The south of France, perhaps?

Secret plans for military occupation of the remaining Vichy-controlled inland and southern regions of mainland France had long been drawn up by the German high command, code-named Case Anton, even though this would abrogate the terms of the Franco-German armistice, signed by Maréchal Pétain in 1940, following the French surrender. Case Anton would ensure the whole of the Mediterranean coast of France would be secured by German and Italian forces, if there were impending signs of an Allied invasion. But what of French Northwest Africa, where there were still only a handful of German officials?

Goebbels, like Hitler, discounted the notion. In the United States, in the October run-up to the November 3 congressional elections, press hostility to the President was rising — "a pretty tough and massive critical mass," Goebbels recorded with satisfaction.[3] All the President seemed to have to offer were words. Not only were the Germans "stealing food from the rest of Europe," the President had warned listeners in his Fireside Chat, but there had been an increase in the "fury" of German "atrocities" in Europe that would not be overlooked — or forgiven, he'd stated. The United Nations, the President declared, "have decided to establish the identity of those Nazi leaders who are responsible for the innumerable acts of savagery. As each of these criminal deeds is committed," he'd emphasized, "it is being carefully investigated; and the evidence is being relentlessly piled up for the future purposes of justice. We have made it entirely clear that the United Nations seek no mass reprisals against the populations of Germany or Italy or Japan. But the ringleaders and their brutal henchmen must be named, and apprehended, and tried in accordance with the judicial processes of criminal law."[4]

Reading the transcript in Berlin, Dr. Goebbels had known far better than the President what "atrocities" were being perpetrated, and on what a sickening scale. He'd put little credence in the President's warnings, however: neither the "additional places" where the Allies would strike, nor the justice that would be meted out for German — and Japanese — "criminal deeds." A Second Front in France that year, or the next, would never succeed, Goebbels reckoned — the British failure at Dieppe having demonstrated the impregnable nature of the Westwall defensive line from Norway to the Spanish border. "The English know as well as we do that they are in no position to launch even a modest start in that direction," he noted contemptuously a few days after the President spoke.[5]

The threat of postwar American justice tribunals Goebbels treated with the same contempt. "One can just dismiss such things with a shrug of one's shoulders," he'd added to his daily diary[6] — secure in his conviction that German victory in the war would make the notion of criminal trials thereafter ridiculous. Even if the renewed German assault on the Eastern Front were to grind to a halt at Stalingrad and in the Caucasus, forcing the Wehrmacht onto the defensive for another winter, German forces had seized a prodigious amount of Russian territory, which could be used to feed the Third Reich rather than its own people. "We have a swath in our possession that will allow us to develop our potential in undreamed of ways," he encouraged himself to believe,[7] recalling Hitler's grand design: German warrior-farmers, controlling a vast eastern border of the Reich in which those Slavs who were allowed to survive (half of all Russian captives were, in reality, killed or starved to death) would be kept as illiterate slaves of their German masters.

The propaganda minister, whose genius had been to manipulate and orchestrate the entire output of German newspapers, radio, film, theater, and publishing, thus dismissed the President's broadcast as hot air, glorying in Roosevelt's "democratic" difficulties with Congress, his embarrassment at Wendell Willkie's almost hysterical calls for a Second Front during a recent visit to Moscow, and declining support for the President in American public opinion polls — with anxious voices saying "We could lose the war!" or "We will lose the war!" Goebbels noted.[8]

But if so, how was it the President sounded so confident in his Fireside Chats, Goebbels wondered[9] — as did Churchill in the English Parliament, too? What were they up to?[10]

On October 6, 1942, Goebbels admitted that, in terms of a possible American or Allied offensive in or around the periphery of Europe, "absolutely nothing is known."[11]

With the Russians continuing their "infernal resistance" at Stalingrad, all eyes were on the Eastern Front, where snow would soon fall. Weather in the English Channel would, by the same token, surely make a cross-Channel attack impossible, whatever reports Wendell Willkie might be taking back to the President from Stalin in Moscow. On October 17, 1942, however, Goebbels noted there were "rumors that the Allies are preparing for something in West Africa. Apparently such plans are quite advanced. It's possible that the British and the Americans are trying to get clear of their commitment [to the Russians], and pretending to Stalin this would be a second front."[12]

No countermeasures were taken by Hitler's headquarters, however, and by October 20, Goebbels was noting that the French — a nation on the down and out, in his view — were getting worried about their hitherto undisturbed colonial territories in North Africa. "Sooner or later the British and above all the Americans will appear there," Goebbels accepted—not as a springboard to attack Italy and Germany so much as for reasons of imperial design, he thought. "The Americans without doubt intend one way or another to inveigle themselves into this war and do everything they can to pick up what's going free, so to speak"—colonies. He and the Führer therefore contented themselves with the assumption that the French could be relied upon to defend their colonial territories with the substantial naval, air, and land forces they had in Morocco, Algeria, Tunisia, and Dakar.

At the White House, meanwhile, the President, reading Ultra decrypts of German signals and hearing from his OSS chief, "Wild Bill" Donovan, could hardly believe the reports from Germany and North Africa. Could the Führer *really* have no idea of the magnitude of what was going to hit him?

That a thousand things could go wrong with the Torch invasion, Roosevelt was well aware, from the notorious autumnal surf off Casablanca to Axis identification of the approaching fleets from America and the British Isles. The Vichy French, too, might react as they were doing on the island of Madagascar still—defending their colonies with everything they had. Would the fact that the "invaders" were American not make a big difference, though? To make absolutely sure, the President prepared special printed leaflets addressed to the people of French North Africa, to be distributed and dropped from the air once the landings took place. He also made a specially recorded audio message to be broadcast on radio—in the President's best French.[13]

How the French would respond thus remained for Roosevelt the biggest question. General Weygand, the pro-American commander in chief of French forces in Africa, had been fired by Marshal Pétain, under pressure of the Nazis, and his successor, Admiral Darlan, was fiercely anti-British and not necessarily pro-American—though he had assured Admiral Leahy that an American landing in overwhelming force would be enough to get the French to agree to a cease-fire, after perhaps token resistance "*pour l'histoire.*" There was also the possibility being explored by Robert Murphy, the former chargé d'affaires at the U.S. Embassy to Vichy France and currently the President's special emissary in Vichy-administered North Africa, that General Henri Giraud, a brave and popular warrior who had escaped to France from a prisoner-of-war camp in

Germany, where he'd been held since 1940, could be used to rally French forces in North Africa to the American cause. For this to happen, however, he would have to be sprung from Vichy France by submarine, and attached to General Eisenhower's headquarters.

At the end of the day, however, the President had as little confidence that the Vichy French wished to be "liberated" as Hitler did. Or Churchill. Life in French Northwest Africa — save for French Jews — had been remarkably easy, and made all the easier by Roosevelt's decision to continue sending food. Resistance to an American invasion might therefore be weak — but the President doubted whether the French would resist a German invasion, either, if Hitler chose to contest the Allied campaign. It wouldn't matter, however. In fact the President was *counting* on German opposition, which would ensure that American troops learn on the field of battle the same military skills that the Russians had had to learn on the Eastern Front — but on ground of America's choosing, at the very limit of German communications, and with the strategic goal of gaining a secure steppingstone, on the threshold of Europe, that could be relentlessly reinforced from the United States, and lead to the elimination of Italy as an Axis belligerent — as Stalin had so swiftly understood. Roosevelt thus remained quietly optimistic, even as his secretary of war slipped into an ever deepening funk.

The Dieppe fiasco, paradoxically, frightened Stimson more as an example of what might befall Torch than it did in relation to his support for a premature cross-Channel invasion. No matter how much the President pointed out the difference — especially the fact that there were still only a handful of Germans in Morocco and Algeria — Stimson continued to argue and conspire to cancel the project. Dimly, though, he became aware that he was testing the President's legendary patience. When he insisted on lowering the age for the draft, or Selective Service, claiming it was immediately necessary for manpower reasons, *before* the congressional elections in November, the President was furious — but agreed to ask Congress for the bill — knowing it would dent his Democratic majority in the House and in the Senate. It did not endear Stimson to the Boss, however — the President refusing to invite Stimson to the White House.

"I have not been seeing as much of the President lately as I used to," Stimson acknowledged in his diary, but ascribed it erroneously to the President's increasing trust in General Marshall's judgment and advice. "That is a good result," he wrote, pretending to welcome the change, "but

I shall have to look out to be sure that it does not cut me out of situations where I have duties and responsibilities as constitutional adviser of the President and will be criticized if I do not present my views to him. I have been much worried lately over the African situation and our variances of views there, but I have presented my own views very clearly to him both verbally and in writing."[14] Reflecting on where it had gotten him, the aging war secretary at last faced up to his failure, however. "I have decided thus far," he noted, "that it would be unwise to do it again."[15]

Stimson, ever the lawyer, wanted a paper trail of protest if the Torch invasion proved a disaster, but steered away from dismissal for being defeatist. On September 17, 1942, only seven weeks before the projected invasion, he finally abandoned his incessant carping — having been told by Marshall to cut it out: by trying to stop the switch of U.S. bomber forces from Bolero to Torch he was only sabotaging General Eisenhower's chances of success.[16] "We are embarked on a risky undertaking but it is not at all hopeless," he acknowledged, "and, the Commander in Chief having made the decision, we must make it a success."[17]

Stimson's reluctant acquiescence in the Torch operation was a relief to the President, but it did not make the path toward victory certain by any means. The speed with which the army and navy operations staffs had to work was phenomenal — and interservice rivalry and disagreement became rife.

For his part General Patton, commanding the Western Task Force that would be setting out on its epic venture from the Chesapeake Bay, became more and more determined to smash Vichy French opposition if it came — but less and less inclined to work with his naval counterpart, Admiral H. Kent Hewitt.

Refusing to micromanage, Roosevelt declined to meddle in operational matters, once his strategy had been laid down and accepted: a confidence in his chosen theater and field commanders that was beginning to mark his leadership style as U.S. commander in chief. On occasion, however, he had no option but to intervene — as he did on hearing of the Patton-Hewitt feud. Learning from Admiral King that Hewitt and Patton had almost come to blows (Patton reported to have unleashed a "torrent of his most Rabelaisian abuse" that had made Hewitt's staff flee in "virtual panic, convinced they could never work with a general so crude and rude as Patton"),[18] the President flatly turned down Admiral King's recommendation that Patton be fired from the invasion lineup. Instead, both officers were summoned to the White House at 2:00 p.m. on October 21, 1942.

"Come in, Skipper, and Old Cavalryman," the President welcomed his two field commanders, "and give me the good news."[19]

Patton entered the Oval Office wearing his ivory-handled pistols, and holding under his arm his helmet with his two oversize major general's stars. Roosevelt had taken a personal interest in the swashbuckling tank commander as far back as 1933, when Patton commanded the cavalry at Fort Myer, and seemed genuinely delighted by his swagger and pugnacious attitude—which contrasted greatly with so many of the War Department personnel he saw. "He was one of the earliest Cavalry officers to shift to tanks," the President later wrote of Patton. "He came to see me two weeks before the American expedition started for Casablanca and I asked him whether he had his old Cavalry saddle to mount on the turret of a tank and if he went into action, with his saber drawn," he recalled with a chuckle. "Patton is a joy."[20]

For his part, now that Torch was definitely "on," the cavalry general was determined to make the American landings a success. But if Admiral Hewitt had hoped the President would intercede over complications with the British over destroyers, and if Patton hoped he would intercede by giving an order to Hewitt that the landings must take place whatever the weather conditions, they were disappointed. The Commander in Chief refused to get involved. "Of course you must," he responded to Patton, declining to be drawn over the matter of touchdown conditions; meanwhile to Hewitt he advised temporizing with the British Admiralty, whose suggestion of switching around British destroyers with American warships threatened to compromise clear American fire control in support of their troops. "I never say no [to the British]," the President confided mischievously to Admiral Hewitt, "but we can stall until it is too late."[21]

"A great politician is not of necessity a great military leader," Patton left the White House thinking[22]—unaware that the very operation he was tasked with was "the President's great secret baby." Unaware, moreover, that to get the landings mounted at all the Commander in Chief had had to fight a far, far more prolonged and arduous battle than the brave tanker was having with his naval counterpart.

From the President's point of view, though, the White House meeting on October 21, two weeks before Torch, was exhilarating. In his characteristic manner Roosevelt had with charm and goodwill gotten his two commanders to stop squabbling and recognize they were on the same side, in a momentous enterprise that would alter the course of the war.

Patton's parting words, spoken in his high falsetto voice as he left the President's office, said it all:

"Sir, all I want to tell you is this. I will leave the beaches either a conqueror or a corpse."[23]

32

Alamein

THE PRESIDENT, TRAVELING to spend the last days of October at Hyde Park, had to pretend to reporters that nothing was afoot. This was so even as the White House Map Room became the focal point for a veritable fusillade of secret communications — and even as, in the midst of a world war, Americans went to the polls on November 3, 1942, to elect a new Congress.

The results were dismal — the Democrats retaining control over both chambers, the Senate and the House, though with sorely diminished majorities.[1]

How different might have been the outcome, the President reflected, if the Torch invasion had been mounted, as he had hoped, on October 30. But war was war, and the lives of the assault troops were too important to be risked without the extra week's training and issue of armaments that General Marshall had deemed necessary, when he asked for — and got — a week's extension of D-day.

Besides, the omens for Torch looked good. On his way to Moscow in August, Churchill had felt compelled to fire General Auchinleck, and appoint a new British Eighth Army commander. The man he'd chosen, Lieutenant General Richard Gott, was yet another of Churchill's poor selections, and would, in the view of almost all Eighth Army veterans, have lost the battle for Egypt. But Gott had been shot down by a flight of Messerschmitts while flying back to Cairo for a bath, and had been burned to death.[2] His replacement, General Bernard Montgomery, had dealt with Rommel's August 31 panzer offensive without turning a hair, and had then set about remaking the Eighth Army into a professional modern force of all arms, working together, in preparation for what he described, in a "Personal Message to be read out to the troops on the morning of D-day" as "one of the decisive battles of history."[3] On the night of October 23, 1942, over a thousand artillery guns of the British Eighth Army opened fire at El Alamein, and an all-out assault had begun in the Egyptian desert—Montgomery hoping to smash the German-Italian African Panzer Army at the very moment when Rommel was away in Berlin, officially receiving his baton as a field marshal in person from the Führer, as well as a standing ovation from his fellow Nazis at the Sportpalast.

Rushing back to Egypt on October 25, Rommel had found his vaunted Panzerarmee Afrika facing disaster—the British not only having attacked in the middle of the night, but having breached the minefields he had ordered to be sewn with half a million land mines. They had killed the acting German Army commander, General Stumme, and broken through with a massive combined force of infantry, tanks, and artillery onto higher ground in the north of the Alamein line, forcing Rommel to counterattack them. The battle had then become a desperate struggle of attrition, causing the Prime Minister to become frantically nervous, and the President—worried lest a British failure prejudice the responses of Vichy officials and military commanders to an American invasion of Morocco and Algeria on November 8—to wish the British could have postponed their offensive for a week, as he had wanted, to synchronize with Marshall's delay of Torch.

Lieutenant General Montgomery—who had ordered the entire Eighth Army to undertake training in night fighting—had refused, however, to order his men into minefield combat (crossing the largest sewn minefield in military history to that date) without at least the light of the full October moon.

By November 2, 1942, when the Alamein battle was still not won, after eleven days, there was growing apprehension. Yet as the disappointing results of the American congressional election came in on the night of November 3, so too did Ultra intercepts of Rommel's desperate appeals to Hitler to be allowed to retreat, taking with him what was left of his once-victorious Afrika Korps.

The tide, then, *was* turning in the Middle East.

Churchill's mood—which had dipped during the last days of October, when victory was still not won[4]—changed from depression to elation.

Hitler's negative response—"siegen oder sterben" (win or die)—marked a turning point in World War II. Not even the famed Field Marshal Rommel, though, could halt the exodus of his mobile units as they fled the battlefield, leaving behind even their own Afrika Korps commander, General von Thoma, and tens of thousands of abandoned, battle-weary German and Italian troops, to surrender to Montgomery.

It was victory, at last, for the British, after three long years of defeat—and the President was as excited as Churchill—for the great pincers he had planned for over a year could soon be applied, if all went well.

Hitler, having so recently entertained and extolled Rommel in Berlin, was mortified. The Führer had moved his headquarters back to East Prussia—chagrined that, having come so close to victory in Russia and in Egypt that summer, victory seemed now to be slipping away from him by the hour.

His troops had not succeeded in swiftly capturing Stalingrad. Nor had they quite breached the vast mountain chain of the Caucasus, despite reaching the peak of Mount Elbrus. Now, with Rommel's Panzer Army in Egypt in full retreat, the possibility of a double German envelopment of the oil fields of the Caucasus from north *and* south became a chimera—indeed, so worried was Hitler by the military situation that Goebbels was told the Führer might not be able to travel to Munich for his annual get-together with Nazi Party stalwarts on November 9, 1942.[5]

Goebbels, for his own part, remained sure that Rommel, the star performer of the German Wehrmacht, would spring something out of the bag to confound the British, as he had so often done before.[6] A German internal Security Service reported on November 4, 1942, that "the people are breathlessly following the battle for North Africa. Their trust in Rommel is so high, they cannot imagine a crisis there. The situation at Stalingrad is murky, but it's hoped the city will be in our hands before the onset of winter."[7]

It was not to be. At his Wolf's Lair headquarters in Rastenburg, "huge problems" were awaiting Hitler. "He still isn't eating either lunch or dinner with his staff," Goebbels noted in his diary on November 4. "Actually this is quite good for his health," Goebbels added, "since that way he saves four or five hours a day, hours he would otherwise need for conversation." Nevertheless, "the way things are going in North Africa are getting to one's nerves—even the Führer's. I'm getting reports he's more and more anxious about the outcome. The whole afternoon is consumed by concern. It's absolutely dreadful we have to wait so long for news. But it's the same for the Führer. Rommel doesn't send much. In these critical hours he'll have other things to do than constantly send us dispatches. We'll have to wait patiently for tomorrow, when things will be clearer."[8]

The next day's news was "somewhat bleak," however, Goebbels noted — plotting how, as Reichsminister for propaganda, he could turn defeat into a story of "calculated withdrawal" to better positions in Egypt.[9] American losses in the naval and land battle for Guadalcanal, as well as the Democratic Party's losses in the congressional elections, would keep American focus on the Far East, surely.[10] Yet even Goebbels had to wonder when he received reports, on the night of November 6, 1942, that Allied warships and troop transports were passing through the Straits of Gibraltar, and entering the Mediterranean Sea — ships whose destination was "still unknown."[11]

Was it a British relief convoy for Malta? Or an attempt to land British forces behind Rommel's front, in Libya? "We're going to do everything in our power to smash it" using air and naval forces, Goebbels recorded. "There are reconnaissance reports of three aircraft carriers and a battleship among them. If we can lure them into a naval battle, we could reverse the defeat we've suffered. . . . Everything else that's going on in the world is being overshadowed by North Africa."[12] He lamented Hitler's increasing reluctance to broadcast or be filmed for weekly newsreels, but hoped the Führer's forthcoming trip to Munich, if he went ahead with it, would give an opportunity for rabble-rousing rhetoric.

Instead he got Torch.

33

First Light

RETURNING FROM HYDE PARK on November 5, 1942, the President stayed briefly in Washington. There he finally revealed to the secretary of state, Cordell Hull, the still-secret details of Operation Torch — tasking him to do his best to keep the Spanish government from interfering from Spanish Morocco, or offering free access across Spain to the Mediterranean to the Germans, once the landings commenced. Then the President formally opened the new White House wing, where Admiral Leahy was now installed as his military chief of staff.

"At 2:20, the President laid a corner stone at the NorthEastern [*sic*] corner of the new addition to the Executive Mansion which contains my office," Leahy recorded in his diary, noting that the ceremony "involved no speeches and no formality other than the usual taking of photographs"[1] — for the press were to be given no chance to get too close to the President, or his staff, on the eve of such a momentous military undertaking, with tens of thousands of American lives at risk.

"Of course we hear no word from the great convoys that are converging on the rendezvous," Secretary Stimson noted that same day, "because they are all under radio silence. But the fact that they are coming is already foreshadowed by messages which are coming out of Germany." Enemy reconnaissance planes had "evidently spotted them in the Gibraltar Channel. Today word came through that the Germans had asked Spain for permission to go through."[2]

Thanks to Montgomery's great victory in Egypt, it seemed unlikely the Spanish, who had remained neutral for so long since 1939, would agree. "The news from Egypt is getting better and better. Rommel is in complete retreat, has lost a large number of tanks and a considerable number of prisoners and, as the day wore on today, the news indicated he was running faster and faster, and the British becoming more and more jubilant. For once matters have been timed admirably for our own action," Stimson — the former refusenik — confessed, "for Hitler's main forces are still tied up in Russia and now Rommel's force seems to have been pretty effectively smashed in the eastern Mediterranean."[3]

So far, so good. General Marshall had warned the war secretary that the President "was very snappy today and was biting off heads," but when Stimson sought an interview at the White House he found Roosevelt "in very good humor," in fact was as "amiable as a basket of chips," as he recorded afterwards with relief[4] — the President even agreeing with Stimson's solution to the problem of competition between the nation's need for soldiers and for specialists in the war industries: namely a presidential appointee to arbitrate between the Manpower Commission and the Selective Service authorities, both of which were working remarkably well.

The tension in the War Department and at the White House was, however, growing by the hour.

The next day, Friday, November 6, 1942, the President addressed the cabinet at 2:00 p.m. Stimson had a new bee in his bonnet, this time about a "military school at Charlottesville." It was designed by the secretary of war "to train officers for proconsular duties after the war was over." As the war was by no means over, and since the cabinet had been divided over the subject at its last meeting, the matter threatened to become a controversial red herring.[5]

Once again Roosevelt had to beat back the temptation to silence his secretary of war — resorting instead to his usual tactic when he didn't want a particular issue to be debated, or another to be raised. As he confided several weeks later to the Canadian premier, Mackenzie King, "I adopt the policy when asked a question that is embarrassing, of stalling to tell a story, and after a time, others forget and lose interest in the question they have asked."[6]

The tactic worked — for the most part. "I had all the typical difficulties of a discussion in a Roosevelt Cabinet," Stimson lamented after the cabinet meeting, once back at the War Department. "The president was constantly interrupting me with discursive stories which popped into his mind while we were talking, and it was very hard to keep a steady thread through, but I kept my teeth in the subject and think I finally got it across."[7]

Roosevelt knew exactly what he was doing, however; the matter was deferred. No sooner was the cabinet meeting over, then, than the President—anxious to evade the press, who might pick up rumors of the impending landings—set off for Shangri-la. With him went Harry Hopkins, Hopkins's new wife, Louise, Grace Tully, Daisy Suckley, and several more guests.

"Quite cold—large fires in the fireplaces," Daisy noted of their arrival. "There was a feeling of excitement. Telephone calls now & then, but the P. keeps conversation light—teases everyone."[8]

They went to bed early. There were "flashing lights" in the woods and some commotion, but eventually she fell asleep.[9]

It was the eve of the largest amphibious invasion launched in American history—an armada of over a hundred ships approaching French Northwest Africa and about to land more than a hundred thousand men on the shores of Morocco and Algeria: Torch.

How such a huge invasion, dispatched from two continents, could be kept so secret and timed to arrive synchronously was nothing short of miraculous.

Daisy Suckley knew something was up, but what it was even she had no idea. "For weeks," she noted in her diary afterwards, the President "has had something up his sleeve." Only a handful of people knew of the operation, she recorded, "though everything, down to the very date, has been planned since July. He spoke of an egg that was about to be laid—probably over the weekend." They might have to return to Washington on Saturday "if the hen laid an egg!" she recalled the President's warning with amusement.[10]

In the event, they remained at Shangri-la. There was little the Commander in Chief could now do. For good or ill, the invasion must go ahead—indeed, on November 1, while still in Hyde Park, Roosevelt had had to crush urgent recommendations from Robert Murphy, his presidential representative in Algiers, who pleaded for the invasion to be postponed for two weeks.

Poor Murphy had run up against a dire difficulty. "Kingpin"—i.e., General Giraud, the "hero" who had escaped from a German prison camp—had been duly contacted in Vichy, on the grounds that he was, in contrast to General de Gaulle in England, the best prospective military leader who might persuade Vichy forces in Morocco and Algeria to lay down their arms, once U.S. troops landed. However, to preserve secrecy, the President's emissaries had declined to tell Giraud in advance when exactly the invasion would take place. Once informed, the brave Frenchman had proven nothing less than a thorn in General Eisenhower's side — insisting he needed more time if he was to be the savior of France.

Murphy, hearing this at his office in Algiers on November 1, had suddenly lost faith in the whole Torch enterprise. "I am convinced that the invasion of North Africa without favorable French High Command will be a catastrophe," he had wired to the President in Washington. "The delay of two weeks, unpleasant as it may be, involving technical considerations of which I am ignorant, is insignificant compared with the result involving serious opposition of the French Army to our landing."[11]

There can have been few more "ridiculous" cables sent by a career diplomat in the days before a major amphibious invasion involving more than a hundred and thirty thousand soldiers, sailors, and airmen in its first wave — as Murphy, to his credit, admitted in retrospect. "The intricate movement of vast fleets from the United States as well as England was already under way, and a delay of even one day would upset the meticulous plans which had been meshed into one master plan by hundreds of staff officers of all branches of the armed forces of both Allied powers," the diplomat afterward reflected — complaining that no one had "briefed" him on the complexity of such an operation of modern war.[12]

General Eisenhower, who had never commanded in combat, nor been responsible for such a vast operation of war involving army, navy, and air components operating simultaneously from two continents, had been rocked by his copy of Murphy's cable, and waited for the President to decide what to do. "Of course it was a preposterous proposal," Secretary Stimson noted in his diary, "but strangely enough" Giraud and his conspirators in Vichy France had "won over 'McGowan' [Murphy's code name] to support it."[13]

In Washington, Secretary Stimson called Giraud's plea to postpone the invasion by two weeks "as impossible as a flight to the moon."[14] Roosevelt felt the same. Receiving Murphy's personal recommendation, the President had been contemptuous. Oh, the French! he'd mused — hearing that General Giraud not only wanted to delay the invasion, but had announced he wished to be made commander in chief of the entire Allied invasion forces, including the Americans, once it took place!

Admiral Leahy, in Washington, spoke with Marshall and King. All were agreed: the Frenchman was mad, and must be dumped if he did not comply with the President's wishes. "The decision of the President," Leahy signaled in an uncompromising cable sent to Murphy from the White House on November 2, "is that the operation will be carried out as now planned and that you will do your utmost to secure the understanding and cooperation of the French officials with whom you are now in contact."[15]

"I personally don't expect much enthusiasm on the part of the French African Army in opposing American troops," Leahy noted in his diary that night, "although the coast defenses of the Navy may be expected to oppose the landings."[16] Even Secretary Stimson, who was feeling daily more confident in the enterprise, called the Murphy cable in his diary "one of the crises which inevitably occur in military operations, particularly in such long and complicated ones as the one we are now launching."[17]

The invasion was still on. The Germans remained unaware of what exactly was coming. The Vichy French were either blissfully ignorant, or squabbling over who would wield power in the aftermath. General Patton, commanding the Western Task Force, thirsted for glory. General Eisenhower cursed at the complexity of dealing with French colonial defenders he needed to befriend in order to fight the real enemy, the Nazis.

And more than a hundred thousand trained assault troops fought seasickness and fear, as "D-day" and "H-hour"—the moment of touchdown—approached.

November 7, 1942, in the Catoctin Mountains, in north-central Maryland, dawned chilly. "One can hear every sound from one room to another," Daisy noted. Even the "doors themselves creak & snap & groan!"[18]

She lit the fire in her bedroom from the paper and kindling outside her door. The camp's staff brought her breakfast. She read the newspaper, then took Fala, the President's dog, for a walk around the grounds. "When I got back to Shangri-La I found the P. sitting in the enclosed porch. He told me that one of the guards last night had challenged a dozen or so men with guns, in the dark. The dozen men refused to stop or answer, so he reported to headquarters. All available soldiers & S.[ecret] S.[ervice] were called out, beat the woods, etc. Much excitement!" Daisy recorded. It turned out that there were only two intruders: "two boys were looking for skunks, & being 'natives' & independent, saw no reason for answering the challenge! All's well!"[19]

The President chuckled—wondering, however, whether this was a microcosm of what would be happening in Northwest Africa. Reports had been confirmed that ships sailing from Great Britain through the Straits of Gibraltar "have been spotted by Spanish and Italian observers," as Secretary Stimson, at the War Department, noted.[20]

Torch was now in the lap of the gods.

The President remained serenely optimistic.

This, the President reflected, was the great virtue of the Torch he was lighting: that whatever transpired on the battlefield—however mixed up the invasion forces, however chaotic the scenes in North Africa, however conflicted the Vichy French defenders of France's African colonies—Torch simply could not fail.

There were still no Germans in French Northwest Africa—and given the sheer size of the secret invasion forces poised to descend on Algeria and Morocco, there was nothing the Germans, or the Vichy French, could do to stop it. Within days of the first landings there would be almost a quarter million American troops, backed by British units, established on the Atlantic and Mediterranean shores, with airfields and seaports to receive reinforcements, drawn from the vast U.S. military arsenal the President had created that year. There was no way Hitler could evict them.

How different a U.S. invasion of France, across the English Channel, would have fared that year—or, without combat experience, the next. The United States had never mounted such an amphibious operation in its history, and there was so much still to learn—even before U.S. troops actually met Germans on the field of battle. The British had been fighting Germans since the spring of 1940, when the "phony" war ended and Hitler launched his massive attack on the Western Front; it had taken them more than two long and unhappy years of combat to win a single battle. How long would it take the U.S. Army?

It didn't matter.

This, again, was something which even the smartest brains in the War Department had been unable to accept, the President reflected: namely, the time it would take for a formerly isolationist, pacifist nation not only to gird itself up for foreign war, but to learn how to fight it there, on the battlefield.

American observers had returned from the British front in Libya and

North Africa that summer with many lessons and recommendations based on desert fighting: especially the need to deal with the Wehrmacht's dreaded 88mm antiaircraft gun used in its lethal mobile antitank role; also the seamless cohesion between Rommel's infantry, panzers, artillery, and his Luftwaffe air support. But until American commanders and their units were tested in battle, such observations were simply theory. Even after years of British combat in North Africa, it had taken a commander as ruthlessly professional as General Montgomery to kick out the duds in the British Eighth Army and recast the way the citizen army of a democracy *should* fight, if it ever hoped to defeat the indoctrinated, disciplined warriors of Hitler's brutal Third Reich: a nation where killing had become the be all and end all of German vengeance for defeat in World War I; where butchering one's own people, of the wrong creed or faith, was accepted by the masses — merciless slaughter, carried out without public protest, and without even a semblance of collective conscience. That stain, that genocidal distortion of humanity, had to be brought to an end, the President was utterly determined — and by putting his first major American army into battle in an area where it could win its spurs, and hearten the free world as well as occupied nations, seemed to him a noble, realistic, and achievable aim.

It was this that caused the President to feel so confident in the days and hours leading up to the Torch invasion, as all around him attested. When Prime Minister Mackenzie King called him on November 6 — his call put straight through to Shangri-la — the Canadian premier had been amazed and delighted to hear the President's voice, immediately. "Said he was feeling very well," and was speaking "from the top of a hill" seventy miles from Washington, and almost two thousand feet "high." The President seemed untroubled by the congressional election results. In fact, "everything considered," Roosevelt remarked, "to still have control of both Houses of Congress in a third term was not too bad." Winston Churchill had called him the night before, "very pleased" with the victory at Alamein; and there would be "other things very soon," the President cautiously assured King — who'd been informed of the Torch operational details both by Churchill and by his minister of defense, since many Canadian naval vessels would be taking part in the invasion armada.

Prime Minister King had consistently argued against a cross-Channel Second Front invasion — and had been proven tragically right when so many of his compatriots were senselessly mown down at Dieppe on August 19. Torch, however, was different—and the Canadian premier was "looking forward" to it. "He said he had been so glad to hear my voice again," King described the President. "He sounded very cheerful"—so much so, in fact, that he "said he wanted to tell me a joke about some of the Italians and the Germans who had been captured in Egypt. He said that they were in terrible shape, Italians were black with dirt, and the Germans in a positively filthy condition . . ."

Mackenzie King didn't get, or lost, the point. But where, in the spring and summer that year, his own mind had been disturbed by the collapse of British forces in the Far East, Indian Ocean, and Middle East, his Canadian heart was now filled with hope.

As to the Pacific, "about all we can do there is to hold our own," Roosevelt confided, "and we are doing that"—the situation at Guadalcanal "much better."[21] But the major blow, as the President had always insisted, was now to be in the West.

The President's telephone rang constantly—Hopkins ignoring his wife, Louise, as he sought to help field the incoming reports and facilitate the President's responses. "Throughout the week-end, Harry was in & out of F.D.R.'s room from breakfast time on," Daisy noted; "Louise stayed in her room all Sunday morning," November 7.[22]

In Washington the situation was more fraught. General Eisenhower had cabled to "inform us that a landing of the American expedition will be commenced at Oran and Algiers at 1:00 a.m. Greenwich civil time, which is 9:00 p.m. Washington time," Admiral Leahy formally recorded in his diary in his new office in the White House East Wing. "Received radio information that one combat loaded ship in a convoy en route to Algeria was torpedoed before reaching the Straits of Gibraltar," he added—telephoning the news to the President at Shangri-la. "This ship, which probably carried 3000 troops, is reported to be afloat and in tow but definitely out of the operation."[23]

In his diary, Secretary Stimson dismissed all office work but that relating to the impending assault, noting that "the underlying thing in our minds is the approaching offensive in Africa." Telegrams were coming in "thick and fast"—zero hour being "early Sunday morning, November 8th in North Africa, "which means the middle of the evening tonight," November 7, in Washington. He and his wife had kindly invited General

Patton's wife, Beatrice, to dinner, "and she came with great eagerness and we three spent the evening together."[24]

At Shangri-la, the President had still said nothing to his guests — though they "couldn't fail" to have noticed "F.D.R. was on edge," Grace Tully later recalled, "and that there must be some unusual reason."[25]

In the privacy of the President's bedroom that evening, Roosevelt finally took the call from Admiral Leahy.

"Thank God. Thank God. That sounds grand. Congratulations," Roosevelt burst out. "Casualties are comparatively light — much below your predictions," he confirmed his understanding of the message.[26]

With a huge sigh of relief the President then "dropped the phone and turned to us," his secretary recalled — being one of the few to be in on the secret of the invasion. "Thank God. We have landed in North Africa," Roosevelt declared. "We are striking back."[27] He still said nothing to his guests, however.

Daisy Suckley, in the sitting room, was vaguely but distinctly aware that something momentous was taking place. "There was a feeling of suspense through dinner, though the Pres. as always was joking & teasing," she recorded. "About 8.30, we left the dining table after a delicious dinner of which the main course was *musk-ox* — It was like the most tender beef but with a tiny difference in taste," she described — prepared and served by the Filipino staff from the USS *Potomac*. "As we were getting settled in chairs & on the sofa with the P. he suddenly said that at nine 'something will break on the radio.'"[28]

The moment, then, had come.

A "portable radio was brought in, as the huge expensive one doesn't work well (quite usual!) & at nine we got the news of the landing of our troops on North Africa!" Daisy noted in her diary. "Morocco, Algiers and Tunisia — Until quite late we all sat around the P., the radio on, he getting word of dispatches by telephone from the White House. It was terribly exciting."[29]

In Washington there was the same anticipation. "At nine o'clock," having also listened to the radio and heard "the proclamations of the President and General Eisenhower which were delivered to the world coincidentally with the landings," Secretary Stimson was told by telephone from the War Department "that the three assaults were under way and the landings had been made" at Casablanca, Oran, and Algiers — successfully. This was a great relief to Stimson — a perennial worrier — who had been concerned over "the prophecies of bad weather which might prevent the landing and disjoint the whole performance." Worse still, it might have caused General Patton, "who is impulsive and brave," to "take off in an impossible sea and suffer great losses."[30]

Soon after, Secretary Stimson felt pleased enough to call the President to congratulate him.

It was a telling conversation — the man who had opposed the Torch invasion from the start to the bitter end, arguing it was too risky to undertake; a man who the previous day had told a visiting British munitions official that, in contrast to Torch, he preferred the idea of "keeping up the pounding on Germany through the air through the winter," and waiting till 1943 to mount a cross-Channel invasion.[31]

Roosevelt graciously said nothing. For one thing, it was too early to crow. Yet for a president who loved American history it was, nevertheless, a moment to savor — eleven months to the day since the Pearl Harbor defeat.

By the time Daisy wrote her diary entry the following day, November 8, 1942, she was aware the event had already passed into "History, & in the papers." Nevertheless, she reflected, it was "thrilling" to have experienced it in the presence of her cousin, the President she adored. "And for the P. it was a tremendous climax, for he had been planning it, arranging it, for months."[32]

On the way back to the White House the next afternoon, Sunday, November 8, the President was so exhausted, he slept in the car. "I had to wake him as we approached the city," Daisy recorded: "it wouldn't look well for the P. of the U.S. to be seen driving through the streets with his eyes closed and his head nodding!"[33]

34

The Greatest Sensation

TODAY EVERYTHING WAS ENTIRELY overshadowed by what was going on in Africa," Secretary Stimson recorded proudly on November 8, 1942. "The whole affair seems to have gone off admirably in respect to its execution and timing. Coming as it does, and was planned to do, on top of the British victory over Rommel in Egypt, it has taken the Nazi forces both at a surprise and at a great disadvantage, and every reaction today which came from Vichy and from the Berlin radios confirmed this."[1]

It was incredible yet true: the Führer caught with his pants down, traveling by train to Munich to give his traditional annual address to Nazi Party stalwarts in the Löwenbräukeller, when "the reports of the Allied landings in North Africa" came through to him.[2] Hitler, for once, was astounded — and furious at the failure of the Abwehr, the German foreign intelligence department.

German as well as Italian analysts had assured the Führer the Allied troop transports that had recently been reported passing Gibraltar must be on their way to Malta, or possibly a landing in Libya, in Rommel's rear. Dr. Goebbels had noted "the ultimate destination of the great armada" was "still unknown," but once it got closer to Italian- and German-dominated airspace "we will descend on it, and give it all we've got. Among other vessels it's reported to have three carriers and a battleship. If we can give it a real pounding, our poor position in North Africa can be made good again."[3]

It was not to be, however. The armada suddenly switched course, to Oran and Algiers.

Goebbels was dumbfounded. "What will happen?" he asked himself in his diary. "The landings thought to be in Italian territory or Rommel's rear" had unexpectedly "switched course in the night" to become a "vast attempted invasion of French Northwest Africa." Information had only come through at three o'clock in the morning; it was the "greatest sensation in ages" — the Americans and British seeking to "seize the initiative," and declaring "this was now the Second Front."[4]

So *that,* Goebbels recognized in shock, explained the long autumn weeks of silence in the Allied camp! The *Americans,* not the British, were coming!

"The Americans have taken the British completely under their wing," Goebbels noted. It was now "coalition warfare" — with the United States in command, and seizing the initiative on a grand scale. "This is their way to help Russia," he added, wincing at the thought. As master of Nazi deceit, he sneered at Roosevelt's proclamation that the landing of U.S. troops in French territory was merely to forestall German occupation, and that the territory would be restored to French control in due course. Yet even the Mephisto of modern propaganda had to admit "the President's appeal to the population not to counter the invasion" would probably work, "for the Americans were coming as friends and not as enemies."[5]

Marshal Pétain, in Vichy, had immediately declared that France would defend its colonies against such an invasion — yet Goebbels had his doubts. "The situation this morning is still utterly murky, not to speak of how it appeared in the night. When I got the first news at 3:30 in the night, I couldn't make any sense of it. I can't get in touch with the Führer, because he's on his way from his military headquarters to Munich. This gives me a few hours to mull things over. . . . These sensational reports have put events in Egypt completely in the shade."[6]

Suddenly the entire war seemed to have been turned on its head. "What will France do?" Goebbels wondered, "and how will it affect, even disrupt our work" in Europe and Russia, "for example if we have to checkmate the French?"[7]

Plans, he knew, had long been drawn up to abrogate the 1940 armistice and occupy the rest of metropolitan France and Corsica, if the Allies attempted an invasion of the French mainland. No war-gaming had been undertaken for the possibility of an invasion of French Northwest Africa by *Americans,* though!

"France's hour" of destiny had come. The Gallic race stood at the threshold of true greatness, if only they would now actually fight with Hitler and the Third Reich, instead of against it — indeed, Pierre Laval, Pétain's deputy, was soon on his way to Munich to propose an egregious new treaty with the Third Reich, in which France would be a full Nazi partner: a Quadripartite Pact.

Meantime, however, whatever Marshal Pétain might say about Frenchmen defending French territory in Northwest Africa "to the last drop of blood" — and hoping thereby to dissuade the Germans from occupying the entire mainland of France — Goebbels had little confidence such a statement by Pétain would prove to be effective on the field of battle, whether the French were facing the Allies or the Germans. The French were a rotten race; they would be conflicted down to their intestines. Was it not better now to simply go ahead and dump the terms of the 1940 armistice: to ignore Pétain and Laval, and overrun metropolitan Vichy France with German troops — to "have a bird in the hand rather than two in the bush?"[8]

That afternoon, November 8, 1942, the Führer finally reached the Brown House in Munich — national headquarters of the Nazi Party in Germany. There the Reichsminister for propaganda met with him.

The Führer had aged since their last meeting in the summer. Only four months before, Hitler had been in "*bester Laune*" — in fine fettle, brimming with pride.[9] The entire Western world had seemed his oyster — stretching as far as the Urals. He still had Blondi with him, the dog he'd acquired in 1941 as a gift from his loyal deputy, Martin Bormann — a German shepherd "of outstanding racial purity," as Goebbels had recorded, glad that the Führer "has at least one being with whom he can be happy." Successes "in every theater of the war put him in a wonderful mood," Goebbels had noted in the summer — Hitler talking "in the most laudatory way about Rommel, who has become the Marshal of the Desert," a good Nazi and a man to whom the Führer would ultimately entrust command of the entire German army "if things get that far."[10]

But they hadn't. Everything had now been upended — stunning Goebbels, but bearing out the Führer's lingering suspicion that the Allies might yet come up with a surprise that would compromise Operation Blue, his drive deeper into Russia.

Goebbels had not shared the Führer's premonition. How stupid it was of the British to talk of cross-Channel landings, he had sneered, since the idea of a Second Front in France had only served to harden German defenses in the coastal areas. "One should never alert people to what is supposed, later, to be a surprise," Goebbels had observed with contempt, two months before the British fiasco at Dieppe. "As a consequence our troops have been reinforced as never before, and made more mobile."[11] Moreover, if the British or Americans thought the local French population would rise up to help them, they had another think coming, Goebbels had noted, recording with satisfaction the Führer's dismissal of such a possibility. "The Führer thinks the chances of French guerillas or partisans are absolutely nil," he recorded — the Germans having shown on the Eastern Front how they dealt with such resistance, and it was not pretty. In sum, "the Führer has not a moment's doubt that an attempted British invasion, lasting possibly eight or ten days, would be a complete catastrophe," which "might lead to a transformation of the war, perhaps even an end to the war."[12]

And yet . . . As Goebbels had confided in his diary, the Führer *did* worry. Churchill got under his skin. Goebbels had remained confident the British would not undertake something as stupid as a cross-Channel invasion, but the Führer had demurred. "As I noted earlier, the Führer is extraordinarily careful and remarks, in this connection, that no general has ever been criticized for having been too well prepared, only for being insufficiently so. The Führer adds that the British *have* to do something with their forty or fifty intact divisions — they can't be expected to simply capitulate to us, given the number they have. We've no idea where they might invade; but given the characters and temperaments of Churchill and Roosevelt anything is possible. Thank God the Führer is so cautious!"[13] Hitler was even having his "so-called Mountain Nest" military headquarters, situated near the Belgian border — the center from which he'd directed his 1940 invasion of the West — completely renovated, so he could move back at any sign of a major cross-Channel offensive. "He makes provision for every eventuality," Goebbels noted, "and doesn't depend on luck. You can really see how he'd love to renew a battle with the British in the West. Mr. Churchill would probably come off a lot worse than the Führer."[14]

That prediction had been made by Dr. Goebbels on June 24, 1942 — shortly after the fall of Tobruk. It had proven prophetic in terms of the disastrous British-Canadian "raid in force" on Dieppe eight weeks later. But now, on November 8, 1942, a real invasion was taking place — and not where German forces could repel them.

Such a strategic throw of the dice had actually occurred to Hitler in the summer, but the Führer had never imagined it would be an American undertaking.[15] "Whether the British might attempt an invasion of Africa is always a possibility," Goebbels had noted on June 23, but "the Führer thinks it would be pointless. What on earth would they [the British] want there? I suppose they could bring the French Vichy territories under British control; but that wouldn't be decisive, in terms of the war's likely course."[16]

Now, on November 8, 1942, the "invasion of Africa" had started — and it filled Hitler with foreboding, since it was clearly a U.S. undertaking, which the French authorities in North Africa might welcome rather than repell. What should the Germans do? Would Mussolini insist on German help? If so, where? Would Rommel's Panzerarmee Afrika be crushed between two Allied pincers, one American, one British? Should Rommel's army be withdrawn across the Mediterranean? Or would Italy then lose heart and be tempted to sue for peace?

Already the neutral countries such as Sweden were turning hostile toward the Third Reich; Spain, equally, was refusing to cooperate, despite the help Hitler had given General Franco in the Spanish Civil War . . .

Germany, which had looked to be on top of the world only three months before, suddenly looked beleaguered — evil, abandoned, vulnerable. The Führer's vaunted military caution had impressed his propaganda minister, but had not served to warn him of an American rather than British assault in Africa. Further east, Rommel's army had been smashed at Alamein and was retreating into Libya. The Russians were refusing to surrender at Stalingrad, and there was no possibility of getting across the Caucasus, now that snow was falling.

Everything had gone wrong. As U.S. commander in chief, Roosevelt had ignored calls for a cross-Channel Second Front — thus avoiding the catastrophe Hitler had prepared for their arrival. Instead, by landing in Morocco and Algeria, the Americans would now have a secure base from which, with their vast industrial capacity and manpower, they could prosecute the war against the Third Reich, advancing from two possible launch pads: the British Isles and North Africa — and forcing the Germans to defend from both directions.

The Führer had been outwitted. Even the most conservative French, who might have resisted a British invasion of French Northwest Africa, could not be relied upon to offer more than token resistance, since the Americans were clearly uninterested in becoming a colonial power, and were already supplying copious amounts of food to the Vichy authorities.

Hitler's worst nightmare — a major war on two fronts — had now come to pass, with Morocco and most of Algeria too distant for even the Luftwaffe to reach, let alone armored forces.

Hitler's first response was to quash any possibility of panic among his staff. The foreign minister, Joachim Ribbentrop, had, for example, joined the Führer's train at Bamberg, in Bavaria — having received news of the American landings at his office on the Wilhelmstrasse in Berlin and flown down immediately. Not surprisingly, Ribbentrop had clutched at straws. It was he, after all, who had negotiated the infamous Nazi-Soviet peace pact with the Russians in 1939 — allowing Hitler to invade the West with impunity, while the Russians watched; he who had been less than enthusiastic, however, about the Führer's decision to declare war on Russia in the summer of 1941; he who had not followed the Führer's logic about the United States being confined to fighting only in the Pacific after Pearl Harbor; he who had opposed the Führer's decision to therefore declare, with supposed impunity, war on America — a decision Ribbentrop had argued vainly against, he later claimed. In any event, it was Ribbentrop the Nazi diplomat who once again attempted to get the Führer to be sensible — begging the Nazi leader to allow him to put out peace feelers to Stalin via Stockholm, while there was still time to negotiate an end to the war of annihilation in the East.

Hitler had brushed away such a proposition.

"From now on," the Führer had snarled, "there will be no more offer of peace."[17]

Absent a miracle, the people of the Third Reich would no longer be asked to fight for *Lebensraum* — living space. It would be *Todesraum* — room to die.

As Führer of the Third Reich, unaccountable to anyone but his own demonic agenda, Hitler was adamant. General Paulus would be denied permission to pull his almost encircled army back from Stalingrad. And though his "siegen oder sterben" order to Rommel at Alamein had been disobeyed, that did not mean he would allow the Marshal of the Desert to bring his Panzer Army back to Europe or Germany. Rather, it would be made, like Field Marshal Paulus's army, to fight until it was destroyed or surrendered.

Torch had thus ignited a veritable funeral pyre, upon which Hitler would prefer to see his nation immolated rather than that he should seek to negotiate a way out or step down as führer — knowing he himself would be tried as a war criminal and executed.

Thus did the Nazi dream meet reality, at last—Europe's largest nation having willingly followed his banner of anti-Semitism and ruthless conquest, yet now facing on November 8 the stark reality that, despite Germany's triumphant victories in the summer of 1942, it was not going to work; that Winston Churchill had inspired his country to hold out, and that, far from focusing exclusively on dealing with the Japanese in the Pacific, the great United States of America, under President Roosevelt, was moving to Europe to bring the thousand-year Third Reich to an inevitable end.

At the Brown House in Munich, addressing his old Nazi Party colleagues and veterans, Hitler did not even mention the American landings, confining himself to a tirade against the Jews—the source of all Germany's misfortunes. In the meantime, however, he had given orders via his High Command Headquarters that German units were immediately to seize and occupy all of Vichy-controlled France, the Pyrenees, and Corsica, as per Case Anton. More, that a fresh German army be assembled to go to Tunisia under General von Arnim, and deny the Americans easy eviction of Rommel's German or Italian forces in North Africa. It was impossible now for Nazi Germany to win the war, Hitler knew—but with luck and stout German hearts, it might not lose it.

If only, he rued, he had not banked on America concentrating upon the Pacific.

35

Armistice Day

AMERICAN FORCES LAND IN FRENCH AFRICA;
BRITISH NAVAL, AIR UNITS ASSISTING THEM;
EFFECTIVE SECOND FRONT, ROOSEVELT SAYS.

SUCH WAS THE *New York Times* banner headline on November 8, 1942.

"SECRET CLOSELY GUARDED — Reporters Locked in Office in White House to Bar Leak Before Release Hour," another headline ran, much to the President's amusement.

As congratulatory messages from world leaders streamed in to the Oval Office — from Stalin, Churchill, Chiang Kai-shek, Mackenzie King, and dozens of others — the President had good reason to be proud.

At his office in the War Department, Secretary Stimson congratulated General Marshall, telling him that Torch "was the most difficult and complex and large expeditionary plan that the United States had ever undertaken in its history — that it had been planned for execution and carried out in a most wonderful and perfect manner, and that I thought that the chief credit belonged to him. He seemed touched by what I said," Stimson noted — having withheld from Marshall "my very grave misgivings as to the hazards of the whole plan strategically."

Stimson was still tormented by the many things that might yet go wrong — and which had caused him to make "my protest to the President last spring or summer and repeated it again to Marshall several times while they were getting ready until I really think he got rather tired of me. But now when we get it out on maps, the hazard of it seems to be more dangerous than ever," the secretary confided in his diary. "It is a hazard that can be met by good luck and by the superb execution of our own men; but when I look at the map and see how easy it would be for

Germany, if she makes a compromise and arrangement with Spain to come down quickly through Spain and with the aid of the 140,000 men in Spanish Morocco, to pinch off the Straits and cut our lines of communication to the eastward, I shiver. . . ."[1]

Stimson's concerns regarding Spain proved groundless. Given his anxiety, it seemed incredible that he had so doggedly pressed for a cross-Channel assault that year, within swift striking distance of more than twenty-six German divisions, insisting upon such an invasion even to the point of mutiny.

For his own part the President had never credited a "compromise arrangement" between Nazi Germany and Franco's Spain, let alone a German drive through the Iberian Peninsula—an armored offensive that would have added yet another enemy to Hitler's ample roster without actually contesting the landings in Morocco or Algeria. (During the Arcadia Conference in January that year the Joint Planning Committee had, in examining prospects for operations in Northwest Africa, determined that "it would take the Germans six weeks to prepare to invade Spain" and a "further six weeks to become firmly established with land and air forces in the South of Spain after they had crossed the Pyrenees.")[2] Yet the hazards of mounting, with only a few weeks to prepare, the largest amphibious undertaking in human history, had certainly been real—and remained so for several days. In the hours that followed the Torch invasion there would, the President knew, be a thousand mistakes, untold misunderstandings, and awkward negotiations with leading French officials, officers, and insurgents, who were expected to help administer the territories liberated by America's legions.

It didn't matter! That was the beauty of Torch. It could not fail—too far from German forces to be extinguished, as a cross-Channel attack could so easily have been by Hitler's air, naval, and ground forces in France and northern Europe.

Above all, there was the simple, clear, and symbolic message that the Torch invasion would send across the occupied world: *The Americans are coming!*

As the President explained to White House correspondents, off the record, on November 10, 1942, "where hundreds of thousands of lives are involved," it was "a pretty good rule of all wars" that you "couldn't find a second front offensive in a department store, ready made." What he didn't say was that Stimson and Marshall's "ready made" cross-Channel Second

Front would have led to a catastrophe, with untold American casualties, and have helped Hitler win the war. It hadn't been viable, not only because the British wouldn't fight, but because the Germans *would* have—the lunacy of such an operation forcing the President, as commander in chief, to look for something else that *was* possible.[3]

Operation Torch had required, he was willing to tell White House reporters, "a great deal of study, a great deal of coordination, a great deal of preparation of all kinds"—in secret, while half the world was demanding a cross-Channel invasion. "And so in succeeding months both Mr. Churchill and I have had to sit quietly and take with a smile, or perhaps you might say take it on the chin," the President said, as the reporters laughed, "as to what all the outsiders were demanding."[4]

And insiders.

For his part, Secretary Stimson remained a bundle of nerves. During the early hours on November 10, "while I lay sleepless," Stimson confided in his diary, "I had one of my bogy fits. This time it hitched around the situation," he moaned, "which the American army was getting itself into in North Africa."

What if "the Germans should force their way or make any arrangement with Spain to come through without opposition and shut the Straits of Gibraltar on us"? he queried yet again. "It was the old objection which I have always had to the plan showing itself up and it seems worse now, in the light of my examination of the maps yesterday, than ever before. I called in General Handy this morning when I got to the office and went over the matter with him. He is the Chief of the Operations Division. He didn't feel any better than I did and the cold facts and figures showed a very serious situation in case the Germans came through."[5]

General Handy did his best to be respectful, but to calm down the Republican secretary, who was, yet again, full of foreboding. Handy showed him the latest reports "showing the vigor and initiative and general efficiency of our troops." These were, Stimson accepted, enough "to reconcile anybody to hazard. Those men can meet almost any danger," he was assured.[6] Still, Stimson could not refrain from calling the secretary of state, Cordell Hull—pouring out his fears and imploring Hull to speak to the President about his anxieties regarding Spain.

The President duly saw Hull at 3:45 p.m., and telephoned Stimson to reassure him. Spain was *not* going to get involved, the President maintained. The invasion was too big, too sudden, too unrelated to Spain's own interests for General Franco to cooperate with Hitler now, when he had failed to do so for two long years since Hitler's invasion of the West.

Overall, the President told them, Torch had gone rather well — despite some troops being landed on wrong beaches, despite French coastal battery fire, despite French submarines, destroyers, and even cruisers as well as air defense going into action, and despite myriad other problems associated with an invasion. The majority of French officers and administrators seemed worried about their pensions, he heard, if they turned against the Vichy government. It would all work out, however.

Now seventy-five, Stimson was, as he confessed, "feeling very tired. The unconscious strain has been pretty heavy on me."[7]

By contrast, President Roosevelt, aged sixty, was feeling at the top of his form.

Torch — *his* Torch — had been lit, and the United States was established in force on the threshold of Europe, with a Vichy cease-fire order already in effect in Algiers and a general cease-fire applicable to the whole of French Northwest Africa in the works, if all went well. The United States had beaten Hitler to the punch — Torch victorious in a matter of only three days.[8]

Admiral Hewitt had shepherded his armada across the Atlantic and put the men of General Patton's Western Task Force ashore with remarkable precision — the French clearly unaware, until the landing craft appeared, of any threat from the sea. Even the ocean had complied — the usual rough winter sea glacially calm and only the quietest surf, as if by biblical command. (The meteorological officer who predicted this was awarded the Legion of Merit by a grateful General Eisenhower.) Much handwringing had concerned the fifteen-inch guns of the latest French battleship *Jean Bart,* in harbor at Casablanca, but dive-bombers from the USS *Ranger* and shells from the sixteen-inch guns of the USS *Massachusetts* silenced it; when it began firing yet again on November 10, more dive-bombers from the *Ranger* sank it.

Aboard the USS *Augusta* — the heavy cruiser on which President Roosevelt had signed the Atlantic Charter the previous year — General Patton had alternately fumed, sworn, and prayed: amazed to see the shore lights, harbor lights, and even lighthouse lights still burning as the U.S. vessels approached the Moroccan coast after a two-week voyage. It was "almost too good to be true. Thank God. He Stays on our side," he'd jotted in his diary[9] — mortified that the gun blasts from the after-turret of the *Augusta* had blown his thirty-two-foot plywood launch to bits, with all his communications equipment, though not his ivory-handled revolvers, which he retrieved.

Once ashore at Fedala, shortly after midday on November 8, he had earned his monicker "Blood and Guts" — delighted at General Harmon's capture of Safi, and General Truscott's seizure of the port and airfield at Lyautey — leaving only Casablanca to be taken. However, in the confusion of the Torch battle he had seen for himself how fortunate his troops were not to have been fighting Germans, rather than the French. He had spent most of the first two days of the invasion literally "kicking ass" — "The French bombed the beach and later strafed it," he wrote in his diary on November 9, describing operations at Fedala. "One soldier who was pushing a boat got scared and ran onto the beach and assumed the Fields [fetal] position and jiberred. I kicked him in the arse with all my might and he jumped right up and went to work. Some way to boost morale. As a whole the men were poor, the officers worse; no drive."[10]

Patton certainly radiated drive, and when the French commander refused to surrender the city of Casablanca, he arranged for an all-out American naval and air blitz to begin at 7:00 a.m. on November 10 — General Eisenhower having sent a chastening cable telling him the Eastern Task Force landings had meantime met almost no resistance. "Algiers has been ours for two days. Oran crumbling rapidly. The only tough nut left is in your hands. Crack it open."[11]

Patton did so — only calling off the devastating firepower of his naval and air support a mere ten minutes before the attack was due to go in, when the French sent word they would surrender, which their senior commanders subsequently did at Patton's headquarters.

"People say that Army Commanders should not indulge in such practices" as "kicking ass" on invasion beaches, Patton later reflected. "My theory is that an Army commander does what is necessary to accomplish his mission and that nearly 80 per cent of his mission is to arouse morale in his men."[12] He and his chosen field commanders had certainly earned the faith the President had vested in them. Early on November 11, Armistice Day, Admiral Leahy reported to the President he'd received a report from London that "the French Military force in Casa Blanca capitulated at 7 a.m. today and that the city of Oran was occupied by American troops last night."[13] Morocco and Algeria were in the Allied bag; there was nothing the Germans could now do about it.

General Giraud, unfortunately, had proven a great disappointment, despite an American team under General Mark Clark that had brought Giraud by submarine from Vichy France to Eisenhower's advance headquarters at Gibraltar, and then to Algiers, where he was supposed to persuade Vichy forces to lay down arms in order to limit the bloodshed. He had failed abysmally to do so — indeed, whether he would have the authority to order his fellow Frenchmen to cease contesting American landings, and even inspire the 140,000 French troops in North Africa to fight the Nazis rather than the American forces, was now a very open question.

General Eisenhower — badgered day and night by Churchill, who could not resist meddling[14] — could only shake his head at news the Germans were landing troops by air in Tunis without a single shot being fired by the Vichy French to stop them, while the latter were continuing in many places to oppose American forces. The French resident-general, Vice Admiral Jean-Pierre Estéva, had even ordered that "German planes be given a friendly reception in eastern Algerian ports."[15]

It was too awful. "If they would only see reason at this moment, we could avoid many weeks of later fighting," Eisenhower railed in a message to his chief of staff in London from his headquarters still in Gibraltar. Unfortunately, "they are not thinking in terms of a cause, but of individual fortunes and opportunities. Consequently, Darlan, Juin, Giraud and the rest cannot combine to place their composite influence behind any particular project. Right this minute they should all be making it impossible for Admiral Estava," the French commander in chief in Tunis, "to permit the German into Tunisia." Instead the French had virtually welcomed the Germans. "He apparently has the equivalent of three divisions down there, and without the slightest trouble, could cut the throat of every German and Italian in the area and get away with it. . . . A situation like this creates in me so much fury that I sometimes wish I could do a little throat-cutting myself!"[16]

Was General Giraud, though, a better hope for getting a general ceasefire to even hold, let alone ginger French forces to fight the incoming Germans? Admiral Leahy thought not — in fact, for his own money, Leahy thought it a godsend that Admiral Darlan, the right-wing commander in chief of all Vichy French forces under Marshal Pétain, had happened to be in Algiers on the night of the Torch invasion, visiting his disabled son who had polio.

Despite being profoundly anti-British and an appeaser of Hitler, Darlan at least seemed to be pro-American, and have real authority over French forces in Morocco, Algeria, and Tunisia. However, he too proved to be a broken reed — declining to break with his head of state, Marshal Pétain, who nevertheless relieved him of his post as commander in chief and made General Nogues in Morocco, an even more egregious appeaser, Darlan's successor, with orders to continue to repel the American forces — while declining to order Frenchmen to lift a finger to resist the German and Italian forces, even as they invaded the remaining Vichy-controlled region of metropolitan France and Corsica, abrogating the 1940 armistice.

In sum, the indifference of Vichy French officers toward the Nazis, in comparison with their intransigence and hostility toward American forces, would have been comical had it not been so disgraceful for a once-great nation, both Leahy and the President reflected.

It didn't matter, though. That was the vantage of Torch: that American force majeure would carry the day, in spite of Vichy French perfidy, Roosevelt was certain — though it now pained him to watch as the French refrained from firing a single bullet to stop the Germans from occupying Tunisia by air and then by sea, while French Vichy forces continued to kill Americans (over five hundred), even as they listened to news of the rest of their homeland being invaded and occupied by the Nazis.[17]

Such, however, was the darker side of war and of alliances.

By contrast the British, who had initially fought so feebly in the Far East and had shot thousands of their own colonial subjects in India rather than accept them as fellow fighters,[18] seemed to have gotten a sort of second wind, once reinforced in Egypt with American armaments and air force groups. In Egypt, under Montgomery's generalship, they had trounced Erwin Rommel's seemingly invincible Panzerarmee Afrika. And in the Torch invasion under an American supreme commander, they had dovetailed their naval, air, and land contributions in remarkably successful fashion . . .

Coalition warfare, then, might yet work. Over time, perhaps, even the French might rediscover their native courage, and fight alongside American forces to defeat Hitler.

This, in sum, was the other subtext of Torch: the first coalition molding of Allied forces against the Axis powers, on the field of battle. It was no longer simply a political front — Roosevelt's twenty-six free nations, speaking as the United Nations opposed to Axis tyranny — but a new, all-powerful assembly of national military forces, fighting under American supreme leadership and command.

There would doubtless be much to learn, commencing with the blooding of American troops in combat with the professional killers of the Wehrmacht, in Tunisia. But with the successful American invasion of North Africa, a start had been made. The Americans were coming.

Reading the messages and reports that came into the White House Map Room, the President felt humbled by the enormity of what had taken place — thankful it had, in the end, borne out his most fervent hopes. At 11:00 a.m. that day, Armistice Day 1942, he therefore went to Arlington Cemetery, Admiral Leahy accompanying him to the Tomb of the Unknown Soldier, across the Potomac.[19]

"Old General Pershing, although he wasn't really fit to do it, he came along and went [in the car] with the President, while Knox and I followed behind," Secretary Stimson recorded in his diary.[20] General Marshall, Admiral King, and the Commandant of the Marine Corps also attended.

Stimson was feeling a great deal better, having heard that, although the "Germans this morning invaded unoccupied France and are rushing through it towards the south coast in an attempt to get to Marseilles and the French fleet at Toulon,"[21] there was no chance that they could now halt or throw back the American forces in Algeria and Morocco. At last Admiral Darlan, Stimson noted, "has ordered all resistance in North Africa to cease."[22]

Torch was over; the pincer campaign to defeat or evict Axis forces from the southern Mediterranean could now proceed.

In the circumstances, the President's address was deeply moving.

"Here in Arlington we are in the presence of the honored dead," Roosevelt, standing with the aid of his fourteen-pound steel leg braces, reminded his audience. "We are accountable to them — and accountable to the generations yet unborn for whom they gave their lives.

"Today, as on all Armistice Days since 1918, our thoughts go back to the first World War; and we remember with gratitude the bravery of the men who fought and helped to win that fight against German militarism. But this year our thoughts are also very much of the living present, and of the future which we begin to see opening before us — a picture illumined by a new light of hope.

"Today, Americans and their British brothers-in-arms are again fighting on French soil. They are again fighting against a German militarism which transcends a hundred-fold the brutality and the barbarism of 1918.

"The Nazis of today and their appropriate associates, the Japanese, have attempted to drive history into reverse, to use all the mechanics of modern civilization to drive humanity back to conditions of prehistoric savagery.

"They sought to conquer the world, and for a time they seemed to be successful in realizing their boundless ambition. They overran great territories. They enslaved — they killed.

"But, today, we know and they know that they have conquered nothing.

"Today, they face inevitable, final defeat.

"Yes, the forces of liberation are advancing."

The President looked around. "Britain, Russia, China, and the United States grow rapidly to full strength," he stated. "The opponents of decency and justice have passed their peak.

"And — as the result of recent events — very recent — the United States' and the United Nations' forces are being joined by large numbers of the fighting men of our traditional ally, France," he declared — hopefully! "On this day, of all days, it is heartening for us to know that soldiers of France go forward with the United Nations."

Which brought the President to the mission of the United States.

"The American Unknown Soldier who lies here did not give his life on the fields of France merely to defend his American home for the moment that was passing. He gave it that his family, his neighbors, and all his fellow Americans might live in peace in the days to come. His hope was not fulfilled," he declared candidly.

Roosevelt was coming to the crux of his vision of the United States as a global guardian of liberty, in a world where it was too easy for the forces of violence, intolerance, and savagery to get their way unless effectively challenged. "American soldiers are giving their lives today in all the continents and on all the seas in order that the dream of the Unknown Soldier may at last come true. All the heroism, all the unconquerable devotion that free men and women are showing in this war shall make certain the survival and the advancement of civilization."[23]

As the spectators and participants in the little ceremony made their way back from Arlington National Cemetery to their cars in the chilling cold, they recognized that, in a sense, America's new journey had just begun. It would not be an easy road, but it was a noble challenge Roosevelt was setting. Moreover, they could take comfort in the fact that the President, who had saved the nation at a time of the worst economic depression it had ever suffered, was now, on a global stage, proving to be perhaps the greatest commander in chief in American history.

Returning to his office at the White House, Admiral Leahy was certainly proud of his commander in chief—the man who liked to call himself "a pig-headed Dutchman."[24] The President had, he noted in his diary, "made a very impressive five minute address to a large gathering of people," with spectators "seated in an amphitheater and standing about in the clear cold morning. Except the President and myself, all of the official party wore heavy overcoats."[25]

Torch had set the tone and determination of the United States in prosecuting the war against the Axis powers. In overruling his generals, the President had, Admiral Leahy reflected, undoubtedly saved his nation from the military catastrophe that would have awaited them on the shores of mainland France. Instead they could now learn in comparative safety the dark arts of modern war—with every chance of ultimate victory. Vast American forces were, after all, now safely established on the threshold of Europe—and of greatness on behalf of their nation, if they could translate that triumph, step by step, into the defeat of Mussolini's Italy, Hitler's Nazi Germany, and finally of Emperor Hirohito's Japan.

More difficulties would arise with the French, and with a rejuvenated Winston Churchill, now that his British Empire stood to be restored, thanks to American might. And there would be the problem of Stalin, the Russians—and the Chinese.

It would all work out, the President assured Leahy. It had been quite a journey over the past eleven months, since December 7, 1941.

Most moving of all to the widower admiral, though, had been the President's prayer, at the end of his address. "Our thoughts," Roosevelt had proclaimed in his unmistakable, lilting tenor voice, "turn in gratitude to those who have saved our Nation in days gone by. God, the father of all living, watches over these hallowed graves and blesses the souls of those who rest here. May He keep us strong in the courage that will win the war, and may He impart to us the wisdom and the vision that we shall need for true victory in the peace which is to come."[26]

Acknowledgments

The study of leadership — moral, literary, political, and military — has been my abiding interest for almost forty-five years as a biographer and historian.

My particular fascination with FDR goes back to *American Caesars,* a Suetonian-style biography of the last twelve U.S. presidents, which I published in 2010. Researching the opening chapter on President Roosevelt, I found it hard to believe that no military biographer or military historian had tackled his military leadership in World War II as commander in chief in a full-scale work. Once I completed *American Caesars* I was able to examine the literature and original documentation more closely. I became even more intrigued — especially at the difference in command styles adopted by Churchill and Roosevelt in directing World War II.

I knew perhaps more than many people of my generation about Winston Churchill as a military leader, and as a striking personality, for I had stayed with him and Lady Churchill at their home at Chartwell, in Kent, while a student at Cambridge University. Moreover, I had spent many, many hours discussing Churchill's leadership with my quasi godfather, Field Marshal Bernard Montgomery, who revered the former prime minister — but lamented his intrusions into the battlefield, and his failure to understand the principles of effective modern command. Later on, over the period of a decade, I spent yet more time interviewing men and women who had known or served under the Prime Minister in World War II for *Monty,* my official life of Montgomery, published in three volumes in the 1980s — in each of which Winston Churchill played a major part.

On the American side, I was also lucky to have interviewed many of the senior surviving World War II commanders and staff officers, from

General Mark Clark to Generals "Lightning Joe" Collins, Max Taylor, and Jim Gavin; from General Al Gruenther to General Freddie de Guingand. In the course of my work I had also gotten to know many senior American World War II military writers and historians, from Forrest Pogue to Russell Weigley, Steve Ambrose, and Carlo D'Este.

And on the German side I was fortunate, too. Thanks to a semester at Munich University and my first marriage to a German (who died tragically in 1973), I had good command of the German language, and sources not available or translated for use by many British or American writers.

In short, I felt confident enough in 2010 to tackle such a project afresh.

The result, *The Mantle of Command: FDR at War,* presents a very different portrait than the conventional characterization of President Roosevelt as commander in chief in World War II. In this respect I was blessed by being able to interview the last living member of FDR's White House team in World War II, Commander George Elsey, who worked in the Map Room, as well as several members of FDR's family, including his granddaughter Ellie and his step-grandson Tom Halsted. Working my way through the many diaries, memoranda, and correspondence kept by the members of the President's staff and military officials, held in various archives in the United States and United Kingdom, I tried my best to reconstruct the story *wie es eigentlich gewesen ist* — how it really was.

For reasons of length I had decided from the beginning to focus on selected landmark moments or episodes in FDR's performance as commander in chief in World War II that best illustrate his responses both to defeat and to victory in war, for good or ill. Unfortunately, even this attempt at condensation proved a failure. The eleven-month period between Pearl Harbor and the first landings of American troops on the threshold of Europe — Operation Torch — seemed to me too important not to reconstruct and get right, given the many alternative, often misleading, accounts that have been given over the years: in particular, that of Winston Churchill in his monumental opus, *The Second World War.*

Interviewing so many World War II commanders and their staffs, I had learned how much of history, in the end, is dependent on the perspective or point of view of the participant. The main perspective of *The Mantle of Command,* let us be clear, is unabashedly that of Franklin D. Roosevelt and the White House he used as his command post in 1941 and 1942. The story, moreover, is a quite fatal one, in terms of world history. Had FDR, in the first year of America's involvement in World War II, not learned to wear the mantle of command so firmly, and to overrule his generals, it is quite possible Hitler would have achieved his aim when declaring war on the United States on December 11, 1941: winning the war in Europe. It is a sobering reflection.

Naturally, in retelling and recasting this extraordinary story I have subjected certain reputations to revision, from those of Winston Churchill and General George Marshall to General Douglas MacArthur and the war secretary, Colonel Henry Stimson. I hope I am not unsympathetic to their memories, serving to the best of their abilities in a world crisis such as we hopefully will never have occasion to repeat or replicate; nevertheless, it seemed important to me to recount the saga from FDR's perspective with absolute if compassionate honesty, since the President did not live to do so. Every other major military participant managed to impart his own account, either autobiographically or via a chosen plaidoyer; only President Roosevelt's POV as commander in chief has remained dark since his death in 1945.

In researching and writing this account — which will be followed by a concluding work — I was helped by a small but wonderful army of professional colleagues, friends, and family. I'd like first to thank my educator wife, Dr. Raynel Shepard, for her everlasting patience in the book's genesis, research, writing, and preparation. Next: Ike Williams, my literary agent in Boston, who saw immediately the potential importance of the undertaking — and found me a well-tempered, experienced commissioning publisher and editor in Bruce Nichols of Houghton Mifflin Harcourt. Bruce not only cut and clarified my often prolix prose, but recognized the need for two separate books to do justice to FDR's wartime story.

To Ike and his associates Katherine Flynn and Hope Denekamp, therefore, and to Bruce Nichols and Melissa Dobson, my copyeditor, my deep gratitude. In terms of colleagues, I have been fortunate to have been a Senior Fellow in the John W. McCormack Graduate School of Policy and Global Studies of the University of Massachusetts, Boston, for many years, and wish to thank Steve Crosby, his successor, Ira Jackson, the staff, and my colleagues there for their constant support — as also the University Provost, Winston Langley, and the ever-helpful staff of the University Library.

The staff and facilities of the Widener Library and Microfilm Department in the Lamont Library, Harvard University, have also been outstanding, as has been the staff of the Boston University Microfilm Department, and the Franklin D. Roosevelt Presidential Library at Hyde Park.

I'm deeply grateful, too, to those colleagues and friends who were willing to read and offer criticism of sections of the growing manuscript, as it evolved, beginning with my oldest Cambridge University friend, Robin Whitby; Professor Mark Schneider; Lieutenant Colonel Carlo D'Este; Professor David Kaiser; James Scott; and Professor Mark Stoler. I'd also like to record my thanks to members of my Boston club, The Tavern, who listened to my early readings from the manuscript and offered advice and encouragement — especially Stephen Clark, Frinde Maher, Alston Purvis, Ed Tarlov, David Scudder, David Amory, and Clive Foss.

Two conferences at which I gave papers based upon chapters of the manuscript were extremely helpful to my work. They were a Raymond E. Mason Jr. Distinguished Lecture on FDR's "Great Spat" with Winston Churchill over India in 1942, delivered at the National World War II Museum in New Orleans as part of the second annual Winston S. Churchill Symposium in July 2012; and a paper on Torch, given at the invitation of Professor David Reynolds to the Guerre des Sables Conference of international World War II scholars at the École Française de Rome in November 2012. The 2012 International Conference on World War II, held at the National World War II Museum in December 2012, was also fruitful, and I thank the director, Dr. Nick Mueller (and conference organizer Jeremy Collins), for inviting me to speak along with fellow panelists Rick Atkinson, Gerhard Weinberg, Allan Millett, Christopher Browning, Conrad Crane, and Mark Stoler.

In the U.K. I would like to thank Allen Packwood, Director of the Churchill Archives Centre in Cambridge, for his help on my visit there, and especially Professor David Reynolds for his hospitality and intellectual support in reexamining the fateful year, 1942, and the story of the collapse of the British Empire, together with its ramifications for FDR.

In London I would like to thank the wonderful staff of the Imperial War Museum's Department of Documents: the former Curator, the late Rod Suddaby, and current Curator, Anthony Richards, for pointing me to useful Churchill and FDR material; also Phil Reid, Director of the IWM's Cabinet War Rooms below Whitehall, for a wonderful personal tour. Also the Liddell Hart Military History Centre at King's College — and my research assistant in London, Jean Simpson, for her help in obtaining documents.

Back in the U.S., I want to record my thanks to the staff of the Manuscript Division Reading Room at the National Archives in Washington, D.C., especially Jeff Flannery, the Head of Reference and Reader Services Section in the Manuscript Reading Room. Also the staff of the Operational Archives of the U.S. Naval History and Command, Washington Navy Yard, D.C., especially John Greco for his help. In Oakland, California, I'd like to thank the volunteers and staff of the presidential yacht, the USS *Potomac,* for their tour — and cruise in San Francisco Bay — in August 2012. And in Boston, my research assistant, Eric Prileson, a graduate of Northeastern University.

As President of BIO — Biographers International Organization — from 2010 to 2012 I was privileged to work with a wonderful committee of fellow biographers, and to participate in excellent annual conferences in Boston, Washington, D.C., Los Angeles, and New York. Thanks to them, the craft of biography has seemed vastly less isolating than in my earlier years, and I want to thank especially Elizabeth Harris and my fellow members of the Boston Biographers Group (BBG), who meet once a month to share progress on their individual projects. Listening to and comparing the practical challenges of biography of fellow practitioners, working on an extraordinary array of different life stories across different centuries, has been, over the past five years, a veritable lifeline to me, and I cannot too highly recommend joining such an organization to anyone contemplating or already working on a biography.

I'd like finally to acknowledge the memories of two women who died recently: Margery Heffron, who cofounded the Boston Biographers Group, but managed to complete her masterpiece, *The Other Mrs. Adams,* before she passed; and my mother, Olive Hamilton, who first invested me with my love of biography, and wrote many herself before passing in January 2012, at age ninety-six — twenty-two years after my father, Lieutenant Colonel Sir Denis Hamilton, DSO, who landed as a twenty-five-year-old battalion commander on D-day, four months after my birth — and inspired my fascination with leadership.

Photo Credits

The Plan of Escape. FDR in Oval Office, summer 1941: Corbis / © Arthur Rothstein; USS *Potomac* off New England coast, Aug. 1941: Corbis / Bettmann

Placentia Bay. USS *Augusta* before the onset of war: U.S. Navy Official / National Archives; FDR welcomes Winston Churchill (WSC) aboard *Augusta,* Aug. 9, 1941: FDR Library; FDR, WSC, and their staffs on board *Augusta,* Aug. 9, 1941: FDR Library

The Atlantic Charter. FDR walks from USS *McDougal* onto HMS *Prince of Wales,* Aug. 10, 1941: FDR Library; WSC greets FDR aboard *Prince of Wales,* Aug. 10, 1941: FDR Library; Divine service aboard *Prince of Wales,* Aug. 10, 1941: FDR Library

Pearl Harbor. Hopkins and FDR in Oval Study, 1941: FDR Library; Japanese bomber's photo of Pearl Harbor, Dec. 7, 1941: National Archives; Burning U.S. battleships, Dec. 7, 1941: FDR Library

A Date Which Will Live in Infamy. White House, night of Dec. 7, 1941: Getty / © Thomas D. McAvoy; FDR before Congress, Dec. 8, 1941: Corbis / Bettmann; Hitler before Reichstag, Dec. 11, 1941: Getty / Keystone-France

Coalition War. FDR greets WSC in Washington, Dec. 22, 1941: FDR Library; FDR and WSC give White House press conference, Dec. 23, 1941: FDR Library; FDR broadcast to nation from White House, Feb. 23, 1942: FDR Library

Spring of '42. Close-up of Commander in Chief, 1942: FDR Library; the Map Room, White House: FDR Library; communicating with MacArthur, Map Room: FDR Library; "Field Marshal" MacArthur and President Quezon in the Philippines: FDR Library

The Raid on Tokyo. Lt. Col. Jimmy Doolittle flies B-25 off USS *Hornet,* April 18, 1942: FDR Library; FDR decorates Doolittle in Oval Office, May 19, 1942: FDR Library

Hawaii Is Avenged. Amagi—built to replace losses at Midway—sunk July 29, 1945: FDR Library; Marine General Thomas Holcombe shows FDR Japanese flag at White House, Sept. 17, 1942: FDR Library

The Fall of Tobruk. FDR with WSC at Hyde Park, June 20, 1942: FDR Library, Margaret Suckley Collection; Rommel in triumph, British surrender of Tobruk fortress and port, June 21, 1942: Bundesarchiv, Federal Archives of Germany; FDR, with Canadian premier Mackenzie King behind him, at Pacific Council, June 25, 1942: FDR Library

Dieppe. General Marshall and War Secretary Stimson plot against the President to insist on cross-Channel landings in 1942: U.S. Signal Corps / National Archives; Canadian dead litter beaches of Dieppe, Aug. 19, 1942: National Archives of Canada; Admiral Bill Leahy, new chief of staff to the Commander in Chief, July 21, 1942: FDR Library / Life

FDR Inspects the Nation. FDR aboard his touring train, Sept. 17–Oct. 1, 1942: FDR Library, Margaret Suckley Collection; FDR disembarking at Fort Lewis, Sept. 22, 1942: FDR Library; FDR waves from rear platform: FDR Library

Gearing Up for Victory. FDR watches launch of *Joseph N. Teal,* Kaiser Shipyard, Oregon, Sept. 23, 1942: FDR Library; inspects aircraft carrier construction, Bremerton Naval Shipyard, Sept. 22, 1942: FDR Library; receives clip at Federal Cartridge Plant, Minnesota, Sept. 19, 1942: FDR Library; reviews tank unit at Fort Lewis, Washington, Sept. 22, 1942: FDR Library; inspects bomber production, Douglas Aircraft Corp., Long Beach, California, Sept. 25, 1942: FDR Library; inspects Ford bomber production, Willow Run, Michigan, Sept. 18, 1942: FDR Library; inspects army units, Fort Lewis, Washington, Sept. 22, 1942: FDR Library; watches rehearsal landings of U.S. Marines, San Diego: FDR Library

Waiting for Torch. FDR dining with staff at Shangri-la, Aug. 1942: FDR Library, Margaret Suckley Collection; close-up of FDR, fall 1942: FDR Library

Torch. General Patton, on Desert Training Center maneuvers prior to Torch invasion: National Archives; General Eisenhower before appointment as supreme commander, Torch invasion: Eisenhower Library; U.S. troops landing near Surcouf, Algeria, Nov. 8, 1942: National Archives; U.S. troops marching toward Algiers, Nov. 8, 1942: National Archives

Armistice Day. President gives address at Arlington Cemetery on Nov. 11, 1942, accompanied by General Pershing: FDR Library; a wreath is laid, Nov. 11, 1942: FDR Library

Notes

PROLOGUE

1. Ross T. McIntire, *White House Physician* (New York: G. P. Putnam's Sons, 1946), 141.
2. See David Reynolds, "FDR's Foreign Policy and the Construction of American History, 1945–1955," in *FDR's World: War, Peace, and Legacies,* ed. David B. Woolner, Warren F. Kimball, and David Reynolds (New York: Palgrave Macmillan, 2008), 7.
3. Viz. Eric Larrabee, *Commander-in-Chief: Franklin Delano Roosevelt, His Lieutenants, and Their War* (New York: Harper & Row, 1987), in which only one of the work's eleven chapters is devoted to the Commander in Chief. Joseph E. Persico's more chronological and wide-ranging narrative of the generals who served under FDR, and their relations with the Commander in Chief, *Roosevelt's Centurions: FDR and the Commanders He Led to Victory in World War II* (New York: Random House, 2013), was published as *The Mantle of Command* went to press.
4. Alan Brooke, *War Diaries, 1939–1945: Field Marshal Lord Alanbrooke,* ed. Alex Danchev and Daniel Todman (Berkeley: University of California Press, 2001), 247.
5. For a recent brief summary and refutation of the "mythology" surrounding the relationship between FDR and his chiefs of staff, see Mark Stoler, "FDR and the Origins of the National Security Establishment," in *FDR's World,* ed. Woolner, Kimball, and Reynolds, 69–78.
6. Barbara Tuchman, *Stilwell and the American Experience in China, 1911–45* (New York: Macmillan, 1970), 241.
7. Mark Stoler, *The Politics of the Second Front: American Military Planning and Diplomacy in Coalition Warfare, 1941–1943* (Westport, CT: Greenwood Press, 1977), 26.
8. Brooke, *War Diaries,* 273.
9. Kenneth Pendar, *Adventure in Diplomacy: Our French Dilemma* (New York: Dodd, Mead, 1945), 152.

1. BEFORE THE STORM

1. "The White House, Washington: Memorandum of Trip to Meet Winston Churchill, August 1941," August 23, 1941, Franklin D. Roosevelt Presidential Library, Hyde Park, NY.
2. Ian Kershaw, *Hitler 1936–1945: Nemesis* (New York: Norton, 2000), 385.
3. Stetson Conn and Byron Fairchild, *The Framework of Hemisphere Defense* (Washington, DC: Office of the Chief of Military History, Department of the Army, 1960), 98.
4. Ibid.
5. See inter alia Hadley Cantril, ed., *Public Opinion 1935–1946* (Princeton, NJ: Princeton University Press, 1951), 1061, 1128, and 1162; Robert Dallek, *Franklin D. Roosevelt and American Foreign Policy, 1932–1945* (New York: Oxford University Press, 1979), 210 and 289; and James MacGregor Burns, *Roosevelt: The Soldier of Freedom* (New York: Harcourt Brace Jovanovitch, 1970), 98–99.
6. Franklin D. Roosevelt, *The Public Papers and Addresses of Franklin D. Roosevelt*, comp. Samuel I. Rosenman, vol. 9, *War—And Aid to Democracies* (New York: Russell & Russell, 1969), 488.
7. The Neutrality Act forbade American flagged vessels from sailing into war zones; permitted belligerents, of any nationality, to be supplied on a cash-and-carry basis only; and mandated that such supplies be carried in non-American shipping, to avoid the United States being drawn into hostilities. At President Roosevelt's request, the Lend-Lease Act of March 1941 had modified the financial terms of trade to belligerents such as Britain, who had been confronting the Nazi menace to the great profit of the American armament industry. No deeper political commitment to war than that, however, was contemplated by Congress.
8. The U.S. Army Air Forces, amalgamating the Army Air Corps and the General Headquarters (GHQ) Air Force, was created on June 20, 1941.
9. Robert Sherwood, *Roosevelt and Hopkins: An Intimate History* (New York: Harper, 1948), 351.
10. Entry of August 3, 1941, John Colville, *The Fringes of Power: 10 Downing Street Diaries, 1939–1955* (New York: Norton, 1985), 424.
11. Martin Gilbert, *Winston S. Churchill*, vol. 6, *Finest Hour: 1939–1941* (Boston: Houghton Mifflin, 1983), 1148.
12. Ibid., 1155.
13. Conn and Fairchild, *The Framework of Hemisphere Defense*, 124.
14. Ibid., 119.
15. Mark A. Stoler, *The Politics of the Second Front: American Military Planning and Diplomacy in Coalition Warfare, 1941–1943* (Westport, CT: Greenwood Press, 1977), 9.
16. Conn and Fairchild, *The Framework of Hemisphere Defense*, 137; and Mark A. Stoler, *Allies and Adversaries: The Joint Chiefs of Staff, the Grand Alliance, and U.S.*

Strategy in World War II (Chapel Hill: University of North Carolina Press, 2000), 51.

17. Stoler, *Allies and Adversaries,* 52–53.
18. "Our Chiefs of Staff believe that the Battle of the Atlantic is the final, decisive battle of the war and everything has got to be concentrated on winning it. Now, the President has a somewhat different attitude. He shares the belief that British chances in the Middle East are not too good. But he realizes that the British have got to fight the enemy wherever they find him. He is, therefore, more inclined to support continuing the campaign in the Middle East": Hopkins to the Prime Minister and British Chiefs of Staff at 10 Downing Street, July 24, 1941, in Robert Sherwood, *Roosevelt and Hopkins,* 314. See also Mark A. Stoler, *Allies in War: Britain and America Against the Axis Powers, 1940–1945* (London: Hodder Arnold, 2005), 31.
19. Hitler later told Mussolini he would prefer to have "three or four teeth pulled" than endure another nine hours' negotiation with the Spanish generalissimo, Franco: Kershaw, *Hitler: Nemesis,* 330.
20. Stoler, *Allies and Adversaries,* 41.
21. Letter of August 2, 1941, in *F.D.R., His Personal Letters,* ed. Elliott Roosevelt (New York: Duell, Sloan and Pearce, 1947–50), vol. 2, 1197.
22. Geoffrey C. Ward, ed., *Closest Companion: The Unknown Story of the Intimate Friendship Between Franklin Roosevelt and Margaret Suckley* (Boston: Houghton Mifflin, 1995), 140.
23. Theodore A. Wilson, *The First Summit: Roosevelt & Churchill at Placentia Bay, 1941* (Boston: Houghton Mifflin, 1969), 34.
24. Entry of August 9, 1941, Henry Harley Arnold, *American Airpower Comes of Age: General Henry H. "Hap" Arnold's World War II Diaries,* ed. John W. Huston (Maxwell Air Force Base, AL: Air University Press, 2002), 226.
25. Ibid., entry of August 4, 1941, 218.
26. Ibid.
27. Ibid., 61.
28. "Operation Riviera, Atlantic Meeting, August, 1941," entry of August 8, Diary of Ian Jacob, Liddell Hart Centre for Military History, King's College London.
29. Ward, *Closest Companion,* 140.
30. Stark Diary, in Mitchell Simpson III, *Admiral Harold R. Stark: Architect of Victory, 1939–1945* (Columbia: University of South Carolina Press, 1989), 92.
31. Ward, *Closest Companion,* 140.
32. The toadfish proved hard to label definitively and was sent to the Smithsonian for further identification.
33. Henry H. Arnold, *Global Mission* (New York: Harper, 1949), 186.
34. Entry of August 8, 1941, Arnold, *American Airpower Comes of Age,* 221.
35. Ibid., 221–22.
36. George C. Marshall, *George C. Marshall: Interviews and Reminiscences for Forrest*

C. *Pogue,* 3rd ed. (Lexington, VA: George C. Marshall Research Foundation, 1991), 285.

37. Wilson, *The First Summit,* 71.
38. Entry of August 8, 1941, Arnold, *American Airpower Comes of Age,* 223.
39. Ibid., entry of August 10, 1941, 228.
40. Ibid., entry of August 8, 1941, 223.
41. Benjamin Welles, *Sumner Welles: FDR's Global Strategist* (New York: St. Martin's, 1997), 303.
42. Sumner Welles, *Where Are We Heading?* (New York: Harper, 1946), 6. See also Wilson, *The First Summit,* 32 and footnote 65, 275.
43. Delivered to Congress January 6, 1941. See also Wilson, *The First Summit,* 154.
44. Conn and Fairchild, *The Framework of Hemispheric Defense,* 125.
45. Ibid., 126.
46. Wilson, *The First Summit,* 185. See also Forrest Pogue, *George C. Marshall,* vol. 2, *Ordeal and Hope, 1939–1942* (New York: Viking, 1966), 153–54.
47. Wilson, *The First Summit,* 71.
48. Elliott Roosevelt, *As He Saw It* (New York: Duell, Sloan and Pearce, 1946), 22.
49. Ibid.
50. Ibid., 22–23.
51. Ibid., 24.
52. Ibid., 24–25.
53. Ibid., 25.
54. H. V. Morton, *Atlantic Meeting* (London: Methuen, 1943), 90–91.
55. "Operation Riviera, Atlantic Meeting, August, 1941," entry of August 9, Jacob Diary.
56. Ibid.
57. Ibid.
58. Ward, *Closest Companion,* 141.
59. Entry of August 10, 1941, Arnold, *American Airpower Comes of Age,* 224.
60. Ibid.
61. Dallek, *Franklin D. Roosevelt and American Foreign Policy,* 285.
62. Morton, *Atlantic Meeting,* 85.
63. Ward, *Closest Companion,* 141.
64. Sherwood, *Roosevelt and Hopkins,* 365.
65. Alexander Cadogan, "Atlantic Meeting," record of 1962, in *The Diaries of Sir Alexander Cadogan, O.M., 1938–1945,* ed. David Dilks (London: Cassell, 1971), 398.
66. Sherwood, *Roosevelt and Hopkins,* 354.
67. Wilson, *The First Summit,* 163.
68. Ward, *Closest Companion,* 141.
69. Since May 10, 1940, Hopkins — whom FDR had once seen as a possible successor to him as president — had lived at the White House as special assistant to the President. In December 1937 he underwent surgery for stomach cancer, and

though the cancer did not recur, he suffered many postgastrectomy problems and relapses. See James A. Halsted, "Severe Malnutrition in a Public Servant of the World War II Era: The Medical History of Harry Hopkins," *Transactions of the American Clinical and Climatological Association* 86 (1975): 23–32.

70. Cadogan, "Atlantic Meeting," *The Cadogan Diaries*, 398; and David Reynolds, *The Creation of the Anglo-American Alliance, 1937–41* (London: Europa, 1981), 258 and footnote 28, 364.
71. Elliott Roosevelt, *As He Saw It*, 29.
72. Ibid., 28.
73. Morton, *Atlantic Meeting*, 86–87.
74. Elliott Roosevelt, *As He Saw It*, 29.
75. Entry of August 10, 1941, *The Cadogan Diaries*, 397.
76. Arnold, *Global Mission*, 252.
77. Entry of August 10, 1941, Arnold, *American Airpower Comes of Age*, 226.
78. Arnold, *Global Mission*, 252.
79. Entry of August 10, 1941, *The Cadogan Diaries*, 397.
80. Wilson, *The First Summit*, 136–37.
81. Winston S. Churchill, *The Second World War*, vol. 3, *The Grand Alliance* (Boston: Houghton Mifflin, 1950), 386.
82. David Reynolds, *In Command of History: Churchill Fighting and Writing the Second World War* (New York: Random House, 2005), 261.
83. Ibid.
84. Ibid.
85. "Operation Riviera, Atlantic Meeting, August, 1941," entry of August 10, Jacob Diary, 22–23.
86. Wilson, *The First Summit*, 98.
87. "Operation Riviera, Atlantic Meeting, August, 1941," entry of August 10, Jacob Diary, 21.
88. Ward, *Closest Companion*, 141.
89. Morton, *Atlantic Meeting*, 113–14.
90. Ward, *Closest Companion*, 141.
91. "Operation Riviera, Atlantic Meeting, August, 1941," entry of August 10, Jacob Diary, 24.
92. Ward, *Closest Companion*, 141.
93. "Operation Riviera, Atlantic Meeting, August, 1941," entry of August 10, Jacob Diary, 25–26.
94. Ibid., 26.
95. Entry of August 10, 1941, Arnold, *American Airpower Comes of Age*, 228.
96. "Operation Riviera, Atlantic Meeting, August, 1941," entry of August 12, Jacob Diary, 36.
97. Stoler, *The Politics of the Second Front*, 10. American planners felt the British paper amounted to "groping for panaceas," in an effort to "bring the United States into the war at the earliest possible date": ibid.

98. "Operation Riviera, Atlantic Meeting, August, 1941," entry of August 11, Jacob Diary, 27.
99. Ibid.
100. Churchill to Eden, May 24, 1941, regarding "Memorandum by Maynard Keynes on British War Aims," Churchill Papers, 20/36, Churchill College, Cambridge, UK.
101. Elliott Roosevelt, *As He Saw It,* 35.
102. Thomas B. Buell, *Master of Sea Power: A Biography of Admiral Ernest J. King* (Boston: Little, Brown, 1980), 130.
103. Elliott Roosevelt, *As He Saw It,* 35.
104. Ibid., 35–36.
105. Ibid., 36.
106. Ibid., 37.
107. Ibid., 38.
108. Ward, *Closest Companion,* 142.
109. Ibid.
110. Ibid.
111. Elliott Roosevelt, *As He Saw It,* 41–42.
112. "Operation Riviera, Atlantic Meeting, August, 1941," entry of August 12, Jacob Diary, 31.
113. Opinion polls showed no change in American reluctance to intercede in the war in Europe following the Atlantic Charter meeting: Stoler, *Allies in War,* 27.
114. Ward, *Closest Companion,* 142.
115. Ibid.
116. "Operation Riviera, Atlantic Meeting, August, 1941," entry of August 12, Jacob Diary, 33. The departure of the Prime Minister's battleship proved more dramatic than expected. The two accompanying U.S. destroyers failed to notice the *Prince of Wales* slowing down, ahead of them, which the battleship did to pass an anchored American vessel. The destroyers almost collided with the battleship's stern. Then, about an hour and a half into the voyage, one of the destroyers suddenly veered across the bow of the *Prince of Wales.* "Our captain ordered full speed astern and missed the destroyer by about 40 or 50 yards," Colonel Jacob recorded with relief. "Apparently the destroyer's helm had jammed hard over, and she came across quite out of control . . .": Ibid., 34.
117. Ward, *Closest Companion,* 142.

2. THE U.S. IS ATTACKED!

1. Eleanor Roosevelt, *This I Remember* (New York: Harper, 1949), 232.
2. Ibid.
3. See Ronald A. Spector, *Eagle Against the Sun: The American War with Japan* (New York: Free Press, 1985), 93–100; Robert B. Stinnett, *Day of Deceit: The Truth About*

FDR and Pearl Harbor (New York: Free Press, 2000); and Christopher Andrew, *For the President's Eyes Only: Secret Intelligence and the American Presidency from Washington to Bush* (New York: HarperCollins, 1995), 105–22, inter alia.

4. Gordon W. Prange, *At Dawn We Slept: The Untold Story of Pearl Harbor* (New York: McGraw-Hill, 1981), 487.
5. Ibid., 467.
6. Ibid., 468.
7. Ibid., 446.
8. Ibid., 468.
9. Ibid., 446.
10. Ibid., 485.
11. Robert E. Sherwood, *Roosevelt and Hopkins: An Intimate History* (New York: Harper, 1948), 426–27.
12. Ibid., 427.
13. Kenneth S. Davis, *FDR, the War President, 1940–1943: A History* (New York: Random House, 2000), 398.
14. Sherwood, *Roosevelt and Hopkins,* 427.
15. Eleanor Roosevelt, *This I Remember,* 233.
16. Station HYPO or Fleet Unit Radio Pacific was the U.S. Navy signals cryptographic unit in Pearl Harbor, but it had no Purple machine equivalent, and received no Purple signals to decrypt. Instead, the unit worked to break some of the Japanese Imperial Navy two-book code JN-25—which was changed on December 1, 1941, forcing U.S. code breakers in Hawaii to begin from scratch.
17. Prange, *At Dawn We Slept,* 294; Stinnett, *Day of Deceit,* 234.
18. Prange, *At Dawn We Slept,* 294.
19. Ibid.
20. Ed Cray, *General of the Army: George C. Marshall, Soldier and Statesman* (New York: Norton, 1990), 255.
21. Gordon W. Prange, *Dec. 7, 1941: The Day the Japanese Attacked Pearl Harbor* (New York: Wings Books, 1991), 98.
22. Cordell Hull, *The Memoirs of Cordell Hull* (New York: Macmillan, 1948), vol. 2, 1095.
23. Prange, *Dec. 7, 1941,* 28.
24. Signal from Admiral Stark, priority, in Stinnett, *Day of Deceit,* 172–73.
25. Prange, *Dec. 7, 1941,* 247.
26. Ibid., 247–48.
27. Sherwood, *Roosevelt and Hopkins,* 430.
28. B. Mitchell Simpson III, *Admiral Harold R. Stark: Architect of Victory, 1939–1945* (Columbia: University of South Carolina Press, 1989), 114. Hopkins, in a memo that night, noted that it was "at about 1.40 p.m.": Sherwood, *Roosevelt and Hopkins,* 430.
29. Simpson, *Admiral Harold R. Stark,* 114. Hopkins noted: "a radio from the

Commander-in-Chief of our forces there advising all our stations that an air raid attack was on and that it was 'no drill.'": Sherwood, *Roosevelt and Hopkins,* 430–31.

30. See Ted Morgan, *FDR: A Biography* (New York: Simon and Schuster, 1985), 186–90 and 200.
31. Marshall, *George C. Marshall Interviews and Reminiscences for Forrest C. Pogue,* 610–11.
32. "FDR Visits Hawaii," Navy History Hawaii, http://navyhistoryhawaii.blogspot.com/2010/07/fdr-visits-hawaii_28.html, accessed April 29, 2011.
33. Franklin D. Roosevelt, "Remarks in Hawaii," July 28, 1934. John T. Woolley and Gerhard Peters, *The American Presidency Project*, http://www.presidency.ucsb.edu/ws/?pid=14729, accessed April 29, 2011.
34. Sherwood, *Roosevelt and Hopkins,* 431.
35. Chris Bellamy, *Absolute War: Soviet Russia in the Second World War* (New York: Knopf, 2007), 140.
36. Ibid., 107.
37. Ibid., 148.
38. Davis, *FDR, the War President,* 339.
39. Hull, *The Memoirs of Cordell Hull,* vol. 2, 1096.
40. Henry L. Stimson and McGeorge Bundy, *On Active Service in Peace and War* (New York: Harper and Brothers, 1948), 391.
41. Sherwood, *Roosevelt and Hopkins,* 431.
42. Prange, *Dec. 7, 1941,* 383; Linda Lotridge Levin, *The Making of FDR: The Story of Stephen T. Early, America's First Modern Press Secretary* (Amherst, NY: Prometheus Books, 2008), 252.
43. Levin, *The Making of FDR,* 252.
44. "The announcement of the attack was made in a brief statement by President Roosevelt. Naval and military targets on the principal island of Oahu have also been attacked." W. Averell Harriman and Elie Abel, *Special Envoy to Churchill and Stalin, 1941–1946* (New York: Random House, 1975), 111.
45. Martin Gilbert, *Winston S. Churchill,* vol. 6, *Finest Hour, 1939–1941* (Boston: Houghton Mifflin, 1983), 1267.
46. John G. Winant, *A Letter from Grosvenor Square: An Account of a Stewardship* (Boston: Houghton Mifflin, 1947), 198–99.
47. Gilbert, *Finest Hour,* 1268.
48. David Reynolds, *In Command of History: Churchill Fighting and Writing the Second World War* (New York: Random House, 2005), 264.
49. Sherwood, *Roosevelt and Hopkins,* 431.
50. Ibid.
51. Ibid., 432.
52. Grace Tully, *F.D.R., My Boss* (New York: Scribner's, 1949), 254.
53. Maurice Matloff and Edwin M. Snell, *Strategic Planning for Coalition Warfare,*

1941–1942 (Washington, DC: Office of the Chief of Military History, Dept. of the Army, 1953), 18.

54. Simpson, *Admiral Harold R. Stark,* 109.
55. Henry H. Arnold, *Global Mission* (New York: Harper, 1949), 193.
56. Tully, *F.D.R., My Boss,* 255.
57. Ibid.
58. Prange, *Dec. 7, 1941,* 255.
59. Tully, *F.D.R., My Boss,* 255.
60. Ibid., 256.
61. Ibid.
62. Draft No. 1, December 7, 1941, Proposed Message to the Congress, Franklin Delano Roosevelt Library, Hyde Park, NY.
63. Tully, *F.D.R., My Boss,* 256.
64. Magic intercept of Tokyo Foreign Office Purple message to the Japanese Embassy in Washington, D.C., dated January 30, 1941, translated on February 7, 1941. Such intercepts could not, of course, be made public by the Roosevelt administration, but the gist of them was shared among all State Department and military authorities.
65. Hull, *The Memoirs of Cordell Hull,* vol. 2, 1098.
66. Tully, *F.D.R., My Boss,* 256.
67. Sherwood, *Roosevelt and Hopkins,* 432.
68. Levin, *The Making of FDR,* 252.
69. Richard Strout, in *Christian Science Monitor,* December 7, 1951, quoted in Prange, *Dec. 7, 1941,* 385.
70. John Morton Blum, ed., *From the Morgenthau Diaries,* vol. 3, *Years of War, 1941–1945* (Boston: Houghton Mifflin, 1967), 2.
71. Eleanor Roosevelt, *This I Remember,* 234.
72. Prange, *At Dawn We Slept,* 557.
73. Sherwood, *Roosevelt and Hopkins,* 433.
74. Ibid.
75. Prange, *At Dawn We Slept,* 556.
76. Ibid.
77. Ibid., 557.
78. Ibid.
79. Ibid., 557–58.
80. Sherwood, *Roosevelt and Hopkins,* 433.
81. Prange, *At Dawn We Slept,* 558.
82. Frances Perkins, Columbia University Oral History, quoted in Lynne Olson, *Citizens of London: The Americans Who Stood with Britain in Its Darkest Hour* (New York: Random House, 2010), 145; Andrew, *For the President's Eyes Only,* 118–19.
83. Morgan, *FDR: A Biography,* 617.

84. Sherwood, *Roosevelt and Hopkins,* 433.
85. Prange, *At Dawn We Slept,* 559.
86. Morgan, *FDR: A Biography,* 618.
87. Prange, *Dec. 7, 1941,* 389.
88. Prange, *At Dawn We Slept,* 560.
89. Ibid.
90. Prange, *Dec. 7, 1941,* 390.
91. Ibid., 384.
92. Prange, *At Dawn We Slept,* 516.
93. Davis, *FDR, the War President,* 347.
94. Sherwood, *Roosevelt and Hopkins,* 433.
95. Ibid., 434.
96. Stinnett, *Day of Deceit,* 3.
97. A. M. Sperber, *Murrow: His Life and Times* (New York: Freundlich Books, 1986), 207.

3. HITLER'S GAMBLE

1. Robert E. Sherwood, *Roosevelt and Hopkins: An Intimate History* (New York: Harper, 1948), 435–36.
2. Ibid., 437.
3. Franklin D. Roosevelt, *The Public Papers and Addresses of Franklin D. Roosevelt,* ed. Samuel Rosenman, vol. 10, *The Call to Battle Stations, 1941* (New York: Russell and Russell, 1969), 514–16.
4. Martin Gilbert, *Churchill and America* (New York: Free Press, 2005), 246.
5. Alan Brooke, "Notes on My Life," in *War Diaries, 1939–1945: Field Marshal Lord Alanbrooke,* ed. Alex Danchev and Daniel Todman (Berkeley: University of California Press, 2001), 209.
6. Cable C-138x, in Warren Kimball, ed., *Churchill & Roosevelt: The Complete Correspondence,* vol. 1, *Alliance Emerging, October 1933–November 1942* (Princeton, NJ: Princeton University Press, 1984), 283.
7. Entry of Tuesday, December 9, 1941, "Secret Diary" of Lord Halifax, Papers of Lord Halifax, Hickleton Papers, Borthwick Institute of Historical Research, University of York, Yorkshire, England.
8. Letter to the Prime Minister, December 9, 1941, Papers of Lord Halifax, Hickleton Papers, Borthwick Institute of Historical Research, University of York, Yorkshire, England.
9. Ibid.
10. Ibid.
11. Letter to Hiram Johnson Jr., December 13, 1941, in Hiram W. Johnson, *The Diary Letters of Hiram Johnson, 1917–1945,* vol. 7 (New York: Garland, 1983).
12. Letter to the Prime Minister, December 9, 1941, Papers of Lord Halifax.
13. Ibid.

14. Andrew Roberts, *The Holy Fox: A Biography of Lord Halifax* (London: Weidenfeld and Nicolson, 1991), 287.
15. Ibid., 280.
16. Ibid.
17. Cable R-73x, draft A, in Kimball, *Churchill & Roosevelt,* vol. 1, 285.
18. Entry of Wednesday, December 10, 1941, Halifax Diary.
19. David Reynolds, *In Command of History: Churchill Fighting and Writing the Second World War* (New York: Random House, 2005), 266.
20. Winston S. Churchill, *The Second World War,* vol. 3, *The Grand Alliance* (Boston: Houghton Mifflin, 1950), 551.
21. Cable C-139x, in Kimball, *Churchill & Roosevelt,* vol. 1, 284.
22. Cable R-73x, draft B, in Kimball, *Churchill & Roosevelt,* vol. 1, 286.
23. Ibid.
24. Cable R-73x, in Kimball, *Churchill & Roosevelt,* vol. 1, 286.
25. Linda Lotridge Levin, *The Making of FDR: The Story of Stephen T. Early, America's First Modern Press Secretary* (Amherst, NY: Prometheus Books, 2008), 262.
26. Entry of December 3, 1941, Galeazzo Ciano, *Diary 1937–1943* (New York: Enigma Books, 2002), 470.
27. Ibid.
28. Ibid., entry of December 8, 1941, 472.
29. Ian Kershaw, *Hitler 1936–1945: Nemesis* (New York: Norton, 2001), 345.
30. Guderian was dismissed on December 26, 1941.
31. Henry Picker, *Hitlers Tischgespräche im Führerhauptquartier, 1941–42,* ed. Gerhard Ritter (Bonn: Athenäum Verlag, 1951), 75.
32. Ian Kershaw, *Fateful Choices: Ten Decisions That Changed the World, 1940–1941* (New York: Penguin Press, 2007), 417, quoting Walter Warlimont, *Inside Hitler's Headquarters, 1939–45* (Novato, CA: Presidio, 1964), 207–8.
33. Kershaw, *Hitler: Nemesis,* 953.
34. Entry of December 8, 1941, Ciano, *Diary,* 472.
35. Gordon W. Prange, *At Dawn We Slept: The Untold Story of Pearl Harbor* (New York: McGraw-Hill, 1981), 428.
36. Hitler to Franz Halder, March 17, 1941, in Kershaw, *Hitler: Nemesis,* 355.
37. Kershaw, *Fateful Choices,* 401.
38. Nikolaus von Below, adjutant to the Führer, in Kershaw, *Hitler: Nemesis,* 954.
39. Kershaw, *Hitler: Nemesis,* 442.
40. Kershaw, *Fateful Choices,* 418.
41. Max Domarus, ed., *Hitler, Speeches and Proclamations 1932–1945: The Chronicle of a Dictatorship,* vol. 4, *The Years 1941 to 1945* (Wauconda, IL: Bolchazy-Carducci, 1997), 2531–51. Notable Americans had sympathized. "There is no question of the power, unity and purposefulness of Germany," Anne Morrow Lindbergh had written her mother after her first trip to Germany in 1936, "it is terrific. I have never in my life been so conscious of such a directed force. It is thrilling when one sees it manifested in the energy, pride and morale of the people — especially

the young people. But also terrifying in its very unity — a weapon made by one man but also to be used by one man. Hitler, I am beginning to feel, is like an inspired religious leader, and as such fanatical — a visionary who really wants the best for his country." Anne Morrow Lindbergh, *The Flower and the Nettle: Diaries and Letters of Anne Morrow Lindbergh, 1936–1939* (New York: Harcourt Brace Jovanovich, 1976), 100.

42. Domarus, *Hitler, Speeches and Proclamations,* vol. 4, 2531–51.
43. Ibid.
44. Letter to Hiram Johnson Jr., December 13, 1941, in Johnson, *The Diary Letters of Hiram Johnson,* vol. 7.
45. Ibid.

4. THE VICTORY PLAN

1. Eleanor Roosevelt, *This I Remember* (New York: Harper, 1949), 233.
2. Mrs. Roosevelt may have misremembered; as AP's White House reporter, A. Merriman Smith, claimed in 1946, "knowledge that Churchill was en route to the White House was generally known among Washington reporters, but not a word was printed or broadcast until he arrived": A. Merriman Smith, *Thank You, Mr. President: A White House Notebook* (New York: Harper and Brothers, 1946), 129.
3. Eleanor Roosevelt, *This I Remember,* 237.
4. Entry of Sunday, December 21, 1941, "Secret Diary" of Lord Halifax, Papers of Lord Halifax, Hickleton Papers, Borthwick Institute of Historical Research, University of York, Yorkshire, England.
5. Ibid., entry of Monday, December 22, 1941.
6. Mary Soames, ed., *Speaking for Themselves: The Personal Letters of Winston and Clementine Churchill* (New York: Doubleday, 1998), 461.
7. Entry of Monday, December 22, 1941, Halifax Diary.
8. Ibid.
9. Entry of December 22, 1941, in Lord Moran, *Winston Churchill: The Struggle for Survival, 1940–1965* (London: Constable, 1966), 11.
10. Ibid.
11. Ibid.
12. See Mark A. Stoler, *The Politics of the Second Front: American Military Planning and Diplomacy in Coalition Warfare, 1941–1943* (Westport, CT: Greenwood Press, 1977), 23.
13. Winston S. Churchill, *The Second World War,* vol. 3, *The Grand Alliance* (Boston: Houghton Mifflin, 1950), 589.
14. Ibid., 589–60.
15. For an excellent summary see Mark A. Stoler, *Allies and Adversaries: The Joint Chiefs of Staff, the Grand Alliance, and U.S. Strategy in World War II* (Chapel Hill: University of North Carolina Press, 2000). See also Mark A. Stoler, *Allies in War: Britain and America Against the Axis Powers, 1940–1945* (London: Hodder Arnold,

2005); and Christopher Thorne, *Allies of a Kind: The United States, Britain, and the War Against Japan, 1941–1945* (New York: Oxford University Press, 1978).

16. Barbara Tuchman, *Stilwell and the American Experience in China, 1911–45* (New York: Macmillan, 1970), 241; Stoler, *The Politics of the Second Front,* 25.
17. Stoler, *The Politics of the Second Front,* 13.
18. Letter to Hiram Johnson Jr., February 19, 1942, in Hiram W. Johnson, *The Diary Letters of Hiram Johnson, 1917–1945,* vol. 7 (New York: Garland, 1983).
19. Churchill, *The Second World War,* vol. 3, *The Grand Alliance,* 588.
20. Stoler, *Allies and Adversaries,* 47–50.
21. The final report, with appendices, was only formally submitted to the President on September 25, 1941, though still dated September 11. "Joint Board Estimate of United States Over-all Production Requirements, September 11, 1941," "Safe" and Confidential Files, Franklin D. Roosevelt Presidential Library, Hyde Park, NY. See also Kenneth S. Davis, *FDR, the War President, 1940–1943: A History* (New York: Random House, 2000), 295.
22. "Joint Board Estimate," "Safe" and Confidential Files, FDR Library. Author's italics.
23. Ibid.
24. Lord Halifax to Prime Minister, October 4, 1941, Papers of Lord Halifax, Hickleton Papers, Borthwick Institute of Historical Research, University of York, Yorkshire, England.
25. "We lunched together, as usual, on his writing table, among his papers and knickknacks, and he talked about everything with great freedom, from Russia to the Philippines": entry of Friday, October 10, 1941, Halifax Diary.
26. Lord Halifax to Prime Minister, October 11, 1941, Papers of Lord Halifax.
27. Churchill, *The Second World War,* vol. 3, *The Grand Alliance,* 482–83.
28. Entry of December 18, 1942, Stimson Diary, Henry L. Stimson Papers, Yale University Library, New Haven, CT.
29. Ibid.
30. Ibid.
31. Jon Meacham, *Franklin and Winston: An Intimate Portrait of an Epic Friendship* (New York: Random House, 2003), 141.
32. Michael F. Reilly and William Slocum, *Reilly of the White House* (New York: Simon and Schuster, 1947), 125.
33. President's Press Conference, White House, December 23, 1941, FDR Library.
34. Smith, *Thank You, Mr. President,* 67.
35. President's Press Conference, White House, December 23, 1941.
36. Smith, *Thank You, Mr. President,* 262.
37. "Operation Arcadia: Washington Conference, December 1941," entry of December 23, Diary of Ian Jacob, 11, Liddell Hart Centre for Military History, King's College London.
38. Ibid., 13.
39. Entry of December 23, 1941, Stimson Diary, 140.

40. Ibid., 140–41.
41. "Operation Arcadia: Washington Conference, December 1941," entry of December 23, Jacob Diary, 15–16.
42. Ibid., 17.
43. "What India was for England, the eastern territory will be for us": Hitler monologues, August 8–11, 1941, quoted in Ian Kershaw, *Hitler 1936–1945: Nemesis* (New York: Norton, 2000), 402.
44. Charles Wilson, Moran Papers, Wellcome Library, London, typescript notes.
45. Ibid., typescript notes, marked "29."
46. Entry of Monday, November 10, 1941, Halifax Diary.
47. Charles Wilson, Moran Papers, typescript notes.
48. Hankey Papers, Churchill College, Cambridge, quoted in David Irving, *Churchill's War,* vol. 2, *Triumph in Adversity* (London: Focal Point, 2001), 339.
49. Thorne, *Allies of a Kind,* 116.
50. Ibid.
51. Andrew Roberts, *Eminent Churchillians* (New York: Simon and Schuster, 1994), 205.
52. Charles Wilson, Moran Papers, typescript notes.
53. Ibid.
54. Ibid.
55. Ibid.
56. Ibid.
57. "Operation Arcadia: Washington Conference, December 1941," entry of December 25, Jacob Diary, 15–16.
58. Ibid.
59. Ibid.
60. Entry of December 25, 1941, Stimson Diary, 145–46.
61. Ibid.
62. Ibid., 145.
63. Ibid., 146.
64. Ibid., 147.
65. Stephen Ambrose, *Eisenhower,* vol. 1, *Soldier, General of the Army, President-Elect, 1890–1952* (New York: Simon and Schuster, 1985), 137–38.
66. Eleanor Roosevelt, *This I Remember,* 242–43.
67. Meacham, *Franklin and Winston,* 144.
68. Ibid., 145.
69. Eleanor Roosevelt, *This I Remember,* 243–44.
70. John Morton Blum, ed., *From the Morgenthau Diaries,* vol. 3, *Years of War, 1941–1945* (Boston: Houghton Mifflin, 1967), 122.
71. Entry of Monday, December 24, 1941, Halifax Diary.
72. Entry of December 27, 1941, Stimson Diary, 144–45.
73. Entry of December 25, 1941, Halifax Diary.
74. Charles Wilson, Moran Papers, typescript notes.

75. Vivian A. Cox, *Seven Christmases*, ed. Nick Thorne (Seven Oaks, Kent, UK: Nickay Associates, 2010), 117.
76. Carlo D'Este, *Warlord: A Life of Winston Churchill at War, 1874–1945* (New York: Harper, 2008), 560.
77. Charles Wilson, Moran Papers, typescript notes.
78. Entry of Monday, December 26, 1941, Halifax Diary.
79. David Lilienthal, *The Journals of David E. Lilienthal* (New York: Harper and Row, 1964), vol. 1, 418.
80. Entry of Monday, December 26, 1941, Halifax Diary.
81. Winston S. Churchill, *The Unrelenting Struggle: War Speeches by the Right Hon. Winston S. Churchill, C.H., M.P*, comp. Charles Eade (Boston: Little, Brown, 1942), 337.
82. Ibid.
83. Smith, *Thank You, Mr. President*, 67.
84. Charles Wilson, Moran Papers, typescript notes.
85. Copy in Stimson Papers, Yale University Library.
86. "Report on the meetings of the nucleus of America First," December 17, 1941, in the home of Edwin S. Webster, copy in Stimson Papers, Yale University Library.
87. Rudolf Schröck, with Dyrk Hesshaimer, Astrid Bouteuil, and David Hesshaimer, *Das Doppelleben des Charles A. Lindbergh* [The double life of Charles A. Lindbergh] (Munich: Heyne Verlag, 2005). See also Joshua Kendall, *America's Obsessives: The Compulsive Energy That Built a Nation* (New York: Grand Central, 2013), 184–94.
88. Cordell Hull, *The Memoirs of Cordell Hull* (New York: Macmillan, 1948), 1117 et seq.

5. SUPREME COMMAND

1. Eleanor Roosevelt, *This I Remember* (New York: Harper, 1949), 243.
2. "Operation Arcadia: Washington Conference, December 1941," "Some Personalities," Diary of Ian Jacob, 26 and 24, Liddell Hart Centre for Military History, King's College London.
3. Ibid., 24.
4. Entry of December 27, 1941, "Secret Diary" of Lord Halifax, Papers of Lord Halifax, Hickleton Papers, Borthwick Institute of Historical Research, University of York, Yorkshire, England.
5. Doris Kearns Goodwin, *No Ordinary Time: Franklin and Eleanor Roosevelt; The Home Front in World War II* (New York: Simon and Schuster, 1994), 320. See also Alonzo Fields, *My 21 Years in the White House* (New York: Coward-McCann, 1961), 82, 88–89.
6. "The President's Secretary to the Secretary of State," with attachment, in Documents and Supplementary Papers, The First Washington Conference, in U.S. Department of State, *Foreign Relations of the United States, The Conferences at*

Washington, 1941–1942, and Casablanca, 1943 (Washington, DC: US Government Printing Office, 1941–43), 375.

7. Kenneth S. Davis, *FDR: The War President, 1940–1943: A History* (New York: Random House, 2000), 372.
8. "Operation Arcadia: Washington Conference, December 1941," "Some Personalities," Jacob Diary, 37–38. President Roosevelt later verified the story to Canadian prime minister Mackenzie King: entry of December 5, 1942, Diaries of William Lyon Mackenzie King, Library and Archives Canada, Ottawa, ON.
9. War Cabinet verbatim report, January 18, 1942, Papers of Lawrence Burgis, in Andrew Roberts, *Masters and Commanders: How Four Titans Won the War in the West, 1941–1945* (New York: Harper, 2009), 87.
10. Forrest Pogue, *George C. Marshall*, vol. 2, *Ordeal and Hope, 1939–1942* (New York: Viking, 1966), 276.
11. Entry of December 27, 1941, Stimson Diary, Henry L. Stimson Papers, Yale University Library, New Haven, CT, 148.
12. "Operation Arcadia: Washington Conference, December 1941," "Sunday, December 28th to Wednesday, December 31st," Jacob Diary, 20.
13. Ibid.
14. Alan Brooke, *War Diaries 1939–1945: Field Marshal Lord Alanbrooke*, ed. Alex Danchev and Daniel Todman (Berkeley: University of California Press, 2001), 215.
15. Ibid.
16. Cable of December 28, 1941, quoted in Winston S. Churchill, *The Second World War*, vol. 3, *The Grand Alliance* (Boston: Houghton Mifflin, 1950), 598.
17. Ibid.
18. Brooke, *War Diaries*, 215.
19. Ibid.
20. "Address to the Congress on the State of the Union," January 6, 1942, Franklin D. Roosevelt, *The Public Papers and Addresses of Franklin D. Roosevelt*, vol. 11, *Humanity on the Defensive 1942* (New York: Russell and Russell, 1969), 32–42.
21. Entry of January 4, 1942, Halifax Diary.

6. THE PRESIDENT'S MAP ROOM

1. Richard Holmes, *Churchill's Bunker: The Secret Headquarters at the Heart of Britain's Victory* (London: Profile Books; Imperial War Museum, 2009), 55.
2. Ibid., 72.
3. Ibid., 83.
4. "Re: Air Raid Shelter," December 15, 1942, Morgenthau Office Diaries, Franklin D. Roosevelt Presidential Library, Hyde Park, NY.
5. Winston S. Churchill, *The Second World War*, vol. 2, *Their Finest Hour* (Boston: Houghton Mifflin, 1949), 331.
6. "Re: Air Raid Shelter," Morgenthau Office Diaries.
7. Ibid.

8. Ibid.
9. Vivian A. Cox, *Seven Christmases,* ed. Nick Thorne (Seven Oaks, Kent, UK: Nickay Associates, 2010), 127–28.
10. Ibid., 284.
11. Ibid., letter of 24.1.41, 134.
12. Ibid., letter of 30.1.41, 285.
13. Ibid., 134.
14. Commander George Elsey, interview with author, September 10, 2011. All Elsey quotes in chapter are from this interview.
15. Cox, *Seven Christmases,* 130.
16. Sublieutenant Cox stayed in Washington until early February 1942. His report to the British Admiralty, detailing the site, the room, its construction, the charts, the personnel, information displayed, manner of display, etc., is reproduced in his memoir: Cox, *Seven Christmases,* 311–19. Cox was distressed, initially, by the failure of the U.S. Navy to cooperate in providing "anything like the complete picture of the [war] situation" that the President "could and should have had": ibid., 314.

7. THE FIGHTING GENERAL

1. Alonzo Fields, *My 21 Years in the White House* (New York: Coward-McCann, 1961), 52–53.
2. *Baltimore Sun,* February 1, 1942, Office of the Coordinator of Information (COI), Press Excerpts, January 27–February 1942, Papers of General Douglas MacArthur, Record Group (RG) 2, MacArthur Memorial Archives [hereafter MMA]; Richard Connaughton, *MacArthur and Defeat in the Philippines* (Woodstock, NY: Overlook Press, 2001), 258.
3. February 2, 1942, COI Press Excerpts, January 27–February 1942, MacArthur Papers, RG 2, MMA; Connaughton, *MacArthur and Defeat in the Philippines,* 258.
4. Connaughton, *MacArthur and Defeat in the Philippines,* 257.
5. Ibid.
6. January 27, 1942, COI Press Excerpts, January 27–February 1942, MacArthur Papers, RG 2, MMA; Connaughton, *MacArthur and Defeat in the Philippines,* 257.
7. January 27, 1942, COI Press Excerpts, January 27–February 1942, MacArthur Papers, RG 2, MMA; Connaughton, *MacArthur and Defeat in the Philippines,* 257.
8. Douglas MacArthur, *Reminiscences* (New York: McGraw-Hill, 1964), 93.
9. William Manchester, *American Caesar: Douglas MacArthur 1880–1964* (Boston: Little, Brown, 1978), 152, quoting Rexford Tugwell, *The Democratic Roosevelt* (Garden City, NY: Doubleday, 1957), 348–51.
10. MacArthur, *Reminiscences,* 101.
11. Ibid.
12. Tugwell, *The Democratic Roosevelt,* 349–50.

13. One of the star performers in establishing the Conservation Corps was Lieutenant Colonel George C. Marshall. Instead of making Marshall a brigadier general, as General Pershing, Marshall's former commanding officer in World War I, had requested, MacArthur — who was known to be pathologically jealous of able subordinates — made him an instructor with the National Guard, without promoting him.
14. Michael Schaller, *Douglas MacArthur: The Far Eastern General* (New York: Oxford University Press, 1989), 13–14 and 18–20; Manchester, *American Caesar,* 156.
15. MacArthur, *Reminiscences,* 96.
16. Carlo D'Este, *Eisenhower: A Soldier's Life* (New York: Holt, 2002), 238.
17. Connaughton, *MacArthur and Defeat in the Philippines,* 109.
18. James Leutze, *A Different Kind of Victory: A Biography of Admiral Thomas C. Hart* (Annapolis, MD: Naval Institute Press, 1981), 218.
19. Schaller, *Douglas MacArthur,* 26.
20. War Department 749 to MacArthur: "Reports of Japanese attacks show that numbers of our planes have been destroyed on the ground. Take all possible steps at once to avoid such losses in your area, including dispersion to maximum possible extent, construction of parapets and prompt take-off on warning": Records of Headquarters, United States Army in the Far East (USAFFE), years 1941–42, USAFFE, Chief of Staff and Commanding General, Radios and Letters Dealing with Plans and Policies, MacArthur Papers, RG 2, MMA. MacArthur also had his own code-breaking unit that had broken the Japanese diplomatic codes (Magic) and a number of Japanese naval codes.
21. William Hassett, *Off the Record with F.D.R., 1942–1945* (New Brunswick, NJ: Rutgers University Press, 1958), 88.
22. Leutze, *A Different Kind of Victory,* 212.
23. Theodore Friend, *Between Two Empires: The Ordeal of the Philippines, 1929–1946* (New Haven, CT: Yale University Press, 1965), 207.
24. Manchester, *American Caesar,* 215; D. Clayton James, *The Years of MacArthur,* vol. 2, *1941–1945* (Boston: Houghton Mifflin, 1975), 33.
25. Telegram of December 22, 1941, MacArthur Papers, RG 2, MMA.
26. James, *The Years of MacArthur,* vol. 2, 54.
27. Schaller, *Douglas MacArthur,* 72.
28. Connaughton, *MacArthur and Defeat in the Philippines,* 225.
29. James, *The Years of MacArthur,* vol. 2, 89.
30. Hassett, *Off the Record with F.D.R.,* 88.
31. Entry of January 29, 1942, in Dwight D. Eisenhower, *The Eisenhower Diaries,* ed. Robert H. Ferrell (New York: Norton, 1981), 46.
32. Letter of November 9, 1941, in Leutze, *A Different Kind of Victory,* 218.
33. Ibid., 212.
34. Telegram of December 13, 1941, MacArthur Papers, RG 2, MMA.
35. Ibid., telegram of December 22, 1941.

36. Ibid., telegram of January 2, 1942.
37. Ibid., telegram of January 7, 1942.
38. Leutze, *A Different Kind of Victory,* 265.
39. "Message from General MacArthur, To All Unit Commanders" (100 copies), MacArthur Papers, RG 2, MMA.
40. Manchester, *American Caesar,* 237.
41. Telegram of January 17, 1942, to General Marshall, MacArthur Papers, RG 2, MMA.
42. See Marshall telegrams to MacArthur 913, 917, 855, 949, 991, inter alia, MacArthur Papers, RG 2, MMA.
43. Entry of January 29, 1942, *The Eisenhower Diaries,* 46.
44. Ibid., entry of February 3, 1942.
45. Telegram 855, December 22, 1941, Records of Headquarters, USAFFE, MacArthur Papers, RG 2, MMA.
46. Telegram 1024, February 8, 1942, MacArthur Papers, RG 2, MMA.
47. Manuel Quezon, *The Good Fight* (New York: D. Appleton-Century, 1946), 261–62.
48. Robert E. Sherwood, *Roosevelt and Hopkins: An Intimate History* (New York: Harper, 1948), 492.
49. Leutze, *A Different Kind of Victory,* 240.
50. General George McClellan was dismissed as general in chief of the Union Army by President Lincoln on November 13, 1862, having failed to exploit his victory at Antietam, and having referred, it was said, to the President and Commander in Chief as a "gorilla." General McClellan later stood against Lincoln in the 1864 presidential election, where he "met with no better success as a politician than as a general": James M. McPherson, *Tried by War: Abraham Lincoln as Commander in Chief* (New York: Penguin, 2008), 141.
51. Sherwood, *Roosevelt and Hopkins,* 492.
52. Ibid.
53. Hassett, *Off the Record with F.D.R.,* 8.
54. Ibid., 14.
55. Ibid., 17.
56. "We've got to go to Europe and fight, and we've got to quit wasting resources all over the world, and still worse, wasting time," Eisenhower noted in his diary on January 27, 1942 — reversing his earlier views. "If we're to keep Russia in, save the Middle East, India, and Burma, we've got to begin slugging with air at West Europe. To be followed by a land attack as soon as possible": Entry of January 27, 1942, *The Eisenhower Diaries,* 43.
57. Ibid., entry of January 23, 1942, 44.
58. Connaughton, *MacArthur and Defeat in the Philippines,* 236.
59. James, *The Years of MacArthur,* vol. 2, 66.
60. Leutze, *A Different Kind of Victory,* 265.
61. Telegram of February 4, 1942, in MacArthur Papers, RG 2, MMA.
62. "Another long message on 'strategy' from MacArthur. He sent one on extolling

the virtues of the flank offensive. Wonder what he thinks we've been studying for all these years. His lecture would have been good for plebes": Entry of February 8, 1942, *The Eisenhower Diaries*, 47.

63. Entry of December 31, 1941, 161, Stimson Diary, Henry L. Stimson Papers, Yale University Library, New Haven, CT.
64. Quezon, *The Good Fight*, 248.
65. MacArthur, *Reminiscences*, 136; full text in MacArthur Papers, RG 4, MMA.
66. Transcript copy of telegram of January 30, 1942, MacArthur Papers, RG 2, MMA. The original wording is different from the paraphrase published in Quezon, *The Good Fight*, 261–63.
67. Telegram to General Marshall of January 31, 1942, MacArthur Papers, RG 2, MMA.
68. MacArthur, *Reminiscences*, 137.
69. Telegram of February 8, 1942, to General Marshall "from President Quezon for President Roosevelt," MacArthur Papers, RG 2, MMA.
70. Entry of February 8, 1942, Stimson Diary.
71. Ibid.
72. Ibid., entry of January 5, 1942.
73. Leutze, *A Different Kind of Victory*, 209.
74. Entry of January 5, 1942, Stimson Diary, 12.
75. Ibid.
76. *Congressional Record*, March 5, 1942. Several congressmen urged the appointment of Congressman John Wadsworth of New York—who wrote to Stimson blaming the *Washington Times Herald* for stoking the Washington "hot bed of rumors... It is a dirty business": Letter to Stimson of March 5, 1942, Stimson Papers.
77. Entry of February 9, 1942, Stimson Diary.
78. Ibid.
79. Ibid.
80. Telegram of February 8, 1942, to General Marshall "from President Quezon for President Roosevelt," in MacArthur Papers, RG 2, MMA.
81. Ibid.
82. Entry of February 9, 1942, Stimson Diary.
83. Forrest Pogue, *George C. Marshall*, vol. 2, *Ordeal and Hope, 1939–1942* (New York: Viking, 1966), 247–48.
84. Entry of February 9, 1942, *The Eisenhower Diaries*, 47.
85. Stimson's earlier draft for a cable to President Quezon gave an unfortunate indication of the war secretary's unprepossessing language. "I am much distressed ... we endeavored to defeat the aggression of Japan ... we have been marshaling our forces.... These difficulties have been accentuated ... The British have been most co-operative ... we have every hope ... our plans are comprehensive but must not be jeopardized by reckless or hasty steps ... You must rest assured that we shall proceed continuously and with all possible speed....": "Draft for

president's reply to Quezon, prepared by HLS February 9, 1942," Stimson Papers. In his diary on February 13 Stimson was still worried about "the President's rather severe telegram of two or three days ago" and "the President's castigation": Stimson Diary, February 13, 1942.

86. Telegram 1029 of February 10, 1942, from President Roosevelt to General MacArthur, MacArthur Papers, RG 2, MMA.
87. Ibid.
88. Ibid.
89. Ibid.
90. Ibid.
91. Arguably, World War II had begun not in Europe but in Asia, in 1937, when Japan invaded China.
92. Telegram 1029 of February 10, 1942, to Commanding General, USAFFE, Fort Mills, February 10, 1942, MacArthur Papers, RG 2, MMA.
93. Entry of February 9, 1942, *The Eisenhower Diaries*, 47.
94. Telegram of February 10, 1942, MacArthur Papers, RG 2, MMA.
95. Ibid., telegram 1037.
96. Ibid., telegram 1031, February 10, 1942.
97. Entry of January 13, 1942, Papers of Paul P. Rogers, Corregidor Diary and Selected Letters, RG 46, 1941–1989, 20 October 1941–11 March 1942, MMA.
98. Ibid., entry of February 12.
99. Telegram from General MacArthur to AGWAR, "For President Roosevelt," February 11, 1942, MacArthur Papers, RG 2, MMA.
100. Ibid.
101. Ibid.
102. Ibid. In fact, General Marshall had already twice raised the issue of MacArthur's own evacuation, cabling first on January 13, then on February 4, 1942, stating that "under these conditions the need for your services there [in the Philippines] might be less pressing than at other points in the Far East": see Connaughton, *MacArthur and Defeat in the Philippines*, 259. Jean MacArthur was adamant she would not leave without her husband, declaring, "We have drunk from the same cup, we three shall stay together": Manchester, *American Caesar*, 249.
103. Manchester, *American Caesar*, 249.
104. James K. Eyre, *The Roosevelt-MacArthur Conflict* (Chambersburg, PA: printed by author, 1950), 40.
105. Ibid., 41.
106. Ibid.
107. Carlos P. Romulo, *I Walked with Heroes* (New York: Holt, Rinehart, and Winston, 1961), 219.
108. Ibid., 223.
109. Eyre, *The Roosevelt-MacArthur Conflict*, 40.
110. Ibid., 42.

111. Ibid.
112. "Secret Priority" Telegram No. 262 to General George C. Marshall, War Department, February 12, 1942, MacArthur Papers, RG 2, MMA.
113. Quezon, *The Good Fight*, 274.
114. Ibid., 275.
115. The matter of MacArthur's controversial payment, solicited from President Quezon, was first raised by Carol M. Petillo in the *Pacific Historical Review* in 1979, later republished as "Douglas MacArthur and Manuel Quezon: A Note on an Imperial Bond" in William M. Leary, ed., *MacArthur and the American Century: A Reader* (Lincoln: University of Nebraska Press, 2001), 52–64.
116. Paul P. Rogers, *The Good Years: MacArthur and Sutherland* (New York: Praeger, 1990), 165; entry of February 13, 1942, Rogers Diary, in Papers of Paul P. Rogers, Corregidor Diary and Selected Letters, RG 46, 1941–1989, 20 October 1941–11 March 1942, MMA. In a memorandum describing his diary, Rogers noted that the money was ordered to be given to MacArthur in a Philippine Presidential "Executive Act Number One," backdated on February 13, 1942, to January 1, 1942. In retrospect Rogers — who admired MacArthur — wondered if possibly he had that day typed an earlier version of the executive order, which was later amended, and the numbers increased — for he specifically recalled MacArthur "at the time saying to Sutherland that the amounts hardly compensated for the salaries they had lost by serving on the Military Mission" — even though MacArthur had been the highest-paid military officer in the world. There was "uproar in Washington," Rogers later learned, for the "radio went from Marshall to Stimson. After some discussion the request was transmitted to Chase National Bank for action, and a copy was sent to the Department of the Interior. [Secretary] Ickes apparently refused to sanction the transfer, and the action seems to have been taken over his head": Rogers, *The Good Years*, 166. For forty years the matter was then hushed up, but looking back, long after the general's death, Rogers, in his note accompanying the donation of his wartime diary to the MacArthur Memorial Archives, was minded to believe that he had "intentionally changed the dollar amount in the diary entry" to protect MacArthur's reputation. "I am sure," he wrote, "in making the entry I changed the $500,000 to $50,000; then I changed [Lieutenant General] Sutherland's figure to $45,000 [instead of $75,000] to keep it in line with MacArthur. The amounts given to [Brigadier General Richard] Marshall and [Lieutenant Colonel Sidney] Huff, being relatively minor as compared with $500,000, were recorded without change": Notes on "The Diary" filed in Papers of Paul P. Rogers, Corregidor Diary and Selected Letters, RG 46, 1941–1989, 20 October 1941–11 March 1942, MMA. Rogers published his account of the matter in *The Good Years*, 165–69.
117. MacArthur later removed his copy of the cable — MacArthur to Adjutant-General, War Department, Washington, #285, February 15, 1942 — but a copy was kept by President Roosevelt in his "Safe" and Confidential Files.
118. Rogers, *The Good Years*, 166.

8. SINGAPORE

1. "Bring Gen MacArthur Home," Speech by Wendell Willkie at the Lincoln birthday dinner of the Middlesex Club in Boston, February 12, 1942, *Vital Speeches of the Day,* vol. 10, no. 4. 297–99.
2. Ibid.
3. Stimson Diary, Henry L. Stimson Papers, Yale University Library, New Haven, CT, entry of February 12, 1942.
4. Ibid., entry of February 13, 1942.
5. Eight Hundred and Sixth Press Conference, February 17, 1942, *The Public Papers and Addresses of Franklin D. Roosevelt,* vol. 11, *Humanity on the Defensive* (New York: Russell and Russell, 1969), 103.
6. Article 7 of the "Preliminary Agreement Between the United States and the United Kingdom" called for the "elimination of all forms of discriminatory treatment in international commerce, and to the reduction of tariffs and other trade barriers; and, in general, to the attainment of all the economic objectives set forth in the Joint Declaration made on August 14, 1941, by the President of the United States of America and the Prime Minister of the United Kingdom."
7. Cable C-25, February 7, 1942, Warren F. Kimball, ed., *Churchill & Roosevelt: The Complete Correspondence,* vol. 1, *Alliance Emerging, October 1933–November 1942* (Princeton, NJ: Princeton University Press, 1984), 351.
8. Correlli Barnett, *The Collapse of British Power* (London: Eyre Methuen, 1972), 107.
9. Winston S. Churchill, *The Second World War,* vol. 4, *The Hinge of Fate* (Boston: Houghton Mifflin, 1950), 88.
10. Cable C-141x, Kimball, *Churchill & Roosevelt,* vol. 1, 287.
11. Cable C-145x, "Part II — The Pacific Front," Kimball, *Churchill & Roosevelt,* vol. 1, 299.
12. Cable C-25, Kimball, *Churchill & Roosevelt,* vol. 1, 349.
13. Alan Brooke, *War Diaries 1939–1945: Field Marshal Lord Alanbrooke,* ed. Alex Danchev and Daniel Todman (Berkeley: University of California Press, 2001), 225.
14. Ibid., entry of February 2, 1942, 226.
15. Ibid., entry of February 3, 1942, 226.
16. Ibid., annotation by Lord Alanbrooke, 226.
17. Martin Gilbert, *Winston S. Churchill,* vol. 7, *Road to Victory: 1941–1945* (London: Heinemann, 1986), 47.
18. Entry of February 9, 1942, Brooke, *War Diaries,* 228.
19. Ibid.
20. Ibid., entry of February 11, 1942, 228–29.
21. Ibid., entry of February 13, 1942, 229.
22. Alexander Cadogan, *The Diaries of Sir Alexander Cadogan,* ed. David Dilks (London: Cassell, 1971), 433.
23. Ibid., entry of February 12, 1942.
24. Entry of February 13, 1942, Brooke, *War Diaries,* 229.

25. Clifford Kinvig, *Scapegoat: General Percival of Singapore* (London: Brassey's, 1996), 208.
26. Sugata Bose, *His Majesty's Opponent: Subhas Chandra Bose and India's Struggle Against Empire* (Cambridge, MA: Belknap Press of Harvard University Press, 2011), 242. There had already been a mutiny in 1940 by Sikh artillerymen tasked with defending Hong Kong: Lawrence James, *The Rise and Fall of the British Empire* (London: Little, Brown, 1994), 492.
27. Cable C-25, Kimball, *Churchill & Roosevelt,* vol. 1, 355.
28. "On the State of the War" (British Library of Information title), or "Through the Storm: A Broadcast Survey of the War Situation," February 15, 1942, in *War Speeches by the Right Hon. Winston S. Churchill, C.H., M.P.*, comp. Charles Eade, vol. 3, *The End of the Beginning* (London: Cassell, 1942), 201–7.
29. Ibid.
30. Ibid.
31. Ibid.
32. Robert Sherwood, *Roosevelt and Hopkins: An Intimate History* (New York: Harper, 1948), 501.
33. "On the State of the War" (British Library of Information title) or "Through the Storm: A Broadcast Survey of the War Situation," February 15, 1942, *The War Speeches of Winston S. Churchill,* vol. 2, 204.
34. In the *Sook Ching,* General Yamashita's order to "clean up" anti-Japanese Chinese captured in Singapore, military cordons were erected around Chinese residential areas to stop Chinese males aged between twelve and fifty from escaping. "The Japanese admitted to responsibility for 6,000 deaths. The figure most commonly quoted by the Chinese community was 40,000": Peter Thompson, *The Battle for Singapore: The True Story of Britain's Greatest Military Disaster* (London: Portrait, 2005), 375. Another source claimed 100,000 deaths, a figure Thompson, comparing it with other Japanese massacres, found "quite credible." "Chinese were beheaded or shot," wrote Brian Farrell in his 2005 account of the fall of Singapore. "The Japanese later admitted to killing at least 6,000. Singapore Chinese claims after the war ranged up to 50,000. An accurate figure might be near the 25,000 [Colonel] Sugita [Head of Intelligence, Japanese Twenty-fifth Army] supposedly admitted to a Japanese reporter": Brian P. Farrell, *The Defence and Fall of Singapore, 1940–1942* (Stroud, Gloucestershire, UK: Tempus, 2005), 385.
35. "Reliable Information from Manila shows American and British civilians subjected to concentration. All have been removed from their homes and families have been separated. All women and children confined in one place and men in another. Living conditions severe. MacArthur": Headquarters, SAFFE to AGWAR, Washington, January 18, 1942, RG 4, MMA. "All reports confirm my previous statements as to the extremely harsh and rigid measures taken against American and English in occupied areas in the Philippines. Such steps are not only unnecessary but are unquestionably dictated by the idea of abuse and special humiliation. I earnestly recommend that steps be taken

through the State Department to have these conditions alleviated. The negligible restrictions apparently applied in the United States to the many thousands of Japanese nationals there can easily serve as the lever under the threat of reciprocal retaliatory measures to force decent treatment for these interned men and women. The only language the Japanese understands [*sic*] is force and it should be applied mercilessly to his nationals if necessary . . . I urge this matter be handled immediately and aggressively through the proper diplomatic channels MacArthur": Cable 179 to AGWAR, Washington, February 1, 1942, RG 4, MMA. General John Dewitt, commander of U.S. Army forces on the West Coast, pressed the secretary of war for evacuation of all Japanese from the vulnerable California defense-industry areas on February 3, and after several weeks of discussion and legal consultation, the evacuation and internment of some 110,000 Japanese Americans from California was ordered under Executive Order 9066 on February 19, 1942.

36. See "1981 Report of the Presidential Commission on the Wartime Location and Internment of Civilians," which concluded that a "grave injustice was done to American citizens who, without individual review or any probative evidence against them, were excluded, removed and detained by the United States during World War II": Roger Daniels, *Prisoners Without Trial: Japanese Americans in World War II* (New York: Hill and Wang, 1993), 3–4.
37. Memorandum of February 16, 1942, in Sherwood, *Roosevelt and Hopkins*, 502–3.
38. "At the Atlantic meeting," Churchill stated before Parliament in London on September 9, 1941, "we had in mind, primarily, the restoration of the sovereignty, self-government, and national life of the [white] States and nations of Europe now under the Nazi yoke . . . So that is a quite separate problem from the progressive evolution of self-governing institutions in the regions and peoples which owe allegiance to the British crown": Richard Toye, *Churchill's Empire: The World That Made Him and the World He Made* (New York: Henry Holt, 2010), 214.

9. THE MOCKERY OF THE WORLD

1. Entry of 14.2.1942, Joseph Goebbels, *Die Tagebücher von Joseph Goebbels* [The diaries of Joseph Goebbels], ed. Elke Fröhlich (Munich: K. G. Saur, 1995), Teil II, Band 3, 308. Quotes from this source have been translated by the author.
2. Entry of 15.2.1942, Goebbels, *Die Tagebücher*, Teil II, Band 3, 321.
3. Ibid., 314.
4. Ambassador Winant to Roosevelt, February 17, 1942, "Safe" and Confidential Files, in Warren F. Kimball, ed., *Churchill & Roosevelt: The Complete Correspondence*, vol. 1, *Alliance Emerging, October 1933–November 1942* (Princeton, NJ: Princeton University Press, 1984), 362.
5. See Carlo D'Este, *Warlord: A Life of Winston Churchill at War, 1874–1945* (New York: Harper, 2008), inter alia.
6. In Harold Nicolson, *Diaries and Letters 1939–1945* (London: Collins, 1967), 211.

7. Ibid., 211.
8. Entry of February 9, 1942, Alexander Cadogan, *The Diaries of Sir Alexander Cadogan, O.M., 1938–1945*, ed. David Dilks (London: Cassell, 1971), 433.
9. Ibid., entry of February 15, 1942, 434.
10. Entry of February 27, 1942, Nicolson, *Diaries and Letters*, 214.
11. Entry of 16.2.1942, Goebbels, *Die Tagebücher*, Teil II, Band 3, 326.
12. Ibid., 326, 325.
13. Ibid.
14. Ibid., entry of 19.2.1942, 340.
15. Cable R-106, February 18, 1942, in Kimball, *Churchill & Roosevelt*, vol. 1, 362–63.
16. Ibid., Cable C-30, February 20, 1942, 364.
17. Entry of 16.1.1942, Goebbels, *Die Tagebücher*, Teil II, Band 3, 120.
18. Secret File, 2-3-42, RG 2, MacArthur Memorial Archives, in Paul P. Rogers, *The Good Years: MacArthur and Sutherland* (New York: Praeger, 1990), 183.
19. See Douglas Lockwood, *Australia's Pearl Harbour: Darwin, 1942* (Melbourne: Cassell Australia, 1966).
20. W. Averell Harriman and Elie Abel, *Special Envoy to Churchill and Stalin, 1941–1946* (New York: Random House, 1975), 126.

10. THE BATTLEGROUND FOR CIVILIZATION

1. Samuel I. Rosenman, *Working with Roosevelt* (New York: Harper, 1952), 1–2.
2. Ibid., 4.
3. Ibid., 330.
4. Ibid., 5.
5. Ibid., 329.
6. Ibid., 5.
7. Ibid., 6.
8. Ibid., 7.
9. Washington's Birthday was on February 22, but since the 22nd fell on a Sunday, the talk was given on February 23, 1942.
10. The Japanese submarine, I-17, under command of Nishino Kozo, shelled oil installations ten miles north of Santa Barbara, California, as well as aimlessly firing shells inland. The attack lasted only fifteen minutes. No one was killed or injured.
11. "We Must Keep On Striking Our Enemies Wherever and Whenever We Can Meet Them"—Fireside Chat on Progress of the War, February 23, 1942, in Franklin D. Roosevelt, *The Public Papers and Addresses of Franklin D. Roosevelt*, comp. Samuel I. Rosenman, vol. 11, *Humanity on the Defensive* (New York: Russell and Russell, 1969), 105–17.
12. Kenneth S. Davis, *FDR: The War President, 1940–1943* (New York: Random House, 2000), 435.

13. Entry of February 24, 1942, Galeazzo Ciano, *Diary 1937–1943* (New York: Enigma Books, 2002), 497.
14. Ibid.
15. Ibid.

11. NO HAND ON THE WHEEL

1. "Safe" and Confidential Files, DC 740.0011, Franklin D. Roosevelt Presidential Library, Hyde Park, NY. See also "India" annotation in Warren F. Kimball, ed., *Churchill & Roosevelt: The Complete Correspondence*, vol. 1, *Alliance Emerging, October 1933–November 1942* (Princeton, NJ: Princeton University Press, 1984), 373.
2. Ibid.
3. Alexander Cadogan, *The Diaries of Sir Alexander Cadogan, O.M., 1938–1945*, ed. David Dilks (London: Cassell, 1971), 438.
4. Ibid.
5. Cable C-34, March 4, 1942, in Kimball, *Churchill & Roosevelt*, vol. 1, 374.
6. Entry of March 5, 1942, *The Cadogan Diaries*, 440.
7. W. Averell Harriman and Elie Abel, *Special Envoy to Churchill and Stalin, 1941–1946* (New York: Random House, 1975), 126–27.
8. Ibid.
9. Letter of March 18, 1942, in Richard Toye, *Churchill's Empire: The World That Made Him and the World He Made* (New York: Henry Holt, 2010), 223.

12. LESSONS FROM THE PACIFIC

1. Entry of January 2, 1942, McCrea Diary, Papers of Captain John McCrea, Box 2, Library of Congress. McCrea had been assistant to Admiral Stark in 1941, and was then appointed naval secretary for the Joint Chiefs of Staff conversations with the British in January 1942, meeting Churchill and President Roosevelt for the first time on January 4, 1942.
2. Ibid., entry of January 16, 1942. McCrea relieved Captain Beardall, who had been made commandant of the U.S. Naval Academy.
3. Ibid., entry of January 18, 1942.
4. Ibid., entry of March 31, 1942.
5. Entry of February 6, 1942, Hart Diary, Papers of Admiral Thomas Hart, Operational Archives Branch, Naval Historical Center, Washington, DC.
6. Ibid.
7. The Battle of the Java Sea took place between February 27 and March 1, 1942. The Combined (Dutch, British, and U.S.) Striking Force, under Rear Admiral Doorman, was ordered into combat by Admiral Helfrich without air cover and with minimal intership communication; in suicidal combat it was annihilated,

suffering the loss of ten warships, including no fewer than five Allied cruisers sunk, while the Japanese Navy did not lose a single warship, or delay its invasion of Java by a single day. Admiral Ernest King called it "a magnificent display of very bad strategy": Samuel Eliot Morison, *History of United States Naval Operations in World War II*, vol. 3, *The Rising Sun in the Pacific, 1931–April 1942* (Boston: Little, Brown, 1948), 132. See Samuel Eliot Morison, *The Two-Ocean War: A Short History of the United States Navy in the Second World War* (Boston: Little, Brown, 1963), 88–98; Ian W. Toll, *Pacific Crucible: War at Sea in the Pacific, 1941–1942* (New York: Norton, 2012), 255–63.

8. Entry of March 10, 1942, Hart Diary.
9. Ibid.
10. Ibid., entries of February 18–19, 1942.
11. Ibid.
12. Ibid.
13. Ibid., entries of March 8 and 9, 1942.
14. "The Secretary, clearly having my appearance before his weekly press conference very much on his mind, sent for me even before I had finished my screed, and I read it to him. He said 'fine' . . . He obviously was much relieved. So I became then and there very decidedly committed to a line as regards participation in publicity" — in print, on radio, and in filmed interviews. Ibid., entry of March 11 et seq.

13. CHURCHILL THREATENS TO RESIGN

1. Entry for March 9, 1942, in Breckinridge Long, *The War Diary of Breckinridge Long: Selections from the Years 1939–1944*, ed. Fred L. Israel (Lincoln: University of Nebraska Press, 1966), 253.
2. United States Department of State, *Foreign Relations of the United States, Diplomatic Papers 1942* [hereafter *FRUS 1942*], vol. 1, *General, The British Commonwealth, The Far East* (Washington, DC: Government Printing Office, 1942), 606.
3. Ibid.
4. Ibid.
5. Christopher Thorne, *Allies of a Kind: The United States, Britain, and the War Against Japan, 1941–1945* (New York: Oxford University Press, 1978), 235.
6. Memorandum of a Conversation by Mr. Calvin H. Oakes of the Division of Near Eastern Affairs, May 26, 1942, *FRUS 1942*, vol. 1, 660. The American ambassador to China in Chungking, after a visit to India, reported the same to the U.S. State Department, recording that "there is bitterness against the Churchill Government for having sabotaged the Cripps Mission" — thus confirming "reports from other sources that there was sabotaging": M. S. Venkataramani and B. K. Shrivastava, *Quit India: The American Response to the 1942 Struggle* (New Delhi: Vikas, 1979), 143.

7. Cable R-116, Warren Kimball, ed., *Churchill & Roosevelt: The Complete Correspondence,* vol. 1, *Alliance Emerging, October 1933–November 1942* (Princeton, NJ: Princeton University Press, 1984), 403–4.
8. Sir Ian Jacob, unpublished autobiography, 73a and 73b, Churchill College Archives, Cambridge.
9. Alexander Cadogan, *The Diaries of Sir Alexander Cadogan, O.M., 1938–1945,* ed. David Dilks (London: Cassell, 1971), 437.
10. "Churchill and Amery had disavowed the possibility of a Cabinet convention. They had reacted against the term 'National Government,' . . . They recoiled from the full Indianization of the Executive . . . Finally they denied Cripps the status of a negotiator": R. J. Moore, *Endgames of Empire: Studies of Britain's Indian Problem* (New York: Oxford University Press, 1988), 97. See also Sarvepalli Gopal, *Jawaharlal Nehru: A Biography* (Cambridge, MA: Harvard University Press, 1976), vol. 1, 279–84; M. S. Venkataramani and B. K. Shrivastava, *Roosevelt, Gandhi, Churchill: America and the Last Phase of India's Freedom Struggle* (New Delhi: Radiant, 1983), 26–28. Most British historians blamed Churchill. "A Viceroy bitterly distrustful of the Congress leadership and lacking political finesse joined forces with a Prime Minister who had no wish to see Indian independence, and for whom the mission was primarily a device to deflect American criticism": Judith M. Brown, *Nehru: A Political Life* (New Haven, CT: Yale University Press, 2003), 148.
11. Cable C-62, Kimball, *Churchill & Roosevelt,* vol. 1, 438.
12. Sir Ian Jacob, unpublished autobiography.
13. Ian W. Toll, *Pacific Crucible: War at Sea in the Pacific, 1941–1942* (New York: Norton, 2012), 255.
14. The Personal Representative of the President in India to the Secretary of State, *FRUS 1942,* vol. 1, 627.
15. The Acting Secretary of State to the Officer in Charge at New Delhi, Personal for President's personal representative, *FRUS 1942,* vol. 1, 627–28.
16. Venkataramani and Shrivastava, *Quit India,* 108–9.
17. Alan Brooke, *War Diaries 1939–1945, Field Marshal Lord Alanbrooke,* ed. Alex Danchev and Daniel Todman (Berkeley: University of California Press, 2001), 245.
18. Telegram C-65, Kimball, *Churchill & Roosevelt,* vol. 1, 443.
19. The most succinct account of the viceroy's obstructionism, the secretary of state for India's change of mind, and Churchill's sheer perfidy (including Churchill's insincerity in claiming Britain would withdraw from overall control of the Raj) is to be found in Moore, *Endgames of Empire,* 94–105.
20. Venkataramani and Shrivastava, *Quit India,* 117.
21. Moore, *Endgames of Empire,* 96–97.
22. Cordell Hull, *The Memoirs of Cordell Hull* (New York: Macmillan, 1948), vol. 2, 1484. See also Venkataramani and Shrivastava, *Quit India,* 44.
23. Winston S. Churchill, *The Second World War,* vol. 4, *The Hinge of Fate* (Boston: Houghton Mifflin, 1950), 185–96. David Reynolds, in his dissection

of Churchill's Second World War memoirs, called Churchill's account of the episode "breathtakingly disingenuous": David Reynolds, *In Command of History: Churchill Fighting and Writing the Second World War* (New York: Random House, 2005), 337.

24. Sir Ian Jacob, unpublished autobiography.
25. Brown, *Nehru: A Political Life,* 148.
26. Cable R-132, Kimball, *Churchill & Roosevelt,* vol. 1, 445.
27. The Personal Representative of the President in India to the Secretary of State, *FRUS 1942,* vol. 1, 631.
28. Venkataramani and Shrivastava, *Quit India,* 141. On April 1, 1942, Halifax predicted that the Cripps mission would fail. When Welles asked Halifax what he thought would happen, in that case, the ambassador had said, "Nothing"— an example of fatal British complacency in view of the riots and deaths that followed: Memorandum of Conversation by the Acting Secretary of State, *FRUS 1942,* vol. 1, 623.
29. Warren Kimball, *Forged in War: Roosevelt, Churchill, and the Second World War* (New York: Morrow, 1997), 140; Warren Kimball, *The Juggler: Franklin Roosevelt as Wartime Statesman* (Princeton, NJ: Princeton University Press, 1991), 134. See also Venkataramani and Shrivastava, *Quit India,* 142.
30. Letter of July 26, 1941, Papers of Lord Halifax, Hickleton Papers, Borthwick Institute of Historical Research, University of York, Yorkshire, England.
31. Ibid., Letter of May 25, 1942.
32. Kimball, *Churchill & Roosevelt,* vol. 1, 446.
33. Winston S. Churchill, *The Second World War,* vol. 4, *The Hinge of Fate,* 219.
34. Cable R-132, Kimball, *Churchill & Roosevelt,* vol. 1, 446.
35. Ibid.
36. Ibid., 446–47.
37. Robert Sherwood, *Roosevelt and Hopkins: An Intimate History* (New York: Harper, 1948), 512.
38. Entry of April 22, 1942, Stimson Diary, Henry L. Stimson Papers, Yale University Library, New Haven, CT.
39. Sherwood, *Roosevelt and Hopkins,* 512.
40. Ibid.
41. Cable C-68, draft A, not sent, April 12, 1942, Kimball, *Churchill & Roosevelt,* vol. 1, 447.
42. Stanley Wolpert, *Nehru: A Tryst with Destiny* (New York: Oxford University Press, 1996), 308–9.
43. Cable C-68, draft A, not sent, April 12, 1942, Kimball, *Churchill & Roosevelt,* vol. 1, 447.
44. Ibid., 447–48.
45. Entry of April 22, 1942, Stimson Diary.

46. Personal Representative of the President in India to the Acting Secretary of State. For the President and Acting Secretary Welles, *FRUS 1942*, vol. 1, 635–37.
47. Ibid.
48. Cable C-68, Kimball, *Churchill & Roosevelt*, vol. 1, 448.
49. Lord Ismay, *The Memoirs of General Lord Ismay* (New York: Viking, 1960), 249.
50. Andrew Roberts, *Masters and Commanders: How Four Titans Won the War in the West, 1941–1945* (New York: Harper, 2009), 152.

14. THE WORST CASE OF JITTERS

1. Entry of Wednesday, April 15, 1942, 308, Diaries of William Lyon Mackenzie King, Library and Archives Canada, Ottawa, ON.
2. Entry of April 1, 1942, McCrea Diary, Papers of Captain John McCrea, Box 2, Library of Congress.
3. Entry of Wednesday, April 15, 1942, 309, King Diary.
4. Ibid., 310.
5. Ibid., 310(b).
6. Ibid., 310(e) and (f).
7. Ibid.
8. Ibid., 309.
9. Churchill himself recounted the "disconcerting" episode in his memoirs: Winston S. Churchill, *The Second World War*, vol. 3, *The Grand Alliance* (Boston: Houghton Mifflin, 1950), 617. See also Jon Meacham, *Franklin and Winston: An Intimate Portrait of an Epic Friendship* (New York: Random House, 2003), 161.
10. Entry of Wednesday, April 15, 1942, 309, King Diary.
11. Ibid.
12. Ibid., 310.
13. Ibid., Memorandum of conversation Mr. King had with President Roosevelt, White House, Washington, DC, 311–12.
14. Ibid.
15. Ibid.
16. Ibid., entry of April 16, 1942, 317.
17. Ibid., 323.
18. Ibid., 318.
19. Ibid.
20. Ibid., 325.
21. Ibid., 315.
22. Ibid.
23. Ibid., 319.
24. Cable C-69, Warren F. Kimball, ed., *Churchill & Roosevelt: The Complete Correspondence* (Princeton, NJ: Princeton University Press, 1984), vol. 1, 454.

25. Ibid., 452–53.
26. Entry of April 16, 1942, 235, King Diary.
27. Ibid, 326.
28. Cable R-134, Kimball, *Churchill & Roosevelt,* vol. 1, 455.
29. Entry of April 16, 1942, 326, King Diary.
30. Ibid.

15. DOOLITTLE'S RAID

1. James H. Doolittle, with Carroll V. Glines, *I Could Never Be So Lucky Again: An Autobiography* (New York: Bantam Books, 1991), 265–66.
2. Entry of April 18, 1942, Stimson Diary, Henry L. Stimson Papers, Yale University Library, New Haven, CT.
3. "Notes written by VADM John L. McCrea, USN (Ret.), Naval Aide to President Roosevelt from 16 Jan. 1942 to 3 Feb. 1943," Papers of John L. McCrea, Library of Congress.
4. Werner Gruhl, *Imperial Japan's World War Two, 1931–1945* (New Brunswick, NJ: Transaction, 2007), 79.
5. Ian W. Toll, *Pacific Crucible: War at Sea in the Pacific, 1941–1942* (New York: Norton, 2012), 296.
6. Entry of April 18, 1942, Stimson Diary.
7. Toll, *Pacific Crucible,* 299.
8. Ibid., 272.

16. THE BATTLE OF MIDWAY

1. Geoffrey C. Ward, ed. *Closest Companion: The Unknown Story of the Intimate Friendship Between Franklin Roosevelt and Margaret Suckley* (Boston: Houghton Mifflin, 1995), diary entry of May 12, 1942, 159.
2. Ibid., 158.
3. Ibid., entry of May 10, 1942, 158.
4. Ibid., entry of June 1, 1942, 164.
5. Commander George Elsey, interview with the author, September 10, 2011.
6. Ibid.
7. Entry of April 12, 1942, Stimson Diary, Henry L. Stimson Papers, Yale University Library, New Haven, CT.
8. Ian W. Toll, *Pacific Crucible: War at Sea in the Pacific, 1941–1942* (New York: Norton, 2012), 370.
9. Cable C-86 of May 7–8, 1942, Warren Kimball, ed., *Churchill & Roosevelt: The Complete Correspondence,* vol. 1, *Alliance Emerging, October 1933–November 1942* (Princeton, NJ: Princeton University Press, 1984), 483.
10. Entry of April 12, 1942, Stimson Diary.

11. Eric Larrabee, *Commander-in-Chief: Franklin Delano Roosevelt, His Lieutenants, and Their War* (New York: Harper & Row, 1987), 360–67; Toll, *Pacific Crucible,* 387.
12. Entry of May 26, 1942, in Ward, *Closest Companion,* 159.
13. William D. Hassett, *Off the Record with F.D.R., 1942–1945* (New Brunwick, NJ: Rutgers University Press, 1958), 54.
14. Entry of May 26, 1942, in Ward, *Closest Companion,* 1159.
15. On April 8, 1942, General Wainwright had wired that exhaustion and malnutrition in his beleaguered forces would, after almost four months of combat, make it impossible to break out of Bataan, with no hope of relief or evacuation. There was great concern in Washington lest the Japanese, who had refused to sign the Geneva Agreements on conduct in war, would massacre the U.S.-Filipino troops, if they did surrender. Roosevelt delegated the decision to General Wainwright — who was all for fighting to the last man standing. Not even he could control events, however, when General King, commanding the U.S. and Filipino troops on the Bataan mainland, fatefully decided his starving, exhausted soldiers could no longer fight, and commenced surrender negotiations, without Wainwright's authority. "Apparently the decision of bitter-end fighting with a possible massacre had been taken out of our hands," Secretary Stimson noted the next morning, when he went in person to Pennsylvania Avenue to tell the President, "who was still asleep. I got in about nine o'clock when he had just awakened and told him the news." The number of troops surrendering totaled 36,853, as of last count taken, two days before, Stimson recorded. At the press conference that the President asked him to give, "I took occasion to point out that there were Navy troops there that had behaved very well; also the fine behavior of the Filipino scouts and the fighting together with the Philippine Army and ourselves in a common cause, as well as a pledge that we would come back and drive the invaders out eventually": entry of April 9, 1942, Stimson Diary.
16. D. Clayton James, *The Years of MacArthur,* vol. 2, *1941–1945* (Boston: Houghton Mifflin, 1975), 169.
17. James MacGregor Burns, *Roosevelt: The Soldier of Freedom* (New York: Harcourt Brace Jovanovitch, 1970), 226.
18. Hassett, *Off the Record with F.D.R.,* 57.
19. Ibid.
20. E. B. Potter, *Nimitz* (Annapolis, MD: Naval Institute Press, 1976), 90–91.
21. Toll, *Pacific Crucible,* 409.
22. Potter, *Nimitz,* 107.
23. Larrabee, *Commander-in-Chief,* 394. The Grumman F6F Hellcat, six months later, provided the answer.
24. Toll, *Pacific Crucible,* 447.
25. Ibid., 446.

26. Ibid., 462.
27. Ibid., 455.
28. Ibid., 478.
29. Ibid., 462.
30. Frank Knox, Memo for the President, enclosing letter to the Attorney-General, June 9, 1942, Franklin D. Roosevelt Presidential Library, Hyde Park, NY.
31. James, *The Years of MacArthur,* vol. 2, 170.

17. CHURCHILL'S SECOND COMING

1. Entry of June 11, 1942, Diaries of William Lyon Mackenzie King, Library and Archives Canada, Ottawa, ON.
2. Ibid.
3. Ibid.
4. Ibid.
5. Ibid.
6. Ibid.
7. Ibid.
8. Cable C-101, June 13, 1942, in Warren Kimball, ed., *Churchill & Roosevelt: The Complete Correspondence,* vol. 1, *Alliance Emerging, October 1933–November 1942* (Princeton, NJ: Princeton University Press, 1984), 510.
9. Carlo D'Este, *Eisenhower: A Soldier's Life* (New York: Henry Holt, 2002), 303.
10. Ibid., 307.
11. Entry of June 17, 1942, Stimson Diary, Henry L. Stimson Papers, Yale University Library, New Haven, CT.
12. Ibid.
13. Ibid.
14. Entry of Thursday, June 18, 1942, "Secret Diary" of Lord Halifax, Papers of Lord Halifax, Hickleton Papers, Borthwick Institute of Historical Research, University of York, Yorkshire, England.
15. Ibid.
16. Ibid.
17. Entry of June 13, 1942, in Alan Brooke, *War Diaries, 1939–1945: Field Marshal Lord Alanbrooke,* ed., Alex Danchev and Daniel Todman (Berkeley: University of California Press, 2001), 265.
18. Prime Minister's Personal Minute to Lord Linlithgow, June 13, 1942, in Martin Gilbert, *Winston S. Churchill,* vol. 7, *Road to Victory, 1941–1945* (Boston: Houghton Mifflin, 1986), 123.
19. Entry of June 19, 1942, Halifax Diary.
20. Entry of June 19, 1942, Stimson Diary.
21. Ibid.
22. Ibid., entry of June 20, 1942.
23. Ibid., entry of June 21, 1942.

24. Ibid., entry of June 20, 1942.
25. Ibid.
26. Ibid.
27. Ibid.
28. Ibid.
29. Ibid., 3.
30. "Safe" and Confidential Files, undated, but probably June 13, 1942, Franklin D. Roosevelt Presidential Library, Hyde Park, NY.
31. Ibid.
32. Diary entry of June 20, 1942, in William D. Hassett, *Off the Record with F.D.R., 1942–1945* (New Brunswick, NJ: Rutgers University Press, 1958), 68.
33. Ibid.
34. Hugh L'Etang, *Fit to Lead?* (London: Heinemann Medical, 1980).
35. Franklin D. Roosevelt Day by Day, Pare Lorentz Center, FDR Library. www.fdrlibrary.marist.edu/daybyday.
36. Winston S. Churchill, *The Second World War*, vol. 4, *The Hinge of Fate* (Boston: Houghton Mifflin, 1950), 338–39.
37. Hassett, *Off the Record with F.D.R.*, 67.
38. Letter to the President, June 19, 1942, para 81, Henry L. Stimson Papers, Yale University Library.
39. Churchill, *The Second World War*, vol. 4, *The Hinge of Fate*, 342.
40. Ibid.
41. Ibid., 343.
42. Hassett, *Off the Record with F.D.R.*, 68.
43. Ibid.

18. THE FALL OF TOBRUK

1. Entry of June 20, 1942, "Secret Diary" of Lord Halifax, Papers of Lord Halifax, Hickleton Papers, Borthwick Institute of Historical Research, University of York, Yorkshire, England.
2. Martin Gilbert, *Winston S. Churchill*, vol. 7, *Road to Victory, 1941–1945* (Boston: Houghton Mifflin, 1986), 123.
3. Robert Sherwood, *Roosevelt and Hopkins: An Intimate History* (New York: Harper, 1948), 589–90.
4. Entry of June 22, 1942, Breckinridge Long, *The War Diary of Breckenridge Long: Selections from the Years 1939–1944*, ed. Frank L. Israel (Lincoln: University of Nebraska Press, 1966), 274.
5. Winston S. Churchill, *The Second World War*, vol. 4, *The Hinge of Fate* (Boston: Houghton Mifflin, 1950), 344.
6. Entry of June 20, 1942, Halifax Diary.
7. Entry of June 21, 1942, Alexander Cadogan, *The Diaries of Sir Alexander Cadogan, O.M., 1938–1945*, ed. David Dilks (London: Cassell, 1971), 458.

8. Ibid.
9. W. Averell Harriman and Elie Abel, *Special Envoy to Churchill and Stalin, 1941–1946* (New York: Random House, 1975), 144.
10. Lord Ismay, *The Memoirs of General Lord Ismay* (New York: Viking, 1960), 254.
11. Alan Brooke, *War Diaries, 1939–1945: Field Marshal Lord Alanbrooke,* ed. Alex Danchev and Daniel Todman (Berkeley: University of California Press, 2001), 269.
12. Ibid.
13. Charles Moran, *Winston Churchill: The Struggle for Survival, 1940–1965* (Boston: Houghton Mifflin, 1966), 37–38.
14. Ibid.
15. Ibid.
16. Entry of 22.6.1942, Joseph Goebbels, *Die Tagebücher von Joseph Goebbels* [The diaries of Joseph Goebbels], ed. Elke Fröhlich (Munich: K. G. Saur, 1995), Teil II, Band 4, 569–70. Quotes from this source have been translated by the author.
17. Ibid., 571.
18. Ibid., entry of 23.6.1942, 575.
19. Ibid., entry of 21.6.1942, 565.
20. Ibid., entry of 23.6.1942, 575, 581, and 589.
21. Ibid., 582 and 581.
22. Ibid., 577.
23. Ibid., entry of 21.6.1942, 563.
24. Ian Kershaw, *Hitler 1936–1945: Nemesis* (New York: Norton, 2000), 515–17.
25. Entry of 24.5.1942, Goebbels, *Die Tagebücher,* Teil 2, Band 4, 357.
26. Kershaw, *Hitler: Nemesis,* 516–17.
27. Entry of 24.5.1942, Goebbels, *Die Tagebücher,* Teil 2, Band 4, 356.
28. Ibid., 357.
29. Ibid., entry of 24.6.1942, 605.
30. Ibid.
31. Ibid., entry of 24.5.1942, 362.
32. Ibid., 363.
33. Ibid., 361–64.
34. Ibid., 359.
35. Ibid., 354.
36. Ismay, *The Memoirs of General Lord Ismay,* 255.
37. Entry of 25.6.1942, Goebbels, *Die Tagebücher,* Teil II, Band 4, 614.

19. NO SECOND DUNQUERQUE

1. Forrest Pogue, *George C. Marshall,* vol. 2, *Ordeal and Hope, 1939–1942* (New York: Viking, 1966), 337.
2. Entry of June 23, 1942, Diaries of William Lyon Mackenzie King, Mackenzie King Papers, Library and Archives Canada, Ottawa, ON.

3. Ibid., entry of June 25, 1942.
4. Ibid.
5. Ibid.
6. Ibid.
7. Ibid.
8. Entry of Saturday, June 13, 1942, "Secret Diary" of Lord Halifax, Papers of Lord Halifax, Hickleton Papers, Borthwick Institute of Historical Research, University of York, Yorkshire, England.
9. Entry of June 25, 1942, King Diary.
10. Ibid.
11. Ibid.
12. Ibid.
13. Ibid., entry of April 16, 1942.
14. Meeting of Pacific War Council at Washington, D.C., Wednesday, April 15, 1942, Mackenzie King Papers.
15. Entry of June 25, 1942, King Diary.

20. AVOIDING UTTER CATASTROPHE

1. Entry of 4.7.1942, Joseph Goebbels, *Die Tagebücher von Joseph Goebbels* [The diaries of Joseph Goebbels], ed. Elke Fröhlich (Munich: K. G. Saur, 1995), Teil II, Band 5, 53. Quotes from this source have been translated by the author.
2. Ibid., entry of 21.7.1942, 160.
3. Ibid.
4. As far back as March 9, 1942, the President had met with his U.S. ambassador to Spain, Alexander Weddell, asking him to consider the disadvantages versus advantages "were an expeditionary force sent to effect a landing 'either in Algeria or on the northwest coast of Africa,'" as Weddell quoted Roosevelt's instructions. There were the four "Favorable results" to be had, Weddell acknowledged when responding on March 24, 1942 — two of which were: "to provide a base from which Europe might be invaded" and "To place hostile elements [Rommel's army] between United Nations' forces in East and West": Weddell Memorandum for the President, March 24, 1942, in "Safe" and Confidential Files, Franklin D. Roosevelt Presidential Library, Hyde Park, NY.
5. William D. Leahy, *I Was There* (New York: Whittlesey House, McGraw-Hill, 1950), 116.
6. The final vote, taken on July 1, 1942, was 475 to 25 in favor of Churchill's government.

21. CITIZEN WARRIORS

1. "Shangri-la," Papers of Captain John McCrea, Box 11, Library of Congress.
2. Ibid.

3. Ibid.
4. Geoffrey C. Ward, ed., *Closest Companion: The Unknown Story of the Intimate Friendship Between Franklin Roosevelt and Margaret Suckley* (Boston: Houghton Mifflin, 1995), diary entry of July 5, 1942, 168.
5. Ibid.
6. Ibid., undated entry, week of June 21, 1942, 167.
7. Ibid.
8. Ibid.

22. A STAGGERING CRISIS

1. Cable C-107, July 8, 1942, Warren Kimball, ed., *Churchill & Roosevelt: The Complete Correspondence,* vol. 1, *Alliance Emerging, October 1933–November 1942* (Princeton, NJ: Princeton University Press, 1984), 520–21.
2. Ibid.
3. Ibid.
4. Ibid., 520.
5. Maurice Matloff and Edwin M. Snell, *Strategic Planning for Coalition Warfare, 1941–1942* (Washington, DC: Office of the Chief of Military History, Dept. of the Army, 1953), 268.
6. Michael Schaller, *Douglas MacArthur: The Far Eastern General* (New York: Oxford University Press, 1989), 70–71.
7. Ibid.
8. Ibid.
9. "Memorandum for the President, July 10, 1942, Washington D.C.," *The Papers of George Catlett Marshall,* vol. 3, *The Right Man for the Job, December 7, 1941–May 31, 1943,* ed. Larry Bland (Baltimore: Johns Hopkins University Press, 1991), 271.
10. Ibid.
11. William D. Hassett, *Off the Record with F.D.R., 1942–1945* (New Brunswick, NJ: Rutgers University Press, 1958), 87.
12. Entry of July 10, 1942, p. 2, Stimson Diary, Henry L. Stimson Papers, Yale University Library, New Haven, CT.
13. "Memorandum for the President, July 10, 1942," *The Papers of George Catlett Marshall,* vol. 3, 271.
14. Matloff and Snell, *Strategic Planning for Coalition Warfare,* 270.

23. A ROUGH DAY

1. Entry of July 12, 1942, Stimson Diary, Henry L. Stimson Papers, Yale University Library, New Haven, CT.
2. Ibid.
3. Ibid., 2.
4. Ibid.

5. Ibid.
6. Leonard Mosley, *Marshall: Organizer of Victory* (London: Methuen, 1982), 151.
7. George C. Marshall, *George C. Marshall: Interviews and Reminiscences for Forrest Pogue*, rev. ed. (Lexington, VA: G. C. Marshall Research Foundation, 1991), 546.
8. Carlo D'Este, *Patton: A Genius for War* (New York: HarperCollins, 1995), 416.
9. Marshall, *George C. Marshall: Interviews and Reminiscences*, 546.
10. Entry of July 12, 1942, 2, Stimson Diary.
11. "Memorandum for the President, Subject: Pacific Operations," July 12, 1942, Stimson Papers.
12. Entry of July 13, 1942, Stimson Diary.
13. Ibid., handwritten annotation on typed Marshall-King Memorandum to the President of July 12, 1942.
14. Andrew Roberts, *Masters and Commanders: How Four Titans Won the War in the West, 1941–1945* (New York: Harper, 2009), 233. FDR was also aware that in the North Atlantic, with almost two million tons a month of merchant shipping being sunk globally (a figure that barely diminished that fall), sinkings were exceeding Allied ship-construction, and would make a cross-Channel Second Front that year impossible to support logistically: B. J. C. McKercher, *Transition of Power: Britain's Loss of Global Pre-eminence in the United States, 1930–1945* (Cambridge: Cambridge University Press, 1999), 317. The President declined to use that argument, however, lest it lend credence to Marshall and King's case for a switch to the Pacific.
15. Maurice Matloff and Edwin M. Snell, *Strategic Planning for Coalition Warfare, 1941–1942* (Washington, DC: Office of the Chief of Military History, Dept. of the Army, 1953), 272.
16. Ibid.
17. Ibid., 272–73.
18. Entry of July 15, 1942, Stimson Diary.
19. Ibid., entry of July 13, 1942.
20. Ibid., entry of July 15, 1942.
21. Ibid., pp. 1–2.
22. Henry Stimson, *On Active Service in Peace and War* (New York: Harper, 1948), 425.
23. Entry of July 15, 1942, Stimson Diary.
24. Ibid.
25. Forrest Pogue, *George C. Marshall*, vol. 2, *Ordeal and Hope, 1939–1942* (New York: Viking, 1966), 346; Mark A. Stoler, *George C. Marshall: Soldier-Statesman of the American Century* (Boston: Twayne, 1989), 101.

24. STIMSON'S BET

1. Maurice Matloff and Edwin M. Snell, *Strategic Planning for Coalition Warfare, 1941–1942* (Washington, DC: Office of the Chief of Military History, Dept. of the Army, 1953), 272–73.

2. Viz. Andrew Roberts, *Masters and Commanders: How Four Titans Won the War in the West, 1941–1945* (New York: Harper, 2009).
3. Entry of July 18, 1942, Leahy Diary, William D. Leahy Papers, Library of Congress.
4. Ibid., entry of July 22, 1942.
5. Entry of July 20, 1942, Stimson Diary, Henry L. Stimson Papers, Yale University Library, New Haven, CT.
6. Ibid., entry of July 23, 1942.
7. Ibid.
8. Ibid.
9. Ibid.
10. Personal and Confidential letter of July 23, 1942, Stimson Papers.
11. Entry of July 24, 1942, Stimson Diary.
12. Entry of July 24, 1942, Leahy Diary.
13. Entry of July 24, 1942, Stimson Diary.
14. Ibid.
15. Ibid.
16. Ibid., entry of July 25, 1942.
17. Ibid.
18. Ibid.
19. Matloff and Snell, *Strategic Planning for Coalition Warfare,* 282.
20. Entry of July 25, 1942, Stimson Diary.
21. Ibid.
22. Ibid.
23. Ibid.
24. "Memorandum for the President: My Views as to the Proposals in Message 625 [message from General Marshall in London]," attached to Stimson Diary, July 25, 1942.
25. Entry of July 25, 1942, Leahy Diary.
26. Forrest Pogue, *George C. Marshall,* vol. 2, *Ordeal and Hope, 1939–1942* (New York: Viking, 1966), 347.
27. Henry Stimson, *On Active Service in Peace and War* (New York: Harper, 1948).
28. Annotations to entry of July 27, 1942, Stimson Diary.
29. Ibid., entry of July 27, 1942.

25. A DEFINITE DECISION

1. Forrest Pogue, *George C. Marshall,* vol. 2, *Ordeal and Hope, 1939–1942* (New York: Viking, 1966), 348.
2. According to American concentration camp historian Roger Daniels, "more than 120,000 individuals were held in relocation centers at one time or another," and the "camp population peaked at a little over 107,000" by early 1943. "By the beginning of the following year it was down to 93,000," and by January 1, 1945, "it had dropped to just 80,000," ending the war at 58,000. By the end of 1945, "every

camp but Tule Lake had been emptied," leaving 12,545 detainees: Roger Daniels, *Prisoners Without Trial: Japanese Americans in World War II* (New York: Hill and Wang, 1993), 73. President Roosevelt opposed the idea of such camps, reasoning that it was far better to arrange that Japanese immigrants and Japanese Americans "whose loyalty to this country has remained unshaken through the hardships of the evacuation which military necessity made unavoidable" should be dispersed and distributed across the country: "75,000 families scattered around on the farms and worked into the community" were "not going to upset anybody"; while Roosevelt declared himself deeply impressed by "the very wonderful record that the Japanese in that battalion in Italy have been making in the war. It is one of the outstanding battalions that we have": White House Press Conference of November 21, 1944, in *Complete Presidential Press Conferences of Franklin D. Roosevelt* (New York: Da Capo Press, 1972), vol. 24, 245–47.

3. Stimson Diary, Henry L. Stimson Papers, Yale University Library, New Haven, CT. President Roosevelt had authorized establishment of the Uranium Committee in October 1939 to coordinate research on nuclear fission, and on October 9, 1941, he created the Top Policy Group comprising himself, Vice President Wallace, Dr. Vannevar Bush, Dr. James Conant, General George Marshall, and War Secretary Henry Stimson to expedite development of an atomic bomb. Concerned lest the Germans outrace the Allies, the President demanded tougher supervision and on September 18, 1942, Colonel Leslie Groves of the Corps of Engineers was promoted to the rank of lieutenant general and formally placed in command of the Manhattan Project, with Secretary Stimson having overall responsibility under the President.
4. "Safe" and Confidential Files, Franklin D. Roosevelt Presidential Library, Hyde Park, NY.
5. Ibid.
6. Stimson Diary.
7. Entry of July 28, 1942, Leahy Diary, Papers of Admiral William D. Leahy, Library of Congress.
8. Ibid.
9. Maurice Matloff and Edwin M. Snell, *Strategic Planning for Coalition Warfare, 1941–1942* (Washington, DC: Office of the Chief of Military History, Dept. of the Army, 1953), 283.
10. Ibid.
11. Ibid.

26. A FAILED MUTINY

1. Robert Sherwood, *Roosevelt and Hopkins: An Intimate History* (New York: Harper, 1948), 615.
2. Forrest Pogue, *George C. Marshall*, vol. 2, *Ordeal and Hope, 1939–1942* (New York: Viking, 1966), 349.

3. Sherwood, *Roosevelt and Hopkins,* 612.
4. Pogue, *Ordeal and Hope,* 349.
5. Entry of August 7, 1942, Stimson Diary, Henry L. Stimson Papers, Yale University Library, New Haven, CT.
6. Ibid.
7. Ibid.
8. Ibid.
9. Ibid., 4.
10. Ibid., entry of August 7, 1942.
11. "DRAFT," Letter to the President, August 10, 1942, Stimson Papers.
12. Entry of August 7, 1942, Stimson Diary.
13. Ibid., entry of August 10, 1942.
14. Ibid., entry of August 11, 1942.

27. STALIN'S PRAYER

1. Winston S. Churchill, *The Second World War,* vol. 4, *The Hinge of Fate* (Boston: Houghton Mifflin, 1950), 428.
2. Robert Dallek, *The Lost Peace: Leadership in a Time of Horror and Hope, 1945–1953* (New York: Harper, 2010), 31. A full account of Churchill and Harriman's trip to Moscow is in Martin Gilbert, *Winston S. Churchill,* vol. 7, *Road to Victory, 1941–1945* (Boston: Houghton Mifflin, 1986), 172–208.
3. Samuel I. Rosenman, *Working with Roosevelt* (New York: Harper, 1952), 349.
4. Ibid.
5. Robert Sherwood, *Roosevelt and Hopkins: An Intimate History* (New York: Harper, 1948), 62; W. Averell Harriman and Elie Abel, *Special Envoy to Churchill and Stalin, 1941–1946* (New York: Random House, 1975), 168.
6. Ibid., 169.
7. Entry of September 3, 1942, Stimson Diary, Henry L. Stimson Papers, Yale University Library, New Haven, CT.
8. "Meeting at the Kremlin on Wednesday, August 12, 1942," in Gilbert, *Road to Victory,* 181.
9. Harriman and Abel, *Special Envoy,* 169.
10. Entry of October 21, 1942, Diaries of William Lyon Mackenzie King, Library and Archives Canada, Ottawa, ON.
11. "Meeting at the Kremlin on Wednesday, August 12, 1942," in Gilbert, *Road to Victory,* 182.
12. Ibid., 181.

28. A TRIP ACROSS AMERICA

1. Cable R-180, Draft A, not sent, initialed GCM, in Warren Kimball, ed., *Churchill & Roosevelt: The Complete Correspondence,* vol. 1, *Alliance Emerging,*

October 1933–November 1942 (Princeton, NJ: Princeton University Press, 1984), 581.

2. Ibid., Cable R-180, 584.
3. Ibid., 583.
4. August 31, 1942, quoted in Stephen Ambrose, *Eisenhower: Soldier, General of the Armies, President-elect* (New York: Simon and Schuster, 1983), 191.
5. Cable R-183, in Kimball, *Churchill & Roosevelt,* vol. 1, 590.
6. Mark W. Clark, *Calculated Risk* (New York: Harper, 1950), 36; Carlo D'Este, *Patton: A Genius for War* (New York: HarperCollins, 1995), 419; Piers Brendon, *Ike, His Life and Times* (New York: Harper and Row, 1986), 86.
7. Arthur L. Funk, *The Politics of TORCH: The Allied Landings and the Algiers Putsch, 1942* (Lawrence: University Press of Kansas, 1974), 100.
8. Cable C-144 of September 5, 1942, in Kimball, *Churchill & Roosevelt,* vol. 1, 591.
9. Ibid., Cable R-185 of September 5, 1942, 592.
10. Franklin D. Roosevelt, *The Public Papers and Addresses of Franklin D. Roosevelt,* comp. Samuel I. Rosenman, vol. 11, *Humanity on the Defensive, 1942* (New York: Russell and Russell, 1969), 350–51.
11. Linda Lotridge Levin, *The Making of FDR: The Story of Stephen T. Early, America's First Modern Press Secretary* (Amherst, NY: Prometheus, 2008), 294–97.
12. Geoffrey C. Ward, ed., *Closest Companion: The Unknown Story of the Intimate Friendship Between Franklin Roosevelt and Margaret Suckley* (Boston: Houghton Mifflin, 1995), 174.
13. Ibid., 175.
14. James MacGregor Burns, *Roosevelt: The Soldier of Freedom* (New York: Harcourt Brace Jovanovitch, 1970), 268.
15. Ward, *Closest Companion,* 175.
16. Arthur Herman, *Freedom's Forge: How American Business Produced Victory in World War II* (New York: Random House, 2012), 200.
17. Ibid.
18. Ward, *Closest Companion,* 175.
19. Ibid.
20. Ibid.
21. Ibid., entry of September 23, 1942, 179.
22. Herman, *Freedom's Forge,* 67.
23. Ibid., 200.
24. Entry of September 29, 1942, in Ward, *Closest Companion,* 182.
25. Ibid., 183.

29. THE PRESIDENT'S LOYAL LIEUTENANT

1. Geoffrey C. Ward, ed., *Closest Companion: The Unknown Story of the Intimate Friendship Between Franklin Roosevelt and Margaret Suckley* (Boston: Houghton Mifflin, 1995), entry of September 29, 1942, 183.

2. Cable R-187, September 26, 1942, in Warren Kimball, ed., *Churchill & Roosevelt: The Complete Correspondence,* vol. 1, *Alliance Emerging, October 1933–November 1942* (Princeton, NJ: Princeton University Press, 1984), 613.
3. Ibid., 645.
4. W. Averell Harriman and Elie Abel, *Special Envoy to Churchill and Stalin, 1941–1946* (New York: Random House, 1975), 166.
5. Arthur Herman, *Gandhi & Churchill: The Epic Rivalry That Destroyed an Empire and Forged Our Age* (New York: Bantam, 2008), 495. For an account of Churchill's use of propaganda in the United States to cast the Quit India movement in the worst possible light, see Auriol Weigold, *Churchill, Roosevelt, and India: Propaganda During World War II* (New York: Routledge, 2008), 140–60.
6. Harriman and Abel, *Special Envoy,* 172.
7. Cable C-148, September 14, 1942, in Kimball, *Churchill & Roosevelt,* vol. 1, 594.
8. Harriman and Abel, *Special Envoy,* 172.
9. Cable C-148, September 14, 1942, in Kimball, *Churchill & Roosevelt,* vol. 1, 594.

30. A CANADIAN BLOODBATH

1. "Meeting at the Kremlin on Wednesday, August 12, 1942," in Martin Gilbert, *Winston S. Churchill,* vol. 7, *Road to Victory, 1941–1945* (Boston: Houghton Mifflin, 1986), 181.
2. Ibid.
3. "Cairo, August, 1942," in Lord Moran, *Winston Churchill: The Struggle for Survival, 1940–1965* (Boston: Houghton Mifflin, 1966), 66.
4. For a full account, see Nigel Hamilton, *The Full Monty,* vol. 1, *Montgomery of Alamein, 1887–1942* (London: Allen Lane, 2001), 427–73.
5. Entry of 20.7.1942, Joseph Goebbels, *Die Tagebücher von Joseph Goebbels* [The diaries of Joseph Goebbels], ed. Elke Fröhlich (Munich: K. G. Saur, 1995), Teil II, Band 5, 348. Quotes from this source have been translated by the author.
6. Ibid., 349.
7. Ibid., 352.
8. Ibid.
9. Ibid.
10. Ibid., 353.
11. Ibid., 371.
12. Ibid., 354.
13. Ibid.
14. Entry of August 19, 1942, Diaries of William Lyon Mackenzie King, Library and Archives Canada, Ottawa, ON.
15. Ibid., entry of August 21, 1942.
16. Ibid., entry of August 24, 1942.

31. SOMETHING IN WEST AFRICA

1. Franklin D. Roosevelt, *The Public Papers and Addresses of Franklin D. Roosevelt,* comp. Samuel I. Rosenman, vol. 11, *Humanity on the Defensive* (New York: Russell and Russell, 1969), 417–18.
2. Entry of 8.9.1942, Joseph Goebbels, *Die Tagebücher von Joseph Goebbels* [The diaries of Joseph Goebbels], ed. Elke Fröhlich (Munich: K. G. Saur, 1995), Teil II, Band 5, 460. Quotes from this source have been translated by the author.
3. Ibid., 464.
4. *The Public Papers and Addresses of Franklin D. Roosevelt,* vol. 11, 417–18.
5. Entry of 16.10.1942, Goebbels, *Die Tagebücher,* Teil II, Band 6, 133–34.
6. Ibid., 134.
7. Ibid., entry of 15.10.1942, 125.
8. Ibid., entry of 26.9.1942, 571–72.
9. Ibid., entry of 9.9.1942, 464.
10. Ibid., 465.
11. Ibid., entry of 17.10.1942, 138.
12. Ibid.
13. William D. Leahy, *I Was There* (New York: Arno Press, 1979), 116.
14. On February 26, 1942, the President had written to Stimson regarding Stimson and General Marshall's order concerning the "reorganization of the Army," but wanted it reworded to "make it very clear that the Commander-in-Chief exercises his command function in relation to strategy, tactics and operations directly through the Chief of Staff. You, as Secretary of War, apart from your administrative responsibilities, would, of course, advise me on military matters": "Safe" and Confidential Files, Franklin D. Roosevelt Presidential Library, Hyde Park, NY.
15. Entry of September 17, 1942, Stimson Diary, Henry L. Stimson Papers, Yale University Library, New Haven, CT.
16. "My own feeling," Stimson confided in his diary, "is that Eisenhower has acted rather precipitately and inquiry should be made as to whether they [U.S. bomber forces] cannot carry on a while longer [in Britain] and perhaps carry on with some bombing right through. It is just another part of the price we are having to pay for this expedition to the south," he railed, "and to pay out of coin most precious to us": entry of September 9, 1942, Stimson Diary.
17. Ibid., entry of September 8, 1942.
18. Carlo D'Este, *Patton: A Genius for War* (New York: HarperCollins, 1995), 421.
19. Martin Blumenson, ed., *The Patton Papers* (Boston: Houghton Mifflin, 1972–74), vol. 2, 94.
20. Eric Larrabee, *Commander in Chief: Franklin Delano Roosevelt, His Lieutenants, and Their War* (New York: Harper & Row, 1987), 486.
21. Blumenson, *The Patton Papers,* vol. 2, 95.

22. Ibid., vol. 2, diary entry of October 21, 1942, 94.
23. Ladislas Farago, *The Last Days of Patton* (New York: McGraw-Hill, 1981), 191–92.

32. ALAMEIN

1. In the House of Representatives the Democrats lost forty-five seats, and in the Senate, eight seats.
2. Nigel Hamilton, *The Full Monty,* vol. 1, *Montgomery of Alamein, 1887–1942* (London: Allen Lane, 2001), 494.
3. Nigel Hamilton, *Monty: The Making of a General, 1887–1942* (New York: McGraw-Hill, 1981), 770.
4. Ibid., 744.
5. Entry of 4.11.1942, Joseph Goebbels, *Die Tagebücher von Joseph Goebbels* [The diaries of Joseph Goebbels], ed. Elke Fröhlich (Munich: K. G. Saur, 1995), Teil II, Band 6, 230. Quotes from this source have been translated by the author.
6. Ibid., 230–31.
7. Ibid., 233.
8. Ibid., 236.
9. Ibid., entry of 6.11.1942, 242–43.
10. Ibid., 244.
11. Ibid., entry of 7.11.1942, 246.
12. Ibid., 246–47.

33. FIRST LIGHT

1. Entry of November 5, 1942, Leahy Diary, William D. Leahy Papers, Library of Congress.
2. Entry of November 5, 1942, Stimson Diary, Henry L. Stimson Papers, Yale University Library, New Haven, CT.
3. Ibid.
4. Ibid.
5. Ibid., entry of November 6, 1942.
6. Entry of December 4, 1942, Diaries of William Lyon Mackenzie King, Library and Archives Canada, Ottawa, ON.
7. Entry of November 6, 1942, Stimson Diary.
8. Entry of November 8, 1942, in Geoffrey C. Ward, ed., *Closest Companion: The Unknown Story of the Intimate Friendship Between Franklin Roosevelt and Margaret Suckley* (Boston: Houghton Mifflin, 1995), 184.
9. Ibid.
10. Ibid.
11. Robert Murphy, *Diplomat Among Warriors* (Garden City, NY: Doubleday, 1964), 121.

12. Ibid.
13. Entry of November 2, 1942, Stimson Diary.
14. Ibid., entry of November 5, 1942.
15. Cable of November 2, 1942, in Murphy, *Diplomat Among Warriors,* 121.
16. Entry of November 2, 1942, Leahy Diary.
17. Entry of November 2, 1942, Stimson Diary.
18. Entry of November 7, 1942, in Ward, *Closest Companion,* 184.
19. Ibid.
20. Entry of November 6, 1942, Stimson Diary.
21. Entry of November 6, 1942, King Diary.
22. Entry of November 8, 1942, in Ward, *Closest Companion,* 185.
23. Entry of November 7, 1942, Leahy Diary.
24. Entry of November 7, 1942, Stimson Diary.
25. Grace Tully, *F.D.R., My Boss* (New York: Scribner's, 1949), 264.
26. Ibid.
27. Ibid.
28. Entry of November 8, 1942, in Ward, *Closest Companion,* 185.
29. Ibid., 186.
30. Entry of November 7, 1942, Stimson Diary.
31. Ibid., entry of November 6, 1942.
32. Entry of November 8, 1942, in Ward, *Closest Companion,* 185–86.
33. Ibid.

34. THE GREATEST SENSATION

1. Entry of November 8, 1942, Stimson Diary, Henry L. Stimson Papers, Yale University Library, New Haven, CT.
2. Ian Kershaw, *Hitler 1936–1945: Nemesis* (New York: Norton, 2000), 541.
3. Entry of 11.7.1942, Joseph Goebbels, *Die Tagebücher von Joseph Goebbels* [The diaries of Joseph Goebbels], ed. Elke Fröhlich (Munich: K. G. Saur, 1995), Teil II, Band 5, 346. Quotes from this source have been translated by the author.
4. Ibid., entry for "Yesterday," drawn up November 9, 1942, Teil II, Band 6, 254.
5. Ibid.
6. Ibid., 256.
7. Ibid., 257.
8. Ibid.
9. Ibid., entry of 23.6.1942, Teil II, Band 4, 594.
10. Ibid., entry of 24.6.1942, 604.
11. Ibid., entry of 23.6.1942, 591.
12. Ibid., entry of 24.6.1942, 605.
13. Ibid.
14. Ibid., 610.

15. Jacques Belle, *L'Opération Torch et la Tunisie* (Paris: Economica, 2011), 65–72.
16. Entry of 23.6.1942, Goebbels, *Die Tagebücher,* Teil II, Band 4, 592.
17. Kershaw, *Hitler: Nemesis,* 539.

35. ARMISTICE DAY

1. Entry of November 9, 1942, Stimson Diary, Henry L. Stimson Papers, Yale University Library, New Haven, CT.
2. Annex 2, Project Gymnast, U.S. Serial ABC-42, in United States Department of State, *Foreign Relations of the United States: The Conferences at Washington, 1941–1942, and Casablanca, 1943* (Washington, DC: Government Printing Office, 1941–1943), 240–43.
3. Eight Hundred and Fifty-Ninth Press Conference, November 10, 1942, in *The Public Papers and Addresses of Franklin D. Roosevelt,* comp. Samuel I. Rosenman, vol. 11, *Humanity on the Defensive* (New York: Russell and Russell, 1969), 462–63.
4. Ibid.
5. Entry of November 10, 1942, Stimson Diary.
6. Ibid.
7. Ibid., entry of November 9, 1942.
8. Allied casualties in the three days, November 8–11, 1942, were 530 killed, 887 wounded, and 52 missing — far lower than the Dieppe fiasco, despite Dieppe having been a vastly smaller operation: George F. Howe, *Northwest Africa: Seizing the Initiative in the West* (Washington, DC: Office of the Chief of Military History, Dept. of the Army, 1957), 173.
9. Entry of November 7, Martin Blumenson, ed, *The Patton Papers* (Boston: Houghton Mifflin, 1974), vol. 2, 102.
10. Ibid., 108.
11. Ibid., 109.
12. Carlo D'Este, *Patton: A Genius for War* (New York: HarperCollins, 1995), 437.
13. Entry of November 11, 1942, Leahy Diary, William D. Leahy Papers, Library of Congress.
14. On November 11, 1942, Eisenhower informed General Marshall of Churchill's "extraordinary impatience," but insisted, "I am not repeat not allowing anything to interfere with my clean cut line of subordination to the Combined Chiefs of Staff": Cable #320, in *The Papers of Dwight David Eisenhower,* ed. Alfred D. Chandler, vol. 2, *The War Years* (Baltimore: Johns Hopkins University Press, 1970), 691.
15. Ibid., 685.
16. Ibid., Cable to Major-General Walter Bedell Smith, November 11, 1942, 693.
17. For a French version of events, see Jacques Belle, *L'opération Torch et la Tunisie* (Paris: Economica, 2011), 44–53.

18. Sugata Bose, *His Majesty's Opponent: Subhas Chandra Bose and India's Struggle Against Empire* (Cambridge, MA: Belknap Press of Harvard University Press, 2011), 223.
19. Entry of November 11, 1942, Leahy Diary.
20. Entry of November 11, 1942, Stimson Diary.
21. Ibid.
22. Ibid.
23. "The Forces of Liberation Are Advancing" — Armistice Day Address, November 11, 1942, *The Public Papers and Addresses of Franklin D. Roosevelt,* vol. 2, 468–70.
24. William D. Leahy, *I Was There* (New York: Arno Press, 1979), 136.
25. Entry of November 11, 1942, Leahy Diary.
26. "The Forces of Liberation Are Advancing," *The Public Papers and Addresses of Franklin D. Roosevelt,* vol. 2, 468–70.

Index